**13TH EDITION**

# The Marriage and Family Experience

## Intimate Relationships in a Changing Society

**Bryan Strong**
University of California, Santa Cruz

**Theodore F. Cohen**
Ohio Wesleyan University

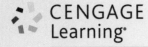
CENGAGE
Learning·

Australia • Brazil • Mexico • Singapore • United Kingdom • United States

## CENGAGE
### Learning®

*The Marriage and Family Experience:*
*Intimate Relationships in a Changing*
*Society,* **Thirteenth Edition**
Bryan Strong and Theodore F. Cohen

Product Director: Marta Lee-Perriard

Product Manager: Elizabeth
Beiting-Lipps

Content Developer: Trudy Brown

Product Assistant: Chelsea Meredith

Marketing Manager: Quynton
Johnson

Content Project Manager: Cheri
Palmer

Art Director: Vernon Boes

Manufacturing Planner: Judy Inouye

Production Service, Composition, and
Illustration: MPS Limited

Photo and Text Researcher: Lumina
Datamatics

Copy Editor: Heather McElwain

Text Designer: Diane Beasley

Cover Designer: Irene Morris

Cover Image: Gillian Laub/Stone/
Getty Images

For product information and technology assistance, contact us at
**Cengage Learning Customer & Sales Support, 1-800-354-9706**

For permission to use material from this text or product,
submit all requests online at **www.cengage.com/permissions**
Further permissions questions can be emailed to
**permissionrequest@cengage.com**

Library of Congress Control Number: 2015960735

Student Edition:
ISBN: 978-1-305-50310-6

Loose-leaf Edition:
ISBN: 978-1-305-67713-5

**Cengage Learning**
20 Channel Center Street
Boston, MA 02210
USA

Cengage Learning is a leading provider of customized learning solutions with
employees residing in nearly 40 different countries and sales in more than
125 countries around the world. Find your local representative at
**www.cengage.com**

Cengage Learning products are represented in Canada by Nelson Education, Ltd.

To learn more about Cengage Learning Solutions, visit **www.cengage.com**

Purchase any of our products at your local college store or at our preferred
online store **www.cengagebrain.com**

Printed in the Canada
Print Number: 01     Printed Year: 2016

*I dedicate this edition to my mother, Eleanor Schoenberg Cohen, who passed away in May 2015. In her 65-year-long marriage to my father, Kalman, she demonstrated what it means to give of oneself lovingly and unconditionally. Throughout my life, she was a role model of how to be a loving and devoted parent. She is greatly missed.*

# Brief Contents

# Contents

# 6

# Understanding Sex and Sexualities 191

# 9
# Unmarried Lives: Singlehood and Cohabitation  323

# 10
# Becoming Parents and Experiencing Parenthood  355

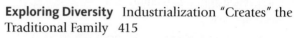

# 14

# New Beginnings: Single-Parent Families, Remarriages, and Blended Families  526

# Boxes

## Real Families

## Public Policies, Private Lives

# Preface

This edition is the 13th in the long literary lifetime of *The Marriage and Family Experience*. Stretching across more than three decades, its contents have changed greatly in keeping with the immense social, cultural, and familial shifts that have occurred since Bryan Strong wrote the first edition. We have witnessed considerable change in definitions of who and what counts as a family, including most recently with the legal recognition of same-sex marriage in 2015. The expectations and experiences people have of their intimate relationships, their marriages, and their relationships with their parents and their children continue to change, alongside shifts in the economy, advances in technology, and changes in the culture, perhaps most notably around issues related to gender, sexuality, and intimacy. The book you have before you is a product of and reflects those changes.

However, in its objectives, much remains the same. From its first to its present edition, *The Marriage and Family Experience* has sought to engage students from a range of academic and applied disciplines across a number of different types of institutions, and to stimulate their curiosity about families. The present edition retains that mission by characterizing and conveying the rich diversity of family experience, the dynamic nature of both the institution of family and of individual families, and the many ways in which experiences of relationships, marriages, and families are affected by the wider economic, political, social, and cultural contexts in which we live.

My personal involvement with *The Marriage and Family Experience* has a shorter history. By the time I entered its life, it was a successful textbook some seven editions old. Now, for the sixth time, I have had the opportunity to revise and update the text. Each time, I have incorporated the latest available research and official statistics on subjects such as sexuality (sexual orientation and expression), marriage, cohabitation, childbirth, child care, divorce, remarriage, blended families, adoption, abuse, the division of housework, and connections between paid work and family life. Once again, there are hundreds of new references in this edition, drawn mainly though not exclusively from research in sociology, psychology, and family studies. I have again tried to feature some of the most interesting issues, controversies, and real-life examples, sometimes drawn straight from recent news stories, popular culture, or narrative accounts, to give readers a better appreciation for how the more academic content applies to real life and to stimulate their fascination with families.

Thinking about my own many years of involvement with *The Marriage and Family Experience*, I marvel at how much has changed, both in the wider society and in my own family. I have been reminded, on a profoundly personal level, of the range of family experiences people have and of the dynamic and unpredictable quality of family life. When I first began working on the eighth edition of this book, I was more than 20 years into a stable marriage and had no reason to imagine ever being single again or remarrying. My wife and I had two young teenagers who formed the center of our too-hectic life together. I was a husband and father, two roles that I valued above all others and that I juggled along with my career as a sociologist and teacher. In the years since, I have been a full-time caregiver when my wife became ill, a widower after her passing, a single parent, a partner in a long-distance relationship, a remarried husband, a stepfather, and an ex-spouse. Both my son, Dan, now 30 and living more than 2,000 miles away with his girlfriend, Marissa, and my daughter, Allison, now married and living with her husband, Joe, and their two cats, have wonderful and busy lives. Most important, both my kids and their partners seem happy. The two stepsons and stepdaughter that I gained when married to their mom have reached their own milestones: Daniel has graduated college, Molly is about to enter college, and the youngest, Brett, is finishing his first year of high school. During my involvement with this book I have seen what a rollercoaster ride family life can feel like, with its many ups and downs. Just in the past year, I have had the joy of witnessing my daughter's wedding and the sadness of being at my mother's funeral. None of this is unique to my life. If anything, my experiences of marriage, fatherhood, caregiving, widowerhood, single parenting, remarriage, stepfatherhood, separation, divorce, and parental loss all just serve to heighten my sensitivity to and appreciation of the many twists and turns that families take and the various roles and relationships covered in this book. They also are constant reminders to me of how—whether in a single lifetime or across a society—we can neither completely anticipate nor fully control the directions our families may take.

## New to This Edition

The changes returning users will see in this edition are mostly content related. In updating the text, I have drawn heavily from reports by such sources as the Pew Research Center, the National Center for Family and Marriage Research, the National Council on Family Relations, the Council on Contemporary Families, or from official sources, such as the U.S. Census Bureau, the Bureau of Labor Statistics, the Centers for Disease Control and Prevention, the National Institute of Justice, the World Health Organization, and many others. These, along with published research from books and journals, are incorporated, where relevant, throughout this revision. Furthermore, this edition continues to make great use of data from such national surveys as the National Survey of Family Growth, the National Survey of Sexual Health and Behavior, the Global Study of Sexual Attitudes and Behavior, the National Survey of Adoptive Parents, the National Intimate Partner and Sexual Violence Survey, and the National Survey of Children's Exposure to Violence. As with previous editions, the 13th edition attempts to capture and characterize the current state of marriage and the family experience.

Second, attention to diversity remains one of the central themes of the book. Therefore, substantial and repeated attention is paid to how our experiences of intimate relationships, marriage, parenthood, work and family, divorce, remarriage, abuse, and so on, are differently experienced across lines of class, gender, race, ethnicity, and sexuality. What is perhaps most noteworthy is the enlarged and more sustained attention to gender and sexuality issues, most evident in Chapters 1, 4, 6, and 9. There is also increased attention to racial and ethnic diversity (including greater coverage of multiracial family experience), and continued attention to religion as it shapes people's attitudes, values, and experiences of many of the topics covered.

Third, I have attempted to reflect wider economic and technological changes as they impact family experiences. Thus, the recession and its aftermath are mentioned in a number of chapters. Even more notably, numerous examples throughout the text illustrate the impact of technological innovations on aspects of people's family experiences, including how people meet and form relationships, communicate with loved ones, and monitor or care for family members.

Fourth, I have made a number of additions to the features of the text that I hope will capture students' interest and engage their curiosity. Roughly two dozen of the almost 60 features are either new to this edition of significantly updated or enlarged. The *What Do You Think?* self-quiz at the start of each chapter has been extensively revised with new true/false questions that follow the content order of the chapter. The true/false quiz questions are treated almost like learning objectives, and instead of providing an answer key close to the quiz, the answers are now provided within the body of the text to highlight the key points made by each question. More specific additions and changes are as follows.

### Content Changes by Chapter

The most notable changes in **Chapter 1**, "The Meaning of Marriage and the Family," include a new section, *"Dramatic Changes, Increasing Diversity, and Continuing Controversy"* addressing the challenges inherent in studying families. Other additions include coverage of the U.S. Supreme Court decision in *Obergefell v. Hodges,* new material addressing cross-cultural data on marriage and extended families, and more attention to gender, sexuality, and race as sources of diversity in attitudes about family issues and in effects on families. I have updated statistics on marital status and household composition in the United States. Once again, I have changed or added to the chapter opening examples of controversial and contested family issues. The examples used in the new edition include a court case over ownership of frozen embryos, an updated discussion of the Kody Brown suit challenging Utah's antipolygamy law, and the domestic violence cases of National Football League players Ray Rice and, especially, Adrian Peterson. As in the past, these are designed to reflect the chapter's continued emphasis on different and competing viewpoints about the meaning of family and the interpretation of changing family patterns. The chapter also contains an up-to-date discussion of the increase in multigenerational households. Changes and additions have been made to some of the boxed features. There is a new *Public Policies, Private Lives* feature on the Obergefell decision, an updated *Issues and Insights* box, "Red and Blue Families," which includes recent research (Wilcox and Zill) on the "reddest" and "bluest" states and on red and blue counties, and a new *Popular Culture* feature on a possible *Modern Family* effect on acceptance of gay marriage.

**Chapter 2,** "Studying Marriages and Families," contains updated data on exposure to popular culture, especially television, new examples of "reality television" programs on families, updated examples of the advice and information genre online, on air, and in print. In discussions of theories, there is a new example illustrating a functionalist approach to wedding rituals, and discussion of intersectionality in the section on feminist perspectives. In discussing research, there is a section on demography—what it is and why it is useful in studying families. Using comments by sociologist Paul Amato, the chapter concludes with more explicit mention of why it is impossible to formulate "universal laws" that apply to everyone's experience of family life.

**In Chapter 3,** "Variations in American Family Life," the coverage of American families across history now includes material from Andrew Cherlin's *Labor's Love Lost*, a history of working-class families in the United States, as well as two new sections—"Late Twentieth-Century Families" and "Families Today"—to better reflect the extent and nature of changes in family life over the past four decades. The section on social class variations now includes material on problems faced by affluent youth, neighborhood effects on opportunities for mobility, and effects of the recession on marriage and divorce, births, and multigenerational families. Data on poverty, the working poor, and children in poverty have all been updated with the latest data available. Material on racial and ethnic variations now includes a more detailed discussion of how the census has defined and measured race, a greatly enlarged discussion of multiracial families, and more attention to diversity of experiences within racial or ethnic groups. In discussing multiracial families,

attention is paid to racial socialization and to experiences of microaggressions, sometimes within one's own extended family. On diversity within groups, there is material differentiating experiences of African Americans and Caribbean black immigrants, and new material on diversity among Asian American groups in their educational attainment, life goals, and where marriage and parenthood rank in their priorities.

**Chapter 4,** "Gender and Family," is the most substantially changed chapter, so as to capture and characterize the recent and ongoing social and cultural changes in how we think about gender. In discussing the concept of gender, there are now sections addressing "what gender is" and "what gender isn't." These are offered as ways to address possible misconceptions as well as to show the breadth of how gender affects our lives. These sections reflect challenges to binary conceptualizations of gender, consideration of gender as a spectrum, and include considerable attention to transgender experience. The new material on transgender experience includes two new features and a later discussion of survey data on transgender family relationships and experiences. The remainder of the chapter has been updated with more recent data, including sections on gender inequality; gender, sexuality, and bullying; media as socialization; gender and religiosity; data on housework and child care; and data on attitudes in support of greater familial gender equality.

**Chapter 5,** "Intimacy, Friendship, and Love," includes much new and/or updated material on the use of websites, smartphones, and texting in initiating, maintaining, and/or ending dating relationships. Additionally, there is new material on women and emotion work; love and sexual intimacy among same-sex and heterosexual couples; friends with benefits relationships; "churning" or relationship cycling; dating in older adulthood; and recent data on breakups and their consequences. In talking about popular cultural emphasis on romantic love, there is also updated data on the romance fiction literary genre, and new popular culture references to love themes in film, using both 2013's *Her*, and 2014's *The Fault in Our Stars* as recent examples.

**Chapter 6,** "Understanding Sex and Sexualities," continues to look at recent data on sexual expression across the life span. It has been updated with data from more recent waves of the National Survey of Family Growth (2011–2013) and the Youth Risk Behavior Survey (2013) in discussing adolescent and young adult sexual experience, as well as more recent

General Social Survey data (on attitudes about different types of sexual expression), Pew Research Center data (survey of LGBT Americans), and Centers for Disease Control and Prevention data on such issues as STI's, including HIV/AIDS. The chapter also contains significantly expanded coverage of LGB sexual issues and experiences, including population estimates, coming out experiences, experience of sexual stigma (including mention of monosexism and biphobia). The new boxed feature, "The Good, the Bad, and the Ugly: Trends in the Status of the LGBT Population in the U.S. and Abroad," focuses on positive indicators suggesting greater acceptance as well as negative indicators such as continued inequality and harassment/violence directed at the LGBT population. The boxed feature on sexting has been updated.

**Chapter 7,** "Communication, Power, and Conflict," has new material on each of the topics in the chapter title. New or updated material on communication includes discussions of sexual communication, aging, and the use of demand-withdraw communication, the question of problems in too much communication, and consideration of positive communication strategies (such as "intentional dialogue"). Material on conflict and conflict management has been updated, with specific sections continuing to focus on conflicts about sex, money, and housework. Material on destructive conflict management and on conflict in same-sex and heterosexual relationships has been updated. The Popular Culture feature, "Staying Connected with Technology," has been updated with data from the Pew Research Center's survey, "Couples, the Internet, and Social Media," as well as other recent research. The new feature, "Should I Stay or Should I Go? Should We Try or Should We Stop?" addresses a recent therapeutic strategy of discernment counseling.

In **Chapter 8,** "Marriages in Societal and Individual Perspective," the most notable changes result from keeping up to date with data on changing marriage rates and shifting attitudes about marriage. The chapter has moved from a consideration of "the marriage debate," to a discussion that highlights the ambiguous status of marriage in the United States, which includes special attention to attitudes and outlooks of millennials. There is new consideration of earlier historical fluctuations in marriage rates, new material on weddings and their costs, new data on marriage and social ties (including to family and in volunteering and charitable giving). The discussion of religion and marriage has been broadened, and the data on racial homogamy versus intermarriage (and roles played by

education and income), religious homogamy, and age-discrepant marriages have all been updated. In the section on who we can marry, the attention to same-sex marriage now includes the *Obergefell* decision, and recent estimates of the numbers of married lesbian or gay male couples. The section on marriage typologies now also includes a typology from the work of John Gottman, and the chapter closing section on the future of marriage now includes reference to Cherlin's *Labor's Love Lost*. The new *Public Policies, Private Lives* feature, "Will You Marry Us?" examines the use of friends and family members as wedding officiants.

In **Chapter 9,** "Unmarried Lives: Singlehood and Cohabitation," data on numbers of singles and the extent of cohabitation again have been updated. Pew Research Center data on why unmarried women and men haven't married are included. The chapter has updated discussions of both premarital and postmarital (prior to remarriage) cohabitation. There is updated and/or enlarged discussion of cohabitation and remarriage, pooling of finances among cohabiting couples, relationship satisfaction among cohabiting couples, and the impact of cohabitation and serial cohabitation on marriage. The material on same-sex cohabitation has been updated, and where available comparisons are made between same-sex and heterosexual married and cohabiting couples. The features titled "Living Apart Together," "Elective Co-Parenting by Heterosexual and LGB Parents," and on "Heterosexual Domestic Partnerships" all have been updated.

**Chapter 10,** "Becoming Parents and Experiencing Parenthood," once again contains updated statistics on fertility, births, unmarried childbirth, infant mortality, pregnancy, mistimed or unwanted pregnancies, pregnancy loss, adoption, voluntary childlessness, and infertility. Updated estimates are given from the U.S. Department of Agriculture on the costs associated with raising children. New data from the third wave of the "Listening to Mothers" survey are used to address women's experiences giving birth. The chapter also includes consideration of competing mothering ideologies ("intensive mothering" versus "extensive mothering"), comparisons of employed versus at-home mothers, and updated data on the wage impact of motherhood for women. More recent data are included on fathers, especially regarding housework and time spent with children. There are also updated discussions of single fathers and at-home fathers. Using the National Survey of Children's Health and the National Survey of America's Families, the chapter consideration of the pleasures and pains of parenthood has been updated. New data or discussions about parents' self-assessments, contact between adults and aging parents, parenting adult children, grandparents raising children, and on nonparental households are included. The section on gay or lesbian parents has been updated and enlarged. A new *Popular Culture* feature looks at research on the potential effects of MTV's *16 and Pregnant* and *Teen Mom* on teen pregnancy and childbearing.

**Chapter 11,** "Marriage, Work, and Economics," contains updated employment and labor force participation data along with data on women's and men's work experiences and dual-earner households. In addressing how work impacts family life, we present 2015 Pew survey data of parental time strains, updated discussions of work-family conflict, and parental guilt by gender; American Time Use Survey data on time spent in housework; and a 2015 comparison of 50 years of time use data from 14 countries. We also update with 2014–15 data the costs of outside child care, and consider trends in unemployment, telecommuting, and flextime. Data on availability of family supportive policies have been updated.

**Chapter 12,** "Intimate Violence and Sexual Abuse," has much new material. This includes new examples to open, and later throughout the chapter reflecting the breadth of family violence and intimate partner violence. We include newer data from the National Intimate Partner and Sexual Violence survey, estimating the prevalence of "minor" and "more severe" intimate partner violence, emotional and psychological abuse (including threats, insults, and excessive efforts to monitor and control), and the impact of abusive behavior on recipients. In addressing dating violence and date rape, there is a new discussion of the concept of "affirmative consent." We have updated the data and discussion on child maltreatment, and consider age, race, parental age, and type of maltreatment. We include new data on sibling violence and on the estimated economic impact of family violence. The discussion of policies to address family violence now better reflect both the advocacy for and the criticisms of mandatory arrest and no-drop prosecution.

**Chapter 13,** "Coming Apart: Separation and Divorce," has updated data on divorce, custody, child support, and alimony, and enlarged coverage of these issues. This is accompanied by a brief discussion of the limitations of divorce data, due to incomplete reporting across the United States (data on divorce does not include data from all 50 states). The chapter uses 2013–14 data to illustrate the different measures of divorce rates. New to the chapter are discussions of the trend in "gray divorce," the risks involved in marrying either too young *or* too old, and the economic impact of divorce. New or updated box features include "Divorcing in Iran and India, but NOT the Philippines," "Making Personal Trouble Public: Sharing One's Divorce Online or in Print," and "Covenant Marriage as a Response to Divorce."

**Chapter 14,** "New Beginnings: Single-Parent Families, Remarriages, and Blended Families," offers updated discussions of trends in single parenting and remarriage, and of the economic status and diversity of living arrangements of single parents. The variations in single-parent households and in remarriage, especially by gender, race/ethnicity, and poverty status, are highlighted. The "benefits" of remarriage are considered, especially as they compare to the benefits of first marriage. In addition, the chapter pays more attention to stepfamilies, including new material on the effects of stepfamily life on marital quality, age differences in children's adjustment to stepfamily life, and the different ways children refer to stepfathers. Data on remarriage and stepfamily life include estimates of how many U.S. marriages are remarriages, how many adults have at least one step-relative, and how that varies along with education, age, and ethnicity.

## Features

### What Do You Think?

Self-quiz chapter openers let students assess their existing knowledge of what will be discussed in the chapter. We have found these quizzes engage students, drawing them into the material and stimulating greater interaction with the course.

### Chapter Outlines

Each chapter contains an outline at the beginning of the chapter to allow students to organize their learning.

### Public Policies, Private Lives

These 12 boxed features focus on legal issues and public policies that affect how we think about and/or experience family life. Among them are new features on the lack of adequate language and policies regarding transgender identities, the Supreme Court decision in *Obergefell v. Hodges*, and the trend toward having friends or family conduct one's wedding, as well as updated features on sexting, the Family and Medical Leave Act, adoptions that dissolve, covenant marriage, and spanking.

## Exploring Diversity

These 11 boxes let students see family circumstances from the vantage point of other cultures, other eras, or within different lifestyles in the contemporary United States. New to this edition are boxes on cross-cultural research on kissing; race, class, and the maintenance of kin ties; and positive and negative trends in the status of LGBT population, both domestically and abroad. Among returning features, the box on divorce in India and Iran now also looks at the lack of divorce in the Philippines. Other retained features address arranged marriage, collectivist versus individualistic cultural constructions of love, dating violence cross-culturally, and the phenomenon of posthumous marriage.

## Issues and Insights

These 14 boxes once again focus on current and high-interest topics. They address such issues as virginity loss; gender, sexuality, and bullying; "living apart together"; and differences in obligations felt toward biological and stepfamily members. The two new Issues and Insights features focus on cross class marriage and discernment counseling for troubled couples. Two returning features on the uses and abuses of technology in families and relationships have been updated, as have the boxes on "red and blue" families, stepfather-stepchild relationships, and living apart together.

## Popular Culture

These 11 features discuss the ways family issues are portrayed through various forms of popular culture. Topics new to this edition include boxes on the possible effects and implications of certain television portrayals, including features on a "*Modern Family* effect" on attitudes about gay marriage, race and class as portrayed in *Blackish*, and whether and how teen pregnancy rates may be affected by such programs as *16 and Pregnant* and *Teen Mom*. Another new feature, "Transgender Faces," looks at popular media attention on Caitlyn Jenner, Jazz Jennings, Chaz Bono, and Laverne Cox, and their possible influence on attitudes toward trans individuals. There is also a new feature,

"Making Personal Trouble Public: Sharing One's Divorce Online and in Print," on some ways in which divorced individuals choose to share their story.

## Real Families

These 10 features give up-close, sometimes first-person, accounts of issues raised in the text as they are experienced by people in their everyday lives. In this edition, there are updated boxes on elective co-parenting by heterosexual and LGB parents, middle-class parenting, and heterosexual domestic partnerships. Returning features include those on blending and unblending families, family caregivers, and a feature on men and childbirth.

## End-of-Chapter Features

Each chapter also has a *Chapter Summary* and a list of *Key Terms,* all of which are designed to maximize students' learning outcomes. The chapter summary reviews the main ideas of the chapter, making review easier and more effective. The key terms are boldfaced within the chapter and listed at the end, along with a page number where the term was introduced. Both chapter summaries and key terms assist students in test preparation.

## Glossary

A comprehensive glossary of key terms is included at the back of the textbook.

# Instructor and Student Resources

*The Marriage and Family Experience,* 13th edition, is accompanied by a wide array of supplements prepared for both instructors and students. Some new resources have been created specifically to accompany the 13th edition, and all of the continuing supplements have been thoroughly revised and updated.

# Resources for Instructors

### Instructor's Resource Center

Available online, the Instructor's Resource Center includes an instructor's manual, a test bank, and PowerPoint slides. The instructor's manual will help instructors organize the course and captivate students' attention. The manual includes a chapter focus statement, key learning objectives, lecture outlines, in-class discussion questions, class activities, student handouts, extensive lists of reading and online resources, and suggested Internet sites and activities. The test bank includes multiple-choice, true/false, short answer, and essay questions, all with answers and text references, for each chapter of the text. The PowerPoints include chapter-specific presentations, including images, figures, and tables, to help build your lectures.

*Cengage Learning Testing Powered by Cognero*
Cognero is a flexible, online system that allows you to:

- Import, edit, and manipulate test bank content from *The Marriage and Family Experience* test bank or elsewhere, including your own favorite test questions.
- Create multiple test versions in an instant.
- Deliver tests from your LMS, your classroom, or wherever you want.

## Resources for Students and Instructors

### MindTap for *The Marriage and Family Experience*, 13th Edition

- MindTap engages and empowers you to produce their best work—consistently—by seamlessly integrating course material with videos, activities, apps, and much more, MindTap creates a unique learning path that fosters increased comprehension and efficiency.
- MindTap delivers real-world relevance with activities and assignments that help students build critical thinking and analytical skills that will transfer to other courses and their professional lives.
- MindTap helps students stay organized and efficient with a single destination that reflects what's important to the instructor, along with the tools students need to master the content.
- MindTap empowers and motivates students with information that shows where they stand at all times—both individually and compared with the highest performers in class.

  Additionally, for instructors, MindTap allows you to:

- Control what content students see and when they see it with a learning path that can be used as is or matched to your syllabus exactly.

- Create a unique learning path of relevant readings and multimedia and activities that move students up the learning taxonomy from basic knowledge and comprehensions to analysis, application, and critical thinking.
- Integrate your own content into the MindTap Reader using your own documents or pulling from sources like RSS feeds, YouTube videos, websites, Google Docs, and more.
- Use powerful analytics and reports that provide a snapshot of class progress, time in course, engagement, and completion.

## Acknowledgments

This book remains the product of many hands. Bryan Strong and, later, Christine DeVault, created a wonderful book from which to teach or study families and relationships. I hope that once again I have retained their emphasis on the meaning and importance of families, along with their effort to engage students' curiosity and interest. I am gratified to continue their efforts.

A number of people at Cengage Learning deserve thanks. Elizabeth Beiting-Lipps, sociology editor, showed considerable enthusiasm, consistent faith, and continued support for this book. I owe her much thanks and appreciation. My developmental editor, Trudy Brown, was truly outstanding. She provided encouragement, reminded me of deadlines (and helped me meet them), offered thoughtful suggestions and wise commentary as she read through the drafts of each chapter, and assisted in the selection of photos used throughout the text. This book has been made stronger, and the processes of writing and revising have been made easier and more gratifying because of her involvement.

I want to extend my thanks to Cheri Palmer, the senior production project manager at Cengage, who oversaw the complex production process with great skill. As always, with patience and flexibility, Jill Traut, project manager at MPS Limited, did an outstanding job on all phases of production. Heather McElwain was tremendously helpful and highly competent in the copyediting. The text looks and reads better because of their involvement. My appreciation also goes to Lumina Datamatics, for finding such good examples of what were occasionally vaguely requested subjects.

Once again, I wish to express deep appreciation to my colleagues and friends at Ohio Wesleyan University for the support they provided me. My

Ohio Wesleyan colleagues, Mary Howard, Jim Peoples, John Durst, Paul Dean, Alper Yalcinkaya and Pam Laucher make me very fortunate to have spent more than 30 years as a member of such a supportive department. They have been exceptional colleagues and remain always treasured friends. The many enthusiastic and curious students I have had in classes make me realize how very fortunate I have been to spend my academic career in Ohio Wesleyan classrooms. Their interest and curiosity about matters of families and relationships helps sustain my own.

I want again to express my appreciation to my family: my parents, Kalman and the late Eleanor Cohen, and my sisters, Laura Cohen and Lisa Merrill, who always formed an especially supportive group. Most importantly, they have been there for me through many life changes and challenges. I cannot adequately thank them.

Finally, my son Dan and daughter Allison will always be in the center of my heart. They have brought more joy to my life than I ever could have expected. As they move through their now adult lives, they continue to make me incredibly proud and remind me how immensely fortunate I am to be their dad and their friend. They are wonderful legacies to their beautiful mother, the late Susan Jablin Cohen, who, in sharing a quarter century of her life with me, shared too in the pride and joy of raising two such incredible people.

# The Meaning of Marriage and the Family

## What Do You Think?

*Are the following statements True or False? You may be surprised by the answers as you read this chapter.*

**T** **F** **1.** Now, same-gender couples may legally marry anywhere in the United States.

**T** **F** **2.** Though many allow polygamy, all cultures throughout the world prefer monogamy—the practice of having only one husband or wife.

**T** **F** **3.** Families are easy to define and count.

**T** **F** **4.** Being related by blood or through marriage is not always sufficient to be counted as a family member or kin.

**T** **F** **5.** Most families in the United States are traditional nuclear families in which the husband works and the wife stays at home caring for the children.

**T** **F** **6.** All cultures traditionally divide at least some work into male and female tasks.

**T** **F** **7.** The number of multigenerational households in the United States is increasing.

**T** **F** **8.** There is widespread agreement about the nature and causes of change in family patterns in the United States.

**T** **F** **9.** African Americans tend to express more conservative views on such family issues as premarital sex, divorce, and gay marriage.

**T** **F** **10.** Researchers agree that when parents divorce, children inevitably suffer long-term trauma.

## Chapter Outline

A course in marriage and the family is unlike almost any other course you are likely to take. At the start of the term—before you purchase any books, attend any lectures, and take any notes—you may believe you already know a lot about families. Indeed, each of us acquires much firsthand experience of family living before being formally instructed about what families are or what they do. These experiences and the relationships in which we have had them are likely among the most important experiences and closest relationships we have known. Whether with parents and siblings; past, present, or future partners and spouses; wider kin or even close nonkin who are "just like family," we are, in part, products of those relationships.

Furthermore, each of us comes to this subject with some pretty strong ideas and personal opinions about families: what they're like, how they should live, and what they need. Our personal beliefs and values shape what we think we know as much as our experiences in our families influence our thinking about what family life is or should be like. But if pressed, how should we describe family life in the United States? Are our families "healthy" and stable? Is marriage important for the well-being of adults and children? Are today's fathers and mothers sharing responsibility for raising their children? How many spouses cheat on each other? Are same-sex couples and heterosexual couples similar or different in how they structure and experience their lives together? What happens to children when parents divorce? Do stepfamilies differ from biological families? How common are abuse and violence in families? Questions such as these will be considered throughout this book. In looking them over, consider not only what you believe to be correct but also why you believe what you do. In other words, think about what we know about families and where our knowledge comes from.

In this chapter, we examine how individuals and society define marriage and family, paying particular attention to the existence of different viewpoints and assumptions about families and family life along with the discrepancies between the realities of family life as uncovered by social scientists and the impressions we may have formed elsewhere. We then look at the functions that marriages and families fulfill and examine extended families and kinship. We close by introducing the themes that will be pursued in the remaining chapters.

## ◾ Personal Experience, Social Controversy, and Wishful Thinking

As we begin to study family patterns and issues, we need to understand that our attitudes and beliefs about families may affect and distort our efforts. In contemplating the wider issues about families that are the substance of this book, it is likely that we will consider our own households and family experiences along with those of people closest to us. How we respond to the issues and information presented throughout the chapters that follow may be influenced by what we have experienced, seen firsthand, and come to believe about families.

### Experience versus Expertise

For some of us, family experiences have been largely loving ones, and our family relationships have remained stable. For others, family life has been characterized by conflict and bitterness, separations and reconfigurations. Most people experience at least some degree of both sides of family life, the love and the conflict, whether or not their families remain intact.

The temptation to draw conclusions about families from personal experiences of particular families is understandable. Thinking that experience translates into expertise, we may find ourselves tempted to generalize from what we experience to what we assume others must also encounter in family life. The dangers of doing that are clear; although the knowledge we have about our own families is vividly real, it is also both highly subjective and narrowly limited.

We "see" things, in part, as we *want* to see them. Likewise, we overlook some things because we don't want to accept them. Our own family members are likely to have different perceptions and attach different meanings to even those same experiences and relationships. Thus, the understanding we have of our families is very likely a somewhat distorted one.

Furthermore, no other family is exactly like one's own family. We don't all live in the same places, and we don't all possess the same financial resources, draw from the same cultural backgrounds, face the same circumstances and build on the same sets of experiences. These make our families somewhat unique. No matter how well we might think we know our own families, they are poor sources of more general knowledge about the wider marital or family issues that are the focus of this book.

## Dramatic Changes, Increasing Diversity, and Continuing Controversy

Learning about marriage and family relationships can be challenging for other reasons. Family life continues to undergo considerable social change. As we will begin to explore in more detail in Chapter 3, for a variety of reasons and in response to a number of influences, the contours and characteristics of U.S. families are in flux.

The rise in cohabitation, the increase in the never-married and formerly married populations, the prevalence of dual-earner couples and single-parent households, and the legalization of same-sex marriage, are some of the more notable examples of how families have changed in recent decades and where we continue to see quite dramatic change. Hence, talking about "marriage and family" as well as writing and reading about them can be difficult given the pace and extent of change. For example, when the previous edition of this textbook went to press, some nine states had legalized same-sex marriage. As these words were first being typed for this edition, same-sex couples could marry legally in 36 states. Then on June 26, 2015, the U.S. Supreme Court rendered a decision in the case of *Obergefell v. Hodges*, which legalized same-sex marriage throughout the United States.

Similarly, technology continues to contribute to changes in the ways we meet potential partners, interact with loved ones, bear and later monitor and raise

our children, and manage our home and work lives. Communications technology has enabled a level of access and interaction between romantic partners or spouses, parents and children, and other family members previously not possible. This raises new questions about such things as how much access we should expect and how frequent our communication should be.

Advances in reproductive science have enabled some individuals and couples who previously would have been infertile to bear children. Equally true, same-sex couples can, if they so choose, use surrogates and sperm or egg donors to have children who are biologically related to at least one of the partners. In the past year the United Kingdom legalized an in vitro fertilization technique that could help prevent children from being born with mitochondrial disease. The process uses the genetic material of three people (by mixing the mother's egg nucleus, with a donor's mitochondria, and then fertilizing the egg with sperm from the father). Reaction to news of such a procedure led some to fear that such "three-parent babies" could be a first step toward "designer babies" (Gallagher 2015).

In part as a by-product of changes such as these and in part as a reflection of the considerable cultural, ethnic, racial, economic, sexual, and religious diversity of the wider population, "the marriage and family experience" differs greatly, even within the United States. Commencing with Chapter 3 but extending throughout the remainder of the text, we strive to capture and convey some of the richly different ways family life is experienced and expressed. The reality of such diversity, however, makes it difficult to capture all the different ways things such as marriage, parenting, and divorce are experienced within a single population, and limits many generalizations, even if they illustrate how most people experience things.

Finally, few areas of social life are more controversial than family matters. Just consider the following recent examples of some family matters. What underlying issues can you identify? What is your position on such issues?

- The practice of polygamy, in which one has more than one spouse at a time, has been illegal in the United States since a U.S. Supreme Court decision in 1879, because it was considered a potential threat to public order (Tracy 2002). Despite this, over the past decade many Americans became more aware of the existence of polygamous families

**True** **1.** Now, same-gender couples may legally marry anywhere in the United States.

living openly in parts of the southwestern United States, especially among some fundamentalist Mormon groups. One of the most well-known examples is the Brown family, of TLC's television series, *Sister Wives*, consisting of Kody Brown, his four wives, Meri, Janelle, Christine, and Robyn, and their 17 children. The Browns successfully challenged part of the Utah law banning bigamy, and asked specifically that the prohibition against unmarried people living together and having sexual relations together be overturned. On August 27, 2014, U.S. District Court Judge, Clark Waddoups, issued a ruling that struck down part of Utah's antipolygamy law, contending that its provision prohibiting cohabitation violated the Browns' freedom of religion. The ruling made it legal for Utah residents to be legally married to one spouse but live with others they also consider to be their spouses (Whitehurst 2014a, 2014b). Yet polygamy remains illegal in Utah and the other 49 states in the United States. Thus, Kody Brown can be married legally to only one of his wives. In February 2015, he divorced Meri, his first wife, and married Robyn, his most recent wife, to provide her children with certain protections. The Browns, along with perhaps 30,000 to 40,000 other individuals living "polygamist lifestyles" in the United States exemplify what legal scholar Ashley Morin characterizes as an "illogical middle ground," in which polygamy laws are only selectively enforced and "even when polygamists openly display their lifestyle," law enforcement generally ignores the practice (Morin 2014). Although most polygamist families reside in Utah and other western states, there are also polygamous Muslim families living elsewhere in the United States, such as Philadelphia, Pennsylvania (Dobner 2011; Morin 2014; Whitehurst 2014a; Young 2010).

- When couples with children separate or divorce, decisions about child custody loom large. For same-gender couples with children, decisions to separate or divorce often take on additional complexity. So it was for a lesbian couple in Florida who had separated after more than a decade together. Years into their loving, committed relationship, they'd decided to have a child together. Because one of the women was infertile, her partner donated the egg that was fertilized with sperm from an anonymous donor and then implanted into the womb of the infertile partner. Their daughter was born in January 2004, given a hyphenated version of both women's last names, and came to consider both women as her parents. Unfortunately, after the couple split up, in keeping with Florida law, only the woman who gave birth to the girl was legally considered the mother, and, therefore, was awarded custody. However, on December 23, 2011, an appeals court overturned the initial ruling and ruled that both women had parental rights to the child. In its decision, the appellate court asked the Florida Supreme Court to consider and clarify the following issue, "Does a woman in a lesbian relationship who gives her egg to her partner have no legal right to the child it produces?" (Stutzman 2011). On November 12, 2013, the Florida Supreme Court ruled that, in fact, both women had parental rights to the child (Farrington 2013).

- Other legal complexities arise from advances in reproductive medicine. On June 12, 2015, a Chicago appeals court ruled that Dr. Karla Dunston could use embryos that she and her ex-boyfriend, Jacob Szafranski, created. Dr. Dunston was receiving cancer treatment when she and Mr. Szafranski reached an agreement for him to donate sperm to create embryos that could be used once her cancer treatment ended. Because they broke up while she was in treatment and before the embryos could be used, Mr. Szafranski was denying her permission to use them. After three court cases, the embryos were awarded to Dr. Dunston, though Mr. Szafranski is again appealing. According to *New York Times* journalist Tamar Lewin, throughout the United States, hundreds of thousands of embryos "in storage" are left over from in vitro fertilization (Lewin 2015).

- Decisions to get or stay married are assumed to be decisions based on falling in or out of love. Sometimes, though, as was the case for Bo and Dena McLain of Milford, Ohio, such decisions are also heavily influenced by much more practical and mundane motives, such as the need to attain or retain health insurance. The McLains married so that Dena could be added to Bo's health insurance plan and thus meet the requirement for insurance imposed by her nursing school. Likewise, many couples whose marriages have effectively ended may stay married to retain health insurance coverage and other benefits that they would lose if they divorced. Most such couples do separate and, though they may live apart, remain married, sometimes for years. Journalist Pamela Paul called them "the un-divorced" (Paul 2010), while Juliet

Bridges, writing in *The Telegraph* in the United Kingdom, called them "not quite married." Much like the McLains' decision to marry, the decision to remain less-than-happily married often partly reflects the privileges found in marriage. Health insurance, pensions, tax advantages, eligibility for Social Security benefits—all may be among the practical matters that sustain such marriages. In the words of couples therapist Toni Coleman, such couples ". . . enjoy the benefits of being married: the financial perks, the tax breaks, the health care coverage. . . . [T]hey just feel they can't live together" (Paul 2010; Sack 2008).

- During the 2014 National Football League season, the league was rocked by arrests of some of its star players for sexual and/or domestic violence. Baltimore Raven running back Ray Rice was suspended after video evidence surfaced revealing him assaulting his fiancé in a hotel elevator and dragging her unconscious body from the elevator. Minnesota Viking, Adrian Peterson, was indicted by a Texas grand jury on charges of reckless or negligent injury to a child after he used a tree branch to spank his four-year-old son, causing "cuts and bruises to the child's back, buttocks, ankles, legs, and scrotum, along with defensive wounds to the child's hands" (Boren 2014). These and other cases led to much public discussion and scrutiny of the National Football League's handling of acts of violence perpetrated by current and former players. As part of its response, the NFL suspended the players involved and formed a special committee of four women with expertise on issues related to sexual and domestic violence. The league ultimately reformed its personal conduct policy to reflect a strengthened stance against sexual assault and domestic violence. While the Rice case was met by fairly uniform condemnation of Ray Rice's behavior, the Peterson case triggered somewhat more divided discussions about corporal punishment, race, and parenting, even among those who agreed that Peterson had crossed the line in the discipline of his young son.

Stories such as these illustrate just some of the kinds of topics and issues raised throughout the remainder of this book. They also raise interesting questions that frequently lack clear answers. For example, how much should the state restrict people's marriage choices? How do policies that privilege married couples influence decisions to enter, exit, or remain in a marriage? How do wider economic conditions influence the internal dynamics of and decisions made by families? Has family law kept pace with advances in reproductive technology, and is it adequate to address diverse sexual lifestyles? At what point should the protection of children take precedence over the privacy of family life? As a society, we are often divided, sometimes strongly and bitterly, on many such family issues. That we are so deeply invested in certain values regarding family life makes a course about families a different kind of learning experience than if you were studying material to which you, yourself, were less connected or invested. Ideally, as a result, you will find yourself more engaged, even provoked, to think about and question things you take for granted. At minimum, you will be exposed to information that can help you more objectively understand the realities behind the more vocal debates.

At a hearing on charges of reckless or negligent injury to a child, Adrian Peterson of the National Football League's Minnesota Vikings consults his attorney, Rusty Hardin. Peterson's case was one of a number of high profile cases that led the league to form a special committee to deal with players charged with family violence and abuse.

David J. Phillip-Pool/Getty Images

## What Is Marriage? What Is Family?

To accurately understand marriage and family, it is important to define these terms. Before reading any further, think about what the words *marriage* and *family* mean to you. As simple and straightforward as this may seem, you may be surprised at the greater complexity involved as you attempt to define these words.

## Defining Marriage

Globally, there is much variation in the percentage of adults who are married and what marriage is like. Sociologists Laura Lippman and W. Bradford Wilcox, reporting on the prevalence of marriage across 43 different countries, state that adults 18 to 49 are most likely to be married in countries in Asia and the Middle East and least likely to be married in Central and South America. Countries in Africa, Europe, North America, and Oceania are said to fall in between. More than 60 percent of adults in South Korea and Malaysia, and more than 70 percent of adults in Indonesia and India are currently married. Among the Middle Eastern countries in their sample, the percentage of adults who are married ranges from 55 percent in Israel to over 60 percent in Turkey and Jordan, to a high of 80 percent in Egypt. At the other end of the spectrum, at 20 percent married, Colombia represents the worldwide low.

As shown in Figure 1.1, slightly over half (53.1 percent) of all adults in the United States, age 18 and older, are married (including those married and living apart). If one includes those currently separated but not divorced, the percentage reaches 55.3 percent. Among males, 54.9 percent are currently married, living with or apart from their spouse. Another 1.9 percent are separated but not divorced and, all told, 68.7 percent have at least experienced marriage (this is, are married, divorced, separated, or widowed). Although a smaller percentage of females is currently married (51.5 percent) or separated (another 2.5 percent), 74.9 percent of females 18 and older are or have at some time been married (U.S. Census Bureau 2014).

Family relationships are often the focus of popular movies. In 2014, *This is Where I Leave You*, featured and exposed the tensions resulting from the coming together of adult siblings and their widowed mother after the death of their father.

**Figure 1.1 Marital Status, U.S. Population 18 and Older**

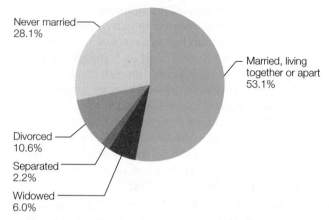

Never married 28.1%

Married, living together or apart 53.1%

Divorced 10.6%

Separated 2.2%

Widowed 6.0%

SOURCE: U.S. Census Bureau, Families and Living Arrangements in the United States, Table 1A.

What is it that these many men and women have at some point entered and experienced? As one goes about trying to define marriage, one might proceed in a number of different directions. Thinking mostly about marriage in the 21st-century United States, for example, might lead one to emphasize marriage as a deeply emotional, sexually intimate, and highly personal relationship between two people in love. Given the past two decades worth of effort expended on marriage equality for gay men and lesbians, one might be inclined to emphasize the legal recognition and more than a thousand rights and protections that accompany marriage in the United States. Still others might approach marriage as a religiously sanctioned relationship. Fans of television programs such as *Say Yes to the Dress* or *Bridezillas* might even associate marriage mostly with the ceremonial celebrations and rituals accompanying weddings. In some ways, all of these have merit, as they reflect the multiple dimensions of marriage.

Anthropologists James Peoples and Garrick Bailey point out that there is so much cultural diversity in how societies define marriage that it is difficult to arrive at a single comprehensive definition that includes all the meanings marriage conveys. Perhaps minimally, **marriage** is a socially and legally recognized union between two people, in which they are united sexually, cooperate economically, and may give birth to, adopt, or rear children. The union is assumed to be permanent, as in "till death do we part," though it may be and often is dissolved by separation or divorce. As simple as such a definition may make marriage seem, it differs among cultures and has changed considerably in our society.

With one exception, the Na of China, marriage has been a universal institution throughout recorded history (Peoples and Bailey 2014). Despite the universality of marriage, widely varying rules across time and cultures dictate whom one can, should, or must marry; how many spouses one may have at any given time; and where married couples can and should live—including whether husbands and wives are to live together or apart, whether resources are shared between spouses or remain the individual property of each, and whether or not children are seen as the responsibility of both partners (Coontz 2005). Among non-Western cultures, who may marry whom and at what age varies greatly from our society. In some areas of India, Africa, and Asia, for example, children as young as six years may marry other children (and sometimes adults), although they may not live together until they are older. In many cultures, marriages are arranged by families who choose their children's partners. In many such societies, the "choice" partner is a first cousin. And in one region of China as well as in certain parts of Africa (e.g., the Nuer of Sudan) and Europe (e.g., France), marriages are sometimes arranged in which one or both parties are deceased.

Considerable cultural variation exists in what societies identify as the essential characteristics that define couples as married. In many societies, marriage entails an elaborate ceremony, witnessed and legitimated by others, which then bestows a set of expectations, obligations, rights, and privileges on the newly married. Far from this relatively familiar construction of marriage, historian Stephanie Coontz notes that in some "small-scale societies," the act of eating together alone defines a couple as married. In such instances, as found among the Vanatinai of the South Pacific, for example, dining together alone has more social significance than sleeping together (Coontz 2005). Anthropological study of Sri Lanka revealed that when a woman cooked a meal for a man, this indicated that the two were married. Likewise, if a woman stopped cooking for a man, their marriage might be considered a thing of the past.

Sociologists Laura Lippman and W. Bradford Wilcox, authors of *The World Family Map 2014*, acknowledge that "across time and space, in most societies and cultures, marriage has been an important institution for structuring adult intimate relationships and connecting parents to one another and to any children that they have together" (Lippman and Wilcox 2014, 14)." Although cultural and historical variation abounds,

the following seem to be shared among all arrangements defined as marriages (Coontz 2005):

- Marriage typically establishes rights and obligations connected to gender, sexuality, relationships with kin and in-laws, and legitimacy of children.
- Marriage establishes specific roles within the wider community and society. It specifies the rights and duties of husbands and wives, as well as of their respective families, to each other and makes such duties and responsibilities enforceable by the wider society.
- Marriage allows the orderly transfer of wealth and property from one generation to the next.
- Additionally, as anthropologists James Peoples and Garrick Bailey (2015) note, marriage assigns the responsibility of caring for and socializing children to the spouses or their relatives.

Many Americans believe that marriage is divinely instituted; others assert that it is a civil institution involving only the state. The belief in the divine institution of marriage is common to religions such as Christianity, Judaism, and Islam, and to many tribal religions throughout the world. But the Christian church only slowly became involved in weddings; early Christianity was at best ambivalent about marriage, despite being opposed to divorce (Coontz 2005). Over time, as the church increased its power, it extended control over marriage. Traditionally, marriages had been arranged between families (the father "gave away" his daughter in exchange for goods or services); by the tenth century, marriages were valid only if they were performed by a priest. By the 13th century, the ceremony was required to take place in a church. As states competed with organized religion for power, governments began to regulate marriage. In the United States today, a marriage must be validated through government-issued marriage licenses to be legal, regardless of whether the ceremony is officiated by legal or religious officials.

## Who May Marry?

Who may marry has changed over the past 150 years in the United States. Laws once prohibited enslaved African Americans from marrying because they were regarded as property. Marriages between members of different races were illegal in more than half the states until 1967, when the U.S. Supreme Court declared, in *Loving v. Virginia*, that such prohibitions

**Ghost or Spirit Marriage**

Looking across cultures, many marriage customs may strike most Americans as unusual. Few, if any, can rival the custom of a marriage where one or both spouses are deceased. A number of versions of so-called ghost, spirit, or posthumous marriages are found among some African countries, in parts of rural China, and in France. In Sudan, among the Nuer, a dead groom can be replaced by a male relative (e.g., a brother) who takes his place at the wedding. Despite being deceased, he—not the living substitute—is considered the husband. Any children born subsequently will be considered children of the deceased man who is recognized socially as the father. In this way, a man who died before leaving an heir can have his family line continue. Among the Iraqw of Tanzania, the "ghost" groom could be the imagined son of a woman who never had a son.

In some parts of rural China, parents of a son who died before marrying may "procure the body of a (dead) woman, hold a 'wedding,' and then bury the couple together," in keeping with the Chinese tradition of deceased spouses sharing a grave (*Economist* 2007). As reported in the *New York Times*, the custom of "minghun" (afterlife marriage) follows from the Chinese practice of ancestor worship, which holds that people continue to exist after death and that the living are obligated to tend to their wants—or risk the consequences. Traditional Chinese beliefs also hold that an unmarried life is incomplete, which is why some parents worry that an unmarried dead son may be an unhappy one (Yardley 2006).

Parents whose daughters had died might sell their daughter's body for economic reasons but also are motivated by the desire to give such daughters a place in Chinese society. As stated by sociologist Guo Yuhua, "China is a paternal clan culture. . . . A woman does not belong to her parents. She must marry and have children

of her own before she has a place among her husband's lineage. A woman who dies unmarried has no place in this world" (Yardley 2006).

In France, in 1959, Parliament drafted a law that legalized "postmortem matrimony" under certain circumstances. These included proof of the couple's intention to marry before one of the partners died and permission from the deceased's family. After a request is submitted to the president, it is passed to a justice minister and ultimately to the prosecutor who has jurisdiction over the locality in which the marriage is to occur. It is then the prosecutor's responsibility to determine whether the conditions have been met and the marriage is to be approved. In June 2011, 22-year-old Frenchwoman Karen Jumeaux sought and received permission from President Nicolas Sarkozy to marry Anthony Maillot, her deceased fiancé and father of her 2-year-old son. Maillot had been killed in an accident, two years earlier at age 20. In a 2009 article in *The Guardian*, it is reported that French government figures estimate that "dozens" of such marriages occur each year (Davies 2009). Such posthumous marriages are largely for sentimental reasons. In fact, French law prevents spouses from any inheritance. Nonetheless, the marriages are retroactive to the eve of the groom's demise. They allow the woman to "carry her husband's name and identify herself as a widow" on official documents. If the woman is pregnant at the time of the man's death, the children are considered legitimate heirs to his estate (Smith 2004). As Jumeaux reported after the posthumous wedding ceremony, "He was my first and only love and we were together for four years. We expected to bring up our son together. I never wanted to do it alone, but fate decided otherwise. Now I am his wife and will always love him" (*Daily Mail Reporter* 2011). ●

were unconstitutional. Each state enacts its own laws regulating marriage, leading to some discrepancies from state to state. For example, in some states, first cousins may marry; other states prohibit such marriages as incestuous.

Of course, the greatest controversy regarding legal marriage over the past two decades has been over the question of same-sex marriage.

During the revision of this text, on June 26, 2015, the United States Supreme Court ruled that based on the 14th Amendment of the United States Constitution, all 50 U.S. states were required to recognize and

license marriages between same-sex couples and to recognize all marriages that were lawfully performed out of state. This decision allowed same-sex couples to legally marry, with all the rights, benefits, and privileges marriage entails. We more fully explore legal aspects of marriage (such as the age at which one can marry, whom one may marry, and so on) in Chapter 8. For now, though, it should be noted that legal marriage bestows literally hundreds of rights, privileges, and protections on couples who marry. Cohabiting couples, whether heterosexual or same-gender couples, do not automatically acquire those same benefits.

"No union is more profound than marriage, for it embodies the highest ideals of love, fidelity, devotion, sacrifice, and family. In forming a marital union, two people become something greater than once they were . . . marriage embodies a love that may endure even past death." (Schwartz 2015)

Those are the eloquent words of Supreme Court Justice Anthony Kennedy, writing for the majority in the historic decision that legalized same-sex marriage. The case, known as *Obergefell v. Hodges*, also included suits brought by couples in Michigan, Kentucky, and Tennessee against their state's gay marriage bans. But it will forever be known as *Obergefell*, and it will reflect the very personal and poignant struggle James Obergefell and John Arthur faced.

Heterosexuals rarely stop to think about the privileges that their sexual orientation confers. One such privilege had long been the right to marry. Those couples who do marry receive many more rights and protections than couples who don't marry. For heterosexuals, marriage versus cohabitation is a matter of choice. Heterosexual couples who choose cohabitation may do so because they prefer the more informal arrangement. They, too, will lack the protections and privileges that accompany marriage, but they elect to cohabit anyway. For many same-sex couples, the historical *inability* to marry has cost them many protections, including the following examples:

- The right to enter a premarital agreement
- Income tax deductions, credits, rates, exemptions, and estimates
- Legal status with one's partner's children
- Partner medical decisions
- Right to inherit property
- The right to a divorce
- Award of child custody in divorce proceedings
- Payment of worker's compensation benefits after death of spouse
- Right to support from spouse

There are also potential personal and emotional benefits related to the right to marry. Knowing that the wider society recognizes, accepts, or respects a relationship may cause feelings of greater self-validation and comfort within the relationship. On the other hand, knowing that people do not respect, accept, or recognize a commitment may cause additional emotional suffering and personal anguish for the partners involved. So it was for James Obergefell and John Arthur of Cincinnati, Ohio.

James Obergefell and his partner, John Arthur, were together more than 20 years. When John became ill with amyotrophic lateral sclerosis, an incurable progressive neurological disease, Jim reacted much as loving spouses

or partners do; he stayed by John's side, even as John's mobility and speech grew weaker. When the U.S. Supreme Court ruled the Defense of Marriage Act to be unconstitutional in 2013, the couple wanted to celebrate by getting married. As residents of Ohio, where same-sex marriage was not legally recognized, they ultimately decided to go out of state to marry. Given John's poor health, travel would not be easy and would require a kind of medical transport that was not inexpensive. They decided to fly to Maryland to get married. They were married in a seven-and-a-half-minute ceremony on the plane by John Arthur's aunt, Paulette Roberts, who had been ordained with the hope of someday marrying Jim and John (Zimmerman 2014). Now legally married in Baltimore, Maryland, they returned home to Cincinnati, Ohio, where their marriage would not be recognized. Not even on John Arthur's death certificate would there be any indication that he and Obergefell were wed. Under Ohio law, Jim Obergefell would not be listed as John Arthur's surviving spouse (Lerner 2015). This motivated Obergefell to bring suit against the state of Ohio, in the case *Obergefell v. Hodges* (Richard Hodges was the director of the Ohio Department of Health).

Obergefell said, "They were going to say, 'No, you don't exist.' It ripped our hearts out. So we filed suit against the state of Ohio" (Ziv 2015).

In authoring the majority opinion, Kennedy spoke about the meaning of marriage:

Marriage responds to the universal fear that a lonely person might call out only to find no one there. It offers the hope of companionship and understanding and assurance that while both still live there will be someone to care for the other. ●

Holding a photo of himself and his late husband, John Arthur, James Obergefell filed suit against the Ohio Attorney General to have his name listed on his spouse's death certificate. The case became the U.S. Supreme Court case, *Obergefell v. Hodges*, which resulted in the Court's decision in June 2015, to legalize same-sex marriage throughout the United States.

The Washington Post/Getty Images

## Forms of Marriage

In Western cultures such as the United States, the only legal form of marriage is **monogamy**, the practice of having only one spouse at one time. Thus, the fundamentalist Mormon polygamists depicted earlier as well as Muslim polygamists living in the United States are in violation of state marriage laws.

Monogamy is also the only form of marriage recognized in *all* cultures. Interestingly, and possibly surprisingly, it is not always the *preferred* form of marriage. The most commonly preferred marital arrangement in many Middle Eastern societies as well as some indigenous African, Southeast Asian, and Melanesian populations is **polygamy**, specifically polygyny—the practice of having two or more wives. One study of 850 non-Western societies found that 84 percent of the cultures studied (representing, nevertheless, a minority of the world's population) practiced or accepted **polygyny**, the practice of having two or more wives. In fact, in more than a quarter of these societies, more than 40 percent of the marriages were polygamous (Ember, Ember, and Low 2007). James Peoples and Garrick Bailey report that prior to colonialism's impact, 70 percent of all societies allowed men to marry more than one wife. Where polygyny is allowed, it tends to be the preferred or most highly valued form of marriage (Peoples and Bailey 2014). Conversely, **polyandry**, the practice of having two or more husbands, is actually quite rare: Where it does occur, it often coexists with poverty, a scarcity of land or property, and an imbalanced ratio of men to women.

Even within polygynous societies, monogamy is the *most widely practiced* form of marriage (Ember, Ember, and Low 2007). In such societies, plural marriages are in the minority, primarily for simple economic reasons: They are a sign of status that relatively few people can afford, and they require wealth that few men possess. As we think about polygyny, we may imagine high levels of jealousy and conflict among wives. Indeed, problems of jealousy may and do arise in plural marriages—the Fula in Africa, for example, call the second wife "the jealous one." Based on data from 69 polygynous societies (56 percent of which were in Africa), Jankowiak, Sudakov, and Wilreker (2005) suggest that co-wife conflict and competition for access to the husband is common, especially in situations where wives are materially dependent on the husband, but there are also circumstances that can reduce conflict (e.g., when the wives are sisters, or when one is fertile and one barren or postmenopausal). Even though conflict and competition among co-wives is often found in polygynous societies, the level is probably a good deal less than would result if our monogamous society was to suddenly allow people multiple spouses. For both the men and the women involved, polygyny brings higher status.

To many in the United States, the idea of being married to more than one spouse or sharing one's spouse with co-wives or co-husbands may seem strange or exotic. However, it may not seem so strange if we look at actual marital practices in the United States. Considering the high divorce and remarriage rates in this country, monogamy may no longer be the best way of describing our marriage forms. In fact, among 40 percent of couples married in 2013, one or both spouses had been married before (Livingston 2014a). Thus, for many, our marriage system might more accurately be called **serial monogamy** (or even *modified polygamy*), a practice in which one person may have several spouses over his or her lifetime despite being wed to no more than one at any given time.

**False** 2. Although monogamy is the only recognized form of marriage in all cultures, it is not always the preferred form.

## Defining Family

As contemporary Americans, we live in a society composed of many kinds of families—married couples, stepfamilies, single-parent families, multigenerational families, cohabiting adults, child-free families, families headed by gay men or by lesbians, and so on. With such variety, how should we define family? What are the criteria for identifying these groups as families?

In its efforts to count and characterize families in the United States, the U.S. Census Bureau defines a **family** as "a group of two people or more (one of whom is the householder) related by birth, marriage, or adoption and residing together; all such people (including related subfamily members) are considered as members of one family" (U.S. Census Bureau 2015a). A distinction is made between a family and a **household**. A household consists of "all the people who occupy a housing unit," whether or not related

**False** 3. Families are not easy to define and count.

(U.S. Census Bureau 2015a). Single people who live alone, roommates, lodgers, and live-in domestic service employees are all counted among members of households, as are family groups. *Family households* are those in which at least two members are related by birth, marriage, or adoption, though unrelated individuals who reside in a household along with the householder and his or her family are counted as family household members (U.S. Census Bureau 2015a). Table 1.1 contains Census data on numbers and kinds of family and non family households. Thus, the U.S. Census reports on characteristics of the nation's households *and* families (see Figure 1.2). Of the 123,229,000 households in the United States in 2014, 81,353,000 or 66 percent were family households. Among family households, 73.3 percent (59,204,000) consisted of married couples, either with or without children. Married couples made up less than half (48.4 percent) of all households in the United States in 2014, and married couples with children under 18 years of age represented 29.4 percent of family households and just under one-fifth (19.4 percent) of all U.S. households (U.S. Census Bureau 2015a).

In individuals' perceptions of their own life experiences, family has a less precise and more varying definition. For example, when we have asked our students whom they include as family members, their lists have

**True** **4.** Being related by blood or through marriage is not always sufficient to be counted as family member or kin.

included such expected relatives as mother, father, sibling, and spouse. Most of those designated as family members are individuals related by descent, marriage, remarriage, or adoption, but some are **affiliated kin or fictive kin**—unrelated individuals who feel and are treated as if they were relatives, such as the following:

best friend
boyfriend
girlfriend
godchild
lover
minister
neighbor
pet
priest
rabbi
teacher

### Figure 1.2 Household Composition, 2010

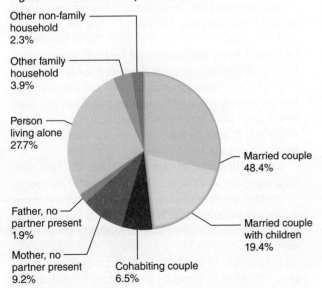

Other non-family household 2.3%
Other family household 3.9%
Person living alone 27.7%
Father, no partner present 1.9%
Mother, no partner present 9.2%
Cohabiting couple 6.5%
Married couple with children 19.4%
Married couple 48.4%

SOURCE: U.S.Census, 2012 Statistical Abstract, Table 59, Households, Families, Subfamilies, and Married Couples

### TABLE 1.1  U.S. Households, 2014

| | Households (in thousands) | Percent of Households |
|---|---|---|
| **Total Households** | **123,229** | **100%** |
| Family households | 81,353 | 66.0% |
| Married couples | 59,629 | 48.4% |
| With children under 18 | 23,933 | 19.4% |
| Cohabiting couples | 8,046 | 6.5% |
| With children under 18 | 2,961 | 2.4% |
| Mother, no partner present | 11,365 | 9.2% |
| With children under 18 | 7,041 | 5.7% |
| Father, no partner present | 2,367 | 1.9% |
| With children under 18 | 1,245 | 1.0% |
| Other family households | 4,769 | 3.9% |
| Female, living alone | 19,034 | 15.4% |
| Male, living alone | 15,151 | 12.3% |
| Other, nonfamily households | 2,869 | 2.3% |

SOURCE: U.S. Census Bureau, Families and Living Arrangements in the United States, Table 1A.

The Meaning of Marriage and the Family     **11**

Furthermore, being related by blood or through marriage is not always sufficient to be counted as a family member or kin. Some individuals consider stepparents or stepsiblings to be family members and even extend that status to *former* stepparents or stepsiblings. Others may draw different distinctions and differentiate between steprelationships and their "real families," thus not counting stepfamily members as part of their families. For individuals, emotional closeness may be more important than biology or law in defining family.

There are also some ethnic differences as to whom people consider to be family. Among Latinos, for example, *compadres* (or godparents) are considered family members. Similarly, among some Japanese Americans, the *ie* (pronounced "ee-eh") is the traditional family. The *ie* consists of living members of the extended family (such as grandparents, aunts, uncles, and cousins) as well as deceased and yet-to-be-born family members (Kikumura and Kitano 1988). Among many traditional Native American tribes, the **clan**, a group of related families, is regarded as the fundamental family unit (Yellowbird and Snipp 1994). Among many African Americans, **fictive kin** are considered to be like family and are treated and expected to act as such (Taylor et al. 2013).

> **False**
> **5.** Most families in the United States are not traditional nuclear families.

To reflect the diversity of family types that coexist within the wider society, the definition of family needs to be expanded beyond the boundaries of the "official" census definition. A more contemporary and inclusive definition describes family as "two or more persons related by birth, marriage, adoption, *or choice* [emphasis added]. Families are further defined by socioemotional ties and enduring responsibilities, particularly in terms of one or more members' dependence on others for support and nurturance" (Allen, Demo, and Fine 2000). Such a definition more accurately and completely reflects the diversity of contemporary American family experience.

## What Families Do: Functions of Marriages and Families

Whether it is a mother/father/child nuclear family, a married couple with no children, a single-parent family, a stepfamily, a dual-worker family, or a cohabiting family, the family generally performs important societal functions and meets certain needs of individuals. Sociologists have identified the following four functions as ways families contribute to social stability as well as to societal and individual well-being: (1) Families provide a source of intimate relationships; (2) families act as units of economic cooperation and consumption; (3) families may produce and socialize children; and (4) families assign social statuses and roles to individuals. Although these are the basic functions that families are "supposed" to fulfill, families do not have to fulfill them all (as in families without children), nor do they always fulfill them well (as in abusive families).

### Intimate Relationships and Family Ties

Intimacy is a primary human need. Chapter 5 will explore in more detail how intimacy is expressed and experienced in friendship and love relationships, and how such experience varies, especially by gender. In addition, research consistently indicates that human companionship strongly influences rates of illnesses, such as cancer or tuberculosis, as well as suicide, accidents, and mental illness. Thus, it is no surprise that studies consistently show married couples and adults living with others to be generally healthier and have lower mortality rates than divorced, separated, and never-married individuals. Although some of this difference results from what is known as the *selection* factor—wherein healthier people are more likely to marry or live with someone—both marriage and cohabitation yield benefits to health and well-being.

Marriage and the family usually furnish emotional security and support. In our families, we generally seek and find our strongest bonds. These bonds can be forged from and sustained by love, attachment, loyalty, obligation, or guilt. The need for intimate relationships, whether or not they are satisfactory, may hold unhappy marriages together indefinitely. Loneliness may be a terrible specter. Among the newly divorced, it may be one of the worst aspects of the marital breakup.

As we will detail in Chapter 3, since the 19th century, marriage and the family have become ever more important sources of companionship and intimacy. As society became more industrialized, bureaucratic, and impersonal, individuals increasingly sought and expected to find intimacy and companionship within their families. In the larger world around us, we are generally seen in terms of some more formal status. A professor may see us primarily as students, a used car salesperson relates to us as potential buyers, and

A major function of marriage and family is to provide us with intimacy and social support, thus protecting us from loneliness and isolation.

Corbis

a politician views us as voters. Only among our intimates are we seen on a personal level, as Maria or Matthew. Before marriage, our friends are our intimates. Upon marrying or cohabiting, our spouse or partner is expected to be the one with whom we are *most* intimate. Especially with our spouse, we are thought to be able to disclose ourselves most completely, share our hopes, express our fears, rear our children, and hope to grow old together.

## Economic Cooperation

The family is a unit of economic cooperation and interdependence. Traditionally, heterosexual families divide responsibilities along gender lines—that is, between males and females, thus fostering interdependence. Although a division of labor by gender is characteristic of virtually all cultures, the work that males and females perform varies from culture to culture (see Chapter 4). Among the Nambikwara in Africa, for example, the fathers take care of the babies and clean them when they soil themselves; the chief's concubines, secondary wives in polygamous societies, prefer hunting over domestic activities. In U.S. society, from the late 19th century through much of the 20th, men were expected to work away from home, whereas women were to remain at home caring for the children and house. Such gendered tasks are assigned by culture, not biology. Only a man's ability to impregnate and a woman's ability to give birth and produce milk are biologically determined.

We commonly think of the family as a consuming unit, but it also continues to be an important producing unit. Family members often are sources of

**True**  **6.** All cultures traditionally divide at least some work into male and female tasks.

goods and services that if produced or provided by outsiders would require expenditure of considerable sums of money. Within families, one is not paid for building a shelf or bathing the children, fixing a leaky faucet or cooking.

Although children contribute to the household economy by helping around the house, they generally are not paid (beyond an "allowance") for such things as cooking, cleaning their rooms, or watching their younger brothers or sisters. Yet they are all engaged in productive, sometimes essential, labor (Dodson and Dickert 2004).

More often women perform the family's role as a service unit. Because work at home is unpaid, the productive contributions of at-home spouses, typically at-home wives, have been overlooked. Furthermore, when employed, women find that greater needs for their "services" await them when they return from their paid jobs. Because family power is partly a function of who earns the money, stay-at-home partners who get no paycheck may have less power because their financial contribution to the family is less visible, whereas their economic dependence is more easily recognized.

## Reproduction and Socialization

The family makes society possible by producing (or adopting) and rearing children to replace the older members of society as they die off. Traditionally, reproduction has been a unique function of the married family. But single-parent and cohabiting families also perform reproductive and socialization functions. Technological advances in assisted reproductive techniques such as artificial insemination and in vitro fertilization also have separated reproduction from sexual intercourse and now allow for the participation of others (e.g., sperm or egg donors, surrogate mothers, and so on) in the reproductive process.

Innovations in reproductive technology also permit many otherwise infertile couples to give birth. Such techniques have also made it possible for lesbian couples and single women without partners to become parents.

The family traditionally has been responsible for **socialization**—the shaping of individual behavior to conform to social or cultural norms. Children are helpless and dependent for years following birth. They must learn how to walk and talk, how to take care of themselves, how to act, how to love,

In talking with sociologist David Karp, 37-year-old Angie reflected on her relationship with her brother who has mental illness. Her words convey her effort to determine where her obligations to care for her family members begin and end:

> It's kind of hard to put . . . into words. I mean I love my parents dearly and I love my brother, and I think you're raised to know what's right and wrong. And I do feel like your family comes first. But by the same token, how much . . . is realistic for a sibling to give up? Are you supposed to give up your life . . . your career . . . your hopes? . . . Just where do you draw the line? Do you do what's right for your family and just do it unselfishly? It's a hard thing. It's easy to say, "Yeah, I'd do anything for my family" until you really have to, until you are faced with it (Karp 2001, 130).

Karp interviewed 60 people with family members who were suffering from diagnosed mental illness. He spoke with parents dealing with a child's mental illness as well as "children of emotionally sick parents, spouses with a mentally ill partner, and siblings of those suffering from depression, manic depression, or schizophrenia" (p. 14). His 60 interviewees each presented a story that is somewhat distinctive. Yet his sociological approach sought to detail "the consistencies and uniformities" that surfaced (p. 24). He raises the following provocative questions—"What do we owe each other?" "What are the moral boundaries of family relationships?" and "To what extent are we bound to care for each other?" (p. 30)—and speaks of "the extraordinary power of love" displayed by his interviewees:

> Even when an ill person treated them with anger and disdain, denied that they were sick, completely disrupted the coherence of everyday life, and did things that were incomprehensible, distressing beyond measure, socially repugnant, or downright dangerous, love kept caregivers caring (p. 16).

Sandra Dorne, 57, would most likely affirm Karp's assertion through her own display of "the extraordinary power of love" for her physically ill sister. Ms. Dorne left behind her life in Orlando, Florida, bought a one-way plane ticket to Allentown, Pennsylvania, and then boarded a bus to Brooklyn, New York, to care for her 69-year-old sister, Patricia Trivisani. Ms. Trivisani suffers from a host of physical ailments, including hypertension, fibromyalgia, osteoporosis, and congestive heart failure. She had been taking 50 pills a day and was hospitalized after having been found on the floor by a neighbor after accidentally overdosing. Now, Ms. Dorne is seeing to her sister's well-being. "Me and Pat's always been close," she stated to reporter Anastasia Economides. "She took care of me because I was the baby sister." With plans to never leave her sister, she claims, "If I caught anyone hurting her, I'd beat them senseless" (Economides 2011).

David Karp found a hierarchy of caring, in which obligations to siblings fall behind those to our spouse, children, or parents. However, as Angie and Sandra illustrate, siblings can and sometimes do step in when their brothers or sisters are in need. Karp suggests that families have been "abandoned" by U.S. society, left on their own without social supports to solve any problems individual members may face. Still, his interviews with caregiving spouses, parents, children, and siblings reveal an "extraordinary reservoir of love, caring, and connection that holds families together, even at a time when family life is so meagerly supported" (Karp 2001, 26). ●

**Although families remain responsible for much of early childhood socialization, preschools and day care centers also often play a large role.**

Robert Kneschke/Shutterstock.com

and how to touch and be touched. Teaching children how to fit into their particular culture is one of the family's most important tasks.

This socialization function, however, is shared by agents and caregivers outside the family. The involvement of nonfamily in the socialization of children need not indicate a lack of parental commitment to their children or a lack of concern for the quality of care received by their children. The most common family among those raising children is a dual-earner couple. Single parents—both mothers and fathers—are likely employed. Thus, many infants, toddlers, and small children are under the care of nonfamily members, thus broadening the caregiving and socialization roles of others, such as neighbors, friends, or paid caregivers. Additionally, since the rise of compulsory education in the 19th century, the state has become responsible for a large part of the socialization of children older than age 5. In Chapter 10, we will address parenting and the care and socialization of children.

## Assignment of Social Statuses and Roles

We occupy various social statuses or positions as family members and we play multiple roles. These statuses and roles provide us with much of our identities. During our lifetimes, most of us will belong to at least two families: the family of orientation and the family of procreation. The **family of orientation** (sometimes called the *family of origin*) is the family in which we grow up, the family that orients us to the world. The family of orientation may change over time if the marital status of our parents changes. Originally, it may be an intact **nuclear family** or a single-parent family; later, it may become a stepfamily. We can even speak of **binuclear families** to reflect the experience of children whose parents separate and divorce. With parents maintaining two separate households and one or both possibly remarrying, children of divorce are members in two different, parentally based nuclear families (Ahrons 1994, 2004).

The common term for the family we form through marriage and childbearing traditionally has been the **family of procreation** (Parkin 1997). Given some of the major changes in family patterns, "the traditional definition of the family must be expanded beyond marriage" (Wu 2003, 173). Because so many families have stepchildren, adopted children, or no children, *and* given the dramatic increases in couples living together outside of marriage, we might use some other term, such as *family of cohabitation,* to refer to the family we form through living or cohabiting with another person, whether we are married or unmarried. Most Americans will form such families sometime in their lives.

Much of our identity is formed in the crucibles of families of orientation, procreation, and cohabitation. In a family of orientation, we are given the roles of son or daughter, brother or sister, stepson or stepdaughter. We internalize these roles until they become a part of our being. In each of these roles, we are expected to act in certain ways. For example, children are expected to obey their parents, and siblings are expected to help one another. Sometimes our feelings fit the expectations of our roles; other times they do not. We may not wish to follow our parents' suggestions or loan money to an unemployed sister and yet feel compelled to do so because of the role expectations we face.

Our family roles as offspring and siblings are most important when we are living in a family of orientation. After we leave home, these roles may gradually diminish in everyday significance, although they continue throughout our lives. In relation to our parents, we never cease being children; in relation to our siblings, we never cease being brothers and sisters. The roles simply change as we grow older.

As we leave a family of orientation, we usually are also leaving adolescence and entering adulthood. Adulthood is defined in part by entering new family roles—those of spouse, partner, or parent. These roles typically take priority over the roles we had in a family of orientation. In our nuclear family system, when we marry, we transfer our primary loyalties from our parents and siblings to our partners. Later, if we have children, we form additional bonds with them. When we assume the role of spouse or bonded partner, we assume an entirely new social identity linked with responsibility, work, and parenting. In earlier times, such roles were considered lifelong. Because of divorce or separation, however, these roles today may last for considerably less time.

Our families influence the status or place we are given in society. Our families place us in a certain socioeconomic class, such as blue collar (working class), middle class, or upper class. We learn the ways of our class through identifying with our families. As shown in Chapter 3, people in different classes experience the world differently. These differences include the ability to satisfy our needs and wants but may extend to how we see men's and women's roles, how we value education, and how we bear and rear our children (Edin and Kefalas 2005; Lareau 2003). Our families also give us our ethnic identities as African American, Latino, Asian American, Italian American, and so forth. Families also commonly provide us with a religious tradition as Protestant, Catholic, Jewish, Greek Orthodox, Islamic, Hindu, or Buddhist—as well as agnostic, atheist, or New Age. These identities help form our cultural values and expectations. These values and expectations may then influence the kinds of choices we make as partners, spouses, or parents.

## Why Live in Families?

As we look at the different functions of the family, we can see that most of them can be fulfilled outside the family. For example, artificial insemination permits a woman to be impregnated by a sperm donor, and embryonic transplants allow one woman to carry another's embryo. Children can be raised communally, cared for by foster families or child care workers, or sent to boarding schools. Most of our domestic needs can be satisfied by microwaving prepared foods or going to restaurants, sending our clothes to the laundry, and hiring help to clean our bathrooms, cook our meals, and wash the mountains of dishes accumulating (or growing new life forms) in the kitchen. Friends can provide us with emotional intimacy, therapists can listen to our problems, and sexual partners can be found outside marriage. Communications technology allows us to meet and friend many people in or well outside the geographic areas in which we live. These individuals can become important partners in day-to-day interaction. With the limitations and stresses of family life, why bother living in families at all?

Sociologist William Goode (1982) suggests several advantages to living in families:

- *Families offer continuity as a result of emotional attachments, rights, and obligations.* Once we choose a partner or have children, we do not have to search continually for new partners or family members who can better perform a family task or function, provide companionship, or bring home a paycheck. We expect our family members—whether partner, child, parent, or sibling—to participate in family tasks over their lifetimes. If at one time we need to give more emotional support or attention to a partner or child than we receive, we expect that the other person will reciprocate at another time. We count on our family members to be there for us in multiple ways. We rarely have the same extensive expectations of friends.

- *Families offer close proximity.* Although communication technologies and social media have reduced the importance of shared space, the fact that family members are physically close by is still of importance. We do not need to travel across town or the country for conversation or help. With families, we do not even need to leave the house; a husband or wife, parent or child, or brother or sister is often at hand (or underfoot). This close proximity facilitates a level of cooperation and communication from which individuals may draw great comfort.

- *Families offer intimate awareness of others.* Few people know us as well as our family members because they have seen us in the most intimate circumstances throughout much of our lives. They have

seen us at our best and our worst, when we are kind or selfish, and when we show understanding or intolerance. This familiarity and close contact teach us to make adjustments in living with others. As we do so, we expand our knowledge of ourselves and others.

- *Families provide many economic benefits.* They offer us economies of scale. Various activities, such as laundry, cooking, shopping, and cleaning, can be done almost as easily and with less expense for several people as for one. As an economic unit, a family can cooperate to achieve what an individual could not. It is easier for a working couple to purchase a house than an individual, for example, because the couple can pool their resources. These are only some of the theoretical advantages families offer to their members. Not all families perform all these tasks or perform them equally well. But families, based on mutual ties of feeling and obligation, offer us greater potential for fulfilling our needs than do organizations based on profit (such as corporations) or compulsion (such as governments).

## Extended Families and Kinship

The **extended family** consists not only of a cohabiting or married couple and their children but also of other relatives, especially in-laws, grandparents, aunts and uncles, and cousins. In most non-European countries, the extended family is often regarded as the basic family unit. Extended families are more common in countries in Asia, the Middle East, Central and South America, and sub-Saharan Africa. According to sociologists Laura Lippman and W. Bradley Wilcox, in most countries in these regions, at least 40 percent of children live in households containing adults other than their parents (Lippman and Wilcox 2014). In certain countries in Africa (such as South Africa, Tanzania, and Nigeria), Asia (e.g., India), South America (e.g., Colombia and Nicaragua), and the Middle East (e.g., Turkey), more than half of children live with and are likely affected by relations with extended kin such as grandparents, aunts and uncles, and cousins (Lippman and Wilcox 2014).

For many Americans, especially those with strong ethnic identification and those in certain groups (discussed further in Chapter 3), the extended family also often takes on great importance (Sarkisian and

Gerstel 2012). Census data estimate that there were over 3.7 million multigenerational family households in the United States, consisting of three or more generations residing together in 2012. This amounted to 4.6 percent of all U.S. family households, though the prevalence of such extended families varied among different groups. Among Hispanics and Blacks, such multigenerational families represent over 8 percent of their family households, followed by Asians at 6.3 percent and non-Hispanic whites at 3.0 percent (Vespa, Lewis, and Kreider 2013). Data from the U.S. Census Bureau's 2009–2011 American Community Survey reveals that multigenerational households were reported most commonly among Native Hawaiian and Pacific Islander families, followed by American Indian and Alaskan Native families (Lofquist 2012).

The Census Bureau's Current Population Survey indicates that more than four million U.S. households included both grandchildren and grandparents in 2012. An estimated 10 percent of all children (about seven million children) in the United States lived in households that contained at least one grandparent. Of those children who lived with a grandparent, 20 percent had no parent present in the households (Ellis and Simmons 2014).

Sometimes, especially among those who don't have firsthand experience living in such households, extended kin may be left out of how they describe and/or define family, because of an assumption that when we speak of *family,* we really mean the nuclear family. That is the model that has come to epitomize what we mean by family. Thus, when someone asks us to name our family members, if we are unmarried, many will probably name our parents, brothers, and sisters. If we are married, we will probably name our spouses and, if we have any, our children. If we cohabit, we are likely to include our partners. Only if questioned further will some bother to include grandparents, aunts or uncles, cousins. Others may go on to include friends or neighbors who are "like family." We may not name all our blood relatives, but we will probably name the ones with whom we feel emotionally close, as shown earlier in the chapter.

Even among those who don't experience living in multigenerational *households,* many Americans maintain what have been called **modified extended families,** which are extended families in which members share contact, care, and support even though they

The numbers of three-generational households has increased in recent years.

Kayte Deioma/PhotoEdit

don't share a residence. Think about your own family. What, if any, role or roles have your grandparents played in your life? Did they babysit for you when you were younger? Did you visit them regularly? Talk on the phone? Exchange gifts? The point is that, even in the absence of sharing a household, grandparents and other extended kin may be important figures in your life and, hence, broaden and enrich your family experiences beyond the nuclear households in which you may live or have lived.

Journalist Tamar Lewin suggests that "in many families, grandparents are the secret ingredients that make the difference between a life of struggle and one of relative ease." They may provide assistance that allows their grandchildren to go to camp, get braces for their teeth, go on vacation, and get music lessons or necessary tutoring, all of which enrich their grandchildren's lives beyond what parents alone could manage. Sociologist Vern Bengtson compiled 20 years of data that he gathered from his undergraduates about how they finance their college educations. Bengtson found that among his own students, grandparents were the third most frequently mentioned source, behind parents and scholarships but ahead of both jobs and loans. And the importance of grandparents includes but goes well beyond those instances in which they either share the households of or provide child care for their young grandchildren. We should note that there are also many instances in which adults help their elderly parents. In a survey conducted by the American Association of Retired Persons (AARP), 25 percent of "baby boomers"

> **True** 7. The number of multigenerational households in the United States is increasing.

expected to have their parents move in with them at some point in time (Green 2009). In either direction, such assistance and support remind us that extended families are important sources of aid and support for one another.

## Kinship Systems

The **kinship system** is the social organization of the family. It is based on the reciprocal rights and obligations of the different family members, such as those between parents and children, grandparents and grandchildren, and mothers-in-law and sons-in-law. Nuclear family roles (such as parent, child, husband, wife, and sibling) combine with extended family roles (such as grandparent, aunt, uncle, cousin, and in-law) to form the kinship system.

### Conjugal and Consanguineous Relationships

Family relationships are generally created in two ways: through marriage and through birth. Family relationships created through marriage are known as **conjugal relationships.** (The word *conjugal* is derived from the Latin *conjungere,* meaning "to join together.") In-laws, such as mothers-in-law, fathers-in-law, sons-in-law, and daughters-in-law, are created by law—that is, through marriage. **Consanguineous relationships** are created through biological (blood) ties—that is, through birth. (The word *consanguineous* is derived from the Latin *con-*, "joint," and *sanguineous*, "of blood.") Relationships between adopted children and parents, though not related by blood, might be considered "fictive consanguineous" relationships in that they are culturally treated as having the same kinds of ties and obligations.

### Kin Rights and Obligations

In some societies, mostly non-Western or nonindustrialized cultures, kinship obligations may be more extensive than they are for most Americans in the 21st century. In cultures that emphasize wider kin groups, close emotional ties between a husband and a wife may even be considered a threat to the extended family.

In a marriage form found in Canton, China, women do not live with their husbands until at least three years after marriage, as their primary obligation remains with their own extended families. Under the traditional marriage system among the Nayar of India, men had a number of clearly defined obligations toward the children of their sisters and female cousins, although they had few obligations toward their own children (Gough 1968).

In U.S. society, the basic kinship system consists of parents and children, but as we have seen, it may include other relatives as well, especially grandparents. Each person in this system has certain rights and obligations as a result of his or her position in the family structure. Furthermore, a person may occupy several positions at the same time. For example, an 18-year-old woman may simultaneously be a daughter, a sister, a cousin, an aunt, and a granddaughter. Each role entails different rights and obligations. As a daughter, the young woman may have to defer to certain decisions of her parents; as a sister, to share her bedroom; as a cousin, to attend a wedding; and as a granddaughter, to visit her grandparents during the holidays.

In U.S. culture, the nuclear family has many norms regulating behavior, such as parental support of children and sexual fidelity between spouses, but the rights and obligations of relatives outside the basic kinship system are less strong and less clearly articulated. Because neither culturally binding nor legally enforceable norms exist regarding the extended family, some researchers suggest that such kinship ties have become voluntary. We are free to define our kinship relations much as we wish. Like friendship, these relations may be allowed to wane (Goetting 1990).

Despite the increasingly voluntary nature of kin relations, our kin create a rich social network for us. Adult children and their parents often live close to one another, make regular visits, and/or help one another with child care, housework, maintenance, repairs, loans, and gifts. The relations among siblings often remain strong throughout the life cycle. In fact, as vividly illustrated by sociologist Karen Hansen's research on "networks of care," kin are frequently essential supports in the ever-more complicated tasks associated with raising children in dual-earner households or single-parent households. Although they are invisible when we focus so intensively on nuclear families, to effectively raise children may require the help of "'other mothers,' aunties, grandmothers and child care workers (as well as) . . . uncles, grandfathers and male friends" (Hansen 2005, 215). Where kin are unavailable or where certain family members are either uncooperative or deemed to be unsuitable, these networks might expand to include neighbors, friends, and paid caregivers.

## Cyber Caregiving and Technological Togetherness

The maintenance of wider kin ties has been enhanced by technological innovations from bicycles, through telephones and automobiles, and most recently through advances made in information technology. Just as wheels allowed family members to visit kin some distance away, and telecommunications allowed people to speak with relatives nearly anywhere in the world, more recent innovations make it possible to have "family get-togethers," monitor each others' actions, and even engage in caregiving activities, despite being separated by great distance.

Via software that allows group video calling, three or more family members can be "brought together" online. For example, you have children or grandchildren off at different colleges or universities, or family members in different parts of the country. Just put them all in a group on Skype and click the call button, and you'll have yourself a big old family reunion that's the next best thing to being face-to-face in the same room!" (https://techboomers.com/t/skype-video-chat). Though limited to four hours per video call, ten hours of video calls per day, and 100 hours per month, group video calls can allow family members to "visit" even when scattered across a number of countries. Google also offers voice and video chat, with similar features. Thus, a man in Ohio could "get together" with his daughter and her husband in Portland, Oregon, his son and son's girlfriend in Las Vegas, his sisters—one in New York and one in New Jersey, where his newly widowed 86-year-old father also lives. People with aging parents have an increasing number of ways to do more than visit. They can oversee their parents' care and ensure their well-being, even from hundreds or thousands of miles away. Consider the following family: Widowed, 70-year-old Elizabeth Roach lives in her home in Virginia. Her son, Michael, lives in Denver. Through a system marketed as GrandCare, movement sensors placed around Ms. Roach's house record and relay her movements, her exact weight, and blood pressure readings to Michael. Michael receives detailed information sent to him via email, text message, or voice mail about what time his mother got into and out of bed, when she opened or closed certain doors, and whether, when, and for how long she opened her refrigerator. In this way, he can determine whether Elizabeth has taken her medications and eaten her meals. In addition to ensuring his mother's well-being, GrandCare helps Michael cope with his feelings about being so far away from his mother. As he told journalist Hilary Stout, "I have a large amount of guilt. I'm really far away. I'm not helping to take care of her, to mow her lawn, to be a good son" (Stout 2010).

Stout proceeds to talk about additional systems that help bridge distance and tend to caregiving activities. For example, there is a medication management system called MedMinder that operates as a computerized pillbox. Once the proper doses of one's medications have been arranged into boxes, the system provides beeps and flashes to remind one to take his or her medication. More than that, the system also alerts others to whether or not a person took the needed medicines. Rachel Meyers receives a tape-recorded phone call at her Brooklyn home notifying her whether her 84-year-old mother, Harriet, has taken her medication. Her brother in Australia gets email notification. BeClose is a wireless monitoring system that, through sensors placed in her widowed aunt's bed, notifies Susan Oertle via email and phone that her 83-year-old aunt went to sleep (Stout 2010).

Innovations such as these have obvious appeal for those who wish to bridge physical distance to maintain a sense of togetherness that transcends geographic boundaries. They can be used to connect spouses separated for work assignments or military deployment, parents and their college-age children, noncustodial divorced parents and their children, or adults and their aging but distant parents. In fact, in providing a kind of "technological togetherness" and "cyber caregiving," these new technologies have the potential to "re-extend" families and allow for the maintenance of ever-closer contact between kin living in geographically dispersed households.

# Multiple Viewpoints of Families

As we noted earlier, marriage and family issues inspire much debate. For instance, those who believe that families of male providers, female homemakers, and their dependent children living together, 'til death do they part, are what families *should be* would not be encouraged by the continued high rates of divorce, increases in cohabitation, or the declining rates of marriage or full-time at-home motherhood. Those on the "other" side who claim that there are basic inequities within the traditional family, especially regarding the status of women, will not mourn the diminishing numbers of breadwinner–housewife families. Similarly, the question of gay marriage divides those who believe that marriage *must* be a relationship between a man and a woman from those who believe that we *must* recognize and support all kinds of families and provide equal marriage rights to all people.

Divisiveness on family issues is neither new nor unique to the United States. In the early 20th-century U.S., there was considerable pessimism about whether families would survive the changing and liberalizing culture of sexuality, the increasing numbers of women delaying marriage for educational or occupational reasons, and the declining birthrate and increases in divorce. In considering the same sorts of changes, others advocated that these trends were positive signs of families adapting to changes in the wider society (Mintz and Kellogg 1988).

In recent years, many other countries have faced similar cultural clashes over trends and changes in family life. In Spain, for example, there has been a dispute pitting the Spanish socialist government against the Catholic Church, as governmental initiatives to legalize same-sex marriage and make abortion and divorce easier or quicker were met with strong and vocal opposition from the church. Whereas some in the Spanish Socialist Workers' Party or among its allies such as the United Left Party believed that Spain had not gone far enough in recognizing and embracing change, organizations aligned with the church, such as the Institute for Family Policy, considered the climate in Spain "family-phobic" (Fuchs 2004).

In Poland, "stormy parliamentary debate" ensued over whether to endorse and sign into law a Council of Europe convention on fighting domestic violence and violence against women, which presses governments and organizations to penalize such violence, help victims, and teaches about tolerance. In the Polish debate, more socially and politically conservative and Catholic lawmakers argued that the regulations went against traditional Polish gender and family roles. A statement on behalf of Poland's Catholic bishops claimed that the lawmakers don't see the "good of the marriage, of the family, of Poland's future demography as their priority." On the other side, Deputy Parliament Speaker and women's rights activist, Wanda Nowicka, contested such accusations and emphasized that the regulation is badly needed to protect Polish women (Scislowska 2015).

**False**

**8.** There is not widespread agreement about the nature and causes of change in family patterns in the United States.

Within the United States two of the more prominent sources of differences in viewpoints on family issues and family change are one's racial or ethnic background and one's religion. With regard to racial and ethnic differences, African Americans have tended to express somewhat more conservative attitudes than whites on a number of family issues such as premarital sex, cohabitation, spanking, gay marriage, and gay parenting. For example, using data from the 2004 to 2014 General Social Survey reveals that compared to whites, blacks are 20 percent more likely to say that homosexual relations are "always wrong" (60.8 percent of blacks versus 40.5 percent of whites) and are 8.5 percent more likely to disagree or strongly disagree that a lesbian couple could raise a child as effectively as a heterosexual couple (47.5 percent vs. 39 percent), and 11 percent more likely to disagree or strongly disagree that a gay male couple could. In addition, Blacks are twice as likely as whites to say that the most important quality in a child is to obey (23.6 percent among blacks, 11.7 percent among whites) and are 14.5 percent more likely to strongly agree (36.7 percent vs. 22.2 percent among whites) that spanking is an effective way to discipline a child. Regarding cohabitation, blacks were 16 percent less likely to agree or strongly agree that living together was an acceptable option (44.7 percent among blacks vs. 60.6 percent among whites). On a number of other family issues, such as attitudes toward extramarital sex, divorce, whether it is better for men to be breadwinners and females to tend home, and whether one approves of supplying birth control to 14- to 16-year-olds, there were very small to no real race differences.

Attitudes toward homosexuality and gay marriage have been among the largest and more consistent areas of racial and ethnic differences. In comparison with whites and other ethnicities, African Americans have been more opposed to same-sex marriage. Although African Americans may be no less supportive of extending "basic civil rights to speech and employment" to gay, lesbian, bisexual, and transgendered people, their opposition to gay marriage remained strong, even as whites and others became more supportive of extending marriage rights to same-sex couples. More recently, though the attitudes of African Americans have also changed, they are still less accepting than those of whites (Sherkat, de Vries, and Creek 2010).

In explaining African Americans' positions, one can point to the second of the sources previously noted above, the influence of religion. Darren Sherkat, Kylan de Vries, and Stacia Creek contend that compared to more secular influences, African Americans' high rates of church attendance and their membership in more conservative Protestant denominations shape their attitudes toward gay marriage (Sherkat, de Vries, and Creek 2010). This influence of religion is not unique to African Americans; across racial lines, membership in more conservative sectarian Protestant denominations has been associated with a greater likelihood to believe that homosexuality is wrong and to be opposed to extending marriage rights to same-gender couples. Members of mainline Protestant denominations, along with Catholics, Jews, and those without religious affiliation, have been more likely to support the legalization of same-gender marriage.

More generally, religious affiliation and participation surface as prominent influences on attitudes toward a host of family issues, including divorce, cohabitation, premarital sex, and gay marriage. For example:

**True** **9.** African Americans tend to express more conservative views on such family issues as premarital sex, divorce, and gay marriage.

- Individuals who report being affiliated with any religious group are more likely to hold more conservative attitudes about family issues than are those with no religious affiliation.
- Conservative religious beliefs are often accompanied by more traditional gender and family attitudes. As expressed by sociologist Jennifer Roebuck Bulanda, ". . . conservative religious traditions espouse the idea of 'traditional' family structure and complementary roles for men and women in family life" (Bulanda 2011, 180).
- Americans who more frequently attend religious services or who believe the Bible to be the Word of God are more likely than their less religious or nonreligious counterparts to support stricter divorce laws (Stokes and Ellison 2010).
- Like the differences between Evangelical Christians and mainline Protestants, differences surface *within* other faiths as well. For example, Reformed Jews tend to be among the most liberal on "family issues," whereas Orthodox Jews are much more conservative. Traditional Catholics hold more conservative views on such issues as divorce, cohabitation, and same-gender marriage than do liberal Catholics. In fact, differences within any particular

faith may be greater than differences between faiths. So, for example, in the Public Broadcasting System's Faith and Family Survey of 2005 (Sherkat, de Vries, and Creek 2010), Evangelical Christians (at 92 percent) and traditional Catholics (at 91 percent) were equally and overwhelmingly in agreement with the statement, "God's plan for marriage is one man, one woman, for life." Much smaller and nearly identical majorities of liberal Catholics and mainline Protestants (60 percent and 62 percent, respectively) agreed. Such differences are often obscured when we look at either overall attitudes of Americans or at differences between Protestants, Catholics, Jews, and others.

- Those who identify themselves as having no religious preference (or as atheists or agnostics) tend to express the most liberal attitudes on family issues.

Jennifer Roebuck Bulanda offers a cautionary note to those who would simply infer that the traditional family attitudes that accompany any particular set of religious beliefs translates into traditional family behavior, or that it does so among all members of those faiths. She suggests that researchers resist the assumption that those who are members of conservative faiths or denominations will all behave in accordance with the ideology of their faiths. She points out that other factors—including gender, race, ethnicity, and social class—likely moderate the relationship between religion and family (Bulanda 2011).

Ultimately, the ways we view families depend on what we conceive of *as* families. Such disagreements, then, reflect both different definitions of family and different values regarding particular kinds of families. Often the product of personal experience as much as of religious background, personal values reflect what we want families to be like and, thus, what we come to believe about the kinds of issues that are raised throughout this book.

## Half Full versus Half Empty

With so much "noise" in the wider society around what family life is and should be like, how families are changing, and whether those changes are good or bad, you may find it difficult to know what conclusions to draw about family issues. Given the lack of societal consensus, it is easy to become confused or be misled about what American families are really like. To some, contemporary family life is weaker because of cultural and social changes and is now, to some extent, endangered. More optimistic interpretations of changing family patterns celebrate the increased domestic diversity of numerous family types and the richer range of choices now available to Americans. Like the proverbial glass, some see the family as "half empty," whereas others see it as "half full." What makes the "half full, half empty" metaphor so apt is that even when looking at the same phenomenon or the same trend, some interpret it as evidence of the troubled state U.S. families are in, and others see today's families as different or changing. So, for instance, although the rates of divorce and marriage, the numbers of children in nonparental child care, or the extent of increase in cohabitation can be objectively measured, like the volume of liquid in a partially filled glass, the meaning of those measures can vary widely, depending on perspective.

## Conservative, Liberal, and Centrist Perspectives

In the wider, societal discourse about families, we can identify opposing ideological positions on the well-being of families. The two extremes, which sociologist Norval Glenn called conservative and liberal, are like the half empty–half full disagreement, a difference between pessimistic and optimistic viewpoints. Conservatives are fairly pessimistic about changes in family life and the state of today's families. To **conservatives**, cultural values have shifted from individual self-sacrifice toward personal self-fulfillment. This shift in values is seen as an important factor in accounting for some of the major changes in family life that occurred beginning with the last three or four decades of the 20th century (especially higher divorce rates, more cohabitation, and more births outside marriage). Conservatives further believe that because of such changes, today's families are weaker and less effective, especially when it comes to raising and meeting the needs of children. Thus, conservatives often recommend policies and programs to reverse or reduce the extent of such changes (repeal of no-fault divorce, introduction of covenant marriage, and programs to promote marriage are all examples).

Compared with conservatives, liberals are more optimistic about the status and future of family life in the United States. **Liberals** tend to believe that the changes in family patterns are just

that—changes—and should not be viewed as signs of familial decline (Benokraitis 2000). The liberal position also typically portrays changing family patterns as products of and adaptations to wider social and economic changes rather than a shift in cultural values (Benokraitis 2000; Glenn 2000). Such changes in family experience lead to a wider range of contemporary household and family types and require greater tolerance of such diversity. Placing great emphasis on economic issues, liberal family policies are often tied to the economic well-being of families, such as the increasing numbers of employed mothers and two-earner households.

According to Glenn, there is yet a third position in the discourse about families. **Centrists** share aspects of both conservative and liberal positions. Like conservatives, they believe that some familial changes have had negative consequences. Like liberals, they identify wider social changes (e.g., economic or demographic) as major determinants of the changes in family life, but they assert greater emphasis than liberals do on the importance of cultural values. They note that too many people are too absorbed in their careers or too quick to surrender in the face of marital difficulties (Benokraitis 2000; Glenn 2000).

## Attitudes Toward Changes in Family Living: Accepters, Skeptics, and Rejecters

A similar three-way division emerges when examining attitudes regarding various trends in the structure of U.S. family life. A Pew Research Center survey of a nationally representative sample of 2,691 adults asked respondents to assess the following trends as either "good for society," "bad for society," or "makes no difference": more unmarried couples raising children; more same-sex couples raising children; increased numbers of single women having children and raising them on their own; more couples living together without getting married; more mothers of young children employed outside the home; increasing numbers of interracial marriages; and increases in numbers of women choosing to not have children. Roughly a third (31 percent) of respondents can be considered **accepters** who see the trends as making no difference to society or as good for society. Almost the same percentage of respondents (32 percent) has been labeled **rejecters,** in that they tend to see these changes as bad for society. The remaining 37 percent has been identified as **skeptics** who share in the relatively tolerant views of the accepters but do express concern about the potential impact of the trends. Typically, accepters are the most tolerant, skeptics less so, and rejecters tend to disapprove of most of the changes about which they were asked.

On many of the trends, accepters and skeptics were much more similar than different. The one major difference between them was their view of the increase in single motherhood. Whereas 99 percent of skeptics saw this increase as bad for society, close to 90 percent of accepters said it had made no difference (74 percent) or has been a good thing (13 percent). On most other issues, both accepters and skeptics were fairly similar and differed greatly from rejecters. Close to or more than 90 percent of rejecters saw the upward trends in couples living together without getting married, unmarried couples raising children, and gay and lesbian couples raising children as "bad for society." Conversely, more than half of accepters and skeptics saw these same trends as "making no difference" (Morin 2011).

There were other interesting differences in the Pew data. When asked which type of marriage— breadwinner/housewife or dual-earner—provided a "more satisfying way of life," half of rejecters favored the male breadwinner household. Nearly three-quarters (73 percent) of accepters favored a dual-earner model, as did 70 percent of skeptics. There were also noteworthy differences in whether different kinds of household arrangements were considered to be families. Although the vast majority of accepters (84 percent) and skeptics (75 percent) both considered a same-sex couple with children to be a family, less than a third (31 percent) of rejecters shared this viewpoint. Just slightly over half of rejecters (55 percent) claimed to consider an unmarried couple with children to be a family. In contrast, 96 percent of accepters and 87 percent of skeptics agreed that such groups were indeed families. Accepters are also more likely to consider unmarried childless couples (60 percent) and same-sex couples without children (68 percent) as families. This same attitude was true for less than half of skeptics and a fifth or less of rejecters.

Looking closer at the composition of each of these clusters exposes demographic differences that may help explain the difference in attitudes. Older people, people who are more religious, whites, married adults, and Republicans tend to have the more pessimistic views that place them among rejecters. Younger people, less religious people, unmarried adults, Hispanics, Democrats, and women express attitudes that place them among accepters or skeptics.

The assumptions within and the differences between these positions are more important than they might first appear to be. The perceptions we have of what accounts for the current status of family life or the directions in which it is heading influence what we believe families *need*. These, in turn, influence social policies regarding family life. As Nijole Benokraitis (2000) states, "Conservatives, centrists, liberals, and feminists who lobby for a variety of family-related 'remedies' affect our family lives on a daily basis" (p. 19).

## Disagreement Among Family Scientists

It should be noted that social scientists are similarly divided in how they perceive contemporary families. In other words, changing family patterns and trends in marriage, divorce, parenting, and child care are explained and interpreted differently even by the experts who study them. For example, in considering the effects of divorce on children, one can find a range of viewpoints, each of which comes from published research rather than personal opinion. The differences between such viewpoints are hard to reconcile.

Constance Ahrons, a noted scholar, member of the Council on Contemporary Families, and author of numerous articles and books on divorce, offers a somewhat encouraging point of view. She contends that "The good news about divorce is that the vast majority of children develop into reasonably competent individuals, functioning within the normal range. . . . (I)t is not divorce per se, but the quality of the relationship between divorced parents that has an important long-term impact on adult children's lives."

In contrast to Ahrons's comments, David Popenoe, also a well-known sociologist, author, and/or editor of numerous books about contemporary American families, and codirector of the National Marriage Project, provides a different perspective. He states, "Divorce increases the risk of interpersonal problems in children." He contends that research suggests that many such problems are long-lasting and may become worse in adulthood. More emphatically, he suggests that, "except in the minority of high-conflict marriages, it is better for the children if their parents stay together and work out their problems than if they divorce." Although these statements are not entirely irreconcilable, they do reveal and may themselves contribute to different overall perspectives about marriage, divorce, and the well-being of children. Thus, it is important to realize that, just as the wider society and culture are fraught with conflicting opinions and

Law professors Naomi Cahn and June Carbone offered another interesting way to look at some of the opposing positions on family issues. In their book, *Red Families v. Blue Families: Legal Polarization and the Creation of Culture,* Cahn and Carbone (2010) built upon the popular political distinction often drawn between conservative "red states" that tend to vote Republican and more liberal "blue states" that more often vote Democratic to look at some major differences in family values and experiences that separate the lives of people in different parts of the United States.

Using statewide comparisons, they described patterns of family life found more frequently in "red states," in the Mountain West, the rural plains, and the South. These include younger ages at marriage and at childbirth, higher teen pregnancy rates, greater emphasis on abstinence outside of marriage and less support for contraceptive availability, greater opposition to abortion and gay marriage, and higher divorce rates. Red families advocate a more traditional family system, where traditional gender roles are "critical to marital stability" (2007, 3). They stress the "unity of sex, marriage, and procreation," and promote delaying becoming sexually active until marriage (Cahn and Carbone 2010).

In contrast, "blue states," such as those on both U.S. coasts, endorse a family model that "celebrates more egalitarian gender roles," delays childbearing, and invests in educational and career goals for both genders. Contraception is "morally compelled," abortion is "the necessary (and responsible) fallback," and acceptance of gay marriage "a matter of basic equality" (Cahn and Carbone 2007, 2). Blue states tend to have lower rates of divorce, lower teen birthrates, lower percentages of teen births occurring within marriage, and higher use of abortion. Compared to their red counterparts, blue families tend to be wealthier, more educated, less religiously active, and have fewer children.

Two of the biggest factors separating the red and blue family models are age and religion. Red families are characterized by younger marriages and younger entries into parenthood. States with the lowest ages at marriage tend to be red, whereas states with the highest median age of marriage are blue (Cahn and Carbone 2007). Populations in red states are more likely to be religious fundamentalists, who believe literally in the words of the Bible. In red states, nearly half of voters identified themselves as fundamentalists compared to just over a fourth of voters in blue states (Cahn and Carbone 2010, 70).

Cahn and Carbone are careful to note that comparisons of rates of family behavior between states may obscure the diversity of experiences and attitudes within states. Although they have been criticized for skewing their analysis in favor of the blue family model, Cahn and Carbone point out problems characteristic of each family model. Despite their emphasis on tradition, red states have the highest divorce rates in the United States, and "their teens are also more likely to become pregnant and to give birth to children the parents are ill-equipped to raise" (Cahn and Carbone 2007, 3). Blue states have much higher numbers of women and men who will never marry, declining fertility rates, and high percentages of people living alone.

In continuing the "red" versus "blue" comparison, sociologist W. Bradley Wilcox and psychologist Nicholas Zill have pointed out that some of the most stable families can be found in some of the bluest (e.g., Massachusetts and Minnesota) and some of the reddest (e.g., Utah and Nebraska) states. They note that both the bluest and the reddest states are most likely to offer high levels of family stability (and low levels of nonmarital births), though the explanations for this may differ between red and blue states (Wilcox and Zill 2015). Wilcox and Zill suggest that low levels of nonmarital childbearing in the reddest states can best be explained by red state family culture (what they call "deep normative and religious commitments to marriage and to raising children within marriage"), whereas in the bluest states a key factor is higher levels of education. At least at the state level, "the bluest and the reddest states in America register the highest levels of family stability in the nation" (2015).

In a follow-up article, Wilcox moves from state- to county-level comparisons in recognition that "there are plenty of blue states with lots of red counties (think Pennsylvania), and vice versa (think Texas)," (2015, 2). He contends that counties that "gave a higher share of their vote to the Republican presidential candidate (Mitt Romney) have a higher share of their population that is married" (2015, 3). The same is said to hold for levels of nonmarital childbearing (lower in "red counties") and the likelihood that children live with their biological parents (higher in red counties). At the same time, divorce is more common in the southern United States, which as Wilcox notes is a "region that tilts red" (Wilcox 2015, 5). Wilcox also claims that one factor explaining why blue counties are more likely to have lower levels of marriage, more family instability, and higher levels of nonmarital childbearing is race: They have more African Americans. As we will examine in Chapter 3, patterns of marriage, childbearing, and child rearing vary across racial lines, though some of that variation is likely tied to class differences between the races. ●

# Has There Been a "Modern Family Effect"?

Any analysis of changing attitudes toward same-sex marriage would show how dramatic a change took place over the past 15 years. In 2001, data from the Pew Research Center showed that a majority of Americans were opposed to same-sex marriage, by a margin of 57 percent to 35 percent. For nearly a decade, from 2001 to 2009, there was small upward movement in the percentages of people who said they were in favor of allowing gay and lesbian couples to marry legally. In April 2009, 35 percent said they favored or strongly favored allowing gays and lesbians to legally marry. By 2015, even before the Supreme Court decision in *Obergefell v. Hodges,* 57 percent of Americans said they supported same-sex marriage, while 39 percent remained opposed. And although some groups remain opposed to gay marriage (e.g., white evangelical Protestants, blacks, Republicans, Conservatives) among all groups, there has been an increase since 2009 in those who say they are in favor of same-sex marriage. So what happened? What changed Americans' minds?

Although acknowledging that demographic shifts and political organizing efforts are likely part of the answer, journalist Spencer Kornhaber, writing in *The Atlantic* in the wake of the Obergefell decision, also raises the possibility that there was a *"Modern Family* effect" in which the successful sitcom's portrayal of the relationship between Mitchell and Cam may have helped to influence attitudes (Kornhaber 2015). Premiering in 2009, approximately ten million viewers watched weekly as the fictional Cam and Mitch were shown "navigating the challenges of being in a long-term relationship" and raising their adopted daughter Lily (Kornhaber 2015). Kornhaber, a staff writer at *The Atlantic* who covers pop culture, notes that in a 2012 *Hollywood Reporter* poll 27 percent of likely voters said that they'd become more pro–gay marriage from the ways in which gay characters were depicted on television. He also notes that "there are news accounts of people crediting their newfound sympathy toward gay people to *Modern Family*" (Kornhaber 2015).

Of course, it is risky to assume that any particular television series or set of characters is responsible for changing public opinion (and thus shaping social reality). As used by sociologists and other analysts of media content, the **reflection hypothesis,** suggests that *media content reflects the values and ideals of the audience.* Media images are said to reflect what people want to see or what already exists. Yet it is also well known that media content *can* shape the values and beliefs of the audience that consumes its content. In that vein, perhaps Cam and Mitch, by being "about as tame as anyone could ask . . . they rarely touch, never talk about sex, and make a big deal over kissing in public," have made gay couples less threatening "and more normal" to those who might have otherwise opposed gay marriage (Kornhaber 2015). They are not the only recent or current popular culture examples, nor are popular culture examples the only or most important factor in raising support for gay marriage. However, they likely factor into an explanation somewhere. ●

Cameron and Mitchell (Eric Stonestreet and Jesse Tyler Young) of the ABC situation comedy, *Modern Family*, at their wedding.

Peter 'Hopper' Stone/Disney ABC Television Group/Getty Images

values about marriage and family relationships, the academic disciplines that study family life occasionally contain a similar lack of consensus.

As we set off on our exploration of marriage and family issues, it is important to realize that many of the topics we cover are part of similar ongoing debates about families. As you try to make sense of the

**False**
**10.** Researchers do not agree about the impact of divorce on children.

material we introduce throughout this book, we require you not to take a particular viewpoint but rather to keep in mind that multiple interpretations are possible. Where different interpretations are particularly glaring (as, for example, in the many issues surrounding divorce), we present them and allow you to decide which better fits the evidence presented.

# The Major Themes of This Text

Throughout the many chapters and pages that follow, as we examine in detail intimate relationships, marriage, and family in the United States, we will introduce a range of theories, provide much data, and look at a number of family issues and relationships in ways you may never have considered before. As we do so, we will visit and revisit the following points.

## Families Are Dynamic

As noted earlier, the family is a dynamic social institution that has undergone considerable change in its structure and functions. Similarly, values and beliefs about families have changed over time and continue to do so. We are more accepting of divorce, employed mothers, and cohabitation. We expect men to be more involved in hands-on child care. We place more importance on individual happiness than on self-sacrifice for family.

In Chapter 3, we explore some of the major changes that have occurred in how Americans experience families. Then, throughout the text, as we address topics such as marriage, divorce, cohabitation, raising children, and managing employment and family, we ask the following: In what ways have things changed, and why? What consequences and implications result from these changes? Because familial change is often differently perceived and interpreted, we also present different possible interpretations of the meaning of change. Are families merely changing, or are they declining?

Throughout much of the text, we also look at how individual family experience changes over time. Families are ever changing—from the formation of love relationships to the entry into marriage or intimate partnerships, from the bearing, raising, and aging of children to the aging and death of parents and spouses.

## Families Are Diverse

Not all families experience things the same way. Beginning with Chapter 3, we look closely at a variety of factors that create differences in family experience. We consider, especially, the following major sources of patterned variation in family experience: social class, race and ethnicity, gender and sexuality, and lifestyle choice.

## Social Class

Different social classes (categories of individuals and families that share similar economic positions in the wider society) have different experiences of family life. Because of both the material and the symbolic (including cultural and psychological) dimensions of social class, our chances of marrying, our experiences of marriage and parenthood, our ties with kin, our experience of juggling work and family, and our likelihood of experiencing violence or divorce all vary. And this is but a partial list of major areas of family experience that differ among social classes.

## Race and Ethnicity

More than 240 different native cultures lived in what is now the United States when the colonists first arrived (Mintz and Kellogg 1988). Since then, U.S. society has housed immigrant groups from the world over who bring with them some of the customs, beliefs, and traditions of their native lands, including those about families. Thus, we can speak of African American families, Latino families, Asian families, Native American families, European families, and so on. In Chapter 3, we provide a brief sketch of the major characteristics of the family experiences of each of these racial or ethnic groups. As we proceed from there, we compare and contrast, where relevant and possible, major differences in family experiences across racial and ethnic lines, and consider the social, cultural, and economic sources of such differences.

## Gender and Sexuality

To understand intimate relationships, marriage and family life requires us to pay sustained attention to gender (the attitudes and behavior expected of individuals because of the sex category into which they have been assigned, socialized, and/or with which they identify). Our cultural ideas and understanding of **gender** continue to change. Long characterized by a strict binary cultural construction of gender that emphasized and sometimes exaggerated differences between male and female experience, more recent ideas about gender contain a broader and more nuanced understanding of gender identities and variations within gender categories. This includes an increasing acceptance and visibility of **transgender** individuals, whose gender identities develop and are expressed in ways that differ from what their biological sex would otherwise predict, as well as a greater understanding of how gender intersects with other social statuses (such as race, class, age, religion, and sexuality).

Gender affects many of the areas of family experience on which we touch in this book. Throughout the text, as we examine such topics as love and friendship, sexual freedom and expression, marriage responsibilities and gratifications, involvement with and responsibilities for children, experience of abuse, consequences of divorce and becoming a single parent, and chances for remarriage, we will identify where women's and men's experiences differ and where they don't. Although gender differences loom large in some areas of family experience, research suggests that on many characteristics and attributes "men and women, as well as boys and girls, are more alike than they are different" (Hyde 2005, 581).

Examining whether and how experiences of such things as intimacy, sexual expression, parenting, abuse, and separation do and don't differ among heterosexual, gay, and lesbian individuals and couples will further our attention to and understanding of the effects of gender and gender difference on relationships.

### Diversity of Chosen Lifestyles

A striking difference between 21st-century families and earlier American families is the diversity of family lifestyles that people choose or experience. There is no family form that encompasses most people's aspirations or experiences. Statistically, the dual-earner household is the most common form of family household with children, but there is considerable variation among dual-earner households and between such households as traditional or single-parent families.

Increasingly, people are choosing to cohabit, either before or instead of marrying. Increasing numbers of couples choose not to have children, and increasing numbers of others choose expensive procedures to assist their efforts and enable them to bear and rear children. This diversity of family types and lifestyles will not soon abate. In the chapters that follow, specific attention is directed at singles (with and without children), cohabiters, childless or child-free couples, and role-reversed households.

### Outside Influences on Family Experience

This book takes a mostly sociological approach to relationships, marriage, and families in that we repeatedly stress the outside forces that shape family experiences. The family is one of the core social institutions of society, along with the economy, religion, the state, education, and health care. As such, the shape and substance of family life is heavily affected by the needs of the wider society in which it is located. In addition, other social institutions influence how we experience our families.

Similarly, cultural influences in the wider society, such as the values and beliefs about what families are or should be like and the norms (or social rules) that distinguish acceptable from unacceptable behavior, guide how we choose to live in relationships and families. Thus, although each of us as an individual makes a series of decisions about the kinds of family life we want, the choices we make are products of the societies in which we live.

In addition, options available to each of us may not reflect what we would freely choose if we faced no constraints on our choices. So, for example, parents who might prefer to stay at home with their children might find such a choice impractical or impossible because economic necessity forces them to work outside the home. Working parents may find the time they spend with their children more a reflection of the demands of their jobs and the inflexibility of their workplaces than of their own personal preferences, just as some at-home parents might prefer to be employed but find that their children's needs, the cost and availability of quality child care, the jobs available to them, and the demands and benefits contained in those jobs push them to stay home.

Our familial life reflects decisions we face, the choices we make, and the opportunities and/or constraints we confront. In the wider discourse about families, we tend to encounter mostly individualistic explanations for what people experience, focusing sometimes exclusively on personal choices. Throughout this text, we examine the wider environments within which our family choices are made and the ways in which some of us are given more opportunities whereas others face limited options.

### The Interdependence of Families and the Wider Society

Following the prior theme, we indicate throughout the book how societal support is essential for family well-being. Equally true, healthy, well-functioning families are essential to societal well-being. To function effectively, if not optimally, families need outside assistance and support. Better child care, more flexible work environments, economic assistance for the neediest families, protection from violent or abusive partners or parents, and a more effective system for collecting child support are just some examples we consider in later chapters of where families clearly have needs for greater societal or institutional support.

In turn, the health and stability of the wider society depend largely on strong and stable families. Families are the sources from which most of the social skills, personality characteristics, and values of individual members of society are formed. Ideally, successful families produce and nurture hope, purpose, and general attitudes of commitment, perseverance, and well-being. Indeed, even the rudimentary maintenance and survival care provided by families make significant contributions to the well-being of a community.

When families fail, individuals must turn elsewhere for assistance; social institutions must be designed to fill the voids left by failing families, as the pathologies created by weak family structures make society a less livable place. Ultimately, there are enormous costs that result from neglecting the needs of families and children in the United States.

Some of the services provided by families are such a basic part of our existence that we tend to overlook them. These include such essentials as the provision of food and shelter—a place to sleep, rest, and play—as well as caregiving, including supervision of health and hygiene, transportation, and the accountability of family members involving their activities and whereabouts. Without families, communities would have to provide extensive dormitories and many personal care workers with different levels of training and responsibility to perform the many activities in which families are engaged.

On a more emotional level, without families, individuals must look elsewhere to satisfy basic needs for intimacy and support. We marry or form marriage-like cohabiting relationships, have children, and maintain contact with other kin (adult siblings, aging parents, and extended kin) because such relationships retain importance as bases for our identities and sources of social and emotional sustenance. We bring to these relationships high affective expectations. When our intimacy needs are not met (in marriage or long-term cohabitation), we terminate those relationships and seek others that will provide them. We believe, however, that those needs are best met in families.

As you now begin studying marriage and the family, it is hoped that you will see that such study is both abstract and personal. It is abstract insofar as you will learn about the general structure, processes, and meanings associated with marriage and the family, especially within the United States. In the chapters that follow, the things that you learn should also help you better understand your own family, how it compares to other families, and why families are the way they are. In other words, as we address family more generally, you will be studying *your* present, *your* past, and *your* future in some ways. By providing a wider sociological context to marriage, family, and intimate relationships, we will show you how and where your experiences fit and why.

## Summary

- Our experiences in our families and relationships affect the kinds of ideas about families and intimate relationships that we bring to a course such as this.

- Family life has changed greatly as can be seen in such things as the increase in cohabitation, singlehood, divorce, dual-earner couples and single-parent households, as well as the legalization of same-sex marriage.

- Technological innovation in communication and reproduction has influenced contemporary family roles and relationships.

- Countries vary widely in the prevalence of marriage. Within the United States a majority of adult women and men have been and are married.

- There is considerable cultural diversity in how societies define marriage and who may marry. At minimum, marriage is a socially and legally recognized union between two people that establishes rights and obligations connected to gender and sexuality, raising children, and relating to the wider community and society.

- In Western cultures, the preferred form of marriage is *monogamy*, in which there are only two spouses. *Polygyny*, the practice of having two or more wives, is preferred throughout many cultures in the world.

- In June 2015, the United States Supreme Court legalized same-sex marriage throughout the United States.

- Defining the term *family* is complex. Most definitions of family include individuals related by descent, marriage, remarriage, or adoption; some also include affiliated kin.

- Four important family functions are (1) the provision of intimacy, (2) the formation of a cooperative economic unit, (3) reproduction and socialization, and (4) the assignment of social roles and status, which are acquired both in a *family of orientation* (in which we grow up) and in a *family of cohabitation* (which we form by marrying or living together).

- Advantages to living in families include (1) continuity of emotional attachments, (2) close proximity, (3) familiarity with family members, and (4) economic benefits.

- The *extended family* consists of grandparents, aunts, uncles, cousins, and in-laws. It may be formed *conjugally* (through marriage), creating in-laws or stepkin, or consanguineously (by birth) through blood relationships.

- The *kinship system* is the social organization of the family. It includes our nuclear and extended families. Kin can be *affiliated*, as when a nonrelated person is considered "kin," or a relative may fulfill a different kin role, such as a grandmother taking the role of a child's mother.

- There are a range of viewpoints about the meaning and implications of various trends underway in family life in the United States. Race and religion are two of the prominent sources of differences in viewpoints. Even social scientists who study families often disagree about what the trends show about the state of contemporary family life.

## Key Terms

accepters  22
affiliated kin  11
binuclear families  15
centrists  22
clan  12
conjugal relationships  18
consanguineous relationships  18
conservatives  22
extended family  16
family  10
family of orientation  15
family of procreation  15
fictive kin  11
gender  26
household  10
kinship system  18
liberals  22
marriage  6
modified extended families  17
monogamy  10
nuclear family  15
polyandry  10
polygamy  10
polygyny  10
reflection hypothesis  25
rejecters  22
serial monogamy  10
skeptics  22
socialization  13
transgender  26

# 2

# Studying Marriages and Families

*Are the following statements True or False? You may be surprised by the answers as you read this chapter.*

**T** **F** 1. When attempting to gain an understanding about families, we need to rely most on our "common sense."

**T** **F** 2. On average, young adults watch more television than children or older adults.

**T** **F** 3. "Everyone should get married" is an example of an objective statement.

**T** **F** 4. We tend to exaggerate how much other people's families are like our own.

**T** **F** 5. In conducting research on families, one always begins with a clearly stated hypothesis.

**T** **F** 6. Many researchers believe that conflict is a normal feature of families.

**T** **F** 7. According to some scholars, only in impersonal, nonfamily relationships do we weigh the costs against the benefits of the relationship.

**T** **F** 8. Every method of collecting data on families is limited in some way.

**T** **F** 9. Much family research proceeds by using data that others have already collected.

**T** **F** 10. It is impossible to observe family behavior.

## Chapter Outline

P erhaps we should offer a word of warning: The subjects covered in this book come up often and unexpectedly in everyday experience. You may be reading the paper, surfing the Web, or watching television and come upon some news about new research on the effects of divorce or day care on children. You might be having lunch with friends or dinner with your parents, and before you know it, someone may make claims about what marriages need or lack or how some kinds of families are just better or stronger than others. Even more personally, you may find yourself experiencing some unanticipated changes in your own family relationships and wondering about the causes, consequences, or commonality of such change. In fact, although hypothetical, the following situation is not an altogether uncommon or unrealistic one.

Imagine you are out having coffee with a close friend. You can see that she is troubled, and after your urging she confides that she is worried about her two-year-old relationship now that she and her partner are separated by nearly 600 miles while at different colleges. You feel for your friend, think hard about her predicament, and, wanting to be supportive, smile reassuringly. She shares the following: "I don't know, I guess maybe I'm worrying too much. After all, how many times have I heard, 'absence makes the heart grow fonder'? Everyone knows that, right?" As you open your mouth to respond, she continues: "But then I guess I do think too much. Don't they say, 'Out of sight, out of mind'? I wonder if I should just prepare myself, even start looking for someone new. Like now." In obvious distress and confusion, she looks to you for advice.

How would you answer your friend? "Absence makes the heart grow fonder" and "out of sight, out of mind" cannot both be true. Moreover, how can "everyone know" one thing even though "they" say the opposite? Surely, there must be a way to resolve such a contradiction.

In this chapter, we examine how family researchers attempt to explore issues such as the hypothetical

> **False**
> **1.** When attempting to gain an understanding about families, we cannot rely simply on our "common sense."

one posed here. In that sense, this chapter differs from the other 13 chapters. Instead of focusing on material about different aspects of the marriage and family experience, it explains and illustrates how we learn the information about relationships and families found in the rest of the book. However, it will enable you to understand better and appreciate more how much our knowledge and understanding of families is enriched by the theories and research procedures we introduce. In learning *how information is obtained and interpreted*, we set the stage for the in-depth exploration of family issues in the chapters that follow.

## ▌ How Do We Know about Families?

As sociologist Earl Babbie (2007) suggests, social research is one way we can learn about things. However, most of what we "know" about the social world, including about families, we have "learned" elsewhere through other less systematic means (Babbie 2007; Neuman 2006). In the previous chapter, we noted the dangers inherent in generalizing from personal experiences. We all do this. If you or someone you know had an unfavorable experience with a long-distance relationship, you probably favor the "out of sight, out of mind" response more than the optimistic one your friend is hoping to hear.

The opening scenario illustrates the difficulty involved in relying on what are often called common sense–based explanations or predictions (Neuman 2006). Commonsense understanding of family life may be derived from "tradition," what everyone knows because it has always been that way or been thought to be that way; from "authority figures," whose expertise we trust and whose knowledge we accept; or from various media sources. Cumulatively, these alternatives to research-based information are typically less than ideal sources of accurate and reliable knowledge about social and family life. Often, what we consider and accept as common sense is fraught with the kinds of contradictions depicted previously (or, for example, "birds of a feather flock together" but "opposites attract"). Even in the absence of contradiction, many commonsense beliefs are simply untrue. For example, it may seem like "common sense" to assume that living together before getting married would better prepare a couple for sharing a household and their lives with each other and thus assure them greater likelihood of a successful marriage. Reality, however, as uncovered through repeated research, turns out to be more complicated. Early research, beginning in the late 1980s, identified a "**cohabitation effect**," wherein couples who lived together and then married were *more* likely to divorce than were couples who married without first living together. Over time, this effect was observed in the United States, Canada, Australia, Sweden, Britain, and a number of other European countries (Kulu and Boyle 2010). Although it may seem counterintuitive, it appears as though, for some couples, living together before marriage has been associated with a greater risk that their subsequent marriages would end in divorce (Kulu and Boyle 2010). Recently, the cohabitation effect lessened, and presently seems mostly not to hold true. Through careful research efforts our understanding of the circumstances under which it does and doesn't operate have increased. We will examine this in more detail in Chapter 9. For now, though, our point is that if we "really want to know" about how families work or what people in different kinds of family situations or relationships experience, we would be better informed by seeking and acquiring more trustworthy information.

**False** 2. On average, young adults do not watch more television than children or older adults.

## How the Media Misrepresent Family Life

Popular culture, in all its forms, is a key source of both information and misinformation about families. Popular culture conveys images, ideas, beliefs, values, myths, and stereotypes about every aspect of life and society, including the family. Because so much of the day-to-day stuff of family life (e.g., caring for children, arguing, dividing chores, and engaging in sexual behavior) takes place in private and behind closed doors, we do not have access to what really goes on. But we

are privy to those behaviors on television, online, and in movies and magazines. Thus, those depictions can influence what we *assume* happens in real families. The various mass media are so pervasive that they become invisible, almost like the air we breathe. Their presence in daily life is extensive, and advances in technology have only increased the use and potential impact of media (Hertlein 2012). The media images of social life to which we are exposed are often exaggerated and unrealistic, and it is likely that the images and stories presented can affect beliefs and attitudes about life, including family life. Researchers from a number of social science disciplines, including sociology, psychology, economics, education, and communications, have undertaken the study of potential media effects on behavior (Coyne, Bushman, and Nathanson 2012). Much of that attention has been directed at the impact of exposure to violent images, whether on television, in movies, music videos, or video games. Additional attention has been directed at stereotypical media portrayals of such topics as gender, sexuality, or race. The amount and breadth of media exposure makes it likely that there are effects, some of which might be considered negative, others positive (Price and Dahl 2012). The distorted or stereotypical portrayals often found in television programming, as well as the *absence of portrayals* of various issues or groups dictate whether potential effects are more positive or negative.

Cumulatively, television, popular music, the Internet, magazines, newspapers, and movies help shape our attitudes and beliefs about the world in which we live. As suggested in the following data, television plays a particularly prominent role in our lives.

It is estimated that, as of 2014, 96 percent of U.S. households own at least one television and that more than half own at least three televisions (Nielsen Company 2015). That translates to an estimated 116.3 million households (Nielsen Company 2014). In terms of *how much* television is being watched, it is estimated that as of 2015, the average adult watches almost 39 hours per week of traditional or "time-shifted" television, that is, viewer-recorded content or programming accessed on an Internet site and watched at a later time (Nielsen Company 2015). Furthermore, with the range of platforms from which to view media content, including multimedia devices, Blu-Ray devices, DVD players, and DVRs, the average U.S. adult, 18 or older, watches almost six hours a day of media content. The group with the greatest number of hours of viewing continues to be those 65 and older, who watch an estimated 249 hours per month (roughly 8 hours a day of traditional or "time-shifted" television) (Nielsen Company 2015). The extent of exposure to television allows for it to have a powerful effect on our values and beliefs, including and especially our beliefs about families.

Prime-time television, in both dramas and situation comedies, unrealistically depicts married life. It minimizes the difficulties couples face in dividing housework and child care and balancing work and family life; understates the unique issues various ethnic families face; inaccurately depicts single-parent family life; inaccurately portrays the relative sexual activity levels of marrieds and singles; and portrays conflict as something easily resolved within 20 or so minutes, often with humor. It also is not inclusive in portraying the family experiences of nontraditional families.

The combined portrayal of family life on daytime television that results from the few remaining soap operas and the variety of talk shows is unrealistic and highly negative. Daytime soap operas depict extremely high rates of conflict, betrayal, infidelity, and divorce that afflict soap opera families; the stereotyping of women as starry-eyed romantics or scheming manipulators; and the distorted and exaggerated portrayal of sex and sexual infidelity.

Daytime talk shows are even worse. From *Jerry Springer* and *Maury* to *Dr. Phil*, talk shows contribute to the idea that American family life is deeply dysfunctional, that parents are anything from "irresponsible fools" to "in-your-face monsters" (Hewlett and West

1998); that spouses and partners routinely cheat on each other and often strike each other; and that teenagers are recklessly out of control. Half-naked fisticuffs on *Jerry Springer* and contested allegations and paternity tests on *Maury* are especially distasteful distortions of families and relationships.

As you will see throughout this book, although families are not without their share of serious problems, daily family life is as poorly represented by daytime television as by prime-time programming.

## (Un)reality Television

The newest genre of television programming is what has been termed *reality television*. Operating without scripts or professional actors, reality television *purports to put* "ordinary people" into situations or locations that require them to meet various challenges, or to expose viewers to lived reality of those individual men, women, and children featured. Much of what is considered reality television (e.g., *Survivor*, *Naked and Afraid*, *The Apprentice*) has nothing to do with relationships, marriages, or families. However, a number of reality programs have focused directly on relationships and family life, including the following current or canceled shows: *Wife Swap*, *Who Wants to Marry a Multi-Millionaire?*, *The Bachelor*, *The Bachelorette*, *Teen Mom*, *Supernanny*, *Say Yes to the Dress*, *Bridezillas*, *Dad Camp*, *Sister Wives*, and the assortment of *Real Housewives* shows. Among the more recent additions one can point to *Married at First Sight*, *Marriage Boot Camp*, and *Surviving Marriage*. Note that these represent just a fraction of the reality genre. Whether these shows match and/or marry people or showcase aspects of families or relationships, they hardly represent what their genre claims as its name. By highlighting extreme cases (e.g., *Bridezillas*) or introducing competitive goals (*The Bachelor*, *The Bachelorette*), and/or artificial circumstances (*Married at First Sight* or *Surviving Marriage*), these shows are no more representative of the reality of relationships or families than the daytime talk shows. It would be dangerous to draw generalizations from shows such as *Supernanny* and conclude that "kids today" are disrespectful or out of control, or to think that the way to fix a broken marriage is to spend five days "stranded" on an island together, undertaking various tasks and surviving assorted tests, even if under the supervision of two professional therapists. Although you may consider yourself too sophisticated to make generalizations from these programs, millions of

Television's depiction of families includes some diverse family types and issues, such as the relationship between single mother Bonnie, and her single-mother adult daughter Christy (Allsion Janney and Anna Farris) of CBS's *Mom*. Such programs can help shape our attitudes and beliefs about families.

Robert Voets/CBS/Getty Images

others watch programs such as these. Are all of them equally sophisticated?

Alongside and increasingly an alternative, even competing presentations of images and ideas about families and relationships can be found on the Internet. Adults (18 and over) are estimated to have spent over 17 hours per week (in the first quarter of 2015) using the Internet on a computer (1 hour and 7 minutes per day) or using a smartphone (1 hour and 27 minutes per day) (Nielsen Company 2015). On a positive note, one can locate and use websites to access and assess good, sound objective research that can fill the gap between what knowledge we have otherwise obtained about families and a wider reality, whether in the United States or across cultures. On the negative side, one can also access distorted, biased, misleading, and inaccurate content. Unfortunately, it is not always evident which type of information one has obtained.

## Advice, Information, and Self-Help Genres

Although you may suspect that most people who watch television comedies, dramas, talk shows, or reality programs realize that they are not the best sources for accurate information about families, there is one genre of the media where such realization may be absent. A veritable industry exists to support what's called "the advice and information genre." It produces self-help and child-rearing books, advice columns, radio and television shows, and numerous articles in magazines and newspapers. In fact, as it transmits information, it also conveys values and norms—cultural rules and standards—about relationships, marriage, and family, often intended as entertainment.

In newspapers in the past, this genre was long represented by such popular "advice columnists" as Abigail Van Buren (whose real name was Pauline Esther Friedman and whose column "Dear Abby" is now written by her daughter, Jeanne Phillips); Dan Savage (whose sex advice column "Savage Love" is widely syndicated in numerous newspapers), and the late Ann Landers (Abby's twin sister, Esther Pauline Friedman, whose daughter, Margo Howard, now writes her own advice column, "Dear Margo").

Web-based columnists, such as Alison Blackman Dunham and her late twin sister Jessica Blackman Freedman, the self-proclaimed "Advice Sisters" (or "Ann and Abby for the new millennium"), helped carry this genre to the Internet. Allison Blackman

Dunham continues to give relationship advice online at her "Great Relationships" website (advicesisters. net). Numerous other online sites offer relationship advice, including sites run by Jeanne Phillips as Dear Abby, and by Margo Howard (now as "Dear Margo") where they respond to readers' online questions. Emily Yoffe writes her "Dear Prudence" column for Slate.com (www.slate.com/authors.emily_yoffe.html). Carolyn Hax writes her daily advice column for the *Washington Post*, and hosts a weekly online discussion and a Facebook page (www.washingtonpost.com/pb /people/carolyn-hax/).

Other advice sources can be found on radio and television. For many years, radio therapists, such as Dr. Joy Browne and Dr. Laura Schlessinger, had daily callers seeking advice or information about relationships, family crises, and so on, and the program *Loveline* with a number of different hosts offered listeners medical and relationship advice since 1983. Presently there are radio call-in shows hosted by clinical psychologists and marriage and family therapists (e.g., Dr. Jenn Berman's, "The Love and Sex Show with Dr. Jenn," Dr. Michelle Cohen's, "On the Couch with Dr. Michelle").

On television, Dr. Phil McGraw's *Dr. Phil* has long been a ratings success. McGraw, a psychologist of close to 30 years, was featured often on *The Oprah Winfrey Show* before landing his own talk show in 2002. His shows cover a range of personal and family issues. Some recent episodes included the following: "I've Divorced My Ex Twice but I'm Still in Love," "My 21-Year-Old Daughter Is Obsessed with Her Controlling, Jealous, Heroin-Addicted Boyfriend," "Parenting While Intoxicated," "Was It Child Abuse, or Was It an Accident?" and "In Sickness and in Health: Should I Get a Divorce from My OCD Wife?"

Dr. Phil is also the author of a number of best-selling books, including *Family First: Your Step-by-Step Plan for Creating a Phenomenal Family* and *Relationship Rescue: A Seven-Step Strategy for Reconnecting with Your Partner*, and has a website from which visitors can obtain a variety of suggestions for how to deal with the kinds of relationship and personal issues featured on his show or in his books.

Of all media genres, the advice and information genre can be especially misleading. After all, it offers advice to and depicts the experiences of "real people." It uses experts, cites studies, and may sometimes sound scientific. However, as the following Popular Culture feature suggests, there are good reasons to be skeptical of the information such sources provide or the advice they offer.

By now, between the previous chapter and the present one, we hope you are aware of the limitations of both personal experience and popular culture as sources of reliable information about familial reality. That doesn't mean that we cannot learn from our own experiences or become more aware of various aspects of family life because of how they are portrayed or presented in the mass media. However, if one's goal is to discover what family life looks and feels like for most people or for people unlike ourselves, we need to turn to more trustworthy sources of information.

## ◼ Researching the Family

Before we examine the specific theories and research techniques that family researchers use, it is important to emphasize that the attitudes of the researchers (or you as you read research) are important. To obtain valid research information, we need to keep in mind the rules of critical thinking. The term *critical thinking* is another way of saying "clear and unbiased thinking."

### The Importance of Objectivity

We all have perspectives, values, and beliefs regarding marriage, family, and relationships. These can create blinders that keep us from accurately understanding the research information. Instead, we need to try to develop a sense of **objectivity** in our approach to information—to suspend the beliefs, biases, or prejudices we have about a subject until we understand what is being said (Kitson et al. 1996). We can then take that information and relate it to the information and attitudes we already have. Out of this process, a new and enlarged perspective may emerge.

One area in which we may need to be alert to maintaining an objective approach is that of family lifestyle. The values we have about what makes a successful family can cause us to decide ahead of time that certain family lifestyles are "abnormal" because they differ from our experience, preference, or our sense of what is "right," "moral," or "best." For example, we may refer to single-parent families as "broken" or say that adoptive parents are "not the real parents."

**False** 3. "Everyone should get married" is an example of a value judgment, not an objective statement.

**True** 4. We tend to exaggerate how much other people's families are like our own.

A clue that can sometimes help us "hear" ourselves and detect whether we are making value judgments or **objective statements** is as follows: A **value judgment** usually includes words that mean "should" and imply that our way is the correct way. An example is, "Everyone should get married." This text presents information based on scientifically measured findings—for example, concluding that "between 85 percent to 90 percent of Americans will likely marry in their lifetimes."

Opinions, biases, and stereotypes are ways of thinking that lack objectivity. An **opinion** is based on our experiences or ways of thinking. A **bias** is a strong opinion that may create barriers to hearing anything contrary to our opinion. A **stereotype** is a set of simplistic, rigidly held, and overgeneralized beliefs about the personal characteristics of a group of people. Stereotypes are fairly resistant to change. Furthermore, stereotypes are often negative. Common stereotypes related to marriages and families include the following:

- Children raised in single-parent households suffer serious disadvantages.
- Stepfamilies are unhappy.
- Lesbians and gay men cannot be good parents.
- Women are instinctively nurturing.
- People who divorce are selfish.

We all have opinions and biases; most of us, to varying degrees, think stereotypically. But the commitment to objectivity requires us to become aware of these opinions, biases, and stereotypes and to put them aside in the pursuit of knowledge.

**Fallacies** are errors in reasoning. These mistakes come as the result of errors in our basic presuppositions. Two common types of fallacies that especially affect our understanding of families are egocentric fallacies and ethnocentric fallacies. The **egocentric fallacy** is the mistaken belief that everyone has the same experiences and values that we have and therefore should think as we do. The **ethnocentric fallacy** is the belief that our ethnic group, nation, or culture is superior to others. In the next chapter, when we consider the differences and strengths of families from different ethnic and economic backgrounds, you need to keep both of these fallacies from distorting your understanding.

# Evaluating the Advice and Information Genre

The various radio or television talk shows, columns, articles, websites, and advice books have several things in common. First, their primary purpose is to sell books or periodicals or to raise program ratings. They must capture the attention of viewers, listeners, or readers. In contrast, the primary purpose of scholarly research is the pursuit of knowledge.

Second, although disseminating information about relationships and families, the media must entertain. Thus, the information and advice must be simplified. Complex explanations and analyses must be avoided because they would most likely undermine the entertainment purpose. Furthermore, the media rely on high-interest or shocking material to attract readers or viewers. Consequently, we are more likely to read or view stories about finding the perfect mate or protecting our children from strangers than stories about new research methods or the process of gender stereotyping.

Third, the advice and information genre focuses on "how-to" information or morality. The how-to material advises us on how to improve our relationships, sex lives, child-rearing abilities, and so on. Advice and normative judgments (evaluations based on norms) are often mixed together. Advice columnists act as moral arbiters, much as do ministers, priests, rabbis, and other religious leaders.

Dr. Phil McGraw is a licensed clinical psychologist who in addition to his television program has authored six *New York Times* no. 1 best-selling books.

Everett Collection

To reinforce their authority, the various media often incorporate statistics, which are key features of social science research. But not all statistics are of equal value. Unless the research from which the statistics are drawn is sound, statistics have little usefulness. They may even misrepresent family reality by poor sampling techniques, inappropriately worded questions, or unwarranted generalizations.

With the media awash in advice and information about relationships, marriage, and family, how can we evaluate what is presented to us? Here are some guidelines:

- Be skeptical. Remember: Much of what you read or see is meant to entertain you. Consider the reliability of the sources and the representativeness of the people interviewed.
- Search for biases, stereotypes, and lack of objectivity. Is conflicting information omitted? Are diverse family forms and experiences represented?
- Look for moralizing. Many times, what passes as fact is disguised moral judgment. What are the underlying values of the article or program?
- Go to the original source or sources. The media simplify, so find out for yourself what the studies said and how well or poorly they were done.
- Seek additional information. The whole story is probably not told, so seek out additional information, especially information in scholarly books and journals, reference books, or college textbooks.

Throughout this book, you will be exposed to a variety of information or data about families. This information may or may not reflect your experiences, but its value is that it will enable you to learn about how other people experience family life. This knowledge and the results of different kinds of responses to family situations enable a more informed understanding of families in general and of yourself as an individual.

Television talk shows, such as *The View*, occasionally discuss topics and issues related to relationships and families.

Lou Rocco/Disney ABC Television Group/Getty Images

Despite the fact that many people believe stereotypes of gay men and lesbians and have biases against same-sex couples as parents, research reveals that heterosexuals, lesbians, and gay men can all be equally effective and loving parents.

From the day of your birth, you have been forming impressions about human relationships and developing ways of behaving based on these impressions. Hence, you might feel a sense of "been there, done that" as you read about an aspect of personal development or family life. However, your study of the information in this book will provide you the opportunity to reconsider your present attitudes and past experiences and relate them to the experiences of others. As you do this, you will be able to use the logic and problem-solving skills of critical thinking so that you can effectively apply that which is relevant to your life.

## The Scientific Method

Family researchers come from a variety of academic disciplines—from sociology, psychology, and social work to communication and family studies (sometimes known as "family and consumer sciences"). Although these disciplines may differ in terms of the specific questions they ask, the data collection techniques they use, or the objectives of their research, they are unified in their pursuit of accurate and reliable information about families through the use of social scientific theories and research techniques. Scholarly research about the family brings together information and formulates generalizations about certain areas of experience. These generalizations help us to better understand the factors that shape family experience and to predict what happens when certain conditions or actions occur.

Family science researchers use what is often referred to as the **scientific method**—well-established

procedures used to collect and analyze information about family experiences. This rigorous approach allows other people to know the source of the information and to be confident of the accuracy of the findings. Much of what the family research scientists do is shared in specialized journals (e.g., the *Journal of Marriage and Family* and the *Journal of Family Issues*), on websites, or in book form. By communicating their results through such channels, other researchers can build on, refine, or further test research findings. Much of the information contained in this book originally appeared in scholarly journals or government reports.

## Concepts, Variables, Hypotheses, and Theories

One of the most important differences between the knowledge about marriage and family derived from family research and that acquired elsewhere is that family research is influenced or guided by concepts and theories. **Concepts** are abstract ideas that we use to represent the concrete reality in which we are interested. We use concepts to focus our research and organize our data. Many examples of concepts—for example, nuclear families, monogamy, and socialization—were introduced in the previous chapter. Family research involves the processes of **conceptualization**, the specification and definition of concepts researchers use, and of **operationalization**, the identification and/or development of research strategies to observe or measure concepts. For example, to study the relationship between social class and child-rearing strategies, we need to define and specify how we are going to identify and measure a person's social class position as well as what we mean by and how we'll measure their child-rearing strategies.

**Theories** are sets of general principles or concepts used to explain a phenomenon and to make predictions that then may be tested and verified empirically. Although researchers collect and use a variety of kinds of data on marriages and families, these data alone do not automatically convey the meaning or importance of the information gathered. Concepts and theories supply the "story line" for the information we collect.

In **deductive research**, the key concepts are turned into **variables**, concepts that can vary in some meaningful way. Marital status is an example of a variable family researchers use. In studying the causes or consequences of different marital statuses, we will have to decide how we wish to differentiate among them. We could elect to compare those who are currently married with all those who are not currently married, or we may seek a finer, more nuanced measure of marital status. After all, we may be married, separated, divorced, widowed, or never married. As researchers explore the causes and/or consequences of marital status, they may formulate **hypotheses**, or predictions, about the causes and/or consequences of marital status and about relationships between marital status and other variables. For example, we might hypothesize that race or social class influences whether or not someone is married. In such an example, race is an **independent variable** and marital status the **dependent variable** in that race is thought to influence the likelihood of becoming or staying married. On the other hand, marital status may be examined as a causal or independent variable in a hypothesized relationship between being married and life expectancy or satisfaction. Finally, marital status might be hypothesized as an **intervening variable**, affected by the independent variable, race, and in turn affecting the dependent variable, life expectancy. In that instance, the hypothesis suggests that race differences in marital status account for race differences in life expectancy (see Figure 2.1).

Rarely do researchers construct theories with only two or three variables. They may hypothesize multiple independent and intervening variables and seek to identify those having the greatest effect on the dependent variable (Neuman 2006). In Figure 2.1, panel (d) is an illustration of this, with regard to the

**False** **5.** Family research generally begins with a topic of interest, rather than a hypothesis.

race–marital status example. There, race is hypothesized to have both direct and indirect effects on marital status. Race is alleged to have effects on both income and education, which in turn are hypothesized to affect marital status. And, finally, race, income, and marital status are all hypothesized to have independent effects on life expectancy.

**Inductive research** is not hypothesis-testing research. Instead, it begins with a topic of interest to researchers (e.g., what happens when couples reverse traditional roles) and perhaps some vague concepts. As researchers gather their data, typically in the form of field observations and/or interviews, they refine their concepts, seek to identify recurring patterns out of which they can make generalizations, and, perhaps, end by building a theory (or asserting some hypotheses) based on the data collected. Theory that emerges in this inductive fashion is often referred to as **grounded theory** in that it is *grounded* or "rooted in observations of specific, concrete details" (Neuman 2006).

## ■ Theoretical Perspectives on Families

On a more abstract level of theory, we can identify major theoretical frameworks or perspectives that guide much of the research about families. These perspectives (sometimes also called *paradigms*) are sets of

**Figure 2.1** The four examples illustrate different ways researchers might hypothesize about the causes or effects of the variable marital status.

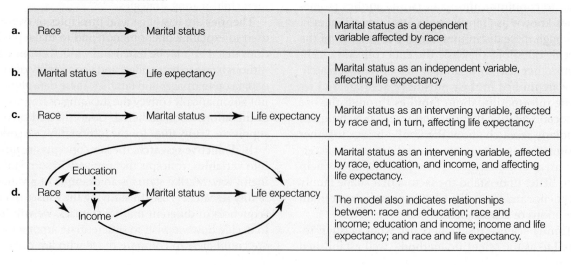

| | | |
|---|---|---|
| **a.** | Race → Marital status | Marital status as a dependent variable affected by race |
| **b.** | Marital status → Life expectancy | Marital status as an independent variable, affecting life expectancy |
| **c.** | Race → Marital status → Life expectancy | Marital status as an intervening variable, affected by race and, in turn, affecting life expectancy |
| **d.** | Education / Race → Marital status → Life expectancy / Income | Marital status as an intervening variable, affected by race, education, and income, and affecting life expectancy. The model also indicates relationships between: race and education; race and income; education and income; income and life expectancy; and race and life expectancy. |

concepts and assumptions about how families work and/or how they fit into society. Theoretical frameworks guide the kinds of questions we raise, the types of predictions we make, where we look to find answers, and how we construct explanations (Babbie 2007).

In this section, we briefly discuss several of the most influential theoretical perspectives sociologists and psychologists have used to study families, including ecological, symbolic interaction, social exchange, developmental, structural functional, conflict, feminist, and family systems theory. Fuller discussions of these and other perspectives are available in a number of books on family theories. Some are organized around the different theoretical perspectives that we discuss in the following sections (e.g., Smith and Harmon 2012; White and Klein 2014), while others are organized more around topics in family studies (e.g., Bengtson et al. 2005; Fine and Fincham 2013). As you examine the summaries that follow, you should notice how the choice of a theoretical perspective can influence the way data are interpreted. Furthermore, as you read this book, you should ask yourself how different theoretical perspectives might lead to different conclusions about the same material.

## Macro-Level Theories

Theories tend to either focus on the family as a social institution or the family as a group of individuals, albeit a unique kind of group. The following four theories are what we call **macro-level theories** in that they typically focus on the family as a **social institution** (see Table 2.1). A social institution is the organized pattern of statuses and structures, roles, and rules by which a society attempts to meet certain of its most basic needs. For example, "the family" refers to groups and relationships through which a society regulates reproduction, socializes children, and provides us with

**TABLE 2.1  Macro- and Micro-Level Theories that Guide Research on Families**

| Macro-Level Theories | Micro-Level Theories |
|---|---|
| Family ecology theory | Symbolic interactionism |
| Structural functionalism | Social exchange theory |
| Conflict theory | Developmental theory |
| Feminist theory | Family systems theory |

© Cengage Learning

emotional support. The economy is the way a society organizes the production and distribution of essential goods and services. Other institutions include religion, politics, and education. Macro-level theories examine how the family as a social institution is affected by the wider society, including the interconnections and influences between families and other social institutions. Sometimes these same perspectives are also applied to analyzing the internal structure of relationships within families. Still, when they do, they typically connect that structure to wider social, economic, and political influences.

### Family Ecology Theory

The emphasis of **family ecology theory** is on how families are influenced by and in turn influence the wider environment. Ecological theory was introduced in the late 19th century by plant and human ecologists. German biologist Ernst Haeckel first used the term *ecology* (from the German word *oekologie*, or "place of residence") and placed conceptual emphasis on **environmental influences**. This focus was soon picked up by Ellen Swallow Richards, the founder and first president of the American Home Economics Association (now known as the American Association of Family and Consumer Sciences), who believed that scientists needed to focus on home and family, "for upon the welfare of the home depends the welfare of the commonwealth" (quoted in White and Klein 2002).

The core concepts in ecological theory include **environment** and **adaptation**. Initially used to refer to the adaptation of plant and animal species to their physical environments, these concepts were later extended to humans and their physical, social, cultural, and economic environments (White and Klein 2002). As applied to family issues, the family ecology perspective asks: How is family life affected by the environments in which families live? We use the plural *environments* to reflect the multiple environments that families encounter.

In psychologist Urie Bronfenbrenner's ecologically based theory of human development, development occurs as individuals engage in reciprocal interaction, on a regular basis over extended periods of time, with "the persons, objects and symbols in [one's] immediate external environment" (Bronfenbrenner and Morris 1998, 96). Through such interactions, which Bronfenbrenner called proximal processes, individuals learn "to make sense of their world and understand their place in it" (Tudge et al. 2009, 200).

For many of us, even so many years later, the awful images are still vividly with us and recalled anytime we hear the name "Katrina." Hurricane Katrina was a true tragedy, a category 5 hurricane that struck the Gulf Coast states of Mississippi, Louisiana, Alabama, and Florida in August 2005. Approximately 1.3 million people were evacuated from Gulf Coast areas in Katrina's path, some of whom were sent "to distant out-of-state shelters" in evacuations that lasted weeks or months (Rendall 2011).

Disasters such as Hurricane Katrina in 2005 often throw families into extreme situations of ambiguous loss.

Michael Ainsworth/Dallas Morning News/Corbis News/Corbis

With winds that occasionally reached 170 miles an hour, Katrina devastated the region. Hardest hit was New Orleans, where 80 percent of the city was submerged underwater. Even a year after Katrina hit, more than half the population of New Orleans, as many as 247,000 people, were still living outside the city (Rendall 2011).

The human cost of Katrina was enormous. More than 1,300 people in five states died from Hurricane Katrina. Suddenly, thousands of people lost spouses, parents, siblings, and children. In addition, many thousands of families, for at least a time, were left in limbo, without news about the whereabouts of missing loved ones or knowledge of whether they had survived. Months after Hurricane Katrina, more than 3,000 of the nearly 11,500 people reported missing were still missing (Roberts 2006). What must families feel in such situations?

Pauline Boss has spent more than 30 years studying families dealing with either physically missing or psychologically missing members. Beginning in the early 1970s, by looking at *psychological father absence* (fathers who were present but distant), Boss broadened her interest to include situations in which any family member might be said to be "there but not there." She labeled such circumstances **ambiguous loss**, "a situation of unclear loss resulting from not knowing whether a loved one is dead or alive, absent or present" (Boss 2004, 554). She suggested that such loss is the most stressful because it remains unresolved and creates lasting confusion "about who is in or out of a particular family" (p. 553). There is no death certificate, no funeral, and no opportunity to

The environment to which individuals adapt as they develop consists of four levels: (1) *microsystem*, (2) *mesosystem*, (3) *exosystem*, and (4) *macrosystem*. Cumulatively, these levels make up the environments in which we live. The *microsystem* contains the most immediate influences with which individuals have frequent contact (e.g., our families, peers, schools, and neighborhoods). Each of these exerts influence over how we develop. The *mesosystem* consists of the interconnections between microsystems (e.g., school experiences and home experiences) and ways they influence each other. The *exosystem* consists of settings in which the individual does not actively participate but that nonetheless affect his or her development (e.g., one's parents' work experiences, salaries, and schedules). Finally, the *macrosystem* operates at the broadest

level, encompassing the laws, customs, attitudes, and belief systems of the wider society, all of which influence individual development and experience (Rice and Dolgin 2002).

In constructing an ecological framework to better understand marriage relationships, sociologist Ted Huston illustrated how marital and intimate unions are "embedded in a social context" (Huston 2000). This social context includes the *macroenvironment*—the wider society, culture, and physical environment in which a couple lives—and their particular *ecological niche*—the behavior settings in which they function on a daily basis (e.g., a poor, urban neighborhood as opposed to a small town or suburb). Also included in the social context is the marriage relationship itself, especially as it is affected by a larger network honor the deceased or bury remains. It prevents family members from reaching psychological closure. As a result, families are immobilized, roles are confused, and tasks remain undone.

Boss considers two situations of ambiguous loss. First is the ambiguous loss of "there but not there," of "physical presence and psychological absence" mentioned previously and applicable in unexpected situations, such as when a family member suffers from dementia (including Alzheimer's disease), autism, depression, or addictions, and in more common situations, such as preoccupation with work, obsessive involvement with the Internet, or divorce followed by remarriage. In the second form of ambiguous loss, members remain psychologically present despite physical absence. This "not there (physically) but there (psychologically)" version of ambiguous loss can be found in tragic situations of war (for families of soldiers missing in action), among families of incarcerated inmates, in families where a member deserts, and in such events as occurred on 9/11 or in the Gulf Coast states, especially if no body is recovered. In addition, researchers have identified situations of ambiguous loss in a variety of cultural contexts beyond the United States and the industrialized West, such as in Nepal and Sudan (Luster et al. 2008, Robins 2010). Refugees, especially children who are separated from family members and resettled, suffer similar uncertainty as they flee or are evacuated and are left not knowing the fate of their parents and siblings (Luster et al. 2008). More common versions of "not there but there" can occur after divorce or adoption, work relocations, and children leaving home and the "nest" emptying (Boss 2004).

Not all situations of ambiguous loss result in the same outcomes or suffering. Some families manage to redraw otherwise ambiguous boundaries (such as when an aunt or uncle steps in and is viewed as a parent). It appears as if some people have a higher tolerance for ambiguity and therefore may be more resilient in instances such as Hurricane Katrina. Other individuals may suffer emotional or psychological wounds after such tragedies. Although surviving family members may also experience post-traumatic stress disorder (PTSD), ambiguous loss is not the same as PTSD. Treatment of PTSD focuses on individuals, not families as whole systems. In addition, PTSD is a pathology, a psychological illness. Ambiguous loss is a situation of stress that can lead to individual suffering but needs to be understood *on the familial level* (Boss 2004).

More than 8,000 of those reported missing in the Gulf Coast after Katrina have been found or their bodies identified. Still, 3,200 or more families struggle to find closure and come to terms with what the storm took from them. Using concepts such as *ambiguous loss* enables us to better understand what they suffer from and why. Such understanding won't eliminate their suffering or reduce the pain of their losses, but it may make it possible to be more effective in any efforts to help them move on. Toward that end, in a recent book, *Loss, Trauma, and Resilience: Therapeutic Work with Ambiguous Loss* (2006), Dr. Boss offers guidelines for practitioners—therapists and teachers—who are working with people suffering from ambiguous loss. ●

of relationships. The final key element in Huston's ecological approach contains the physical, psychological, and social attributes of each spouse, including attitudes and beliefs about their relationship and each other. Each of these environments influences and is influenced by the others. His ecological approach suggests that to fully understand marriage, we must explore the interconnections among these three elements.

More recently, Heather Helms, Andrew Supple, and Christine Proulx advocated using Huston's ecological model specifically to study Mexican-origin married couples. They contended that to more completely understand how Mexican-origin couples' marriages fare during the early years of parenthood, one must pay attention to the ecological contexts in which such marriages are embedded. They suggested that without assessing the macroenvironmental context, which includes their country of origin, their adaptation to U.S. culture, their socioeconomic conditions, and their more immediate social and physical environments, one couldn't fully understand couples' experiences (Helms, Supple, and Proulx 2011).

Ecological theory has been and continues to be a popular perspective among family researchers. In recent years, it has been used broadly in the study of a wide range of family issues. For example, in recent years, it has been applied to such topics as sibling relationships (Whiteman, McHale, and Soli 2011), how children of incarcerated mothers maintain family relationships (Poehlman et al. 2008), the effects of economic restructuring on rural women and their families (Ames, Brosi, and Damiano-Teixeira 2006), the mothering behavior of women involved in street-level prostitution (Dalla 2004), and changes in the quality of lesbian relationships as couples become parents (Goldberg and Sayer 2006). It has been especially prominent in studying low-income families and in exploring the multiple influences on children's and adolescents' behavior and school performance (e.g., Bowen et al. 2008).

Regardless of the specific topic, ecological approaches examine how the broader social environments affect family experience. In many ways, much of what we examine in subsequent chapters shares at least this level of ecological focus. We cannot understand what happens within families without considering the wider cultural, social, and economic environments within which family life takes place.

**Critique.** There have been a variety of criticisms of ecological theory (White and Klein 2002). It is not always entirely clear which system best accounts for the behavior we attempt to explain (e.g., microsystem, mesosystem, exosystem, or macrosystem), how exactly that outcome results, or how the different systems influence one another. A second criticism contends that the theory has been more effectively applied to individual or familial development and growth. Families are prone to degeneration or decline as much as they are to development and growth, yet ecological theory often fails to address this. Finally, some suggest that the theory may not apply as well to a range of diverse, especially nontraditional, families.

## Structural Functionalism Theory

**Structural functionalism theory** was for a long time the dominant perspective from which sociologists studied society and the family. In the latter two or three decades of last century it saw its acceptance and use diminish for reasons that we will note. Although many now reject the functionalist perspective, it still surfaces, especially in certain areas of sociology, such as research on aging (Connidis 2012).

The structural functionalist perspective seeks to explain how society works, how families work, and how families relate to the larger society and to their own members. The theory has been used largely in sociology and anthropology, disciplines that focus on the study of society rather than of individuals. When structural functionalists study the family, they typically look at three aspects: (1) what functions the family as an institution serves for society (discussed in Chapter 1), (2) what functional requirements family members perform for the family, and (3) what needs the family meets for its individual members.

Structural functionalism was deeply influenced by biology. It treats society as if it were a living organism, like a person, animal, or tree. The theory sometimes uses the analogy of a tree in describing society. In a tree, there are many substructures or parts, such as the trunk, branches, roots, and leaves. Each of these structures has a function in keeping the tree alive. The roots gather nutrients and water from the soil, the leaves absorb sunlight, and so on. Society is like a tree insofar as it has different structures that perform functions for its survival. These structures are called **subsystems,** each of which is necessary for the survival of the wider societal system.

The subsystems of society are the major social institutions, including the family, religion, government, education, and the economy. Each of these performs functions that help maintain society, just as the different parts of a tree serve a function in maintaining the tree. Religion gives spiritual support, government ensures order, and the economy produces goods. The family provides new members for society through procreation and socializes its members so that they fit into society. In functionalist theory, all institutions work in harmony for the good of society and one another. In fact, an underlying concern characteristic of functionalism is "in what way(s) does the phenomenon being studied contribute to the survival and stability of the wider society (or group)?"

**The Family as a System.** Families themselves may also be regarded as systems. When looking within "the family," structural functionalists examine how the family organizes itself for survival and what functions the family as a whole performs for its members. For the family to survive, its members must perform certain functions that traditionally were divided along age and gender lines. During the heyday of the functionalist approach, men and women in the United States were expected to perform different roles and were assigned different tasks: Men worked outside the home to provide an income, whereas women performed household tasks and child rearing. To functionalists, the male breadwinner–female homemaker nuclear family "was ideally adapted to the needs of industrial society, was essential for marital stability, and overall satisfaction was said to be maximized when men work and women don't" (Ruggles 2014).

According to structural functionalists, the family molds the kind of personalities it requires to carry out its functions and meet its needs. It encourages different personality traits for men and women as a way to ensure its survival. Men develop instrumental traits, and women develop expressive traits. *Instrumental traits* encourage competitiveness, coolness, self-confidence, and rationality—qualities that will help a person succeed in the world of work. *Expressive traits* encourage warmth, emotionality, nurturing, and sensitivity—qualities appropriate for someone responsible for taking care of a family and managing a household.

Such a division of labor and differentiation of temperaments is seen as efficient because it allows each spouse to specialize, thus minimizing competition, creating interdependence between spouses, and reducing ambiguity or uncertainty over such things as who should work outside the home or whose outside employment is more important. For these reasons, such role allocation may be deemed functional, at least for heterosexual couples.

Although the structural functionalist perspective is not nearly as prominent as it was in the mid-20th century when it dominated American sociology, other functionalist analyses can still be found in contemporary scholarship on families. Without valorizing a particular form of family, some researchers do assess family phenomena by examining their functional contributions. For example, a 2014 article by sociologist Medora Barnes looked at the functions of traditional wedding rituals among a small sample of white, middle-class young women who had recently had a formal or traditional wedding. Noting that functionalist theory stresses that rituals, including wedding rituals, exist because they fulfill specific functions, Barnes examined the functions of wedding rituals especially in the context of today's patterns of marriage and divorce.

Traditionally, wedding rituals performed the following functions: serving as a formal rite of passage from one life stage to another; providing reassurance that one's choice of partner has the approval of family and friends; and uniting two families and fostering emotional connections between members. Today, formal, traditional weddings, along with some newer wedding rituals such as Unity Candles, Rose Ceremonies, and Family Medallion Services, continue to serve as ways to acknowledge and strengthen family bonds. Additionally, having a traditional wedding means that couples successfully plan together and hold a large and very public ceremony acknowledging their commitment to each other. Although women do more of the wedding planning work, brides gain reassurance that they and their future husbands can work together and overcome life's challenges in navigating the wedding planning process. In a climate in which marriage rates have declined, divorce rates remain high, and women's and men's roles in and outside of marriage have changed, these two functions become potentially even more important (Barnes 2014).

**Critique.** Although structural functionalism was an important and widely used theoretical approach to the family, it has declined in significance in recent decades for several reasons.

First, how do we know which family functions are vital? The family, for example, is supposed to socialize children, but the schools, peer groups, and the media have taken over much socialization. Is this "functional"? Does it diminish the importance of families?

Second, structural functionalism typically looks at the family abstractly, from a distance far removed from the daily lives and struggles of men, women, and children. It views the family in terms of functions and roles. Family interactions, the lifeblood of family life, are absent, making structural functionalism of little relevance to real families in the real world. On a related note, historical demographer Steven Ruggles points out that empirical evidence for the efficiency of or satisfaction within breadwinner–housewife families was typically absent from functionalist analysis, even when structural functionalist theorists were extolling its virtues.

Third, it is not always clear what function a particular structure serves. For example, what is the function of the traditional division of labor along gender lines—efficiency, survival, or the subordination of women? If interdependence, specialization, and clarity of role responsibilities made breadwinner–homemaker households most "functional," those same objectives can be met by household arrangements wherein women earn incomes and men stay home, rear kids, and tend house. In some relationships, these role reversals might be even more functional. But structural functionalism has a conservative bias against change. Aspects that reflect stability are called functional, and those that encourage instability (or change) are called dysfunctional. By that logic, traditional roles would be described as functional, and nontraditional roles would be more likely seen as dysfunctional, deviant, or harmful.

Functionalism has been especially criticized for disregarding inequality, power, and conflict in families. Even within the idealized breadwinner–housewife family, not all members perceived such an arrangement as ideal (Ruggles 2014). These criticisms are the very focal concerns of the next perspective, conflict theory.

## Conflict Theory

The basic premise of conflict theory sets it apart from structural functionalism. Where structural functionalists assert that existing structures benefit society, conflict theorists ask, "Who benefits?" **Conflict theory** holds that life involves discord and competition. Conflict theorists see society not as basically cooperative but as divided and unequal, with individuals and groups in conflict with one another. Given the scarcity of society's resources, such as wealth, income, prestige, or power, groups are in competition for those resources. As a consequence of such competition, inequality results. Conflict theory directs attention to how the wider gender, racial, and economic inequalities that result in society influence families.

**True** 6. Many researchers believe that conflict is a normal feature of families.

Conflict theory is sometimes applied to the conflict and inequality that exists within families between members over such things as the division of responsibilities, allocation of resources, and levels of commitment.

Harry Stewart/Alamy Stock Photo

**Sources of Conflict in Families.** We can also analyze marriages and families in terms of internal conflicts and power struggles. Conflict theorists agree that love and affection are important elements in marriages and families, but they believe that conflict and power are also fundamental. Marriages and families are composed of individuals with different personalities, ideas, values, tastes, and goals. Each person is not always in harmony with every other person in the family.

Conflict theorists do not believe that conflict is bad; instead, they think that it is a natural part of family life. Families always have disagreements, from small ones, such as what movie to see, to major ones, such as how to rear children. Families differ in the number of underlying conflicts of interest, the degree of underlying hostility, and the nature and extent of the expression of conflict. Conflict can take the form of competing goals or differences in role expectations and responsibilities. For example, an employed mother may want to divide housework 50/50, whereas her husband insists that household chores are not his responsibility.

**Sources of Power.** When intrafamily conflict occurs, who wins? Family members have different resources and amounts of power. There are four important sources of power: legitimacy, money, physical coercion, and love. When arguments arise in a family, a man may want his way "because I'm the head of the household" or a parent may argue "because I'm your mother." These appeals are based on legitimacy—that is, the belief that the person is entitled to prevail by right. Money is also a potentially powerful resource within marriages and families, much as it is in the wider society. "As long as you live in this house . . ." is a parental directive based on the power of the purse. Because men have tended to earn more than women, they have had greater economic power, which, in part, can translate into marital power. More than money is involved, however, because when wives outearn husbands, they don't reap the same benefits or enjoy the same amount of power. Physical coercion, threats, or force make up another important source of power. Spanking children and physical abuse of spouses are examples of this. Finally, there is the power of love. Love can be used to coerce someone emotionally, as in "If you really loved me, you'd do what I ask."

**Critique.** A number of difficulties arise in conflict theory. First, conflict theory derives from politics and economics, in which self-interest, egotism, and competition are dominant elements. Yet, is such a harsh judgment of human nature justified? People's behavior is also characterized by self-sacrifice and cooperation.

Love is an important quality in relationships. Conflict theorists do not often talk about the power of love or bonding, yet the presence of love and bonding may distinguish the family from all other groups in society. We often will make sacrifices for the sake of those we love. We will defer our wishes to another's desires; we may even sacrifice our lives for a loved one. Second, conflict theorists assume that differences lead to conflict. Differences can also be accepted, tolerated, or appreciated. Differences do not necessarily imply conflict. Third, conflict in families is not easily measured or evaluated. Families live much of their lives privately, and outsiders are not always aware of whatever conflict exists or how pervasive it is. In addition, much overt conflict is avoided because it is regulated through family and societal rules. Most children obey their parents, and most spouses, although they may argue heatedly, do not employ violence.

## Gender and Feminist Perspectives

Thanks to the feminist movement of the 1960s and 1970s, new questions and ways of thinking about society and social life, including the meaning and characteristics of families, arose. Although there are some notable distinctions between different kinds of **feminist perspectives**, when applied to families, they tend to broadly share certain concerns: seeing gender as a social construct as opposed to a biological given; recognizing the extent and seeking to explain gender differences and inequalities in social, economic, and familial roles; focusing on and making visible women's family experiences; acknowledging and appreciating family diversity; and advocating for change (Humble et al. 2006). As mentioned last chapter and addressed more in Chapter 4, gender theorists in recent years have paid increasing attention to those who are transgendered and whose gender identities and expression fall outside of binary constructions of gender.

Feminists critically examine the ways in which family experience is shaped by gender—the social and psychological aspects of being female or male. This is the orienting focus that unifies most feminist writing, research, and advocacy. Feminists maintain that family and gender roles have been constructed by society and do not derive from biological or absolute conditions. They further tend to believe that men have created family and gender roles to maintain power over women. Basically, the goals of the feminist perspectives are to work to accomplish changes and create conditions in society that remove barriers to opportunity and oppressive conditions and are, instead, "good for women" (Thompson and Walker 1995).

Although that characterization applies to the multiple feminist perspectives, there are differences of emphasis between them (Budig 2004). For example, *liberal feminism* emphasizes the acquisition and protection of equal legal rights and equal economic opportunities as essential for women to achieve a greater quality of life. *Radical feminism* tends to depict the male-dominated family as a major source of the oppression of women through either men's control of women's household labor or their sexuality and reproduction. *Socialist feminism* asserts that women's status is a product of both capitalism and patriarchy, which exploit women's roles in both production and reproduction, whereas *Marxist feminism* focuses on how capitalism exploits women, who—especially as wives and mothers—provide essential yet unpaid care (Budig 2004). Sociologist Michelle Budig asserts that there are also pro-family versions of radical and socialist feminism, both of which contend that it is women's family roles and "exclusion from politics and economics" that lead women to develop "superior female virtues," such as cooperation, caring, and protecting. Beginning in the 1990s, greater attention was raised to "multiple forms of oppression" that one could experience as a result of one's gender, race, class, sexual orientation, physical ability, age, or religion, and how these statuses intersect (Hill Collins 1998, 2000). Known as **intersectionality**, this became a central concept in *Black feminism* and *multiracial feminism* (Smith and Hamon 2012). It has since become a common way in which family researchers understand family diversity.

**Gender and Family: Concepts Created by Society.** Who or what constitutes a family cannot be taken for granted. The "traditional family" is no longer the predominant family lifestyle. Today's families have great diversity. What we think family should be is influenced by our own values and family experiences. Research demonstrates that couples actually may construct gender roles in the ongoing interactions that make up their marriages (Zvonkovic et al. 1996).

Are there any basic biological or social conditions that require the existence of a particular form of family? Some feminists would emphatically say no. Some object to efforts to study the family because to do so accepts as "natural" the inequalities built into the traditional concept of family life. Feminists urge an extended view of family to include all kinds of sexually interdependent adult relationships regardless of the legal, residential, or parental status of the partnership. For example, families may be formed of committed relationships between lesbian or gay individuals, with children obtained through adoption, from previous marriages, or through artificial insemination.

**From Analysis to Action.** Feminists have an action orientation alongside their analytical one as they strive to raise society's level of awareness regarding the oppression of women. Furthermore, some feminists make the point that all groups defined on the basis of age, class, race, ethnicity, disability, or sexual orientation are oppressed; they extend their concern for greater sensitivity to all disadvantaged groups as well as to the intersections of gender with race, class, sexual orientation, age, and other statuses. Feminists assume that the experiences of individuals are influenced by the social system in which they live. Therefore, the experiences of each individual must be analyzed to form the basis for political action and social change. The feminist agenda is to attend to the social context as it affects personal experience and to work to translate personal experience into community action and social critique.

Feminists share a belief in the need to challenge and change the system that exploits and devalues women. This being the case, some feminist researchers have described themselves as having "double vision"—the ability to be successful in the existing social system and simultaneously work to change oppressive practices and institutions.

**Gender Includes Men.** Inspired and influenced by the writing and research of feminist scholars, many social scientists focus on how men's experiences are shaped by cultural ideas about masculinity and by their efforts to either live up to or challenge those ideas (Kimmel and Messner 2007). Instead of assuming that gender only matters to or includes women, this perspective looks at men as men, or as "gendered beings," whose experiences are shaped by the same kinds of forces that shape women's lives (Kimmel and Messner

2007). We now have a greatly enlarged and still growing body of literature about men as husbands, fathers, sexual partners, ex-spouses, abusers, and so on (e.g., see Cohen 2001; Gerson 1993, 2009; Jacobson and Gottman 2007; LaRossa 1997, 2011; Marsiglio 1998, 2004a, 2004b; Marsiglio and Roy 2012; Miller 2011). Throughout this book, we explore how gender shapes women's and men's experiences of the family issues we examine.

**Critique.** There is no single feminist perspective. Rather than a unified theory, there are a variety of perspectives representing thinking across the feminist movement. All of the viewpoints share an integrating focus relating to inequality between women and men in society and, for present purposes, especially in family life.

Some family scholars criticize feminists' focus on power and inequality as a description of how people experience and construct their family relationships. So, for example, the ways in which feminists address the inequities that befall full-time housewives or women in traditional families fails to recognize the degree to which some women desire and choose to be housewives. Although feminist perspectives are also sometimes criticized for ignoring the experiences of men, it is also true that much of the research on men and masculinity was inspired by work done by feminist researchers. Finally, some criticize feminist family theorists' emphasis on actively promoting change.

## Micro-Level Theories

The four **micro-level theories** discussed next look at families from a different angle. They emphasize what happens within families, looking at everyday behavior, interaction between family members, patterns of communication, and so on. Rather than attempting to analyze "the family," they are more useful for examining what happens between individuals in "families" that accounts for the relationships we form and maintain. Whereas macro-level theories focus on families at the institutional level, addressing how they are related to other institutions and how they vary from one society to another and/or one historical period to another, micro-level theories focus their attention at how and why members of families interact, communicate, make decisions, divide responsibilities, and raise children—in other words, how they experience family life.

## Symbolic Interaction Theory

**Symbolic interaction theory** looks at how people interact with one another. An **interaction** is a reciprocal act, the everyday words and actions that take place between people. For an interaction to occur, at least two people must both act and respond to each other. When you ask your sister to pass the potatoes and she does it, an interaction takes place. Even if she intentionally ignores you or snaps at you, "I'm not your maid, get the potatoes yourself," an interaction occurs (even if it is not a positive one). Such interactions are conducted through symbols, words, or gestures that stand for something else.

When we interact with others, we do more than simply react to others; we interpret or define the meaning of their words, gestures, and actions. If your sister did not respond to your request for the potatoes, what did her nonresponse mean or symbolize? Hostility? Rudeness? A hearing problem? We interpret the meaning of her action and, *on the basis of that interpretation,* act accordingly. If we interpret the nonresponse as not hearing, we may repeat the request. If we believe that it symbolizes hostility or rudeness, we may become angry.

The notion of "meaning" is central to symbolic interaction theory. We interpret or attach meanings to interactions, situations, roles, relationships, and other individuals whenever we encounter them. Moreover, for interaction to occur, these meanings must, to some degree, be shared. Symbolic interactionists study how we arrive at or construct such shared meanings and how they affect relationships and the roles we play in them.

**Family as the Unity of Interacting Personalities.**
In the 1920s, sociologist Ernest Burgess (1926) defined the family as a "unity of interacting personalities." This definition has been central to symbolic interaction theory and in the development of marriage and family studies. Marriages and families consist of individuals who interact with one another over time. Such interactions and relationships define the nature of a family: a loving family, a dysfunctional family, a conflict-ridden family, an emotionally distant family, a high-achieving family, and so on.

In marital and family relationships, our interactions are partly structured by **social roles**—established patterns of behavior that exist independently of a person, such as the role of wife or husband existing independently of any particular husband or wife.

Each member in a marriage or family has one or more roles, such as husband, wife, mother, father, child, or sibling. These roles help give us cues as to how we are supposed to act. When we marry, for example, these roles help us "become" a wife or a husband; when we have children, they help us "become" a mother or father.

Symbolic interactionists study how one's sense of self develops and how it is maintained in the process of acquiring these roles. We are, after all, more than simply the roles we fulfill. There is a core self independent of our being a husband or wife, father or mother, or son or daughter. Symbolic interactionists ask how we fulfill our roles and continue to be ourselves and, at the same time, how our roles contribute to our sense of self. Our identities as humans emerge from the interplay between our unique selves and our social roles.

Only in the most rudimentary sense are families created by society. According to symbolic interactionists, families are "created" by their members. Each family has its own unique personality and dynamics created by its members' interactions. Classifying families by structure, such as nuclear family, stepfamily, and single-parent family, misses the point of families. Structures are significant only insofar as they affect family dynamics. It is what goes on inside families—the construction, communication, and interpretation of shared meanings—that is important.

A study exploring how parents differently define the idea of "quality time" with their children nicely illustrates the symbolic interactionist emphasis on meanings and definitions of social and familial reality. Sociologist Karrie Ann Snyder analyzed interview transcripts of 110 married, mostly dual-earner, middle- and upper-middle-class couples with teenage children and identified the different meanings that parents attach to the concept of "quality time" with their children. Assessing what parents said about the purpose of spending quality time, the activities associated with quality time, their main role or responsibility during quality time, and how much time parents thought was necessary to qualify as "quality" time, Snyder (2007) identified three categories of parents:

- *Structured planning parents* perceived that they had an overall shortage of time with their children and believed that only through efforts to specifically set aside time from their normal, hectic family and work lives for special and carefully scheduled family activities could they experience "quality time" with their teenagers.

- *Child-centered parents* also felt a deficit in the amount of time spent with their children but defined quality time in terms of having intimate, heart-to-heart conversations with their teenagers about their (children's) needs and interests *whenever and wherever such conversations occurred.*

- *Time-available parents* felt that all time spent at home with their families, good or bad, was quality time. They emphasized the amount of time together more than what occurred during that time and felt that both structured planning parents and child-centered parents use their claims of "quality time" to ease their guilt for not spending enough time with their families. Time-available parents were somewhat skeptical of the culturally understood version of quality time.

Snyder (2007) suggests that parents' definitions of quality time may differ between mothers and fathers in the same households as well as vary across different families. She further suggests that the versions of quality time that parents construct and convey may be strategic, designed to fit their parenting behaviors, especially their efforts to juggle their job demands and time with their families. In this way, they can "make sense of and explain their actions to outsiders and themselves by drawing on socially acceptable words and images to talk about their behaviors" (p. 322).

In symbolic interactionist terms, we arrive at definitions of parenting responsibility and family needs that then guide our behavior. But, at the same time, we employ a definition that serves to justify our behavior and allows us to maintain a positive sense of self. We then communicate that definition to others in the hope of influencing how they will define and assess our parenting behavior.

In general, whenever we assess our relationships, when we feel that our partner does (or does not) understand us, that we communicate well (or poorly), or that our relationship can (or cannot) withstand the difficulties created by long distance, we are illustrating dynamics that are at the heart of symbolic interaction research. The quality and stability of our relationships will depend on how we characterize such features of our relationships. Accurate or not, these subjective interpretations influence how we feel about, think about, and act within our intimate and family relationships.

**Critique.** Although symbolic interaction theory focuses on the daily workings of the family, it suffers from several drawbacks. First, the theory tends to

How family members interact with one another is partly determined by how they define their roles and by the meanings they attach to such behaviors as housework and child care.

Discpicture/Alamy Stock Photo

minimize the role of power in relationships. A partner with greater power may well have a greater chance to influence how the other person comes to define social reality. If a conflict exists, it may reveal more than differences in meaning, and it may take more than simply communicating to resolve it. If one partner strongly wants to pursue a career in Los Angeles and the other just as strongly wants to pursue a career in Boston, no amount of communication and role adjustment may be sufficient to resolve the conflict. The partner with the greater power in the relationship may prevail.

Second, symbolic interaction does not fully account for the psychological aspects of human life, especially personality and temperament. It sees us more in terms of our roles, thus neglecting the self that exists independently of our roles and limiting our uniqueness as humans.

Perhaps most important, the theory does not place marriage or family within a larger social context. It thereby disregards or minimizes the forces working on families from the outside, such as economic or legal discrimination against minorities and women.

## Social Exchange Theory

According to **social exchange theory,** we measure our actions and relationships on a cost–benefit basis, seeking to maximize rewards and minimize costs by employing our resources to gain the most favorable outcome. An outcome is basically figured by the equation *Reward − Cost = Outcome*. However, one's assessment of the outcome also takes into account what he or she expects or believes one should get from a relationship and what one believes he or she could get from available alternatives.

**How Exchange Works.** At first glance, exchange theory may be the least attractive theory we use to study marriage and the family. It seems more appropriate for accountants than for lovers. But the use of a cost–benefit analysis to measure our actions and assess our relationships is commonplace.

One reason many of us may not recognize our use of this interpersonal accounting is that we do much of it almost automatically. If a friend is unhappy with a partner, you may ask, "What are you getting out of this relationship? Is it *worth it?*" To determine the worth of his relationship, your friend will start listing pluses and minuses: When the emotional costs outweigh the benefits of the relationship, your friend will probably end it (or you will probably recommend taking such action). This weighing of costs and benefits is social exchange theory at work.

We may tend to think of concepts like rewards and costs as tangible objects, like money. In personal relationships, however, resources, rewards, and costs are more likely to be things such as love, companionship, status, power, fear, and loneliness. As people enter into relationships, they have certain resources—either tangible or intangible—that others consider valuable, such as intelligence, warmth, good looks, or high social status. People consciously or unconsciously use their various resources to obtain what they want, as when they "turn on" the charm.

Two other considerations that exchange theory incorporates are the expectations we bring to social situations and relationships and the availability of better (i.e., more rewarding) alternatives. If, for example, we find ourselves unfulfilled by a relationship, a situation most likely experienced as a cost, how we respond is influenced by what we expected to experience and whether we have more attractive alternatives. If one enters marriage with low expectations, anticipating that fulfillment would have to be found in other aspects of life, one might remain in a marriage in which the levels of intimacy or communication are low, because we expected nothing more. Conversely, if we expected high levels of closeness, physical intimacy, and sharing, even a marriage with average levels of such characteristics will be perceived as lacking.

**False** 7. According to some scholars, we weigh the costs against the benefits of all relationships.

We also compare what we have to available alternatives. If we are unhappy in our relationship but we have no better alternative or all alternatives (including being on one's own) are perceived as less attractive, we may well stay in the relationship. On the other hand, we might be moved to leave even a relatively satisfying relationship if something more attractive to one comes along.

**Equity.** A corollary to exchange is **equity**: Exchanges that occur between people have to be fair, to be balanced. We are always exchanging favors: You do the dishes tonight, and I'll take care of the kids. Often we do not even articulate these exchanges; we have a general sense that they will be reciprocated. If, in the end, we feel that the exchange was not fair, we are likely to be resentful and angry. Some researchers suggest that people are most happy when they get what they feel they deserve in a relationship. Oddly, both partners (the one feeling deprived and the one feeling "overbenefited") may well feel uneasy in an inequitable relationship. When partners recognize that they are in an inequitable relationship, they try to restore equity in one of three ways:

- They attempt to restore actual equity in the relationship.
- They attempt to restore psychological equity by trying to convince themselves and others that an obviously inequitable relationship is actually equitable.
- They decide to end the relationship.

Society regards marriage as a permanent commitment. Because marriages are expected to endure, exchanges take on a long-term character. Instead of being calculated on a day-to-day basis, outcomes are judged over time.

An important ingredient in these exchanges is whether the relationship is fundamentally cooperative or competitive. In cooperative exchanges, both husbands and wives try to maximize their "joint profit" (Scanzoni 1979). These exchanges are characterized by mutual trust and commitment. Thus, a husband might choose to work part-time and care for the couple's infant so that his wife may pursue her education. In a competitive relationship, however, each is trying to maximize individual profit. If both spouses want the freedom to go out whenever or with whomever they wish, despite opposition from the other, the relationship is likely to be unstable.

**Applying Exchange Theory: Does Sexual Frequency Matter?** Exchange theory has been applied to a number of areas of marriage and family, including mate selection or partner choice, the transition to parenthood, caregiving by aging spouses and adult children, and decisions to divorce. Two recent applications of exchange theory applied it to examine whether and when the frequency of sexual relations affects marital and/or cohabiting relationships. Interested in long-term marital or cohabiting relationships where couples engage in little to no sexual activity, sociologists Denise Donnelly and Elisabeth Burgess (2008) studied 352 people who were involuntarily celibate (i.e., having desired but not having had sex for at least six months prior to being interviewed). The social exchange perspective was applied specifically to a subset of 77 people (51 percent males) who were either married or partners in cohabiting relationships of one year or more. Although they asked other questions as well (e.g., How do relationships become involuntarily celibate?), here we consider their analysis regarding why partners in involuntarily celibate relationships stay in their relationships:

> [They] weigh the rewards and costs of staying in their relationships, keeping in mind past experiences and available alternatives, investments of resources such as time, money, and psychic energy, and social prescriptions regarding partnerships, marriages, and families (Donnelly and Burgess 2008, 529).

Although the assessment of rewards and costs differs depending on gender as well as on one's personal attitudes and beliefs, couples were more likely to stay together when the following occurred:

- They were happy with other, nonsexual aspects of the relationship.

- They perceived a lack of alternatives, believing that the probability of finding another partner is not worth the risks associated with leaving their current partner. Women were more likely to come to such a conclusion.

- They perceived that, for the sake of their families, they had a responsibility to stay in their relationships.

- They emphasized the commitment that marriage entails and expressed awareness of social expectations.

Summarizing their results, Donnelly and Burgess contend that "people stay in sexless relationships when they perceive both the benefits of staying and the costs of leaving as high" (p. 530). It isn't until they assess the sexual situation—along with other aspects of the relationship—as "unbearable" that they consider leaving.

Sociologists Scott Yabiku and Constance Gager (2009) also used social exchange theory concepts to examine how the frequency of sexual relations might affect the stability of couple relationships. Using data from two waves of the National Survey of Families and Households, they hypothesized that marriages and cohabiting relationships will be differently affected by the frequency of their sexual relations. Specifically, they theorized that cohabitors have greater expectations for sexual activity than do married women and men, experience lower barriers (i.e., lesser costs) of ending unsatisfying relationships, and, given their relative youth and lesser time away from the dating market, benefit from a higher availability of sexual alternatives to their current relationships. Their results, drawn from a sample of 5,440 married couples and 462 cohabiting couples, showed that sexual frequency was negatively associated with dissolution for both types of couples, but significantly more so for cohabitors (Yabiku and Gager 2009). Where married individuals could draw upon a wider range of potential "rewards" of their relationships even in situations of low sexual frequency, cohabitors were less able to do so. Additionally, married individuals faced a wider range of potential costs to ending their relationships than did cohabitors.

**Critique.** Social exchange theory assumes that we are all rational, calculating individuals, weighing the costs and rewards of our relationships and making cost–benefit comparisons of all alternatives. In reality, sometimes we are rational, and sometimes we are not. Sometimes we act altruistically without expecting any reward. This may be most true of love relationships and parent–child interactions. Social exchange theory also has difficulty ascertaining the value of costs, rewards, and resources, as such values may vary considerably from person to person or situation to situation.

### Family Development Theory

Of all the theories discussed here, **family development theory** is the only one exclusively directed at families (White and Klein 2002). It emphasizes the patterned changes that occur in families through stages and across time. In its earliest formulations, family development theory borrowed from theories of individual development and identified a set number of stages that all families pass through as they are formed: growth with the birth of children, change during the raising of children, and contraction as children leave and spouses die. Such stages comprised the *family life cycle*. Eventually, other concepts were introduced to replace the idea of a family life cycle. Roy Rodgers (1973) and Joan Aldous (1978, 1996) proposed the notion of the *family career*, which was said to consist of subcareers like the marital or the parental career, which themselves were affected by an educational or occupational career. Most recently, the idea of the *family life course* has been used to examine the dynamic nature of family experience.

The family life course consists of "all the events and periods of time (stages) between events traversed by a family" (White and Klein 2002). Because all these concepts emphasize the change and development of families over time, they are complementary and overlapping.

Family development theory looks at the changes in the family that typically commence in the formation of the premarital relationship, proceed through marriage, and continue through subsequent sequential stages. The specification of stages may be based on family economics, family size, or developmental tasks that families encounter as they move from one stage to the next. The stages are identified by the primary or orienting event characterizing a period of the family history. An eight-stage family life cycle might consist of the following: (1) beginning family, (2) childbearing family, (3) family with preschool children, (4) family with schoolchildren, (5) family with adolescents, (6) family as launching center, (7) family in middle years, and (8) aging family. As we grow, each of us responds to certain universal developmental challenges. For example, all people encounter **normative age-graded influences**, biological or social influences that are clearly correlated with age, such as the biological processes of physical maturation, puberty, and menopause, or typical events, such as the beginning of school, death of parents, and the advent of retirement, that are linked to age. **Normative history-graded influences** come from historical facts that are common to a particular generation, such as the political and economic influences of wars and economic depressions, and that are similar for individuals in a particular age group (Santrock 1995).

The life course perspective emphasizes the dynamic nature of families and the changes in family relationships that accompany life transitions. The impact of life

transitions (e.g., getting married, experiencing parental divorce or remarriage, transitioning to parenthood, etc.) is in part dependent on the *timing* of such transitions. For example, an alteration in family structure will have different effects if experienced in early childhood than in adolescence or young adulthood (Heard 2007). Also central to the life course perspective is the idea of interdependent or **linked lives**. Stressing the reciprocal influences family members have on one another, the concept emphasizes how the shape of an individual's life course is influenced by the shape of the life courses of others. Speaking specifically of parents and children, sociologists Melissa Milkie, Alex Bierman, and Scott Schieman point out that "events, transitions, and values of each generation have the potential to affect the other in powerful ways" (2008, 86). As articulated by Emily Greenfield and Nadine Marks, this concept suggests that "circumstances in adult children's lives would have implications for the lives of their parents and, accordingly, that adult children's problems would affect their parents' psychological and relational well-being" (2006, 443). There are also a variety of developmental theories that more narrowly examine stages involved in specific family phenomena, such as falling in love, choosing a spouse, or experiencing divorce. Instead of attempting to depict all stages families might encounter, these theories look at the unfolding of specific aspects of family life across stages and through various life transitions. You will find such approaches in a number of later chapters.

**Critique.** An important criticism sometimes made of family development theory is that it assumes the sequential processes of intact, nuclear families. It further assumes that all families go through the same process of change across the same stages. Thus, the theory downplays both the diversity of family experience and the experiences of those who divorce, remain childless, or bear children but never marry (Winton 1995). For example, stepfamilies experience different stages and tasks (Ahrons and Rodgers 1987). Nevertheless, the universality of the family life cycle may transcend the individuality of the family form. Single-parent and two-parent families go through many of the same development tasks and transitions. They may differ, however, in the timing and length of those transitions.

A related criticism points out that gender, sexual orientation, race, ethnicity, and social class all create variations in how we experience family dynamics. The very sequence of stages may reflect a middle- to upper-class family reality. Many lower- and working-class families do not have lengthy periods of early childless marriage. The transitions to marriage and parenthood may be encountered simultaneously or in reverse of what the stages specify. In neglecting these sorts of variations, the developmental model can appear overly simplistic.

## Family Systems Theory

**Family systems theory** combines two of the previous sociological theories, structural functionalism and symbolic interaction, to form a more psychological—even therapeutic—theory. Mark Kassop (1987) notes that family systems theory creates a bridge between sociology and family therapy.

**Structure and Patterns of Interaction.** Like functionalist theory, family systems theory views the family as a structure of related parts or subsystems: the spousal subsystem, the parent–child subsystem, the parental subsystem (spouses relating to each other as co-parents), the sibling subsystem, and the personal subsystem (the individual and his or her relationships). Each part carries out certain functions. One of the important tasks of these subsystems is maintaining their boundaries. For the family to function well, the subsystems must be kept separate (Minuchin 1981). Husbands and wives, for example, should try to prevent their conflicts from spilling over into the parent–child subsystem. Sometimes a parent will turn to the child for the affection that he or she ordinarily receives from a spouse. When the boundaries of the separate subsystems blur, as in incest, the family becomes dysfunctional.

As in symbolic interaction theory, interaction is important in systems theory. A family system consists of more than simply its members. It also consists of the pattern of interactions of family members: their communication, roles, beliefs, and rules. Marriage is more than a husband and wife; it is also their pattern of interactions, how they act in relation to each other over time (Lederer and Jackson 1968). Each partner influences and in turn is influenced by the other partner. And each interaction is determined in part by the previous interactions. This emphasis on the pattern of interactions within the family is a distinctive feature of the systems approach.

Any change in the family—such as a child leaving the family, family members forming new alliances, and hostility distancing the mother from the father—can create disequilibrium, which often manifests itself in emotional turmoil and stress. The family may try to restore the old equilibrium by forcing its "errant" member to return to his or her former position, or it may adapt and create a new equilibrium with its members in changed relations to one another.

**Analyzing Family Dynamics.** In looking at the family as a system, researchers and therapists believe the following:

- *Interactions must be studied in the context of the family system.* Each action affects every other person in the family. The family exerts a powerful influence on our behaviors and feelings, just as we influence the behaviors and feelings of other family members.
- *The family has a structure that can be seen only in its interactions.* Each family has certain preferred patterns of acting that ordinarily work in response to day-to-day demands. These patterns become strongly ingrained "habits" of interactions that make change difficult.
- *The family is a purposeful system; it has a goal.* In most instances, the family's goal is to remain intact as a family, achieving **homeostasis**, or stability. This makes change difficult, for change threatens the old patterns and habits to which the family has become accustomed.
- *Despite resistance to change, each family system is transformed over time.* A well-functioning family constantly changes and adapts to maintain itself in response to its members and the environment. The family changes as partners age and as children are born, grow older, and leave home. The family system adapts to stresses to maintain family continuity while making restructuring possible.

Although it has been applied to a variety of family dynamics, systems theory has been particularly influential in studying *family communication* (White and Klein 2002). As applied by systems theorists, interaction and communication between spouses are the kinds of systems wherein a husband's (next) action or communication toward his wife depends on her prior message to him. But through research in family communications, we recognize that marital communication is more complex than a simple quid pro quo or reciprocity expectation, such as, "If she is nasty, he is nasty." Well-known marriage researcher John Gottman has explored marital communication patterns that differentiate distressed from nondistressed couples. He identifies the importance of nonverbal communication over that of verbal messages spouses send (White and Klein 2002).

**Critique.** It is difficult for researchers to agree on exactly what family systems theory is. Many of the basic concepts are still in dispute, even among the theory's adherents, and the theory is sometimes accused of being so abstract that it loses any real meaning (White and Klein 2002).

Family systems theory originated in clinical settings in which psychiatrists, clinical psychologists, and therapists tried to explain the dynamics of dysfunctional families. Although its use has spread beyond clinicians, its greatest success is still in the analysis and treatment of dysfunctional families. However, the basic question is whether its insights apply to healthy families as well as to dysfunctional ones. Do healthy families, for example, seek homeostasis as their goal, or do they seek individual and family well-being?

## Applying Theories to Family Experiences

Although the preceding theories were illustrated with numerous examples, it is worthwhile to stress that each will raise different questions and suggest different explanations for family phenomena. For instance, returning for a moment to the chapter opening scenario, each of the theories would pose different questions about long-distance relationships, given their major assumptions. This would be equally true for any other research topic, such as how couples divide housework and child care, why family violence and abuse occur, or how unemployment affects marital stability. Table 2.2 gives an example of the kind of question each might raise about long-distance relationships.

### TABLE 2.2 Applying Theories to Long-Distance Relationships

| Theory | Assumptions about Families | Applied to Long-Distance Relationships |
|---|---|---|
| Ecological | Families are influenced by and must adapt to environments. | How do the characteristics of each partner's different living environments affect their abilities to maintain their commitments to the relationship? |
| Symbolic Interaction | Family life acquires meaning for family members and depends on the meanings they attach. | What meaning do couples attach to being separated? How does this alter their perceptions of the relationship? |
| Social Exchange | Individuals seek to maximize rewards, minimize costs, and achieve equitable relationships. | How do both partners define the costs and rewards associated with their relationship? |
| Family Development | Families undergo predictable changes over time and across stages. | What are the stages or phases that couples encounter as they adjust to being separated? |
| Structural Functionalism | The institution of the family contributes to the maintenance of society. On a familial level, roles and relationships within the family contribute to its continued well-being. | How does physical separation function to maintain or threaten the stability of the relationship? |
| Conflict | Family life is shaped by social inequality. Within families, as within all groups, members compete for scarce resources (e.g., attention, time, power, and space). | To what extent does one partner benefit more from being apart? |
| Family Systems | Families are systems that function and must be understood on that level. | How does being physically separated make it difficult for the couple to communicate effectively or maintain the equilibrium of the relationship? |
| Feminist | Gender affects our experiences of and within families. Gender inequality shapes how women and men experience families. Families perpetuate gender difference. | How are women and men differently affected by separation (i.e., do women carry more of the burden of managing and maintaining the relationship)? |

© Cengage Learning

It is also worth noting that much of the theorizing that family researchers do is more narrowly focused on given family phenomena. For example, there are assorted theories of how we choose our spouses and life partners, how families cope with stressful situations, how gender socialization in families occurs, why relationships might turn violent, and so on. The theories that one finds tested and put forth in these and other areas are often called "middle-range" theories, or even "mini-theories." They are specific to the phenomena under study and are not put forward to explain broader or more general family phenomena. This is true of much of the literature published in academic journals (White and Klein 2008). Throughout the rest of this text, you will come across a number of such theories. Occasionally, it will be clear what broader paradigm or perspective such theories seem to be guided by. However, at times, it may appear as if these narrower theories are disconnected to any of the major perspectives or, conversely, are influenced by multiple perspectives.

Keep in mind that whether theory guides research or is the product of research, theories alone will not give us the kind of understanding we want about relationships and families. For that, we need to collect information, or data, from or about real people and real families. The remainder of this chapter focuses on the research process and techniques of data collection.

## ■ Conducting Research on Families

In gathering their data, researchers use a variety of techniques. Some researchers ask the same set of questions of great numbers of people. They collect information from people of different ages, sexes, living situations, and ethnic backgrounds. This is known as "representative sampling." In this way, researchers can discover whether age or other background characteristics influence people's responses. This approach to research is called **quantitative research** because it deals with large quantities of information that is analyzed and presented statistically. Quantitative family research often uses sophisticated statistical techniques to assess the relationships between variables. Survey research and, to a lesser extent, experimental research (discussed in the following sections) are examples of quantitative research.

Other researchers study smaller groups or sometimes individuals in a more in-depth fashion. They may place observers in family situations, conduct intensive interviews, do case studies involving information provided by several people, or analyze letters, diaries, or other records of people whose experiences represent special aspects of family life. This form of research is known as **qualitative research** because it is concerned with a detailed understanding of the object of study.

In addition to using information provided specifically by people participating in a research project, researchers use information from public sources. This research is called **secondary data analysis**. It involves reanalyzing data originally collected for another purpose. Examples might include analyzing U.S. Census data and official statistics, such as state marriage, birth, and divorce records. Secondary data analysis also includes content analysis of various communication media, such as newspapers, magazines, letters, and television programs.

Social science research on families often becomes incorporated into public policy discussions or debates and influences the outcomes of such discussions. Thus, research on such topics as the effects of divorce on children, the causes and consequences of family and intimate partner violence and abuse, the effects of different parenting situations on children's development (e.g,, gay parents, single parents, teen parents), as well as research on families of different economic statuses (especially those in poverty) or racial or ethnic backgrounds often influences policy recommendations and decisions.

One of the most well-known examples of this is Daniel Patrick Moynihan's 1965 study, *The Negro Family: The Case for National Action* (Moynihan 1965), widely known as "the Moynihan Report." As Stephanie Coontz, professor of history and family studies and director of Research and Public Education for the Council on Contemporary Families, contends that for nearly a decade, there was little academic or policy discussion about Moynihan's claims about the connections between family structure and poverty. As economic circumstances worsened, Moynihan's research "got a lot more traction, especially among politicians" and "by the 1990s Moynihan's theme had been adopted with a vengeance by Democrats and Republicans alike" (Coontz 2015). Though Coontz points out that much of Moynihan's dire claims about the connections between family structure (i.e., being raised by single parents) and poverty, crime, violence, and educational achievement were not "borne out," his research had great influence.

## Ethics in Family Research

Family science researchers conduct their investigations using **ethical guidelines** agreed on and shared by professional researchers. These guidelines protect the privacy and safety of people who provide information in the research. For example, any research conducted with college students requires investigators to present the plan and method of the research to a review committee. This ensures that participants' involvement is voluntary, that their privacy is protected, and that they will be free of harm. Professional organizations in each of the disciplines that research family experience publish a code of ethics that their members are expected to follow. There is considerable overlap in the guidelines provided by such organizations as the American Sociological Association, the American Psychological Association, the National Council on Family Relations, the American Association of Family and Consumer Sciences, and the National Association of Social Workers. Although the language may differ, the following are common concerns.

To protect the privacy of participants, researchers promise them either anonymity or confidentiality. **Anonymity** insists that no one, including the researchers, can connect particular responses to the individuals who provided them. Much questionnaire research is of this kind, providing that no identifying information is found on the questionnaires. According to the rules of **confidentiality**, the researchers know the identities of participants and can connect what was said to who said it but promise not to reveal such information publicly.

To protect the safety of research participants, researchers design their studies with the intent to minimize any possible and controllable harm that might come from participation. In social research on relationships and families, such harm is not typically physical harm but rather embarrassment or discomfort. Much of what family researchers study is ordinarily kept private. Talking about personal matters with an interviewer or even answering a series of survey questions may create unintended anxiety on the part of the participants. At best, researchers carefully design their studies to minimize the likelihood and reduce the extent of such reactions. Unfortunately, they cannot always be completely prevented (Babbie 2007).

> **True**
> **8.** Every method of collecting data on families is limited in some way.

Research ethics also require researchers to conduct their studies and report their findings in ways that assure readers of the accuracy, originality, and trustworthiness of their reports. Falsifying data, misrepresenting patterns of findings, and plagiarizing the research of others are all unethical.

The data about marriage and the family that researchers gather come from four basic research methods: survey research, clinical research, observational research, and experimental research. No single method is best for studying marriage and the family. Each technique has its strengths and weaknesses, and each may provide important and useful information that another method may not (Cowan and Cowan 1990). That's why many advocate and practice **triangulation**. Although this term is used in psychology and family therapy to refer to a particular form of family dysfunction, as used in discussions of social research, triangulation refers to the use of multiple methods in undertaking any particular study. Rather than "just observe," researchers might also interview people or distribute questionnaires to get a sense of how people think or feel about behavior of interest to researchers.

## Survey Research

The **survey research** method, using questionnaires or interviews, is the most popular data-gathering technique in marriage and family studies. Surveys may be conducted in person, over the telephone, or by written questionnaires. Typically, the purpose of survey research is to gather information from a smaller, representative group of people and to infer conclusions valid for a larger population. Questionnaires offer anonymity, may be completed fairly quickly, and are relatively inexpensive to administer.

Quantitative questionnaire research is an invaluable resource for gathering data that can be generalized to the wider population. Because researchers who use such techniques typically draw or use *probability-based random samples,* they can estimate the likelihood that their sample data can be safely inferred to the population in which they are interested. Furthermore, preestablished response categories or existing scales or indexes used by all respondents allow more comparability across a particular sample and between the sample data and related research.

## Questionnaires and Interviews

**Questionnaires** are documents consisting of a series of questions or statements to which a researcher wants research participants to answer or react. Typically consisting mostly of closed-ended questions, in which the researcher has supplied all possible answers from which a participant is to choose, questionnaires usually do not allow in-depth responses. Instead, a person must respond with a short answer, a *yes* or *no*, or a choice on a scale of, for example, 1 to 10, from *strongly agree* to *strongly disagree,* from *very important* to *unimportant,* and so on. Unfortunately, marriage and family issues are often too complicated for questionnaires to explore in depth.

Interview techniques avoid some of these shortcomings of questionnaires because interviewers are able to probe in greater depth and follow paths suggested by the interviewees. They are also typically better able to capture the particular meanings or the depth of feeling that people attach to their family experiences.

Consider the following example from sociologist Sharon Hays's interview study of 38 mothers of 2- to 4-year-olds. In describing how priorities are restructured when a woman becomes a mother, one of Hays's informants offered this comment on the life changes associated with being a mother:

> I think the reason people are given children is to realize how selfish you have been your whole life—you are just totally centered on yourself and what you want. And suddenly here's this helpless thing that needs you constantly. And I kind of think that's why you're given children, so you kinda think, okay, so my youth was spent for myself. Now, you're an adult, they come first. . . . Whatever they need, they come first.

Narrative data of this kind convey much about the experience of parenthood, including a depth of feeling and degree of nuance that quantitative questionnaire data cannot provide. By having respondents circle or check the appropriate preestablished response categories to a researcher's questions, we may never identify what that response truly means to respondents or how it fits within the wider context of their life. However, interviewers are less able to determine how commonly such experiences or attitudes are found. Interviewers may also occasionally allow their own preconceptions to influence the ways in which they frame their questions and to bias their interpretation of responses.

There are problems associated with survey research, whether done by questionnaires or interviews. First, how representative is the sample (the chosen group) that volunteered to take the survey? In the case of a probability-based sample, this is less of a concern. However, self-selection (volunteering to participate) also tends to bias a sample. Second, how well do people understand their own behavior? Third, are people underreporting undesirable or unacceptable behavior? They may be reluctant to admit that they have extramarital affairs or how often they yell at their children, for example. If people are *unable* or *unwilling* to answer questions honestly for any reason, the survey technique will produce incomplete or inaccurate data. Nevertheless, surveys are well suited for determining the incidence of certain behaviors or for discovering traits and trends.

Much of the research that family scientists conduct and use—on topics as far-reaching as the division of housework and child care, the frequency of and satisfaction with sex, or the effect of divorce on children or adults—is derived from interview or questionnaire data. Sociologists more commonly use surveys than psychologists because they tend to deal on a more general or societal level rather than on a personal or small-group level. But surveys are less able to measure well how people interact with one another or what they actually do. For researchers and therapists interested in studying the dynamic flow of relationships, surveys are not as useful as clinical, experimental, and observational studies.

## Time Use Surveys

**Time use surveys** represent an interesting application of survey research. Also sometimes known as the **time use diary** method (Robinson 1999), this technique uses self-administered surveys in which people record their activities at designated points in time (e.g., specific hours of the day) or report how much time they spend or spent in activities of interest to researchers, such as in housework, child care, and/or commuting to and from their place of employment. Typically, a time diary is a chronological record of one's activities, and includes one's primary activity, the time one started, the location of one's activity, and the presence of others. Additionally, time diary research often includes reports of "secondary" or simultaneous activity. Prior time diary research indicates that individuals spend perhaps three to four hours a day in simultaneous activities (e.g., folding laundry while helping a child with homework). In fact, this may be especially true of time

Time use surveys and time diaries are often used to determine how much time individuals spend in housework and child care.

Dash/Hero Images/Corbis/Alamy Stock Photo

people spend in household tasks and/or child care (Paolisso and Hames 2010). Much of what we know about who does what in families and households comes from time use surveys such as the American Time Use Survey, administered by the U.S. Bureau of Labor Statistics. For example, based on findings from the 2003 survey, whereas nearly 100 percent of women engaged in some form of domestic work on a daily basis, 85 to 90 percent of men engaged in such work on any given day (Robinson and Martin 2009). Throughout the remaining chapters, we will incorporate data from time use surveys.

## Secondary Analysis of Existing Survey Data

Secondary analysis is one of the techniques family researchers use most frequently. Because of the various costs associated with conducting surveys on large,

nationally representative samples, researchers often turn to one of the many available survey data sets, such as the General Social Survey (GSS) conducted by the National Opinion Research Center at the University of Chicago. The GSS includes many social science variables of interest to family researchers. Family researchers also often use data issued by the U.S. Census Bureau, the National Center for Health Statistics, the Bureau of Labor Statistics, and other government agencies that include many descriptive details about the U.S. population, including characteristics of families and households; birth, marriage, and divorce rates; and employment and time use.

Additional examples of available survey data of particular value to family researchers include the National Survey of Families and Households (NSFH), the National Health and Social Life Survey (NHSLS), the National Longitudinal Study of Adolescent Health (Add Health), and the Fragile Families and Child Well-being Study. The NSFH has provided much information about a range of family behaviors, including the division of housework, the frequency of sexual activity, and the relationships between parents and their adult children. The NHSLS is based on a representative sample of 3,432 Americans, aged 18 to 59, and contains much useful data about sexual behavior. Add Health is the largest longitudinal study ever (following a sample of respondents over a period of time) of more than 90,000 adolescents, from 80 high schools and more than 50 middle schools. They were initially studied in 1994, and then again in 1996, 2001–2002, and 2007–2008. The study examines the "social, economic, psychological, and physical well-being," and includes data on family, friendships, peer groups, and romantic relationships (www.cpc.unc.edu/projects/addhealth).

The Fragile Family and Child Wellbeing Study follows almost 5,000 children born between 1998 and 2000, three-quarters of whom were born to unmarried parents (i.e., "fragile families" who are at "greater risk of breaking up"). In addition to studying children's cognitive and emotional development, physical health, and home environment, the project contains interview data with parents on their attitudes, relationships, parenting behavior, health, economic situation, and employment status (www.fragilefamilies.princeton.edu/about.asp). In recent years at least a dozen articles in the *Journal of Marriage and Family* were based on data from the Add Health survey, the Fragile Families Study, and the National Survey of Families and Households.

**True** 9. Much family research proceeds by using data that others have already collected.

Some of the most useful and influential research on families, like both the National Survey of Families and Households and The National Survey of Family Growth, has come from the field of **demography**, often defined as "the study of the size, distribution, and composition of the population, changes therein, and the demographic processes underlying demographic change, including fertility, mortality, and migration" (Lichter 2013, F2). Changes in family structure can be better understood by identifying changes in characteristics of populations. Data collected by demographers help "provide the big picture, identify emerging trends, and set an empirical baseline that invites other fine-grained, nuanced, or specialized empirical analyses" (p. F1).

The major difficulty associated with all secondary data analysis is that the material collected in the original survey may "come close to" but not be exactly what you wanted to examine. Say, for example, you wanted to use the General Social Survey to look at attitudes toward marriage and divorce or different kinds of sexual behaviors. Perhaps, if designing your own research instrument, you would have worded the items differently to capture the essence of what you are most interested in. Likewise, perhaps you would have asked additional questions to further or more deeply explore your topical interest (Babbie 2007). This disadvantage, although real, does not negate the enormous benefits associated with secondary analysis, the greatest of which is the availability of and easy access to broad sets of data on large, representative samples from which a host of family-related issues can be quickly and inexpensively examined.

## Clinical Research

**Clinical research** involves in-depth examination of a person or a small group of people who come to a psychiatrist, psychologist, or social worker with psychological or relationship problems. The **case-study method**, consisting of a series of individual interviews, is the most traditional approach of all clinical research; with few exceptions, it was the sole method of clinical investigation through the first half of the 20th century (Runyan 1982).

Clinical researchers gather a variety of additional kinds of data, including direct, firsthand observation, or analysis of records. Rather than a specific technique of data collection, clinical research is distinguished by its examination of individuals and families that have sought some

kind of professional help. The advantage of clinical approaches is that they offer long-term, in-depth study of various aspects of marriage and family life. The primary disadvantage is that we cannot necessarily make inferences about the general population from them. People who enter psychotherapy are not a representative sample. They may be more motivated to solve their problems or have more intense problems than the general population (Kitson et al. 1996).

One of the more widely cited and celebrated clinical studies is Judith Wallerstein's longitudinal study of 60 families who sought help from her divorce clinic. Wallerstein published three books—*Surviving the Breakup: How Children and Parents Cope with Divorce; Second Chances: Men, Women, and Children a Decade after Divorce;* and *The Unexpected Legacy of Divorce: The 25 Year Landmark Study*—following the experiences of most of the children in these families (she has retained 93 of the original 131 children whom she first interviewed in 1971) at 5, 10, and 25 years after divorce (Wallerstein 1980, 1989, 2000). All three books are sensitively written and richly convey the multitude of short- and long-term effects of divorce in the lives of her sample. Her critics have questioned whether findings based on such a clinically drawn sample (60 families from Marin County, California, who sought help as they underwent divorce) apply to divorced families more generally (Coontz 1998).

Clinical studies, however, have been fruitful in developing insight into family processes. Such studies have been instrumental in the development of family systems theory, discussed earlier in this chapter. By analyzing individuals and families in therapy, psychiatrists, psychologists, and therapists such as R. D. Laing, Salvador Minuchin, and Virginia Satir have been able to understand how families create roles, patterns, and rules that family members follow without being aware of them.

## Observational Research

Observational research and experimental studies (discussed in the next section) account for a small minority of published research articles. In **observational research**, scholars attempt to study behavior systematically through direct observation while remaining as unobtrusive as possible. To measure power in a relationship, for example, an observing researcher may sit in a home and videotape exchanges between a husband and a wife or bring couples into a lab situation.

**False** **10.** There are various ways to observe family behavior.

Some observational research involves family members being given structured activities to carry out. These activities involve interaction between family members (e.g., cohabiting couples) that can be observed and analyzed. They may include problem-solving tasks, putting together puzzles or games, responding to a contrived family dilemma, or managing conflict. Different tasks are intended to elicit different types of family interaction, which then provide the researchers with opportunities to observe behaviors of interest. For example, researchers have used observational techniques designed to elicit conflict between couples. How couples manage the conflict, what kind of negative affect they express, and how they interact are all observed and recorded.

A notable example of this kind of observational research followed 100 couples over a 13-year period from their premarital period through the period of greatest risk of divorce. Multiple sources of data were used, including the videotaping of couple interactions. The research revealed that premarital interaction patterns can help predict relationship outcomes some ten or more years later (Clements, Stanley, and Markman 2004).

In more "naturalistic" observation, researchers either observe people in their homes and/or as they go about their daily routines, or observe in more public settings where people remain unaware that they are being watched. Researchers Belinda Campos, Anthony Graesch, Rena Repetti, Thomas Bradbury, and Elinor Ochs undertook a naturalistic observation study of 30 dual-earner, heterosexual couples with two or more children to explore the challenges dual-earner parents face in their attempts to forge and enhance high-quality relationships. Each of the 30 families was observed and video-recorded in their own homes over two weekdays and one weekend. During the weekday observations, families were recorded in the mornings between waking up and when children and parents left the household for school and work. Recording resumed once children and/or parents returned home, and continued until children went to bed for the night. Researchers focused their analysis on *reunions*—behavior toward a parent returning from work, and on the *physical proximity* of family members throughout the evening (Campos et al. 2009). Their observational data revealed gender differences in reunions, with women tending to return home earlier, being greeted with positive reactions, and receiving information reports from spouses and children. Men tended to return later and were either received positively or ignored as family members were engaged in other activities. Their data also showed that during evenings, mothers spent more time with children than did fathers, while fathers spent more time alone. Meanwhile, couples were infrequently observed together without their children (Campos et al. 2009).

The most obvious disadvantage of observational research is that when individuals know they are being observed, they may hide or suppress unflattering or inappropriate behaviors, such as dealing with decisions through threats of violence, when the observer is present. Individuals within families, as well as families as groups, are concerned with appearances and the impressions they make.

Researchers may bring their own biases into what they see and how they interpret what they see. Some of that is unavoidable and could affect any social research. But observational research is the most subjective of the techniques. It is the hardest in which to remain objective, and, as a consequence, research is often hard to replicate.

A third problem that observational researchers encounter involves the essentially private nature of most family relationships and experiences. Because we experience most of our family life "behind closed doors," researchers typically cannot see what goes on "inside" without being granted access. For more public family behavior (e.g., caring for children in public places), observational data can be effectively used. However, some critics may point out the ethical dilemma of observing people when they don't know they are being studied. This puts researchers who use naturalistic observation in the position of having to justify their research strategy as necessary despite it falling outside the standard of "voluntary participation."

There are aspects of family life that can be easily observed, such as care for children in public.

Jose Luis Pelaez In./Blend Images/Alamy Stock Photo

> What goes on at home, behind closed doors, may not be easily accessible to observational researchers.

Jonathan Nourok/PhotoEdit

## Experimental Research

In **experimental research**, researchers isolate a single factor under controlled circumstances to determine its influence. Researchers are able to control their experiments by using *variables*, aspects or factors that can be manipulated in experiments. Recall the earlier discussion of types of variables, especially independent and dependent variables. In experiments, independent variables are factors manipulated or changed by the experimenter; dependent variables are factors affected by changes in the independent variable.

Because it controls variables, experimental research differs from the previous methods we have examined. Clinical studies, surveys, and observational research are correlational in nature. Correlational studies measure two or more naturally occurring variables to determine their relationship to one another. Because correlational studies do not manipulate the variables, they cannot tell us which variable causes the others to change. By design, in experimental studies researchers manipulate the independent variables, so that they might determine whether and which variables affect the other variables.

Experimental findings can be powerful because such research gives investigators control over many factors and enables them to isolate variables. For example, psychologists Michael Wohl and April McGrath designed a study to test the notion that "time heals all wounds." Specifically, they were interested in the effect of "temporal distance" (i.e., the passage of time) on the willingness people have to forgive someone who committed an interpersonal transgression against them. Using three different experimental designs and both real and hypothetical transgressions, they were able to determine that the more time that passed since a transgression, the more likely individuals were to forgive their transgressor. In addition, one's subjective sense of time affects one's ability to forgive. They note that "the more a person feels temporally removed from the transgression, regardless of elapsed clock time, forgiveness becomes a more likely response" (Wohl and McGrath 2007, 1032).

The obvious problem with such experimental studies is that we may well respond differently to people in real life than we do in controlled situations, especially in paper-and-pencil situations.

Experimental situations are often faint shadows of the complex and varied situations we experience in the real world. Thus, even when researchers can design and carry out a carefully controlled experiment, "the artificial setting . . . and its accompanying controls still can limit the researcher's ability to generalize . . . to naturalistic settings" (i.e., real life) (Small 2005, 324).

## Applied Family Research

Despite our earlier lengthy consideration of theories, a good amount of family research is what we call **applied research**. The focus of such research is more practical than theoretical. It tends to be less concerned with formulating theories, generating concepts, or testing hypotheses. Data are gathered in an effort to solve problems, evaluate policies or programs, or estimate the outcome of some proposed future change in policy. For example, an applied researcher might study the effectiveness of a new mandatory arrest policy in reducing domestic violence incidents or the success of abstinence-only education in reducing teen pregnancy rates, the likely outcomes of new welfare policies on poor single mothers and their children, or whether and how repealing no-fault divorce laws might influence divorce rates.

Data from applied research tend to be of less interest to academic family researchers who are engaged in what is called basic or pure research than to policy makers, program directors, and heads of agencies. Such individuals may not even publish the findings of their research in journals or books but rather disseminate them in organizational reports read by

**Researching Dating Violence Cross-Culturally**

Family researchers use surveys to examine all sorts of family issues, but few aspects of intimate and family relationships are as disturbing as the issues of violence and abuse. They represent the worst of family relationships and the opposite of what we believe such relationships ought to be like. We will look at many of the issues that surround family and intimate partner violence in some detail in Chapter 12. For now, we look at comparative survey data on one form of intimate partner violence—dating violence—because such data are representative of the kind of data that family researchers can obtain through survey instruments.

The following reported findings are from the International Dating Violence Study, a remarkable effort by a consortium of researchers to examine partner violence among samples of college students at 68 different colleges and universities in 32 different countries (including 2 in Africa, 7 in Asia, 13 in Europe, 4 in Latin America, 2 in North America, 2 in the Middle East, plus Australia and New Zealand). A total of 13,701 college students were surveyed (71 percent female), most of whom were enrolled in sociology, psychology, family studies, or criminology classes. The surveys explored a number of interesting issues, including respondents' relevant attitudes (e.g., toward intimate partner violence and corporal punishment) and experiences (had they ever assaulted a partner, injured a partner, been spanked as a child, and so on). Researchers used the revised Conflict Tactics Scales (CTS2), which measures both more "minor" and more severe assaults. As described by sociologist and principal investigator Murray Straus (2008):

> The CTS2 items to measure "minor" assault are: (1) pushed or shoved, (2) grabbed, (3) slapped, (4) threw something at partner and (5) twisted arm or hair. The items in the "severe" assault scale are: (1) punched or hit a partner, (2) kicked, (3) choked, (4) slammed against a wall, (5) beat up, (6) burned or scalded and (7) used a knife or gun on partner.

Considering only students who had been in a relationship of at least a month, the following findings are based on reports of 4,239 respondents. If a respondent reported that one or more of these acts had occurred (either as perpetrator or victim) in the prior 12 months, an assault was recorded. Among the key findings that Straus reports, the following stand out:

- The level of self-reported assaults against a dating partner ranged from a low of 16.6 percent (Portugal) to a high of 77.1 percent (Iran) with a median of 31.2 percent. The rate in the United States was 30 percent.
- The percentage of students reporting severely assaulting a partner ranged from 1.7 percent (Sweden) to 23.2 percent (Taiwan), with a median of 10.8 percent. The U.S. rate was 11.0 percent.
- In every one of the 32 national settings, the largest category of partner violence was "bidirectional" (i.e., both are violent). In the United States, nearly 70 percent of the self-reported partner violence was bidirectional.
- When looking at "severe violence," again the largest category is "both violent" (56.6 percent in the United States).
- In only four of the 32 national settings was the percentage of unilateral male-only violence greater than the percentage of unilateral female-only violence (Iran, Tanzania, Greece, and Brazil).
- The second largest category of both overall and severe violence was "female-only," wherein the female was the only partner to use violence.

Straus is careful to caution against overgeneralizing from these findings. He points out that because the sample is a *convenience sample of college students*, the findings cannot be extrapolated either to the general populations of those countries (college students are not representative of national populations) or to college students in the settings included (convenience samples of students in a handful of social science courses do not allow us to generalize to all college students). Still, the data are consistent with much additional research in indicating high levels of violence among dating partners.

These results document internationally what has been known for a long time—that physical assaults against partners in dating and marital relationships are by far the most prevalent type of violent crime (Straus 2004). In Chapter 12, we explore the issues addressed here in much greater detail. ●

only a small number of others. Although they may recognize the need for careful attention to detail in data collection and analysis, the practical needs that motivate them also push applied researchers to make "trade-offs" and to "compromise scientific rigor to get quick, usable results" (Newman 2004, 11).

## How to Think about Research

The last chapter discussed research into the effects of divorce on children and demonstrated that researchers occasionally arrive at very different findings and draw even contrary conclusions. You may have been left wondering how research results can be so different. This will occur frequently throughout the next 12 chapters. Differences in sampling and methodological techniques help explain why studies of the same phenomenon may arrive at different conclusions. They also help explain a common misperception many of us hold regarding scientific studies. Many of us believe that because studies arrive at different conclusions, *none* are valid. What conflicting studies may show us, however, is that researchers are constantly exploring issues from different perspectives and with different techniques as they attempt to arrive at a consensus.

In addition, researchers may discover errors or problems in sampling or methodology that lead to new and different conclusions. They seek to improve sampling and methodologies to elaborate on or disprove earlier studies. In fact, the very word *research* is derived from the prefix *re-*, meaning "over again," and *search*, meaning "to examine closely." And that is the scientific endeavor: searching and re-searching for knowledge.

It is also important to recognize that there are always going to be exceptions to any pattern of findings that family scientists identify. Even sound sampling and rigorous research strategies should not be expected to yield either identical results between studies or results that apply to every family. As articulated by sociologist Paul Amato, writing as then president of the National Council on Family Relations, "People and families do not react to a given stimulus (or a change in an independent variable) in identical ways. Instead a wide variety of outcomes usually occurs. Given the extreme variability that exists in the social world, it is not possible to formulate universal laws that apply to everyone" (Amato 2014, 3).

This is important for two reasons. First, don't assume that the patterns reported in this book *will happen* in your life. Your experience may constitute an exception to the more general pattern. All human beings don't behave the same way even when faced with very similar circumstances or constraints. Second, and equally important, don't dismiss findings reported here because they don't fit your experiences or those of people you may know. Instead, try to account for why your experience departs from the more generally observed social regularities.

By using critical thinking skills and by understanding something about the methods and theories family researchers use, we are in a position to more effectively evaluate the information we receive about families. We are also better able to step outside our personal experience, go beyond what we've always been told, and begin to view marriage and family from a sounder and broader perspective. In Chapters 3 and 4, we take such steps and explicitly examine the factors and forces that create differences in family experience.

## Summary

- What we think we know about families comes from many sources. Our commonsense understanding of families is influenced by personal experience, tradition, and media representations.

- The average adult in the United States views almost six hours a day of media content.

- We need to be alert to maintain *objectivity* in our consideration of different forms of family lifestyle. *Opinions, biases,* and *stereotypes* are ways of thinking that lack objectivity.

- *Fallacies* are errors in reasoning. Two common types of fallacies are *egocentric fallacies* and *ethnocentric fallacies*—the belief that all people are or should be the same as we are or that our way of living is superior to all others.

- *Theories* attempt to provide frames of reference for the interpretation of data. Macro-theories of marriage and families include family ecology, structural functionalism, conflict theory, and feminist theories. Micro-theories include symbolic interaction, social exchange, family development, and family systems. Middle-range theories are empirically derived theories, more narrowly designed to address specific family phenomena.

- *Conceptualization* is the process of identifying and defining the concepts we are studying, and *operationalization* is the development of research strategies to observe our concepts.

- *Deductive research* tests hypotheses are statements in which we turn our concepts into variables and specify how variables are related to each other.

- *Inductive research* does not test hypotheses. It begins with a more general interest, and, as data are collected, concepts are specified in more detail, leading to the development of hypotheses and to grounded theory.

- Family researchers apply the *scientific method*—well-established procedures used to collect information.

- Professional family researchers follow ethical principles to protect participants from having their identities revealed and to minimize the discomfort the subjects experience from their participation in the research.

- Research data come from surveys, clinical studies, and direct observation, in which naturally occurring variables are measured against one another. Data are also obtained from experimental research.

- Frequently, family researchers make use of the availability of existing survey data, gathered by others and made available for secondary analysis.

- To overcome limitations with any particular method, researchers often engage in triangulation—the use of multiple methods and/or multiple sources of data.

- Family researchers strive to identify and account for patterns of behavior. There will be exceptions to all patterns. Exceptions do not negate the importance or validity of research conclusions.

## Key Terms

adaptation 39
ambiguous loss 40
anonymity p. 55
applied research 60
bias 35
case-study method 58
clinical research 58
cohabitation effect 32
concepts 37
conceptualization 37
confidentiality 55
conflict theory 44
deductive research 37
demography 58
dependent variable 38
egocentric fallacy 35

environment 39
environmental influences 39
equity 49
ethical guidelines 55
ethnocentric fallacy 35
experimental research 60
fallacies 35
family development theory 50
family ecology theory 39
family systems theory 52
feminist perspectives 45
grounded theory 38
homeostasis 52
hypotheses 38
independent variable 38
inductive research 38
interaction 47
intersectionality 45
intervening variable 38
linked lives 51
macro-level theories 39
micro-level theories 46
normative age-graded influences 51
normative history-graded influences 51
objective statements 35
objectivity 35
observational research 58
operationalization 37
opinion 35
qualitative research 54
quantitative research 54
scientific method 37
secondary data analysis 54
social exchange theory 48
social institution 39
social roles 47
stereotype 35
structural functionalism theory 42
subsystems 42
survey research 55
symbolic interaction theory 47
theories 37
time use diary 56
time use survey 56
triangulation 55
value judgment 35
variables 37

# 3

# Variations in American Family Life

*Are the following statements True or False? You may be surprised by the answers as you read this chapter.*

T F **1.** Compared with contemporary families, colonial family life was considered much more of a private matter.

T F **2.** Slavery destroyed the African American family system.

T F **3.** Industrialization transformed the role families played in society as well as the roles women and men played in families.

T F **4.** Compared with what came both before and after, families of the 1950s were unusually stable.

T F **5.** Within upper-class families, husbands and wives are relatively equal in their household roles and authority.

T F **6.** Lower-class families are the most likely to be single-parent families.

T F **7.** Marital relationships can suffer as a result of either downward or upward mobility.

T F **8.** Compared with Caucasian families, relationships between African American husbands and wives are more traditional.

T F **9.** Asian American or Latino families show much variation within each group, depending on the country from which they came, why they left, and when they arrived in the United States.

T F **10.** European ethnic groups are as different from one another as they are from African Americans, Latinos, Asian Americans, or Native Americans.

Dorothy Littell Greco/The Image Works

## Chapter Outline

One thing you can almost always count on is that sometime during the term or semester, whether in class or in conversation, someone will make the oft-heard statement, "Well, all families are different." There is a lot of truth to that sentiment. For example, your family is not like your best friend's family in every way. Furthermore, assuming your best friend is someone a lot like you (which, as you've probably noticed, is common among people who become best friends), the differences between your families likely understate how richly variable family experience actually is.

It is true that in some ways family researchers are interested in discovering, describing, and explaining patterns in family experience. Of course, *every family is different,* and we will look closely at some patterned variations that separate and diversify family experiences over the next few chapters. Although we could include a number of factors as sources of such variation, the current chapter is concerned with the following four: time, social class, race, and ethnicity. In subsequent chapters, we also look at how gender (Chapter 4) and sexual orientation (Chapter 6) shape people's experiences of relationships and families. Then, throughout the remainder of the book, we draw comparisons and make contrasts among different types of households and families—singles, cohabiting and married couples, parents and nonparents, single-parent households and two-parent households, dual earners, male breadwinner–female homemaker households and "role reversers," first marriages and remarriages, and step-relationships in blended families and blood relationships in birth families. Therefore, the task we start here won't end until you finish this book.

In this chapter, we begin by detailing the historical development of the kinds of families that predominate in the United States today, noting key transformations and the forces that created them. This accomplishes two things: It gives you a better sense of where today's American families have come from, and it enables you to see how different family life has been across generations, even within the same families. We then shift our attention to some major economic, ethnic, and racial variations that diversify contemporary families in the United States.

# ■ American Families across Time

In Chapter 1, we noted that marriages and families in the United States are dynamic and must be understood as the products of wider cultural, demographic, and technological developments. Although more attention tends to be paid to changes that have occurred in the post–World War II era, those changes represent only some of the more recent instances of more than 300 years of change that make up the history of American family life from the colonial period on into the 21st century. Armed with this brief history, we can recognize and make connections between changes in society and changes in families. In addition, we will be better positioned to assess the meaning of some of the more dramatic changes that have occurred recently in American family life. Finally, on a more personal level, you can better understand your own genealogies and family histories by recognizing the shifting stage on which they were played out.

## The Colonial Era (1607–1776)

The colonial era is marked by differences among cultures, family roles, customs, and traditions. These families were the original crucible from which our contemporary families were formed.

### Native American Families

The greatest diversity in American family life may have existed during our country's earliest years, when it was populated by 2 million Native Americans, representing more than 240 groups with distinct family and kinship patterns. Many groups were **patrilineal:** Descent was traced, and rights and property flowed from the father. Others, such as the Zuni and Hopi in the Southwest and the Iroquois in the Northeast, were **matrilineal:** Rights and property descended from the mother.

Native American families tended to share certain characteristics, although it is easy to overgeneralize. Most families were small. There was a high child mortality rate, and mothers breast-fed their infants; during breast-feeding, mothers abstained from sexual intercourse. Children were often born in special birth huts. As they grew older, the young were rarely physically disciplined. Instead, they were taught by example, praised when they were good and publicly shamed when they were bad.

Children began working at an early stage. Their play, such as hunting or playing with dolls, was modeled on adult activities. Ceremonies and rituals marked transitions into adulthood. Girls underwent puberty ceremonies at first menstruation. For boys, events such as growing the first tooth and killing the first large animal when hunting signified stages of growing up. A days-long initiation ritual under the supervision of elders known as a vision quest often marked the transition to manhood.

Marriage took place early for girls, usually between 12 and 15 years; for boys, it took place between 15 and 20 years. Again, though, the range of diversity was striking. Some tribes arranged marriages; others permitted young men and women to choose their partners. Most groups were monogamous, although some allowed two wives. Some tribes permitted men to have sexual relations outside of marriage when their wives were pregnant or breast-feeding.

## Colonial Families

From earliest colonial times, America has been an ethnically diverse country. In the houses of Boston, the mansions and slave quarters of Charleston, the mansions of New Orleans, the haciendas of Santa Fe, and the Hopi dwellings of Oraibi, Arizona (the oldest continuously inhabited place in the United States, dating back to AD 1150), American families have provided emotional and economic support for their members.

**The Family.** Colonial America was initially settled by waves of explorers, soldiers, traders, pilgrims, servants, prisoners, farmers, and slaves. In 1565, in St. Augustine, Florida, the Spanish established the first permanent European settlement in what is now the United States. In 1607, America's first permanent English colony was founded in Jamestown, Virginia. Some 13 years later, Plymouth colony was founded in what is now Plymouth, Massachusetts. But the members of these first groups came as single men—as explorers, soldiers, and exploiters.

In 1620, the leaders of the Jamestown colony, hoping to promote greater stability, began importing English women to be sold in marriage. The European colonists who came to America attempted to replicate their familiar family system. This system, strongly influenced by Christianity, emphasized **patriarchy** (rule by father or eldest male), the subordination of women, sexual restraint, and family-centered production. In the North, the colonial family was basically a multifunctional social and economic institution, the primary unit for producing most goods and caring for the needs of its members. The family planted and harvested food, made clothes, provided shelter, and cared for the necessities of life. As a social unit, the family reared children and cared for the sick, infirm, and aged. Its responsibilities included teaching reading, writing, and arithmetic because there were few schools. The family was also responsible for religious instruction: It was to join in prayer, read scripture, and teach the principles of religion.

Unlike in New England, the plantation system that came to dominate the Southern colonies did not give the same priority to family life. Hunting, entertaining, and politics provided the greatest pleasure. The elite plantation owners continued to idealize gentry ways until the Civil War destroyed the slave system on which they based their wealth.

**Marital Choice.** Romantic love was not the decisive factor in choosing a partner; one practical 17th-century marriage manual advised women that "this boiling affection is seldom worth anything" (Fraser 1984). Because marriage had profound economic and social consequences, parents often selected, or had the right to refuse, their children's choices of mates. In the 17th century, 8 of the 13 colonies had laws *requiring parental approval* and imposed sanctions as harsh as imprisonment or whipping on men who "insinuated" themselves into a woman's affections without her parents' approval (Coontz 2005). Even in instances without such restrictions, in which individuals were "free to choose," children rarely went against their parents' wishes. If parents disapproved, their children typically gave up out of fear of the social and financial consequences of defying their parents (Coontz 2005). Love was not irrelevant but came after marriage. It was a person's duty to love his or her spouse. The inability to desire and love a marriage partner was considered a defect of character.

Although the Plymouth colonists prohibited premarital intercourse, they were not entirely successful. **Bundling,** the New England custom in which a young man and woman spent the night in bed together, separated by a wooden bundling board, provided a courting couple with privacy; it did not, however, encourage restraint. By the latter part of the 18th century, between 30 percent and 40 percent of New England brides were pregnant by the time they married (Godbeer 2004).

**Family Life.** The colonial family was strictly patriarchal, and both the church and the community reinforced such paternal authority (Mintz 2004). Unlike the present where we consider our families to be private matters into which outsiders should refrain from intervening or interfering, colonial family life was much more of a public phenomenon.

> **False**
>
> **1.** Unlike contemporary families, colonial family life was much more public.

Considering its patriarchal quality, historian Steven Mintz (2004) describes the range of fathers' influence. Fathers were

> responsible for leading their households in daily prayers and scripture reading, catechizing their children and servants, and teaching household members to read so that they might study the Bible. . . . Child rearing manuals were thus addressed to men, not their wives. They had an obligation to help their sons find a vocation or calling, and a legal right to consent to their children's marriage. Massachusetts Bay Colony and Connecticut underscored the importance of paternal authority by making it a capital offense (i.e., punishable by death) for youths sixteen or older to curse or strike their father. (p. 13)

The authority of the husband/father rested in his control of land and property. In a society dependent on agriculture and farming, such as colonial America, land was the most precious resource. The manner in which the father decided to dispose of his land affected his relationships with his children. In many cases, children were given land adjacent to the father's farm, but the title did not pass into their hands until the father died. This power gave fathers control over their children's marital choices and kept them geographically close.

This strongly rooted patriarchy called for wives to submit to their husbands. The wife was not an equal but was a helpmate. This subordination was reinforced by traditional religious doctrine. Like her children, the colonial wife was economically dependent on her husband. On marriage, she transferred to her husband many rights she had held as a single woman, such as the right to inherit or sell property, to conduct business, and to attend court.

For women, marriage marked the beginning of a constant cycle of childbearing and child rearing. On average, colonial women had six children and were consistently bearing children until around age 40. In addition to their maternal responsibilities, colonial women were expected to do a wide range of chores from cooking and cleaning to spinning, sewing, gardening, keeping chickens, and even brewing beer (Mintz and Kellogg 1988).

**Childhood and Adolescence.** The colonial conception of childhood was radically different from ours. First, children were believed to be evil by nature. The community accepted the traditional Christian doctrine that children were conceived and born in sin.

Second, childhood did not represent a period of life radically different from adulthood. Such a conception is distinctly modern (Aries 1962; Mintz 2004). In colonial times, a child was regarded as a small adult. From the time children were 6 or 7 years old, they began to be part of the adult world, participating in adult work and play.

Third, children between the ages 7 and 12 were often "bound out" or "fostered" as apprentices or domestic servants (Mintz 2004). They lived in the home of a relative or stranger where they learned a trade or skill, were educated, and were properly disciplined. **Adolescence**—the separate life stage between childhood and adulthood—did not exist. They went from a shorter childhood (than what we are accustomed to) to adulthood (Mintz 2004; Mintz and Kellogg 1988). Thus, our contemporary notions of a rebellious life stage filled with inner conflicts, youthful indiscretions, and developmental crises do not fit well with the historical record of Plymouth colony (Demos 1970; Mintz 2004).

### African American Families

In 1619, a Dutch man-of-war docked at Jamestown in need of supplies. Within its cargo were 20 Africans who had been captured from a Portuguese slaver. The captain quickly sold his captives as indentured servants. By 1664, when the British gained what had been Dutch-governed New Amsterdam, 40 percent of the colony's population consisted of African slaves. During the 17th and much of the 18th centuries, enslaved Africans and their descendants faced difficulty forming and maintaining families. It was hard for men, who often outnumbered women 40 percent to

60 percent or worse, to find wives. Enslaved African Americans were more successful in continuing the traditional African emphasis on the extended family, in which aunts, uncles, cousins, and grandparents played important roles. Although slaves were legally prohibited from marrying, they created their own marriages.

Childhood experience was often bitter and harsh. It was common for children to be separated from their parents because of a sale, a repayment of a debt, or a plantation owner's decision to transfer slaves from one property to another (Mintz 2004). Despite the hardships placed on them, enslaved Africans and African Americans developed strong emotional bonds and family ties. Slave culture discouraged casual sexual relationships and placed a high value on marital stability. On the large plantations, most enslaved people lived in two-parent families with their children. To maintain family identity, parents named their children after themselves or other relatives or gave them African names. In the harsh slave system, the family provided strong support against the daily indignities of servitude. As time went on, the developing African American family blended West African and English family traditions (McAdoo 1996).

**False** **2.** Slavery did not destroy the African American family system.

**True** **3.** Industrialization transformed the role families played in society as well as the roles women and men played in families.

## Marriages and Families in the 19th Century

In the 19th century, the traditional colonial family form gradually vanished and was replaced by the modern family. In this transition, families became more egalitarian, less patriarchal, and more affection based.

### Industrialization and the Shattering of the Old Family

In the 19th century, the industrialization of the United States transformed the face of America. It also transformed American families from self-sufficient farm families to wage-earning, increasingly urban families. As factories began producing gigantic harvesters, combines, and tractors, significantly fewer farm workers were needed. Looking for employment, workers migrated to the cities, where they found employment in the ever-expanding factories and businesses. Because goods were now bought rather than made in the home, the family began its shift from being primarily a production unit to being more of a consumer- and service-oriented unit.

With this shift, a radically new **division of labor** arose in the family. Men began working outside the home in factories or offices for wages they then used to purchase the family's necessities and other goods. Men became identified as the family's sole provider or breadwinner. Their work was given higher status than women's work because it was paid in wages. Men's work began to be increasingly identified as "real" work, distinct from the unpaid domestic work women did.

Sociologist Andrew Cherlin details how "the growth of wage labor profoundly affected family life" (Cherlin 2014, 11). First, the primary setting for men's work was moved out of the home, which meant, too, that the work worlds of husbands and wives were separated. Industrial workplaces became mostly worlds of men. Married women, especially those with children, tended not to work outside the home, unless and except for when their families needed the additional income. However, working-class women contributed to their family's income by doing such things as taking in boarders or engaging in piecework—"small-scale assembly work that could be done at home . . ." (p. 27).

In addition, as Cherlin points out, relations between parents and children were transformed, as industrialization led to a decline in parental authority. Compulsory schooling laws and laws against child labor gradually reduced parents' ability to send their

Strong family ties endured in enslaved African American families. The extended family, important in West African cultures, continued to be a source of support and stability.

children to work. Also, industrialization changed how workers saw themselves. Men took pride in their ability to endure the demanding and often repetitive industrial jobs to support their families. They also felt as though they fulfilled their responsibility *to* their families by what they did *for* their families, not by their participation *in* their families (Cherlin 2014).

## Marriage and Families Transformed

Without its central importance as a work unit and less and less the source of other important societal functions (e.g., education, religious worship, protection, and recreation), the family became the focus and abode of feelings. The emotional support and well-being of adults and the care and nurturing of the young became the two most important family responsibilities.

**The Power of Love.** This new affectionate foundation of marriage brought love to the foreground as the basis of marriage and represented the triumph of individual preference over family, social, or group considerations. Stephanie Coontz (2005) reports that "by the middle of the nineteenth century there was near unanimity in the middle and upper classes throughout western Europe and North America that the love-based marriage, in which the wife stayed home and was protected and supported by her husband, was a recipe for heaven on earth" (p. 162).

Women now had a new degree of power: They were able to choose whom they would marry. Women could rule out undesirable partners during courtship; they could choose mates with whom they believed they would be compatible. Mutual esteem, friendship, and confidence became guiding ideals. Without love, marriages were considered empty shells. Although working-class men and women only gradually adopted such middle- and upper-class ideas of marital love and intimacy, the working class more readily accepted middle-class ideas about gender roles.

**Changing Roles for Women.** The two most important family roles for middle-class women in the 19th century were those of housewife and mother. As there was a growing emphasis on domesticity in family life, the role of housewife increased in significance and status. Home was the center of life, and the housewife was responsible for making family life a source of fulfillment for everyone. For many women, especially middle-class women, this "doctrine of separate spheres" was wholeheartedly accepted and

enthusiastically embraced (Coontz 2005). But even many among the working and lower classes ultimately came to recognize "good economic sense" in having a wife at home. With what women could contribute to their families through their performance of such tasks as making clothing, preparing food, growing vegetables, and so on, families benefited more than they would from the contribution of women's wages, estimated at only about one-third of the wages men earned. Although many working-class and poor women did work outside the home for large stretches in their lives, "the ideology of male breadwinning and female homemaking became entrenched in working-class aspirations" (Coontz 2005, 175).

Women also increasingly focused their identities on motherhood. People in the 19th century witnessed the most dramatic decline in fertility in American history. Between 1800 and 1900, fertility dropped by 50 percent. Whereas at the beginning of the 19th century, an American mother typically gave birth to between 7 and 10 children, beginning "in her early twenties and (giving birth) every two years or so until menopause," the average number of births had fallen to just three by 1900 (Mintz 2004).

Women reduced their childbearing by insisting that they, not men, control the frequency of intercourse. Child rearing rather than childbearing became one of the most important aspects of a woman's life. Having fewer children and having them in the early years of marriage allowed more time to concentrate on mothering and opened the door to greater participation in the world outside the family. This outside participation manifested itself in women's heavy involvement in abolition, prohibition, and women's emancipation movements.

**Childhood and Adolescence.** A strong emphasis was placed on children as part of the new conception of the family. A belief in childhood innocence replaced the idea of childhood corruption. A new sentimentality surrounded the child, who was now viewed as born in total innocence. Protecting children from experiencing or even knowing about the evils of the world became a major part of child rearing.

People in the latter part of the 19th century also witnessed changes in the lives of young people that eventuated in the "beginning" of adolescence (Kett 2003). In contrast to colonial youths, who participated in the adult world of work and other activities, young people in the late 19th and early 20th centuries were kept economically dependent and separate from adult activities, remained in school until their mid-teens, and often felt apprehensive when they entered the adult world. This apprehension sometimes led to the emotional conflicts associated with adolescent identity crises. In a 1904 book, psychologist G. Stanley Hall described the "emotional upheaval and fluctuating emotions, . . . contradictory tendencies toward hyperactivity and inertness, selfishness and altruism, bravado and a sense of worthlessness," that characterized this stage of development (Mintz 2004, 196).

Education also changed as schools, rather than families, became responsible for teaching reading, writing, and arithmetic as well as educating students about ideas and values. Conflicts between the traditional beliefs of the family and those of the impersonal school were inevitable. At school, the child's peer group increased in importance.

## The African American Family in Slavery and Freedom

Although there were large numbers of free African Americans—100,000 in the North and Midwest and 150,000 in the South—most of what we know about the African American family before the Civil War is limited to the slave family.

**The Slave Family.** By the 19th century, the slave family had already lost much of its African heritage. Under slavery, the African American family lacked two key factors that helped give free African American and Caucasian families stability: autonomy and economic importance. Slave marriages were not recognized as legal. Final authority rested with the owner in all decisions about the lives of slaves. The separation of families was a common occurrence, spreading grief and despair among thousands of slaves. Furthermore, slave families worked for their masters, not themselves. It was impossible for the slave husband/father to become the provider for his family. The slave women worked in the fields beside the men. When an enslaved woman was pregnant, her owner determined her care during pregnancy and her relation to her infant after birth.

Slave children endured deep and lasting deprivation. Often shoeless, sometimes without underwear or adequate clothing, hungry, underfed and undernourished, and forced into hard physical labor as young as age 5 or 6, slave children suffered considerably. Rates of illness and death in infancy and childhood were high. Furthermore, family life was fragile and often disrupted. Steven Mintz reports that separation of children from parents, especially fathers, was so common that at least half of all enslaved children experienced life separate from their father because he died, lived on another plantation, or was a white man who declined to acknowledge that they were his children. By their late teens, either temporary or permanent separation from their parents was something virtually all slave children had suffered (Mintz 2004).

Still, it is important to reiterate that slavery did not destroy all aspects of slave families. Despite the intense oppression and hardship to which they were subjected, many slaves displayed resilience and survived by relying on their families and by adapting their family system to the conditions of their lives (Mintz and Kellogg 1988). This included, for example, relying on extended kinship networks and, where necessary, on unrelated adults to serve as surrogates for parents absent because of the forced breakup of families.

Furthermore, enslavement did not forever destroy the African American family system. In no way does saying this diminish the horrors of slavery. Instead, it acknowledges the resilience of those who survived enslavement, and it illustrates how family systems may be pivotal sources of support and key mechanisms of surviving even the most extraordinary distress.

**After Freedom.** In 1865, with the ratification of the Thirteenth Amendment to the U.S. Constitution, slavery was outlawed. When freedom came, the formerly enslaved African American families had strong emotional ties and traditions forged from slavery and from their West African heritage (Guttman 1976; Lantz 1980). Because they were now legally able to marry, thousands of former slaves formally renewed their vows. The first year or so after freedom was marked by what was called "the traveling time," in which African Americans traveled up and down the South looking for lost family members who had been sold. Relatively few families were reunited, although many continued the search well into the 1880s.

African American families remained poor, tied to the land, and segregated. Despite poverty and continued exploitation, the Southern African American family usually consisted of both parents and their children. Extended kin continued to be important.

## Immigration: The Great Transformation

In the 19th and early 20th centuries, great waves of immigration swept over America. Between 1820 and 1920, 38 million immigrants came to the United

States. Historians commonly divided them into "old" immigrants and "new" immigrants.

**The Old and the New Immigrants.** The old immigrants, who came between 1830 and 1890, were mostly from western and northern Europe. During this period, Chinese also immigrated in large numbers to the West Coast. The new immigrants, who came from eastern and southern Europe, began to arrive in great numbers between 1890 and 1914 (when World War I virtually stopped all immigration).

Japanese also immigrated to the West Coast and Hawaii during this time. Today, Americans can trace their roots to numerous ethnic groups.

As the United States expanded its frontiers, surviving Native Americans were incorporated. The United States acquired its first Latino population when it annexed Texas, California, New Mexico, and part of Arizona after its victory over Mexico in 1848.

**The Immigrant Experience.** Most immigrants were uprooted; they left only when life in the old country became intolerable. The decision to leave their homeland was never easy. It was a choice between life and death and meant leaving behind ancient ties.

Most immigrants arrived in America without skills. Although most came from small villages, they soon found themselves in the concrete cities of America. Again, families were key ingredients in overcoming and surviving extreme hardship. Because families and friends kept in close contact even when separated by vast oceans, immigrants seldom left their native countries without knowing where they were going—to the ethnic neighborhoods of New York, Chicago, Boston, San Francisco, Vancouver, and other cities. There they spoke their own tongues, practiced their own religions, and ate their customary foods. In these cities, immigrants created great economic wealth for America by providing cheap labor to fuel growing industries.

In America, kinship groups were central to the immigrants' experience and survival. Passage money was sent to their relatives at home, information was exchanged about where to live and find work, families sought solace by clustering together in ethnic neighborhoods, and informal networks exchanged information about employment locally and in other areas.

The family economy, critical to immigrant survival, was based on cooperation among family members. For most immigrant families, as for African American families, the middle-class idealization of motherhood

Except for Native Americans, most of us have ancestors who came to America—voluntarily or involuntarily. Between 1820 and 1920, more than 38 million immigrants came to the United States.

and childhood was a far cry from reality. Because of low industrial wages, many immigrant families could survive only by pooling their resources and sending mothers to work and even sending their children to work in the mines, mills, and factories.

Most groups experienced hostility. Crime and immorality were attributed to the newly arrived ethnic groups; ethnic slurs became part of everyday parlance. Strong activist groups arose to prohibit immigration and promote "Americanism." Literacy tests required immigrants to be able to read at least 30 words in English. In the early 1920s, severe quotas were enacted that slowed immigration to a trickle.

## Marriages and Families in the 20th Century

By the beginning of the 20th century, the functions of American middle-class families had been dramatically altered from earlier times. Families had lost many of their traditional economic, educational, and welfare functions. Food and goods were produced outside the family, children were educated in public schools, and the poor, aged, and infirm were increasingly cared for by public agencies and hospitals. The primary focus of the family was becoming even more centered on meeting the emotional needs of its members. In time, cultural emphasis would shift from self-sacrificing familism (i.e., making sacrifices in one's own pursuit of happiness or satisfaction for the well-being of one's family) to more self-centered individualism (wherein one's family's well-being was less important than one's own), and individuals' sense of their connections and obligations to their families would be greatly transformed.

## The New Companionate Family

Prior to the 20th century, romantic love was considered an unwise, unsafe basis for selecting one's spouse; by the first decades of the 20th century, it was seen as essential (Cherlin 2004). This new ideal family form rejected the "old" family based on male authority and sexual repression. This new family form was based on what's called the **companionate marriage**.

There were four major features of this companionate family: (1) Men and women were to share household decision making and tasks; (2) marriages were expected to provide romance, sexual fulfillment, and emotional growth; (3) wives were no longer expected to be guardians of virtue and sexual restraint; and (4) children were no longer to be protected from the world but were to be given greater freedom to explore and experience the world; they were to be treated more democratically and encouraged to express their feelings (Mintz and Kellogg 1988). Most critically, spouses were supposed to not just love each other but also be *in love* (Cherlin 2009).

## Through the Depression and World Wars

The history of 20th-century family life cannot be told without considering how profoundly family roles and relationships were affected by the Great Depression and two world wars. Although many different connections could be drawn, two seem particularly significant: changes in the relationship between the family and the wider society, and changes in women's and men's roles in and outside of the family.

**Linking Public and Private Life.** The economic crisis during the Depression was staggering in its scope. Unemployment jumped from less than 3 million in 1929 to more than 12 million in 1932, and the rate of unemployment rose from 3.2 percent to 23.6 percent.

Over that same span of time, average family income dropped 40 percent (Mintz and Kellogg 1988). To cope with this economic disaster, families turned inward, modifying their spending, increasing the numbers of wage earners to include women and children, and pooling their incomes. Often it was a broadened "inward" to which they turned because people often

took in relatives or relied on kinship ties for economic assistance (Mintz and Kellogg 1988).

Ultimately, these more personal, intrafamilial efforts proved insufficient. President Franklin Roosevelt's New Deal social programs attempted to respond to the social and economic despair that more localized efforts were unable to alleviate. Farm relief, rural electrification, Social Security, and a variety of social welfare provisions were all implemented in the hope of doing what local communities and individual families could not. Such federal initiatives reflected a dramatic ideological shift wherein government now bore responsibility for the lives and well-being of families (Mintz and Kellogg 1988).

Precipitated by the mass entrance into the workforce of millions of previously unemployed women, including many with young children, there was a clear need and opportunity for public resources to be committed to child care. Unfortunately, the federal government's response was slow and inadequate given the sudden and dramatic increase in need and demand (Filene 1986; Mintz 2004; Mintz and Kellogg 1988). Most mothers who entered the labor force had to rely on neighbors and grandparents to provide child care. When such supports were unavailable, many had no choice but to turn their children into "latchkey" kids fending for themselves (Mintz 2004). Unlike some of our European allies who invested more heavily in policies and services to accommodate employed mothers (Mintz and Kellogg 1988), it took the federal government two years to "appropriate funds to build and staff day-care centers, and the funds were sufficient for only one-tenth of the children who needed them" (Filene 1986, 164). Despite having engineered a propaganda campaign to entice women into jobs vacated by the 16 million men who entered the service, the government remained ambivalent about welcoming mothers of young children into those positions. However inadequate or slow government efforts were, they were still greater than what followed for most of the rest of the century.

**Gender Crises: The Great Depression and World Wars.** Both the Depression and the two world wars (especially World War II) reveal much about the gender foundation on which 20th-century families rested. During the Depression, it was men whose gender identities and family statuses were threatened by their lost status as providers. During each world war, women were the ones who faced challenges that required them to abandon their gender socialization

During World War II, women were urged to enter the labor force and especially to enter nontraditional occupations left vacant by the deployment of men overseas. The images here illustrate the kinds of messages women received and the kinds of jobs they helped fill.

J. Howard Miller/Fine art/Corbis

Harold M. Lamber/Archive Photos/Getty Images

and step into roles and situations that fell outside their traditional familial roles. In each instance, the familial gender roles and identities had to be altered to match extraordinary circumstances.

What is especially striking about men's reactions to their job loss is their internalization of fault for what was a society-wide economic crisis. Given how widespread unemployment was, one might think that men would at least take some comfort in knowing that the predicaments they faced were not of their own making. Yet they had so deeply internalized their sense of themselves as providers that their identities, family statuses, and sense of manhood were all invested in wage earning and providing. When unable to provide, many men were deeply shaken. Some were even driven to the point of emotional breakdown or suicide by their sense of economic failure (Filene 1986).

For many families, survival depended on the efforts of wives or the combination of women's earnings, children's earnings, assistance from kin, or some kind of public assistance. For those who depended at least somewhat on women's earnings, there were other gender consequences of running the household. Sometimes, men were pressed by their wives to contribute domestically in the women's "absence."

Although some did, many others resisted (Filene 1986). Sometimes women displayed ambivalence about the meaning of male unemployment and male housework. Whereas 80 percent of the women who were surveyed in 1939 by the *Ladies' Home Journal* thought an unemployed husband should do the domestic work in the absence of his employed wife, 60 percent reported they would lose respect for men whose wives outearned them (Filene 1986).

If the Depression revealed male anxiety about their familial roles as providers, we see in women's experiences during World Wars I and II that gender crises were not limited to men. During both wars, the absence of millions of men meant that women were pressed to step into their vacant shoes and participate in wartime production. During World War I, 1.5 million women entered the wartime labor force, many in jobs previously held largely by men (Filene 1986). During World War II, the number of employed women rose even more dramatically. Between 1941 and 1945, the numbers of employed women increased by more than 6 million to a wartime high of 19 million (Degler 1980; Lindsey 1997). Furthermore, "nearly half of all American women held a job at some time during the war" (Mintz and Kellogg 1988).

Whereas single women had long worked and poor or minority women had worked even after marriage, the biggest changes in women's labor force participation during World War II were the increased employment among married, middle-class women, and the types of jobs into which women were being hired. Despite the widely held cultural emphasis on the special nurturing role of women and the belief that the home was a woman's "proper place," American society needed women to take over for the absent men. Once enticed into nontraditional female employment, women received both material and nonmaterial benefits that were hard for many to surrender once the war ended and men returned.

Materially, women in traditionally male occupations received higher wages than they had in their past, more sex-segregated work experiences. As important, they also found a sense of gratification and enhanced self-esteem that were often missing from the jobs they were more accustomed to. However reluctant they may have been to take on such work, many were clearly more than a little ambivalent to leave it. To assist women in their departures from these jobs, pro-family rhetoric and a new ideology extolling the value and importance of women's roles as mothers and caregivers were broadly conveyed by a variety of sources (e.g., popular media, social workers, and educators).

### Families of the 1950s

In the long history of American family life, no other decade has come to symbolize so much about that history despite actually representing relatively little of it (Coontz 1997; Mintz and Kellogg 1988). In many ways, the 1950s appear to be a period of unmatched family stability, during which marriage and family seemed to be central to American lives. It was a time of youthful marriages, unusually high marriage rates and birthrates, stable and uncharacteristically low divorce rates, and economic growth. With a prosperous economy, many couples were able to buy homes on the income of only one wage earner—the husband. This was the period during which the breadwinner–homemaker family, with its traditional gender and marital roles, reached its peak. Man's place was in the world, and woman's place was in the home. Women were expected to place motherhood first and to sacrifice their opportunities for outside advancement to ensure the success of their husbands and the well-being of their children.

> **True**
> **4.** Compared with what came both before and after, families of the 1950s were unusually stable.

Given the meaning often invested in this era, it is important to understand that the 1950s were atypical. Compared with both what came before and what followed, families of the 1950s were unique. This is important: It means that anyone who uses this decade as a baseline against which to compare more recent trends in such family characteristics as birth, marriage, or divorce rates starts with a faulty assumption about how representative it is of American family history. Looking at those same trends with a longer view reveals that the changes that followed the 1950s were more consistent with some familial patterns evident in the 19th and earlier part of the 20th centuries (Mintz and Kellogg 1988). For example, during the 1950s, the divorce rate increased less than in any other decade of the 20th century. Similarly, after more than 100 years of declining birthrates and shrinking family sizes, during the 1950s, "women of childbearing age bore more children, spaced . . . closer together, and had them earlier and faster" than had previous generations (Mintz and Kellogg 1988, 179).

It cannot be emphasized enough how much familial experience of the 1950s was created, sustained by, and depended on the unprecedented economic growth and prosperity of the postwar economy (Coontz 1997). The combination of suburbanization and economic prosperity, supplemented by governmental assistance to veterans, allowed many married couples to achieve the middle-class family dream of home ownership while raising their children under the loving attention of full-time caregiving mothers. We must be careful, though, not to oversimplify the family experience of the 1950s. Americans did not all benefit equally from the economic prosperity and opportunity of the decade. Thus, overgeneralizations would leave out the experiences of poor and working-class families and racial minorities for whom neither full-time mothering nor home ownership were common (Coontz 1997). In addition, many, especially women, found that the ideal lifestyle of the period left them longing for something more (Friedan 1963).

### Late Twentieth-Century Families

When we look at family changes that occurred in subsequent decades, we need to recognize that economic factors, again, were among the most important determinants of some more dramatic departures from the 1950s model. This especially pertains to the emergence of the dual-earner household. As Stephanie Coontz (1997, 47) points out, "By the mid-1970s, maintaining the prescribed family lifestyle meant for many couples giving up the prescribed family form. They married later, postponed children, and curbed their fertility; the wives went out to work." They did this not in rejection of the family lifestyle of the 1950s, but in the pursuit of central features of that lifestyle, such as home ownership, no longer as attainable on the earnings of one wage earner.

The latter decades of the last century saw certain trends emerge and/or continue to grow. Most notably, increasing numbers of dual-earner couples, increases in cohabitation, delaying entry into marriage, and a leveling off and decline in divorce continued. Certain legal developments, such as the passage and gradual widespread adoption of no-fault divorce, and the wider use of joint custody of children of divorce, changed the process and outcome of divorce for many couples and their children. Revolutions in reproductive medicine changed the possibilities for previously infertile couples to have children. Wider availability of more effective contraception along with access to legal abortion meant that one could more effectively prevent or safely and legally terminate unwanted pregnancies, thus freeing sexual intimacy from the fear of pregnancy. The discovery in the early 1980s of HIV/AIDS had a somewhat chilling effect on gay men and eventually wider cultural expectations about safe sex. The passage of the Family and Medical Leave Act brought job-protected family leave to many in the United States. Though this legislation paled in comparison with the policies and protections afforded new parents elsewhere, it was an improvement over

the even more precarious and vulnerable position new parents found themselves in prior to its passage.

The remaining 11 chapters of this book look closely at families and family issues in the first decades of the current century. The characteristics these families display did not emerge suddenly but were established over years. Beginning with the latter years of the 1950s and escalating through and then beyond the 1960s and 1970s, some striking family trends surfaced. These trends persisted through and beyond the end of the 20th century, leaving marriages and families reshaped, and the meaning and experience of family life significantly altered.

Tables 3.1, 3.2, and 3.3 depict a variety of such changes. As Table 3.1 shows, beginning in the 1960s,

**TABLE 3.1  Median Age at First Marriage, 1960–2011**

| Year | Males (age) | Females (age) |
|------|-------------|---------------|
| 1960 | 22.8 | 20.3 |
| 1970 | 23.2 | 20.8 |
| 1980 | 24.7 | 22.0 |
| 1990 | 26.1 | 23.9 |
| 2000 | 26.8 | 25.1 |
| 2008 | 27.4 | 25.6 |
| 2011 | 28.7 | 26.5 |
| 2014 | 29.3 | 27.0 |

SOURCE: U.S. Census Bureau, 2011a, "Current Population Survey, March and Annual Social and Economic Supplements, 2011."

**TABLE 3.2  Trends in Marriages, Divorces, and Births: 1970–2010**

|  | 1970 | 1980 | 1990 | 2000 | 2010 |
|--|------|------|------|------|------|
| Marriages | 2,159,000 | 2,390,000 | 2,443,000 | 2,329,000 | 2,096,000 |
| Marriage rate | 10.6% | 10.6% | 9.8% | 8.3% | 6.8% |
| Divorces | 708,000 | 1,189,000 | 1,182,000 | 944,000 | 872,000 |
| Divorce rate | 3.5% | 5.2% | 4.7% | 4.1% | 3.6% |
| Births | 3,731,000 | 3,612,000 | 4,158,000 | 4,063,000 | 3,999,386 |
| Birthrate* | 18.4 | 15.9 | 16.7 | 14.5 | 13.0 |

NA means data not available.
*Rate per 1,000 people.
SOURCES: Munson and Sutton 2005; Tejada-Vera and Sutton 2009.

**TABLE 3.3 Couples and Children: 1970–2014**

|  | 1970 | 1980 | 1990 | 2000 | 2010 | 2014 |
|---|---|---|---|---|---|---|
| Married couples | 44, 728,000 | 49,112,0000 | 52,317,000 | 55,311,000 | 58,410,000 | 59,630,000 |
| Unmarried cohabiting couples | 523,000 | 1,600,000 | 2,900,000 | 4,500,000 | 7,500,000 | 7,913,000 |
| Percent of children living with two parents | 85.4% | 76.6% | 72.5% | 69.1% | 65.7% | 64.4% |
| Percent of children living with one parent | 12.0% | 19.7% | 24.7% | 26.7% | 26.6% | 27.5% |
| Births to unmarried mothers | 399,000 | 666,000 | 1,165,000 | 1,308,000 | 1,693,658 | 1,604,495 |
| As percent of all births | 11% | 18% | 28% | 33% | 41% | 40.3% |

SOURCES: U.S. Census Bureau, "Current Population Survey Reports, America's Families and Living Arrangements, 2008"; National Center for Health Statistics, "Births: Final Data, National Vital Statistics Reports," vol. 57, #7, 2010.

Americans increasingly delayed marriage. By the mid-1990s, the median age at marriage for both genders was higher than it had been in more than a century. This trend has continued, and as can be seen in Table 3.1, American marriages now start later than ever before.

Careful inspection of the trends in Tables 3.2 and 3.3 shows that for much of the time period that is portrayed, an increasing divorce rate joined the falling marriage rates, and that a dramatic increase in cohabitation occurred alongside a similarly marked increase in births outside marriage and in single-parent families. But family trends are neither always linear nor always unambiguous. They may rise, unexpectedly stabilize, or even reverse direction. This is evident in the divorce rates as seen in Table 3.2. Similarly, the birthrate dropped before rising again and ultimately falling even lower. Thus, it appears that, even in the short term, the only constant in family life is change (Mintz and Kellogg 1988).

To some, many of the changes depicted in Tables 3.2 and 3.3 might appear to signal family decline, implying especially a diminishing importance or appeal of marriage, leaving the future of the family as most Americans perceived it and marriage, in particular, seemingly in some doubt (Popenoe 1993). However, one could also look at these same trends more optimistically, taking a more liberal position that change, in and of itself, is not a bad thing. In fact, with these changes come more choices for people about the kinds of families or lifestyles they wish to create and experience (Coontz 1997; Mintz and Kellogg 1988).

## Families Today

Certainly, today's families do reflect considerable diversity of structure. In painting a picture of today's families, we would include many categories: the "traditional" breadwinner–homemaker families with children, at-home dad families with children, two-earner couples with children, single-parent households with children, marriages without children, cohabiting couples with or without children, blended families, and gay and lesbian couples with or without children. Even within some of these categories one finds variation. For example, two-earner couples may set up their work schedules such that they work different shifts and can take care of their children with minimal involvement of outside caregivers. Single-parent households, despite being structurally the same will have different experiences and face different circumstances if they are products of divorce, death of a parent, or births to single mothers. These are the families that we will address throughout the remainder of this book.

Of the many trends that characterize families in the early 21st century, the following merit special comment:

- *Cohabitation.* As commonly used, **cohabitation** refers to unmarried couples sharing living quarters and intimate and sexual relationships. Few trends rival cohabitation in terms of the extent of change.

In 1970, cohabitation was a relatively uncommon and socially questionable lifestyle. Many cohabitors hid the lifestyle from their families and, sometimes, from their friends. Today, cohabitation is widely and more openly practiced by heterosexual and same-sex couples, by young people, by middle-age or even elderly couples, and by people of all races and classes. If one includes same-gender unmarried cohabiting couples, the total number of couples living together without marrying exceeds 8 million. Nearly half of U.S. women 15 to 44 have cohabited, and one-fourth of never-married 25- to 34-year-olds are currently cohabiting (Wang and Parker 2014). More than 3 million cohabiting couples are raising children together. Although cohabiting couples share the characteristics of living together in sexually and emotionally intimate relationships, couples who cohabit do so for many different reasons, and these different reasons lead to other differences among cohabitors. As we shall see in Chapter 9, some cohabitors live together as an alternative to marriage, others as a test of their compatibility and suitability for marriage, and still others for convenience. Many same-sex cohabitors live together because they are unable to legally marry where they live. Cohabitation and its effect on marriage are to be considered.

- *Marriage.* Although more than 80 percent of the U.S. population is predicted to marry at least once in their lifetimes, marriage has undergone significant changes in recent decades. The expectations people bring, the kinds of relationships they attempt to construct, the roles they play as spouses, and the levels of satisfaction they express all have changed. The struggle waged for marriage equality that culminated in the June 2015 U.S. Supreme Court's decision in *Obergefell v. Hodges* to make same-sex marriage fully legal led to more public discourse about marriage in legal, social, and religious contexts. Overall, marriage rates have declined, especially among women and men of lower income and for those with less than a college education. Presently, married couples head less than half of U.S. households, the first time such households have dropped below 50 percent of all U.S. households.

  However, even among the population least likely to marry, marriage remains a highly valued life goal. At the "other end," marriage rates for college-educated women have actually increased. Thus, as will be explored further in Chapter 8, the status of marriage is more complicated than meets the eye.

- *Divorce, remarriage, and blended families.* After a period of escalating rates, divorce rates appear to have declined and stabilized over the past few decades. Unfortunately, we don't know how much of the decline reflects an undercount of divorce or a real reduction in the rate because the data available to track marriages and divorces have become less trustworthy, especially because the federal government stopped funding the collection of marriage and divorce data (Amato 2015). We do know that divorce remains a traumatic disruption in the lives of millions of U.S. couples and children each year. The effects of divorce are not the same on women and men, adults and children, or children of varying ages, as we shall explore further in Chapter 13. Furthermore, the social acceptability of divorce and the legal processes that surround divorce remain controversial public issues.

  The prevalence of divorce is also associated with both remarriages and blended families. Around half of men and women over 25 who have ever divorced are currently remarried. Nearly 20 percent of currently married women and 21 percent of currently married men have been married at least twice (Kreider and Ellis 2011a). These subsequent marriages face stresses beyond those of first marriages, especially when children are involved. As a result, they fail at a rate higher than first marriages. An estimated 9 percent of U.S. children who lived in a two-parent household lived in a household with a stepparent. About 70 percent of such children lived with their biological mothers and a stepfather (www.childstats.gov/americaschildren/family1.asp). **Blended families,** or families containing either a stepparent, a step-sibling, or half sibling, are even more common. Roughly 14 percent of all children younger than 18 years old were estimated to live in families with at least one step-sibling, half sibling, or adopted sibling (Kreider and Ellis 2011a). Relationships in blended families are more complicated than those in "intact" families in part because of divided loyalties to one's "real" (i.e., original) family.

- *Unmarried motherhood and single-parent families.* The increase in births outside of marriage, though not quite as dramatic as the increase in cohabitation, is nonetheless among the more striking changes in family patterns. Roughly 40 percent of births in 2014 were to unmarried women. Among some racial and ethnic groups—African Americans, American Indian and Alaska Natives, and Puerto Ricans, for example—the percent of all births that are to unmarried mothers exceeds 60 percent. Although more than a quarter of such women live with the fathers of their children, unmarried childbearing, like divorce, has helped drive the increase in single parent–headed households. In fact, the proportion of single mothers who never married (as opposed to having been married and divorced) has become increasingly large. As of 2014, more than a quarter of U.S. children under 18 lived in households headed by either an unmarried mother or an unmarried father, though the vast majority of the more than 20 million children in single-parent households lived with their mothers without their fathers (U.S. Census Bureau, Current Population Survey, 2015a). Female-headed households typically face economic hardships, and more children in single-parent versus two-parent households experience a variety of social, psychological, and educational difficulties.

- *A steady increase in the numbers of U.S. adults who are unmarried, including especially, increasing numbers who intend never to marry.* There are currently more than 100 million unmarried adults in the United States, representing more than 48 percent of the U.S. population over 18. Of that group, 62 percent had never married, while 14 percent were widowed and 24 percent were divorced. More than one of every four U.S. households consist of people who live alone (U.S. Census Bureau, "Facts for Features: Unmarried and Single Americans Week" 2014b).

## Factors Promoting Change

A number of different forces in society drive trends in marriages and families. In looking over the major changes to families in the United States, we can identify four especially important factors that initiated these changes: (1) economic changes, (2) technological innovations, (3) demographics, and (4) gender roles and opportunities for women.

### Economic Changes

As noted earlier, over time, the family has moved from being an economically productive unit to a consuming, service-oriented unit. Where families once met most needs of their members—including providing food, clothing, household goods, and occasionally surplus crops that it bartered or marketed—most of today's families must purchase what they need.

Economic factors have been responsible for major changes in the familial roles women and men play.

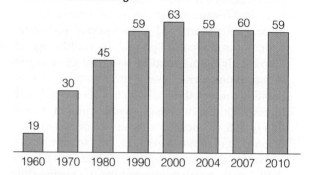

**Figure 3.1** Percentage of Married Women Employed Outside the Home Who Have Children 6 Years Old or Younger

SOURCE: U.S. Census Bureau 2006: Table 586; U.S. Department of Labor, Bureau of Labor Statistics, "Employment Status of Parents by Age of Youngest Child and Family Type, 2009–2010."

Inflation, economic hardship, and an expanding economy led to married women entering the labor force in unprecedented numbers. More than half of married women with preschool-aged children are typically employed outside the home (see Figure 3.1). As a result, the dual-earner marriage and the employed mother have become common features of contemporary families. As women have increased their participation in the paid labor force, other familial changes have occurred. For instance, women are less economically dependent on either men or marriage. This provides them greater legitimacy in attempts to exercise marital power. It has also increased the tension around the division of household chores and raised anxiety and uncertainty over who will care for the children. The recent recession affected families at all economic levels, albeit in different ways and to greater or lesser extents.

### Technological Innovations

The family has been affected by most major innovations in technology—from automobiles, telephones, cell phones, televisions, and microwaves to personal computers and the Internet. These devices were designed or invented not to transform families but to improve transportation, enhance communication, expand choices and quality of entertainment, and maximize efficiency. Nevertheless, they have had major repercussions in how family life is lived and experienced.

For example, older devices, such as automobiles and telephones, as well as more recent innovations, such as personal computers and cell phones, have aided families in maintaining contact across greater distances, thus allowing extended families to sustain closer relations and nuclear family members to stay available to one another through school- and job-related travel or relocation. The proliferation of automobiles also changed the residential and relationship experiences of many Americans. It became possible for people to live greater distances from where they work—thus contributing to the suburbanization of the United States—and to experience premarital relationships (i.e., "dating relationships") away from more watchful adult supervision.

Televisions and, more recently, the Internet have altered the recreation and socialization activities in which families engage, with both beneficial and negative consequences. As important as the entertainment function of both television and the Internet are, they also operate as additional socialization agents beyond parents and other relatives. What we watch on television or view and read on the Internet has the potential to shape our values and beliefs about the world around us. As shown in a subsequent chapter, the Internet has also greatly expanded our options for meeting friends, potential partners, and spouses. Finally, a host of innovations in communications, including cell phones, email, instant messaging, and texting, have altered the ways in which parents monitor children and how family members remain in contact with one another despite geographic distance.

The range of domestic appliances—from washing machines and dishwashers to microwaves—has altered how the tasks of housework are done. Although we might be tempted to conclude that such devices free people from some time- and labor-intensive burdens associated with maintaining homes, historical research has shown that this is not automatically so. For instance, as technology made it possible to more easily wash clothes, the standards for cleanliness increased. In the case of microwaves, the time needed for tasks associated with meal preparation has been reduced, freeing people to spend more time in other activities (not necessarily as families and often away from their families—at work, for example).

Finally, revolutions in contraception and biomedical technology have reshaped the meaning and experience of sexuality and parenthood. Much of what we call the "sexual revolution" in the 1970s and beyond was fueled partly by safer and more reliable methods of preventing pregnancy, such as the birth control pill. Regarding parenthood, people who in the past would have been unable to become parents have the opportunity to enjoy childbearing and child rearing as a result of **assisted reproductive technologies**—including medical advances such as in vitro fertilization as well as surrogate motherhood and sperm donation. Such developments have thus altered the meaning of parenthood, as multiple individuals may be involved in any single conception, pregnancy, and eventual birth. Sperm and/or egg donors, surrogate mothers, and the parent or parents who nurture and raise a child all can claim in some way to have reared the child in question. Such changes have complicated the social and legal meanings of parenthood, as they have opened the possibility of parenthood to previously infertile couples or same-sex couples.

## Demographics

The family has undergone dramatic demographic changes in areas that include family size, life expectancy, divorce, and death. Three important changes have emerged:

- *Increased longevity.* As people live longer, they are experiencing aspects of family life that few experienced before. In colonial times, because of a relatively short life expectancy, husbands and wives could anticipate a marriage lasting 25 years. Today, couples can remain married 50 or 60 years. Today's couples can anticipate living more years together after their children are grown; they can also look forward to grandparenthood or great-grandparenthood. Because men tend to marry women younger than themselves and on average die younger than women do, U.S. women can anticipate a prolonged period of widowhood.

- *Increased divorce rate.* Even with the more recent decrease, the long-term increase in the divorce rate, beginning in the late 19th century (even before 1900, the United States had the highest divorce rate in the world), led to the rise of single-parent families and stepfamilies. In this way, it has dramatically altered the experience of both childhood and parenthood and has altered our expectations of married life.

- *Decreased fertility rate.* As women bear fewer children, they have fewer years of child-rearing responsibility. With fewer children, partners are able to devote more time to each other and expend greater energy on each child. Children from smaller families benefit in a variety of ways from the greater levels of parental attention, although they may lack the advantages of having multiple siblings, such as learning to share, to care for, and to negotiate with others (Downey and Condron 2004). From the adults' perspective, smaller families afford women greater opportunity for entering the workforce and enjoying the enhanced economic status that follows.

## Changing Gender Roles and Opportunities for Women

Changes in gender roles are a fourth force contributing to alterations in U.S. marriages and families. The history summarized earlier indicated some major changes that took place in women's and men's responsibilities and opportunities. These gender shifts then directly or indirectly led to changes in both the ideology surrounding and the reality confronting families. The emphasis on child rearing and housework as women's proper duties lasted until World War II, when, as we saw, there was a massive influx of women into factories and stores to replace the men fighting overseas. This initiated a trend in which women increasingly entered the labor force, became less economically dependent on men, and gained greater power in marriage.

The feminist movement of the 1960s and 1970s led many women to reexamine their assumptions about women's roles. Betty Friedan's *The Feminine Mystique* challenged head-on the traditional assumption that women found their greatest fulfillment in being mothers and housewives. The women's movement emerged to challenge the female roles of housewife, helpmate, and mother, appealing to some women as it alienated others.

More recently, the dual-earner marriage made the traditional division of roles an important and open question for women. Today, contemporary women have dramatically different expectations of male–female roles in marriage, child rearing, housework, and the workplace than did their mothers and grandmothers. Changes in marriage, birth, and divorce rates and in the ages at which people enter marriage have all been affected by women's enlarged economic roles.

We also have witnessed changes in what men expect and are expected to do in marriage and parenthood. Although it may still be assumed that men will be "good providers," that is no longer enough. Married men face greater pressure to share housework and participate in child care. Although they have been slow to increase the amount of housework they do, there has been more acceptance of the idea that greater father involvement benefits both children and fathers. New standards and expectations of paternal behavior and more participation by fathers in raising children help explain the ongoing changes—from how dual-earner households function to why we are more accepting of fathers staying home to care for their young children.

The U.S. Census Bureau estimates that, in 2013, 214,000 married stay-at-home dads were home caring for their children while their wives were employed. These at-home dads cared for more than 430,000 children under 15 years of age. In comparison, there were an estimated 5.2 million at-home mothers in 2014. In 2011, fathers provided child care for 17.8 percent of preschool-age children and 29.3 percent of preschool-age children with employed mothers. Finally, there were approximately 2 million single fathers, representing 17 percent of all single parents (U.S. Census Bureau 2011b; U.S. Census Bureau 2011c). Gender issues are so central to family life that they are the subjects of the entire next chapter. It is not an exaggeration to say that we cannot truly understand the family without recognizing the gender roles and differences on which the family rests.

## Cultural Changes

We can, in conclusion, point to a shift in American values from an emphasis on obligation and self-sacrifice to individualism and self-gratification (Amato et al. 2007; Bellah et al. 1985; Coontz 1997; Mintz and Kellogg 1998). The once strong sense of **familism**, in which individual self-interest was expected to be subordinated to family well-being, has given way to more open and widespread individualism in which even marriages and/or families can be sacrificed for individual happiness and personal fulfillment.

This shift in values has had consequences on how people choose between alternative lifestyle paths. For example, complex decisions—about whether and how much to work, whether to stay married or to divorce, and how much time and attention to devote to children or to spouses—are increasingly weighed and made against a backdrop of pursuing self-gratification and individual happiness. We have embraced the idea that what makes individuals happy, in the long run, is what is best for them and their families. Such shifts in values, alone, have not changed families, but they have contributed to the choices people make, out of which new family forms predominate (Coontz 1997).

# How Contemporary Families Differ from One Another

The preceding discussion traced some ways families have changed throughout history and why. In that sense, it has led us to family life of today. But today's families differ from one another, a topic we now explore. We look first at economic factors that differentiate families and then at cultural characteristics, and race and ethnicity.

## Economic Variations in Family Life

A **social class** is a category of people who share a common economic position in the stratified (i.e., unequal) society in which they live. We typically identify classes using economic indicators. One such indicator is wealth, which is the sum value of one's assets such as bank accounts, houses, stock, retirement accounts, and so on, minus the sum of one's debts, such as credit card debt, mortgages, other outstanding loans, and so on. The other major economic indicator of class is the amount of income earned from one's occupation or any other sources of earnings. Social class has both a structural and a cultural dimension. Structurally, social class reflects the occupations we hold (or depend on), the income and power they give us, and the opportunities they present or deny us. The cultural dimension of social class refers to any class-specific values, attitudes, beliefs, and motivations that distinguish classes from one another. Cultural aspects of social class are somewhat controversial, especially when applied to supposed "cultures of poverty"—an argument holding that poor people become trapped in poverty because of the values they hold and the behaviors in which they engage (Harrington 1962; Lewis 1966). What is unclear regarding "cultures of class" is how much difference there is in the values and beliefs of different classes and whether such differences cause or follow the more structural dimensions that separate one class from another.

To an extent, there is also a psychological aspect to social class, an internalization of economic status in the self-images we form and the self-esteem we possess. These may also be seen as consequences of other aspects of class position, such as the self-identity that results from the prestige accorded to work or the respect paid to accomplishments. Like the structural and cultural components of social class, these are brought home and affect our experiences in our families.

The effect of social class is far-reaching and deep. In an article about how social class affects marriage, *New York Times* reporter Tamar Lewin quotes one of her sources, Della Mae Justice, of Pikeville, Kentucky. Justice grew up in the coal-mining world of Appalachia, in a house without indoor plumbing. Having put herself through college and later law school, she is now solidly and unambiguously middle-class. Justice says, "I think class is everything, I really do. When you're poor and from a low socioeconomic group, you don't have a lot of choices in life. To me, being from an upper class is all about confidence. It's knowing you have choices, knowing you have connections" (Lewin 2005).

Clearly, many facets of our lives (often referred to by sociologists as **life chances**) are affected by our **socioeconomic status**, including our health and well-being, safety, longevity, religiosity, and politics. A host of family experiences also vary up and down the socioeconomic ladder. For instance, class variations can be found in such family characteristics as age at marriage, age at parenthood, timing of marriage and parenthood, division of household labor, ideologies of gender, socialization of children, meanings attached to sexuality and intimacy, and likelihood of violence or divorce.

Conceptualizations of social class vary in both how class is defined and how many classes are identified and counted in U.S. society. For example, in Marxian formulations of social class, it is a person's relationship to the means of production—as owner or worker—that defines class position. The capitalist class owns the factories and machinery and employs workers in the processes of production. The workers own nothing but their own ability to work, for which they are paid a wage. In other models, people are grouped into classes because of similar incomes, amounts of wealth, degrees of occupational status, and years of education. Whether we claim that the United States has two (owners and workers), three (upper, middle, and lower), four (upper, middle, working, and lower), six (upper-upper, lower-upper, upper-middle, lower-middle, upper-lower, and lower-lower), or more classes, the important point about the concept of social class is that life is differently experienced by individuals across the range of identified classes and similarly experienced by people within any one of the class categories. As sociologist Dennis Gilbert notes, "defining classes and specifying the dividing lines between them is as much art as science" (Gilbert 2011, 242).

**Marrying across Class**

In examining patterns of marriage in the United States, much scholarly and popular attention has been paid to intermarriage, or marrying outside of one's race or faith. A later section of this chapter looks at data on racial intermarriage and multiracial families. As to religion, although nearly 70 percent of married people responding to a Pew Research Center survey said that their spouse was of the same religious faith as themselves, it appears as though such shared religious background is less important in recent marriages than it was in the past. Nearly 40 percent of those who married since 2010 had a spouse of a different religious faith. Considering those who married before 1960, just under a fifth were in religious intermarriages.

In some interesting recent research, sociologist Jessi Streib looked at another kind of "marrying out," in a study of 32 couples whose marriages crossed social class. Although all 64 respondents were "white, college educated, adults in heterosexual marriages," half had grown up in the working class and were married to spouses who had grown up in the middle class. Streib notes that although respondents minimized the role and relevance of class differences in their marriages, in fact such differences "left a deep imprint" on their lives together.

Streib shows how those who grew up in working-class families develop different outlooks and approaches compared to those who grew up in middle-class homes. She found that her respondents who'd grown up working-class had what she called a *laissez-faire* approach to daily life, a "go with the flow, enjoy the moment," approach in which they "assumed things would work out without their intervention" (2015, 42). Those men and women born and raised in middle-class families took more of

what Streib called a *managerial approach* to daily life, seeing the need to plan, organize, and oversee things in order to have them work out. Streib shows how these different approaches surfaced in multiple areas of couple's lives together—including managing their money, housework, paid work, parenting, leisure, time, and even emotions. Respondents from working-class backgrounds preferred to "let things take their own course," without the need to intervene and oversee. Spouses from middle-class backgrounds wanted instead "to plan, organize, and oversee" (2015, 42).

In an article for *Contexts: Understanding People in Their Social Worlds*, Streib illustrates these concepts by describing differences that surfaced in the marriage of Christie and Mike. Christie, from a working-class family, had a looser attitude toward money, and more relaxed attitude toward paid work and household tasks. Mike from a middle-class background, wanted to take a more hands-on approach to saving money, advancing in their careers, and structuring their division of household tasks. Even in their emotional expression, these differences surfaced. Mike wanted to manage, slowly process, and weigh how to express his feelings; Christie preferred to express emotions "as they were felt and in the way they were felt" (2015, 44).

Streib notes that although the different approaches were common, they were not gendered in the same way for all couples. Instead, it was one's class origin that tended to determine whether one took a laissez faire or managerial approach. Furthermore, most couples she interviewed recognized the differences as valid and understandable, and dealt with them with admiration and respect. However, for a minority of sample couples, such differences were more problematic and divisive. ●

Based on some common sociological models, we can describe these social classes in the United States: the upper-upper class, or "elites"; the lower-upper class, or "everyday rich"; the upper-middle class; the "middle-middle class"; the working class; and the lower class, which can be further subdivided into the "working poor" and the underclass.

## Upper Class(es)

Roughly 1 percent of the U.S. population occupies an "upper-class" position. At the very top of this group are those referred to as "upper-upper class" or sometimes

called the "ruling class" or the "elite." They own a third of all assets, and their wealth is worth more than the "bottom" 90 percent of the population (Henslin 2006). Their "extraordinary wealth" often takes them into the hundreds of millions if not billions of dollars (Curry, Jiobu, and Schwirian 2002). They own stocks and bonds, prosperous businesses, and commercial real estate, allowing them to earn great amounts of income whether or not they are employed (Gilbert 2011). In fact, this class is defined by their dependence on income generated from what they own and/or inherited. Many among the upper-upper class were born into their fortunes, and others joined the "superrich" through the information technology boom or in entertainment or sports (Kimmel and Aronson 2011). Sociologist Dennis Gilbert suggests that an income of $2 million a year is "typical" for this upper 1 percent.

The rest of the upper class live on yearly incomes ranging from hundreds of thousands to a million dollars, own substantial amounts of wealth, and enjoy much prestige. The members of this "lower-upper class" might be considered the "everyday rich" (Kimmel and Aronson 2011). They are professionals and corporate executives and managers who tend to have degrees from prestigious universities (Kimmel and Aronson 2011). They live well in private homes in exclusive communities and enjoy considerable privilege. In some formulations of class, a distinction is drawn between the elite and the lower-upper class in terms of "old" versus "new" money (Langman 1987; Steinmetz, Clavan, and Stein 1990). In other words, some separate the upper class based on how they achieved and how long they have enjoyed their affluence.

## Middle Classes

In some analyses, the middle class is considered the largest class, representing between 45 percent and 50 percent of the population (Gilbert 2011; Henslin 2006). Often, the middle class is subdivided into two groupings: the upper-middle class and the middle-middle class.

The **upper-middle class** consists of highly paid professionals (e.g., lawyers, doctors, accountants, and engineers), upper managers, and small business owners, who have annual incomes that may reach into the hundreds of thousands of dollars. They are typically college educated, although they may not have attended the same elite colleges as the upper-upper class, and often possess advanced degrees (Henslin 2006). Women and men of the upper-middle class have incomes that allow them luxuries such as home ownership, vacations, and college educations for their children. Roughly 15 percent of the population is upper-middle class (Gilbert 2011; Henslin 2006).

The **middle class** comprises a larger portion of the population, typically estimated at around 30 percent (Gilbert 2011) to 35 percent (Henslin 2015). Although it is impossible to specify an exact income threshold that separates the middle (or middle-middle class) from the upper-middle class, sociologist Michael Kimmel and co-author Amy Aronson (2011) suggest that the upper-middle class has household incomes above $80,000, while the middle-middle class is

between $40,000 and $80,000. Dennis Gilbert (2014) suggests that the "typical" upper-middle class household income may be $150,000, as compared to what he calls the "middle class" at around $70,000. Again, as Gilbert notes, there is more art than science to these depictions.

Of course, in areas of the country with higher costs of living, such incomes will not bring you the same levels of comforts associated with being middle class. The middle class is comprised of white-collar service workers, technicians, educators, salespeople, and nurses. Additionally, many "high-demand" service personnel (e.g., police or firefighters) and some blue-collar workers do reach the income threshold as indicated. Generally, those in the middle class possess less wealth, live on less income, and have less education (or less prestigious degrees) and social standing than their professional and managerial counterparts (e.g., physicians, attorneys, and managers) in the upper-middle class. They own or rent more modest homes and purchase more affordable automobiles than their upper-middle-class counterparts, and they hope, but with less certainty, to send their children to college. Importantly, they do not have the same sorts of safety nets protecting them from sliding "down the ladder" with periods of increased and prolonged unemployment as occurred in the United States during and following the recession of 2007–2009.

## Working Class

About 30 percent of the U.S. population can be considered **working class,** though the line separating them from the middle class is not always clear (Gilbert 2014). Members of this class tend to work in low-level white-collar or blue-collar occupations (as more likely unskilled or semiskilled laborers) and typically have high school or vocational educations, though some have attended college. Factory workers, clerical workers, retail salespeople, and custodians are all examples of the working class. Typically, they have jobs that offer little opportunity to move up and generate low levels of job satisfaction. Household incomes in the working class range, though $40,000 is seen as a typical working-class household income, depending on whether they are two-earner or one-earner households (Kimmel and Aronson 2011). Members of the working class live somewhat precariously, with little savings and few liquid assets should illness or job loss occur. They also have difficulty buying their own homes or sending their children to college (Curry, Jiobu, and Schwirian 2002). Significantly, as we shall see, the extended family is of great importance to many in the working class.

## Lower Classes: The Working Poor and the Underclass

If by lower class one means those who fall just above or beneath the poverty threshold, the lower classes can be divided into two—the working poor and the underclass. Poverty is consistently associated with marital and family stress, increased divorce rates, low birth weight and infant deaths, poor health, depression, lowered life expectancy, and feelings of hopelessness and despair. Poverty is a major contributing factor to family dissolution. Poor families are characterized by irregular employment or chronic underemployment. Individuals work at unskilled jobs that pay minimum wage and offer little security or opportunity for advancement. Although many lower-class individuals rent substandard housing, homelessness is a problem among poor families. Karen Seccombe (2000) effectively describes the problems: "Poverty affects one's total existence. It can impede adults' and children's social, emotional, biological, and intellectual growth and development" (p. 1,096). She further notes that over a year, most poor families experience one or more of the following: "eviction, utilities disconnected, telephone disconnected, housing with upkeep problems, crowded housing, no refrigerator, no stove, or no telephone" (Seccombe 2000, 1,096).

According to the U.S. Census Bureau, the official poverty rate in the United States in 2013 was 14.5 percent, down slightly and for the first time since 2006 from the previous year's rate (15 percent). This translates to over 45 million Americans (DeNavas-Walt and Proctor 2014).

Age and sex make a difference in the amounts of individuals in poverty. Children (those under 18) have the highest poverty rate of any age group. In 2013, just under 20 percent of children were in poverty. Among 18- to 64-year-olds, the rate dropped to 13.6 percent, and among those over 65, the rate was 9.5 percent.

Women are more likely than men to be poor; in 2013, 15.8 percent of females were poor compared to 13.1 percent of males. Among those over 65, the size of the gender gap widens, as 11.6 percent of women and 6.8 percent of men are below the poverty threshold. Large gender differences exist among women and men as heads of households. Female-headed families are almost twice as likely as families headed by single males to be poor; 30.6 percent of families with female heads and 15.9 percent of families with male heads

### TABLE 3.4 Poverty Guidelines, 2015

The 2015 Poverty Guidelines for the 48 Contiguous States and the District of Columbia

| People in Family | Poverty Guideline |
|---|---|
| 1 | $11,770 |
| 2 | 15,930 |
| 3 | 20,090 |
| 4 | 24,250 |
| 5 | 28,410 |
| 6 | 32,570 |
| 7 | 36,730 |
| 8 | 40,890 |

For families with more than eight people, add $4,160 for each additional person.
SOURCE: U.S. Department of Health and Human Services, "2015 Poverty Guidelines," aspe.hhs.gov/2015-poverty-guidelines.

were below poverty. Among married couple–headed families, the poverty rate dropped considerably, to 5.8 percent, representing just under 3.5 million families (DeNavas-Walt and Proctor 2014).

The Department of Health and Human Services provides poverty guidelines to reflect who is eligible for governmental assistance. The 2014 poverty guidelines are illustrated in Table 3.4.

African Americans and Hispanics, or Latinos, along with American Indians and Alaskan Natives, are more likely to experience poverty than are Caucasians or Asian Americans (MacArtney, Bishaw, and Fontenot 2013). As reflected previously, those living in poverty, like their upper- and middle-class counterparts, can be subdivided.

**Working Poor.** Since 1979, there have been large increases in the proportion of the population who, despite paid employment, live in poverty. The label *working poor* refers to people who spent at least 27 weeks in the labor force, working or looking for a job, but whose incomes fell below the poverty threshold (U.S. Bureau of Labor Statistics 2015i). They work in low-skilled and low-paid occupations that require less than college educations, putting them very close to the "poverty line" established by the U.S. government. They may rely on food stamps, food pantries, and welfare even when employed and on unemployment assistance should they lose their jobs. The vulnerability of the working poor cannot be overstated. They live from paycheck to paycheck, week to week, with no opportunity to save money. Their jobs (e.g., delivering pizza, cleaning homes, migrant farmwork, day labor, low-pay factory work) don't typically offer benefits such as health insurance or sick pay, making them very vulnerable to illness or accidents (Henslin 2006; Kimmel and Aronson 2011). Many single mothers and their children fall into this class, but even a family headed by two minimum wage–earning spouses can find themselves among the working poor. Factors such as low wages, occupational segregation, and the dramatic rise in single-parent families account for why having a job and an income may not be enough to keep people out of poverty.

In 2013, more than 10.5 million people were classified as working poor, a rate of 7 percent. This includes over 5 million women (5.4 million) and men (5 million), with women's rate (7.8 percent) higher than men's rate (6.3 percent). Blacks and Hispanics (at 13.3 percent and 12.8 percent, respectively) were more likely than whites (6.1 percent) and Asians (4.5 percent) to be among the working poor. As to families, in 2013, 5.1 million families, down from 5.5 million a year earlier, lived below the poverty line even with at least one member in the labor force for at least half the year. As with the earlier poverty profile, families maintained by women were more likely (at 26.7 percent) than either male-headed families (15.1 percent) or families headed by married couples (8.8 percent) to be among the working poor. Families with children were much more likely to be working poor than were families without children. (U.S. Bureau of Labor Statistics, "A Profile of the Working Poor," July 2015i). Although their family members may be working or looking for work, these families cannot earn enough to raise themselves out of poverty. An individual working full time at minimum wage simply does not earn enough to support a family of three. Thus, this kind of poverty results from problems in the economic structure—low wages, job insecurity or instability, or lack of available jobs.

### The Underclass or Ghetto Poor.

Finally, at the very bottom, we find what some researchers call the "underclass" (Gilbert 2011). The underclass tends to be concentrated in the inner cities, in substandard housing, if not homeless. They lack education and job skills and have inadequate nutrition and no health care (Kimmel and Aronson 2011). When employed, they typically do low-paying, menial work but are often dependent on government assistance to survive. Some engage in criminal activities. Their lives are bleak at best.

Disproportionately African Americans and Latinos, the underclass represents a deeply disturbing counterpoint to wider cultural values and beliefs that are definitive features of American life. Their lifestyles and circumstances challenge cherished images of wealth, opportunity, and economic mobility. Their behaviors, actions, and problems are often responses to lack of opportunity, urban neglect, and inadequate housing and schooling.

Because the label "underclass" has connotations that many consider to be derogatory and/or racially specific, other researchers suggested alternative terms to refer to this segment of the U.S. class structure. For example, notable sociologists such as William Julius Wilson and Herbert Gans rejected the "underclass" label in favor of such alternatives as "ghetto poor" (Wilson 1991) or "undercaste" (Gans 1993). In both cases, their objective was to move away from a label whose popular and scholarly use was directed mostly at poor Latinos and African Americans or had come to emphasize lifestyle choices made by those surrounded by persistent and profound poverty, and to instead move toward a term with a more structural and race-neutral emphasis. With the export of manufacturing, few job opportunities exist in the inner cities; the jobs that do exist are usually service jobs that fail to pay their workers sufficient wages to allow them to rise above poverty. Schools are substandard. The infant death rate approaches that of third world countries. The housing projects are infested with crime and drug abuse, turning them into kingdoms of despair. Estimating the size of this portion of the lower class is difficult; however, estimates range from around 5 percent of the population (Henslin 2006) to 12 percent (Gilbert 2011).

### Spells of Poverty.

Most of those who fall below the poverty threshold tend to be there for spells of time rather than permanently. In typical times, poverty lasts less than a year for most poor people. About a quarter of the U.S. population may require some form of assistance at one time during their lives because of changes in families caused by divorce, unemployment, illness, disability, or death. About half of our children are vulnerable to poverty spells at least once during their childhood. Many families who receive assistance are in the early stages of recovery from an economic crisis caused by the death, separation, divorce, or disability of the family's major wage earner. Many who accept government assistance return to self-sufficiency within a year or two. Most children in these families do not experience poverty after they leave home.

Two major factors are related to the beginning and ending of spells of poverty: changes in income and changes in family composition. Many poverty spells begin with a decline in earnings of the head of the household, such as a job loss or a cut in work hours. Other causes include a decline in earnings of other family members, the transition to single parenting, the birth of a child to a single mother, and the move of a youth to his or her own household.

**Poor Women and Children.** The **feminization of poverty** is a painful fact that has resulted primarily from fairly high rates of divorce, increasing numbers of unmarried women with children, and women's lack of economic resources. When women with children divorce, their income and standard of living fall, occasionally dramatically. By family type, 41 percent of single-mother families with children under 18 were below the poverty line in 2013. In families with younger children, the rate was even higher—55 percent of single mother–headed families with children under age 6 were living in poverty (DeNavas-Walt and Proctor 2014).

Looking at poor children in 2013, 14.7 million children, down from 16.1 million the year before, were poor. The child poverty rate was 19.9 percent, meaning one out of five children under 18 in the United States was living below the poverty threshold in 2013. Child poverty rates decreased from 2012 to 2013 for non-Hispanic whites, Hispanics, and Asian American children, but did not change for black children (Renwick 2014). The rate is higher among younger children—25 percent of children under age 6, living in families, were poor. Among children 6 to 17, 21 percent were poor (children in poverty by age group, kidscount.org).

Like their parents, poor children often move in and out of spells of poverty, depending on major changes in family structure, employment status of family members, or the disability status of the family head. These variables affect ethnic groups differently and help account for differences in child poverty rates. African Americans, for example, have significantly higher unemployment rates and greater numbers of never-married single mothers than do other groups. As a result, their child poverty rates are higher.

Close to 40 percent (38.3 percent) of black children lived in poverty in 2013, more than three times the rate for white children, at 10.7 percent. Among Hispanics, 30.4 percent of children lived in poverty, and among Asian children the 2013 rate was 10.1 percent (Tavernise 2015). Being poor puts the most ordinary needs—from health care to housing—out of reach.

These statistics may understate the extent of economic hardship faced by children in the United States. Many researchers recommend using a standard other than the official poverty threshold to gauge the breadth of childhood economic vulnerability. With evidence that families typically need incomes that amount to twice the official poverty threshold to make ends meet, if we use 2015 levels, a family of four would need $48,500 to avoid being characterized as low income. In 2013, nearly 32 million children, the equivalent of 44 percent of U.S. children, could have been classified as low-income.

On the other hand, to get a sense of those who are most vulnerable, 6 percent of the population, over 20 million people, had incomes less than half of the poverty threshold. They lived in extreme (or "deep") poverty. Nine percent of children lived in families with incomes below 50 percent of the poverty threshold in 2013. Racially, 19 percent of black, non-Hispanic children, 13 percent of Hispanic children, and 5 percent of white, non-Hispanic children lived in families with incomes below one-half of the poverty threshold in 2013 (www.childstats.gov/americaschildren/eco1.asp).

## Class and Family Life

Working within this framework, we can note some ways in which individuals in different classes experience family life. Although we could consider a number of family characteristics (including divorce, domestic violence, and the division of labor), we look briefly at class-based differences in marriage relationships, parent–child relationships, and ties between nuclear and extended families.

### Marriage Relationships

Within upper-class families, we tend to find sharply sex-segregated marriages in which women are subordinated to their husbands. Upper-class women often function as supports for their husbands' successful economic and political activities, thus illustrating the **two-person career** (Papanek 1973).

> **False**
> **5.** In upper-class families, marriages are sharply sex-segregated.

Family experiences are affected by such variables as social class and ethnicity.

Although a wife's supportive activities may be essential to her husband's success, such wives are neither paid nor widely recognized for their efforts. Rather than having their own careers, they often volunteer within charitable organizations or their communities. They are free to pursue such activities because they have many servants—from cooks to chauffeurs to nannies—who do the domestic work and some child care or supervision.

Middle-class marriages tend to be *ideologically* more egalitarian and are often two-career marriages. In fact, middle-class lifestyles increasingly *require* two incomes. This creates both benefits and costs for middle-class women. The benefits include having more say in family decision making and greater legitimacy in asking for help with domestic and child-rearing tasks. The costs may include the failure to receive the help they request. Because most employed wives earn less than their husbands, the strength of their role in family decision making may still be less than that of their husbands. Middle-class marriages are "ideologically" more egalitarian because middle-class couples more highly value and more readily accept the ideal of marriage as a sharing, communicating relationship in which spouses function as "best friends."

Once more explicitly traditional, working-class marriages are becoming more like their middle-class counterparts. Whereas such marriages in the past were clearly more traditional in both rhetoric and division of responsibilities, they have moved toward a model of sharing both roles and responsibilities in recent years. The sharply segregated, traditional marriage roles evident even just two decades ago have given way to two-earner households, increasingly driven by the need for two incomes.

Especially among those working-class couples who work "opposite" shifts, higher levels of sharing domestic and child care responsibilities are often found, as is greater male involvement in home life (Rubin 1994). The reality of being the only parent home may force men to take on tasks that women might otherwise have. In such instances, it is necessity more than ideology that influences men's participation. Such involvement may be more valued in the circles in which middle-class men live and work but be more of a practicality or necessity for working-class men. Thus, working-class men may understate and middle-class men may exaggerate what and how much they do, making class comparisons that much more difficult to draw.

Marriage is least common and least stable among the lower classes. Men are often absent from day-to-day family life. Resulting from the combination of high divorce rates and widespread nonmarital childbearing, almost a third of female-headed households are poor. The poverty rate for families with a female householder (30.6 percent) was nearly twice the poverty rate for single male-headed households (15.9 percent). There were 4.6 million female-householder families in poverty and 1 million male-householder families in poverty in 2013. In comparison, 3.5 million, or 5.8 percent of married-couple families were poor in 2013 (DeNavas-Walt and Proctor 2014).

The cultural association of men's wage earning with fulfillment of their family responsibilities subjects lower-class men to harsher experiences within families. They are less likely to marry. If married, they are less likely to remain married, and when married, they derive fewer of the benefits that supposedly accrue in marriage.

When marriages cross class lines, other problems can arise. People may find themselves feeling out of step, as if they are in a world where there are different, perhaps dramatically different, assumptions about how to discipline and raise children, where to go and what to do on vacation, and how to save or spend money (Lewin 2005). Marriage across class lines is more difficult to measure than interracial marriage or religious intermarriage, but using education as an indicator of class, there appear to be fewer cross-class marriages than in the past. Most of those marriages that do cross class lines are now between women with more education marrying men with less. This combination does not bode especially well for the future stability of the marriages (Lewin 2005).

## Parents and Children

The relationships between parents and children vary across social lines, but most research has focused on the middle and working classes (Kohn 1989; Lareau 2003). Among some upper-class families, nannies or au pairs may do some hands-on child rearing. Certainly, mothers are involved, and relationships between parents and children are loving, but parental involvement in economic and civic activities may sharply curtail time with children. For upper-class parents, an important objective is to see that children acquire the appropriate understanding of their social standing and that they cultivate the right connections with others like themselves. They may attend private and exclusive boarding schools and later join appropriate clubs and organizations. Their eventual choice of a spouse may receive especially close parental scrutiny.

Addressing the experiences of *affluent* families (upper-middle- to upper-class), Megan Haselschwerdt describes some of the unique pressures affluent parents and children face. Extreme pressure on children to be the best at something (or on parents to have their children be the best at something), academic pressure, extracurricular pressure, and pressure to get into the right colleges are all said to have the potential to take a toll on young women and men of the upper-middle and upper classes. Citing research by Suniya Luthar and colleagues, Haselschwerdt notes that affluent youth have reported at least as much "disturbance across a variety of domains," as do extremely poor urban teens (Haselschwerdt 2012, F15). High levels of female anxiety and depression and high levels of using cigarettes, alcohol, marijuana, and hard drugs are among the potential negative consequences

of "extremely heightened achievement pressure" and low levels of parental availability. Of course, young people reap benefits from their family's affluence but also more negatives than we might typically think.

A considerable amount of research indicates that working- and middle-class parents socialize their children differently and have different objectives for child rearing (Hays 1996; Kohn 1989; Lareau 2003; Rubin 1994). Although all parents want to raise happy and caring children, middle-class parents tend to emphasize autonomy and self-discipline, and working-class parents tend to stress compliance (Hays 1996; Kohn 1989). Whereas both middle- and working-class mothers value education, working-class mothers in these studies saw and therefore stressed education as essential for their children's later life chances. Middle-class mothers took for granted that their children would receive good-quality educations, and emphasized instead the importance of building children's self-esteem. And although both classes of mothers acknowledged using spanking to discipline their children, middle-class mothers spanked more selectively and favored other methods of discipline (e.g., "time-out") (Hays 1996).

One of the more fascinating and influential class comparisons is Annette Lareau's *Unequal Childhoods* (2003). Lareau contends that "social class does indeed create distinctive parenting styles . . . that parents differ by class in the ways they define their own roles in their children's lives as well as in how they perceive the nature of childhood" (p. 748). Lareau introduces the concepts of *concerted cultivation* and *accomplishment of natural growth* to represent class-based differences in philosophy of child rearing.

Middle-class families engage in concerted cultivation. Parents enroll their children in numerous extracurricular activities—from athletics to art and music—that come to dominate their children's lives as well as the life of the whole family. Through these activities, however, children partake in and enjoy a wider range of outside activities and interact with a range of adults in authoritative positions, giving them experiences and expertise that can serve them well later. Because of the way household life tends to center around children's schedules and activities, the other members of these middle-class families (parents and siblings) are forced to endure a frenzied pace and a shortage of family time.

Working-class parents, lacking the material resources to enroll their children in such activities, tend to focus less on developing their children and more on letting them grow and develop naturally, play freely in unsupervised settings, and spend time with relatives and in the neighborhood. For a sense of how children may have experienced these two approaches, see the Real Families feature in this chapter.

Lower-class families are the most likely of all families to be single-parent families. Single parents, in general, may suffer stresses and experience difficulties that parents in two-parent households do not, but this situation is exacerbated for low-income single parents (McLanahan and Booth 1989). Parent–child relationships suffer from a variety of characteristics of lower-class life: unsteady, low-pay employment; substandard housing; and uncertainty about obtaining even the most basic necessities (food, clothing, and so forth). All of these can affect the quality of parent–child relationships and the ability of parents to supervise and control what happens to and with their children.

**True** **6.** Lower-class families are the most likely to be single-parent families.

### Extended Family Ties

Links between nuclear family households and extended kin vary in kind and meaning across social class. By some measures, the least closely connected group may be the middle class, as families may find themselves the most physically removed from their kin because of the geographic mobility that accompanies their economic status. As Matthijs Kalmijn (2004) observed among the upper-middle-class families he studied in the Netherlands, they live almost three times as far from their siblings and more than three times as far from their parents and their (adult) children as do families of the lowest-educated class. Similar class differences can be observed in the United States.

Middle-class families do visit kin or phone regularly and are available to exchange aid when needed. Still, the emphasis is on the conjugal family of spouses and children. Closer connections may be found among both the working and the upper classes, although the reasons differ. In the case of working-class families, there are often both the opportunity and the need for extensive familial involvement. Opportunity results from lesser levels of geographic mobility, which results in closer proximity and allows more continuous contact to result. The need for involvement is created by pooling resources and exchanging services (e.g., child care) that often result between adults and their parents or among adult siblings. Intergenerational upward mobility may lessen the reliance on extended families (see discussion later in this chapter).

Upper-class families, especially among the "old-money" upper class, highly value the importance of family name and ancestry. They tend to maintain strong and active kinship groups that exert influence in the mate selection processes of members and monitor the behavior of members. Inheritance of wealth gives the kin group more than symbolic importance in their ability to influence behavior of individual members.

Among the lower class, kin ties—both real and fictive—may be essential resources in determining economic and social survival. Grandparents, aunts, and uncles may fill in for or replace absent parents, and multigenerational households (for example, children living with their mothers and grandmothers) are fairly common. **Fictive kin ties** refer to the extension of kinship-like status to neighbors and friends, thus symbolizing both an intensity of commitment and a willingness to help one another meet needs of daily life (Liebow 1967; Stack 1974).

### The Dynamic Nature of Social Class

Like other aspects of family life, social class position is not set in stone. Individuals may experience **social mobility**, movement up *or* down the social class ladder. Either kind of social mobility can affect family relationships, especially, although not exclusively, intergenerational relationships (Kalmijn 2004; Newman 1988; Sennett and Cobb 1972). For example, children who see their parents "fall from grace" through job loss and dwindling assets may start to look differently at those parents. Fathers who once seemed heroic may become the source of concern and even resentment as their job loss threatens the lifestyle of the family on which children depend (Newman 1988). Children who in adulthood climb *upward* occasionally find their relationships with their parents suffering as a result. As they are exposed to new values and ideas that differ from those their parents hold, generational tension and social distance may follow. Furthermore, as they move into a new social circle, parents (as well as less mobile siblings) may appear to fit less well with their new life circumstances. The more they strive to fit into new circles and circumstances accompanying their increased social standing, the less well they may fit comfortably within their ongoing family relationships.

In a tightly constructed and concise presentation, Natalia Sarkisian and Naomi Gerstel (2012) explore the role of race and class (as well as gender) in maintaining active ties with extended family. Based on the data available to them as they wrote *Nuclear Family Values, Extended Family Lives: The Power of Race, Class, and Gender*, they examine racial/ethnic patterns of marriage, single motherhood, and extended family relationships as they differ between whites, blacks, and Latinos/as. They note, for example, that although among all three groups there has been a decline in the percent who are married, compared to whites, both blacks and Latinos/as have considerably lower rates of marriage and much higher proportions of childbirths outside of marriage. Using national survey data, they then address striking differences in wider kin relationships, noting that blacks and Latinos/as are more likely to live in family households with extended kin and are more likely to help and rely on help from their wider families. Smaller proportions of whites (less than a fifth) than Latinos/as (approximately a third) or blacks (roughly 40 percent) shared households with relatives. Extending the analysis beyond shared households, they note that where only a third of whites live within two miles of kin, over half of blacks and Latinos/as live within such proximity.

Their data on support are especially striking. Whites are more likely than blacks or Latinos/as to exchange (give or receive) large amounts of money, and white women are more likely than their minority counterparts to provide or receive emotional support (share problems, offer or receive advice). Black and Latino/a kin are more likely to provide practical assistance, providing assistance with child care and housework, providing rides or running errands when needed.

Significantly, in accounting for differences across race and ethnic lines, they turn to social class as the major causal factor. Whites are more likely to give financial assistance to their kin because they are more able financially to afford to do so. Blacks and Latinos/as come to need and receive more practical help (child care, rides, household tasks) because they tend to have lower incomes. In fact, as they report, "Simply put, Whites, Blacks, and Latinos/as with the same amount of income and education have similar patterns of involvement with their extended families" (2012, 31). They conclude that "routine engagement with extended family" is a strategic response to difficult economic realities, a "class-based strategy" (p. 31). Here, then, as with the earlier feature on Annette Lareau's research on class differences in parenting and childhood, in accounting for differences "class trumps race" (Sarkisian and Gerstel 2012). ●

Aside from the difficulty of fitting parents and siblings into a new social standing, practical considerations imposed by a job may create obstacles that prevent individuals from maintaining closer relationships. As is true elsewhere, ascending the ladder to a higher rung may require geographic relocation. Such jobs may also impose greater demands on the individual's time. To these constraints of time and distance, one can add that as someone establishes new friendships and participates in leisure activities, further reductions in opportunity and availability to family may result (Kalmijn 2004).

Marital relationships, too, may be altered by either downward or upward mobility. Research indicates that some men who lose their jobs and "slide downward" react to their economic misfortune by abusing their spouses, turning to alcohol or other substances, withdrawing emotionally, or leaving the home (Newman 1988; Rubin 1994). Changes in the marriage are not entirely of men's doing; after an initial period of sympathy and support, wives may grow impatient with their husbands' unemployment or alter their positive views of the husbands' dedication as a worker or job seeker. In addition, as couples are forced to scale back their accustomed lifestyle, tensions may rise, and resentment and distance may grow.

Upward mobility may transform marriage relationships as well. Some women face situations where, after sacrificing to help launch their husbands' careers by supporting them through school, they are left by those same husbands once they have achieved their career goals. With their own increasing economic opportunity, some women find that marriage becomes less desirable because of the constraints it continues to impose on their career development.

**True**   **7.** Marital relationships can suffer as a result of either downward or upward mobility.

Louise and Don Tallinger are proud parents of three boys: 10-year-old Garrett, 7-year-old Spencer, and 4-year-old Sam. They are also busy professionals: Louise is a personnel consultant, and Don is a fundraising consultant. Between them, they earn $175,000, making them comfortably upper-middle class. They travel a lot for work; Don is out of town an average of three days a week, and Louise flies out of state early in the morning four or five times a month, returning sometime after dinner. Don often doesn't return home from work until 9:30 p.m. This middle-class family of five is one of the families Annette Lareau studied in her fascinating class comparison, *Unequal Childhoods* (2003).

With 10-year-old Garrett's involvement in baseball, soccer, swim team, piano, and saxophone lessons, his schedule dictates the pace and routines of the household. Lareau (2003) describes what life is like for Don and Louise:

Rush home, rifle through the mail, prepare snacks, change out of . . . work clothes, make sure the children are appropriately dressed and have the proper equipment for the upcoming activity, find the car keys, put the dog outside, load the children and equipment into the car, lock the door and drive off. (p. 42)

The Tallingers epitomize a middle-class child-rearing strategy that Lareau called "concerted cultivation." This lifestyle, dedicated as it is to each child's individual development and enrichment, is exhausting just to read about. It may be familiar to you as an extreme example of your experiences; Lareau argues that it is not uncommon among the middle class.

The Tallinger brothers hardly go long stretches without some scheduled activity, most of which require adult-supplied transportation, adult supervision, and adult planning and scheduling. Rarely can they count on playing outside all day like Tyrec Taylor, a working-class child from Lareau's study. Like other children in working- and lower-class families, Tyrec's daily life is less structured, more flexible and free-flowing, more oriented around kin and neighborhood peers, and more suited to developing such skills as negotiation, mediation, and conflict resolution.

Of course, working- and lower-class children like Tyrec could not participate in all the activities that the Tallinger boys do. By the Tallingers' estimate, Garrett's activities alone cost more than $4,000 a year, which at the time of Lareau's research was an even more sizable sum than it might sound, and Garrett was just the oldest of three. Assuming his brothers would be engaged in comparable levels of such outside involvement, the costs would multiply and become even more striking. Travis Dorsch, a professor specializing in the psychology of sports and the founding director of the

Families in Sports Lab at Utah State University, estimates that the amount families are spending on sports has grown to over 10 percent of their gross income, and that with that degree of investment the potential for kids to feel pressured or parents to become overinvolved and behave in a negative way increases (Sullivan 2015).

Although Lareau is careful to illustrate what middle-class children like the Tallinger boys miss out on—free play, closer connections to relatives, more time for themselves away from adult supervision and control, comfort with and ability to amuse themselves, and less fatigue—she also illustrates the many benefits they receive beyond involvement in activities that they enjoy. The Tallingers believe that all the activities that their boys participate in teach them to work as part of a team, to perform on a public stage and in front of adults, to compete, to grow familiar with the many performance-based assessments that will come at them through school and work experiences, and to prioritize. The children travel to tournaments, eat in restaurants, and stay in hotels; they may fly to summer camps or special programs out of state or overseas. Indeed, Lareau suggests that children like the Tallinger boys may travel more than working-class and poor *adults*. These experiences, combined with what the Tallingers teach the boys at home, promote skills that enhance their chances of staying or even moving higher up in the middle class.

Research by sociologist Hilary Levey Friedman extends Lareau's analysis by examining the competitive nature of the kinds of activities in which parents enroll their children. In her book, *Playing to Win: Raising Children in a Competitive Culture*, Friedman details what parents hope their children learn from their involvement in such activities as soccer, chess, and competitive dance. Such lessons include the importance of being able to bounce back and persist in the face of failure, and the ability to perform within time limits, in public, and in stressful situations. She calls such lessons "competitive kid capital," and examines how middle-class parents perceive its importance in raising their children to succeed and maintain or improve upon a middle-class life (Friedman 2013). In lifestyles such as those both Friedman and Lareau describe, family life is often organized and ruled by large calendars that detail the children's sports, play activities, music, tournaments, and scouting events. It then falls on the parents to see that their children arrive at these activities, often directly from one to another. As Lareau (2003) somberly puts, "At times, middle-class homes seem to be little more than holding places for the occupants during the brief periods when they are between activities" (p. 64).

## Constraints on Mobility

Social mobility is differently experienced by Americans of different races and by women and men. A report by economist Julia Isaacs (2007) for the Brookings Institution compared adult children's incomes in the late 1990s and early 2000s to their parents' incomes in the 1960s. She found that though close to two-thirds of both white and African American children exceed the incomes of their parents; if one looks specifically at middle-income families, blacks are *much less likely* to experience upward intergenerational mobility. Among middle-income families, the proportion of white children who have family incomes greater than that of their parents was more than twice the size of the proportion of middle-income blacks who exceeded their parents' incomes. In addition, nearly half (47 percent) of middle-income African American children ended up *falling to the bottom* of the income distribution at some point in their lives. Only 16 percent of white children experience this same sort of fall.

A report on downward mobility by the Pew Charitable Trusts identified some factors that increased one's chances of falling out of the middle class (Acs 2011). Defining the middle class as those whose incomes ranged between the 30th and 70th percentiles in income, the report showed the prevalence of downward mobility and the effects of race, gender, and marital status. Among the findings, More than one-fourth (28 percent) of adults from middle-income families fell out of the middle, and 19 percent had incomes at least 20 percent less than that of their parents. Black men raised in middle-income families were 17 percent more likely than were white men to fall out of the middle class (data for Hispanic men were not statistically significant). Almost 40 percent of black men from middle-income families experienced downward mobility out of the middle class. There was no real difference between white, African American, and Hispanic women.

Race is not the only factor affecting mobility. Gender, marital status, and even the neighborhood in which one is raised also make a difference. In the aforementioned report on downward mobility, gender mattered more among whites than blacks, as 30 percent of white women "fell" compared to 21 percent of white men. Regarding marital status, married women were considerably less likely to fall than were divorced, separated, widowed, or never-married women. Divorced, widowed, or separated women were 31 percent to 36 percent more likely than married women to be downwardly mobile. Never-married women were 16 percent to 19 percent more likely than married women to fall. Similar but smaller differences surfaced among men as well.

Neighborhood effects are also evident in one's chances of upward mobility (Chetty and Hendren 2015). Some residential areas provide better opportunity for social mobility than others. Areas with lower levels of income inequality, better schools, lower levels of violent crime, less racial segregation, and larger proportions of two-parent households offer the best chances of moving up. "Geographic immobility," or the inability to relocate, can impede upward social mobility. As stated by historian Stephanie Coontz, "People live in different social locations. Some offer clear views and easily navigable paths to desired outcomes. Others are marked by constricted views, blocked exits, and narrow alleys that often lead to dangerous curves or dead ends" (Coontz 2015).

## Structural Downward Mobility: U.S. Families during the Recession

The previous examples depict consequences of individual upward or downward mobility. Similarly, families can see their circumstances dramatically altered by **structural mobility,** which refers to when large segments of a population experience upward or downward mobility resulting from changes in the society and economy. The recent state of the U.S. economy since the beginning of the recession in 2007 through its official end in June 2009 and beyond, made downward mobility a harsh and painful reality for large numbers of U.S. families across the social class spectrum. Widespread job losses, increases in home foreclosures, crises on Wall Street, and the failures of many big employers caused many families to suffer a variety of familial consequences along with their economic losses. With increased debt, lower incomes, and dwindling savings, many families were pushed beyond their limit.

For some couples, one consequence was a reallocation of household roles and responsibilities in two-earner families as husbands who suffered job losses became dependent on their wives' incomes. Although some couples adapted well to such change, others had to endure more and more frequent tense conversation and negotiation, as they faced hard financial decisions and alterations in their accustomed lifestyles. For others, the consequences were more visibly dramatic, including increased risk of suicide, domestic violence, or homelessness. Looking at domestic violence, the National Domestic Violence Hotline (National Conference of State Legislatures 2009) reported that calls to the hotline had increased 21 percent between the third quarter of 2007 and 2008, as the recession took hold. As the poor economy lingered, increasing rates of domestic abuse seemed to follow. In 2009, for example, the city of Philadelphia saw a decrease in the overall rate of violent crime but a 67 percent increase in domestic homicides. This was not unique to Philadelphia as there were reports of increases in domestic abuse nationwide. Perhaps even more telling, these increases occurred after a 15-year period of declining rates of domestic abuse. In May 2009, a study put out by the Mary Kay Ash Charitable Foundation showed that 75 percent of the nation's domestic violence shelters experienced an increase in women seeking help since September 2008. According to the report, 73 percent of these shelters attributed the increase to financial issues (Urbina 2009). In April 2012, a survey conducted by the Police Executive Research Forum found that more than half (56 percent) of the 700 agencies that responded reported an increase in the numbers of domestic violence incidents, and indicated that they believed the poor economy was to blame for the rise. This reflects an increase of about 15 percent from a similar survey in 2010 (Johnson 2012). Although suggestive, some sociologists warn about making sweeping generalizations, unsupported by more scientific research (Cohen 2011; Peterson 2011).

The impact of the recession on divorce is even less certain. Although mounting economic pressures certainly strained many marriages to or beyond their breaking points, the wider economy may have forced spouses to stay married even if they had intended to divorce. Although data show that the divorce rate decreased through the recession, divorce rates had been declining since before the recession, so it is hard to determine whether further declines reflect the economy's impact on marriage and divorce (Cohn 2012). Summing up his analysis of the recession's impact on divorce, sociologist Phillip Cohen suggests "it is quite likely that the economic crisis both caused some divorces and prevented some divorces. However, the balance of the evidence suggests it prevented more than it caused" (Cohen 2014).

Childbearing and child rearing were also affected. Couples who were planning to have or adopt children may have had to consider delaying their entry into parenthood or downsizing the number of children they hoped to have. Birthrates dropped during the recession, for both married and unmarried women. Although married women's birthrates started to increase in the aftermath of the recession (i.e., 2011 and 2012), this has not yet been the case for unmarried women (Cohen 2014).

The long-term effects on children are potentially chilling. It is estimated that the economic crisis may have pushed an additional 3 million children into poverty, costing the United States a yearly loss of $35 billion over the lifetimes of these children. More than half of children who fell into poverty with their families may suffer longer term. An estimated 25 percent will remain there through half of their remaining childhoods. It has been estimated that in adulthood, those who spend most of their childhoods in poverty

Families are affected by wider economic conditions, as was evident during the Great Depression and is also the case during the recent economic recession.

Marion Post Wolcott/Historical/Corbis

Kristoffer Tripplaar/Alamy Stock Photo

will earn, on average, 39 percent less than the median income, while those who spend between one-quarter to one-half of their childhoods in poverty are projected to suffer about half that level of reduced earnings (Linden 2008). Even young adults who graduated from college during the recession are vulnerable. Faced with reduced career prospects, many of them may get stuck in lower-prestige jobs even after the economy recovers (Shellenbarger 2009).

A less ominous effect of the recession on children was the increase in numbers of children living with a grandparent, during and after the recession. The onset of the recession caused a "surge" in both the numbers of children living with a grandparent, and those whose primary caregiver was a grandparent. About 10 percent of U.S. children (7.7 million) were living in the same home with a grandparent in 2011, two years after the recession's official end. For 3 million of those children, the grandparent was their primary caregiver. In 80 percent of cases of living with a grandparent, at least one parent is also present in the household, creating a multigenerational household. These households are, at least in part, products of people having lost jobs, homes, or both during the recession (DeSilver 2013).

## ◼ Racial and Ethnic Diversity

The United States is a richly diverse society. This is not likely to be news to you: Americans have long taken pride in the multicultural mix of groups, whether conceptualized as "melting" together into one large pot or envisioned more like a salad bowl, with each group retaining its uniqueness even when tossed all together. In the remainder of this chapter, we look at some racial and ethnic variations in family experience. Before we do, we need to first note how it is that we know about the multiplicity of different groups that make up the U.S. population.

Every ten years from 1790 through 2010, the decennial census has asked about race to provide a count of the different racial groups that comprise the U.S. population. Across that span of time, there have been a number of revisions to the methodology used in the census, especially with regard to the number and nature of categories from which individuals choose or into which they are placed. Originally, census takers were instructed to identify an individual's color. In 1880, the instructions were modified to ask about "color or race." The 1950 census dropped the term "color," but it was brought back for the 1970 census. Neither the 1960 nor 1980 census used the term "race"; instead, the census taker was instructed to place the individual in the appropriate racial category among those listed (Cohn 2015).

The 2000 census was the first that enabled individuals to self-identify by indicating more than one race category in completing the census questionnaire. This was significant both for individuals of multiple-race backgrounds and for those who desired a more complete and accurate count of the breadth of racial diversity within the wider population. The 2010 census provided another opportunity to measure the multiracial population. According to data from the 2000 census and the 2010 census, the number of respondents who reported multiple races increased from 6.8 million to 9 million, representing an increase from 2.4 percent to 2.9 percent of the U.S. population. Using a different and broader definition to measure the multiracial population, the Pew Research Center included people who indicated that one of their parents or one of their grandparents was of a different race. This yielded an estimate of 6.9 percent of the U.S. adult population. If one considered "Hispanic" as a race category, the estimate increases to 8.9 percent of the population as multiracial (Wang 2012). In a later section of this chapter, we will look more closely at some characteristics and experiences found in research on multiracial families.

In recognition of the complexity and potential confusion generated in asking about race, the Census Bureau is considering dropping terms such as *race* and *origin* entirely from the 2020 census and asking simply that people choose and check all "categories" that describe them. Such an item would be followed by categories that would likely include the following eight choices: White; Hispanic, Latino, or Spanish origin; Black or African American; Asian; American Indian or Alaskan Native; Middle Eastern or North African; Native Hawaiian or other Pacific Islander; and some other race, ethnicity, or origin. In response, people would then choose the category or categories that best fit from those provided.

### Defining Race, Ethnicity, and Minority Groups

Before we begin to look more closely at ethnic or racial diversity in family experience, we need to define several important terms. A **race** or **racial group** is a group of people, such as whites, blacks, and Asians, classified according to their **phenotype**—their anatomical and physical characteristics. Racial groups share common phenotypical characteristics, such as skin color and facial structure. The concept of race is often misused and misunderstood to the extent that some sociologists refer to it as a myth (Henslin 2006). This refers specifically to any assumption of racial purity or homogeneity within racial groupings (in skin color, facial features, and so on), and any belief that certain racial groups are superior or inferior in comparison to one another. Socially, however, we perceive or identify ourselves and others within racial classifications, and we are treated and act toward others on the basis of race. As a consequence, race becomes a highly significant and very real factor in shaping our life experiences. Although its biological importance may be doubtful, its social significance remains great.

An **ethnic group** is a set of people distinct from other groups because of cultural characteristics. Such things as language, religion, and customs are shared within and allow us to differentiate between ethnic groups. These cultural characteristics are transmitted from one generation to another and may then shape how each person thinks and acts—both inside and outside families.

Either a racial or an ethnic group can be considered a **minority group**, depending on social experience, not numerical size. Minority groups are so designated because of their status (position in the social hierarchy), which places them at an economic, social, and political disadvantage (Taylor 1994). Thus, African Americans are simultaneously an ethnic, a racial, and a minority group in the United States (as well as an *ancestry category* as shown previously).

As we will soon see, ethnic and/or racial differences are related to and often difficult to untangle from social class differences because there are differences in the overall economic standing of the various racial and ethnic groups in the United States. Some differences in family patterns reflect cultural background factors or distinctive values characteristic of different groups; others are shaped by social experiences—including experiences of discrimination—had by members of different racial groups. However, some ethnic or racial differences in family patterns, such as in nonmarital childbearing or relations with extended kin, also reflect the different social and especially economic circumstances under which different groups live (Hattery and Smith 2012; Sarkisian, Gerena, and Gerstel 2007; Wildsmith and Raley 2006).

The best estimates of the racial composition of the U.S. population come from data gathered by the U.S. Census Bureau's decennial census of the population. Complying with the constitutional mandate, a census of the population is undertaken in each year ending in a zero. The 2010 census data revealed that more than 40 percent of the U.S. population consisted of ethnic or racial minorities: Of these, 14 percent were African American (alone or in combination with another race), 16 percent were Hispanic, 7 percent were Asian/Pacific Islander, and 2 percent were Native American. Another 3 percent are biracial or multiracial. A more recent census estimate compares the U.S. population characteristics for 2014 to what is projected for 2060. According to those data, the population is projected to change as illustrated in Figure 3.2.

By 2060, more than half of the U.S. population is expected to be made up of minorities, with Hispanics projected to be almost 30 percent of the overall population; non-Hispanic blacks, 14.3 percent; Asian Americans, 9 percent; and American Indians and Alaska Natives, 1 percent. Native Hawaiians and Pacific Islanders are projected to represent less than one percent of the 2060 population. As presently projected, 6 percent of the U.S. population will be biracial or multiracial. Non-Hispanic whites are expected to represent 43.6 percent of the total 2060 U.S. population (Colby and Ortman 2015, Table 2).

Until the latter decades of the 20th century, most research about marriages and families in the United States tended to focus on white, middle-class families. The nuclear family was the norm against which all other families, including single-parent and stepfamilies, were evaluated. A similar distortion unfortunately influenced our understanding of African American, Hispanic or Latino, Asian American, and Native American families. Instead of recognizing the strengths of diverse ethnic family systems, misguided researchers viewed these families as "tangles of pathology" for failing to meet the model of the traditional nuclear family (Moynihan 1965). Part of this distortion resulted from the long-term scarcity of studies on families from African American, Latino, Asian American, Native American, and other ethnic groups. Furthermore, many earlier studies focused on weaknesses rather than strengths, giving the impression that all families from a particular ethnic group were riddled with problems.

For example, what's known as the **"culture of poverty"** approach depicts African American families

**Figure 3.2** Current and Projected Racial and Ethnic Composition of U.S. Population, 2015 and 2060

### 2015 Population

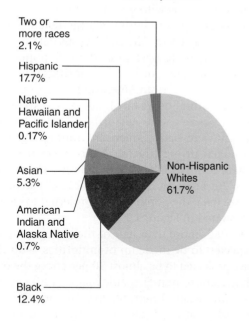

Two or more races
2.1%

Hispanic
17.7%

Native Hawaiian and Pacific Islander
0.17%

Asian
5.3%

American Indian and Alaska Native
0.7%

Non-Hispanic Whites
61.7%

Black
12.4%

### 2060 Projected Population

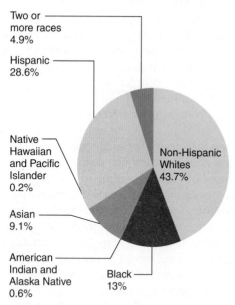

Two or more races
4.9%

Hispanic
28.6%

Native Hawaiian and Pacific Islander
0.2%

Asian
9.1%

American Indian and Alaska Native
0.6%

Non-Hispanic Whites
43.7%

Black
13%

NOTE: Percentages may not add up to 100 percent.
SOURCE: U.S. Census Bureau Population Division, Percent Distribution of the Projected Population by Hispanic Origin and Race for the United States: 2015-2060 (NP2014-T11).

as being deeply enmeshed in illegitimacy, poverty, and welfare as a result of their slave heritage. As one scholar notes, the culture of poverty approach "views black families from a white middle-class vantage point

and results in a pejorative analysis of black family life" (Demos 1990). This approach ignores most families that are intact or middle class. It also fails to see African American family strengths, such as strong kinship bonds, role flexibility, love of children, commitment to education, and care for the elderly.

The United States is a pluralistic society. Thus, it is important that students and researchers alike re-examine diversity among our different ethnic groups as possible sources of strength rather than pathology (DeGenova 1997). For instance, cultures may vary widely in how to define the best interests of children (Murphy-Berman, Levesque, and Berman 1996). Differences may not necessarily be problems or signs of weakness but may, instead, be solutions to problems; in fact, they may more accurately be signs of adaptation rather than weakness. As two family scholars pointed out, "Whether a phenomenon is viewed as a problem or a solution may not be objective reality at all but may be determined by the observer's values" (Dilworth-Anderson and McAdoo 1988).

As we embark on our discussion of race and ethnicity in family life, it is especially important to try to avoid ethnocentric fallacies (a term introduced in Chapter 2)—beliefs that your ethnic group, nation, or culture is innately superior to others. It is also important to remember that some family patterns that differ between different ethnic or racial groups may reflect, at least in part, differing socioeconomic circumstances experienced by these groups. For example, American Community Survey data from the U.S. Census show higher levels of poverty among American Indians/Alaskan Natives, blacks, and Hispanics, and lower levels among whites and Asians. Thus, some family characteristics that are associated with poverty or economic hardship may show up more prominently for certain racial/ethnic groups though their origin may be at least partially economic in nature. In the following sections, we consider briefly some distinctive characteristics and strengths of families from various ethnic and cultural groups.

## Racial and Ethnic Groups in the United States

The following brief profiles highlight some of the key features of family life for the major ethnic and racial minorities in the United States. As shown in Figure 3.3, family patterns vary both between and within these groups, with economic factors helping to account for the extent and nature of the differences.

**Figure 3.3** Percent Distribution of Hispanic Population, 2015 to 2060

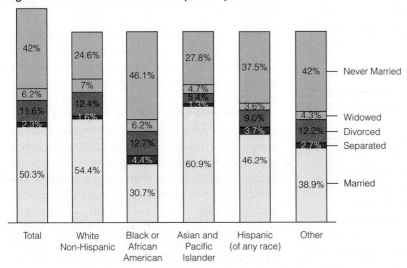

SOURCE: U.S. Census Bureau, Population Division, Table 11, Percent of the Projected Population by Hispanic Origin and Race for the United States: 2015 to 2060 (NP2014-T11).

## Black and African American Families

According to the Census Bureau in 2010, there were nearly 39 million blacks in the United States, representing 13 percent of the U.S. population. If we were to include those who consider themselves biracial—in this case, black combined with one or more other races—the 2010 total reached 42 million people, or 14 percent of the U.S. population (Rastogi et al. 2011). The current census estimate is that in 2060, 60 million people, 14 percent of the population, will be black. Again, if we added in those who consider themselves multiracial—black in combination with one or more other races—the proportion is expected to reach 17.9 percent (Colby and Ortman 2015).

It is worth noting that the Black population contains more diversity than is often realized or studied (Taylor et al. 2013). Recent increases in the population of black immigrants from Caribbean countries (such as Haiti and Jamaica) as well as from Africa, have resulted in sizable portions of the black populations of many U.S. cities now being foreign born. Monica Anderson of the Pew Research Center reports that there are more than four times as many black immigrants in the United States today as in 1980, making them almost 9 percent of the black population. In some cities, such as New York, Miami, Fort Lauderdale, and Washington, DC, they represent a much more sizable share of the black population, reaching around a quarter of the black population in New York and Boston, more than a third in Miami, and 44 percent of the black population in Fort Lauderdale, Florida (Anderson 2015; Taylor et al. 2013).

Caribbean blacks make up more than half of the foreign-born black population. Between 2000 and 2013, the number of black Africans in the United States increased to a high of 1.4 million, making them more than a third of the total foreign-born black population. By 2060, foreign-born blacks are expected to be a sixth of the black population (Anderson 2015).

A comparison of foreign-born and U.S.-born blacks shows that foreign-born blacks tend to be older, have higher household incomes ($43,800 vs. $33,500), are more likely to have a college degree, and are less likely to live in poverty. Research into marriage and marital satisfaction as well as into support given to and received from extended family, fictive kin, friends, and church congregations shows that black Caribbeans' and African Americans' family experiences differ and, as Chalandra Bryant and colleagues point out, one should not assume similar family experiences or meanings just because both groups can be classified as black (Bryant et al. 2010).

There is also economic diversity within the black population that is sometimes overlooked in strict racial comparisons or in characterizations of family patterns and experiences. In identifying racially specific patterns, it is important to keep in mind the influence that economic realities have in shaping those patterns and that the experiences of middle-class and higher-status blacks may get somewhat lost in the following overall profiles or comparisons between races.

Compared with the total U.S. population, African Americans are less likely to be married (see Figure 3.2). Although they are about as likely to be divorced or widowed, a much greater percentage of blacks than whites have never married (46.1 vs. 24.6 percent). Blacks are much more likely to bear children outside of marriage and more likely to live in single-parent, mostly mother-headed families. More than 70 percent (71 percent) of all births to black women in 2013 were to unmarried women, compared to a rate nationwide of 40.6 percent, and a rate of 29.3 percent among non-Hispanic whites, and of 53.2 percent among Hispanics (Martin et al., 2015).

Although African Americans are as likely as the general population to live in family households, their households differ from the family households in the general population. In 2013, there were 9.8 million black family households, representing 61.8 percent of all black households. Close to half of black *family* households consisted of married couples. Married couples headed 28.2 percent of all black households as compared to almost half (48.4 percent) of households in the wider population. Because of high rates of divorce and of births to unmarried women, an estimated 39 percent of African American children lived in households with two parents in 2014. Meanwhile, 54.9 percent lived with either single mothers (50.8 percent) or single fathers (4.2 percent), and 6.1 percent lived with neither parent (U.S. Census Bureau, 2014a, Current Population Survey, America's Families and Living Arrangements: 2014, Table C9, Children by Presence and Type of Parents, Race, and Hispanic Origin).

Considering families rather than households, data from 2014 reveal that 44.1 percent of African American families were married-couple families, 27.8 percent were single mothers living with their children under 18, and 3.2 percent were single fathers and their children under 18. The equivalent percentages for non-Hispanic whites show more than three-fourths (79 percent) of white families headed by married couples with or without children, 7.5 percent by unmarried women with children, and 2.1 percent by single fathers and their children. African American families are also more likely to consist of grandchildren living with a grandparent (3.2 percent) or adult children living with their parents (12.1 percent) than are white, Asian, or Hispanic families (U.S. Census Bureau, Current Population Survey, 2014a, Social and Economic Supplement, Table FG10).

A variety of measures indicate that as a group, blacks experience greater economic hardship than do whites. Unemployment rates, income differentials, and poverty rates all illustrate this. Compared to the general population, African Americans are less likely to have completed college, are more likely to be unemployed, and are more likely to be in poverty. Black unemployment (as of June 2015) is twice the rate as white unemployment (10.7 percent vs. 4.8 percent). Additionally,

- Median household income for African Americans in 2013 was $34,598. This was 59.4 percent of the median income for white households, and 66 percent of the $51,939 median for all U.S. households.

- Data from 2013 reveal racial aspects of poverty. According to U.S. Census Bureau data, 27.2 percent of blacks fell below the poverty line. The percentage of impoverished blacks is close to three times the percentage of poor whites (9.6 percent) and nearly twice the percentage of the general population who are poor (14.5 percent). Among families, similar race patterns surface. In 2013, 25.8 percent of African American families were below poverty. This was slightly higher than the percentage of Hispanic families in poverty (22.3 percent), but notably higher than the percent of Asians (7.7 percent) and whites (6.9 percent) who were poor. Among female-headed black households, 42.5 percent were below poverty in 2013 (U.S. Census Bureau, 2012c, Statistical Abstract, Table 716; DeNavas-Walt and Proctor 2014). Nationwide, roughly 20 percent of children under 18 live in poverty. The poverty rate for black children is nearly twice as high at 38.3 percent, and black children have three times the rate of poverty as white children (10.7 percent) (Patten and Krogstad 2015).

Along with a high value placed on children, there are several other noteworthy features of African American families, some of which can be related to the aforementioned economic facts. First, marriages tend to be more egalitarian in their sharing of family roles and division of labor. Black men have more positive attitudes toward working wives, take on a slightly larger share of household labor, and spend more time on domestic tasks and child care activities (McLoyd et al. 2000; Sarkisian and Gerstel 2012). Both middle-class and working-class African American women identify breadwinning as a part of their role as mothers (Sarkisian and Gerstel 2012). African American families, in contrast to Caucasian families, have a long history of being dual-earner families, in part as a result of economic need. As a consequence, employed women have played important roles in the African American family.

Second, marital relations more often show signs of greater distress than is true of the general population (Bryant et al. 2010). Some evidence indicates African Americans have a greater likelihood of spousal violence, lower levels of reported marital happiness, and a greater tendency

**False** **8.** Compared with Caucasian families, relationships between African American husbands and wives are less traditional.

## *Blackish*, Race, and Class

In the fall of 2014, ABC premiered a new situation comedy, *Black-ish*, portraying an upper-middle-class, black family with two economically and professionally successful parents and their four children. As introduced in the premier episode and central to the entire series, the show raises the question of whether and how growing up in affluence and economic prosperity weakened the racial awareness and identity of the children in the family. This becomes a major source of anxiety for Dre, the father, though it is less so for Rainbow, his biracial spouse, and is the driving concern in multiple episodes, throughout its successful first year (it was renewed for at least a second season).

Between Dre's successful advertising career (as a senior vice president) and Rainbow's career as an anesthesiologist, they have put their family in a much more privileged class position than he grew up in, and the show focuses attention onto whether and how his upward mobility has left his children, especially his son Andre, Jr., unable to appreciate in the same way as his father, what it means to be black in the United States.

As critic Mary McNamara wrote in the *Los Angeles Times*, "The primary defining force of the Johnson kids is money, not color. They are children of privilege, and like all children, they do not understand what that privilege represents. Dre is caught in the contradiction felt by parents throughout the ages: They want a better life for their children, but not at the price of historical amnesia" (McNamara 2014).

The show has received both considerable critical acclaim and some fairly harsh criticism. Reviewers have praised the show for featuring a stable upper-middle-class black family, and for bringing up important issues related to race in the United States in ways that are "mild and entertaining," while offering valuable "political and social commentary" (www.commonsensemedia.org/tv-reviews/black-ish).

Creator Kenya Barris, in an interview with Melinda Sylvester (http://faithandsoul.net/Film-Entertainment-BLACKish.html), offered the following:

This is a black family that we're using as a lens to show a picture to America about culture and identity, and to look at how there are generational differences within a black family in the modern world that aren't necessarily what we're used to.

Writing in a more critical way, Pastor Frances Cudjoe Waters contends,

What *Black-ish* misses . . . is that one of the primary things that unite black people is this country is that regardless of socioeconomic status, skin complexion, and other life choices, black folks in America have a shared history and current reality of struggling against stereotypes, institutional and legislative racism, and continued barriers precisely because we continue to be judged by the color of our skin, more than by the content of our character and the uniqueness of our journey.

Despite criticism such as this, the show essentially depicts how challenging racial socialization can be, especially in multiracial families and those whose generations have faced such significantly different socioeconomic realities.

---

to consider ending their marriages. Researchers have connected these characteristics to the financial strains and stressful life circumstances that more often characterize daily life for many African Americans. These stresses include continued economic hardship, racial discrimination, adverse work conditions, and the responsibilities of parenthood that many African Americans carry into marriage with them (Bryant et al. 2010; McLoyd et al. 2000).

Economic disadvantage also increases the likelihood that African Americans find housing in more distressed, if not impoverished, urban areas. Such neighborhoods may impose additional stresses on couples and families, including heightening the concerns parents have about the well-being and safety of their children, while at the same time lack the kinds of services and resources that might promote more positive family experiences (Bryant et al. 2010).

Third, kinship bonds—whether blood-related or fictive/affiliated kin—are especially important because they can provide economic assistance and emotional support in times of need (Taylor 1994; Taylor et al. 1991). African Americans have a strong tradition of familism (emphasis on family and family loyalty), with an important role played by intergenerational ties. This sense of familial obligation creates a willingness to help and seek help from kin as needed. Some young African American couples may be expected to help other kin, even when their own resources are limited (Bryant et al. 2010).

In raising their children, black parents have been found to approve of and use corporal punishment more than do white parents. We will return to this in Chapter 12, but for now it merits mention that this "racial characteristic" may, itself, be an outcome of class more than of race. Sociologists Angela Hattery and Earl Smith make this quite explicit in their assessment that rather than being a product of race, blacks may be more likely to favor or approve of spanking because they are "more likely to be poor, young parents, live in the South, and hold conservative religious views, all of which are highly correlated with both attitudes toward spanking and actual use of corporal punishment" (2012, 26).

Still on the subject of parenting, blacks, like other racial minorities (including multiracial individuals), engage in a process of ethnic or **racial socialization**, to prepare their children to live in a society where they are likely to encounter racism and racial inequality. Parents' racial socialization includes teaching their children about their heritage and history, thus promoting their familiarity with customs and traditions and enhancing their pride. In addition, ethnic-racial socialization may also include attempts to prepare children for discrimination by raising their awareness of it. Messages emphasizing hard work, self-acceptance, and equality are also prevalent both in black families and across different ethnic minorities (Hughes et al. 2006).

Finally, African Americans are much more likely than Caucasians to live in **extended households**, households that contain several different generations. Black children are more likely than other children to live in a household with a grandparent. In many of these households, the grandparent has assumed responsibility for her grandchild(ren) (typically, this grandparent is a grandmother) (U.S. Census Bureau 2008).

Many of these and other characteristics are often associated with harsher economic circumstances and thus may not be features inherent in African American families (Hattery and Smith 2012). For example, when race differences are examined according to income, racial differences become smaller (see Table 3.5). Poor African Americans have divorce rates more similar to poor Caucasians, and middle-class African Americans have divorce rates more similar to middle-class Caucasians. Thus, understanding socioeconomic status, especially poverty, is critical in examining African American life (Hattery and Smith 2012). Combined with the tendency of upper-status African American families (i.e., middle- and upper-middle-class) to be as stable as Caucasian families of comparable status, these economic indicators suggest that much of what we may assume to be race differences are confounded by economic differences. Said differently, maybe social class differences sometimes *masquerade as race ones*.

This more economic argument may pertain especially well to an understanding of race differences in marriage rates, divorce rates, and the numbers of single mother–headed families. The most widely applied argument is that blacks' "marital prospects" have shifted dramatically, especially among the poor (Aponte, Beal, and Jiles 1999). Wilson's notion of the "male marriageable pool index" emphasizes the importance of male employment to their "marriage-ability" (Wilson 1987). Downward shifts in male employment patterns would then account for some of the decline in marriage rates and the increase in single mother–headed families. As expressed by sociologists

**TABLE 3.5  Race, Ethnicity, and Socioeconomic Status**

|  | Total | White | African American | Hispanic | Asian American |
|---|---|---|---|---|---|
| Median household income (2013) | $51,939 | $58,270 | $34,598 | $40,963 | $67,065 |
| Percent unemployed (as of July 2015) | 5.3 | 4.6 | 9.1 | 6.8 | 4.0 |
| Percent in poverty (2013) | 14.5 | 9.6 | 27.2 | 23.5 | 10.5 |
| Percent of families with children in poverty (2013) | 12.4 | 6.9 | 25.8 | 22.3 | 7.7 |
| Percent of children in poverty (2013) | 19.9% | 10.7 | 38.3 | 30.4 | 10.1 |
| Median wealth (2013) | $81,400 | $141,900 | $11,000 | $13,700 | $91,440 |

SOURCES: Bureau of Labor Statistics (2015b); DeNavas-Walt, Carmen, and Proctor (2014); Kochhar and Fry (2014); and U.S. Census Bureau, Current Population Survey (2014a).

Angela Hattery and Earl Smith, "poverty, not race, seems to be the key contributor to declining marriage rates" (Hattery and Smith 2012, 23).

Not only are African Americans unlikely to devalue marriage, they may actually more highly value marriage than do other groups. Asked about the importance of marriage when a man and woman have decided to spend their lives together, African Americans were more likely (77 percent) than either whites (69 percent) or Hispanics (71 percent) to say "very important" or "somewhat important." African Americans are also most likely to claim that divorce should be avoided except in an "extreme situation" (Pew Research Center 2007a). Thus, despite being the least likely to be married, African Americans may be most likely to value marriage.

Despite the benefits of linking class and race in our efforts to understand family diversity, we cannot simply reinterpret all race differences as economic in nature. Don't forget that a major feature of race in U.S. society is that it determines much about the kind of treatment we receive from others. Thus, the opportunities we are offered or refused and whether others insult, avoid, or think less of us are all affected by race. An interpretation of race differences as only (or even largely) class differences unfortunately minimizes or ignores such expressions of racism and discrimination, and at the same time fails to acknowledge patterns that may have cultural origins to them—such as greater emphasis on extended family ties or gender equality. Racial discrimination, including that which results in restricted economic opportunities as well as that which is experienced day-to-day, amount to stressors that can spill over into color, and eventually even influence or determine marital and familial outcomes (Bryant et al. 2010).

## Hispanic Families

Hispanics (or Latinos) are now the largest ethnic group in the United States. In fact, more than half the growth of the U.S. population between 2000 and 2010 was from an increase in the Hispanic or Latino population. Additionally, though presently the third fastest growing group, between 2014 and 2060, the Hispanic population is expected to increase 115 percent. By 2060, it is projected that three out of ten Americans will be Hispanic. These increases result from both immigration and relatively high birthrates. In the 2013 U.S.

**True** **9.** Latino and Asian American families show much variation within each group, depending on the country from which they came, why they left, and when they arrived in the United States.

Hispanic population, 64.1 percent of Hispanic Americans were of Mexican descent, 9.5 percent were Puerto Rican, and another 3.7 percent were Cuban. The remaining 24 percent included almost 9 percent from Central American countries and 6 percent from South American countries. Nearly two-thirds of the Hispanic population was born in the United States, 35 percent were foreign born.

Compared to the population overall, the population of non-Hispanic whites and the other ethnic or racial minorities discussed in this chapter, as a whole, Hispanics have the youngest median age: 27.4. Socioeconomically, Hispanics are less well off than the general population. The median family income for Hispanics in 2012 was $40,963—lower than the income of Asians or whites but higher than the median incomes for African American or American Indian families. Whereas the overall poverty rate was 14.5 percent in 2013, 23.5 percent of Hispanics were poor. Roughly 22 percent of Hispanic families were below poverty, and—at 33 percent—a greater proportion of Hispanic children live in poverty than either Asian (14 percent) or white children (14 percent). Still higher child poverty rates are found among African Americans.

Familial characteristics tell a similar story. Generally, Hispanics are less likely than either Asians or non-Hispanic whites to be married. They are more likely to be divorced than Asian Americans but less likely than either African Americans or non-Hispanic whites to be divorced. Nearly a third of Hispanic children live with one parent, with 90 percent of those children living with their mothers.

It is important to note that, in addition to these sorts of differences between Hispanics and non-Hispanics, there is considerable diversity *among Hispanics* in terms of ethnic heritage (such as Mexican, Cuban, Puerto Rican, or Central or South Americans), socioeconomic status, and family characteristics. For example, although 13 percent of Hispanics had bachelor degrees or more education in 2010, educational attainment ranged from a low of 7 percent among Salvadorans to highs of 32 percent among Colombians and 30 percent among Peruvians (Motel and Patten 2012). One-quarter of those of Cuban descent were college educated compared to 16 percent of Puerto Ricans and 9 percent of Mexican Americans. It is worth noting that among each of the ten largest Hispanic population groups in the United States, the percent with college degrees increased between 2000 and 2010 (Motel and Patten 2012).

Looking at 2010 incomes among these same ten Hispanic groups shows a range from $34,000 among Dominicans to $50,000 among Ecuadorians. Among all Hispanics, median household income was $40,000. Using 2010 data on poverty rates, they ranged from a low of 13 percent among Colombians and 14 percent among Peruvians, to highs that were roughly twice that, of 27 percent among Hondurans, Puerto Ricans, and Mexicans (with Guatemalans and Dominicans at 26 percent) (Motel and Patten 2012). For comparison, for this same period, the overall Hispanic poverty rate was 25 percent. Among Hispanics, there are some interesting connections between the aforementioned familial and economic characteristics. Research indicates that the percentage of children born to either teenage or unmarried mothers varies between Hispanic groups, with Puerto Ricans (at 11.8 percent) and Mexicans (at 11.1 percent) having the highest percentages of births to teenage mothers; Puerto Ricans also had the highest percentage of births to unmarried mothers (64.6 percent). Among Hispanic women, the lowest percentage of births to unmarried mothers is among Cuban women and women from Central or South America (50.1 percent in 2013). In 2013, births to teenage mothers ranged from 4.6 percent among Cubans to 11.8 percent among Puerto Ricans (Martin et al. 2014). This diversity is partly but not entirely because of economics. It is further accentuated by the varying proportions of U.S.-born and foreign-born Latinos in each group.

Substantial Latino diversity also surfaces in differences between immigrant women and U.S.-born Latinas. For example, there was a 7 percent difference in the percentage of births to unmarried mothers between U.S.-born women of Mexican descent (53 percent) and women born in Mexico (46 percent). Among Puerto Rican women, the gap was 9 percent, with 65 percent of births to U.S.-born Puerto Rican women occurring outside of marriage, whereas the figure was 56 percent among foreign-born Puerto Rican women (Hummer and Hamilton 2010).

Finally, avoid overgeneralizing characterizations of Mexican, Puerto Rican, Cuban, or any other Latino family types. None of these groups have a singular family system. More specifically, there is much diversity in Mexico, Cuba, or Puerto Rico, some of which results from socioeconomics, some from rural versus urban living, some from religion, and so on (Aponte, Beal, and Jiles 1999).

Traditional Mexican and Puerto Rican families can be characterized by a few distinctive cultural traits: high value placed on children and childbearing, devotion to family (*i.e., familismo*), and male dominance (*i.e., machismo*). Children are especially important. Birth and fertility rates are still higher among Hispanics than among the general U.S. population, non-Hispanic whites, African Americans, or Asian Americans. Hispanics are also more likely to have larger families. More than a quarter of Hispanic family households contain five or more people. This is higher than the percentage of comparable households among whites (10.1 percent), blacks (14.4 percent) or Asian Americans (17.2 percent). As with many other household and family characteristics, there is a large difference between U.S.-born and foreign-born Hispanics, with more than 30 percent of foreign-born and just under 20 percent of U.S.-born Hispanics having family households of five or more. The larger family size may in part reflect the fact that Hispanics are significantly more likely than whites or African Americans to identify the purpose of marriage as having children, and much less likely to see marriage as for mutual happiness and fulfillment (Pew Research Center 2007a).

*La familia* is based on the nuclear family, but it also includes the extended family of grandparents, aunts, uncles, and cousins. All tend to live close by, often on the same block or in the same neighborhood. There is close kin cooperation and mutual assistance, especially in times of need, when the family bands together. Family unity and interdependence, sometimes extended to include fictive kin (e.g., Cuban *compadres* and *comadres*—godparents), reflect the importance of extended kin ties.

Children are highly valued in Latino culture, and the family is understood to be the basic source of emotional support for children.

Monkey Business Images/Shutterstock.com

One indicator of the importance Hispanics place on family ties can be seen in answers to a national survey conducted by the Pew Hispanic Center. When asked about the importance of family over friends, 90 percent of the 2,000 Hispanics surveyed "strongly agreed" (78 percent) or "agreed somewhat" (12 percent) that relatives are more important than friends. Additionally, 82 percent agreed strongly (68 percent) or somewhat agreed (14 percent) that it is better for children to live in their parents' home until they get married. Yet, even the greater integration of Hispanic extended families (living nearby, staying in touch, and providing assistance) cannot be entirely disconnected from social class, and—among Mexican Americans, for example—may account for their greater tendency to live with or close to extended kin and provide household and practical help to extended kin (Sarkisian, Gerena, and Gerstel 2007).

Male dominance, as suggested, although often exaggerated in the misuse or misunderstanding of *machismo*, is part of traditional Latino family systems but has declined, especially among dual-earner couples. Migration and mobility disrupt traditional Latino family forms and lead to change. This change can be seen as part of a wider process of "convergence," in which distinctive ethnic traits diminish over time (Aponte, Beal, and Jiles 1999). Although there has been a tradition of male dominance, current day-to-day living patterns suggest that noteworthy change has occurred. Women have gained power and influence in the family as they have increased their participation in paid employment. When wives are co-providers, Hispanic men spend more time on household tasks (Aponte, Beal, and Jiles 1999; McLoyd et al. 2000).

## Asian American Families

Asian Americans are the fastest-growing minority in the United States. In 2010, the more than 19 million Asian Americans and Pacific Islanders represented 5.6 percent of the U.S. population. Of this population, 14.7 million people (85 percent) identified themselves as "Asian alone," while an additional 2.6 million reported themselves as "Asian in combination" with some other racial group. More than half of those indicating "Asian in combination" identified themselves as "Asian white" (Hoeffel et al. 2012).

Similar to Hispanics, Asian Americans are richly diverse, comprising Chinese, Filipino, Asian Indians, Japanese, Vietnamese, Koreans, Thai, and other groups. The six largest Asian American groups—Chinese Americans, Filipino Americans, Asian Indians, Vietnamese Americans, Korean Americans, and Japanese Americans—each number a million people or more. Cumulatively, three groups make up 60 percent of the Asian American population: Chinese at 23 percent, Filipino Americans at 19.7 percent, and Asian Indians at 18.4 percent. By adding Vietnamese, Americans, Korean Americans, and Japanese Americans, more than 80 percent of the Asian American population is accounted for (pewsocialtrends.org/asianamericans).

General comparisons show that, in many key ways, Asian Americans are unlike other racial or ethnic minorities in that, on a variety of social, educational, and economic indicators, Asian Americans stand out from the general population and the population of non-Hispanic whites for the levels of success they enjoy. Asian Americans are less likely than the societal population to be unemployed or to fall beneath the poverty level. Given these economic characteristics, it is perhaps unsurprising that Asians are also more likely to be married, less likely to be divorced, and less likely to bear children outside of marriage, as these familial characteristics often vary by income and education.

However, the diversity among Asian Americans is considerable, and lost in a "general comparison" is the fact that certain Asian American groups, such as Hmong, Laotians, and Cambodians, fare far less well than is suggested by a less nuanced racial comparison. For example, consider education. Although Asian Americans as a group are much more likely to obtain higher levels of education than the general public, there is quite considerable variation among Asian American subgroups as suggested in Table 3.6.

In 2012, just over half of all Asian Americans 25 and older had bachelor degrees or higher compared to 29.1 percent among the general population. Additionally, more than 20 percent of Asian Americans had graduate or professional degrees compared to 10 percent of the general population. But as the data in Table 3.6 show, educational attainment varies considerably (Jaschik 2013).

Similarly, when looking at data on income, one must keep in mind the wide range that characterizes the Asian American population. For example, among the Asian alone population, the 2013 median income was $72,472, which made it higher than that of the general population and higher than among whites. However, Asian American median incomes ranged from a low of $51,331 among Bangladeshis to a high of $100,547 among Asian Indians (U.S. Census Bureau 2014a).

## TABLE 3.6 Educational Attainment of Asian American Subgroups

| Group | Less than High School Diploma | Bachelor's Degree or Higher |
|---|---|---|
| Hmong | 37.9% | 14.7% |
| Cambodian | 37.4% | 14.1% |
| Laotian | 33.8% | 12.4% |
| Vietnamese | 29.4% | 25.8% |
| Chinese | 19.3% | 51.5% |
| Thai | 16.8% | 43.8% |
| Bangladeshi | 16.6% | 49.9% |
| Pakistani | 13.4% | 53.9% |
| Indian | 8.8% | 71.1% |
| Korean | 8.3% | 52.7% |
| Filipino | 7.9% | 48.1% |
| Sri Lankan | 7.6% | 57.4% |
| Indonesian | 7.3% | 48.7% |
| Japanese | 5.3% | 47.7% |
| Taiwanese | 4.8% | 74.1% |

SOURCE: Jaschik 2013.

Also, notable differences exist between those Asian Americans born in the United States and those who are immigrants. Thus, it is fitting that an analysis of the Asian American population would be called, "A Community of Contrasts: Asian Americans in the United States, 2011" (Asian Pacific American Legal Center and Asian American Justice Center 2011). Where Asian Indians and Chinese, Filipino, Korean, and Japanese Americans tend to have higher incomes and greater levels of education than the overall population, certain other groups, such as Hmong, Laotian, and Cambodian Americans, have much lower levels of income and education and higher rates of poverty and unemployment. Although Asian American unemployment overall is lower than other groups, unemployment among Hmong, Laotian, and Cambodian Americans is higher than average U.S. unemployment (Asian Pacific American Legal Center and Asian American Justice Center 2011). Overall poverty or unemployment

data masks some considerable variation within the Asian American population.

Familial characteristics also differ between Asian populations. In marital status, for example, although Asians were overall more likely than the general population to be married, this was not uniformly true. Although 59 percent of Asian Americans were married in 2010, the percentage among groups ranged from 48 percent for Cambodians to 71 percent for Asian Indians and Bangladeshis. (www.pewsocialtrends.org/2012/06/19/chapter-1-portrait-of-asian-americans/). Similarly, divorce rates (per 1,000 people), ranged considerably, from lows among Bangladeshis (4.6 percent), Asian Indians (5.3 percent), and Pakistanis (7.1 percent) to highs among Cambodian (24.2 percent), Laotians (26.3 percent), and Thai Americans (28.1 percent) (familyinequality.wordpress.com/2014/11/13/asian-divorce-rate/).

Clearly, much diversity can be observed within Asian American families based on where they're from, time of arrival in the United States, and reasons for coming to this country (e.g., political vs. economic). More recent immigrants retain more culturally distinct characteristics, such as family structure and values, than do older groups, such as Chinese Americans and Japanese Americans. Asian American families tend to be slightly larger than the average U.S. family (averaging 3.5 compared to 3.2 members), although there is wide variation between older and more recent immigrants. Among the more assimilated Japanese, average family size is just under three members. Among more recent Asian immigrants (e.g., Cambodians, Laotians, Vietnamese, and Hmong), families average closer to or over four members. The greater family size reflects the presence of extended kin.

Values that continue to be important to Asian Americans in general include a strong sense of importance of family over the individual, self-control to achieve societal goals, and appreciation of cultural heritage. A Pew Research Center report on Asian Americans, *The Rise of Asian Americans* (2013), included survey data on a number of relevant opinions and values, and illustrates the importance placed on family. Although the report received some criticism for overemphasizing Asian American social and economic success and downplaying the considerable differences across Asian groups, the report also contained a breadth of data on many aspects of Asian American experience and attitudes. For example, when asked to consider the importance of each of the following possible goals, including being a good parent, having a successful marriage, owning one's own home, being successful in a high-paying career or profession, living a very religious life, helping people in need, and having a lot of free time to relax and do what they wanted to, being a good parent and having a successful marriage were the two deemed most important. Two-thirds of Asian Americans ranked "being a good parent" as one of the most important things in their lives. More than half (54 percent) responded that "having a successful marriage" was one of the most important things. Comparable data for the United States overall found that although the same two goals were considered most important, the percentages designating these family goals as among "the most important things" in their lives were smaller. Being a good parent was identified by 50 percent, and having a successful marriage was identified by 34 percent as "one of the most important things in life." Interestingly, in these attitudes, too, there was some variation across different Asian groups (see Figure 3.4).

Chinese Americans place a great deal of importance on treating one's parents and elders with respect and looking to them for guidance (Benner and Kim 2009). Asian, especially Chinese, parents tend to exercise strong parental control while encouraging their children to develop a sense of independence and strong motivation for achievement (Ishii-Kuntz 1997; Lin and Fu 1990). In the aforementioned Pew Research Center survey, Asian Americans indicated seeing a gap between their parenting and what they saw of most American parents. More than three-fifths responded that American parents do not put enough pressure on their children to do well in school.

Parents' own experiences of discrimination or prejudice can influence how they go about socializing their children, especially given the strong ties between parents and children characteristic of Asian culture more generally (Benner and Kim 2009). Parents' experiences of discrimination may influence them to use ethnic or racial socialization practices, which in turn may contribute to greater stress in their children and a greater sense of their own discrimination (Benner and Kim 2009).

Migration and assimilation alter many traditional Asian family patterns. For example, among Japanese families, there are considerable differences among the *Issei* (immigrant generation), the *Nisei* (first-generation U.S.-born), and the *Sansei* and subsequent generations on such family characteristics as the relative importance of marriage over extended kin ties, the role of love in the choice of a spouse, and the relationship between the genders (Kitano and Kitano 1998). Similarly, we can draw distinctions between traditional Vietnamese families and U.S.-born Vietnamese. Attitudes toward marriage and family, changes in familial gender roles, increased prevalence of divorce, and single-parent households all separate the generations. We can also see marked change between parents' and children's attitudes about individualism and self-fulfillment versus family obligation and self-sacrifice (Tran 1998).

The most dramatic change affecting Chinese Americans has been their sheer increase in numbers over the past half century. The Chinese American population increased from 431,000 in 1970 to over 3.8 million (with an additional 215,000 Taiwanese) in 2010 (Hoeffel et al. 2012). Nearly 90 percent of Chinese

**Figure 3.4** U.S. Asians: Life Goals and Priorities, Value of Parenthood and Marriage

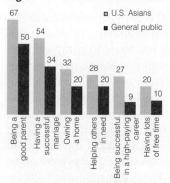

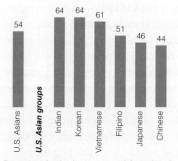

SOURCE: Pew Research Center (2012).

immigrants in 2010 were born in mainland China (McCabe 2012). Because of the large numbers of new immigrants, it is important to distinguish between U.S.-born and foreign-born Chinese Americans; less research is available concerning the latter. Contemporary U.S.-born Chinese families continue to emphasize familism, although filial piety and strict obedience to parental authority have become less strong. Chinese Americans tend to be better educated, have higher incomes, and have lower rates of unemployment than the general population. Their sexual values and attitudes toward gender roles tend to be more conservative. Chinese American women are expected to be employed and to contribute to the household income. More than 2.8 million people 5 and older speak Chinese at home, making Chinese the second most widely spoken non-English language in the United States after Spanish. Korean, Vietnamese, and Tagalog were each spoken by over a million people (U.S. Census Bureau, 2012a).

## Native American Families

Some 5.2 million Americans identify themselves as being of native descent, that is, as American Indian or Alaska Native, either alone or in combination with another ethnic group. According to the U.S. Census, just under half (49 percent) of this population are American Indian/ Alaska Native only, and 51 percent were American Indian or Alaska Native in combination with one or more other groups (U.S. Census Bureau, "American Indian and Alaska Native Heritage Month: November 2014"). Cumulatively, this population represented roughly 2 percent of the 2013 U.S. population. It is also a population that is expected to increase to more than 11 million by the year 2060.

Those who continue to be deeply involved with their own traditional culture give themselves a tribal identity, such as Diné (Navajo), Lakota, or Cherokee, rather than a more general identity as Indians or Native Americans (Kawamoto and Cheshire 1997). There are 566 federally recognized Indian tribes, the largest of which include the Cherokee, Navajo, Blackfeet, Latin American Indian, Choctaw, Sioux, and Chippewa. If looking at tribal groupings with the largest number of people identifying with it in combination with multiple other races, the largest, by far, is Cherokee, with a population of more than 819,000 (based on 2010 decennial census data). The Navajo is the next largest,

numbering over 330,000 (Norris, Vines, and Hoeffel 2012). The Navajo are also the tribal group with the largest population of individuals identifying with it and no other group. Those who are more acculturated, such as urban dwellers, tend to give themselves an ethnic identity as Native Americans or Indians. Although there has been an increase in the numbers of Americans identifying themselves as American Indian, the proportion reporting a tribal identity has been decreasing. This is truer of multiracial than of single-race American Indians (Cohn 2015). Economic indicators reveal that the American Indian/Alaska Native population is a relatively disadvantaged group. They are more likely to be unemployed, to be below poverty, and have substantially lower median household and family incomes. The 2013 median household income for American Indian and Alaska Native households was $36,252, well below the $52,176 for U.S. households overall.

In 2012, two-thirds of American Indian and Alaska Native households were family households, with an average size of 3.7 people. In comparison, families overall were smaller, with an average of 3.25 people. A third of Native American/Alaska Native families contained children under 18. Decennial census data indicated that, in 2010, 40 percent of American Indian and Alaska Native families were headed by married couples. An additional 21 percent was headed by single women, and 8.9 percent by single men. According to 2012 American Community Survey data, 37.4 percent of American Indian and Alaska Native males and 35.5 percent of females were married; 45.9 percent of males and 37.9 percent of females had never married (http://libguides.cmich.edu/c .php?g=104348&p=677782).

Other social, economic, and familial characteristics include the following:

- Based on data from the 2013 American Community Survey, the American Indian and Alaska Native populations were less likely to complete high school and college than was the general population. Whereas 86 percent of the general population age 25 and older had completed high school, 82 percent of the American Indian and Alaska Native populations had. Of the American Indian and Alaska Native populations, 17.6 percent had completed college, compared to 29.1 percent of the general population. (U.S. Census Bureau Facts for Features: American Indian and Alaska Native Heritage Month: November 2014)

- Based on 2013 data supplied by the Bureau of Labor Statistics, American Indians and Alaska Natives had the lowest labor force participation (59.2 percent), the lowest proportion of the population employed (51.6 percent), and an unemployment rate (12.8 percent) just under that of the 2013 rate for blacks (13.1 percent) (U.S. Bureau of Labor Statistics, August 2014a).

- Of the 45,991 births to American Indian or Alaska Native mothers, two-thirds were to unmarried mothers, second only to the rate among blacks, but exceeding the rate among whites, Hispanics, and Asian Americans. American Indian or Alaska Native mothers had the youngest average age at first birth (22.9) and the highest rate of teenage childbirth (12.6 percent).

Although there is considerable variation among different tribal groups and hence no single type of American Indian or Alaska Native family, three aspects of Native American families are important: First, extended families are significant. These extended families may be different from what the larger society regards as an extended family. They often revolve around complex kinship networks based on clan membership rather than birth, marriage, or adoption. Concepts of kin relationships may also differ. A child's "grandmother" may be an aunt or great-aunt in a European-based conceptualization of kin (Yellowbird and Snipp 1994).

Second, increasingly large numbers of Native Americans marry non–Native Americans. Among married Native Americans, more than half have non-Indian spouses. A Pew Research Center survey of multiracial Americans found the two largest multiracial groups to be white/American Indian, representing 50 percent of multiracial adults, and black/American Indian, representing another 12 percent (Cohn 2015). United States Census data also found the largest multiracial group to be non-Hispanic white and American Indian, though the percentage reporting that combination was much smaller. This may result from the different ways the U.S. Census and the Pew Research Center define and count the multiracial population. These different methodologies will be addressed in the section on multiracial families.

With such high rates of intermarriage, a key question is whether Native Americans can sustain their ethnic identity. Writing more than two decades ago, Michael Yellowbird and Matthew Snipp (1994, 236) wondered if "Indians, through their spousal choices,

may accomplish what disease, Western civilization, and decades of federal Indian policy failed to achieve" (p. 236). In a Pew Research Center survey, multiracial American Indians are said to have few connections with the Native American population; only a fifth (22 percent) describe themselves as having a lot in common with American Indians, whereas more than 60 percent say they have a lot in common with whites. Just under a fifth (19 percent) report having a lot of contact with their American Indian relatives.

Third, family characteristics are affected by the economic status of American Indians and Alaska Natives. Given the higher levels of unemployment and poverty and lower overall earnings and educational attainment, social class may once again be confounding our attempt to look at patterns of family living.

## European Ethnic Families

The sense of ethnicity among Americans of European descent varies considerably. Much of this variation is generational, and as among immigrants and their children, ethnic identification may be much greater than it is among subsequent generations. In fact, across and despite ethnic differences, many families experience similar pressures toward and processes of assimilation and acculturation. Historically, members of European ethnic groups sought to assimilate—to adopt the attitudes, beliefs, and values of the dominant culture, especially because assimilation seemed necessary to achieve upward mobility. Most white ethnic groups have assimilated to a considerable degree—they have learned English, moved from their ethnic neighborhoods, and married outside their group, but many continue to be bound emotionally to their ethnic roots. These roots are psychologically important, giving them a sense of community and a shared history. This common culture is manifested in shared rituals, feast days, and saint's days, such as St. Patrick's Day.

Except for some West Coast enclaves, such as Little Italy in San Francisco, white ethnicity is strongest in the East and Midwest. The Irish neighborhoods of Boston, the Polish areas of Chicago, and the Jewish sections of Brooklyn, for example, have strong ethnic identities. Common languages and dialects are spoken in the homes, stores, and parks. Traditional holidays are celebrated; the foods are prepared from recipes passed down through generations. Elders speak of the old country and their villages—even if it was their parents or grandparents who immigrated.

As is true of some non-European ethnic groups, as children grow up and move from their neighborhoods, their ethnic identity often becomes weaker in terms of language and marriage to others within their group—but they may retain some elements of ethnic pride. Their ethnicity is what Herbert Gans (1979) calls *symbolic ethnicity*—an ethnic identity that's used only when the individual chooses. Symbolic ethnicity has little effect on day-to-day life. It is not linked to neighborhoods, accents, the use of a foreign language, or working life. Others cannot easily identify the person's ethnicity; he or she "looks" American. Nevertheless, for many Americans, ethnicity has emotional significance. A person is Irish, Jewish, Italian, or German, for example—not only an American.

European ethnic groups differ from one another in many ways. However, a major study of contemporary American ethnic groups (Lieberson and Waters 1988) found that European ethnic groups are more similar to one another than they are to African Americans, Latinos, Asian Americans, and Native Americans. The researchers concluded that a European–non-European distinction remains a central division in our society. There are several reasons for this. First, most European ethnic groups no longer have **minority status**—that is, unequal access to economic and political power. Some scholars suggest that what separates ethnic groups into distinctive lifestyles is their social placement. As groups become more similar in their access to opportunities, their family lifestyles may "converge" toward a common pattern, one that includes smaller families, increased divorce, less interdependent ties with extended families, and less male dominance (Aponte, Beal, and Jiles 1999). Second, because most European ethnic groups are not physically distinguishable from other white Americans, they are not discriminated against racially.

## Multiracial Families

The population of multiracial families is diverse and growing. In fact, according to a Pew Research Center report, it is growing at three times the general population's rate of growth (Munguia 2015). The Pew estimates include both those who self-identified as multiracial *as well as* those who said they had at least one parent or grandparent of a different race, even if the latter did not label themselves multiracial. Using this methodology, just under 7 percent of U.S. adults could be considered multiracial. Only a little more than a fifth (22 percent) of those counted as multiracial would apply such a label to themselves (Munguia 2015).

The multiracial population is made up of many different combinations, the four largest being white *and* black, white *and* some other race, white *and* Asian, and white *and* American Indian or Alaskan Native (Jones and Bullock 2013). Each of these combinations represents over a million people, including 1.8 million identifying as white *and* black, 1.7 million as white *and* some other race, 1.6 million as white *and* Asian, and 1.4 million as white *and* American Indian or Alaskan Native. Cumulatively, these four groups represent just under three-quarters of those who identified themselves as being of two or more races. Twenty percent of the multiracial population identified themselves as white *and* black, 19 percent as white *and* some other race, 18 percent as white *and* Asian, and 16 percent as white *and* American Indian or Alaskan Native. Three of the four saw considerable increases in their size, with only the white *and* some other race category diminishing in size between 2000 and 2010. The white *and* black population experienced the biggest increase, growing by 134 percent or more than a million people (Jones and Bullock 2013).

Unsurprisingly, the increases in the multiracial population are tied to increases in racial and ethnic intermarriage. Approximately 15 percent of all new marriages in the United States in 2010 were between spouses of different racial or ethnic backgrounds, more than double the proportion in 1980, when it was less than 7 percent (Wang 2012). According to the Pew Research Center's analysis of census data and additional nationwide telephone surveys, in 2010, 28 percent of Asians, 26 percent of Hispanics, 17 percent of blacks, and 9 percent of whites married out of their race or ethnic group (Wang 2012). Along with the increases in intermarriage, there has been an increase in the wider public's acceptance of marrying outside of one's race or ethnic group. Whereas 65 percent of Americans described marriage between people of different races as either "unacceptable" (28 percent)

**False** **10.** European ethnic groups are more similar to each other than they are to African Americans, Latinos, Asian Americans, or Native Americans.

or as "acceptable for others but not for themselves" (37 percent), just under two-thirds of Americans now say that "it would be fine with them" if a member of their own family married outside their own racial or ethnic group (Wang 2012). Such an increase in public acceptance likely contributes to and follows from the increases in interracial and interethnic marriages. Further discussion of interracial marriage is contained in Chapter 8.

Just as the multiracial population contains much diversity in the combinations of races that comprise their families, so, too, do the experiences of multiracial individuals and families differ considerably, making it impossible to identify a single shared experience of being multiracial (Munguia 2015). For example, although both white–Asian and black–American Indian individuals are multiracial, the former are among the most likely to consider themselves multiracial whereas the latter were among the least (Hunguia 2015). When multiracial individuals who don't identify as such are asked why they don't identify as multiracial, their answers include because they look like one race, were raised as one race, or closely identify with a single race. Also, never having known the ancestor who was a different race lessens the likelihood of identifying oneself as multiracial. Among multiracial individuals, racial identity may change over the course of their lifetimes, and about a fifth of mixed-race adult respondents acknowledge having been pressured by family (11 percent), friends (9 percent), or "society in general" (15 percent) to adapt a single race identity (Pew Research Center 2013).

The issue of racial socialization of children can become more complicated for multiracial families, simply because parents do not share the same racial backgrounds and experiences (Rollins and Hunter 2013; Schlabach 2013). Families may consist of a parent whose own socialization and life experiences prepared them for more privileged status and a parent whose experiences have been more marginalized. Black parents in multiracial families have been found more likely to try to prepare their children for potential experiences of racism or discrimination, whereas white parents have been more likely to try to teach about racial equality and may teach their biracial children that the best way to cope is by ignoring racist comments. Research on 73 biological mothers of multiracial children revealed that mothers used one of three approaches to the racial socialization process: *promotive*, *protective*, or *passive* (Rollins and Hunter 2013). *Promotive socialization* (used by 37 percent of sample mothers) attempts to strengthen one's child's sense of self, self-confidence in one's abilities, and cultural traditions and values. It "encourages a color-blind worldview" and does not directly teach or prepare one's child for possible discrimination. *Protective socialization* (used by 27 percent of sample mothers) explicitly prepares children for potential discriminatory experiences, and also pushes one's child to stand up for one's rights, exercise self-control, and identify with one's sense of self as part of a racial group. *Passive racial socialization* involves doing nothing to prepare one's child for the possibility of discrimination. These approaches did not vary by mother's race, though their use did vary by their children's biracial heritage (Rollins and Hunter 2013). Parents of black/white children and black/other minority children were more likely to engage in *promotive socialization*. Almost three-quarters of mothers of white/other minority children engaged in *passive racial socialization*, whereas few mothers with a black/white child did so (Rollins and Hunter 2013). Sociologist Sarah Schlabach's study of multiracial adolescent well-being shows that the gender of the minority parent may make a difference in the well-being of children. She found that for social and emotional well-being, having a minority mother is disadvantageous. Multiracial children with a white father and minority mother were found to have the worst outcomes when compared to monoracial counterparts (Schlabach 2013).

With regard to both friends and family, multiracial adults may feel as though they "straddle two (or more) worlds," and their interactions with family and friends may show greater closeness and acceptance in one world than in the other(s). Black/white biracial adults feel as though they have more in common with, feel closer to, and more accepted by their black relatives and are much more likely to have friendship networks that are predominantly black than predominantly white. Biracial adults who are white and Asian are more likely to have developed and sustain closer ties to whites than to Asians. They are also almost twice as likely to say they have a lot in common with white people (60 percent) as with Asian people (33 percent). Regarding closeness and contact with family, more biracial adults who are black and white have had a lot of contact with their black family members than with their white family members. Almost 70 percent say that in their lifetimes they have had a lot of contact with black relatives. Conversely, a quarter of them report having had little contact, and 40 percent no contact with their white relatives (Pew Research Center 2013). Black/white multiracial adults are more than twice as likely as multiracial respondents in general (21 percent vs. 9 percent) to say they have been treated badly by a relative because of their mixed racial background (Pew Research Center 2013).

Related to the previously mentioned point, multiracial individuals may be more likely to experience intrafamilial racial **microaggressions**. The concept of microaggressions refers to "brief, commonplace daily verbal, behavioral, or environmental indignities that communicate hostile, derogatory or negative slights and insults toward people of color" (Nadal et al. 2013). Although monoracial blacks, Hispanics, or Asians and multiracial people report experiencing microaggressions, multiracial Americans are more vulnerable than single-race minorities to encountering such slights and insults *within their families* from their monoracial relatives (Nadal et al. 2013). The kinds of within-family microaggressions reported by multiracial adults match those found in studies of women and of lesbian, gay, and bisexual people (Nadal et al. 2013). The within-family microaggressions lead authors, psychologist Kevin Nadal, and colleagues to suggest families prepare their children and promote healthy multiracial identities via racial socialization (Nadal et al. 2013).

This chapter has covered much ground. As we have now seen, American families are diverse in a host of ways. They vary across time and, within any given period, between racial, ethnic, and socioeconomic groups. Family diversity is reflected throughout subsequent, more specialized chapters as relevant variations by race, class, or ethnicity are discussed. Thus, our goal of understanding American families will be made more complete and representative.

Acknowledging the diversity that exists across families has personal consequences as well. It ought to make us a bit more cautious in generalizing from our particular set of family experiences to what others "must also experience." In addition, in noting how historical, economic, and cultural factors shape our families, we link our personal experiences to broader societal forces. In that way, we are better able to apply "sociological imaginations" to family experiences, identifying how our private and personal family worlds are largely products of when, where, and how we live (Mills 1959). Simply put, if we come of age during a period of great economic upheaval, we may put off marrying, bearing children, or divorcing because of the opportunities and constraints we face. Similarly, the kinds of family experiences we are able to have are limited or enhanced by the economic resources at our disposal, regardless of what we might otherwise choose to do.

Despite the extent to which the factors discussed in this and the next chapter may limit your opportunities or narrow your range of choices, remember that you do and will make choices about what kind of family you wish to create. You decide whether to marry, whether to bear children, how to rear your children, whether to stay married, and so on. A major goal of this book is to equip you with a foundation of accurate information about family issues from which you can make sound choices more effectively.

## Summary

- In the early years of colonization, there were 2 million Native Americans in what is now called the United States. Many families were *patrilineal;* rights and property flowed from the father. Other tribal groups were *matrilineal.* Most families were small.

- In colonial America, marriages were arranged. Marriage was an economic institution, and the marriage relationship was *patriarchal.*

- Colonial American family life was more public and subject to more scrutiny from members of one's community.

- In the early 17th century, African American families in the United States continued the African tradition

  that emphasized kin relations. Most slaves lived in two-parent families that valued marital stability.

- In the 19th century, industrialization revolutionized the family's structure and transformed gender roles, childhood was sentimentalized, and adolescence was invented. Marriage came to be based more on emotional bonds. Enslaved families were broken up by slaveholders, and marriage between slaves was not legally recognized. African American families formed solid bonds nevertheless.

- Beginning in the 20th century, *companionate marriage* became the marital ideal. During the Great Depression and the two world wars, women and men were faced with challenges to their traditional gender roles and identities. The Depression also initiated an enlarged role for the government in protecting poor families.

- Family life in the 1950s was an exception to more general trends of rising divorce and nontraditional gender roles. Changes that followed in subsequent decades and into the 21st century include increases in cohabitation, unmarried motherhood, single-parent families, and blended families, along with changes in the numbers of people who enter or forego marriage, and those who reenter marriage following divorce.

- The movement for same-sex marriage drew attention to the legal rights and protection marriage provides, and culminated in the June 2015 United States Supreme Court decision, in *Obergefell v. Hodges* to legalize same-sex marriage.

- Family life varies greatly across social class, including marital relationships, parent–child relationships, and relationships with extended kin, which all show differences across and between social classes.

- Family experience differs between African Americans, Hispanics, Asians, and Native Americans. *Socioeconomic status* is an important element in understanding diversity between and within ethnic and racial groups.

- The increase in interracial marriage has led to increasing numbers of multiracial families. Although their experiences vary depending on the specific combination of races involved, multiracial families must deal with a greater complexity of racial socialization and with the possibility of experiencing racial microaggressions within their own families.

- For many Americans of European descent, ethnicity is symbolic and has little effect on day-to-day life. Most members of European ethnic groups are physically indistinguishable from other white Americans and no longer have minority status.

## Key Terms

adolescence  67
assisted reproductive technologies  79
blended families  77
bundling  66
cohabitation  76
companionate marriage  72
culture of poverty  95
division of labor  68
ethnic group  95
extended households  100
familism  80
feminization of poverty  86
fictive kin ties  89
life chances  81
matrilineal  65
middle class  83
minority group  95
minority status  108
patriarchy  66
patrilineal  65
phenotype  95
race  94
racial group  94
racial socialization  100
social class  81
social mobility  89
socioeconomic status  81
structural mobility  92
two-person career  86
upper-middle class  83
working class  83

# 4

# Gender and Family

## What Do You Think?

*Are the following statements True or False? You may be surprised by the answers as you read through this chapter.*

T F **1.** The only universal feature of gender is that all societies sort people into only two categories.

T F **2.** Gendered roles are unaffected by ethnicity.

T F **3.** Gender differences reflect the instinctive nature of males and females.

T F **4.** Parents are not always aware that they treat their sons and daughters differently.

T F **5.** Both boys and girls suffer from gender-related problems in school.

T F **6.** During adolescence, the influence of peers on gender greatly diminishes.

T F **7.** Research shows it is possible for women and men to establish work or family roles that are counter to their socialization.

T F **8.** Compared with traditional roles, contemporary male gender roles place more emphasis on the expectation that men will be actively involved with their children.

T F **9.** Among African American couples, the gender division of labor tends to involve more sharing of wage earning and domestic work.

T F **10.** There are heterosexual couples whose relationships are not based on gender.

Tetra Images/Alamy Stock Photo

112

## Chapter Outline

I f you travel south on Brice Road out of the central Ohio town of Reynoldsburg, you'll find that the road name changes, and for the next sizable stretch you will be driving down Gender Road. In this area south of Columbus, Ohio, Gender Road will take you to the town of Canal Winchester, where it ends. By the time you will have gone from one endpoint to the next, you will have been on Gender Road for more than 7 miles. From birth on, as each of us travels through life, including into or out of family roles and relationships, the road we take—how it twists or turns—and the places it takes us is also a gender road. Measured in years rather than miles, it lasts as long as we live.

Think about your own life. Have you ever stopped to consider how similar or different your life might be if you hadn't been born into and raised in the gender you presently occupy? Would you be the kind of person you are? Would you have the same roles and relationships within your families or have the same friends? Would your goals be the same as they are now? Would you envision and later experience the same familial future? Asking ourselves such questions reminds us that gender places us on a path and, in so doing, helps shape much of what we do, who we are, what we expect, and what happens to us. It is not the only such influence, but avoiding it may be nearly impossible.

In this chapter, we examine interconnections between gender and our experiences in and of families. In truth, one cannot fully understand families without taking gender into account. Like social class and race, it is one of the major influences over the kind of intimate and family relationships we experience. We begin the chapter by discussing the concept of gender, what it is, what it isn't, and how our understanding of it continues to change. As we proceed, we consider some gender and socialization theories to illustrate how much our families influence how we learn to act masculine and feminine. Along with the family, we briefly highlight other sources of gender socialization, how our peers, the various media, and our educational and religious experiences all contribute to our ideas

about gender. We then briefly describe how gender influences some areas of family experience, a theme that we will then carry throughout the remaining chapters. Finally, we discuss efforts to change gender roles on both an individual and a collective level.

## What Gender Is, What Gender Isn't

Each of us occupies many different social statuses and attempts to perform multiple roles associated with those statuses. In other words, each of us is many things to many people—sons or daughters, sisters or brothers, friends and/or lovers, spouses or partners, mothers or fathers, as well as students, workers, perhaps athletes and/or actors, and so on. Of the many different statuses we hold, few have the potential importance or influence that gender has in shaping the lives we come to lead. For many, it will be the central status. For others, it may share center stage with other statuses, such as your race or social class, but it will likely be on that stage somewhere. For our purposes, it is especially important to consider as gender is one of the most significant factors in differentiating our family experiences. Before beginning to more closely examine some of the interconnections between gender and family experience, let's examine the concept itself (Figure 4.1).

### What Gender Is

Gender is one of the most basic components of who we are, how we see ourselves and how others see and treat us. Each of us possesses a **gender identity**, a sense of ourselves with regard to socially established categories such as male or female. *Most* individuals identify themselves as *either* male or female, *usually*—but not always—in accord with their biological sex characteristics. We say "most," "either," and "usually," because not everyone identifies within a binary construction of gender (i.e., in which you are either male or female), nor does gender identity always follow

## Figure 4.1  What Gender Is/What Gender Isn't

**Gender *is*:**

- One of the most basic components of who we are and how we see ourselves.
- A social status with differential access to society's rewards, opportunities, and obstacles.
- A means for classifying people and then expecting them to act accordingly.
- A phenomenon that exists on a psychological, social, economic, and political level.
- A source of both difference and inequality.
- A culturally variable, historically dynamic phenomenon.
- A socially acquired, socially reproduced status.
- A performance.
- A set of differences that are exaggerated in importance.

**Gender *isn't*:**

- The same as sex.
- Synonymous with women.
- The same as sexuality/sexual orientation.
- Always dichotomized.
- "Polarized," despite the way it is culturally constructed in the United States.
- Simply and easily categorized into singular versions of masculinity and femininity.
- Inevitable. Socialization may be inescapable but no gender outcome or arrangement is inevitable.

from and match one's biological sex, the physical characteristics—such as chromosomes, hormones, anatomical structure, and reproductive organs—associated with being female or male.

Typically today, the term *transgender* is used to refer to any and all individuals who develop gender identities that differ from their biological sex characteristics. Increasingly, the term **cisgender** is used to refer to those whose gender identities match their biological sex. Some transgender individuals may identify as *neither* male nor female, others *as both* male and female, and still others as *alternating from* male to female, feeling, living, and acting sometimes as one and at other times as the other gender. Also, some transgender individuals do accept a binary view of gender, develop a stable gender identity, but identify as male despite being biologically female, or as female despite being biologically male. They just feel and believe that they were "born into the wrong body."

Among those who believe that they were meant to be the opposite sex, some may opt for reconstructive surgery in an attempt to bring their biology more into line with the identity they have developed. Others reject such surgeries and simply live as the gender with

which they identify. Such was the case with transgender male, Thomas Beatie, formerly Tracy, who in 2008 was hailed in the press as the "first man to give birth." As Beatie explained, "for as long as I can remember, and certainly before I knew what this meant, I wanted to live my life as a man" (Beatie 2008, 6). With the aid of testosterone to stimulate muscle development and the growth of facial hair, exercise, and surgery to remove his breasts, Tracy became Thomas. But by choosing not to have his female reproductive organs removed, Thomas retained the ability to conceive, carry, and—in 2008—give birth to daughter, Susan. Though he identified and lived as a married male, Thomas proceeded to give birth to two more children, Austin in 2009 and Jensen in 2010.

The transgender category includes those who, like Beatie, alter their social but not necessarily all their physical characteristics, as well as those who undergo sex change procedures to attempt to bring their physical characteristics fully in line with their identities. In light of increasing awareness and attention paid to and by transgender individuals, the **gender binary**—a dichotomous understanding of gender—is being replaced by a more nuanced understanding of a wider range of identities and ways of expressing those identities, what frequently has come to be known as a **gender spectrum**. As evidence of this, consider the much wider range of terms from which one can choose to designate one's gender. This is discussed in the Public Policies, Private Lives Box, *What Should We Call Each Other?*

We are in the midst of a rather remarkable transformation in how we think, talk about, and act regarding gender. Transgender individuals, though not very numerous, have become quite visible, especially in popular culture (see Popular Culture Box: *Television's Transgender Faces*). Chaz Bono, Laverne Cox, costar of the television series, *Orange Is the New Black*, Caitlyn Jenner, and transgender teen star of the TLC series, *I am Jazz*, Jazz Jennings, have all become public faces of transgender reality, and are raising both public awareness and acceptance of the transgender population.

Estimates of the size of the transgender population are hard to come by. There are no national population-based figures available to use. Based on research by the Pew Research Center, 4.3 percent of its nearly 1,200 LGBT (lesbian, gay, bisexual, transgender) respondents identified as transgender. Demographer Gary Gates estimates that this would translate to 0.15 percent of the adult population. Gates notes that a Massachusetts-based survey found 0.5 percent of adults identifying as transgender. Attempting to apply these estimates to the wider U.S. population would translate to between 360,000 and 1.2 million adults who identify themselves as transgender (Gates 2015b).

The key aspect of gender for those who identify themselves as transgender is that of gender identity. But for transgender and cisgender individuals alike, gender identity encompasses perhaps the most personal experience we have of being gendered. At the same time that gender defines how we see ourselves, gender is also a means for classifying others and being so classified ourselves. Expectations for how we are to behave are shaped by those classifications. Such expectations comprise what is usually meant by the concept of **gender roles**, a term used to refer to expectations attached to being identified as male or female. Although a familiar concept, it is not especially useful to talk of *a male gender role* or *a female gender role*. Instead, it may be preferable to talk about **gendered roles**, because rather than a singular female role or male role, multiple roles can be assigned to and performed by those identifying or being identified as female more often or instead of by people identifying or identified as male and vice versa.

As much of this chapter will illustrate, gender is a socially acquired, socially reproduced status. In thinking about how gender is acquired and reproduced, how we come to learn the expectations in those gendered roles, we'll need to consider **gender socialization**, the process by which we come to learn what behaviors, demeanor, and temperament are expected of us, by virtue of our assigned gender. In later sections of this chapter, we will review some of the sources (or agents) of gender socialization and the messages they convey. But socialization need not be assumed to explain everything about gender that is social in origin; social situations also matter. So, although some gendered behavior is an attempt to perform within the expectations we come to learn, some is more a product of the different structural positions and social situations in which males and females find themselves. If, for example, one is responsible for the care of infants, from their birth through early childhood, one will develop a greater understanding of and sensitivity to the needs of infants and young children. If women more than men have been so situated, women will come to possess and display a greater awareness of and sensitivity to how to care for young children than will men, at least in part because of the situational circumstances and demands they have had to meet.

Gender is one of the most influential social statuses we occupy, in that it determines, in part, our access to and receipt of sought-after societal rewards. The range of opportunities we have available to us may be broadened or narrowed by the gender with which we are identified. In that way, gender, like our earlier discussions of race and social class, is a source of inequality as well as of difference, and **gender stratification (or inequality)** can be found to a greater or lesser degree in all but the least complex societies, such as among foraging people. Like social class, gender affects our life chances.

Most societies, both past and present, can be considered patriarchies, or societies dominated by males, where economic and political institutions are controlled by men and in which males tend to exercise greater power in interpersonal relationships, Although there are also societies identified as more **egalitarian**, in which females and males possess similar amounts of power and where neither dominates the economic or political institutions and systematically restricts the other, true **matriarchies**—meaning societies where women rule over men and where men are denied the right to political office or are excluded from religious rituals—have not been documented.

Globally, male dominance is commonplace. A variety of practices illustrate extreme male dominance and severe female subordination. Some of these have quite profound, even life-and-death, consequences. These include:

- *Sex selection and selective abortion.* As explained in The World Fact Book (CIA 2015), a country's "sex ratio at birth" is used as an indicator of "the existence of certain kinds of sex discrimination. . . . For instance, high sex ratios at birth in some Asian countries are now attributed to sex-selective abortion and infanticide due to a strong preference for sons." This has been especially true in India and China, where there continue to be gaps in the sex ratio of the number of girls to the number of boys. In India, the gaps are partly the products of the use of ultrasound technology to determine the sex of one's offspring in utero, followed for many by the selective abortion of female fetuses. Decried as a "national shame" by the prime minister of India, the practice there led to a substantial decline in the sex ratio of the number of girls for every 1,000 boys. Recent sex ratio data shows that for every 1,000 males, 943 females were born in India in 2015 (indiaonlinepages.com). In 2011, India had 7 million

**What Should We Call Each Other?**

There has been a variety of consequences accompanying the many changes in our understanding and acceptance of those who question and challenge prevailing sexual and gender categories. With wider recognition of those who are *transgender,* many previously taken-for-granted practices and policies have had to be reconsidered and changed. Organizations and institutions, including especially educational institutions, have had to alter their housing policies, their facilities, and their official documents. One such change is in how we speak of ourselves, what labels we choose and use, and what choices are made available to us. Official governmental and institutional policies have had to be adapted, as increasing numbers of people wish not to be assigned to categories that fail to capture the ways in which they see and understand themselves. Consider the following examples.

In 2014, Facebook offered its users ways to identify themselves outside of the gender binary of male or female. A few years earlier, Google had made it possible for subscribers to Google+ to select their genders from among male, female, or other. As reported by *New York Times* journalist Jennifer Conlin, this was especially appealing and important to young people who were among "a growing number of high school and college students who are questioning the gender roles society assigns individuals simply because they have been born male or female" (Conlin 2011). Though not Google's intent, the "other" option appealed to those who advocated giving people their choice of "preferred gender pronouns," which could include using "they," "them," and "theirs" instead of either he or she, him or her, his or hers. For those for whom the plural dictionary pronouns don't seem to work, "there are numerous made-up ones now in use, including "ze," "hir," and "hirs," which are intended to be inclusive of both genders (Conlin 2011).

Facebook took the issue of gender identities a good deal further, offering users as many as 71 different gender options (Williams 2014). Of course, this was not meant to suggest that there are more than 70 genders, as understood by Facebook, but that there were multiple ways to describe oneself outside of gender binary categories. Among the Facebook-provided options, one could find the following: androgyne, asexual, bigender, female-to-male transman, gender fluid, gender neutral, genderqueer, hermaphrodite, intersex person, male-to-female transgender woman, nonbinary, pangender, polygender, transgender, transman, transwoman, two-spirit person, Cis Female or Cis Male, Cis Man or Cis Woman, transsexual man, transsexual woman, or transsexual person (Dewey 2014; Williams 2014). Both the *Washington Post* in 2014 (Dewey 2014) and the *New York Times* in 2015 (Scelfo 2015b), published lists and definitions of some of the terms now more frequently or centrally in use. The list in the *Washington Post* had 21 of the identity options. The *Times,* calling its a "gender-neutral glossary," provided a much shorter list of terms but included along with their definitions, the terms *transgender, gender nonconforming, genderqueer, pangender, cisgender,* and *trans* (Scelfo 2015b). Although most of the terms are intended to be used to refer to people whose identities do not match the gender category into which they had been assigned at birth, in addition, the term *cisgender* has come to be used as a way of referring to those who are "non-trans," or for whom there is "a match between the gender they were assigned at birth, their bodies, and their personal identity" (Schilt and Westbrook 2009).

As to official documents or institutional policies, in 2010, Australia granted recognition to Norrie May-Welby, a Scottish-born Australian citizen, as the first legally genderless person. May-Welby was born male, underwent gender reassignment surgery in 1989, but ultimately decided that being female felt no more right than being male. Having experienced unhappiness both as a male and a female, May-Welby decided to become neither. "The concepts of man or woman don't fit me. The simplest solution is not to have any sex identification" (Fugate 2010). In 2010, May-Welby was issued a revised birth certificate, indicating "no specific sex" rather than male or female. The New South Wales Registry of Births, Deaths, and Marriages eventually revoked its earlier approval of May-Welby's application, and thus set in motion a prolonged legal process. In 2014, the High Court of Australia ruled that New South Wales and the Australian Capital Territory must recognize a third gender on official documents. The ruling was seen as likely to apply to most of the country, as five of the seven Australian states and territories have comparable language in their legislation (Baird 2014). Along with Australia, at least six other countries—Nepal, India, Pakistan, Bangladesh, Germany, and New Zealand—also offer an alternative option to the male/female binary on some legal documents. Nepal is said to have been the first to do so (Basu 2015), and Germany the first European country to do so (Pasquesoone 2014). Journalist Jacinta Nandi notes that Germany allows both for individuals to change their legal gender if they can show medical diagnosis of transsexualism and have been living as their preferred sex for three or more years. German parents of intersex children are also given the option of checking a third box, X, instead of either M or F (Nandi 2013).

*Intersexuality,* making up a variety of different conditions, is estimated to occur in between one in 1,500 to one in 2,000 births (Intersex Society of America, www.isna.org/faq/frequency). At birth, intersex infants cannot be easily labeled as females or males. As described by the Intersex Society of North America:

> a girl may be born with a noticeably large clitoris, or lacking a vaginal opening, or a boy may be born with a notably small penis, or with a scrotum that is divided so that it has formed more like labia. Or a person may be born with mosaic genetics, so that some of her cells have XX chromosomes and some of them have XY. (www.isna.org/faq/what_is_intersex)

Within the United States, educational institutions have grappled with policies to address the needs and preferences of their transgender students. For example, after a decade of action taken by students, staff, and faculty, and six months spent adjusting university software, the University of Vermont has adopted policy changes that allow transgender students "to select their own identity—a new first name, regardless of whether they've legally changed it, as well as a chosen pronoun—and records these details in the campus-wide information system so that professors have the correct terminology at their fingertips" (Scelfo 2015a). Journalist Julie Scelfo mentions that about 100 other schools have contacted the University of Vermont, in an effort to devise a strategy to accommodate transgender students who "are laying claim to a degree of identity freedom nearly unimaginable when the first L.G.B.T student centers were established" (Scelfo 2015a).

Meanwhile, students are also taking action themselves, "organizing identity conferences and inventing new vocabularies, which include pronouns like 'ze' and 'xe,'

and pressing administrations to make changes that validate, in language, the existence of a gender outside the binary" (Scelfo 2015a).

In so doing, they are bringing to their local environments the same understanding about gender that Norrie May-Welby brought to attention in Australia: "Maybe people will understand there's more options than the binary," And, in so doing, they are raising the same question Norrie raises: "It's important for people to have equal rights in society. Why should people be left out because they're seen as not male or female? They should be recognised wherever they are and allowed to participate in society at an equal level" (Davidson 2014). ●

Some individuals identify themselves as of both genders or neither gender. Australian Norrie May-Welby is the first person to be officially and legally regarded as genderless.

Norrie May-Welby/AFP/Getty Images/Newscom

**Television's Transgender Faces**

The growing visibility and increasing acceptance of the transgender population can be seen in many ways, not the least of which is in the way the media, especially television, have covered and featured various transgender individuals. Caitlyn Jenner, Laverne Cox, Chaz Bono, and Jazz Jennings are four examples of transfemales or transmales who have been prominently featured and mostly positively presented. Likewise, Jill Soloway's critically acclaimed and award-winning Amazon television series, *Transparent*, starring Jeffrey Tambor, sensitively and often humorously depicts the transition of Mort, divorced father of three adult children, into Maura. Both the series and Tambor received critical praise and awards, including Golden Globe Awards for outstanding comedy or music series and for Best Actor in a comedy or musical series. *Transparent* also received awards from the Director's Guild, and Tambor was awarded the Critics Choice Award for Best Actor. As of the writing of this edition, *Transparent* has ten pending nominations for Emmy Awards, given for excellence in television. Though the show received some criticism for casting a cisgender male as Maura, the quality and poignancy of the series have been widely recognized. In preparation for a second season of *Transparent*, Soloway has added a transgender writer to the staff of writers.

What is striking about the individual examples noted here is their widely positive reception from the viewing public. Chaz Bono is an author, actor, advocate, and activist for social justice and LGBT rights. Born female to famous parents, singing duo Sonny and Cher, the 46-year-old Bono came out as a lesbian in 1995, and more than a dozen years later transitioned from Chastity into Chaz. Laverne Cox is an Emmy-nominated actress and one of the stars of the Netflix series, *Orange Is the New Black*, in which she plays Sophia Burset, a transfemale inmate in the show's Litchfield Penitentiary. As her website notes, Cox is the first transwoman of color to have a leading role on a mainstream scripted television show. Along with her acting, writing, and filmmaking, Cox has "travelled the country speaking about moving beyond gender expectations to live more authentically" (www.lavernecox .com/events/).

Jazz Jennings is a 14-year-old transfemale who has been in the spotlight since she was a preschooler. Diagnosed at the age of 3 with "gender identity disorder" (now known as gender dysphoria), Jennings was one of the youngest individuals to be so diagnosed. From age 6, when she and her parents Jeanette and Greg Jennings first appeared on a *20/20* segment with Barbara Walters, through return interviews with Walters as well as with Katie Couric, Rosie O'Donnell, and others over the years, she and her family have been unusually open to sharing her and their story of her transition from Jared to Jazz. In documentaries on the Oprah Winfrey Network (which also has shown Chaz Bono's *Being Chaz)*, and currently in the TLC series *I Am Jazz*, Jennings has given audiences a look inside the life of a transgender teen (and her family) and has spoken openly and poignantly about her relationships with classmates and peers, including her desire to date and her frustrations at her inability to as of yet succeed in that aspect of adolescence. She has served as a Human Rights Campaign Foundation youth ambassador and became a model in Clean & Clear's *See the Real Me* campaign (www.oprah.com/inspiration/ Jazz-Jennings-Transgender-Equality).

---

fewer girls than boys age 6 or younger, a gap that has increased from 6 million a decade earlier (E. Brown 2011). Similar practices in China have led to a comparable demographic imbalance (Hesketh, Lu, and Xing 2011). In China, about 118 boys are born for every 100 girls (Powledge 2015). As a consequence, the Chinese government is enacting some strict limits as to how and when sex-determination tests can be conducted and under what circumstances women can have legal abortions (IANS 2014).

- *Honor killings.* The United Nations estimates that as many as 5,000 women and girls are killed each year in what are known as "honor killings."

Honor killings reportedly occur most often in the Middle East and South Asia, but have also been reported in countries such as Brazil, Canada, Ecuador, Egypt, Italy, Sweden, Uganda, the United Kingdom, and the United States. In one well-publicized Canadian case, three young Afghan sisters, ages 19, 17, and 13, along with a fourth woman, were drowned by members of their family, their bodies later discovered in a submerged car. Mohammad Shafia, 58, his wife Tooba Yahya, 42, and their son Hamed, 21, were each found guilty of four counts of first-degree murder. According to prosecutors, the women were killed because they

Jennings has also written a book (*I am Jazz*) and started a foundation for transgender youth, The Transkids Purple Rainbow Foundation. According to its website, the foundation is "committed to the premise that Gender Dysphoria is something a child can't control and it is society that needs to change, not them. Families need to support their children and be encouraged to allow them to grow-up free of gender roles." The foundation "is committed to enhancing the future lives of TransKids by educating schools, peers, places of worship, the medical community, government bodies, and society in general, in an effort to seek fair and equal treatment and of all transyouth" (www .transkidspurplerainbow.org).

In 2014, Jazz Jennings was named one of *Time* magazine's 25 Most Influential Teens of 2014 (www.tlc .com/tv-shows/i-am-jazz/).

Caitlyn Jenner may be the most visible of these four transpeople. Formerly Bruce Jenner, a gold medal Olympian in the 1976 summer Olympics, long-time face on Wheaties cereal boxes, and one of the stars of the successful reality series, *Keeping Up with the Kardashians*, Jenner has been married three times and is the father of six kids from his three marriages. He is also the stepfather of four from his marriage to Kris Kardashian. Jenner's transition from Bruce to Caitlyn has been a greatly detailed and well-documented journey, which is now the focus of her new reality series, *I Am Cait*. Talking about her transition with journalist Diane Sawyer in April 2015, Jenner revealed, "For all intents and purposes, I am a woman. I was not genetically born that way . . . as of now I have all the male parts. As of now we're different, but we

still identify as female" (Hopkins 2015). Acknowledging a lifelong struggle with gender identity issues, Jenner expressed the hope that by being public in her personal transition, that perhaps she could change attitudes and perceptions people have of those who are transgender (Hopkins 2015). Although hers is an atypical profile for one who would be seen as a gender pioneer (wealthy, politically conservative, Republican), she aspires to have influence on the wider public. On July 15, 2015, Caitlyn Jenner was awarded the Arthur Ashe Courage Award by the sports network ESPN. At a film showing that told Jenner's story and her later speech, an audience that frequently teared up gave Caitlyn Jenner a standing ovation and the desired apparent acceptance.

These four featured individuals have each become important symbols in a movement toward greater transgender acceptance and support. They may be especially helpful in making transgender seem less exotic or unusual and in making transpeople seem just like everyone else. After all, as transgender author Jennifer Finney Boylan, author of the memoir, *She's Not There*, said, "We don't want anything other than our humanity." It is also likely though that Chaz Bono, Jazz Jennings, Caitlyn Jenner, and Laverne Cox have each endured criticism, occasional ridicule, and, at times, hostility or rejection. These, however, may be somewhat more benign when compared to the plight of many transwomen, transmen, and especially transkids, for whom acceptance, approval, and support often may be hard to come by, and for whom "hatred, discrimination, and violence" are potentially predictable parts of their journeys (Haberman 2015). ●

brought dishonor to the family by the way they dressed, who they dated, and with whom they socialized (Austen 2012).

- *Forced marriage of girls.* In Yemen, in 2008, 10-year-old Nojud Muhammed Ali went to a courthouse seeking a divorce from her husband, Faez Ali Thamer, age 30. Forced to marry at age 8, Nojud wanted freedom from her marriage, claiming that her husband beat and raped her. Pulled out of the second grade by her father and then married off to her husband, Nojud's case became an international news story. Although she is only one of thousands of young girls forced into such circumstances—a

2006 study estimated that 52 percent of rural Yemeni girls were married by age 15—she drew considerable acclaim for her successful determination to escape her marriage. The attention Nojud received is credited with inspiring others—including an 8-year-old Saudi girl married off by her father to a man in his 50s—to seek annulments or divorces (Ali 2010; BBC News 2009; Kristoff 2010).

- *Female circumcision and genital mutilation.* Undertaken largely to control female sexuality, the procedure "involves partial or total removal of the external female genitalia," provides no health benefits, and carries considerable risk for the girls and young women

In successfully divorcing her 30-year-old husband, 10-year-old Nojud Mohammed Ali became an inspiration to many young girls whose families had forced them to marry men much older.

Khaled Fazaa/AFP/Getty Images

Generally, the only limit on the jobs that women or men hold is social custom, not biology or individual ability. Even sex-segregated jobs such as firefighting can be performed by either gender.

Tom Carter/PhotoEdit

who undergo such operations. Yet the World Health Organization estimates that somewhere between 100 million and 140 million girls and young women worldwide have been "circumcised," including more than 90 million females age 10 and older in Africa, where, every year, approximately 3 million girls are at risk of undergoing these procedures (World Health Organization 2012). Female circumcision is not limited to developing nations; an estimated 100,000 women have undergone female circumcision in Britain, despite it being illegal and potentially fatal (Owen 2012). In 2010, the World Health Organization, along with other international organizations, published a "global strategy" to address female circumcision, and strongly recommended that health care professionals stop performing such procedures (World Health Organization 2012b). The Centers for Disease Control and Prevention estimate that in the United States, between 150,000 and 200,000 girls are "at risk of being forced to undergo cutting" (Roberts and Smith 2014).

In addition to examples such as these, one can find numerous other indicators of gender inequality in many parts of the world. Looking more broadly at the institutional realms of politics and economics allows us to recognize the most persistent elements of male dominance in the United States, where men continue to dominate political office, at the municipal, state, and especially national levels, and where women continue to lag behind men in such work-related matters as wages, mobility, authority, and autonomy.

### Politics

In U.S. politics, women have been and continue to be underrepresented. At the highest level of elected office, no woman has been elected president or vice president of the United States, though in the elections of 2008, 2012, and 2016, women have been serious contenders or been nominated for these positions. By the time you are reading this, the United States may have elected Hillary Clinton its first female president. Internationally, as of January 2015, 10 women were serving as heads of state and 14 as heads of government. As of 2015, women make up just about a fifth of the U.S. Senate and House of Representatives, hold less than 25 percent of statewide and state legislative office, and were mayors in only 12 of the largest U.S cities. Only 6 of the 50 U.S. governors are female; in fact, only 37 women have ever served as state governors in the United States, and 27 states have never had a female governor (www.cawp.rutgers.edu /history-women-governors).

## Economics

Economically, although women continue to make gains, men in the United States have enjoyed long-standing advantage in the workplace and the boardroom. Less than 5 percent of CEOs of Fortune 500 companies are female, and women hold 17 percent of the board positions of those companies (www .pewsocialtrends.org/2015/01/14/chapter-1-women -in-leadership/). In other economic measures, men tend to earn greater amounts of income, experience more upward mobility, exercise more authority and enjoy more autonomy in their jobs, and are given more prestige for the work they do. Given the inequality in wages, women are more likely than men to be poor, and individuals in female-headed households are especially prone to poverty (see Figure 4.2). Such gender inequalities are not the same across all racial and ethnic groups, as is evident from data on the wage gap, where, when compared to white males, white females fare better than either African American or Hispanic males (see Table 4.1).

## Families

Within male-dominated societies, families tend to be male-dominated as well. In many societies, men have the right to have more than one spouse, descent is

**TABLE 4.1** Median Usual Weekly Earnings, Full-Time Wage and Salary Workers, 2nd Quarter, 2015

|  | Male | Female | Female as % of Male |
|---|---|---|---|
| All races | $886 | $726 | 82% |
| White, non-Hispanic | $914 | $742 | 82% |
| African American | $696 | $615 | 88% |
| Asian American | $1,085 | $836 | 77% |
| Hispanic | $619 | $572 | 92% |

SOURCE: U.S. Department of Labor, Bureau of Labor Statistics, www.bls.gov /news.release/pdf/wkyeng.pdf, July 2015.

traced through males, property is passed from fathers to sons, and men exercise authority. In the United States, the division of responsibilities may be the most visible familial advantage that men hold. Women carry the vast bulk of responsibility for domestic work and child care. Even women who are employed full time come home to what sociologist Arlie Hochschild called a **second shift** of housework and child care at the end of their paid workday. Whether one uses absolute (i.e., numbers of hours per day or week) or relative measures (percent of total housework that women and men do), and whether one takes men's or women's estimates of how they divide housework, women do considerably more (Bianchi, Robinson, and Milkie 2006; Lee and Waite 2006). How much more varies some from study to study, but equal sharing of housework is unusual regardless of how one chooses to measure it. We return to the division of housework and child care later both in this chapter and later ones. Especially for employed women with children, having more of the responsibility for household and children translates into having more to think about, more to worry about, and less free time. It is likely a contributing factor in accounting for why women's levels of happiness declined relative to men's happiness from 1972 to 2006 (Stevenson and Wolfers 2007).

Using decision making as a measure of power or dominance among

**Figure 4.2** Percentage of People in Poverty by Type of Family, 2013

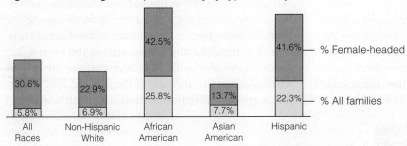

SOURCE: DeNavas-Walt, Carmen and Bernadette D. Proctor, U.S. Census Bureau, Current Population Reports, P60-249, Income and Poverty in the United States: 2013, U.S. Government Printing Office, Washington, DC, 2014, Table B-1.

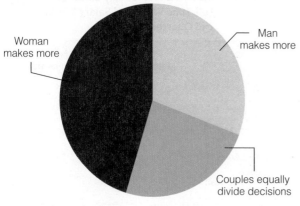

**Figure 4.3 Who Makes the Decisions at Home?**

Woman makes more

Man makes more

Couples equally divide decisions

SOURCE: Morin and Cohn 2008.

heterosexual cohabitants and married couples tells a different story. Survey data by the Pew Research Center indicated that women have more domestic decision-making input than do men (Morin and Cohn 2008). Looking at decisions of four different kinds—choosing what to do on the weekend, buying major purchases for the home, deciding what to watch on television, and managing household finances—in 43 percent of couples, women make more decisions than men, whereas respondents indicate that men make more of the decisions in only 26 percent of couples (see Figure 4.3). Furthermore, in 43 percent of the couples, men have final say in *none* of the four areas.

Decision-making data of this kind are not ideal. They are based on what people say, without opportunity for behavioral confirmation. Additionally, in some ways, decision making can be another responsibility that one carries in the household rather than an indication of how much power one has. Nevertheless, data such as these prevent us from claiming that U.S. men dominate families via their role in decision making.

## Across Cultures

Like many other social phenomena, gender is culturally variable and dynamic. Ideas and expectations associated with gender differ across cultures, sometimes immensely, and they may change greatly over time. Thus, how gender is understood and the characteristics and attributes expected of females or males vary from one society to another and, within any society, over time. Even the most basic assumptions we have had about gender—that there are only two

**False** **1.** In many cultures, the idea of only two categories of gender is absent.

genders and that one's gender doesn't change over time—were not shared universally and have recently come to be questioned. Few areas of social experience have undergone as rapid and dramatic a shift as have those accompanying gender. As a result, even introducing basic concepts is more challenging here than has been the case in previous editions of this textbook.

Rejecting the binary view of gender in favor of a broader and more fluid notion of gender may be newer and still foreign to many in the United States, but in many other cultures, the idea of a gender binary has been absent. We find instances of more than two genders into which people can be placed or place themselves among a number of South Asian societies, including India, Pakistan, Nepal, and Bangladesh; a variety of Native American cultures, including the Navajo, Sioux, Zuni, Mohave, and Lakota; as well as cultures of the South Pacific islands of Samoa, Tonga, and the Hawaiian islands (Kimmel 2011; Peoples and Bailey 2006; Roscoe 2000). Although the vast majority of the populations of these societies identify themselves as either males or females, there are culturally accepted categories for people who see themselves as other than female or male.

For example, anthropologist Serena Nanda described the hijra of India as "*neither* man nor woman" (Nanda 1990). Most hijras are biological males who voluntarily undergo an emasculation procedure, not for the purpose of becoming women but specifically to become hijras. The hijras often occupy a ceremonial and ritualized role; they may perform for a fee by singing and/or dancing at weddings or births, while others may turn to begging or prostitution to survive. Hijra communities are also found in other South Asian societies, though some, such as in Bangladesh, are much less tolerant and accepting of their presence (Khan and Hussain 2009).

Among Native Americans, there are numerous examples of third and even occasionally fourth gender categories. For example, the *nadleehi* among the Navajo, the *winkte* among the Lakota, the *hamana* among the Zuni, *berdache* among the cultures of the Great Plains, as well as the *alyha* and *hwame* among the Mojave, are all examples of gender categories that are other than male or female. In the case of the Mojave, after undergoing specific initiation procedures, both males and females could transition from their gender statuses, males becoming alyha and females becoming hwame. Whereas the term *berdache* has been used as a kind of umbrella category covering the range of alternative gender constructs among Native Americans, one increasingly finds the term *two-spirit* being used instead. Two-spirit individuals are said to have both a masculine and feminine spirit. As hinted at in the previous examples, they are "found in every region of the continent, in every kind of society, and among speakers of every major language group" (Roscoe 2000, 7).

There are also other, less extreme, cultural variations in conceptualizations of gender. As anthropologist Margaret Mead studied, among the Arapesh of New Guinea, both males and females possess what we consider feminine traits. Men and women alike tend to be passive, cooperative, peaceful, and nurturing. The father is said to "bear a child," as well as the mother; only the father's continual care can make a child grow healthily both in the womb and in childhood. Only 80 miles away, the Mundugumor live in remarkable contrast to the peaceful Arapesh. About the Mundugumor, Mead (1975) offered this description:

> Both [Mundugumor] men and women are
> expected to be violent, competitive, aggressively
> sexed, jealous and ready to see and avenge insult,
> delighting in display, in action, in fighting.

Given the great differences between the two cultures regarding gender, Mead concluded:

> Many, if not all, of the personality traits which we
> have called masculine or feminine are as lightly
> linked to sex as are the clothing, the manners,
> and the form of headdress that a society at a given
> period assigned to either sex.

## Within Cultures

Although a society may embrace a particular version of masculinity or femininity as more appropriate, every society actually has *multiple* masculinities and femininities, out of which a version emerges that is more widely accepted and expected (Kimmel 2011). These models, what have been called **hegemonic masculinity and emphasized femininity**, are then held up as the standards for all women and men to emulate. Such standards are dynamic and culturally variable, changing over time, and—within a given time and place—being challenged for cultural dominance by those who advocate other versions of masculinity or femininity (Connell 1995; Gilmore 1990; Kimmel 2011).

In helping to account for much of one's tastes, style of interpersonal communication, how one presents oneself to others, and how one sees oneself, gender also limits or expands opportunities in life and gives shape to the distribution of such societal resources as income, power, respect, and safety throughout society. Whereas one's sexual identity and orientation determine how one sees oneself in sexual terms and who one is attracted to for sharing sexual intimacies, it, too, is related to social resources such as those previously noted. In fact, both gender and sexuality, like social class, can affect one's life chances. We will look closely at sexuality and sexual identity in Chapter 6.

Gender is important on both a micro—that is, personal and interpersonal—and macro—societal—level. As conceptualized by many gender theorists, gender is, itself, a *social structure* that constrains behavior by the opportunities it offers or denies us (Connell 1987; Lorber 1994; Risman 1998). The consequences of the different opportunities afforded women and men can be seen at the *individual* level in the development of gendered selves, at the *interactional* level in the cultural expectations and situational meanings that shape how we "do gender," and at the *institutional* level in such things as sex-segregated jobs, a wage gap, and other economic and institutional realities that differentiate women's and men's experiences (Risman 1998). Although we may more often focus on individuals making choices that reflect their internalization of gender expectations, situations and institutions also shape behavior.

On both the individual and societal levels, gender is dynamic and highly variable. On an individual level, this means that our understanding of what it means to be and behave like a male or female likely changes as we age and varies some from person to person and/or from one group to another. Thus, what a child or teenager thinks about gender will likely differ from and become broader in adulthood. Similarly, what parents believe a female or male ought to do or not do or be or not be will likely differ from what friends or classmates believe. On a more macro or societal level, this means how gender is understood and what characteristics and attributes are expected of males and females vary from one society to another, as do the opportunities afforded to and the extent of inequality between women and men in the distribution of desired roles, responsibilities, and resources. Within a society, these characteristics typically vary over time or from one (class, racial, or ethnic) group to another. Again, among the familiar notions most Americans learn, that there are only two genders and that your gender is unchangeable, are not shared universally.

## Doing Gender

An additional insight about gender is the recognition that much of our behavior inside and out of families is a kind of gender performance. In other words, gender is a performance, constructed and/or reproduced in everyday social interactions and encounters. As first articulated by gender theorists Candace West and Don Zimmerman, more than what we *are*, gender is something we *do* (West and Zimmerman 1987; Risman 1998). This pertains to familial as well as nonfamilial behavior. As Greer Fox and professor of family and child development, Velma Murry (2000, 1,164), explain it, "Men and women not only vary in their degree of masculinity or femininity but have to be constantly persuaded or reminded to be masculine and feminine. That is, men and women have to 'do' gender rather than 'be' a gender." We "do gender" whenever we take into account the gendered expectations in social situations and act accordingly. We don't so much perform an internalized role as tailor our behaviors to convey our suitability as a woman or a man in the particular situation in which we find ourselves (West and Zimmerman 1987). To fail to conform to the expectations for someone of our gender in a given situation exposes us to potential criticism, ridicule, or rejection as an incompetent or immoral man or woman (Risman 1998). Much of our day-to-day behavior—such as the way we walk, talk, sit, and dress—is a means of communicating our gender to others. But in living up to or within those social expectations, we help create and sustain the idea of gender difference. According to Michael Kimmel (2011, 135), "Successfully being a man or a woman simply means convincing others that you are what you appear to be."

Although we see the social construction or "doing" of gender in all kinds of social settings, the family is a particularly gendered domain (Risman 1998). There are cultural expectations about how wage earning, housework, child care, and sexual intimacy should be allocated and performed between women and men. Thus, much of the experience that people have in their families is understandable as both an exercise in and a consequence of how they and others "do gender."

## What Gender Isn't

Although related, gender is not the same as sex, nor is it the same as sexuality or sexual orientation. As already noted, sociologists and other social scientists tend to differentiate sex from gender, using sex to refer to physiological differences between females and males, and gender to refer to just about everything else. People sometimes mistakenly assume a connection between gender and sexuality, that one's sexuality will determine and be revealed by whether one behaves in gender-appropriate or gender-inappropriate ways. In other words, they link characteristics of gender expression or display with assumptions of sexual preference. For example, women who depart from the variety of behavioral norms associated with femininity and female roles might be assumed to be lesbians, and men who depart from masculinity and reject male roles ("feminine" men) must be gay men. Neither is true. Sexual preference cannot be "read" by demeanor or role behavior. Men who fit within norms of "masculine behavior" may be heterosexual, but they could be gay or bisexual (or asexual or pansexual). Men whose behavior seems "feminine" by wider cultural standards may be gay, but they could be bisexual, or heterosexual (or, again, pansexual or asexual). The same holds true for women.

On a second level, some make connections between gender and sexual orientation by raising doubts and articulating suspicions about the sexual orientation of those who depart from gender expectations. In this way, gender norms are bolstered and reinforced. Men, especially, may monitor and restrict their behavior so as to avoid the disparaging and unwanted sort of question, "What are you anyway, a fag?" These potential doubts accomplish the feat of keeping people conforming to gender roles and expectations. Sociologist C. J. Pascoe's qualitative study of a California high school (featured in the book *Dude, You're a Fag*) demonstrates how the use of such sexualized labels is geared more toward identifying and policing "failed masculinity" (i.e., gender nonconformity) than failed heterosexuality (Pascoe 2007).

Sometimes people assume that talking and studying about gender means talking about and studying women. In other words, women are thought to be the ones for whom gender matters, as though they have genders and men don't. Of course this is completely erroneous. Gender is not synonymous with women, and gender issues affect males as well as females. There is much good research about masculinity and how men's lives are shaped by expectations that they behave in socially acceptable, appropriately masculine ways. What exactly it means to be masculine varies, as does the meaning of femininity.

Within a given society one is likely to find *multiple* masculinities and femininities, with expectations and ideals varying from one (class, racial, or ethnic) group to another. In other words, gender is not the same for all women and men, nor is it alone in shaping our life experiences. In Chapter 2, the idea of *intersectionality* was introduced. It is one of the most important and influential contributions to any understanding of gender. The critical idea in intersectional analysis is that social experience is not the same for all women just because they are female, or for all men just because they are male. Instead, gender is seen as intersecting other statuses that one occupies or identities one possesses. Initially formulated with race and then class in mind, intersectional analysis has been extended to include sexuality, and in some instances disability, nationality, age, and ethnicity.

**False** **2.** Gendered roles are affected by ethnicity.

In studying women's experiences, intersectional theorists assert that one needs to always consider the question: *Which* women's experiences? To talk about gender as though it is the only or even always the most meaningful identity or status is misleading and incomplete.

Psychologist Tiffany Townsend provides the following example: "African American women may share a history of racial oppression with African American men or a similar understanding of gender discrimination with European American women, but the impact of both racism and sexism adds a unique perspective for African American women, which helps to shape their world view, and ultimately their behavior" (Townsend 2008, 432). The same is clearly true of men's lives, where age, ethnicity, class, and sexuality add considerable diversity to men's experiences as men (Hurtado and Minha 2008, 338).

Traditionally, male and female, masculinity and femininity, and men and women have been assumed to be as different as night and day. Familiar gender stereotypes fit this same pattern of assuming polar differences: If men are aggressive, women must be easygoing; if men have **instrumental traits** (task-oriented), women must have **expressive traits** (emotion-oriented); if men are rational, women must be irrational; if women are expressive, men are expected to be stoic (Kimmel 2011; Lips 1997). The popular books by author John Gray suggested, if "men are from Mars, women are from Venus." Sociologist Michael Kimmel refers to this as the "interplanetary theory of gender," with its core assumption that females and males are polar opposites and fundamentally different (Kimmel 2011). Psychologist Sandra Bem (1993) calls this assumption of male and female as opposites **gender polarization**, and suggests that it is the fundamental assumption that we tend to make about gender. Our entire society is organized around such supposedly polar differences (Kimmel 2011).

Viewing masculinity and femininity as opposites has several implications. First, if a person differs from the male or female stereotype, he or she is seen as being more like the other gender. If a woman is sexually assertive, for example, she not only is less feminine but also is believed to be more masculine. Similarly, if a man is nurturing, he not only is seen as less masculine but also as more feminine. Second, because males and females are perceived as opposites, it is thought that they cannot share the same traits or qualities. A "real man" possesses exclusively masculine traits and behaviors, and a "real woman" possesses exclusively feminine traits and behaviors. A man is assertive, and a woman is receptive; in reality, both men and women are often both assertive and receptive. Third, because males and females are viewed as opposites, they are believed to have little in common with each other. Men and women are viewed as unable to understand each other, nor can they expect to do so. Difficulties in their relationships are attributed to their "oppositeness."

Although our culture encourages us to look for differences and, when we find them, to exaggerate their degree and significance, men and women are significantly more alike than we are different. We have tended to ignore the most important fact about males and females: that we are both human. As humans, we are significantly more alike biologically and psychologically than we are different (Hyde 2005). As men and women, we share similar respiratory, circulatory, neurological, skeletal, and muscular systems. (Even the penis and the clitoris evolve from the same undifferentiated embryonic structure.) Hormonally, both men and women produce androgens and estrogen (but in different amounts). Where men and women biologically differ most significantly is in terms of their reproductive functions: Men impregnate, whereas women menstruate, gestate, and nurse. Beyond these reproductive differences, biological differences are not great. In terms of social behavior, studies suggest that men are more aggressive both physically and verbally than women; the gender difference, however, is not large. Most differences can be traced to culturally specific expectations, male–female status, and gender stereotyping. We are capable of displaying both masculine and feminine qualities as is suggested by the notion of **androgyny**.

Although we are more similar than different in our attributes and abilities, large and meaningful differences do exist in the *statuses* we occupy and the privileges and responsibilities these carry. Although either gender may have the *ability* to nurture children, support families, clean, or cook, these tasks are assumed to be more appropriate for one gender than the other. Although women and men may possess the *ability* to do many kinds of jobs, as noted earlier, the labor force is gender segregated into jobs that are disproportionately male or female. Men's jobs typically carry more prestige, earn higher salaries, and offer more opportunity for advancement than do women's jobs. We often refer to these differences as "gaps." The "wage gap" refers to the difference between what men tend to earn and what women tend to earn (see Table 4.1). We can also speak of "prestige gaps" or "mobility gaps." Jobs that tend to be among the most highly respected jobs (typically, jobs such as physician, attorney, and engineer) tend to be held disproportionately by men. Jobs that are held largely by women (such as types of clerical work, elementary and preschool teaching, household service, and nursing) are often undervalued (and, not surprisingly, underpaid). Gaps such as these illustrate the persisting gender stratification in the United States. In much the same way that we conceptualize the upper and middle classes as being "above" the working and lower classes, we can consider the genders to be stratified.

> **False** **3.** Gender differences are largely the product of socialization.

At the same time that we see more people more widely questioning formerly automatic assumptions about gender and revising cultural conventions, within relationships and families, gender still matters. Said differently, our family experiences *are* highly "gendered" (i.e., differently experienced for women and men). For example, women and men tend to have different ideas and experiences of love. In heterosexual marriages, roles and responsibilities differ greatly between the genders. As parents, caregiving activities tend to be unequally divided by gender. The difficulties of juggling family responsibilities and paid employment are not typically the same for both genders. Research has shown that women and men seek divorces for different reasons, play different roles in the divorce process, and suffer different outcomes when their marriages end. The causes, context, and consequences of intimate violence between spouses or partners differ between women and men. In short, as we examine both in this chapter and throughout the remaining ones, gender affects nearly all aspects of families and intimate relationships.

# ▌Gender Socialization

To most family scientists, gender differences are mainly products of differently socializing males and females. Even if they recognize the potential effects of some biological characteristics or physical processes in contributing to gendered behavior, most psychologists and sociologists tend to emphasize the critical importance of socialization. In doing so, they may take different approaches to accounting for how we are socialized to be females or males. Two prominent theories used to explain how we learn what is expected of us are social learning theory and cognitive development theory.

Many theorists see gender like any other socially acquired role. They stress that we have to be socialized to act according to the expectations attached to our status as female or male. The emphasis on socialization has been considerable, although consensus on the process of socialization has not. In other words, although there is considerable agreement that we undergo gender socialization, there are different theories of how such socialization proceeds.

## Socialization through Social Learning Theory

**Social learning theory** is derived from behaviorist psychology and its emphasis on observable events and their consequences rather than internal feelings and drives. According to behaviorists, we learn attitudes and behaviors as a result of social interactions with others (hence, the term *social learning*).

The cornerstone of social learning theory is the belief that consequences control behavior. Acts regularly followed by a reward are likely to occur again; acts regularly punished are less likely to recur. Girls may be rewarded for playing with dolls, but most boys are not.

This behaviorist approach has been modified recently to include **cognition**—that is, mental processes (such as evaluation and reflection) that intervene between stimulus and response. The cognitive processes involved in social learning include our ability to use

**The Work Daughters Do to Help Families Survive**

Listen as two of the teenage girls Lisa Dodson and Jillian Dickert (2004) studied describe their contributions to their families:

> I have to take care of the house and take care of the kids and I don't go outside. I have to stay home. They have to work and so I take over.
>
> *15-year-old Ella*

> I have to clean up the kitchen in the morning before school and then do whatever shopping or whatever on the way home. I cook for the kids [younger sister and cousin] before I start my [home]work.
>
> *16-year-old Anita*

Ella and Anita carry heavy family responsibilities, yet because of a tendency to focus either on middle-class families or on younger children, the importance of children's contributions to household labor has not always been recognized and understood. In many families, however, especially low-income or single-parent families, the contributions of children, particularly daughters, have become part of a "survival strategy" without which their families would suffer greatly (Dodson and Dickert 2004).

Research that has looked into the effects of children's housework and child care contributions reveals both potentially positive and potentially negative consequences of such participation. Among the positive outcomes, children's sense of empathy, self-esteem, self-reliance, and maturity can all be enhanced by making meaningful contributions to their families. Interestingly, and perhaps especially pertinent for teens like Ella and Anita, low-income, inner-city young people who participate meaningfully in the assistance of their families may gain self-confidence and a "sense of mattering" that they otherwise may not experience (Burton 2007).

Regarding negative consequences, disproportionate caretaking can "deprive children of their own developmentally appropriate experiences and harm their health and well-being" (East 2010). As Patricia East reviews, especially in the case of larger amounts and/or long durations of caretaking, stress, academic difficulties, and depression have been found to be associated with teens' involvement and responsibility for care work (East 2010).

Although both sons and daughters often contribute labor to the household, what they do, how much they do,

and the consequences of their labor—both for themselves and for their families—greatly differ (East 2010; Gager, Cooney, and Call 1999). Typically, research finds that daughters are more likely to be pressed to provide care, do more, and start younger. They also may, themselves, seek out ways to help and participate (East 2010). Using data on 825 high school students who were part of the larger Youth Development Study, Constance Gager and colleagues (1999) compared the household labor of sons and daughters when they were in 9th and later 12th grade. Daughters spent more time in household tasks than sons, with a *growing gap* between 9th and 12th grades. Boys spent their household time on "male tasks" such as taking out trash or doing yard work. Female tasks, such as cooking, laundry, and caring for other family members, are more repetitive and time consuming. Girls also devoted more time to homework, paid work, and volunteering than boys did, resulting in an adolescent version of a leisure gap between the genders. Summing up their findings, Gager, Cooney, and Call (1999) report, "When we consider all household tasks, teenage girls are more likely to pick up the slack when the need arises" (p. 993).

Daughters' contributions to their households become even more evident in the research reported by Dodson and Dickert (2004). Girls like Ella and Anita are not merely helping out; they are indispensable ingredients in their families' survival. Driven by economic necessity, low-income parents, especially single mothers, are pushed to depend on their daughters to do what they themselves are unavailable to do. This includes caregiving and domestic work. In caring for younger siblings, girls may feed and wash them, help them with schoolwork, monitor their activities, and put them to bed. Household chores might include cooking, cleaning, laundry, shopping, and even household maintenance. In short, daughters do what mothers are unable to do, either because of employment-induced absence (e.g., entering the labor force, working increased hours, or commuting greater distances) or because of familial circumstance (e.g., birth, adoption, maternal illness, or illness of a former child care provider) (Dodson and Dickert 2004).

Sons may also help, but daughters are perceived as more responsible and more "naturally inclined" to provide effective care for home and siblings. As Dodson and Dickert (2004, 326) put it bluntly, many daughters "lose the opportunity to focus on their own young lives." ●

language, anticipate consequences, and make observations. These cognitive processes are important in learning gender roles. By using language, we can tell our daughter that we like it when she does well in school and that we don't like it when she hits someone. A person's ability to anticipate consequences affects behavior. A boy does not need to wear a pink princess gown in public to know that such dressing may lead to negative consequences. Finally, children observe what others do. A girl may learn that she "shouldn't" play certain video games by seeing that the players of those same games in video arcades are almost always boys.

We also learn gender roles by imitation, according to social learning theory. Learning through imitation is called **modeling**. Most of us are not even aware of the many subtle behaviors that, as females or males, shape the roles we play and how we play them—for example, the ways in which men and women use different mannerisms and gestures, speak differently, and so on. We don't "teach" these behaviors by reinforcement. Children tend to model friendly, warm, and nurturing adults; they also tend to imitate adults who are powerful in their eyes—that is, adults who control access to food, toys, or privileges. Initially, the most powerful models that children have are their parents.

As children grow older and their social world expands, so do the number of people who may act as their role models: siblings, friends, teachers, media figures, and so on. Children sift through the various demands and expectations associated with the different models to create their unique selves.

## Cognitive Development Theory

In contrast to social learning theory, **cognitive development theory** focuses on children's active interpretation of the messages they receive from the environment. Whereas social learning theory assumes that children and adults learn in fundamentally the same way, cognitive development theory stresses that we learn differently, depending on our age. Swiss psychologist Jean Piaget (1896–1980) showed that children's abilities to reason and understand change as they grow older.

Lawrence Kohlberg (1969) took Piaget's findings and applied them to how children assimilate gender-role information at different ages. At age 2, children can correctly identify themselves and others as boys or girls, but they tend to base this identification on superficial features, such as hair and clothing. Girls have long hair and wear dresses; boys have short hair and never wear dresses. Some children even

Playing "dress up" is one way children model the characteristics and behaviors of adults. It is part of the process of learning what is and isn't appropriate for those of their gender.

Mims/Shutterstock.com

Natalie Kauffman/Design Pics Inc/Alamy Stock Photo

believe that they can change their sex by changing their clothes or hair length. They don't identify sex in terms of genitalia, as older children and adults do. No amount of reinforcement will alter their views because their ideas are limited by their developmental stage.

When children are 6 or 7 years old and capable of grasping the idea that basic characteristics do not change, they begin to understand that gender is permanent. A woman can be a woman even if she has short hair and wears pants. Interestingly enough, although children can understand the permanence of sex, they tend to insist on rigid adherence to gender-role stereotypes. Even though boys can play with dolls, children of both sexes often believe they shouldn't because "dolls are for girls." Researchers speculate that children exaggerate gender roles to make the roles "cognitively clear."

According to social learning theory, children learn appropriate gender-role behavior through reinforcement and modeling. But according to cognitive development theory, once children learn that gender is permanent, they independently strive to act like "proper" girls or boys. They do this on their own because of an internal need for congruence, the agreement between what they know and how they act. Also, children find that performing the appropriate gender-role activities is rewarding. Models and reinforcement help show them how well they are doing, but the primary motivation is internal.

## Learning Gender Roles and Playing Gendered Roles

Although biological factors, such as hormones, clearly are involved in the development of male and female differences, the extent of biological influences is not easily discerned. Moreover, it is difficult to analyze the relationship between biology and behavior because learning—or gender socialization—begins at birth. In this section, we explore gender-role learning from infancy through adulthood, emphasizing the influence of our families in the construction of our ideas about gender. It should be stressed that gender socialization is *lifelong;* it begins at birth or even before we are born and continues throughout our lives. Along the way, numerous **agents of socialization**, the variety of individuals, groups, and institutions that help shape us, contribute to our development as males and females. First and perhaps foremost is the family.

## Childhood and Adolescence

Parents are major influences in our gender development. Even *before* children are born, parents begin to build ". . . the foundation for the gendered interests and tendencies they expect those children to have but also sharpening their sense of themselves as gendered persons, through the connections they anticipate sharing with their future children" (Kane 2009, 372).

As early as the first day after birth, parents tend to describe their daughters as soft, fine featured, and small, and their sons as hard, large featured, big, and attentive. Fathers tend to stereotype their sons more extremely than mothers do, but fathers and mothers are reinforced by others in relating differently to female and male infants (Kimmel 2011). Although it is impossible for strangers to know the gender of a diapered baby, once they learn the baby's gender, they respond "accordingly"; in other words, they describe and react to babies based on social and cultural ideas about gender.

In the United States, after the first few months of our lives, infant girls and boys are treated and talked to differently (Clearfield and Nelson 2006; Lippa 2005). In fact, as psychologists Melissa Clearfield and Naree Nelson report, research, including their own, repeatedly has revealed differences in the amount and nature of parental verbal interaction and play with male and female infants and toddlers. They point out that these differences are evident well before children are able to speak and even before they become independently mobile (Clearfield and Nelson 2006). Mothers of 2-year-olds ask daughters more questions than they ask sons, including asking more interpretive questions to daughters (e.g., "You like that, don't you?"), whereas they issue more directives to their sons (e.g., "Come here") (Clearfield and Nelson 2006; Lippa 2005).

Girls are usually held more gently and treated more tenderly than boys, who are initially the recipient of more holding, kissing, rocking, and touching. By 6 months of age, girls are held, talked to, and soothed more than boys and receive more parental engagement. Clearfield and Nelson (2006) suggest that one implication of such differences is that infant daughters are being encouraged to seek help, whereas infant sons are encouraged to be independent. As they grow, boys are ordinarily subjected to rougher forms of play, and both girls and boys begin to absorb expectations about appropriate behavior. In many ways, socialization into masculinity is more rigid, and gender-inappropriate behavior is less tolerated among boys than among girls (Kane 2012). Boys who behave in any way that appears effeminate are often stigmatized as "sissies" or "fags," whereas girls who behave in more masculine ways may receive the more ambiguous label of "tomboy." The connotations of the terms are quite different.

Such gender-role socialization occurs throughout our lives. By middle childhood, although conforming to gender-role behavior and attitudes becomes increasingly important, there is still considerable flexibility (Absi-Semaan, Crombie, and Freeman 1993). By late childhood and adolescence, that conformity becomes most characteristic. The influence of peers, discussed later, increases as they become important participants in policing gender-appropriate behavior. They are joined by teachers and the media as increasingly influential in shaping our understanding and performance of gender. However, the primary agents of our early gender socialization are parents.

## Parents as Socialization Agents

During infancy and early childhood, a child's most important source of learning is the primary caretaker—often both parents or just the mother, father, grandmother, or someone else. Many parents may be unaware of how much their words and actions contribute to their children's gender-role socialization, nor are they always aware that they treat their sons and daughters differently because of their gender. Although parents may recognize that they respond differently to sons than to daughters, they usually have a ready explanation—the "natural" differences in the temperament and behavior of girls and boys. Parents may also believe that they adjust their responses to each child's unique personality. In an everyday living situation that involves changing diapers, feeding babies, stopping fights, and providing entertainment, it may be difficult for harassed parents to recognize that their own actions may be at least partly if not largely responsible for some of the very differences that they attribute to nature or personality.

The role of nature cannot be ignored completely, however. Temperamental characteristics may be present at birth. Also, many parents who have conscientiously tried to raise their children in a nonsexist way have been frustrated to find their toddler sons shooting each other with carrots or trucks, or daughters primping in front of the mirror. Indeed, it is increasingly

**True** **4.** Parents are not always aware that they treat their sons and daughters differently.

likely that hormones and/or chromosomes influence some gender differences. At the same time, it is undeniable that children are socialized differently based on their gender.

**How Parents Shape Gender Differences.** Childhood gender socialization occurs in many ways, with parents the strongest influencers. Parents give shape to gender differences in the clothing, toys, and books they buy for their children; the activities in which they involve them; the environments in which they raise them; and the behavioral expectations they convey to them.

Females are socialized with more attention placed on the importance of physical appearance. Women's studies professor Lori Baker-Sperry and sociologist Liz Grauerholz examined one particular genre of children's literature—classic fairy tales—and determined that such tales tended to emphasize the importance of physical appearance for female characters, especially young female characters, and associated a character's "goodness" with her beauty and attractiveness. The tales that have been most popular, republished in books and/or remade into films, are the tales that most stress the importance of female beauty (Baker-Sperry and Grauerholz 2003). In the more than 4,000 children's books published annually, females are rarely portrayed as brave or independent and are typically presented in supporting roles (Renzetti and Curran 2003).

A more recent study of award-winning children's picture books examined books that won the prestigious Caldecott Medal or received Caldecott Honors between 1990 and 2009. Concentrating on the visual messages conveyed via illustrations, researchers Peter Crabb and Deb Martino examined 490 illustrations from 68 different books. They found that somewhat traditional gender messages were still being conveyed through the depiction of the gendered use of household versus production artifacts. Household objects used for food preparation, cleaning, family care, repair, home crafts, and gardening included such things as pots and pans, knitting needles, brooms, and washing machines. Production artifacts were objects used in agriculture, manufacturing, construction, transportation, and defense, such as hammers, tractors, cars, and guns. Crabb and Marciano found that females were much more likely than were males to be depicted with household items, little having changed from the

1990s through the first decade of this century. Males were much more likely than females to be depicted with "production artifacts." That these gender patterns showed little change from 1990 to 2009 suggests that children are being exposed to visual images that continue to reinforce a more traditional gender division of labor (Crabb and Marciano 2011).

Children's toys and clothing also reflect and reinforce gender differences. Boys prefer such toys as blocks, trucks and trains, guns and swords, and tool sets. Girls prefer to play with dolls and domestic toys and to dress up. At least some, if not most, of the preference appears to be the result of parental preferences, peer influence, and positive reinforcement (Lippa 2005). Some stores and toy companies have been targets of criticism and complaint for reinforcing gender stereotypes. Recently, the retail store Target announced a change in how it would display toys and children's bedding, by removing gender labels. Instead of "boy's toys" and "girl's toys," the company will just sell toys as "toys." As reported by Jessica Contrera in *The Washington Post*, "Along with grouping all toys together, the aisles will no longer have colored backdrops to indicate gender, such as pink and yellow for girls or blue and green for boys" (Contrera 2015). This change was met with some mixed reaction, applauded by some for breaking down stereotypes and criticized by others for "political correctness" (Contrera 2015).

Parents construct the physical environments in which they raise their children in very gender-differentiated ways. More than 30 years ago, research established that parents furnished and decorated their children's rooms in ways that reflect traditional notions and expectations about gender. Psychologists Erin Sutfin, Megan Fulcher, Ryan Bowles, and Charlotte Patterson (2007) found that the "great majority" of children in their study were being raised similarly in gender-differentiated settings (i.e., their bedrooms). Interestingly, in comparing heterosexual parents to lesbian parents, they reported that lesbian mothers were less likely to create highly stereotypical environments for their children. This is in keeping with their finding that lesbian mothers had more liberal attitudes and egalitarian beliefs than heterosexual parents expressed.

In general, children are socialized by their parents through four subtle processes: manipulation, channeling, verbal appellation, and activity exposure (Oakley 1985):

- *Manipulation.* From an early age, parents treat daughters more gently (e.g., telling them how beautiful they are or advising them that nice girls do not fight) and sons more roughly (telling them how strong they are or advising them not to cry). Eventually, children incorporate such views as integral parts of their personalities. Differences in girls' and boys' behaviors may result from parents expecting their children to behave differently.

- *Channeling.* Children are directed toward specific objects and activities and away from others. Toys, for example, are differentiated by gender and are marketed with gender themes. Parents purchase different toys for their daughters and sons, who—influenced by advertising, the reinforcement by their parents, and the enthusiasm of their peers—are attracted to gendered toys (Kimmel 2011).

- *Verbal appellation.* Parents use different words with boys and with girls to describe the same behavior. A boy who pushes others may be described as "active," whereas a girl who does the same may be called "aggressive."

- *Activity exposure.* Both genders are usually exposed to feminine activities early in life, but boys are discouraged from imitating their mothers, whereas girls are encouraged to be "mother's little helpers." Chores are categorized by gender (Dodson and Dickert 2004; Gager, Cooney, and Call 1999). Boys' domestic chores take them outside the house, whereas girls' chores keep them in it—another rehearsal for traditional adult life.

Although it is generally accepted that parents socialize their children differently according to gender, there are differences between fathers and mothers. Fathers pressure their children more to behave in gender-appropriate ways. Fathers set higher standards of achievement for their sons than for their daughters, play more interactive games with their sons, and encourage them to explore their environments (Renzetti and Curran 2003). Fathers emphasize the interpersonal aspects of their relationships with their daughters and encourage closer parent–child proximity. Mothers also reinforce the interpersonal aspect of their parent–daughter relationships, typically engaging in more "emotion talk" with their daughters than with their sons, and— unsurprisingly—girls are more adept at monitoring emotion and social behavior as early as first grade (Renzetti and Curran 2003).

There are also differences among parents that, based on her study of more than 40 parents of preschoolers, sociologist Emily Kane sorted into five different patterns. These patterns were based on such things as whether parents believe gender to be largely biological or mostly social in origin, and whether parents strive to reproduce or resist gendered behavior in their children (Kane 2012). Kane identified the following five types of parents:

- *Naturalizers.* These parents see gender as mostly biological and act in ways to reproduce the gender structure, in part out of heightened awareness of how others would react to their children's nonconformity.

- *Cultivators.* They see the origin of gender behavior as largely social, act in ways that will reproduce gender conformity in their children, but are not motivated to do so by concern for how others will judge them or their children.

- *Refiners.* They acknowledge both biological and social influences on their children's gender behavior, are aware of the reactions and potential judgments of others and act with a mixture of both resistance and conformity.

- *Innovators.* These parents actively resist gender structures and do so without concern for how they or their children will be judged.

- *Resisters.* These parents are opposed to gendered patterns but express concern about the reactions and judgments of others.

In addition to differentiating among parents, Kane notes that though parents are major influences over their children's development of gendered selves, they "do not act alone" (Kane 2012, 2). By this she means two different things: Parents are joined by other socialization influences in the shaping of their children, and children, themselves, are not passive participants in this process. Children act in ways to "achieve a 'normally sexed identity' as part of navigating their peer cultures and the wider social world" (Kane 2012, 21). Kane also notes that parents commonly assumed that their children were/would be heterosexual (by referring to their kids' boyfriends or girlfriends) and that parents of sons expressed less approval of their sons crossing the line of gender-appropriate behavior than did parents of daughters.

Both parents of teenagers and the teenagers themselves believe that parents treat boys and girls differently. It is not clear, however, whether parents are reacting to existing differences or creating them. It is probably both, although by that age, gender differences are fairly well established in the minds of adolescents.

Although both females and males are given responsibilities in the household, girls' chores are more often daily domestic duties whereas boys' are often of the occasional or seasonal nature.

AJA Productions/The Image Bank/Getty Images

Jose Luis Pelaez Inc/Getty Images

As we saw in the previous chapter, ethnicity and social class are important in how parents interact with and socialize their children. Among Caucasians, working-class families tend to differentiate more sharply than middle-class families between boys and girls in terms of appropriate behavior; they tend to place more restrictions on girls. African American families tend to socialize their children, especially their daughters, toward more egalitarian gender roles (Hill 2002). African American families socialize their daughters to be more independent and assertive than Caucasian families do. Indeed, among African Americans, the "traditional" female role model may never have existed. The African American female role model, in which the woman is both wage earner and homemaker, is more typical and more accurately reflects the African American experience (Lips 1997). It is less clear whether and how much African American males' socialization departs from more traditional male socialization (Hill 2002).

In Latino families as well as in Asian families, gender socialization is frequently more traditional, especially with regard to the kind of play parents encourage, the domestic responsibilities they assign, and the restrictions they impose on their children, especially their daughters (Raffaelli and Ontai 2004; Yee et al. 2007). Traditionally, Latinos and Latinas learned what was expected of them within the gender ideals of **marianismo**, **machismo**, and **caballerismo**, constructs that specify the expectations for how a Latina female and Latino male should act. Marianismo includes such qualities as submissiveness, selflessness, nurturance, sexual purity, self-sacrifice (especially for one's family), and the ability to endure suffering (Pina-Watson et al. 2014). Machismo includes the idea of male superiority and dominance over women, while caballerismo emphasizes chivalry and respecting and protecting the honor of women (Pina-Watson et al. 2014).

Of course, parents don't socialize their children in a vacuum, and the home is not an isolation chamber. Two points follow from this fairly obvious realization: First, no matter what parents intend for their children, they have limited control over the entirety of the gender socialization to which their children are exposed. Parents who have downplayed the importance of feminine appearance norms or masculine aggressiveness may be disappointed to find that, while at the babysitter's, their daughter got to play dress up with accompanying makeup and lipstick, or that their son had his choice of a well-stocked arsenal of toy guns from which to choose his weapon. Other children,

other adults in or outside the family, the various media, and eventually school experience will also be part of how one's child is socialized.

In a related way, other people will bear witness to the ways in which parents go about socializing their children. Though most of family life is lived in relative private for most people, when unconventional socialization occurs and becomes known, parents and their children can be the recipients of public reaction. Two recent examples illustrate this well. In a 2011 ad for J. Crew emailed to subscribers, J. Crew president Jenna Lyons was featured along with her young son, Beckett, with his toenails painted pink. Beneath the photo of Lyons and her son is copy that read, "Lucky for me I ended up with a boy whose favorite color is pink. Toenail painting is way more fun in neon" (Moss 2011). When the ad appeared, it generated considerable backlash in various media outlets and was the subject of numerous talk show commentaries in which concern was expressed for the "damage" that might be done to Beckett by having pink toenails. Psychiatrist Dr. Keith Ablow claimed of the ad, "This is a dramatic example of the way that our culture is being encouraged to abandon all trappings of gender identity." Likewise, Erin Brown of the conservative Media Research Center characterized the ad as "blatant propaganda celebrating transgendered children" and claimed that there was a trend in which "gender-confused boys wanting to dress and act like girls . . . [was] . . . seeping into mainstream culture" (E. Brown 2011; Moss 2011).

Similar public outrage resulted after an article appeared in the *Toronto Star* newspaper about a couple who had decided to keep the gender of their child a secret, even from grandparents and other family and friends. Motivated by a desire to "mitigate at least some of the gendered messages children are blitzed with," and paying "tribute to freedom and choice in place of limitation, a stand-up to what the world could become . . . ," Storm's parents Kathy Witterick and David Stocker made the decision to keep Storm's sex a mystery to all but a handful of people. When news of their decision was featured in the *Toronto Star*, a "storm" of controversy and reaction engulfed the family. Even more intense than what followed Jenna Lyons and her son, Beckett, Witterick and Stocker were praised by some but heavily and harshly criticized by others, including child psychiatrists, psychologists, and other professionals for their parenting decision and its potential impact on their child (Newton 2011; Poisson 2011).

## Other Sources of Socialization

Although primary, both in importance and in exposure, families are not the only influences on the ideas we acquire about gender. Our early lives are lived in the company of many others who shape our ideas about men and women, femininity and masculinity. As children grow even just a little older, their social world expands, and so do their sources of learning.

**School.** Around the time children enter preschool or kindergarten, teachers (and peers, discussed next) become important influences and potential role models in their overall development and also their gender socialization. Because most child care providers, as well as preschool, kindergarten, and elementary schoolteachers are women, children's interactions with adults are primarily with women. Teachers and peers also monitor children's behavior, reinforcing gender differences along the way.

As children grow and pass through elementary, middle, and high school, perhaps going on to college, they are the recipients of an education that is itself "gendered" and are being shaped to conform to gender-specific expectations. We might say, then, as sociologist Michael Kimmel (2011) does, that schools are like gender factories. We are the products.

In what ways are school experiences "gendered"? There are at least two different stories to tell here: one older, one newer, one female, one male. Through much of the 1990s, researchers emphasized the following picture of gendered school experience. Classroom observations documented that boys were louder and more demanding and received a disproportionate amount of the teacher's attention. Teachers called on boys more often, were more patient with boys in their explanations, and were more generous toward them with their praise. The curriculum was more male centered, often leaving out issues of importance to females and ignoring accomplishments of women. Such curricular materials were joined by teacher expectations and classroom interaction patterns in devaluing women and discouraging female students. Girls, praised for their appearance and the neatness of their work more than its substance or quality, grew more tentative and hesitant as they approached and entered middle school.

By high school, they suffered drops in their self-esteem and self-confidence, prefacing their answers with disclaimers: "I'm probably wrong, but . . ." or "I'm not sure, but. . . ." Intelligent girls often found that they were devalued by boys and often overlooked by teachers. Only in all-girl schools, argued Myra and David Sadker (1994), did female students assert themselves vigorously in class. The Sadkers believed that girls benefited from gender-segregated schools and classes by not having to compete with boys for teachers' attentions, not becoming overly concerned with their appearance, and not having to fear that their intelligence would make them undesirable as dates. The picture in coeducational settings was bleaker; coed schools had "failed at fairness," and girls suffered the harsher consequences (Sadker and Sadker 1994).

Fast-forward through the last 15 to 20 years and on into the present. From kindergarten through high school, we are increasingly finding that the performance of boys, not girls, most lags. Girls generally excel over boys in all areas during grade school. They have less difficulty learning to read, learn to read earlier, score higher on fourth-grade standardized reading and writing tests, and are less likely to be diagnosed with learning or speech problems or to repeat a grade. Boys are twice as likely to be diagnosed with learning disabilities, prescribed medication, or placed in special education classes. In middle school and high school, girls score higher than boys on standardized reading and writing tests, are more likely to plan on attending college, and are less likely to drop out. Increasingly, more and more boys say they dislike school (Tyre 2006).

As a result of the variety of trends noted here, attention to and concern about how gender shapes educational experience and how schools reinforce gender differences have been broadened and even shifted. Researchers became as concerned with what boys experience in school, why, and with what consequences (Gurian and Stevens 2005; Pollack 1998; Tyre 2006). For example, the attention boys receive from teachers is not always positive—they are subject to more discipline and receive more of the teacher's anger than do girls, even when the disruptiveness of their behavior is similar. Furthermore, their academic performance often suffers, as indicated by their rates of failing, acting up, and/or dropping out. To some, including family counselor and author Michael Gurian, this constitutes a crisis, requiring institutional and curricular intervention (Gurian and Stevens 2005).

True **5.** Both boys and girls suffer from gender-related problems in school.

As school curricula become more rigid and more focused on assessment and demonstrating proficiency, teachers have less leeway to teach to students' strengths or needs and less tolerance for the typically boy style of learning—disorganized, distracted, high energy, and potentially disruptive. Boys are also often unwilling to seek help and admit weakness. As much as some saw the earlier call for all-girl schools as a remedy for girls' school problems, some are now embracing it as a solution for what ails boys (Tyre 2006).

One indicator of the difficulties boys face can be seen in data on educational attainment. In 2012–13, females earned 57 percent of bachelor degrees, 60 percent of all master's degrees, and a little over half (51.4 percent) of all doctorates (Digest of Education Statistics 2014). Gender doesn't operate alone in shaping school experiences. Race and class matter, too. In fact, much of the gender issues noted previously are more prominent among minority students. For example, the gender gap in higher education turns out to be smaller among white students than among African Americans, Hispanics, or Native Americans (Snyder and Dillow 2011, Table 235).

In actuality, there is a third gender-related story worth telling. In recent years, much attention has been paid to bullying behavior, much of which varies by and/or is tied in with gender and sexual identity. From elementary school on, children and adolescents are potential targets of bullying behavior. Though we will look more closely at this in the discussion of peers as socialization agents, it merits mention here because being bullied at school is associated with greater likelihood of missing school, dropping out, and performing poorly.

As shown in Figure 4.4, data from the National Center for Education Statistics reveal that more than a fifth (22 percent) of students ages 12 to 18 report having been bullied at school during the school year. One can see in the bar graph that more females than males acknowledge being bullied. More specifically, females were more likely than males to report that they had been called names, insulted, or made fun of; had rumors spread about them; or were purposely excluded from activities. More males than females had been physically bullied (pushed, shoved, tripped, or spit on) and threatened with harm.

Gender expression and sexuality can increase one's susceptibility to victimization. Findings from the 2013 National School Climate Survey (Kosciw et al. 2014),

**Figure 4.4 Bullied at School: Results from National Center of Education Statistics**

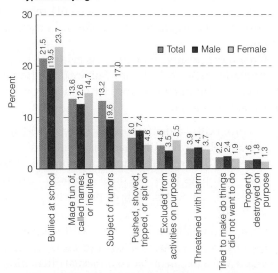

SOURCE: National Center for Education Statistics. "Fast Facts: Bullying," 2015. https://nces.ed.gov/fastfacts/display.asp?id=719.

a survey of nearly 8,000 LGBT students, ages 13 to 21, include the following:

- Nearly three-quarters were verbally harassed because of their sexual orientation, and 55 percent said they'd been verbally harassed because of their gender expression.

- More than a third (36 percent) were physically harassed (having been pushed or shoved) in the past year because of their sexuality, and 23 percent because of their gender expression.

- Around one out of six (16.5 percent) were physically assaulted because of their sexual orientation, and 11 percent because of their gender expression.

- Approximately half (49 percent) of LGBT respondents had been targets of cyberbullying or electronic harassment.

- Among transgender students specifically, 32 percent had been prevented from wearing clothes that were deemed inappropriate for someone of their legally recognized sex, 42 percent had been prevented from using their preferred name, and almost 60 percent had been made to use a bathroom or locker room of their legal sex.

Again, such patterns have consequences for one's overall well-being as well as one's experience of education and schooling. Due to their feeling unsafe or uncomfortable, a majority of the LGBT sample reported that they avoided extracurricular activities and school functions, over a third avoided such gender-segregated spaces as locker rooms and bathrooms, and nearly a third (30.3 percent) reported missing at least an entire day of school in the past month.

On a more positive note, there is evidence in the latest National School Climate Survey that, though hostile, the school climate for LGBT students is improving. For example:

- In 2013, LGBT students reported a lower incidence of homophobic remarks than was found in all previous NSCS surveys. The most common, "that's so gay," continues to be the most common form of "anti-LGBT language," but its prevalence too has declined.

- Compared to all previous years, LGBT students in the 2013 survey reported lower levels of physical and verbal harassment based on sexual orientation. They also report experiencing a level of sexual orientation based physical assault.

- Harassment and assault based on gender expression had similarly declined.

**Peers.** A child's age-mates, or **peers**, become especially important when the child enters school. By granting or withholding approval, friends and playmates have great influence on us. They may affect what games we play, what we wear, what music we listen to, what television programs we watch, and even what cereal we eat or beverage we drink. Peer influence is so pervasive that it is hardly an exaggeration to say that in some cases, children's peers tell them what to think, feel, and do.

Peers also provide standards for gender-role behavior in several ways, such as through the play activities they engage in, the toys with which they play, and the approval or disapproval they display, verbally or nonverbally, toward others' behavior. Children's perceptions of their friends' gender-role attitudes, behaviors, and beliefs encourage them to adopt similar ones so that they are accepted. If a girl's female friends play soccer, she is more likely to play soccer. If a boy's male friends display feelings, he is more likely to display feelings.

During adolescence, peers continue to have a strong influence, one that often leaves parents feeling helpless and as though their importance has

**False** **6.** During adolescence, the influence of peers on gender remains strong.

been reduced in guiding or shaping their sons and daughters. Much of that influence may be a kind of "boundary maintaining," or what sociologist Barrie Thorne called "border work," reinforcing the limits around acceptable and appropriate gender behavior and pointing out and sanctioning transgressions. This may be more obvious among males, whose peers may more openly and harshly stigmatize and penalize them for displays of femininity. Boys who don't sufficiently display masculine qualities are often recipients of severe peer harassment and potential ostracism. This is often sexualized into allegations or intimations that a nonmasculine boy is gay (Pascoe 2007).

The line leading from so-called border work into the kind of harassment or bullying addressed previously is often a fine one, particularly for children and teens whose gender behavior or sexuality fall outside what peers deem appropriate. Beginning in elementary school, children who do not conform to traditional gender norms are at greater risk of bullying and name-calling than are their cisgender peers, and as a consequence, they report feeling less safe at school.

See the Issues and Insights: *Gender and Bullying* for more discussion of peer harassment and bullying.

**Popular Culture and Mass Media.** In terms of both quantity of exposure and content of images, the various mass media are important contributors to gender socialization. In sheer quantity of exposure, media content is prominent in the lives of children and teenagers, and in terms of amount of exposure, television is dominant in children's media use. Data from a Nielsen report, *Kid's Audience Behavior across Platforms,* demonstrates the extent to which children from 2 to 18 years old are exposed to mass media messages. According to the report, in 2014, more than 95 percent of children between ages 2 to 17 spent at least some time watching either live or on-demand programs during the month of the survey. Looking just at television, children 2 to 11 were either watching traditional TV, watching time-shifted TV, or watching video on the Internet a total of almost 27 hours per week, whereas 12- to 17-year-olds were watching close to 23 hours a week (Nielsen Company 2015).

Looking specifically at children "0 to 8," Common Sense Media and Victoria Rideout report that almost 60 percent of children watch television at least once every day, making television the medium used most frequently. Children 8 or younger average

**Gender and Bullying**

Phoebe Prince had just recently turned 15. She was a "new girl" at South Hadley High School in western Massachusetts, having emigrated from Ireland with her family in 2009. After having had a "brief fling" with a senior football player, Phoebe became a target for a group of her classmates, including an ex-girlfriend of the senior male. Phoebe's tormentors engaged in prolonged, relentless harassment and bullying. Persistent name-calling, stalking, threats, and physical intimidation were directed toward her. She was called an "Irish slut," had her books knocked out of her hands, received threatening text messages, and on January 14, 2010, after being tormented in the school library, cafeteria, and hallways, had a canned drink thrown at her as she walked home from school. Sometime later that afternoon, Phoebe went up to her bedroom and hung herself from a bar in her closet.

Phoebe Prince's suicide led prosecutors to bring charges against nine teenagers. Six of those teens—four girls and two boys—faced charges that included statutory rape, criminal harassment, stalking, and assault. Three younger girls faced charges of delinquency. None of those charged were given jail sentences. Instead, they were placed on probation that, if completed successfully, would eventuate in their having a clean criminal record. Some were also assigned to community service. Eventually, the town of South Hadley reached a settlement of nearly a quarter of a million dollars with Phoebe's parents, for the failure of the school to act and bring an end to the mistreatment she was enduring. The Prince case captured attention of many well beyond South Hadley, including people across the United States and in Ireland, where Phoebe was taken home to be buried. At least part of the attention was the result of the merciless teasing and harassment she suffered at the hands of other females.

Nine months after Phoebe Prince's death, in a three-week span in September 2010, three boys—two 13-year-olds and a 15-year-old—committed suicide after enduring unrelenting harassment by classmates, in other words, bullying—because each was believed to be gay. Billy Lucas, Asher Brown, and Seth Walsh didn't know one another; they lived hundreds of miles apart, but each was targeted

for mistreatment by peers at school (Connolly 2012). A published review of recent research identified 19 studies that connected suicides among lesbian, gay, and bisexual adolescents to bullying experienced at school. This was found to be especially true for those who displayed cross-gender behavior, demeanor, or appearance (Haas et al. 2011).

Suicide is an extreme and relatively uncommon reaction to bullying; the vast majority of bullied children and teens do not take their lives. However, among young people who do commit suicide, many had prior experience of being bullied. A published review of relevant research suggests that victims of bullying have suicide rates two to nine times the rate of suicide among nonvictims (Kim and Leventhal 2008). Bullying is also related to depression, which itself is associated with elevated risk of suicide. In fact, the group of young people at the greatest risk of suicide may be those who are both victims and perpetrators of bullying (Suicide Prevention Resource Center 2011).

Although long associated with male-on-male (and male-on-female) behavior, bullying is the monopoly of neither gender. However, there are notable differences in how females and males bully and in what leads them to be bullied. Males report experiencing more physical bullying than do females. Females report more bullying by rumors and sexual comments, and are more likely to report being teased or joked about. Females are also more likely to be targets of cyberbullying, being bullied via email and text messages. Again, their harassers are most commonly other females. Consequences of such "indirect" bullying are themselves direct and potentially severe. Victimization is associated with psychological distress, such as depression, which may lead to more severe health problems. Repeated indirect bullying is also associated with a significant decrease in girls' self-esteem.

Albeit different, bullying is done by both genders, to both genders. A substantial amount of bullying is "about gender," as bullies target for harassment, ridicule, exclusion, or threats, those whose behavior or demeanor depart from expected and accepted gender styles. ●

nearly two hours a day of total "screen media" use, with the average increasing to 2:21 among 5- to 8-year-olds. More than a third (36 percent) have a television in their bedrooms. When asked whether media contributes to having more or less face-to-face family

time, most parents said it had neither effect. However, more than a quarter (28 percent) said that media contributed to less family time.

In an earlier 2010 report by the Kaiser Family Foundation on media use and exposure among children

8 to 18, Victoria Rideout, Ulla Foehr, and Donald Roberts (2010) provided the following information:

- Three out of four 8- to 18-year-olds said they had a television in their bedrooms; almost half of the sample had cable or satellite in their bedrooms. A third of the sample had a computer with Internet access in their bedrooms.

- Just under two-thirds of sampled 8- to 18-year-olds said that the television is usually on during dinner; close to half (45 percent) said that it was on "most of the time."

- Eight- to 18-year-olds spent more than seven and a half hours a day using various media and were exposed to ten hours and 45 minutes a day of media content. For more than three hours a day, children "multitasked," and used two media simultaneously.

- "Screen time" (television, video games, computer, and movies) took up over seven and a half hours a day; two and a half hours spent listening to music (i.e., radio, tapes, CDs, and MP3s); and, among those who reported reading, less than an hour a day was spent reading. Of course, the exact mix of these varies by age, race, sex, and personal interest.

With such heavy exposure, the content of media messages becomes even more significant. In all its forms, electronic as well as print, the mass media depict females and males quite differently. Women are underrepresented across a range of media content. For example, research on prime-time television, music videos, G-rated movies, and video games finds substantial to severe underrepresentation of females relative to males. This disparity has characterized television for more than three decades (Collins et al. 2011). When they are portrayed, women are often sexually objectified, trivialized, stereotyped, or they are ignored (Collins et al. 2011). Beyond underrepresentation, researchers suggest that gender and sexual stereotypes characterize television (Ward and Friedman 2006), advertising, popular music lyrics (Bretthauer, Zimmerman, and Banning 2007), music videos (Bell, Lawton, and Dittmar 2007; Wallis 2011), and video games (Burgess, Stermer, and Burgess 2007; Williams et al. 2009).

The women depicted on television have been less representative of the range of women than the men depicted have represented men. A 2003 study of gender and age of characters revealed that female characters continue to be younger than male characters. The largest percentage of female characters was in the 20- to 29-year-old range, and the largest age range among male characters was 30 to 39. Within age groups, such as adolescence, females are underrepresented and stereotyped (Walsh and Ward 2008). As summarized by researchers Jennifer Walsh and Monique Ward, young female characters "groomed, whined, shopped, and did chores," while males worked, fought, and rebelled (Walsh and Ward 2008, 137). Across media, heavy emphasis is placed on female appearance, especially thinness, and on females as sex objects. Male characters typically have been shown as more aggressive and constructive than female characters. They solve problems, take action, and rescue others from danger, often by resorting to violence.

Do such gendered portrayals matter? Research indicates that they do. For example, after viewing select music videos, daytime talk shows, and prime-time programs, individuals are more likely to endorse stereotypical gender roles, emphasize female thinness, and accept myths about women and sexual assault (Kahlor and Morrison 2007). Even exposure to images of Barbie dolls can affect attitudes and self-esteem of young girls. Adolescent girls who viewed just three music videos featuring thin and glamorous women showed an increase in their own level of dissatisfaction with their bodies (Bell, Lawton, and Dittmar 2007).

Research on attitudes toward LGBT individuals and issues suggests that media images can have a positive effect. In addition to the so-called "*Modern Family* effect," featured in Chapter 1, research has indicated that the more accepting attitudes of young people toward LGBT issues has been influenced in part by increased exposure to gay and lesbian characters on television. Using exposure to an ABC sitcom, *Grace Under Fire*, which ran from 1993 to 1998 and in 1995 introduced a "gay-related storyline" (Gupta 2014), Jeremiah Garretson found that those respondents who were familiar with *Grace Under Fire* increased their positive attitudes toward gays and lesbians. Among younger respondents, the increase was especially noteworthy (Gupta 2014).

**Religion.** Studies indicate multiple connections between religion and gender. First, one's religious background can be a force in shaping one's attitudes about gender issues (Whitehead 2012). According to sociologist Andrew Whitehead, religious women and men are much more likely to support a more traditional gender ideology, wherein "men are supposed to inhabit the public sphere while women are meant to take care

More frequent attendance at religious services is one way in which women appear more religious than men.

Syracuse Newspapers/S Cannerelli/The Image Works

of the domestic or private sphere" (Whitehead 2012). Various aspects of religion appear to affect one's gender ideology, including one's religious affiliation, the frequency with which one attends religious services, whether one believes that the Bible is the literal Word of God, and whether one perceives God to be male. Conservative Protestants tend to endorse more traditional gender roles for both females and males, as do those who attend services more often, consider their scriptures to be the Word of God, and carry a masculine image of God (Whitehead 2012). Religiosity has also been found to contribute to more likely opposition to gay marriage (Kimmel 2011).

There is a second connection between religion and gender. By a host of measures, women are more religious than are men. Seeking to explain why women are more—or why men might be less—religious, some emphasize different socialization experiences and others look at sex-based biological dispositions (Bradshaw and Ellison 2009; Miller and Stark 2002). However divided researchers might be as to why, there is little disagreement that women show higher levels of religiosity than men. Using data from the General Social Survey collected between 2004 and 2014, we can identify the following statistically significant gender differences on a variety of indicators of religious belief and behavior. The data indicated that:

**True** 7. Research shows it is possible for women and men to establish work or family roles that are counter to their socialization.

- 21 percent more women (68 percent) than men (47 percent) said that they pray at least once a day; men were twice as likely as women to say that they pray less than once a week or never (33 percent vs. 16 percent);

- 21 percent more women (57 percent) than men (36 percent) said they ask for God's help once to several times a day;

- 16 percent more women (67.4 percent) than men (51.5 percent) said they "know that God exists";

- 17 percent more women (61 percent) than men (44 percent) say they feel "very close" to God or "as close as possible";

- 19 percent more women than men (66.5 percent vs. 47.1 percent) say they feel God's guidance most days to many times a day;

- 7 percent more women (29 percent) than men (22 percent) said they attend religious services every week or more than once a week, and 7 percent more men than women said they attend religious services once a year or never; and

- 13 percent more women (65 percent) than men (52 percent) said that they consider themselves to be very religious or moderately religious.

Finally, it should be noted that in some of the fundamental beliefs that characterize Christianity, Islam, Judaism, and their institutional policies, especially regarding women's access to positions of religious leadership, religion has been criticized for perpetuating and justifying gender inequality. Religious doctrine can be said to reinforce gender inequality by making gender differences appear to be divinely created (Kimmel 2011). Recall, though, that religious beliefs don't automatically or inevitably translate to behavior.

## Continued Gender Development in Adulthood

Although more attention is directed at early experiences and socialization in childhood and adolescence, gender development doesn't stop there. Many life experiences that we have in adulthood alter our ideas about and actions as males and females. Again, families loom large in reshaping our gendered ideas and behaviors.

Research offers examples in which adult life experiences transform how we act as males or females. This is especially evident in the work of sociologists like Kathleen Gerson and Barbara Risman (Gerson 1985, 1993, 2010; Risman 1986, 1987, 1998), whose more structural analyses showed how adult life experiences both inside and outside of families have the potential to restructure our identities, redefine our role responsibilities, and take us in directions quite different from those suggested by our early gender socialization. The life one leads is often different from what one was raised to lead or expected to lead (Gerson 2010; Risman 1987, 1998).

Focusing on adulthood is important because it reveals the gaps that often exist between earlier gender socialization and adult experiences. To some scholars, this diminishes the importance of socialization and discredits theories that deterministically link early socialization to later life outcomes (Gerson 1993). In some ways, those socialization-based theories may be no better than *biological determinism,* in which we are seen as limited to those behaviors that our genetic or hormonal characteristics allow. They simply substitute socialization for biology (Risman 1989). Socialization *is* important, especially in affecting our expectations and offering us role models for lives we might live. But life is more circuitous than linear. Unanticipated twists and turns often take us in directions we neither expected nor intended. Research on women's and men's career and family experiences bear this out. For example, in Kathleen Gerson's research on women's and men's career and family choices, many people developed commitments to either careers or parenting that were unexpected and grew out of their experiences in jobs and relationships (Gerson 1985, 1993). Some women and men who anticipated "traditional" adult outcomes moved in nontraditional directions based on the levels of fulfillment and opportunity at work, the experiences and aspirations of their partners, and their experiences with children. Similarly, men and women who aspired to nontraditional outcomes (career attachment for women and involved fatherhood for men) "reluctantly" abandoned those directions as a result of firsthand experiences at home and work.

Gerson's book, *The Unfinished Revolution: How a New Generation is Reshaping Family, Work, and Gender in America* (2009), is a study of the family backgrounds and future hopes and plans of 120 young women and men. In recounting their earlier family lives, many report having experienced significant change in how their families were structured or in how their parents managed employment and parenting. According to Gerson, the key to successful family outcomes appeared to be gender flexibility in responding to the circumstances families find themselves facing.

In articulating their plans for the future they hoped they'd have, both Gerson's female and male interviewees expressed a desire for an equal partnership and a sharing of employment and responsibility for caring for home and children. Only 15 percent of the women and less than 30 percent of the men expressed a desire for strict, clearly defined gender boundaries. However, despite their expressed ideals, they showed an awareness that an egalitarian family life would be hard to create and maintain, especially given the inflexibility of paid employment. Both women and men tended to offer "fallback plans," should life's circumstances force a change in plans. However, men's and women's fallbacks were different. Men saw a more traditional division of responsibilities, where they would support their families financially while their spouses took responsibility for caring for their children, as the next best outcome. Women saw a more independent and self-reliant lifestyle as the best alternative should their ideals remain beyond their grasp.

Barbara Risman's research on single custodial fathers pointed to the ability of fathers to develop nurturing qualities that departed from their prior socialization. Men who unexpectedly found themselves as lone custodial parents developed nurturing abilities that their socialization had not included. More important than how they were raised was how they interacted with their children as well as the lack of a female in their lives to whom caregiving tasks could be assigned. Thus, these single fathers "mothered" their children in ways that were more like women's relationships with children than what one would predict (Risman 1986). Importantly, socialization contributes to but neither guarantees nor restricts us to any particular family outcome.

In addition, as one moves through adulthood, one has new experiences or encounters different sources of gender-role learning, both of which can alter one's sense of their gendered roles. Marriage and parenthood can alter one's understanding of what being a female or male means, as can experiences in the workplace and/or attending college.

## College

Within the college setting, many young adults learn to think critically, to exchange ideas, and to discover the bases for their actions. There, many young adults first encounter alternatives to traditional gender roles, either in their personal relationships, their cocurricular or extracurricular activities, or in their courses. Research indicates that college liberalizes people's attitudes about a host of issues, including ideas about gender, attitudes toward women's employment opportunities, and family roles and responsibilities (Cotter, Hermsen, and Vanneman 2011, 2014). Looking across educational levels, compared to those with less than high school, high school, and those with some college, college graduates were found to hold more egalitarian attitudes on a number of gender-related issues (Cotter, Hermsen and Vanneman 2014).

This effect of college may be especially so for students who major in the humanities, social sciences, or arts; live on campus; have jobs while attending school; socialize with others of different racial and ethnic backgrounds; and take women's studies courses. Both women's and men's attitudes tend to become more liberal, though in general, at both entry and exit from school, women's attitudes are more liberal than men's (Bryant 2003).

## Marriage

People's experiences in marriage, parenthood, and the workplace can also lead to changes in gender attitudes, beliefs, and behavior. Marriage is an important source of gender-role learning because it creates the roles of husband and wife, which are not merely the same as male and female. Our spouses have expectations of how one should act as a husband or wife, and these expectations are important in shaping behavior. Such expectations may require significant adjustments to one's prior expectations of what a male or female does in marriage.

Husbands tend to believe in innate gender roles more than wives do. This should not be especially surprising because men tend to be more traditional and less egalitarian about gender roles. Husbands stand to gain more in marriage by believing that women are "naturally" better at cooking, cleaning, shopping, and caring for children.

Household work can affect women psychologically in many of the same ways that paid work affects them in female-dominated occupations, such as clerical and service jobs. Women in both situations feel greater levels of frustration because of the repetitive nature of the work, time pressures, and being held responsible for things outside their control. Such circumstances do not encourage self-esteem, creativity, or a desire to achieve. Moreover, housework *on top of paid work* is a reality with which many women but much fewer men contend.

**True** **8.** Compared with traditional roles, contemporary male gender roles place more emphasis on the expectation that men will be actively involved with their children.

How couples divide housework can affect the quality of their relationships. This will be addressed in a later chapter, but consider just one example. Earlier research on couple relationships found that heterosexual couples who split and tried to share domestic chores had less satisfying sexual relationships than couples who divided responsibilities in more traditional ways. Such sharing of housework in heterosexual-couple households has been shown to have less effect on recent married and cohabiting couples than on those who met and married three decades ago. Research by sociologists Daniel Carlson, Amanda Miller, Sharon Sassler, and Sarah Hanson (2014) found that, with the exception of heterosexual couples in which the male became the one responsible for the bulk of housework, couples who departed from traditional roles by sharing domestic chores and responsibilities were as satisfied with their sexual lives and had sex as frequently as did those with more traditional divisions of labor (Carlson et al. 2014).

## Parenthood

For most men and women, life is dramatically transformed by parenthood. The concept **traditionalization** describes how becoming parents may lead even egalitarian-minded couples to more traditional outcomes, especially if new mothers take extended leave to care for their infants (Dew and Wilcox 2011; Fox 2009).

Motherhood typically alters life more significantly and visibly for women than fatherhood does for men. For some men, fatherhood may still mean little more than providing financially for their children. One is unlikely to find many who would associate motherhood only with providing. As parents, mothers do more, are expected to do more, and are expected to juggle what they do with paid employment. As a consequence, fatherhood has not typically created the same degree of work–family conflict that motherhood has, nor does it add as many new responsibilities onto men's shoulders. However, fathers are finding themselves more stressed, more concerned about whether they are spending enough time with their children, and more critical of their performances as fathers. Fathers who strive to be fully or nearly equal co-parents, especially, may discover the ways in which the demands of parenthood clash with demands of the workplace. Whereas a man's work role traditionally allowed him to fulfill much of his perceived parental obligation, we now expect more out of fathers and fathers expect more out of themselves.

Not only have our expectations shifted toward more nurturing versions of fatherhood, but whereas traditional fatherhood was tied to marriage and often to a breadwinner–homemaker marital structure, approximately 40 percent of all current births occur outside of marriage, and nearly half of all current marriages end in divorce. In most two-parent family households with children, both parents are employed. What, then, is the father's role for a man who is not married to his child's mother, who is divorced and does not have custody, or who shares wage earning with his spouse? What are his role obligations as a single father as distinguished from those of married fathers? For many men, the answers to such questions are not altogether clear and must be actively constructed in response to circumstances.

Women today have somewhat greater latitude as wives and mothers. It is now both accepted and expected that women will work outside the home, at least until they become mothers, and, more than likely, continue or return to paid employment sometime after they have children. Even with increases in the numbers of women who remain childless, women may be expected to become mothers and be subjected to social pressure toward motherhood. As noted previously, once children are born, roles tend to become more traditional, even in previously nontraditional marriages. Often, the wife remains at home, at least for a time, and the husband continues full-time work outside the home. The woman must then balance her roles as wife and mother against her needs and those of her family.

### The Workplace

Experiences in the workplace can also lead to changes in one's attitudes and expectations. Both men and women are psychologically affected by their occupations. Work that encourages self-direction, for example, makes people more active, flexible, open, and democratic; restrictive jobs tend to lower self-esteem and make people more rigid and less tolerant. If we accept that sex-segregated female occupations are often of lower status with little room for self-direction and less opportunity for much advancement, we can anticipate that some women may become less achievement oriented, and suffer when competing with and compared to men. With different opportunities for promotion, men and women may express different attitudes toward achievement. Women may downplay their desire for promotion, suggesting that promotions would interfere with their family responsibilities. But this really may be related to a need to protect themselves from frustration because many women are in jobs where promotion to management positions is somewhat less likely.

An especially interesting analysis of the ways gender affects how one is perceived and treated at work is Kristen Schilt's, *Just One of the Guys? Transgender Men and the Persistence of Gender Inequality* (2011). Interviews with 54 female-to-male transgender men, ages 18 to 64, whose first experiences of work were as females before becoming men, reveal how profoundly their experiences of and at work were altered by their becoming male. Schilt spoke with both white-collar and blue-collar workers, as well as both "open transmen" (who acknowledged their transition to male either through remaining in the same job or in a new job), and "stealth transmen" who got new jobs as men and kept their transition secret from coworkers and employers. What Schilt shows is that as (trans)men, her interviewees become recipients of work-related rewards and privileges that they did not get when they worked as women. In fact, they must "negotiate being treated as not just *different from* women but *better than* women." Fully two-thirds of the transgender men she interviewed experienced rewards such as new opportunities, greater authority, and economic gains (Schilt 2011).

## Gendered Family Experiences

Characterizing contemporary gendered family experience is difficult for a number of reasons. *First*, there is variation across the vast array of households and families. Some characteristics that may accurately depict gendered roles and relationships among white, middle-class, married heterosexual couples do not accurately represent roles and relationships in households and families across lines of race, class, or sexuality. Even religious differences may make characterizations difficult, as there are some notable differences across religious groups. Those from more conservative religions, such as Mormons and conservative Protestants, have continued to adhere more strongly to beliefs in more traditional gendered family roles than have others (Ammons and Edgell 2007). Such beliefs aren't always expressed in choices and/or behaviors (Ammons and Edgell 2007), but they make it somewhat more problematic to describe changes in attitudes or expectations regarding gender in family life.

Class and race diversity are more familiar to us. Middle-class women and men have tended to be more liberal and egalitarian than working-class or upper-class women and men, at least in their stated attitudes or beliefs. *In practice*, working-class women and men may behave in more equal ways, even if they do so more out of circumstance-induced necessity than because of an egalitarian ideology. Their professed attitudes and beliefs may remain more traditional than their behavior suggests.

> **True**
> **9.** Among African American couples, the gender division of labor tends to involve more sharing of wage earning and domestic work.

Due to a combination of factors, including African heritage, legacy of slavery (which subjected women to the same labor and hardships as men), or economic discrimination and hardship, African American women have long been less likely to occupy the status of full-time housewife, even during times when greater numbers of white women were at home full time than are today. As a consequence, the gender division of labor in black households has been characterized as more flexible, with more involvement of women in more instrumental tasks and more men sharing domestic work. African American men are generally more supportive than Caucasian or Latino men of egalitarian gender roles.

As we noted, in traditional Latina gender roles, the notions of *marianismo, machismo, and caballerismo* have been said to influence gender roles and relationships. Drawn from the Catholic ideal of the Virgin Mary, *marianismo* stresses women's roles as self-sacrificing mothers suffering for their children and subordinating themselves to males (McLoyd et al. 2000; Vasquez-Nuttall et al. 1987). But such subordination is based more on respect for the male's role as provider and on mothers being the pillars of their families than it is on subservience (Pina-Watson et al. 2014). Latina increasingly adopted values incompatible with a belief in male dominance and female subordination.

*Second,* because gendered family roles and relationships are in such a state of flux, characterizations that applied even in the recent past may have undergone or be undergoing change that makes such characterizations now limited, at best. Related to this point, some changes in both attitudes about and behaviors within families have undergone change that may be best appreciated by taking a longer view, whereas shorter-term

views might show stalled progress or even suggest reversals in movements toward greater gender equality (Sayer 2015; Sullivan, Gershuny, and Robinson 2015). In the late 1990s and early 2000s, there appeared to be a slowing in the trend toward greater sharing of household labor (Sullivan, Gershuny, and Robinson 2015) and in the attitudinal acceptance of new, more equal, gender roles (Cotter, Hermsen, and Vanneman 2014). These trends, especially the attitudinal changes, have once again begun to move in the direction of more equal gender roles and relationships.

*Third,* in speaking about gendered family roles and relationships, we are talking about both beliefs and behaviors, attitudes and actions. In some instances, driven by necessity, for example, actions may change before we see appreciable attitude change. For example, the more substantial male job loss of the recent recession may have pushed men to become more involved in or take responsibility for more housework and child care. On a more micro level, a male who is the spouse or partner of a female whose job requires her to work second shift (i.e., 4 P.M. to midnight) may take on more housework and child care (even if managed by his female partner or spouse), not because his attitudes about gender have become more egalitarian but rather because he is home, she is not, and someone must do what needs to be done.

Within the past generation, there has been a significant shift from traditional toward more egalitarian gender roles (Brewster and Padavic 2000; Lang and Risman 2007). Women have changed more than men, but men, too, have changed. These changes seem to have had effects across classes, although not to the same extent or at the same pace. As illustrated previously, college-educated women and men, especially, are considerably less likely to hold traditional ideas about gender, work, and family roles (Brewster and Padavic 2000; Bryant 2003; Gerson 2010).

Within the family, attitudes toward gender roles have become more liberal, moving in the direction of more sharing of housework and child care. Researchers estimate that from the 1960s to the 21st century, men's share of housework had doubled from about 15 percent to about 30 percent. Moreover, this trend has been observed in industrialized countries throughout the world. A 20-country comparison showed an overall increase of men's share of housework and child care from less than a fifth of all housework in the

1960s to more than a third by 2003 (Sullivan and Coltrane 2008). This trend still places women at a disadvantage, especially by making them responsible for more of the housekeeping and child care activities, but it also gives one reason to assume that movement in the direction of greater sharing will continue.

Further reason to suspect additional movement in the direction of more sharing can be found in national public opinion data. A report by the Pew Research Center (2007b) reveals the extent to which "sharing household chores" has come to be seen as a vital ingredient in achieving marital success. More than 60 percent of respondents identified such sharing as "very important," while only 7 percent considered it "not very important." This placed sharing housework third in importance behind faithfulness (93 percent) and a happy sexual relationship (70 percent), but made it more important than adequate income, good housing, shared religious beliefs, shared tastes or interests, children, and agreement on politics. There was minimal difference between the percent of women (61 percent) and men (64 percent) or between fathers (66 percent) and mothers (62 percent) in stressing the importance of sharing housework. Of course, how people answer in a survey may be neither an indicator of their true feelings nor a predictor of their behavior.

## Women's and Men's Roles in Families and Work

Traditionally, women's lives centered around marriage and motherhood. When a woman left adolescence, she was expected to either get a job, go to college, or marry and have children. Although a traditional woman might work before marriage, she was not expected to defer marriage for work goals and was expected to be "expecting" soon after marriage. Within the household, she was expected to subordinate herself to her husband. Often this subordination was sanctioned by religious teachings. Circumstances permitting, this may still reflect the intentions and preferences of women with more traditional values. However, there were class and racial differences in women's traditional work and family roles.

Even though the traditional roles for white women have typically been those of wife and mother, increasingly over the past few decades these roles were joined by that of employed worker or professional. Attitude changes have occurred as well. Despite a brief period between the late 1990s and early 2000s, in which opinion data suggested a slip in support of gender

equality, attitudes supportive of gender equality in home and work have rebounded (Cotter, Hermsen, and Vanneman 2014). Consider these indicators:

- An all-time high of 68 percent *disagree* that it is better for men to earn the money and women to tend the home.
- Roughly two-thirds of those surveyed *disagreed* that preschool children suffered if their mother was employed outside the home.
- Nearly three-fourths (72 percent) *agree* with the notion that an employed mother can have as warm a relationship with her children as a stay-at-home mother.

It is now generally expected that most women will be employed at various times in their life. According to data from the Pew Research Center, less than a fifth (18 percent) adults surveyed agreed that "women should return to their traditional roles in society." Just under eight of ten adults (79 percent) disagreed with this prescription for what role adult women should play (Wang, Parker, and Taylor 2013).

Other opinion data from Pew suggests that a majority of those surveyed (62 percent) consider dual-earner marriages in which both spouses take care of the home and children more likely to be a more satisfying lifestyle than the traditional arrangement of male breadwinner and female housewife. Almost three out of four respondents say they consider the increase in women working outside the home a positive change (Parker 2012a). It is also typically economically necessary. This is understood and recognized as a benefit of increases in the numbers of women working outside the home. When asked how the increasing number of women working for pay outside the home has affected family life, two-thirds (67 percent) say this change has made it easier for families to earn enough money to live comfortably (Wang, Parker, and Taylor 2013). However, data also show that the increased employment of women is something that many see as having been a mixed blessing. Three-quarters of adults surveyed responded that more women in the workplace has made it harder for parents to raise children, and half said it has made it harder for marriages to succeed (Wang, Parker, and Taylor 2013).

One definite consequence of increases in women's employment is the issue of how to manage the competing demands of paid employment and caring for family, a situation that forces women more than men to "juggle" responsibilities. How they respond to such circumstances depends in part on their relative commitment to work and family, what sociologist Mary Blair-Loy (2003) calls their *schemas of devotion* or what Arlie Hochschild (1989) called their *gender ideologies*. Those women who are more "family-committed" often attempt to reduce the conflict between work and family roles by giving family roles precedence. As a result, they tend to work outside the home in greatest numbers before motherhood and after divorce, when single mothers generally become responsible for supporting their families. After marriage, most women are employed even after the arrival of the first child. Regardless of whether a woman is employed full time, she almost always continues to remain the one with primary responsibility for housework and child care.

Cultural expectations impose high standards of devotion and labor-intensive self-sacrifice on women who become mothers, what is often described as the **intensive mothering ideology**—the belief that children need full-time, unconditional attention from mothers to develop into healthy, well-adjusted people. Those mothers who embrace such an ideology find themselves in a demanding position, trying to meet the expectations attached to intensive mothering, but mothers who also choose or need to work outside the home are faced with a particularly difficult dilemma (Hays 1996). We will explore mothering ideologies more in Chapter 10, but for now it should be noted that, as an ideal, intensive mothering is something more attainable by economically privileged women, and may not even reflect the perspectives of women of color whose lives have long consisted of integrating motherhood with employment (Hill Collins 2000; Walls, Helms, and Grzywacz 2014).

Sociologist Karen Christopher has detailed an alternative to the *intensive mothering ideology* that is especially relevant for employed mothers, which she calls **extensive mothering** (Christopher 2012). Extensive mothering entails delegating much daily child care to others while still seeing oneself as ultimately responsible for the development and well-being of their children and, instead, emphasizing "being in charge" of the care one's child(ren) receive and enhancing ones financial status. Christopher interviewed 40 employed mothers about their attempts to resolve the competing and conflicting expectations associated with trying to live up to intensive mothering demands and being seen as an ideal worker.

Married mothers in this sample construct scripts of "extensive mothering," in which they delegate substantial amounts of the day-to-day child care to others, and reframe good mothering as being "in charge" of and ultimately responsible for their children's well-being (Christopher 2012).

There is some evidence that employed mothers have adopted a standard of mothering that fits more closely with the extensive than intensive mothering ideology. Asked to rate their performance as mothers, more than three-fourths (78 percent) of employed mothers reported that they are doing an excellent or good job raising their children. Two-thirds of at-home mothers rated themselves at a similar level (Parker 2013).

Although husbands were once the final authority, wives have greatly increased their power in decision making. Between the late 1970s and mid-1990s, data on attitudes about marriage indicated increasing acceptance of more gender-equal relationships, where women contribute to decision making, share wage earning, and have assistance with family caregiving. After a slight drop in such egalitarian attitudes in the late 1990s, research indicates that they are rising again, especially regarding the acceptance of women's employment and, as we saw earlier, men's involvement in domestic work.

In fact, some couples develop and act on an ideology of sharing and fairness, valuing and pursuing such relationship characteristics as equality and equity (Risman and Johnson-Sumerford 1998; Schwartz 1994). Although these "peer" and **postgender relationships** (i.e., relationships lived outside the constraints of gender expectations) cannot yet be considered as the norm, they reflect the most concerted efforts to establish greater equality in marriage. Research also indicates that couples who share housework and wage earning are less likely to divorce than couples in which men earn all the income and women do all the housework (Cooke 2008).

## Men's Roles in Families and Work

Central features of the traditional male role—whether among Caucasians, African Americans, Latinos, or Asian Americans—stressed male dominance and men as breadwinners. Males are generally regarded as being more power oriented and more competitive than females, even at a young age (Gneezy and Rustichini 2004; Stewart and

**True** **10.** There are heterosexual couples whose relationships are not based on gender.

McDermott 2004). Statistically, men demonstrate higher degrees of aggression, especially violent aggression (such as assault, homicide, and rape). They are more often arrested for violent crimes and are more common victims of violent offenses other than rape and sexual assault (Kimmel and Messner 2010). Aggressiveness is one of a handful of attributes on which consistent moderate to large differences between the genders have been documented (Hyde 2005). Although aggressive traits may be useful in the corporate world, politics, and the military, such characteristics are rarely helpful to a man in fulfilling marital and family roles that are more likely to require understanding, cooperation, communication, and nurturing.

Traditionally, across ethnic and racial lines, male roles centered on providing, and the centrality of men's work identity affected their family roles as husbands and fathers. Men's identity as providers took precedence over all other family functions, such as nurturing and caring for children, doing housework, preparing meals, and being intimate. Because of this focus, men who retain traditional attitudes may become confused if their spouses expect greater sharing or intimacy; they believe that they are good husbands when they are good providers. When circumstances render them unable to provide, the blow to their self-identity can be quite powerful (Rubin 1994). However, because race, ethnicity, and economic status often overlap, certain categories of men face more difficulty meeting the expectations of the traditional provider role. Because African Americans and Latinos often fare less well economically, men often are left unable to lay claim to the household status and power that traditional masculine roles promise. Both African American and Latino men have more positive attitudes toward employed wives. Ethnic differences in traditional notions of masculinity and men's roles are more evident among older and less-educated African Americans and among Mexican Americans not born in the United States (McLoyd et al. 2000).

Because the key assumption about male gender roles has been the centrality of work and economic success, many earlier researchers failed to look closely at how men interacted *within their families*. Over the past three decades, there has been considerable popular and scholarly attention to whether and how men's family lives have changed. Researchers began to ask some of the same questions about men's lives that were previously asked about women's, looking at whether and how men juggle paid work and family, whether they maintain sufficient involvement in each, and how they feel about their efforts. Although we may not yet see employed fathers talked about with the same levels of concern that have been directed toward working mothers, more are questioning whether and how men experience conflicts between work and parenting. Sociologists Oriel Sullivan and Scott Coltrane note that national public opinion data indicate increasing acceptance of gender equality and, along with that, the expectation that spouses will share both wage earning and family responsibilities, as does Kathleen Gerson's more recent interview study (Gerson 2010). Additionally, men appear to be more involved in caring for their children than ever before (Sullivan and Coltrane 2008). Later chapters look at men's experiences of marriage, parenting, and the division of household labor.

Still, alongside enlarged emphasis on men's more nurturing qualities, men continue to be expected to work and to support or help support their families.

As contemporary male gender roles allow increasing expressiveness, men are encouraged to nurture their children.

Myrleen Pearson/PhotoEdit

Although women's financial contributions may be no less essential in maintaining their family's standard of living or even keeping them out of poverty, women are not judged as successful wives and mothers based on whether they succeed at paid employment. As a result, most men have less role freedom *to choose* whether to work than do women. Of course, women's choices are limited as well, but more by economic reality and necessity than gender expectations. Those men who aspire toward equal involvement in child care are especially likely to discover that the complexity of juggling work and family is not restricted to women. Men who attempt the same juggling act often experience similar role strain and role overload (Gerson 1993; Hansen 2005). In fact, recent Pew Research Center survey data indicate that half of dads surveyed say it is difficult to balance work and family. In addition, almost as many employed fathers (34 percent) as employed mothers (40 percent) reported that they "always feel rushed." Employed dads were much more likely than employed moms to say that they spent too little time with their kids (Parker 2015).

## Continued Constraints of Contemporary Gendered Roles

Despite the limitations that traditional roles may place on us, changing them is not and has not been easy. Gendered roles are closely linked to self-evaluation. Our sense of adequacy often depends on our gender performance as defined by parents and peers in childhood ("You're a good boy" or "You're a good girl"). Because gendered roles often seem to be an intrinsic part of our personality and temperament, we may defend these roles as being natural, even if they are destructive to a relationship or to ourselves. To threaten an individual's gender role is to threaten his or her gender identity as male or female because people do not generally make the distinction between gendered roles and gender identity. Such threats are an important psychological mechanism that can keep people in traditional roles.

Even though substantially more flexibility is offered to men and women today, contemporary gendered roles and expectations continue to limit our potential. Indeed, there is evidence that some stereotypes about gender traits are still very much alive. Asked in a national survey to identify whether certain traits were truer characteristics of men, women, or both men and

women, respondents answered in ways that reinforced gender stereotypes. Both women and men associated arrogant, stubborn, and decisive with men more than with women. Traits identified as truer of women than of men included compassionate, emotional, creative, intelligent, honest, and manipulative. Interestingly, men said two traits—ambitious and hardworking—were truer of men than of women, whereas women said the reverse (Taylor et al. 2008).

The situation of contemporary women and men in dual-earner, heterosexual households imposes its own constraints. Because they continue to shoulder more of the responsibility for housework and child care *on top of full-time jobs*, women often experience fatigue, stress, resentment, and a lack of leisure. However, men are not immune to these same stresses. Whereas 56 percent of "working moms" report difficulty trying to balance competing demands of work and family, so did 50 percent of working dads (Parker 2015).

The social structure also reinforces traditional gender norms and behaviors and makes some changes more difficult. The workplace, especially, helps enforce traditional gender roles and can make change more difficult. Despite its shrinking magnitude, the continued wage disparity between men and women is a case in point. Such a significant difference in income makes it "rational" for the man's paid work role in many heterosexual couples to take precedence over the woman's paid work role. If someone needs to remain at home to care for children or an elderly relative, it makes more "economic sense" for a heterosexual woman to quit her job because her male partner probably earns more money. This, too, may be changing, as more women in dual-earner households are becoming near equal, equal, or even primary wage earners (Mundy 2012). For now, however, wage disparities represent yet another way in which outside institutions influence gendered roles in families.

### Transgender Family Experience

As we saw earlier, ideas about gender have been broadened by the increased visibility and acceptance of transgender individuals. Our knowledge about how transpeople experience marriages, families, and intimate relationships has only recently begun to grow after being virtually nonexistent. Using the major journals in the marriage/couple and family therapy fields and adding in journals such as *The Journal of Marriage and Family, Journal of Family Issues, The Family Journal: Counseling and Therapy for Couples and Families,* and the *Journal of Counseling Psychology,* Markie Blumer, Mary Green, Sarah Knowles, and April Williams undertook a content analysis of 10,739 articles published in 17 different journals between 1997 and 2009, and found that only 9 (0.0008 percent) dealt with transgender issues or included gender variance as a variable. Writing in 2012, they asserted that "transgender issues are ignored and marginalized by marriage/couple and family therapy scholars and researchers alike" (2012, 1), and advocated for "increased attention to trans issues" within these fields (2012, 11).

Data gathered for the National Transgender Discrimination Survey (NTDS) provide some greater information about the family experience of those who are transgender (Grant et al. 2011). Based on the responses of a sample of 6,450 transgender and gender-nonconforming respondents who come "from all 50 states, the District of Columbia, Puerto Rico, Guam, and the U.S. Virgin Island," the survey revealed both positive and negative family relationships and experiences. Of the negative findings, almost three-fifths (57 percent) of respondents experienced some form of family rejection. Among those who had transitioned (and were living in their preferred gender), more than half (55 percent) saw their relationships end. Among those who had come out to their partners, almost half (45 percent) of those who came out to their partners saw their relationships ended because of their transgender identity or gender nonconformity. Of the sample respondents, 40 percent indicated that their parents or other family members elected not to speak or spend time with them because of their gender identity or gender expression.

Among those respondents who were parents, about half (49 percent) said their relationships with their children "remained the same" or were "in some ways better, some ways worse," after coming out as transgender. Another 22 percent said relationships were somewhat (12 percent) or much (10 percent) improved. Thus, more than 70 percent had experienced either neutral or somewhat positive outcomes as parents. Their children continued to speak and/or spend time with them. Of respondents, 29 percent said their situations as parents had become somewhat worse (16 percent) or much worse (13 percent). Relationships with children were subject to disruption or interference from other people. Of those with children, 29 percent experienced limited contact with their children due to actions by their ex-partner. For 13 percent of those who were parents, courts limited or stopped relationships with their children.

More than 40 percent of the sample (45 percent) reported that their families "were as strong today as before they came out." Similarly, 43 percent reported that they maintained the majority of their family bonds. Taken altogether, these and other data from the National Transgender Discrimination Survey showed a combination of familial and relationship outcomes or effects of coming out as transgender. These family outcomes led to other consequences, as "acceptance was strongly connected with a range of positive outcomes while family rejection was connected with negative outcomes" (p. 88). Those who were rejected by family had almost twice the rate of suicidality and three times the frequency of homelessness compare to those who were recipients of family acceptance (Grant et al. 2011).

As transgender and gender nonconforming lifestyles gain and retain public awareness and acceptance, the family experiences revealed in the NTDS data are likely to improve as well. The survey results were published in 2011. This is well before *Orange Is the New Black* or *Transparent* aired on television, before Caitlyn Jenner transitioned from Bruce to Caitlyn, and before Jazz Jennings and her family opened their lives so widely to the general public. These and other kinds of recent developments may well help improve the familial circumstances as depicted previously.

## Gender Movements and the Family

Gender issues have been the source of much collective action and the focus of a number of social movements that press for change. These movements include the range of perspectives within the contemporary women's movement and also various "men's movements" that, although less visible, have organized to change aspects of men's lives. We look briefly here at some of the ways these movements have framed and acted on family matters.

### Women's Movements

A complete history of feminism in the United States is well beyond the scope of this book. In the 18th, 19th, and 20th centuries, women organized around issues such as economic justice, abolition of slavery, temperance, and women's suffrage. In their antislavery activity during the 19th century, many women were sensitized to the extent of their own oppression and disadvantage, which helped energize their pursuit of voting rights (Lindsey 1997; Renzetti and Curran 1999). After gaining the right to vote with the passage of the Nineteenth Amendment in 1919, many women withdrew from active feminist involvement because they thought they had reached equality with men (Renzetti and Curran 1999).

During the 1960s, feminism resurfaced dramatically. Catalyzed by the publication of Betty Friedan's *The Feminine Mystique*, many women began to look critically at the sources of their "problems with no names," and the family was seen as a major culprit. In addition, wage inequality was made a public issue through President John F. Kennedy's Commission on the Status of Women in 1961 and the passage of the Equal Pay Act of 1963. Then, in 1966, the National Organization for Women (NOW) was established. Over the past 40-plus years, this liberal, reform-oriented feminist organization has grown to include more than a half million members in its more than 500 chapters throughout the United States. It is the largest, although not the only, organized plank of the women's movement, and its philosophy represents one of a number of "feminisms" (Lindsey 1997; Renzetti and Curran 2003). Like NOW, the Feminist Majority Foundation has, since its founding in 1987 and through its education, research and action programs, and policy development, striven to "empower women economically, socially, and politically" (feminist.org/welcome/index.html). Both organizations have among their concerns women's reproductive rights and health, safe and affordable contraception, and the protection of women's abortion rights. NOW also focuses on family issues such as paid parental leave and work–family balance, mothers' and caregivers' rights, and support for equal marriage rights.

As we noted in Chapter 2, there are multiple feminist theoretical positions. So, too, are there multiple feminist political ideologies. Contemporary feminist positions range across a spectrum including liberal, socialist, radical, lesbian, multiracial, and postmodern feminism (Lorber 1998; Renzetti and Curran 1999). Each has a specific emphasis on issues, and advocates different strategies to improve women's lives.

Judith Lorber sorts these various feminist perspectives into three broader categories: **Gender-reform feminism** is geared toward giving women the same rights and opportunities that men enjoy; **gender-resistant feminism** advocates more radical, separatist strategies for women out of the belief that their subordination is too embedded in the existing social system; and **gender-rebellion feminism** tends to emphasize overlapping and interrelated inequalities of gender, sexual orientation, race, and class (Lorber 1998; Renzetti and Curran 1999).

Given this diversity of opinion, it is difficult to characterize *one* "feminist" position on families. Furthermore, such attempts occasionally exaggerate or simplify complex positions. Thus, one can find multiple feminist positions on family issues. Although family responsibilities may well have been downplayed or criticized by many feminist activists in the 1960s and 1970s, much of the activism on a number of family issues today has and continues to come from feminists. American feminists have emphasized issues such as abortion rights, reproductive freedom, ending bigotry against gays and lesbians, and ending violence against women, but they have also been active at the forefront of pushing for parental leave, child care, and so forth.

Various attempts have been made to document how many women in the United States consider themselves feminists. Such estimates have varied, though most place the figure at 30 percent or less (McCabe 2005; Reid and Purcell 2004; Roy, Weibust, and Miller 2007). Data from large national surveys, such as the General Social Survey, suggest that around a quarter of female respondents consider themselves feminists, with better-educated women, white women, and lower-income women being more likely to identify as feminists, and more religiously active women less likely (Peltola, Milkie, and Presser 2004). Roy, Weibust, and Miller (2007) found that the willingness to identify oneself as a feminist was affected by stereotypes about feminists and could be altered by exposing young women to positive descriptions of feminism and feminists. Complicating the situation, Alyssa Zucker (2004) found that some women express feminist attitudes but distanced themselves from a feminist label. Consistent with Zucker's claim, a 2009 CBS poll found that roughly two-thirds of the women who responded credited the women's movement with "improving their lives," yet just 27 percent said they consider themselves feminists. In an earlier 2008 poll, when given the dictionary definition of a feminist as "someone who believes in the social, political, and economic equality of the sexes," two-thirds of the women who responded identified themselves as feminists (Alfano 2009). Thus, more women support feminist ideas and appreciate the contributions feminism has made than are willing to call themselves feminists.

When asked by sociologist John Durst to reflect on her situation, then 32-year-old Karen Wilson described having what she considered an almost ideal life. "Oh, how much time do you have? . . . Do you have like three hours? I love it! I love it! God, there's just so many things about it." Karen is a success in her career in sales and promotions for a communications company. She has a husband she loves and two children she adores, a 3-year-old daughter and a 4-month-old son. As the full-time breadwinner for her family, she describes how she and Kevin reached the arrangement they have and what she most enjoys about it:

> You know what, we talked about this before we got married. . . . I planned on working after I had kids. [I told Kevin] . . . "if you want to stay home, great, but I can't stand it!" And he said, "Yeah, we can do that." Then, BAM! This opportunity came along for me to make more money than he was and I said, "You wanna do it? You wanna live the dream?"

Pushed to identify what she sees as the biggest positives of her lifestyle, she enthusiastically replied:

> I like being able to get away between 8 and 5 and to have a lot more control over my life without having to worry about two other responsibilities (son and daughter) and Kevin, too. I should say all three of them. I like that. I like being able to turn it off and just go, but I like coming back and having my daughter's little face pressed against the window (waiting), Kevin standing there with a beer in his hand, the dog running around me, it's really nice to come home to. . . . I love bringing home the paycheck and telling Kevin, "Here, honey, split it up. . . ." I love that. I love contributing; I just think it's the ultimate.
>
>  I love not having all the responsibilities he has. I hated cooking. I hated the dishes, the laundry—I felt like it was the least rewarding job anybody could have because you never get any pats on the back. I like having a title and being able to say, "This is what I do. I'm contributing to my family." . . . And I like that Kevin's just so calm and relaxed and really laid-back. The kids keep him moving constantly yet at the end of the day he's still relaxed enough to talk to me. I think it's been really wonderful.

Research on intimate relationships, marriage, and family consistently reveals the importance of gender in dividing up domestic responsibilities and shaping personal and familial experiences. Typically, women perform two to three times as much housework as men, and employed wives experience greater stress and enjoy less leisure than their husbands (Coltrane 2000). The consistency with which such inequalities are reported may give the impression of inevitability, that they are somehow unavoidable parts of marriage and parenthood, but couples such as the Wilsons offer a more hopeful scenario to those who might wish to someday depart from the norm, whether to create more equal partnerships or, more dramatically, reverse roles.

Sociologists Barbara Risman and Danette Johnson-Sumerford interviewed their own sample of 15 "postgender" couples who explicitly reject conventional conceptions of gender, opting instead for more gender-neutral relationships. That is, they carefully and intentionally share responsibility for paid work and share responsibility as caregivers for their children.

At a minimum, they "changed how gender works in their families." Furthermore, "in the negotiation of marital roles and responsibilities, they have moved beyond using gender as their guidepost" (Risman and Johnson-Sumerford 1998, 24).

Regardless of the route couples took to arrive at their postgender family arrangements, they used criteria other than gender to organize their daily activities. They have rejected the ideas that "wifehood involves a script of domestic service or that breadwinning is an aspect of successful masculinity" (Risman and Johnson-Sumerford 1998, 38). Such couples are still rare, and their lifestyles may require high levels of female income and professional autonomy if women are to be able to move beyond male dominance or privilege.

The significance of couples like the Wilsons or Risman and Johnson-Sumerford's postgender couples is that they reveal a wider range of possible marital outcomes than most literature reports. There is no inevitable inequality that engulfs married couples. Equality and fairness take work and persistence but are possible for those who seek them. ●

## What about Men?

Just as there is no one perspective on how women should be or what they should do, neither is there unanimity about men's lives. Just as there are multiple feminisms, each with its own agenda, there are different viewpoints on whether, in what direction, and how men ought to change (Clatterbaugh 1997; Messner 1997). Similarly, academic studies of men and masculinity have seen the recent emergence of "male studies," a more conservative approach to studying men than the approach typical in more pro-feminist gender studies or men's studies curricula.

Organizationally, the closest parallel to NOW among men has been the National Organization for Men Against Sexism (NOMAS). NOMAS is a pro-feminist, gay-affirmative, antiracist men's organization founded in 1975 by a group of male students in a women's studies course at the University of Tennessee. Among the issues NOMAS emphasizes and acts on are child custody, fathering, ending male violence, gay rights, and reproductive rights (www.nomas.org/history). Central to the pro-feminist men's movement is the issue of fairness. Pro-feminist men believe that men ought to share responsibilities within their households and that women and men ought to be equal partners. Also, pro-feminists argue that both men and children would benefit from closer connections between fathers and their children.

It is interesting to note the different positions feminist and pro-feminist activists take on the family. Where the resurgent women's movement of the 1960s grew partly out of the articulation of discontent with women's domestic roles and responsibilities, male activists often pressed for expanding and enlarging men's family roles. Where early "second-wave" feminists (1960s to 1970s) attempted to sever the automatic assumptions and connections people typically made among women, children, and families as a way of liberating women to pursue other aspirations, pro-feminist male activists speak of such involvement as potentially "liberating" for men because it reconnects them to their emotional sides and broadens their lives beyond wage earning. Looked at more closely, these positions are understandable. They are also not as different as they seem. What earlier feminists railed against was not "the family" but the gendered family. They were less antagonistic to what women felt toward and did in the family than they were toward what men did not do. Because of the differential burden women carried in households, family life imposed constraints on women's opportunities for outside involvements in ways it did not on men's. More recently, pro-feminist male activists have acknowledged men's lack of involvement or weaker family commitments, and opposed defining men solely in terms of what they do away from the family. They, too, object to the way family life has been gendered.

Contemporary gender roles are still in flux. Few men or women are entirely egalitarian or traditional. Few with egalitarian attitudes, for example, divide all labor along lines of ability, interest, or necessity rather than gender. Also, marriages that claim to be traditional rarely have wives who submit to their husbands in all things. Furthermore, some who express egalitarian attitudes, especially males, may be more traditional in their behaviors than they realize, while others share more equally without being ideologically driven. Within marriages and families, the greatest areas of gender inequality continue to be the division of housework and child care. But change continues to occur in the direction of greater gender equality, and this equality promises greater intimacy and satisfaction for both men and women in their relationships.

## Summary

- Gender is one of the most meaningful statuses we occupy and one of the most basic components of who we are, how we see ourselves, and how we are seen and treated by others.

- Gender identity is one's sense of self within gender categories, and usually, but not always fits, with the gender into which we've been assigned based on apparent biological characteristics.

- Transgender individuals develop gender identities that differ from their biological sex characteristics. Increasing attention to transgender experience has challenged binary conceptualizations of gender.

- Despite popular beliefs, men and women are actually more similar than different. Innate gender differences are generally minimal; differences are encouraged by socialization.

- Across societies, much variation exists in how gender is perceived, including the perception of how many gender categories there are. Within societies, one finds multiple versions of masculinity and femininity, and gender experience intersects with and varies across race, class, sexuality, and other social statuses.

- *Gendered roles* are those roles a person is expected to perform as a result of being male or female in a particular culture. *Gender identity* refers to the sense one has of *being male or* female.

- *Gender* is a source of both inequality and difference. A majority of societies are *patriarchies,* societies in which males dominate and control economic and political institutions. *Egalitarian societies* are those in that neither gender possesses, such dominance, and in which males and females have similar amounts of power. *Matriarchies,* in which women dominate institutionally, have not been found.

- Symbolic interactionists view gender as something we actively create or "do" in everyday situations and relationships, not an internalized set of behavioral and personal attributes.

- Two important socialization theories are social learning theory and cognitive development theory.

- Parents, teachers, and peers (age-mates) are important agents of socialization during childhood and adolescence. Similarly, religion and mass media help shape our ideas about gender.

- Marriage, parenthood, and the workplace also influence the development of adult gender roles.

- The gendered roles we play in adulthood are affected by situations, opportunities, and constraints that can alter the path established by socialization.

- Traditional male roles emphasize dominance and work. Traditional female roles have greater role diversity according to ethnicity.

- Contemporary gender roles are more *egalitarian* than the traditional ones of the past. They reflect (1) the acceptance of women as workers and professionals, (2) increased questioning of motherhood as a core female identity, (3) greater equality in marital power, and (4) the expansion of male family roles.

- Changing gender-role behavior is often difficult because we evaluate ourselves in terms of fulfilling gender-role concepts. Also, gender roles have become an intrinsic part of us and our roles, and social structure reinforces traditional roles.

- There have been various social movements dedicated to challenging or changing women's or men's roles, including various feminisms and various "movements" and perspectives on men and masculinity.

## Key Terms

agents of socialization  129
androgyny  126
caballerismo  133
cisgender  114
cognition  126
cognitive development theory  128
egalitarian  115
expressive traits  125
extensive mothering  145
gender binary  114
gender identity  113
gender polarization  125
gender role  115
gendered roles  115
gender-rebellion feminism  149
gender-reform feminism  149
gender-resistant feminism  149
gender socialization  115
gender spectrum  114
gender stratification  115
hegemonic masculinity and emphasized femininity  123
instrumental traits  125
intensive mothering ideology  145
machismo  133
marianismo  133
matriarchies  115
modeling  128
peers  136
postgender relationships  145
second shift  121
social learning theory  126
traditionalization  141

# Intimacy, Friendship, and Love

# 5

## What Do You Think?

*Are the following statements True or False? You may be surprised by the answers as you read through the chapter.*

**T** **F** **1.** Those who lack positive intimate relationships are at greater risk of illness, depression, and accidents.

**T** **F** **2.** Friendship and love share many characteristics.

**T** **F** **3.** Americans have always chosen their spouses on the basis of romantic love.

**T** **F** **4.** Men have a more instrumental style of expressing love than do women.

**T** **F** **5.** Love is more important to heterosexual women than to lesbians, gay men, or heterosexual men.

**T** **F** **6.** Love and commitment are inseparable.

**T** **F** **7.** It is common for people in the United States to form couple relationships with people from similar backgrounds.

**T** **F** **8.** In hooking up, common on college campuses, there is no sexual double standard.

**T** **F** **9.** A high degree of jealousy is a sign of true love.

**T** **F** **10.** Although painful, breaking up can induce positive changes that can improve later relationships.

Jose Luis Pelaez, Inc/Blend Images/Getty Images

153

# Chapter Outline

"Love doesn't make the world go round, love is what makes the ride worthwhile," or so said American businessman Franklin P. Jones. From a variety of indicators, it appears as though Americans share that sentiment. We are, it seems, in love with love. We can see this in the ways we live our daily lives, especially in the kinds of relationships we want, seek, and make, and in the steps we take to find and keep them. It is also evident in the popular culture that we produce and consume. There, we can see our love affair with love in everything from the music we choose to the things we read and watch.

Love is *the* dominant theme of popular music, where song titles and lyrics are typically testimonies to the power, pleasure, and pain associated with falling in and out of love. In the books we read, there is a whole genre devoted to love and "romance," which, based on 2013 data, is the largest in the consumer book market. Romance fiction makes up 13 percent of adult fiction, and in 2013 generated an estimated $1.08 *billion* in sales. According to the Romance Writers of America, nearly 65 million Americans read romance novels, over 84 percent of whom are female (Romance Writers of America 2013).

However, even more than in our music or books, our devotion to love stands out especially well in movies. Romantic movies, *love stories* as they are often appropriately called, provide us with vivid scenes and memorable lines filled with heartfelt, often poignant declarations of the depth of a character's love. Often specific love scenes stay with us, even coming to symbolize our very idea of true love. Some of the more successful, acclaimed, and/or award-winning movies of the last decade had love stories at their core, albeit in very different ways. In 2005's *Brokeback Mountain*, the love was between two men, ranch hand Ennis Del Mar and rodeo cowboy Jack Twist, across 20 years, two marriages, children, and divorces, until Jack's tragic death. In 2008's critically acclaimed *WALL-E*, the love relationship was between two robots, but whatever other themes characterized the movie, WALL-E's feelings for EVE and his determined desire to just hold her hand are front and center. In *Twilight*, the first in the series of films based on the immensely successful four-book series by Stephenie Meyers, the love story is between a vampire and a human (who later becomes a vampire herself), and the passionate yet dangerous love that leads them to take great risks to be together. And in 2013's *Her*, the love is between Theodore, a lonely man and Samantha, his computer's operating system. Cutting across and unifying these very different stories is the fact that that they each depict the overpowering, life-shaping experience of falling and being in love.

One of the top grossing movies of 2014 was *The Fault in Our Stars*. Based on the popular 2012 novel by John Green, the film tells the story of two teenagers, Hazel Grace Lancaster and Augustus Waters, who meet in a cancer support group and ultimately fall in love. Although both are dying, the film is a testament to the power of love. In one scene, Gus sums up one of the central themes of the story—despite all the bad things that can and often do happen, love is worth sharing:

> I am in love with you. And I know that love is just a shout into the void, and that oblivion is inevitable, and that we're all doomed. And that one day all our labor will be returned to dust. And I know that the sun will swallow the only earth we will ever have. And I am in love with you.

Much like the wider culture that surrounds them, families in the United States place high value on love. Decisions about entering or exiting a marriage, assessments of the quality and success of any particular marriage, and devotion between spouses or between parents and children all come down to love. On both an individual level and a familial level, then, it is important to consider the role love plays in our lives. This chapter is devoted to such consideration, exploring the role love plays in our personal and familial lives, what it is, and the processes through which we go about finding it. However, before we turn to love, we ought to consider the broader phenomenon of intimacy, including the intimacy of friendship.

# The Need for Love and Intimacy

Humans require other humans with whom we feel close and to whom we can commit. We need to form relationships in which we can share ourselves with others, exchange affection, and feel connected. In the developmental model psychologist Erik Erikson formulated, this was the great task facing us in young adulthood—intimacy versus isolation; either we satisfy our need for intimacy, or we remain socially and emotionally isolated (Hook et al. 2003). In psychologist Abraham Maslow's **hierarchy of needs**, after meeting our physiological needs and needs for safety, our social needs—for intimacy and love—are the most fundamental of human needs (Maslow 1970). Just what exactly is intimacy, and why is it so important?

In its most general sense, **intimacy** refers to closeness between two people. Sometimes we associate "intimacy" or "being intimate" with sexual relations. Certainly, sexual relations are part of physical intimacy, as are kisses, caresses, and hugs. However, it is more the emotional intimacy, having someone to talk to, to share ourselves with that is such an important part of our social and psychological well-being.

Reviewing research and theory on intimacy, psychologist Misty K. Hook and colleagues (2003) suggest that intimacy consists of four key features: the presence of *love and/or affection*, *personal validation*, *trust*, and *self-disclosure*. The more we feel as though another person likes or loves us, the more comfortable we will be sharing our innermost feelings and revealing our most personal thoughts. When we feel as though we are understood and appreciated, we feel more accepted and safer to freely open up to another person without the fear of being judged or betrayed. Finally, being intimate entails **self-disclosure**, sharing both the facts of our lives and our deeper feelings (Hook et al. 2003).

Intimate relationships provide us with a variety of benefits. They buffer us against loneliness, provide us with positive feelings about ourselves and others, give us confidence that our needs will be fulfilled in the future, and enhance our self-esteem.

> **True**
> **1.** Those who lack positive intimate relationships are at greater risk of illness, depression, and accidents.

Intimate relationships are connected to happiness, contentment, and a sense of well-being. They also offer protection from some stress-related symptoms and reduce our likelihood of illness, depression, and accidents. People who lack satisfying, positive intimate relationships are at greater risk of illness; once ill, they recover more slowly and have higher susceptibility to relapse or recurrence of their illness.

In a relationship, intimacy can be expressed in a variety of ways—talking together, listening to each other, making time for each other, being open and honest with each other, and trusting each other. The importance of intimacy in defining relationship quality cannot be stressed strongly enough. It is a more important factor in relationship satisfaction than independence (autonomy, individuality, and freedom), agreement (harmony and few quarrels), or sexuality (sexual harmony and satisfaction and physical contact) (Hassebrauck and Fehr 2002). This is not only true in the United States but in other countries as well. In comparative research using German and Canadian samples, intimacy was the factor most highly correlated with relationship satisfaction in both countries and for both males and females (although it was somewhat more strongly correlated with women's than with men's relationship quality and may have different meanings for females and males) (Hassebrauck and Fehr 2002; Hook et al. 2003).

## The Intimacy of Friendship and Love

Although both have proved difficult to define with precision or consistency, friendship and love are the two most important sources of intimacy people have. They help preserve both our physical and mental well-being. The loss of a friend and especially a loved one can lead to illness and even suicide.

Friendship can supply the foundation for a strong love relationship. Shared interests and values, acceptance, trust, understanding, and enjoyment are at the root of friendship and form a basis for love. As much as they may provide similar benefits, love and friendship are not the same thing. Even though individuals want and value many of the same qualities in friends and lovers—such as trust, acceptance, kindness,

> **True**
> **2.** Friendship and love share many characteristics.

and warmth—they exercise more selectivity in choosing romantic partners than they do friends.

Research by Susan Sprecher and Pamela Regan suggests that warmth, kindness, openness, expressiveness, and a sense of humor are considered the most desirable and important qualities in both friends and partners. However, romantic partners were subjected to higher standards for these attributes, suggesting that such qualities are more important in romantic partners than in friends. Romantic partners are also evaluated on the basis of attributes such as appearance or social status; friends are not (Sprecher and Regan 2002).

Potential friends may be deemed desirable on the basis of their specific combination of unique attributes and how those attributes match our needs and wants at a given point in time. Romantic partners, on the other hand, are more carefully selected, and their desirability is more carefully evaluated on the basis of their possession of certain qualities or attributes that might indicate their commitment to the relationship, their potential reproductive success, and their eventual attachment to offspring. In terms John Scanzoni and colleagues introduced, romantic partners are selected on the basis of their seeming ability to satisfy multiple needs that are products of the multiple "interdependencies" two people share. Interdependencies consist of shared activities, statuses, and patterned exchanges between two people. Romantic partners are expected to be able to satisfy four types of interdependencies: intrinsic (e.g., emotional support), extrinsic (e.g., money or services), sexual (sexual activity), and formal (shared legal status). Friends, however, typically provide only intrinsic resources (Scanzoni et al. 1989; Sprecher and Regan 2002). Research on college students' friendships and romantic relationships by psychologists Robert Fuhrman, Dorothy Flannagan, and Mike Matamoros found that romantic partners were held to higher expectations than were friends, even when the romantic relationships were of much shorter duration than the friendships. They suggest that this may result from the importance of romantic partners as attachment figures, the exclusivity that characterizes romantic relationships, and the promise of long-term commitment that often characterizes developing romantic relationships (Fuhrman, Flannagan, and Matamoros 2009).

## Why It Matters: The Importance of Love

Love is both a feeling and an activity. One feels love for someone, and one acts in a loving manner. Love is essential to our lives. Love binds us together as

Love stories, even those with sad endings, are popular in literature and film, such as 2014's "The Fault in Our Stars," based on the successful and critically acclaimed novel by John Green.

Photos 12/Alamy Stock Photo

partners, spouses, parents and children, and friends and family. Individuals make major life decisions, such as marrying, on the basis of love. Love creates bonds that we hope will enable us to endure the greatest hardships, suffer the severest cruelty, and overcome any distance.

We can look at love in many ways besides through the eyes of lovers, although other ways may not be as entertaining. Whereas love was once the province of lovers, madmen, poets, and philosophers, social scientists have also taken a look at love. Although there is something to be said for the mystery of love, understanding how love works in the day-to-day world may help us keep our love vital and growing.

## Love and Families in the United States

Romantic love is the primary basis for family formation in the United States, as it has been for most of the last two centuries (Coontz 2004). Although marriages in the United States were never quite as formally arranged as they have been in other places in the world, they were guided by more practical considerations and subject to more parental, especially paternal, control throughout the 18th century. By the end of the 19th century, however, most active parental involvement in their children's marriage choices had dissipated (Coontz 2004; Mintz 2004). Economic developments had decreased the dependency of adult children on their parents; increasingly, economic opportunity could be found without parental assistance, which freed people from worrying about the consequences of parental disapproval of their choice of mate. With increasing economic activity among women, a spread of legal and social recognition of women's rights, and enhanced opportunity for young people to meet and mingle, American courtship was further transformed (Mintz and Kellogg 1988; Murstein 1986). Love, as individuals experience, perceive, and pursue it, became the vehicle that drove mate selection.

In the early decades of the 20th century, new ideals about marriage and family emerged. Although American family life had already shifted from a largely economic to a mostly emotional emphasis, this was extended even further with the emergence and celebration of *companionate marriage*, wherein spouses were to be each other's best friends, confidants, and romantic partners (Mintz and Kellogg 1988). Where once such love would have been considered a "risky basis for marriage," it became the foundation on which marriage was to be built and the criterion by which spouses were to be chosen (Cherlin 2009).

Selecting a spouse on the basis of romantic love has consequences. It may lead to a greater tendency to idealize the partner, display affection toward the partner, and attach more importance to sexual intimacy (Medora et al. 2002). Ironically, perhaps, the high emphasis placed on love as the basis for spousal choice contributes to the patterns of divorce and remarriage in the United States. The qualities that people "fall in love" with may not be easy to sustain across the lifelong duration of a marriage. Thus, marriages are more likely to be perceived as "failures" when individuals sense that those qualities are gone or diminished. They may then seek those same idealized qualities from subsequent marriage partners.

## The Culture of Love

Within our marriage practices, one finds a number of distinct but related cultural beliefs about the character and place of love. However, the prevalence and extent of **homogamy**—the tendency for people to marry others much like themselves—casts some doubt on some of our ideas about the blind, irrational qualities of love and how such an emotion carries us into marriage. Perhaps love is more controllable and rational than presumed (and therefore not truly blind) because people seem to fall in love with people like themselves. On the other hand, perhaps love *is* blind

> **False** **3.** Americans have not always chosen their spouses on the basis of romantic love.

(i.e., uncontrollable and irrational) but choosing a partner or spouse is not. Love is not the only determinant of mate selection, and people don't necessarily marry simply because they've fallen in love. Some "loves" are recognized as unwise marriages. Finally, the social circles within which a person lives and moves limit love. Thus, our "one and only" may be drawn from a smaller pool than what the romantic mystique surrounding love suggests. With the emergence of the Internet and the proliferation of websites devoted to finding and forming relationships, this is likely changing for many people. As we will see in a later section, such sites widen the pool of potential partners with whom to form a relationship as people become less dependent on finding someone from among those they see or meet face-to-face on a daily basis. Even with the tendencies toward falling in love with and/or marrying individuals similar to ourselves, it is worth remembering that most Americans who marry say they are marrying because they are in love.

In addition, there are other beliefs making up the ideology of romanticism in U.S. culture (see the Exploring Diversity feature on cultural differences of love). Many Americans believe that love strikes powerfully on first sight, that each of us has one, and only one, "one and only," and that as long as we love each other, everything else will work out. As we will see, these beliefs are not as widely shared in other cultures as they are in the United States and western European societies.

Neither "falling in love" nor the experience of romantic love are unique to Americans; 90 percent of the 166 societies William Jankowiak and Edward Fischer (1992) examined recognize and value love as an important element in building intimate relationships. In fact, scholars now consider love to be universal, though there is variation in how commonly it is experienced or expressed (Hatfield and Rapson 2000). But love appears to play an especially significant role in American mate selection, perhaps even more so than in other Western societies (Peoples and Bailey 2006). It fits well with and helps reinforce other features of American families and society. Love-based marriage validates the importance of individual autonomy and freedom from parental intervention and control, establishes the relative independence of the **conjugal family** from the extended family, and fits with the wider social freedoms granted to adolescents and

Although most cultures recognize and value love, the meanings and expectations attached to love vary, sometimes greatly. In individualistic cultures, such as the United States, people value passionate love, the kind experienced as an "intense longing for another," a "lovesickness" that often takes us on a roller coaster of "elation and despair, thrills and terror" (Kim and Hatfield 2004, 174). If reciprocal, passionate love brings us ecstasy and fulfillment; if unrequited, it can bring us emptiness and sadness. In individualistic cultures, it is expected that people marry out of such an intense love, which is to be the most important factor in finding a spouse. This is part of the greater romanticism found in such societies, where prime importance is given to the emotional element of relationships and there is a stronger belief in each of the following components associated with romanticism:

1. Love conquers all.
2. For each person there is "one and only one" romantic match.
3. Our beloved should and will meet our highest ideals.
4. Love can and often most powerfully does strike "at first sight."
5. We should follow our hearts and not our minds when choosing a partner. (Medora et al. 2002)

In collectivist cultures, including many Asian societies such as Japan, China, India, and Korea, individual happiness is subordinated to group well-being. Loyalty, especially to the wider kin group and extended family, dictates decisions people make about entering marriage and who they shall marry. Higher value is placed on what Elaine Hatfield and Richard Rapson (1993) call companionate love, a less intense emotion in which warm affection and tenderness is felt and expressed toward those to whom our lives are deeply connected. Importance is placed on shared values, commitment, intimacy, and trust. In collectivist cultures, passionate love and marriage based on romantic love are seen negatively as potential threats to family-approved and/or arranged marriages, associated with sadness and jealousy, and thought to interfere with family closeness and kin obligations (Kim and Hatfield 2004). More traditional and less developed collectivist Eastern cultures, such as China and India, are reported to attach the least importance to romantic love. The idea of baring the soul, sharing, or confiding innermost and heartfelt feelings to a partner receives more cultural

validation in the United States and other individualistic cultures than in collectivist cultures (Kito 2005).

Additional cross-cultural research compared the attitudes toward romantic love of college undergraduates from the United States, Turkey, and India. In the United States, romanticism is idealized, and topics such as love, dating, and the media openly and frequently discuss or cover finding a partner. India is a sexually conservative and more collectivist society in which family stability is valued above individual gratification, and autonomy and marriages are frequently arranged. Turkey is a society "in transition." The ideal of romantic love was introduced as part of the processes of Westernization and secularization. Comparing the attitudes of college undergraduates from the three countries, researchers found the students from the United States to be most romantic and the Indian students to be least romantic (Medora et al. 2002). The study used a 29-item, five-point scale (from 1 = strongly disagree to 5 = strongly agree), wherein individuals could score between a low of 29 and a high of 145, with higher scores indicating more romanticism. Items included statements such as the following:

> "Somewhere there is an ideal mate for most people. The problem is just finding that one."
> "Love at first sight is often the deepest and most enduring type of love."

The average scores were as follows:

|  | United States | Turkey | India |
| --- | --- | --- | --- |
|  | N = 200 (86 male, 114 female) | N = 223 (114 male, 114 female) | N = 218 (98 male, 120 female) |
| Mean score | 86.09 | 74.92 | 70.33 |
| Standard deviation | 15.6 | 13.6 | 14.4 |

*From Medora et al. (2002).*

In all three national subsamples, females scored higher than males. Overall, the gender difference was as shown here:

|  | Male | Female |
| --- | --- | --- |
| Mean score | 74.63 | 79.81 |

*From Medora et al. (2002).* ●

Although it is difficult to come up with a formal definition of love, we usually know what we mean when we tell someone we love them. Such feelings are important at the individual, relationship, and institutional level.

young adults (Goode 1982). All of these make romantic love functional in industrial societies (Goode 1977). Conversely, in societies in which nuclear families are deeply embedded in extended families or in which it is important for economic or political reasons to create alliances and exchanges through marriages, romantic love is not the central factor in mate selection. In such societies, it may be entirely irrelevant (Medora et al. 2002).

Love reflects the positive factors, such as caring and attraction, that draw two people together and sustain them in a relationship. Related to love, **commitment** reflects the stable factors, including not only love but also obligations and social pressure that help maintain a relationship, for better or for worse. Although love and commitment are related, they are not inevitably connected. It is possible to love someone without being committed, without making the sacrifices and adjustments needed to sustain the relationship. It is also possible to be committed to someone without loving that person. Even in the absence (or disappearance) or love, we might remain in a relationship such as marriage because of perceived obligation, for the sake of the children, or because of fear of how other aspects of our life might be negatively affected. Yet, when all is said and done, most of us long for a love that includes commitment and a commitment that encompasses love.

**True** **4.** Men have a more instrumental style of expressing love than do women.

# Gender and Intimacy: Men and Women as Friends and Lovers

As shown in the previous chapter, many areas of our lives are gendered, meaning that males and females experience them differently. Intimacy, friendship, and love are among such areas, as is sexual intimacy, the topic of Chapter 6. In much social scientific literature, there is a recurring theme highlighting men's supposed shortcomings as friends and partners when compared to women. Unlike women, who are said to relate more easily and deeply with others and who develop and display a greater capacity for disclosing and sharing their inner selves, men maintain greater emotional distance, even as they experience their closest relationships.

Of note, sociologist Francesca Cancian (1985) argued that there is a gender bias in our cultural constructions of love that likely distorts our understanding of how both men and women love. She referred to this bias as the **feminization of love**. By defining or "seeing" love in largely expressive terms (telling each other how you feel), important qualities or aspects of *both* women's and men's intimacy are ignored or overlooked. For example, much of what women do as expressions of love (for spouses and children, especially) consists of **instrumental displays**, via tasks associated with nurturing and caregiving, more than **expressive displays**, such as telling others how much we care about or love them. Although done *out of love*, such activities may not be seen as displays of love. Likewise, because men may believe they "show" or express love by *what they do* more than by *what they say*, conceptualizing or recognizing love largely in terms of things said renders men's sincere attempts to show intimacy invisible and leaves them looking especially inadequate as intimate partners (Hook et al. 2003).

Hook and colleagues (2003) note the following gender differences in intimacy. To women, intimacy means sharing love and affection and expressing warm feelings toward someone. To men, being intimate may mean engaging in sexual behavior and being physically close. For women, positive sexual interaction tends to require an emotional closeness (Peplau 2001). Women display intimacy in their verbal exchanges, express more empathy, and are more likely than men are to come to an understanding of what others are feeling. In heterosexual relationships, they have been found to work harder than their partners to generate and sustain the intimacy in their relationships both by urging their male partners to share their feelings and by occasionally repressing their own feelings to enhance their partner's well-being (Umberson, Thomeer, and Lodge 2015).

Men are more likely to react to disclosures of negative or problematic emotions by trying to solve a supposed problem. Men also associate intimacy with "doing" things together or for another person and often find women's need or desire to "talk things through" puzzling. Although men may feel as though they show intimacy by sharing activities and interests, telling stories, and even sitting together in silence, women associate intimacy with being together and sharing themselves with another (Hook et al. 2003).

## Gender and Friendship

The critique of the cultural feminization of love applies to friendship as well. We tend to conceptualize "real" or "true" friendship by such qualities as emotional support and self-disclosure—telling each other innermost feelings and sharing personal experiences (Sprecher and Hendrick 2004). We expect friends to share their inner lives with one another, including how they feel about each other. The closer the friend, the more personal and more frequent the disclosures. This conceptualization measures friendship against a standard more consistent with female friendships and may underestimate the "real" intimacy that men's friendships contain, especially if such closeness is expressed in other, more covert ways, such as through joking behavior and shared athletic activity (Hook et al. 2003; Swain 1989).

Indeed, there are gender differences in disclosure in same-sex friendships. If intimacy means self-disclosure, then as early as age 6, female friendships are more intimate. This gender difference is accentuated in adolescence and persists into and through adulthood (Benenson and Christakos 2003; Fuhrman, Flannagan, and Matamoros 2009). Women experience and express "closeness" with each other through conversation, disclosing more of both a positive and a negative nature (Hook et al. 2003; Sprecher and Hendrick 2004). They also expect their female friends to provide greater emotional closeness than males expect of their male friends (Fuhrman, Flannagan, and Matamoros 2009).

Female friendships typically involve more personal disclosure than do male friendships.

Simon Winnall/Taxi/Getty Images

There are other noteworthy differences between male and female friendships. In childhood and adolescence, boys spend more of their time in groups and in group activities, especially physical activities, games, and sports; girls spend more time in dyads (groups of two) and engage in more mutual disclosure. In fact, research on children and adolescents by psychologist Amanda Rose reveals that although girls expect benefits from sharing their problems with one another, such as feeling less alone and more cared for, boys said that such sharing would feel like "a waste of time" (Nauert 2011). When conflict arises between close friends, males may have an easier time reaching resolution. Within a group context, we can draw others in, drawing on third parties to act as mediators, serve as allies, or even become alternate partners. With more loyalty to the larger group, one-on-one conflict may be kept to a lower level (Benenson and Christakos 2003).

Boys share less with one another, spend less time sharing, and engage in less **co-rumination** (excessive discussion of personal problems). Co-rumination shares characteristics of self-disclosure but entails a mutual "excessive focus on problems and concerns" (Rose 2002; Rose, Carlson, and Waller 2007). Although co-rumination may be an aspect of close friendships and has been linked to higher levels of friendship satisfaction, it is also related to higher instances of anxiety and depression symptoms (Calmes and Roberts 2008). As a consequence, where boys are less likely to enjoy the stronger sense of closeness or deepened friendship that are increased by co-ruminating, they are also spared some of the emotional fallout that results from dwelling on problems (Rose 2002; Rose, Carlson, and Waller 2007).

In addition, men display less affection, using either words or touch, than women do toward their friends (Dolgin 2001). Yet female friendships appear to be more fragile. With increasingly intense sharing comes more opportunity for misunderstanding or even for conflict. Furthermore, when females' closest friendships end, they are more likely to "find themselves alone" (Benenson and Christakos 2003).

Heterosexual men are more open and intimate in cross-sex relationships than in their friendships with other men (Dolgin 2001). Wives or romantic partners are often the closest confidants in men's lives. In those relationships, men find themselves reaping the benefits that come from greater disclosure, even if such disclosure may need to be drawn out of them by wives or female partners and the levels at which they disclose don't match what their partners desire. Certainly, the tendency to funnel their intimacy into one relationship, especially marriage, is consistent with the cultural expectations of companionate marriage, where one's spouse is expected to be one's closest confidant. But even outside marriage, the depth of men's disclosure to women stands in contrast to the male–male style, suggesting not so much inability as unwillingness at or discomfort with male–male intimacy.

Experiences and expectations of cross-sex friends seem to vary depending on whether one or both friends is also in a romantic relationship. Research on college students suggests that for those women and men who are *not* in romantic relationships, expectations are higher for cross-sex friendships than for same-sex friendships. Those who are in romantic relationships have similar expectations for both their same-sex and cross-sex friendships (Fuhrman, Flannagan, and Matamoros 2009).

A unique cross-sex friendship type is known as the "friends with benefits" (FWB) relationship. Such relationships emerge out of cross-sex friendships and incorporate sexual intimacy into those relationships without the expectation or desire for greater commitment or pushing the relationship into a love relationship (Hughes, Morrison, and Asada 2005). These people retain the benefits of friendship, such

as mutual understanding, support, and companionship, *along with* physical intimacy more typical of romantic relationships, yet with neither the responsibilities nor the expected commitment of a romantic sexual relationship. Friends with benefits may be most common on college campuses, where studies have reported that 48 percent to 62 percent of college student respondents have experienced such relationships (Hughes, Morrison, and Asada 2005). In their study of 888 young adults, Jesse Owen and Frank Fincham found that 54.3 percent of males but only 42.9 percent of females reported having at least one FWB relationship and such relationships were deemed to have led to more positive than negative emotional reactions (Owen and Fincham 2011). Research by Paul Mongeau and colleagues found considerable diversity among friends with benefits relationships in terms of the nature of the intimacy and depth of emotional investment, eventually identifying multiple types of such relationships. At one end were the true friends type, while at the other were "just sex partners" (Mongeau et al. 2013).

## Gender and Love

With regard to love, the genders have been found to differ in a number of ways. Men fall in love more quickly than women, describe more instrumental styles of love (i.e., love as "doing"), and are more likely to see sex as a way to express love, yet more easily separate love and sex. Interestingly, they are also more likely to subscribe to romantic beliefs about love and to believe in love at first sight. Because men have fewer deeply intimate, self-disclosing friendships, when they find this quality in a relationship, they are more likely to perceive that relationship as special. Having more intimates with whom they can share their feelings, women are less likely to be as quick to characterize a particular relationship as love. In addition, women traditionally could do so less safely unless other economic criteria were also met. Thus, men could afford to be more romantic, and women needed to be more realistic, taking into account a man's suitability as a stable provider along with his ability to satisfy needs for intimacy (Knox and Schacht 2000).

Gender differences may be more exaggerated in *what people say* than in *what they do*. This may be true of both friendship and love, where bigger differences show up in how the genders talk *about* relationships than in what they actually do or experience (Eastwick and Finkel 2008; Walker 1994). Despite the gendered expectations and definitions of love, significant gender differences in self-disclosure are absent within the context of heterosexual, romantic love relationships; men and women disclose similarly (Sprecher and Hendrick 2004).

In identifying the factors that shape men's and women's intimate relationships and *could* contribute to differences between the genders, most researchers point in some way to lessons of gender socialization. Within dominant cultural constructions of masculinity and femininity, males are inexpressive, competitive, rational, and uncomfortable with revealing their innermost feelings, especially of vulnerability or of affection toward other males. Females are allowed and encouraged to express a wider range of feelings without concern for the consequences.

Other researchers suggest that gender-specific relationship styles emerge because of differences in how males and females resolve the developmental task of early childhood identity formation (Chodorow 1978; Rubin 1985). As a result of being "mothered" and having the closest early relationship be with a female, the genders develop different ways of relating. Females develop "permeable ego boundaries" open to relationships with others, and they retain a strong connection with their mothers. Males are forced to separate from their mothers, identify with absent or less present fathers, and build boundaries around themselves in relation to their most nurturant caregivers. This haunts them throughout their later relationships because it makes them less able to "connect" intimately with others (Rubin 1985). Women experience themselves in the context of relationships, whereas men—depicted as "selves in separation"—remain oriented more toward independence and task completion (Kilmartin 1994).

Even without attention to the complexity associated with personality formation during early development, one could also emphasize the likely role model consequences of being "mothered" but not "fathered." Without a loving, attentive, nurturing presence from fathers or other male role models, boys might come to inhibit their own emotional expressiveness, identifying such behavior as typical of mothers (and women in general) and to be avoided. Because of the still greater involvement of mothers versus fathers in caring for most young children and the greater prevalence of single-mother over single-father households, boys may well still have fewer available role models for intimacy. Furthermore, what role models they have are themselves the products of gender socialization and

carry a style of relating that results from that socialization. Girls have the opportunity to observe up close a caring, loving female role model from which they learn how to relate and express love.

Finally, still others stress evolutionary explanations for gender differences. Beginning with the idea that each gender has different "reproductive strategies," differences in intimacy are linked to such sex-specific goals. For males, the objective is to reproduce as widely as possible, seeing that their genetic material is spread widely in multiple offspring. For women, the objective is to see that each child they bear successfully survives to a healthy adulthood. Such a difference is said to explain numerous other differences, especially in areas of intimacy, love, and sexuality. For example, from an evolutionary perspective on qualities desired in romantic and sexual partners, females will desire males of high status who are ambitious and dependable. Males will desire physically attractive females who can bear them healthy children (Sprecher and Regan 2002).

## Showing Love: Affection and Sexuality

Within relationships based on romantic or passionate love, the emotional connection between partners is expressed in many ways, typically through displays of affection and through sexual desire and activity. Most people assume the state of "being in love" to include sexual desire. Two people in a relationship lacking sexual desire are assumed to not be in love (Regan 2000). Psychologist Lisa Diamond (2004) challenges this assumption, noting that sexual desire often occurs in the absence of romantic or passionate love and, more controversially, that romantic love, even in its earliest and most passionate stage, does not require sexual desire. She offers the examples of prepubertal children who have yet to undergo the hormonal changes necessary for adult levels of sexual desire who, nonetheless, experience intensely romantic infatuations as well as instances wherein people fall in love with others of the wrong gender (heterosexuals with someone of the same sex and gay men and lesbians who fall in love with someone of the other gender). Regarding the latter, Diamond notes that both cross-cultural and historical evidence documents such romantic relationships, even in the absence of sexual desire.

**A Kiss Is Just a Kiss? It Depends**

One of the most widely accepted and expected ways that lovers, partners, and spouses express affection, love, and passion is by kissing. Couples recall their "first kiss" as a turning point in their relationships, and one or both partners approach that moment with anticipation and occasionally anxiety. Until we kiss, we're not sure of the nature of our relationships. Are we more than friends? Kisses become markers of deepening intimacy and connection. And, according to researchers, we spend roughly 20,000 minutes of our lives (two weeks) sharing kisses. The centrality of kissing as a display of affection, a precursor to further intimacy, and an expression of romantic love is widespread in the United States. As anthropologists are moved to do, William Jankowiak, Shelly Volsche, and Justin Garcia (2015) wanted to know just how common "romantic-sexual kissing" was across cultures. Was it a nearly universal custom as the academic and popular literature had contended?

Conducting the first cross-cultural survey to establish the presence or absence of romantic-sexual kissing, they used data from the Standard Cross-Cultural Sample and the Human Relations Area Files World Cultures to draw a sample of 168 cultures from nine cultural areas: Africa, Asia, Europe, Middle America, Caribbean, Middle East, North and South America, and Oceania. They further contacted 88 ethnographers, asking whether in their research they observed or heard of people kissing on the mouth "in a sexual, intimate setting." Of the 28 who responded, only 14 had studied couple intimacy. Of these 14, 12 "had never observed a single instance of romantic-sexual kissing within their studied culture" (Jankowiak, Volsche, and Garcia 2015). Six additional ethnographers

whose work was with foraging, horticultural, or agricultural groups were contacted, and they provided data on 13 cultures not in the initial sample.

They employed what they consider a conventional definition of romantic-sexual kissing, "lip-to-lip contact that may or may not be prolonged. In other words, the romantic-sexual kiss is not a passing glance of the lips, but rather the intentional touching of the lips that is more focused and thus potentially more prolonged" (Jankowiak, Volsche, and Garcia 2015). They also categorized their sample societies in terms of their levels of societal complexity, labeling them as either "egalitarian," "simple," or "complex."

Although other researchers such as Rafael Wlodarski and Robin Dunbar (2014) had claimed that kissing was "a near-ubiquitous custom among human cultures," Jankowiak, Volsche, and Garcia found that romantic-sexual kissing was found in only 77 of their sample cultures (46 percent), but was absent in 91 of the 168 (54 percent) cultures in their sample. Furthermore, romantic-sexual kissing was present most frequently in complex societies:

> Significantly, no ethnographer working with sub-Saharan African, New Guinea, or Amazonian foragers or horticulturalists reported having witnessed any occasion in which their study populations engaged in a romantic-sexual kiss. . . . Like other romantic and sexual behaviors, while kissing may be a way to communicate intimacy in some societies or may function as a specific eroticized activity in others, . . . for quite a few kissing is seen as unpleasant, unclean, or simply unusual. ●

Although love and sex are separate phenomena, recent research shows that for both men and women, sex often includes intimacy and caring, key aspects of love, and love is most often expected to include sexual desire. Men and women who feel the greatest sexual desire for dating partners are also likely to report the strongest feelings of passionate love. Interestingly, sexual *activity* (as measured by the average weekly number of "sexual events" in which partners engaged) is not as strongly associated with amount or depth of passionate love (Regan 2000).

Besides sexual intimacy, we show intimacy and love in many other ways. Back rubs, caresses, cuddling, holding hands, hugging, and kissing—either on the lips or face are all ways to display affection (Kent and El-Alayli 2011). Such displays have been found to enhance relationships and contribute to further relationship development or maintenance. Research on both heterosexual and same-sex relationships has found that such displays are connected to the degree to which partners are satisfied with and feel intimacy within their relationship (Kent and Al-Alayli 2011). Some of these displays occur openly, in public, as we say, and show others that we are a couple. Holding hands, being out together alone, and telling others that we are a couple are all examples of public displays of affection and couple status (Vaquera and Kao 2005). More privately, we may exchange presents,

Interracial relationships are increasing in number and acceptability.

Bob Pool/Photographer's Choice/Getty Images

tell each other how we feel (saying that we love each other), and just think about ourselves as a couple. Finally, the physical acts, from kissing to touching under clothes or with no clothes on, touching each other's genitals, and having sexual intercourse, are all "intimate displays" (Vaquera and Kao 2005).

For some types of couples, such as interracial and same-sex couples, public displays of affection may be deliberately kept to a minimum or avoided because of the potential for social disapproval. Using data drawn from the 1994–1995 National Longitudinal Study of Adolescent Health with its large, nationally representative sample of high school students, Elizabeth Vaquera and Grace Kao (2005) examined how displays of affection varied between intraracial and interracial couples and whether the potential "stigmatizing" of such relationships led to different ways of behaving as a couple. They found that interracial couples displayed lower levels of affection in terms of both public and private displays. However, when it comes to intimate displays of affection (e.g., kissing, touching one's genitals, and sexual intercourse), no difference was found between interracial and intraracial couples. Erin Kent and Amani El-Alayli compared public displays of affection among women in

committed heterosexual relationships and those in lesbian relationships and found that although private displays of affection did not differ between women who were partners in same-sex and different-sex relationships, women in same-sex relationships displayed significantly less physical affection in public settings (Kent and El-Alayli 2011). Vaquera and Kao (2005) also reported differences in the displays of affection across racial groups (i.e., among intraracial couples of different racial backgrounds). They determined that compared with Caucasians, African American couples displayed less public affection but more intimate affection. Hispanics, Asians, and Native Americans displayed lower levels of intimate affection in public than African American or Caucasian couples. All minority couples displayed less public affection than white couples did. In terms of "private displays," no statistically significant differences were found among racial groups (Vaquera and Kao 2005).

## Gender, Love, and Sexual Activity

For both women and men, sexual desire—but not sexual activity—is associated with passionate love (Regan 2000). Yet gender differences have been observed in the relationship between love and sex. Men and women who are not in an established relationship have different expectations. Men are more likely than women to more easily separate sex from affection, thus feeling freer to engage in sex in the absence of a close emotional connection whereas women attach greater importance to relationships as the "context" for sexual expression (Diamond 2004; Laumann et al. 1994). Lisa Diamond suggests a number of possible reasons for this gender difference. First, men are more likely than women to first experience sexual arousal "in the solitary context of masturbation," whereas women are more likely to experience sexual arousal for the first time within a heterosexual relationship. Second, as shown in the next chapter, women and men have been differently socialized about the legitimacy of sexual expression. Women have been expected and encouraged to restrict sexual desire and activity to intimate relationships in which they find themselves. Men have been raised with more "license" regarding casual sexual relationships. Finally, Diamond notes that the gender difference may partly be explained by biological factors; specifically, certain neurochemicals, such as oxytocin, that mediate bonding also mediate sexual behavior. As much as oxytocin might be associated with caregiving, it is also released in greater amounts in women than in men during sexual activity. Oxytocin is also associated with orgasmic intensity (Diamond 2004). In their study of lesbian, gay, and heterosexual couples, researchers Debra Umberson, Mieke Beth Thomeer, and Amy Lodge found that half of their couples across relationship types reported that, over time, emotional intimacy increased while sexual frequency declined. As they note, this suggests that emotional intimacy and sexual expression may become less connected over time (Umberson, Thomeer, and Lodge 2015).

## Sexual Orientation and Love

We need to make three important points about how love is experienced by people of different sexual orientations. Love is neither more nor less important for heterosexuals, gay men, lesbians, and bisexuals (Aron and Aron 1991; Patterson 2000; Peplau and Cochran 1988). Given that men, in general, are more likely than women to separate love and sex, it is unsurprising that gay men are especially likely to make this separation. Although gay men value love, they also tend to value sex as an end in itself. Furthermore, they place less emphasis on sexual exclusiveness in their relationships (Patterson 2000). However, researchers suggest that heterosexual males are not very different from gay males in terms of their acceptance of casual sex. Lesbians and heterosexual couples tend to be more supportive than gay men of monogamy and sexual fidelity. This is probably because of gender more than sexual orientation; heterosexual males would be as likely as gay males to engage in casual sex if women were equally interested. Women, however, are not as interested in casual sex; as a result, heterosexual men do not have as many willing partners available as do gay men (Hendrick and Hendrick 1995).

Second, love may have special significance in the formation and acceptance of a same-sex or bisexual identity. Although significant numbers of women and men have had sexual experiences with members of the same sex or both sexes, relatively few identify themselves as lesbian or gay. One important element in solidifying such an identity is falling and/or being in love with someone of the same sex. Love signifies a commitment to being gay or lesbian by unifying the emotional and physical dimensions of a person's sexuality (Troiden 1988). For a gay man or lesbian,

Love transcends sexual orientation. Heterosexual and same-gender couples value and seek loving relationships.

Queerstock, Inc./Alamy Stock Photo

love marks the beginning of sexual wholeness and acceptance. Some researchers believe that the ability to love someone of the same sex, rather than having sex with him or her, is the critical element that distinguishes being gay or lesbian from being heterosexual.

Third, some of what have been reported as gender differences in experiencing and expressing intimacy may vary depending on the relational context (i.e., whether the couple is a same-sex couple or a male–female couple). Although there may be differences in general between women's and men's experiences, intimacy "is enacted and experienced by men and women in different ways depending on whether they are in a relationship with a man or a woman" (Umberson, Thomeer, and Lodge 2015, 542). Behavior that women display in love relationships with men may well differ from behavior they display in a lesbian relationship. Likewise, how men address issues in their intimate heterosexual relationships may differ from how those same things play out between gay male partners. This would indicate that gender of one's partner is also an important factor in shaping experiences of love and intimacy.

In their in-depth interviews with 50 couples (15 lesbian, 15 gay, and 20 heterosexual), Debra Umberson, Mieke Beth Thomeer, and Amy Lodge found that, indeed, on some issues "gender trumps relational context," whereas on other issues the gendered relational context is more influential. So, for example, women did more emotion work in their relationships

**False** 5. Love is not more important to heterosexual women than to lesbians, gay men, or heterosexual men.

than men did to reduce boundaries and minimize emotional distance between themselves and their partners, regardless of whether those partners were female or male. Other relationship issues, such as creating and respecting boundaries and providing emotional space, played out differently depending on whether one was in a same-sex as opposed to a heterosexual relationship. Thus, both women in heterosexual and men in gay male relationships—both being partnered with males—reported doing more of this type of emotion work (Umberson, Thomeer, and Lodge 2015).

## Love, Marriage, and Social Class

In many ways, the romantic view of love-based marriage represents a middle-class version of marriage. Among upper-class families, there is a greater sense of urgency in ensuring that their children marry the "right kind" of people because considerable wealth and social position may be at stake. Furthermore, upper-class families have more ability to exercise such control by the threat of withholding inheritance from the maverick child who dares act without consideration of parental preference. Among the working class, marriage was often entered as a means to escape economic instability and parental authority and to be seen as an adult. Eventually, working-class couples embraced many of the ideals of "companionate marriages" (e.g., expecting more sharing and communication). More recently, aspects of what is called "individualized marriage," in which dual-earner couples "follow independent paths to growth and change over their married lives, sharing their feelings, supporting each other, and engaging in the joint project of raising children," has come to characterize marriage among the college-educated middle class (Cherlin 2014, 163). It is not entirely clear to what extent such a view has also become the normative expectation among less educated men and women of the working class. However, as Cherlin notes, "Such a movement could reduce class differences in the conduct of family life" (Cherlin 2014,164).

What is unmistakably clear about class differences in intimate relationships and marriage is the substantial gap that has come to separate the family lives of college-educated middle-class couples from their less educated peers. In fact, the "retreat from marriage" that has been a source of great public discussion turns out to be a class-based phenomenon. Therefore, even if expectations and aspirations for intimate and marital relationships were the same across classes, how

those intimate relationships are constructed and experienced are not the same (Cherlin 2009, 2014). The economic circumstances that define someone's life may induce different ways of linking love and marriage.

## ◾ What Is This "Crazy Little Thing Called Love"?

Despite centuries of discussion, debate, and complaint by philosophers and lovers, no one has succeeded in finding a single definition of love on which all can agree. Ironically, such discussions seem to engender conflict and disagreement rather than love and harmony. Some researchers wonder whether such definitions are even possible (Myers and Shurts 2002).

We may not so much have formal definitions of love as we do **prototypes** of love (i.e., models of what we mean by love) stored in the backs of our minds. Some researchers suggest that instead of looking for formal definitions of love, it is more important to examine people's prototypes, that is, to consider what people mean by the concept of love when they use it. When someone says "I love you," he or she is referring to a prototype of love rather than its definition. If people think about their partner all the time, feel happy when with that partner and sad (or less happy) when apart, and spend all or most of their available time together, these thoughts, feelings, and behaviors are compared against a mental model or prototype of love (Regan 2003). If people's experiences match the different characteristics of love, individuals then define themselves as in love (Regan, Kocan, and Whitlock 1998). By thinking in terms of prototypes, we can study how people actually use the word *love* in real life and how the meanings they associate with love help them define and assess the progress of their intimate relationships.

In an effort to discover people's prototypes, researcher Beverley Fehr (1988) asked respondents to rate the central features of love and commitment. There were 68 different features of love mentioned by at least two respondents. In order, the 12 *central* attributes of love they listed are as follows:

- Trust
- Caring
- Honesty
- Friendship

- Respect
- Concern for the other's well-being
- Loyalty
- Commitment
- Acceptance of the other the way he or she is
- Supportiveness
- Wanting to be with the other
- Interest in the other

Many other characteristics were also identified as features of love (euphoria, thinking about the other all the time, butterflies in the stomach, and so on). These, however, tend to be *peripheral*. As relationships progress, the central aspects of love become more characteristic of the relationship than the peripheral ones, and it is on the basis of those aspects that we assess our relationships, moving toward feeling increased love or moving out of the relationship. Violations of central features of love are considered more serious than violations of peripheral ones. A loss of caring, trust, honesty, or respect threatens love, whereas the disappearance of butterflies in the stomach does not. Psychologist Pamela Regan, along with coauthors Elizabeth Kocan and Teresa Whitlock, undertook a prototype analysis of the more specific concept of *romantic love*. When they asked their respondents to list all the features or characteristics of "being in love," participants identified 119 different features. Of these, the central features of romantic love were trust, honesty, sexual attraction, acceptance and tolerance, happiness, spending time together, and sharing thoughts and secrets. There were few negative aspects noted, and those that were (e.g., anger, depression, and fear) were seen as peripheral characteristics (Regan, Kocan, and Whitlock 1998).

Love is also expressed behaviorally in several ways, with the expression of love often overlapping thoughts of love:

- Verbally expressing affection, such as saying "I love you"
- Self-disclosing, such as revealing intimate personal facts
- Giving nonmaterial evidence, such as offering emotional and moral support in times of need and showing respect for the other's opinion
- Expressing nonverbal feelings such as happiness, contentment, and security when the other is present

- Giving material evidence, such as providing gifts or small favors or doing more than the other's share of something
- Physically expressing love, such as by hugging, kissing, and other sexual activity
- Tolerating and accepting the other's idiosyncrasies, peculiar routines, or annoying habits, such as forgetting to put the cap on the toothpaste

These behavioral expressions of love are consistent with the prototypical characteristics of love. In addition, research supports the belief that people "walk on air" when they are in love. Researchers have found that those in love view the world more positively than those who are not in love (Hendrick and Hendrick 1988).

## Studying Love

Scholarly study of love was first pioneered by sociologists, and later—from the mid 1970s on—became a "respectable" topic for psychological study. Presently, the interest in studying love is multidisciplinary:

> Today, scholars from a wide variety of theoretical disciplines—anthropologists, clinical psychology, communication studies, developmental psychology, evolutionary psychologists, historians, neurobiologists, neuroscientists, primatologists, social psychologists, and sociologists, among others—are attempting to understand the nature of love. (Hatfield, Bensman, and Rapson 2012)

A thorough review of the research on love finds a number of definitions that are tied to a variety of research instruments developed to measure love. Elaine Hatfield, Lisamarie Bensman, and Richard Rapson (2012) identified 33 different scales used to measure passionate or romantic love. Next, we look briefly at some of the more influential approaches to studying love.

### Styles of Love

Hendrick and Hendrick's Love Attitudes Scale is a 42-item instrument based on and designed to measure sociologist John Lee's (1973, 1988) six styles of love:

- *Eros*. Romantic or passionate love. Erotic lovers delight in the tactile, the sensual, and the immediate; they are attracted to beauty (although beauty may be in the eye of the beholder). They love the lines of the body and its feel and touch. They are fascinated by every detail of their beloved. Their love burns brightly but soon flickers and dies.

- *Ludus.* Playful or game-playing love. For ludic lovers, love is a game, something to play at rather than to become deeply involved in. Love is ultimately ludicrous. Love is for fun; encounters are casual, carefree, and often careless. "Nothing serious" is the motto of ludic lovers.

- *Storge* (pronounced STOR-*gay*). Love between companions. It is, writes Lee, "love without fever, tumult, or folly, a peaceful and enchanting affection." It usually begins as friendship and then gradually deepens into love. If the love ends, it also occurs gradually, and the couple often becomes friends once again.

The first three are "primary" styles, which can be combined to generate the following secondary styles:

- *Mania.* Coming from the Greek word for "madness," mania is an obsessive love, a combination of ludus and eros, characterized by an intense love–hate relationship. For manic lovers, nights are marked by sleeplessness and days by pain and anxiety. The slightest sign of affection brings ecstasy briefly, only to have it disappear. Satisfactions last but a moment before they must be renewed. Manic love is roller-coaster love.

- *Agape* (pronounced ah-GAH-pay). A combination of eros and storge, agape is altruistic love. Agape is love that is chaste, patient, selfless, and undemanding; it does not expect to be reciprocated. Agape emphasizes nurturing and caring as their own rewards. It is the love of monastics, missionaries, and saints more than that of worldly couples.

- *Pragma.* A combination of storge and ludus, pragma is a practical, pragmatic style of love. Pragmatic lovers are primarily logical in their approach toward looking for someone who meets their needs. They look for a partner who has the background, education, personality, religion, and interests compatible with their own. If they meet a person who meets their criteria, erotic or manic feelings may develop.

These styles, Lee cautioned, are relationship styles, not individual styles. The style of love may change as the relationship changes or when individuals enter different relationships. In addition to these pure forms, there are mixtures of the basic types: storgic–eros, ludic–eros, and storgic–ludus. According to Lee, a person must thus find a partner who shares the same style and definition of love to have a mutually satisfying love affair. The more different two people are in their styles of love, the less likely it is that they will understand each other's love. Of these six basic types, psychologists Bianca Acevedo and Arthur Aron (2009) suggest that *romantic love* is best seen as eros, without the obsessiveness of mania and as distinct from the more friendship-like attachment of storge.

Love styles are also linked to gender and ethnicity. Research indicates that heterosexual and gay men have similar attitudes toward eros, mania, ludus, and storge, and that gay male relationships have multiple emotional dimensions (Adler, Hendrick, and Hendrick 1989). Women endorse the storgic and pragma styles and outscore men on measures of the manic style (Hendrick and Hendrick 1995). As to cultural differences, different styles tend to characterize Asians, African Americans, Latinos, and Caucasians. Asian Americans have a more pragmatic style of love than do Latinos, African Americans, or Caucasians, and they place a high value on affection, trust, and friendship (pragma and storge). Latinos often score higher on the ludic characteristics (Regan 2003).

## Hatfield and Sprecher's Passionate and Companionate Love

Hatfield and Sprecher divide love into two types: passionate and companionate. As shown earlier in the Exploring Diversity feature, **passionate love** is "an intense longing for union with another" and is familiar to us because it most fits our ideas of romantic love, or the state of being in love (Kim and Hatfield 2004). Passionate love can be seen through cognitive, emotional, and behavioral indicators. **Companionate love** refers more to the warm and tender affection we feel for close others. It includes friendship, shared interests and activities, and companionship. It is milder and less intense, may lack sexual attraction or desire, and produces less of the extreme highs and lows people experience from passionate love (Acevedo and Aron 2009; Kim and Hatfield 2004).

## Sternberg's Triangular Theory of Love

Sternberg's Triangular Love Scale is based on his **triangular theory of love**. According to the theory, love is composed of three elements that can be visualized as the points of a triangle: intimacy, passion, and decision or commitment. The *intimacy* component refers to the warm, close feelings of bonding you experience when you love someone. It includes such things as giving and receiving emotional support to and from your partner, being able to communicate with your partner about intimate things, being able to understand each other, and valuing your partner's presence in your life. The *passion* component refers to the elements of romance, attraction, and sexuality in a relationship. These may be fueled by the desire to increase self-esteem, to be sexually active or fulfilled, to affiliate with others, to dominate, or to subordinate. The decision or *commitment* component consists of two parts, one short term and one long term. The short-term part refers to your decision that you love someone. You may or may not make the decision consciously, but it usually occurs before you decide to make a commitment to that person. The commitment represents the long-term aspect; it is the maintenance of love, but a decision to love someone does not necessarily entail a commitment to maintaining that love.

Each of these components can be enlarged or diminished in the course of a relationship, and their changes will affect the quality of the relationship. They can also be combined in different ways in different relationships or even at different times in the same love relationship, creating one of the following eight different outcomes:

- Liking (intimacy only)
- Romantic love (intimacy and passion)
- Infatuation (passion only)
- Fatuous love (passion and commitment)
- Empty love (decision or commitment only)
- Companionate love (intimacy and commitment)
- Consummate love (intimacy, passion, and commitment)
- Nonlove (absence of intimacy, passion, and commitment)

These types represent extremes that probably few of us experience. Not many of us, for example, experience infatuation in its purest form, in which there is no intimacy. The categories are nevertheless useful for examining love (except for empty love, which is not really love).

Romantic love combines intimacy and passion. It is similar to liking, but it is more intense and includes a physical or emotional attraction. It may begin with an immediate union of the two components—with friendship that intensifies with passion or with passion that develops intimacy. Although commitment is not an essential element of romantic love, it may

develop. Sternberg also believed that throughout the duration of even successful relationships, passion lessens, while intimacy and commitment increase (Acevedo and Aron 2009).

What then results is companionate love. Companionate love often begins as romantic love, but as the passion diminishes and the intimacy increases, it is transformed. Some couples are satisfied with such love; others are not. Those who are dissatisfied in companionate love relationships may seek extrarelational affairs to maintain passion in their lives. They may also end the relationship to seek a new romantic relationship in the hope that it will remain romantic.

## Love and Attachment

The **attachment theory of love** maintains that the degree and quality of attachments one experiences in early life influence one's later relationships. It examines love as a form of attachment that finds its roots in infancy (Hazan and Shaver 1987; Shaver, Hazan, and Bradshaw 1988). Phillip Shaver and his associates (1988) suggested that "all important love relationships—especially the first ones with parents and later ones with lovers and spouses—are attachments." On the basis of classic infant–caregiver work by John Bowlby (1969, 1973, 1980), some researchers suggest numerous similarities between attachment and romantic love (Bringle and Bagby 1992; Downey, Bonica, and Rincon 1999; Shaver, Hazan, and Bradshaw 1988). These include the following:

| Attachment | Love |
| --- | --- |
| Attachment formation and quality depend on attachment object's responsiveness, interest, and reciprocation. | Feelings of love are related to lover's feelings. |
| When attachment object is present, infant is happier. | When lover is present, person feels happier. |
| Infant shares toys, discoveries, and objects with attachment object. | Lovers share experiences and goods and give gifts. |
| Infant coos, talks baby, and "sings." | Lovers coo, talk baby talk, talk, and sing. |
| There are feelings of oneness with attachment object. | There are feelings of oneness with lover. |

According to research by Downey and Feldman (1996), rejection by parents of their children's needs can lead to the development of **rejection sensitivity**, or the tendency to anticipate and overreact to rejection.

Individuals who develop rejection sensitivity seek to avoid rejection by their partners and closely monitor, even overanalyze, the relationship dynamics for signs of potential rejection. As Pamela Regan (2003) notes, even "minimal or ambiguous" rejection cues may lead to feelings of rejection and to anger, jealousy, and despondency. Rejection-sensitive people tend to be less satisfied with their relationships and more likely to see them end.

Based on studies conducted by Mary Ainsworth and colleagues (1978, cited in Shaver, Hazan, and Bradshaw 1988), there are three styles of infant attachment: (1) secure, (2) anxious or ambivalent, and (3) avoidant. In *secure attachment*, an infant feels secure when the mother is out of sight. He or she is confident that the mother will offer protection and care. In *anxious or ambivalent attachment*, the infant shows separation anxiety when the mother leaves. He or she feels insecure when the mother is not present, and this insecurity results from her being inconsistently available, leaving the infant afraid to leave her side. In *avoidant attachment*, the infant senses the mother's detachment and rejection when he or she desires close bodily contact. The infant shows avoidance behaviors with the mother as a means of defense. In Ainsworth's study, 66 percent of the infants were secure, 19 percent were anxious or ambivalent, and 21 percent were avoidant.

Some researchers believe that the styles of attachment developed during infancy continue through adulthood. Others, however, question the validity of applying infant research to adults as well as the stability of attachment styles throughout life (Hendrick and Hendrick 1994). Still others have used attachment styles as a way to understand people's experiences of relationship satisfaction as well as how people cope with a breakup (Fagundes 2012).

## Secure Adults

Secure adults find it relatively easy to get close to others. They are comfortable depending on others and having others depend on them. They believe that they are worthy of love and support and expect to receive them in their relationships (Regan 2003). They generally do not worry about being abandoned or having someone get too close to them. More than avoidant and anxious or ambivalent adults, they feel that others generally like them; they believe that people are generally well intentioned and good-hearted. In contrast to others, secure adults are less likely to believe in media images of love and are more likely to believe that

romantic love can last. Their love experiences tend to be happy, friendly, and trusting. They are more likely to accept and support their partners. Reportedly, compared to others, secure adults find greater satisfaction and commitment in their relationships (Pistol, Clark, and Tubbs 1995).

## Anxious or Ambivalent Adults

Anxious or ambivalent adults feel that others do not or will not get as close as they themselves want. They feel unworthy of love, need approval from others, and worry that their partners do not really love them or that they will leave them (Regan 2003). They also want to merge completely with the other person, which sometimes scares that person away. More than others, anxious or ambivalent adults believe that it is easy to fall in love. Their experiences in love are often obsessive and marked by a desire for union, high degrees of sexual attraction and jealousy, and emotional highs and lows.

## Avoidant Adults

Avoidant adults feel discomfort in being close to others; they are distrustful and fearful of becoming dependent (Bartholomew 1990). Thus, to avoid the pain they expect to come from eventual rejection, they maintain distance and avoid intimacy (Regan 2003). More than others, they believe that romance seldom lasts but that at times it can be as intense as it was at the beginning. Their partners tend to want more closeness than they do. Avoidant lovers fear intimacy and experience emotional highs and lows and jealousy.

In adulthood, the attachment styles developed in infancy combine with sexual desire and caring behaviors to give rise to romantic love. Comparing across these three types of attachment styles indicates that women and men alike tend to prefer a romantic partner with secure attachment styles. They also tend to find more satisfaction in their relationships, experience more happiness, hold more positive views of their partners, and display fewer negative emotions (Regan 2003).

Psychologists Michelle Drouin and Carly Landgraff (2012) studied how attachment style was associated with texting and sexting behavior among a sample of 744 college students in committed relationships. Texting and sexting were both significantly related to attachment style, with text messaging more common among those with secure attachments (i.e., those with less attachment avoidance), and sexting (both texts and pictures) more common among those with insecure attachments.

## Love and Commitment

We expect our romantic partner to be there for us through "thick and thin." Later, should we enter into marriage, we pledge our love "for better, for worse, for richer, for poorer, in sickness and in health, *till death do us part.*" In other words, we expect that, along with loving us, our partners will be and stay committed to us and to our relationship. Although we generally make commitments to a relationship because we love someone, love alone is not sufficient to make a commitment last. Our commitments seem to be affected by several factors that can strengthen or weaken the relationship. Ira Reiss (1980) believes that there are three important factors in commitment to a relationship that interact to increase or decrease the commitment:

**False** **6.** Love and commitment are not inseparable.

1. *The balance of costs and benefits.* Whether or not we like it, humans have a tendency to look at romantic and marital relationships from a cost–benefit perspective. Most of the time, when we are satisfied, we are unaware that we judge our relationships in this manner. But as shown in our discussion of social exchange theory in Chapter 2, when there is stress or conflict, we might ask ourselves, "Just what am I getting out of this relationship?" Then we add up the pluses and minuses. If the result is on the plus side, we are encouraged to continue the relationship; if the result is negative, we are more likely to discontinue it.

2. *Normative inputs.* Normative inputs for relationships are the values that you and your partner hold about love, relationships, marriage, and family. These values can either sustain or detract from a commitment. How do you feel about a love commitment? A marital commitment? Do you believe that marriage is for life? Does the presence of children affect your beliefs about commitment? What are the values that your friends, family, and religion hold regarding your type of relationship?

3. *Structural constraints.* The structure of a relationship will add to or detract from commitment. Depending on the type of relationship—whether it is dating, living together, or marriage—different roles and expectations are structured. In marital relationships, there are partner roles (husband–wife) and economic roles (employed worker–homemaker). There may also be parental roles (mother–father).

Commitments are more likely to endure in marriage than in cohabiting or dating relationships, which tend to be relatively short lived. The reason that commitments tend to endure in marriage may or may not have anything to do with a couple being happy. Marital commitments may last because norms and structural constraints compensate for the lack of personal satisfaction.

For most people, love seems to include commitment and commitment seems to include love. Beverly Fehr (1988) found that if a person violated a central aspect of love, such as caring, that person was also seen as violating the couple's commitment. If a person violated a central aspect of commitment, such as loyalty, it called love into question. Because of the overlap between love and commitment, we can mistakenly assume that someone who loves us is also committed to us. Or a person can intentionally mislead the partner into believing that there is a greater commitment than there actually is. Even if a person is committed, it is not always clear what the commitment means: Is it a commitment to the person or to the relationship? Is it for a short time or a long time? Is it for better and for worse?

## ■ Finding Love and Choosing Partners

You probably know what the following paragraph represents:

> I have been searching the world over, looking for my true love, my soul mate, someone that I can't wait to see or talk to at the end of the day or wake up next to every morning. I am a friendly, compassionate, hardworking female who enjoys music, dancing, travel, and the beach. Looking for someone who wants to share a movie, dinner, a laugh, and maybe a lifetime. I know you're out there somewhere.

Of course, most people will recognize that this is a personal ad, one of the many that increasingly can be found each day on multiple sites on the Internet. There are many such "dating sites," some of which cater to the general population (e.g., Match.com or eHarmony.com), whereas others are specific to certain populations based on age (e.g., SeniorPeopleMeet.com and OurTime.com), race or ethnicity (e.g., AsiaFriendFinder.com, JapanCupid.com, BlackPeopleMeet .com, LatinAmericanCupid.com, or InterracialMatch.com), religion (Muslima.com, ChristianMingle.com, and JDate.com), sexual orientation (e.g., GayDating.com or Grindr.com), and other specialized sites (e.g., SingleParentMeet.com, MillionaireMatch.com, MilitaryCupid.com, EquestrianCupid.com, Purrsonals.com, VeggieDate.com, and HotSaucePassions.com). There are sites that are based on one's self-assessed looks or appearance (e.g., LargeFriends.com, BBPeopleMeet.com, UglySchmucks.com,), as well as sites that bring together people with various ailments or allergies (e.g., GlutenfreeSingles .com and SingleswithFoodAllergies.com). In addition to such online sites, one can still also find print versions of ads such as the previous one in many newspapers or magazines.

In the past two decades, reality television programs have been added into the mix, pushing such mate-seeking attempts into previously uncharted water. On February 15, 2000, the Fox network aired *Who Wants to Marry a Multi-Millionaire?*, among the first of what would become a wave of reality shows to capitalize on our age-old fascination with how people get together. In subsequent years, shows such as *The Bachelor* and *The Bachelorette*, *Joe Millionaire*, *Date My Mom*, *Are You the One?*, *Married by America*, and *Dating Naked* have aired, along with numerous others on multiple networks, offering viewers variations on how people meet and choose a romantic partner and, in some instances, start what they hope will be a long-term relationship. One of the more recent and most interesting of such shows is *Married at First Sight*, in which a panel of relationship experts, including psychologist Dr. Joseph Cilona, sexologist Dr. Logan Levkoff, sociologist Dr. Pepper Schwartz, and spiritual adviser Greg Epstein, use "scientific principles" to match and create three couples. The catch is that the partners in these couples meet for the first time as they are about to exchange their vows. Strangers till they wed, the marriages are fully legally binding marriages. As described on the show's website,

> Over the course of several weeks, episodes capture each couple's journey as they go from wedding, to honeymoon, to early nesting, to the daily struggle of working on their marriage. After several weeks together, each couple must make a decision: do they remain together or decide to divorce?

(www.fyi.tv/shows/married-at-first-sight)

As of this writing, two of the two of three couples from the initial season were still together but all three from the second season had filed for divorce (Morabito 2015).

Aside from popular interest and curiosity, there is also considerable social science interest in understanding how people find and select their spouses or partners. In addition, researchers have studied *who* we choose and *why* we choose those particular individuals. Although love is the major criterion used to select a spouse, and most people in the United States who marry likely would say they are doing so out of love, many factors operate alongside and on love. In theory, most of us are free to select as partners those people with whom we fall in love, but other factors enter the process, and our choices become somewhat limited by societal norms and rules of mate selection. Once you understand some principles of mate selection in our culture, without ever having met a friend's new boyfriend or girlfriend, you can deduce many things about him or her. For example, if a female friend at college has a new boyfriend, you would be safe in guessing that he is about the same age or a little older, probably taller, and a college student. Furthermore, he is probably about as physically attractive as your friend (if not, their relationship may not last); his parents are more likely than not of the same ethnic group and social class as hers; and he is probably about as intelligent as your friend. If a male friend has a new girlfriend, many of the same things apply, except that she is probably the same age or younger and shorter than he is. Obviously, some relationships will depart from such conventions, and many will have one or two characteristics on which the partners differ (or differ more), but you will probably be correct in most instances. These are not so much guesses as deductions based on the principle of *homogamy*, the pairing of people with similar characteristics or from similar backgrounds. This is discussed more fully in Chapter 8. For now, it bears mentioning that same-gender couples are less homogamous than heterosexual couples, with gay males being least likely to be homogamous (Schwartz and Graf 2009). The reasons for this may reflect opportunity, choice, or a combination of both. The

**True** **7.** It is common in the United States for people to form couple relationships with people from similar backgrounds.

opportunity one has to meet a potential partner is more limited for gay men and lesbians, given the smaller population from which to find a partner with whom one can form a relationship. In a sense, then, gay men and lesbians are pushed to "cast a wider net" than are heterosexuals (Schwartz and Graf 2009). In addition, gay men and lesbians may be more liberal and willing to accept or even desire to find a less traditional relationship, qualities that could increase the likelihood of choosing someone different from oneself.

## The Relationship Marketplace

The process of choosing partners is affected by bargaining and exchange. People select each other in a kind of **marketplace of relationships**. The notion of a "marketplace" effectively conveys that, as in a commercial marketplace, people enter exchange relationships when they form relationships, much as when they exchange goods. Unlike a real marketplace, however, the "relationship marketplace" is more of a process, not a place, in which *people* are the goods exchanged. Each of us has certain resources—such as socioeconomic status, looks, and personality—that determine our marketability. As Matthijs Kalmijn (1998) puts it, "Potential spouses are evaluated on the basis of the resources they have to offer, and individuals compete with each other for the spouse they want most by offering their own resources in return." Individuals bargain with the resources they possess. They

People tend to choose partners who are about as attractive as themselves.

Claudia Kunin/The Image Bank/Getty Images

size themselves up and rank themselves as a good deal, an average package, or something "available at a discount" because of lack of interested "buyers"; they do the same with potential dates and, ultimately, mates. Such "exchanges" are more often between equally valuable goods. In other words, people tend to seek others about *as attractive* or *as intelligent* as they perceive themselves (Sprecher and Hatfield 2009).

## Physical Attractiveness: The Halo Effect, Rating, and Dating

Pretend for a moment that you are at a party, unattached. You notice that someone is standing next to you as you reach for some chips or a drink. He or she says hello. In that moment, you have to decide whether to engage him or her in conversation. On what basis do you make that decision? Is it looks, personality, style, sensitivity, intelligence, or something else? Or imagine yourself searching online on one of the many dating sites. Would you select someone who didn't post a photograph? What about someone whose photograph was not especially attractive? Would you "swipe right" or rule such a person unacceptable and "swipe left?"

In face-to-face situations, most people consciously or unconsciously base such initial decisions at least in part on appearance. If you decide to talk to or contact the person, you probably formed a positive opinion about how he or she looked. In other words, he or she looked "cute," "hot," or looked like a "fun person." In the party setting, perhaps the person gave a "good first impression," or seemed "interesting." Physical attractiveness is particularly important during the initial meeting and early stages of a relationship.

In the online relationship marketplace, appearance may well be decisive. In fact, on many sites people's profiles will receive little attention unless accompanied by photos. Furthermore, users frequently stress how they will only respond to people who post multiple recent photos. Although the sites themselves may base decisions on what profiles to recommend to a user on the basis of compatibility of backgrounds or interests, individual users place great decision-making weight of the attractiveness of one's photos.

### The Halo Effect

Most people would deny that they are interested in others *just* because of their looks. However, individuals tend to infer qualities in others based on looks. This inference is called the **halo effect**—the assumption that good-looking people possess more desirable social characteristics than unattractive people. In a well-known experiment (Dion et al. 1972), students were shown pictures of attractive people and asked to describe what they thought these people were like. Attractive men and women were assumed to be more sensitive, sexually responsive, poised, and outgoing than others; they were assumed to be more exciting and to have better characters than "ordinary" people. Furthermore, attractive people are preferred as friends, candidates, and prospective employees, and they even receive more leniency if they are defendants in court (Ruane and Cerulo 2004). Research indicates that, overall, the differences between perceptions of attractive and average people are minimal. When attractive and average people are compared to those considered to be unattractive, there are pronounced differences, with those perceived as unattractive being rated more negatively (Hatfield and Sprecher 1986).

Research has demonstrated that good-looking companions increase our status. In one study, men were asked their first impressions of a man seen alone, arm in arm with a beautiful woman, and arm in arm with an unattractive woman. The man made the best impression with the beautiful woman. He ranked higher alone than with an unattractive woman. In contrast to men, women do not necessarily rank as high when seen with a handsome man. A study in which married couples were evaluated found that it made no difference to a woman's ranking if she was unattractive but had a strikingly handsome husband. If an unattractive man had a strikingly beautiful wife, it was assumed that he had something to offer other than looks, such as fame or fortune.

## Trade-Offs

In mixing with and meeting people, one doesn't typically gravitate to the *most attractive* person in the room but rather to someone from among those one considers to be about as attractive as oneself. Sizing up someone at a party or dance, a man may say, "I'd have no chance with her; she's too good looking for me." Even if people are allowed to specify the qualities they want in a date, they are hesitant to select anyone notably different from themselves in social desirability.

Susan Sprecher and Elaine Hatfield suggest that in situations where rejection is a possibility, people are more likely to make realistic choices and approach someone of their same perceived level of attractiveness, regardless of their actual preference. They further note that in research on actual couple relationships,

both partners typically tend to be judged to be of similar levels of attractiveness. Additionally, there is more "complex matching" that goes on as well. Physical attractiveness is only one potential resource one has to offer a potential partner, and one can "compensate for lack of physical attractiveness" by offering other desirable traits, such as status, money, kindness, or a charming personality (Sprecher and Hatfield 2009). A woman who values status, for example, may accept a lower level of physical attractiveness in a man if he is wealthy or powerful.

## Are Looks Important to Everyone?

For the wider majority of people who are more ordinary looking, it will come as a relief to know that looks aren't everything. Looks are most important to certain types or groups of people and in certain situations or locations (e.g., in classes, at parties, and in bars, where people do not interact with one another extensively on a day-to-day basis). Looks are less important to those in ongoing relationships and to older rather than younger adults. Those who interact regularly—as in working together—put less importance on looks (Hatfield and Sprecher 1986). In adolescence, the need to conform and the impact of peer pressure make looks especially important, as we may feel pressured to go out with handsome men and beautiful women.

Men tend to care more about how their partners look than do women (Buunk et al. 2002; Regan 2003). This may be attributed to the disparity of economic and social power. Because men tend to have more assets (such as income and status) than women, they can afford to be less concerned with their potential partner's assets and can choose partners in terms of their attractiveness. Because women lack the earning power and assets of men, they may have to be more practical and choose a partner who can offer security and status. Unsurprisingly, then, women are more likely than men to emphasize the importance of socioeconomic factors (Regan 2003).

Most research on attractiveness has been done on first impressions or early dating. At lower levels of relationship involvement, physical attractiveness is more important. As relationship involvement increases, status and personality become more important, and appearance less. For long-term relationships (e.g., marriage), women and men prefer mates about as attractive as themselves. For short-term, less involved relationships, both men and women prefer more attractive mates. Bram Buunk and colleagues (2002) interpret this pattern to reflect potential costs of having someone to whom others are strongly attracted as a long-term partner. Physical attractiveness continues to be important throughout marriage. It is, however, joined by other qualities, and these other attributes are deemed more important.

## Bargains and Exchanges

Likening relationships to markets or choosing partners by an exchange may not seem romantic, but both are deeply rooted in marriage and family customs. In some cultures, for example, arranged marriages take place only after extended bargaining between families. The woman is expected to bring a dowry in the form of property (such as pigs, goats, clothing, utensils, or land) or money, or a woman's family may demand a bride-price if the culture places a premium on women's productivity. Traces of the exchange basis of marriage still exist in our culture in the traditional marriage ceremony when the bride's parents pay the wedding costs and "give away" their daughter.

## Gender Roles

Traditionally, relationship exchanges have been based on gender. Men used their status, economic power, and role as protector in a trade-off for women's physical attractiveness and nurturing, childbearing, and housekeeping abilities; women, in return, gained status and economic security in the exchange.

The terms of bargaining have changed some, however. As women experience more career success and become less economically dependent, what do they ask from men in the marriage exchange? Clearly, many women expect men to bring more expressive, affective, and companionable resources into marriage. An independent woman does not have to "settle" for a man who brings little more to the relationship than a paycheck; she wants a man who is a partner, not simply a provider.

But even today, a woman's bargaining position may not be as strong as a man's. Women earn between 75 percent and 80 percent of what men earn, are still significantly underrepresented in many professions, and have seen many of the things women traditionally used to bargain with in the marital exchange—such as children, housekeeping services, and sexuality—become devalued or increasingly available outside of relationships. A man does not have to rely on a woman to cook for him, sex is often accessible in the singles world, and someone can be paid to do the laundry and clean the apartment.

Women are further disadvantaged by the **double standard of aging**. Physical attractiveness is a key bargaining element in the marital marketplace, but the older a woman gets, the less attractive she is considered. For women, youth and beauty are linked in most cultures. Furthermore, as women get older, their field of potential eligible partners declines because men tend to choose younger women as mates.

## Going Out, Hanging Out, and Hooking Up

If relationships are like products in a marketplace, how do most young people do their shopping? Some contend that this process has become much more casual and informal than in the past. Rather than *dating*, young people refer instead to terms like *getting together*, *hanging out*, or *hooking up*. Although getting together and hanging out are intended to capture and reflect increasing informality in romantic relationships, **hooking up** is a particular type of relationship, prominent on college campuses, that we consider in a later section.

Dating practices have become more diverse, but "its purposes and activities are similar to when it first emerged" (Surra et al. 2007). In fact, young people who date do so for the predictable reasons—intimacy, companionship, and support—and tend to adhere to certain gendered scripts, especially at the beginning of dating relationships. Furthermore, given increases in divorce and cohabitation, later ages at marriage, and increased prevalence of singlehood, individuals may date more people and date for more of their adult lives, making dating more important than ever (Surra et al. 2007).

Although there are some general rules of mate selection that are important in the abstract, they do not tell us how relationships begin. The actual process of beginning a relationship is discussed in the sections that follow.

### Meeting Face-to-Face

On a typical day, one may see dozens, hundreds, or even thousands of men and women. But seeing isn't enough; one must become aware of someone for a relationship to begin. It may take only a second from the moment of noticing to meeting, or it may take days, weeks, or months. Sometimes "noticing" occurs between two people simultaneously; other times it may take considerable time, and sometimes it never happens.

The setting in which you see someone can facilitate or discourage meeting each other. **Closed fields**, such as small classes or seminars, dormitories, parties, and small workplaces, are characterized by a small number of people who are likely to interact whether or not they are attracted. In such settings, you are likely to "see" and interact simultaneously. In contrast, **open fields**, such as beaches, shopping malls, bars, amusement parks, and large university campuses, are characterized by large numbers of people who do not ordinarily interact. Common venues for first encountering a dating partner are social gatherings, such as bars and parties.

How is a meeting initiated, and who tends to do the initiating? Among heterosexuals, does the man initiate it? On the surface, the answer appears to be yes. A recent qualitative study of cohabiting working-class heterosexual couples revealed that the most common pattern was for the men to have initiated the relationships, and for women to "demonstrate their receptiveness," thus adhering to fairly traditional gender expectations (Sassler and Miller 2011b, 490). With an enlarged sample of both working-class and middle-class couples, male initiation of dating relationships was especially true of middle-class couples (Sassler, Miller, and Green 2011). In reality, however, the woman often contributes or even covertly initiates a relationship by communicating indirectly or even nonverbally her availability and interest (Metts and Cupach 1989). A woman will glance at a man once or twice and catch his eye; she may smile or flip her hair. If the man moves into her physical space, the woman then relies on nodding, leaning close, smiling, or laughing.

Regardless of who initiates contact, a variety of verbal and nonverbal signals are used to convey attraction and interest to a potential partner. Smiling, moving closer to, gazing at, laughing, and displaying "positive facial expressions" are all gestures to convey interest or "flirt" (Regan 2003). Touch is also an important element in flirting, whether the touch consists of lightly touching the arm, hand, face, or hair of the target of interest or rubbing fingers across the other's arm (Regan 2003).

If a man believes that a woman is interested, he often initiates a conversation using an opening line. The opening line tests the woman's interest and availability. According to women, the most effective are innocuous, such as "I feel a little embarrassed, but I'd like to meet you" or "Are you a student here?" The least effective are sexual come-ons, such as "You really turn me on. Do you want to have sex?" or corny lines such as "If I had a nickel for every time I met someone as beautiful as you, I'd have a nickel." Women, more than men, prefer direct but innocuous opening lines over cute, flippant ones, such as "What's a good-looking babe like you doing in a college like this?" Much as they are more likely than women to use "a line," men are more likely to initiate a meeting directly, whereas women are more likely to wait for the other person to introduce himself or herself or to be introduced by a friend. An introduction has the advantage of a kind of prescreening, as the mutual acquaintance may believe that both may hit it off. Parties, classes, workplaces, bars and clubs, or events centered on hobbies are all fairly common settings in which young adults meet.

## Meeting Online

The Internet continues to gain enormous popularity as a major way for people to "meet" a potential partner. Online, people can introduce themselves in fantasy-like images. A growing number of people first "meet" in cyberspace, find common interests, and form relationships that develop and intensify rapidly, even before they ever actually meet face-to-face. What might take months to develop in traditional dating can develop online within weeks or even days. Where dating typically proceeds predictably from two people seeing, meeting, evaluating each other's attractiveness and personal appeal, sharing personal interests and information, and—if compatible and trusting—moving toward deeper and more meaningful disclosure, online dating relationships are characterized by earlier and more deeply personal sharing and self-disclosing, perhaps even in the first few emails. Many couples have already developed a strong emotional bond by the time they meet face-to-face (Rosen et al. 2008). In fact, one might say that they fall in love "from the inside" through heartfelt disclosures online and/or on the phone.

What was once stereotyped and even stigmatized as a last resort taken by desperate people has become commonplace. In research about online dating by the Pew Research Center's Internet and American Life Project beginning in 2005 and more recent research in 2013, 11 percent of all Internet-using adults in the United States reported having gone on an Internet dating site for the purpose of meeting a potential partner (Madden and Lenhart 2006; Smith and Duggan 2013).

Moreover, as Mary Madden and Amanda Lenhart asserted, "In a remarkably short time, online dating has revolutionized how people seek romantic partners and initiate relationships with them." In 2013, Aaron Smith and Maeve Duggan's reported the following:

- Eleven percent of Internet users and 7 percent of cell phone apps users have used an online dating site or a dating app on their cell phone.
- More than a third of adults who are single and seeking partners (38 percent), have visited online sites or used cell phone dating apps.
- An estimated 42 percent of Americans report knowing someone who has used an online dating site or app, up from 31 percent a decade earlier. Just under 30 percent (compared to 15 percent in 2005) say they know someone who met their partner or spouse via an online site or cell phone app.
- Of those people who use online sites or dating apps, two-thirds report having gone on a date arranged through such sites (up from 43 percent in 2005). Nearly a fourth (23 percent) report having embarked on a long-term relationship or marrying someone they met via an online site or app.
- Five percent of Americans who are married or in long-term partnerships say they met their partner or spouse online.

Along with increased use and familiarity with online dating websites and apps, online dating has become more accepted. The percentage of Internet users who thought online dating was a good way to meet people increased 15 percent between 2005 and 2013, from 44 percent to 59 percent. There was also a 6 percent increase, from 47 percent to 53 percent, in the percentage of Internet users who agreed that online dating allowed one to get to know a lot more people and therefore to find a better match. However, online dating continues to generate mixed reactions:

- Looking specifically at those who have used online dating, 79 percent agreed that online dating is a good way to meet people, and 70 percent felt it increased one's chances of finding a more suitable partner.
- Although less than in 2005, there was still some skepticism expressed about online dating in the Pew research. One-fifth (21 percent) of Internet users agree with the statement that people who use such sites "are desperate" (down from 29 percent in 2005). In addition, 33 percent agree that online

dating keeps people from settling down because it affords them more options for people to date. Among online daters, 13 percent agree that those who use such sites are desperate, and 29 percent agree that online dating discourages settling down.

- Furthermore, among online daters, 54 percent said that they felt that someone had seriously misrepresented themselves in their profile, and 28 percent had been contacted by someone who made them feel harassed or uncomfortable. More female (42 percent) than male (17 percent) online daters had experienced such uncomfortable contact.

Social networking sites (SNS) such as Facebook or Twitter are also becoming a more prominent part of dating. A third of all SNS users have gone online to check up on former dates or partners. Among those 18 to 29, nearly half had used social networking sites for such reasons. In addition, 30 percent of SNS users who are looking for partners or who have recent dating experience have used their social networking sites to get information about potential dates.

Social networking sites are also sometimes the venue in which one sees "relationship drama" played out (Smith and Duggan 2013). Around a quarter of SNS users have blocked or unfriended someone who was making them uncomfortable by flirting with them. Former dating partners are also players in such online drama; 22 percent of SNS users have unfriended or blocked a former relationship partner, and 17 percent have either deleted or untagged photos of themselves with someone with whom they used to be in a relationship. Among 18- to 29-year-olds, such steps are even more commonly used.

In the United States, close to 4,000 businesses are considered dating services, and they cumulatively generate revenue in excess of $2 billion (Picker and Sherman 2015). These, along with some of the more prominent social networking sites, provide individuals with ways to initiate and develop relationships. At least a dozen different online dating services claim a million or more members. These include Match.com and eHarmony.com, the two biggest such sites, as well as OKcupid.com, Date.com, Zoosk.com, and PerfectMatch.com. Industry estimates in 2012 claimed more than a quarter of a million marriages a year are between people who met on an online dating site (Online Dating Magazine 2012).

According to statistics reported in the *Wall Street Journal*, U.S. dating sites are expected to earn $1.17 billion in 2015, up from $1.08 billion in 2014. Dating

Increasingly, people are turning to the many online dating sites to meet potential partners.

apps, too, are expected to see an increase of close to $60 million, from $572 million to $628.8 million (Wells 2015).

Single men and women also continue to make use of print-based personal ads, matchmaking services, and so on. Both online and in print, ads often reflect stereotypical gender roles. Men advertise for women who are attractive and de-emphasize intellectual, work, and financial aspects. Women advertise for men who are employed, financially secure, intelligent, emotionally expressive, and interested in commitment.

For lesbians and gay men, the problem of meeting others in person is exacerbated because they cannot necessarily assume that the person in whom they are interested shares their orientation. Instead, they must rely on identifying cues, such as meeting at a gay or lesbian bar or events, wearing a gay or lesbian pride button, or being introduced by friends to others identified as being gay or lesbian (Tessina 1989). Once a like orientation is established, gay men and lesbians usually engage in nonverbal processes to express interest. Both lesbians and gay men tend to prefer innocuous opening lines. To prevent awkwardness, the opening line usually does not make an overt reference to orientation unless the other person is clearly lesbian or gay.

Online services, such as GayDating.com, CompatiblePartners.net (part of eHarmony), and OutPersonals.com are among the more popular online services specifically designed for people seeking same-gender partners. Many of the most popular traditional dating sites also offer options for those seeking same-gender relationships. Online dating sites or apps may have

particular appeal for gay men and lesbians. According to sociologist Michael Rosenfeld, among romantic relationships formed between 2007 and 2009, one out of five heterosexual couples and three out of five same-sex couples met online (Rosenbloom 2011). Such online dating sites as Grindr for gay men and Her (formerly Dattch) or Brenda for lesbians may be attractive to those who otherwise have "thin markets" of potential partners. Although one can search for same-sex partners on the major dating sites, those sites are mostly populated by and designed for males seeking females, or females seeking males.

## Dating

In a social and family system such as we find in the United States and societies like it, with the emphasis on romantic love and love-based marriage mentioned earlier, dating is the process through which we interact with potential partners and eventually select our eventual partners and spouses. Stated a little differently, "dating is a publicly expressed practice undertaken by romantically interested partners for the purpose of getting to know one another better" (Eaton and Rose 2011, 843). In addition, especially for adolescents and young adults, dating provides opportunities to learn about intimate relationships, experience closeness, develop skills in communication and conflict management, and explore ones identity. At the same time, the process can have such negative consequences as anxiety, depression, jealousy, and even suicidal ideation. It also can expose one to risks associated with becoming and being sexually active (e.g., pregnancy, sexually transmitted diseases) and to the possibility of experiencing dating violence (Kerpelman 2014).

For many of us, asking someone out for the first time is not easy. Shyness, fear of rejection, and traditional gender roles, in which women are expected to wait to be asked may fill us with anxiety and nervousness. (Sweaty palms and heart palpitations are not uncommon when asking someone out the first time.) Both men and women contribute, although sometimes differently, to initiating a first date. Men are more likely to ask directly for a date: "Want to go see a movie?" Women are often more indirect. They hint or "accidentally on purpose" run into the other person: "Oh, what a surprise to see you *here* studying for your marriage and family midterm!" Although women may initiate dates, they do so less often than do men (Laner and Ventrone 2000).

In addition, research indicates that both women and men believe in certain gendered roles regarding cross-gender dating, especially as pertains to first dates, and that these expectations are fairly similar to those obtained in research over the past 25 years. These include the idea that men *should* initiate first dates, pick up the woman, pay for the date, and walk or drive the woman home. Men also have a higher expectation for sexual intimacy to occur.

Men also express a desire for women to more actively participate in initiating relationships, either by asking directly for a date or by at least hinting. Interesting results from research by communications researchers Mary Morr Serewicz and Elaine Gale show that men expect more sexual intimacy on first dates when initiated by women than when initiated by men. Women, however, do not develop that same expectation. Serewicz and Gale found that men expect "more than kissing" on a female-initiated date more than on a male-initiated date. Women thought "a good night kiss" was more likely on a male-initiated date than on a female-initiated date. When it comes to physical intimacy, men interpret first dates in more sexual terms, whereas women interpret the first date in more social or romantic terms (Serewicz and Gale 2008).

Although both women and men have ideas about what behaviors are most likely of men and of women on first dates, their ideas don't always match. In one study of 103 males and 103 females, women were slightly more egalitarian than men; almost twice as many women as men thought that either gender could do the inviting or initiating, and 22 percent of women compared to only 9 percent of men thought that either person could pick up the bill (Laner and Ventrone 2000). Gendered dating scripts introduce potential problems for men and women. A woman who wants to see a man again faces a dilemma: how to encourage him to ask her out again without engaging in more sexual activity than she really wants. Men often felt that they didn't know what to say, or they felt anxious about the conversation dragging. Communication may be a particularly critical problem for men because traditional gender roles do not encourage the development of intimacy and communication skills among males. A second problem, shared by almost identical numbers of men and women, was where to go. A third problem, named by 20 percent of the men but not mentioned by women, was shyness. Although men can take the initiative to ask for a date, they also face the possibility of rejection. For shy men, the fear of rejection is especially acute. A final problem—and, again, one not shared by women—was money, cited by 17 percent of the men. Men apparently accept the idea that they are the ones responsible for paying for a date.

In addition to potential gender differences, cultural differences have also been identified. For example, as reported by K. Anh Do and Yan R. Xia (2014), dating is much more restricted in collectivist countries such as China, Israel, Egypt, and India. For example, where a majority of young people in countries like the United States and Canada are likely to report being in a relationship, the comparable figure in China is roughly a third (Do and Xia 2014). As noted earlier in this chapter, in collectivist countries one's loyalty is to one's family of origin. Marriage is the bringing together of two families and attention is paid to such things as dowries, arranged marriages, and bride-prices. Dating then is either very limited if it occurs at all (Do and Xia 2014).

Research by Aaron Smith and Maeve Duggan of the Pew Research Center showed that for people who own cell phones and are single and looking for a partner or are in relationships of ten years or less, technology has entered the process. Smith and Duggan report that though more people indicate that they have asked someone out on a date by using their cell phones and calling (52 percent), young adults, 18 to 29 years of age, are more likely than older adults to say they have asked someone out on a date by text messaging; 47 percent of 18- to 29-year-olds compared to 33 percent among 30- to 49-year-olds, and 21 percent among those 50 to 64 have texted an invitation for a date (Smith and Duggan 2013). Texting has also entered the process of ending dating relationships, as 17 percent of the Internet or cell phone users surveyed by Smith and Duggan have broken up with someone via text, online message, or email, and 17 percent have had someone break up with them via text, an online message, or an email. Looking at young adults, 18 to 29, shows that they have had more such experiences: 22 percent have broken up with someone via an electronic method, and 28 percent have had someone break up with them, via email, text message, or online message (Smith and Duggan 2013).

## Problems in Dating

Dating is often a source of both fun and intimacy, but a number of problems may be associated with it. In dating relationships, disagreements inevitably occur. When disagreements do occur, who generally wins? Does it depend on the issue? When one person wants to go to the movies and the other wants to go to the beach, where do they end up going? If one wants to engage in sexual activities and the other doesn't, what happens?

Chapter 7 examines such questions in more detail. For now, it is sufficient to note that typical of much relationship research, the female demand–male withdraw pattern found in many marriages is common among heterosexual dating couples. Of the 108 participants in dating relationships studied by David Vogel, Stephen Wester, and Martin Heesacker (1999), 51 percent reported having a female demand–male withdraw communication pattern. Another 28 percent described their communication as male demand–female withdraw, and 21 percent had no pronounced pattern. The female demand–male withdraw pattern was more often the style couples engaged in "difficult discussions" used. David Vogel and colleagues suggest that either version of the demand–withdraw pattern may prove to be a problem for dating couples as far as their relationship satisfaction and cohesion are concerned (Vogel, Wester, and Heesacker 1999). They recommend reduction of the overall level of demand–withdraw behavior as an important step toward enhancing the quality of relationships.

## Costs and Benefits of Romantic Relationships

As anyone who has had a romantic relationship can attest, relationships bring both positive and negative experiences. In other words, when asked about their romantic relationships, people identify both rewards (companionship, sexual gratification, feeling loved and loving another, intimacy, expertise in relationships, and enhanced self-esteem) and costs (loss of freedom to socialize or date, investment of time and effort, loss of identity, feeling worse about oneself, stress and worry about the health or durability of the relationship, and other nonsocial costs, like lower grades) (Sedikides, Oliver, and Campbell 1994, cited in Regan 2003).

Males and females differ some in what costs and rewards they identify. More males than females identify sexual gratification as a benefit of romantic relationships, and women are more likely than men to identify the benefit of enhanced self-esteem. More women than men mention loss of identity, feeling worse about themselves, or growing too dependent on their partners as relationship costs. Males, on the other hand, stress perceived loss of freedom (to socialize or date) and financial costs more than women do (Regan 2003).

## Dating in Older Adulthood

Although discussions about dating may typically lead one to think about teens and young adults, dating relationships are experienced across the life span among unmarried women and men. Approximately a third of baby boomers are unmarried, and a majority is unmarried among older women and men (adults over 65) (Brown and Shinohara 2013). As Susan Brown and Sayaka Shinohara report in their portrait of dating patterns among older adults, with increases in the population of older unmarried women and men, there is an increase in the pool of potential dating partners. Using data from the 2005–2006 National Social Life, Health, and Aging Project's sample of 3,005 individuals ages 57 to 85, Brown and Shinohara construct a profile of older adult daters and the role dating plays in their lives, with special concern for differences between women and men.

Based on the sample data, Brown and Shinohara estimate that one out of seven (14 percent) of older singles were in dating relationships. Among this population:

- Fourteen percent were in dating relationships, with 27 percent of men and 7 percent of women reporting dating partners. All told, nearly two-thirds of unmarried older adult daters were men. More than 25 percent of men but less than 10 percent of women were dating.

- Dating was more common, at 18 percent, among those 57 to 64, declining to 14 percent among those 65 to 74, and 9 percent among those 75 to 85. At each of these three older age levels, men were more likely to be dating than women: Of men 57 to 64, 32 percent reported they were dating, 27 percent of those 65 to 74, and 24 percent of those 75 to 85. For women, the comparable percentages were 11 percent, 7 percent, and 3 percent among those 75 to 85.

- Comparing daters and nondaters showed the following differences. Daters were younger than nondaters by an average of about 3 years (68 vs. 71 years old). Where most nondaters were widowed (56 percent), most daters were divorced or separated (57 percent). Daters were also more likely than nondaters to be working (40 percent vs. 25 percent), to have college degrees (37 percent vs. 16 percent), and to be wealthier. Daters were also more likely than nondaters to be more socially connected.

- Looking within gender revealed the following: Younger women were more likely to be dating than older women, and divorced women were more than 50 percent more likely to be dating than were widowed women. Among men, never-married men were less likely to be dating than divorced men, black men were 2½ times more likely to be dating than white men, and college-educated men were more than twice as likely to be dating as those with less education. Being socially connected was positively associated with women's dating but was not related to dating for men. Wealth and health were both positively associated with men's dating but were unrelated to dating for women.

Brown and Shinohara note that the sample included people who identified having a partner, thus it misses those who are "in the market searching for a partner" (p. 1201). Also, the role of choice or desire to date could not be determined. Some nondaters may have refrained from dating by choice whereas others faced unfavorable circumstances. They urge further research for dynamics of later-life dating and the consequences of dating in the lives of daters (Brown and Shinohara 2013).

## Hooking Up

As sociologist Kathleen Bogle (2005) points out, as far back as the 1970s, social scientists were forecasting the end of *dating* as we knew it, especially on college campuses. Nonetheless, researchers continued to use the term *dating* to refer to behaviors that seemed less and less to fit what dating meant. Bogle, author of the book *Hooking Up: Sex, Dating, and Relationships on Campus* (2008), differentiates between dating and hooking up, the latter being the pattern she says is dominant among college students today.

What is "hooking up"? Bogle (2005, 2) states that it "generally involves a college man and a college woman pairing off at the end of a party or evening at a bar to engage in a physical or sexual encounter." The term is used differently by different individuals, but all uses imply some degree of physical intimacy, be it "having sex," "making out," "fooling around," or even "everything but intercourse" (Bogle 2008). Unlike the physical intimacy that *might* occur in dating, couples who hook up understand and expect that some level of physical intimacy will be shared. Furthermore, sex between dating partners typically might occur after the couple had dated (at least once), whereas in hookups, partners "become sexual first and then *maybe* someday

A pattern prevalent on college campuses, "hooking up" differs in a number of important ways from more typical "dating" relationships.

Hans Neleman/The Image Bank/Getty Images

go on a date." In addition to occurring earlier, sex in hookups is unrelated to the degree of emotional closeness between the partners and implies no commitment beyond the encounter itself. As articulated by both women and men Bogle studied, the most likely outcome of a hookup is nothing. Once they part, neither pursues a relationship with the other.

Although they don't include the same gendered role expectations that dating does, women and men experience hookups differently. Especially as they move beyond the freshman year, men continue to want hookups with "no strings attached" (i.e., no expectation for anything further developing), whereas many women begin to desire "some semblance of a relationship" (Bogle 2008, 96). Men have hardly any restrictions imposed on them in the hookup culture, whereas women can experience reputational damage and labeling if they hook up with too many men; hook up with two men who knew each other well, especially if there was little time between the encounters; or hang around a specific fraternity too much or if their behavior (e.g., how much they drink, how they dress, or whether they behave "wildly") is deemed excessive. Hooking up comes with a sexual double standard even as it "frees" both genders to engage in more sexual encounters (Bogle 2008).

**False** **8.** Hooking up comes with a sexual double standard.

Bogle suggests that the hookup culture makes relationship building more difficult, as there is less opportunity to get to know each other or develop feelings for each other. In other words, hooking up is a less effective method for "finding love" than conventional dating has been. The hookup model makes the transition to the "real world" beyond college more problematic. Away from college campuses, dating, not hooking up, is the norm, and dating continues to be characterized by gendered scripts regarding such tasks as initiating the date, driving, paying, and so on.

## Jealousy: The Green-Eyed Monster

In addition to bringing great joy, love relationships are often the source of painful insecurities and jealousy. What exactly is jealousy? As studied by researchers, **jealousy** is an aversive response that occurs because of a partner's real, imagined, or likely involvement with a third person (Bringle and Buunk 1985; Sharpsteen 1993). Jealousy sets the boundaries for what individuals or groups feel are important relationships; others cannot trespass these limits into other emotional and/or sexual relationships without evoking jealousy.

**False** **9.** A high degree of jealousy is not a sign of true love.

Some people may think that jealousy proves love and, by flirting with another person, may try to test their partner's interest or affection by attempting to make him or her jealous. If one's date or partner becomes jealous, the jealousy is taken as a sign of love. But making jealousy a litmus test of love is dangerous because jealousy and love are not necessarily related. Jealousy may be a more accurate yardstick for measuring insecurity or possessiveness than love (for a discussion of changing cultural attitudes toward jealousy, see Mullen 1993).

Social psychologists suggest that there are two types of jealousy: suspicious and reactive (Bringle and Buunk 1991). **Suspicious jealousy**, which generally occurs in the early stages of a relationship, is jealousy that occurs when there is either no reason to be suspicious or only ambiguous evidence to suspect that a partner is involved with another. **Reactive jealousy** is jealousy that occurs when a partner reveals a current, past, or anticipated relationship with another person. It is a more intense jealousy. Basic trust is questioned, and the damage can be irreparable.

### Gender Differences in Jealousy

Both men and women are susceptible to jealous fears that their partner might be attracted to someone else because of dissatisfaction with the relationship, attractiveness of a rival, or the desire for sexual variety. Women feel especially vulnerable to losing their partner to a physically attractive rival, whereas jealousy in men is evoked more by a rival's status (Buunk and Dijkstra 2004). Furthermore, men and women become jealous about different matters. Men tend to experience more jealousy when they feel their partner is sexually involved with another man. Women, by contrast, tend to experience jealousy over intimacy issues, when they feel their partner is too emotionally close to someone else (Buunk and Dijkstra 2004; Cramer et al. 2001–2002). This gender difference has been found in research in the United States as well as in China, Germany, Japan, Korea, the Netherlands, and Sweden (Cramer et al. 2001–2002).

In multiple studies, psychologist Robert Cramer and colleagues asked samples of undergraduate women and men to indicate whether emotional or sexual infidelity would distress or upset them more—if their partner was to become emotionally unfaithful (by forming a deep emotional attachment to another) or sexually unfaithful (by enjoying passionate sexual intercourse with another person). Others were asked to imagine that *both* infidelities had occurred and indicate which would distress them more (Cramer et al. 2001–2002).

Like much jealousy research, Cramer and colleagues have used evolutionary theory to account for these gender differences. They suggest that emotional infidelity is more distressing for women than for men because, at least in theory, it threatens a romantic partner's commitment and, therefore, continued access to material resources and economic stability needed to ensure the healthy growth and development of offspring. Men, on the other hand, are more distressed by sexual infidelity than women are because it decreases their "paternity certainty" through the loss of sexual exclusivity (Cramer et al. 2001–2002).

In a more recent study, Robert Cramer, Ryan Lipinski, John Meteer, and Jeremy Houska (2008) looked at these gender patterns of jealousy among a sample of 189 male and female California undergraduates. They asked participants to imagine a situation where either (but not both) sexual or emotional infidelities were occurring, or to imagine that both sexual and emotional infidelities were occurring and report which would be more distressing to them. In line with their expectations, again more women than men were disturbed by their partner becoming deeply emotionally attached but not sexually involved with someone else. This distressed 74 percent of the females and 45 percent of the males. More men (55 percent) than women (26 percent) found the possibility of their partner enjoying a sexual but not emotionally fulfilling relationship distressing. In response to hypothetical infidelity of both kinds, "more women (81 percent) than men (45 percent) were distressed by the emotional component of the combined infidelity and more men (55 percent) than women (19 percent) were distressed by the sexual component" (Cramer et al. 2008).

Both men and women react to jealousy with a host of emotions. Betrayal, anger, rejection, hurt, distrust, anxiety, worry, suspicion, and sadness are all possible. The kind of emotional reaction appears to depend on the type of infidelity that provokes it. Following emotional infidelity, such feelings as anxiety, suspicion, worry, distrust, and threat are more common. Bram Buunk and Pieternel Dijkstra call this type of jealousy *preventive* jealousy. Following sexual infidelity, jealousy was expressed more through anger, sadness, a sense of betrayal, hurt, and rejection. Buunk and Dijkstra (2004) label this *fait accompli* (after the fact) jealousy. Further differentiating the genders, following emotional infidelity, jealousy was evoked in men by a rival's dominance and was experienced mostly as a sense of threat. Following sexual infidelity, men's jealousy was evoked by his rival's physical attractiveness, not his dominance, and was experienced as betrayal or anger. For women, after emotional infidelity, a rival's physical attractiveness evoked a sense of threat, whereas after sexual infidelity, women's jealousy responses were unaffected by any particular characteristics of her rival (Buunk and Dijkstra 2004).

A recent study by psychologists Rosanna Guadagno and Brad Sagarin (2010) examined whether this same gender pattern would surface in perceptions of online infidelity. They asked a sample of 332 heterosexual undergraduate students to choose whether they would be more upset or distressed by a partner forming a deep emotional attachment to someone who they only interact with online or engaging in cybersex with someone who they only interact with online. They found the same gender difference in jealousy as had been found in research on "offline" interaction. In other words, men were more distressed than were women at the notion of their partners having online sexual encounters with someone else than they were at the thought of their partners forming deep emotional bonds with someone they met online. Women were more jealous at the thought of their partners developing an emotional attachment (Guadagno and Sagarin 2010).

## Managing Jealousy

Jealousy can be unreasonable or a realistic reaction to genuine threats. Unreasonable jealousy can become a problem when it interferes with an individual's well-being or that of the relationship. Dealing with irrational suspicions can often be difficult because such feelings touch deep recesses in us. As noted earlier, jealousy is often related to personal feelings of insecurity and inadequacy. The source of such jealousy lies within a person, not within the relationship.

If we can work on the underlying causes of our insecurity, then we can deal effectively with our irrational jealousy. Excessively jealous people may need considerable reassurance, but they must at some point confront their own irrationality and insecurity. If they do not, they may emotionally imprison their partner. Their jealousy may destroy the very relationship they were desperately trying to preserve.

Managing jealousy requires the ability to communicate, the recognition by each partner of the feelings and motivations of the other, and a willingness to reciprocate and compromise (Ridley and Crowe 1992). If the jealousy is well founded, the partner may need to modify or end the relationship with the "third party" whose presence initiated the jealousy. Modifying the third-party relationship reduces the jealous response and, more important, symbolizes the partner's commitment to the primary relationship. If the partner is unwilling to do this—because of lack of commitment, unsatisfied personal needs, or other problems in the primary relationship—the relationship is likely to reach a crisis. In such cases, jealousy may be the agent for profound change.

It's important to understand jealousy for several reasons. First, jealousy is a painful emotion filled with anger and hurt. Its churning can turn one inside out and make one feel out of control. If one can understand jealousy, especially when it is irrational, then one can eliminate some of its pain. Second, jealousy can help cement or destroy a relationship. It helps maintain a relationship by guarding its exclusiveness, but when in its irrational or extreme forms, it can destroy a relationship by its insistent demands and attempts at control. There is a need to understand when and how jealousy is functional and when and how it is not. Third, jealousy is often linked to violence. It is a factor in precipitating violence or emotional abuse in dating relationships among both high school and college students; among marital partners, abusive partners often use it to justify their violence (Jewkes 2002). It has been linked to abuse and violence in other countries, such as China (Wang et al. 2009). Rather than being directed at a rival, jealous aggression is often used against a partner, especially when accompanied by alcohol (Foran and O'Leary 2008).

## Breaking Up

"Most passionate affairs end simply," Elaine Hatfield and G. William Walster (1981) noted. "The lovers find someone they love more." Love cools; it changes to indifference or hostility. Perhaps the relationship ends because one partner shows a side that the other partner decides is undesirable. Or couples disclose *too much*, revealing negative feelings or ideas that lead to unhappiness and the demise of the relationship (Regan 2003).

Relationships outcomes are also influenced by outside influences. Perhaps some new opportunity for greater fulfillment appears in someone else or in a return to a more autonomous and independent state. Even satisfying relationships may end under these circumstances (Regan 2003). Over the course of their lifetimes, most people will experience multiple relationships and endure numerous breakups (Tashiro and Frazier 2003).

Breaking up is typically painful because few relationships end by mutual consent. The extent of distress caused by breakups is revealed by research indicating that many people include breaking up among the "worst events" they can experience; they are also among the biggest risk factors for adolescent and young adult depression (Graham, Keneski, and Loving 2014a; Tashiro and Frazier 2003). The most

common depressive symptoms accompanying breakups are those associated with grief and bereavement, including insomnia and intrusive thoughts. One may even experience physiological reactions including sore throats, sneezing and other upper respiratory symptoms, as well as elevated blood pressure, and chest pain that is indicative of *heartbreak syndrome* (Graham, Keneski, and Loving 2014b).

For college students, breakups are more likely to occur during vacations or at the beginning or end of the school year. Such timing is related to changes in the person's daily living schedule and the greater likelihood of quickly meeting another potential partner. Holidays, including—ironically—Valentine's Day, also may increase the risk of breakup (see the Popular Culture feature).

Research into the process of breaking up indicates that relationships often begin to sour as one partner grows quietly dissatisfied. Psychologist Steve Duck (1994) called this the *intrapsychic* phase, and sociologist Diane Vaughan (1990) talked about "keeping secrets" (Duck 1994; Vaughan 1990). One partner decides that something is wrong with the relationship, considers the possibility of ending the relationship, weighs the likely outcomes associated with being out of the relationship, and begins to build an identity as "single." All of this may happen before the other partner learns what has happened. By the time the "initiator" informs the partner, the partner is forced to play "catch-up" in that the initiator is a few steps ahead in the exiting process. This is further discussed in Chapter 13.

Because of the variety of problems that plague relationships, many couples break up. In the process of breaking up, both the initiator and the rejected partner suffer.

Radius Images/Alamy Stock Photo

**Chocolate Hearts, Roses, and . . . Breaking Up? What about "Happy Valentine's Day"?**

Every year on February 14, millions of Americans exchange tokens of love and affection. As the day approaches, post office branches fill with Hallmark cards. Florists take orders and send roses around the country. Chocolate hearts show up on store shelves. Jewelry is bought for and given to those we love. Millions of dollars are spent in efforts to show and tell our "one and only" how much we love them, and, collectively, the country celebrates love and romance in the name of Valentine's Day. You may be familiar with such rituals as both a giver and a recipient. You may be wondering what if anything is noteworthy about such rituals. One of the lesser-known aspects of this holiday devoted to love is the effect it can have on ongoing love relationships. This effect was provocatively captured by exploratory research undertaken by Katherine Morse and Steven Neuberg (2004) in their study following the relationship outcomes for 245 undergraduate students (99 male and 146 female, mean age of 19.5 years) from the week before to the week after Valentine's Day. The average relationship length across all research participants was 18 months, suggesting that these were meaningful relationships. The results may surprise you, especially if you consider yourself a romantic at heart.

It seems that "Valentine's Day is harmful to many relationships" (Morse and Neuberg 2004, 509). Although this may seem hard to imagine at first, Morse and Neuberg remind us that, although limited, research has shown that holidays can affect behaviors. The best illustration of this is the effect major holidays (e.g., Christmas, Thanksgiving, and New Year's Eve and Day) have on suicide rates: Essentially, they postpone such acts from before to after the holiday.

Morse and Neuberg (2004) predicted that during the two-week period straddling Valentine's Day (from one week before the holiday to one week after the holiday), there would be more breakups than in comparison periods from other nonholiday times of year, and, indeed, Valentine's Day posed relationship hazards. The overall odds of breaking up were 5.49 times greater during the Valentine's Day period than during the comparison months (which did not differ from one another). They further determined that the effect of the holiday on breakups was the result of a catalyst effect. The holiday had no effect on breakups among high-quality or improving relationships but did affect breakups among those in moderately strong and weak relationships if they were encountering relationship downswings. Already suffering from diminishing expectations and unfavorable comparisons to other relationships or potential partners, such relationships might be deemed not worth the effort and expenses associated with trying to successfully play out the Valentine's Day script, thus "making the option of relationship dissolution more attractive" (Morse and Neuberg 2004, 512).

They suggest that the catalyst effect may in part be a favor to troubled relationships in that it facilitates a breakup that was likely anyway, and hence "saved these couples the psychological stress, wasted time, and wasted resources that result from perpetuating a doomed relationship" (Morse and Neuberg 2004, 524). However, they also admit that because many long-term relationships go through periods of ups and downs, it is "at least plausible that a good number of our couples might have otherwise survived the downward blip in relationship expectations and quality had it not been for the catalytic effects of Valentine's Day" (p. 524). ●

SOURCE: Morse and Neuberg (2004, 509–527).

Breaking up is rarely easy, regardless of whether one is on the initiating or "receiving" end. Adolescents and young adults are more likely to have more negative emotional reactions and to struggle more in the aftermath of a breakup. Also, the longer the relationship was, the more invested the individual was in it, the more satisfied and committed she or he was, and the more difficult one believes it will be to find another relationship, the worse the breakup consequences are likely to be (Graham, Keneski, and Loving 2014a; Regan 2003). Social support and self-esteem appear to be important factors in helping someone recover more quickly and completely (Regan 2003).

Also important in shaping the impact of breakups are the **attributions** individuals make and use ("What happened?" "Why did it end?") to account for the demise of a relationship. Attributions may be important factors in efforts to avoid such problems in later relationships, shielding people from experiencing the heartache that accompanies a breakup. Ty Tashiro and Patricia Frazier (2003) suggest that there are four such attributions:

- *Personal*. Personal traits and characteristics are identified as causes of relationship failure ("If only I hadn't been so jealous . . .").
- *Other*. Personal traits and characteristics of the partner are seen as the causes of relationship failure ("He or she was always so insensitive . . .").
- *Relational*. The unique combination of one's own traits and characteristics with those of the other partner is perceived as the cause of the breakup ("We just wanted different things . . .").
- *Environmental*. The social environment is identified as the cause of the breakup. It comprises many things, from familial pressure and disapproval of the relationship to work pressures to "alternative romantic partners."

According to Tashiro and Frazier (2003), those who construct accounts to explain why their relationships failed usually cite "relational" attributions. These are followed by "other" attributions, "personal" attributions, and "environmental" attributions. Although environmental attributions are quite uncommon, environmental factors weigh heavily on relationships. Ironically, environmental factors may be the "real cause" of a breakup incorrectly attributed to something else.

Attributions are also related to how distressing a breakup is felt to be. People who apply relational attributions are happier, more confident, and more socially active. "Other" attributions are associated with greater distress, including sadness, lack of self-confidence, and greater pessimism. Research on personal attributions is mixed, with some showing that it is related to less and some suggesting that it is associated with more distress.

Psychologists Erica Slotter, Wendi Gardner, and Eli Finkel contend that a major source of distress surrounding breakups is the influence they have on individuals' self-concepts. In addition to sadness at the ending of a valued relationship, breakups force former partners to redefine themselves, and this redefinition is part of a process of restructuring of the self. Individuals may experience changes in their self-concepts and must renegotiate their sense of self without the relationship or partner's input. Individuals may experience self-confusion. Having constructed an identity that was shaped by being in the relationship, having

merged "activities, social networks, goals, and even aspects of their self-concept," they must now experience a painful process of self-change (Slotter, Gardner, and Finkel 2010).

Research demonstrates that, alongside pain and distress, breakups *can* induce positive changes that improve the quality of subsequent relationships. One might expect that the degree to which breakups are associated with positive rather than negative outcomes (e.g., growth rather than distress) would depend on such things as who initiated the breakup (noninitiators suffer greater distress), one's gender, and/or one's personality (people high in "neuroticism" are more likely to suffer from distress). Of these, gender differences occur in "stress-related growth," with women reporting more growth following breakups than men. In addition, people high in "agreeableness" reported more post-breakup growth (Tashiro and Frazier 2003).

> **True** **10.** Although painful, breaking up can induce positive changes that can improve later relationships.

In their study of 92 undergraduates (75 percent female ranging from 18 to 35, with a mean age of 20 years), Ty Tashiro and Patricia Frazier found that positive change, such as personal growth, following breakups was common, with respondents reporting positive changes that they believed would strengthen their future relationships and the chances for success in those relationships. Using the same four categories of attributions discussed earlier, the most commonly reported positive changes were "personal related" (e.g., "I learned not to overreact."), followed by "environmental" (e.g., improved family relationships and increased success in school), "relational" (e.g., better communication), and "other" (remember, "other" refers to characteristics of the other with whom we have a relationship). Individuals who use environmental attributions were most likely to report both distress from a breakup and having experienced growth as a result of the breakup. Potentially, those who can explain the failure of their relationship in terms of changeable environmental factors are in a better position to learn from and implement such changes in future relationships (Tashiro and Frazier 2003).

## Breakups among Gay and Lesbian Couples

As we will see in Chapter 6, there are both similarities and differences between same-sex and heterosexual couples. Couples in a relationship, especially those that share a household, encounter many of the same day-to-day issues (e.g., housework, money management, and the effects of outsiders, such as family and friends, on relationships). Furthermore, all couples need to manage issues such as communication and conflict management. Given these similarities, how do same-sex couples and heterosexual couples compare in terms of susceptibility to breaking up?

Research has shown that same-sex couples tend to be more likely than heterosexual couples to break up (Wagner 2006). Citing research findings by Lawrence Kurdek (1994), Cynthia Wagner reports that married heterosexuals—in a three-way comparison of married and cohabiting heterosexuals, gay, and lesbian couples—had the lowest rate (4 percent) of breaking up within 18 months of getting together ("relationship dissolution") and that lesbian couples had the highest (18 percent). However, the cause of such differences is more likely the result of *marriage* than of *sexuality*. All cohabiting couples had similar "dissolution rates," and all were significantly higher than the rate found among married heterosexuals. Marriage is more likely to be associated with cultural acceptance and social support. Furthermore, once married, it is more difficult to simply walk away or to separate easily. Using comparative data from Sweden and Norway, Kurdek illustrates that state-sanctioned and recognized legal unions between gay men or lesbians lowered the rates at which such relationships broke up, even though they still did so at levels greater than among married heterosexuals (Wagner 2006).

Research that has compared the rate of breakups between same-sex couples and married heterosexuals finds that the differences become strikingly similar when one limits the comparison to same-sex couples in legally recognized relationships (Rosenfeld 2014). A study that compared Vermont gay and lesbian couples who were in civil unions to those not in civil unions found that although relationships were similar regarding divisions of family labor and social support from family and friends, three years later those in civil unions were more likely to still be together than were those not in civil unions. Those without civil unions had rates of breaking up (9.3 percent) two and a half times the rate among those in civil unions (3.8 percent), whereas same-sex couples in civil unions had rates much more similar to rates among married heterosexual couples (2.7 percent) (Balsam et al. 2008). The most recent research following passage of gay marriage laws and civil union legislation in many states shows it does appear as though breakups and dissolutions among gay and lesbian couples have occurred at levels more similar to

those among married heterosexuals when relationships are formalized (Rosenfeld 2014).

What becomes of relationships once couples break up? How do former romantic partners relate to each other? A study of 131 predominantly female (81 percent), mostly Caucasian (75 percent) undergraduates who had broken up with a recent romantic partner found that friendship status was significantly correlated with romantic satisfaction during the now-ended relationship. Those individuals who were satisfied with the past relationship were more likely to be friends with their former partners (Bullock et al. 2011). Research on 298 individuals from same-sex and 272 individuals from heterosexual romantic relationships reveals some interesting similarities in "postdissolution relationships" (Lannutti and Cameron 2002). Many gay men and lesbians, as well as many heterosexuals, report remaining (or becoming) friends with former partners, especially following the "let's just be friends" type of breakup. Those friendships are different, however, from friendships in which two people have no shared romantic past. In comparing characteristics of postdissolution relationships, Pamela Lannutti and Kenzie Cameron found that heterosexuals reported moderate amounts of satisfaction and emotional closeness and low levels of interpersonal contact and sexual intimacy with former partners. Gay and lesbian respondents revealed high levels of satisfaction, moderate levels of emotional intimacy and personal contact, and low levels of sexual intimacy in their postdissolution relationships. For both same-sex and heterosexual former partners, postdissolution relationships also differ from intact or ongoing romantic relationships (Lannutti and Cameron 2002).

## Churning and Relationship Cycling

One factor that differentiates among breakups is whether couples reconcile and reunite. Research estimates indicate that it is not uncommon for young adult relationship breakups to be followed by reconciliations with the same partners. Between 30 and 50 percent of young adults in dating relationships and perhaps a third of cohabiting couples are estimated to have experienced at least one breakup with a current partner (Vennum and Johnson 2014).

As reviewed by Amber Vennum and Matthew Johnson, characteristics associated with such couples are suggestive of less than optimal outcomes, as couples who engage in **churning**, or **relationship cycling**, tend to have more doubts about the future of their relationships and lower levels of commitment, express less relationship satisfaction, and report poorer communication and higher levels of abuse in their relationships. Only a third of couples who have a trial separation later reconcile, and half of those separate again within three years (see Vennum and Johnson 2014). Those who not only reconcile but also later marry have been found to be at a higher risk of marital distress, instability, and the negative consequences of such for their own and children's well-being. Among the 170 "cyclical couples" in the larger sample of 564 couples Vennum and Johnson studied, premarital cycling had "small but robust" effects on their marriages over the first five years, including lower levels of satisfaction, less closeness, and more conflict and uncertainty in the relationship.

## Lasting Relationships through the Passage of Time

As discussed earlier, romantic love occupies a particularly valued place in societies such as the United States. Its emergence and presence can move two people into marriage, and its absence or disappearance can lead them to terminate their relationship. However, there is also the fairly widespread belief that romantic love cannot be sustained in relationships that last many years. Ultimately, it is believed that romantic love will be transformed or replaced by a quieter, more lasting love, and that such a transformation is probably beneficial. However, there is evidence that should cause one to question the inevitability of such a transformation (Acevedo and Aron 2009).

### Changes in Intimacy, Passion, and Commitment over Time

Many of the major theories of love predict that the transformation from romantic to companionate love is inevitable. For example, returning to Robert Sternberg's triangular theory of love, love has three components: passion, intimacy, and commitment. A love that contains all three, what Sternberg called **consummate love**, or complete love, is idealized and sought after but difficult to find and even harder to sustain (Sternberg 2007). The passage of time affects our levels of intimacy, passion, and commitment though not at the same pace or in the same direction.

**Intimacy over Time.** When two people first meet, intimacy increases rapidly as we make critical discoveries about each other, ranging from our innermost

thoughts of life and death to our preference for strawberry or chocolate ice cream. As the relationship continues, the rate of growth decreases and then levels off. After the growth levels off, the partners may no longer consciously feel as close to each other. This may be because they are beginning to drift apart, or it may be because they are becoming intimate at a different, less conscious, even deeper level. This kind of intimacy is not easily observed. It is a latent intimacy that nevertheless is forging stronger, more enduring bonds between the partners.

**Passion over Time.** Passion may be subject to habituation. What was once thrilling—whether love, sex, or roller coasters—becomes less so the more we get used to it. Once we become habituated, more time with a person (or more sex or more roller-coaster rides) does not increase our arousal or satisfaction.

If the person leaves, however, we experience withdrawal symptoms (fatigue, depression, and anxiety), just as if we were addicted. In becoming habituated, we have also become dependent. We fall beneath the emotional baseline we were at when we met our partner. Over time, however, we begin to return to that original level.

**Commitment over Time.** Unlike intimacy and passion, time does not necessarily diminish, erode, or alter commitments. Our commitment is most affected by how successful our relationship is. Even initially, commitment grows more slowly than intimacy or passion. As the relationship becomes long term, the growth of commitment levels off. Our commitment will remain high as long as we judge the relationship to be successful. If the relationship begins to deteriorate, the commitment will probably decrease after a time. Eventually, it may disappear, and an alternative relationship may be sought.

### Disappearance of Romance as Crisis

The disappearance (or transformation) of passionate love is often experienced as a crisis in a relationship. But intensity of feeling does not necessarily measure depth of love. Intensity often diminishes over time. It is then that we begin to discover if the love we experience for each other is one that will endure.

Our search for enduring love is complicated by our contradictory needs. Elaine Hatfield and William Walster (1981) offer this observation:

> What we really want is the impossible—a perfect mixture of security and danger. We want someone who understands and cares for us,

someone who will be around through thick and thin, until we are old. At the same time, we long for sexual excitement, novelty, and danger. The individual who offers just the right combination of both ultimately wins our love. The problem, of course, is that, in time, we get more and more security—and less and less excitement—than we bargained for.

The disappearance of passionate love, however, can also enable individuals to refocus their relationship. They are given the opportunity to move from an intense one-on-one togetherness that excludes others to a togetherness that includes family, friends, and external goals and projects. They can look outward on the world together.

### The Reemergence of Romantic Love

Contrary to what pessimists believe, many people find that they can have both love and romance and that the rewards of intimacy include romance. Romantic love may be highest during the early part of romantic relationships and decline as stresses from child rearing and work intrude on the relationship. Most studies suggest that marital satisfaction proceeds along a U-shaped curve, with highest satisfaction in the early and late periods. Romantic love may be affected by the same stresses as general marital satisfaction. Romantic love begins to increase as children leave home. In later life, romantic love may play an important role in alleviating the stresses of retirement and illness.

So it is that, among those whose marriages survive, passion and romance do not necessarily decline over time. In fact, Bianca Acevedo and Arthur Aron contend that long-term romantic love is very real and that individuals who are in such relationships are likely to have higher levels of satisfaction with their relationships, better mental health, and a better sense of overall well-being. They suggest that awareness of the existence of long-term romantic relationships might give couples in committed relationships something toward which to strive. Long-term marriages or couple relationships don't inevitably lose the intensity, engagement, or sexual interest that characterize romantic love (Acevedo and Aron 2009).

Although we increasingly understand the dynamics and varied components of love, the experience of love itself remains ineffable, the subject of poetry rather than scholarship. A journal article is not a love poem, and romantics should not forget that love exists in the everyday world. Researchers have helped us increasingly understand love in the light of day—its nature, its development, and its varied aspects—so that we may better be able to enjoy it in the moonlight.

## Summary

- Intimate relationships such as love and friendship satisfy our basic need for closeness with others and offer numerous emotional, psychological, and health benefits.

- Love is considered to be a cultural universal, though it varies in how it is defined and experienced. Americans value passionate, romantic love. Within the ideology of romanticism, love is seen as blind, irrational, uncontrollable, and likely to strike at first sight. In addition, it is believed that there is a "one and only" for each of us.

- Cultural expectations surrounding friendship and love in the United States place heavy emphasis on self-disclosure. Such expectations are more compatible with females' experiences and expectations.

- *Prototypes* of love and commitment are models of how people define these two ideas in everyday life.

- Commitment is affected by the balance of costs to benefits, normative inputs, and structural constraints.

- The *marketplace of relationships* refers to the selection activities of men and women when sizing up someone as a potential date or mate. In this marketplace, each person has resources, such as social class, status, age, and physical attractiveness.

- Initial impressions are heavily influenced by physical attractiveness. A *halo effect* surrounds attractive people, from which we infer that they have certain traits, such as warmth, kindness, sexiness, and strength.

- The setting in which you see someone can facilitate or discourage a meeting; *closed fields* allow you to see and interact simultaneously, whereas an *open field* makes meeting more difficult.

- Increasingly, Americans are turning to online sites and smartphone apps to meet potential partners.

- *Dating scripts* prescribe certain behavior as expected of each gender. For women, problems in dating include sexual pressure, communication, and where to go on the date; for men, problems include communication, where to go, shyness, and money.

- *Hooking up* is a fairly widespread phenomenon on college campuses. It differs from conventional dating in its focus on sexual intimacy early, before building a relationship, and in its lack of longer-term objectives.

- About a third of baby boomers and a majority of those over 65 are unmarried, which has led to an increase in the desire to date and in the formation of dating relationships among older women and men. Men make up the majority of older adult daters.

- Breakups are commonplace. In accounting for breakups, we attribute the cause to one of the following: our own or our partner's personal characteristics, characteristics within the relationship, or environmental influences.

- *Relationship cycling* or *churning* refer to relationships that reconcile after breakups, sometimes more than once. The outcomes for such relationships are not encouraging.

- Gay and lesbian individuals report higher levels of satisfaction with post-breakup friendships with former partners than do heterosexuals.

- *Jealousy* is an aversive response that occurs because of a partner's real, imagined, or likely involvement with a third person. Jealousy acts as a boundary marker for relationships.

- There are gender differences in what causes partners to become jealous.

- Time affects romantic relationships. Intimacy may level off, and commitment tends to increase, provided that the relationship is judged to be rewarding.

- The transformation from romantic to more companionate love is not inevitable. There are long-term relationships that retain the intensity and sexual interest that are characteristic of passionate love.

## Key Terms

attachment theory of love  169
attributions  185
churning  187
closed fields  176
commitment  159
companionate love  168
conjugal family  157
consummate love  188
co-rumination  160
double standard of aging  175
expressive displays  159
feminization of love  159
halo effect  173
hierarchy of needs  155
homogamy  157
hooking up  175
instrumental displays  159
intimacy  155
jealousy  182
marketplace of relationships  173
open fields  176
passionate love  168
prototypes  166
reactive jealousy  182
rejection sensitivity  169
relationship cycling  187
self-disclosure  155
suspicious jealousy  182
triangular theory of love  168

# Understanding Sex and Sexualities

6

MoMo Productions/The Image Bank/Getty Images

191

I t is now time to consider sex. For many of you, that must seem like a silly statement. Quite apart from this book, we often consider, think about, or take steps to pursue—or avoid—sexual encounters. Our popular culture is heavily sexualized. Advertising, in particular, uses sexual innuendo and imagery to sell us any number of products. Furthermore, being sexual is an essential part of being human. Through our sexuality, we are able to connect with others on the most intimate levels, revealing ourselves and creating deep bonds and relationships. Sexuality is a source of great pleasure and profound satisfaction. It is the typical means by which we reproduce, transforming ourselves into mothers and fathers. Paradoxically, sexuality also can be a source of guilt and confusion, a pathway to infection, and a means of exploitation and aggression. Examining the multiple aspects of sexuality helps us understand our sexuality and that of others. It provides the basis for enriching our relationships.

In this chapter, we offer an overview of sexuality and sexual issues, especially as they are interconnected with relationships, marriages, and family life. We begin by considering the sources of our sexual learning and proceed through sexual development and expression in young, middle, and later adulthood, including the gay, lesbian, or bisexual identity process. We consider the shifts from traditional to modern sexual scripts, and the social control of sexuality. When we examine sexual behavior, we cover the range of activities and relationships in which people engage. Ultimately, we look at nonconsensual sexual relations, sexual problems, and dysfunctions; birth control; sexually transmissible diseases, human immunodeficiency virus (HIV), and acquired immunodeficiency syndrome (AIDS); and sexual responsibility. We hope that this chapter will help you make sexuality a positive element in your life and relationships.

## Sexual Scripts

Although we might consider sexual behavior one of the most natural parts of being human, what we do, think about, and feel regarding sexual activity is the product of our social learning. Organizing and directing our sexual impulses are culturally shared sexual scripts, which we learn and act out. A **sexual script** consists of expectations of how to behave sexually as a female or male and as a heterosexual, bisexual, lesbian, or gay male. Like a sexual road map or blueprint offering us general directions, a sexual script enables each individual to organize sexual situations and interpret emotions and sensations as sexually meaningful (Hynie et al. 1998). It may be more important than our own experiences in guiding our actions. Over time, we may modify our scripts, but we will not throw them away.

In looking at sexual scripts, we see how society influences our sexual attitudes and behavior. This is not unique to the United States; every society regulates and controls the who, what, when, where, and why of sexuality:

*Who.* We are taught to have sex with people who are unrelated, around our own age, and of the other sex (heterosexual). Having sex with yourself (masturbation), with members of the same sex (gay or lesbian sexuality), and with relatives (incest) are typically less socially acceptable choices. In most societies, extramarital relationships are prohibited, even though they might frequently occur. Examining the "*who*s" in sexual scripts raises the following question: With whom is it or would it be acceptable to engage in sexual behaviors?

*What.* Society classifies various sexual acts as good or bad, moral or immoral, and appropriate or inappropriate. Although these designations may seem absolute, they, like all aspects of sexual scripts, are culturally and historically relative (Kimmel 2011).

*When.* For teenagers and young adults, "when" might mean when parents are out of the house. If one is a parent, "when" may entail waiting until one's children are asleep or one's teens are out of the house. Usually, such timing is related to privacy, but "when" may also be related to the age at which sexual activity is expected to start and stop, how often people are expected to engage in sexual relations, and when sex should begin in a relationship. Finally, it may pertain to times when sex is considered appropriate or inappropriate. Some societies frown on a woman engaging in sex during her menstrual flow, for a period after the birth of a child, or while nursing (Miracle, Miracle, and Baumeister 2003).

*Where.* Where do sexual activities occur with society's approval? In the United States, they usually occur in a bedroom, where a closed door signifies privacy. For adolescents, automobiles, fields, beaches, and motels may be identified as locations for sex; churches, classrooms, and front yards usually are not. "Where" may also extend to where it might be considered appropriate or inappropriate to discuss sex or to expose parts of your body.

*Why.* There are many reasons for having sex: procreation, love, passion, revenge, intimacy, exploitation, fun, pleasure, relaxation, boredom, achievement, relief from loneliness, exertion of power, and on and on. Some of these reasons are approved by society; others are not. Some we conceal; others we do not.

## Gender and Sexual Scripts

Our gender socialization is also in part a socialization into sexuality as we learn what sexual behavior is appropriate, legitimate, and acceptable for each gender. Sexual scripts are most powerful during adolescence, when individuals are first learning to be sexual. Gradually, as one gains experience, one may modify these sexual scripts. As children and adolescents, we learn our sexual scripts primarily from our parents, siblings, peers, and the media. As we get older, interactions with our partners become increasingly important.

Traditional heterosexual sexual scripts are clearly "gendered," prescribing different roles and responsibilities to females and males. The traditional male script casts men as the initiators of sexual encounters. Men are expected to be assertive, confident, and knowledgeable about sexual matters. They are supposed to know how to please their partners as well as how to coax their partners to share sexual intimacy. Sex is a goal to be achieved, and each sexual encounter is almost like a "conquest," enhancing the male's self-esteem and reputation.

The male script places performance expectations on men. The orgasm is the proof in the pudding; the more orgasms, the better the sex. If a woman does not have an orgasm, the male feels that he is a failure because he was not good enough to give her pleasure. Researchers who have studied sexual stereotypes observe that men's sexual identity may depend heavily on a capricious physiological event: getting and maintaining an erection. They note that the following traits are associated with the traditional male role: (1) sexual competence, (2) ability to give partners orgasms, (3) sexual desire, (4) prolonged erection, (5) being a good lover, (6) fertility, (7) reliable erection, and (8) heterosexuality. Common to all these beliefs, sex is seen as a performance in which men are both the directors and the principal actors.

The traditional female script prescribes females a more passive role in sexual relations. Females are expected to wait for and comply with the male's initiation of sexual activity and to be pleased with how each sexual encounter progresses. Where sex is the goal to be achieved in the male sexual script, the female script focuses more on feelings than on sex, more on love than on passion. When not sanctioned by love or marriage, sexually active women risk their reputations. The traditional female sexual script left many women unable to talk about sex easily because they were not expected to have strong sexual feelings. Some women may feel comfortable enough with their partners to have sex with them but not comfortable enough to communicate their needs to them. Under more traditional expectations, to keep her image of sexual innocence and remain pure, a woman does not tell a man what she wants.

Women often "learn" that the right way to experience orgasm is from penile stimulation during sexual intercourse. But there are many ways to reach orgasm: through oral sex; manual stimulation before, during, or after intercourse; masturbation; and so on. Women who rarely or never have an orgasm during heterosexual intercourse may be deprived by not sexually expressing themselves in other ways.

These gendered sexual scripts affect the ways in which many people look at a range of sexual and sexually related behaviors. For example, by portraying males as active, sexually aggressive initiators whose sexual response cannot be easily controlled or constrained and females as sexually passive, innocent, and even nonsexual, traditional scripts ignore the possibilities of sexually reluctant males or sexually coercive females. This, in turn, obscures the phenomenon of female sexual offending, especially with male victims, and leaves police, victims, and helping professionals in the dark about what may be a more common phenomenon than is understood.

## Contemporary Sexual Scripts

As gender roles have changed, so have sexual scripts. To a degree, traditional sexual scripts have been replaced by more liberal and egalitarian ones. Contemporary sexual scripts include the following elements for both sexes:

- Sexual expression is positive.
- Sexual activities are a mutual exchange of erotic pleasure.
- Sexuality equally involves both partners, and the partners are equally responsible.
- Legitimate sexual activities are not limited to sexual intercourse but also include masturbation and oral–genital sex.
- Either partner may initiate sexual activities.
- Both partners have a right to experience orgasm, whether through intercourse, oral–genital sex, or manual stimulation.
- Nonmarital sex is acceptable within a relationship context.
- Gay, lesbian, and bisexual orientations and relationships are increasingly open and accepted or tolerated, especially on college campuses and in large cities.

Contemporary scripts both give greater recognition to female sexuality and are relationship centered rather than male

centered. As we said earlier, traditional scripts have been replaced *to a degree*. Certainly, the kind of sexual encounters that are common in "hooking up" are less gendered in that either party can initiate and both parties can expect sexual gratification. However, women who have several concurrent sexual partners or casual sexual relationships, for example, are still more likely to be regarded as more promiscuous than are men in similar circumstances. Women often carry out the "suppression of female sexuality," whether through maternal influence or the judgments of female peer groups (Miracle, Miracle, and Baumeister 2003).

From their study of 698 heterosexual couples from a northeastern college campus and surrounding community, researchers Kathryn Greene and Sandra Faulkner (2005) contend that the contemporary sexual scripts for females and males are more alike, especially within committed heterosexual relationships where there is much overlap in the behavior of women and men. However, they also suggest that the "rules and roles" regarding sexual relationships in some ways still point to a "double standard." This is most evident on issues such as how many lifetime partners are "acceptable" (as opposed to "too many"), who in the relationship suggests using condoms or other methods of contraception, who should initiate sexual activity, and how openly one should discuss sexual matters. Greene and Faulkner found that couples who expressed less belief in the sexual double standard and other traditional sexual attitudes also reported more sexual communication and sexual self-disclosure. Such communication is further associated with higher relationship satisfaction (Greene and Faulkner 2005).

## How Do We Learn about Sex?

Children and adolescents are subjected to gender-specific messages about sexuality as well as both subtle and explicit socialization into heterosexuality (and away from homosexuality and bisexuality). Sources of influence include parents, siblings, peers, media, and school.

### Parental Influence

Children learn a great deal about sexuality from their parents. They learn both because their parents set out to teach

**True** **1.** Expectations for women's and men's behavior in sexual relationships have become more similar.

them and because they are avid observers of their parents' behavior. Even in families where open and active discussion of sexual topics is avoided, lessons about sex are taking place. When silence surrounds sexuality, it suggests that one of the most important dimensions of life is off-limits, bad to talk about, and dangerous to think about.

Parents convey sexual attitudes to their children in a number of ways. What parents say or do, for example, to children who touch their "private parts" or try to touch either their mother's or some other woman's breasts conveys meanings about sex to a child. Parents who overreact to children's curiosity about their bodies and others' bodies may create a sense that sex is wrong. On the other hand, parents who acknowledge sexuality rather than ignoring or condemning it help children develop positive body images, comfort with sexual matters, and higher self-esteem (Miracle, Miracle, and Baumeister 2003). Research suggests that what parents teach their children about sex and sexuality is shaped by what they themselves were taught and reflects what they wish they had received from their parents (Byers, Sears, and Weaver 2008). Existing research also indicates that most parent–child discussion about sex is really mother–daughter discussion about sex, that sons receive much less parental insight and information than daughters do, and that sons tend to receive what information they do learn from their mothers, not their fathers (Lehr et al. 2005).

As young people enter adolescence, they are especially concerned about their own sexuality, but they may be too embarrassed or distrustful to ask their parents directly about these "secret" matters. Meanwhile, parents may also experience embarrassment when discussing sex with their adolescents. That embarrassment along with the fear that they might not be able to answer their adolescents' questions are among the biggest reservations mothers express regarding talking about sexual matters with their teens (Byers, Sears, and Weaver 2008). Furthermore, many parents are ambivalent about their children's developing sexual nature. They are often fearful that their children (daughters especially) will become sexually active if they have too much information. They tend to indulge in wishful thinking: "I'm sure Jessica's not really interested in boys yet" or "I know Tyrese would never do anything like that." As a result, they may put off talking seriously with their children about sex, waiting for the "right time," or they may bring up the subject once, say their piece, breathe a sigh of relief, and never mention it again. Sociologist John Gagnon called this the "inoculation" theory of sex education: "Once is enough." But children may need frequent "boosters" where sexual knowledge is concerned.

Research is somewhat mixed regarding the consequences of parent–child communication about sexuality (Bersamin et al. 2008). There is evidence that parents' attitudes toward premarital sex and sexuality can strongly influence adolescent sexual behavior. For example, mothers' disapproval of "risky sexual behavior" delays adolescents' initial experiences of sexual intercourse and reduces the frequency with which adolescents engage in sexual intercourse. The more positive, effective, and comprehensive the communication about sexuality, the more adolescents delayed their initial sexual experiences. Some research suggests that "early, clear communication" between parents and their teenage children leads to lower levels of teen sexual activity and, for those who become sexually active, to greater understanding and use of safe-sex practices (Lehr et al. 2005; Dilorio, Kelly, and Hockenberry-Eaton 1999). In addition, research cited by Lehr and colleagues indicates that mothers who discuss sex-related issues influence their children's later protective sexual behaviors, including condom use (Dittus, Jaccard, and Gordon 1999; Miller et al. 1998). Other researchers (Newcomer and Udry 1985; O'Sullivan et al. 1999) report little to no effect of mother–child sexual communication on subsequent teen sexual behavior, and still other research suggests that it is associated with greater involvement in sexual activity, although this may be as much a consequence as a cause—in other words, parents of teens who are sexually active or expected to soon be sexually active may then engage in more discussion of sexual matters with their adolescents (Bersamin et al. 2008; Paulson and Somers 2000).

The effectiveness of parental supervision is less equivocal than the consequences of communication. Parental monitoring—knowledge of who their children are with, where they are, and how they spend their time—is inversely associated both with adolescents becoming sexually active and with the total number of sexual partners teens have; it is positively related to adolescents' use of contraceptives. For example, high school students who received high levels of parental supervision (i.e., were unsupervised less than five hours a week) were less likely to have had sexual intercourse. They also had fewer lifetime sexual partners than teens who were unsupervised more than five hours a week (Bersamin et al. 2008).

Parents, especially mothers, are important sources of information and advice about sexuality. Although both sons and daughters speak more to mothers than to fathers about sexual issues, most parent–child sexual communication is really between mothers and daughters.

Steve Skjold/Alamy Stock Photo

Because parents assume that their children are (or will be) heterosexual, they may avoid—intentionally or merely without thinking it relevant—discussion about sexual orientation. In so doing, they leave their children less aware of homosexuality and bisexuality, which, as a result, become almost invisible. In addition, in imagining their young children's futures, many parents envision that their children will someday enter a heterosexual marriage (Kane 2012; Shibley-Hyde and Jaffe 2000).

## Siblings

Parents are not the only family members that help shape one's sexual attitudes and behavior. Siblings, specifically older siblings, are influential sources of sexual socialization, and by their own behavior may become role models for their younger siblings' sexual

attitudes and behaviors (Kowal and Blinn-Pike 2004). In fact, siblings may in some instances be more influential sources of sexual socialization than parents, though their influence can be positive or negative. Adolescents may feel less reluctance to seek information about sex or contraception from an older sister or brother than from a parent because siblings are assumed to be less judgmental, less embarrassed, and less punitive than parents. One study that followed 297 midwestern high school students over 42 months found that sibling conversation about safe(r) sexual practices was associated with less risky attitudes about sexual behavior when it is "in concert with parent–adolescent conversations about sex" (Kowal and Blinn-Pike 2004, 382).

Certain characteristics of sibling relationships appear to influence the frequency of sibling sexual communication. Where adolescents report positive relationships with older siblings, communication is more likely regardless of whether the older sibling is perceived to be sexually conservative or to take risks. Girls with older sisters described their relationships as closer, perceived their sisters to have safer attitudes about sex, and engaged in more conversations about sexual matters than any other sibling dyad (i.e., brother–brother or brother–sister) (Kowal and Blinn-Pike 2004).

## Peer Influence

Adolescents garner a wealth of information, as well as much misinformation, from one another about sex. They often put pressure on one another to carry out traditional gender roles. Boys encourage other boys to become and be sexually active. Those who are pressured must camouflage their inexperience with bravado, which increases misinformation; they cannot reveal sexual ignorance. Even though many teenagers find their earliest sexual experiences less than satisfying, many still seem to feel a great deal of pressure to conform, which may mean becoming or continuing to be sexually active. As we will explore in a later section, virginity for many young people may be experienced as a stigma, whereas virginity loss is seen either as a way to shed the stigma (more true for males) or merely as part of growing up (Carpenter 2002).

Evidence suggests that more teens are delaying their initial experience of sexual intercourse, thus creating a somewhat different peer culture. Findings from the National Survey of Family Growth show that from 1995 through 2010, the percentages of females and

**TABLE 6.1** Percentage of Never-Married Teens, Ages 15 to 19, Who Ever Had Intercourse by Age and Sex, United States 1995 to 2011–13

|  | 1995 | 2002 | 2006–2010 | 2011–2013 |
|---|---|---|---|---|
| **Female** | | | | |
| 15–19 years old | 49.3% | 45.5% | 42.6% | 44.1% |
| 15–17 years old | 38.0% | 30.3% | 27% | 30.2% |
| 18–19 years old | 68.0% | 68.8% | 62.7% | 64.4% |
| **Male** | | | | |
| 15–19 years old | 55.2% | 45.7% | 41.8% | 46.8% |
| 15–17 years old | 43.1% | 31.3% | 28% | 34.4% |
| 18–19 years old | 75.4% | 64.3% | 63.9% | 64.0% |

SOURCE: NSFG, Series 23, No. 31, Table 1; Martinez and Abma (2015).

males who had experienced sexual intercourse had declined. In the more recent period, from the 2006–2010 to 2011–2013 surveys, there was a slight uptick in the percentages of both female and male teens who had experienced sexual intercourse (see Table 6.1). The 2011–2013 data are still lower than the 1995 data (and for females, lower than the 2002 data as well). We look further at trends in sexual involvement in a later section of this chapter.

## Media Influence

Although many may believe that the media have no effect on them, the various mass media profoundly affect our sexual attitudes and may also affect our sexual choices and behaviors (Brown et al. 2006; Collins et al. 2011; Strasburger 2005). In fact, "private electronic media [have] become the primary sexuality educator of youth," and "young teens (ages 10–15) consider the mass media as more important sources of information about sex and intimacy than parents, peers, and sexuality education programs" (Allen et al. 2008, 518).

Certain media—most notably television and popular music—have been shown to have an effect on attitudes about sex and on sexual behavior. For example, research undertaken by social psychologist Rebecca Collins and colleagues (Collins et al. 2004) found that both teenagers' sexual beliefs and their behavior are affected by their exposure to sexual content on

television. Sexual content is common in prime-time programming. After reviewing a sample of almost 1,000 programs, 7 out of 10 were found to have sexual content and convey messages about sex. Talking about sex was more common than was portraying sexual activity (Kunkel et al. 2005).

Extensive exposure to television portrayals of sexual behavior, televised talk about sex, and talk about contraception can all influence the ways teens think about sex and how they act. Collins and associates (2004) found that extensively exposed viewers of sexual content were twice as likely as those who viewed the lowest level of sexual content to initiate sexual intercourse or progress from less (e.g., kissing) to more advanced levels of sexual expression (e.g., oral sex). Other research, stretching back from the 1980s to the 2000s, finds similar associations between viewing sexual content on television and earlier involvement in sexual activity.

What we listen to has its own supposed effects. Physician and researcher Brian Primack and colleagues (2009) examined the effect of exposure to music

Popular magazines such as *Seventeen* and *Cosmopolitan* are part of the sexual socialization that many young women in the United States experience. Although both are widely read, they convey different messages about sexuality.

**Sex, Teens, and Television**

"Forgive me, Father, for I have sinned. It's been . . . a while since my last confession."

"What troubles you, my child?"

"After being broken up with my boyfriend for exactly 20 minutes, I succumbed to inebriation, performed at a speakeasy, and surrendered my virtue to a self-absorbed ass. The only good news is that he's a total pig who'll act like it never happened, thank God. . . . Sorry. Truthfully, I'm not even Catholic."

This exchange, between a priest and Blair Waldorf, the main character of the recent and popular television program, *Gossip Girl*, is from a season 1 episode. Based on the successful book series by author Cecily von Ziegesar, *Gossip Girl* tells the story of a group of girls at an exclusive private high school in New York City. Sex is central to the story line of many episodes, and characters plot to have sex, steal each other's boyfriends, and cheat on their own.

Teenage sex is often featured prominently on other widely watched series, including ABC Family's, *Pretty Little Liars*, another program based on a successful 16-book series, this by author Sara Shepard. In what follows from a first-season episode of the popular series, two of the main characters, Mona and Hanna, are discussing whether Hanna and her boyfriend will have sex:

Mona: So, the pressure's on.

Hanna: What do you mean?

Mona: Not all of us have a Sean to wear to that party, and I'm not gonna spend the night guarding the bushes so you can jump each other's bones.

Hanna: Okay, we're not gonna be doing it in the bushes.

Mona: Whatever. Have you guys even done it yet?

Hanna: It's not a race, Mona.

Mona: Okay mom, seriously. No one's pushing you to be a natty ho, but you guys have been going out for months. If you're not together in that way, how do you know you're together-together? How long can you wait before you lose him?

*Gossip Girl* portrays teenage sex as part of a lifestyle filled with risk taking that also includes underage drinking and illegal drug use. Girls flaunt their sexuality, and use sex to further their popularity and betray their friends. The show occasionally blurs the line between consensual sex and sexual exploitation. From her content analysis of *Gossip Girl*, media researcher Elke Van Damme contends that in many story lines, males do not respect females' sexual boundaries, females are degraded as sexual objects, and females' bodies and their sexuality are treated as part of the consumer culture ("goods" you pay for) or as prizes to be claimed by males. On the positive side, she also notes that female characters are occasionally portrayed as taking sexual initiative (Van Damme 2010).

*Pretty Little Liars* features more diversity of sexual issues, choices, and lifestyles, ranging from sex between a high school teacher and one of his students, teens who remain virgins out of religious conviction, and female characters who more freely engage in sexual relations. It is noteworthy, too, for featuring main characters in same-sex

lyrics, especially those that are sexually degrading (sex based only on physical characteristics and in which consent is not mutual but reflects a power imbalance). In Primack et al.'s study, it wasn't just the level of exposure but also the nature of the sexual content that mattered. High levels of exposure to lyrics describing degrading sex were associated with higher levels of sexual behavior. Where nearly a third of the teens overall had had intercourse, 45 percent of those with the greatest amount of exposure to sexually degrading lyrics had sexual intercourse. This effect was found for both genders. Exposure to lyrics describing nondegrading sex did not have the same effect. Only 21 percent of those teens least exposed to sexually degrading lyrics but exposed to nondegrading sexual lyrics had had intercourse (Primack et al. 2009).

Electronic media, as well as magazines, newspapers, and print advertising can also affect what young people learn about and learn to expect from sex. Sometimes the messages are mixed. For example, research on the effects of magazines on young women suggests that print media can affect women's attitudes and beliefs about sex. The more "teen-focused magazines" expose young women to a contradictory message that encourages them to be sexually provocative in their demeanor and dress but that discourages them from being sexually active. Encouraged to devote much time and effort toward making themselves physically appealing to boys and to presenting themselves as sexual objects, girls are at the same time discouraged from and warned about pursuing sexual relationships. Males are negatively portrayed as "either emotionally

Television programs popular with teenagers, such as *Gossip Girl*, often focus on and prominently feature sexual issues.

Kristin Callahan/Everett Collection

relationships. Both shows are examples of programs directed toward teens where sex is central to the plot. As research by the Rand Corporation reveals, heavy viewing of programs with sexualized content appears to influence teens who watch them. Interestingly, teens themselves agree that television affects the sexual behavior of *other teens* but do not believe that they themselves are affected. A 2002 survey of 15- to 17-year-olds posed the following among its questions:

How much, if at all, do you think the sexual behaviors on TV influence the sexual behaviors of teens your age? How much, if at all, do you think the sexual behaviors on TV influence your own behavior?

| | Influence Others | Influence Own Behavior |
|---|---|---|
| A lot | 32% | 6% |
| Somewhat | 40% | 16% |
| A little | 22% | 28% |
| Not at all | 6% | 50% |

SOURCE: Kaiser Family Foundation (2002).

Where only 6 percent of teens said that their own behavior was influenced a lot by sexual content they saw on television, an identically small percentage thought that television had no influence on other teens. Also, where almost 75 percent of teens thought that other teens were influenced at least somewhat, less than 25 percent thought that they themselves were influenced at least somewhat.

inept . . . or as sexual predators" (Kim and Ward 2004, 49), neither of which are flattering to males or encouraging for females entering the world of heterosexual relationships. More "adult-focused magazines," such as *Cosmopolitan*, convey a different message. Sexually aggressive women are portrayed positively, almost in the same way as a stereotypical male is portrayed. College-age women who more frequently read magazines such as *Cosmopolitan* were less likely to perceive sex as risky or dangerous and "more likely to view sex as a fun, casual activity and to be supportive of women taking charge in their sexual relationships" (Kim and Ward 2004, 53).

Using a variety of designs and questions, research strongly suggests that teenage behavior is influenced by the amount of exposure teens have to sexual content in the media. A 2006 study looking at the effects of television along with popular music, movies, and magazines on the behavior of more than 1,000 North Carolina middle school students from more than a dozen middle schools found that white students who were in the top 20 percent of exposure to sexual content at ages 12 to 14 were twice as likely to have engaged in sexual intercourse by ages 14 to 16 as those in the bottom 20 percent of exposure. Among African Americans, media exposure had less effect. Instead, sexual experience was influenced more by their friends' sexual behavior and by their perceptions of what their parents expected of them (Brown et al. 2006).

A second study examined six different media, adding newspapers and Internet sites to the four media already mentioned, assessing both the amount of sexual content and the effects of that content. Communications researchers Carol Pardun, Kelly Ladin L'Engle, and Jane Brown (2005) found that popular music and movies have the largest percentage of sexual content, with television third. In terms of impact, movies and music were more strongly associated with both "light" sexual behavior (e.g., being alone with a romantic partner, light kissing, or "French kissing") and "heavy" sexual behavior (breast touching, penis or vagina touching, oral sex, or sexual intercourse) and with one's intentions to have sex than were the other media, including television. This study also indicated that it was really how much sexual content, not what kind of sexual content, one was exposed to that mattered in affecting one's behavior and expectations.

As we get older, parents, peers, and the media eventually become less important in our sexual learning. As we experience interpersonal sexuality, our sexual partners become the most important sources of modifying traditional sexual scripts. In relationships, men and women learn that the sexual scripts and models they learned from parents, peers, and the media may not necessarily work in the real world. They adjust their attitudes and behaviors in everyday interactions. If they are married, sexual expectations and interactions become important factors in their sexuality.

## A Caution about Data on Sex

Readers should keep in mind that the data reporting estimates of the frequency or prevalence of various sexual behaviors presented here and throughout this chapter are subject to potential response biases on the part of survey participants. In general, survey respondents may be affected by what researchers refer to as a **social desirability bias**, leading them to answer researchers with more acceptable responses. Although this is true in all survey research, questions about sexual relations may be especially susceptible to either exaggeration or understatement, depending on the behavior in question and the type of person responding. Therefore, estimates of number of partners and certain behaviors—such as masturbation, oral sex, anal sex, same-sex relations, and infidelity—may be lower than the "real incidence" of such behavior. The same may be true of estimates of sexual intercourse, though that may be so more for women than for men (who, in fact, may exaggerate about the

numbers of partners they've had or other aspects of their sexual experience) and for more religious than less religious people.

Unfortunately, there is little way to get around this problem. Sexual behavior is the most private of behaviors. If one wishes to study what people do sexually, who they do it with, how often, and in what settings or situations, one can only ask about it. Asking about sex, even with the promise of anonymity for participants, risks receiving answers that may be less than truthful. Some people will likely answer truthfully and completely. Regrettably, there is no way to know how many are truthful and how many aren't.

Similarly, sampling biases can lead to estimates that are far from accurate reflections of true population characteristics or behaviors. A noteworthy example of this is Shere Hite's research on infidelity among married women. In her book *Women and Love*, Hite claimed that on the basis of her sample findings, 70 percent of U.S. women married five years or more had had extramarital relationships. This estimate was more than *four times higher* than most other scientific estimates. How did she arrive at such a result? Hite sent 100,000 questionnaires to women reached through a variety of women's groups and organizations. She did not draw a scientifically representative random sample of U.S. women. Thus, the experiences of her sample were not likely to reflect the experiences or attitudes of most women. Furthermore, of her initial questionnaires, she received 4,500 back, a response rate of 4.5 percent, much lower than accepted standards in social research. Given that 95 percent of her sample did not respond, any estimate based on the small percentage who did respond will be suspect. The research cited throughout this chapter and the wider book is more scientifically rigorous research. Still, in encountering the material that follows, we recommend that readers keep in mind the difficulty in surveying people about sexual matters as well as the importance of adhering to methodological rigor.

## Attitudes about Sex

As we begin to look more closely at sex and sexuality, it is worth pointing put that there have been notable changes in American's attitudes toward sexual behaviors and lifestyles. Generally speaking, Americans are much more accepting of premarital and same-sex sexual activity

> **True** **2.** Data on sexual behavior are prone to a social desirability bias that may lead people to exaggerate some but understate other sexual experiences.

(Twenge, Sherman, and Wells 2015). Using data from the General Social Survey, Jean Twenge, Ryne Sherman, and Brooke Wells detail increasing acceptance of premarital sexual relations, sex among teenagers, and same-sex sexual relations. They point out that in that same time span, attitudes toward extramarital sex grew less accepting. Whereas less than a third of Americans responded that premarital sex was "not wrong at all" in the early 1970s, 55 percent expressed such sentiments in the most recent surveys. Regarding same-sex sexual relations, we see highest levels of approval or acceptance among especially younger women and men. Among 18- to 29-year-olds in particular, more than half, 56 percent, described same-sex relations as "not wrong at all" in the 2010s. In comparison, among 18- to 29-year-olds in the early 1990s, 26 percent believed same-sex relations were not wrong (Twenge, Sherman, and Wells 2015). Generally, blacks and women were less accepting of each sexual lifestyle or behavior item than were whites or men. In a later section we'll look more closely at attitudes toward nonmarital sexual relations.

# Sexuality in Adolescence and Young Adulthood

Several tasks challenge young adults as they develop their sexuality:

- *Establishing a sexual orientation.* Children and adolescents may engage in sexual experimentation such as playing doctor, kissing, and fondling members of both sexes without such activities being associated with sexual orientation. By young adulthood, a heterosexual, gay, lesbian, or bisexual orientation emerges. Most young adults develop a heterosexual orientation. Others find themselves attracted to members of the same sex and begin to develop a gay, lesbian, or bisexual identity.
- *Integrating love and sex.* As we move into adulthood, we need to develop ways of uniting sex and love.
- *Forging intimacy and commitment.* Young adulthood is characterized by increasing sexual experience. Through dating, cohabitation, and courtship, we gain knowledge of ourselves and others as potential partners. As relationships become more meaningful and intimate, sexuality can be a means of enhancing intimacy and self-disclosure as well as a means of obtaining physical pleasure.

- *Making fertility or childbearing decisions.* Childbearing is socially discouraged during adolescence, but fertility issues become critical, if unacknowledged, for young single adults. If sexually active, how important is it for them to prevent or defer pregnancy? What will they do if the woman unintentionally becomes pregnant?
- *Developing a sexual philosophy.* As we move from adolescence to adulthood, we reevaluate our moral standards, using our personal principles of right and wrong and of caring and responsibility. We develop a philosophical perspective to give coherence to our sexual attitudes, behaviors, beliefs, and values. Sexuality must be placed within the larger framework of our lives and relationships, integrating our personal, religious, spiritual, or humanistic values with our sexuality.

## Adolescent Sexual Behavior

If we take a long view of changes in teenage sexual attitudes and behavior, today's teenagers are more tolerant of premarital sex and more likely to engage in it than their parents' generation did.

Yet, as noted earlier and revealed in what follows, teenage and young adult sexuality have undergone some pronounced changes in more recent years. Using data from the 2011–2013 National Survey of Family Growth and from the 2013 Youth Risk Behavior Survey (Kann et al. 2014), we can draw the following sketch of adolescent and young adult sexual experiences:

- Over 40 percent of never-married 15- to 19-year-old females (44 percent) and close to half of 15- to 19-year-old males (47 percent) reported that they had experienced sexual intercourse at least once. Thirty percent of 15- to 17-year-old females and 34 percent of 15- to 17-year-old males had sexual intercourse experience. Among 18 and 19-year-old never-married teens, 64 percent of both females and males had had sexual intercourse.
- For both males and females in the National Survey of Family Growth, the median age at first sexual intercourse was around 17 (17.2 for females, 16.8 for males).
- Among the high school students in the 2013 Youth Risk Behavior Study, nearly 6 percent (5.6 percent) had their first sexual intercourse before turning 13 years of age. There were noteworthy race differences regarding such experiences: 3.3 percent of whites,

6.4 percent of Hispanics, and 14 percent of black high school student respondents said their first experience of sexual intercourse occurred before they were 13. Looking more closely at race and sex, 24 percent of black male respondents and 14 percent of black female respondents reported having had intercourse by the time they were 12 (Kann et al. 2014).

- Again, drawing from the Youth Risk Behavior Study, in 2013, among females, there was almost no difference between non-Hispanic whites (45.3 percent) and Hispanics (46.9 percent), with larger differences between white and Hispanic females and black females (53.4 percent) in the percentage reporting themselves as having ever had sexual intercourse. Among males, the differences were larger: 42.2 percent of white males, 51.7 percent of Hispanic males, and 68.4 percent of African American males reported that they had had sexual intercourse at least once (see Figure 6.1). Overall, 46 percent of 15- to 19-year-old females and 47.5 percent of males had experienced sexual intercourse.

- As reflected in Figure 6.2, there are differences in the percentage of high school students reporting themselves as sexually active. Defining "sexually active" as having had sexual

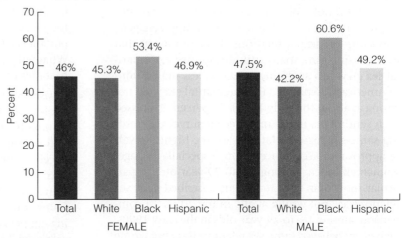

**Figure 6.1** Percentage of High School Students Who Ever Had Sexual Intercourse, by Sex and Race

SOURCE: Table 63, Kann, L., S. Kinchen, S. Shanklin, et al.; Youth Risk Behavior Surveillance—United States, 2013. MMWR 2014; 63(no.4), June 13, 2014.

intercourse with at least one person in the three months prior to the survey, whites (32.8 percent) and Hispanics (34.7 percent) were less likely than blacks (42.1 percent) to be sexually active. Of note, this difference was mostly a consequence of a much higher percentage of sexually active black *males* (estimated at 47 percent) than white or Hispanic males. Among females, race differences were between 2 to 3 percent.

- Among 15- to 19-year-olds, three-fourths of those who had had sexual intercourse experienced it with someone with whom they were going steady. A sixth of 15- to 19-year-olds reported that they first had

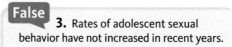

**False**

**3.** Rates of adolescent sexual behavior have not increased in recent years.

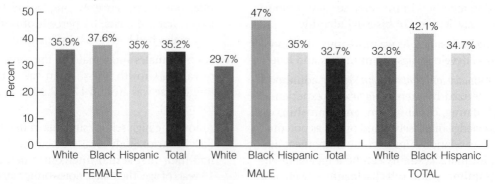

**Figure 6.2** Percentage High School Students Who Were Currently Sexually Active, by Sex and Race/Ethnicity 2013

SOURCE: Kann, Kinchen, Shanklin, et al., MMWR 2014:63, Table 65.

sexual intercourse with someone they had just met or with whom they were just friends.

- Fifteen percent of the high school students in the Youth Risk Behavior Surveillance Survey reported having had sexual intercourse with four or more people during their life. This was more common among older high schoolers, as almost one-fifth (18.5 percent) of 11th graders and nearly a fourth (23.4 percent) of 12th graders responded that they'd had such experience. Overall, males were somewhat more likely to report having had intercourse with at least four different partners (16.8 percent of males, 13.2 percent of females). Among sexually active teens, only about one out of seven (13.7 percent) reported that the last time they had sexual intercourse, neither they nor their partner had used any method of contraception to prevent pregnancy. Both genders reported using condoms more than any other method of contraception, both the first time they had sexual intercourse and the most recent time. Both sexes report birth control pills as the next most widely used contraceptive method.

| TABLE 6.2 Never-Married Adolescent and Young Adult Lifetime Sexual Experiences, from National Survey of Sexual Health and Behavior | | | |
| --- | --- | --- | --- |
| **Male** | 14–15 | 16–17 | 18–19 |
| Masturbation | 67.5% | 78.9% | 86.1% |
| Mutual masturbation | 5.7% | 20.3% | 49.3% |
| Received oral sex from a female | 13% | 34.4% | 59.4% |
| Gave oral sex to a female | 8.3% | 20.2% | 60.9% |
| Vaginal intercourse | 9.9% | 30.3% | 62.5% |
| **Female** | 14–15 | 16–17 | 18–19 |
| Masturbation | 43.3% | 52.4% | 66% |
| Mutual masturbation | 9.0% | 19.7% | 38.8% |
| Received oral sex from a male | 10.1% | 25.8% | 62% |
| Gave oral sex to a male | 12.8% | 29.1% | 61.2% |
| Vaginal intercourse | 12.4% | 31.6% | 64% |

SOURCE: Herbenick et al. (2010).

Additional details about adolescent and young adult sexuality can be gleaned from the National Survey of Sexual Health and Behavior (NSSHB), a large study of sexual experiences of a nationally representative sample of 5,865 women and men ages 14 to 94. This survey asked about a wide range of sexual behaviors and whether respondents had experienced them in the prior month, prior 12 months, or at some earlier point in their lifetimes (Herbenick et al. 2010). Although we will examine some of the findings about older adults in a later section, here we look at some of the findings about the sexual experiences of 14- to 19-year-olds in the NSSHB. Table 6.2 shows the extent of and variation in sexual experiences of 14- to 15-year-old, 16- to 17-year-old, and 18- to 19-year-old females and males in this sample.

Of note, given the small numbers of gay, lesbian, and bisexual respondents that are included in studies based on nationally representative samples, estimates of same-gender sexual experience are not included here. As the researchers remind, without deliberately oversampling males and females who report gay, lesbian, or bisexual identities, the numbers included are too small for accurate statistical projections (Herbenick et al. 2010).

As the data reveal, adolescents and young adults become increasingly sexually experienced, reflecting an apparent "developmental trajectory of sexual expression" (Herbenick et al. 2010, 261). Although a minority of 14- to 17-year-olds report having had partnered sexual experience, by 18 to 19, close to or over 60 percent of females and males have had oral sex and sexual intercourse, and roughly half of males and almost 40 percent of females report having experienced

By ages 18 to 19, a majority of U.S. teens have experienced both oral sex and sexual intercourse.

Jacob Lund/Shutterstock.com

For many people, the 2008 suicide of 18-year-old Jessica Logan first caught their attention and introduced them to what since has come to be known as "sexting." Jesse Logan, a high school senior who was planning to go to college, got caught up in a nightmarish situation resulting from her decision to send nude pictures of herself, by cell phone, to her boyfriend. After the two broke up, her ex-boyfriend sent some of the photos to other girls at their Ohio high school. Before long, the pictures were more widely distributed among students. Jesse Logan became an easy target for harassment and name-calling, had things thrown at her, and was pretty badly bullied. She even went on a Cincinnati television station to tell her story, saying that she wanted to warn others about the ramifications of sending such images and spare them the same sort of consequences that she was experiencing. Instead of attending college, she committed suicide, hanging herself in her bedroom closet.

This truly tragic story is extreme, and the outcome is mostly uncommon. However, Jesse Logan's shame- and despair-driven suicide is far from the only such incident. In a riveting and deeply disturbing 2013 article in *Rolling Stone*, journalist Nina Burleigh recounts the suicide of 15-year-old Audrie Pott. On September 10, 2012, Audrie Pott hung herself after facing the humiliation of having images of her, seminude and passed out, distributed online. After getting drunk at a party, she woke the next day to find her body covered with scrawled images and words made with a Sharpie marker. Three males at the party had taken Pott to an upstairs bedroom where she was sexually assaulted and photographed while unconscious. When interviewed by police, the boys who assaulted her admitted that, in addition to coloring half her face black, they had pulled down her bra to expose her breasts, on which they drew lines and circles, pulled down her shorts, and one of them had written "anal" above her buttocks with an arrow pointing downward. There were drawings all over her body, "including her genital area." One of the three males who participated tried to sell images to a pornographic website. Two of the boys were found to have additional nude photos of other girls on their cell phones (Burleigh 2013). In the course of her article, Burleigh refers to other such cases, including one just a couple of years earlier involving a popular cheerleader at a neighboring California high school (Burleigh 2013).

Although suicide is an uncommon result following having images of oneself sexted, instances of teenagers exchanging nude or seminude pictures of themselves via cell phone have become increasingly common, and their stories ever more familiar to us. Called "sexting" to emphasize the use of text messaging for sharing sexually explicit material, the phenomenon began to be addressed in the popular press around 2007. By 2009, "sexting" was so well known that it was a finalist for the *New Oxford American Dictionary*'s "word of the year" (Judge 2012). The behaviors that are associated with sexting have left parents, educators, and lawmakers in a quandary. What, if anything, can or should be done?

Efforts to answer that question often lead to other questions, including the question of how common or uncommon sexting is among teens. Two recent studies have examined college students' experiences with sexting. Heidi Strohmaier, Megan Murphy, and David DeMatteo found that among the 175 students at a northeastern university, more than a quarter had sent photographic sext messages while they were minors. Altogether, more than half had experience sending, receiving, or forwarding sext messages (Strohmaier, Murphy, and DeMatteo 2014). Kathy Martinez-Prather and Donna Vandiver surveyed 378 college freshmen at a southeastern university and found that as high schoolers, almost a third had sent a sext containing an image of themselves, while more than half said that they had received such messages. Of students who had not sent photographic sext messages, 27 percent had considered doing so. In addition, 18 percent of the sample reported photographing their own naked genitals, breasts, or buttocks for their personal viewing (Martinez-Prather and Vandiver 2014).

An earlier, widely quoted statistic of sexting cases in the media coverage suggests that around 20 percent of teens have sent or posted nude or seminude photographs or videos of themselves. This statistic is the product of a survey cosponsored by *CosmoGirl* and the National

mutual masturbation (Herbenick et al. 2010). Also evident, the most common sexual activity for both females and males is masturbation, and by each of the ages indicated, greater percentages of males than females report having masturbated.

## Unwanted, Involuntary, and Forced Sex

Discussions of sexual experience are incomplete unless one acknowledges that some sexual experiences are coercive. Using 2011–2013 data from the

Campaign to Prevent Teen and Unplanned Pregnancy. Of the nearly 1,300 13- to 26-year-olds who were asked a series of questions about sending or receiving sexually suggestive material, girls were a little more likely than boys (22 percent vs. 18 percent) to have sent nude or seminude photos or videos of themselves.

Other research, such as the Associated Press/MTV survey of more than 1,200 14- to 24-year-olds, found that 29 percent of the young people sampled had received either text messages or images that contained sexual content. Of those, 17 percent forwarded the images to another person. One in ten had shared a nude image of themselves, with females (13 percent) being more likely to do so than males (9 percent). Seventeen percent of sexually active respondents and 8 percent of non–sexually active respondents had shared nude images of themselves (Lounsbury, Mitchell, and Finkelhor 2011). It is worth noting that teenagers are not the only ones who engage in sexting. According to 2013 survey data from the Pew Research Center, 9 percent of adult cell phone owners reported sending a nude or nearly nude image of themselves to someone, up from 6 percent in 2012. Twenty percent reported that they had received such messages, where just a year earlier 15 percent had been recipients of sext messages. Among 18- to 24-year-olds, the percentages were higher, with 22 percent having sent, and 44 percent having received sexts. According to the Pew data, adults who own smartphones are more likely than other cell phone owners to send and receive sext messages (Pew Research Center 2014).

Regardless of how common sexting is or how abnormal and unexpected we should consider it, for minors especially, it carries potential legal risks as it is against the law. Criminal justice authorities thus face the problem of how to police teens who consensually send and receive "sext messages" and whether and how to punish those who are caught. Large U.S. cities, such as San Diego, California, now have special investigative teams within their police departments to address sexting and other Internet crimes. The San Diego unit investigates as many as 70 cases at a time (Wilson 2008). Punishment is a more vexing matter. Simply stated, sexting is a crime, and in many states, those who send sexually suggestive photos of someone under age 18 can be arrested and charged with distributing child pornography, even if the images are of oneself and the exchange is consensual. Consider these examples:

- A 17-year-old Wisconsin boy was arrested and charged with a variety of offenses, including sexual exploitation of a child and distributing child pornography, because he posted nude pictures of his 16-year-old ex-girlfriend on his MySpace page.
- An Orlando, Florida, 18-year-old is now registered as a felony sex offender, charged with child pornography and sentenced to five years probation for sending a nude photo of his 16-year-old girlfriend to her family and friends after the two had had an argument. As a convicted sex offender, he has been dismissed from his college and is having trouble finding a job.
- A 14-year-old New Jersey girl was arrested for child pornography for posting more than two dozen nude pictures of herself on her MySpace page.

Aware that the legal consequences of sexting are potentially steep and long-term, some state legislatures have explored and taken steps to lessen the punishment for offenders who get caught. In the absence of new laws, sexting cases will continue to fall under child pornography statutes, which make it "illegal to possess, distribute, or manufacture pornography involving anyone less than 18 years of age." Punishments for those convicted of child pornography can be as much as ten years in prison (Wilson 2008). Even in the event states reduce the penalties associated with sexting, many feel that sexting is less a crime than an indication of a more personal problem. Maureen Kanka—whose 7-year-old daughter Megan was murdered by a sex offender, prompting the creation and passage of "Megan's Law" (requiring released sex offenders to register with local authorities and mandating notification of the community)—reacted to the arrest of the 14-year-old New Jersey girl by advocating that such teens need counseling and intervention rather than arrest. "The only person she exploited," Kanka said, "was herself." ●

National Survey of Family Growth (NSFG), one out of five women, 18 to 44, report ever having been forced by a male to have sexual intercourse. Nearly 7 percent (6.7 percent) responded that their first sexual intercourse was "not voluntary."

Among 18- to 44-year-old males, almost 5 percent (4.6 percent) say they've ever been forced by a female to have sexual intercourse or by a male to have oral or anal sex (www.cdc.gov/nchs/nsfg/key_statistics/f.htm).

Looking specifically at teenagers, according to the 2013 Youth Risk Behavior Surveillance Survey, 7 percent of high school students—10.5 percent of high school females and 4.2 percent of high school males—responded that they had ever been physically forced to have sexual intercourse (Kann et al. 2014, Table 19).

Whether one considers "involuntary," "unwanted," or "forced" sexual intercourse, it is clear that many young people first become sexually active as a result of pressure and/or force. Using 2006–2010 NSFG data, slightly more than 1 out of 10 (10.8 percent) females 18 to 24 years old who had sexual intercourse before age 20 described their first vaginal intercourse as "not voluntary" in that they "really didn't want it to happen at the time." Another 48 percent reported that they had "mixed feelings" ("Part of me wanted it to happen at the time, part of me didn't"). Among 18- to 24-year-old males, 38 percent "either didn't want it to happen at the time" or had "mixed feelings" (Martinez et al. 2011).

Racial and age differences surface here. Combining those who said they didn't want it to happen and those who expressed ambivalence (i.e., had mixed feelings), white females and males were less likely than their African American or Hispanic counterparts to describe their first intercourse as other than wanted. Considering age, 70 percent of 18- to 24-year-old females who were 14 or younger the first time they had sexual intercourse either "didn't want it to happen" or "had mixed feelings" when it happened. Those young women who were 18 to 19 when they first experienced sexual intercourse were less likely to characterize it as unwanted (8.9 percent) or themselves as ambivalent (39.5 percent). Still, that suggests nearly half of those who were 18 or 19 the first time they had sexual intercourse characterized the experience as something other than "really wanted" (Martinez et al. 2011).

There are still other indicators of teens suffering sexual pressure, engaging in behaviors when they didn't want to, or engaging in certain sexual acts to avoid having to engage in others. Among those who were sexually active, 33 percent felt that "things were moving too fast sexually" in their relationship, 24 percent reported having done something sexually that they didn't really want to, and 21 percent had participated in oral sex to avoid sexual intercourse (Kaiser Family Foundation 2003). Furthermore, Youth Risk Behavior Surveillance data show that one out of ten high school students (14 percent of high school females and 6 percent of high school males) responded that they had experienced sexual dating violence, defined as kissing, touching, or being physically forced to have sexual intercourse when they didn't want to by someone they were going out with or dating (Kann et al 2014).

## Virginity and Its Loss

Despite the attention paid previously to teenage sexual relationships, a majority of high school students report never having had sexual intercourse. Technically, they are virgins. But what makes one a virgin? This fairly simple and straightforward-sounding question is a little more complicated. Is virginity more broadly the preservation of "innocence" through the lack of sexual *experience,* or is it more narrowly the lack of *sexual intercourse experience?* Most people agree that we maintain **virginity** as long as we refrain from sexual (vaginal) intercourse. But we occasionally hear people speak of "technical virginity" to refer to people who have had a variety of sexual experiences but have not had sexual intercourse. Such individuals are hardly sexually naive and lack some other connotations associated with the concept of virginity (e.g., innocence and purity). Data indicate that a "very significant proportion of teens has had experience with oral sex, even if they haven't had sexual intercourse, *and may think of themselves as virgins*" (Lewin 2005, emphasis added). Research findings from the 2006–2008 National Survey of Family Growth released by the National Center for Health Statistics reveal that roughly half of all 18- to 24-year-olds who had ever had oral sex indicated that they had done so prior to ever having had sexual intercourse.

Sociologist Laura Carpenter, author of *Virginity Lost: An Intimate Portrait of First Sexual Experiences* (2005), acknowledges that losing virginity has different meaning for males and females (see the Issue and Insights feature). Women are more likely to be worried about negative outcomes of their first experience of intercourse. In addition, women are more worried about pregnancy, more likely to be nervous, more likely to be in pain, and less likely to experience orgasm. They are also more likely to experience postcoital guilt and express with regret the wish that they had waited.

## Converging Patterns for Women and Men

As recently as the 1980s, young women were more likely to value virginity and to contemplate its loss primarily within committed romantic relationships, and men welcomed opportunities for casual sex and expressed disdain for virginity. Research in the 1990s revealed increasing similarities between women and men. More young men than before were expressing pride and happiness about being virgins. Growing numbers of young women were perceiving virginity in neither a positive nor a negative light, with a minority eagerly anticipating "getting it over with." By the 1990s, gender differences in the age at which one first engaged in intercourse had all but disappeared. By 1999, age at first vaginal sex was between 16 and 17 for both females and males (Carpenter 2002). Today, it is around 17 for both sexes.

**True** **4.** The most common reason teens give for abstaining from sexual intercourse is that such behavior is against their religion or morals.

With so much attention paid to sexuality within the media and peer culture, what motivates some young people to retain virginity or maintain abstinence? There is evidence suggestive of both moral and more pragmatic reasons. Among teens who have never had sexual intercourse, the most common reason given by both females (41 percent) and males (31 percent) in the NSFG survey is that to do so is against their religion or morals. Other reasons females gave included not having "found the right person yet" (18.7 percent) and not wanting to get pregnant (17.6 percent), whereas answers for males included not having found the right person (29.4 percent) followed by "not wanting to get a female pregnant"(12.6 percent) (Martinez et al. 2011). What the 2006–2010 NSFG survey also showed is that a teen's family background and circumstances were associated with whether they had or hadn't had sexual experiences. Females and males who lived with both parents were less likely to have become sexually experienced. Slightly more than a third of teenage girls who lived with both parents were sexually experienced. In comparison, more than half (54 percent) of those who lived in some other parental arrangement were sexually experienced. Parental characteristics also seemed to matter. Smaller percentages were sexually experienced for both males and females whose mothers had graduated from college, and/or whose mothers gave birth to their first child at or over age 20 (Martinez et al. 2011).

# Gay, Lesbian, and Bisexual Identities

In the United States, people are most commonly classified as involved in **heterosexuality** (sexually attracted to members of the other gender), **homosexuality** (sexually attracted to members of the same gender), or **bisexuality** (attracted to both genders). These categories reflect both **sexual identity** and **sexual orientation.** Sexual identity refers to whether one perceives oneself as heterosexual, gay, lesbian, bisexual, **pansexual** (someone who is physically or romantically attracted to others, regardless of their gender identity or biological sex) or **asexual** (one who lacks sexual attraction to anyone). Sexual orientation on the other hand often includes sexual identity and also refers to the sex to whom one is attracted and with whom one engages in sexually intimate relations (APA 2011).

Although these categories are familiar to us today, such acceptance has not always been the case, and the categories do not necessarily reflect reality. As late as the 19th century, there was no concept of "homosexuality." Both the label *homosexual* and the label *heterosexual* first appeared in print in the United States in a medical journal in 1892 (Katz 2004).

The interaction of numerous factors—social, biological, and personal—leads to the unconscious formation of sexual orientation. The two most important components of sexual orientation are the gender of our sexual partner and how we label ourselves (e.g., heterosexual, gay, lesbian, bisexual, pansexual, asexual). Finally, our sexual orientation may change over time. Thus, what was true of past relationships or attractions may not fit with the present or may differ from what we envision for our future (Klein 1990; Miracle, Miracle, and Baumeister 2003).

The familiar threefold categorization of sexual orientation so commonly used today may no longer accurately depict the range that exists in how people construct their sexual identities or experience their sexual orientations—who we are attracted to, who we have relations with, who we fantasize about, the type of lifestyle we live, and how we identify ourselves. On any of these items, we may be *exclusively* oriented toward the other sex or our sex, *mostly* drawn to the other sex or our sex, or oriented to both sexes *about*

## Issues and Insights — The Different Meanings of Virginity Loss

Researchers looking into the meanings people attach to the loss of their virginity and how they negotiate the transition from virgin to nonvirgin find that virginity has different meanings for women of different ages (Carpenter 2002; Houts 2005). As Leslie Houts (2005, 1,097) says, "the meaning of being a virgin at 14 is very different than at age 24, or at age 34." At younger ages, virginity may be culturally expected; at a somewhat later age, it may be respected and celebrated; and at "too old an age," it may be viewed with curiosity or suspicion. In each of these scenarios, the meaning of virginity loss is different (Houts 2005).

Laura Carpenter's qualitative interview research with 61 women and men suggests that people draw on three themes to make sense of their lost virginity: virginity as a gift, virginity as a stigma, and virginity loss as part of the transition to adulthood. Although many individuals indicated more than one of the following categories, the following patterns of response were revealed:

- *Virginity as a gift.* Half of Carpenter's informants recalled that at some point in their lives, they had thought of virginity as a gift that they were giving to someone, ideally to someone they loved, and to which the recipient would give enhanced love and commitment in return.
- *Virginity as a stigma.* More than a third of Carpenter's sample saw their virginity as something to hide and something they wished to shed as soon as possible ("at the first available opportunity, often with relatively casual partners, such as friends or strangers"). The sexual double standard of even contemporary sexual scripts made it easier for women both to hide and to shed their virgin status.
- *Virginity loss as part of growing up.* More than half of Carpenter's interviewees thought that the loss of virginity was inevitable and desirable, "just another experience" in the process of becoming an adult, with minimal gender differences in the interpretation of the experience. Where gender did surface prominently was in how much physical pleasure or enjoyment was experienced with the loss of virginity. For a majority in this group, including 75 percent of the women and 60 percent of the men, virginity loss was not physically enjoyable.

Sexual orientation also colored people's interpretations of their loss of virginity. Gay men and lesbians were more likely than heterosexuals to have seen the loss of virginity as a step in the process of growing up (73 percent vs. 46 percent). Heterosexual women and men were more likely to have perceived virginity as a gift than were gays or lesbians (54 percent vs. 31 percent). Interestingly, among gay men, lesbians, bisexuals, and heterosexuals who shared an interpretive framework, experience of virginity loss was quite similar. ●

---

*equally.* We may also be *asexual* and lack sexual attraction to anyone sexually, or *pansexual,* with the ability to be sexually attracted to male or female partners (as with bisexuality), as well as to androgynous and/or transgendered partners. Shortly before this book went to press, singer and actress Miley Cyrus announced that she was "pansexual" and attracted to individuals regardless of their sex or gender identities.

Because *homosexual* obscures the differences between what women and men experience, here we refer to lesbians and gay men, the L and G of the LGBT acronym (which is now often extended as far as LGBTQIA to include lesbian, gay, bisexual, transgender, queer, intersex, and asexual). In addition, replacing the term *homosexual* may help us see individuals more as whole people; sexuality is not the only significant aspect of the lives of gay men, lesbians, bisexuals, or heterosexuals. Love, commitment, desire, caring, work, possibly children, religious devotion, passion, politics, loss, and hope are also, if not more, important.

At different times, especially in the past, those with lesbian or gay orientations have been the recipients of much social stigma and negative social reaction. They might have been called sinful, sick, perverse, or deviant, reflecting traditional religious, medical, and psychoanalytic approaches. With such prejudices as underpinnings, LGBT individuals have also been victims of discrimination and targets of hate-motivated violence, second only to race-based hate crime.

Current attitudes about the LGBT population shows increased social acceptance. One of the most telling indications of this can be seen in the following data from the Pew Research Center. Asked what their reaction would be if they had a child who told them he or she was gay or lesbian, in 1985, 89 percent said they would be somewhat or very upset. Only 9 percent

**TABLE 6.3** Projected Reactions to a Child Revealing He or She was Gay or Lesbian: 1985–2015

| Year | % Very or Somewhat Upset | % Not Upset |
|------|--------------------------|-------------|
| 1985 | 89% | 9% |
| 2000 | 73% | 23% |
| 2004 | 60% | 36% |
| 2013 | 40% | 55% |
| 2015 | 39% | 57% |

SOURCE: *Los Angeles Times* Surveys (1985–2004 data); Pew Research surveys, conducted May 1–5, 2013, and May 12–18, 2015.

said they would not be upset. Fifteen years later, in 2000, nearly three-fourths of those surveyed said they would be somewhat or very upset, and 23 percent said they would not be upset. By 2013, a majority said they would not be upset (see Table 6.3).

By 2015, almost three-fifths of those surveyed said it would not upset them to learn their child was gay or lesbian, whereas 30 years ago, nine out of ten surveyed said they would be upset at such a reality (Gao 2015). Contemporary thinking in sociology and psychology has rejected the biased and unscientific labels and focused, instead, on how women and men come to identify themselves as lesbian, gay, bisexual, how their sexual identities affect their relationships and experiences, and what effect society has on them.

**True** 5. The actual percentage of the population that is lesbian, gay, or bisexual is not known.

## Counting the Gay, Lesbian, and Bisexual Populations

The actual percentage of the population that is lesbian, gay, or bisexual is not known. According to the 2011–2013 National Survey of Family Growth data, about 17 percent of women 18 to 44 years of age report having ever had sexual experience with another woman at some point in their lives. Among males, almost 6 percent (5.7) report having ever had oral or anal sex with another male at some point in their lives, and 3.1 percent report having done so within the prior 12 months of the survey (www.cdc.gov/nchs/nsfg/key_statistics/s.htm#sexualmales). However, does one or even ten experiences with a same-sex partner make a person gay or bisexual?

Considering self-designated sexual orientation, demographer Gary Gates reports estimates from three different surveys—the General Social Survey, the Gallup Daily tracking survey, and the National Survey of Family Growth. Across the three surveys, between 3 to 4 percent identified as lesbian, gay, or bisexual. In both the General Social Survey and the National Survey of Family Growth, nearly 9 percent said they had had a same-sex sexual experience since turning 18. The National Survey of Family Growth, the only one of the three that asks about same-sex sexual attraction, reported that 11 percent of respondents acknowledged some same-sex sexual attraction. Translating these sample percentages to population estimates, Gates reports that these would yield estimates of between 8.2 and 8.7 million Americans who self-identify as lesbian, gay, or bisexual, 20.2 to 20.4 million adults who have had a same-sex sexual experience, and more than 26 million Americans who have a same-sex sexual attraction (Gates 2015b).

Averaging results from seven large surveys conducted in the United States, Gates had earlier estimated that nearly 9 million Americans self-identified as lesbian, gay, bisexual, or transgendered. In terms of sexual experience, Gates had estimated that 19 million people reported having participated in some same-gender sexual behavior, and 26 million acknowledged some same-gender sexual attraction. Putting it in a somewhat different light, he noted that about as many Americans self-identified as lesbian, gay, bisexual, or transgendered as there are people living in New Jersey, and as many people self-reported same-sex sexual attractions as there are individuals living in the state of Texas (Gates 2011).

Population estimates depend on the methodologies, interviewing techniques, sampling, or definitions that researchers use. Furthermore, sexuality is more than simply sexual behaviors; it also includes attraction and desire. One can be a virgin or celibate and still be gay, lesbian, bisexual, transgender, or heterosexual. Finally, sexuality is varied and changes over time; its expression at one time is not necessarily its expression at another.

## Identifying Oneself as Gay or Lesbian

Identifying oneself as lesbian or gay may include several phases, usually beginning in late childhood or early adolescence. Sexual attraction to members of the same gender almost always precedes gay or lesbian activity, sometimes by several years.

## Stages in Acquiring a Lesbian or Gay Identity

The first stage in acquiring a lesbian or gay identity has been characterized as containing a mixture of feelings, many of which are clear reflections of the ways in which LGB sexualities were perceived and LGB people were treated. These reflections include the fear, confusion, denial, and the perception that one's sexual desires mark one as different from others. The person may find it difficult to label the emotional and physical desires for the same sex. Adolescents especially may fear their family's discovery of their attraction to others of the same gender.

In the second stage, if these feelings recur often enough, the person recognizes the attraction, love, and desire as homoerotic.

Two significant factors in identifying sexual orientation are (1) the gender of one's partner, and (2) the label one gives oneself (lesbian, gay, bisexual, or heterosexual).

Image Source/Getty Images

The third stage includes the person's self-definition as lesbian or gay. This may entail a considerable struggle because it means accepting a label that historically has been considered deviant. Perhaps, with the increasing acceptance of LGB lifestyles, the legalization of same-sex marriage, and the visibility of same-sex and transgender issues and people in popular culture, this will be easier for individuals to navigate and accept. In the Pew Research Center's report, *A Survey of LGBT Americans,* 92 percent of all LGBT adults surveyed said that compared to ten years ago, society has become more accepting of LGBT issues and individuals. Gallup's public opinion polls have tracked attitudes toward homosexuality. As of 2014, almost three-fifths of the U.S. population (58 percent) now views homosexuality as "morally acceptable" (Riffkin 2014).

## Coming Out

Being lesbian or gay is often associated with a total lifestyle and way of thinking. In making the gay or lesbian orientation a lifestyle, **coming out**—publicly acknowledging one's gayness—has become especially important as an affirmation of sexuality. Coming out may jeopardize many relationships, but it is also an important means of self-validation. By publicly acknowledging a gay or lesbian orientation, a person begins to reject the stigma and condemnation associated with it. Generally, coming out occurs in stages, first involving family members, especially the mother and siblings and later the father. Gender, age, race, and ethnicity may all interact with and color the coming-out process. Reportedly, the process may be more difficult for ethnic minority families who tend to be more reluctant to talk about sexuality issues. Younger women and men are coming out at younger ages than older cohorts did, and women tend to come out at somewhat later ages than do men (Grov et al. 2006). There is no ideal time or sequence in telling others of one's sexuality (Fisher and Lerner 2005).

Despite a clear trend toward greater acceptability and tolerance, there is still potential uncertainty as to how specific individuals might react to one's sexual disclosure, as a sizable portion of the population still does not perceive LGBT lifestyles as acceptable. Questions then arise about whether to hide one's sexual identity, or make the identity known ("to come out of the closet") and how widely to make such disclosure.

The Pew Research Center survey of LGBT Americans reports that the median age at which those who revealed their sexual orientation or gender identity to a friend or family member was 20, with gay men reporting a slightly younger age (18) than either lesbians (21) or bisexuals (20). Bisexuals (at 24 percent) were more likely than either gay men or lesbians (4 to 5 percent) to have told neither friends nor family about their sexual orientation. Younger gay males and lesbians report disclosing their sexuality at earlier ages (Mezey 2015). Among those in the Pew survey who were younger than 30, the median age at which they report having told a friend or family member was 17. One-fourth of those under 30 came out to a friend or family member before they were 15 years old.

In her book *LGBT Families*, sociologist Nancy Mezey reminds that the coming-out process can be difficult and is frequently accompanied by considerable emotional distress. Citing a number of studies, she notes that LGBT youth experience more self-harm, depression, emotional distress, suicidal ideation, and suicide attempts, and are more likely to report higher rates of alcohol or substance abuse than do their heterosexual counterparts. Most of the studies that report such problems are between 2006 and 2012. With hope and greater levels of even wider acceptance, this picture will improve.

In looking at coming out to parents and family, LGBT adults in the Pew survey are most likely to have told a close friend of their sexual orientation or gender identity, with nearly nine out of ten (86 percent) having disclosed to a friend. Lesbians and gay men are more likely than bisexuals to have told their parents about their sexual orientation, and they are more likely to have told their mothers than their fathers. Seventy percent of gay men, 67 percent of lesbians, and 40 percent of bisexuals reported having told their mothers. Over half (53 percent) of gay men, 45 percent of lesbians, and 24 percent of bisexuals reported having told their fathers (Pew Research Center 2013). Bisexual women were two to three times more likely to have told their parents than were bisexual men.

Although around two-thirds of gay men and lesbians reported that telling their parents was difficult, the vast majority said that it either improved or didn't change their relationship with their parents. As Nancy Mezey reports, research on parents' reactions to their LGBT daughters and sons suggests that positive parental reactions can lead to fewer problems and better support systems for LGBT youth, whereas negative reactions can have negative effects on their children and the relationships they have with them (Mezey 2015).

Some gay men and lesbians may go through two additional stages. One stage is to enter the gay subculture. A gay person may begin acquiring exclusively gay friends, going to gay bars and clubs, or joining gay activist groups. In the gay world, gay and lesbian identities incorporate a way of being in which sexual orientation is a major part of one's identity as a person. The final stage begins with a person's first lesbian or gay affair. This marks the commitment to unifying sexuality and affection. Sex and love are no longer separated.

Gay men and women are often "out" to varying degrees. Some may be out to no one, some to their romantic partners, and others to close friends and romantic partners but not to their families, employers, associates, or fellow students. Still others may be out to everyone. Due to fear of reprisal, dismissal, or public reaction, lesbian and gay schoolteachers, police officers, members of the military, politicians, and members of other such professions are rarely out to their employers, coworkers, or the public.

**Outing** refers to the practice of publicly identifying "closeted" gays or lesbians. Some claim that outing is politically justified, rationalizing that if gays and lesbians stay quiet about their sexual orientation, negative stereotypes about homosexuals remain unchallenged. They reason that as heterosexuals discover that some of their friends and family members or even public figures with whom they are familiar and who they respect are gay or lesbian, they may modify their attitudes about homosexuality in a more accepting direction (Miracle, Miracle, and Baumeister 2003).

How does one "become" gay, lesbian, bisexual, or even heterosexual for that matter? Such a question is neither easily answered nor inconsequential. If sexual orientation is biologically based, discrimination against gay men, lesbians, or bisexual women and men is especially unjustified. It becomes no different than discriminating against someone because of their age, their gender, or their race, all statuses over which we exercise no control.

A Pew Research Center "Survey of LGBT Americans," asked sample respondents at what age they first felt that they might not be straight or heterosexual. Though the median age among all LGB adults was 12, gay men reported that as early as 10 years of age they felt that they might be gay, whereas among lesbians and bisexuals, the median age at which such early feelings were experienced was 13. Almost 40 percent of gay men (compared to 23 percent of lesbians and 18 percent of bisexuals) said that before they were 10, they were questioning their sexuality. By the time they were 14, 84 percent of gay men, 61 percent of lesbians, and 57 percent of bisexuals said they felt as though they were not heterosexual. Among those who claimed to know for sure that they were lesbian, gay, bisexual, or transgender, 17 is the median age at which they say they knew this. Here too there were some differences among LGB respondents; among gay men, 15 was the median age at which they knew for sure that they were gay. For lesbians and bisexuals, the median ages were 18 and 17 (Pew Research Center 2013).

Research on the self-identification process suggests that among those who report never questioning that they were heterosexual until later in their lives, such as college age or even middle age, most men and many of the women attribute this delayed identification to denial. However, many women (but not many men) reject the idea that they were driven by uncontrollable or irresistible desires, saying instead that they "chose" to become involved with a same-sex partner and that their choice was a political one, associated with their particular feminist politics. For others, it was a choice motivated out of the desire for more equal, more intimate relationships than they believed they could have with men (Butler 2005).

## Sexual Frequency and Exclusivity

When it comes to how couples structure and experience their sexual relationships, gay male and lesbian couples differ in two important aspects: the frequency with which they engage in sexual relations and the meanings they attach to sexual exclusivity. Within their relationships, sexual intimacy is less frequent among lesbian couples than among gay male or heterosexual couples, although nongenital or nonsexual affection (e.g., cuddling, kissing, and hugging) is reportedly more common. Of course, as is true among heterosexuals, there is wide variation among gay couples and among lesbian couples in terms of how often they share sexual intimacies. Also like heterosexuals, both gay male and lesbian couples tend to see a decline in sexual frequency over time. Comparatively, across couple types, lesbians have less frequent sex, and gay male couples have sex most frequently. This difference is most apparent in the early stages of couples' relationships (Peplau and Fingerhut 2007).

Research has consistently indicated that sexual exclusivity is less characteristic of gay male couples than either heterosexual or lesbian couples. It also appears to be of less importance to gay male couples—especially those not in civil unions—than to lesbians or heterosexual couples (Peplau and Fingerhut 2007; Solomon, Rothblum, and Balsam 2005). Psychologist and professor of sexuality studies, Colleen Hoff, the principal

Research shows that gay male couples are less likely to remain monogamous, and their relationships are less threatened by extrarelational sex.

EpicStockMedia/Shutterstock.com

investigator in a study of 556 male couples, characterizes the difference as follows: "With straight people, it's called affairs or cheating, but with gay people it does not have such negative connotations" (James 2010). In a qualitative study of 39 gay male couples that Hoff conducted with sexuality researcher Sean Beougher, partners in 64 percent of the relationships had reached an open agreement that sex with other people was acceptable, though some restrictions were imposed or expected. For example, some who arrived at such agreements specified that they only pertained to threesomes, and then only when done together. A third of the sample couples had closed agreements, choosing to remain monogamous "in the classic sense" (Hoff and Beougher 2010).

Of course, research done prior to the legal recognition of gay and lesbian couple relationships, first via civil unions or domestic partnerships and gradually through marriage, shows that tolerance for extrarelational sex is lower among those in legally recognized unions. Monogamy and romantic love are more important to lesbians and to heterosexual women than to men in heterosexual or gay relationships (Spitalnick and McNair 2005).

## Anti-LGBT Prejudice and Discrimination

**Antigay prejudice** is a strong dislike, fear, or hatred of lesbians, gay men, bisexuals, and transgender people because of their sexuality or gender expression. **Homophobia** is an irrational or phobic fear of gay men and lesbians, whereas **heterosexism** is bias and/or discrimination in favor of heterosexuals. Like other forms of bias (e.g., racism and sexism), there are both

**Figure 6.3** Attitudes on Morality of Gay/Lesbian Relations, 2001–2012

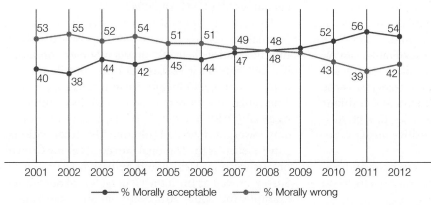

Do you personally believe gay or lesbian relations are morally acceptable or morally wrong?

institutionalized forms of heterosexism—such as the continued lack of antidiscrimination protection for those who are LGBT—and individualized expressions of heterosexism—as in telling jokes or making disparaging comments about sexual minorities. Prior research has indicated that whether in the workplace or in academic environments, such mistreatment has negative consequences for one's health, performance, and well-being. Outcomes have included diminished self-esteem, higher rates of depression, lower levels of life satisfaction, and more substance abuse (Silverschanz et al. 2008). Encouragingly, data suggest that attitudes toward the morality of gay and lesbian relations have improved, as can be seen in Figure 6.3. Along with the data in Table 6.3, this indicates a shift toward much greater acceptance of same-sex sexuality. As we shall see later in Chapter 8, the same trend has occurred regarding attitudes toward gay marriage.

Psychologist Perry Silverschanz and colleagues used Gregory Herek's definition of heterosexism—actions that deny, denigrate, or stigmatize nonheterosexual behavior, relationships, identity, or community—to assess how commonly heterosexism was experienced by a sample of more than 3,000 college students of varying sexual orientations at a northwestern U.S. university. Silverschanz et al. wanted to assess the impact of heterosexism on those who were exposed to either direct and personal harassment (e.g., being called a "fag" or a "dyke" or being called perverted) or "ambient" heterosexist harassment (e.g., overhearing offensive jokes, crude remarks, or name-calling in one's presence). Of their sample, 41 percent had some

experience with heterosexist harassment, including 39 percent of heterosexual students and 57 percent of sexual minorities. The impact of heterosexist harassment was worse when one experienced both personal and ambient harassment, and such effects affected both sexual minorities and heterosexuals (Silverschanz et al. 2008).

Other research confirms the continued experience of harassment and mistreatment by sexual minorities.

Gregory Herek, Regina Chopp, and Darryl Strohl (2007) note characteristics more consistently associated with heterosexuals who have negative attitudes toward gay males and lesbians. Typically, males harbor more prejudice and express more negative attitudes toward gays and lesbians than women do. This is addressed in some detail by sociologist Meredith Worthen (2012), who documents some of the gender pattern in attitudes toward both LGBT women and men. Reviewing existing research, she notes that in comparison to heterosexual women, heterosexual men report lower levels of "support of homosexuals." Also, heterosexual men report more negative attitudes toward gay men and more positive attitudes toward lesbians, whereas heterosexual women tend to report similar attitudes toward gay men and lesbians (Worthen 2012). Of potential reasons for this gender difference, Worthen points out that heterosexual men have more negative attitudes toward "gender nonconformity," especially from other men. She also notes that men may fear the possibility of sexual advances from gay men, which could threaten their sense of "heteromasculinity."

Aside from gender, negative attitudes toward gay men and lesbians have been found to be more common among heterosexuals who are older, less well educated, more frequent attendees at religious services, believers in the literal truth of the Bible, more supportive of traditional gender roles, and less sexually permissive. They are more likely to be Republican, politically conservative, and live in areas where hostility toward gays and lesbians is common. They are also more likely to believe that homosexuality is a choice, and are less likely to have close friends or family who are openly gay or lesbian (Herek, Chopp, and Strohl 2007). On the other hand, both long-term personal contact with well-liked gay men or lesbians, and even brief positive media exposure to admired and counterstereotypical gays and lesbians (e.g., singers such as Melissa Etheridge or Tracy Chapman, writer Alice Walker, former Congressman Barney Frank) can reduce the amount of implicit prejudice (i.e., unconscious attitudes) toward gay men and lesbians. Those with little personal contact with gays and lesbians maintain more implicit antigay attitudes, though such attitudes could be improved by positive media exposure (Dasgupta and Rivera 2008).

As shown in the research by Silverschanz and colleagues, antigay prejudice can have adverse effects on heterosexuals, too. It can:

- create fear and hatred—aversive emotions that cause distress and anxiety;

- alienate heterosexuals from gay family members, friends, neighbors, and coworkers;

- limit the expression of a range of behaviors and feelings, such as hugging or being emotionally intimate with same-sex friends for fear that such intimacy may be "homosexual";

- and lead to exaggerated displays of masculinity by heterosexual men trying to prove they are not gay.

## Bisexuality

As we noted earlier, bisexuals are individuals attracted to members of both genders. Also, as mentioned, some use the term *pansexual* to avoid assumptions of either a gender or sexual binary. Asked what their bisexual identities meant to them, most of Paula Rust's respondents said it meant that they had "the potential to be sexually, emotionally, and/or romantically attracted to members of both sexes or genders" (Rust 2004, 216). For many, it is the capacity or potential, not necessarily the actual experience, that makes them identify themselves as bisexual. For some, bisexuality is expressed in alternating relationships with women and men. Others have concurrent sexual relationships with women and men (e.g., "I have a girlfriend, with whom I have sex often and am very attracted to, but I am still attracted to men, with whom I also have sexual relations") (Rust 2004, 218). Still others base their self-definitions more on feelings than on any actual relationships, past or

**False**  **6.** Bisexuality is less widely accepted than male homosexuality or lesbianism.

present (e.g., "over 99 percent of my sexual interactions have been heterosexual. But I fantasize about women a great deal and enjoyed the one-on-one encounter I had") (Rust 2004, 218).

Trying to count or estimate the size of the bisexual population has proven to be difficult. Using data from the 2013 National Health Interview Survey, conducted by the Centers for Disease Control and Prevention and the National Center for Health Statistics, less than 1 percent of the more than 34,000 adults identified as bisexual (Dahlhamer et al. 2014). This is a considerably lower estimate than obtained by other national surveys, such as the National Survey of Family Growth or the National Health and Nutrition Examination Survey, as well as from the General Social Survey. For example, the 2006–2010 National Survey of Family Growth found 2.6 percent of respondents identifying themselves as bisexual, 3.9 percent of women, and 1.2 percent of men. The National Health and Nutrition Examination Survey found almost these same percentages: 2.6 percent of adults, 3.9 percent of women, and 1.4 percent of men. In the 2013 National Health Interview Survey, 0.5 percent of men and 1.1 percent of women identified themselves as bisexual.

Complicating the situation, should one use self-described sexual orientation, sexual attraction toward both women and men, or sexual experiences with both women and men to define one as bisexual? Demographer Gary Gates illustrates the nature of this difficulty. If one uses General Social Survey items asking about one's sexual identity, 1.2 percent report being bisexual. If one defines being bisexual solely on behavioral measures, the proportions change considerably. Gates reports that since age 18, nearly 7 percent (6.8) of adults report having had both same-sex and different-sex partners. In comparison, just 1 percent say they have had only same-sex partners. Using sexual experience in the previous five years, as opposed to lifetime experience, 1.5 percent say they have had sex with both same-sex and different-sex partners. Just under 2 percent have had exclusively same-sex partners. If one also considers gender, among LGB-identified adults, bisexuals represent a majority among women, but less than a third (30 percent) among men (Gates 2012).

Becoming bisexual requires the rejection of two recognized categories of sexual identity: heterosexual and homosexual. Within the past two decades, researchers began looking at both the attitudes held about

# The Good, the Bad, and the Ugly: Trends in the Status of the LGBT Population in the U.S. and Abroad

As we saw in Chapter 1, in the summer of 2015, the United States Supreme Court legalized same-sex marriage. This was, without a doubt, a major step in the provision of equal rights and protections for gay male and lesbian couples and one more sign of increased acceptance of same-sex relationships. With U.S. legalization, there are (as of October 2015), nearly two dozen countries in which same-sex couples can or will soon be able to legally marry (Argentina, Belgium, Brazil, Canada, Denmark, England, Finland, France, Greenland, Iceland, Ireland, Luxembourg, the Netherlands, New Zealand, Norway, Portugal, Scotland, South Africa, Spain, Sweden, the United States, and Uruguay. Same-sex marriage is also legally recognized in parts of Mexico).

Additionally, according to a United Nations Human Rights Council report from May 2015, between 2011 and 2015, 14 countries adopted or strengthened antidiscrimination and hate crime laws by extending protections for sexual orientation and/or gender identity, three countries abolished criminal penalties for homosexuality, and dozens of countries have introduced special gender and sexual sensitivity training for a host of medical, educational, and criminal justice personnel. These developments are further positive steps in a movement for greater equality, stronger protections from discrimination, and more concerted attempts to improve the quality of life for LGBT individuals.

In the last decade in the United States, support for same-sex marriage rights and the rights of gay and lesbian couples to adopt rose steadily and dramatically, even well before the landmark *Obergefell v. Hodges* decision. A host of public opinion polls found that between 55 percent and 60 percent of those sampled supported gay marriage rights, with similar percentages favoring the right of gay or lesbian couples to adopt. For example, according to a Public Religion Research Institute survey of more than 4,500 American adults, from November and December 2013 (more than 18 months before the *Obergefell* decision), solid majorities expressed support for gay marriage, rights of gay and lesbian couples to adopt, and protection from employment discrimination. Similar findings emerge from other public opinion surveys, whether conducted by the Pew Research Center, Gallup, Quinnipiac University, ABC News/*Washington Post*, NBC News/*Wall Street Journal*, or CBS News/*New York Times* (www.pollingreport.com/civil.htm).

The picture is a little less rosy when one looks at whether LGBT individuals are protected from discrimination. In more than half of all U.S. states, LGBT women and men have no protection from discrimination in housing, employment, and public accommodation. Writing prior to the *Obergefell* decision, National Public Radio journalist Tanya Ballard Brown quotes David Stacy, government affairs director for the Human Rights Campaign, "With limited or no federal protections, an LGBT person can get legally married in most states, but then be evicted from an apartment and denied a home loan" (Brown 2015, www.npr.org/sections/itsallpolitics/2015/04/28/402774189/activists-urge-states-to-protect-the-civil-rights-of-lgbt-people).

Outside the United States, similar "good news/bad news" pertains. Alongside the positive developments described in the opening paragraph, the United Nations Human Rights Council notes that the global picture still reflects ". . . continuing, pervasive, violent abuse, harassment and discrimination affecting LGBT and intersex persons in all regions" (UN Human Rights Council 2015, 20). The Human Rights Council report notes that at least 76 nations/states continue to have laws that criminalize consensual adult same-sex relationships; typically, such laws prohibit "certain types of sexual activity or any intimacy between persons of the same sex" (p. 13). In fact, in Iran, Saudi Arabia, the Sudan, Yemen, Mauritania, and parts of Nigeria and Somalia, consensual same-sex relations can be punished with the death penalty.

Aside from state-sanctioned violence and punishment, homophobic and transphobic violence occurs in all regions of the world. Such violence includes sexual assaults, assaults, beatings, kidnappings, and murders. The UN report indicates that such hate-motivated violence "is often particularly brutal, and in some instances characterized by levels of cruelty exceeding that of other hate crimes" (p. 8). Examples given include deep knife cuts, genital mutilation and anal rape, stoning and dismemberment. LGBT hate-motivated assaults and murders in numerous countries in the Americas, Africa, Europe, and the Middle East are noted, alongside nonlethal violence, such as assaults and rapes of lesbians and kidnappings and beatings of gay men. A 2013 survey of more than 90,000 LGBT individuals throughout Europe found that a quarter of respondents had been threatened or attacked within the prior five years. Within the United States, sexually motivated bias or hate crimes are second only to race-based hate violence. In the United States in 2013, 18 hate-violence homicides and 2,001 instances of anti-LGBT violence were documented (Mastropasqua 2015).

Clearly, the picture that emerges is a highly mixed one. Positive developments are evident and noteworthy. So, too, do we see areas of continued discrimination, and—ugliest of all—violence. ●

and the prejudices directed at bisexuals. Studies fairly consistently show that, like gays and lesbians, bisexual people are often the targets of hostility, harassment, and discrimination. Herek, Gillis, and Cogan (1999) reported that 15 percent of bisexual women and 27 percent of bisexual men in their sample had experienced a property or violent crime. These rates are similar to what lesbians (19 percent) and gay men (28 percent) reported. A Kaiser Family Foundation survey of 405 lesbians, gay men, and bisexuals found that 60 percent of bisexuals had experienced some form of discrimination, 52 percent had suffered verbal abuse, and 26 percent felt that their families did not accept them because of their sexual orientation. These percentages were lower than the comparable percentages for lesbians and gay men, suggesting that bisexuality may be somewhat less stigmatized than homosexuality (Herek 2002).

Harboring negative attitudes toward bisexual men and women seems to be associated with many of the same individual characteristics that were linked to antigay sentiments, such as frequent attendance at religious services, a conservative political ideology, and having had minimal prior contact with bisexual men or women. Interestingly, heterosexuals tend to report more negative attitudes toward bisexuals than toward gays and lesbians, suggesting that **biphobia** is a separate and distinct phenomenon from homophobia (Worthen 2012). Other similar concepts of relevance include notions such as *monosexism*, wherein only heterosexuality or homosexuality are recognized as valid sexual orientations, and *bisexuality erasure* or *bisexual invisibility*, terms that refer to the subsequent lack of recognition and respect for bisexuality as a legitimate sexual orientation. Perceived and portrayed as either a phase or not a legitimate sexual orientation, bisexual erasure is associated with both a lack of acknowledgment of bisexuality and a lack of institutional support for bisexuality (Raymond 2015).

Because they might be perceived as rejecting both heterosexuality and homosexuality, bisexuals can also be stigmatized by gay men and lesbians who might view bisexuals as "fence-sitters" not willing to admit their homosexuality, as "traitors to the gay and lesbian movement," or as people simply "playing" with their orientation (Herek 2002; Worthen 2012). Thus, bisexuality may not be taken seriously by either group. Furthermore, bisexuals may separate themselves from gay men and lesbians in an attempt to strengthen their political influence and visibility, leading at times to "a GL vs. B scenario" (Worthen 2012).

### Becoming Bisexual

In 1994, the first model of bisexual identity formation was developed (Weinberg, Williams, and Pryor 1994). According to this model, bisexual women and men go through four stages in developing their identity:

1. *Initial confusion.* This may last years. People may be distressed by being sexually attracted to both sexes, may believe that their attraction to the same sex means an end to their heterosexuality, or may be disturbed by their inability to categorize their feelings as either heterosexual or homosexual.

2. *Finding and applying the bisexual label.* For many, discovering there is such a thing as bisexuality is a turning point. Some find that their first heterosexual or same-sex experience permits them to view sex with both sexes as pleasurable; others learn of the term *bisexuality* from friends and are able to apply it to themselves.

3. *Settling into the identity.* At this stage, bisexuals begin to feel at home with and accept the bisexual label.

4. *Continued uncertainty.* Bisexuals don't have as active or visible a community or social environment that reaffirms their identity. Despite being settled in, many feel persistent pressure from gay men and lesbians to relabel themselves as homosexual and to engage exclusively in same-sex activities.

Online research with 157 bisexual-identified adults sought to identify what bisexuals perceive to be positive aspects of being bisexual. The mostly female, cross-national sample identified 11 specific aspects of being bisexual as positive. The following five were the most commonly mentioned: freedom to love without regard for sex/gender; freedom to explore diverse relationships and experiences; freedom from labels, roles, and social rules; honest and/or authentic self; and the achievement of a unique perspective on gender and sexuality (Rostosky et al. 2010).

## ■ Sexuality in Adulthood

Psychosexual development and change does not end in young adulthood. It continues throughout our lives. In middle age and old age, our lives, bodies, sexuality, relationships, and environment continue to change. New tasks and new satisfactions arise to replace or supplement older ones.

## Developmental Tasks in Middle Adulthood

In the middle adult years, some tasks of psychosexual development begun but only partly completed or deferred in young adulthood (e.g., issues surrounding intimacy or childbearing) may continue. Because of separation or divorce, we may find ourselves facing the same intimacy and commitment tasks at age 40 that we thought we completed 15 years earlier (Cate and Lloyd 1992). But life does not stand still; it moves steadily forward, whether or not we're ready. Other developmental issues appear, including the following:

- *Redefining sex in marital or other long-term relationships.* In new relationships, sex is often passionate and intense and may be the central focus. But in long-term marital or cohabiting relationships, the passionate intensity associated with sex is often eroded by habituation, competing parental and work obligations, fatigue, and unresolved conflicts. Sex may need to be redefined as a form of intimacy and caring. Individuals may also need to decide how to deal with the possibility, reality, and meaning of extramarital or extrarelational affairs.

- *Reevaluating sexuality.* Single men and women may need to weigh the costs and benefits of sex in casual or lightly committed relationships. In long-term relationships, sexuality often becomes less central to relationship satisfaction. Nonsexual elements such as communication, intimacy, and shared interests and activities become increasingly important to relationships. Women who have deferred their childbearing begin to reappraise their decision: Should they remain child-free, "race" against their biological clocks, or adopt a child?

- *Accepting the biological aging process.* As we age, our skin wrinkles, our flesh sags, our hair grays (or falls out), our vision blurs—and we become less attractive and less sexual in the eyes of society. By our 40s, our physiological responses have begun to slow noticeably. By our 50s, society begins to "neuter" us, especially if we are women who have gone through menopause. The challenges of aging are to accept its biological mandate and to reject the stereotypes associated with it.

## Sexuality and Middle Age

Men and women view and experience aging differently. As men approach their 50s, they fear the loss of their sexual capacity but not their attractiveness; for women, the reverse is true. As both age, purely psychological stimuli, such as fantasies, become less effective for arousal. Physical stimulation remains effective, however.

Among American women, sexual responsiveness continues to grow from adolescence until it reaches its peak in the late 30s or early 40s; it is usually maintained near the same level into the 60s and beyond. Data from both the United States and elsewhere have yielded inconsistent research findings on women's sexuality at midlife. Some studies suggest that rates of sexual intercourse, levels of sexual interest, frequency of orgasm, extent of sexual fantasizing, vaginal lubrication, and satisfaction with a partner all decline in midlife. Others show no decline in sexual interest, responsiveness, or "functioning." About the only thing that can be safely concluded is that considerable variability occurs in midlife women's sexuality.

Having emotional and psychological needs met (feeling attractive, appreciated, independent, understood, and productive) is related both to feeling attractive and to satisfaction with one's sex life. Frequency of intercourse and orgasm and finding sex pleasant, enjoyable, and satisfying are associated with higher levels of marital adjustment and contentment, although it is not clear whether marital quality causes or follows sexual satisfaction (Fraser, Maticka-Tyndale, and Smylie 2004). Data from the United States, Great Britain, and France indicate age differences that may be the result of cohort differences (based on differences in sexual socialization and changing cultural attitudes) or possible effects of aging. Around the age of 50, the average American woman begins **menopause**, which is marked by a cessation of the menstrual cycle and an end to fertility. Menopause is not a sudden event. Usually, the menstrual cycle becomes increasingly irregular for several years preceding menopause. Menopause does not end interest in sexual activities. The decrease in estrogen, however, may cause thinning and dryness of the vaginal walls, making intercourse painful. The use of vaginal lubricants will remedy much of the problem.

There is no male equivalent to menopause. Male fertility slowly declines, but men in their 80s are often fertile. Men's physical responsiveness is greatest in late adolescence or early adulthood; beginning in their 20s, men's responsiveness begins to slow imperceptibly. Changes in male sexual responsiveness become apparent only when men are in their 40s and 50s. As a man ages, achieving erection requires more stimulation and time, and the erection may not be as firm.

## Psychosexual Development in Later Adulthood

As we leave middle age, new tasks confront us, especially dealing with the process of aging itself. Our health and the presence or absence of a partner are key aspects of this time in our lives.

Many of the psychosexual tasks older Americans must undertake are directly related to the aging process:

- *Changing sexuality.* As physical abilities change with age, sexual responses change as well. A 70-year-old person, although still sexual, is not sexual in the same manner as an 18-year-old. Sexuality tends to be more diffuse, less genital, and less insistent. Chronic illness and increasing frailty understandably result in diminished sexual activity and desire. These considerations contribute to the ongoing evolution of the individual's sexual philosophy.

- *Loss of partner.* One of the most critical life events is the loss of a partner. After age 60, there is a

Sexuality among the aged tends to be sensual and affectionate. Older couples may experience an intimacy forged by years of shared joys and sorrows that is as intense as the passion of young love.

Volt Collection/Shutterstock.com

significant increase in spousal deaths. As having a partner is the single most important factor determining an older person's sexual interactions, the death of a partner signals a dramatic change in the survivor's sexual interactions.

The developmental tasks of later adulthood are accomplished within the context of continuing aging. Their resolution helps prepare us for acceptance of our own eventual mortality.

# ◼ Adult Sexual Behavior

In this section, we examine various sexual behaviors.

## Autoeroticism

**Autoeroticism** consists of sexual activities that involve only the self such as sexual fantasies, masturbation, and erotic dreams. Autoeroticism is one of our earliest and most universal yet also less accepted expressions of sexual stirrings. By condemning it, our culture sets the stage for the development of deeply negative inhibitory attitudes toward sexuality.

### Sexual Fantasies

Erotic fantasizing may be the most universal of all sexual behaviors, but because such sexual fantasies may touch on feelings or desires considered personally or socially unacceptable, they are typically not widely discussed. Although they are normal and serve certain functions (such as escape or rehearsal for later sexual behavior), fantasies may also interfere with an individual's self-image, causing a loss of self-esteem as well as confusion.

Various studies report that between 60 percent and 90 percent of respondents fantasize during sex—the percentage depending on gender, age, and ethnicity (Knafo and Jaffe 1984; Miracle, Miracle, and Baumeister 2003). A large-scale study (Michael et al. 1994) found that 54 percent of the men and 19 percent of the women thought about sex daily.

Women and men have sexual fantasies, although their fantasies often differ. Can you tell the gender of the individuals who supplied the following fantasies?

- "It's evening time, the sun is setting, I'm on a tropical island, a light breeze is blowing into my balcony doors, and the curtains [white] are fluttering lightly in the wind. The room is spacious, and there is white everywhere, even the bed. There are flowers of all kinds, and the light fragrance fills the room."

- "Have sex on the beach."

If you guessed that the first of these fantasies was from a woman and the second was from a man, you guessed correctly. These are real examples that Michael Kimmel and Rebecca Plante received from undergraduates at three New York colleges or universities. Men's and women's fantasies contained similarities (e.g., in the acts they described), but the differences were more striking: Women's fantasies were longer and more vivid, using more emotional and sensual imagery, especially in describing the setting; men more often fantasized about *doing something sexual to someone,* whereas women's fantasies were often more passive and gentler, of *having something sexual done to them,* and women's fantasies tended to have more emotional and romantic content (47 percent of women described their fantasy partners as boyfriends or husbands; only 15 percent of men depicted their fantasy partners as "significant others"). Women's fantasies were also often romantic stories of love and affection, whereas men's had less romance and less emotional language or context.

## Masturbation

**Masturbation** is the manual stimulation of one's genitals. Individuals masturbate by rubbing, caressing, or otherwise stimulating their genitals to bring themselves sexual pleasure. Masturbation is an important means of learning about our bodies. Girls, boys, women, and men may masturbate during particular periods or throughout their entire lives. However, it is among the most sensitive topics to study and tends to be viewed somewhat negatively by many. Thus, data on prevalence and frequency of masturbation are especially prone to underreporting (Robbins et al. 2011).

Research on gender and sexual behavior typically finds male–female differences in masturbation. Males typically have had significantly more masturbatory experiences than females. Data from the National Survey of Sexual Health and Behavior indicate that more than 20 percent of women from each age group from 20 through over 70 reported having masturbated within the month prior to the survey, and more than 40 percent of women 14 to 70 reported having masturbated within the prior year. Among men, close to or over 60 percent of those ages 16 to 59 reported having masturbated within the prior month, and more than 70 percent reported having done so within the

preceding 12 months. For males 16 and older, nearly 80 percent to well over 90 percent reported having masturbated in their lifetimes. Among females, the percentages reporting ever masturbating ranged from 66 percent to 80 percent.

Although the rate is significantly lower for those who are married, many people, especially men, continue to masturbate even after they marry. There are many reasons for continuing the activity during marriage: Masturbation is a pleasurable form of sexual excitement, a spouse may be unavailable or unwilling to engage in sex, sexual intercourse may not be satisfying, the partners may fear sexual inadequacy, or one partner may want to act out fantasies. In marital conflict, masturbation may act as a distancing device, with the masturbating spouse choosing masturbation over sexual intercourse as a means of emotional protection (Betchen 1991).

Cohabitation has a different effect than marriage on frequency of masturbation. Many cohabiting men masturbate often despite the presence or availability of a sexual partner. Thus, social factors other than the presence of a partner affect masturbation. In citing reasons for why they masturbate, only a third of women and men list an unavailable partner (Laumann et al. 1994).

## Interpersonal Sexuality

We often think that sex is sexual intercourse and that sexual interactions end with orgasm (usually the male's). But sex is not limited to sexual intercourse. Heterosexuals engage in a variety of sexual activities, which may include erotic touching, kissing, and oral and anal sex. Except for sexual intercourse, gay and lesbian couples engage in sexual activities similar to those experienced by heterosexuals.

### Touching

Because touching, like desire, does not in itself lead to orgasm, it has largely been ignored as a sexual behavior. Sex researchers William Masters and Virginia Johnson (1970) suggest a form of touching they call **pleasuring**—nongenital touching and caressing. Neither partner tries to stimulate the other sexually; the partners simply explore each other. Such pleasuring gives each a sense of his or her own responses; it also allows each to discover what the other likes or dislikes. We cannot assume that we know what any particular individual likes because there is too much variation among people. Pleasuring opens the door to communication; couples discover that the entire body rather than just the genitals is erogenous.

As we enter old age, touching becomes increasingly significant as a primary form of erotic expression. Touching in all its myriad forms—ranging from holding hands to caressing, massaging to hugging, and walking with arms around each other to fondling—becomes the touchstone of eroticism for the elderly. Touching tends to be the primary form of erotic expression for married couples more than 80 years old.

### Kissing

Kissing as a sexual activity is probably the most common and acceptable of all premarital sexual activities. The tender lover's kiss symbolizes love, and the erotic lover's kiss simultaneously represents passion. Both men and women in one study regarded kissing as a romantic act, a symbol of affection and attraction (Tucker, Marvin, and Vivian 1991). A cross-cultural study of jealousy found that kissing is also associated with a couple's boundary maintenance: In each culture studied, kissing a person other than the partner evoked jealousy (Buunk and Hupka 1987).

The lips and mouth are highly sensitive to touch. Kisses discover, explore, and excite the body. They also involve the senses of taste and smell, which are especially important because they activate unconscious memories and associations. Often we are aroused by familiar smells associated with particular sexual memories: a person's body smells, perhaps, or perfumes associated with erotic experiences. In some cultures—among the Borneans, for example—the word *kiss* literally translates as "smell." Among traditional Eskimos and Maoris, there is no mouth kissing, only the nuzzling that facilitates smelling.

Although kissing may appear innocent, it is in many ways the height of intimacy. The adolescent's first kiss is often regarded as a milestone, a rite of passage, the beginning of adult sexuality (Alapack 1991). Philip Blumstein and Pepper Schwartz (1983) report that many of their respondents found it unimaginable to engage in sexual intercourse without kissing. They found that those who have a minimal (or nonexistent) amount of kissing feel distant from their partners but engage in coitus nevertheless as a physical release.

The amount of kissing differs according to orientation. Lesbian couples tend to engage in more kissing than heterosexual couples, and gay male couples kiss less than heterosexual couples. As many as 95 percent of lesbian couples, 80 percent of heterosexual couples, and 71 percent of gay couples engage in kissing

**Kissing is probably the most acceptable premarital sexual activity.**

Jim Arbogast/Photodisc/Getty Images

whenever they have sexual relations (Blumstein and Schwartz 1983).

## Oral–Genital Sex

In recent years, oral sex has become an increasing part of our sexual scripts. Heterosexuals, bisexuals, gay men, and lesbians engage in it. In recent years, it has increased among teens and young adults in part as an alternative to sexual intercourse. The two types of oral–genital sex are cunnilingus and fellatio. **Cunnilingus** is the erotic stimulation of a woman's vulva by her partner's mouth and tongue. **Fellatio** is the oral stimulation of a man's penis by his partner's sucking and licking. Cunnilingus and fellatio may be performed singly or simultaneously. Oral sex is an important and healthy aspect of adults' sexual selves.

According to survey data, oral–genital sex is fairly common, especially among whites. Ninety percent of 25- to 44-year-old males and 88 percent of 25- to 44-year-old females have engaged in heterosexual oral sex. Among 15- to 44-year-olds, the percentages are somewhat smaller and reveal notable race differences. Where 88 percent of 15- to 44-year-old white females and 87 percent of white males have had heterosexual oral sex, the percentages among African Americans and Hispanics are lower.

### Vaginal Intercourse

**Sexual intercourse** or **coitus**—the insertion of the penis into the vagina and subsequent stimulation—is a complex interaction. As with many other types of activities, the anticipation of reward triggers a pattern of behavior. The reward may not necessarily be orgasm, however, because the meaning of sexual intercourse varies considerably at different times for different people. There are many motivations for sexual intercourse; sexual pleasure is only one. Other motivations include showing love, having children, gaining power, ending an argument, demonstrating commitment, seeking revenge, proving masculinity or femininity, or degrading someone (including oneself).

Looking at adults ages 25 to 44, data indicate that nearly all have had a sexual intercourse experience. The National Survey of Family Growth data indicate that (as of 2002), 97 percent of 25- to 44-year-old males and 98 percent of 25- to 44-year-old females report having had sexual intercourse at least once in their lives (Mosher, Chandra, and Jones 2005). The National Survey of Sexual Health and Behavior found that more than 85 percent of women, from age 20 on, reported having had sexual intercourse in their lifetimes. Among men, 86 percent to 93 percent of men from age 25 and older have had vaginal intercourse in their lifetimes (see Table 6.4)

Yet a third study, the Global Study of Sexual Attitudes and Behaviors (GSSAB), sampled 27,500 women and men, ages 40 to 80, in 29 countries. Among the U.S. adults, approximately 80 percent of men and 70 percent of women reported having had sexual intercourse within the prior year, and more than a third of the men (35.4 percent) and more than a quarter of the women (27.8 percent) reported engaging in sexual intercourse more than once a week (Laumann et al. 2009).

Men tend to be more consistently orgasmic than women in sexual intercourse. Part of the reason may be that the clitoris often does not receive sufficient stimulation from penile thrusting alone to permit orgasm. Many women need manual stimulation during intercourse to be orgasmic. They may also need to be more assertive. A woman can manually stimulate herself or be stimulated by her partner before, during, or after intercourse. But to do so, she has to assert her own sexual needs and move from the idea that sex is centered around male orgasm.

## Anal Eroticism

Sexual activities involving the anus are known as **anal eroticism**. The male's insertion of his erect penis into his partner's anus is known as **anal intercourse**. Both heterosexuals and gay men may participate in this activity. For heterosexual couples who engage in it, anal intercourse may be an experiment or occasional activity rather than a common mode of sexual expression. This is suggested by data from the National Survey of Men. Data from the NSSHB reveal that from age 25 through 59,

**TABLE 6.4** Lifetime Estimate of Men's and Women's Sexual Experiences, National Survey of Sexual Health and Behavior

| Men | 20–24 | 25–29 | 30–39 | 40–49 | 50–59 | 60–69 | 70+ |
| --- | --- | --- | --- | --- | --- | --- | --- |
| Masturbation | 91.8 | 94.3 | 93.4 | 92 | 89.2 | 90.2 | 80.4 |
| Received oral | 73.5 | 91.0 | 89.7 | 86.2 | 82.7 | 75.3 | 57.6 |
| Gave oral | 70.9 | 85.6 | 88.2 | 84.4 | 77.3 | 72.5 | 61.6 |
| Vaginal | 70.3 | 89.3 | 92.6 | 89.3 | 85.8 | 86.9 | 88.1 |
| Anal (inserter) | 23.7 | 45.2 | 44.5 | 43.1 | 40.4 | 26.7 | 13.8 |
| Women | 20–24 | 25–29 | 30–39 | 40–49 | 50–59 | 60–69 | 70+ |
| Masturbation | 76.8 | 84.6 | 80.3 | 78 | 77.2 | 72 | 58.3 |
| Received oral | 79.7 | 88.1 | 82 | 86.3 | 83.4 | 79 | 47.4 |
| Gave oral | 77.6 | 89 | 80.1 | 83.1 | 80 | 73.1 | 42.7 |
| Vaginal | 85.6 | 90.7 | 88.7 | 94.5 | 94 | 92.4 | 89.2 |
| Anal | 39.9 | 45.6 | 40.4 | 40.6 | 34.6 | 29.8 | 21.2 |

© Cengage Learning

more than 40 percent of males report having ever had heterosexual anal intercourse. Similar estimates were obtained among women from age 20 to 49.

Other data on sexual behavior can be examined in the following from the National Survey of Family Growth (Mosher, Chandra, and Jones 2005). Within the sample of 12,000, among 25- to 44-year-old women and men, 35 percent of women and 40 percent of men had engaged in heterosexual anal intercourse at least once. Data in Figure 6.4 illustrate that there are only modest racial differences in the percentages of 15- to 44-year-old men who have engaged in heterosexual anal sex; among females, larger race differences surface. White males and females are more likely to have had heterosexual anal sex than either African American or Hispanic males and females (Mosher, Chandra, and Jones 2005).

From a health perspective, anal intercourse is the riskiest form of sexual interaction and the most prevalent sexual means of transmitting the HIV among both gay men and heterosexuals. Because the delicate rectal tissues are easily torn, HIV (carried within semen) can enter the bloodstream. (HIV is discussed later in this chapter.) Using a lubricated condom significantly decreases the risk of transmitting HIV during anal sex (Bell 1999).

# Sexual Expression and Relationships

Sexuality exists in various relationship contexts that may influence our feelings and activities. These include nonmarital, marital, and extramarital contexts.

**Figure 6.4** Sexual Experiences of 15- to 44-Year-Old Males and Females, by Race

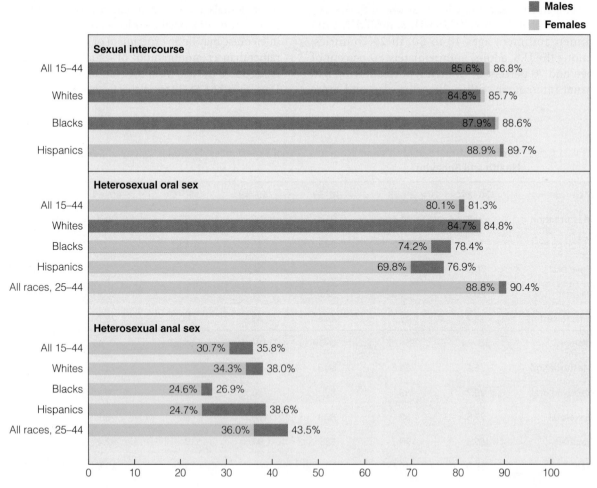

SOURCE: Chandra, et al. 2011, Table 6.

## Nonmarital Sexuality

**Nonmarital sex** encompasses sexual activities, especially sexual intercourse, that take place outside of marriage. We use the term *nonmarital sex* rather than **premarital sex** to describe sexual behavior among unmarried adults in general. Among forms of nonmarital sex, only **extramarital sex**—sexual interactions that take place outside the marital relationship between at least one married partner—continues to be consistently frowned upon. General Social Survey data from 1990 to 2008 show the relative reactions to extramarital and premarital sex. More than 90 percent of males and females said that having sex with someone other than one's spouse while married is always or almost always wrong. However, when asked about sex before marriage, only 31 percent of males and 39 percent of females responded that premarital sex was always or almost always wrong. Only when asked specifically about teenagers 14 to 16 years old having sexual relations did respondents express strong opposition to sex before marriage, with 84 percent of males and 90 percent of females saying that such sexual activity was "always" or "almost always wrong" (see Figure 6.5).

Looking at behavior, among never-married men and women, 89.1 percent of females aged 15 to 44 and 91.8 percent of males aged 20 to 44 had sexual intercourse before marriage. However, because increasing numbers of never-married adults are over 30, "premarital sex" does not seem to adequately describe the nature of their sexual activities.

**Figure 6.5** Attitudes about Nonmarital Sex, by Gender

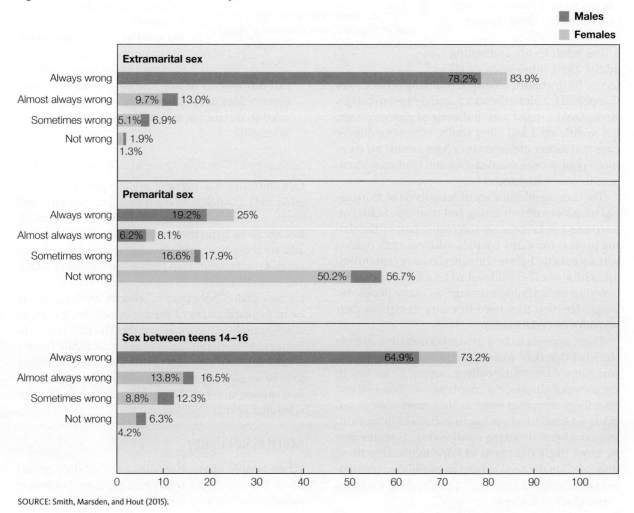

SOURCE: Smith, Marsden, and Hout (2015).

Second, at least 10 percent of adult Americans will never marry; it is misleading to describe their sexual activities as "premarital." Third, many adults are divorced, separated, or widowed; 30 percent of divorced women and men will never remarry. Fourth, between 3 percent and 10 percent of the population is lesbian or gay, and gay and lesbian sexual relationships can only recently be categorized as "premarital," now that gays and lesbians are given the right to marry.

### Sexuality in Dating Relationships

Over the past several decades, there has been a remarkable increase in the acceptance of premarital sexual intercourse, a decline in the numbers of people who believe that premarital sex is "always wrong," and an increase in the percentages who feel it is "not wrong at all." This trend has been interpreted as a shift toward "moral neutrality" regarding intercourse before marriage (Christopher and Sprecher 2000).

For adolescents and young adults, the combination of effective birth control methods, changing gender roles that permit females to be sexual, and delayed marriages have played a major part in the rise of premarital sex. For middle-aged and older adults, increasing divorce rates and longer life expectancy have created an enormous pool of once-married men and women who engage in nonmarital sex.

The increased legitimacy of sex outside of marriage has transformed both dating and marriage. Sexual intercourse has become an acceptable part of the dating process for many couples, whereas only petting was acceptable before. This increased acceptance of premarital sex is considered to be a factor in people delaying or foregoing marriage, as many people no longer feel that they need to marry to express their sexuality in a relationship.

There appears to be a general expectation among students that they will engage in sexual intercourse sometime during their college careers. As we saw in the previous chapter, the emergence of "hooking up" on college campuses suggests that many college students no longer feel sex has to occur within an existing emotional or loving relationship. Females may be more likely than men to hope that such a bond develops, but the need to have an established committed relationship is no longer a prerequisite for many sexually active students.

**Directing Sexual Activity.** As we begin a sexual involvement, we have several tasks to accomplish:

1. *We need to practice safe sex.* Ideally, we need information about our partners' sexual history and whether he or she practices safe sex, including the use of condoms. Unlike much of our sexual communication, which is nonverbal or ambiguous, we need to use direct verbal discussion in practicing safe sex.

2. *Unless we are intending a pregnancy, we need to discuss birth control.* Condoms alone are only moderately effective as contraception, although they help prevent the spread of sexually transmitted diseases. To be more effective, they must be used with contraceptive foam or jellies or with other devices.

3. *We need to communicate about what we like and need sexually.* What kind of **foreplay** or afterplay do we like? Do we like to be orally or manually stimulated during intercourse? What does each partner need to be orgasmic? Many of our needs and desires can be communicated nonverbally by our movements or other physical cues. But if our partner does not pick up our nonverbal signals, we need to discuss them directly and clearly to avoid ambiguity.

> **True** **7.** Partners in cohabiting relationships are more likely to be sexually unfaithful than are married couples.

### Sexuality in Cohabiting Relationships

Cohabitation has become a widespread phenomenon in U.S. culture. In contrast to married men and women, cohabitants have sexual intercourse more often, are more egalitarian in initiating sexual activities, and are more likely to be involved in sexual activities outside their relationship (Waite and Gallagher 2001). The higher frequency of intercourse, however, may be because of the "honeymoon" effect: Cohabitants may be in the early stages of their relationship, the stages when sexual frequency is highest. The differences in frequency of extrarelational sex may result from a combination of two factors: Norms of sexual fidelity may be weaker in cohabiting relationships, and men and women who cohabit tend to conform less to conventional norms.

## Marital Sexuality

When people marry, they discover that their sexual life is different than it was before marriage. Sex is now morally and socially sanctioned. Most heterosexual interactions take place in marriage, yet we as a culture seem ambivalent about marital sex. On the one hand, marriage is the only relationship in which sexuality is fully legitimized. On the other hand, marital sex is an endless source of humor and ridicule: "Marital sex? What's that?"

## Sexual Interactions

A variety of large-scale studies report consistent findings in regard to how often married couples engage in sexual intercourse and in how sexual frequency changes over the course of a marriage. Married couples report engaging in sexual relations about once or twice a week, or about six to seven times a month (Christopher and Sprecher 2000).

Sexual intercourse tends to diminish in frequency the longer a couple is married. For newly married couples, the average rate of sexual intercourse is about three times a week. As shown in Table 6.5, as couples get older, sexual frequency drops. In early middle age, married couples make love an average of 1.5 to 2 times a week. After age 50, the rate is about once a week or less. Among couples 75 and older, the frequency is a little less than once a month (Christopher and Sprecher 2000).

Table 6.5 shows reported frequency of sexual intercourse among married women and men at different ages. As is apparent and expected, older women and men are more likely to report little to no sexual intercourse, and less likely to report frequent intercourse. Note, however, that although the data show age patterns, they do not display duration of marriage factors.

> **False**
>
> **8.** A decline in the frequency of intercourse does not always indicates problems in the marital relationship.

Within marriage, decreased frequency does not necessarily mean that sex is no longer important or that the marriage is unsatisfactory. For dual-worker families and families with children, fatigue and lack of private time may be the most significant factors in the decline of frequency. Couples also report "being accustomed" to each other. In addition, activities and interests other than sex engage them. The decline in interest and frequency of sex may begin within the first two years of marriage (Christopher and Sprecher 2000).

## Bringing New Meanings to Sex

Sex within marriage is significantly different from nonmarital sex in at least three ways: It is expected to be monogamous, procreation is a legitimate goal, and such sex takes place in the everyday world. These differences present each person with important tasks.

**Monogamy.** Before marriage or following divorce, a person may have various sexual partners, but within marriage, all sexual interactions are expected to take place between the spouses. Approximately 90 percent of Americans believe that extramarital sexual relations are "always" or "almost always" wrong (Christopher and Sprecher 2000; Miracle, Miracle, and Baumeister 2003). This expectation of monogamy lasts a lifetime; a person marrying at 20 commits to 40 to 60 years of sex with the same person. Within a monogamous relationship, each partner must decide how to handle fantasies, desires, and opportunities for extramarital sexuality. Does one tell one's spouse about having

### TABLE 6.5  Reported Frequency of Sexual Intercourse by Married Women and Men, by Age

|         | 18–24 | | 25–29 | | 30–39 | | 40–49 | | 50–59 | | 60–69 | | 70+ | |
|---------|------|------|------|------|------|------|------|------|------|------|------|------|------|------|
|         | F | M | F | M | F | M | F | M | F | M | F | M | F | M |
| None    | 11.8 | 4.2 | 3.5 | 1.6 | 6.5 | 4.5 | 8.1 | 9.1 | 22 | 20.6 | 37.9 | 33.9 | 54.2 | 53.5 |
| Few/yr  | 14.7 | 12.5 | 11.6 | 9.3 | 16.3 | 15.6 | 21.7 | 16.2 | 23.7 | 25 | 20 | 21.2 | 25.4 | 24.2 |
| Few/mo  | 14.7 | 16.7 | 47.7 | 46.3 | 50.2 | 47.3 | 46.6 | 51 | 36.2 | 38.3 | 35.9 | 35.4 | 18.3 | 15 |
| 2–3/wk  | 35.3 | 45.8 | 35.2 | 37.1 | 21.9 | 26.8 | 19.9 | 20.8 | 16.9 | 15 | 6.2 | 9.5 | 1.4 | 5.8 |
| >/=4/wk | 23.5 | 20.8 | 2 | 5.9 | 5.1 | 5.8 | 2.7 | 3.7 | 1.1 | 1.1 | 0 | 0 | 1.4 | 0 |

SOURCE: National Survey of Sexual Health and Behavior (2010).

fantasies about other people? Does one have an extramarital relationship and, if so, tell one's spouse? How should one handle sexual conflicts or difficulties with his or her partner?

**Socially Sanctioned Reproduction.** Even with more than 40 percent of births being to unmarried women, marriage in most segments of society remains the more socially desirable and approved setting for having children. In marriage, partners are confronted with one of the most crucial decisions they will make: the task of deciding whether and when to have children. Having children will profoundly alter a couple's relationship. If they decide to have a child, sexual activity may change from simply an erotic activity to an intentionally reproductive act as well.

**Changed Sexual Context.** Because married life takes place in a day-to-day living situation, sex must also be expressed in the day-to-day world. Sexual intercourse must be arranged around working hours and at times when the children are at school or asleep. One or the other partner may be tired, frustrated, or angry.

In marriage, some emotions associated with premarital sex may disappear. For many, the passion of romantic love, especially as experienced in the earliest period of a relationship, eventually disappears as well, to be replaced with a love based on intimacy, caring, and commitment.

## Celibate Marriages

The discussion of social exchange theory in Chapter 2 used the example of involuntarily celibate couples to illustrate how such couples make the decision to remain together (Donnelly and Burgess 2008). Such relationships, which may amount to around 14 to 15 percent of marriages, can also be instructive illustrations of the role that sex plays in marriage, and the causes and consequences of celibacy.

Using six months of desiring but not having any form of sexual contact as their measure of involuntary celibacy, sociologists Denise Donnelly and Elisabeth Burgess identified a number of factors that contribute to declines in sexual relations. In addition to the impact of the passage of time and the disappearance of novelty, they point to the following stressors that affect sexual activity:

- Late-term pregnancy and/or early postpartum adjustments

- Competing time demands, especially for dual-earning couples, opposite-shift couples, or couples who are either raising children or caring for aged or ill parents

- Chronic illness, disability, or mental illness (e.g., depression)

- Guilt or conflict resulting from one's religious beliefs

The potential consequences resulting from sexual inactivity threaten the quality or stability of a relationship. More specifically, Donnelly and Burgess identify decline in relationship quality, such as lower levels of satisfaction or happiness; sexual dissatisfaction and frustration; infidelity; decreased mental health; and relationship instability.

## Relationship Infidelity and Extramarital Sexuality

As we noted, a fundamental assumption in our culture is that marriages are sexually and emotionally monogamous. This assumption is not unique to the United States. Eric Widmer, Judith Treas, and Robert Newcomb (1998) undertook comparative research using a 24-country sample of more than 33,000 respondents and found strong and widespread disapproval of extramarital sex, although people in different countries varied some in their levels of disapproval, with some (e.g., those in Russia, Bulgaria, and the Czech Republic) being more tolerant than the majority. In the United States, roughly 80 percent of Americans believe extramarital sex is "always wrong." It has been estimated that between a quarter and half of divorces in Western countries cited a spouse's cheating as the major complaint (Mark, Janssen and Milhausen 2011). Research by Paul Amato and Denise Previti (2003) identified infidelity as the most commonly reported cause of divorce. Nearly a third of men (31 percent) and close to half of women (45 percent) have cited extramarital affairs as a reason for having separated (Atwood and Seifer 1997; DeMaris 2013). Even after controlling for the quality of the marriage, if extramarital sex has transpired, the risk of marital disruption is increased by two-thirds. Indeed, "regardless of how well adjusted the marriage appears to be, extramarital sexual involvement extracts a cost on marital stability" (DeMaris 2013, 21). However, divorce proneness also predicts infidelity (Nowak et al. 2014).

As sociologist Al DeMaris points out, there are gender differences worth noting: First, men are more

Infidelity is a major contributor to relationship problems and divorce.

*Larry Dale Gordon/The Image Bank/Getty Images*

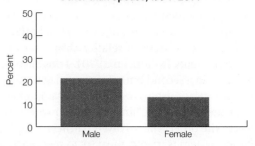

Figure 6.6  Percent of Ever-Married People Who, While Married, Had Sex with Someone Other than Spouse, 1994–2014

SOURCE: Smith, Marsden, and Hout (2015).

likely to engage in infidelity than women, as suggested by various estimates of prevalence that will be noted further in the following material. Second, women pursue extramarital relations more out of unhappiness with their spouse and inequity in their relationships. Women may be seeking emotional more than sexual fulfillment. Men may be seeking sexual excitement and variety. Third, men have more difficulty forgiving sexual infidelity than emotional affairs. Women's adultery puts marriages at greater risk of disruption (DeMaris 2013).

## How Much Infidelity and Extramarital Sex Is There?

Although we sometimes overstate the amount of "cheating" that goes on, it is neither an isolated phenomenon nor restricted to married couples. As we reported previously, there is more nonmonogamy among cohabiting couples than among married couples and among gay male couples than among lesbian or heterosexual couples. There are varying estimates of how prevalent extramarital sex is in the United States:

- Looking at General Social Survey data from 1994 to 2014, among respondents who were or had been married (thus including divorced, separated, and widowed respondents), 21.2 percent of the males and 13 percent of the females admitted to having had sex outside of their marriages (see Figure 6.6).

- Examining only a more recent period (2006 to 2014), among survey respondents who were married at the time they were surveyed, 20.2 percent of

the males and 13.5 percent of the females said that they had had sex with someone other than their spouse during the time they had been married.

Researchers Kristen Mark, Erick Janssen, and Robin Milhausen note that in surveys using nationally representative samples of U.S. women and men, 20 percent to 25 percent of men and 10 percent to 15 percent of women reported having participated in extramarital sex during their marriages (Mark, Janssen, and Milhausen 2011).

## Types of Infidelity

We tend to think of extramarital involvements as being sexual, but they may actually assume several forms (Moore-Hirschl et al. 1995; Thompson 1993). They may be (1) sexual but not emotional, (2) sexual and emotional, or (3) emotional but not sexual (Thompson 1984). Less is known about extramarital relationships in which the couple is emotionally but not sexually involved. People who engage in extramarital affairs have a number of different motivations, and these affairs satisfy a number of different needs (Adler, Hendrick, and Hendrick 1996; Moultrup 1990).

## Characteristics of Extramarital Sex

Most extramarital sexual involvements are sporadic. Most extramarital sex is not a love affair; it is generally more sexual than emotional. Affairs that are both emotional and sexual appear to detract more from the marital relationship than do affairs that are only sexual or only emotional (Thompson 1984). More women than men consider their affairs emotional; almost twice as many men as women consider their affairs only sexual. About equal percentages of men and women are involved in affairs that they view as both sexual and emotional. As we saw in the last chapter, research suggests that men are more bothered by the sexual nature of a partner's infidelity, whereas women are disturbed more by the emotional aspect (Christopher and Sprecher 2000; DeMaris 2013).

An emotionally significant extramarital affair creates a complex system of relationships among the three individuals (Moultrup 1990). Long-lasting affairs can form a second but secret "marriage." In some ways, these relationships resemble polygamy, in which the outside person is a "junior" partner with limited access to the other. The involved partners, who know that their system is triadic, must try to meet each other's needs for time, affection, intimacy, and sex while taking the uninvolved partner into consideration. Such extramarital systems are stressful and demanding. Most people find great difficulty in sustaining them. If both people involved in the affair are married, the dynamics become even more complex.

One factor consistently found to be associated with lower levels of infidelity is religiousness. Interestingly, David Atkins and Deborah Kessel analyzed General Social Survey data and determined that of the many possible measures of religiousness (e.g., feeling close to God, praying, having faith, experiencing a religious turning point, and attending religious services), only attendance at religious services was associated with lower levels of infidelity. Reporting religion to be important in one's life without attending religious services is actually positively associated with infidelity (Atkins and Kessel 2008). With attendance at religious services comes ties to a community (i.e., a congregation) from which one can not only draw support but also receive scrutiny and criticism should one's behavior stray from shared norms. That such attendance is typically joint attendance *with one's spouse* may make one feel closer to one's spouse, less likely to feel disengaged or to drift apart, either of which could prompt infidelity (Atkins and Kessel 2008).

Although much research finds that more men than women engage in extrarelational sex, the gender gap has been narrowing. If one looks at acts other than intercourse, women acknowledge as much infidelity as do men (Mark, Jannsen, and Milhausen 2011). Other factors associated with higher rates of infidelity include marital status (though more the case for women than for men), income (higher-income individuals are more prone to infidelity), and employment (due to increased opportunity). According to economist Masanori Kuroki, the risk of an extramarital affair rises along with the numbers of opposite sex co-workers (Kuroki 2013). Socioeconomic status appears to affect men's infidelity but does not have the same effects on women's. Relationship variables such as satisfaction or happiness with one's relationship and perceived sexual incompatibility may more strongly affect women's likelihood of cheating than men's (Mark, Jannsen, and Milhausen 2011).

## Sexual Nonmonogamy and Sexual Orientation

Among heterosexual couples, lesbian couples, and gay male couples, gay male couples have been and continue to be more likely to both accept and experience sexual nonmonogamy. Furthermore, in comparing monogamous and nonmonogamous (i.e., "faithful" and "unfaithful") gay male couples, research often finds no differences in relationship adjustment or satisfaction (LaSala 2004). However, more recent research examining whether and how gay male couples in civil unions and those not in civil unions differ finds that gay male couples in civil unions are less accepting of nonmonogamy than those not in civil unions. However, even among gay males in civil unions, a higher percentage expressed openness about nonmonogamy than among either lesbians or heterosexuals.

Some clinicians have gone as far as to deem those who condemn nonmonogamy as dysfunctional "heterocentrist," meaning that they are applying standards that may pertain to heterosexual relationships too broadly. Given that males think about and act differently regarding sexual relationships, we might assume that gay male couples would display these tendencies even more than heterosexual couples and lesbians (each of whom has at least one female partner). Michael LaSala (2004) reminds us that research has established that, compared to women, men are more likely to separate sex and love, to engage in sexual relationships in the absence of emotional involvement, to engage in sexual relations within even "casual relationships," and to consider having sex with strangers.

Just as a portion of gay men maintains nonexclusive relationships, many gay couples, especially those who are able to marry or enter civil unions, construct relationship boundaries that prohibit sex with others. Like heterosexual and lesbian couples, such men come to see infidelity as a breach of trust. In LaSala's (2004) sample of 121 gay male couples, 60 percent described their "relationship agreements" as assuming

**True** **9.** The gender gap in infidelity is narrowing.

monogamy (40 percent were in sexually open relationships). However, among these 73 couples, 33 had breached this expectation and broken their monogamous agreement. This latter group had the lowest scores on satisfaction, expression of affection toward one's partner, and relationship adjustment. Interestingly, however, when those who engaged in nonmonogamous sex in the prior 12 months were removed from the analysis, there appeared to be no real difference between those whose monogamous expectations had been upheld and those whose expectations had been violated.

## Sexual Enhancement

Sexual behavior cannot be isolated from our personal feelings and relationships. Sometimes dissatisfaction arises because the relationship itself is unsatisfactory; at other times, the relationship itself is good but the erotic fire needs to be lit or rekindled. Such relationships may grow through **sexual enhancement**—improving the quality of a sexual relationship.

Being aware of our sexual needs is often critical to enhancing our sexuality. Gender-role stereotypes and negative learning about sexuality often cause us to lose sight of our sexual needs. Sex therapist Bernie Zilbergeld (1993) suggests that to fully enjoy our sexuality, we need to explore our "conditions for good sex," those things that make us "more relaxed, more comfortable, more confident, more excited, more open to your experience."

Different individuals report different conditions for good sex. More common conditions include the following:

- *Feeling intimate with your partner.* Emotional distance can take the heart out of sex.

- *Feeling sexually capable.* Generally, this relates to an absence of anxieties about sexual performance.

- *Feeling trust.* Both partners may need to know that they are emotionally safe with the other and confident that they will not be judged, ridiculed, or talked about.

- *Feeling aroused.* A person does not need to be sexual unless he or she is sexually aroused or excited. Simply because your partner wants to be sexual does not mean that you have to be.

- *Feeling physically and mentally alert.* Both partners should not feel particularly tired, ill, stressed, preoccupied, or under the influence of excessive alcohol or drugs.

- *Feeling positive about the environment and situation.* A person may need privacy, to be in a place where he or she feels protected from intrusion.

# ■ Sexual Problems and Dysfunctions

Many of us who are sexually active may experience sexual difficulties or problems. Recurring problems that cause distress to an individual or his or her partner are known as **sexual dysfunctions**. Although some sexual dysfunctions are physical in origin, many are psychological. Some dysfunctions have immediate causes, others originate in conflict within the self, and still others are rooted in a particular sexual relationship.

In estimating the prevalence of sexual problems, we draw on data from the U.S. sample of the Global Study of Sexual Attitudes and Behaviors (GSSAB), a 29-country study of over 27,000 women and men ages 40 to 80. Both men and women may suffer from hypoactive (low or inhibited) sexual desire or a lack of sexual interest, though this sexual problem was reported by a much higher percentage of women (33 percent) than men (18 percent), and was the most frequently reported female sexual problem. Other dysfunctions women experienced are arousal difficulties (the inability to become erotically aroused, also sometimes called lubrication difficulties), orgasmic dysfunction (the inability to attain orgasm), finding sex not pleasurable, and **dyspareunia** (painful intercourse). In research reported by Laumann et al. (2009), arousal difficulties, orgasmic difficulties, and finding sex not pleasurable were reported by roughly 20 percent of the U.S. sample of women ages 40 to 80. Pain during sex was reported by 12.7 percent of female respondents.

Other large-scale surveys on women's experiences of sexual problems or difficulties have reached estimates of slightly more than 40 percent of U.S. women, 18 and older. As would be expected, age was a factor in how common sexual problems were. Considering low desire, low arousal, and orgasmic difficulties, 27 percent of women 18 to 44 reported experiencing such problems. Among women 45 to 64, the percentage increased to 45 percent, and among women 65 and over, 80 percent reported any of the three problems. Interestingly, older women were the least likely to become personally distressed about sexual problems, despite being most likely to actually experience such problems (Shifren et al. 2008).

Among male respondents to the GSSAB, the most commonly reported dysfunctions included early ejaculation (26 percent), **erectile dysfunction** (the inability to achieve or maintain an erection) (22.5 percent), lack of sexual interest (18 percent), delayed orgasm or difficulty in ejaculating (12.4 percent), and finding sex not pleasurable (11.2 percent). An additional 3 percent of male respondents reported pain during sex as a problem (Laumann et al. 2009).

## Causes of Sexual Problems

It is generally believed that between 10 percent and 20 percent of sexual dysfunctions are structural in nature. Physical problems may be *partial* causes in another 10 percent or 15 percent (Kaplan 1983; LoPiccolo 1991). Various illnesses may have an adverse effect on a person's sexuality (Wise, Epstein, and Ross 1992). Alcohol and some prescription drugs, such as medication for hypertension, may affect sexual responsiveness (Buffum 1992).

Among women, diabetes, hormone deficiencies, and neurological disorders, as well as alcohol and alcoholism, can cause orgasmic difficulties. Problems of low or absent desire may be elevated by such physical factors as thyroid problems and incontinence. Arousal problems are more likely among women with thyroid problems, arthritis, urinary incontinence, and inflammatory or irritable bowel disease (Shifren et al. 2008). Painful intercourse may be caused by an obstructed or thick hymen, clitoral adhesions, a constrictive clitoral hood, or a weak pubococcygeus muscle. Coital pain caused by inadequate lubrication and thinning vaginal walls often occurs as a result of decreased estrogen associated with menopause. Lubricants or hormone replacement therapy often resolve the difficulties.

Among males, diabetes and alcoholism are among the leading physical causes of erectile dysfunctions; atherosclerosis is another important factor (LoPiccolo 1991; Roenrich and Kinder 1991). Smoking may also contribute to sexual difficulties (Rosen et al. 1991). Prostate disease and a lack of physical activity are also potential contributing factors for lack of sexual interest (Laumann et al. 2009).

## Psychological or Relationship Causes

Among the prominent nonphysical factors associated with sexual dysfunctions are depression, performance anxiety, and conflicts within the self. In the GSSAB, diagnosed depression was associated with erectile difficulties in men, arousal and orgasmic difficulties in women, and a lack of sexual interest in both women and men (Laumann et al. 2009). Men's depression-related erectile difficulties could actually be consequences of antidepressant medications, as such sexual side effects are common of many antidepressants (Laumann et al. 2009). In a study of more than 31,500 adult women in the United States, depression was associated with a doubling of the odds of distressing sexual problems (Shifren et al. 2008). **Performance anxiety**—the fear of failure—is also among the most important immediate causes of erectile dysfunctions and, to a lesser extent, of orgasmic dysfunctions in women (Kaplan 1979). If a man does not become erect, anxiety is a fairly common response. Some men experience their first erectile problem when a partner initiates or demands sexual intercourse. Women are permitted to say no, but many men have not learned that they too may say no to sex. Women suffer similar anxieties, but they tend to center around orgasmic abilities rather than the ability to have intercourse. If a woman is unable to experience orgasm, a cycle of fear may arise, preventing future orgasms. A related source of anxiety is an excessive need to please one's partner.

The relationship itself, rather than either individual, sometimes can be the source of sexual problems. Disappointment, anger, or hostility may become integral parts of a deteriorating or unhappy relationship. Such factors affect sexual interactions because sex can become a barometer for the whole relationship.

Sexual problems can become self-fulfilling, as they can cause performance anxiety or a fear of future failures to achieve erections or experience orgasms.

Mauro Speziale/The Image Bank/Getty Images

Understandably, relationship issues can affect our sexuality in several ways, such as through poor communication that inhibits our ability to express our needs and desires, power struggles in which sexuality becomes a tool in struggles for control, and sexual sabotage where partners ask for sex at the wrong time, put pressure on each other, and frustrate or criticize each other's sexual desires and fantasies. People most often do this unconsciously. Conversely, couples who effectively communicate their sexual needs and show that they care about each other are likely to have a lower risk of sexual problems. Uncertainty about the future of the relationship and/or one's partner's commitment have been linked to erectile problems in men and orgasmic difficulties in women.

## Resolving Sexual Problems

Sexual problems can be embarrassing and emotionally upsetting. Perhaps the first step in dealing with a sexual problem is to turn to immediate resources. Talking about the problem with one's partner, finding out what he or she thinks, discussing specific strategies that might be useful, and simply communicating feelings and thoughts can sometimes resolve the difficulty. That is often a first step, and is the most common help-seeking behavior of both men and women; 43 percent of women and men who reported having at least one sexual problem reported talking to their partner (Laumann et al. 2009).

**False** **10.** The most common strategy for resolving sexual problems is to talk with one's partner, not a physician.

One can also go outside the relationship, seeking medical advice or talking with friends with whom one can share feelings and anxieties, asking whether they have had similar experiences, and learning how they handled them. Aside from one's circle of friends and intimates, there are ever-increasing, additional resources on which one can draw. A growing number of self-help books dealing with sexuality and relationship issues line the shelves in bookstores and libraries. An online search for books dealing with "sexual difficulties" generated almost hundreds of titles from major online bookstores. There are also numerous websites one can access and consult. Not all of these will be sites offering help or advice. Some may be pornographic, and others may carry exaggerated claims designed to sell products, but many websites offer information compiled or overseen by medical, psychiatric, psychological, nursing, or educational specialists.

Cumulatively, partners, friends, websites, and books may provide information and grant individuals needed "permission" to engage in sexual exploration and discovery by making such inquiries normal. From these sources, we may learn that our sexual issues, problems, fantasies, and behaviors are not unique. Such methods are most effective when the dysfunctions arise from a lack of knowledge or mild sexual anxieties.

If one remains unable to resolve his or her sexual difficulties, despite conversation with one's partner, consultation with friends, and/or reading books, magazines, or other resources, seeking professional assistance is a logical next step. It is important to realize that seeking such assistance does not signal personal weakness or failure. Rather, it demonstrates an ability to reach out and a willingness to change. It is a sign of caring for one's partner, one's relationship, and oneself. To determine whether one's problems have medical explanations or treatments, one will need to consult an outside professional. In the Global Study of Sexual Attitudes and Behaviors, of those respondents reporting at least one sexual problem, 22 percent of men and 16 percent of women reported talking to medical doctors. However, 76 percent of men and 80 percent of women reported seeking no help from a health professional. When questioned about reasons for not seeking medical help, the most prominent reasons for both men and women were as follows:

| Reason | % men | % women |
|---|---|---|
| Did not think it serious/waiting for it to go away | 36.3 | 38.1 |
| Problem is normal with aging/I am comfortable the way I am | 25.4 | 28.2 |
| Doctor cannot do much/do not think it is a medical problem | 14.9 | 13.5 |

An additional 3 percent of men and 5 percent of women said they were not comfortable talking to an MD, the MD is a friend, or the MD is of the wrong gender. Majorities of both women and men, not just those experiencing sexual problems, thought that doctors should routinely ask about a patient's sexual health (Laumann et al. 2009).

For those whose problems stem mostly from psychological or relationship causes, therapists can help deal with sexual problems on several levels. Some focus directly on the problem, such as lack of orgasm, and suggest behavioral exercises, such as pleasuring and masturbation, to develop an orgasmic response. Others focus on the couple relationship as the source of difficulty. If the relationship improves, they believe that sexual responsiveness will also improve. Still others work with individuals to help them develop insight into the origins of the problem to overcome it. Therapy can also take place in a group setting. Group therapy may be particularly valuable for providing partners with an open, safe forum in which they can discuss their sexual feelings and experience and discover commonalities with others. However, only 3.7 percent of men and 6.3 percent of women who had acknowledged having a sexual problem reported talking to a psychiatrist, psychologist, or marriage counselor (Laumann et al. 2009).

## ▶ Issues Resulting from Sexual Involvement

Most of us think of sexuality in terms of love, passionate embraces, and entwined bodies. Sex involves all of these, but what we so often forget (unless we are worried) is that sex is also a means of reproduction. Whether or not we like to think about it, many of us (or our partners) are vulnerable to unintended pregnancies. Not thinking about pregnancy does not prevent it; indeed, not thinking about it may increase the likelihood of its occurring. Sex can also involve transferring sexually transmitted infections from one partner to another. Unless we practice **abstinence**, refraining from sexual intercourse, we need to think about unintended pregnancies and sexually transmitted infections and then take the necessary steps to prevent them.

### Sexually Transmitted Infections, HIV, and AIDS

Americans continue to experience an epidemic of sexually transmitted infections (STIs). There are an estimated 20 million new cases of STIs in the United States each year. Counting both new and existing infections, the Centers for Disease Control estimates that there are more than 110 million infections nationwide. Half of all infections occurring among 15- to 24-year-olds, even though they are only a quarter of

those who are sexually experienced. College students are among the population at highest risk of contracting an STI. Risks to females may be more severe than risks to males. Undiagnosed or left untreated, chlamydia and gonorrhea can lead to infertility, ectopic pregnancy, and chronic pelvic pain.

Most people might wince if asked on a first date, "Do you have chlamydia, herpes, HIV, HPV (human papillomavirus), or any other sexually transmissible disease that I should know about?" However, given the risks of contracting an STI and the consequences associated with infection, it is a question whose answers one ideally should know before becoming sexually involved. Just because a person is "nice" or good-looking or available and willing is no guarantee that he or she does not have one of the STIs discussed in this chapter. One can become infected through such sexual contact as sexual intercourse, oral sex, or anal sex. If diagnosed early, chlamydia, gonorrhea, syphilis and trichomoniasis can be easily treated and cured. Unfortunately, no one can tell by a person's looks, intelligence, or demeanor whether he or she has contracted an STI. The costs of becoming sexually involved with a person without knowing about the presence of any of these diseases are potentially steep.

The most prevalent STIs in the United States are trichomoniasis, chlamydia, gonorrhea, HPV, genital herpes, syphilis, hepatitis, and HIV and AIDS. Table 6.6 briefly describes the symptoms, exposure intervals, treatments, and other information regarding the principal STIs. According to data from the Centers for Disease Control (CDC), in 2013, there were 1.4 million reported cases of chlamydia, a slight decrease (1.5 percent) from 2012 but more than four times the number of gonorrhea cases. It is likely that the real annual incidence of chlamydia is perhaps twice that, with half or more cases going undiagnosed (Chlamydia, CDC Fact Sheet). Chlamydia is more common among females than males and is highest among 15- to 19-year-olds followed by 20- to 24-year-olds. Racially, the rate of reported cases of chlamydia among black women was almost six times (5.8) the rate among white women, and the rate among black men was almost eight times the rate among white men. That said, between 2009 and 2013, the rate of chlamydia infections increased among all racial groups except for blacks; the rate of infection among blacks decreased 5.2 percent, whereas the rates increased by 6 to 7 percent among Hispanics and Asians, 23.9 percent for American

**TABLE 6.6 Principal Sexually Transmitted Disease**

| STD and Infecting Organism | Time from Exposure to Occurrence | Symptoms | Medical Treatment | Comments |
|---|---|---|---|---|
| Chlamydia (*Chlamydia trachomatis*) | 7 to 21 days | Women: 80% asymptomatic; others may have vaginal discharge or pain with urination. Men: 30% to 50% asymptomatic; others may have discharge from penis, burning urination, pain and swelling in testicles, or persistent low fever. | Doxycycline, tetracycline, erythromycin. | If untreated, it may lead to pelvic inflammatory disease (PID) and subsequent infertility in women. |
| Gonorrhea (*Neisseria gonorrhoeae*) | 2 to 21 days | Women: 50% to 80% asymptomatic; others may have symptoms similar to chlamydia. Men: itching, burning, or pain with urination; discharge from penis ("drip"). | Penicillin, tetracycline, or other antibiotics. | If untreated, it may lead to PID and subsequent infertility in women. |
| Human papillomavirus (HPV) | 1 to 6 months (usually within 3 months) | Variously appearing bumps (smooth, flat, round, clustered, fingerlike white, pink, brown, and so on) on genitals, usually penis, anus, vulva. High-risk HPV can lead to cervical cancer. | Surgical removal by freezing, cutting, or laser therapy. Virus remains in the body after warts are removed. Chemical treatment with podophyllin (80% of warts eventually reappear). | HPV vaccine was developed and licensed for use in 2006. |
| Genital herpes (*Herpes simplex virus*) | 3 to 20 days | Small, itchy bumps on genitals, becoming blisters that may rupture, forming painful sores; possibly swollen lymph nodes; flulike symptoms with first outbreak. | No cure, although acyclovir may relieve symptoms. Nonmedical treatments may help relieve symptoms. | Virus remains in the body, and outbreaks of contagious sores may recur. Many people have no symptoms after the first outbreak. |
| Syphilis (*Treponema pallidum*) | Stage 1: 1 to 12 weeks Stage 2: 6 weeks to 6 months after chancre appears | Stage 1: Red, painless sore (chancre) at bacteria's point of entry. Stage 2: Skin rash over body, including palms of hands and soles of feet. | Penicillin or other antibiotics. | Easily cured, but untreated syphilis can lead to ulcers of internal organs and eyes, heart disease, neurological disorders, and insanity. |
| Hepatitis (hepatitis A or B virus) | 1 to 4 months | Fatigue, diarrhea, nausea, abdominal pain, jaundice, and darkened urine due to impaired liver function. | No medical treatment available; rest and fluids are prescribed until the disease runs its course. | Hepatitis B is more commonly spread through sexual contact and can be prevented by vaccination. |
| Urethritis (various organisms) | 1 to 3 weeks | Painful and/or frequent urination; discharge from penis. Women may be asymptomatic. | Penicillin, tetracycline, or erythromycin, depending on organism. | Laboratory testing is important to determine appropriate treatment. |

SOURCE: Strong and DeVault (1997).

Indians/Alaska Natives, and 32.7 percent for whites (CDC 2013).

Left untreated, chlamydia can lead to pelvic inflammatory disease (PID), which can then cause damage to the uterus, fallopian tubes, and surrounding tissue, causing infertility, ectopic pregnancy, and an elevated risk of HIV infection. The CDC estimates that 10 percent to 15 percent of women with untreated chlamydia infections develop PID (Centers for Disease Control and Prevention 2012c).

Gonorrhea, the second most common STI, remained stable between 2012 and 2013. In 2013, more than 333,000 cases were reported to the Centers for Disease Control and Prevention, though the CDC estimates that more than 800,000 new infections actually occurred. Gonorrhea infections are a major cause of PID, and the Centers for Disease Control and Prevention express the concern that gonorrhea may become resistant to the most effective treatment. Females have a slightly higher rate of developing gonorrhea than males. Rates of gonorrhea are also highest among adolescents and young adults, and among racial and ethnic minorities. African Americans accounted for 58.4 percent of reported gonorrhea cases and a rate of infection more than 12 times that of whites; American Indians have four times the rate of whites; and Hispanics have nearly twice the rate of whites. In turn, whites have a gonorrhea rate nearly twice the rate among Asians.

After reaching an all-time low in 2000, the number of syphilis cases has increased 36 percent since 2006. Among almost all subpopulations, syphilis increased for each of the following groups: Rates grew 18 percent among males, 10 percent among females, 25 percent among African Americans, 23 percent among Hispanics, 5 percent among whites, and 6 percent among Native Americans. Only among Asians did the rate remain unchanged. Two-thirds of syphilis cases are the result of male same-sex relations.

HPV is a sexually transmitted infection that is passed through genital contact. Twenty million Americans are currently affected. At some point in their lives, it is estimated that 50 percent of sexually active women and men acquire HPV, and that at any one point, 1 percent of sexually active adults have genital warts (a possible consequence of HPV infection). More serious consequences include certain cancers, including cervical cancer and cancers of the vagina, anus, or penis. Recently, a new vaccine has been developed to prevent the types of HPV that lead to most cervical cancer and genital warts. The vaccine is recommended

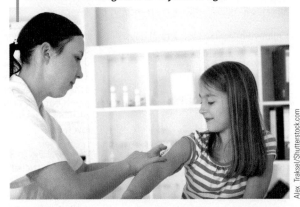

Gardasil, a vaccine for human papillomavirus (HPV), protects recipients against most of the HPV strains that are linked to 70 percent of cervical cancers, and is recommended for girls 11 to 12 years of age.

Alex_Traksel/Shutterstock.com

for girls 11 and 12 years old and for 13- to 26-year-old females who have not been vaccinated.

## HIV and AIDS

The **human immunodeficiency virus (HIV)** is the virus that causes **acquired immunodeficiency syndrome (AIDS)**. The disease is so termed because of its characteristics: It is acquired—because people are not born with it; it manifests as an immunodeficiency—because the disease relates to the body's immune system, which is lacking in immunity; and it is a syndrome—because the symptoms occur as a group.

Overall, the effects of HIV have been devastating, with the worst effects happening not in the United States but in other parts of the world, especially sub-Saharan Africa and the Caribbean. According to a UNAIDS World Aids Day 2011 Report, through 2010, more than 60 million people globally have been infected with the HIV virus, and roughly 30 million have died of AIDS-related causes (UNAIDS World AIDS Day Report 2011). The same report estimates that approximately 34 million people are living with HIV worldwide, a 17 percent increase from a decade ago. This is not a reflection of increases in the rate of infection, as annual infections declined 21 percent between 1997 and 2010. Instead, it is due largely to improved treatment and medications that have allowed more people with HIV infections to live more years after being infected (UNAIDS World AIDS Day Report 2011). Worldwide, sub-Saharan Africa accounts for more than two-thirds of people living with HIV and nearly half of all deaths from AIDS-related illnesses. Additionally, in the Caribbean and sub-Saharan Africa, more than half of all people living with HIV infections are women (UNAIDS World AIDS Day Report 2011).

In the United States, it is estimated that 13,712 people with AIDS died in 2012, and more than 658,000 people have died from HIV or AIDS since the beginning of the epidemic (CDC 2015b).

Furthermore, approximately 50,000 new HIV infections are diagnosed per year. In 2013, an estimated 47,352 people in the United States received HIV diagnoses. The Centers for Disease Control and Prevention further estimates that 1.2 million people age 13 and older in the United States who have been infected and are living with HIV are unaware that they have been infected (CDC 2015b). Men who have sex with men (MSM), though only approximately 4 percent of the U.S. population, continue to bear the brunt of HIV and AIDS. In 2010, they accounted for more than three-fourths of new HIV infections among men and nearly two-thirds (63 percent) of all new infections. An additional 3 percent of infections occurred among men who have sex with men with a history of injection drug use (Kaiser Family Foundation 2014).

In 2010, injection drug users accounted for 8 percent of new HIV infections and represented 15 percent of those living with HIV. Over 186,000 deaths have occurred among injection drug users since the beginning of the epidemic, including over 3,500 in 2012. A total of 92,613 people infected with AIDS through heterosexual sex have died since the epidemic began, including an estimated 4,550 in 2012.

A look at HIV and AIDS among U.S. women shows that more than 9,000 women, 13 and older, were newly diagnosed with HIV in 2013, making them almost one-fourth (20 percent) of those newly infected with HIV. Of the total number of U.S. women newly diagnosed in 2009, 57 percent were African American, 21 percent were white, and 16 percent were Hispanic. Women accounted for just under a quarter of the estimated 26,680 AIDS diagnoses in 2013. At a total of 242,178, women are also 24 percent of the 1,184,618 total AIDS diagnoses in the United States through 2013. Since the beginning of the AIDS epidemic through 2012, an estimated 117,797 women diagnosed with AIDS have died, including more than 3,500 women who died during 2012. The vast majority of HIV infections in women are the result of heterosexual contact with infected partners. Black heterosexual women are the group with the fourth largest number of new HIV infections, after white, black, and Hispanic

men who have sex with men. In 2010, black women accounted for almost two-thirds (64 percent) of newly infected women, followed by white women at 18 percent and Hispanic women at 15 percent (CDC 2015a). HIV and AIDS cases have hit African Americans and Latinos especially hard, with each group infected at disproportionate rates. African Americans suffered more than 40 percent of new infections in 2009 and 2010, and accounted for 41 percent of people living with an HIV infection in 2011; Hispanics accounted for an estimated 21 percent of new HIV infections in 2010, and made up 20 percent of those living with HIV. From the beginning of the epidemic through 2012, more than 100,888 Hispanics and 270,726 blacks with AIDS have died (CDC 2015b).

Although AIDS was initially discovered in gay men and was thought of early on as a "gay disease" or the "gay plague," sexually transmitted cases among heterosexuals account for more than a quarter of new HIV infections. Without discounting or diminishing the devastation that the gay community suffered from AIDS and HIV or signs of resurging rates of infection, it is important to keep in mind that heterosexuals and bisexuals are also at risk and also become infected. Virtually all adults in the United States are or will soon be related to, personally know, work with, or go to school with people infected with HIV or will know others whose friends, relatives, or associates test HIV-positive.

As these words are being typed, there is still no surefire vaccine to prevent HIV, nor is there a cure for those who are or will become infected. Significant strides have been made in fighting the disease, suppressing its symptoms, and prolonging life for those who are infected. Hence, advances in treatment have reduced AIDS-related morbidity and mortality rates. Many who are infected have had their lives extended by such advances. Routine HIV testing, especially with the wider availability of rapid testing, can lead to earlier diagnoses which can lead to earlier and more effective intervention (Kaiser Family Foundation 2014).

In addition to new treatments that can lengthen the life span of an AIDS-infected person as much as 15 years (Fallon 2005), we have considerable knowledge about the nature of the virus and how to reduce the likelihood of infection:

- *HIV attacks the body's immune system.* HIV is carried in the blood, semen, and vaginal secretions of infected people. A person may be HIV-positive (infected with HIV) for years before developing AIDS symptoms.

- *HIV is transmitted only in certain clearly defined circumstances.* It is transmitted through the exchange of blood (as by shared needles or transfusions of contaminated blood), through sexual contact involving semen or vaginal secretions, and from an infected woman to her fetus through the placenta. Infected mothers may also transmit the infection during delivery or through breast milk (Miracle, Miracle, and Baumeister 2003).

- *All those with HIV (whether or not they have AIDS symptoms) are HIV carriers.* They may infect others through unsafe sexual activity or by sharing needles; if they are pregnant, they may infect the fetus.

- *Heterosexuals, bisexuals, gay men, and lesbians are all susceptible to the sexual transmission of HIV.* No group owns HIV, and no group is immune from the possibility of infection.

- *There is a definable progression of HIV infection and a range of illnesses associated with AIDS.* HIV attacks the immune system. AIDS symptoms occur as opportunistic diseases—diseases that the body normally resists—that infect an individual. The most common opportunistic diseases are pneumocystis carinii pneumonia and Kaposi's sarcoma, a skin cancer. It is the opportunistic disease, rather than the HIV, that kills people with AIDS.

- *The presence of HIV can be detected through various kinds of antibody testing.* Though the most common method of testing for HIV is through blood tests, including one that tests dried blood, there are also tests to be used on oral fluid or urine (U.S. Food and Drug Administration 2015a).

Anonymous testing is available at many college health centers and community health agencies. In addition, there are now test kits in which one collects a sample at home and sends it to a lab to be evaluated by health professionals, and still other test kits in which one collects and tests the sample at home. Different tests vary in sensitivity, specificity, time it takes to obtain results, and whether there is need for a follow-up test to confirm the results. They also differ in terms of whether blood or oral fluid is to be tested (U.S. Food and Drug Administration 2015b).

HIV antibodies develop between one and six months after infection. Antibody testing should take place one month after possible exposure to the virus and, if the results are negative, again six months later. If the antibody is present, the test will be positive. That means that the person has been infected with HIV and

that an active virus is present. The presence of HIV does not mean, however, that the person necessarily will develop AIDS symptoms in the near future; symptoms generally occur 7 to 10 years after the initial infection.

## Protecting Yourself and Others

As with avoiding unintended pregnancies, the safest practice to avoid STIs is *abstinence,* forgoing sexual relations. There is no chance of contracting STIs without sexual contact, although HIV infection can and does occur through nonsexual transmission (e.g., intravenous drug use with shared needles). If one is sexually active, however, the key to protecting oneself and others is to talk with one's partner about STIs in an open, nonjudgmental way and to use condoms. Because many people are uncomfortable asking about STIs, one can open the topic by revealing one's anxiety: "This is a little difficult for me to talk about because I like you, and I'm embarrassed, but I'd like to know whether you have herpes, or HIV, or whatever." If *you* have an STI, you can say, "Look, I like you, but we can't hook up right now because I have a herpes infection, and I don't want you to get it."

Remember, however, that not every person with an STI knows she or he is infected. Women with chlamydia and gonorrhea, for example, generally don't exhibit symptoms. Both men and women infected with HIV may not show any symptoms for years, although they are capable of spreading the infection through sexual contact. If you are or are planning to be sexually active but don't know whether your partner has an STI, use a condom. Even if you don't discuss STIs, condoms are simple and easy to use without much discussion. Both men and women can carry them. A woman can take a condom from her purse and give it to her partner. If he doesn't want to use it, she can say, "No condom, no sex." Because HIV can be transmitted through semen and vaginal secretions, when engaging in oral sex, it is also recommended to use a condom (on a man) and a dental dam (on a woman).

# ▌ Sexual Responsibility

Because we have so many sexual choices today, we need to be sexually responsible. Sexual responsibility includes the following:

- *Disclosure of intentions.* Each person needs to reveal to the other whether a sexual involvement indicates love, commitment, recreation, and so on.

- *Freely and mutually agreed-upon sexual activities.* Each individual has the right to refuse any or all sexual activities without the need to justify his or her feelings. There can be no physical or emotional coercion.

- *Use of mutually agreed-upon contraception in sexual intercourse if pregnancy is not intended.* Sexual partners are equally responsible for preventing an unintended pregnancy in a mutually agreed-upon manner.

- *Use of "safer sex" practices.* Each person is responsible for practicing safer sex. Safer sex practices do not transmit semen, vaginal secretions, or blood during sexual activities and guard against STIs, especially HIV and AIDS.

- *Disclosure of infection from or exposure to STIs.* Each person must inform his or her partner about personal exposure to an STI because of the serious health consequences, such as infertility or AIDS, that may follow untreated infections. Infected individuals must refrain from behaviors—such as sexual intercourse, oral–genital sex, and anal intercourse—that may infect their partner. To help ensure that STIs are not transmitted, a condom and/or dental dam should be used.

- *Acceptance of the consequences of sexual behavior.* Each person needs to be aware of and accept the possible consequences of his or her sexual activities. These consequences can include emotional changes, pregnancy, abortion, and STIs.

Responsibility in many of these areas is facilitated when sex takes place within the context of an ongoing relationship. In that sense, sexual responsibility is a matter of values. Is responsible sex possible outside an established relationship? Are you able to act in a sexually responsible way? Sexual responsibility also leads to the question of the purpose of sex in your life. Is it for intimacy, erotic pleasure, reproduction, or other purposes?

As we consider the human life cycle from birth to death, we cannot help but be struck by how profoundly sexuality weaves its way through our lives. From the moment we are born, we are rich in sexual and erotic potential, which begins to take shape in our sexual experimentations of childhood. As children, we are still unformed, but the world around us haphazardly helps give shape to our sexuality. In adolescence, our education continues as a mixture of learning and yearning. But as we enter adulthood, with greater experience and understanding, we undertake to develop a mature sexuality: We establish our sexual orientation as heterosexual, gay, lesbian, or bisexual; we integrate love and sexuality; we forge intimate connections and make commitments; we make decisions regarding our fertility and sexual health; and we develop a coherent sexual philosophy. Then, in our middle years, we redefine sex in our intimate relationships, accept our aging, and reevaluate our sexual philosophy. Finally, as we become elderly, we reinterpret the meaning of sexuality in accordance with the erotic capabilities of our bodies. We come to terms with the possible loss of our partner and our own end. In all these stages, sexuality weaves its bright and dark threads through our lives.

## Summary

- Our sexual behavior is influenced by *sexual scripts:* the acts, rules, stereotyped interaction patterns, and expectations associated with male and female sexual expression.

- Traditional sexual scripts are gendered, prescribing different sexual roles for women and men. Contemporary sexual scripts are more egalitarian, though evidence suggests that there is still a sexual double standard in which different sexual behaviors are accepted and expected of men and women. Even amid a longer-term trend toward more open acceptance of nonmarital sexual behavior, a recent trend seems to point toward a decline in sexual activity among teenagers.

- Estimates of sexual behaviors and experiences are prone to either exaggeration or understatement, depending on the behavior.

- Estimates of sexual behaviors of adolescents and young adults show large gender differences in sexual experience. More females than males report their first experience of intercourse as unwanted. More males than females report masturbating.

- The actual percentage of the U.S. population that is lesbian, gay, or bisexual remains unknown. Estimates vary depending on whether one considers self-identity, sexual attraction, or sexual experience.

- Identifying oneself as gay or lesbian occurs in stages.

- A continuing difference between gay male couples and both heterosexual and lesbian couples is in the acceptance of extrarelational sex. Gay couples in civil unions are more likely to expect and experience monogamy.

- Lesbian, gay, and bisexual individuals may confront prejudice or hostility, including verbal abuse, discrimination, or violence. Negative attitudes toward bisexuals may be even harsher than attitudes toward gay men and lesbians, and may come from heterosexuals, gay males, or lesbians.

- Developmental tasks in middle adulthood include (1) redefining sex in marital or other long-term relationships, (2) reevaluating one's sexuality, and (3) accepting the biological aging process.

- *Autoeroticism* consists of sexual activities that involve only the self. It includes sexual fantasies and masturbation, both of which are very common, and erotic dreams.

- In the United States, the most common and acceptable of all premarital sexual activities is kissing. As seen in Chapter 5, in many cultures romantic kissing is not found.

- *Oral–genital sex*, which includes *cunnilingus* and *fellatio*, is practiced by heterosexuals, gay men, and lesbians. Data indicate both increasing rates of oral sex among teenagers and young adults and persisting race differences in rates of oral sex.

- *Anal eroticism* is practiced by both heterosexuals and gay men. From a health perspective, *anal intercourse* is the most common means of sexually transmitting HIV. Among females, there are notable differences in rates of heterosexual anal sex.

- *Nonmarital sex* includes all sexual activities, especially sexual intercourse, that take place outside of marriage. *Premarital sex* has gained in acceptability, whereas *extramarital sex* has not. Extrarelational sex occurs among heterosexual cohabitants, gay male couples, and lesbian couples at higher rates than among married couples.

- Sexual dysfunctions (such as orgasmic or arousal difficulties in women or erectile dysfunction or premature ejaculation in men) may be physiological or psychological in origin. Both genders may suffer from reduced sexual interest as well as orgasmic difficulties. Men may experience erectile difficulties and women may experience arousal difficulties.

- *Sexually transmitted infections (STIs)*, especially chlamydia and gonorrhea, are epidemic. *Acquired immunodeficiency syndrome (AIDS)* is caused by the *human immunodeficiency virus (HIV)*, which attacks the body's immune system. HIV is carried in the blood, semen, and vaginal fluid of infected people.

## Key Terms

abstinence 232
acquired immunodeficiency syndrome (AIDS) 234
anal eroticism 221
anal intercourse 221
Antigay prejudice 212
asexual 207
autoeroticism 218
biphobia 216
bisexuality 207
coitus 220
coming out 210
cunnilingus 220
dyspareunia 229
erectile dysfunction 230
extramarital sex 223
fellatio 220
foreplay 224
heterosexism 212
heterosexuality 207
homophobia 212
homosexuality 207
human immunodeficiency virus (HIV) 234
masturbation 219
menopause 217
nonmarital sex 223
outing 211
pansexual 207
performance anxiety 230
pleasuring 219
premarital sex 223
sexual dysfunctions 229
sexual enhancement 229
sexual identity 207
sexual intercourse 220
sexual orientation 207
sexual script 192
social desirability bias 200
virginity 206

# Communication, Power, and Conflict

7

Are the following statements True or False? You may be surprised by the answers as you read this chapter.

T F **1.** The meaning conveyed through nonverbal communication is precise and unambiguous.

T F **2.** Although it is often an expression of closeness or intimacy, touch can also be a sign of dominance.

T F **3.** Female partners tend to give more positive and fewer negative messages than husbands.

T F **4.** The demand–withdraw pattern of communicating has been observed widely among heterosexual couples but seems absent among same-gender couples.

T F **5.** In couple relationships, the partner with more power may vary from one situation to another.

T F **6.** The party with the least interest in continuing a relationship generally has the power in it.

T F **7.** A woman's power in marriage inevitably increases when she earns as much or more than her husband.

T F **8.** Heterosexual, gay, and lesbian couples are similar in the amount of conflict they experience.

T F **9.** Among married couples, money-related conflicts are the most likely to remain unresolved.

T F **10.** The healthiest relationships are those that avoid conflict.

Jupiterimages/Stockbyte/Getty Images

# Chapter Outline

"Your mother called again."

That seems like such a simple, ordinary statement. It hardly seems like the kind of comment that would provoke an argument, nor does it appear particularly revealing about the tone or quality of a marriage or relationship. In fact, it sounds so routine, so "matter-of-fact," that we might overlook its significance and potential effect on married or coupled life.

Of course, we only have the four words; we don't know *how* they were said. What was the tone of voice, the cadence or rhythm of speech? Was it, "Your mother called again" or "Your mother called. Again." Or, combining tone and cadence, "Your mother called. Again!" We also have no information about the nonverbal signs. What was the expression on the face of the speaker—say a wife to a husband—when the statement was made? Did she smile? Roll her eyes? Frown? Shake her head? Stand with her arms folded in front of her? All these aspects of nonverbal communication help reveal more of the meaning and significance of such a statement. Clearly, seemingly simple comments such as this may have greater importance than whatever words they otherwise convey.

Finally, of potentially even greater significance is how the other person responds to a statement such as this one. Whether he or she responds with "an irritable groan," a laugh (as if to say, "What? Again!"), a defensive explanation of the frequency of such maternal phone conversations, an expression of concern ("I hope she's all right"), or a positive discussion of his or her mother tells us a lot. A nonresponse may tell us yet more. It may suggest indifference and lack of interest in talking with the partner. Exchanges surrounding statements such as this one, "mundane and fleeting" as they may appear to be, can build and, in the process, greatly affect the quality of a relationship, the amount and nature of conflict, and the feelings of closeness and romance among partners or spouses (Driver and Gottman 2004).

Thinking about the kinds of relationships that are the focus of this book, what is it you most want or expect from your intimate relationships, especially from lovers, spouses, family members, and other intimate relationships? Chances are, if you list the many characteristics or qualities you desire in such relationships, somewhere on that list will be "communication." We want our loved ones to share their feelings and ideas with us and to understand the ideas or feelings that we voice to them. After all, as shown in Chapter 5, that is how we expect to share intimacy. We want to be able to communicate effectively.

Chances are that "conflict" will not be included among your most desired relationship characteristics. After all, who wants to argue? We tend to see and experience conflict as a negative to be avoided. Yet conflict is as much a feature of intimate relationships as are love and affection. As long as we value, care about, and live with others, we will experience occasions when we disagree, when disagreements lead to conflict, and when we find ourselves in the middle of arguments. An absence of conflict not only is unrealistic but would be unhealthy as well. *How* we resolve our disagreements tells us much about the health of our relationships and the power dynamics that exist within them.

Both communication and conflict are inextricably connected to intimacy. When we speak of communication, we mean more than just the ability to relay information (e.g., "Your mother called"), discuss problems, and resolve conflicts. We also mean communication for its own sake: the pleasure of being in each other's company, the excitement of conversation, the exchange of touches and smiles, and the loving silences. Through communication, we disclose who we are and what we feel, and from such self-disclosure, intimacy grows.

One of the most common complaints of married partners, especially unhappy partners, is that they don't communicate. But it is impossible not to communicate—a cold look may communicate anger as effectively as a fierce outburst of words. What these unhappy partners mean by "not communicating" is that their communication is somehow driving them apart rather than bringing them together, feeding and creating conflict rather than resolving it. Among married couples, communication patterns are strongly associated with marital satisfaction (Noller and Fitzpatrick 1991).

In this chapter, we explore patterns and problems in communication in marital and intimate relationships. We also examine the role of power in marital relationships, where it comes from, and how it is expressed in both the styles of communication and the outcomes of conflict. Finally, we look at the relationship between conflict and intimacy, exploring different types of conflict and approaches to conflict resolution. We look especially at three of the more common areas of relationship conflict: conflicts about sex, money, and housework.

# Verbal and Nonverbal Communication

When we communicate face-to-face, the messages we send and receive contain both a verbal and a nonverbal component. Verbal communication expresses the *basic content* of the message, whereas **nonverbal communication** reflects more of the *relationship* part of the message. The relationship part conveys the feelings and attitudes of the speaker (friendly, neutral, or hostile) and indicates how the words are to be interpreted (as a joke, request, or command). To understand the full content of any message, we need to understand both the verbal and the nonverbal parts.

For a message to be most effective, both the verbal and the nonverbal components should be in agreement. If you are angry and say "I'm angry," and both your facial expression and your voice show anger, the message is clear and convincing. But if you say "I'm angry" in a neutral tone of voice and a smile on your face, your message is ambiguous. More commonly, if you say "I'm not angry" but clench your teeth and use a controlled voice, your message is also unclear. Your tone and expression make your spoken message difficult to take at face value.

The ability to interpret nonverbal communication correctly appears to be an important ingredient in successful relationships. The statement "What's wrong? *I can tell something is bothering you*" reveals the ability to read nonverbal clues, such as body language or facial expressions. This ability is especially important in ethnic groups and cultures that rely heavily on nonverbal expression of feelings, such as Latino and Asian American cultures. Although the value placed on nonverbal expression may vary among groups and cultures, the ability to communicate and understand nonverbally remains important in all cultures.

In addition to both verbal and nonverbal communication, people are increasingly communicating with one another via technologies that allow some of the same qualities of verbal communication minus any of the information conveyed nonverbally or via tone of voice. This occasionally leaves the recipient of a text message, an email, or some other written message uncertain about the intention and meaning of the message. However, in the same way as we rely on nonverbal cues to aid in interpreting the emotions behind someone's spoken words, we often incorporate emoticons and emojis in cybercommunications to clarify how to "read" the feelings behind one's typed words (Lo 2008). Such electronically mediated communication is rapidly increasing and, along with an increase in its volume of use, has become a unique tool in maintaining relationships (see the Popular Culture feature in this chapter).

## The Functions of Nonverbal Communication

Whenever two or more people are together and aware of each other, it is impossible for them *not to communicate*. Even when you are not talking, you communicate by your silence (e.g., an awkward silence, a hostile silence, a tender silence, or an indifferent silence). You communicate by the way you position your body and tilt your head, your facial expressions, your physical distance from the other person or people, whether and how you touch or avoid touch, and so on. Take a moment, right now, wherever you are reading this, and look around you. If there are other people in your presence, how and what are they communicating nonverbally?

Research supports the idea that nonverbal communication has important consequences that extend beyond the message and the moment. For example, parents can affect their children's physical and mental health by their nonverbal communication. The same is true of marital partners' effects on each other.

Supportive nonverbal behavior can benefit relationship partners by actually affecting

> **False**
> **1.** The meaning conveyed through nonverbal communication is often imprecise and ambiguous.

Suppose you received the following text message while in class or at work or while otherwise engaged in some activity away from your significant other:

"Luv u. Miss u. C u later."

Despite your physical separation, it is likely that you would recognize such a message as a warm and loving gesture, a means of staying connected and feeling close, despite being in different places and engaged in different activities. Such messages, made possible by the immense spread of cell phone technology, add a dimension of closeness to our romantic and familial relationships.

A 2014 report from the Pew Research Center, "Couples, the Internet, and Social Media," examined how couples in serious relationships, including marriages, use the Internet and mobile phones. Calling the devices "key actors in the life of many American couples," cell phones, social media, and the Internet have had both positive and negative effects on how couples experience their intimate relationships. Sampling 2,252 adults, 18 and older, 1,428 of whom were either married, cohabiting, or in a committed relationship, the survey detailed how respondents used and assessed the impact of technology on their relationships. Especially among younger adults, technology has enabled couples to use text messages and online communication in ways that can enhance close feelings to their partners. Among 18- to 29-year-olds in serious relationships, just over 40 percent responded that they have felt closer to their partner because of text messages received or online conversations had. Almost a fourth (23 percent) reported that they had used technology to resolve an argument that they were having trouble resolving in person. A third of 18- to 29-year-old text message users had exchanged text messages with a partner or spouse while both were in the home together.

The Pew survey showed substantial differences in how individuals use and assess technology's impact based on their age. Only one out of ten adults 65 and over say that technology has had some relationship impact. Not all relationship impact has been positive, however. Among younger adults and those who had been in their relationships ten years or less, almost a fifth (18 percent) had argued with their partner about how much time one of them was spending online. This was even more the case among those who owned cell phones. More than 40 percent (42 percent) of 18- to 29-year-olds in serious relationships (including marriages) have felt as though their partner/spouse was distracted by their cell phone. Almost 30 percent (29 percent) of 30- to 49-year-olds had the same complaint.

A study of 276 young adults suggestively indicated that text messaging may have gender-specific meanings (Schade et al. 2013). For example, women who texted their partner a lot considered their relationship to be more stable. On the other hand, men who receive or send a lot of texts reported being less satisfied with their relationship. Researchers Lori Cluff Schade and colleagues also found that women were more likely than men to use text messages to manage their relationships, work out issues, or make decisions (Schade et al. 2013). Another 2012 study, using data from 354 undergraduates, found that nearly half had received bad news via a text message; almost half (46 percent) had been broken up with electronically, via a text message or email, and 39 percent had ended a relationship electronically. More than 40 percent had been told of infidelity via some electronic message, and 25 percent had communicated their own infidelity to a partner via message.

Texting technology creates the expectation of *perpetual contact*, encouraging partners to believe that their loved one should be accessible 24 hours a day, 7 days a week (Katz and Aakhus 2002). As David Knox describes, it gives partners "an instant and continuous connection throughout the day so that they are, in effect, 'together all the time'" (Knox 2014, F14). Texting also creates a need for clarity and shared understanding of the "rules" couples wish to enact (Miller-Ott, Kelly, and Duran 2012). Communications scholars Aimee Miller-Ott, Lynne Kelly, and Robert Duran (2012) investigated the importance of cell phones and cell phone rules in relationship satisfaction with romantic relationships among their sample of 227 undergraduate students. They found that rules regarding the use of cell phones to manage relationship issues (e.g., agreements not to fight or discuss interpersonal issues over the phone via talk or text) helped predict relationship satisfaction. The *absence* of rules restricting how often they could text or call each other and whether it was all right to check each other's phone and text logs were associated with greater relationship satisfaction. Partners felt more satisfied when there were no limits on when and how often they could call or text each other, and when they were not prohibited from checking each other's phone usage (Miller-Ott, Kelly, and Duran 2012).

Cell phone technology has significantly altered communication, making it possible for spouses and partners to keep in touch, and for parents to monitor and even supervise children from afar. Within couple relationships, cell phones can also be a source of texting-related troubles, by facilitating arguing, cheating, or even breaking up via exchanges of texts. ●

their immune system and their overall health. Negative nonverbal communication can negatively impact individual (marital) partners as well as threaten the stability of the relationship between them (Giles and LePoire 2006). As first suggested by psychologist Nancy Henley (1977), nonverbal communication can also reveal much about power in relationships. People with more power use various nonverbal methods to express and maintain their dominance.

One of the problems with nonverbal communication, however, is the imprecision of its messages. Is a person frowning or squinting? Does the smile indicate friendliness or nervousness? A person may be in reflective silence, but we may interpret the silence as disapproval or distance. We may incorrectly infer meanings from expressions, eye contact, stance, and proximity that are other than what is intended. Even if we correctly read the tone of a partner's nonverbal behavior we may incorrectly attribute it to relational causes when the cause may in fact be nonrelational. Accurately decoding nonverbal messages is important in the success of a relationship. For example, if one incorrectly attributes a spouse's or partner's negative affect to the relationship, it can cause that individual to assume falsely that he or she is responsible and is negatively affecting one's partner. Husbands' abilities to recognize nonrelational negative affect in their wives' nonverbal communication is of particular importance for the well-being of a relationship (Koerner and Fitzpatrick 2002). However, by acting on the meaning we read into nonverbal behavior, we give it more weight and make it of greater consequence than it initially might have had.

Despite the potential for misinterpretation or ambiguity, nonverbal communication can be a means to convey our attitudes, form and make impressions, express our emotions, and handle an ongoing interaction.

## Conveying Interpersonal Attitudes

Nonverbal messages are used to convey attitudes. Holding hands can suggest intimacy; sitting on opposite sides of the couch can suggest distance. Not looking at each other in conversation can suggest discomfort or lack of intimacy. Rolling eyes at another's statement conveys a negative attitude or reaction to what's being said or the person saying it, even if the eye-rolling culprit claims, "What? I didn't say *anything.*" Indeed, she or he said quite a bit.

## Expressing Emotions

Our emotional states are expressed through our bodies, our gestures, and our expressions. Smiles, frowns, furrowed brows, tight jaws, tapping fingers—all express emotion. Expressing emotion is important because it lets our partner know how we are feeling so that he or she can respond appropriately. It also allows our partner to share our feelings, whether that means to laugh or weep with us. It is this feature of nonverbal communication that is most lacking from phone conversations and electronic communication. Without those emotional cues that we read and come to depend upon, it is sometimes a challenge to know just what the person on the other end of the phone is "really saying."

## Handling the Ongoing Interaction

Nonverbal communication helps us handle the ongoing interaction by indicating interest and attention. Posture and eye contact are especially important. Are you leaning toward the person with interest or slumping back, thinking about something else? Do you look at the person who is talking, or are you distracted, glancing at other people as they walk by or watching the clock?

## The Importance of Nonverbal Communication

According to psychologist John Gottman (1994), even seemingly simple acts, such as rolling one's eyes in response to a statement or complaint made by a spouse, can convey **contempt**, a feeling that the target of the expression is undesirable. Contempt can be displayed verbally as well, through such things as insults, sarcasm, and mockery. Along with contempt, there are three other negative behaviors that indicate particularly troubled and vulnerable relationships. These others are criticism (especially when it is overly harsh), defensiveness, stonewalling (or avoiding). Together, these four behaviors made up what Gottman called the "four horsemen of the apocalypse," spelling potential for eventual divorce (Gottman 1994). Eventually, Gottman added a fifth—belligerence. Gottman suggested that all these are warning signs of serious risk of eventual divorce (Gottman 1994; Gottman et al. 1998). Conversely, couples who communicate with affection and interest and who maintain humor amid conflict can use such a *positive affect* to diffuse potentially threatening conflict (Gottman et al. 1998).

As you think about Gottman's danger signs, consider how easily they can be expressed and conveyed via nonverbal communication as well as by things we say to one another. For example, failing to make eye contact is a way of avoiding or stonewalling. The common gesture of raising your hands in front of yourself and "pushing at the air" communicates defensiveness to those you are interacting with; it is as if you were saying "back off." In fact, nonverbally, you *are* saying just that. When one's nonverbal communication differs from one's verbalized messages—as in a parent denying being angry at her teenage son as she stands before him with her arms crossed in front of her—we consider the nonverbal behavior a truer indication of the person's real feelings (Smith et al. 2011).

On the positive side, such nonverbal behaviors as touch, proximity (i.e., physical closeness), smiling, and gazing help define the intimacy of an interaction. Such behaviors also differentiate closer from more casual relationships and, within relationships, more satisfied from less satisfied partners. In the next section, we will look specifically at some important means of communicating nonverbally.

## Proximity, Eye Contact, and Touch

Three forms of nonverbal communication that are especially interesting are proximity, eye contact, and touch. Awareness of the ways in which such forms of nonverbal communication convey intimacy may enable one to observe interactions and get a fairly good sense of the closeness and warmth within that relationship.

Even the act of gently touching hands can communicate closeness between two people.

Nicolesa/Shutterstock.com

### Proximity

Nearness, in terms of physical space, time, and so on, is referred to as **proximity**. Where we sit or stand relative to another person can signify levels of intimacy or the type of relationship. In a social situation, the face-to-face distances between people when starting a conversation are clues to how the individuals wish to define the relationship. A distance of 0 to 18 inches is considered an **intimate zone** not typically found among people interacting in public settings and typically reserved for one's intimate relationships (e.g., romantic partners, close friends, parents, and young children).

The intimate zone is followed by a space of 1.5 to 4 feet that Hall (1966) considered one's **personal space**. Within this area, one can access a variety of kinds of sensory information, though not with the same detail as in the intimate zone. One is close enough to touch the person, one can see details of a person's appearance, and one can still obtain some (though not as much) sense of the person's scent. It is the more common distance at which people interact in public that allows them to move closer (as intimacy might dictate) or move even farther away from each other (Altman and Chemers 1984).

Although we speak here of the most personal zones as understood in the United States, different cultures have their own standards of spatial norms (Hogh-Olesen 2008). In most cultures, decreasing the distance signifies an invitation to greater intimacy or a threat. Moving away denotes the desire to "distance oneself," and possibly even terminate the interaction. For example, when standing at an intermediate distance from someone at a party, one sends the message that intimacy is not encouraged. If either party wants to move closer, however, he or she risks the chance of rejection. Therefore, they must seek out and exchange cues, such as eye contact, laughter, or small talk, before moving closer to avoid facing direct rejection. If the person moves farther away during this exchange or, worse, leaves altogether ("Excuse me, I think I see a friend"), he or she is signaling disinterest. But if the person moves closer, there is the "proposal" for greater intimacy.

As relationships develop, couples do more than just narrow the spatial distance that partners maintain while interacting. They may also engage in close gazing into each other's eyes, holding hands, and walking with arms around each other. These require closer proximity and signal greater intimacy.

However, because of cultural differences, there can be misunderstandings about the nature or propriety of an interaction. The neutral distance for Latinos, for example, is much closer than for Caucasians, who may misinterpret the distance as close and may respond by withdrawing or, at the other extreme, by flirting. Anthropologist Henrik Hogh-Olesen looked at individuals and their spatial patterns in six different cultures—Greenland, Finland, Denmark, Italy, India, and Cameroon. He reported that "the participants from northern countries such as Greenland, Finland, and Denmark systematically kept a significantly larger distance to strangers than the participants from Italy, India, and Cameroon" (Hogh-Olesen 2008). Evidently, standards of acceptable physical closeness vary across cultures as well as within cultures, according to one's ethnic background, the nature of the relationship between participants, and the social context in which interaction is occurring.

There is more than just distance to consider. In addition to the actual space between two people, one can identify other physical signs that reveal aspects of the closeness or quality of their relationship. Two such signs are body orientation and lean. Face-to-face body orientation is associated with greater intimacy, as is a forward lean (Anderson, Guerrero, and Jones 2006).

## Eye Contact and Facial Expressions

In looking for signs of the degree of closeness or intimacy shared between two people, eye contact is an important component. Much can be discovered about a relationship by watching whether, how, and how long people look at one another. Making eye contact with another person, if only for a split second longer than usual, is interpreted as a signal of interest. You can often distinguish people in love by their prolonged looks into each other's eyes. In addition to eye contact, dilated pupils may be an indication of sexual interest (or poor lighting).

As with proximity, the level of eye contact may differ by culture. For example, research reveals that compared to Americans, Arabs gaze longer and more directly at their partners. Furthermore, in countries where physical touch or contact during interaction is accepted and common, individuals engage in more gazing with those with whom they are interacting than is observed among those in noncontact cultures (see Matsumoto 2006). In the United States, African Americans display less eye contact than do whites (Dovidio et al. 2006).

Eye contact may be best understood as part of a broader category of information and emotions conveyed and communicated with one's face. Via various facial expressions, the face may be the most important conveyor of the level of intimacy (or, conversely, animosity) shared between people in social interaction.

"Pleasant" facial expressions, especially smiles, help convey warmth and display a sense of comfort.

Interesting research by psychologists Masaki Yuki, William Maddux, and Takahiko Masuda identified cultural differences between Japanese and Americans in the importance paid to the eyes as opposed to the mouth in conveying and interpreting emotions. They suggest that the eyes are more of a diagnostic cue for Japanese, whereas the mouth (e.g., smiles, frowns, and so on) is a more influential facial cue for Americans. When subjects were shown photographs of faces in which competing emotions were displayed (e.g., happy eyes, sad mouth), the eyes were the more influential cue for Japanese, whereas the mouth was more influential for Americans (Yuki, Maddux, and Masuda 2007).

## Touch

Touch is the most basic of all senses; it contains receptors for pleasure and pain, hot and cold, and rough and smooth. Touch is a life-giving force for infants. If babies are not touched, they may fail to thrive and may even die. We hold hands with small children and those we love.

A review of the research on touch finds it to be extremely important in human development, health, and sexuality (Hatfield and Rapson 1994). Touch is associated with intimacy across many different types of relationships, from close friends to romantic partners to family members. Indeed, one might state that without touch, intimacy is nearly impossible. When it is welcome, touch of the face or torso is experienced as especially intimate (Anderson, Guerrero, and Jones 2006).

We convey feelings via a variety of nonverbal means—proximity, touch, and eye contact.

Epic Stock Media/Shutterstock.com

People vary in their responsiveness and receptiveness to touch, with some people being more "touch avoidant" (Andersen, Andersen, and Lustig 1987; Guerrero and Andersen 1991). For some, touch can be experienced as a violation. A stranger or acquaintance may touch you in a way that is too familiar. Your date or partner may touch you in a manner you don't like or want. Sexual harassment often consists of unwelcome touching.

Touching is a universal part of social interaction, but it varies in both frequency and meaning across cultures and between women and men (Dibiase and Gunnoe 2004). Often, touch has been taken to reflect social dominance. Based largely and initially on research by Nancy Henley (1977) in which men were found to touch women more than women touched men, the generalization was drawn that touch is often a privilege that higher-status, more socially dominant individuals enjoy over lower-status, more subordinate individuals. This generalization was further modified some by research that revealed that when individuals were of close but different statuses, the lower-status person often strategically used touch as a means of "making a connection" with the higher-status person. Status differences also determined the *type of touch*; lower-status individuals were more likely to initiate handshakes, and higher-status individuals were more likely to initiate somewhat more intimate touching, such as placing a hand on another's shoulder (Dibiase and Gunnoe 2004).

In couple relationships, touch may be used to display closeness, support, warmth, as well as dominance. This is especially true of hand touches, such as hand on arm, hand on shoulder, hand on hand, and so on. Communications research suggests that such touch can be a direct attempt to exert or signal power over the person being touched (Smith et al. 2011).

What about culture? Differences surface in a number of interesting ways. For example, people in colder climates use relatively larger distance and hence relatively less physical contact when they communicate, whereas people in warmer climates prefer closer distances. Latin Americans are comfortable at a closer range (have smaller personal space zones) than North Americans. Middle Eastern, Latin American, and southern European cultures can be considered "high-contact cultures," where people interact at closer distances and touch one another more in social conversations than people from noncontact cultures, such as those of northern Europe, the United States, and Asia (Dibiase and Gunnoe 2004). In so-called high-contact cultures, the kind of touch used in greetings is more intimate, often consisting of hugging or kissing, whereas a firm but more distant handshake is an accepted greeting in noncontact cultures.

Comparing women and men in the United States, Italy, and the Czech Republic, Rosemarie Dibiase and Jaime Gunnoe found that gender differences in touch varied across the three cultures. Men tended to engage in more "hand touch" than women, and women engaged in more "non–hand touch" in all three cultures, though the extent of gender difference varied some in the three countries observed (Dibiase and Gunnoe 2004).

**True** **2.** Although it is often an expression of closeness or intimacy, touch can also be a sign of dominance.

Touch may well be the most intimate form of nonverbal communication. Touching seems to go hand in hand with self-disclosure. Those who touch seem to self-disclose more; touch seems to be an important factor in prompting others to talk more about themselves (Heslin and Alper 1983; Norton 1983).

The amount of contact, from almost imperceptible touches to "hanging all over" each other, helps differentiate lovers from strangers. How and where a person is touched can suggest friendship, intimacy, love, or sexual interest.

Sexual behavior relies above almost all else on touch: the touching of self and others and the touching of hands, faces, chests, arms, necks, legs, and genitals. Sexual behavior is skin contact. In sexual interactions, touch takes precedence over sight, as we may close our eyes to caress, kiss, and enjoy sexual activity. When we shut our eyes to focus better on the sensations aroused by touch, we shut out visual distractions to intensify the tactile experience of sexuality.

## Gender Differences in Communication

Compared with men's nonverbal communication patterns, women smile more; express a wider range of emotions through their facial expressions; occupy, claim, and control less space; and maintain more eye contact with others with whom they are interacting (Smith et al. 2011; Lindsey 1997). In mixed-gender conversations, men tend to display a more "open and relaxed posture" and use more hand touches. In their use of language and their styles of speaking, further differences emerge

(Lakoff 1975; Lindsey 1997; Tannen 1990). Women have been found to use more qualifiers ("It's *kind of* cold out today"), use more tag questions ("It's kind of cold out today, *don't you think?*"), use a wider variety of intensifiers ("It was *awfully* nice out yesterday; now it's kind of cold out, don't you think?"), and speak in more polite and less insistent tones. Male speech contains fewer words for such things as color, texture, food, relationships, and feelings, but men use more and harsher profanity ("It's so damn cold out!") (Lindsey 1997). In cross-gender interaction, men talk more and interrupt women more than women interrupt men. In same-gender conversation, men disclose less personal information and restrict themselves to safer topics, such as sports, politics, or work (Lindsey 1997).

There are two things to note about such differences between female and male styles of both verbal and nonverbal communication. The more male style fits with positions of dominance, whereas the more female style is often found among people in subordinate positions. At the same time, women's style of communicating is characterized more by cooperation and consensus seeking; thus, it is also situationally appropriate and advantageous to relationship building and maintenance (Lindsey 1997; Tannen 1990). In light of these facts, researchers differ in their interpretations of these gender patterns: those who see women's style more as a reflection of subordination versus those who see gender patterns as just reflecting difference.

## Gender Differences in Partner Communication

In addition to overall gender differences in communication noted earlier, researchers have identified several

Touching the face of one's partner is an example of especially intimate nonverbal communication.

Geoff Manasse/Queerstock, Inc./Alamy Stock Photo

**False** **3.** Female partners tend to give more positive and fewer negative messages than male partners.

gender differences in how heterosexual spouses or partners communicate (Klinetob and Smith 1996; Noller and Fitzpatrick 1991).

First, wives tend to *send clearer messages* to their husbands than their husbands send to them. Female partners are often more sensitive and responsive to their partner or spouse's messages both during conversation and during conflict. They are more likely to reply to either positive messages (e.g., compliments) or negative messages (e.g., criticisms) than are their husbands, who may not reply at all.

Second, female partners tend to *give more positive or negative messages;* they tend to smile or laugh when they send messages, and they send fewer clearly neutral messages. Thus, among heterosexual couples, men's neutral responses make it more difficult for women to decode what their partners are trying to say. If a wife asks her husband if they should go to dinner or see a movie and he gives a neutral response, such as "Whatever," does he really not care, or is he pretending he doesn't care to avoid possible conflict?

Third, although communication differences in marital arguments between husbands and wives are usually small, they nevertheless follow a typical pattern. Wives tend to set the emotional tone of an argument. They escalate conflict with negative verbal and nonverbal messages ("You're not even listening to me!") or de-escalate arguments by setting an atmosphere of agreement ("I understand your feelings"). Husbands' input is less important in setting the climate for resolving or escalating conflicts. Wives tend to *use emotional appeals more than husbands*, who tend to reason, seek conciliation, and find ways to postpone or end an argument. A wife is more likely to ask, "Don't you love me?," whereas a husband is more likely to say, "Be reasonable."

## ■ Communication Patterns in Marriage

Communication occupies an important place in marriage. Research in the United States as well as in Europe (e.g., Italy), Asia (Taiwan), and Latin America (Brazil) indicates that partners' satisfaction with relationships is affected by the quality of their communication (Christensen, Doss, and Atkins 2006). When couples have communication problems, they often fear that their marriages are seriously flawed. In addition, negative communication is associated with both less relationship satisfaction and greater instability. As shown in a subsequent section, one of the most common complaints of couples seeking therapy is about their communication problems (Burleson and Denton 1997).

There continues to be a substantial amount of research examining premarital and marital communication. Researchers are finding significant correlations between the nature of communication and satisfaction as well as differences in male versus female communication patterns in marriage.

## Premarital Communication Patterns and Marital Satisfaction

"Drop dead, you creep!" is hardly the thing someone would want to say when trying to resolve a disagreement in a dating relationship. But it may be an important clue as to whether such a couple should marry. Many couples who communicate poorly before marriage are likely to continue the same way after marriage, and the result can be disastrous for future marital happiness. Researchers have found that how well a couple communicates before marriage can be an important predictor of later marital satisfaction (Markman et al. 2010). If communication is poor before marriage, it is not likely to significantly improve after marriage—at least not without a good deal of effort and help. On the other hand, *self-disclosure*—the revelation of our own deeply personal information—before or soon after marriage is related to relationship satisfaction later. Talking about your deepest feelings and revealing yourself to your partner builds bonds of trust that help cement a marriage.

Whether a couple's interactions are basically negative or positive can also predict later marital satisfaction and divorce. In a notable experiment by John Gottman (1979), 14 premarital couples were evaluated using "table talk," sitting around a table and simply engaging in conversation. Each couple talked about various topics. Using an electronic device, each partner electronically recorded whether the message was positive or negative. Gottman found that the negativity or positivity of the couple's communication pattern barely affected their marital satisfaction during their first year. This protective quality of the first year is known as the **honeymoon effect**—which means that you can say almost anything during the first year, and it will not seriously affect marriage (Huston, McHale, and Crouter 1986). But after the first year, couples

with negative premarital communication patterns were less satisfied than those with positive communication patterns.

More recent research by psychologist Howard Markman and colleagues indicates that early risk factors, observable in couples' communication prior to marriage, were associated with lower levels of marital adjustment over the first five years of marriage. Observed and self-reported negative premarital communication more clearly predicted later marital quality than did positive communication. They also found significant association between self-reported negative communication and later divorce (Markman et al. 2010). Clearly, other factors arise during a marriage such as infidelity and/or falling out of love that can have a more immediate effect in causing marital decline in couples even when their premarital communication patterns are more positive. Nevertheless, premarital communication quality can reveal couples who may be at risk for lower-quality marriages and higher likelihood of divorce (Markman et al. 2010).

## Sexual Communication

To have a more satisfying sexual relationship, a couple must be able to communicate effectively with each other about expectations, needs, attitudes, and preferences (Byers 2011; Regan 2003). Couples' sexual communication might address sexual health, sexual pleasure, sexual limits, and partners' sexual histories (Greene and Faulkner 2005). The frequency with which couples engage in sexual relations, the types of sexual behaviors they engage in, and the quality of their sexual involvement depend on such communication. The more effective the communication, the more satisfied partners will be with their sexual relationship. Unfortunately, rates of sexual self-disclosure even among partners in long-term committed relationships are discouragingly low (Byers 2011).

As addressed in Chapter 6, in both married and cohabiting heterosexual relationships, women and men often follow sexual scripts. Traditional sexual scripts leave the initiation of sex (i.e., the communication of desire and interest) to men, with women then in a position of accepting or refusing men's overtures. Reviewing the literature on sexual communication, Pamela Regan observes that regardless of who takes the role of initiating, the efforts are usually met with positive responses. Both attempts to initiate and positive responses are rarely communicated explicitly and verbally (Regan 2003, 84):

> A person who desires sexual activity might turn on the radio to a romantic soft rock station, pour his or her partner a glass of wine, and glance suggestively in the direction of the bedroom. The partner . . . might smile, put down his or her book, and engage in other nonverbal behaviors that continue the sexual interaction without explicitly acknowledging acceptance.

Interestingly, lack of interest or refusal of sexual initiations is more likely to be communicated directly and verbally (e.g., "Not tonight, I have a lot of work to do"). By framing refusal in terms of some kind of account, the refusing partner allows the rejected partner to save face (Regan 2003).

According to psychologist E. Sandra Byers, intimate partners have difficulty communicating to each other what pleases and displeases them sexually, and yet effective sexual communication may be difficult, but it is important if couples hope to construct and keep mutually satisfying sexual relationships. Well beyond whatever gratification partners experience from a sexual encounter, effective sexual communication can have a range of longer-term, positive outcomes such as greater relationship satisfaction and positive sexual health and well-being over the life span (Byers 2011). A study of 608 mostly young (18 to 30), heterosexual couples who had been dating for an average of two years, found that couples' sexual communication and sexual assertiveness were both associated with greater relationship satisfaction. In addition, couples who endorsed less traditional sexual attitudes, less gendered sexual scripts, and less belief in the sexual double standard reported more sexual self-disclosing and partner communication. Even when their sexual communication failed to lead to the desired outcome (e.g., a particular kind of sexual intimacy, more frequent sex, etc.), the sexual communication ultimately had beneficial effects on couple relationships. We must trust our partner enough to express our feelings about sexual needs, desires, and dislikes, and we must be able to hear the same from our partner without feeling judgmental or defensive (Regan 2003).

## Demand–Withdraw Communication

One prominent type of familial communication, especially evident in conflict communication, is referred to as **demand–withdraw communication**—a pattern in which one person makes an effort to engage the other person in a discussion of some issue of importance to the person initiating the conversation. The one raising the issue may criticize, complain, or suggest a need for change in the other's behavior or in the relationship. In the demand–withdraw pattern, the other party, in response to such overtures, withdraws by either leaving the discussion, failing to reply, or changing the subject (Klinetob and Smith 1996).

In many ways, the demand–withdraw pattern is an understandable outcome of differences in what each partner in a relationship wants. One partner, often the wife in heterosexual marriages, wants something different from the status quo. She is then left with a choice between doing nothing and confronting her husband. If she resists or avoids bringing up the topic, nothing will change, a situation she may deem unacceptable. Instead, she raises the subject, voices her complaint, and presses for change. The other partner, more often the husband in heterosexual marriages, may want no change, as he is satisfied with the way things are (or otherwise not interested in the change his partner wishes to discuss). By agreeing to discuss the subject, there is a chance that tension will rise and conflict will result. He also runs the risk of having to ultimately agree to change in the way his wife desires. Wanting neither to argue nor to agree with his partner's wishes, he withdraws.

As the preceding illustrates, the person desiring the change and thus making the demand is in a potentially vulnerable, less powerful position than the person withdrawing from the interaction. The latter can choose to change or not. By withdrawing, he or she maintains the status quo. Withdrawal has other consequences. Just as it keeps the conflict from escalating, it may prevent the resolution of the conflict by curtailing needed communication and necessary relationship adjustment (Holley, Sturm, and Levenson 2010; Papp, Kouros, and Cummings 2009).

The demand–withdraw pattern is common in the United States and cross-culturally (Christensen, Doss, and Atkins 2006). Although most often studied within heterosexual couples, it has also been observed in same-gender couples (Holley, Sturm, and Levenson 2010), and in conflict between parents and their adolescents (Caughlin and Malis 2004). In both versions of parent–child demand–withdraw (i.e., instances of parents demanding and adolescents withdrawing, or instances of adolescents demanding and

**False** **4.** The demand–withdraw pattern of communicating has been observed widely among both heterosexual and same-gender couples.

parents withdrawing), both adolescents' and parents' relationship satisfaction can be negatively affected.

Research on heterosexual couples has often shown a tendency for the demand–withdraw pattern to be associated with gender. As in the earlier illustration, women were more often found to "demand" and men to "withdraw." Researchers have considered a variety of explanations for this more common gender pattern, from a more biologically based explanation about how the genders physiologically respond to conflict, to more socialization-based explanations about how the genders learn to act within relationships. Men and women may well have different physiological responses to conflict that theoretically could help produce the familiar male withdrawal that is part of the female demand–male withdraw pattern of communication. With greater tolerance for physiological arousal, women can maintain the kinds of high levels of engagement that conflict contains. Compared to women, men show different and highly unpleasant physiological reactions to disagreements—more rapid heartbeat, quickened respiration, and the release of higher levels of epinephrine in their endocrine systems. Withdrawal from potential or likely conflict may be a means of avoiding these reactions (Gottman and Levenson 1992; Levenson, Carstensen, and Gottman 1994).

Those who emphasize socialization note that females are socialized to seek greater connectedness with others, whereas males are socialized to seek greater independence and autonomy. Theoretically, this could motivate women to desire greater closeness, which in turn could put them in the position of making demands of male partners. Relationship dynamics are also important. Women are more often the partners seeking to initiate some sort of change in the relationship, whether that be increased intimacy, increased male participation in housework, or increased male involvement in child care (Christensen and Heavey 1993; Christensen, Doss, and Atkins 2006).

Psychologists Sarah Holley, Virginia Sturm, and Robert Levenson sought to determine the extent to which the common gender pattern in demand–withdraw communication results from essential differences between the genders or from power differences that surface in the form of females pressing for change and males resisting their overtures by avoiding, curtailing, or withdrawing from discussion. Holley, Sturm, and Levenson (2010) compared conflict behaviors in 21 heterosexual couples, 21 gay male couples, and 21 lesbian couples. If the demand–withdraw pattern was in fact a product of gender, it should have been displayed in heterosexual couples in the form of females demanding and males withdrawing, and lesbian couples should have shown higher overall levels of demand behaviors, whereas gay male couples would show higher overall levels of withdrawing behavior. Instead, there were *no differences* between couple types in their display of demand–withdraw behavior, either by their own self-reported accounts or in ratings by researchers observing their interactions. The results support an emphasis on differential power across couple types—the more change in aspects of the relationship a person desires relative to his or her partner, the more that person will demand and the more the partner will withdraw, regardless of their genders (Holley, Sturm, and Levenson 2010). Although this study looked at only one discussion of an ongoing, unresolved relationship problem, other research comparing same-gender couples and heterosexual couples has reached the same conclusion: There is a lack of significant differences in demand–withdraw behavior between heterosexual and same-gender couples (Baucom, McFarland, and Christensen 2010).

In addition to gender, there is also research that connects the roles in demand–withdraw communication more to whomever's issue is being discussed. Women have been found to withdraw when men raise issues or about which they are concerned or with which they are unhappy. In fact, same-sex couples have been found to show a demand–withdraw pattern. Across couple types, the partner who is seeking change in some aspect of their partner's behavior or

The demand–withdraw style of communication is common among couples, but it is considered ineffective, even dysfunctional, in pursuing the resolution of a disagreement.

George Doyle/Stockbyte/Getty Images

the relationship raises an issue ("demands") while the partner being asked to make changes may withdraw (Holley, Sturm, and Levenson 2010). It is important to note that there is also research that has shown more similarity between females and males in the roles they take in the demand–withdraw pattern.

Although the demand–withdraw pattern is fairly common, it is neither a healthy nor an effective style of communication or conflict resolution (Papp, Kouros, and Cummings 2009). It is associated with less marital satisfaction and higher likelihood of partners' experiencing anger and other negative emotions, relationship failure or divorce, and may even be a predictor of violence within the couple relationship, especially among couples with high levels of husband demand–wife withdraw (Papp, Kouros, and Cummings 2009; Sagrestano, Heavey, and Christensen 1999). Psychologist John Gottman and colleagues contend that the demand–withdraw pattern is "consistently characteristic of ailing marriages," while researchers Lauren Papp, Chrystyna Kouros, and Mark Cummings indicate that demand–withdraw has been linked to both relationship dysfunction and "individual maladjustment" (Gottman et al. 2002, in Christensen, Doss, and Atkins 2006; Papp, Kouros, and Cummings 2009).

Sarah Holley, along with Claudia Haase and Robert Levenson, has also looked into whether couples display age-related changes in demand–withdraw behavior. Observing 127 middle-aged and older couples at three different points over 13 years, they found that there was an increase in the use of avoidance behaviors (such as changing topics, delaying the discussion, hesitating) by both husbands and wives. Other behaviors that characterize demand–withdraw communication were stable over that same time period. They suggest that where avoidance behavior among young couples would be problematic, among long-term married couples, the same behaviors may be neutral or even adaptive, in that it can move the discussion in a more benign direction and onto potentially neutral or even pleasant topics. It also allows them to avoid more toxic areas (Holley, Haase, and Levenson 2013).

## Communicating Too Much?

In an article for the National Council on Family Relations' Summer 2015 Report, John Caughlin and Erin Basinger pose the question of whether people want—or should want—"open and honest communication" about everything in our relationships and families. As we saw in Chapter 5, intimacy in the United States is built on an assumption of personal self-disclosure of one's innermost feelings. This was shown to characterize both constructions of close friendship as well as of showing love. As Caughlin and Basinger refer to it, there is an "ideology of openness," in which being open is crucial to the success of relationships.

Clearly, self-disclosure is an important part of growing and being close to another person. It is central to our notions of love and intimacy. Avoiding disclosures can lead spouses, partners, or family members to be unaware of each other's desires or needs. However, Caughlin and Basinger (2015) raise the issue of whether moderation is a wiser standard to use than the idea that one should share everything. They note that proponents of *relational dialectics theory* and *communication privacy management theory* both recognize the opposing impulses in close relationships, and contradictions between complete openness and privacy.

Believing in the value of complete openness can lead one to disappointment or dissatisfaction at anything less than that. It may be more important to have a sense that one *can* be open than it is to actually share everything. This should not be taken as a justification for avoiding communication, Caughlin and Basinger note, but the line between enough and too much sharing is not always clear (Caughlin and Basinger 2015).

## ■ Other Problems in Communication

Studies suggest that poor communication skills precede the onset of marital problems (Gottman 1994; Markman 1981; Markman et al. 1987). Even family violence has been seen by some as the consequence of deficiencies in the ability to communicate (Burleson and Denton 1997). In this section, we consider some additional sources of communication difficulties and suggest ways to develop better communication.

### Topic-Related Difficulty

Some communication problems are topic dependent more than individual or relationship based. That is, some topics are more difficult for couples to talk about. As Keith Sanford (2003, 98) states, "It would seem easier to resolve a disagreement about what to do on a Friday night than a disagreement about whether one spouse is having an affair." If some topics are more difficult to discuss than others, couples are likely to display poorer communication when discussing those topics.

**TABLE 7.1  Ten Topics that Are Most Difficult for Couples to Discuss**

| Topic | Difficulty Score* |
|---|---|
| Relationship doubts (possibility of divorce) | 4.58 |
| Disrespectful behavior (lying, rudeness) | 4.50 |
| Extramarital intimacy boundary issues (use of pornography, jealousy) | 4.42 |
| Excessive or inappropriate display of anger (yelling, attacking) | 4.25 |
| Sexual interaction | 4.17 |
| Lack of communication (refusal to talk) | 4.00 |
| In-laws and extended family | 3.83 |
| Confusing, erratic, emotional behavior | 3.75 |
| Criticism | 3.58 |
| Poor communication skills (being unclear or hard to understand) | 3.46 |

*1 = 5 extremely easy; 5 = extremely difficult
SOURCE: Sanford (2003).

In an attempt to determine the difficulty of different topics, Sanford gave a sample of 12 licensed PhD psychologists a list of topics and asked them to provide their best guesses as to how difficult each topic is for couples to discuss and resolve (from 1 being extremely easy to 5 being extremely difficult). The list consisted of 24 topics, generated from a sample of 37 couples who were asked to identify two unresolved issues in their relationships. The 10 topics to which the psychologists assigned the highest "difficulty scores" are listed in Table 7.1.

Other familiar relationship trouble spots and their assigned ratings include child-rearing issues (3.42), finances (3.42), lack of listening (3.08), household tasks (2.33), and not showing sufficient appreciation (2.25). Although the scores demonstrate differences in the degree of sensitivity of different marital issues, these differences do not themselves appear to determine how couples communicate about them (Sanford 2003).

## Barriers to Effective Communication

We can learn to communicate, but it is not always easy. Traditional male gender roles, for example, work against the idea of expressing feelings. Traditional masculinity calls for men to be strong and silent, to ride off into the sunset alone. If men talk, they talk

about things—cars, politics, sports, work, or money—but not about feelings. In addition, both men and women may have personal reasons for not expressing their feelings. They may have strong feelings of inadequacy: "If you really knew what I was like, you wouldn't like me." They may feel ashamed of or guilty about their feelings: "Sometimes I feel attracted to other people, and it makes me feel guilty because I should only be attracted to you." They may feel vulnerable: "If I told you my real feelings, you might hurt me." They may be frightened of their feelings: "If I expressed my anger, it would destroy you." Finally, people may not communicate because they are fearful that their feelings and desires will create conflict: "If I told you how I felt, you would get angry."

Earlier, we mentioned John Gottman's "four horsemen of the apocalypse," his emphasis on especially damaging or dangerous communication issues. To Gottman, *criticism, contempt, defensiveness,* and *stonewalling* present particularly problematic communication and conflict management styles. *Criticism* refers to verbal attacks in which the other person's character is made the issue, as in "Why do you always do that?!" or "You never try to see my side of things!" *Contempt,* addressed earlier in the nonverbal display of rolling one's eyes, conveys a sense of superiority, and can be expressed via sarcastic tones of voice, mocking one's partner, sneering, name-calling, or using hostile humor. The Gottman Institute (2013) lists contempt as the greatest predictor of divorce. *Defensiveness* in response to a criticism from one's partner is a way of blaming one's partner for even raising their issue, and makes oneself out to be a victim of an unfair or undeserved attack. In fact, it conveys the sense that "It's not me that's the problem, it's you." Last, *stonewalling* occurs when one avoids issues by withdrawing from the conversation. Such a withdrawal may be needed to calm oneself down and engage in *physiological self-soothing* so as to avoid exploding in anger or imploding. The Gottman Institute recommendation is to take at least a 20-minute time-out, during which the person spends time doing something soothing and distracting, like listening to music or exercising (Gottman Institute 2013).

Health and Human Sciences professor Kevin Zimmerman offers the following to be used in lieu of the four horsemen: instead of criticizing one's partner, use "I statements" to convey how one feels; rather than gestures of contempt, communicate appreciation and respect for one's partner even while engaged in uncomfortable dialogue; rather than reacting defensively and dismissing as unfair one's partner's criticism, accept part of the responsibility for the difficulties being experienced; and instead of stonewalling, remain in or return to the conversation after a short break (Zimmerman 2015).

## ▌ Positive Communication Strategies

Researchers have found a number of patterns that distinguish the communication patterns in satisfied and dissatisfied marriages (Gottman 1995; Hendrick 1981; Noller and Fitzpatrick 1991; Schaap, Buunk, and Kerkstra 1988). The following characteristics tend to be found among couples in satisfying marriages:

- Willingness to accept conflict but to engage in conflict in nondestructive ways.

- Less frequent conflict and less time spent in conflict. Both satisfied and unsatisfied couples, however, experience conflicts about the same topics, especially about communication, sex, and personality characteristics.

- The ability to disclose or reveal private thoughts and feelings, especially positive ones, to a partner. Dissatisfied spouses tend to disclose mostly negative thoughts to their partners.

- Expression by both partners of equal levels of affection, such as tenderness, words of love, and touch.

- More time spent talking, discussing personal topics, and expressing feelings in positive ways.

- The ability to encode (send) verbal and nonverbal messages accurately and to decode (understand) such messages accurately. This is especially important for husbands. Unhappy partners may actually decode the messages of strangers more accurately than those from their partners.

Kevin Zimmerman describes an effective couple communication strategy developed by Harville Hendrix, known as intentional dialogue. The approach "slows the process and makes both speaker and listener more deliberate about how they interact" (Zimmerman 2015, F13). The process has three stages: mirroring, validation, and empathy.

In the *mirroring stage,* on receiving "small chunks of a sentence or two at a time" from the speaker, the listener mirrors back by paraphrasing or restating what she or he has heard, and then asking "Did I get it?" This allows the speaker to correct anything the listener may have misheard. This stage continues until the speaker is talked out on the topic.

The *validation stage* allows the listener to communicate back that the speaker's point of view is legitimate even if the listener disagrees with it. As Zimmerman describes, validation is potentially the most difficult part of the process, the objective of which is to convey that if the situation were reversed, the partner would probably have similar feelings.

At the *empathy stage,* the listener tries to imagine what the partner is feeling, saying things like, "I imagine you might be feeling . . . Is this accurate?" If corrected by the partner, the listener then tries to mirror the correction. Otherwise, the listener moves on to ask whether there are additional feelings to share and address.

Kevin Zimmerman draws from Jeffrey Foote, Carrie Wilkens, and Nicole Kosanke's *Beyond Addiction: How Science and Kindness Help People Change* (2013), and their list of the following "positive communication strategies," that can improve couple communication:

1. Be positive: Use words and tone of voice that are respectful and affirming. Also, describe what one wants not what is unwanted.

2. Be brief: Clarify and stick to one's central message or request.

3. Be specific: Referring to specific behaviors or examples makes it easier for the listener to understand and try to implement in his or her behavior.

4. Label your feelings: By stating ones feelings in a calm and nonaccusatory manner, it will be easier to gain empathy from the listener.

5. Offer an understanding statement: Comments that indicate that you can understand why the listener might feel as she or he does or have done what she or he did will reduce defensiveness and enhance cooperation and collaboration.

6. Accept some responsibility: Even just saying something like "I make such mistakes too sometimes" communicates the idea that the two are "in this together."

7. Offer to help. Asking "How can I help?" conveys nonjudgmental support. (Zimmerman 2015, F14)

The effects of communication between spouses on couples' satisfaction with their marriages can be found in many cultures, though not necessarily to the same extent found among marriages in the United States. Given the premium placed on intimacy and more romantic conceptualizations of love, marriages in the United States are especially susceptible to the effects of positive and negative communication on their sense of marital well-being. However, even in cultures where marriage is more "practical," arranged, and based on matching partners on a host of characteristics (e.g., Pakistan), marital satisfaction is affected by the nature of marital communication (Rehman and Holtzworth-Munroe 2006).

# ■ Power, Conflict, and Intimacy

Although we may find it unusual or even unpleasant to think about family life in these terms, day-to-day family life is highly politicized. By that we mean that the politics of family life—who has more power, who makes the decisions, and who does what—are complex and can be a source of conflict between spouses or intimate partners. Like other groups, families possess structures of power. As used here, **power** is the ability or potential ability to influence another person or group, to get people to do what you want them to do, whether or not they want to.

Often, we are unaware of or, if aware, deny the power aspects of our intimate relationships.

> **True**
> **5.** In couple relationships, the partner with more power may vary from one situation to another.

We want to believe that intimate relationships are based on love alone. Furthermore, the exercise of power is often subtle, even at times nonverbal, as in the use of hand touches that are often used as a part of a strategy to get one's partner to comply with one's wishes (Smith et al. 2011). On the other hand, when we think of power, we tend to think of coercion or force; as we show here, however, power in relationships between spouses or intimate partners takes many forms and is often experienced as neither coercion nor force. A final reason we are not always aware of power is that power is not constantly exercised. It is most likely to come into play only when an issue is important to both people and they have conflicting goals. Furthermore, power may be situational in couple relationships and change from situation to situation depending on the goals each partner is seeking (Smith et al. 2011).

As a concept, power in marital and other relationships has been said to consist of power bases, processes, and outcomes. *Power bases* are the economic and personal assets (such as income, economic independence, commitment, and both physical and psychological aggression) that make up the source of one partner's control over the other. *Power processes* are the "interactional techniques" or methods that partners or spouses use to try to gain control over the relationship, the partner, or both, such as persuasion, problem solving, or demanding. *Power outcomes* can be observed in such things as who has the final say and who determines—or potentially could determine and control—the outcome of attempted decision making (Byrne, Carr, and Clark 2004; Sagrestano, Heavey, and Christensen 1999).

## Power and Intimacy

The problem with power imbalances and the blatant use of power is the negative effects they have on intimacy. If partners are not equal, self-disclosure may be inhibited, especially if the powerful person believes that his or her power will be lessened by sharing feelings. Genuine intimacy is maximized when there is equality in power relationships. Decision making in the happiest marriages seems to be based not on coercion or tit for tat but on caring, mutuality, and respect for each other. Women or men who feel vulnerable to their mates may withhold feelings or pretend to feel what they do not. Unequal power in marriage may

Nonverbal displays of power in relationships include hand touches by the more powerful to the body of the less powerful partner.

Image Source/Corbis

encourage power politics. Each partner may struggle with the other to keep or gain power.

It is not easy to change unequal power relationships after they become embedded in the overall structure of a relationship, yet they can be changed. Talking, understanding, and negotiating are the best approaches. Still, in attempting changes, a person may risk estrangement or the breakup of a relationship. He or she must weigh the possible gains against the possible losses in deciding whether change is worth the risk.

## Sources of Marital Power

Traditionally, husbands have held authority over their wives. In Christianity, the subordination of wives to their husbands has its basis in the New Testament. Paul (Colossians 3:18–19) states, "Wives, submit yourselves unto your husbands, as unto the Lord." Such teachings reflected the dominant themes of ancient Greece and Rome. Western society continued to support wifely subordination to husbands. English common law stated, "The husband and wife are as one and that one is the husband." A woman assumed her husband's identity, taking his last name on marriage and living in his house.

The U.S. courts formally institutionalized these power relationships. The law, for example, supported the traditional division of labor in many states by making the husband legally responsible for supporting the family and the wife legally responsible for maintaining the house and rearing the children. She was legally required to follow her husband if he moved; if she did not, she was considered to have deserted him. But if she moved and her husband refused to move with her, she was also considered to have deserted him (Leonard and Elias 1990).

Legal and social support for the husband's control of the family declined through most of the 20th century, especially the latter decades. Still, even into the 1970s, judicial discourse reflected an assumption of a "unitary spousal identity" wherein the husband and wife were one, represented by the husband. Ultimately, this was replaced by a more egalitarian model in which marriage was a partnership between equals, each of whom retained an independent legal existence, enjoyed the same rights, and held mutual responsibilities (Mason, Fine, and Carnochan 2001). Especially through employment and wage earning, wives have gained more power in the family, increasing their influence in deciding such matters as family size and how money is spent.

Sociologist Jessie Bernard (1982) drew an important distinction between authority and power in marriage. Authority is based in law, but power can be derived from personality. A strong, dominant woman may, in her relationship, be as likely to exercise power over a more passive man as vice versa simply through the force of personality and temperament.

Among heterosexual dating couples, power imbalances are common, whether they are measured by who makes decisions or who is perceived to be more powerful. Such imbalances tend to favor males over females. An interesting gender pattern finds men perceiving themselves to be more powerful in decision making, whereas women are more likely to characterize decision making as equal (Sprecher and Felmlee 1997).

The relationship among gender, power, and violence is complex. Although some research suggests that men's violence is an expression of men's power over their wives or female partners (and of women's powerlessness), research also asserts that men who themselves feel powerless may use violence. Framed in this way, violence can be a method through which men who lack power or have a need for power attempt to control their wives. Even the threat of violence can be an assertion of power because it may intimidate women into complying with men's wishes even against their own (Kimmel 2011; Levitt, Swanger, and Butler 2008; Sagrestano, Heavey, and Christensen 1999).

If we want to see how power works in marriage, we need to look beneath gender stereotypes and avoid overgeneralizations. Women have considerable power in marriage, although they often feel that they have less than they actually do. They may fail to recognize the extent of their power; because cultural norms traditionally put power in the hands of their husbands, women may look at norms rather than at their own behavior, failing to recognize the degree to which they wield power. A woman may feel that her husband holds the power in the relationship because she believes that he is supposed to be dominant. Similarly, women who may believe their relationships to be egalitarian nevertheless may exercise control over domains of day-to-day life. For example, linguist Alexandra Johnston illustrated the process of **gatekeeping** in her analysis of "manager–helper interactions" between a couple she calls Kathy and Sam in their distribution of caregiving responsibility for their daughter Kira. Despite their commitment to and endorsement of shared parenting, their roles are not as equal as they believe or would like. Although they spend equal amounts of time and participate in the same caregiving tasks, Kathy is the primary decision maker for many of the issues concerning the care of 2-year-old Kira (Johnston 2007). Finally, husbands may believe that they have more power in a relationship than they actually do because they see only traditional norms and expectations.

Power is not a simple phenomenon. Researchers generally agree that family power is a dynamic, multidimensional process (Szinovacz 1987). Generally, no single individual is always the most powerful person in every aspect of the family. Nor is power always based on gender, age, or relationship. Power often shifts from person to person, depending on the issue.

## Explanations of Marital Power

**Relative love and need theory** explains power in terms of an individual's involvement and needs in a relationship. Each partner brings certain resources, feelings, and needs to a relationship. Each may be seen as exchanging love, companionship, money, help, and status with the other. What each gives and receives, however, may not be equal. One partner may be gaining more from the relationship than the other. The person gaining the most from the relationship is the most dependent.

Love itself is a major power resource in a relationship. Those who love equally are likely to share power equally (Safilios-Rothschild 1976). Such couples are likely to make decisions according to referent, expert, and legitimate power.

> **True**
> **6.** The party with the least interest in continuing a relationship generally has the power in it.

> **False**
> **7.** A woman's power in marriage does not automatically increase when she earns as much or more than her husband.

### Principle of Least Interest

Akin to relative love and need as a way of looking at power is the **principle of least interest**. Sociologist Willard Waller (Waller and Hill 1951) coined this term to describe the situation in which the partner with the least interest in continuing a relationship enjoys the most power in it.

Quarreling couples may unconsciously use the principle of least interest to their advantage. The less involved partner may threaten to leave as leverage in an argument: "All right, if you don't do it my way, I'm going." The threat may be extremely powerful in coercing a dependent partner. It may have little effect, however, if it comes from the dependent partner because he or she has too much to lose to be persuasive. Knowing this, the less involved partner can easily call the other's bluff.

In their study of 101 heterosexual dating couples, sociologists Susan Sprecher and Diane Felmlee found that Waller's principle of least interest described the power imbalances in their sample couples. They found that the partners who perceived themselves to be more emotionally involved and invested in the relationship also perceived themselves to have less power than their partners. This pattern held true for both women and men, but men were significantly more likely to perceive themselves as the less involved partner. Women's perceptions echoed men's, as they saw their male partners as less invested in their relationships than they perceived themselves to be (Sprecher and Felmlee 1997).

### Resource Theory of Power

In 1960, sociologists Robert Blood and Donald Wolfe studied the marital decision-making patterns as revealed by their sample of 900 wives. Using "final say" in decision making as an indicator of relative power, Blood and Wolfe inquired about a variety of decisions (e.g., whether the wife should be employed, what type of car to buy, and where to live), and who "ultimately" decided what couples should do. They noted that men tended to have more of such decision-making power and attributed this to their being the sole or larger source of the financial resources on which couples depended. They further observed that as wives' share of resources increased, so did their roles in decision making (Blood and Wolfe 1960).

This **resource theory of power** has been met with both criticism and some empirical support. By focusing so narrowly on resources, the theory overlooks other sources of gendered power. Specifically, it fails to explain the power that many men continue to enjoy when their wife outearns them, or why a woman's power doesn't automatically increase when she earns more than her husband. In fact, sociologist Karen Pyke contended that when husbands perceive their wife's employment and earnings as threats to their status and identities rather than as gifts, wives derive less power from their earnings and employment (Pyke 1994). It appears that marriages in which couples are relatively equal in their earnings are the most egalitarian couples (Tichenor 2005a). As sociologist Veronica Tichenor asserted, "Money, then, is still linked to power, but only for husbands" (Tichenor 2005a, 202). We will return to this shortly.

The resource theory has also been criticized for equating power with decision making and for ignoring that having power may sometimes mean that one is freed from having to make decisions. In still other circumstances, one may exercise power by influencing how others make decisions, by forcing them to consider the possible consequences of making a decision about which one disapproves. This may be best understood using Steven Lukes's three-dimensional view of power. To Lukes, power may be *overt*, *latent*, or *hidden*. This is nicely depicted in the following example:

> If a husband and wife struggle over domestic labor, and the husband successfully resists the greater participation the wife seeks, he has exercised overt power. If his wife then accepts the situation and avoids raising the issue again out of fear of renewed conflict, he has exercised latent power. But . . . if this issue is never raised . . . because the wife accepts it as her duty to bear the domestic labor burden even when she is employed outside the home . . . the husband has benefited from the hidden power in prevailing gendered practices and ideology. (Tichenor 2005a, 194)

Despite continuing disagreement among researchers about how to measure marital power, the most commonly used method still relies on determining who has "final say" in decision making (Amato et al. 2007).

## Feminist or Gender Perspective

Even though women have considerable power in marriages and families, it would be a serious mistake to overlook the inequalities between husbands and wives. As feminist scholars have pointed out, major aspects of contemporary households and families point to important areas in which women are often clearly subordinate to men: Examples are the continued female responsibility for housework and child rearing, inequities in sexual gratification (sex is often over when the male has his orgasm), the extent of violence against women, and the sexual exploitation of children.

Feminist scholars suggested several areas that required further consideration (Szinovacz 1987). First, they believed that too much emphasis had been placed on the marital relationship as the unit of analysis. Instead, they believed that researchers should explore the influence of society on power in marriage—specifically, the relationship between social structure and women's position in marriage. Researchers could examine, for example, the relationship of women's socioeconomic disadvantages, such as lower pay and fewer economic opportunities than men, to female power in marriage. However, if left at that, such a focus would be consistent with major assumptions of resource theory—because men commonly bring in (and traditionally have brought in) more income, men have derived power in heterosexual relationships from their contributions. What is not consistent with resource theory but is central to a gender theory of power is the finding that when women outearn their spouses, the power outcome is quite different. As sociologist Veronica Tichenor's research illustrated:

> I am not arguing that these women *prefer* to be dominated. Rather, these women are afraid that their tremendous resources will *make them look powerful*, or that their husbands will experience their resource disadvantage as domination—or worse, as emasculation. Therefore these wives defer to their husbands in order to prove that they are not trying to dominate them, and are therefore "real women" and "good wives." (Tichenor 2005b, 201)

Tichenor notes that even though the higher-earning wives don't derive the benefit of greater power due to earning more money, the money = power formula may still work for husbands. In other words, "the gendered expectation of men's dominance overrides the potential power in being the major breadwinner" (Tichenor 2005b, 204). Gender has influence on the balance (or imbalance) of power that is somewhat immune to the influence of spouses' earnings (Tichenor 2005b). Second, these scholars argued that many of the decisions that researchers study are trivial or insignificant in measuring "real" family power. Researchers cannot conclude that marriages are becoming more egalitarian on the basis of joint decision making about such things as where a couple goes for vacation, whether to buy a new car or appliance, or which movie to see. The critical decisions that measure power are such issues as how housework is divided, who stays home with the children, and whose job or career takes precedence. Consider, for example, Shirley Hsiao-Li Sun's study of 16 married Chinese immigrant couples in Canada. She found that in a substantial number of sample couples, wives more than husbands made decisions about such things as day-to-day expenses, whether the wives were earning less than their husbands or even unemployed. Furthermore, husbands deliberately avoided making decisions about household expenditures as a way of maintaining their masculinity. In other words, by treating such matters as too trivial, unimportant, or as "women's business," the men were "doing gender." Finally, Sun notes that when examined closely, couples who reported "joint decision making" on important financial matters often consisted of women seeking input from husbands so that the decisions would be acceptable to the men in the end (Sun 2010).

Some scholars suggested that we shift the focus from marital power to family power. Researcher Marion Kranichfeld (1987) called for a rethinking of power in a family context. Even in instances where women's *marital power* may not be equal to men's, a different picture of women in families may emerge if we examine power within the entire family structure, including power in relation to children. The family power literature has traditionally focused on marriage and marital decision making. Kranichfeld, however, felt that such a focus narrows our perception of women's power. Marriage is not the same thing as family, she argues, and in the wider family context, women often exert considerable power. Their power may not be the same as male power, which tends to be primarily economic, political, or religious. But if by *power* we mean having the ability to change the behavior of others intentionally, women have a good deal of power in their families.

Research on marital violence suggests that the level of absolute power has violent consequence for couples. In relationships that are *either* male dominated *or* female dominated, we find the highest levels of violence. In relationships that are "power divided," there is less violence, and in egalitarian relationships, we see the lowest levels of violence (Sagrestano, Heavey, and Christensen 1999). Aside from violence, egalitarian

heterosexual couples also report the highest levels of relationship satisfaction, whereas couples in which the female has more power than the male have been found to have the lowest levels of satisfaction for both partners (Sprecher and Felmlee 1997).

One difficulty with discussions of "egalitarian relationships" is the question of whether such relationships are truly equal. Feminist research has revealed that power processes often seem to favor men, even among self-professed equal couples. Carmen Knudson-Martin and Anne Rankin Mahoney's (1998) study of equal couples—in which each spouse perceives the relationship to be characterized by mutual accommodation and attention and each spouse has the same ability to receive cooperation from the other in meeting needs or wants—is a case in point. Although couples described their relationships as equal and their roles as "non–gender specific," men wielded more power than women. Wives made more concessions to fit their daily lives around their husbands' schedules than husbands did to fit their lives around the schedules of their wives. Women were also more likely than their husbands to report worrying about upsetting or offending their spouses, to do what their spouses wanted, and to attend to their spouses' needs (Fox and Murry 2000).

It appears as if characterizing an unequal marriage as equal allows a couple to ignore real if covert power differences that might otherwise threaten their relationships (Fox and Murry 2000). As seen in sociologist Arlie Hochschild's classic study of the "second shift," the housework and child care that awaits couples at the end of their work days, spouses may construct family myths in which they redefine unequal divisions of responsibilities as fair, and in this way preserve their relationships (Hochschild 1989).

In their qualitative interview study of 15 middle-class African American couples with young children, researchers Randi Cowdery and colleagues illustrate how power may in some ways be bestowed upon husbands by wives who are especially sensitive to their husbands' lack of societal status and power. In deferring to their husbands and "'letting' the men have more power at home," women were trying to protect their husbands from "indignities experienced in the larger society" (Cowdery et al. 2009, 35). They also note that some women expressed fear that unless they deferred to their husbands, the men might just leave. In that way, "some men seemed to have power just by their presence" (p. 35). Once again, gender issues help determine the exercise of marital power.

# Intimacy and Conflict

Conflict between people who love each other may initially seem to be a mystery. The simultaneous coexistence of conflict and love has puzzled human beings for centuries. An ancient Sanskrit poem reflected this dichotomy:

> In the old days we both agreed
> That I was you and you were me.
> But now what has happened
> That makes you, you
> And me, me?

We expect love to unify us, but often it doesn't. Two people don't really become one when they love each other, although they may have this feeling at first. In reality, they retain their individual identities, needs, wants, and pasts while loving each other—and it is a paradox that the more intimate two people become, the more likely they may be to experience conflict. But conflict itself is not dangerous to intimate relationships; it is the manner in which the conflict is handled. Conflict itself is natural. Coupled life introduces multiple and frequent occasions for conflict. Heterosexual, gay, and lesbian couples are similar in the amount of conflict they experience and in many of the issues that provoke it (Wright and Loving 2011). If this is understood, the meaning of conflict changes, and it will not necessarily represent a crisis in the relationship.

David and Vera Mace, prominent marriage counselors and founders of the Association for Couples in Marriage Enrichment, observed that on the day of marriage, people have three kinds of raw material with which to work. First, they have things in common—the things they both like. Second, they have the ways in which they are different, but the differences are complementary. Third, unfortunately, they have differences between them that are not complementary and that cause them to meet head-on with a big bang. In every relationship between two people, there are a great many of those kinds of differences. So when they move closer to each other, those differences become disagreements (Mace and Mace 1979).

## Experiencing Conflict

The presence of conflict within a relationship, marriage, or family doesn't automatically suggest trouble or indicate that love is going or gone; it may mean quite the opposite. In the healthiest of relationships, it is common and normal for couples to have disagreements or conflicts. All couples have disagreements, with an estimated average of one to two arguments a month. Thus, the average married couple would experience between 84 and 168 arguments in seven years (Wright and Loving 2011). The important factor is not *that* they have differences or even how often or what in particular they fight about but *how* constructively or harmfully they resolve their differences. By using occasions of conflict to implement mutually acceptable behavior changes or to decide that the differences between them are acceptable, couple relationships may grow and solidify as a product of their differences. Couples who resolve conflict with mutual satisfaction and who find ways to adapt to areas of conflict tend to be more satisfied with their relationship overall and are less likely to divorce.

It seems that in couples where one or both partners keep quiet and don't vent their frustrations or express their feelings, problems may result for either the individual partners or the longer-term stability of the relationship. One study of almost 4,000 women and men found that during spousal arguments, nearly a third of men and a quarter of women said that they usually kept their feelings inside during an argument (Parker-Pope 2007). Such behavior, called **self-silencing**, had particularly harsh effects on women; women who kept their feelings to themselves during marital arguments were four times as likely to die during the ten-year span of the research compared to women who said that they always expressed their feelings. Men's health

**True** 8. Heterosexual, gay, and lesbian couples are similar in the amount of conflict they experience.

Conflict is an inevitable and normal part of being in a relationship. How people express conflict is more important than the fact that they have conflict.

Wavebreakmedia/Shutterstock.com

was not measurably affected by whether they did or didn't express themselves during a fight.

If we handle conflicts in a healthy way, they can help strengthen our relationships. But conflicts can go on and on, consuming the heart of a relationship, turning love and affection into bitterness and hatred. In the following section, we look at ways of resolving conflict in constructive rather than destructive ways. In this manner, we can use conflict as a way of building and deepening our relationships.

Most research has revealed no significant differences between cohabiting and married couples in their frequency of conflict over such relationship aspects as time spent together, in-laws, money, sex, decisions about childbearing, or the division of responsibilities. Nor have most studies found cohabitants to differ significantly in the likelihood of heated arguments or the level of open disagreements. Based on a recent study of more than 1,200 people who were either in dating relationships (220), cohabiting (231), or married (801), psychologists Annie Hsueh, Kristen Rahbar Morrison, and Brian Doss established that compared to married individuals, those in cohabiting relationships were more likely to report problems associated with relationship conflict, including disagreement about values and goals for the future and an inability to resolve conflicts (Hsueh, Morrison, and Doss 2009). Cohabitants were also more likely than individuals in dating relationships to report problems with arguments and with conflict resolution.

## Dealing with Anger

Differences can lead to anger, and anger transforms differences into fights, creating tension, division, distrust, and fear. Most people have learned to handle anger by either venting or suppressing it. As indicated earlier, suppressing anger is unhealthy, especially for women. It also can be dangerous to the relationship because it is always there, simmering beneath the surface. It leads to resentment, that brooding, low-level hostility that poisons both the individual and the relationship.

Anger can be dealt with in a third way: when conflict escalates into violence. Especially in a culture that cloaks families in privacy, surrounds people with beliefs that legitimize violence, and gives them the sense that they have a right to influence what their loved ones do, escalating anger can result in assault, injury, and even death. Given the relative power of men over women and adults over children, threats against one

person's supposed advantage may provoke especially harsh reactions. We look closely at the causes, context, and consequences of family violence in Chapter 12.

Finally and most constructively, anger can be recognized as a symptom of something that needs to be changed. If we see anger as a symptom, we realize that what is important is not venting or suppressing the anger but finding its source and eliminating it.

Not all conflict is overt. Some conflict can go undetected by one of the partners. As such, it will have minimal effect on him or her and is not likely to lead to anger. In addition, not all "conflicts" (i.e., of interest, goals, wishes, expectations, and so on) become *conflicts*. Spouses and partners can approach their differences in many ways short of overt conflict (Fincham and Beach 1999).

## How Women and Men Handle Conflict

In keeping with observed gender differences in communication, research has identified differences in how men and women approach and manage conflict. As summarized by Rhonda Faulkner, Maureen Davey, and Adam Davey (2005), we can identify the following gender differences:

- As we saw earlier in discussing the demand–withdraw pattern, women are more likely than men are to initiate discussions of contested relationship issues. Where men have been found to be more likely to withdraw from negative marital interactions, women are more likely to pursue conversation or conflict.

- Typically, women are more aware of the emotional quality of and the events that occur in the relationship.

- In the course and processes of conflict management and resolution, men take on instrumental roles, and women take on expressive roles. Men approach conflict resolution from a task-oriented stance, as in "problem solving"; women are more emotionally expressive as they pursue intimacy.

We need to bear in mind that the research designs used to study patterns of interaction in conflict management may have exaggerated the gender connection by commonly asking couples to engage in discussion of topics of greater importance to females than to males (e.g., intimacy and child-rearing practices). When both partners in heterosexual couples were required to discuss an area in which they would like their partner to make changes, gender patterns were more

"Did you bring me to this country for exploitation?" Such is the plaintive appeal of 41-year-old Yong Ja Kim, a Korean immigrant, to her husband, Chun Ho Kim. What is it she is objecting to? In what way does she feel exploited? Sociologist Pyong Gap Min (2001) researched the consequences of immigration for marital relations among Korean immigrant couples. Existing research indicated that marital conflicts had emerged among Korean immigrants to the United States because of women's increased role in the economic support of families without concurrent changes in their husbands' gender attitudes or marital behavior. Min sought to delve more deeply into such conflicts.

Among Min's interviewees were Yong Ja Kim and Chun Ho Kim, husband and wife who work together at their retail store six days a week from 9:30 A.M. to 6:00 P.M. On returning home, he watches Korean television programs and reads a Korean daily newspaper while she prepares dinner. Defensively, he retorts,

> It makes no sense for her to accuse me of not helping her at home at all. In addition to house maintenance, I took care of garbage disposal more often than she and helped her with grocery shopping very often. I did neither of the chores in Korea.

To his wife, however, the comparison is not between what he did in Korea and what he does in the United States but between what *he does* and what *she does*:

> I work in the store as many hours as you do, and I play an even more important role in our business than you. But you don't help me at home. It's never fair. My friends in Korea work full-time at home, but don't have to work outside. Here, I work too much both inside and outside the home.

Culturally, there are noteworthy differences between the traditional status of husbands in Korea and the situations of most immigrant Korean men in the United States. Traditionally, Korean husbands were breadwinners and patriarchal heads of their families. Wives and children were expected to obey their husbands and fathers. Women were further expected to bear children and cater to their husbands and in-laws. Although the traditional South Korean family system has been "modified," it remains a patriarchal system, justified by Confucian ideology. As they have immigrated to the United States, Korean women's involvement in paid employment has increased radically. In the process, traditional gender attitudes and male sense of self as patriarch and provider have been undermined.

Jose Maciel, Zanetta Van Putten, and Carmen Knudson-Martin's qualitative study of 12 couples in which at least one spouse immigrated to the United States introduces the notion of a *gender line* to help make sense of some of the conflicts that immigrant couples experience (Maciel, Van Putten, and Knudson-Martin 2009). The gender line is a way to visualize the division of power and amount of equality between husbands and wives. As part of the adjustment to their new cultural surroundings, women push this gender line. This is illustrated in the following comment from a woman they refer to as Maria as she compares gender expectations in the United States to those in her homeland of Puerto Rico:

> Here in the United States, I learned to confront conflict by speaking out. In my native country I am not supposed to speak up. (Maciel, Van Putten, and Knudson-Martin 2009, 17)

In pushing the gender line, women try new ways of acting and communicating that were unacceptable and/or avoided in their native country. Where the initial gender line reflects the gender expectations that shaped couple interactions prior to exposure to a new cultural environment, women directly and indirectly push their spouses to accept their influence in decision making and to listen to their opinions. Although the women are the ones pushing toward change, men are said to set the line. Wanting to keep the women satisfied, the men ultimately "give in and 'allow' the women to have influence" (Maciel, Van Putten, and Knudson-Martin 2009, 19).

Referring to the Korean immigrants he studied, Pyong Gap Min notes that with immigration to the United States, most Korean immigrant men encountered significant downward occupational mobility. This, in turn, resulted in further "status anxiety." They compensated by seeking ways to assert their authority in the household, only to find that their wives and children no longer granted them such status automatically. Min states that Mr. Kim "could not understand much and how fast his wife had changed her attitudes toward him since they had come to the United States. He did not remember her talking back to him in Korea." ●

varied. Significantly more woman demand–man withdraw behavior occurred when couples addressed the woman's top issue, but there was also more man demand—woman withdraw behavior during discussions of issues most important to the man. Thus, it is crucial to avoid overgeneralizing gender patterns in partners' conflict styles; importance of the issue to each party also affects conflict behavior.

## Conflict Resolution and Relationship Satisfaction

How couples manage conflict is one of the most important determinants of their satisfaction and the well-being of their relationships (Greeff and deBruyne 2000). Happy couples are not conflict free; instead, they tend to act in positive ways to resolve conflicts, such as changing behaviors (putting the cap on the toothpaste rather than denying responsibility) and presenting reasonable alternatives (purchasing toothpaste in a dispenser). Unhappy or distressed couples, in contrast, use more negative strategies in attempting to resolve conflicts ("If the cap off the toothpaste bothers you, then *you* put it on").

Thus, we can talk of "constructive and destructive" or "helpful and unhelpful" conflict management (du Plessis and Clarke 2008; Greeff and deBruyne 2000; Noller 2015). Constructive conflict management is characterized by flexibility, a relationship rather than individual (self-interest) focus, an intention to learn from their differences, and cooperation. Additional characteristics of "helpful" conflict management include compromise, negotiation, turn taking, calm discussions, careful listening, and trying to understand the other's perspective.

Destructive conflict management includes the following:

- Mutual avoidance: Couples ignore the issue, acting as though it doesn't matter. Issues that aren't addressed cannot be resolved. As such, resentment can build, and distance between partners may grow.

- Coercion: Here, one partner verbally attacks the other over some issue, serious or trivial. The attacked partner "counterattacks." Not only is the issue likely to go unresolved, but both partners may also conclude that there's no point in raising and discussing issues.

- Demand–withdraw: Discussed earlier, this style is one in which one partner raises an issue of concern and the other partner withdraws rather than discussing or addressing it. The demander is left frustrated, the withdrawer may feel frustrated, and the issue remains unresolved. Unresolved issues are more likely to recur or even escalate.

- Obliging: One partner agrees to meet the request or demand of the other, even though she or he disagrees with it. It produces a short-term peace but can lead to resentment over the absence of compromise. If the same partner consistently obliges the other, she or he may grow frustrated and resentful,

the other partner may come to believe she or he can bully the other into compliance. (Noller 2015).

Additional characteristics of "unhelpful" conflict management include confrontation, competition, complaining, criticizing or insulting, defensiveness, and displaying contempt.

One of the strongest predictors of marital unhappiness and of the possibility of eventual divorce is whether couples engage in hostile conflict. Hostile conflict is a pattern of negative interaction wherein couples engage in frequent heated arguments, call each other names and insult each other, display an unwillingness to listen to each other, and lack emotional involvement with each other (Gottman 1994; Topham, Larson, and Holman 2005). Once such patterns become the normative pattern in a relationship, they are difficult to change.

Research is mixed as to the effectiveness of humor in conflict management. More satisfied couples display higher levels of nonsarcastic humor and, during discussions of problems, share laughter. The type of humor appears to make a critical difference. *Affiliative humor*, where one says funny or witty things or tells jokes in an effort to reduce tension and enhance the relationship, may in fact lead to less distress and assist discussions of and facilitate resolution of problems. *Aggressive humor*, which is used to tease, ridicule, or disparage the other, may impede such efforts and lead to greater distress (Campbell, Martin, and Ward 2008).

## What Determines How Couples Handle Conflict?

Many factors might affect how couples approach and attempt to manage the inevitable conflict that relationships contain. Among these, premarital variables, including carryover effects of upbringing, may be particularly influential. Glade Topham, Jeffry Larson, and Thomas Holman (2005) suggest that such influence may be conscious or unconscious; may affect behaviors and patterns of interaction as well as attitudes, beliefs, and self-esteem; and may remain even in the absence of contact with the family of origin.

Family-of-origin factors can be explained by social learning theory or attachment theory. Learning theory suggests that by observing parents and how they interact with each other, we develop a **marital paradigm**: a set of images about how marriage ought to be done, "for better or worse" (Marks 1986). When we fail to experience a positive model of marriage as children, we may develop ineffective communication or conflict resolution skills. Attachment theory suggests that our attachment style influences the way conflict is expressed in relationships (Pistole 1989). Secure parent–child relationships lead us to be more self-confident and socially confident, more likely to view others as trustworthy and dependable, and more comfortable with and within relationships. Individuals who had insecure parent–child attachments are more demanding of support and attention, more dependent on others for self-validation, and more self-deprecating and emotionally hypersensitive (Topham, Larson, and Holman 2005).

In contrast to anxious or ambivalent and avoidant adults, secure adults are more satisfied in their relationships and use conflict strategies that focus on maintaining the relationship. Helping the relationship stay cohesive is more important than "winning" the battle. Secure adults are more likely to compromise than are anxious or ambivalent adults, and anxious or ambivalent adults are more likely than avoidant adults to give in to their partners' wishes, whether or not they agree with them.

Although either husbands' or wives' family-of-origin experiences *could* negatively affect marital quality and conflict management, the influences are not equivalent. Wives' family-of-origin experiences—including the quality of relationships with their mothers, the quality of parental discipline they received, and the overall quality of their family environments—are more important than husbands' experiences in predicting hostile marital conflict (Topham, Larson, and Holman 2005).

There are two "analytically independent" dimensions of behavior in conflict situations: assertiveness and cooperativeness (Greeff and deBruyne 2000). *Assertiveness* refers to attempts to satisfy our own concerns; *cooperativeness* speaks to attempts to satisfy concerns of others. With these two dimensions in mind, we can identify five conflict management styles:

- *Competing.* Behavior is assertive and uncooperative, associated with "forcing behavior and win–lose arguing." This style can lead to increased conflict as well as to either or both spouses feeling powerless and resentful (Greeff and deBruyne 2000).
- *Collaborating.* Behavior is assertive and cooperative; couples confront disagreements and engage in problem solving to uncover solutions. Collaborative conflict management may require relationships that are relatively equal in power and high in trust. Using this style then accentuates both the trust and the commitment that couples feel.

- *Compromising.* This is an intermediate position in terms of both assertiveness and cooperativeness. Couples seek "middle-ground" solutions.
- *Avoiding.* Behavior is unassertive and uncooperative, characterized by withdrawal and by refusing to take a position in disagreements.
- *Accommodating.* This style is unassertive and cooperative. One person attempts to soothe the other person and restore harmony.

Research has yielded inconsistent ("diverse") results about the relationship outcomes of each of these styles. Some studies favor one style—collaboration—over all others as the only style displayed by satisfied couples. There is research suggesting that avoidance is dysfunctional and antisocial, yet there is research that finds avoidance associated with satisfied, nondistressed couples. Finally, although some research suggests that husbands and wives have happier marriages when they agree on how to manage conflict, other findings indicate that discrepancies in spouses' beliefs about conflict are not predictive of how satisfied they are (Greeff and deBruyne 2000).

## Conflict Resolution across Relationship Types

All couple relationships experience conflict. Using self-reported and partner-reported data, Lawrence Kurdek (1994) explored how 75 gay, 51 lesbian, 108 married nonparent, and 99 married parent couples handled conflicts. Essentially, the differences across couple type were less impressive than were the similarities. The four types of couples did not significantly differ in their level of ineffective arguing, and there were no noteworthy differences in their styles of conflict resolution as measured by the Conflict Resolution Styles Inventory (CRSI). The CRSI includes four styles of conflict resolution: (1) *positive problem solving* (including negotiation and compromise), (2) *conflict engagement* (such as personal attacks), (3) *withdrawal* (refusing to further discuss an issue), and (4) *compliance* (such as giving in). Ratings were obtained from both partners about themselves and the other partner. There was little indication that the frequency with which conflict resolution styles were used varied across couple type. As Kurdek (1994) notes, there is similarity in relationship dynamics across couple types.

Others have reported similar comparisons, showing that same-sex and heterosexual romantic relationships are "extremely similar" on many characteristics (Markey et al. 2014). Same-sex and heterosexual couples report similar levels of conflict, intimacy, satisfaction and relationship commitment. Randi Hennigan and Linda Ladd report that looking at conflict and conflict resolution across couple types has revealed that same-sex and heterosexual couples are more alike than they are different in how they go about experiencing and handling conflict (Hennigan and Ladd 2015). Specifically, they note that heterosexual and same-sex couples use similar strategies to resolve conflicts. The most common among these are "to communicate, negotiate or compromise, understand, and problem solve" (Hennigan and Ladd 2015, F8).

Social worker and therapist Arlene Istar Lev suggests that same-sex couples have the advantage of having had similar (gender) socialization. The similarity may lead them to develop more similar approaches to conflict, a greater attunement during conflict, and more similar approach to how best to resolve it (Lev 2015).

A qualitative study of problem areas reported by a sample of more than 1,200 individuals who were either cohabiting, dating, or married (231 cohabiting, 220 dating, or 801 married) uncovered both similarities and differences (Hsueh, Morrison, and Doss 2009). Cohabiting and dating individuals were similar in most regards, with two exceptions. Those in dating relationships were more than twice as likely to report problems with relationship commitment or security, but less than half as likely to report problems with arguments. In comparison to married individuals, cohabiters were much more likely to report problems with arguments, relationship commitment and security, and problems with a previous relationship. Married individuals were significantly more likely than cohabiters to identify emotional affection and distance and lack of physical affection or sex as problems (Hsueh, Morrison, and Doss 2009).

## Common Conflict Areas: Sex, Money, and Housework

Even if, as the Russian writer Leo Tolstoy suggested, every unhappy family is unhappy in its own way, marital and intimate couple conflicts still tend to center on certain recurring issues, especially communication, children and parenting, sex, money, personality differences, how to spend leisure time, in-laws, infidelity, and housekeeping. In this section, we focus on three areas: sex, money, and housework. Then we discuss general ways of resolving conflicts.

## Conflict about Sex

Fighting and sex can be intertwined in several ways (Strong and DeVault 1997). A couple can have a specific disagreement about sex that leads to a fight. One person wants to have sexual intercourse, and the other does not, so they fight. A couple can have an indirect fight about sex. The woman does not have an orgasm, and after intercourse, her partner rolls over and starts to snore. She lies in bed feeling angry and frustrated. In the morning, she begins to fight with her partner over his not doing his share of the housework. The housework issue obscures why she is angry. Sex can also be used as a scapegoat for nonsexual problems. A husband is angry that his wife calls him a lousy provider. He takes it out on her sexually by calling her a lousy lover. They fight about their lovemaking rather than about the issue of his provider role. A couple can fight about the wrong sexual issue. A woman may berate her partner for being too quick during sex, but what she is really frustrated about is that he is not interested in oral sex with her. She, however, feels ambivalent about oral sex ("Maybe I smell bad"), so she cannot confront her partner with the real issue. Finally, a fight can be a cover-up. If a man feels sexually inadequate and does not want to have sex as often as his male partner, he may pick a fight and make his partner so angry that the last thing he would want to do is to have sex with him.

It is hard to tell during a fight if there are deeper causes than the one about which a couple is fighting. Is a couple fighting because one wants to have sex now and the other doesn't? Or are there deeper reasons involving power, control, fear, or inadequacy? If they repeatedly fight about sexual issues without getting anywhere, the ostensible cause may not be the real one. If fighting does not clear the air and make intimacy possible again, they should look for other reasons for the fights. It may be useful for them to talk with each other about why the fights do not seem to accomplish anything. In addition, it would be helpful if they step back and look at the circumstances of the fight, what patterns occur, and how each feels before, during, and after a fight.

Sexual tensions and strains can arise because of these other conflicts that happen to play

themselves out in the physical relationship. Sexual conflicts can also arise because of differences in partners' expectations and desires regarding the frequency of sexual relations. Researchers Brian Willoughby, Adam Farero, and Dean Busby found that higher "individual sexual desire discrepancy" (or the difference between partners in their desired frequency of sexual behavior) was associated with more frequent conflict and with negative outcomes for both relationship stability and satisfaction (Willoughby, Farero, and Busby 2014).

Lauren Papp, Marcie Goeke-Morey, and E. Mark Cummings used diary reports of relationship conflicts among a hundred married couples. Intimacy conflicts, which included conflicts with both sexual intimacy and verbal or physical displays of affection, represented approximately 8 percent of the 748 recorded instances of marital conflict. Papp and colleagues note that such conflicts were likely to be recurrent and be of both immediate and longer-term importance. In fact, for the more distressed couples, disagreements about intimacy can affect partners deeply, provoking feelings of inadequacy or shame, and affecting self-esteem (Papp, Goeke-Morey, and Cummings 2013).

With a more "positive, respectful, affirming process of conflict resolution," partners may deepen the respect and admiration they feel for each other, develop a greater level of trust and of self-esteem in their relationship, and grow more confident that the relationship can withstand and grow through future conflict. These can create positive feelings and comfort with each other that facilitate sexual desire (Metz and Epstein 2002). Although the conflicts being resolved need not be sexual, positive and constructive relationship conflict resolution may provide affirmation of the love and intimacy two people share, bring emotional relief, and even serve as a sexual stimulant (Metz and Epstein 2002). Thus, the intensity of pleasure supposedly accompanying "makeup sex" is another reminder of how conflict and its resolution can affect sex regardless of whether it is about sex.

## Money Conflicts

Money is a major source of marital conflict in families in the United States and abroad. This is especially evident in the first year of marriage (Wright and Loving 2011).

Money conflicts are likely to be more contentious and are the most likely of conflict areas

**True** **9.** Among married couples, money-related conflicts are the most likely to remain unresolved.

to continue longer without resolution. They also pre-dict divorce more effectively than other conflict areas (Dew 2011; Dew, Britt, and Huston 2012). Among cohabiting couples, relative to other problem areas such as conflicts over household chores, sex, time to-gether, and parents, financial conflict was the only one to predict the eventual dissolution of the relationship. Sociologist Jeffrey Dew suggests that financial conflicts may well be qualitatively different than other conflict areas couples face (Dew 2011).

According to Lauren Papp, E. Mark Cummings, and Marcie Goeke-Morey (2009), though conflicts over money were not the most frequent of those they studied, money-related conflicts were more stressful, problematic, and recurrent. They have been character-ized as more "intense and significant," longer lasting, and of greater importance. In part, it is more difficult to avoid such conflicts, as decisions about money and how it is to be or needs to be spent confront couples on a regular basis (Dew 2015). In research by Papp et al., couples experienced and expressed more distress in conflicts about money, including sadness, fear, anger, and withdrawal. Compared to other areas studied, these conflicts were more likely to remain unresolved.

Couples disagree or fight over money for a number of reasons. One of the most important has to do with power. Earning wages has traditionally given men power in families. Work in the home has not been rewarded by wages. As a result, full-time homemakers have been placed in the position of having to depend on their husbands for money. In such an arrange-ment, if there are disagreements, the woman is at a disadvantage, and the old cliché "I make the money but she spends it" has a bitter ring to it. As women in-creased their participation in the workforce, however, power relations within families have shifted some. Studies indicate that women's influence in financial and other decisions increases if they are employed outside the home.

Another major source of monetary conflict is al-location of the family's income. Not only does this involve deciding who makes the decisions, but it also includes setting priorities. Is it more important to pay a past-due bill or to buy a new television set to re-place the broken one? Is a dishwasher a necessity or a luxury? Should money be put aside for long-range goals, or should immediate needs be satisfied? Set-ting financial priorities plays on each person's values and temperament; it is affected by basic aspects of an individual's personality. Jeffrey Dew points out that any couple, regardless of their income, may experience money matters having a negative impact on their re-lationships (Dew 2015).

Dating relationships are a poor indicator of how a couple will deal with money matters in marriage. Dating has clearly defined rules about money: Either the man pays, both pay separately, or they take turns paying. In dating situations, each partner is financially independent of the other. Money is not pooled, as it usually is in a committed partnership or marriage. Power issues do not necessarily enter spending deci-sions because each person has his or her own money. Differences can be smoothed out fairly easily. Both in-dividuals are financially independent before marriage but financially interdependent after marriage. Even cohabitation may not be an accurate guide to how a couple would deal with money in marriage. Cohabi-tators generally do not pool all of their income, but neither do they keep their money totally separate (Dew 2015). Nevertheless, it is the working out of financial interdependence in marriage that is often so difficult.

Why do we find it difficult to be financially interde-pendent and talk about money? There may be several reasons. First, we don't want to appear to be unroman-tic or selfish. If a couple is about to marry, a discussion of attitudes toward money may lead to disagreements, shattering the illusion of unity or selflessness. Second, gender roles make it difficult for women to express their feelings about money because women are tra-ditionally supposed to defer to men in financial mat-ters. Third, because men tend to make more money than women, women feel that their right to disagree about financial matters is limited. These feelings are especially prevalent if the woman is a homemaker and does not make a financial contribution, but they devalue her child care and housework contributions.

In studying different areas of conflict among married couples, including money, spending time together, in-laws, housework, and sex, Jeffrey Dew, Sonya Britt, and Sandra Huston found financial conflicts were the most predictive of a couple's future together. As Dew reports, "For husbands, financial conflicts were the only type of conflict that was associated with future divorce. For wives, financial conflicts were the type of conflict that most strongly related to divorce" (Dew 2015, F11).

## Housework and Conflict

The division of responsibility for housework can be one of the most significant issues couples face. Expressed well by Shannon Davis and Theodore Greenstein, "we can frame housework as something that people are un-willing to do, which is under the purview of one actor who then has to figure out, perhaps through the use of power, how to get another actor to participate" (Davis and Greenstein 2013, 67). It can become a source of tension and conflict within marriage, especially for dual-earner couples (Davis and Greenstein 2013; Dew, Britt, and Huston 2012; Hochschild 1989). Part of this is an understandable consequence of the inequality in each spouse's contribution; compared to most women, most men do not do much housework. As we will explore in more detail in Chapter 11, whether or not they are employed outside the home and whether or not there are children in the home, wives bear the bulk of housework responsibility in marriages. Among cohabiting heterosexual couples, women do more housework than do men. Although there is variation among them, same-sex couples tend to divide housework in a more equal manner than heterosexual couples (Goldberg 2013). When same-sex couples have less equal divisions of housework, the partner who earns less or has less job prestige or greater job flexibility tends to be the partner who does more housework (Goldberg 2013). Interestingly, because of the expectation that housework will be divided more fairly and equally, partners in same-sex couples may find themselves more distressed by less-than-equal arrangements. Should they compare their arrangements to heterosexual couples they know, they may be more satisfied. When they compare their situations to other same-sex couples or to the ideal of equality, they may find themselves less content (Goldberg 2013).

Among heterosexual couples, a husband or partner's lack of involvement can create resentment and affect the levels of both conflict and happiness in a marriage. Longitudinal research on married couples reveals that husbands whose wives perceived that the division of housework was unfair report higher levels of marital conflict over time (Faulkner, Davey, and Davey 2005). Similarly, in her acclaimed study of the division of housework among 50 dual-earner couples, Arlie Hochschild (1989) argued that men's level of sharing "the second shift" (i.e., unpaid domestic work and child care) influenced the levels of marital happiness couples enjoyed and their relative risk of divorce. This held true whether couples were traditional or egalitarian in their views of marriage.

In a comparative study, sociologist Leah Ruppaner analyzed housework conflict cross-nationally by comparing 25 countries. Ruppaner showed that the level of reported housework conflict varied by country depending on the level of gender equality and full-time female labor force participation, thus suggesting that household decisions are affected by the wider social context. Men and women in countries with high levels of female labor force participation and high rates of gender egalitarianism reported the least housework conflict. Looking across households, both women and men reported less conflict over housework when men increased their housework hours. On the other hand, couples in which women did all the housework reported the greatest amount of conflict (Ruppaner 2010).

How much each spouse contributes to the household is only the more observable aspect of the "politics of housework." In addition, couples must reach agreements about standards, schedules, and management of housework. Conflicts about standards are struggles over *whose* standards will predominate: Who decides whether things are "clean enough"? Similarly, disputes about schedules reflect *whose* time is more valuable and which partner works around the other's sense of priorities. Finally, arguments about who bears responsibility for organizing, initiating, or overseeing housework tasks are also disputes about who will have to ask the other for help, carry more responsibility in his or her head, and risk refusal from an uncooperative partner.

Thus, housework conflicts have both practical and symbolic dimensions. Practically, there are things that somehow must get done for households to run smoothly and families to function efficiently. Couples must decide who shall do them and how and when they should be done. In this way, they are more similar to money conflicts (Dew, Britt, and Huston 2012). On a more symbolic level, disputes over housework may be experienced as conflicts about the level of commitment each spouse feels toward the marriage. Because marriage symbolizes the union of two people who share their lives, work together, consult each other, and take each other's feelings and needs into consideration, resisting housework or doing it only under duress may be seen as a less-than-equal commitment. Indeed, studying the division of responsibility for and performance of housework provides us "a window into power in couples" (Davis and Greenstein 2013).

Even in the absence of overt conflict over the allocation of tasks and time, one cannot assume that there is no conflict. It means only that the conflict is not openly expressed. Wives in more traditional marriages are more likely than wives in egalitarian relationships to avoid conflict over housework even if they are dissatisfied with their domestic arrangements. They may withdraw from discussions of the division of labor as a way of avoiding the issue. Because egalitarian couples may engage in more open discussion and conflict over housework responsibilities, such conflict gives them more opportunity to establish a solution (Kluwer, Heesink, and Van de Vliert 1997).

## When the Fighting Continues

The end of a relationship does not necessarily indicate the end of conflict between the former spouses or partners. This is especially so for couples who have children while together. For such couples, it will be difficult to avoid contact, and with contact there will be ample opportunities for continued conflict. In fact, the Personal Responsibility and Work Opportunity Reconciliation Act of 1996 mandates that children need regular contact with both parents, so long as such contact serves the good of the child (Yeager 2009). Even when a parent has a history of abusing the child(ren), after participating in mandated counseling, he or she may be awarded supervised visitation that the custodial parent is not to prevent.

Custody, visitation, and child support are all potential opportunities for conflict and confrontation, as disagreements about co-parenting are said to be major sources of postdivorce conflict. Sociologist Erica Owens Yeager notes that, out of a desire to see children retain meaningful parental contact with both parents, such well-intended policies as joint custody and shared parenting, along with generous visitation, all serve to maximize the opportunities to experience conflict.

In addition to these issues surrounding children and parenting, conflict over financial matters such as spousal or child support, and concern over the child's well-being while with one's former spouse may arise, as "the existence of a mutual child maintains a bond between two people who have gone to considerable trouble and likely expense to terminate their voluntary bond" (Yeager 2009, 680). Furthermore, even in the absence of children, former partners—whether or not formerly married—may carry lingering hostilities. What having children together does is ensure that those former spouses have occasions and issues conducive to conflict.

## ◼ Consequences of Conflict

Although conflict is a normal part of marriages and relationships, excessive conflict can have negative personal or relationship consequences. Among couples who engage in frequent conflict, spouses can suffer negative consequences to their physical and mental health and to their overall well-being (Choi and Marks 2008). Spouses in high-conflict marriages may engage in more behaviors that negatively affect their physical health, such as smoking or drinking alcohol, while at the same time lacking the benefit of social and emotional support that accompanies happier marriages. Marriages assessed as low-quality marriages expose spouses to higher risk of depression, which can have physical health consequences. It is even possible, health researchers suggest, that repeated exposure to marital conflict and tension can induce physiological effects that eventually affect one's physical health. All in all, marital conflict has effects on a host of outcomes related to individual mental and physical health, family health, and child well-being.

### Mental Health

There are links between experiencing marital conflict and suffering from depression, eating disorders, being physically and/or psychologically abusive of partners, and male alcohol problems (including excessive drinking, binge drinking, and alcoholism). There is less evidence connecting marital conflict to elevated levels of anxiety.

### Physical Health

Marital conflict is associated with poorer overall physical health, as well as certain specific illnesses. These include cancer, heart disease, and chronic pain. Cross-national research has shown that marital and relationship conflict can affect cardiovascular, endocrine, and immune system functioning. Conflict also can affect asthma symptomology, how one experiences physical pain, and the body's ability to heal wounds. Distress from such conflict also predicts an increased risk of early mortality (Wright and Loving 2011).

Psychologists Brittany Wright and Timothy Loving note that it isn't just the fact of conflict but the way the conflict is expressed that most matters. For example, husbands and wives in hostile relationships show an increase in blood pressure and vascular resistance (how much blood vessels resist the flow of blood throughout the vascular system, with greater resistance associated with problematic effects on health). Wright and Loving note as well that when spouses perceive that their partners are generally supportive, they tend to experience lower blood pressure and vascular resistance both before and during conflicts (Wright and Loving 2011). In similar fashion, the extent of hostility surrounding conflict affects hormone levels, and this effect tends to be stronger for wives than for husbands, especially when they engage in the wife demand–husband withdraw pattern of interaction. Married men's endocrine responses appear more affected by their own behavior than by the behavior of their wives. However, regardless of the style of expression, simply disagreeing with someone has effects on the body (Wright and Loving 2011).

## Familial and Child Well-Being

Marital conflict may disrupt the entire family, especially if the conflict is frequent, intense, and unresolved. Marital conflict has been shown to be connected to poorer parenting, problematic parent–child attachments, and greater frequency and intensity of parent–child or sibling–sibling conflict. Consequences for children can be particularly harmful when the conflict centers on issues about the children and child rearing. The most destructive form of marital conflict appears to be when couples engage in attacking and withdrawing (hostility and detachment). In addition, when marriage is characterized by the absence of or low levels of warmth, mutuality, and harmony between parents, along with the presence of high levels of competitiveness and conflict, children develop more externalizing and peer problems (Katz and Woodin 2002). When parental marriages lack relationship cohesiveness, are devoid of playfulness and fun, and yet have high degree of conflict, children miss out on the warmth, intimacy, and security that healthy families can provide (Katz and Woodin 2002).

Research reveals numerous problematic effects of marital conflict on children, including health problems, depression, anxiety, conduct problems, and low self-esteem. When marital conflict is frequent, intense, and child centered, it has especially negative consequences for children. Peer relations also suffer when children are exposed to early and prolonged high levels of parental conflict. This is especially severe when children have insecure parental attachment, and can be observed in children as young as 3 years old (Lindsey, Caldera, and Tankersley 2009).

How do children react to marital conflict? Research indicates that children are distressed by both verbal and physical conflict but are reassured by healthy conflict resolution. Witnessing threats, personal insults, verbal and nonverbal hostility, physical aggressiveness between parents or by parents toward objects (e.g., breaking or slamming things), and defensiveness all can give rise to "heightened negative emotionality" (Cummings, Goeke-Morey, and Papp 2003). When parental conflict leads one parent (or both) to withdraw as a means of dealing with the differences between them, children's distress is also worsened (Goeke-Morey, Cummings, and Papp 2007).

Conversely, when parents engage in calm discussion and display affection and continued support even while engaged in conflict, children react positively. Conflict resolution lessens the negative effects of parental conflict on children, especially when what children see is parents compromising with each other so as to resolve their differences (Goeke-Morey, Cummings, and Papp 2007). Parents' displays of support, including providing validation to one another and affection during conflict, may reassure children that the marital relationship remains strong and loving even though parents disagree (Cummings, Goeke-Morey, and Papp 2003). However, the absence or failure of resolution causes anger, sadness, and distress. A frequently posed question, one that we consider in Chapter 13, is whether the effects of conflict on children are worse than the effects of divorce.

Sociologist Erica Owens Yeager points out that research on children's reactions to parental conflict shows that parents do not typically downplay their conflict because of the presence of children, and that conflicts children witness are as negative if not more negative as conflicts the parents privately engage in. When children witness a high level of conflict, they tend to view conflict episodes in light of what they have witnessed in the past. Conflicts will then seem more severe through the eyes of such children (Yeager 2009). Yeager also notes that even if parents think that the relationship has been "repaired," children's perceptions may be otherwise (Yeager 2009).

Children react to parental conflict in a variety of ways, depending on how the parents handle themselves. Although children can be hurt by outward displays of anger and especially by witnessing violence, "healthy conflict management" may be beneficial for children to witness.

Corbis

## Can Conflict Be Beneficial?

As we noted earlier, conflict is a normal and predictable part of living with other people, especially given the intensity of emotions that exist within marriage. Conflict itself is not necessarily damaging; there may be benefits of conflict in which spouses' "conflict engagement" (especially that of husbands) predicts positive change in husbands' and wives' satisfaction with marriage. It appears as though some negative behavior—such as conflict—may be both healthy and necessary for long-term marital well-being. Like too much conflict, too little conflict (suggestive of avoidance) may lead to poorer outcomes. However, the outcome of conflict varies, along with the meaning and function of conflict behavior. It can as easily reflect engagement with a problem as it can suggest withdrawal from the problem (Christensen and Pasch 1993). Furthermore, it may be part of an effort to maintain the relationship or conversely indicate that one or both partners have given up on the relationship (Holmes and Murray 1996). Thus, as Frank Fincham and Steven Beach (1999, 54) suggest, "We have to identify the circumstances in which conflict behaviors are likely to result in enhancement rather than deterioration of marital relationships."

# Resolving Conflicts

There are a number of ways to end conflicts and solve problems. You can give in, but unless you believe that the conflict ended fairly, you are likely to feel resentful. You can try to impose your will through the use of power, force, or the threat of force, but using power to end conflict leaves your partner with the bitter taste of injustice. Less productive conflict resolution strategies include *coercion* (threats, blame, and sarcasm), *manipulation* (attempting to make your partner feel guilty), and *avoidance* (Regan 2003).

More positive strategies for resolving conflict include *supporting your partner* (through active listening, compromise, or agreement), *assertion* (clearly stating your position and keeping the conversation on topic), and *reason* (the use of rational argument and the consideration of alternatives) (Regan 2003). Finally, you can end the conflict through negotiation. In negotiation, both partners sit down and work out their differences until they come to a mutually acceptable agreement. Conflicts can be solved through negotiation in three primary ways: agreement as a gift, bargaining, and coexistence.

## Agreement as a Gift

If you and your partner disagree on an issue, you can freely agree with your partner as a gift. If you want to go to the Caribbean for a vacation and your partner wants to go backpacking in Alaska, you can freely agree to go to Alaska. An agreement as a gift is different from giving in. When you give in, you do something you don't want to do. When you agree without coercion or threats, the agreement is a gift of love, given freely without resentment. As in all exchanges of gifts, there will be reciprocation. Your partner will be more likely to give you a gift of agreement. This gift of agreement is based on referent power, discussed earlier.

## Bargaining

Bargaining in relationships—the process of making compromises—is different from bargaining in the marketplace or in politics. In relationships, you want what is best for the relationship, the most equitable deal for both you and your partner, not just the best deal for yourself. During the bargaining process, you need to trust your partner to do the same. In a marriage, both partners need to win. The result of conflict in a marriage should be to solidify the relationship, not to make one partner the winner and the other the loser. Achieving your end by exercising coercive power or withholding love, affection, or sex is a destructive form of bargaining. If you get what you want, how will that affect your partner and the relationship? Will your partner feel that you are being unfair and become resentful? A solution has to be fair to both, or it won't enhance the relationship.

## Coexistence

Although unresolved conflict may, over time, wear away at marital quality, sometimes differences simply cannot be resolved. In such instances, they may need to be lived with. If a relationship is sound, often differences can be absorbed without undermining the basic ties. All too often, we regard a difference as a threat rather than as the unique expression of two personalities. Rather than being driven mad by the cap left off the toothpaste, perhaps we can learn to live with it.

## Forgiveness

Related to the issues of conflict and its resolution is the topic of *forgiveness*. Conceptualized as a reduction in negative feelings and an increase in positive feelings toward a "transgressor" after a transgression, an attitude of goodwill toward someone who has done us harm, and showing compassion and forgoing resentment toward someone who has caused us pain, research has determined that forgiveness has long-term physical and mental health benefits for the person forgiving. Forgiveness is associated with enhanced self-esteem, positive feelings toward the transgressor, and reduced levels of negative emotions, such as anger, grief, revenge, and depression. In a relationship context, forgiveness has been defined as "the tendency to forgive partner transgressions over time and across situations" (Fincham and Beach 2002).

Forgiveness has been found to be a crucial element of married life. It is an important aspect of efforts to restore trust and relationship harmony after a transgression. Most "forgiveness narratives" mention motivations such as a partner's well-being, restoration of the relationship, and love (Fincham and Beach 2002). Forgiveness has been shown to resolve existing difficulties and prevent future ones. It also enhances marital quality, as can be seen in the positive association between forgiveness and marital satisfaction and longevity (Kachadourian, Fincham, and Davila 2004).

Research has identified both personal and relationship qualities associated with the ability or tendency to forgive. Qualities such as agreeableness, religiosity, humility, emotional stability, and empathy are associated with forgiveness. Pride and narcissism are associated with decreased tendencies to forgive. Individuals who are more accommodating within their relationships, are more securely attached, and have more positive models of self and others are also more likely to be forgiving toward partners who have committed transgressions.

Not all relationship transgressions are equivalent. The ability to forgive relatively minor transgressions doesn't automatically guarantee forgiveness of more major transgressions. In heterosexual couples, wives who display tendencies to forgive seem able to do so in both minor and major transgressions. For husbands, on the other hand, tendencies to forgive apply more to major transgressions. It appears as though men may not consider minor transgressions important enough to warrant either receiving apologies or granting forgiveness (Kachadourian, Fincham, and Davila 2004).

## Helping Yourself by Getting Help

Despite good intentions and communication skills, we may not be able to resolve our relationship problems on our own. Accepting the need for professional assistance may be a significant first step toward reconciliation and change. Experts advise counseling when communication is hostile, conflict goes unresolved, individuals cannot resolve their differences, and/or a partner is thinking about leaving.

Marriage and partners counseling are professional services whose purpose is to assist individuals, couples, and families in gaining insight into their motivations and actions within the context of a relationship while providing tools and support to make positive changes. Skilled counselors offer objective, expert, and discreet help. Much of what counselors do is crisis or intervention oriented.

It may be more valuable and perhaps more effective to take a preventive approach and explore dynamics and behaviors before they cause more significant problems. This may occur at any point in relationships: during the engagement, before an anticipated pregnancy, or at the departure of a last child.

Each state has its own degree and qualifications for marriage counselors. The American Association for Marriage and Family Therapy is one association that provides proof of education and special training in marriage and family therapy. Graduate education from an accredited program in social work, psychology, psychiatry, or human development, coupled with a license in that field, ensures that the clinician has received necessary education and training.

It may be necessary to seek outside assistance to resolve conflicts effectively and preserve one's relationship.

Mangostock/Shutterstock.com

What should one do when one's marital conflicts remain unresolved or communication problems seem unavoidable no matter what one is discussing with one's spouse? At what point does one begin to consider the possibility that things just might not work out? Additionally, at what point, if ever, does one's spouse reach the same conclusion? These are the kinds of questions that plague troubled relationships. Typically, they may be what leads couples to therapy and couples counseling. A relatively new, innovative approach in the field of marital counseling and couples therapy is an approach known as "discernment counseling." Developed by long-time marriage and family therapist, William Doherty, founder of The Doherty Relationship Institute, discernment counseling is designed to assist those "mixed-agenda couples," couples in which one spouse is "leaning out" while the other is still "leaning in" and hoping that the marriage can be saved. According to Doherty's estimate, this is the situation for perhaps 30 percent of couples in marriage counseling.

Discernment counseling is a method of assisting troubled couples in making the decision of whether to stay in and work on their marriages or to divorce. It is designed to help the "leaning out spouse" decide whether to leave the marriage, while helping the other spouse cope without "pleading, threatening, or otherwise turning off the already irritated spouse" (Bernstein 2012). As journalist Diane Mapes quotes Doherty, the aim is to help couples decide whether "to improve the marriage or let it go" (Mapes 2012).

Discernment counseling is characterized by three features that differentiate it from more traditional marriage counseling:

1. The goal is not to solve the couple's problems but help them to figure out whether problems can be solved.

2. The process involves mainly individual conversations with each spouse.
3. It is always short term, consisting of between one and five hour-and-a-half to two-hour sessions. (discernmentcounseling.com/)

Sessions are described as follows: "Both spouses come in and there's a check-in (with a counselor) then you meet for part of the session with one (spouse) and part with the other." This is followed by "a check-out where you meet with both and summarize what each is taking out with them."

Doherty reports that up to 40 percent of divorced people have regrets about their divorce, often because they feel that they and their spouse didn't try hard enough to save the marriage. He offers, "It's almost always a good idea to slow the process down and look at the marriage from five different angles, including what your role in it was. . . . If people end a marriage without looking at their own contributions to the problems, they are leaving with a big blind spot" (Mapes 2012). The process is considered successful "when people have clarity and confidence in their decision and more fully understand what's happened to their relationship" (discernmentcounseling.com).

Of those who go through discernment counseling, about half choose to attempt a reconciliation and enter couples therapy. Most of the rest, though they may proceed directly to divorce, do so only after having carefully considered the options.

"You may end up with a realistic plan to restore your marriage to health. Or you may end up with a decision to divorce that you will be less likely to regret in years to come, and with learning about yourself that you can carry into new relationships" (discernmentcounseling.com /leaning-out-spouse/). ●

---

It also offers consumers recourse if questionable or unethical practices occur. However, this recourse is available only if the practitioner holds a valid license issued by the state in which he or she practices. Mental health workers belong to any one of several professions:

- *Psychiatrists* are licensed medical doctors who, in addition to completing at least six years of postbaccalaureate medical and psychological training, can prescribe medication.

- *Clinical psychologists* have usually completed a PhD, which requires at least six years of postbaccalaureate course work. A license requires additional training and the passing of state boards.

- *Marriage and family counselors* typically have a master's degree and additional training to be eligible for state board exams.

- *Social workers* have a master's degree requiring at least two years of graduate study plus additional training to be eligible for state board exams.

- *Pastoral counselors* are clergy who have special training in addition to their religious studies.

Financial considerations may be one consideration when selecting which one of the preceding to see. Typically, the more training a professional has, the more he or she will charge for services.

A therapist can be found through a referral from a physician, school counselor, family, friend, clergy, or the state department of mental health. In any case, it is important to meet personally with the counselor to decide if he or she is right for you. Besides inquiring about his or her basic professional qualifications, it is important to feel comfortable with this person, to decide whether your values and belief systems are compatible, and to assess his or her psychological orientation. Shopping for the right counselor may be as important a decision as deciding to enter counseling in the first place.

Marriage or partnership counseling has a variety of approaches: Individual counseling focuses on one partner at a time, joint marital counseling involves both people in the relationship, and family systems therapy includes as many family members as possible. Regardless of the approach, all share the premise that, to be effective, those involved should be willing to cooperate. Additional logistical questions, such as the number and frequency of sessions, depend on the type of therapy.

At any time during the therapeutic process, one has the right to stop or change therapists. Before doing so, however, one should ask oneself whether his or her discomfort is personal or has to do with the techniques or personality of the therapist. This should be discussed with the therapist before making a change. Finally, if one believes that therapy is not benefiting him or her, a change in therapists seems necessary.

If we fail to communicate, we are likely to turn our relationships into empty facades, with each person acting a role rather than revealing his or her deepest self. But communication is learned behavior. If we have learned *how not to* communicate, we can learn *how to* communicate. Communication will allow us to maintain and expand ourselves and our relationships.

## Summary

- Communication includes both verbal and nonverbal communication. For the meaning of communication to be clear, verbal and nonverbal messages must agree.

- Much nonverbal communication, such as levels of touching, varies across cultures and between women and men. The genders also have been found to differ in verbal communication.

- In marital communication, wives send clearer, less ambiguous messages; send more positive, more negative, and fewer neutral messages; and take more active roles in arguments than husbands do.

- *Demand–withdraw communication* is common among couples. One partner, more often the woman, will raise an issue for discussion, and the other partner, more likely the man, will withdraw from the conversation instead of attempting to communicate.

- Demand–withdraw patterns may be a reflection of the relative power of each partner, of gender socialization, or even of biological differences between women and men in their reactions to conflict.

- Rather than gender, which partner is seeking change in some aspect of the relationship and which partner withdraws varies depending on whose issue is being discussed.

- *Power* is the ability or potential ability to influence another person or group. There are six types of marital power: coercive, reward, expert, legitimate, referent, and informational.

- There are a variety of explanations for relationship power. One prominent idea, the *principle of least interest,* is that the person who has the least invested in the relationship is, as a result, more powerful.

- Resource-based theories of power fail to account for why women don't gain power as much as men do, even when they bring in more of the financial resources that couples require. Theories that focus on decision making as indicators of power may miss more covert expressions of power.

- Feminist insight into relationship power raises awareness of the gender dynamics that contribute to the distribution of power. Even self-described equal (or egalitarian) couples often still reveal power differences and inequalities that more often favor men.

- Conflict is natural in intimate relationships. *Basic conflicts* challenge fundamental rules; *nonbasic conflicts* do not threaten basic assumptions and may be negotiable.

- Conflict is natural and common among all types of couples. Major sources of conflict include sex, money, and housework. Conflict can have effects on the mental and physical health of spouses or partners, the health of the relationship, and the well-being of children.

- *Forgiveness* is an important part of efforts to restore trust and rebuild relationship harmony. It is positively associated with both relationship satisfaction and stability (i.e., longevity).

## Key Terms

contempt 243
demand–withdraw communication 249
gatekeeping 256
honeymoon effect 248
intimate zone 244
marital paradigm 262
nonverbal communication 241
personal space 244
power 254
principle of least interest 256
proximity 244
Relative love and need theory 256
resource theory of power 257
self-silencing 259

# Marriages in Societal and Individual Perspective

# 8

Inti St. Clair/Blend Images/Photolibrary

**What Do You Think?**

*Are the following statements True or False? You may be surprised by the answers as you read through this chapter.*

T  F  **1.** Trends in cohabitation and divorce clearly indicate a decrease in the importance of marriage in the United States.

T  F  **2.** Most Americans now agree with the view that marriage is an outdated institution.

T  F  **3.** Couples who frequently attend religious services together have a lower risk of domestic violence, infidelity, and divorce.

T  F  **4.** Once same-gender couples marry, their marriage must be recognized as legal by all 50 U.S. states.

T  F  **5.** Interracial marriages between African American men and Caucasian women, and between African American women and Caucasian men are at greater risk of divorce than are marriages between Caucasian men and women.

T  F  **6.** Married people report greater happiness and better health than unmarried people, but only if their marriages are happy.

T  F  **7.** Couples who are unhappy before marriage significantly increase their happiness after marriage.

T  F  **8.** Married people are less likely to socialize with friends and neighbors than are never-married or previously married women and men.

T  F  **9.** Marital relationships appear to be more affected by minor stresses than by major stresses such as unemployment or serious illness.

T  F  **10.** After losing a spouse, women are more likely than men are to remarry.

## Chapter Outline

… to have and to hold from this day on, for better or for worse, for richer, for poorer, in sickness and in health, to love and to cherish; until death do us part.

As you probably realize, those words are a traditional version of wedding vows that, in some similar form or fashion, are exchanged between two people as they enter marriage. Some may add more religious language, some may be less traditional, and some may be longer or more personal, as couples write their own versions of vows. It is likely, however, that most will convey an intention to share life's ups and downs, to be as one *together*, and to so commit for as long as both people live.

When two people exchange wedding rings and vows, they make a public declaration, in a ceremony overseen by a legally recognized officiate, usually a clergyperson or justice of the peace. Once they do, each newly married couple embarks on a journey that is simultaneously intensely personal, inarguably public, and in recent years, highly politicized. Our goals in this chapter are to examine marriage in all three ways—as a relationship between spouses, as a commitment certified by the state and celebrated by one's wider circle of friends and kin, and as a legal relationship undergoing dramatic changes and challenges.

The chapter begins by considering the current status and direction of marriage in the United States today. Although most Americans will marry at some point, fewer enter and stay in marriage today than did in the recent past. Is marriage less valued than it was in the past? As a society, are we less committed to marriage as a central life goal, and are those who marry less willing or able to work hard to make their marriages work? Will more and more people decide to forego marriage, deeming it less important for achieving meaningful life goals or is marriage more likely to remain central among the most highly valued accomplishments, the "gold standard" in the universe of intimate relationships (Amato, Booth, and Johnson

2007)? In describing the ambiguous status of marriage in the United States, we will look at the role of socioeconomic status and race. Is marriage increasingly limited to certain populations more than others?

We then shift to the experiences associated with getting and being married. We describe how people choose their spouses and who they tend to choose and examine issues that confront couples as they enter marriage and attempt to share a lifetime together. Along the way, we identify some factors that predict marital success, discuss marital roles and boundaries, and look at how having and raising children affects married couples. Finally, we turn to middle-aged and later-life marriages and the end of marriage with the death of one's spouse, and we survey the different patterns and factors that characterize lasting marriages.

## ■ Marriage in American Society

Marriage has long been the foundation on which American families are constructed. Although we have always recognized and valued our ties connecting us with our wider families, marriage has been the centerpiece of family life in the United States. As we saw in Chapter 1, within the United States, the relationship with our spouse in our nuclear family system tends to be more important than our relationships with our extended families. The person we marry is expected to be someone with whom we will share everything, a soul mate and partner "for as long as we both shall live."

As central as marriage has been to our family system, it can be difficult and confusing to reach conclusions about either the current status or the future direction of marriage in the United States. Even the "marriage experts" cannot always agree about whether marriage is or isn't "endangered," or whether it has retained or lost its appeal and its meaning as a major life goal to which people aspire. Writing a decade ago, sociologist Paul Amato (2004b) referred to this situation as the **marriage debate**, noting that, despite their expertise, even respected scholars could not agree; whereas some portrayed marriage as weaker and "in decline," others portrayed it as dynamic, changing, and resilient. How is this possible? Why couldn't they agree?

## Behavior Trends

Let's consider some of the indicators of what is a mixed portrait of marriage in the United States today, looking first at a number of trends in behavior, before turning to survey data on attitudes and values:

- Looking at the current marital status of the U.S. population shows that slightly more than half, approximately 53.1 percent of U.S. adults *18 and older*, are currently married. Another 18.8 percent are formerly married, being either widowed (6.0 percent), separated (2.2 percent), or divorced (10.6 percent). Although separation and divorce are certainly not the best news about marriage, these data show that, cumulatively, nearly 73 percent of adults 18 and older are currently or have been married. (As we proceed, some data are presented for those 18 and older, other data for those 15 and older.)

- In numerical terms, an estimated 127 million Americans 18 years and older are currently married (U.S. Census Bureau, America's Families and Living Arrangements, 2014, Table A1). It is projected that between 75 to 80 percent of never-married Americans will have married by the time they reach their 50s (Wang and Parker 2014). This is a higher percentage than what we observe in other Western societies (Cherlin 2009).

- Looking specifically at young adults (those 18 to 29), researchers Wendy Wang and Kim Parker report that a record high share of young adults is likely never to marry. They estimate that around 25 percent of never-married young adults will still be unmarried by the time they reach their mid-40s to mid-50s (ages by which the vast majority of those who will ever marry have married) (Wang and Parker 2014).

- The median ages at first marriage for U.S. men and women are at their highest recorded point. In 2013 the median age at which men entered their first marriage was 29.0; among women, the median age at first marriage was 26.6.

**False** **1.** Trends in cohabitation and divorce do not indicate a decrease in the importance of marriage in the United States.

- The marriage rate in the United States is at the lowest point in over a century's worth of data. Tracking the marriage rate from 1890 to the present shows earlier periods of ups and downs, yet since 1970 the rate declined steadily, reaching a low of between 31 and 32 per 1,000 women *15 and older* in 2010 and remaining there between 2010 and 2014.

- With a 2012 divorce rate of 17.3 per 1,000 women 18 and older in first marriages, and with 13.4 percent of Americans 15 and older currently divorced, marriage appears more fragile in the United States than in other Western societies (Stykes, Payne, and Gibbs 2014; www.nationmaster.com/country-info/stats/People/Divorce-rate ). More than 40 percent of new marriages are projected to end in divorce, which pessimistically might be taken to indicate a lack of commitment in marriage (Gibbs and Payne 2011).

- Although many divorced women and men eventually do reenter marriage, the rate of remarriage after divorce has declined. In the 1990s, most divorced women (69 percent) and men (78 percent) remarried, and the remarriage rate was 43 per 1,000 formerly married women and men. One could interpret these statistics to mean that marriage remained highly valued, even if one's own marriage or marriage partner was deemed to have failed to meet one's expectations. Since the 1990s, the remarriage rate, now at 28 per 1,000 previously married people, has declined, dropping more than 40 percent since 1990 and 16 percent just since 2008 (Payne 2015).

- As reported by the Pew Research Center, whereas less than one in ten Americans over age 25 had never married in 1960, in 2012, 23 percent of men and 17 percent of women ages 25 and older had never been married.

- Along with singlehood, cohabitation and births to unmarried mothers (either single or cohabiting) also increased across the last three decades of the 20th century and first decade of this century. Together, one might ask whether these trends mean that marriage has become less attractive, less valued, and is perceived as less essential, even as a prerequisite for having and raising children.

In thinking about what these indicators might suggest, it is worth pointing out the following. As noted, marriage rates have fluctuated before, often dipping and rising at different points throughout the 20th century. For example, in 1920, the U.S. marriage rate was at its highest, at 92 per 1,000 unmarried women 15 and older. After having dropped to slightly under 68 (67.8) just a decade later in 1930, the rate climbed back to 90 in 1950. By 1970 the rate was down to 76 per 1,000 unmarried women over 15. Since 1970, as noted previously, it has dropped to the current rate of 31 to 32 per 1,000 unmarried women 15 and older. Given the extent and length of time associated with the most recent decline, it is unlikely that it will return to anywhere near those previously reached rates.

It is also worth noting that though the current declines in getting, being, and staying married cuts across racial and ethnic lines, they have not been equally steep for all groups. The decline in marrying has been greatest among blacks. Larger proportions of Asian, white, and Hispanic women than black women are presently married (Lamidi 2015).

## Attitudes about Marriage

At present, we can see a mixed picture of how valued and important marriage is. Marriage remains highly valued, even with increased acceptance of divorce and nonmarital lifestyles. For example, June 2013 survey data collected by the Gallup organization, found that across different ages and races, majorities of those sampled were either married or had never married but want to someday marry. Notably, *among 18- to 34-year-olds,* the age range that includes those most likely to support unmarried lifestyles, 28 percent were married and another 56 percent had yet to marry but reported wanting to someday marry. Among 35- to 54-year-olds, 65 percent were married, and 12 percent more were never married but wanted to marry. Among the older respondents (55 and older) 64 percent were married and 2 percent had never married but hoped or wanted to someday marry (Newport and Wilke 2013).

Among 18- to 34-year-old non-Hispanic whites, 87 percent had either married (34 percent) or said they wanted to someday marry (53 percent). Among nonwhites, 81 percent were either already married (20 percent) or had not married but wanted to eventually marry. Similar patterns were found across income levels and educational differences. In fact, the percentage of people who had not married and don't want

to marry tended to be quite low, across different ages, races, income levels, and education, usually less than 10 percent (Newport and Wilke 2013).

**False** **2.** Most Americans do not agree with the view that marriage is an outdated institution.

Similarly, data reported by the Pew Research Center found that only 13 percent of never-married American adults reported that they don't ever want to marry. In the Pew survey, more than half (53 percent) of those who had never married said they wanted to marry eventually. However, nearly a third said they were unsure.

Past data, drawn from the World Values Surveys of respondents from more than 60 countries, found that fewer Americans (10 percent) agreed with a statement suggesting that marriage was an "outdated institution" as compared with adults in the other Western countries (Cherlin 2009). Compared to countries such as Canada, the United Kingdom, France, Germany, Italy, and Sweden, the U.S. data expressed stronger support for marriage in the United States (see Figure 8.1). By 2010, however, when asked in a Pew Research Center survey whether they agreed that marriage as an institution was becoming obsolete, almost 40 percent of those surveyed agreed (see Figure 8.2). In the same survey, nearly half (47 percent) of never-married respondents who agreed that marriage was becoming obsolete also expressed a desire to someday marry (Cohn et al. 2011).

Although we don't have more recent comparable data on the exact same question, it does appear, in other ways, that perhaps Americans' attitudes about marriage have become somewhat more mixed. For example, in 2013, when Gallup asked how important it was for two people to marry if they want to spend the rest of their lives together or if they have a child together, almost two-thirds (64 percent) felt it was somewhat or very important for both questions. Just seven years earlier (2006), more than three-fourths (76 percent) felt it important to marry if one was raising a child together and nearly three-fourths (73 percent) felt it important if a couple planned to spend their lives together. Even among those who weren't married but want to marry, in 2013, 57 percent thought it at least somewhat or very important for a couple who want to spend their lives together to get married, and 55 percent thought that for a couple raising a child (Newport and Wilke 2013).

According to Harris poll data, 70 percent of women and men in the United States think marriage is important to Americans in general, and 75 percent

## Figure 8.1  Belief that Marriage Is an Outdated Institution, International Comparison

**Marriage Is an Outdated Institution**

Italy
Spain
United States
Canada
Norway
Germany
Sweden

NOTE: Selected countries/samples: Canada (2006), France (2006), Germany (2006), Great Britain (2006), Italy (2005), Norway (2007), Spain (2007), Sweden (2006), United States (2006)
SOURCE: World Values Survey, www.worldvaluessurvey.org.

### Figure 8.2  Percent of U.S. Population that Believes Marriage Is Obsolete

**Is Marriage Becoming Obsolete?**

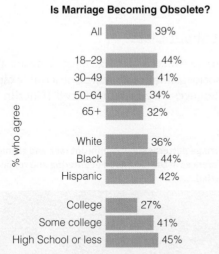

% who agree

| | |
|---|---|
| All | 39% |
| 18–29 | 44% |
| 30–49 | 41% |
| 50–64 | 34% |
| 65+ | 32% |
| White | 36% |
| Black | 44% |
| Hispanic | 42% |
| College | 27% |
| Some college | 41% |
| High School or less | 45% |

NOTE: Whites and blacks include only non-Hispanics. Hispanics are of any race.
SOURCE: Pew Research Center (2010).

(72 percent of men and 79 percent of women) say marriage is important or very important to them personally. At the same time, almost three out of four surveyed (67 percent of men and 76 percent of women) in the Harris survey agreed that marriage was less important today than in the past. Asked whether they agreed with the statement that "Marriage is not necessary," 39 percent agreed (strongly or somewhat) while 61 percent disagreed (with 34 percent disagreeing strongly). Among the youngest respondents (i.e., "Millennials"), more than half agreed that marriage was not necessary. At the opposite generational end, among those 68 and older, less than a fifth (19 percent) agreed (www.theharrispoll.com/health-and-life/The_State_of_Marriage_in_the_U_S____It_is_Very_Important_to_Half_of_Americans.html).

In discussing marriage, to this clearly mixed assortment of behavioral and attitudinal data already noted, we should also consider the decade-plus long struggle for marriage equality. Whether one assumes that the push for marriage equality was more about a claim for *equality* than about a reverence for *marriage*, it does serve as a reminder of the special qualities assumed of marriage. In seeking and gaining the legal right to marry, the movement for same-sex marriage was a reminder of the status marriage holds, and of the rights, recognition, and respect that relationships gain with Justice Anthony Kennedy's explanation for the Supreme Court's *Obergefell vs. Hodges* decision. Declaring in his closing paragraph that "No union is more profound than marriage," Justice Kennedy affirmed the central and special tie that marriage creates and represents:

> (Marriage) embodies the highest ideals of love, fidelity, devotion, sacrifice, and family. In forming a marital union, two people become something greater than once they were. As some of the petitioners in these cases demonstrate, marriage embodies a love that may endure even past death. It would misunderstand these men and women to say they disrespect the idea of marriage. Their plea is that they do respect it, respect it so deeply that they seek to find its fulfillment for themselves. Their hope is not to be condemned to live in loneliness, excluded from one of civilization's oldest institutions. They ask for equal dignity in the eyes of the law. The Constitution grants them that right.

Given the variety of trends and attitudes acknowledged, what are we to conclude about the status and future of marriage? In the first decade of the 21st century, considering mostly demographic trends, some scholars contended that a retreat from marriage had occurred or was under way in the United States and would likely continue. In making that claim, they used and emphasized as evidence some of the kinds of indicators already noted—trends such as older ages at first marriage for both women and men, more people never marrying, significant increases in cohabitation and nonmarital births, and continued high divorce rates. In fact, some asserted that marriage had actually "been in retreat for more than a generation," as fewer men and women "ever marry," and that the "U.S. withdrawal from marriage" persisted at least into the 21st century. Some suggested that the retreat from marriage was associated with increases in employment of women, smaller gender wage gaps in earnings, and persistent economic inequality between racial groups (Schoen and Cheung 2006).

## The Economic and Demographic Aspects Discouraging Marriage

Closer inspection of marriage trends indicates that whatever "retreat from marriage" has occurred has not occurred equally among all social groups. Instead, racial, economic, and educational differences can be seen. As noted previously and shown earlier in Chapter 3, there are considerable differences in marital status for different racial, ethnic, and economic groups. Looking again using 2014 census data reveals the differences as detailed in Figure 8.3.

Whereas majorities of Caucasians and Asian Americans are married, less than half of Hispanic Americans are married, and less than 30 percent of African Americans are married (Lamidi and Payne). Adding the widowed, separated, and divorced to the portion married, 77 percent of whites, 74 percent of Asians, and 65 percent of Hispanics *are or have been* married compared to 57 percent of African Americans (see Figure 8.3). Some other racial differences to note: Although young people in general expect to marry someday, fewer young African Americans express an expectation to ever marry, and those who do report an older desired age at marriage than whites (Crissey 2005; Manning, Longmore, and Giordano, 2007). African Americans are more likely to divorce than Caucasians. Divorced African Americans are less likely than divorced Caucasians to remarry. Blacks are also much more likely to bear children outside of marriage. Although roughly 41 percent of all children born in the United States in 2010 were born to unmarried mothers, the racial differences were pronounced: Approximately 29 percent of all births to Caucasian women compared to more than 70 percent of births to African Americans, and more than half of births to Hispanic women were to unmarried mothers (www.childtrendsdatabank.org). Of births to Asian American women in 2009, 17 percent occurred outside of marriage (Hamilton, Martin, and Ventura 2010).

## What about Class?

Within the shifts in marriage rates, there are also notable socioeconomic differences, observable especially in differences by educational level (Cherlin 2014).

Marriage patterns show significant race and ethnic differences in the likelihood of entering and remaining married.

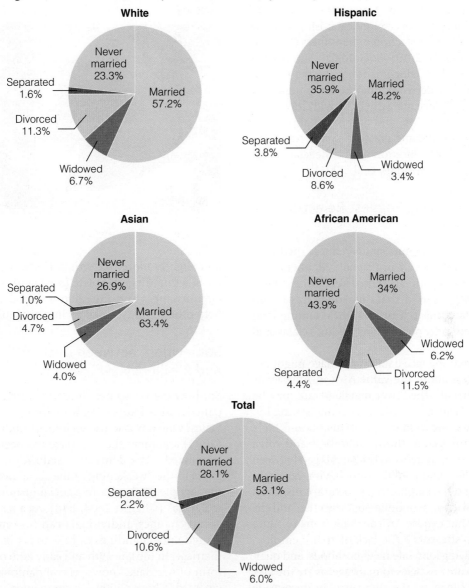

Figure 8.3 Marital Status, U.S. Population 18 and Older, by Ethnicity

**White**
- Never married 23.3%
- Married 57.2%
- Separated 1.6%
- Divorced 11.3%
- Widowed 6.7%

**Hispanic**
- Never married 35.9%
- Married 48.2%
- Separated 3.8%
- Divorced 8.6%
- Widowed 3.4%

**Asian**
- Never married 26.9%
- Married 63.4%
- Separated 1.0%
- Divorced 4.7%
- Widowed 4.0%

**African American**
- Never married 43.9%
- Married 34%
- Widowed 6.2%
- Separated 4.4%
- Divorced 11.5%

**Total**
- Never married 28.1%
- Married 53.1%
- Separated 2.2%
- Divorced 10.6%
- Widowed 6.0%

SOURCE: U.S. Census, America's Families and Living Arrangements: 2014, Table A1.

For example, although lifetime marriage rates among women have dropped by 5 percent in the United States, they have declined by 30 percent for women without a high school diploma (Gibson-Davis, Edin, and McLanahan 2005). Among college-educated white women, the prospect of marrying has *grown greater*, whereas it has decreased among those without college degrees (Huston and Melz 2004).

For both women and men, educational attainment, particularly obtaining a college degree is positively associated with the likelihood of marriage (Cherlin 2014; Schoen and Cheng 2006). In addition, in the 1980s and 1990s, marriages among college-educated women became *more* stable—that is, less likely to end in divorce—than they had been in the previous decade; among women at the bottom of the educational distribution, marriage became less stable (Edin, Kefalas, and Reed 2004). In discussions of a retreat from marriage among Hispanics, Oropesa and Landale (2004) emphasized how limited economic opportunities may be major barriers or disincentives to marriage.

Once married, socioeconomic factors influence marital quality and stability. A number of studies detail the harmful impact of economic hardship and low income on marital satisfaction or happiness (Conger, Conger, and Martin 2010). Sociologist Paul Amato and colleagues (2007) document how higher rates of marital problems, decreased marital happiness, and greater risk of divorce are associated with lower levels of income, education, and occupational prestige. Meanwhile, higher income, greater financial resources, and higher levels of education have beneficial consequences for marriage (Amato et al. 2007).

## Does Not Marrying Suggest Rejection of Marriage?

Even if low socioeconomic status reduces the likelihood of marriage, it may not signal a deliberate rejection of marriage. In fact, Edin, Kefalas, and Reed (2004, 1,008) asserted that "marriage has by no means lost its status as a cultural ideal among low-income and minority populations" (Edin, Kefalas, and Reed 2004).

Despite what the race data on marriage might appear to suggest about the value African Americans place on marriage, the lower marriage rate may be misleading. African Americans remain "strong believers in the value of marriage" (Huston and Melz 2004). Psychologist Anthony Chambers and Aliza Kravitz argue that, in trying to bridge this gap between supposed values and patterns of behavior, researchers have most often emphasized structural constraints that limit social and economic opportunities, and circumstances that expose African Americans to more "relationship stressors." The lack of secure employment, well-paying jobs, safe neighborhoods, and quality education are obstacles to marriage, as are the lack of "marriageable men," the high rate of births outside of marriage (affecting the "marriageability" of many women), and some of the interpersonal dynamics within relationships (e.g., gendered power struggles, difficulties co-parenting, and betrayal and trust issues). Among many African American heterosexual couples, the women have higher earnings and more education than do the men, increasing the probability of conflict over the division of housework or the roles each should play in decision making. Taken together, these produce "a unique set of contextual challenges" that help explain why the African American marriage rate is so low (Chambers and Kravitz 2011). Although blacks are somewhat more likely than other whites

African Americans are less likely than other groups to marry.

Ariel Skelley/Blend Images/Getty Images

and slightly more likely than Hispanics to say they consider marriage obsolete (see Figure 8.2), external circumstances and the relationship dynamics they help produce must be taken into account.

## Somewhere between Decline and Resiliency

So, how *are* we to best understand what has happened and is happening to marriage in the United States? One way is to use sociologist Andrew Cherlin's (2004) argument that marriage has been "deinstitutionalized." The deinstitutionalization of marriage refers to the "weakening of the social norms that define people's behavior in a social institution such as marriage" (Cherlin 2004, 848). As a result of wider social changes, individuals can no longer rely on shared understandings of how to act in and toward marriage. In the late 19th and early 20th centuries, the form of marriage known as companionate marriage emerged (Cherlin 2009). In companionate marriages, spouses were expected to supply each other with companionship, friendship, romantic love, and mutually gratifying sexual intimacy. Held together by love and friendship between spouses rather than social obligations, characterized by egalitarian as opposed to the earlier patriarchal ideals for marriage, and encouraging spouses to focus on self-development and expression, the companionate marriage was the widely shared cultural ideal by the 1950s.

Beginning in the 1960s and accelerating in the 1970s, the companionate marriage began to lose ground to a form of marriage Cherlin (2009) calls the **individualized marriage**. Partly the product of "cultural upheavals of the 1960s and 1970s," individualized marriages emphasize personal fulfillment and personal growth in marriage and expect that our spouses will facilitate such growth and be sources of unprecedented support (Amato 2004a). In an individualized marriage, emphasis is placed on self-development, flexible and negotiable roles, and openness and communication in problem solving. In this newer form of marriage, "spouses are free to grow and change. . . . [W]hat matters most is not merely the things they jointly produce—well-adjusted children, nice homes—but also each person's own happiness" (Cherlin 2009, 90).

This is where the marriage debate centers. Some scholars see the changes and trends described here as worrisome because they undermine marriage as an institution that meets the needs of society. They believe that we have become too individualistic and too focused on personal happiness and have less commitment to making our marriages work. Such attitudes help explain the increases in cohabitation, single parenthood, and divorce, as individuals pursue what they most want regardless of their effects on others. To proponents of this viewpoint, we need to enact policies to reinstitutionalize marriage, to restrict and decrease divorce, and to strengthen values such as marital commitment, obligation, and sacrifice.

Others put more emphasis on marriage as a relationship between two individuals and stress the value of such characteristics of contemporary marriage as self-development, freedom, and equality between spouses. Rejecting the idea that we have grown too individualistic or selfish, they also challenge the idea that ongoing trends should be seen with such negativity. Even the increase in divorce may be seen as an opportunity for happiness for adults and a means of escape for children from dysfunctional or dangerous environments.

As articulated by sociologist Paul Amato, neither the marital decline perspective (the belief that marriage is endangered) nor the marital resilience perspective (the belief that marriage is changing though still highly valued) is consistently or uniformly supported by the variety of available data on marriage. Along with David Johnson, Alan Booth, and Stacey Rogers, Amato compared two national surveys of married women and men in the United States, one from 1980 the other from 2000 (Amato et al. 2007).

As expected, given some trends we have already discussed, the demographics of marriage had changed considerably. Age at first marriage had increased, as had the proportion of remarried individuals and couples marrying after first cohabiting, the proportion of wives in the labor force, and the share of household income that married women contributed. Gender relations had changed in less traditional directions. Couples also became more religious and expressed greater support for the norm that marriage was for life.

Linking these sorts of changes to shifts in marital quality, data appear to partially support both the marital decline perspective and the marital resilience perspective. In other words, some changes in late 20th-century marriage led to declines in marital happiness and interaction, and were associated with increases in likelihood of divorce. Yet other changes were associated with improved marital quality, such as an improved economic standing of married couples, the adoption of less traditional gender roles, and an increase in the belief in marriage as a lifelong relationship (Amato et al. 2007). And the overall effect? Although the average level of marital interaction declined significantly (couples became less likely to eat dinner together, go shopping together, visit friends together, and share downtime together), as Amato expresses, "In general, these changes tended to offset one another, resulting in little net change in mean levels of happiness and divorce proneness in the U.S. population" (2004b, 101).

## Religion and Marriage

Part of the supposed retreat from marriage consists of the delayed age at which women and men first enter marriage. Along with race and social class, religious affiliation and religiosity (how religious one is) are among the factors that may influence whether and when people choose to marry. Religious differences have been identified in mate choice, childbearing, and child rearing; the division of housework; domestic violence; marital quality; and divorce (Xu, Hudspeth, and Bartkowski 2005).

Within marriage, shared religious participation is associated with higher marital quality such that couples who share core religious beliefs and values tend to report themselves as being more satisfied and engaging in less conflict (Ellison, Burdette, and Wilcox 2010).

**True** 3. Couples who frequently attend religious services together have a lower risk of domestic violence, infidelity, and divorce.

Couples in which spouses are of the same faith tend to have somewhat higher levels of happiness and satisfaction with their marriages, and couples who more frequently attend religious services together have a lower risk of divorce. Frequent attenders also report lower levels of domestic violence and infidelity, though the latter outcome is dependent on whether couples report themselves as satisfied with their marriages (Mahoney 2010). This illustrates that religious involvement might both promote or encourage positive behavior and discourage negative behavior. Involvement in religious communities may serve as a barrier to behaviors that would otherwise threaten the stability of marriages (Fincham and Beach 2010).

Frequent attendance at religious services may be considered an indication of both being highly committed to the norms and values of one's faith and of the importance of being part of a religious or spiritual community. In fact, across religious traditions, including Muslims, Jews, Christians, and Latter-day Saints, highly religious couples report that their religious communities help them maintain their families (Mahoney 2010). Frequent attenders may derive support from other church members, benefit from the moral messages communicated through sermons, gain from the exposure to positive role models within their faith community, and take advantage of such potential supports as pastoral counseling or church-sponsored marital enrichment programs (Ellison, Burdette, and Wilcox 2010).

However, religion's effect on marriages and families are neither automatically nor inevitably positive. Wesley Burr, Loren Marks, and Randal Day (2012) describe how religious or sacred matters can occasionally be harmful to individuals or families. Although they do contend that major religious traditions offer "ideas and answers" that are consistent with and helpful to achieving "the basic and enduring goals in family life," they also offer both of the following propositions as core parts of their analysis:

- When one experiences parts of family life as sacred, those parts take on "a unique, unusually powerful, and salient influence in families" (Burr, Marks, and Day 2012, 17).
- How people *act* as a result of their ideals and beliefs about the sacred will determine whether religious matters are helpful or harmful in families.

According to the theory Burr, Marks, and Day constructed, religious or sacred matters can be harmful to families in a number of ways. For example, if one considers marriage as sacred, the pain, guilt, or sense of failure that would follow marital distress or divorce will be greater. A second way in which religious beliefs or ideas can have harmful familial effects is when members take them too far (i.e., get carried away) or misunderstand religious teachings, and as a result behave in ways that are harmful to familial well-being (Burr, Marks, and Day 2012).

## Religion and the Importance of Marriage

Religious traditions and denominations differ in the kinds and degree of emphasis they place on marriage. Although Judeo-Christian religious groups tend to support marriage, uphold marriage and family as desirable and important lifestyles, and discourage both premarital and extramarital sex, there are differences among them, especially in the extent to which they support traditional gender roles and relationships and reject divorce, abortion, and homosexuality (Xu, Hudspeth, and Bartkowski 2005). Conservative Protestant denominations and Latter-day Saints (Mormons) articulate especially strong commitments to marriage, encouraging members to marry and stay married, by portraying marriage as "part of God's plan for self-development . . . in this life, as well as . . . long-term spiritual salvation" (Xu, Hudspeth, and Bartkowski 2005, 589–90).

Fundamentalists and Evangelical Protestants are more likely to advocate and support traditional family roles, emphasizing female submission in marriage and male headship of the household. Although other conservative religious traditions have somewhat different focal concerns (e.g., Catholicism's prohibition of birth control and abortion, and its emphasis on the centrality of the mother role for women), they share an emphasis on female domesticity and a separation of men's and women's roles into a public/private split (Bulanda 2011). However, in an important analysis, sociologist Jennifer Roebuck Bulanda points out that gender ideologies are not uniformly shared among the members of any one religion. For example, an estimated 40 percent of evangelical Protestants do not endorse the more traditional gender ideals characteristic of their religion. As Bulanda states, "rhetoric does not universally translate into belief" (2011, 181).

In addition, the link between belief and behavior is a tenuous one. Despite a stated belief in the idea of husbands as heads of households and wives as their submissive expressive partners, most conservative

In U.S. society, the expectation is that singles find their eventual life partner through the process of dating. Dating, or whatever else it might be called, allows us to test out our suitability for each other, develop stronger and closer relationships, fall in love, and select our life partners. Marriage without love goes against the culture of romantic love and these established patterns of mate selection. Although we might consider marriage without love an exceptional case, anthropologists tell us that most people in traditional cultures do not consider love the basis for their entry into marriage.

Marriage customs vary dramatically across cultures, and marriage means different things in different cultures. If we consider how marriages come about—how they are "arranged"—we find that it is usually not the bride and groom who have decided to marry, as is the case in our own society today. Typically, the parents and elders have done the matchmaking, sometimes relying on intermediaries and matchmakers to locate suitable spouses for their children.

These strategies are still practiced and are not entirely restricted to other countries. *New York Times* journalist Stephen Henderson tells the story of Rakhi Dhanoa and Ranjeet Purewal, and captures some of the motivation behind using others to arrange marriages: "Each wanted a love marriage . . . yet neither would dream of marrying someone who wasn't a Sikh." An immigration lawyer in New York whose parents emigrated from Punjab, India, 27-year-old Dhanoa decided that she wanted to marry someone of the same faith. "I began to appreciate that my religion is based on complete equality of the sexes," she said. At the same time, Purewal was beginning to think about finding a partner. His mother had approached Jasbir Hayre, a Sikh matchmaker, living nearby in New Jersey. So, when it came time to throw a party for her own daughter, Hayre invited both Dhanoa and Purewal. Although he had firmly believed in choosing for himself based on love, like Dhanoa, Purewal came to feel as though there were important issues to take into account. "I was adamant that I'd marry whoever I wanted. . . . But seeing how different cultures treated their families, I realized the importance of making the right match." After two months of mostly covert dating, "their cover was blown, on a double date, [and] the matchmaker was quickly summoned to negotiate marital arrangements" (Henderson 2002).

The story of Dhanoa and Purewal illustrates a variation of a phenomenon common in many parts of the world. In most cultures, marriage matches do not result from individuals meeting and dating; instead, the parents of the bride and groom arrange the marriage of their children. In some cultures, mothers are the primary matchmakers, as in traditional Iroquois culture. In others, fathers have a dominant voice in arranging marriage, as in traditional Chinese society. In still other cultures, the pool of elders involved in matchmaking is more extensive, including grandparents, aunts, uncles, and even local political and religious authorities, such as tribal chiefs and clan leaders. In all these instances, though, marriage is a major event in the life of two families—both the bride's and the groom's—as well as for the clan, tribe, and community to which each family belonged. As such, important matters must be taken into account before agreeing to any particular match. Families must know how a particular marriage affects the family as a whole. The feelings and love between an individual bride and groom are subordinate to the greater interests and welfare of the family, clan, and community. ●

Protestant men and women describe their actual family practices (e.g., decision making) as egalitarian. Despite the fact that most conservative religions stress that the woman is responsible for caring for home, husband, and children, their home lives in reality are often less gendered.

Also of particular interest is the concept of **marital sanctification**, a process through which one's marriage is believed to be sacred, have divine character, and in which God is believed to be an active partner in the relationship (Ellison, Burdette, and Wilcox 2010). This may, for some, suggest that God had a direct hand in bringing the spouses together and foster the view that one's marriage is truly a blessing from God. Sanctification is thought to strengthen a couple's commitment to their marriage, increase their willingness to put forth the effort needed to sustain and nourish their marriage, encourage them to put their spouse's needs ahead of their own, accept the differences between them and their spouse, and foster unconditional forgiveness in and after conflict. Marital sanctification may also strengthen a couple's resilience in dealing with the difficulties and stresses they are likely to face (Ellison, Burdette, and Wilcox 2010).

Indeed, based on interviews with a Texas-based sample of more than 1,200 married individuals, sociologists Christopher Ellison, Andrea Henderson, Norval Glenn, and Kristine Harkrider found that marital sanctification was an important predictor of marital quality, in terms of levels of happiness, commitment, bonding, and positive emotion, and further noted that these positive results were especially evident in couples facing financial strain and/or high levels of general stress (Ellison et al. 2011). Spouses who considered their marriage sacred reported more positive and fewer negative feelings toward their partners, spent and enjoyed more time together, and showed a greater degree of commitment to working on their relationships (Ellison et al. 2011).

Michael Goodman, David Dollahite, Loren Marks, and Emily Layton studied 184 religiously diverse couples from a variety of racial, socioeconomic, and geographic backgrounds to examine how religious faith affects couples' commitment to their marriages and contributes to their approaches to coping with life challenges. They found that for sample couples, their religious beliefs strengthened both their commitments to their relationships and helped them develop ways to more effectively cope with life challenges. They recommend further research to pursue the question of how generalizable their research findings are in the wider population, and whether secular beliefs work similarly for nonreligious couples to strengthen their commitments and assist in their coping abilities (Goodman et al. 2013).

## Who *Can* We Marry?

The majority of U.S. adults can marry the partner of their choice. In Chapter 1, we looked at some restrictions imposed on one's choice of a marriage partner. As we noted then, who we are allowed to legally marry has undergone change and challenge over the past 150 years in the United States over such issues as number of spouses in a lifetime and marriage across racial lines and, most recently, over the question of marriage between two people of the same sex.

What criteria do state marriage laws currently specify regarding eligibility to marry? Each state enacts its own laws regulating marriage, leading to some discrepancies from state to state. Although some restrictions or lack thereof are uniform across all 50 states (e.g., no state prohibits people from marrying someone of another race, no state allows siblings to marry

each other), others may be more variable—such as those regarding marriage among cousins, the minimum age at which people can marry, and the legality of same-sex marriage before the U.S. Supreme Court decision in *Obergefell v. Hodges*.

All states limit people to one living husband or wife at a time, and will not issue a marriage license to anyone with a living spouse. Once an individual is married, the person must be legally released from the relationship either by death, divorce, or annulment before he or she may legally remarry. Limitations that some but not all states prescribe are the requirement of blood tests, and being of good mental capacity.

### Marriage between Blood Relatives

*Nowhere in the United States* is marriage allowed between parents and children, grandparents and grandchildren, brothers and sisters, uncles and nieces, and aunts and nephews. Half siblings (e.g., children who share the same biological mother but different fathers) are similarly restricted.

Perhaps these restrictions are unsurprising. Such blood relations are commonly considered "too close," and marriage within such relationships is seen as incestuous and unacceptable. Some states disallow all "ancestor/descendant marriages," and a handful of states explicitly extend the prohibition against marriage between parents and children to marriages between parents and their adopted children. Although many state marriage statutes articulate very specific restrictions, some states, such as Ohio or Washington, more simply and generally prohibit marriage between relatives "closer than second cousins."

Along with Canada and Mexico, 19 U.S. states and the District of Columbia allow first cousins to marry. Six other states allow such marriages under certain circumstances. The remaining half of U.S. states do not allow first-cousin marriages (National Council of State Legislatures, www.ncsl.org/research/human-services/state-laws-regarding-marriages-between-first-cousi.aspx). Some may find it surprising that so many states allow first cousins to marry, thinking that if they were to have children together they would face risks of passing genetic defects to their children. Furthermore, there is debate about the justification for prohibiting first-cousin marriages, common in many other parts of the world, including the Middle East, Europe, and South Asia. One genetics researcher estimates that as many as 20 percent of marriages worldwide are between first cousins (Willing 2002). As to the risk to offspring of such marriages, there is only a slightly elevated risk (of an additional 2 to 3 percent) of such children inheriting recessive genetic disorders such as cystic fibrosis or Tay-Sachs disease (Willing 2002).

## Age Restrictions

State laws regulate and restrict marriage based on age requirements. Throughout the United States, 49 of 50 states require both would-be spouses to be at least 18 years old to marry without parental consent. In Nebraska, both partners must be at least 19 years old. Some states will waive the age requirement under certain circumstances (e.g., if the woman is pregnant), but in such instances, the couple may need approval from a court. Many states allow couples to marry in their early to mid-teens, provided that they secure parental and/or court consent.

Independent of the legal restrictions on when one can enter marriage, Americans continue to delay the ages at which they marry. As little as one-fourth of Americans now marry before the age of 25, and those who do tend to be more religious, are disproportionately drawn from economically disadvantaged backgrounds, and are less likely to attend or graduate from college (Sassler 2010).

## Number of Spouses

Recall the discussion in Chapter 1 of the controversy over polygamy. No state in the United States allows an individual to *marry legally* if he or she is already married. In other words, all 50 states consider monogamy the only legally accepted form of marriage. If a divorced or widowed man or woman wishes to remarry, he or she must present evidence of the legal termination of the prior marriage or of the death of his or her former spouse.

## ▌ Marriage Equality: The Controversy over Same-Sex Marriage

In Chapter 1, we briefly considered the issue of same-sex marriage, now legal everywhere in the United States. It is worthwhile to review the gradual change in legal recognition and rights that occurred in the United States over the past two decades. Internationally, during the same time frame, gay marriage was legalized in a number of countries, beginning in the Netherlands in 2000. As these words are being typed, gay marriage is now legal in close to two dozen countries.

Domestically, in the 1990s, U.S. courts rendered decisions that appeared to pave the way toward U.S. legalization of same-sex marriage. The two most notable cases were in Hawaii and Vermont. In 1993, the Hawaii Supreme Court ruled that denying gay men and lesbians the right to marry was unconstitutional in that it violated the equal protection clause of the state's constitution. This decision led many to anticipate the eventual legalization of same-sex marriage in the United States. It also caused opponents of gay marriage to take action. A number of state legislatures, along with the federal government, passed laws that declared marriage to be the union of one man and one woman, thus preventing the forced acceptance of gay or lesbian marriages should the Hawaiian decision stand up to an appeal.

In 1996, Congress passed the **Defense of Marriage Act**, and President Bill Clinton signed it into law. This act denied federal recognition to same-sex couples and gave states the right to legally ignore gay or lesbian marriages should they gain legal recognition in Hawaii or any other state. But the earlier Hawaiian decision did not stand. In a November 1998 ballot, 69 percent of Hawaiian voters chose to amend the state constitution, giving lawmakers the power to block same-sex marriage and limit legal marriage to heterosexual couples.

As 1999 drew to a close, the state of Vermont then took a major step toward what some believed, others hoped, and opponents feared would be the eventual legal recognition of gay marriage. There, three same-sex couples filed lawsuits, challenging a 1975 state ruling prohibiting same-sex couples from marrying. On December 20, 1999, the Vermont Supreme Court ruled that the state legislature had to either grant marriage rights to same-sex couples or assure them a legal equivalent to marriage, providing them the same range of state benefits married heterosexuals enjoyed.

On April 26, 2000, Vermont Governor Howard Dean signed into law legislation recognizing same-sex "civil unions." Although they were not marriages, "civil unions" were officially entered, offered the same rights and protections as marriages, and had to be officially dissolved if they failed. As of April 2008, close to 10,000 such civil unions had been recorded in Vermont, nearly 1,500 between state residents and more than another 8,000 involving residents of other states, the nation's capital, and several other countries, including Canada (Vermont Guide to Civil Unions 2005).

With the 5–4 U.S. Supreme Court decision in *Obergefell v. Hodges*, same-gender couples can marry in all 50 U.S. states.

Rob Melnychuk/Digital Vision/Getty Images

In 2001 and 2002, California and Connecticut passed laws granting gay or lesbian domestic partners some of the many benefits that accompany marriage (including tax benefits, stepparent adoption, sick leave, and permission to make medical decisions). Neither was quite as sweeping in scope as Vermont's, but both furthered the extension of relationship recognition to same-gender couples. Of greatest significance, on November 18, 2003, the Massachusetts Supreme Court ruled that the state's ban of same-sex marriage was unconstitutional and gave the state legislature six months to remedy the situation. Although Vermont's response was to create civil unions that provided the same rights and benefits as legal marriage, the Massachusetts court's decision specified the *right to marry*. Although the Massachusetts legislature and governor remained opposed to same-sex marriage, on February 4, 2004, the state supreme court ruled four to three that a "civil union" solution was unacceptable in that it would constitute "an unconstitutional, inferior, and discriminatory status for same-sex couples." Beginning with Massachusetts (on May 17, 2004), marriage between two men or two women became a legal reality for the first time anywhere in the United States.

There then followed a period of more than four years between legalization in Massachusetts and a second state, which turned out to be Connecticut, on November 12, 2008. Slowly and gradually, other states began to pass legislation allowing gay marriage: Iowa (on April 24, 2009), Vermont (on September 1, 2009), New Hampshire (on January 1, 2010), the District of Columbia (on March 3, 2010), New York (on June 24, 2011), Washington (passed February 14, 2012; effective June 7, 2012), and Maryland (passed March 1, 2012; effective January 1, 2013), all joined Massachusetts and Connecticut within the same amount of time that it had taken for Connecticut to join Massachusetts. Meanwhile, other developments of note included the following: In 2011, President Barack Obama directed the United States Justice Department to stop defending the Defense of Marriage Act against challenges to its constitutionality. On May 9, 2012, in an interview with a journalist, the president further reported that he now believed that same-sex couples should have the same right to legally marry as heterosexuals enjoy. Furthermore, in 2013, in a 5-to-4 decision, the U.S. Supreme Court ruled the Defense of Marriage Act provision denying federal benefits to same-sex couples who had married in states recognizing such marriages was unconstitutional.

The pace and breadth of further legal recognition increased. At the end of 2013, gay marriage was legal in 16 states. That number more than doubled in 2014, and by the time the U.S. Supreme Court began reviewing four same-sex marriage cases in January 2015, 37 states and the District of Columbia had legalized same-sex marriages. On June 26, 2015, by a 5-to-4 vote, a decision was reached in the case known as *Obergefell v. Hodges*, legalizing same-sex marriage and providing the right to marry to gay male and lesbian couples throughout the United States.

Across this same span of time, as gay marriage was gaining legal status, one can see pronounced shifts in wider public opinion on same-sex marriage. General Social Survey data, Pew Research Center survey data, or Gallup public opinion data reveals that whatever source one uses to track public opinion, gay marriage went from being widely disapproved of to widely supported. For example, between 2004 and 2014, the General Social Survey shows the percentage agreeing with the statement, "Homosexual couples should have the right to marry" climbing from 30 percent to 56 percent. In fact, if one looks from the first time the question was included in the GSS (1988), the percentage agreeing went from 11 percent to its 2014 level of 56 percent, with 2010 being the first GSS survey showing more people agreeing than disagreeing (Associated Press-NORC Center for Public Affairs Research, Issue Brief, 2015).

Pew Research Center data reflect the same trend, with 2004 data showing 31 percent in favor of and 60 percent opposed to gay marriage, and 2015 data indicate a near reversal, 57 percent were in favor and 39 percent were opposed. Pew data show 2011 to be the first time that

**Figure 8.4** Attitude toward Same-Sex Marriage, 1996–2011

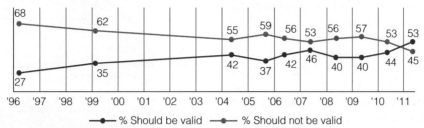

**Do you think marriages between same-sex couples should or should not be recognized by the law as valid, with the same rights as traditional marriages?**

● % Should be valid ━ % Should not be valid

NOTE: Trend shown for polls in which same-sex marriage question followed questions on gay/lesbian rights and relations.
1996–2005 wording: "Do you think marriages between homosexuals…"
SOURCE: From Lydia Saad, "Americans still Oppose Gay Marriage." Copyright © 2012 Gallup, Inc. All rights reserved.
The content is used with permission; however, Gallup retains all rights of republication.

a higher percentage favored (46 percent) than opposed (45 percent) gay marriage (Pew Research Center 2015). Finally, as shown in Figure 8.4, Gallup poll data show Americans favoring gay marriage by 60 to 37 percent as of May 19, 2015 (more than a month before the Supreme Court's decision in *Obergefell v. Hodges*). Beginning in 2011, the poll question asked, "Do you think marriages between homosexuals should or should not be recognized by the law as valid, with the same rights as traditional marriages?"; majorities of those sampled responded that they should be valid (McCarthy 2015).

According to Gary Gates and Frank Newport, there are an estimated 390,000 same-sex married couples in the United States. An additional 600,000 couples, or 1.2 million adults, maintain same-sex domestic partnerships (Gates and Newport 2015). With the right to marry now legal everywhere in the United States, it is expected that some of those couples will likely marry, while yet others who lived in states that offered neither marriage nor domestic partnership opportunities will also add to the numbers of same-sex marriages.

**True**

**4.** Once same-gender couples marry, their marriage must now be recognized as legal by all 50 states.

## ▪ Who *Do* We Marry? The Marriage Market, Who and How We Choose

People usually marry others from within their same large group—such as within the nationality, ethnic group, or socioeconomic status with which they identify—because they share common assumptions, experiences, and understandings. This practice, known as **endogamy**, strengthens group structure. If people already have ties as friends, neighbors, work associates, or fellow church members, a marriage between such acquaintances solidifies group ties. But another, darker force may lie beneath endogamy: the fear and distrust of outsiders, those who are different from ourselves. Both the need for commonality and the distrust of outsiders urge people to marry individuals like themselves.

Conversely, the principle of **exogamy** requires us to marry outside certain groups—specifically, outside our own family (however defined). Exogamy is enforced by taboos deeply embedded within our psychological makeup. The violation of these taboos may cause a deep sense of guilt. A marriage between a man and his mother, sister, daughter, aunt, niece, grandmother, or granddaughter is considered incestuous; women are forbidden to marry their corresponding male relatives. Beyond these blood relations, however, the definition of incestuous relations changes. One society defines marriages between cousins as incestuous, whereas another may encourage or even require such marriages.

### Homogamy

As norms defining the boundaries that determine partner or spouse, endogamy and exogamy interact to *limit* the field of eligibles. The field is further limited by society's encouragement of *homogamy*, the tendency to choose a mate whose personal or group characteristics are similar to ours. This is also known as **positive assortative mating** (Blackwell 1998). **Heterogamy** refers to the tendency to choose a mate whose personal or group characteristics differ from our own. There are strong pressures toward homogamy. We may make homogamous choices regarding any number of characteristics, including age and race, and also such characteristics as height (Blackwell 1998). As a result, our choices of partners tend to follow certain patterns. These homogamous considerations generally apply to heterosexuals, gay men, and lesbians in their choice of partners, though not to the same extent.

The most important elements of homogamy are race and ethnicity, religion, socioeconomic status, age, and personality characteristics. These elements are strongest in first marriages and weaker in second and subsequent marriages. They also influence our choice of sexual partners, as well as those we date and with whom we enter cohabiting relationships (Blackwell and Lichter 2004).

## Race and Ethnicity

Most marriages are between members of the same race. A Pew Research Center analysis of census data shows that in 2013, 6.3 percent of all marriages, and 12 percent of new marriages were interracial (Wang 2015). Although the concept of racial intermarriage is often assumed to refer to black–white marriage, in fact, Figures 8.5 and 8.6 show that American Indian, Hispanic Americans, and Asian Americans are more likely than African Americans to marry interracially (Wang 2015).

Based on an analysis of census data by sociologist Wendy Wang, we can note the following about interracial marriage in the United States:

- Among the newly married in 2013, more than a quarter of Asians (28 percent) and Hispanics (26 percent), and about a fifth (19 percent) of blacks "married out" of their race/ethnicity. The comparable

**Figure 8.5** Intermarriage Rates, by Race and Ethnicity, 2010

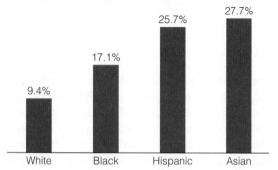

**% of Newlyweds Married to Someone of a Different Race/Ethnicity**

NOTE: Asians include Pacific Islanders. Whites, blacks, and Asians include only non-Hispanics. Hispanics are of any race.
SOURCE: Reprinted with permission. Copyright © Pew Research Center, Social & Demographic Trends Project. "The Rise of Intermarriage." http://www.pewsocialtrends.org/2012/02/16/the-rise-of-intermarriage/

**Figure 8.6** Intermarriage Types, Newly Married Couples in 2010

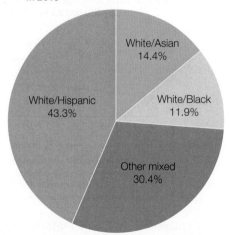

NOTE: Asians include Pacific Islanders. Whites, blacks, and Asians include only non-Hispanics. Hispanics are of any race.
SOURCE: Reprinted with permission. Copyright © Pew Research Center, Social & Demographic Trends Project. The Rise of Intermarriage. http://www.pewsocialtrends.org/2012/02/16/the-rise-of-intermarriage/

figure among whites was 7 percent. In addition to the racial differences that these figures reveal, across the four ethnic/racial categories, around three-fourths or more married homogamously.

- Among both Hispanics and whites, men and women were equally likely to intermarry. Among Asians, women are considerably more likely to marry interracially than are men, 37 percent versus 16 percent. Among blacks, the pattern is reversed; whereas one-fourth of black men married interracially in 2013, half that many (12 percent) black women married a spouse of a different race.

- Although *overall*, couples who marry out are economically and educationally similar to couples who "marry in," as shown in Tables 8.1 and 8.2, when one looks at the range of different combinations,

**TABLE 8.1** Median Incomes of Interracial and Intraracial Newlyweds

| | |
|---|---|
| White/Asian | $70,952 |
| Asian/Asian | $62,000 |
| White/White | $60,000 |
| White/Hispanic | $57,900 |
| White/Black | $53,187 |
| Black/Black | $47,700 |
| Hispanic/Hispanic | $35,578 |

SOURCE: Pew Research Center, Social and Demographic Trends Project, 2012. "The Rise of Intermarriage." www.pewsocialtrends.org/2012/02/16/the-rise-of-intermarriage.

**TABLE 8.2  Share of Interracial and Intraracial Newlyweds with Both College Educated**

| | |
|---|---|
| Asian/Asian | 53% |
| White/Asian | 41% |
| White/White | 23% |
| White/Hispanic | 19% |
| White/Black | 15% |
| Black/Black | 10% |
| Hispanic/Hispanic | 5% |

NOTE: Asians include Pacific Islanders. Whites, blacks, and Asians include only non-Hispanics. Hispanics are of any race.
SOURCE: Pew Research Center, Social and Demographic Trends Project, 2012. "The Rise of Intermarriage." www.pewsocialtrends.org/2012/02/16/the-rise-of-intermarriage.

there are noteworthy income and educational differences between them. Wendy Wang (2012) provided analysis of 2008–2010 data from the Pew Research Center (see Tables 8.1 and 8.2).

Within the broad racial or ethnic categories, one finds variations in rates of intermarriage. For example, among Asian Americans, Japanese and Filipino Americans marry white spouses at a greater rate than do Asian Indian Americans or Southeast Asians (Burton et al. 2010). Among the Hispanic population, Puerto Rican women and men are least likely to marry white spouses and are the most likely to marry African Americans. However, Puerto Ricans are more likely than African Americans to intermarry with whites. Of course, as noted previously, homogamy is the more prevalent pattern for all groups.

**False**

**5.** Interracial marriages between African American men and Caucasian women are at a greater risk of divorce than are marriages between Caucasian men and women, but marriages between African American women and Caucasian men are not.

Interracial marriage varies greatly across different cities and regions in the United States. David Harris and Hiromi Ono assert that without taking into consideration "local" marriage markets, we cannot completely and accurately understand racial marriage patterns. For example, the 2008 racial composition of the U.S. population was as follows: 63.7 percent non-Hispanic whites, 16.3 percent Hispanics, 12.6 percent African Americans, and 4.8 percent Asians. (An additional 6 percent identified some other race.) The racial composition of major cities exhibited substantial deviations from the national pattern, and cities differed from one another in important ways. For example, whites were 45 percent of the population in Philadelphia but only 12 percent in Detroit. Asians were at least 25 percent of the population in San Jose, San Francisco, and Honolulu but no more than 2 percent of the population in Phoenix, San Antonio, and Detroit (Harris and Ono 2005). These differing population compositions matter because where there is greater opportunity to find spouses of the same race, rates of homogamy are higher and the intermarriage rate is less. On the other hand, racial and ethnic heterogeneity are associated with higher levels of intermarriage (Kalmijn 1998).

Matthijs Kalmijn points out that marrying *outside* the group is not the same for all ethnic groups. For example, when Latinos marry "out," they are more likely marrying Latinos of a different cultural origin than they are white, European Americans. Asians, on the other hand, are much less likely to marry Asians of a different background and more likely when "marrying out" to marry whites (Kalmijn 1998). Kalmijn further indicates that the highest rates of homogamy are among blacks. The lowest rates are among European ethnic groups and among American Indians. Hispanics and Asians have intermediate homogamy rates (Kalmijn 1998).

## Black–White Intermarriage

According to Esther Lamidi, Susan Brown, and Wendy Manning, in 2014, 95 percent of black and white marriages were racially homogamous, down from virtually 100 percent in "a slow but steady decline," over the past 50 years (Lamidi, Brown, and Manning 2015a). Among those born between 1983 and 1997, the Millennial generation, the share of black and white couples who are married *intra*racially stands at 94 percent, the lowest level across five generations (Lamidi, Brown, and Manning 2015a). Furthermore, as reflected in public opinion data from July 2013, nearly nine out of ten (87 percent) surveyed by Gallup said they approved of marriages between blacks and whites. Among blacks, 96 percent said they approved, whereas among whites the percentage was lower but still represented more than 80 percent of those whites surveyed (Newport 2013). Among both blacks and whites, the respective percentages are the highest levels of approval Gallup has recorded. In a similar vein, data from the Pew Research Center show that 37 percent of those surveyed in 2014 thought "having more people of different races marrying each other" was a good thing for society, whereas only 9 percent considered it "bad for society." Half of those surveyed suggested it doesn't make much difference, which itself suggests an attitude more accepting than rejecting (Wang 2015).

Among black–white intermarried couples, between 70 to 75 percent of all marriages are between an African American man and a white woman (Qian and Lichter 2011). This same pattern persists among interracial cohabitors and holds regardless of level of education (Batson, Qian, and Lichter 2006). However, the proportion of black–white marriages does vary by education. Whereas 97 percent of black–white couples with less than a high school education marry within their race, the percentage of those married interracially stands at 6 percent among those with at least some college or college degrees (Lamidi, Brown, and Manning 2015a).

Overall, interracial marriages have been found to be at greater risk of divorce than racially homogamous marriages, though the risk is not the same for all interracial pairings. Some research has found interracial marriages where one spouse is Asian to be more stable than marriages of two white spouses. Likewise, marriages between African American women and white men are more stable even than marriages between white husbands and wives (Wang 2012). In contrast, the particular combination of marriage between a black man and a white woman faces the greatest risk of instability (Bratter and King 2008; Zhang and van Hook 2009).

Qualitative research sheds some light on experiences of intermarried African American and Caucasian couples. Based on their study of 76 such intermarriages (52 black male–white female couples and 24 black female–white male couples), Leigh Leslie and Bethany Letiecq (2004) suggested that success in black–white intermarriages may depend on the degree to which the partners possess pride in their race or culture without diminishing other races. This appears to be especially true for the black spouse in such marriages and seems to influence the quality of married life well into the marriage. Those who had resolved issues of racial identity and developed a strong black identity while showing racial tolerance and appreciation of other races more positively evaluated their marriage, felt less ambivalent about it, and/or worked harder to maintain it. However, those who had more negative assessments of either black or white culture experienced lower marital quality (Leslie and Letiecq 2004).

## Religion

Religion has long been a significant factor in marital choice and has been consistently linked to marital quality and stability. Sociologist Scott Myers (2006) summarized some past research as follows:

Only around one out of ten racial intermarriages is between an African American and Caucasian.

Huntstock/Stockbyte/Getty Images

- Religious homogamy has had a stronger effect on marital quality than has the level of religiosity of either spouse or of the couple.

- Marital conflict has been found to be higher among couples with theological differences. The greater the differences in spouses' religious beliefs, the more likely the marriage is to be deemed unhappy.

- Religious homogamy unites spouses who have similar values, thus strengthening their commitment and giving them a unified approach to marital and family matters. Myers found that religious homogamy, especially when expressed in joint church attendance, continues to be associated with higher marital quality, albeit to a lesser extent than in the past.

Most religions still oppose interreligious marriage because they believe that it weakens individual commitment to the faith. Although interreligious dating and marriage have increased, religious homogamy remains evident in American patterns of choosing spouses. Data

**Figure 8.7** Religious Homogamy: Percent of Individuals Living with or Married to a Partner of the Same Religion

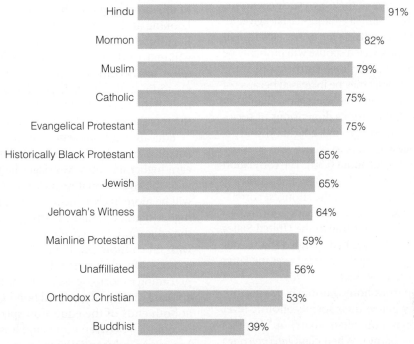

| | |
|---|---|
| Hindu | 91% |
| Mormon | 82% |
| Muslim | 79% |
| Catholic | 75% |
| Evangelical Protestant | 75% |
| Historically Black Protestant | 65% |
| Jewish | 65% |
| Jehovah's Witness | 64% |
| Mainline Protestant | 59% |
| Unaffilliated | 56% |
| Orthodox Christian | 53% |
| Buddhist | 39% |

SOURCE: Pew Research Center. "America's Changing Religious Landscape," May 12, 2015.

from 1972 to 1996 indicate that religious homogamy for Protestants, Catholics, and Jews in the United States continued to occur more often than would be predicted by chance (i.e., if religion made no difference in people's spousal choices) (Bisin, Topa, and Verdier 2004).

According to data collected and analyzed by the Pew Research Center, spouses share the same religion in just under 70 percent of married couples. However, comparing marriages over the past 50 years, where more than eight out of ten (81 percent) of marriages among couples married before 1960 were between spouses of the same religion, that percentage has gradually but steadily declined. Among those married between 2010 and 2014, 61 percent are religiously homogamous. As Caryle Murphy reports, nearly a fifth of all U.S. marriages since 2010 are between a Christian and a religiously unaffiliated spouse (Murphy 2015).

A look at different religious faiths reveals differences in the likelihood of being in a relationship (married or cohabiting) with a spouse or partner who shares your same religion. Data from the Pew Research Center report, "America's Changing Religious Landscape," reveal that where more than nine out of ten Hindus are married or living with a spouse or partner who is also Hindu, less than four in ten Buddhists report being married to or living with a romantic partner who shares

their Buddhist faith. Between these extremes, one can see there is much variation (see Figure 8.7).

Couples from different religious backgrounds who marry do have greater risk of marital unhappiness and divorce than those from similar religious backgrounds (Myers 2006). However, religious homogamy is less of a predictor of later marital quality than it once was (Ellison, Burdette, and Wilcox 2010). For example, in a study of Jewish marriages, what matters more in predicting the amount of conflict and instability is the extent of agreement or disagreement on Jewish issues, not what self-reported labels people use to identify themselves (Chinitz and Brown 2001).

Much of the concern that is expressed about religious intermarriage has to do with the potential effects on children and the conflicts that might ensue over child-rearing issues. With data from the National Survey of Families and Households, sociologists Richard Petts and Chris Knoester found that religious heterogamy is associated with more marital conflict and with less religious participation. Children of intermarried couples are more likely to engage in substance use (e.g., underage drinking and/or marijuana use), though they are no more likely to engage in delinquency, experience academic difficulties, or suffer from reduced self-esteem (Petts and Knoester 2007).

## Socioeconomic Status

Most people marry others of their own socioeconomic status and of the same or similar educational background. Even if a person marries outside his or her ethnic, religious, or age group, the selected spouse will probably be from the same socioeconomic level. Furthermore, some ethnic or racial homogamy may be increased because of tendencies toward socioeconomic homogamy (Bankston and Henry 1999). Of the various dimensions of socioeconomic status (family background, education, and occupation), the weakest appears to be between spouses' class origins (correlation of about 0.30). The correlation between husbands' and wives' occupational statuses is stronger (around 0.40). However, the strongest correlation is between spouses' educational backgrounds (approximately 0.55). This holds true in the United States as well as most other countries. In the United States, educational homogamy "strongly" increased over the latter decades of the 20th century (Kalmijn 1998).

Not everyone marries homogamously. Men more than women marry below their socioeconomic level (**hypogamy**); women more often "marry up" (a practice known as **hypergamy**). When class *intermarriage* occurs, it is rarely a case of spouses from opposite extremes (i.e., paupers and princesses). Both the upper and the lower levels of the class spectrum appear more "closed" than the middle levels (Kalmijn 1998).

Much research reveals a tendency for individuals to select spouses within their same educational level (less than high school, high school, some college, or college degree). In 2014, more than half (56 percent) of heterosexual marriages were between spouses with the same level of education (Lamidi, Brown, and Manning 2015b). Of those that weren't, a slightly higher percentage (23 percent) were between more educated wives and less educated husbands, than between more educated husbands and less educated wives (20 percent). Esther Lamidi, Susan Brown, and Wendy Manning note that, in fact, the percentage of marriages in which women had more education than their husbands has been increasing for more than a quarter century (since the late 1980s).

Lamidi, Brown, and Manning also note that in 2014, a greater percentage of white (56 percent) than black (52 percent) marriages were between spouses with the same levels of education. They further point out that black women more than white women are likely to be in marriages where they are more educated than their husbands. Among white marriages, men are more likely to have more education than their wives (Lamidi, Brown, and Manning 2015b).

In fact, college graduates are more likely to marry other college graduates than they are to marry others of the same religion who don't have college degrees. Looking at those with lower levels of education, the likelihood that people without a high school degree will marry others who are more educated than themselves has become less in recent years (Cherlin 2010). Of additional note, college-educated individuals are overall the most likely to ever marry but they are also the most likely to delay marriage (Cherlin 2010).

Educational homogamy can increase economic inequality, as more highly educated individuals tend to earn higher incomes. Marriages between two spouses with the educational resources to earn high incomes will be more likely to achieve or retain middle-class or higher status. Similarly, marriages at the lower end of the educational spectrum bring together two spouses with much more limited earnings potential, making it more difficult for them to achieve upward mobility. According to sociologists Christine Schwartz and Robert Mare (2005), educational homogamy has increased at both ends of the education spectrum. In fact, the income gap between more- and less-educated couples has grown substantially wider since 1960. Rich Morin of the Pew Research Center reports that in 2005, a couple in which both spouses were high school educated earned on average only 83 percent of the average household income. On the other hand, a couple in which both spouses had college degrees earned, on average, 219 percent of the average income (i.e., more than twice the average income) (Morin 2014).

Occupationally, the biggest divide has been between those in white-collar and those in blue-collar occupations. It appears as though the cultural status, not the economic status, of occupations is a more important factor in determining compatibility and attractiveness for marriage (Kalmijn 1998). A teacher may not find a plumber to be an attractive marriage choice, even if the teacher and the plumber have very similar incomes. More than the economic resources associated with an occupation is often considered in potential partners.

## The Marriage Squeeze and Mating Gradient

An important factor affecting the marriage market is the ratio of men to women. The **marriage squeeze** refers to the gender imbalance reflected in the ratio of available unmarried women and men. Because of this imbalance, members of one gender tend to be "squeezed" out of the marriage market. The marriage squeeze is distorted, however, if we look at overall figures of men and women without distinguishing between age and ethnicity. Overall, among unmarried women and men from ages 25 to 64 years old, the ratio of men to women was nearly equal; there were 97 unmarried men per 100 unmarried women. Combining widowed, divorced, and never-married people, in 2010, there were 113 unmarried men, aged 18 to 44, for every 100 unmarried women (U.S. Census Bureau 2010). But once one takes other variables into account, the pool of potential husbands for heterosexual women is reduced. For example, if one considers *employed never-married* men, the ratio of such potential husbands drops to 91 per 100 never-married women (Wang 2014). Similarly, examining ethnicity, the pool of employed men drops for both white and black women, but the decline has been greater for blacks. Thus, many women, especially African American women and/or women of certain ages, are squeezed out of the marriage market as eligible males are scarcer. There is also variation by location. Washington, DC, is a good, if extreme, example. There are 62 males for every 100 females, dramatically reducing women's opportunities for marrying, and giving men greater bargaining power (Abbott 2010).

"All the good ones are taken" is a common complaint of women in their mid-30s and beyond, even if there are still more men than women in that age bracket. The reason for this is the **mating gradient**, the tendency for women to marry men of higher status. Although we tend to marry those with the same socioeconomic status and cultural background, men tend to marry women slightly below them in age, education, and so on.

## Age

Americans tend to marry those of similar ages. The trend toward age homogamy commenced in the latter decades of the 19th century and continued through the 20th. In 1900, husbands were an average of five years older than wives; by the end of the century, the gap was between two and three years (Rolf and Ferrie 2008).

Defining age heterogamy as marriages in which wives are two or more years older than husbands or husbands are five or more years older than their wives, Esther Lamidi, Susan Brown, and Wendy Manning (2015a) report that over the past 50 years, approximately 40 percent of marriages were age heterogamous. They also offer these further details:

- Among couples who could be considered age heterogamous, there's a continuing trend in which the

share of such marriages with older husbands and those with older wives are becoming more similar.

- In 2014, 15 percent of married couples had older wives, an increase from 9 percent of such marriages over the past 50 years.

- Couples in which wives are college educated are typically age homogamous. Age heterogamy is typically associated with wives who did not complete high school. Both heterogamous marriages with older husbands and those with older wives are more prevalent in marriages where wives have less than high school educations.

Age is important because we view ourselves as members of a generation, and each generation's experience of life leads to different values and expectations. Furthermore, different developmental and life tasks confront us at different ages. A 25-year-old woman wants something different from marriage and from life than a 50-year-old man does. By marrying people of similar ages, we often ensure congruence for developmental tasks. As the gap between grooms' and brides' ages has narrowed in recent years, the ages at which both men and women enter marriage have consistently climbed.

Research suggests that the importance individuals place on age varies *by age* differently for men than for women. As men age, they prefer women progressively younger than themselves. Women, on the other hand, prefer for their partners to be about the same age (ranging from slightly younger through slightly older), up to ten years older than themselves. This does not appear to vary much, even as women age (Buunk et al. 2002). Interesting data from the United States and Australia reveal that the same age preferences that exist among heterosexuals exist among homosexual men and women—men prefer younger partners, and women prefer partners of about the same age (Over and Phillips 1997).

Historically, research tended to portray age-discrepant marriages as problem-filled, fraught with clashing values and a great imbalance of power. These were said to lead to greater dissatisfaction and elevated risks of divorce (Boyd and Li 2003). Research by Karen Pyke and Michele Adams (2010) on American marriages between older husbands and much younger wives, and by Monica Boyd and Anne Li on Canadian "May–December" marriages (2003) indicates that age-discrepant marriages are more common in second or subsequent marriages and in marriages in which other forms of heterogamy (most frequently racial) are also found. Age difference, by itself, appears to have little effect on marital satisfaction.

## Marital and Family History

An interesting application of the concepts of homogamy and heterogamy (intermarriage) can be found with regard to marital history. Essentially, never-married people are more likely to marry other never-married people than they are to "intermarry" by marrying divorced people (Ono 2005). Hiromi Ono questions whether this is a "by-product" of other homogamous patterns (such as age, socioeconomic status, or parenthood status) or a deliberate choice that individuals make to marry someone of similar marital history. For example, a divorced person may believe that only another divorced person will similarly understand and have experience with the lingering ties to prior marriages.

Conversely, never-married individuals who marry divorced partners may find that they have to deal with lower amounts of resources because of the continued demands of former spouses and the needs of children from former marriages. This, in turn, may give rise to jealousy and impede the development of needed levels of trust (Ono 2005). Ono determined that **marital history homogamy** occurs more as a result of deliberate choices. Ono also reasonably speculated that parental status, like marital history, operates in a similar fashion. Parents make lifestyle concessions to their parenting responsibilities that nonparents don't have to make. Where children and their needs become priorities for parents, nonparents can maintain other priorities.

The structure of an individual's family of origin also turns out to be a factor in the process of mate selection. Children of divorced parents often marry other children of divorced parents. Research by Nicholas Wolfinger suggests that coming from a divorced home increases the likelihood of choosing another child of divorce as a spouse by 58 percent. Although homogamy often is associated with a greater chance for marital happiness and stability, family structure homogamy may be a noteworthy exception because marriages in which both spouses are children of divorce face greater odds of marital failure. Marriages in which either spouse comes from a divorced family are twice as likely to fail as those in which neither spouse is a child of divorce. When both spouses are from divorced homes, their marriages face three times the likelihood of failure as marriages between two children of intact parental marriages (Wolfinger 2003).

## Residential Propinquity

An additional, long-observed homogamous factor is based on the principle of **residential propinquity**—the tendency we have to select partners (for relationships and for marriages) from a geographically limited locale. Put differently, the likelihood of marriage decreases as the distance between two people's residences increases. The obvious explanation behind this is one of opportunity. In most instances, to start dating or get together with someone, you have to first meet. Our chances of meeting are greater when our daily activities (shopping, commuting, eating out, and so forth) overlap.

Although it is easy to trivialize this tendency as too obvious to be meaningful, consider the implications it can have for some of the other patterns of homogamy. American communities are often segregated by class, race, or both. In some towns, they may even have religious splits (e.g., the Catholic side and the Protestant side of town or a Jewish neighborhood). Public schools, being neighborhood based, further the tendency for us to associate with others like ourselves. Thus, the types of people with whom we are most likely to come into contact and with whom we might develop intimate relationships or eventually marry are a lot like ourselves. Meeting at school promotes age, educational, and social class homogamy (Kalmijn and Flap 2001).

Thus, within a society somewhat residentially segregated by race or social class, residential propinquity may explain some other homogamous tendencies by how it limits our opportunity. But the story is more complicated than where we live. After all, unmarried people do not simply wander around an area or region looking for a spouse; they spend most of their lives in small and functional places, such as neighborhoods, schools, workplaces, bars, and clubs. Such local marriage markets are often socially segregated, which is why they are important for explaining marriage patterns. In the sociological literature, three local markets have been considered most often: the school, the neighborhood, and the workplace. Of these three, schools are considered the most efficient markets because they are homogeneous with respect to age and heterogeneous with respect to sex (Kalmijn 1998).

## Understanding Homogamy and Intermarriage

Factors in the choice of partner interact with one another. Ethnicity and socioeconomic status, for example, are often closely related because of discrimination. Many African Americans and Latinos are working class and are often not as well educated as Caucasians. Thus, a marriage that is endogamous in terms of ethnicity is also likely to be endogamous in terms of education and socioeconomic status.

Matthijs Kalmijn (1998) identified three social forces that help explain marriage patterns: (1) the preferences of individuals for resources in a partner, (2) the influence of one's peer group, and (3) the constraints of the marriage market. It appears that all three of these combine to produce the tendencies toward homogamy and the patterns of mate choice we observe, but it is difficult to determine the relative strength of the factors or what is "most influential" in shaping mate selection practices. What we can say with more certainty is that the presence of both opportunity constraints and outside influence (or "interference") makes it unwise to conclude that homogamy automatically reflects hostility or animosity toward others unlike oneself. It may not even illustrate an outright preference for people like oneself.

Hiromi Ono (2005) contends that—especially with regard to race, education, and social class patterns but also with reference to marital history—homogamy has the potential to widen social inequality. What about consequences of intermarriage? Kalmijn (1998) argues that intermarriage potentially has the following effects:

- Intermarriage can decrease the importance of cultural differences because the children of mixed marriages are less likely to identify themselves with a single group. Even when mixed couples socialize children into the culture of a single group, the children are less likely to identify with that group when intermarriage in society is common.

- Through intermarrying, individuals may question and lose negative attitudes they have toward other groups. Spouses and their wider networks (of kin and friends) gain the opportunity to get to know people "different" from themselves and question any biases and stereotypes they previously held.

## Theories and Stages of Choosing a Spouse

At most, homogamy narrows the pool from which we might seek a spouse. Furthermore, sharing background characteristics doesn't automatically guarantee that two people will feel close to each other, fall in love, and/or want to share their lives. Homogamy alone isn't sufficient to account for whom we choose. A range of theories has been suggested to address the question of why we select particular individuals. Do "opposites attract"? Do "birds of a feather flock together"? Do we unconsciously select people like our parents? What is more important: finding someone who seems to think as we do about things or finding someone whose behavior fits what we expect in a partner?

Each of the preceding questions illustrates the premise of an existing theory of mate selection. The commonsense notion that "opposites attract" is in keeping with **complementary needs theory**, the belief that people select as spouses those whose needs are different from their own. Thus, an assertive person who has difficulty compromising will be drawn to a less outgoing and highly adaptable person. The complementarity and subsequent interdependence are alleged to strengthen the bond between the two partners.

The notion that "birds of a feather flock together" is more in keeping with theories such as **value theory** or **role theory**, in which gratification follows from finding someone who feels and/or thinks like we do. Having someone who shares our view of what's important in life or who acts in ways that we desire in a partner validates us, and this sense of validation leads to an intensification of what we feel toward that other person.

**Parental image theory** suggests that we seek partners similar to our opposite-sex parent. Some versions of parental image theory draw on Freudian concepts such as the Oedipus complex, whereas others point toward the lasting impressions made by our parents (Eshelman 1997; Murstein 1986).

A variety of **filter theories** suggest that the choice of a partner or spouse is based on a number of different factors that become more or less important as relationships develop, grow, and change (Caughlin and Huston 2010). Bernard Murstein developed a social exchange–based filter theory known as **stimulus–value–role theory** to depict what happens between that "magic moment" with its mysterious chemistry of attraction and the decision to maintain a long-term relationship such as marriage. Murstein's theory identifies three stages of romantic relationships: the stimulus, value, and role stages. At each stage, if the exchange seems equitable, the two will progress to the next stage and ultimately remain together:

- In the *stimulus* stage, each person is drawn or attracted to the other before actual interaction. This attraction can be physical, mental, or social. During the stimulus stage, with little other information on which to evaluate the other person, we make potentially superficial decisions. This is especially evident during first encounters.

- In the next stage, the *value* stage, partners weigh each other's basic values, seeking compatibility. As relationships continue, each person discovers the other's philosophy of life, politics, sexual values, religious beliefs, and so on. Wherever they agree,

As is widely known, marriage proposals are often expressed in such a way that they build to or contain the familiar question, "Will you marry me?" The question may be preceded by a poem, a song, or some sort of stunt, and the proposal may contain much more than the question itself. There are many clever and creative ways in which a hopeful spouse "pops the question" to the other, seeking an affirmative answer on the way to forming a life together as married. Once the proposal recipient says yes, as couples plan their weddings, a second question often gets popped, this one more a "wedding proposal" than a marriage proposal. Increasingly, the recipient is a friend or relative of either or both spouses-to-be, who is asked to officiate at the wedding and marry the couple.

Historically, weddings were performed by clergy or officials of the state. Still today, every U.S. state allows religious clergy or certain city, county, or state officials to officiate at weddings (Hoesly 2015). Doctoral candidate Dusty Hoesly notes that there is variation among the states as far as who is allowed to perform a legally valid wedding ceremony. For example, some states allow individuals to become deputized for the day of the wedding, thus allowing them to perform the wedding ceremony. Some states require that if religious officiants are used that they first officially register (i.e., with a county clerk). Other states don't require such registration. Some states even allow the marrying couple to perform the ceremony. However, the majority of states require either a clergyperson or official to officiate, in order for the marriage to be legally recognized (Hoesly 2015).

The Universal Life Church (ULC), founded by Kirby J. Hensley in 1959 (as Life Church), has provided a means for those who wish to officiate weddings to become ordained as ministers, by offering free, lifetime ordination, via mail order or especially through online application. Dusty Hoesly describes the ULC as among the largest U.S. religions and reports that it has, since1962, ordained more than 20 million people as ministers, thus enabling them to perform wedding ceremonies. In fact, the ability to officiate weddings is said to be the primary reason that individuals seek to be ordained by the ULC, with church leaders estimating that as more than 80 percent of ULC ministers get ordained just for that reason (Hoesly 2015).

The Universal Life Church is not the only online source of ordination. Journalist Elissa Gootman cites Rose Ministries and the Church of Spiritual Humanism, as well as ministries that use the Universal Life Church name (Gootman 2012). The Rose Ministries offers online ordination packages. For $89.95 (as of February 2016), one can be officially ordained, be issued a one-year license, receive a Certificate of Ordination, a clip-on clergy badge, and five ready-to-use wedding certificates. Lesser-priced packages are also available (http://openordination .org/ordination.php?crn=198). The Church of Spiritual Humanism offers an online, completely free, lifetime ordination. The ULC ordination is also without charge. All three prominently mention the right to perform legal wedding ceremonies as among the rights accompanying and the reasons to seek ordination.

For couples who desire to have a friend or relative marry them, the online ordination process offers a way for the preferred officiant to become legally eligible to do so. Though data are hard to come by, there are indications that this is becoming more and more common, especially among Gen Xers and Millennials (Gootman 2012; Hoesly 2015). Older ages at first marriage, and increases in cohabitation, interfaith marriages, the percentages of religiously unaffiliated individuals, and do-it-yourself (DIY) orientations, have greatly increased the popularity of having one's wedding officiated by a friend or relative. According to Dusty Hoesly, the wedding industry itself has caught on and promotes such a choice as part of a personalization of one's wedding. Social historian Stephanie Coontz suggests that there may also be some wishful thinking at work, in the thought that if one's wedding is unique, so too will their marriage be unique (Gootman 2012).

There are those who strike a somewhat cautious note. Judith Johnson (2013), a life coach and interfaith minister, suggests it may not be a good idea. She notes that one needs to be careful in researching state law as to whether clergy with online ordinations are accepted as legally valid officiants in the state in which you are getting married. She also points out that with all that has to be done to design and conduct a wedding ceremony, there is risk in making one's friend or relative responsible for bearing a burden about which they know very little. In addition to subjecting the chosen officiant to potential stress, one also risks the success of one's wedding. ●

SOURCES: Hoesly (2015); Johnson (2013); Gootman (2012).

it is a plus for the relationship. However, if they disagree—for example, on religion—it is a potential minus for the relationship. Each person adds or subtracts the pluses and minuses along value lines. Depending on the outcome, the couple will either disengage or go on to the next stage.

- Eventually, in the *role* stage, each person analyzes the other's behaviors, or how the person fulfills his or her roles as lover, companion, friend, and worker and potential husband or wife and mother or father of potential children. Are the person's behaviors consistent with marital roles? Is he or she emotionally stable? This aspect is evaluated in the eighth and subsequent encounters (Murstein 1986).

Although the stimulus–value–role theory has been one of the more prominent filter theories explaining relationship development, some scholars have criticized it, especially regarding the question of whether we actually test the degree of "fit" between us and our partners. In reality, we might underestimate the importance of certain issues or, conversely, be focused more extensively on others. For example, religious fundamentalists and atheists may sometimes believe that they are compatible. They may not discuss religion; instead, they might focus on the "incredible" physical attraction in their relationship or their mutual desire to raise a family. They may believe that religion is not that important, only to discover after they are married or after they have children that it is important.

None of these theories has been proven to be *the explanation* of how two people select each other for marriage. In fact, it is unlikely that any single theory can account for the broad range of relationships that people construct and enter.

# ▌ Why Marry?

If you stopped each couple just before they exchanged their vows and asked, "Why are you doing this? Why are you getting married?," you might hear many different answers, though somewhere among them, probably early and emphasized, would be because they are in love. The greatest attraction of marriage is probably the love and intimacy that we expect to come with it. A nationally representative sample of 1,003 young adults (20 to 29 years old) demonstrated the extent to which our views about marriage and, perhaps, the appeal of

marriage is rooted in the intimacy and love we hope to find there. More than nine out of ten never-married respondents endorsed the notion that "when you marry, you want your spouse to be your soul mate, first and foremost" (Whitehead and Popenoe 2001, in Cherlin 2006). In addition, more than 80 percent of women surveyed indicated that it was more important to "have a husband who can communicate his deepest feelings" than a husband who is financially successful (Cherlin 2004). Clearly, we are drawn to marriage in pursuit of a level of love and intimacy that we believe may not otherwise be available or possible. As sociologist Paul Amato (2004b) puts it, we tend to see marriage as "the gold standard" for relationships.

Among the many reasons for marriage, we can easily recognize the role of possible economic and social pressures (i.e., "pushes" toward marriage) as well as the strong desires to have and raise children together, which, for many, still seem to be best accomplished in marriage. However, whereas a decade ago sociologist Paul Amato (2004b) suggested, "Most people will continue to see marriage as the best context for bearing and raising children," and will marry if they desire to become parents, today this may be felt less strongly, as suggested by recent opinion data.

In addition to potential "pushes" are the "pulls of marriage," the attractions or draws that marriage presents. For many, marriage symbolizes that two people have reached a stage in their lives as well as in their relationships, and that in it they have attained "a prestigious, comfortable, stable style of life" (Cherlin 2004, 857).

If the practical importance of marriage has diminished, if marriage can no longer be counted on to cement a relationship, allowing spouses to confidently invest themselves in each other without fear, invest their time and energy in raising children together, and invest financially in acquiring such goods as cars and homes, the "symbolic significance" of marriage remains considerable and attractive. It has become less a marker of conformity and more a marker of prestige (Cherlin 2004).

## Benefits of Marriage

In what ways does being married benefit the women and men who marry? We have already seen that marriage confers legal benefits. Quite literally, marriage offers more than a thousand rights, benefits, protections, and/or privileges to couples who marry. There are more than 1,100 federal benefits and protections alone. In addition to these legal benefits, marriage confers benefits in economic well-being (e.g., higher income, greater productivity, and mobility at work), physical and mental health, and personal happiness. Married couples are better off financially than those living in all other types of households (Hirschl, Altobelli, and Rank 2003). Marriage both reduces the risk of poverty and increases the probability of affluence. Defining *affluence* as living in a household that earns ten times the poverty level, Thomas Hirschl, Joyce Altobelli, and Mark Rank conclude that married-couple households are more likely to attain affluence than those living outside of marriage. Women, in particular, face a much greater likelihood of attaining affluence in marriage than outside of marriage (Hirschl, Altobelli, and Rank 2003; Wu and Hart 2002/2003).

The health advantages associated with marriage are considerable. Sociologists Deborah Carr and Kristen Springer report that research, both in the United States and abroad, has shown that marriage has strong beneficial effects on spouses' health and well-being in such areas as psychological distress, suicide, and overall mortality. Additional studies report protective effects of marriage, as indicated by number of illnesses, self-rated health status, and chronic conditions (Carr and Springer 2010). In a study looking across marital statuses at married, cohabiting, and single individuals, Kelly Musick and Larry Bumpass report that married people experienced more beneficial health outcomes than cohabitants, though cohabitants fared better in reported happiness (Musick and Bumpass 2012). The marital health benefits are experienced across all income and age groups and are found among both women and men as well as among whites and African Americans, though the magnitude of the beneficial effects may differ by gender and race (Carr and Springer 2010). Musick and Bumpass (2012) report that the positive effect of marriage on health can be seen both among married couples who cohabited prior to marrying and those who did not. In addition, although both genders derive health-related advantages from marriage, men's morbidity and mortality seem more strongly affected than women's are by their marital *status* (i.e., by being married in comparison to those in other marital statuses), whereas women's health appears to be more dependent on marital *satisfaction* (Fincham and Beach 2010). However, married women report lower levels of stress and substance use than do counterparts who are divorced or separated, so there is still an advantage that appears to accompany their being married (Witters and Sharpe 2014).

Some of the health advantages that accompany marriage come from the fact that married people tend to live healthier lifestyles than unmarried people. Researchers at the Centers for Disease Control and Prevention concluded that married women and men are less likely to smoke, drink heavily, or be physically inactive and are less likely to suffer from headaches and serious psychological distress (Schoenborn 2004; Stimpson and Wilson 2009). The social and emotional support one derives from having a spouse also help improve one's health and well-being. When marriages end, women suffer increased depression, and men suffer poorer physical and mental health.

## Is It Marriage?

In considering the benefits that seem to accompany marriage, researchers have explored whether these benefits truly follow marriage or were instead reflections of differences in the types of people who do and don't marry. Sometimes phrased as a difference between *selection* into marriage and *protection* afforded by marriage, it raises the question of whether there is something unique and beneficial about being married (i.e., marriage protects us from certain unhealthy risks) or whether those who marry are somehow unique compared to those who don't marry (i.e., selection). Deborah Carr and Kristen Springer refer to these as the **social causation** and **social selection** perspectives.

Although we have painted these as alternatives— as *either* social selection *or* social causation—the benefits associated with marriage more accurately reflect both processes. Thus, although healthier and more stable individuals may be more attractive as marriage partners, thus bringing better mental and/or physical health with them into their marriages, a good marriage also has healthful and stabilizing effects on those who marry. Such is the conclusion of research by sociologists Daniel Hawkins and Alan Booth. They contend that happier and healthier people may be more likely to marry, but also that marriage itself is associated with increased psychological well-being and at least half of the observed difference in health between married and unmarried individuals (Hawkins and Booth 2005). Carr and Springer suggest that three primary factors account for the beneficial marriage-health connection: economic resources, social control, and psychosocial support and strain. Of these, the *economic resources* reflect both social selection and social causation. Those who marry are likely to have more economic resources before marriage and/or have qualities that are perceived to make them more likely to achieve greater economic stability if not prosperity. Therefore, they are more attractive as and to potential spouses. The *social control* factor reflects the monitoring that spouses do to ensure their partners are engaging in more healthful and less unhealthy behavior. Finally, the *psychosocial support* that a good marriage might bring may enable married individuals to better handle difficulties and stresses that could otherwise undermine health (Carr and Springer 2010). Kelly Musick and Larry Bumpass (2012) suggest that as cohabitation becomes even more common and accepted, indeed institutionalized, and the boundaries between cohabitation and marriage further blur, one may well see more comparable advantages being enjoyed by cohabiting as by married couples.

## Or Is It a *Good* Marriage?

It should be noted that we said that a *good marriage* has healthful and stabilizing effects; what happens to those in "not-so-good" marriages or marriages in which one or both spouses are unhappy, where there are high levels of conflict and spouses' expectations remain unmet? Much research shows that marital distress puts one at greater risk of poorer physical health. Conversely, high levels of marital quality (as measured in terms of marital happiness or satisfaction) are significantly associated with good physical health. Richard Miller and colleagues suggest that this is explained by the social and emotional support marriage can provide, alongside the monitoring that spouses do of each other's health and habits (Miller et al. 2013).

Research indicates that unhappy, conflict-ridden marriages fail to provide the health, happiness, or other benefits that otherwise appear to separate happily married people from the unmarried. Furthermore, long-term unhappy marriages appear to have negative effects on spouses. Hawkins and Booth (2005) contend that remaining unhappily married actually lowers one's happiness, life satisfaction, and self-esteem and is associated with poorer overall health. As we saw in Chapter 7, negative spousal interactions and marital conflict "can impair immune response, slow wound healing, heighten susceptibility to infectious agents, and increase cardiovascular reactivity, all factors that compromise physical health in the long run"

**True** **6.** Married people report greater happiness and better health than unmarried people, but only if their marriages are happy.

(Carr and Springer 2010). Divorce, though associated with psychological distress, elevated levels of depression, and lower levels of self-esteem and happiness, may nonetheless improve the health and well-being of those in unsatisfying or unhappy marriages. This appears especially true for those divorced women and men who remarry, but even those who remain unmarried after divorce report greater self-esteem, more satisfaction with their lives, and better overall health than do unhappily married people. They assert that unhappily married individuals are *moderately worse off* than those who divorce.

Psychologist Eli Finkel points out that our expectations and ideals for marriage have been raised higher, so high in fact that they require a greater investment of time and energy to reach and realize. When we are unable to invest the needed attention and energy, we may be unable to achieve the kind of marriages we now idealize and the sense that the marriage is a failure (or at least less than we dreamed of), may be greater than when expectations were lower. Finkel refers to "the all-or-nothing" marriage, tracing its emergence to the transition from institutional marriage to companionate marriage and ultimately individualized (or "self-expressive") marriage. This latest type of marriage promises potentially unprecedented benefits, assuming one can and

Although marriage can bring many benefits, unhappy, high-conflict marriages may leave spouses worse off than had they not married or ended their marriage.

Axel Ieschinski/Alamy Stock Photo

does devote time to what it requires, spending time together, talking, sharing activities, and socializing as couples with shared friends. Couples who can do this are considerably more likely to consider themselves very happily married than are spouses who cannot or who don't. Unfortunately, data indicate that, on average, people spend less, not more, time with their spouses due to the demands associated with work and with time-intensive/labor-intensive parenting. Furthermore, for "less wealthy" Americans, these heightened expectations are more likely to be unmet, in the face of unemployment, working, and juggling multiple jobs (Finkel 2014).

# Predicting Marital Success

The period before marriage is especially important because couples learn about each other—and themselves. Courtship sets the stage for marriage. Many of the elements important for successful marriages, such as the ability to communicate in a positive manner and to compromise and resolve conflicts, develop during courtship. They are often apparent long before a decision to marry has been made (Cate and Lloyd 1992). Couples who are unhappy before marriage are more likely to be unhappy after marriage as well, as "troubled courtships generally give way to unhappy or fragile marriages" (Huston 2009).

> **False** **7.** Couples who are unhappy before marriage do not increase their happiness after marriage.

Ted Huston and Heidi Melz (2004, 952) describe three "prototypical courtship experiences," each of which has different likely consequences for couples who marry. Of critical importance in differentiating these courtships are personality characteristics of partners, which affect "both the dynamics of their courtships and the success of their marriages." Some qualities, such as warmheartedness, a caring disposition, or an even temper, are important determinants of whether people create happy and stable marriages (Huston 2009). Other qualities, such as being less stubborn, less independent-minded, and more conscientious, are important factors in determining whether couples stay married. These personality characteristics are associated with the three courtships and marital outcomes that Huston and Melz (2004) identify as follows:

- *Rocky and turbulent courtships.* Such courtships are characterized by periods of upset and anger, distress and jealousy over potential rivals, and uneasiness about placing love in "undeserving hands" (p. 950). They are more typically experienced by "difficult" personalities, people who are exceedingly independent-minded, who lack conscientiousness, and who have high anxiety. If men are excessively independent, they may make poor husbands, and their marriages are likely to be "brittle." If men and women high in anxiety marry each other, their marriages tend to be unhappy but lasting marriages.

- *Sweet and undramatic courtships.* Such partnerships are between people with "good hearts" who are helpful, sensitive to the needs of others, gentle, warm, and understanding. Good-hearted couples find enjoyment and pleasure in each other's company. Their marriages are more likely to be satisfying and enduring.

- *Passionate courtships.* These are characterized by partners "plunging into love, having sex early in the relationship, and deciding to marry one another within a few months" (p. 950). Such couples begin marriage as "star-crossed lovers," sharing far more affection than typical of even newly married couples, "but over the first two years, much of the sizzle fizzles" (p. 950). They are also vulnerable to divorce.

Huston and Melz (2004, 949) contend that we can tell "from the psychological makeup of partners and how their courtships unfolded, whether they would be delighted, distressed, or divorced years later." How couples reach marriage, as well as what types of personal traits they bring into marriage, are important (Huston and Melz 2004).

Whether marriage is an arena for growth or disenchantment depends on the individuals and the nature of their relationship. It is a dangerous myth that marriage will change a person for the better: An insensitive single person likely becomes an insensitive husband or wife. Undesirable traits may become magnified in marriage because we must live with them in close, unrelenting, and everyday proximity.

Family researchers have found numerous premarital factors to be important in predicting later marital happiness and satisfaction. Although they may not necessarily apply in all cases—and when we are in love, we may ignore such factors, believing that we are the exceptions—they are worth thinking about. As enumerated by researchers Jeffry Larson and Rachel

Hickman, these include *background factors* (e.g., age at marriage, level of education, race, parental marital status, and so on), *contextual factors* (e.g., support and approval from friends and freedom from pressures to marry), *individual traits and behaviors* (e.g., level of self-esteem, interpersonal skills, physical health, or illness), and *couple characteristics* (e.g., being from similar backgrounds; possessing similar values, attitudes, beliefs, and gender-role expectations; and communication and conflict management skills) (Larson and Hickman 2004). We now turn to some of these.

## Background Factors

Age at marriage is an especially important, perhaps even the most important, background factor in shaping marital outcomes. People who "marry young" are at greater risk of seeing their marriages fail. Adolescent marriages (where either party is younger than 20) are especially likely to end in divorce. Such young marriages may be more prone to divorce because of the immaturity and impulsivity of the partners (Clements, Stanley, and Markman 2004). In addition, early entry into marriage is associated with reduced educational attainment, which itself can curtail one's later occupational and economic success and thus contribute to additional marital stress.

As we saw earlier, the trend in the United States has been toward delaying marriage; on average, women and men are entering their first marriages at older ages than ever before—at age 29 for men and at age 27 for women. Nevertheless, despite this trend, more than 25 percent of women and 15 percent of men marry before they turn 23 (Uecker and Stokes 2008). Certain characteristics are associated with greater likelihood of marrying in one's early 20s, including one's gender, race, region, socioeconomic status, religion, education, and history of cohabitation. Women are twice as likely as men to marry early. Sociologists Jeremy Uecker and Charles Stokes identify a white, rural southerner from families of lower socioeconomic status (as measured by parents' incomes and educations) as most likely to marry young. Other factors associated with early marriage include having lower educational aspirations and attainment, growing up in a conservative Protestant or Mormon family, having parents who married early (22 or younger), and having cohabited (Uecker and Stokes 2008). Uecker and Stokes point out that although young, disadvantaged women and men in urban areas of the United States are the ones most likely to retreat from marriage, young, disadvantaged women and men from rural and southern areas are likely *to embrace* marriage.

Marriage age seems to have less effect as people age further into adulthood. In other words, differences are slight between those who marry in their mid- to late 20s and those who marry in their 30s and older. Length of courtship is also related to marital happiness. The longer you date and are engaged to someone, the more likely you are to discover whether you are compatible with each other. But you can also date "too long." Those who have long, slow-to-commit, up-and-down relationships are likely to be less satisfied in marriage. They are also more likely to divorce.

Socioeconomic factors, such as education, income, and even occupational prestige are associated with marital quality and stability. Higher levels of education are positively associated with levels of marital satisfaction and marital stability, whereas lower levels of education, income, or prestige are associated with less happiness, more marital problems, and a higher likelihood of divorce (Conger, Conger, and Martin 2010).

Other factors include level of religiousness and family background. Higher religiousness, especially by wives, is associated with greater probability of happy and stable marriages (Clements et al. 2004). Parental divorce may cause someone either to shy from marriage or to marry with the determination not to repeat the parents' mistakes. Parental divorce increases risks to married children; those who grew up in households where parents divorced are considerably more likely to experience a divorce themselves (Amato and De Boer 2001). This pattern has been identified in the United States and at least 15 other countries (Diekmann and Schmidheiny 2004).

## Personality Factors

We bring with us into our marriages personality characteristics, attitudes and values, habits and preferences, and unique personal histories and early experiences. As you can imagine, your partner's personality will affect your life, your relationship, and your marriage considerably. Such personality characteristics are relatively stable and likely exert influence on the quality and outcomes of our marriages (Bradbury and Karney 2004). Personality characteristics may be most significant during courtship, when those with undesirable or incompatible personalities are weeded often out—or ought to be, at least in theory.

Researchers tend to focus more attention on relationship process and change than on personality.

Personality seems fixed and unchanging. Nevertheless, it clearly affects marital processes. For example, a rigid personality may prevent negotiation and conflict resolution, and a dominating personality may disrupt the give-and-take necessary to making a relationship work, whereas warmth, an even temperament, and a forgiving and generous attitude toward one's spouse contribute to happy, stable marriages. In Ted Huston's longitudinal study, following couples from courtship through early marriage and at nearly 14 years after they were wed, there was notable stability to assessments of spouses' personalities made when couples were first married. These early assessments predicted the feelings that couples had and the behavior they displayed *nearly 14 years later* (Huston and Melz 2004). Spouses in happy marriages were higher in *communal expressiveness.* They had kind and caring dispositions, tended to be more affectionate, and managed to put a positive spin on their partner's personality. Conversely, *trait anxiety,* characterized by moodiness and irritability, was associated with more marital antagonism, more demand–withdraw communication, and with "cycles of anger and hurt" (Huston 2009, 318). Individuals high in trait anxiety were generally more "down" about their marriages. Thus, personal attributes and characteristics matter greatly in shaping marital outcomes.

The ability to identify and communicate emotions is also associated with marital satisfaction. Such ability appears to differ by gender, with men having more difficulty identifying emotions expressed by women and with expressing and effectively communicating their own emotional state. Also gendered is the importance of effective emotional communication; women's marital satisfaction is more affected by the effective communication of emotions than is men's. This suggests that the emotional skills one brings with one into marriage are important factors in determining marital quality because they affect the level of shared intimacy (Cordova, Gee, and Warren 2005).

## Relationship Factors

Besides personality characteristics, researchers have also examined aspects of premarital interaction and relationships that might predict marital success. Writing about couples followed across more than a decade, psychologist Ted Huston notes that signs of future marital problems surface during courtship and in the early years of marriage, as "troubled courtships generally give way to unhappy or fragile marriages" (Huston

2009, 323). Other aspects of premarital relationships that are linked to later marital satisfaction and divorce include approval of one's family and friends, premarital pregnancy, and cohabitation (Wilson and Huston 2013).

Not all research substantiates the idea that marital success or failure is determined by how spouses communicate and solve problems (Bradbury and Karney 2004). Problem-solving skills are important but perhaps not as important as the emotional climate within which such skills are implemented. "If spouses have a reservoir of good will and they show their affection regularly, they are more likely to be able to work through their differences, to warm to each other's point of view, and to cope effectively with stress" (Huston and Melz 2004). If couples can maintain humor, express "genuine enthusiasm for what the partner is saying," and convey their continued affection for each other, couples with low levels of problem-solving ability will experience similar outcomes (in terms of shifts in marital satisfaction) as couples more skilled at problem solving (Bradbury and Karney 2004).

The same holds for conflict. As discussed in the previous chapter, the presence of conflict early in marriage does not indicate that the marriage is doomed any more than the absence of conflict guarantees positive feelings of warmth or more affection. April Wilson and Ted Huston do point out that the interplay of personality and relationship factors may foretell later trouble. Premarital couples who are highly critical of each other and make less than adequate effort to work through their problems establish patterns that may lead to less satisfying marriages. They found that a "shared reality," as measured by feeling comparable levels of love and experiencing similar changes in their estimates of the likelihood they would actually marry, were especially important predictors of divorce. Stable marriages are more likely when couples have similar feelings of love, and share beliefs about and progress similarly in their confidence that they will marry (Wilson and Huston 2013).

Psychologist John Gottman, who articulated some of the warning signs of potential future marital distress (e.g., the "four horsemen of the apocalypse") addressed last chapter, reportedly can predict with impressive accuracy, which newly married couples will and won't divorce within the decade after they enter marriage. Gottman's methodology focuses on listening to five minutes of how couples argue and whether they display behaviors indicative of an eventual breakup. What are these behaviors? Based on the Gottman Institute's Gottman Relationship Blog, they consist of the following (Fulwiler 2014):

- *Harsh startups to the discussion:* This pertains to when either partner begins the discussion in an accusatory or negative way, or displays contempt. With such a beginning, the discussions are said to be destined to fail.

- *Displays of the "four horsemen":* Again, these four warning signs consist of criticism (rather than complaining about a particular behavior, the person attacks the character of one's spouse), contempt (sarcasm, cynicism, eye-rolling, mocking, etc.), defensiveness, and stonewalling (tuning out the partner and avoiding discussion).

- *Flooding:* When a spouse's criticism's, contempt, and negativity are overwhelming, one experiences "flooding," a feeling likened to feeling "shell-shocked").

- *Physiological arousal signs:* These are increases in blood pressure, heart rate, and the secretion of adrenalin, which cause one to experience a fight or flight response. In either instance, the discussion fails.

- *Failed repair attempts:* A serious sign of likely instability is when one spouse's attempts to de-escalate the tension and conflict (via humor, a smile, an apology, etc.) and keep the negative emotions from "spiraling out of control" fail, perhaps because one's partner is feeling flooded.

- *Bad memories:* The reported "final sign that divorce is inevitable," results when couple's recall and describe their history with a negative view.

# Engagement, Cohabitation, and Weddings

For married-couple families, the beginning stages of constructing their families unfold as they experience a serious romantic relationship, followed eventually by engagement or cohabitation, and ultimately by a wedding, the ceremony that represents the beginning of a marriage.

## Engagement and Cohabitation

Engagement is the culmination of the premarital dating process. Today, in contrast to the past, engagement retains significance more as a ritual than as a binding commitment to be married. Engagement is losing even its ritualistic meaning, however, as more couples start out in less formal relationships or by living together. These couples are less likely to become formally engaged. Instead, they announce that they "plan to get married." Because it lacks the formality of engagement, "planning to get married" is also less socially binding.

According to three different wedding industry sources (weddingindustrystatistics.com, theweddingreport.com, and xogroupinc.com), the average engagement in the United States in 2011 was between 12 and 15 months. Engagement signifies the depth of the couple's commitment to their relationship, makes their intention to marry known, and helps define the goal of the relationship as marriage. Engagement performs several functions, including the following:

- Engagement prepares couples for marriage by requiring them to think about the realities of everyday married life: money, friendships, religion, in-laws, and so forth. Partners are expected to begin making serious plans about how they will live together as a married couple.

- Engagement is the beginning of kinship. The future marriage partner begins to be treated as a member of the family. He or she begins to become integrated into the family system.

- The engagement period allows the marrying couple to plan the wedding.

- Engagement allows the prospective partners to strengthen themselves as a couple.

Men and women may need to deal with a number of social and psychological issues and even wrestle with potential doubts during engagement, including questions about such matters as their readiness for marriage (e.g., Are we mature enough?), whether their partner is the right person for them (e.g., Is she or he really "the one"?), and how their lives will change (e.g., What am I giving up? What am I getting?).

In addition, couples may discover or experience difficulties with the following:

- *Gender-role conflict*—disagreement over appropriate male and female roles;

- *Idealization and disillusionment*—the tendency to believe that your partner is "perfect" and to become disenchanted when he or she is discovered to be "merely" human;

- *Marital expectations*—beliefs that the marriage will be blissful and free of conflict, and that your partner will be entirely understanding of your needs; and

- *Self-knowledge*—an understanding of yourself, including your weaknesses as well as your strengths.

Psychologists Blaine Fowers and David Olson constructed a fourfold typology of engaged couples using a premarital inventory known as PREPARE. The PREPARE inventory consisted of 125 items designed to indicate couples' strengths and work areas across 11 different relationship areas (including among them, for example, communication, conflict resolution, sexual relationship, realistic expectations, financial management, and leisure). Based on their findings, they identified four types of engaged couples: *vitalized*, *harmonious*, *traditional*, and *conflicted*.

The rapid and remarkable rise of cohabitation has continued to rewrite the story of contemporary marriages and families, and along the way altered the premarital stage. As we will see in the next chapter, the meaning of cohabitation varies; for some people, cohabitation is an alternative way of *entering marriage*. More than half of first marriages occur among couples who are already cohabiting. For still others, cohabitation is an alternative to marrying.

Although cohabiting couples may be living together before marriage, their relationship is not legally recognized until the wedding, nor is the relationship afforded all of the same social legitimacy. For example, relatives may not consider one's cohabitants as kin. As will be discussed in Chapter 9, the risks previously found for those who marry following cohabitation, especially if they began living together before being formally engaged, have diminished greatly. Such couples were, in the past, found to have a higher divorce rate, report lower levels of positive attributes about their marriages on average, and have more negative marital interactions than did marriages in which couples were engaged before they began living together (Rhoades, Stanley, and Markman 2009; Stanley et al. 2010). Cohabitation does, however, perform some of the same functions as engagement, such as preparing the couple for some realities of marriage and helping them think of themselves as a couple.

## Weddings

There were approximately 2.2 million weddings in the United States in 2014. Data from a variety of industry sources indicate that the nationwide average cost of a wedding was between $28,000 and $30,000 per wedding, though costs vary considerably depending on the kind of wedding one arranges and where one marries. In Manhattan, in New York City, average wedding costs can exceed $70,000, whereas in Utah, reportedly the state with the lowest statewide average cost, the average wedding costs are just over $13,000 (www.weddingstats.org/average-cost-of-a-wedding.html). Of course, wedding expenses will vary as well based on season, style, and even day of the week. It is estimated that approximately half of weddings end up exceeding the budgeted costs, but that only around 16 percent will actually exceed $30,000 (www.weddingstats.org).

Weddings are ancient rituals that symbolize a couple's commitment to each other. The word *wedding* is derived from the Anglo-Saxon *wedd*, meaning "pledge." It included a pledge to the bride's father to pay him in money, cattle, or horses for his daughter (Ackerman 1994; Chesser 1980). When the father received his pledge, he "gave the bride away." The exchanging of rings dates back to ancient Egypt and symbolizes trust, unity, and timelessness because a ring has no beginning and no end. It is a powerful symbol. To return a ring or take it off in anger is a symbolic act. Not wearing a wedding ring may be a symbolic statement about a marriage. Another custom, carrying the bride over the threshold, was practiced in ancient Greece and Rome. It symbolized the belief that a daughter would not willingly leave her father's house. The eating of cake is similarly ancient, representing the offerings made to household gods; the cake made the union sacred (De Coulanges 1980). Many contemporary African Americans have incorporated the African tradition

Weddings carry multiple meanings, both about the individuals marrying and the nature of their commitment.

of jumping the broomstick, carried to America by enslaved tribespeople and into their wedding ceremonies (Cole 1993).

The honeymoon tradition can be traced to a pagan custom for ensuring fertility: Each night after the marriage ceremony, until the moon completed a full cycle, the couple drank honey wine. The honeymoon was literally a time of intoxication for the newly married man and woman. Flower girls originated in the Middle Ages; they carried wheat to symbolize fertility. Throughout the world, gifts are exchanged, special clothing is worn, and symbolically important objects are used or displayed in weddings (Werner et al. 1992).

Wedding ceremonies, celebrations, and rituals such as those described are rites of passage encompassing rites of separation (e.g., the giving away of the bride), aggregation, and transition. It is especially noteworthy as a rite of transition wherein it marks the passage from single to married status. The wedding may also reflect the degree to which both the bride's and the groom's "social circles" are part of the transition into marriage. As such, weddings vary. As Matthijs Kalmijn (2004) describes, they range from highly public large weddings to highly private weddings with just a couple of witnesses. Of course, the aforementioned costs of weddings will range considerably between the two extremes.

Marriage is a major commitment, and entering marriage may provoke considerable anxiety and uncertainty. Is this person right for me? Do I really want to get and be married? What is married life going to be like? Will I be a good wife or husband? These are examples of the kinds of anxieties brides and grooms might feel as they approach marriage. Weddings may increase the commitment by creating and involving an audience that can serve as witnesses to the commitment a marrying couple is making.

Andrew Cherlin (2004, 856) suggests that where weddings had historically been celebrations of a kinship alliance between two kin groups and later a reflection of parental "approval and support" for their child's marriage, today's weddings are more a symbolic demonstration of "the partners' personal achievements and a stage in their self-development." A wedding is, in part, a statement, as is the purchase of a house. It says, "Look at what I have achieved. Look at who I have become." Seen this way, we can understand why, despite the economic obstacles they face, low-income couples can honestly contend that a major barrier preventing them from marrying is

insufficient money to have a "real wedding" (i.e., a church wedding and reception party). For some, "going down to the courthouse" is not a real or sufficient wedding (Smock 2004). A big wedding means that a couple "has achieved enough financial security to do more than live from paycheck to paycheck" (Cherlin 2004, 857). Both "the brides and grooms of middle America" and low-income, unmarried parents alike desire "big weddings," even if the nature of "big" varies between the two (Edin, Kefalas, and Reed 2004). This is all part of the deinstitutionalization of marriage mentioned earlier. Marriage and the wedding that signifies its beginning has become more of a symbol of individual achievement and development. If it is no longer the foundation of adult life, it still serves as a capstone (Cherlin 2004).

To other analysts, weddings are seen as mostly "occasions of consumption and celebrations of romance" (Cherlin 2004, 857). Indeed, weddings of today are big business, and the "bridal business" is estimated to be a $48 billion industry (www.weddingstats.org). Not all couples, however, have elaborate formal weddings. Although no national data are available on the percentage of weddings that are civil as opposed to religious ceremonies, civil ceremonies (typically performed by a judge, notary, or justice of the peace) had increased substantially in the latter decades of the last century (Grossman and Yoo 2003). Because of the expense, some couples opt for civil ceremonies, which may have relatively minimal cost beyond the marriage license.

Whether a first, second, or subsequent marriage, a wedding symbolizes a profound life transition. Most significantly, the partners take on new marital roles. For young men and women entering marriage for the first time, marriage signifies a major step into adulthood. Some apprehension felt by those planning to marry may be related to taking on these important new roles and responsibilities. Therefore, the wedding must be considered a major rite of passage. When they leave the wedding scene, the couple leaves behind singlehood; they are now responsible to each other as fully as they are to themselves and more than they are to their parents.

However, if we focus too much on the ceremonial aspect of marriage, we overlook two important points: First, marrying is a process that begins well before and continues after the couple exchanges their vows. Second, the legal or ceremonial aspect of marrying may not be the most profound part of the transition.

# In the Beginning: Early Marriage

As couples navigate their way through the transition to marriage, they experience changes in a host of areas or dimensions of their day-to-day lives. Past analyses of both divorce and remarriage have used the concept of stations of marriage to represent both the dynamic and the multidimensional nature of transitions out of and back into marriage (Bohannan 1970; Goetting 1982). By people's approach to divorce and remarriage, these analyses imply that marriage itself is multidimensional. Thus, the concept of "stations" should work equally well when used to depict the complex process of marrying (for further discussion of Bohannan's stations of divorce, see Chapter 13). Both Bohannan and Goetting stressed that marital transitions are thick with complexity, and couples experience new emotional, social, economic, and legal realities. They also undergo alterations in their self-identities and ways of understanding social reality. For those who enter as parents or become stepparents upon marriage, the transition to marriage is complicated by the transition to or demands of parenthood.

Ted Huston and Heidi Melz (2004) contend that newly married couples are typically affectionate, very much in love, and relatively free of excessive conflict, a state that might be called "blissful harmony." Within a year, this affectionate climate "melts" into a more genial partnership. As they point out, "One year into marriage, the average spouse says, 'I love you,' hugs and kisses their partner, makes their partner laugh, and has sexual intercourse about half as often as when they were newly wed" (Huston and Melz 2004, 951). Even though conflict is not necessarily more frequent or intense, it is less likely to be embedded in the highly affectionate climate of new marriage when it occurs. Thus, it may feel worse.

Huston and Melz (2004, 952) also found that couples establish a "distinctive emotional climate" from the outset that does not change over the initial two years of marriage; they are either happy or unhappy. Thus, it is not the case that unhappy couples begin on a blissful happy note and see things fail; instead, "most unhappy yet stable marriages fall short of the romantic ideal" from the beginning. All couples, even happy ones, have their ups and downs. Happy couples, however, typically contain two people who are both warm and even-tempered.

## Establishing Marital Roles

The expectations that two people have about their own and their spouse's marital roles are based on gender roles and their own experience. Traditional legal marriage was explicitly gendered and clearly patriarchal. It established the husband as head of the household and imposed other expectations according to gender (Weitzman 1981).

Whether or not the provisions of traditional legal marriage once reflected what many couples actually experienced in marriage, such assumptions no longer fit with contemporary marital reality. For example, the husband traditionally may have been regarded as head of the family, but today power tends to be more shared, albeit perhaps not equally. In dual-earner families, both men and women contribute to the financial support of the family, sometimes with the wife earning a larger share of the couple's income. Although responsibility for domestic work still rests largely with women, men have gradually increased their involvement in household labor, especially child care. The mother is generally still responsible for child rearing, but as we shall examine in Chapters 10 and 11, fathers are participating quite a bit more.

### Marital Tasks

Newly married couples need to begin a number of marital tasks to build and strengthen their marriages. The failure to complete these tasks successfully may contribute to what researchers identify as the **duration-of-marriage effect**—the accumulation over time of various factors such as unresolved conflicts, poor communication, grievances, role overload, heavy work schedules, and child-rearing responsibilities that might cause marital disenchantment. The tasks couples face are primarily adjustment tasks and include the following:

- *Establish marital and family roles.* Spouses need to discuss marital-role expectations for oneself and one's partner and make appropriate adjustments to fit each other's needs and the needs of the marriage.

- *Provide emotional support for the partner.* Couples must learn how to give and receive love and affection, support the other emotionally, and fulfill personal identity as both an individual and a partner.

- *Adjust personal habits.* Each partner faces adjusting to each other's personal ways by enjoying, accepting, tolerating, or changing personal habits, tastes, and preferences, such as differing sleep patterns, levels of personal and household cleanliness, musical tastes, and spending habits.

Spouses also need to do the following:

- Negotiate gender roles.
- Make sexual adjustments.
- Establish family and employment priorities and negotiate a division of labor.
- Develop communication skills and learn how to effectively share intimate feelings and ideas with each other.
- Manage budgetary and financial matters.
- Establish kin relationships, participate in extended family, and manage boundaries between family of marriage and family of orientation.
- Participate in the larger community.

As you can see, newly married couples must undertake numerous tasks as their marriages take form. Marriages take different shapes according to how different tasks are shared, divided, or resolved. It is no wonder that many newlyweds find marriage harder than they expected. But if the tasks are undertaken in a spirit of love and cooperation, they offer the potential for marital growth, richness, and connection. If the tasks are avoided or undertaken in a selfish or rigid manner, however, the result may be conflict and marital dissatisfaction.

## Establishing Boundaries

When people marry, many still have strong ties to their parents. Until the wedding, their family of orientation has greater claim to their loyalties than their spouse-to-be. After marriage, the couple must negotiate a different relationship with their parents, siblings, and in-laws. Loyalties shift from their families of orientation to their newly formed family.

Of course, relationships with parents and siblings continue after marriage and continue to influence and be influenced by marriage. Difficulties between one's spouse and one's family can add considerable strain to a marriage. For the most part, the better one's relationship with one's in-laws, the more likely the marriage is to be happy and the lower the likelihood of divorce. In-laws may supply needed assistance, whether that takes the form of economic assistance, practical

help, emotional support, or advice. Newly married couples may have little money or credit and ask parents to loan money, cosign loans, or obtain credit. But such financial dependence can keep the new family tied to the family of orientation. The parents may try to exert undue influence on their children because their money is being spent. With regard to emotional support and advice, maintaining close relations with family may make parents and in-laws feel entitled to offer unsolicited advice, which can contribute to a feeling of interference and to potential distress or conflict as a result (Orbuch et al. 2013).

The critical task is to form a family that is interdependent rather than independent or dependent. It is a difficult task, as parents and their adult children begin to make adjustments to the new marriage. Research suggests differences by both gender and race in creating and maintaining wider family connections. Women more than men are likely to bear more responsibility and take more action to maintain or strengthen ties with both sides of their new families. A recent analysis of longitudinal data on 373 married couples (199 black and 174 white) explored the connections between closeness to in-laws in early marriage and marital stability across the first 16 years of marriage (Orbuch et al. 2013). Men's feelings of closeness to in-laws reduced the likelihood of divorce over time for both black and white men. Terri Orbuch and colleagues suggest that those early ties may serve both to connect new husbands to their in-laws and strengthen their marital relationships, as wives value their husbands' being able to get along with their families. Interestingly, Orbuch et al. found a similar outcome for black wives but not for whites. Black wives who reported feeling close to their in-laws early in marriage were significantly less likely to divorce. On the other hand, white wives who reported feeling close to in-laws early in marriage were *significantly more likely* to divorce. Orbuch et al. suggest that for white wives, close ties to in-laws may be felt and perceived as interference, whereas for black wives such closeness may provide psychological and practical benefits that are especially helpful in buffering the couple against economic and other kinds of stressors (Orbuch et al. 2013).

When couples marry, they need to maintain bonds with their families of orientation, and to participate in the extended family network, they cannot let those bonds turn

**True** **8.** Married people are less likely to socialize with friends and neighbors than are never-married or previously married women and men.

# Can We Learn Lessons about Marriage from *Wife Swap* and *Trading Spouses*?

Marriage is lived behind closed doors and drawn curtains. Although we see many different married couples out and about in public settings, we really don't know how they live their daily lives. In fact, we have the opportunity to see "up close and in person" very few marriages aside from our own should we, in fact, marry. As a consequence, our understanding of just how diverse marriages really are—how many different ways people construct and experience their marital relationships—is very limited. Although researchers such as John Cuber and Peggy Harroff or Yoav Lavee and David Olson have constructed typologies (discussed later in this chapter) to illustrate different types of marriages, we may still operate with the misconception that there is one right or best way to be married or believe that most marriages are alike.

That's what makes a reality television program like ABC's *Wife Swap* (or Fox's version, *Trading Spouses: Meet Your New Mommy*) potentially so interesting. What the show does, as did its award-winning British predecessor, is take two married women with children and have them "swap lives," living in the other's house with the other's family for a two-week period of time. During the first week, the visiting wife tries to live by the other's rules and standards. During the second week, the families are supposed to adhere to what the visiting wife recommends.

Even acknowledging the distortions likely created by filming and selectively editing people's daily lives, each episode of *Wife Swap* gives us a glimpse at two very different households, often from different socioeconomic circumstances, and how difficult it is for the two women to switch places. Although the hope (and claim) is that the women and their families learn other ways in which they could structure their lives, they also learn to appreciate each other as well. For viewers, however, there is more. As American studies professor Allison McCracken observes,

> *Wife Swap* reveals the specificity of people's lives through attention to the mundane, rather than sensational, details that accompany the "wifely" role: cleaning, cooking, child care, spousal negotiations, religious practices, professional responsibilities. (McCracken 2005)

Writing in the Irish newspaper *The Independent* about the British version of the show, journalist John Masterson offers, "It should be compulsory viewing for anyone thinking of shacking up with anyone. Even before, 'Do you take this man/woman etc.,' the cleric should ask, 'Have you both watched *Wife Swap*?' After they have both solemnly uttered, 'I did,' then the ceremony may continue" (Masterson 2008). While we wouldn't go that far, we recognize the value of such inside views of marriage and family life, especially as experienced by an outsider. •

into chains. It is a potentially delicate balancing act, made more difficult by claims that marriage makes in people's lives.

Sociologists Natalia Sarkisian and Naomi Gerstel used two large national surveys to assess what they call the "greedy" nature of married life. Of interest, they found that married women and men are less involved with their parents and siblings than either the never-married or previously married. They are less likely to visit, call, write, or provide either emotional or practical help to their parents. They are also less likely to see, call, or write their siblings, though they are slightly more likely than the formerly married to offer emotional or practical help to siblings. Interestingly, divorced women and men are less involved with their parents than are never-married people. The pattern of "diminished ties," especially to parents, persists even after marriages end. Even after taking into account such factors as time demands, needs and resources, demographics, or extended-family characteristics, the

impact of marriage on family ties, especially to parents, remains (Sarkisian and Gerstel 2008).

Looking beyond the family of orientation to relationships with friends and neighbors, Gerstel and Sarkisian found that married people are less likely to socialize with friends and neighbors than are either the never-married or the previously married. Once married couples have children, they show more similarity to the never-married or previously married in providing friends and neighbors with practical or emotional help, though they are still less likely to hang out with either neighbors or friends (Gerstel and Sarkisian 2006).

Ideally, both families of orientation and peers (friends and neighbors) will understand, accept, and support these breaks. The new family must establish its own boundaries. Married couples should decide how much interaction with and assistance to others, especially their families of orientation, is desirable and how much influence these others may have.

Looking further outside their personal networks, we find research suggesting that marriage also can affect volunteering and contributing to charity, but differently for women and men (Einolf and Philbrick 2014). Christopher Einolf and Deborah Philbrick report that women and men differ in terms of how much time and money they contribute to charitable organizations. Most research comparing the genders in terms of volunteering and charitable giving finds that women in the United States do more volunteering than do men, though the research on charitable giving has been somewhat inconsistent. Einolf and Philbrick examined what effect marriage seemed to have on women and men's volunteering and giving. Using data from the nationally representative Panel Study of Income Dynamics, they found that newly married men were more likely to give money to charity and to increase the amount they give compared to prior to marriage. Newly married women tended to volunteer less often than prior to marriage and, when they did volunteer, did so for fewer hours than prior to marriage (Einolf and Philbrick 2014).

## Social Context and Social Stress

Even with all the attention paid to the dynamics of spousal relationships, marital success is also affected by things that happen outside of and around married couples (Bradbury and Karney 2004). Marriages are affected by the wider context in which we live, including "the situations, incidents, and chronic and acute circumstances that spouses and couples encounter," as well as the developmental transitions they undertake (Bradbury and Karney 2004). Changes in employment, the transition to parenthood, health concerns, friends, finances, in-laws, and work experiences can all affect the quality of marriage relationships. As Thomas Bradbury and Benjamin Karney (2004, 872) express, "Theoretically identical marriages are unlikely to achieve identical outcomes if they are forced to contend with rather different circumstances."

Similarly, they contend that marriages that are "rather different" in their internal dynamics may reach similar outcomes in quality, depending on whether the wider context is especially healthy or especially "toxic" (Bradbury and Karney 2004). From their research on married couples, Bradbury and Karney offer the following points to consider:

> **True** **9.** Marital relationships appear to be more affected by minor stresses than by major stresses such as unemployment or serious illness.

- Marital quality was lower among couples experiencing higher average levels of stress.
- During times of elevated stress, more relationship problems were perceived, and a partner's negative behaviors were more often viewed as selfish, intentional, and blameworthy.

Incorporating research findings from other studies, they also offer the following especially supportive evidence of the importance of social context on marital interaction and quality:

- Observational research found that because of greater job stress, blue-collar husbands were more likely than white-collar husbands to respond with negative affect to negative affect from their wives in problem-solving discussions.
- Among married male air-traffic controllers, on high-stress days in which they received support from their wives, they expressed less anger and more emotional withdrawal.
- Among a sample of more than 200 African American couples, those living in more distressed neighborhoods (as measured by a composite that included such things as income and the proportion of the neighborhood on public assistance, living in poverty, being unemployed, and living in single-parent households) experienced less warmth and more overt hostility.

Cumulatively, findings such as these remind us that improving the quality of marriage may take more than educating couples about marriage or even providing therapeutic intervention. It may be necessary to attend to and "fix" contextual circumstances also, even if it means "bypassing couples and lobbying for change in environments and conditions that impinge on marriages and families" (Bradbury and Karney 2004, 876). This doesn't lessen the importance or potential benefit of either premarital education or therapeutic assistance for marriage; it simply puts social contexts and circumstances onto an equally significant level of importance.

Although one might expect major stressor events such as unemployment, death of a loved one, accidents, or acute illness to seriously affect couples, recent research points to more "minor," everyday stresses such as those associated with childrearing, conflicts in the workplace, or problems with neighbors as playing a more important role in understanding how couple relationships function (Randall and Bodenmann 2009). These more minor and external stresses are reported to "spill over" into the couple relationship, affecting communication and marital quality. Family researchers Ashley Randall and Guy Bodenmann noted that such everyday stress is often associated with relationship deterioration, and they recommend that researchers consider the intensity, duration, and origin of stress if they hope to understand its effects (Randall and Bodenmann 2009).

## Marital Commitments

What keeps us in a relationship? Is it something internal to an individual (a reflection of attitudes, values, and beliefs), or is it something external (the outcome of constraints)? Just what does the commitment to marriage entail?

Trying to sort out the meaning and experience of **marital commitment**, Johnson, Caughlin, and Huston (1999) identify three major types of commitment, each of which operates within marriage:

- *Personal commitment.* In essence, this is the degree to which one wishes to stay married to his or her spouse. As such, it is affected by how strongly one is attracted to one's spouse, how attractive one's relationship is, and how central the relationship is to one's concept of self.

- *Moral commitment.* This is the feeling of being "morally obligated" to stay in a relationship, resulting from one's sense of personal obligation ("I promised to stay forever and I will"), the values one has about the lifelong nature of marriage (a "relationship-type obligation"), and a desire to maintain consistency in how one acts in important life matters ("I am not a quitter, I have never been a quitter, and I won't quit now").

- *Structural commitment.* This is our awareness and assessment of alternatives, our sense of the reactions of others and the pressures they may put on us, the difficulty we perceive in ending and exiting from a relationship, and the feeling that we have made "irretrievable investments" in a relationship and that leaving the relationship would mean we had wasted our time and lost opportunities all for nothing.

Personal commitment is more a product of love, satisfaction with the relationship, and the existence of a strong couple identity. Moral commitment is the product of our attitudes about marriage and divorce, our sense of a personal "contract" with our spouse, and the desire for personal consistency. Finally, structural commitment is a product of attractive alternatives, social pressures, fear of termination procedures, and the feeling of sacrifices we have made and cannot recover. Johnson, Caughlin, and Huston (1999) suggested that in our efforts to understand why marriages do or don't last, we tend to look mostly at personal commitment. To more accurately understand the fate of a marriage, we need to look at how all three types are experienced and how each influences the outcome and experience of marriage.

## How Parenthood Affects Marriage

Marriages change over time. Circumstances change, individuals age, couples may become parents, and spousal roles and responsibilities are subject to renegotiation. Marital satisfaction may ebb and flow, as marital conflicts may increase or decrease depending on the situations and circumstances couples face. Much research attention has been devoted to examining the effects of children on marriage. The presence of children in the household appears to lower marital satisfaction and increase marital conflict, in part by virtue of the division of housework and child care becoming more traditional (Cowan and Cowan 2011; Hatch and Bulcroft 2004). In fact, there has been considerable research that reveals a "deteriorating of marital functioning" when couples become parents. As Philip Cowan and Carolyn Pape Cowan report, "more than 50 studies" replicate the finding that spouses' satisfaction with their couple relationship declines after the first child is born (Cowan and Cowan 2011). Although it is sometimes unclear how soon such decline begins, how long such deterioration lasts, or whether it is much different than what occurs among couples without children, spouses appear to devote less time and energy to their marital relationships after becoming parents.

The Cowans note that the decline in women's marital satisfaction tends to start sooner than the decline in men's. Through the increasingly traditional division of housework and child care, spouses are more likely to occupy different spheres, and as such, experience loneliness, disappointment, and unhappiness with their relationship. They also note that marital satisfaction among parents continues to decline for the next 15 years or more (Cowan and Cowan 2011).

As a consequence of the demands imposed by new parenthood, the arrival of children is also frequently followed by an immediate decline in both independent and shared leisure activities. Although some of the decrease is gradually reversed, neither spouse fully returns to the level of either the independent or the shared leisure that they maintained before becoming parents. Given that shared leisure time affects the way spouses feel about their relationship and each other, a reduction in those activities may contribute to a lower level of marital satisfaction or happiness after spouses become parents (Claxton and Perry-Jenkins 2008). In fact, with all that accompanies the transition to parenthood (see Chapter 10), it is unsurprising that more frequent conflict and tension ensue, that couples often change the ways in which they handle or resolve conflict, and that marital satisfaction drops.

Interestingly, as mentioned, marital satisfaction declines in the early years of marriage even among nonparents. This duration-of-marriage effect suggests that as couples move beyond the honeymoon phase, as they grow more familiar, they assess each other and their relationship more realistically. Any unrealistic premarital or even newly married expectations about married life or about their partners can translate into disappointment as couples experience the reality of marriage (Van Laningham, Johnson, and Amato 2001).

Marriages are also affected by changes each spouse experiences. Husbands and wives do not carry identical parenting or wage-earning responsibilities and thus may have very different experiences if and as children are born and grow, and as jobs increase or peak in demandingness. In addition, changes in employment, income, or residence can potentially affect how spouses evaluate their marriages (Van Laningham, Johnson, and Amato 2001).

One common outcome of becoming parents is the increased traditionalization of the division of responsibilities between spouses. This too will be addressed in Chapter 10. For now, it is enough to note the following. According to survey data from the Pew Research Center, young adults (under 30) are the most likely of those 18 and older to favor a dual-income marriage, where husband and wife both have jobs and both take care of the house and children, over the idea of a breadwinner–homemaker marriage. Almost three-quarters of 18- to 29-year-olds (72 percent) prefer this dual-income marriage ideal over the male breadwinner–female homemaker model (favored by 22 percent of 18- to 29-year-olds). Yet,

The arrival and presence of children profoundly affect marital relationships.

among married couples where the wife is under 30, we find the highest percentage of breadwinner–homemaker marriages (32 percent). Why? The answer has to do with parenting. More than half of these younger couples have preschool-age children at home. With very young children at home, the dual-income model, though perhaps ultimately their preferred division of roles, is less evident, and the percent of breadwinner–homemaker households is more likely than among couples who are older and whose children are older. Among couples where wives are 30 to 49 years old, there are higher percentages of dual-income marriages (62 percent), and fewer male breadwinner–female homemaker households (27 percent) (Wang 2013).

## ◼ Middle-Aged Marriages

Middle-aged couples in their 40s and 50s frequently have families with adolescents and/or young adults leaving home. Adolescence brings with it a number of cognitive, emotional, physical, and social changes, and adolescents may trigger considerable family reorganization on the part of parents: They stay up late, play loud music, infringe on their parents' privacy, and leave a trail of empty pizza cartons, popcorn, and dirty socks in their wake. Conflicts over tidiness, study habits, communication, and lack of responsibility may emerge. Adolescents want rights and privileges but have difficulty accepting responsibility. Such conflict can and often does spill over into the husband–wife relationship. In finding that marital satisfaction decreased for both mothers and fathers of adolescents, sociologist Ming Cui and psychologist M. Brent Donnellan suggest that "there might be noteworthy challenges involved in the parenting of adolescents that can spill over to the marital relationship" (Cui and Donnellan 2009). Furthermore, the research literature indicates changes in parent–child relationships, including less time spent together, less mutual acceptance, increasing emotional distance, and increasing conflict—particularly between adolescent daughters and their mothers (Whiteman, McHale, and Crouter 2007)—can affect how parents relate as spouses.

Married couples with adolescent children may disagree about such issues as their teens' dating, increasing autonomy, and their assumption of more adult roles and responsibilities. Such disagreements may be at least part of the cause of declines in spouses' reports of marital love and satisfaction during this stage of parenting. Shawn Whiteman, Susan McHale, and Ann Crouter found that among the 188 families that they studied over a seven-year period, both fathers and mothers seemed especially affected by the onset of adolescence of their firstborn children, with both expressing declines in positive assessments and increases in negativity (Whiteman, McHale, and Crouter 2007).

Whiteman, McHale, and Crouter (2007) remind us of the importance of recognizing that other things may occur during the period of parenting adolescents that can also cause marital changes. For parents at midlife, their own aging process may cause them distress. As their hair grays or their vision weakens, they may feel distress that can spill over into their feelings about their marriages.

## Families as Launching Centers

Some couples may be happy or even grateful to see their children leave home, some experience difficulties with this exodus, and, increasingly, many continue to accommodate their adult children under the parental roof. As children are "launched" from the family (or "ejected," as some parents wryly put it), the parental role becomes increasingly less important in daily life. The period following the child's exit is commonly known as the **empty nest**.

Much research has found what appeared to be an increase in marital satisfaction after children leave. Such studies were mostly cross-sectional in design, studying marriages of different durations at the same single point in time. From those studies, it appeared that satisfaction trended upward among couples whose children left home. Longitudinal studies, following a sample of couples over a span of time, find no such bump in satisfaction. Instead, marital

satisfaction appears to be relatively stable over time or declines (Van Laningham, Johnson, and Amato 2001).

Once children are gone, couples must re-create their family, minus their children. Some couples may divorce at this point if the children were the only reason the pair remained together. The outcome is more positive when parents have other, more meaningful roles, such as school, work, or other activities, to which to turn (Lamanna and Riedmann 1997).

## The Not-So-Empty Nest: Adult Children and Parents Together

For an increasing number of middle-age couples, the empty nest stage may be delayed, while many others who had reached that stage may find it disrupted and find their nests refilling with the return of their young adult children. A higher proportion of adult children are living with their parents today than even during the depths of the recession of the last decade. Of adults 18 to 34, 67 percent of 18- to 34-year-olds were living independently during the first quarter of 2015; more than one-fourth (26 percent) were living with parents (Fry 2015a). This represents a slight increase in the percentage living with parents compared to 2010, when 24 percent of 18- to 34-year-olds lived with their parents, and 69 percent lived independently. In 2007, more than 70 percent (71 percent) of 18- to 34-year-olds lived independently, while 22 percent lived with family (Fry 2015a). Young adults at home are such a common phenomenon that one of the leading **family life cycle** scholars suggested an additional family stage: adult children at home (Aldous 1990). Sue Shellenbarger, in the *Wall Street Journal*, referred to this stage as an "open nest" (Shellenbarger 2008), while Katherine Newman (2012) referred to this familial form, in which adult children remain or return to the parental home, as the "accordion family," expanding and contracting as needed. Newman's book, *The Accordion Family: Boomerang Kids, Anxious Parents, and the Private Toll of Global Competition*, examines the global economic circumstances that have created accordion families in many countries, including in her analysis of the United States, Japan, Italy, Spain, and Scandinavian countries such as Denmark and Sweden. We will explore the parent–adult child relationship in Chapter 10. For now, the important point is that the expectation of a marriage after child rearing is becoming less and less likely for many middle-aged couples.

These young adults are sometimes referred to as the **boomerang generation**, though, in truth, their shared household can result from either young adults or their parents moving in with the other, leading some to now also refer to "boomerang parents" (Gross 2008).

Researchers note that there are important financial and emotional reasons for this trend. The high unemployment rates and a tightened job market related to the U.S. recession are among the factors that have contributed to the increase in adult children returning or remaining home. High divorce rates, as well as personal problems, push adult children back to the parental home for social support and child care, as well as for cooking and laundry services.

This new stage generally is not one that parents have anticipated. According to Katherine Newman's analysis, the situation may be a mixed blessing for parents. It allows parents to continue as active parts of their children's lives, which can lead to closer, and more equal, parent–child relationships. She suggests that having young adults at home can also create a more exciting home atmosphere from which parents may gain, as their young adult children introduce them to new popular cultural ideas and trends and shake up what might otherwise have grown quite routine (Newman 2012). The opportunity to get to know their children as adults may be especially important to fathers, who likely have less of a social support system to draw upon. For mothers as well as fathers, the accordion family can make parents feel needed (Newman 2012).

Of course, having one's children move home or remain there beyond what was anticipated can also create problems, in that it delays the transition to the empty nest, a time that commonly allows spouses to focus more on their marriages. It also creates ambiguity in such things as curfews, responsibilities, and financial expectations, all of which may also be experienced as things about which parents, as spouses, disagree with each other. These issues will be discussed more in Chapter 10.

## Reevaluation

Middle-aged people find that they must reevaluate relations with their adult children, those who live with them as well as those who live independently. Many middle-age couples also must now incorporate new family members as in-laws. Some must also begin considering how to assist their own parents who are becoming more dependent as they age. Any and all of these issues may take away from a marriage.

Couples in middle age tend to reexamine their aims and goals. The man may decide to stay at home or not work as hard as before. The woman may commit herself more fully to her job or career, or she may remain

at home, enjoying her new childfree leisure. Because the woman has probably returned to the workplace, wages and salary earned during this period may represent the highest amount the couple will earn.

As people enter their 50s, they probably have advanced as far as they will ever advance in their work. They have accepted their own limits, but they also have an increased sense of their own mortality. They not only feel their bodies aging but also begin to see people their own age dying. Some continue to live as if they were ageless—exercising, working hard, and keeping up or even increasing the pace of their activities. Others become more reflective, retreating from the world. Some may turn outward, renewing their contacts with friends, relatives, and especially their children and grandchildren.

# ▌ Aging and Later-Life Marriages

More than 46 million Americans are over age 65, representing an estimated 14.5 percent of the U.S. population. Nearly 2 percent (1.9 percent) of the population is 85 or older. Of those over 65, 56 percent are females. Of those over 85, 66 percent are females. More than half of those over age 65 are married (57 percent), and a little more than one-fourth (27 percent) are widowed. Among men over 65, 72 percent are married and living with their spouse. Given the gender differences in life expectancy and the tendency for women to marry men who are older than themselves, it is unsurprising that far fewer women over 65 (46 percent) are married and living with a spouse (see Figure 8.8). Americans age 65 can expect to live another 19.2 years on average, 17.6 for males and 20.3 for females (Centers for Disease Control and Prevention 2011).

**Figure 8.8  Marital Status of Persons 65+, 2010**

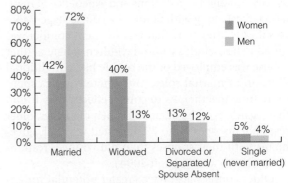

SOURCE: Administration on Aging, U.S. Department of Health and Human Services. "A Profile of Older Americans: 2011." www.aoa.gov/aoaroot/aging statistics /Profile/2011/docs/2011profile.pdf.

Among the elderly, women also are more likely to be poor. In 2013, among those over 65, an estimated 7 percent of men and 12 percent of women were below poverty (Cubanski, Casillas, and Damico 2015). However, at 10 percent overall, a smaller percentage of the elderly are poor than the overall percentage of the population living in poverty.

Beliefs that the elderly are neglected and isolated tend to reflect myth more than reality. Rather than being isolated, those over 65 are more likely to be living with either a spouse or other relatives than alone, though this is more evident among men than among women (72 percent of men and 46 percent of women lived with a spouse). Finally, as individuals age, most appear to remain fairly well connected to their families. More than half of older women and men with children say they are in contact with a son or daughter on a daily basis; 54 percent say they either see, speak to, or email on a daily basis with a son or daughter. Another 40 percent say they are in touch at least once a week. When asked, more than 80 percent say they are very satisfied with their parent–child relationships (Pew Research Center 2015).

## Marriages among Older Couples

Compared with middle-aged couples, older couples engage in less conflict. Laurie Hatch and Kris Bulcroft studied the frequency of marital conflict among women and men ages 20 to 79, and found that older husbands and wives reported less frequent conflict in areas including amount of time spent together, sex, money, and household tasks. They note a number of possible explanations for the difference in frequency of conflict, including effects of aging, stage in the family life course, and birth cohort. They ultimately emphasize cohort effects (i.e., one's age or generation and its effect on one's expectations and experience of marital conflict) along with family life course stage. More specifically, they note that less frequent conflict was found in households without children, where neither spouse was employed or the couple had a traditional division of marital roles, and where couples spent more time together. As to cohort effects, they found that regardless of the duration of their marriages, older respondents, ages 60 to 79, reported fewer disagreements with their spouses than did younger respondents (Hatch and Bulcroft 2004).

Older couples also have greater potential for engaging in pleasurable activities together and separately, such as dancing, travel, or reading (Levenson,

Marriage relationships continue to face new challenges and circumstances as couples age.

AP Images/Shavonne Potts

Carstensen, and Gottman 1993). Research in the 1990s showed that older people without children experienced about the same level of psychological well-being, instrumental support, and care as those who have children (Allen, Bleiszner, and Roberto 2000).

## Widowhood

Marriages are finite; they do not last forever. Eventually, every marriage is broken by divorce or death. Despite high divorce rates, most marriages end with death, not divorce. "Till death do us part" is a fact for most married people.

In 2014, 11 percent of men and 35 percent of women age 65 and older were widowed (Administration on Aging 2014). Using data from 2014, we can see these percentages increase for both genders as they age: 6.3 percent of men and 19.9 percent of women between ages 65 and 74 were widowed. Among those 75 to 84 years old, 15.1 percent of men and 46.6 percent of women were widowed. Finally, 33.1 percent of men and nearly three-fourths (72 percent) of women 85 and older were widowed. Because women live five to seven years longer on average than men, as shown previously, most widowed people are women (U.S. Census Bureau America's Families and Living Arrangements: 2014, Table A1).

Demographic facts of life expectancy yield many more widows than widowers, thus creating for men "many more opportunities to date and remarry should they choose to" (Carr 2004, 1,052). And they take advantage of such opportunities, as "within 6 months of bereavement, men were significantly more likely to express interest in either dating or marrying than were widowed women." By 18 months after spousal loss, only about one in ten (9 percent) widows report having gone on a date compared to nearly one in four (23 percent) widowers (Sassler 2010, 566). Widowhood is often associated with a significant decline in income, plunging the grieving spouse into financial crisis and hardship in the year or so following death.

**False** **10.** After losing a spouse, women are less likely than men are to remarry.

This is especially true for poorer families. Feelings of well-being among both elderly men and elderly women are related to their financial situations. If the surviving spouse is financially secure, he or she does not have the added distress of a dramatic loss of income or wealth.

Recovering from the loss of a spouse is often difficult and prolonged. A woman may experience considerable disorientation and confusion from the loss of her role as a wife and companion. Having spent much of his life as part of a couple—having mutual friends, common interests, and shared goals—a widower suddenly finds himself alone. Whatever the nature of marriage, those women and men whose spouses have died may experience grief, anger, distress, and loneliness. Physical health appears to be tied closely to the emotional stress associated with widowhood.

One common response of widowed women and men is to glorify or "sanctify" their marriages and their deceased spouses. This is especially true shortly after a spouse's death (Carr 2004). One way in which women and men differ in their reactions is that women who had close marriages may feel less open to seeking and forming a new relationship with another man, retaining the feeling that they are "still married" to their late husbands. Men who were in close marriages may be especially motivated to establish another marriage. Having experienced and grown dependent on the emotional support and intimacy of their marriages, they may have few other alternative sources of support to whom they can turn. Thus, they have greater incentive to form a new emotionally supportive marriage or partner relationship (Carr 2004). In addition to the loss of their confidant and chief source of emotional support, they may have limited experience managing households, cooking, and cleaning, and may suffer from poor nutrition and distress over the conditions in which they live as a result (Carr 2004).

Widows certainly suffer, too, although they may be beneficiaries of more practical help from their children and draw on emotional support from a wider and deeper network of friends. Additionally, the complexity of remarriage, along with the possible loss of Social Security, the lack of encouragement or acceptance from one's children, and the challenges created in going from separate to shared living can all affect one's desire to remarry after the loss of one's spouse. There is also evidence that widowed women more than men may come to enjoy the freedom that

The loss of one's spouse confronts women and men with a variety of deep and painful losses. Although both women and men lose their chief source of emotional support, women typically have wider and deeper friendship networks to turn to for support.

Rolf Bruderer/Spirit/Corbis

comes with being able to pursue their own interests and desires (Sassler 2010). Increasing numbers of widowed women and men are choosing to live together rather than formally remarry. There is also evidence that older cohabiting couples consider their relationships happier, more equitable, and more harmonious compared to younger cohabitors (Sassler 2010).

# Enduring Marriages

Examining marriages over time is an important way of exploring the different tasks we must undertake at different times in our relationships. Which marriages last? What researchers find is what many of us already know: Despite what might be more widely believed, we cannot assume that happy marriages will be stable ones, or that lasting marriages are happy ones. Many unhappily married couples stay together, and some happily married couples undergo a crisis and break up. In general, however, the quality of the marital relationship appears to show continuity over the years. Much of the discrepancy between happiness and stability results because happiness or satisfaction is an evaluative judgment of a marriage relative to what we expected from marriage and what better alternatives are available. Stability results more from assessments of the costs and rewards of staying in or leaving a marriage. Unhappy marriages may be enduring ones because there are no better alternatives, because the costs of leaving exceed the costs of staying married, or both.

Long-term marriages are not immune to conflict. Middle-aged and older couples may continue to experience conflict over such issues as children, money, communication, recreation, sex, and in-laws. Surviving together does not require couples to eliminate or avoid conflict.

A study by Robert and Jeanette Lauer (1986) used a somewhat modest definition of *long-term* to look at marriages that last. Their study of 351 couples married at least 15 years (most were married a good deal longer) found the following to be the "most important ingredients" that men and women identified to explain their marital success:

1. Having a spouse who is a best friend and whom you like as a person
2. Believing in marriage as a long-term commitment and sacred institution

3. Consensus on such fundamentals as aims and goals and philosophy of life
4. Shared humor

When assessing marriages, keep in mind that there is considerable diversity in married life. This can be seen in the numerous typologies that researchers have constructed to identify key characteristics or qualities that differentiate marriages. This includes typologies that address specific populations (e.g., Alzheimer patients and their spouses, African Americans, Israeli long-term marriages), as well as typologies that have been applied more generally to marriages (Allen and Olson 2001; Cohen, Geron, and Farchi 2010; Kaplan 2001). Two of the attempts to document and differentiate types of marriages that couples construct are John Cuber and Peggy Harroff's five-fold typology of marriage and Yoav Lavee and David Olson's seven types of marriages (Cuber and Harroff 1965; Lavee and Olson 1993).

In Cuber and Harroff's (1965) five types of marriage, each could either last "till death do us part" or end in divorce. Thus, these are not degrees of marital success but rather different kinds of marriage relationships:

- **Conflict-habituated marriages** are relationships in which tension, arguing, and conflict "permeate the relationship." It may well be that conflict is what holds these couples together. It is at least understood to be a basic characteristic of this type of marriage.

- **Passive-congenial marriages** are relationships that begin without the emotional "spark" or intensity contained in our romantic idealizations of marriage. They may be marriages of convenience that satisfy practical needs in both spouses' lives. Couples in which both spouses have strong career commitments and value independence may construct a passive-congenial marriage to enjoy the benefits of married life and especially parenthood.

- **Devitalized marriages** begin with a high level of emotional intensity that dwindles over time. From the outside looking in, these marriages may closely resemble passive-congenial relationships. What sets them apart is that they have a history of having been in a more intimate, sexually gratifying, emotional relationship that has become an emotional void. Obligation and resignation may hold such couples together, along with the lifestyle they have built and the history they have shared.

- **Vital marriages** appeal more to our romantic notions of marriage because they begin and continue with high levels of emotional intensity. Such couples spend much of their time together and are "intensely bound together in important life matters." The relationship is the most valued aspect of their lives, and they allocate their time and attention on the basis of such a priority. Conflict is not absent, but it is managed in such a manner as to make quick resolution likely.

- **Total marriages** are relationships in which characteristics of vital relationships are present but to a wider and deeper degree, with the "points of vital meshing" extended across more aspects of daily coupled life. Spouses appear to share everything.

Differentiating between these five types, Cuber and Harroff noted that the first three types were more common than the last two. As many as 80 percent of the relationships among their sample were of one of the first three types. Both vital and total marriages (what they called *intrinsic marriages*) were relatively rare. Again, we must remember that the researchers were not sorting relationships into "successful" versus "unsuccessful," or "good" versus "bad." Marriages of all five types were enduring marriages; thus, if one defines marriages as successes by whether they last, all five types were marital successes. Furthermore, they suggested that any of the five types could end in divorce, and none was immune from the possibility of dissolution, although the reasons for divorce would differ.

Yoav Lavee and David Olson (1993) constructed a seven-type marriage typology from an analysis of the marriages of more than 8,000 couples voluntarily in marriage enrichment programs or marital therapy. Although such a sample may be a difficult one from which to generalize, Lavee and Olson suggested that we could differentiate couples into seven different types on the basis of their satisfaction or dissatisfaction with the nine areas of married life: personality issues, conflict resolution, communication, sexual intimacy, religious beliefs, financial management, leisure, parenting, and relationships with friends and family. Of their types, *vitalized couples* (9 percent of the sample) reported themselves satisfied with all nine areas. At the opposite end, *devitalized couples* reported problems in all nine areas. Keeping in mind that the sample was drawn from either clinical or enrichment intervention, the devitalized were by far

the most common type, representing 40 percent of their sample.

The remainder of the sample was relatively evenly divided across the other types: *balanced, harmonious, traditional, conflicted,* and *financially focused.* All types except the vitalized reported problems, although the areas and extent of problems differed across these types. For example, the financially focused (11 percent) had problems in all areas but financial matters. Traditional couples (10 percent) reported problems in handling conflict, communication, sexual intimacy, and parenting. The conflicted (14 percent) reported themselves generally satisfied with only their parenting, leisure activities, and religious beliefs. Even those couples designated as harmonious (8 percent) tended to have difficulties in areas such as religious beliefs, parenting, and relations with family and friends. Balanced couples (8 percent) were generally satisfied with all areas except financial matters.

For different reasons, we need to be cautious about generalizing too far from either Cuber and Harroff or Lavee and Olson. Nonetheless, in both typologies, 75 percent or more of the sample couples were in marriages that many would define as unattractive, when held against our idealized and romanticized expectations of marriage. They are seemingly held together by something other than a deep emotional connection. In addition, both typologies underscore that marriage need not be free of conflict to last.

Similarly, a third, more recent, typology of couples, based on over 40 years of research on couple relationships, was developed by psychologist John Gottman, who categorized couples into five types: *conflict-avoiding, validating, volatile, hostile,* and *hostile-detached.* Of these, the conflict-avoiding, volatile, and validating—though different in how and how much they express emotions, whether they express sufficient positive affect and avoid the more problematic ways of interacting (e.g., criticism, contempt, stonewalling, and defensiveness)—are said to be "happy couple types," and the hostile and "hostile-detached" couples were both considered unhappy. Between the latter two, hostile couples remained married while hostile-detached couples were prone to divorce. Among the factors that are critical to differentiating the happy from the unhappy types is whether couples are able to maintain a ratio of positive-to-negative affect of at least five to one (Fulwiler 2014).

Most obviously, these and any other marital or relationship typologies illustrate that not all marriages are alike. In other words, there is no one way to "do marriage." This is a simple and obvious but potentially important point, as we attempt to study and understand marriage and as we might enter and experience it.

Throughout marriage, from the earliest most hopeful and optimistic beginning until death or divorce do us part, we are presented with opportunities for growth and change as we enter our roles as husbands or wives, become parents or stepparents, and still later become grandparents. As we have seen, marriages and families never remain the same. They change as we change, as we learn to give and take, as children enter and exit our lives, and as we create new goals and visions for ourselves and our relationships. In our intimate relationships, we are offered the opportunity to discover ourselves.

As marriage continues to undergo changes, we are left to wonder about what the future holds. We will not likely see a return to traditional marriages any more than we should expect a total disappearance of marriage. As Paul Amato (2004b) suggested, alternatives to marriage have become increasingly accepted and more widespread. People will continue to have sex prior to marriage, live together without being married, have children outside of marriage, avoid marriage altogether, and divorce if their marriages are flawed. However, in his book *The Marriage-Go-Round*, sociologist Andrew Cherlin (2009) offers a more cautionary note to at least consider. Cherlin, himself, is optimistic about marriage. Like Amato and colleagues, he believes that marriage will continue at least to have symbolic value in the United States, serving as the highest achievement—a capstone—for adult family life, if no longer the foundation of such a life, and acting as a marker to demonstrate that one has acquired the resources needed for marriage—a good education, a decent job, and a partner who is willing to pledge to stay with you indefinitely. However, using sociologist William Ogburn's concept of **cultural lag**, he warns of another, more pessimistic interpretation. Cultural lag is the outcome of rapid social change, when part of the culture (e.g., technology) changes more rapidly than another part (e.g., behavior). In this vein, he acknowledges that it is at least possible that people are entering marriage because that is exactly what their parents and grandparents before them

did. Perhaps they haven't yet realized that they can have most if not all of the benefits of marriage without marrying, that they can cohabit with many of the legal rights and benefits formerly reserved for marriage, that they can enjoy sexual relations without marrying by making use of the availability of more effective birth control, or that they can have and raise children successfully and acceptably outside of marriage. At some point, they may decide that there is no longer a good reason to marry and may stop doing so. Only time will tell.

In his more recent book, *Labor's Love Lost: The Rise and Fall of the Working-Class Family in America* (2014), Andrew Cherlin points out the marriage gap between working-class and college-educated middle-class Americans. Record numbers of those without college degrees now cohabit, have children outside of marriage, and increasingly, inside cohabiting relationships, before ultimately moving on to other cohabiting relationships or to heading single-parent households, whereas college-educated Americans tend to put off childbearing until after they have entered marriage. This illustrates how much the retreat from marriage has to do with economic opportunity and the "hollowing out" of the labor market. In accounting for such changes, he suggests that if the cause of rising rates of serial cohabitation and unmarried parenthood was part of a cultural rejection of marriage, we ought to see it happening more widely among college-educated women and men as well, given that they are exposed to the same wider climate regarding marriage. To date, this has not occurred. Instead, Cherlin points out, we have a historically unique, class-linked split, "between marriage-based families on the top rungs of the social ladder and cohabitation and single parent–based families on the middle and bottom rungs" (Cherlin 2014, 195). Such a split creates a situation where for those, especially children, on the lower rungs, family complexity and instability result, and adults are disconnected from "the institutions that have historically anchored civic life" (p. 195). Arguing that because these are the products of both economic and cultural changes, he reminds us that they will require both economic and cultural responses. In other words, instead of simply urging people to get married and/or to wait longer to have a child, we will need to address the occupational and income-related needs of those whose lifestyles are least likely to include marriage.

# Summary

- There is an ongoing *marriage debate* over the status and future of marriage. The two extreme positions in this debate are the *marital decline* and *marital resilience* positions.

- Behavioral indicators of a *retreat from marriage* include increasing percentages of adults remaining unmarried, living together, having children outside of marriage, and divorcing. The retreat from marriage varies considerably by race, education, and socioeconomic status. However, more than 80 percent of Americans are expected to someday marry.

- The *deinstitutionalization of marriage* refers to weakening of the social norms that define people's behavior in a social institution such as marriage. In *individualized marriage*, new emphases on personal self-fulfillment and freedom of choice become more important than marital commitment and obligation.

- Legal limits imposed on choice of marriage partner include gender, age, family relationship, and number of spouses.

- Restricting people to spouses from within their same group (e.g., tribe, race, or religion) is known as *endogamy*. *Exogamy* compels people to select partners from outside their same group.

- Within the population of eligible potential partners, we tend to marry others like ourselves on criteria such as age, race, religion, social class, and family history, a practice known as *homogamy*. This appears to result from deliberate choice, social pressure, and opportunity.

- Marriage provides various benefits to married people, including economic benefits, health benefits, and psychological benefits. Research supports both a selection effect (healthier and better-adjusted people are more likely to marry) and a protection effect (marriage provides a range of protective resources enabling people to prosper). The benefits depend on the quality of the marriage.

- The relationships that precede marriage often predict marital success because marital patterns emerge during these times. Premarital factors correlated with marital success include (1) background factors, (2) personality factors, and (3) relationship factors.

- Engagements prepare the couple for marriage by involving them in discussions about the realities of everyday life, involving family members with the couple, and strengthening the couple as a social unit. Increasingly, couples are living together prior to getting engaged.

- Marital success is affected by the wider social context and the extent and kind of social stresses couples face.

- A practical task in early marriage is to establish boundaries separating the newly formed family from the couple's families of orientation. Married women and men experience diminished ties to their families of origin.

- The arrival and raising of children leads to a decline in shared leisure activities by spouses, reduced marital satisfaction, and increased marital conflict. Marriages without children also experience a reduction in satisfaction.

- In middle age, many married couples must deal with issues of independence in regard to their adolescent children and with the departure of their young adult children from the home. For many families, there is no *empty nest* because of the increasing presence of adult children in the home.

- Most marriages end with the death of a spouse. Most surviving spouses are women. Widowers have greater opportunity to date, repartner, or remarry than do widows.

- Marriages differ from one another. One popular typology contrasts five types of marriage: *conflict-habituated*, *devitalized*, *passive-congenial*, *vital*, and *total*. These reflect different conceptualizations and experiences of marriage, not different degrees of marital success.

- The future of marriage is unclear. Despite changes in the nature or necessity of marriage, it has retained considerable symbolic value. Most people still value marriage, and most unmarried people expect to marry someday.

## Key Terms

boomerang generation  314
complementary needs theory  297
conflict-habituated marriages  318
cultural lag  320
Defense of Marriage Act  287
devitalized marriages  318
duration-of-marriage effect  308
empty nest  314
endogamy  289
exogamy  289
family life cycle  314
filter theories  297
heterogamy  289

# Unmarried Lives: Singlehood and Cohabitation

# 9

Biocreative/Shutterstock.com

**What Do You Think?**

*Are the following statements True or False? You may be surprised by the answers as you read this chapter.*

T F **1.** Among unmarried adults in the United States today, a majority have never been married.

T F **2.** There are more unmarried women than there are unmarried men.

T F **3.** The increase in the unmarried population appears to be a global phenomenon.

T F **4.** Unmarried women and men often encounter discriminatory treatment.

T F **5.** Compared to their divorced and widowed counterparts, never-married men fare better than never-married women.

T F **6.** For most women, marriage is preceded by cohabitation.

T F **7.** Most cohabiting couples choose to live together because they don't perceive marriage as a desirable goal.

T F **8.** Cohabitation before a remarriage improves one's likelihood of having a successful remarriage.

T F **9.** Only same-gender cohabiting couples can enter into a legal domestic partnership.

T F **10.** Same-gender couples are more likely to be interracial than are heterosexual couples.

T he next time you are in class, take a look around the room. It is likely that most of the people who share that classroom space with you expect or hope to someday marry. After all, this is true of the majority of young women and men. It is also probable that some of your classmates don't intend to marry, and that some who do intend to wed, won't, despite whatever desire or intention they possess. It is further likely that more of them fit into those categories—not wanting or intending to marry, and not marrying despite wanting to someday—than have previously. Despite the title of this book and the emphasis we place on marriage, not everyone is or desires to be part of a couple, and not every couple intends to marry. Some people prefer the freedom and flexibility they associate with being unattached and unmarried. Others might prefer to be part of a couple but are unable to find partners with whom they wish to share their lives. In either of those instances, some of your classmates will find themselves making a life of and on their own. Still others find themselves in a relationship, perhaps even living with a partner but remaining unmarried, either by choice or because they don't feel ready to marry. And some will experience a number of different types of relationships, some more casual, some quite serious, and find themselves with a series of partners as they move through their lives. These are the lifestyles that are the focus of this chapter: singlehood, and heterosexual and same-sex cohabitation.

**True**

**1.** Among unmarried adults in the United States today, a majority have never been married.

## Singlehood

Even a casual inspection of demographics in this country illustrates the striking increases that have occurred in both singlehood and cohabitation. The trends, which have taken root and grown substantially since 1960, include an eclectic combination of divorced, widowed, and never-married individuals. Each year, more and more adult Americans are among the population that is unmarried (see Table 9.1).

According to U.S. Census estimates, in 2014, approximately 106.9 million unmarried Americans

age 18 or older were either never married (67.3 million), divorced (25.3 million), or widowed (14.3 million). Of this group, 63 percent had never married, 24 percent were divorced, and 13 percent were widowed. If one also adds the 5.3 million people who were separated but not divorced, along with the 3.4 million who were married but living apart, 54 million males and 61.6 million females were living without a spouse. They represented 47 percent of all U.S. residents age 18 and over. Of this population, 58 percent had never married, 22 percent were divorced, 12 percent were widowed,

**TABLE 9.1 Percentage of Population 15 and Older Who Are Unmarried**

| Year | Men | Women |
|------|-----|-------|
| 1890 | 48% | 45% |
| 1900 | 47% | 45% |
| 1910 | 46% | 43% |
| 1920 | 42% | 43% |
| 1930 | 42% | 41% |
| 1940 | 40% | 40% |
| 1950 | 32% | 34% |
| 1960 | 30% | 34% |
| 1970 | 34% | 39% |
| 1980 | 37% | 41% |
| 1990 | 39% | 43% |
| 2000 | 42% | 45% |
| 2010 | 46% | 48% |
| 2014 | 46% | 49% |

SOURCE: Marital Status Data for 1890–1970 from U.S. Census Bureau (1989); data for 1980–2000 from U.S. Census Bureau (2001); data for 2010 from U.S. Census Bureau (2011b, Table A-1); data for 2014 from U.S. Census Bureau, 2014, Table A-1.

and just under 8 percent were legally married but separated or living apart.

Even if one considers only those officially unmarried, 45 percent of Americans age 18 and older were either never married, widowed, or divorced. Looking at households instead of individuals, almost 58 million households were headed by unmarried men or women in 2014, representing 47 percent of all households. Another 5.8 million households were headed by individuals who were still married but separated (3.7 million) or who for reasons such as jobs or military deployment were married but living apart from their spouses (2.1 million) (U.S. Census Bureau 2014). Over a quarter (28 percent) of all households in 2014 were one-person households. More than 34 million people, over 19 million women and 15.2 million men, lived alone.

## The Unmarried Population

There are more single women than men; according to the U.S. Census Bureau, among those 18 and older who are currently unmarried, there are 87 men for every 100 women (U.S. Census Bureau 2013). Data from the Pew Research Center allows us to look at how the ratio varies by age and employment status. For example, among unmarried Americans age 25 to 64, there are an estimated 97 men to every 100 women. Among all unmarried adults, there are 65 employed men for every 100 unmarried women (Wang and Parker 2014).

Women make up just over 53 percent of all unmarried Americans. Almost 17 percent (16.8 percent) of the unmarried population, representing nearly 18 million people, are age 65 years old or older (U.S. Census Bureau 2014).

A historical comparison in Table 9.1 reveals a decline in the percentage of the population that was unmarried from the late 19th century until the 1960s. The trend then reversed, and across the past four decades of the 20th century and through the first decade and a half of the 21st, the percentage of women and men age 15 and older who were unmarried grew steadily (we use age 15 and older here because of data available for purposes of historical comparison).

Internationally, similar trends can be observed. For example, as reported by sociologist Eric Klinenberg in his book, *Going Solo: The Extraordinary Rise and*

*Surprising Appeal of Living Alone* (2012), in countries such as Sweden, Denmark, Norway, and Finland, between 40 to 45 percent of all households are one-person households. In Japan, around 30 percent of households are one-person households, whereas countries such as Germany, France, the United Kingdom, Australia, and Canada all have higher percentages of one-person households than does the United States. In other words, the increasing single population is a global phenomenon; between 1996 and 2006, the number of adults living alone worldwide went from 153 million to 202 million, a 33 percent increase (Klinenberg 2012).

The percentage of unmarried Americans (18 or older) varies widely by race and ethnicity. Based on 2014 data, 61.6 percent of African Americans, 35.6 percent of Asians, 48 percent of Hispanics, and 41.2 percent of non-Hispanic whites were unmarried. Another 2.5 percent of whites, 6.5 percent of Hispanics, 4.0 percent of Asians, and 6.5 percent of African Americans were legally married but not residing with a spouse, either because they had separated or were living apart due to factors such as jobs or military deployment (U.S. Census Bureau 2014). The percentage of people who had never married also shows considerable racial or ethnic variation: 23.3 percent of non-Hispanic whites, 43.9 percent of African Americans, 35.9 percent of Hispanics, and 26.9 percent of Asian and Pacific Islanders, age 18 and older, had *never married*. Furthermore, an additional 11.3 percent of non-Hispanic whites, 11.5 percent of African Americans, 8.6 percent of Hispanics, and 4.7 percent of Asian Americans were divorced (U.S. Census Bureau 2014, Table A-1). Thus, the population of singles is both very large and quite diverse. Among African Americans, those who never married and live alone are the fastest-growing segment of the black middle class (Klinenberg 2012).

The varieties of lifestyles that single or unmarried people maintain are really too numerous to fit under one "umbrella," too complex to be understood within any one category, and too diverse to be accounted for by any one explanation. The lifestyles include the never-married and the divorced; young, middle-age, and old single parents; gay men, lesbians, and bisexuals; and widows and widowers. In addition, as noted previously, more than 5.3 million married

True
**2. There are more unmarried women than there are unmarried men.**

True
**3. The increase in the unmarried population appears to be a global phenomenon.**

Throughout much of the United States, by the third week of September, we are surrounded by signs of summer's end. Leaves start to change color, vacations become memories, and students from preschool to professional school have returned to classes. Along with the official arrival of autumn that same week comes a weeklong celebration, now known as Unmarried and Single Americans Week. Unmarried and Single Americans Week acknowledges and recognizes single women and men, celebrates the lifestyles of the unmarried, and acknowledges the contributions unmarried people make to American society.

First started in Ohio by the Buckeye Singles Council in 1982, the week was originally called National Singles Week. Taken over by the American Association for Single People in 2001, the weeklong "celebration" was renamed in recognition that many unmarried people are in relationships or are widowed and don't identify with the "single" label (U.S. Census Bureau 2012b). In 2002, the association changed its name to Unmarried America. Although the designated week has been around now for more than 30 years and is recognized by mayors, city councils, and governors in some 33 states, as of 2005, it had yet to be "legitimized" and incorporated into mainstream U.S. culture, as indicated by both the absence of greeting cards for the occasion and the number of people (including millions of unmarried people) unaware that the weeklong recognition exists (Coleman 2005).

Another sign of increasing attention being paid to the growth in numbers and importance of single women and men can be seen in the efforts to broaden the academic curriculum so as to keep pace with the changing demographics. For example, psychologist Bella DePaulo, law professor Rachel Moran, and sociologist and gender studies scholar E. Kay Trimberger provided interdisciplinary advocacy for a broader and deeper attention to be paid to singles, including formerly married, never-married, gay, lesbian, bisexual, and heterosexual people; single parents; and singles from across ethnic, racial, and class boundaries.

DePaulo, Moran, and Trimberger recommended that scholars apply a "singles perspective" to a host of topics of study (e.g., friendship, stereotyping and discrimination, and laws and policies that privilege married couples) and in disciplines such as law, sociology, psychology, women's and gender studies, and ethnic studies. Such an effort would, they claimed, "broaden and deepen scholarship while enriching the intellectual life of the classroom." DePaulo, who sociologist Eric Klinenberg calls "the minister of truth for the solo nation," has two blogs, *All Things Single (and More)* and, at the Psychology Today website, *Living Single*. The latter is described as "a myth-busting, consciousness-raising, totally unapologetic take on single life." For good reason, alongside and in recognition of domestic and international increases in the unmarried population, singles are currently the recipients of increasing public and academic attention. ●

---

women and men were separated, and another 3.4 million were married but living apart from their spouse in 2014. Some singles live with other people, whereas others live alone. In 2014, 28 percent of the more than 123 million households were men and women living alone (U.S. Census Bureau 2014).

These different populations experience diverse living situations that affect how singleness is experienced. However, when we think about those generally regarded as "single," we often think mostly of young or middle-age heterosexual women and men who are not living with someone, and working rather than attending school or college. Although there are numerous single lesbians and gay men, they have not traditionally been included *as singles* in research on unmarried men and women, in part because of their long-standing inability to marry and because of the

association of marriage with heterosexuals. In coming years, we should expect to have more data that allow us to compare gay men and lesbians who remain unmarried and on their own from those who cohabit with a partner or from those who marry.

## Never-Married Singles in the United States: An Increasing Minority

The growth in the percentage of *never-married adults*, from 20.3 percent in 1980 to 28.1 percent of those 18 and older in 2014, has occurred across all population groups (see Table 9.2). In part, this increase (like the creation of National Unmarried and Single Americans Week) reflects a change in the way in which society views this way of life. Many singles appear to be postponing marriage to an age that makes better

The proportion of the population that is unmarried varies considerably by race.

economic and social sense. Many of these young women and men, technically counted among the *never married*, may more accurately be considered "yet to be married." In 2013 public opinion data from Gallup, 5 percent of those polled were never married and do not want to get married, whereas 21 percent

were never married and want to marry (Newport and Wilke 2013). D'Vera Cohn summarized some Pew Research Center data indicating a majority of never-married respondents would like to marry at some point in their lives. Of those who have never married, 61 percent said they would like to get married, whereas just 12 percent said they do not want to marry (another 27 percent of those who had never married were unsure whether they wanted to marry) (Cohn 2013).

The divorce rate, which peaked in the early 1980s and has since decreased and stabilized, also contributed to the numbers of singles. In 2014, 9.3 percent of men and 11.8 percent of women age 18 and over were divorced (U.S. Census Bureau 2014). In addition to creating more unmarried people, the fear of divorce likely deters some from marrying. The proportion of widowed men and women has declined slightly but remains mostly similar to past numbers. Among older people, singlehood most often occurs because of the death of a spouse rather than by choice. However, there has also been a notable increase in what's being called "gray divorce," or divorce among those over 50 years old. Americans over 50 are twice as likely to get divorced as people over 50 were two decades ago (Jaffe 2014). As society continues to value individualism and choice, the numbers of singles will most likely continue to grow.

In addition to the aforementioned delaying of marriage and persistent divorce rate, other factors are worth addressing:

- *More liberal social and sexual standards.* As we saw in Chapter 6, nonmarital sex became and remains more acceptable than in the past, thus reducing—if not removing—what used to be one of the stronger motives for marrying.

- *Uneven ratio of unmarried men to unmarried women.* As we have touched on previously, the opportunity to marry is not the same for all groups. Some women (e.g., African American women) may find themselves *squeezed out* by the imbalanced sex ratio within their same race. Among blacks, highly educated women greatly outnumber comparably educated men. Among lower-income inner-city populations, there is a deficit in "marriageable" men resulting from a combination of high rates of male joblessness, incarceration, and premature deaths due to violence. Additionally, high rates of births to unmarried women may render many single mothers as less desirably "marriageable."

**TABLE 9.2** **Percentage of Never-Married Women and Men by Age, 1970–2015**

| Age | Male 1970 | Male 2015 | Female 1970 | Female 2015 |
|-----|-----------|-----------|-------------|-------------|
| 20–24 | 35.8 | 90.5 | 54.7 | 84.0 |
| 25–29 | 10.5 | 67.3 | 19.1 | 54.3 |
| 30–34 | 6.2 | 40.6 | 9.4 | 31.1 |
| 35–39 | 5.4 | 25 | 7.2 | 21.1 |
| 40–44 | 4.9 | 20.3 | 6.3 | 15.5 |

SOURCE: Fields and Casper (2001); U.S. Census Bureau (2015, Table A-1).

- *Increasingly expanded educational, lifestyle, and employment options open to women.* Such changes have reduced women's economic need to be married, lessened their need to stay in a bad marriage, and expanded their lifestyle options outside of marriage.

Sociologist Eric Klinenberg (2012) considers the rising status of women, both in the United States and other "advanced nations," as among the driving forces in the global growth of the single population. In addition, he considers the communications revolution and mass urbanization to be pivotal in making unmarried life more comfortable and appealing than in the past. The former makes it possible for people to "experience the pleasures of social life—not to mention vast amounts of entertainment—even when they are home alone," while the latter provides the possibility of "a robust social life" simply by leaving one's home and venturing outside, where we are surrounded by gyms, coffee shops, and other gathering places (p. 15).

The final factor that Klinenberg considers is the lengthening of the life span. This produces more elderly, especially elderly women. Although acknowledging that aging alone can present many challenges, in the United States and elsewhere, many older women are finding it more feasible and appealing to live alone (Klinenberg 2012).

Looking not at these "big-picture" factors but at how never-married people account for their marital status, data from the Pew Research Center point to the importance of economic factors, age, and readiness to marry, and not having found a partner who has the qualities one is seeking in a spouse (Wang and Parker 2014). With regard to the latter, in identifying qualities they were seeking, both women and men stressed a spouse's having a steady job and sharing similar ideas about having and raising children. Although the same two reasons were identified as most important by both genders, where shared ideas about having and raising children was the most important quality men reported seeking in a spouse, for women that was a close second to having a steady job (Wang and Parker 2014).

Racially, blacks and Hispanics were more likely than whites to mention the importance of a potential spouse having a steady job. Blacks and Hispanics were also nearly twice as likely as whites to mention the importance of finding a spouse with at least as much education as themselves (Wang and Parker 2014).

When intentionally single heterosexual people form relationships within the singles world, both the man and the woman tend to remain highly independent. Unmarried women and men are typically employed and, thus, tend to be economically independent of each other. They may also be more emotionally independent because their energy may already be heavily invested in their work or careers. Their relationships consequently tend to more greatly emphasize autonomy and egalitarian roles. Employed single women tend to be more involved in their work, either from choice or from necessity, than are married women, but the result is the same: They are accustomed to living on their own without being supported by a man.

Aside from the attractions of living alone, the personal, social, and economic costs associated with being single have been greatly reduced. Singlehood, especially in one's 20s and 30s, is unlikely to be met with the same kinds or levels of criticism or suspicion as, say, two or three decades ago. Although being unmarried has become less stigmatized, media images (from news stories through entertainment media)

The delay in age at marriage is one of the biggest factors in the increase in both singlehood and cohabitation.

Rob Melnychuk/DigitalVision/Getty Images

With greater economic opportunity and success, women's economic need to marry is less today than in the past.

Novastock/Stock Connection Blue/Alamy Stock Photo

often have cast marriage in a less attractive and more precarious light. Finally, as more people forgo or delay marriage, one can do so with less guilt, greater peer support, and less parental pressure. Cultural expectations to marry and marry young have lessened, sex outside of marriage has grown in acceptability and availability, increasing numbers of women and/or couples have embarked on parenthood outside of marriage, and one is likely to feel less economic pressure to marry because of an increasing ability to find economic security outside of marriage.

Furthermore, we have reduced the need to marry. For example, men can obtain many of the "services" long provided by wives—such as cooking, cleaning, intimacy, and sex—outside of marriage without being tied down by family demands and obligations. Thus, men may not have a strong incentive to commit, marry, or stay married.

## Types of Never-Married Singles

In examining the varied experiences of unmarried women and men, much depends on two factors: *intention*—whether being single is a reflection of choice or circumstances—and the *anticipated duration*—whether being single is thought to be a temporary or permanent lifestyle. More specifically, we can differentiate between the following categories:

**True** 4. Unmarried women and men often encounter discriminatory treatment.

- *Voluntarily and temporarily unmarried.* These are usually younger men and women actively pursuing education, career goals, or taking advantage of some of the perceived advantages of being single. These include the freedom and flexibility to socialize whenever, however, and with whomever they want, including the freedom to have a number of different sexual relationships. By remaining unmarried for a time, they have more opportunity to mature and to establish themselves (Klinenberg 2012). Although they are not actively seeking marital partners, they remain open to the idea of marriage, possibly even expecting to someday marry.

- *Involuntarily and temporarily unmarried.* Women and men in this category would prefer to be married and are actively and consciously seeking marital partners. Their desire to be married rather than single makes it less likely that they will perceive advantages to remaining unmarried.

- *Voluntarily and permanently unmarried.* This refers to individuals who regard themselves as permanently single and are resolved to remain so. They include some never-married single mothers who prefer to raise their children alone, women or men who choose to center their lives around their jobs or careers, as well as more "hard-core" singles who simply prefer to be single and do not intend to marry.

- *Involuntarily unmarried.* Never-married individuals in this category might have wished to be married but, having been unable to find a partner, are otherwise resigned to remaining unmarried. These might include well-educated, high-earning women over age 40 who would have married but faced a shortage of similar men as a result of the marriage gradient, African American women confronted by the shortage of suitably marriageable men, and any other women and men whose circumstances, health, or finances have prevented them from entering marriage.

Individuals may shift from one type to another at different times. All but the voluntarily and permanently unmarried share an important characteristic: They want or wanted to move from a single status to a romantic couple status. The dramatic increase in the single population noted previously is more accurately accounted for by increases in the numbers of those who are voluntarily unmarried, whether temporarily or permanently, those we might also simply call single by choice.

## Singlism and Matrimania

Until perhaps recently there have been many myths and misconceptions about those who are unmarried (DePaulo 2006). As articulated and systematically debunked by psychologist Bella DePaulo, these include the following:

1. All singles, especially single women, want to be coupled.

2. Singles are lonely, perhaps bitter, and envy their coupled friends.

3. Singles are self-centered and immature.

4. Single women will someday regret not having married and not having families. They are alternately and unflatteringly portrayed as either sexually deprived or sexually promiscuous.

5. Single men are frequently portrayed as threatening, irresponsible, sexually obsessed, or gay.

6. Children of single parents are destined to suffer emotionally, socially, academically, financially, and behaviorally.

7. Singles lack a partner and, therefore, lack a purpose.

8. Singles will age and die alone.

9. Married people with families deserve special benefits, perks, higher pay, and other resources that singles don't need.

DePaulo considers such myths and stereotypes a part of the evidence of what she calls **singlism**. Singlism, like racism, sexism, and heterosexism, embodies stereotyping, discrimination, and negative, dismissive treatment, though without the segregation, hatred, and violence that other, more "brutally stigmatized" groups face (DePaulo 2006). According to DePaulo, singles (i.e., unmarried women and men) are stereotyped on the basis of these myths.

Beyond stereotyping, though, singlism also has included discriminatory treatment of unmarried people. They can be asked and expected to stay late at work, travel over the holidays, and pay more (per person) for vacation packages, club memberships, and even meals in restaurants and often must do without discounted health benefits, greater Social Security options, lower tax bills, and higher salaries. According to DePaulo, it is almost as though single people were considered to be worth less than married women and men. She notes that surviving spouses can collect the Social Security benefits of their deceased spouses, including small amounts to assist with funeral expenses. When unmarried women and men die, "no other adult can receive your benefits. Your money goes back into the system" (DePaulo 2006, 6).

Alongside the stereotyping of and discrimination against singles comes the glorifying of couples, especially married couples. DePaulo calls this **matrimania**, "the over-the-top hyping of marriage and coupling" (www.psychologytoday.com/blog/living-single). She notes that the matrimania mythology glorifies marriage as "the gold standard," supposedly offering the power to transform an immature single into a mature spouse—to create commitment, sacrifice, selflessness, intimacy, and loyalty where none existed before (i.e., when single). Marriage is seen as the source of happiness—"not just garden-variety happiness, but deep and meaningful well-being. A sense of fulfillment that a single person cannot even fathom" (DePaulo 2006, 13).

Between the glorification and celebration of marriage and the

**False** 5. Compared to their divorced and widowed counterparts, never-married women fare better than never-married men.

stereotyping and discrimination singles face, many unmarried women and men may be made to suffer needlessly. Their status as unmarried, however, does not by itself cause them suffering and misery. In fact, the health and well-being of single women and men is often misunderstood or mischaracterized in assessments of the effects of different marital statuses. Much like the discussion in Chapter 8 of the well-being of married women and men, the well-being of unmarried women and men is similarly misrepresented when comparisons fail to sufficiently differentiate, in this case between never-married, divorced, and widowed singles. Differences between widowed, divorced, and never-married adults from the "baby boom generation" surface in living arrangements, health, economic resources, and the maintenance of social ties, thus indicating a need to more carefully consider not just whether people are married or unmarried, but what marital history precedes one's unmarried status (Lin and Brown 2012; Musick and Bumpass 2012). This is further complicated by gender differences, as when compared against their divorced and widowed counterparts, never-married women fare better than never-married men (Lin and Brown 2012). On a number of measures, never-married singles fare better than widowed, separated, or divorced singles, and are not far behind married women and men. For example, the mental and physical health of older never-married women are better than those of their formerly married counterparts and equal to their married ones (Carr and Springer 2010).

# ■ Cohabitation

Few changes in patterns of marriage and family relationships have been as dramatic as changes in cohabitation. What in the 1960s was rare and may have been relegated to hushed whispers and secrets kept from families is now a common, normative, perhaps even recommended experience (King and Scott 2005) (see Table 9.3).

## The Rise of Cohabitation

Over the past 40 years, cohabitation has increased from approximately 400,000 to 20 times as many heterosexual couples. It has increased across all socioeconomic, age, and racial groups. Using a combination of census reports, it is estimated that as of 2015, there are more than 8.3 million cohabiting heterosexual couples in

**TABLE 9.3  Percentage of Cohabiting among Women, Ages 19 to 44, by Race, 1987 to 2011–13**

|  | 1987 | 1995 | 2002 | 2006–2008 | 2011–2013 |
|---|---|---|---|---|---|
| Non-Hispanic whites | 32% | 47% | 54% | 59% | 67% |
| African Americans | 36% | 45% | 57% | 61% | 59% |
| Hispanics | 30% | 40% | 52% | 56% | 64% |

SOURCE: Fields and Casper (2001); U.S. Census Bureau (2011b, Table A-1); Manning and Stykes 2015.

the United States, and an additional estimated 600,000 unmarried, cohabiting same-sex couples (Gates and Newport 2015; U.S. Census Bureau 2015a). The estimate of same-sex couples does not include the roughly 400,000 married gay and lesbian couples, bringing the total of same-gender couples to over a million (Gates and Brown 2015). With over 8 million couples living together unmarried, we can see how steep an increase in cohabitation has occurred, especially since 1970 (see Figure 9.1).

Looking at the percentage of people who cohabit at some point in their lives reveals just how much of an increasingly common experience cohabitation has become. The proportion of women who have experienced cohabitation has increased from around a third of women 19 to 44 in 1987 to just under two-thirds (65 percent) of women 19 to 44 between 2011 and 2013. Between 2011 and 2013, three-fourths of women 30 to 34 had ever cohabited. Almost as large a share of 25- to 29-year-old women, and nearly 70 percent of 35- to 39-year-old and 40- to 44-year-old women had ever cohabited (Manning and Stykes 2015).

Looking at households, not individuals, Esther Lamidi reports that in the nearly two decades between 1995 and 2012, cohabiting households more than doubled. In contrast, married-couple households declined by 10.5 percent.

> **True** **6.** For most women, marriage is preceded by cohabitation.

Figure 9.1  Cohabitation, U.S. 1960–2010

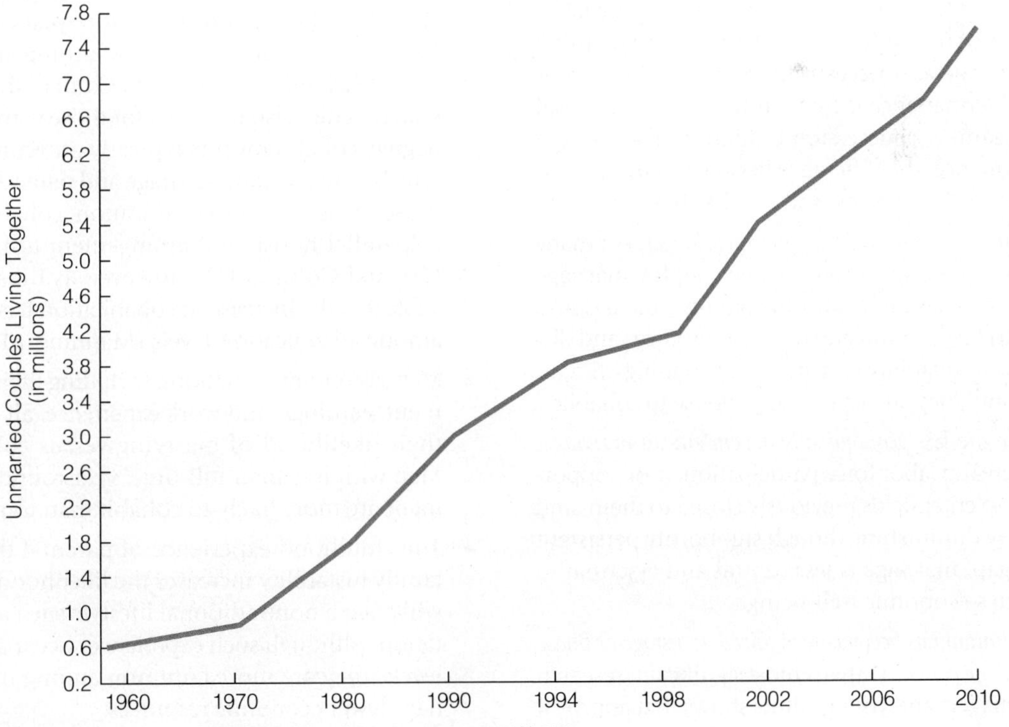

SOURCE: U.S. Census Bureau (2011), Families and Living Arrangements, Table UC1.

These trends occurred across racial and ethnic lines. For example, among whites, blacks, and Asians, there were roughly twice as many cohabiting households in 2012 as in 1995. Among Hispanics, the proportion of cohabiting households had increased fourfold (Lamidi 2014).

The typical marriage is now preceded by cohabitation. As reported by sociologists Wendy Manning and Bart Stykes (2015), since 2000, between 66 and 69 percent of women who married cohabited prior to their first marriages (Manning and Stykes 2015). In 1987, a third of women, 19 to 44, had ever cohabited. Between 2011 and 2013, that percentage had essentially doubled, to 65 percent (Manning and Stykes 2015). The cohabitation increase has occurred across racial lines, as can be seen in the data in Table 9.3 reflecting change over the 20 years between 1987 and 2011–13.

In accounting for the increase in cohabitation, we can return to some of the same factors that helped explain the increase in singlehood (Smock, Casper, and Wyse 2008):

- *The general climate regarding sexuality is more liberal than it was a generation ago.* Sexuality is more widely considered to be an important part of a person's life, whether or not he or she is married. Unmarried couples can more freely engage in an ongoing sexual relationship with less stigma.

- *The meanings of marriage and divorce have changed.* Because of the increase in divorce during the latter decades of the 20th century, many no longer think of marriage as a necessarily permanent commitment. Permanence is frequently replaced by serial monogamy—a succession of marriages—and as a consequence, the difference between marriage and living together has lost some of its sharpness.

- *Men and women are delaying marriage longer.* For many couples, especially lower-income couples, marriage is seen as a desirable outcome but one that requires them to first accumulate sufficient resources and display their suitability to marry. While putting off marriage until they are ready, many choose to cohabit.

- *Women are less economically dependent on marriage.* With greater labor force participation, more opportunity to enter fields previously closed to them, and a slowly diminishing though stubbornly persistent wage gap, marriage is less central and essential to women's economic well-being.

- *Cohabitation has become normalized as a stage or phase in one's life course.* Perhaps most significant, research on teenagers and young adults shows that approval of cohabitation is common and that many young adults expect to someday cohabit with someone, perhaps along the way toward marriage.

Examining these items reveals both economic and cultural factors at work. Many of these same factors help account for other familial changes, including the previously discussed increase in the population of unmarried women and men. According to Smock, Casper, and Wyse (2008), the combination of cultural and economic factors changed the societal context within which people make lifestyle choices. They contend that these same two influences operate on a more individual or micro level in terms of the resources one possesses and the attitudes and values one embraces. Looking first at socioeconomic influences, we can note the following:

- Cohabitation is more likely among those with less education. Individuals with less than a college education are twice as likely to cohabit as individuals with a college degree (Fry and Cohn 2011). Looking at women ages 19 to 44, Wendy Manning and Bart Stykes report that the share who have ever cohabited is greatest among those with less than 12 years of education (76 percent), though in recent years that percentage has leveled off, increasingly only slightly, while there has been a more sizable increase among those with some college (at 64 percent in 2011–13) or four or more years of college (58 percent as of 2011–13) (Manning and Stykes 2015). Richard Fry and D'Vera Cohn of the Pew Research Center also note that for those with a college degree, cohabitation is typically an economically sensible step toward marriage and parenthood. For those without a college education, cohabitation is a "parallel household arrangement to marriage" (Fry and Cohn 2011). However, as illustrated in Table 9.4, the increase in cohabitation has occurred among all education levels (Manning 2010).

- Men's economic situations, including their employment, earnings, and work experience, all influence their likelihood of marrying versus cohabiting. Men with less than full-time, year-round employment are more likely to cohabit than to marry.

- The childhood experience of parental divorce or family instability increases the likelihood that one will enter a nontraditional lifestyle such as cohabitation. Although such experiences occur at all class levels, they are more common among those who have fewer economic resources.

**TABLE 9.4 Percent Cohabiting by Education, 1987 to 2011–2013**

|  | 1987 | 1995 | 2002 | 2006–2008 | 2011–2013 |
|---|---|---|---|---|---|
| < 12 years | 43% | 58% | 64% | 73% | 76% |
| 12 years | 32% | 50% | 63% | 69% | 67% |
| 1–3 years college | 30% | 40% | 49% | 52% | 64% |
| 4 or > years college | 31% | 37% | 45% | 47% | 58% |

SOURCE: Manning (2010), Manning and Stykes 2015.

- Among women 19 to 44, greater shares of white (67 percent) and Hispanic (64 percent) than black women (59 percent) report ever having cohabited.

- There are racial differences that often overlap with class differences regarding the likelihood of *growing up in* cohabiting parent households, with African American and Hispanic children overrepresented in such households. Additionally, black and Hispanic women are more likely to become pregnant while cohabiting and to continue to cohabit rather than marry after their children are born.

Exploring the role of attitudes and values, Smock, Casper, and Wyse (2008) note that cohabitation is more common among

**False** **7. Couples cohabit for many different reasons.**

Hispanic and African American couples are more likely to cohabit than are whites or Asian Americans.

John Lund/Blend Images/Alamy Stock Photo

those who have more liberal attitudes—those who are more likely to support nontraditional families and egalitarian gender roles and who are less religious.

## Types of Cohabitation

Although the picture sketched so far has addressed the overall phenomenon of and increase in cohabitation, there is considerable diversity among cohabitors.

There is no single reason to cohabit, no one type of person who cohabits, and no one type of cohabiting relationship. One popular typology differentiates among four different types of cohabiting relationships: *substitutes or alternatives for marriage, precursors to marriage, trial marriages,* and *coresidential dating* (Casper and Bianchi 2002; Casper and Sayer 2002; Phillips and Sweeney 2005). These can be distinguished by whether partners anticipate a married future, their perceptions of the stability of the relationship, and their general attitudes toward cohabiting relationships:

- *Trial marriage.* In **trial marriages**, the motive for living together *outside marriage* is to assess whether partners have sufficient compatibility to successfully *enter marriage.* They are undecided as to their likelihood of marriage, are uncertain about their suitability for marriage, and, by cohabiting, hope to determine whether they should proceed to marry.

- *Precursor to marriage.* When the relationship is a **precursor to marriage**, partners share an expectation that eventually they will marry. In fact, such couples may get engaged before or while cohabiting.

- *Substitute for marriage.* When the relationship is a *substitute for marriage,* partners are not engaged and have no intention or expectation to marry but they do anticipate staying together.

- *Coresidential dating.* In coresidential dating, the relationship is more like a serious boyfriend–girlfriend relationship, lacking any intention or expectation to marry; cohabitation offers such couples greater convenience than does living apart (Bianchi and Casper 2000).

In neither substitutes for marriage nor coresidential dating is there any expectation of eventually marrying, even though many such cohabitors do ultimately marry. As to the expected duration of the relationship, partners in a coresidential dating situation expect to live together for only a relatively short time. In either a substitute for marriage or a precursor to marriage, couples expect to be together a long time. In the case of trial marriages, couples don't know whether and how long they will stay together (Heuveline and Timberlake 2004).

A second typology separates cohabitation into the following *five types:*

- *Prelude to marriage.* Cohabitation is used as a "testing ground" for the relationship. Cohabitants in this type of situation would likely marry or break up before having children. The duration of this type is expected to be relatively short, and couples should decide to transition into marriage or to part ways relatively soon after beginning to cohabit.

- *Stage in the marriage process.* Unlike the prior type, couples may reverse the order of marriage and childbearing. They cohabit for somewhat longer periods, typically in response to opportunities that they can pursue "by briefly postponing marriage" (p. 1,216). Both partners understand that they intend to eventually marry.

- *Alternative to singlehood.* Considering themselves to be too young to marry and with no immediate intention to marry, such couples prefer living together to living separately. Having a commitment level more like a dating couple than that of a married couple, such relationships will be prone to separation and breaking up.

- *Alternative to marriage.* Couples choose living together over marriage but choose, as married couples would, to form their families. Greater acceptance of unmarried childbirth and child rearing will increase the numbers of couples experiencing this type of cohabitation. Such couples would not likely transition into marriage but would likely build lasting relationships.

- *Indistinguishable from marriage.* Such couples are similar to the previous type but are more *indifferent*

rather than *opposed* to marriage. As cohabitation becomes increasingly accepted and parenting receives support regardless of parents' marital status, couples lack incentive to formalize their relationships through marriage (Heuveline and Timberlake 2004).

Finally, consider a third typology of cohabitation, highlighting the factors that couples consider in deciding to live together, the "tempo of relationship advancement" into cohabitation, and the language used, or story told, by couples in accounting for their cohabiting (Sassler 2004). Sharon Sassler (2004, 498) identifies six broad categories of reasons that couples decide to cohabit—finances, convenience, housing situation, desire, response to family or parents, and as a trial—out of which she constructs a three-category typology:

- *Accelerated cohabitants* decided to move in together quickly, typically before they had dated six months. Emphasizing the strength and intensity of their attraction and their connection and the fact that they were spending a lot of time together and identifying finances and convenience as major reasons for their decision, they contended that moving in together felt like "a natural process."

- *Tentative cohabitants* admitted to some uncertainty about moving in together. Together for 7 to 12 months on average before living together, they typically saw each other less often than the "accelerateds" did before moving in together (e.g., three or maybe four nights a week) or had experienced disruptions in their relationship with one of the partners being gone for a period, slowing their progression into cohabitation. They often mentioned "unexpected changes in their residential situation" as a reason for their decision. Absent such a situation, they might not have moved in together when they did.

- *Purposeful delayers* were the most deliberate in the decision-making process. Their relationships progressed more gradually, taking more than a year before they decided to live together and allowing them an opportunity to discuss future plans and goals. They most often mentioned housing arrangements and finances as the reasons they moved in together.

As the three typologies reveal, not all cohabitors desire, intend, or expect to marry. Although cultural attitudes and values—as well as ideas about singlehood, dating, marriage, and cohabitation—somewhat determine whether someone expects to marry, socioeconomic criteria are also of importance. Also, cohabitation is changing and, since 2000, has become less likely to transition to marriage. So although cohabitation continues to increase, it has become somewhat less selective and, among more recent cohabiting unions, more "delinked" from marriage (Guzzo 2014).

## What Cohabitation Means to Cohabitors

The meaning of cohabitation varies for different groups. As discussed last chapter, there is a significant class divide that can be seen in the different rates of cohabitation, serial cohabitation, and in the likelihood of having children while cohabiting, by individuals with different amounts of education. Although cohabitation is common and experienced by a majority of young college-educated adults before they enter marriage, couples typically delay childbearing until after marriage (Cherlin 2014).

There are also notable race and ethnic differences. For African Americans, cohabitation is more likely to be a substitute for marriage than a trial marriage, and blacks are more likely than whites to conceive, give birth, and raise children in a cohabiting household. Indirectly, this implies that cohabitation may be a more committed relationship for blacks than for whites, a more acceptable family status, and a more acceptable family form within which to rear children—even though blacks are no more likely than whites to say that they approve of cohabitation (Phillips and Sweeney 2005).

The same appears to be true among Hispanics. The idea of "consensual unions" outside of marriage goes back a long way in Latin America, especially among the economically disadvantaged. More than among whites, cohabitation for Hispanics is more likely to become an alternative to marriage. Again, we draw this conclusion from rates of nonmarital pregnancy and childbearing. Julie Phillips and Megan Sweeney (2005, 299) report that cohabitation for Hispanics "may be a particularly important context for *planned* childbearing" (emphasis added).

One of the most notable social effects of cohabitation is that it delays the age of marriage for those who live together. *Ideally*, then, should cohabitors go on to marry, cohabitation could encourage more stable marriages because the older people are at the time of marriage, the less likely they are to divorce. However, as we shall soon see, cohabitation neither ensures nor precludes more stable marriages.

In the United States, cohabiting couples still lack most of the rights that married couples enjoy, a topic to which we will return shortly. According to Judith Seltzer (2000), children of cohabiting couples may also be disadvantaged unless they have legally identified fathers. For example, in the absence of legally established paternity, children are not guaranteed financial support from nonresidential fathers. This situation differs greatly from what exists in many other countries. In Sweden, for instance, the law treats unmarried cohabitants and married couples the same in such areas as taxes and housing. In many Latin American countries, cohabitation has a long and socially accepted history as a substitute for formal marriage (Seltzer 2000).

There are additional disadvantages. Cohabiting couples may also find that they cannot easily buy a house together because banks may not count their income as joint; they also usually cannot and don't automatically qualify for health insurance benefits. If one partner has children, the other partner is usually not as involved with the children as he or she would be if they were married. Cohabiting couples may find themselves pressured to marry if they have a child, though again, this may be a diminishing likelihood and may vary across race and class. Finally, as Andrew Cherlin points out, in the United States cohabiting relationships generally don't last more than two years; couples either break up or marry (Cherlin 2009, 2014).

## Cohabitation and Remarriage

Roughly half of those who remarry after a divorce cohabit before formally remarrying, and postdivorce cohabitation is now commonplace (Xu, Bartkowski, and Dalton 2011; Xu, Hudspeth, and Bartkowski 2006). In fact, one reason that remarriage rates have declined is because postmarital cohabitation has increased (Seltzer 2000). At minimum, the prevalence of postdivorce/pre-remarriage cohabitation delays remarriage. For some, it may well replace remarriage or at least enable some couples to avoid remarriages that carried greater risk of later instabilities of separation and divorce.

Living together takes on a different quality among those who have been previously married. Some postdivorce cohabitors may choose to live together because of their experiences in their earlier failed marriage. Sociologist Xiaohe Xu and colleagues (2006) suspect that this may be especially true if those postdivorce cohabitors didn't cohabit prior to their first marriage. They may see cohabitation as a way to test the relationship that they missed—and suffered the consequences—in their earlier marriage. Additionally, postdivorce cohabitation has likely delayed the process of remarrying, much as premarital cohabitation leads to a delay of first marriage (Xu, Bartkowski, and Dalton 2011). Cohabitation after divorce has become fairly common, and has become a period more like extended courtship or dating. It is not typically used as a permanent substitute for remarriage but rather part of a prolonged path leading to a remarriage (Xu, Bartkowski, and Dalton 2011).

Marital quality and happiness appear to be lower among postdivorce (pre-remarriage) cohabitors. Happiness in remarriage is nearly 30 percent less for couples who cohabited before remarrying than for those who did not live together before remarriage. Similar outcomes resulted regarding the effects of postdivorce cohabitation on the stability of remarriages (Xu, Hudspeth, and Bartkowski 2006). Remarriage instability (divorce or separation) rates are higher for postdivorce cohabitors. According to Xiaohe Xu, John Bartkowski, and Kimberly Dalton (2011), the odds of a second divorce or separation are more than twice as high for those who cohabit following divorce (and prior to remarrying) as for those who remarry without first cohabiting.

**False** **8.** Cohabitation before a remarriage does not improve one's likelihood of having a successful remarriage.

## Cohabitation and Marriage Compared

Perhaps the major difference between cohabitation and marriage results from the fact that marriage is institutionalized and cohabitation isn't. Even with all the continuing changes to marriage discussed in the previous chapter, there are still more "shared understandings" of what marriage means—its rules and roles and rights and responsibilities—than there have been about cohabitation. Of course, this may be changing, as cohabitation has become a normative, expected part of the adult life course. Still, there are other important differences that have been found between marriage and cohabitation to which we now turn.

### Different Commitments

Marriages begin with spouses anticipating a lifelong commitment to each other. A lesser level of commitment characterizes the majority of cohabiting couples when compared to married couples. Generally, when a couple lives together, their primary commitment is to each other, but it is a more transitory commitment. As long as they feel they love each other, they will stay together. For as long as they feel gratified and benefitted by the relationship, they will continue to cohabit. In marriage, the couple makes a commitment not only to each other but also to their marriage. Articulated in vows that promise "till death do we part," or "as long as we both shall live," marriages start with a pronouncement of lifelong commitment, even though more than 40 percent of marrying couples end their marriages through divorce. Cohabitants make no such declaration and frequently are less committed to the certainty of a future together (Waite and Gallagher 2001).

However, there is a wider range in how much commitment one finds among cohabiting couples compared to married couples, with some cohabiting couples being as committed as are most married couples. Michael Pollard and Kathleen Mullan Harris (2013) used survey data from 18- to 26-year-olds who were part of the National Institute of Child Health and Human Development's National Longitudinal Study of Adolescent Health (Add Health) to explore the dynamics of cohabiting relationships. They looked specifically at differences among heterosexual cohabiting couples and between cohabitors and married women and men on such relationship characteristics as commitment and closeness. In general, cohabitors were somewhat less likely than married respondents to claim that they were "completely committed" to their relationships, but there was interesting variation between types of cohabiting relationships. People who were in a *precursor to marriage relationship* were as likely to report themselves as "completely committed" to their partners as were married respondents, and were more likely to characterize their relationships at the maximum level of closeness. On the other hand, those in more casual *coresidential dating* relationships were less than half as likely as either married or other cohabitors to express the same level of commitment (see Figure 9.2).

Living together tends to be a more temporary arrangement than marriage, and heterosexual cohabiting relationships in recent years have become both more short term and less frequently linked to eventual marriage (Seltzer 2000; Smock, Casper, and Wyse 2008). A man and woman who are living together

**Figure 9.2 Type of Relationship and Couples' Reports of Closeness and Commitment**

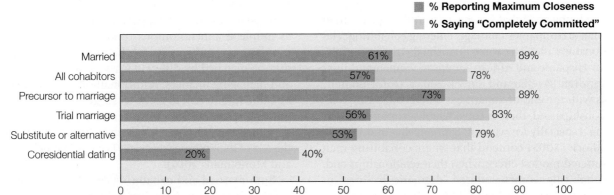

■ % Reporting Maximum Closeness
□ % Saying "Completely Committed"

|  | | |
|---|---|---|
| Married | 61% | 89% |
| All cohabitors | 57% | 78% |
| Precursor to marriage | 73% | 89% |
| Trial marriage | 56% | 83% |
| Substitute or alternative | 53% | 79% |
| Coresidential dating | 20% | 40% |

SOURCE: Pollard and Harris (2007).

may not work as hard to save their relationship. Less certain of a lifetime together, they are more likely to continue to live somewhat more autonomous lives. In marriage, spouses will typically do more to save their marriage, giving up dreams, work, ambitions, and extramarital relationships for marital success.

Unmarried couples are also less likely than married couples to be encouraged by others to make sacrifices to save their relationships. Parents may even urge their children who are "living together" to split up rather than give up plans for school or a career. If a cohabiting couple encounters sexual difficulties, it is more likely that they will split up. It may be easier to abandon a problematic relationship than to change and fix it. Again, among cohabitants who intend to marry, relationships—including the aspect of commitment—are not significantly different from marriages. Cohabitating partners who are highly educated and employed and who enjoy higher levels of income are more likely to express an intention to marry, are more likely to marry, and are less likely to break up (Brown and Booth 1996; Smock, Casper, and Wyse 2008).

## Sex

There are differences in the sexual relationships and attitudes of cohabiting and married couples. Linda Waite and Maggie Gallagher (2001) suggest that married couples experience more fulfilling sexual relationships because of their long-term commitment to each other and their emphasis on *exclusivity*. Because they expect to remain together, married couples have more incentive to work on their sexual relationships and discover what most pleases their partners.

Heterosexual cohabitants, however, have more frequent sexual relations. Whereas 43 percent of married men reported that they had sexual relations at least twice a week, 55 percent of cohabiting men said that they had sex two or three times a week or more. Among married women, 39 percent said that they had sex at least twice a week, compared with 60 percent of never-married cohabiting women. Sex may also be more important in cohabiting relationships than in marriages. Waite and Gallagher (2001) go as far as calling it the "defining characteristic" of cohabitants' relationships.

Married couples are also more likely to be sexually monogamous. According to data from the National Health and Social Life Survey, 4 percent of married men said that they had been unfaithful in the 12 months prior to the survey; four times as many, 16 percent, of cohabitants reported infidelity. Among women, the equivalent comparison shows that 1 percent of married women compared with 8 percent of cohabiting women admitted to having had sex outside their relationship. Judith Treas and Deirdre Giesen (2000, 59) obtained similar findings, even when they controlled for how permissive individuals were toward extramarital sex: "This finding suggests that cohabitants' lower investments in their unions, not their unconventional values, accounted for their greater risk of infidelity." Importantly, however, they find that in the vast majority of cohabiting relationships, sexual exclusivity *is* expected. In fact, 94 percent of cohabiting heterosexuals expected sexual exclusivity in their relationships (Treas and Giesen 2000).

## Finances

Overall, cohabiting women and men have more precarious economic situations than married couples. The latter have higher personal earnings and higher household incomes and are much less likely to live in poverty. There is also evidence that cohabitation carries an "economic premium" over remaining single and living apart that is *comparable* to though less than what accompanies marriage; entering cohabiting relationships alleviates some financial distress, especially for Hispanic and African American women and their children (Avellar and Smock 2005). Unfortunately, as with the end of a marriage, when cohabiting relationships end, there is considerable economic suffering, especially for women. Sarah Avellar and Pamela Smock (2005) contend that where cohabiting men suffered modest effects when their relationships end, cohabiting women suffered "dramatic declines" in their standards of living. Men suffered declines of roughly 10 percent in their household income. For women, there is a more notable loss of household income (33 percent) and a striking spike in the level of poverty (reaching nearly 30 percent in poverty) following breakups.

Perceptions of current and future financial stability are more instrumental in decisions that couples make to marry than in decisions to live together. In ongoing relationships, partners' relative financial contributions differently affect married and cohabiting couples. For cohabitors, relative equality in partners' earnings seems to stabilize relationships. Conversely, among married couples in which each spouse contributes equally to the family income, the risk of divorce is heightened (Raley and Sweeney 2007; Weisshaar 2014).

Cohabiting and married couples also differ greatly in terms of whether and how they pool their money, which is typically a symbol of commitment (Cannon 2009; Hamplova, Le Bourdais, and Lapiere-Adamcyk 2014). People generally assume that a married couple will pool their money, as it suggests a basic trust or commitment to the relationship and a willingness to sacrifice individual economic interests to the interests of the relationship. Additionally, married couples who do not pool their money are more likely to be dissatisfied with their marriages than married couples who do (Dew 2009). Data from the United States but also from many European countries (including France, Britain, Spain, Norway, Denmark, Sweden, and the Czech Republic) indicate that cohabiting couples are much less likely than married couples to pool their money (Hamplova, Le Bourdais, and Lapiere-Adamcyk 2014). In fact, one of the reasons couples cohabit rather than marry, or maintain committed noncohabiting partnerships (increasingly referred to as "living apart together" or LAT relationships, and discussed later in the chapter) rather than cohabit, is to maintain a sense

of financial independence (Lyssens-Danneboom and Mortelmans 2014). The distinction between how married versus cohabiting couples handle money is not as simple as a dichotomous "pooled versus separate monies," as the diversity in cohabiting relationships may be matched by different ways of thinking about and managing finances. Additionally, "pooling" may be done to different degrees, as suggested by "partial pooling systems," in which partners pool some of their incomes to be used for shared expenses but keep the rest of their monies separate (Lyssens-Danneboom and Mortelmans 2014).

Both married and cohabiting couples experience conflict about money and spending, but such conflicts affect couples differently. Sociologist Jeffrey Dew (2009) studied factors associated with breakups among cohabiting and married couples. Although disagreements about sex, housework, or time spent together did not account for breakups among cohabiting couples, fights about money and feelings of unfairness in how money was handled did. Dew contends that conflicts about money can damage either a

A major difference between cohabitors and married couples can be seen in how they manage finances and whether they pool their money.

PhotoAlto/John Dowland/PhotoAlto Agency RF Collections/Getty Images

marriage or a cohabiting relationship, but estimates the increased risk of breaking up to be much greater among cohabiting couples (70 percent increase) than among married couples (25 percent increase in risk of divorce). "Differing evaluations of how to equitably split expenses can be a potential minefield. . . . Compared to married couples, cohabiting couples tend to be less financially intertwined and more concerned about fairly sharing the costs of the relationship" (Cannon 2009).

Finally, cohabitation can bring financial benefits that result from our tax system and Social Security policies, though it limits the protection one can expect should the relationship end. When both partners earn similar amounts, by being legally single and filing their taxes as such, they enjoy the benefit of larger standard deductions than they would if they were married. Regarding Social Security, some, especially elderly, men and women might decide to live together instead of marry because if they were to marry, they would lose some of their Social Security benefits (Willetts 2003). Unlike the ending of a marriage, when cohabiting couples break up, there is no legal requirement as to how money and property are to be divided. Unless they have drawn up and entered a legal contract, there is no obligation for one partner to financially support the other after a breakup regardless of how support might have been provided during the relationship. Support of children is a different matter; *if paternity has been confirmed*, a father has the obligation to contribute to the support of any children born during the cohabiting relationship, even after such a relationship ends (http://family.findlaw.com/living-together).

## Children

The arrival of children tends to stabilize marriages, lowering the likelihood that couples will divorce. The birth of the first child and the presence of young children are especially associated with reductions in the likelihood of separation and divorce. Stepchildren have the opposite effect; their presence increases the risk of divorce (Manning 2004). How are cohabitors affected by the arrival and presence of children? Wendy Manning found that births during cohabitation do not seem to significantly affect—either positively or negatively—the cohabiting relationship. The odds of marriage are 72 percent greater and the odds of separation 17 percent lower for cohabiting women who conceive a child during cohabitation. Especially for white women, conceiving a child while cohabiting promotes increased stability of the cohabiting

relationship and increases the likelihood of marriage (Manning 2004).

Perhaps of greater import is how children fare in cohabiting as opposed to married households. Some of that will be addressed in the next chapter, when we look at the diversity of parenting situations. For now, drawing from Wendy Manning's review of how cohabitation affects children's well-being, we note the following:

- Cohabiting families are more likely to be poor, and poverty exposes children to less favorable outcomes.

- Cohabiting families are more unstable, break up more often and more quickly, and such instability can negatively impact children. Children born to cohabiting parents experience almost three times as many family transitions as children born to married parents.

- Cohabiting families lack the legal protections marriage brings, and children in cohabiting families may lack the institutional and social supports that children in married-parent households receive.

- Children born into cohabiting households have been found to be at greater risk of health disadvantages, negative psychological and social effects, and poorer academic outcomes.

Of course, given the higher levels of instability and lower socioeconomic circumstances associated with cohabiting parents, one cannot separate the child outcomes from that wider socioeconomic and familial context.

## Health and Mortality

You will recall from the previous chapter that marriage bestows health benefits on the married. Generally, married people live longer and healthier lives and suffer from fewer chronic or acute health concerns than the single, divorced, separated, or widowed. Although there is a likely selection factor that partly accounts for the comparative advantages married people enjoy, some of the benefits are products of healthier lifestyles, evident in such things as lower rates of alcoholism and problem drinking and healthier body weights. Do similar health benefits apply to cohabitants? After analyzing Canadian data on the health status of 6,494 women and 5,368 men, Zheng Wu and colleagues conclude that married people have somewhat better general health than cohabitants, who, in turn, have better health than the separated and divorced, widowed, and never married. The difference between cohabitants and married people loses statistical significance once researchers control for other factors. By providing us with the social support of a loving partner, both marriage and cohabitation appear to "protect" the health of those in stable unions compared to those who lack such relationships. One thing to keep in mind: Because cohabitation is typically of shorter duration and more likely to fail or end, cohabitants are at a disadvantage compared to married people and may find that as their relationships end, their health slides (Wu et al. 2003).

Hui Liu and Corinne Reczek (2012) explored the relationship between cohabitation and adult mortality. The mortality of married white men and women was lower than that of their cohabiting counterparts, whereas for black women and men there were no significant differences between marriage and cohabitation. Compared to their unpartnered counterparts, cohabiting white and black men had lower mortality rates. Interestingly, though higher mortality was found for single compared to married or cohabiting men, the same did not hold for single women when compared to cohabiting women (Liu and Reczek 2012).

## Relationship Quality and Mental Health

Research across three decades (Brown and Booth 1996; Brown, Manning, and Payne 2014; Burgoyne 2012) has indicated that heterosexual cohabiting couples, especially those without plans to marry, have poorer relationship quality than do married couples. On average, compared to individuals who are married, cohabitors report lower levels of happiness with their relationships, more fighting, more violence, and lower confidence in their relationship stability, and they perceived their relationships to be less fair (Brown, Manning, and Payne 2014). Recent research continues to support that characterization, but with the following, more specific, qualification: As reported in research by sociologists Susan Brown, Wendy Manning, and Krista Payne, those who "marry directly," *without first cohabiting* report the highest levels of marital quality when compared to married individuals who first cohabited, cohabitors with plans to marry, and cohabitors with no plans to marry. At the other extreme, cohabitors with no plans to marry reported the lowest relationship quality of the four groups. Those who are cohabiting but have plans to marry their partners report relationship quality much like married individuals who cohabited before marrying. As Brown, Manning, and Payne's findings emphasize, cohabitors are not all the same with regard to whether they expect and plan

to marry, and their relationship quality varies accordingly. Similarly, among those who are married, we find those who did and didn't cohabit. As Brown, Manning, and Payne note, their data reveal the importance of differentiating both among marrieds in terms of their cohabitation experience and among cohabitors in terms of their intentions to marry (Brown, Manning, and Payne 2014).

A report by Sarah Burgoyne using data from 724 married couples and 316 cohabiting couples revealed that in 57 percent of married couples and 36 percent of the cohabiting couples, both partners reported themselves as "very satisfied." Among married couples, those who had "married directly" (i.e., *not cohabited*) agree more often that they are very satisfied with their marriages compared to those married couples who cohabited prior to marrying (61 percent vs. 53 percent). Among cohabiting couples in which both partners report an intention to marry, there is almost twice the proportion of partner agreement that they are very satisfied (47 percent) than found among cohabiting couples with at least one partner who does not have marriage plans (25 percent) (Burgoyne 2012). Thus, once again, the *type* of cohabiting relationship makes a difference.

Cohabitors are positioned somewhere between married and those never married, widowed, or divorced individuals who are unpartnered (Liu and Reczek 2012) in terms of their receipt of social support (e.g., love, care, advice) and social integration (feeling connected). Cohabitation is frequently short term, and cohabitants are less likely to receive the same levels and kinds of social support from their families and friends. They are also more likely to worry about their relationships and experience more psychological distress than those who are married (Liu and Reczek 2012).

## Work

Traditional marital roles call for the husband to work; it is left to the discretion of the couple whether the woman works. Contemporary families often cannot afford the luxury of a one-income household. Still, gender roles in marriage have emphasized men's economic provision as a major component of men's family responsibilities. In cohabiting relationships, the man is not expected to support his partner (Blumstein and Schwartz 1983). If the woman is not in school, she is expected to work. If she is in school, she is nevertheless expected to support herself. With less certainty about the future of their relationships, cohabiting women may be less willing to restrict their outside employment or to spend time and energy on housework that could be spent on paid work.

Married couples often disagree about the division of household work. Both married and cohabiting women tend to do more of the domestic work than their male partners (Seltzer 2000; Shelton and John 1993; Waite and Gallagher 2001). But cohabiting women spend about five to six fewer hours per week on housework than do married women (Ciabattari 2004). Cohabiting women who are not employed or who have children in the home tend to do more housework. Whether women intend to marry their partners does not significantly affect their time spent on housework. However, marital intentions loom large in influencing men's housework performance. Men who intend to marry *someone other than their current partner* (i.e., intend to marry "someday" but not the woman they are living with) do 8 fewer hours of total housework and 4.4 fewer hours of core housework (housecleaning, cooking, laundry, shopping, and dishes) than men who definitely plan to marry their cohabiting partners. Cohabiting men with stronger commitments to their partners do more housework than men who are least committed to their relationships (Ciabattari 2004).

## Effect of Cohabitation on Later Marriage

The effects of premarital cohabitation on marriage have been the focus of considerable research and popular attention over a number of years. Although it may seem surprising and goes against the logic of cohabiting couples who think that cohabitation helps prepare them for marriage, thus reducing their risk of divorce if and when they marry, three decades worth of research indicated the opposite. Addressing that research, sociologist Arielle Kuperberg reports that couples who lived together before marriage and then married were found, on average, to have a 33 percent higher likelihood of divorcing when compared to couples who married without first living together (Kuperberg 2014). Meanwhile, little research supported the idea that cohabitation led to beneficial marital outcomes when compared against couples who didn't cohabit before marriage (Stanley and Rhoades 2009). This came to be known as the *cohabitation effect* and was found repeatedly in a variety of different studies, inspiring researchers to consider what it was—either about cohabitation or about cohabitors—that accounted for the greater likelihood that their marriages would end in divorce.

In marriages that were previously cohabiting relationships, research has documented higher levels of disagreement and instability, lower levels of commitment, and greater likelihood of divorce. Family researchers Scott Stanley and Galena Rhoades (2009) summarized some "negative effects" that were found to follow cohabiting couples into marriage. They note that research has associated cohabitation with the following:

- More negative marital communication
- Lower levels of satisfaction with marriage
- Lower levels of male commitment to one's spouse
- Gradual "erosion" of the value of marriage and child rearing
- Greater likelihood of divorce

However, more recent and current research have failed to find such outcomes, prompting sociologists to question and reject a "one-size-fits-all" approach to the question of how cohabitation affects cohabitors who later marry. Research by Wendy Manning and Jessica Cohen found that among couples who married since the middle 1990s, the cohabitation effect (i.e., the increased risk of divorce for couples who married following cohabiting) did not pertain. In fact, as reported by sociologist Arielle Kuperberg, for some women whose risk of divorce was high, such as women raised in stepfamilies or single-parent families, women who had had a child premaritally, and women who had had above average numbers of past sex partners, cohabitation while engaged to marry actually was associated with *lowered risk* of divorce (Kuperberg 2014). Let's look at the issue a bit more closely.

First, the effect cohabitation has on subsequent marriage has not been the same for all groups. For example, Julie Phillips and Megan Sweeney report that for Caucasian women, there was a 9 percent difference between those who had and had not cohabited, whereas 37 percent of those who cohabited before marriage saw their marriages end within 10 years compared to 28 percent who did not cohabit. The cohabitation effect is much smaller among African Americans and Hispanics. Among African American women, 51 percent of those who had cohabited before marriage and then married saw their marriages fail within 10 years compared to 48 percent of those who had never cohabited, only a 3 percent difference. Among Mexican American women, the difference was 6 percent: 32 percent who had cohabited experienced

Imagine that two friends of yours who are a couple are trying to decide what to do about their relationship now that both are about to graduate from college. Having been together for the past 17 months, they love each other and hope to have a long future together. You may be wondering, "So, what's the problem?" The problem is that neither partner feels ready to take the "big step" and get married. At the same time, they don't want to break up. Although they could continue their relationship as dating partners, seeing each other as often as they can and/or choose, a relationship such as that fails to reflect the intensity of their feelings or the strength of their desire to be together. They are now seriously considering moving in together. What advice do your friends need as they weigh their options?

According to the American Bar Association's *Family Legal Guide,* your friends should think about reaching a formal cohabitation agreement that spells out how they will share expenses, manage finances, and share any property that they acquire together. Of course, a more romantic way of thinking may lead them to think that simply declaring their love for and to each other and promising to care for each other is enough for them. If they stay together, it may well be enough. If they part, as most cohabiting couples ultimately do, having a written agreement that covers a variety of details will spare them much difficulty.

Some other recommendations, courtesy of Thomson Reuters's Findlaw.com website, include the following:

1. Partners should keep their finances separate and maintain accurate records of what each contributed to property held by the other. In the case of major purchases, joint purchases should be in both names. Purchases where only one partner is making the payments (including the down payment) should be in the name of the person actually paying.
2. It is best that cohabitors avoid making joint purchases, combining their money in joint bank accounts, or entering debt together. It is also recommended that one avoid being completely dependent financially on the other. Unlike marriage, where a financially dependent spouse may receive certain consideration and compensation if the marriage fails, former cohabitors are more vulnerable.
3. It is strongly recommended that cohabiting partners not present themselves as husband and wife, use the same last name, or allow others to refer to them as "Mr. and Mrs.," or they risk acquiring a married status via common law marriage in those states where

such marriages are still recognized. Even without common law marriage, partners who act as married are vulnerable to being sued for "palimony" by their ex should the relationship end. **Palimony** is like alimony but applies to cohabiting couples. Palimony is an award of support in which the couple were not married but lived together for a long period and then terminated their relationship. A determining factor in awarding such support is whether there was an agreement that one partner would support the other in return for the second making a home and performing other domestic duties beyond sexual pleasures.

However, an ex–cohabiting partner does not automatically have the right to property or support as might a divorcing spouse simply based on his or her status as a spouse. Instead, cohabitors must prove that there was an agreement—written, oral, or implied—that one partner would support the other through and beyond the relationship and in return that the other would take responsibility for maintaining the household. An implied agreement is implied from how couples behave or have behaved in the past and consists of "unspoken 'understandings'" between partners. If one partner had consistently supported the other financially, that may persuade a jury that there was an implied agreement that such behavior would continue, even if the relationship ended (USLegal.com 2010). An oral agreement (i.e., "I will always take care of you") without some substantiating documentation may be difficult to prove, as either party can claim not to remember or to have been misunderstood or simply lie and deny having come to any agreement.

Ironically, though many cohabitors may choose to live together rather than marry because they think it unnecessary to demonstrate their commitment via a "piece of paper" and want to avoid the personal and financial costs associated with divorce if things don't work out, the absence of some kind of formal written agreement (a "piece of paper") or of clear-cut policies regarding custody and support leaves cohabitors vulnerable.

Finally, your friends need to be aware of the following. No matter how devoted or dedicated they are to each other, most states do not automatically give cohabitors the right to help each other in medical emergencies, to inherit property, or to be beneficiaries in each other's pension plans. Even property bought while together may not be divided equally if the relationship ends (Hartwell-Walker 2008). ●

marital failure compared to 26 percent who hadn't co-habited. Among *foreign-born* Mexican Americans, there were more marital failures among women who *had not* cohabited than among those who had (Phillips and Sweeney 2005). Thus, we can say that the negative effect of cohabitation on later marriage has been greatest among whites and may not even hold true for African Americans or Mexican Americans (Smock, Casper, and Wyse 2008).

Like race or ethnicity, age also matters. Older co-habitants are more likely than their younger counterparts to view their relationship as an alternative to marriage. Younger cohabitants more likely see their relationship as a prelude to marriage. For older men and women, decisions to cohabit *rather than* marry may well be motivated by financial considerations. Fears of losing benefits, such as a deceased spouse's pension, or of seeing savings intended for their children's inheritance spent or becoming shared may prompt older couples to live together. If they do marry instead, such concerns may lead them to craft carefully negotiated prenuptial agreements (Levaro 2009). Nonfinancial incentives to live together rather than marry include the fear of losing one's spouse (i.e., experiencing widowhood again) or of becoming caregivers for a new spouse. These are more likely and strongly felt among older women than among older men. Another concern that some older women may have that would deter them from remarrying more than from cohabiting is the desire to retain some of the greater freedom and flexibility that they perceive to accompany cohabitation but that might be lost in marriage (Levaro 2009).

Within cohabiting relationships, older cohabitants also report higher levels of relationship quality on numerous aspects of their relationships—fairness, having fewer disagreements, spending more time alone together, being less likely to argue heatedly, and being less likely to think that their relationship is in trouble or may end. For older cohabitants, cohabiting relationships tend to last longer, be less likely to transition into marriages, and tend to be more harmonious than cohabiting relationships among younger cohabitors (Brown, Bulanda, and Lee 2012). Older cohabitants seem less negatively affected than younger cohabitants by the absence of plans to marry. Clearly, as Valarie King and Mindy Scott (2005, 283) suggest, "cohabiting relationships are indeed different for older and younger adults."

Social demographer Sharon Sassler suggests that, in addition to and related to age, the length of time one is involved in a romantic relationship before moving in together is important. Sassler notes that there are notable differences by education in how quickly couples move in together. Based on her interviews with over 150 cohabitors, the majority of those with less than a college degree began cohabiting within six months of the start of their relationships. Among those with college degrees, the move to cohabitation was much more gradual. Half had been involved for more than a year, while more than a third were romantically involved for more than two years before beginning to live together (Sassler 2015).

Of still additional significance is the type of cohabitation. The negative impact of cohabitation on marriage is not found among cohabitors who begin living together already engaged or with definite plans to get married at some later time (Manning and Cohen 2012; Smock, Casper, and Wyse 2008). Conversely, those who enter into cohabiting relationships before getting engaged and then go on to marry are more likely to experience lower levels of marital quality than either couples who had not cohabited or cohabitors who were engaged first. As expressed by family researchers Scott Stanley and Galena Rhoades (2009, F3), "The greatest risk is for those who do not have mutual clarity about the future together because they are increasing the likelihood of marriage [by cohabiting] before clarifying . . . important matters of fit, intention, and commitment." Research by Wendy Manning and Pamela Smock found that for over 50 percent of the cohabiting couples they studied, cohabitation "just sort of happened. One thing led to another and bingo, the couple was living together" (Stanley and Rhoades 2009, F3).

The most recently available research suggests that the negative cohabitation effect has weakened as more and more couples cohabit and cohabitation as a lifestyle becomes more widely accepted. Among younger couples who marry after living together, the risk of divorce may be no greater than the risk for couples who marry without first living together. In fact, according to sociologists Wendy Manning and Jessica Cohen, "since the mid-1990s, *whether men or women cohabited with their spouse is not related to marital stability*" (2012, 384, emphasis added). There is also evidence in Manning and Cohen's research to support the idea that, for some women, especially those engaged at the outset of cohabiting, cohabitation was associated with significantly lower odds of marital instability (Manning and Cohen 2012).

The risk of marital instability may still be greater among those couples who "slide" into cohabitation

As this chapter has addressed, an increasing number of couples in long-term, loving, and committed relationships forgo marriage in favor of living together. In recent years, there also has been an increase in the number of couples in long-term, loving, and committed relationships who forgo living together and instead maintain their relationships across different households. Characterized by labels such as "home alone together," "dual-dwelling duos," and, most commonly, "living apart together" (or LAT), such arrangements are left invisible when researchers discuss and differentiate between single, cohabiting, and married individuals. Though unmarried and not residing together, such couples consider themselves partnered. Thus, considering them "single" is misleading; they consider themselves couples. To the individuals involved, these are much more than casual relationships.

LAT relationships have become more common, both in the United States and abroad. They have been longer recognized as distinctive types of relationships in Scandinavian countries such as Sweden and Norway, and are currently on the rise in many European countries, including France, Germany, Spain, and the United Kingdom, as well as in Australia, Canada, and Japan (Reimondos, Evans, and Gray 2011). Survey data indicate that 10 percent of men and 11 percent of women, ages 18 to 79, were maintaining such noncohabiting relationships. They are more common in western than eastern Europe, reaching close to 10 percent of the population of 18- to 79-year-olds in western Europe and 5 percent or less in eastern Europe (Leifbroer, Poortman, and Seltzer 2015).

Estimates of LATs in Australia suggest between 7 to 9 percent of adults have partners they do not live with, and in Canada, it is estimated that 1 in every 12 adults age 20 and older, and 56 percent of Canadians in their 20s, are in LAT relationships (Milan and Peters 2003; Reimondos, Evans, and Gray 2011). Based on data gathered in Australia's Household, Income and Labour Dynamics in Australia (HILDA) Survey, some 1.1 million Australians, 24 percent of the unmarried population, could be considered "living apart together" (Reimondos, Evans, and Gray 2011).

Like marriage and cohabitation, main features of LAT relationships include a long-term commitment, public identification as a couple, open acknowledgment of love for one's partner, and involvement in regular sexual relations with one's partner (Connidis 2006). Unlike marriage and cohabitation, these features are shared and expressed while living apart from each other. Although LAT couples share the fact of living apart, there are differences among them based on intent. According to sociologist Sasha Roseneil, one can distinguish between those who are "regretfully apart," at one end of a continuum, while at the other are those who can be considered "gladly apart." In between, we would put couples who consider themselves "undecidedly apart" (Roseneil 2006).

LAT relationships can be the choice of couples of a variety of ages and statuses. For younger couples, LAT relationships might be chosen to allow one or both partners to continue to live at home, supported in part or full by parents, while maintaining a committed,

rather than actively decide to live together with the intent to ultimately marry. Once a couple is living together, they may find themselves on a path to marriage simply because it is harder to end their relationship than if they were just dating. Such things as "the idea of moving out, splitting things and friends up, and finding another place to live" may lead many cohabitors to remain together, and remaining together increases the likelihood that they will eventually marry. Once married, such "sliders" may still experience marital difficulties that increase their likelihood of divorce (Stanley and Rhoades 2009).

Related to the matter of *type of cohabitation* is the question of one's past history of cohabiting relationships. Individuals who cohabit with only their future

spouse are unlikely to experience the negative outcomes otherwise associated with cohabitation. On the other hand, **serial cohabitation**, where individuals have cohabited with more than one partner, has more problematic outcomes. Research by Daniel Lichter and Zhenchao Qian revealed that female cohabitors who had prior cohabiting relationships were much less likely to marry their partner than were women who had cohabited only with their current partner. Those serial cohabitors who did enter marriage experienced more than double the risk of divorce compared to women who cohabited only with their future husband (Lichter and Qian 2008).

In their consideration of factors that help account for negative effects of cohabitation on marriage, Laura monogamous, couple relationship. Young adults may face housing and financial constraints that delay their ability to establish a common residence with their chosen partners. At the same time, they may not feel ready to commit to living together. For those who are older, such as in midlife, a variety of practical reasons might motivate them to maintain separate residences while, at the same time, considering their lives to be joined together. For example, for those who have children, living apart together can allow them to focus on the things that need to be done for and with their children without the added complication of a live-in partner. In instances where one or both partners own their residence, this arrangement allows them to maintain the value of two residences. For those who have been married and divorced, the LAT arrangement allows them to avoid subsequent marital failure.

There are aspects of LAT relationships that make them especially attractive to older women and men. For older women, a LAT relationship allows them to maintain the freedom and independence that accompanies having one's own space. It also frees them from the potential obligation to provide long-term care for a partner whose health diminishes. Also potentially important to older couples, a LAT relationship can more clearly protect their financial resources for future inheritance to be left to one's children and grandchildren. In marriage or cohabitation, such money might be contested by a surviving spouse or cohabiting partner.

Research by Aart Liefbroer, Anne-Rigt Poortman, and Judith Seltzer (2015) examined LATs in ten European countries. They suggest that such relationships are chosen for both practical and ideological reasons (i.e., as an alternative to cohabitation or marriage), but more often for practical reasons, or as a result of practical constraints to establishing a coresidence. They also note that living apart together appears to be more of a stage than an alternative to other relationship situations, pointing out that in all the countries they examined, majorities of respondents expect to be living with their partners (married or cohabiting) within three years (Liefbroer, Poortman, and Seltzer 2015).

Of course, across all age groups, by living together apart, partners can maintain their independence and autonomy, avoid the need for frequent compromise, and live life "on one's own terms." Similarly, living apart together can be an attractive option for couples where individual partners wish to remain in their apartments, homes, or neighborhoods or to retain their affordable rent-controlled apartments or expensive real estate (Rosenblum 2013). Advocates contend that living apart together can be the best of both worlds, as partners can both "be themselves" and "be a couple." As sociologist Ingrid Arnet Connidis (2006) suggested, LAT arrangements can allow one to simultaneously satisfy one's need for intimacy and autonomy, and companionship and independence.

More negative assessments cast the choice to enter and maintain a LAT relationship as self-centered and indicative of an inability or unwillingness to compromise (Hart 2006). Like so many other issues that surface throughout this text, the ultimate meaning and significance of living apart together can be interpreted in more than one way.

Tach and Sarah Halpern-Meekin raise the possibility that there may be another important but often overlooked factor at work. Using data from the National Longitudinal Survey of Youth, their analysis indicates that premarital childbearing explains much of the lower marital quality found among former cohabitors who married. Such couples were more likely to be parents and to be experiencing the downswing that parenthood induces in marital quality (Tach and Halpern-Meekin 2009).

Is there anything about the *types of people* who choose to live together before marrying that might also make them more likely to divorce should they face marital difficulties? Susan Brown and Alan Booth (1996) suggested that the characteristics of people who cohabit are more influential than the cohabiting experience itself. People who live together before marriage tend to be more liberal, more sexually experienced, and more independent than people who do not live together before marriage. They also tend to have slightly lower incomes and are slightly less religious than noncohabitants (Smock 2000). As cohabitation has become so much more common in the years since these comparisons were drawn, it is likely that there are smaller differences between those who do and don't cohabit.

At the same time, there have been suggestions that cohabitation itself may affect individual partners and their relationships. Compared with married couples, cohabiting partners tend to have more similar incomes and divide household tasks more equally. Egalitarian arrangements may be harder to sustain once married, especially once couples have children, and strain or conflict may occur. Sociologist Judith Seltzer (2000) suggests that marriage presses couples toward a gendered division of labor. As cohabitation continues to increase among people from varying backgrounds, we will be better positioned to see whether and how the cohabitation effects on later marital quality and stability have changed and, if so, for which groups of cohabiting couples. If, as some report, the effect is now "trivial" at best, is it similarly slight for all cohabitors (Smock, Casper, and Wyse 2008)? Will it become so? Finally, it stands to reason that at least some poorly chosen relationships will break up at the cohabitation stage and never become marriages or divorces. Thus, even if cohabitation does not guarantee marital success, it does show some high-risk couples that, in fact, they were not meant for each other. This will at least spare them the later experience of divorce (Seltzer 2000).

## ■ Common Law Marriages and Domestic Partnerships

Before the 19th century, U.S. couples who lived together without marrying would, after a short period of living together, enter what is known as **common law marriage**. A couple who "lived as husband and wife and presented themselves as married" was considered married. Originating in English common law, as practiced in the United States, common law marriage was seen as a practical way to enable marriage for couples who wanted to be married but were too geographically removed from both an individual with the authority to marry them and a place where they could obtain a marriage license (Willetts 2003). Common law marriage became less necessary in the 19th century as the availability of officials who could perform marriage ceremonies grew. Although most states no longer allow or recognize common law marriage, according to the National Conference of State Legislatures information about common law marriage in the United States in 2015, the following states and the District of Columbia still do:

Colorado
District of Columbia
Iowa
Kansas

Montana
South Carolina
Texas
Utah

In addition, in Alabama, Oklahoma, and Rhode Island, common law marriages are recognized in case law. In New Hampshire, common law marriages are recognized for inheritance purposes, meaning only after one partner dies. Additional states, such as Georgia, Pennsylvania, and Ohio, for example, "grandfathered" common law marriages established prior to some designated date. They no longer allow new common law marriages (National Conference of State Legislatures, www.ncsl.org).

If you happen to be reading this in one of these states and you meet the requirements described earlier, congratulations, you have just been pronounced married! Although we are being facetious, common law marriage does unite into legal marriage two people who never sought and never obtained a marriage license. Once in such a marriage:

> if you choose to end your relationship, you must get a divorce, even though you never had a wedding. Legally, common law married couples must play by all the same rules as "regular" married couples. If you live in one of the common law states [and live together] and don't want your relationship to become a common law marriage, you must be clear that it is your intention not to marry. (Solot and Miller 2005)

Benefits associated with common law marriage include the right to spousal support and an equitable division of property in the event that the relationship ends, and the right to inherit property or monies upon the death of one's partner. In states that recognize common law marriages, the amount of time a couple must live together before being considered married varies. What is common among them and essential to the meaning of common law marriage is that they have presented themselves as if married by acting like they are married, telling people they are married, and doing the things married people do (including referring to each other as "husband" and "wife"). In states that don't recognize common law marriages, no matter how long you live together or how married you act, you are not married. Of note, *no state* in the United States recognizes same-sex common law marriage.

> **False**
> **9.** All cohabiting couples can enter into a legal domestic partnership.

In **domestic partnerships**, cohabiting heterosexual, lesbian, and gay couples in committed relationships obtain certain legal rights and protections of marriage. In some ways, domestic partnerships can be considered alternative forms of cohabitation with certain formal rights and protections. Civil unions, which as of this writing are available to committed couples in New Jersey, Illinois, Hawaii, and Colorado, are more like alternative versions of marriage.

Domestic partnerships can be offered and regulated by a state or municipality or be recognized by private employers in the provision of employee benefits such as medical coverage and other benefits. Domestic partnership benefits are more limited than civil unions, but can include a variety of the following:

- Medical, dental, and vision insurance
- Sick leave
- Bereavement leave
- Death benefits
- Accident and/or life insurance
- Parental leave
- Tuition reduction or remission at universities where a partner is employed
- Housing rights
- Access to recreational facilities (http://family .findlaw.com/living-together)

In 1982, the *Village Voice* newspaper (based in New York City) became the first private business to offer domestic partnership benefits to employees and their partners. Two years later, in 1984, Berkeley, California, was the first U.S. city to enact a domestic partnership ordinance and extend it to both heterosexual and same-sex couples (Willetts 2003). In 1997, San Francisco extended health insurance and other benefits to their employees' domestic (which includes same-sex) partners, requiring all companies contracting with the city or county to offer the same benefits to same-sex couples that they provided married couples. Individual employers, such as the Gap, Levi Strauss, and Walt Disney Company, soon followed suit, introducing domestic partner policies that have now become fairly common in the private sector as well as in many local and state governments, colleges, and universities. Journalist Tara Siegel Bernard reports that, according to the Human Rights Campaign, approximately two-thirds of Fortune 500 companies offer domestic partner benefits to employees with same-sex partners. Three-fifths of those companies offer those same benefits

to workers with opposite-sex partners. Bernard also points out that based on data from the Bureau of Labor Statistics, National Compensation Survey, slightly more than a third of all those employed in the private sector had access to domestic partner benefits for same-sex domestic partners.

At present, it is unclear as to whether and how many companies will continue to provide those benefits now that same-sex couples can legally marry in every state in the United States (Bernard 2015). A poll taken by the consulting firm, Mercer Signal, found that of the more than 150 employers surveyed, of those firms that provide domestic partnership coverage, 4 percent had already dropped such coverage in the wake of the Supreme Court ruling on gay marriage. Fifteen percent said they will be dropping such coverage at the next open enrollment period, and another 28 percent are considering dropping such coverage. Just over half said they were not considering dropping such coverage. Of employers whose domestic partner benefits also cover heterosexual domestic partners, three-fifths (62 percent) said they had no plans to drop the coverage (Umland 2015).

The following states recognize domestic partnerships: California, Oregon, Washington, Maine, Hawaii, Nevada, and Wisconsin, along with the District of Columbia. Meanwhile, Colorado, Delaware, Hawaii, Illinois, New Jersey, and Rhode Island offered civil union protection for both opposite-sex and same-sex cohabiting couples. Upon statewide legalization of gay marriage, Delaware and Rhode Island replaced their civil unions with same-sex marriages. In addition to the states mentioned, many city governments throughout the United States recognize and register domestic partnerships, including Albuquerque, Atlanta, Baltimore, Boston, Chicago, Denver, Minneapolis, New Orleans, New York, Philadelphia, Phoenix, Portland (Oregon), St. Louis, and Tucson, as well as a large number of smaller cities throughout the United States (family.findlaw.com/marriage/state -and-local-provisions-regarding-cohabitation.html)

Domestic partners, whether heterosexual, gay, or lesbian, still lack some of those legal rights and benefits that come automatically with marriage. These include the right to do the following:

- File joint tax returns
- Automatically make medical decisions if your partner is injured or incapacitated
- Automatically inherit your partner's property if he or she dies without a will

"I get kind of upset when people say that a domestic partnership is an alternative to marriage. . . . That's not what it is. . . . It's a different approach to looking at partnerships in sort of a legal sense."

The preceding comment is from 26-year-old Marie as she talked to sociologist Marion Willetts (2003) about her four-year-long licensed relationship. Willetts interviewed 22 other licensed heterosexual domestic cohabitants in the first study that attempted to uncover and document the motives for embarking on a domestic partnership instead of marriage. Although the rights and benefits bestowed by domestic partnership recognition could be obtained by marrying, some couples opt instead to enter licensed domestic partnerships. Typically, they must sign an affidavit declaring that they are not married to someone else and that they are not biologically or legally related to each other. They further pledge to be mutually responsible for each other's well-being and to report to authorities any change in their relationship—either marriage or dissolution. Willetts notes that motives behind heterosexual couples' choice to *cohabit* rather than marry may include economic benefits for some couples. This is especially true if partners have similar incomes and can take larger standard deductions on their income taxes when each files as single than they would as a married couple. Older women and men may avoid remarriage and cohabit instead, thus retaining their individual Social Security benefits. Other cohabiting heterosexual couples are motivated by more personal and philosophical benefits, such as rejecting the assumptions that are part of legal marriage, not wanting the state to intervene in their relationships, and wanting to avoid past marital failures. But what about motives to license partnerships?

Economic benefits, including health insurance coverage, access to university-owned family housing or in-state tuition benefits, or access to family membership rates in outside organizations, made up the motives most cited by Willetts's interviewees with domestic partners. For formerly married cohabitants, licensed partnerships allowed them to avoid reentering marriage yet to obtain the protection and recognition of documentation. For others, such as 31-year-old Leslie, obtaining a domestic partnership license with her partner, Alan, was a means to obtain recognition in the eyes of others that they had made a deep and meaningful commitment to each other, even in the absence of a wedding: "I guess [we wanted] to sort of be counted. There's [sic] relationships that mean a lot that aren't recognized by law and to sort of be counted in that count in the city."

When Willetts posed the question of why that mattered, Leslie continued:

It's difficult to be in a relationship where people are like, "Oh, aren't you married?" or "Are you not married?" . . . It's like an issue all the time. "Why aren't you married, you've been together for 10 years?" . . . so we were like, "We'll get a domestic partnership [to have some sort of documentation in response to these questions]." But it wasn't really something that meant a great deal to us. . . . It wasn't a big deal.

Although Leslie and Alan desired recognition, they wished to avoid too much interference, such as what accompanies a marriage license: "We didn't want to have any law interfere in our relationship, or we didn't feel we needed to have a legal stamp on our relationship."

Other respondents stressed wanting to avoid the trappings of the patriarchal institution that they perceived marriage to be or wanting to demonstrate support for friends whose same-sex relationships were long denied the right to marry. Licensed partnerships did not, however, give heterosexuals the same recognition and support with their families or friends that they would have had if they had married. Marie comments on what her four-year-long licensed partnership lacked:

With a marriage license, there's that sort of social and economic and political legitimacy involved in it. . . . With our domestic partnership, nobody gave us any sort of crockery, nobody bought us a house, nobody sends us anniversary cards, and nobody sort of celebrated, or has celebrated that, you know, that special day [when she and her partner obtained their certificate].

Marie did not feel that legally defining licensed partners as though they were married was desirable: "Once the court says, 'Well, we're going to define this as marriage' . . . once you start having courts that intervene in using words that this is like a marriage, it takes away from, once again, the legitimacy of these other sorts of different types of families that can come about."

As noted in the chapter, legalization of same-sex marriage may well lead states and municipalities that have domestic partnership ordinances to deem them no longer necessary and abandon them. Now that same-sex couples can also legally marry, why continue to offer this other legal category? Of course, for heterosexual domestic partners, the right to legally marry was not what kept them from marrying. If domestic partner policies and protections become scarcer, and if employers start eliminating domestic partner benefits such as health coverage, those heterosexuals who have chosen domestic partnerships will find themselves once again in more vulnerable positions. ●

- Collect unemployment benefits if you quit your job to move with a partner who has obtained a new job
- Live in neighborhoods zoned "family-only"
- Obtain residency status for a noncitizen partner to avoid deportation

Keep in mind that heterosexual domestic partnerships and same-sex domestic partnerships and civil unions resulted from different motivations. Among heterosexuals, domestic partnership has been a deliberately chosen alternative to marriage, as can be seen in the Real Families feature. For at least some gay and lesbian couples, domestic partnerships or civil unions were selected as the closest approximation to legal marriage available to them. A great many same-sex couples who entered domestic partnerships would have married if marriage had been an available and/or accessible option. It is also important to note that with the legalization of same-sex marriage, many of the corporate, municipal, and statewide policies are in flux. With marriage now available to same-sex couples, many employers are reconsidering their policies; states that recognize civil unions may convert them to marriages, as Rhode Island and Vermont did, and domestic partner policies may change. As stated on the legal website, Findlaw.com, "As of September 2015, the status of domestic partner benefits (in light of the *Obergefell* decision) remains unclear" (family .findlaw.com/domestic-partnerships/domestic -partners.html). Unmarried cohabiting couples are advised to consult the state and local laws to determine what rights and protections are currently available outside of marriage.

## Gay and Lesbian Cohabitation

Based on 2014 census data, there were 783,100 same-sex couple households in the United States, an increase of more than 20 percent since 2010, and more than double the number of such households (358,390) in 2000 (www.census.gov/hhes/samesex/; Gates 2012). Gary Gates and Taylor N. T. Brown of the Williams Institute at the UCLA School of Law supplement the 2014 census data with 2015 Gallup data to estimate that there are more than a million same-sex couple households in

the United States, of which they project that 486,000 same-sex couples are married couples. Who are these couples, and how do they compare to heterosexual couples?

- Lesbian couples make up slightly more than half (51 percent) of all same-sex couple households. They also represent just over half (53 percent) of married same-sex couples (Gates 2015a).

- Gary Gates and Taylor N. T. Brown note that LGB individuals are less likely than heterosexuals to be in married or cohabiting relationships. Roughly 40 percent of LGB adults live with a partner or spouse and, among LGB individuals, lesbians are most likely to be in cohabiting relationships at rates similar to that of heterosexual women (Gates and Brown 2015).

- Among all same-sex couples, 27 percent of female couples and 8 percent of male couples are raising a child under 18. Among married same-sex couples, the percentages rise to 36 percent among females and 17 percent among males. Among unmarried cohabiting same-sex couples, 24 percent of female couples and 6 percent of male couples are raising at least one child under 18 in the household (Gates 2015a).

- On average, partners in same-sex couple households are five years younger than women and men in different-sex couples (44.8 vs. 49.7 years old). Among both same-sex and different-sex couples, married partners tend to be older than unmarried partners, though the average difference is much greater among different-sex couples (13 years) than among same-sex couples (less than a year) (Gates 2015a).

- Unmarried same-sex couples have a median income 27 percent lower than that of their married same-sex counterparts. Among different-sex couples, the gap between unmarried and married is larger (46 percent). Income differences can be seen in Figure 9.3. Similar differences can be seen in poverty rates. Whereas 4 percent of married same-sex couples were in poverty in 2013, the rate was 18 percent among cohabiting same-sex couples. A similar but larger marital status gap can be seen among different-sex couples, where 6 percent of married couples but 30 percent of unmarried couples were in poverty in 2013 (Gates 2015a).

**True** **10.** Same-gender couples are more likely to be interracial than are heterosexual couples.

**Figure 9.3** Median Incomes for Same-Sex and Heterosexual Unmarried and Married Couples

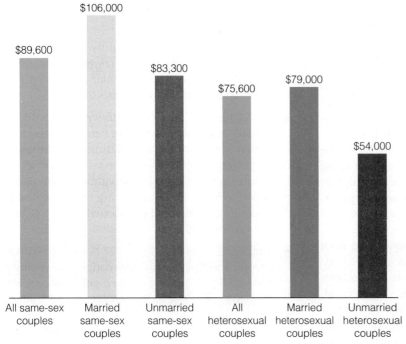

SOURCE: Gates (2015b).

In many ways, same-sex couples want, experience, and struggle with many of the same things as heterosexual couples. They seek similar qualities in their partners, valuing affection, dependability, shared interests, and similarity of religious beliefs (Peplau and Fingerhut 2007). Comparative research has indicated that gay men, lesbians, and heterosexuals report similar levels of attraction and love for their partners, though same-sex couples have been found to report higher levels of relationship quality and satisfaction; greater affection, joy, and humor; and less whining, belligerence, and negative feelings (Balsam et al. 2008; Gottman et al. 2003). It remains unclear as to whether the measures of higher relationship quality result from

The following draws upon census data from the 2014 American Community Survey:

- Same-sex couples are more likely to be interracial couples than are heterosexual couples. Of same-sex couples, 14.8 percent are interracial compared to 6.8 percent of married heterosexual couples and 13.4 percent of different-sex cohabiting couples. Among same-sex couples, male couples were more likely than female couples (17.3 vs. 12.4 percent) to be interracial.

- In a larger percentage of same-sex couples than different-sex couples, both partners are college educated. In 31 percent of same-sex couples (31.4 percent of male couples, and 30 percent of female couples), both partners are college educated. Among different-sex couples, both spouses are college educated in 23.7 percent of married couples and 12.5 percent of unmarried couples.

In terms of relationships themselves, we can note the following. The existing research comparing same-sex and heterosexual relationships is mixed as to how similar or different they are from each other.

Same-sex cohabitors are more likely than married heterosexual couples to be dual earners.

Jack Slomovits/Photodisc/Getty Images

unique advantages and benefits of having a partner of the same gender who has experienced more similar socialization and shares a more similar communication style, or if it is a consequence of the more binding ties of marriage that might keep more unhappily married heterosexuals together. Among most same-sex couples, relationships are held together more by choice (Balsam et al. 2008).

Among all three types of couples, conflict arises over many of the same issues with about the same frequency. Although among couples together two years or less, same-sex couples were more likely to break up (22 percent for lesbian couples, 16 percent for gay male couples, and 4 percent for married couples), some of this difference is due to the legal binds of marriage, not sexual orientation (over the same period, 17 percent of heterosexual cohabitants together less than two years broke up). Indeed, among long-term couples—such as those together ten years or more—there were only very modest differences in their rate of breaking up (6 percent among lesbians and 4 percent among gay couples and among married heterosexuals).

Same-sex couples who desire to marry can now marry throughout the United States. Previously, it was observed that same-sex couples who legalized their relationships with either civil unions or where available, legal marriage, were less likely to end their relationships than were same-sex couples not in legally recognized relationships. The difference in termination rates suggests that legal ties may solidify relationships either by introducing more difficulty ending the relationship or by serving to legitimize and validate the nature of the couple's commitments.

Gay, lesbian, and heterosexual couples struggle over the same sorts of issues: money, housework, power, and abuse. When relationships end—because of breakup or death—partners all suffer similarly. However, gay and lesbian couples often lack the supportiveness of family, friends, and others that married heterosexual couples can mostly take for granted. Thus, when relationship issues arise, conflict occurs, and losses result, gay men and lesbians may not receive the encouragement, support, advice, and sympathy that heterosexuals receive (Peplau, Veniegas, and Campbell 2004).

Major areas of difference have been identified in the importance attached to gender and gender role behavior (as expected, greater among heterosexuals than in gay or lesbian relationships), in the presence or absence of role models for healthy relationships and for resolution of difficulties (scarcer for gay and lesbian couples), and in sexual behavior. Same-sex couples place greater emphasis on fairness in shouldering domestic responsibilities and sharing power within the relationship (Peplau and Fingerhut 2007; Solomon, Rothblum, and Balsam 2005).

Lesbians and gay men who are neither married nor in civil unions also have fewer barriers than do married heterosexuals to ending their relationships once troubles surface. This makes it unlikely that lesbians and gay men will live in long-term, dissatisfying, "miserable and deteriorating" relationships, but perhaps more gay and lesbian relationships than heterosexual relationships will end that could have been saved or improved with patience and effort. In addition, gay and lesbian couples must deal with disagreements about how much they wish to disclose their sexuality to others. Such disagreements may lead a more open partner to pressure a less open partner with the threat of disclosure or leave the more open partner feeling as though the less open partner is less committed to the relationship (Peplau, Veniegas, and Campbell 2004).

Regardless of their sexual orientation, most people want a close, loving relationship with another person. For same-sex and heterosexual couples alike, intimate relationships provide love, romance, satisfaction, and security.

## Same-Sex Couples: Choosing and Redesigning Families

In addition to the couple relationship itself, gay men and lesbians actively construct their wider families in ways that broaden our conception of family life. Lesbian, gay, bisexual, and transgendered women and men are "acting in creative ways to plan families, keep them together, and redefine what it means to 'do family'" (Allen 2007a, 175). As masterfully introduced by anthropologist Kath Weston in her book *Families We Choose*, gay men and lesbians construct **families of choice**, whose boundaries are fluid and cross household lines. They are "people who are there for you" and who can be counted on to provide support and assistance, whether of an emotional or a material nature. In addition to one's partner, such chosen families include varying combinations: children (biological or adopted), a lover's children and biological or adopted kin, friends, and ex-lovers. They can coexist simultaneously with or function in the absence of one's biological or adopted family.

Psychologist Valory Mitchell (2008, 309) suggests that LGBT families "have both the freedom and the need to define their commitment boundaries and expectations of one another." Speaking of lesbians, though true of same-sex couples more broadly, Mitchell asserts the following:

- Lesbians *make families*, independent of legal and blood ties, that are enduring and satisfying.
- Lesbians blend members of their families of origin (families into which they were born), families of procreation (families they make via marriage, cohabitation, and childbearing), and families of choice *to create* a "fabric of family life" that extends over one's life span and across generations.
- Many lesbians are parents, raising children who do at least as well as children of heterosexual parents.

As emphasized previously, lesbians and gay men construct and count on their families of choice, which may become the only family or the more important ones they have.

Although we look somewhat more closely at gay men's and lesbians' experiences as parents in Chapter 10, it is instructive to at least consider the range of households and relationships that same-sex couples with children construct to become and be parents. As summarized by journalist John Bowe:

> For many gay parents, the family structure is more or less based on a heterosexual model: two parents, one household. . . . Then there are families . . . that from the outset seek to create a sort of extended nuclear family, with two mothers and a father who serves, in the words of one gay dad, as "more than an uncle and less than a father." How does it work when [a child] has two mommies, half a daddy, two daddies, or one and a half daddies? (2006, 69)

Sociologist Judith Stacey suggested that the gay men and lesbians who construct such innovative family forms are motivated by a strong desire to have children, "drawing from all kinds of traditional forms [of family], but at the same time . . . inventing new ones" (Bowe 2006, 69). The potential complexity of such new forms can be considerable, as illustrated in the Real Families feature titled "Elective Co-Parenting by Heterosexual and LGB Parents."

## When Friends Are Like Family

Sociologist Margaret K. Nelson authored a sensitive and thoughtful account of the illness and later death of a close friend and colleague (whom she calls "Anna Meyers"). Noteworthy in Nelson's account is how a group of friends came together to become Anna's support system. Without a lover, partner, or spouse; having had no children; and with a biological family of orientation that "was so dysfunctional . . . that their input was not only inappropriate but irrelevant," Anna's care was left in the hands of Nelson, a second colleague, and later a "small group of others whom [we] consulted before making decisions because they also cared about and took good care of Anna" (Nelson 2006, 78).

Nelson (2006, 78) uses Anna's illness and death to ask two fundamental questions: "Were [we] supposed to act as if we were family?" and "Just what does acting like family mean?" Ultimately, Nelson's account is testament to the important, family-like role friends often play. She cites the work of anthropologist Carol Stack, whose urban ethnography *All Our Kin* revealed the extent of interdependence among low-income, inner-city African American single mothers and the **fictive kin** relationships they developed to survive. Recalling our earlier discussion (Chapter 1) of sociologist Karen Hansen's work (*Not So Nuclear Families*), individuals often construct networks of care to assist with day-to-day tasks associated with parenting. Although these networks typically include some members of their biological families, they also typically include unrelated others, such as neighbors and, especially, friends. Nelson's essay "Caring for Anna" really illustrates the vital role that friends play for each other—how they may come together "like family" to assist, support, and care for each other.

Ideally, the material in this chapter has made our consideration of family life more complete. Most Americans, including most of your classmates, expect to and are likely to marry at least once. Still, failure to include those who will make other choices—to remain unmarried, to live together without marrying, or to create one's own supportive network and "do family" in less conventional ways—would have left out a sizable and growing portion of men and women. Harkening back to Chapter 1, our current discussion reminds us of the diversity of family experience and how much our families depend on societal recognition, legal protection, and support.

Pride Angel is an "online connection website" designed to bring together women and men who wish to have a child and have no partner with whom they can do so (Jadva, Freeman, Tranfield, and Golombok 2015). As such, it is among those online sites intended to facilitate a process that has come to be known as "elective co-parenting." Members of Pride Angel are individuals who are seeking sperm and egg donors or recipients as well as those looking for potential co-parents (Jadva et al. 2015). At the time that psychologist Vasanti Jadva and colleagues embarked on their survey of Pride Angel members, the site had nearly 28,000 members, with more than 60 percent registered as sperm recipients and nearly 20 percent as sperm donors. Thirteen percent registered as "co-parents."

Although elective co-parenting has only recently begun to be used by heterosexuals wishing to experience parenting, it has been more prevalent among gay men and lesbians. While detailing the experiences of a number of gay men and lesbians who had cooperated in bearing children together, journalist John Bowe (2006) introduced readers to three men: Mark, David, and R. Mark, 48, is the biological father of two boys born to lesbian partners Jean and Candi, each of whom gave birth to one of the boys. David, 43, decided to help his friends, P. J. and Vicki, have children by fathering two children with Vicki. Along with his partner Bobbie, he shares in the raising of sons Eli and Wyatt. R. is a gay man in his mid-40s who agreed to donate his sperm to an interracial lesbian couple so that they could have a biracial child. R. is white and upper class, characteristics he shares with the nonbiological partner who sought his assistance.

The roles and relationships of the women and men involved in each of these situations vary greatly. Mark sees the children once a week, "sometimes alone, sometimes with his long-term partner Jeffrey, . . . and sometimes with one or both mothers" (Bowe 2006, 70). He feels as though the relationships that have developed between the four adults—the fathers and mothers—have created a " brother–sister feeling." "People make a lot out of it," he told Bowe, "but it's really quite simple: You've got four parents now instead of two. And they're all together."

David, Bobbie, P. J., and Vicki weathered a serious medical crisis when their son Eli had a cancerous tumor removed from his abdomen. The successful surgical removal of the tumor was then followed by chemotherapy, a stem cell transplant, and radiation treatment. For a time, David found himself more extensively involved in caring for Eli than he had anticipated or intended. However, this was followed by a period where Vicky felt David wasn't doing enough. Bobbie, on the other hand, has felt himself to be the "fourth wheel," called out for being either too involved or too aloof. Even with such complexities, the whole family went away for a week together to visit David's mother.

Initially, R. has felt that the arrangement he agreed to left him more removed from his daughter than he would like. "I was always kept at a distance. I was never brought in in a way where I felt like I was being acknowledged as really more than just a friend." Now, with his daughter having turned 6, he is being encouraged by her birth mother to develop a relationship with their daughter.

As each example reveals, gay men and lesbians who wish to become parents can enter into a variety of co-parenting arrangements or avoid co-parenting altogether and opt, instead, for surrogates or anonymous sperm donors. When they do construct an arrangement to bear a child together, other choices await. Sometimes they formalize their arrangements with agreements covering everything from visitation schedules to the what-ifs of partners moving away, starting new relationships, or even dying. Drafting such documents doesn't guarantee that judges will abide by what individuals have indicated should parties wind up in disputes in court, but they do help outline expectations and boundaries. ●

Lesbians and gay men who have children often raise them in arrangements in which up to four partners act as parents.

Steven C. De La Cruz/Image Source/Getty Images

# Summary

- Because of delayed marriage, increased economic and educational opportunities and commitments for women, increased divorce, and liberal social and sexual standards, there has been a dramatic increase in the unmarried population (including both formerly married and never married).

- The population of women and men who are unmarried is diverse; it varies considerably across racial and ethnic lines, and contains a variety of lifestyles and situations, making it difficult to characterize or account for them with a single generalization or explanation.

- The unmarried may be differentiated on the basis of their desire to be married or unmarried and whether they believe that their single status is temporary or permanent.

- Singlism refers to the stereotyping and discrimination directed at the unmarried. Matrimania is the glorification and hyping of marriage. Both contribute to negative treatment and experiences faced by the unmarried.

- Cohabitation has become increasingly accepted for many of the same reasons as has singlehood: a more liberal sexual climate; the changed meaning of marriage, divorce, and cohabitation; delayed marriage; and reduced economic dependence of women on marriage.

- Cohabitation is more common among African Americans and Hispanic Americans than among whites or Asian Americans, those with less education, those whose parents divorced, and those who are less religious, who are more likely to favor egalitarian gender roles, and who have less traditional attitudes about family. Men who work less than full time, year-round, are more likely to cohabit.

- Reasons for and types of cohabitation vary. Decisions to cohabit may be driven by finances, convenience, and housing situations as well as by more relationship-relevant factors. It may be a substitute or alternative to marriage, a precursor to marriage, a trial marriage, or a convenient alternative to dating or to singlehood.

- Gender and race affect the meanings attached to cohabitation. Women are more likely than men to view cohabitation as a step toward marriage. African Americans and Hispanic Americans are more likely to view cohabitation as an alternative to marriage.

- Domestic partnership laws and policies provide some legal rights or workplace benefits to cohabiting couples, including gay and lesbian couples.

- Compared with marriage, cohabitating relationships are more transitory, have different commitments, and lack economic pooling and social support. They also differ in sexual relationships, finances, health benefits, relationship quality, and household responsibilities. Perhaps most important, whereas marriage is "institutionalized," cohabitation is not.

- Cohabitants who later marry tend to be more prone to divorce because of both selection factors (the type of people who cohabit) and experiential factors (consequences of cohabitation itself). Among those who begin cohabiting with an intent to marry, such outcomes are much less likely.

- Factors such as the age or race of partners and prior history of cohabitation affect the outcomes of the cohabiting relationships and subsequent marriages.

- The Census Bureau reports almost 650,000 households headed by married or cohabiting same-sex couples.

- On average, same-sex cohabiting couples tend to be younger, more racially diverse, and dual earners, and to have higher incomes. Same-sex couples who are raising children tend to be less well off than their married heterosexual counterparts.

- Beyond the couple relationship, gay men and lesbians create or construct families of choice, which may consist of combinations of one's partner, members of one's biological or adopted family, one's children, one's partner's children, members of a partner's biological or adopted family, friends, and ex-lovers. These families of choice are the people to whom one turns for emotional or material support or assistance. When raising children, same-sex couples often form families of choice that bring together gay men and lesbians in co-parenting relationships.

- Regardless of sexual orientation, for some people, "family" consists of close friends who, though they lack social recognition, provide emotional, social, and practical support sought from families.

## Key Terms

common law marriage  346
domestic partnership  347
families of choice  351
matrimania  330
palimony  342
precursor to marriage  333
serial cohabitation  344
singlism  330
trial marriages  333

# Becoming Parents and Experiencing Parenthood

# 10

Ron Chapple/Photodisc/Getty Images

355

"It's unbelievable.". . . There's really no way a nonparent can think like a parent. It's really knocked me for a loop. And in my wildest dreams, I never thought of it. . . . Something just creeps into your life and all of a sudden it dominates your life. It changes your relationship with everybody and everything, you question every value and every belief you ever had. And you say to yourself, "This is a miracle." It's like you take your life, open up a drawer, put it all in a drawer, and close the drawer.

These comments convey a 33-year-old man's thoughtful reactions to becoming a first-time father. As he reflects on it, becoming a parent is life defining and life altering. He is not alone. Along with new and demanding responsibilities, becoming a parent introduces profound changes in how we see ourselves, how we live, what we think about, and how we feel. Simultaneously, parents experience changes in their social relationships and how they are viewed by others. These changes are neither minor nor temporary. Becoming a parent is as profound a life change as any other we make.

Not everyone decides to become a parent. With widespread availability of effective contraception and access to legal abortion, women and men can decide whether and when to have children. The first part of this chapter focuses on the choices people make whether to have children and the range of factors that figure into the decision-making process. We examine the characteristics of those who decide to or are forced to forgo parenthood. But those who embark on parenthood face other choices. *How* should they become parents? For some, bearing a child is difficult or impossible, leading them to attempt to adopt or take advantage of the ever-expanding options presented by advances in reproductive technology. And *when* should they become parents? Is there an optimal time or age for

entering motherhood or fatherhood? Throughout this chapter, we explore these choices.

Our focus then shifts to how women and their partners experience pregnancy, how they navigate the transition to parenthood, and the changes parenthood introduces into parents' lives. Special attention is paid to the responsibilities and challenges confronting mothers and fathers in the United States today and the diversity of parent–child relationships.

# Fertility Patterns and Parenthood Options in the United States

There were just under four million births (3,985,924) in the United States in 2014, an increase of 1 percent from the previous year but a drop of 7.7 percent from the all-time highest total of 4,317,119 births recorded in 2007. From 2007 to 2010, the U.S. birthrate steadily declined. The decline slowed between 2010 and 2013, and had its first increase in 2014. As shown in Table 10.1, between 2013 and 2014, there was a 1 percent increase in total births, birthrate (births per 1,000 population),

> **False** **1.** The number of births in the United States declined from 2007 to 2010.

TABLE 10.1 **Births, Birthrates, and Fertility Rates, United States, 2013 and 2014**

|  | 2013 | 2014 |
| --- | --- | --- |
| Total births | 3,932,181 | 3,985,924 |
| Birthrate | 12.4 | 12.5 |
| Fertility rate | 62.5 | 62.9 |
| Total fertility rate | 1,857.5 | 1,861.5 |

SOURCE: Martin, et al. (2015), Table 1.

general fertility rate (births per 1,000 women, age 15 to 44), and total fertility rate (the estimated total number of births a hypothetical group of 1,000 women would have over their lifetimes).

Of course, a long-term view would show that in the midst of the baby boom, the U.S. fertility rate was considerably higher. In 1960, for example, the fertility rate was 118 (Martin, Hamilton, Ventura, et al. 2012).

Fertility rates and birthrates vary considerably according to social and demographic characteristics, such as age, race and ethnicity, income, and marital status. Between 2013 and 2014, there were decreases for age groups under 25. Racially, rates increased among Asian or Pacific Islander women and among white women. Among American Indian and Alaska Natives, Hispanic women, as well as for unmarried women, rates declined slightly from 2013 to 2014 (Hamilton et al. 2015a). In addition to the 2013–2014 percentage change, it is worth noting that, considered in a longer-term historical context, these birthrates are relatively low. Table 10.2 and Figures 10.1 and 10.2 show variation in birth and fertility rates by age and ethnicity for 2013.

**Figure 10.1  Fertility Rates, Race, and Ethnicity, 2013**

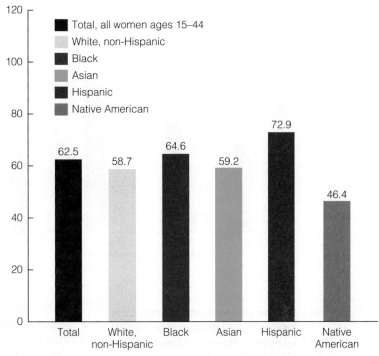

SOURCE: Martin et al. (2015).

Figures 10.1 and 10.2 show birth and fertility rates by race and ethnicity. Although not reflected in the figures, recall from Chapter 3, there is much diversity across Hispanic ethnicities. Within the Hispanic population, birthrates vary considerably, from a high among Central and South Americans (22.3 in 2012) to a low among Cuban Americans (8.9 in 2012). Cultural, social, and economic factors play a significant part in influencing the number of children a family has. Because of a combination of higher fertility rates and continuing immigration patterns, Hispanics have become our nation's largest minority group, having surpassed African Americans.

Fertility rates also vary by education and income. In 2012, women with the highest level of education (e.g., graduate or professional school) had the highest fertility rate (66.0 births per 1,000 women ages 15 to 50). High school graduates had the next highest rate (57.7), followed by women with college educations (56.2). Women with less than a high school education (46.8) and women with some college or associate degrees (51.6) had the lowest levels (Monte and Ellis 2014).

Income, too, affects rates of childbearing (see Table 10.3). Birthrates tend to decline, and the

**TABLE 10.2  Birthrates by Age per 1,000 Women, 2013–2014**

| Age | 2013 | 2014 | % change, 2013–2014 |
|---|---|---|---|
| 15–19 | 26.5 | 24.2 | −9% |
| 20–24 | 80.7 | 79.0 | −2% |
| 25–29 | 105.5 | 105.7 | +2% |
| 30–34 | 98.0 | 100.8 | +3% |
| 35–39 | 49.3 | 50.9 | +3% |
| 40–44 | 10.4 | 10.6 | +2% |
| 45–49 | 0.8 | 0.8 | No difference |

SOURCE: Hamilton et al. (2015a).

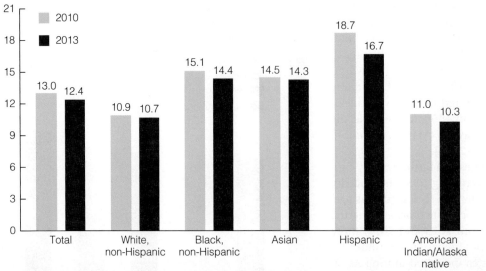

**Figure 10.2  Birthrates, Race and Ethnicity, 2010 and 2013**

SOURCE: Martin et al. (2015, Table 1 and 5).

**TABLE 10.3  Women Who Had a Child in the Last Year per 1,000 Women, Age 15 to 50, by Poverty Status, 2012 (numbers in thousands)**

| Poverty Status | Births per 1,000 Women |
|---|---|
| Below 100% of poverty | 81.7 |
| Below 50% of poverty | 88.5 |
| 100% to 199% of poverty | 62.8 |
| 200% or more above poverty | 44.8 |

SOURCE: Monte and Ellis (2014).

percentage of women who are childless increases as poverty status goes down (i.e., income goes up).

Approximately one in five women in their mid-40s who have a college level or higher degree have not had children. This number is the lowest its been since 1994, declining steadily since 2004. Women in the same age group with less education also saw decreases in childlessness, but have seen less variation in the last 20 years (Livingston 2014).

As will be shown later, in addition to income, being childless (or child-free) also varies by characteristics such as race and education.

## Unmarried Parenthood

In 2014, there were 1.6 million births to unmarried women, an increase from the 1.59 million in 2013, but a drop in the overall percentage of births to unmarried mothers and a drop since the number peaked in 2007 (Hamilton et al. 2015a). However, taking a longer view, the number of births to unmarried women in 2010 was two and a half times higher than the number in 1980, and nearly 19 times higher than the 1940 level. The 2010 birthrate among unmarried women was 49.9 per 1,000 women, making such births 40.8 percent of all births. Contrary to the impressions one may have, more than half (57 percent) of births to unmarried mothers actually were to women in cohabiting relationships (Payne, Manning, and Brown 2012). As such, though the mothers and their partners were unmarried, they did constitute two-parent, couple-headed households.

As reflected in Table 10.4, the percentage of births to unmarried women differs for different racial or ethnic groups. Almost three-fourths of African American and more than half of Hispanic American births in 2014 occurred to unmarried women. Additionally, two-thirds of births to Native American and Alaskan Americans were nonmarital. In comparison, 29.2 percent of births to non-Hispanic whites and 16.4 percent of births to Asian American women were nonmarital births. As noted in Chapter 3, within broad categories such as Asian American or Hispanic American are some notable variations. In 2013, for example, non-marital births made up 53.2 percent

**False** **2.** The number of births to unmarried women has been decreasing since 2007.

## TABLE 10.4  Percent of Births to Unmarried Mothers by Ethnic Origin, 2013–2014

| Ethnic Origin of Mother | 2013 | 2014 |
|---|---|---|
| All ethnic groups | 41% | 40.8% |
| Non-Hispanic whites | 29.3% | 29.2% |
| Non-Hispanic blacks | 71.5% | 70.9% |
| American Native or Alaskan Native | 66.4% | 65.7% |
| Asian or Pacific Islanders | 17.0% | 16.4% |
| Hispanics | 53.2% | 52.9% |

SOURCE: Hamilton et al. (2015a), Table 6.

of all Hispanic births, but ranged from 50.1 percent among Cuban Americans; 51.9 percent among Mexican Americans; 56.1 percent among Central and South Americans; to 64.6 percent among Puerto Rican Americans (Martin et al. 2015).

Although a large majority of births to teenagers are outside of marriage, births to unmarried teens represent less than a fifth of the total number of births to unmarried mothers (17.9 percent) (Hamilton et al. 2015a).

**True** **3.** Births to teenagers have declined substantially over the past 20 years.

On the subject of births to teens, at 26.5 births per 1,000 females ages 15 to 19, the rate of U.S. teen childbirths dropped 10 percent from 2012 to 2013, and has declined substantially, dropping 36 percent between 2007 and 2013, and 57 percent since 1991. The rate has fallen for teenagers across racial or ethnic backgrounds, declining by 12 percent among non-Hispanic white and black teens, 12 percent among Hispanic teens, and 13 percent among Asian American teens. Even with such declines, the 2011 U.S. rate of 26.5 births per 1,000 females ages 15 to 19 remains among the highest teen birthrates found among developed countries (Jayson 2012; Martinez, Daniels, and Chandra 2012).

Once again, it is worthwhile to note the potential ambiguity of estimates of unmarried mothers, in this case, teen mothers. More than 40 percent (44 percent) of births to unmarried teens occurred among teens in cohabiting households. Thus, although legally unmarried, almost half of unmarried teenage mothers were not truly "single mothers," in the sense of bearing and caring for their children without partners (Payne, Manning, and Brown 2012).

## Forgoing Parenthood: "What If We Can't?" "Maybe We Shouldn't"

Increasingly, there are women who don't become mothers, and couples who never become parents. How should we refer to them? In the past, the most common way to describe all adults without children was as *childless*, meaning simply that they had no children. However, the population considered *childless* is more diverse than the term suggests. There are really three different childless populations among women of childbearing age: those who are temporarily childless, involuntarily childless, and childless by choice (Martinez, Daniels, and Chandra 2012). In other words, there are women and couples who don't *yet* have children, *but expect* to, those who desire and have tried to have children but have been unsuccessful, and those who don't want to have children. Calling all three "childless" obscures the differences between them. A further difficulty results from the fact that "childless" carries the connotation that such women or couples are lacking, and that they are "less something" that they wanted or were *supposed to have*. This description simply doesn't fit the experiences of those childless by choice, though it no doubt describes the experiences of those women 15 to 44 years old who have an "impaired ability" to have children or those couples who seek help for infertility. Both are *involuntarily childless*.

Worldwide, it is estimated that over 70 million of the more than 800 million women in married or consensual relationships face an impaired ability to conceive, going more than 12 months without becoming pregnant despite not using contraception. In developed countries, the estimated prevalence of infertility ranges from 3.5 percent to 16.7 percent; in less developed countries, the prevalence ranged from 6.9 percent to 9.3 percent of women. According to a 2007 study, approximately 40 million sought treatment (Boivin et al. 2007).

Based on research by the World Health Organization (WHO), which included data from 277 national surveys in 190 countries, 2 percent of women 20 to 44 were unable to have their first live birth and another 10.5 percent of women who had already given birth at least once were unable to have a baby despite five years of trying (Mascarenhas et al. 2012) The WHO estimates 48.5 million couples were unable to have a child, and that global infertility rates remained relatively stable from 1990 to 2010 (Warren-Gash 2013).

In the United States, an estimated 6 percent of married women, 15 to 44, are infertile. Among married, childless women, the percentage who are infertile climbs to 14 percent (based on the 2006–2010 National Survey of Family Growth). The Centers for Disease Control and Prevention further estimated that 11 percent of U.S. women ages 15 to 44 had **impaired fecundity**, a broader term referring to difficulty conceiving or carrying a pregnancy to term. It includes those who had some physical impairment that prevented them from having a baby or made it physically difficult or dangerous to have a child, and those who were continuously married or cohabiting and had gone three years without using contraceptives but had not gotten pregnant. Among married women, 12 percent were experiencing impaired fecundity. In numbers, 6.7 million women, ages 15 to 44, reportedly experienced impaired fecundity, 3.1 million women with no children and 3.6 million women who had had prior births. Among married women 15 to 44, 3.1 million experienced impaired fecundity and 1.5 million were infertile. Among males 15 to 44, 4 percent were nonsurgically sterile and 5 percent were considered subfertile. Among men 25 to 44, 12 percent experienced some form of infertility (Chandra, Copen, and Stephen 2013).

Approximately 12 percent, representing some 7.4 million U.S. women of reproductive age, reported having ever received infertility services. Among childless women 15 to 44, 6.5 percent had previously received infertility services. Such services included advice, prescription drugs, testing, and artificial insemination.

The majority of women age 15 to 44 without children can be considered *temporarily childless*. They intend and expect to become mothers *at some point* but have not yet had children. Compared to both those who were involuntarily childless as well as the voluntarily "childless by choice," women who were temporarily childless tend to be younger, less likely to have ever married or to be cohabiting, and more likely to have completed college (Martinez, Daniels, and Chandra 2012).

Looking at U.S. women ages 40 to 44 sheds light on those who remain childless throughout their reproductive years. In 2014, 15 percent of 40- to 44-year-old women in the United States had never given birth, the lowest percentage of childlessness since 1994. Examining this group more closely shows that those women with higher levels of education have higher

rates of childlessness. Having said that, it is worth pointing out that among 40- to 44-year-old women with postgraduate degrees, the percentage with no children (22 percent) is down 8 percent from 1994 (Pew Research Center 2015b). Rates vary by race as well, with higher percentages of white women (17 percent) than black (15 percent), Asian (13 percent), or Hispanic women (10 percent) reporting not having had children.

What about those who explicitly *choose* not to have children? How should we differentiate them from their reluctantly childless counterparts? In more recent years, the term *childless* has been joined by *child-free* when referring to those without children. In the United States, we have experienced a cultural and a demographic shift in which there are now more women and couples than previously who *expect and intend* to remain nonparents. They are more likely to consider themselves **child-free**. Not wanting or needing to be seen as objects of sympathy or portrayed as though they lack something essential for personal and relationship fulfillment, they are more likely to consider themselves "free" of the responsibilities of parenting. Their "childlessness" is by choice.

Using data from the 2006–2010 National Survey of Family Growth, it has been estimated that 8.5 percent of 20- to 24-year-old women and 8.4 percent of women ages 25 to 44 expect to have no children in their lifetimes. Among women who were married at the time of the survey, 5.2 percent expected to have no children, while 11.8 percent of never-married women indicated the same expectation (Martinez, Daniels, and Chandra 2012).

Who are the women who remain or expect to remain *child-free*? Research indicates that compared to mothers, the child-free are women with the highest levels of education, those employed in high-status occupations such as managerial and professional occupations, and those with high levels of household income in dual-earner or dual-career marriages. They are also less religious, have less traditional family attitudes, are more likely to be firstborn or only children, and have less traditional ideas about gender.

Historically, white women were more likely than African American women to choose not to have children. Recent research suggests that this race gap has greatly diminished and may no longer hold. The rates of childlessness have increased among both groups, especially among women with higher socioeconomic

> **True** **4.** Among African American women, married women are more likely than unmarried women to be childless.

status and greater educational attainment. Both black and white college-educated women have rates of childlessness 5 to 10 percent higher than the overall rate for women of their same race (Lundquist, Budig, and Curtis 2009). Examining data from the National Survey of Family Growth, sociologists Jennifer Hickes Lundquist, Michelle Budig, and Anna Curtis found an especially striking racial difference in the effect of marital status on the likelihood of childlessness. Among white women, married women had a much reduced risk of childlessness when compared to unmarried women. Among African Americans, married women were *more likely* than unmarried women to be childless (Lundquist, Budig, and Curtis 2009).

Social scientist Rosemary Gillespie (2003, 133) suggests that what has changed most over the past quarter century is the emergence of "a more radical rejection or push away from motherhood." She asserts that an increasing number of women are resisting and rejecting the cultural expectations that automatically associate women with motherhood. Instead, she suggests, "modernity has given rise to wider possibilities for women" (p. 134). One can see evidence of Gillespie's claim in the numbers of books, organizations, and websites—such as www.nokidding.net, www.childfree.net, or www.happilychildfree.com—advocating or defending a child-free lifestyle.

Although most couples have *some* idea before they marry that they will or will not have children, it is not always possible to "disentangle" those whose childlessness is voluntary from those for whom it is involuntary. Even those who "choose" not to become parents may do so after a lengthy process of postponing childbearing. Delaying childbearing has the potential to increase one's risk of age-related infertility. Lundquist, Budig, and Curtis point out that those women or couples who are ambivalent about having children may be indirectly choosing childlessness by "voluntarily postponing parenthood until age-related infertility 'forces' the decision on them." They also point out that some women who are biologically unable to have children, and thus fall into the "involuntarily childless" category, might well have chosen not to have children even if they could have (Lundquist, Budig, and Curtis 2009, 744).

Many studies of child-free marriages indicate a higher degree of marital adjustment or satisfaction than is found among couples with children. Given the time and energy required by child rearing, these findings are not particularly surprising. It has also been observed that divorce is more probable in child-free marriages, perhaps because child-free couples do not stay together "for the sake of the children," as do some other unhappily married couples. That being the case, there will be fewer unhappy childless couples who stay married than there will be unhappily married couples with children.

Today, although greater in number than in the past, child-free women and couples may still find themselves occasionally somewhat stereotyped, perceived as career oriented, materialistic, individualistic, or selfish, with child-free women more negatively perceived than child-free men. Sociologist Kristin Park (2002) identified a variety of strategies the child-free have used to reduce the stigma attached to not wanting children. These strategies included the following:

- *Passing.* This involves pretending to intend someday to become parents.
- *Identity substitution.* This includes feigning an involuntary childless status as well as letting other statuses (e.g., as a voluntary single or an atheist) dominate a social identity.
- *Condemning the condemners.* This may mean suggesting that some people have children for the wrong or for selfish reasons or that they do so without thinking fully about the responsibility.
- *Asserting their "right" to self-fulfillment.* Park (2002) contends that this is a modern type of justification.
- *Claiming a biological "deficiency."* The individual lacks the desire or lacks the nurturing "instinct."
- *Redefining the situation.* This turns potential accusations around by showing how the lifestyle allows nurturing qualities to be used in other ways or allows the individual to be productive. Some also claim that their careers just don't allow for the inclusion of children.

An inspection of the preceding strategies shows that some are more defensive than others, suggesting that would-be parents feel that they must justify their child-free status. Others are more proactive, redefining childlessness as something socially valuable (Park 2002).

## Waiting a While: Parenthood Deferred

Although most women who become mothers still begin their families while in their 20s, the trend toward later parenthood can be expected to continue to grow, especially in middle- and upper-income groups.

Cultura RM Exclusive/Erin Lester/Getty Images

A number of factors contribute to this. More career and lifestyle options are available to single women today than in the past. Marriage and reproduction are no longer economic or social necessities. People may take longer to search out the "right" mate (even if it takes more than one marriage to do it), and they may wait for the "right" time to have children. Increasingly effective birth control (including safe, legal abortion) has also been a significant factor in the planned deferral of parenthood.

Besides giving parents a chance to complete an education, build careers, and firmly establish their own relationship, delaying parenthood can be advantageous for other reasons. Older parents may be more emotionally mature and thus more capable of dealing with parenting stresses (although age isn't necessarily indicative of emotional maturity). Speaking economically, raising children is expensive. Delaying parenthood until one's economic position is more secure makes good sense for many people given the economic effect of parenthood.

## How Expensive Are Children?

Cost estimates of raising a "typical" child, born in 2012, to age 18, range between $216,910 (for those in the lower third of the income range) and $501,250 (for those in the upper third income bracket). These estimates, from the United States Department of Agriculture (USDA), do not include costs associated with prenatal medical care or childbirth, nor do they include costs of college or other postsecondary education (Lino 2013). Furthermore, some, such as author Pamela Paul (2008), contend that the USDA estimates are unrealistically low. The government estimate of child care costs are significantly less than the average obtained from the National Association of Child Care Resource and Referral Agencies (NACCRA), a network of more than 805 child care resource and referral centers. Paul notes that, according to the NACCRA, the average cost of one year of child care for an infant ranges from $3,803 to over $13,000, regardless of how much parents earn.

Paul (2008) also notes how much variation there is throughout the United States in many of the costs associated with raising children, such as the aforementioned cost of child care. In cities such as New York and San Francisco, she suggests that one may have difficulty finding caregivers who will accept less than $500 a week for full-time (40 hours of) child care. That amounts to $26,000 a year just on child care.

Other cost estimates to consider include the cost of higher education. The College Board reports that the average 2015–2016 cost of tuition and fees for a year of college at a private four-year college or university was $32,405. At a public school, the costs are considerably less (in-state tuition costs average $9,410 for a year), but multiplying by four or more years, with likely increases over that time, would still add a significant amount to the USDA estimated costs (College Board 2015). Add to these figures the cost of room and board (between $10,138 and $11,516), and the costs grow significantly. The College Board estimates that the tuition, fees, and room and board costs total between $19,548 for a year of college at an in-state public institution, to $43,921 at a private college or university. Keep in mind that these are averages. They also don't include the cost of books or other materials. Also left off is any additional needed "spending money." Although it is true that many students receive some form of financial aid, college expenses greatly add to the real costs of raising children into adulthood.

For parents, there is also the matter of wages lost or reduced because of cutting back one's time at work. For example, a parent who earned $40,000 a year but reduced to half-time employment until her or his child turned 18 is incurring a cost of at least $360,000 (assuming an unlikely static income and

18 years of steady employment) on top of whatever direct expenses one must assume. Those parents who limit their time out of the labor force to a few years will incur fewer costs but are still encountering some of the financial impact of raising children.

When including costs of higher education, costs for prenatal care and childbirth, as well as estimated wages lost while raising children, the total cost of raising one child from birth to age 22 reaches a much greater, much more daunting total. If one adds those into the total costs, including estimates for inflation, the amounts can easily exceed a million dollars (more than $2,000,000 for the upper third income bracket). The costliness of raising children is certainly one factor in a number of related trends, including the delay of parenthood, the decision to have fewer children, and, for some, the decision not to have children.

## Choosing When: Is There an Ideal Age at Which to Have a Child?

Although we have briefly addressed the question of delayed or deferred parenthood, we should point out that delaying parenthood "too long," like having children "too young," may entail risks and brings costs. For example, research on the health effects for mothers caused by their age at first birth reveals that both "unusually young" and "unusually old" mothers face health risks.

### Teen Mothers

Mothers who bear their first child in their teens face nearly twice the risk of anemia as women who have their first child between ages 30 and 35 (Mirowsky 2002). They also face a greater likelihood of pregnancy-related hypertension, preeclampsia (a medical condition that combines high blood pressure with excess protein in the urine, organ damage, and swelling of a mother's hands and face), and low-birth-weight babies. Aside from potential health problems, teen mothers (and fathers) experience worsened educational outcomes, though some of that may be due to "preexisting socioeconomic and other factors" that make both teen parenthood and dropping out more likely (Mollborn 2007). Once they do become parents, teen parents are likely to have less material resources, given the increased economic needs introduced by the birth of a baby along with their limited resources, which in turn also affects their educational attainments (Mollborn 2007).

Furthermore, children born to teen mothers have been found to be more likely to experience a variety of negative consequences, including delinquency, depression, anxiety, poor performance in school, dropping out of school, higher incarceration rates, and becoming teen parents themselves. They also face greater risks of poverty and unemployment as adults. Although some problems, such as higher levels of drug use, gang membership, and unemployment, were greater for boys than for girls born to women who had their first child while still teenagers, both boys and girls experienced a greater likelihood of becoming teen parents themselves (Campa and Eckenrode 2006; Levine, Emery, and Pollack 2007; Pogarsky, Thornberry, and Lizotte 2006).

What about teen parenting might contribute to these negative outcomes? Teen motherhood may lead many young mothers to end their schooling "early," assume parenting responsibilities before having completed their own social development, delay their entry into the labor market, and be less likely to marry the fathers of their children. These, in turn, reduce the household resources, lessen the father's involvement, and create a poorer context within which to raise one's children. It is possible that such social and economic conditions are among the consequences associated with teen motherhood, but it is also possible that such conditions lead to teen motherhood. In fact, it is quite likely that both are true. The Popular Culture feature, "16 and Pregnant, Teen Mom, and the Reality of Teen Pregnancy and Motherhood," looks at how teen motherhood has been depicted in recent popular television programs.

### Older Mothers

At the opposite end, nearly 40 percent of all births are to women over 30 years old, and nearly 15 percent are to women over 35. Women's studies professor Elizabeth Gregory points out that what is new about this is the postponing of first births into ones 30s, noting that, historically, when women tended to have larger numbers of children, they often had their last child in their later 30s or into their 40s. However, whereas in 1970, 1 percent of *first* births was to a woman 35 or older, in 2010 one of every 12 first births was to women over 35, and a quarter of first births was to women in their 30s (Gregory 2012).

Are there significant risks for pregnancy- and labor-related distress among older first-time mothers? The answer is less definite than one might imagine. For

# 16 and Pregnant, Teen Mom, and the Reality of Teen Pregnancy and Motherhood   *With Bridget McQuaide and Clarissa Roof*

Teen pregnancy is a controversial, often misunderstood topic in the United States, made more complicated by various stereotypes and stigmas that may distort and misrepresent the realities teen mothers face. For example, many associate teen pregnancy and teen mothers only with certain socioeconomic and ethnic groups. Additionally, the trends in and prevalence of both teen pregnancy and teen motherhood are not widely recognized. Both have declined and continue to decline, yet their portrayal in popular culture can leave the opposite impression.

Beginning in 2009, MTV has aired two series, *16 and Pregnant*, and its spinoff, *Teen Mom*, both of which are intended to illustrate the "realities" of what it means to be, first, a pregnant teen and then a young mother. The intention of both programs is to educate viewers about the difficulties faced by the teens and especially the range of struggles inherent in these circumstances. However, the presentation of teen pregnancy and parenthood in these programs is also somewhat problematic.

One of the most important issues that the shows downplay is the range of options for a teen upon finding out she is pregnant. She could keep the baby, carry the baby and then give it up for adoption, or she could get an abortion. However, *16 and Pregnant* really only features the first of the options. Abortion is a rare, nearly nonexistent option in the series, though it was a special feature accompanying a 2010 episode of the

show, and was, in 2012, the featured storyline of another. Adoption, too, is rarely represented. In this way, the series may constrict viewers' ideas about options available to pregnant teens.

Another concern arising from the shows' portrayals of what it means to become a teen parent is the very different depiction of teen fathers and fatherhood in contrast to that of teen mothers and motherhood. Frequently depicted as irresponsible, undependable, and uninvolved in caring for their child, the young fathers are often very negatively portrayed. After all, it is the young women who are "16 and pregnant," and the child's birth leads to "teen moms" not "teen families" or "teen moms and dads." Frequently, the featured actions of the fathers play into and reinforce the impression that they are unable to give up the freedom that comes with being young, and that they only add further stress to the lives of the young women they have impregnated. Of course, the drama surrounding the young women trying to balance pregnancy and eventual motherhood with the wants and demands of the father may be exactly the situations that prove to get higher ratings. Ultimately, most of the couples break up, and many who marry later divorce.

One of the most interesting aspects of the shows, especially *16 and Pregnant*, is the potential effects of the programs on their mostly young viewers. The series are part of a collaboration with the nonprofit National Campaign to Prevent Teen and Unplanned Pregnancy,

example, due to a variety of physical changes, aging reduces the likelihood that one will have a successful pregnancy. As women age, ovaries become less able to release eggs. The eggs themselves are neither as numerous nor as healthy. But caution is needed in reacting to such information, and as Elizabeth Gregory describes, there are certain myths about the risks of older motherhood. For example, one may hear that a 30-year-old woman has only 12 percent of her eggs left. However, this is in comparison to the number of eggs in a *20-week-old fetus*. Even a healthy 25-year-old has only about a fifth of her eggs left (Gregory 2012). Furthermore, the vast majority of 35-year-old women, roughly two-thirds of women through age 39, and about half of 41-year-old women can get pregnant without aid (Gregory 2012). However, the Centers for Disease Control and Prevention continues to recommend that women over 35 who have tried

unsuccessfully for six months or more to become pregnant consult a physician (Centers for Disease Control and Prevention 2012a).

In pregnancy, older mothers (over 35) do have a somewhat elevated risk of a number of complications, including higher rates of pregnancy-related hypertension, fetal mortality, miscarriage, and cesarean-section deliveries. Their offspring are also at somewhat greater risk of birth defects. As to the risk of birth defects, Gregory points out that the increased risk of autism, for example, that is attributable to mothers' age, appears to be less than one percent. In fact, older fathers have been associated with somewhat higher risks. She further contends that not only might the risks for children of older mothers be exaggerated, possible benefits have been overlooked. For example, children's test scores improve along with a delayed entry to motherhood (Gregory 2012).

which was founded in 1996 and receives part of its funding from the Department of Health and Human Services. Many of the commercial breaks for the shows began with a public service announcement directing viewers to itsyoursexlife.com and informing them that "Teen pregnancy is 100% preventable." The DVD releases of these shows even include informational guides, intended to serve as educational resources for young people about the perils of sex and pregnancy. So what are the effects?

Here, recent research suggests two widely different conclusions. Economists Melissa Kearney and Phillip Levine (2014) suggest that *16 and Pregnant* was responsible for a 5.7 percent drop in teen birthrates, representative of a third of the overall reduction in teen birthrates. Kearney and Levine looked at Nielsen ratings, Google searches, and Twitter tweets, and found that areas of high viewership of the show had bigger reductions in teen births. Also, locations in which there were more Twitter tweets and Google searches about the show also had higher levels of searches and tweets about pregnancy prevention. According to Kearney, what teens watch can have a big effect on what they think and how they act (NPR 2014).

Communications researchers Jennifer Stevens Aubrey, Elizabeth Behm Morawitz, and Kyung Bo Kim conducted an online field experiment in an effort to more directly test the effects of watching *16 and Pregnant*. Their study,

"Understanding the Effects of MTV's *16 and Pregnant* on Adolescent Girls' Beliefs, Attitudes, and Behavioral Intentions toward Teen Pregnancy" (2014), had 121 teenage girls watch either an episode of *16 and Pregnant* or an episode of another MTV program, *Made*. They found that those who watched *16 and Pregnant* reported a lower perception of the risk of pregnancy as well as greater perception that the benefits of teen pregnancy outweigh the risks. Jennifer Stevens Aubrey suggests that this reflects the fact that *16 and Pregnant* contains mixed messages: While the young women often made statements suggesting regret and indicating that they would make different choices if they found themselves in such situations again, they are also shown getting lots of attention and in the company of their adorable babies. She suggests that if the goal is to prevent teen pregnancies, *16 and Pregnant* is probably not doing the job (Harwood 2014).

One definite plus conveyed by the two programs is in debunking some stereotypes of teen pregnancy by illustrating that teen pregnancy and young motherhood are not just issues affecting poor young women living in neighborhoods of crime and deep poverty where inhibitions are depicted as running wild, but in the life of the average 16-year-old girl who is juggling high school, cheerleading practice, and homecoming nominations, and who now has the added stress of determining whether and how to care for another life.

Later first births and the care associated with infants can also take a physical toll on women. It has been suggested that for many women, the kind of physical energy required to care for children may decline with age (MacDorman and Kirmeyer 2009; Mirowsky 2002). Here, again, the evidence is mixed. Mothers who start families after age 35 experience a stronger than average increase in happiness, live longer, on average, than other mothers, and avoid some of the "motherhood penalty"—the decline in lifetime income that mothers, in general, have been found to experience (Gregory 2012).

In recent decades, both in the United States and elsewhere, the age of women at first motherhood has increased. In the United States in 2012, the mean age at first childbirth for women was 25.8, up from 25.6 a year earlier. After climbing steadily between 1970 and 2000 (see Table 10.5), the mean age has changed little

### TABLE 10.5  Women's Mean Age at First Childbirth, 1970–2012

| Year | Median Age |
|------|------------|
| 1970 | 21.4 |
| 1980 | 22.7 |
| 1990 | 24.2 |
| 2000 | 24.9 |
| 2009 | 25.2 |
| 2012 | 25.8 |

SOURCES: Mathews and Hamilton (2002); Hamilton, Martin, and Ventura (2010); U.S. Census Bureau News, "Facts for Features," CB14-FF.12, April 16, 2014 (Mother's Day): May 11, 2014.

| TABLE 10.6 | Women's Mean Age at First Childbirth by Race and Ethnicity, United States, 2009 |
| --- | --- |

| Racial or Ethnic Group | Mean Age |
| --- | --- |
| White, non-Hispanic | 26.1 |
| African American | 22.9 |
| Hispanic | 23.3 |
| Asian American | 28.8 |
| Native American | 22.0 |
| All groups | 25.2 |

SOURCE: Martin et al. (2011).

since then, fluctuating between 25.0 and 25.2 between 2000 and 2009 (Hamilton, Martin, and Ventura 2010; Mathews and Hamilton 2002).

There is also ethnic and racial variation in the age at first birth, with Native Americans having the youngest average age (22.0) and Asian Americans the oldest (28.8). Table 10.6 illustrates the age pattern by race.

As suggested previously, father's age is also of some relevance. Children born to fathers over 40 are at a greater risk of autism when compared to children born to fathers younger than 30 years old. Risks of certain birth defects and of cognitive impairments, such as problems with concentration, memory, and reading, also increase among children of older fathers (Harms 2012). Thus, despite the more frequently expressed concern directed at women who delay motherhood, men's age at fatherhood bears its own potential consequences.

Having a child in one's teens or after age 35 can be risky for both mothers and children.

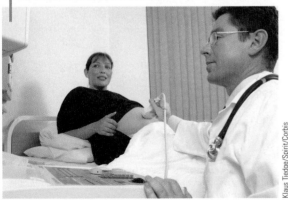

Klaus Tiedge/Spirit/Corbis

# Pregnancy in the United States

Women and men who become parents enter a new phase of their lives. For those who bear their own children, this phase begins with pregnancy. From the moment it is discovered, a pregnancy affects people's feelings about themselves, their relationship with their partner, and the interrelationships of other family members.

According to a 2013 Centers for Disease Control and Prevention report on pregnancy in the United States, there were nearly 6.4 million pregnancies in 2009, down 12 percent from the 1990 peak of 6.8 million, and a continuance of the downward trend since 2007 (Curtin et al. 2013).

Of the nearly 6.4 million pregnancies in the United States in 2009, 4.1 million (65 percent) resulted in births, 1.1 million (17 percent) in induced abortions, and 1.1 million (17 percent) in fetal loss (Curtin et al. 2013). As shown in Figure 10.3, pregnancy outcomes are affected by both marital status and race. In 2009, 86 percent of pregnancies among married women resulted in a live birth, while only 50 percent of pregnancies to unmarried women resulted in births. Unmarried women were much more likely to end their pregnancies via abortion than were married women (Curtin et al. 2013).

There are race differences in the number of expected lifetime pregnancies and in pregnancy outcomes. African American women were less likely than either white or Hispanic women to have their pregnancies result in childbirths. They were nearly one and a half times as likely as Hispanic women and almost three times more likely than white women to end their pregnancies by abortions (see Figure 10.3).

## Being Pregnant

A woman's feelings during pregnancy vary dramatically according to who she is, how she feels about pregnancy and motherhood, whether the pregnancy was planned, whether she has a secure home situation, and many other factors. Her feelings may be ambivalent; they will probably change over the course of the pregnancy.

### Planned versus Unplanned: Was It a Choice?

Based on data from the National Survey of Family Growth 2011–2013, it is estimated that about half of all pregnancies and more than a third (37.1 percent)

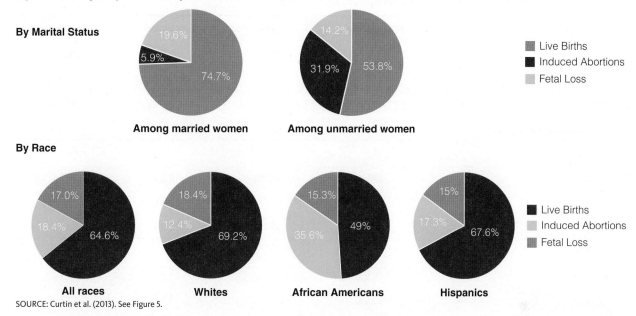

**Figure 10.3 Pregnancy Outcomes by Marital Status and Race**

**By Marital Status**

Among married women

Among unmarried women

Legend: Live Births, Induced Abortions, Fetal Loss

*Among married women:* 74.7%, 5.9%, 19.6%

*Among unmarried women:* 53.8%, 31.9%, 14.2%

**By Race**

All races — 64.6%, 18.4%, 17.0%

Whites — 69.2%, 12.4%, 18.4%

African Americans — 49%, 35.6%, 15.3%

Hispanics — 67.6%, 17.3%, 15%

Legend: Live Births, Induced Abortions, Fetal Loss

SOURCE: Curtin et al. (2013). See Figure 5.

of all births to women ages 15 to 44 were either unwanted (12.9 percent) or mistimed (21.6 percent) at the time of conception. As can be seen in Table 10.7, the experience of unwanted or mistimed pregnancy is more common among unmarried women, minority women, women with lower levels of education, and women who were younger at the time they gave birth (Mosher, Jones, and Abma 2012).

Further details about unintended pregnancies include the following:

- Every year, about 5 percent of reproductive age women (i.e., 15 to 44) have an unintended pregnancy. By age 45, a majority of all women in the United States will have experienced an unintended pregnancy.

- In all 50 states in the United States, at least 38 percent of pregnancies are unintended. In Washington, DC, and 29 states, unintended pregnancies make up more than half of all pregnancies.

- Rates are highest among impoverished women, whose unintended pregnancy rate is five times that of women at the highest income level. As a consequence, poor women also have high rates of unplanned births and abortions, despite the fact that they are less likely than higher-income women to end their unintended pregnancies via abortions. The

rate of unplanned pregnancy among higher-income, white women is one-third the national rate.

- Combining the public expenditures at the state and federal levels, the costs for births resulting from unintended pregnancies is estimated to be in excess of $11 billion.

**True**

**5.** A majority of U.S. women are likely to experience an unintended pregnancy during their childbearing years.

Comparing consequences of planned and unplanned pregnancies reveals the potential risks that are greater for mothers and their children when pregnancy is mistimed or unplanned. Women and couples whose pregnancies were unplanned face greater likelihood of problems associated with lack of readiness or preparedness for parenting. Mothers who give birth to "unintended" babies are more likely to report experiencing psychological problems, such as postpartum depression. Infants born from unintended pregnancies are more likely to suffer both physical and social disadvantages.

### How Pregnancy Affects Couples' Relationships

A woman's first pregnancy is especially important because it has traditionally symbolized the transition to maturity. Even as social norms change and it becomes more common and acceptable for women to defer childbirth until they have established a career or to choose not to have children, the significance of first

## TABLE 10.7  Intendedness of Births at Conception, 2006–2010

| Characteristic | Total Intended | Total Unintended | Unwanted | Mistimed |
|---|---|---|---|---|
| **Total** | 62.9% | 37.1% | 13.8% | 23.3% |
| Married | 76.6% | 23.4% | 7.2% | 16.2% |
| Cohabiting | 49.3% | 50.7% | 20.4% | 30.3% |
| Unmarried, not cohabiting | 33.1% | 66.9% | 27.7% | 39.2% |
| **Race** | | | | |
| Hispanic | 57.1% | 42.9% | 18.1% | 24.8% |
| White | 69.3% | 30.7% | 9.3% | 21.4% |
| Black | 46.5% | 53.5% | 22.9% | 30.6% |
| **Education** | | | | |
| H.S. | 58.9% | 41.1% | 23.2% | 17.9% |
| H.S./GED | 59.9% | 40.1% | 17.3% | 22.8% |
| Some college | 63.3% | 36.7% | 12.6% | 24.1% |
| College (or more) | 83.3% | 16.7% | 4.0% | 12.6% |
| **Age** | | | | |
| 15–19 | 22.8% | 77.2% | 19.3% | 57.9% |
| 20–24 | 49.9% | 50.1% | 16.5% | 33.6% |
| 25–44 | 74.6% | 25.4% | 11.8% | 13.7% |

SOURCE: Mosher, Jones, and Abma (2012).

pregnancy should not be underestimated. It is a major developmental milestone in the lives of mothers—and fathers (Marsiglio 1991; Notman and Lester 1988; Snarey et al. 1987). In fact, a couple's relationship is likely to undergo changes during pregnancy. Communication is particularly important at this time because each partner may have preconceived ideas about what the other is feeling. Both partners may have fears about the baby's well-being, the approaching birth, their ability to parent, and the ways in which the baby will affect their relationship. All these concerns are normal (Kitzinger 1989). If a pregnant woman's partner is not supportive or if she does not have a partner, it is important that she find other sources of support—family, friends, or women's groups—and that she not be reluctant to ask for help.

The first trimester (three months) of pregnancy may be difficult physically and emotionally for an expectant mother. She may experience nausea, fatigue, and painful swelling of the breasts. She may also fear that she will miscarry. Her sexuality may undergo changes, resulting in unfamiliar needs (for more, less, or differently expressed sexual love), which may in turn cause anxiety. (Sexuality during pregnancy is discussed later in this chapter.) Education about the birth process, information about her body's functioning, and support from partner, friends, relatives, and health care professionals are the best antidotes to her fear.

During the second trimester, most nausea and fatigue disappear, and a pregnant woman can feel the fetus move within her. Worries about miscarriage will probably begin to diminish because the riskiest part of fetal development has passed. Pregnant women may look and feel radiantly happy. Some women, however, may be concerned about their increasing size; they may fear that they are becoming unattractive. A partner's attention and reassurance can help ease this fear.

The third trimester may be the time of the greatest difficulties in daily living. The uterus, originally about the size of the woman's fist, has now enlarged to fill the pelvic cavity and is pushing up into the abdominal cavity, exerting increasing pressure on the other internal organs. Water retention (edema) is a fairly common problem during late pregnancy; it may cause swelling in the face, hands, ankles, and feet. It can often be controlled by reducing salt and refined carbohydrates (such as bleached flour and sugar) in the diet. If dietary changes do not help this condition, however, the woman should consult her physician.

Another problem is that the woman's physical abilities are limited by her size. She may be required by her employer to stop working at some point during her pregnancy. A family dependent on her income may suffer hardship. And the woman and her partner may become increasingly concerned about the upcoming birth.

Some women experience periods of depression in the month preceding their delivery; they may feel physically awkward and sexually unattractive. Many, however, feel an exhilarating sense of excitement and anticipation marked by energetic bursts of industriousness. They feel that the fetus is a member of the family. Both parents may begin talking to the fetus and "playing" with it by patting and rubbing the expectant mother's belly.

## Sexuality during Pregnancy

It is not unusual for a woman's sexual feelings and actions to change during pregnancy, although there is great variation among women in these expressions of sexuality. Some women feel beautiful, energetic, sensual, and interested in sex; others feel awkward and decidedly "unsexy." A woman's feelings may also fluctuate during this time. Generally, by the third trimester of pregnancy, approximately 75 percent of first-time mothers report loss of sexual desire, and between 83 percent and 100 percent report reduced frequency of sexual intercourse (De Judicibus and McCabe 2002).

Men may feel confusion or conflicts about sexual activity during pregnancy. They, like many women, may have been conditioned to find the pregnant body unerotic. Or they may feel deep sexual attraction to their pregnant partner yet fear that their feelings are "strange" or unusual. They may also worry about hurting their partner or the baby.

A couple, especially during their first pregnancy, may be uncertain as to how to express their sexual feelings. The following guidelines may be helpful:

- Even during a normal pregnancy, sexual intercourse may be uncomfortable. The couple may want to try positions such as side by side or rear entry to avoid pressure on the woman's abdomen and to facilitate more shallow penetration.

- Even if intercourse is not comfortable for the woman, orgasm may still be intensely pleasurable. She may wish to consider masturbation (alone or with her partner) or cunnilingus.

- Both partners should remember that there are no rules about sexuality during pregnancy. This is a time for relaxing, enjoying the woman's changing body, talking a lot, touching each other, and experimenting with new ways—both sexual and nonsexual—of expressing affection. (Strong and DeVault 1997)

## Men and Pregnancy

Obviously, pregnancy is something men do not experience directly. It is the woman's body that carries the fetus and undergoes profound change along the way. For men, pregnancy is only accessible vicariously. Still, how men navigate the pregnancy process has consequences for their later conceptualization of and involvement in fathering (Marsiglio 1998).

The roles men play in supporting their partners, participating in the preparation for parenthood, and at the birth also are significant. Not all men act in similar ways. Some may be relatively detached, others fully involved, and still others practical in their participation in the pregnancy (Marsiglio 1998). The way men act during pregnancy (reading material, attending prenatal classes, involving themselves in the birth process, and so on) may affect how they later relate with their newborns. When men are involved prenatally—supporting their partner (e.g., by helping with chores, taking her to the doctor, and buying needed items) and experiencing the unborn child (e.g., listening to the child's heartbeat or examining ultrasound images)—they are more likely to be involved postbirth with their partners and their infants (Cabrera, Fagan, and Farrie 2008).

Men's anxieties during pregnancy cover a number of areas, including the health of both fetus and partner, whether they will be good fathers, how fatherhood will affect their lives, and how well they will manage their economic responsibilities, especially given new expenses and reduced spousal income. Although a man's traditional role as father centered on providing, the concern over competence as a provider is not the major source of men's pregnancy anxieties. Men whose employment is unstable or whose income is insufficient will experience more provider anxiety than will men who simply take for granted that they can meet their financial responsibilities (Cohen 1993).

# Experiencing Childbirth

Women and couples planning the birth of a child have decisions to make in a variety of areas—birthplace, birth attendants, medications, preparedness classes, circumcision, and breast-feeding, to name but a few. In the past few decades, there was much criticism directed at what was seen as excessive and intrusive institutionalized control of women's birth experiences.

## The Critique against the Medicalization of Childbirth

The concept of the *medicalization of childbirth* depicts women receiving impersonal, almost assembly line–quality care during labor and delivery and lacking much input or control over their own childbirth experiences. The concept of control is central to the critique of medicalization: Critics contend that women have less say and control over the process than they should and less than the medical professionals whose expertise they seek.

The critique also takes into account the medical procedures used. Such things as routine use of monitoring devices, administering of enemas, rates of episiotomies (a surgical procedure to enlarge the vaginal opening by cutting through the perineum toward the anus) and cesarean-section deliveries, use of forceps or vacuum suction to assist in pulling the fetus from the womb—all of these reflect what critics have suggested is society's increasing dependence on technology and medical control. Critics of medicalization contend that many of the aforementioned procedures are more often employed for the convenience and control of the obstetrician than because of medical necessity. In general, critics recognize and value the potential lifesaving use of such interventions *when needed* but question procedures that seem to place medical convenience above women's interests or needs.

## The Feminist Approach

The question of what women most want or need has been central to what became a feminist critique of contemporary childbirth. Along with consumer advocates and government policy makers, feminists and activists in the "women's health movement" raised concerns and objections about the medicalization of childbirth. Feminists believe women have a right "to be informed, fully conscious, and to experience childbirth as a 'natural' process" (Treichler 1990). Feminists raised questions about how much medical intervention and control are necessary to reduce risks associated with the "normal, natural physiological process" of childbirth. They asserted that most pregnant women are essentially healthy and require minimal medical management during the birth process.

Many of the previous criticisms have been heard and addressed by hospitals and medical practitioners. For example, most hospitals have responded to the need for family-centered childbirth. Fathers and other relatives or close friends typically participate today. "Birthing rooms," with softer lighting and more comfortable birthing chairs, are increasingly common. Today, most hospitals encourage "rooming in" (the baby stays with the mother rather than in the nursery) or a modified form of rooming in. Moreover, women have more choices as to how they wish to give birth. Nevertheless, as the following sections show, the birth process continues to be characterized by considerable use of medical procedures.

## What Mothers Say

Keeping the preceding feminist critique and the critique of medicalization in mind, women's assessments of their experiences and treatment during pregnancy, while in labor and giving birth and after they give birth, show a mixed picture of satisfaction and discontent. This is evident in data collected in the "Listening to Mothers III" surveys, based on telephone interviews and online surveys with a combined sample of 2,400 women who gave birth between July 2011 and June 2012. Some key findings are as follows:

1. According to the survey, almost half of all mothers (47 percent) felt their care providers were "completely trustworthy," and an additional 33 percent thought they were "very trustworthy." Similarly, 36 percent rated their quality of maternity care as excellent, and 47 percent rated it as good.

2. Although women generally felt positive about their birthing experiences, several reported holding back

on asking questions because their care provider seemed rushed (30 percent), she might be viewed as difficult (23 percent), or because she wanted something different from the recommendation of her care provider (22 percent).

3. All of the surveyed women gave birth in a hospital, and obstetricians attended to most (70 percent). Midwives attended around 10 percent of the births, with the remainder divided among family physicians, other physicians, nurses, or physician's assistants.

4. At 99 percent, almost every woman reported receiving "supportive care" or attention during labor and delivery. Most often support came from spouses or partners (77 percent), followed by nursing staff (46 percent), another family member or friend (37 percent), and a doctor (31 percent).

5. Somewhat consistent with the critique of "medicalization," women reported experience with the following medical interventions during labor and delivery:

   • In the survey, 67 percent were given epidural analgesias to relieve pain, 16 percent were given a narcotic pain reliever, and 7 percent were given general anesthesia. Of the sample, 83 percent used some form of pain medication.

   • Some women reported feeling pressured to receive medical interventions including labor induction (15 percent), epidural analgesia (15 percent), or cesarean section (13 percent). Significantly, three times as many mothers who received an induction (25 percent) or cesarean (25 percent) said they received pressure compared with mothers who did not receive pressure (8 percent).

   • Of the sample, one in three women had cesarean sections.

Authors of the report on the survey assert that large segments of women experienced potentially invasive interventions when giving birth, even when they were healthy, their pregnancies were normal, and their labor was uncomplicated. Such interventions may not be medically necessary. Many women did not have the knowledge they needed or were unable to make the choices they desired (e.g., only 41 percent of women who received episiotomies said that they had a choice in the decision).

## Giving Birth

Sociologist Karin Martin (2003) conducted intensive interviews with a small sample of first-time mothers. The 26 mostly white heterosexual women,

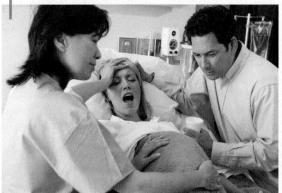

Family-centered childbirth allows fathers to participate alongside mothers in the birth process.

Shotshop GmbH/Classic Collection/Alamy Stock Photo

ranging in age from 20 years to over 40, were interviewed within three months of having given birth. Instead of exploring the macro-level and institutional dimensions of childbirth, Martin wanted to know how women experienced childbirth and how their experiences were shaped by gender identity. Deep within us, even in "seemingly natural experiences like birth," are our culturally constructed gender identities (Martin 2003, 57). Martin found that even during childbirth, women are "doing gender," acting compliant, nice, and kind. Martin's informants recalled trying not to "bother" strangers in adjoining rooms, remembered trying hard to remain attentive during conversations, and described doing things that indicated they were putting the needs of others ahead of their own. Even though they had to impose on others (doctors, nurses, midwives, husbands, and so on) for things (back rubs, quiet, patience, information, and so on), they recalled feeling badly about doing so. They found it hard not to feel "rude" or "selfish" for making the demands and imposing on others.

Martin (2003) suggests that the feminist critique of the medicalization of childbirth may be correct in highlighting how institutional control over birth shapes the experience. But it is only part of the story. She contends that women's birth experiences are also regulated and controlled "from within" by internalized gender identities. Even when "given permission" to depart from gender expectations, to act in gender-deviant ways, they found themselves at odds with such behavior. It was not "how they are" or "who they are" (Martin 2003).

I remember thinking, "Nobody else cares." My wife was knocked out, everybody else in the room was taking care of my wife and the baby, and the baby was wet, cold, red, really an ugly-looking thing. Truthfully, I was instantly bonded; it was like a marriage. She [my daughter] opened her eyes a little bit and I immediately began to relate to what she saw.... I was thrilled. I followed her after she opened her eyes and tried to imagine what she saw and what she might be seeing.... It was real exalting.

*Mark, 37-year-old father of one*

My wife had the shakes and couldn't hold the baby. I held her [my daughter] and sang her a lullaby. She was looking at my face, wasn't focusing, but I could see something going on. She could obviously hear too. I got to hold her for like 15 minutes. It was all so exciting and incredible...and strange.

*Bill, 36-year-old father of one*

The preceding comments are the recollections of two fathers witnessing the births of their daughters. Told to sociologist Theodore Cohen, they reveal how deeply some men are moved by their involvement at birth. In the United States and many other countries, it is now common for fathers to attend, witness, and often even actively assist in the birth of their children. It may be so common that we forget how relatively recent it is for men to enjoy such access. In 1960, only about 15 percent of fathers attended the birth of their child in the delivery room. Although estimates vary, by the first years of the 21st century, between 75 percent and 80 percent of fathers were present at childbirth (Mishori 2006).

We see the same trend elsewhere. In the United Kingdom, fathers are now in attendance at 90 percent of births (Sims 2009). Similar trends have been observed in other European countries and in Canada. Attendance at birth offers fathers an opportunity to feel part of the birth process and to offer support to their partners. Among men in the United Kingdom, the most frequently cited motivations for attendance at birth are out of support for their partners, out of curiosity, or because of pressure. In the United States and Canada, there is a fourth reason: Men often play the role of "coach," assisting their partner to implement what they have been taught in prenatal classes (Johnson 2002). Where once hospital practice and cultural expectations kept men out of the delivery room, now they are expected to be present. To illustrate this coercive element, in Martin Johnson's exploratory study of 53 British fathers, 57 percent of the men said they felt pressured to be there through labor and delivery. For example, "You don't get a choice, not really. It is assumed that you want to be there; I mean I did, but that is not the point. It's like not having a choice."

Finally, in Johnson's (2002) study, men's reactions to what they saw and experienced were both positive and negative.

On the negative side, 56 percent of the men identified as their most overwhelming memory the pain they witnessed their partner suffering. One man, Ben, claimed that he felt as though he ought to be experiencing pain himself: "In a strange way, when she dug her nails into my hands, I wanted to embrace the pain; it was like my share."

On the positive side, and unsurprisingly, men were deeply moved by the birth and awed by their partners' strength and resilience. ●

## Infant Mortality

The **infant mortality rate** in the United States has remained the same or decreased significantly in all but two years between 1958 and 2015. The estimated 2015 rate of 5.87 infant deaths per 1,000 live births is continuing to decline. Even with that trend, the U.S. rate is substantially higher than the rates in most of the developed world. Forty-two other countries had rates at or below 4.0. Bermuda, Norway, Singapore, Japan, Iceland, and Monaco all had rates of 2.5 or less (CIA 2015).

In 2015, the United States estimated infant mortality rate ranked 58th in the world, meaning that 57 countries had *lower* infant mortality rates than the United States (CIA 2015). In 2012, the United States ranked 49th (CIA 2015) and a similar cross-national comparison in 1960 found the United States ranked 12th (MacDorman and Mathews 2008). So although the rate has continued to decline, rates in other countries are declining more rapidly.

Data reveal noteworthy racial and ethnic variations, from rates below 5.0 among Asians, Cuban Americans, Mexican Americans, and Central or South Americans to a high of 11.11 among African Americans (see Figure 10.4). The U.S. targeted goal rate of reducing infant mortality by 2010 to 4.5 was never reached, nor was the determination to eliminate racial and ethnic

**Figure 10.4** Racial and Ethnic Comparison of Infant Mortality Rates, United States, 2008 and 2013

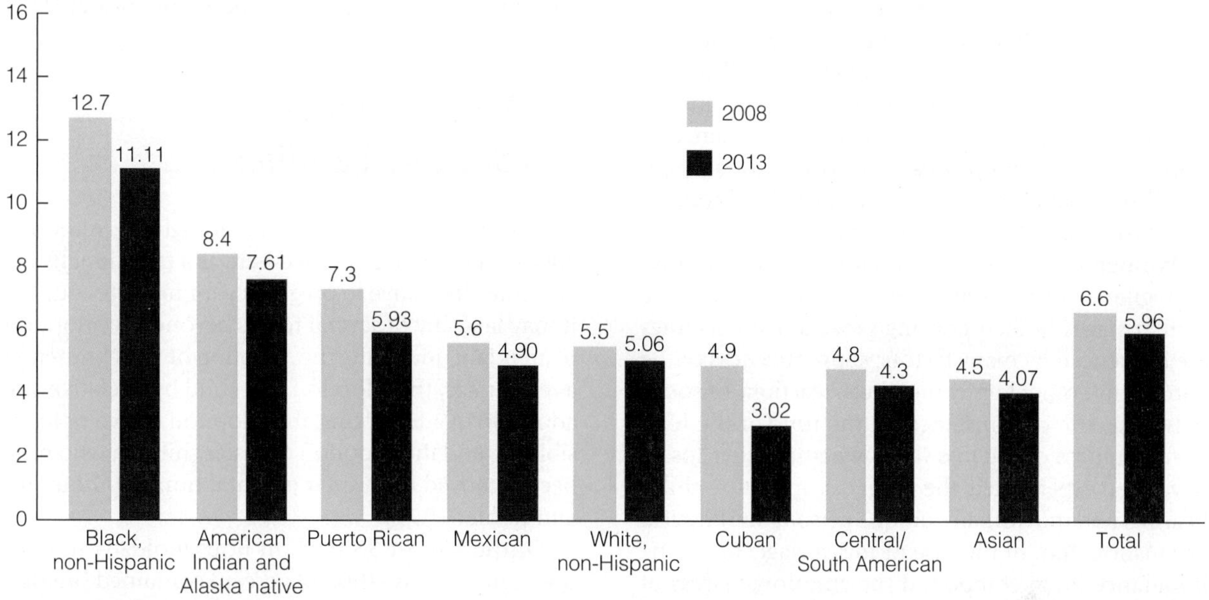

SOURCE: Mathews, MacDorman, and Thoma (2015).

disparities successfully accomplished (Mathews, MacDorman, and Thoma 2015)

Of the thousands of infant deaths, many are associated with poverty, which often hits racial minorities harder. The United States still lags behind many other countries in providing health care for children and pregnant women. In France, Sweden, and Japan, for example, all pregnant women are entitled to free prenatal care. Free health care and immunizations are also provided for infants and young children. One in six children born in the United States is born to mothers who received no prenatal care through the first trimester of pregnancy. Almost one in eight children has no health insurance (Ruane and Cerulo 2004).

Although many infants die of poverty-related conditions, others die from congenital problems (conditions appearing at birth) or from infectious diseases, accidents, or other causes. Sometimes, the causes of death are not apparent. The Centers for Disease Control and Prevention reports that approximately 3,500 infants die unexpectedly in the first year of life. In 2013, of such deaths, around 1,500 were classified as due to **sudden infant death syndrome (SIDS)**, a perplexing phenomenon wherein an apparently healthy infant dies suddenly while sleeping (Centers for Disease Control and Prevention 2015e). SIDS was deemed the cause of 45 percent of all sudden

unexpected infant deaths (SUID) in 2013. An additional 31 percent of such deaths were attributed to unknown causes, and 24 percent to accidental suffocation and strangulation in bed.

There has been substantial progress made in reducing SIDS. In 1990, SIDS-related death rates were 130.3 per 100,000 live births. In 2013, the rates were less than a third of that, at 39.7 per 100,000 live births. Looking at both the broader category of sudden unexpected infant deaths and the more specific category of SIDS, there is noteworthy ethnic/racial variation. Rates were highest among American Indians/Alaska Natives (190.5 per 100,000) and non-Hispanic blacks (171.8). The rate among non-Hispanic whites (84.4) was less than half of those rates, while rates were lowest among Hispanics (50.8) and Asians/Pacific Islanders (34.7) (Centers for Disease Control 2015e).

## Coping with Loss

The depth of shock and grief felt by many who lose a child before or during birth is sometimes difficult to understand for those who have not had a similar experience (Layne 1997). What they may not realize is that most women form a deep attachment to their children even before birth. The loss of the child must be acknowledged and felt before psychological healing can take place. Instead, however, women typically find that friends, relatives, and coworkers want to pretend that "nothing has happened" (Layne 1997).

Equally problematic are the common reactions from medical personnel and midwives. Medical personnel, especially physicians, may perceive pregnancy loss as "medically unimportant" and as evidence of normal and natural processes at work. Not surprisingly, this reaction may exacerbate the pain of couples who lose a child before or during birth.

Women (and sometimes their partners) who lose a pregnancy or a young infant generally experience similar stages in their grieving process. Their feelings are influenced by many factors: supportiveness of the partner and other family members, reactions of social networks, life circumstances at the time of the loss, circumstances of the loss itself, whether other losses have been experienced, the prognosis for future childbearing, and the woman's unique personality. Physical exhaustion and, in the case of miscarriage, hormone imbalance often compound the emotional stress of the grieving mother.

The initial stage of grief is often one of shocked disbelief and numbness. This stage gives way to sadness, spells of crying, preoccupation with the loss, and perhaps loss of interest in the rest of the world. It is not unusual for parents to feel guilty, as if they had somehow caused the loss, although this is rarely the case. Anger (toward the physician, perhaps, or God) is also a common emotion.

Experiencing the pain of loss is part of the healing process (Vredevelt 1994). This process takes time—months, a year, perhaps more for some. Support groups and counseling are often helpful, especially if healing does not seem to be progressing or depression and physical symptoms do not appear to be diminishing. Planning the next pregnancy may be curative, too, although we must keep in mind that the body and spirit need some time to heal.

Men's reactions to pregnancy loss have been largely overlooked. Where understandable attention and effort is directed toward the woman, men are the ones that health care staff, family, and friends expect to make necessary arrangements and decisions while remaining stoic. While trying to care for and support their partners, men are also grieving their own loss. As counselor Celeste Eckman-Himanek suggests, "It appears that when an expectant mother is transformed by miscarriage into a patient, the man is demoted from expectant father to caregiver" (Eckman-Himanek 2011, F-11). Men also face a situation in which, wanting not to worsen their partner's grief by expressing their own sadness, they appear uncaring and unaffected by the miscarriage (Eckman-Himanek 2011).

## Choosing How: Adoptive Families

Parenthood is not only entered biologically. Although adoption is being examined here as a traditionally acceptable alternative to pregnancy for infertile couples, it may take any of several forms beyond the adoption of an infant, including the adoption of stepchildren in a remarriage, the adoption of a child by a relative, the adoption of adolescents, the adoption of two or more siblings, and the adoption of foster children who have been removed from their parental homes (Grotevant and Kohler 1999).

Of the almost 35 million households with children under 18 in 2010, 90 percent contained biological children only, 4 percent included biological and stepchildren, 2 percent had only stepchildren, and the remainder were divided equally between households with both biological and adopted children and households with adopted children only (Kreider and Lofquist 2014).

Many people—married and single, with or without biological children—choose to adopt, not because they are unable to conceive or bear their own children but because they are ideologically committed to adoption. Some have concerns about overpopulation and the number of homeless children in the world. They may wish to provide families for older or disabled children. Thus, the population of adoptive families is diverse in terms of both motivation and circumstances.

There are an estimated 2 million adopted children, representing 2 percent of all children in the United States. Including adoptions of all types from all sources—adoptions from private agencies, foster care, international adoptions, and stepparent adoptions—there were roughly 136,000 adoptions in both 2007 and 2008. The 135,813 adoptions in 2008 represents an 8 percent increase over the 127,985 recorded adoptions in 2000 (Child Welfare Information Gateway 2011b). Approximately 25 percent of adopted children were adopted internationally. Of the nearly 240,000 internationally adopted children under 18 between 2009 and 2011, a quarter of the children were adopted from Europe, with 70 percent of those adopted from Russia. Another 50 percent of international adoptions were from

Asia, with China and Korea being primary source countries. Twenty percent of international adoptions were from Latin America (Kreider and Lofquist 2014).

## Why People Adopt

Based on an analysis of data from the 2007 National Survey of Adoptive Parents, the two most prominent motives adoptive parents gave for their choice to adopt were: to give a child a permanent home (81 percent) and to expand their families (68 percent). Just over half also indicated an inability to have a biological child (52 percent) (Ela 2011). These motivations were differently expressed across adoption types. For example, among those who adopted internationally, the most common reason mentioned was to expand the family (92 percent), followed closely by wanting to give a child in need a permanent home (90 percent). The latter reason was the most commonly cited reason by those who adopted from foster care (86 percent) or via private adoptions (70 percent) (Ela 2011).

Further detail shows how motivations varied by type of adoption. Those who adopted internationally said they did so because they thought a U.S. adoption would be difficult (mentioned by 65 percent of those who adopted internationally), they wanted an infant (63 percent), and/or they preferred a closed adoption with no contact with the birth family (53 percent). Those who adopted from public foster care said they did so because such adoptions were less costly (59 percent) and they thought they would be better able to get a child sooner (27 percent) and/or to adopt a special needs child (24 percent). Among those who adopted through a private agency, reasons given were because they wanted an infant (82 percent), they wanted a healthy child (75 percent), and/or they thought they would get a child sooner (41 percent) (Ela 2011).

## Characteristics of Adoptive Families

The following profile of adopted children and their families draws heavily from the 2007, first-of-its-kind, National Survey of Adoptive Parents (NSAP). This survey was an add-on to the National Survey of Children's Health, and offers representative data on adoptive families, as reported by the parents:

- Overall half of adopted children are female. When looking in more detail by the type of adoption, children adopted internationally are much more likely to be female than male, by a ratio of two to one. If, instead, one looks at children adopted from foster care, an estimated 57 percent are male.

- Although a majority of adoptive parents are white, a minority of adopted children are white. In fact, adopted children are disproportionately African American (23 percent) and Asian (15 percent), when compared to the overall percentages of black (14 percent) and Asian (4 percent) children. Compared to the overall population of children under 18, 37 percent of adopted children but 56 percent of all children in 2007 were white. Furthermore, 40 percent of adopted children are a different race, ethnicity, or culture than their adoptive parents.

- Three-quarters of adoptive families have just one adopted child; another 19 percent have two adopted children, and the remaining 6 percent have three or more adopted children.

- Over 40 percent of adopted children lived with their birth families at some point before being adopted. Almost all (97 percent) of adopted children five years old or older know they were adopted.

- Economically, families with adopted children are somewhat better off than those without. They are less likely to be poor compared to families with children overall, tend to have higher median incomes, and adopted children were as likely as children overall to live with two married parents. Almost 40 percent of adopted children were the only child under 18 living in their adoptive households. This was especially true of children adopted through private adoptions.

- The NSAP reveals that adopted children benefit from enriching family environments. They are more likely to be read to, sung to, or told stories daily, and those who are school age are more likely to participate in extracurricular activities.

- Eight out of ten adopted children have parents who described their relationship with their child as very warm and close. In addition, although 15 percent of adoptive parents reported the relationship with their child to be "more difficult" than they'd anticipated, 42 percent of adoptive parents reported the relationship as "better than ever expected." Almost 90 percent of adopted children have parents who said they would "definitely" adopt their child again. Of children 5 and older, more than 90 percent had parents who described their child's feelings about the adoption as positive or mostly positive. (Vandivere, Malm, and Radel 2009)

The costs of adoption can be quite steep. The Child Welfare Information Gateway provides the following estimates of adoption-specific costs. Such costs vary, depending on the type of adoption, the type of agency used, whether they adopt domestically or internationally, and so on. According to the NSAP, about a third of all adoptions cost nothing. This pertains mostly to adoptions from foster care, though even 22 percent of private adoptions were reported to have no adoption-related cost. On the other hand, 20 percent of private adoptions cost between $10,000 and $20,000, and another 12 percent cost more than $20,000. Half of international adoptions had costs in excess of $20,000 (Vandivere, Malm, and Radel 2009). A Child Welfare Information Gateway "Costs of Adopting" fact sheet (2011a) estimated somewhat higher adoption expenses: up to $2,500 for adoptions from foster care, from $5,000 to $40,000 for private domestic infant adoption, and from $15,000 to $30,000 for international adoptions.

## Open Adoption

Adoption laws vary widely from state to state: Some prohibit private adoption, and other states have laws that are considered quite supportive of it. With confidentiality no longer the norm, the trend toward **open adoption**, in which there is contact between the adoptive family and the birth parents, has been dramatic (Siegel and Smith 2012). The contact can be direct, with birth mother and adoptive parents having contact, or mediated through an agency. Estimates indicate that 55 percent of adoptions are fully open, 40 percent are mediated, and only 5 percent are closed adoptions (Siegel and Smith 2012). The amount, frequency, and nature of contact can vary quite a bit. Mediated adoption may be better considered as "semi-open," with contact occurring through some third party (e.g., an adoption agency staff member). In other cases, contact may consist of exchange of gifts; sharing of photographs; periodic communication by telephone, email, or letters; or even face-to-face contact (Grotevant 2008).

Once somewhat divided over whether open or closed adoption was more in the best interest of the involved parties, consensus has been reached in favor of open and mediated adoptions. Research by Harold Grotevant and Ruth McRoy has tracked the experiences of 169 adopted women and men for more

**False** **6.** Adopted children are less likely to be poor than are children who live with their biological parents.

than 20 years. In addition to the adoptees themselves, Grotevant and McCoy have studied (between 1986 and 1992) adoptive parents, birth mothers, and other family members from an original sample of 190 adoptive families, 171 adoptees, and 169 birth mothers. Current adoption policies favor some form of open adoption as in the best interests of both the child and the birth parents. Regarding the outcomes of open versus closed adoptions, Grotevant (2008) found that adoptees, adoptive parents, and birth mothers all derived benefits from open adoption. Adoptees came to a better understanding of the meaning of adoption and the circumstances surrounding their adoption. Adoptive parents were less fearful of birth parents trying to reclaim their child, while birth mothers were likely to experience less unresolved grief.

Adoptive families face unique problems and stresses. They may struggle with physical and emotional strains of infertility, endure uncertainty and disappointment as they wait for a child, and spend all their savings and then some in the process. They often face insensitivity or prejudice. For example, an adopted child may be asked, "Who is your *real* mother?" or "Are you their *real* daughter?" Adoptive parents may be congratulated by well-meaning folks—"Oh, you're doing such a good thing!"—as though they had made a sacrifice of some kind in choosing to build a family in this way. Even grandparents may reject adopted grandchildren (at least initially), especially if the adoption is interracial.

# ■ Becoming a Parent

The time immediately following birth is a critical period for family adjustment. No amount of reading, classes, and expert advice can fully prepare expectant parents for the real thing. The three months or so following childbirth (the fourth trimester) constitute the **postpartum period**. This is a time of physical stabilization and emotional adjustment.

Many new mothers who may well have lost most of their interest in sexual activity during the last weeks of pregnancy may find themselves slowly returning to pre-pregnancy levels of desire. Both relationship satisfaction and postpartum depression are important predictors of the levels of sexual desire and satisfaction, and of changes in sexual frequency following childbirth. Enjoyment returns gradually. According to

A truism of parenthood that is often invoked as one of the unique and stress-inducing aspects of being a parent is that once you have a child, "you can't give it back." Of course, there are parents who abandon their children, but, for all practical purposes, once you become a parent, you remain a parent. This quality of parenthood does not equally pertain to adoptive parenthood. In fact, experts estimate that between 10 percent and 25 percent of adoptions "fail." Such failures can occur before the adoption is finalized (known as "disruption") or after (known as "dissolution"). Dissolution permanently ends an adoption (Boss 2008). Between 1 and 5 percent of adoptions result in dissolutions (Child Welfare Information Gateway 2012).

A report on adoption dissolutions and disruptions suggests that a number of different factors have been found to be associated with adoption disruptions, including both child factors and adoptive family factors. Child factors include older age at time of adoption, emotional or behavioral problems, continued strong attachment to one's birth mother, and having been subjected pre-adoption to sexual abuse. Adoptive family factors included having been new or matched to the adopted child (as opposed to having first been foster parents to the child), having unrealistic expectations of the adoption outcome, and experiencing a lack of post-adoption social support (Child Welfare Information Gateway 2012).

The consequences of adoption dissolutions are traumatic for both the adoptive parents and the child, and each suffers as a result of the experience. As professor of child development and family studies Pauline Boss (2008) suggests, adoption dissolution is an instance of ambiguous loss (see Chapter 2), a loss without finality or resolution, a loss lacking closure. "Who is in and who is out" of the family is left uncomfortably unresolved.

Boss (2008) suggests that because of the lack of community support, an established ritual, or mourning period, the stigma associated with both the "failure" and having surrendered one's adopted child, and adoption policies that force an absolute cutoff between parent(s) and child, families are left struggling with how to manage. Furthermore, clinicians may not all be familiar with circumstances such as these and therefore provide less effective support. Conversely, Boss offers guidelines to assist families in building resiliency (e.g., to find meaning in the failed adoption, learn to balance what one can and cannot control, reconstruct ones identity and rethink the roles one plays, and recognize that failed adoptions will generate mixed feelings and lingering attachments).

Boss (2008) offers a number of suggested strategies for both prevention of and intervention in failed adoptions. But a more important goal may be to prevent the parenting failures in birth families that lead children into institutional or foster care in the first place. Barring that, whatever steps can be taken to prevent adoptions from dissolving must be taken, so as "to prevent such traumatic loss for the sake of all concerned—the adoptive parents, the birthparents, and, especially, the children" (Boss 2008, F19). ●

---

research by Margaret De Judicibus and Marita McCabe (2002), at two weeks postpartum, few new mothers report sexual intercourse as pleasurable; by 12 weeks, two-thirds of women say that sex is "mostly enjoyable." Even then, however, 40 percent complain of some difficulties. Relationship satisfaction is at its lowest at this point. Nearly six out of seven couples report reduced frequency of intercourse at four months postpartum.

The postpartum period also may be a time of significant emotional upheaval. Even women who had easy and uneventful births may experience a period of postpartum or "baby blues" characterized by periods of crying from time to time, difficulty sleeping, unpredictable mood changes, fatigue, irritability, occasional mild confusion, decrease in focus or concentration,

or lapses of memory. These reactions may begin just three or four days after giving birth, typically worsening until the fifth day and concluding by around the tenth day.

More severe than the "baby blues," postpartum depression can be truly devastating. The Centers for Disease Control and Prevention estimate that between 11 percent and 18 percent of new mothers experience "frequent postpartum depressive symptoms." Women who suffer from postpartum depression may experience combinations of the following: sadness, low self-esteem, greatly disturbed sleep, moderate to severe anxiety, obsessive thoughts, an inability to care for oneself or the baby, panic attacks, and thoughts of harming oneself or the baby (Declercq et al. 2013). These more severe postpartum reactions can leave women feeling as though they are "losing their minds" (Layne 1997). The following are words of women who shared their experiences with actress/author Brooke Shields, herself a survivor of postpartum depression:

"I cried incessantly."

"I wanted to run away and never return."

"I didn't feel attached to my baby at all."

"I couldn't even look at my baby for the longest time."

"I kept waiting for my baby's real mother to arrive."

(Shields 2005)

Infants of women suffering postpartum depression also suffer, as postpartum depression interferes with a mother's abilities to respond to her newborn's needs and may lead to poor emotional and cognitive development (Layne 1997). Many women blame themselves for their fluctuating moods. They may feel that they have lost control over their lives because of the dependency of their newborns.

Biological, psychological, and social factors are all involved in postpartum depression. Biologically, during the first several days following delivery, there is an abrupt fall in certain hormone levels. The physiological stress accompanying labor, dehydration, blood loss, and other physical factors contribute to lowering the woman's stamina. Psychologically, conflicts about her ability to mother, ambiguous feelings toward or rejection of her own mother, and communication problems with the infant or partner may contribute to the new mother's feelings of depression and helplessness. Finally, the social setting into which the child is born is important, especially if the infant represents a financial or emotional burden for the family.

Recent research, published in *The Lancet Psychiatry,* and characterized as "the largest study to date on postpartum depressive symptoms," drew on data from 7 different countries and 19 different institutions; it differentiated between three classes of postpartum depression, based on variation in severity of symptoms. For two-thirds of those women who experienced the most severe instances—with symptoms that included panic, frequent crying, and even suicidal thoughts, the depression symptoms actually began during pregnancy. Women with moderate depression were more likely to experience symptoms after giving birth (Belluck 2015; Postpartum Depression: Action Towards Causes and Treatment (PACT) Consortium 2015; Walsh 2015).

Depression-like symptoms are not limited to biological mothers or even to women. Within even just a few weeks after adoption finalization, some adoptive parents may experience what researchers have called "postadoptive depression syndrome," or PADS. The weight of parental responsibilities, the tedium of caring for a child, difficulties forming an attachment to one's child, and doubts about one's parenting capabilities may surface. Even without the hormonal changes induced by childbirth that contribute to postpartum depression, adoptive parents—especially mothers—struggle with many of the same questions, doubts, and feelings that new birth mothers experience (Child Welfare Information Gateway 2010).

Postpartum counseling before discharge from the hospital may help couples gain perspective on their situation so that they will know what to expect and can evaluate their resources.

Men, too, can get a form of postpartum blues. An estimated 4 percent of new fathers may experience depression during the first year following their child's birth (Centers for Disease Control and Prevention 2012b). When infants arrive, many fathers do not feel prepared for their new parenting and financial responsibilities. Some men are overwhelmed by the changes that take place in their marital relationship. Fatherhood is a major transition for them, but their feelings are overlooked because most people turn their attention to the new mother.

The transition to parenthood can be made somewhat easier if the new parents understand in advance that a certain amount of fatigue and stress is inevitable. They need to ascertain what sources of support will be helpful to them, such as friends or family members who can help out with preparing meals or running errands. They also need to keep their lines of communication open—to let each other know when they are feeling overwhelmed or left out. It's also important that they plan time to be together, alone or with the baby—even if it means telling a well-meaning relative or friend they need time to themselves.

## Taking on Parental Roles and Responsibilities

Despite the many months of pregnancy, the actual transition to parenthood happens in the instant of birth. Such an abrupt transition from a nonparent to a parent may create considerable stress. Parents take on parental roles literally overnight, and the job goes on without relief around the clock. Many parents express concern about their ability to meet all the responsibilities of child rearing.

There have been a number of influential analyses of the transition to parenthood (Cowan and Cowan 1992; Fox 2009; LaRossa and LaRossa 1981; Rossi 1968). Sociologist Alice Rossi (1968) offered a classic analysis of what parents experience as they enter the new social reality of parenthood. According to Rossi, entering parenthood is stressful because of the nature of the role of parent and the characteristics of the parental role transition.

Rossi (1968) singled out the following features of entering parenthood:

- *Irreversibility*. Unlike nearly any other role, once one enters parenthood, one cannot easily leave without incurring significant social or legal repercussions.

- *Lack of preparation*. There is almost nowhere and no way to realistically practice parenting and, furthermore, there has been little formal societal effort to acquaint and equip individuals with what to expect when they become parents.

- *Idealization and romanticization*. Expectations about parenthood are often unrealistic and overly idealized. If and when reality turns out to be less than ideal, it is easy to become frustrated and disappointed.

- *Suddenness*. Despite months of awareness of impending parenthood, one goes from nonparent to parent in the moment of childbirth and assumes all the role responsibilities with that same suddenness.

- *Role conflict*. The parental role affects all the other roles one plays, encroaching on time spent with a spouse or partner and complicating paid employment.

More recently, Carolyn and Philip Cowan (1992, 2000) identified five domains in which new parents experience change as a result of the arrival of children:

- *Identity and inner-life changes*. New parents no longer think of themselves the same way they did before their children were born; their priorities and personal values change.

- *Shifts within the marital roles and relationship*. Parenthood alters how couples divide tasks or allocate responsibilities. A common pattern is for the division of family labor to become more traditionally gendered. As considerable research has indicated, new parents often find that their relationship quality declines.

- *Shifts in intergenerational relationships*. Becoming parents alters—often improving and intensifying, sometimes straining—the relationship between new parents and *their* parents.

- *Changes in roles and relationships outside the family*. New parenthood, especially new motherhood,

Although becoming a parent is stressful, the role of mother or father is deeply fulfilling for many people.

may force changes in other nonfamily roles and relationships, such as at work or in friendships. Although some of these changes may be temporary (e.g., leaving work only for the length of a parental leave), they nonetheless compound other things to which new parents are adjusting.

- *New parenting roles and relationships*. Couples must arrive at an agreeable division of child care. One parent may feel put upon or taken advantage of in the way that the couple allocates their individual time and energy to child care tasks.

The Cowans suggest that the difficulties associated with the parental transition are more difficult for contemporary parents because of some major features of the social climate in which they parent. First, contemporary parenthood is more *discretionary* or optional, making decisions about whether and when to have children subject to more discussion, negotiation, and potential dispute. Second, many new parents, especially middle-class parents, are relatively *isolated* geographically from their wider kin groups and other long-term social supports. Third, changes in women's roles have introduced more *role conflict* for new mothers and have increased women's need and legitimate demand for more sharing by their partners. Fourth, the social policies that address the needs of parents are weak to nonexistent. Fifth, there are few enviable or attractive role models for effective parenting. Sixth, today's families are supposed to fulfill all our emotional needs. Parenting is stressful and requires mutual effort and sacrifice. But effort and sacrifice don't fit compatibly with individual emotional fulfillment. Thus, the difficulties may become sources of resentment and estrangement (Cowan and Cowan 1992, 2000).

## Parenthood and Traditionalization

Many of the stresses new parents feel reflect the different roles and responsibilities women and men acquire when they become parents.

Overall, mothers seem to experience greater stress than fathers. They also show stronger declines in satisfaction with their marriages. Furthermore, even among couples who have relatively egalitarian marriages before the birth of their first child, marriage typically become more traditional once a child is born (Fox 2009). After becoming mothers, women tend to assume responsibility for more of the housework than they did before becoming parents (Kluwer 2010). If the mother continues to be employed outside the home or if the woman is single, she will have a dual role as both caregiver and provider. She will also probably have the responsibility for finding adequate child care, and when circumstances dictate, she will most likely stay home to take care of a sick child. Multiple role demands are the greatest source of stress for mothers.

The traditional division of family work and caregiving surfaces as an important factor in the frequently found decline in marital satisfaction. In her systematic review of the literature on the transition to parenthood, psychologist Esther Kluwer states that women who expect a more equal division of family work after childbirth are likely to experience greater declines in marital satisfaction in early parenthood as their relationships take on more gendered roles (Kluwer 2010).

There are various other sources of parental stress. Fathers often describe severe stress associated with their work. Both mothers and fathers must be concerned about having enough money. Other sources of stress involve infant health and care, infant crying, interactions with the spouse (including sexual relations), interactions with other family members and friends, and general anxiety and depression. Changes in marital quality, increased marital conflict, and decreased interaction between spouses are also common as couples transition to parenthood (Crohan 1996; Kluwer 2010).

As noted in Chapter 8, some of the marital changes, such as increased conflict and declining satisfaction, are also found among couples without children. One notable exception, experienced by new parents but not found among nonparents, is an increase in passive avoidance, defined as "becoming quiet and pulling away" from one's partner or spouse (Kluwer 2010). Though it may be considered a better option than more active confrontation and conflict, passive withdrawal leaves open the possibility that conflict might resurface eventually (Kluwer 2010).

**False** **7.** Egalitarian marriages don't usually remain so after the birth of the first child.

The first years of child rearing are likely to be both stressful and joyful, bringing strains as well as rewards. A variety of factors lessen the stress somewhat. As new parents, the partners experience less stress if they (1) have already developed a strong relationship, (2) are open in their communication, (3) have agreed on family planning, and (4) originally had a strong desire for the child. Despite planning, the reality for most is that this is a stressful time. Accepting this fact while developing time management skills, patience with oneself, and a sense of humor can be most beneficial.

# Parental Roles

By this point one might be wondering, "But what is it like to actually *be a parent* and to raise children?" Having a child is unlike any other experience we undertake. The changes in our lives are wide-ranging and irreversible, the *potential* rewards are great, and the sacrifices are many. Most people continue to decide to embark on this journey and take on the challenging tasks, to which this chapter now turns.

Over the past four decades or so, major changes in society have profoundly influenced parental roles. Parents today cannot necessarily look even to their own parents as models. Although most mothers and fathers of today's children have at least some things in common with mothers and fathers throughout history, such as the desire for their children's well-being, they have had to chart a new course in many areas. Here we briefly review motherhood and fatherhood, highlighting some major changes that have transformed the meaning and experience of each.

## Motherhood

To many, a chapter about parenting might be assumed to be mostly about mothers and children because "parenting" and "nurturing" are treated as though they are synonymous with "mothering." Furthermore, many women see motherhood as their "destiny." Given the choice of becoming a mother or not (with "not" made possible and more controllable through birth control and abortion), most women would choose to become a mother at some point in their lives, and they would make this choice for positive reasons. Some women make no conscious choice; they become a mother without weighing their decision or considering its effect on their own lives and the lives of their children and partners. The potential negative consequences of a nonreflective decision—bitterness, frustration, anger, or depression—may be great. Yet it is also possible that a woman's nonreflective decision will turn out to be "right" and that she will experience unique personal fulfillment as a result.

Although researchers are unable to find any purely instinctual motives for having children among humans, they recognize many social and cultural motives impelling women to become mothers. When a woman becomes a mother, she may feel that her identity as an adult is confirmed. This may be especially true for those women with fewer marketable skills, more limited education, fewer financial resources, and more restricted opportunities. For such women, having a baby is a desirable and achievable goal. In the words of sociologists Kathryn Edin and Maria Kefalas (2005, 46), "Unlike their wealthier sisters, who have the chance to go to college and embark on careers…poor young women grab eagerly at the surest source of accomplishment within their reach: becoming a mother."

Having a child of her own proves her womanliness because, from her earliest years, she has been trained to assume the role of mother. The stories a girl has heard, the games she has played, the textbooks she has read, the religion she has been taught, the television she has watched—all have socialized her for the mother role. The idea of a maternal instinct reflects a belief that mothering comes naturally to women. For women who struggle with the new roles and responsibilities that motherhood brings, such an idea can be frustrating and can produce guilt. Add to this the assumption that mothers instinctively or intuitively "know" how to nurture children and the inherent ability of women to breast-feed, and we can quickly see the enormous pressures that can face new mothers more than new fathers.

Compounding the situation are those ambiguous cultural expectations alluded to earlier. "Too much" mothering? "Not enough" mothering? What do children really need, and what should mothers give and do? Women receive unclear, often contradictory, messages. Furthermore, the standards against which mothers have tended to be judged (and have often come to judge themselves) are often unrealistic and idealized, putting women in a situation of comparing themselves to a model to which it is difficult, if not impossible, to fully "measure up." Recall the earlier discussion (Chapter 4) of sociologist Sharon Hays's analysis of the **ideology of intensive mothering**. This ideology has portrayed mothers as the essential caregivers, who should be child centered, guided by experts, and emotionally absorbed in the labor-intensive and financially demanding task of child rearing. As a result, mothers come to see the child's needs as primary and the child as innocent and pure, and go on to invest much of their time, energy, emotion, and money in a very labor-intensive, self-sacrificing process of parenting their children (Hays 1996). The intensive mothering ideology confronts mothers and women who contemplate motherhood with cultural contradictions. Living up to its standards is difficult even for stay-at-home mothers. For women employed outside the home, the ideology can provoke self-doubt, guilt, and a sense of being negatively judged by others. As Hays (1996) noted, almost no woman can resolve this cultural no-win situation. Women who forgo childbearing may be perceived as "cold" and "unfulfilled." An employed woman with children may be told that she is selfishly neglecting her children. If she scales back her workload but stays in a job, she may be "mommy tracked," put in a less demanding but also less important and less upwardly mobile position. Finally, at-home mothers, in meeting the intensive mothering mandates, will be seen by some as "useless" or "unproductive" (Hays 1996).

In fact, women who aspire to embody the intensive mothering mandates appear to be more likely to suffer as a result (Rizzo, Schiffrin, and Liss 2012). Research by psychologists Kathryn Rizzo, Holly Schiffrin, and Miriam Liss looked at the mothering beliefs and consequences of those beliefs among 181 mothers of preschool-age children. Those who adhered to the intensive mothering beliefs reported higher levels of stress, guilt, and depression, and lower levels of happiness than mothers who did not (Rizzo, Schiffrin, and Liss 2012).

Recall, too, that sociologist Karen Christopher found that among employed mothers she interviewed, many constructed an alternative version of good mothering that she calls *extensive mothering*. Extensive mothering stresses mothers "being in charge," rather than "being the one" to provide all the care for their children. *Extensive mothering* entails entrusting and delegating significant amounts of daily child care to others, sees the benefits to both mothers and children of maternal employment, and has mothers monitoring the care their children are provided by others (Christopher 2012). Research by Jill Walls, Heather Helms, and Joseph Grzywacz (2014) found that their sample of 205 full-time employed mothers of infants did not consistently endorse intensive mothering beliefs. Instead, most of the employed mothers held ideas about good mothering that were achievable in and congruent with their own employment situations. For example, where intensive mothering stresses mothers' availability to their infants and young children, the full-time, employed mothers defined that in terms of being psychologically and emotionally available to their children (Walls, Helms, and Grzywacz 2014).

No matter what she chooses or does, a woman may find that there are some who question her decision and object to her lifestyle. Such mixed messages create a maternal double bind regarding *the degree* and *kind* of involvement mothers should have in their children's lives that has not, as yet, come to define men's

The ideology of intensive mothering casts mothers as essential and defines motherhood as labor intensive and self-sacrificing.

Juergen Hasenkopf/Alamy Stock Photo

experiences of fathering. In addition, the recent popular attention paid to "helicopter moms," as well as the harsh criticism legal scholar Amy Chua received after the publication of her book, *The Battle Hymn of the Tiger Mother,* suggest that mothers will be subject to scrutiny, almost regardless of their chosen parenting style.

Motherhood affects women's employment experiences, as shown in Chapter 11. One notable way that women are affected is in their earnings. Estimates differ, but it is clear that women with children earn less than their counterparts without children. According to a 2013 Bureau of Labor Statistics report on women's 2012 earnings, median weekly earnings for women with children under age 18 were $680, whereas women without children had median weekly earnings of $697 (U.S. Bureau of Labor Statistics, October 2013b). For men, parenthood has the opposite association with earnings; the median weekly earnings for men with children were $946 as compared to $799 for men without children. According to sociologist Michelle Budig, the effects of motherhood on earnings are not restricted to any one group or race. She does note that among white women, the mommy penalty is especially strong for low-wage women. Budig's research also indicates that the "motherhood penalty" increases with each child a woman bears (Linn 2013).

In addition, regardless of their employment status, the responsibilities of parenthood continue to fall more heavily on women than on men, even as children age and move into their teens (Kurz 2002). Such responsibilities give employed mothers more to do and less time to do it. At the same time that women's parental status affects their employment experiences, a mother's employment status is also likely to affect the amount of time she spends with children. Compared to at-home mothers, women who are employed full-time spend less time with their children, approximately eight less hours per week (Bianchi 2011; Kendig and Bianchi 2008). However, it bears mentioning that by 2000, employed mothers were spending more time in primary child care than nonemployed mothers did in 1975 (Bianchi 2011).

There has been an increase in the share of mothers who are not employed outside the home. According to a report by D'Vera Cohn, Gretchen Livingston, and Wendy Wang (2014), some 10.4 million women were "stay-at-home mothers," home so as to care for their families, or as a result of unemployment, disability, or being in school. These mothers represent 29 percent of all mothers living with children 18 or younger, and an increase from a 1999 low of 23 percent. They report that nearly 30 percent, 21.1 million children, are being raised by at-home mothers, an increase from 24 percent (and 17.3 million) in 2000. For purposes of historical comparison, they note that in 1970, just under half (48 percent) of children had a stay-at-home mother.

In discussing at-home mothers, Cohn, Livingston, and Wang (2014) report that approximately two-thirds of these mothers are "traditional" at-home mothers, meaning they are married to full-time wage-earning spouses. They also provide the following details:

- Married at-home mothers are more likely to say they are home to care for their families (85 percent) than are single (41 percent) or cohabiting (64 percent) at-home mothers. All together, 6 percent of at-home mothers said they were home because they couldn't find a job.

- Stay-at-home mothers, regardless of their marital status, tend to be younger and less educated than employed mothers. However, those at-home mothers who are married and have employed husbands tend to be more highly educated and less likely to be in poverty than single or cohabiting at-home mothers.

- Hispanic (38 percent) and Asian (36 percent) mothers were more likely to be at home than black (27 percent) or white mothers (26 percent).

- Stay-at-home mothers report spending an average of 18 hours a week engaged in child care activities and 23 hours a week on housework. In contrast, mothers employed outside the home report spending 11 hours a week in child care and 14 hours a week doing housework. At-home mothers also report five more hours of sleep and nine more hours of leisure per week compared to employed mothers.

- Married at-home mothers spend more time in child care activities (20 hours per week) than either cohabiting (19 hours) or single (15 hours) at-home mothers.

The time mothers spend with their children is affected by more than whether they are employed. A mother's marital status is associated with the amount of time spent with children; single mothers spend less time with their children than married mothers spend, though the gap—estimated at about three to five hours per week by sociologists Sarah Kendig and Suzanne Bianchi (2008)—is not huge (single mothers spend between 80 percent and 90 percent the amount

of time that married mothers spend). It isn't necessarily marital status per se that is operating but rather characteristics or experiences that accompany marital status. For example, because single mothers, especially divorced single mothers, often have the need to spend more hours at work, their time with children is affected.

Still other factors that have been found to *increase* mothers' time with children are level of education, economic status, and characteristics of children (such as the number of children a mother has, their ages, and their genders). College-educated mothers spend more overall time, including both routine time and more interactive activities, with their children. Mothers with preschool-age children in the home spend more time in child care activities. Most (but not all) research also suggests a positive relationship between the number of children a mother has and her time spent in child care (the more children a mother has, the more time she spends on child care) (Kendig and Bianchi 2008).

## Fatherhood

When we speak of "mothering a child," the meaning is fairly clear: Mothering is a process that involves nurturing and caring for the physical and emotional well-being of a child, almost daily, for at least 18 consecutive years. *Fathering* is more ambiguous. Nurturing behavior by a father toward his child has not typically been referred to as *fathering*. As often used today, the term *parenting* is intended to describe the child-tending behaviors of *both* mothers and fathers (Atkinson and Blackwelder 1993). In fact, the popular meaning of *fathering* has at times been reduced to the act of impregnating the child's mother (as in "who *fathered* that child?"). Of course, cultural expectations of fathers are much broader and deeper than that, and fathers today participate more than they did in the past in a broader range of child care activities, and demonstrate multiple types of involvement in raising their children. However, although we have witnessed increases in the numbers of at-home fathers who are primary caregivers, the number of single father–headed households, and the time fathers in two-parent households spend with their children, the cultural expectations of fathers are not the same as the expectations of mothers: Typically, fathers do not participate to the same extent or in the same ways that mothers participate, and fathers' actual involvement in child care does not necessarily match the broader cultural expectations of them.

Beginning in the 1980s and 1990s and continuing through to the present, the depth and kinds of male involvement or absence in the lives of their children became popular topics of scholarly interest and societal concern (Eggebeen and Knoester 2001). In all the commentary and analysis, however, we are still left with something short of a consensus about the state of fatherhood in the United States. As much as we seem to expect, welcome, and celebrate men's increased involvement with children and in child care, we continue to lament men who behave irresponsibly toward their children, as perhaps best epitomized by the image of the divorced and deadbeat dad. Just what are fathers supposed to do, and how well or poorly are they doing it?

Most men expect to someday become fathers, and indeed most do. Although there has been an increase in the percentage of men who remain childless (from 15 percent in 1987 and 1988 to 21 percent between 2006 and 2010), fatherhood is a normative part of men's lives (Stykes 2011; Tichenor et al. 2011). In analyzing today's fathers and today's families, we find diverse opinions and a range of experiences of fathering. There is evidence indicating that fathers have become more emotionally connected to and involved in the lives of their children (Eggebeen and Knoester 2001). Some commentators point proudly to the wider cultural embracing of a "new father" model against which many men now measure themselves (Lamb 1993; Smith 2009). Feminist ideology is often credited with being influential in shifting the emphasis to a more expressive and engaged model of fathering, but many men pursue more involved versions of fatherhood as part of their own quest for deeper relationships with their children than they themselves may have had. When pressed, most men today compare themselves favorably with their own fathers in both the quality and the quantity of involvement they have with their children.

The "nurturant father," as Michael Lamb (1997) referred to him, can participate in virtually all parenting practices (except, of course, gestation and lactation) and experience similar emotional states to those that mothers experience. It is clear that fathers can feel a connection to their infants that men were often thought to lack (Doyle 1994). Furthermore, researchers have reconceptualized *father involvement* in recognition that there are many ways in which fathers are influential participants in their children's development. Fathering activities such as communicating, teaching, caregiving, protecting, and sharing affection are ways in which fathers might be "involved" with their children, and all are viewed as beneficial to the development and well-being of both children and adults (Hawkins and Dollahite 1997; Palkovitz 1997).

Although this new standard of fatherhood has been widely hailed and embraced, it is still somewhat unclear whether, how much, or how widely it reflects actual behavior. As conceptualized by sociologist Ralph LaRossa, the study of fathering needs to separate the **culture of fatherhood** from the **conduct of fatherhood.** Writing in 1988, LaRossa contended that there had been substantial changes in the culture of fatherhood in the more nurturing direction as described here; he noted that it was less clear whether or how much the *conduct of fatherhood* had kept pace. More than 20 years later, we still need to keep LaRossa's distinction in mind. In addition, when we look at how fathers compare to mothers, on average, fathers are neither as involved with nor as close to their children, including their teenaged children, as mothers are (Kurz 2002). This is an important reminder that reality may be different from rhetoric when it comes to what people actually do or believe they should do in their families.

Of course, in trying to determine what and how much fathers and mothers actually do, it is also important to consider the source of those reports. Researchers have found discrepancies between mothers' and fathers' accounts of how involved fathers are and in what aspects of child care and child rearing they are involved. Compared to what fathers report, mothers' reports of father involvement tend to be lower. Although there may be as many reasons for fathers to

Fathers are increasingly involved in parenting roles—not just playing with their children but also changing their diapers, bathing, dressing, feeding, and comforting them.

exaggerate their estimates of how involved they are as there are reasons mothers might underestimate father involvement, the mother–father discrepancy points to the complexity of measuring father involvement and to the need to consider which parent is providing the estimates (Mikelson 2008).

As the preceding examples show, it is difficult to generalize too widely about today's fathers. Although most of today's fathers recognize father involvement as beneficial and aim to be more broadly involved with their children than what they perceive fathers to have been in the past, many fathers may also feel confused and uncertain about what is expected of them. Models of highly involved fathers are still relatively new and men's wage-earning responsibilities are still strongly felt, thus fathers themselves may not know exactly what and how much they're supposed to do. Although it is true that to be a good father today consists of more than "earning as caring," for most men it still means earning. Involvement in child care, sharing of household labor, being more than a "helper" to one's spouse or partner are all among the expectations fathers face and hold for themselves (Rehel and Baxter 2015). Those men for whom being a father is of central importance in how they perceive themselves are more likely to be more highly involved in mental and physical caregiving activities (Tichenor et al. 2011).

As sociologists David Eggebeen and Chris Knoester (2001) found in their research, fathers and nonfathers differed in three areas—social, intergenerational/familial, and occupational. Men living with dependent children were significantly less likely to participate in leisure pursuits or social activities with friends. Men without children, men who lived away from their children, and men who lived with stepchildren attended church much less often than men who lived with their own biological or adopted children. Fathers who lived with their own biological or adopted children were more likely to have regular contact with aging parents and adult siblings than were men without children or men with stepchildren. Even fathers who lived apart from their children had more frequent contact with parents and siblings, suggesting that "fatherhood tightens intergenerational family ties" (Eggebeen and Knoester 2001, 389).

Psychologist David Kerr and colleagues identified additional positive social consequences of fatherhood. They followed a sample of at-risk males from adolescence to young adulthood and found that men engaged in less criminal behavior following the birth of their first child. They also smoked less and less often used alcohol (Kerr et al. 2011). For some men who experience a lack of employment success, fatherhood can provide an opportunity for enhanced self-worth (Tichenor et al. 2011).

In addition to whatever difficulty accompanies measuring father's behaviors with their children and their constructions of fathering, it is important to remember that today's fathers and mothers are still held to different parenting standards or expectations for involvement with their children. The aforementioned ideology of *intensive mothering* has yet to be joined by an ideology of intensive *fathering* or even a somewhat gender-neutral and equal ideology of intensive *parenting* (Hook and Chalasani 2008). We still tend to see father involvement as positive *when possible,* and we tend to view mother's caregiving as *essential and expected.* The cultural stereotype is that mothers are *supposed to be involved.*

Fatherhood has changed and is changing, perhaps more than motherhood. The biggest changes in motherhood are changes in mothers' employment and marital status. Where mothers are more likely to be employed than in prior generations, and more are unmarried, the tasks in which mothers are expected to participate and the relationships mothers attempt to build and maintain with their children are more in line with mothering as it long has been experienced; this is less the case for fathers and fathering. Consider the following:

- Since 1965, fathers have doubled the time spent doing housework and tripled the time they spend with children (Parker 2015). Fathers of preschoolers, ages 3 to 5, read to them an average of six times a week, compared to mothers who do so nearly seven times a week. Seventy percent of fathers report that in a typical week they have dinner with their child every night (Carr, Springer, and Cowan 2010).

- As they attempt to be more involved with their children, employed fathers are now as likely as employed mothers to say that they find it difficult to balance their work and family responsibilities (Parker 2015).

- Almost half of all fathers (46 percent) say they feel as though they don't spend enough time with their children (compared to 23 percent of mothers). Employed fathers are nearly as likely as mothers to say they "always feel rushed" (Parker 2015).

- In addition to the previous item, there are currently 1.9 million single fathers, representing 16 percent of all single parents. According to census data, almost 10 percent of single fathers were in households with three or more children (U.S. Census Bureau 2015b).

- There are now an estimated two million stay-at-home dads, representing 7 percent of fathers with children at home. Like the earlier discussion of stay-at-home moms, at-home fathers are home for a number of reasons, not just to care for their children or families. In a Pew Research Center report on at-home dads, Gretchen Livingston reports that, in 2012, about a third (35 percent) of at-home dads were at home and out of work because of disability or illness. Almost a quarter (23 percent) of fathers were at home because they were unemployed and couldn't find a job. Another 21 percent reported that they were home primarily to take care of their children. This latter category of at-home dads most embodies the idea of cultural change, and is also the type that Livingston reports has increased the most, from 5 percent in 1989 to the 2012 level of 21 percent of all fathers who are at home. Of course, whatever reasons have them at home, while at home with children, they are in a position to supply care and undertake housework.

## What Parenthood Does to Parents

One might wonder about the consequences of parenthood in the lives of fathers and mothers. Much of the research literature focuses attention rather narrowly on negative outcomes associated with becoming and being parents, while wider cultural attitudes and beliefs might be said to overstate the positives and downplay the costs (Simon 2008). After all, the negative outcomes, what might be considered costs of parenthood are easily noted and often quantifiable—less hours of sleep, more physical fatigue, greater financial expense, often a reduced household income, less time to oneself or together with one's spouse or partner, less sexual intimacy, and so on. The positives, what we might consider the rewards, are more intangible and more difficult to articulate or explain. How can something that is associated with so many different "costs" come to be seen as "the greatest thing I ever did," as many parents describe parenthood (Nelson, Kushlev, and Lyubomirsky 2014).

In their review of the literature on the relationship between parenthood and well-being, psychologists S. Katherine Nelson, Kostadin Kushlev, and Sonja Lyubomirsky (2014) note that researchers have tried to get at both the "pains and pleasures" of parenting in a variety of ways, by simple comparisons of parents and nonparents, by exploring changes that occur across the transition to parenthood, and by comparing parents' experiences when with their children against their experiences during other activities. With each approach, findings have been somewhat mixed. For example, in comparative studies of parents and nonparents on such outcomes as happiness, general well-being, life satisfaction, and depression, some studies have found parents with lower levels of well-being, some indicate higher levels of well-being, and some find similar levels of well-being. Research on the transition into parenthood finds increasing well-being during pregnancy and earliest parenthood that tends to be followed by longer-term decline. However, some research suggests an overall increase in positive emotions. Research on parents' rating of their well-being while caring for their children has also led to mixed findings. In time diary rankings of how parents felt during different parts of their day, "child care seems to be about as enjoyable (i.e., as in generating positive affect) as doing housework or surfing the Web, and somewhat less enjoyable than shopping or watching TV" (Nelson, Kushlev, and Lyubomirski 2014, 851). When measured differently, child care has been found to be associated with more positive affect than other daily activities against which it has been compared.

Among the positives, children can produce and deepen one's feelings of joy, strengthen one's social ties (especially to family but potentially to friends, neighbors, and community institutions), increase one's self-esteem, and afford one the opportunity to fulfill an expected role and transition to adulthood. Parenthood can bring social approval, love and affection, and increased marital stability. It initiates a shift in one's priorities and in the importance one attaches to other social roles. It also presents the opportunity for a feeling of **generativity**, of being committed to guiding or nurturing others (Nomaguchi and Milkie 2003). Parents feel a greater sense of purpose, meaning, and life satisfaction than nonparents (Simon 2008; Tichenor et al. 2011).

Among the negatives, research on women and men in the United States indicates an association between being a parent and such negatives as depression, emotional distress, marital strains, higher levels of stress, greater expenses, more work and housework, and depleting demands on one's time and energy (Nomaguchi and Milkie 2003; Simon 2008). Research by David Murphey, Tawana Bandy, Kristin Anderson Moore, and P. Mae Cooper used data from the National Survey of

America's Families (NSAF) and the National Survey of Children's Health (NSCH) to examine "parental aggravation" experienced by U.S. parents of children from birth through 17. To measure aggravation, they looked at how often the child does things that really bother the parent; how often the parent felt their child was harder to care for than most children the same age; and how often the parent felt angry with the child. If a parent expressed having all three feelings, at least "sometimes" (or generated combined responses of three or higher on a scale of 0 to 9), the parent was considered to have experienced parental aggravation. They point out that between 1997 and 2007, the proportion of parents who report aggravation increased from about 20 percent to 35 percent. In 2011–12, a third of parents nationwide reported parental aggravation (Murphey et al. 2014).

## The Effects of Parenthood on Marriage and Well-Being

Early research depicted the transition to parenthood as a crisis leading to a decline in marital quality and satisfaction. More recent research has tied becoming and/or being parents to such outcomes as depression (Evenson and Simon 2005) and, as reported in Chapter 8, traditionalization of marital roles and relationships. So, what generalizations can we draw about how having children affects parents? Based on S. Katherine Nelson, Kostadin Kushlev, and Sonja Lyubomirsky's review of the literature on "the pains and pleasures of parenting," the impact of parenthood is variable. One can say, as family researcher Heather Helms-Erikson (2001, 1100) expressed, parenthood leaves "some couples faring better following the birth of their first child, others worse, and still others seemingly unchanged."

In studies that followed samples of couples from the first trimester of pregnancy into new parenthood, results indicated that although close to half experienced a decline in their marital quality, that outcome was not true of all sampled couples. In fact, in two different studies, a fifth of couples reported increases in couples' well-being (Belsky and Kelly 1994; Belsky and Rovine 1990). Even among couples whose experiences fit the more common pattern of declining relationship quality (i.e., more conflict, less communication, declining feelings of love), such declines varied in magnitude. However, they are likely to persist beyond the transitional period after childbirth (Kluwer 2010).

**False**

**8.** Compared to nonparents, parents are more likely to experience depression.

Given the very mixed picture already described, it is safe to say that rather than an either/or (pleasure or pain) question, the impact of parenthood varies in part based on such things as the context and circumstances surrounding entry into it, parental and child demographics (including age of parents, gender, marital status, social class), and parental and child psychological factors (including child temperament, parenting style, and parental social support). Consider these factor examples from the review by S. Katherine Nelson, Kostadin Kushlev, and Sonja Lyubomirsky (2014):

- Parental age: Whereas young parents are reportedly less happy than their nonparent peers, middle-age and older parents are at least as happy or even happier than their peers without children.

- Child's age: Parenting younger children is more physically demanding and potentially more stressful as one confronts such issues as less sleep, midnight feedings, discipline problems, and so on. When children begin school, add homework battles and other school- and peer-related issues with which parents are confronted.

- Parent's gender: Research has been more consistent in showing benefits of fatherhood to men, where findings about motherhood are more mixed. Compared to nonparent peers, fathers report greater life satisfaction, happiness, meaning, and less depression, whereas mothers report only less depression. The gender difference is likely tied to the differing kinds and degrees of responsibility mothers and fathers carry. Most active, hands-on parenting is still done by mothers. Women also assume more responsibility for thinking about, worrying, and seeking information about children and their needs as well as planning and managing the division of responsibility for child care.

- Social support: Parenthood may induce a decline in time for other relationships. Reductions in the size of one's social network will also affect parental happiness and well-being because such reductions mean one is less likely to receive social support.

Outcomes of parenthood also differ depending on the circumstances under which one parents (whether one is married, never married, divorced; employed or unemployed; the parents' level of economic well-being; and whether parenting by choice or not). A later section looks at the effect of parents' marital status on parent–child relationships. The next chapter examines how employment affects parents. The most obvious difference between the impact of parenthood on women and men is in the child-rearing responsibility that women acquire when they become mothers.

It is interesting to note that in spite of potential difficulties meeting the high and/or ambiguous expectations of contemporary mother or father roles, according to survey data collected by the Pew Research Center, most parents report that they are doing a very good (45 percent) to excellent (24 percent) job raising their children, and only 6 percent rate their performance as a parent as fair or poor (Pew Research Center 2013b). Just under three-quarters (73 percent) of mothers surveyed rated their performance very good or excellent, and just under two-thirds of fathers (64 percent) gave themselves equally "high grades." Married parents were more likely than unmarried parents to say they were doing a very good to excellent job, and among mothers, working mothers rated their parenting more highly than did at-home mothers. Nearly eight of ten (78 percent) employed mothers said they were doing a very good or excellent job raising their children, 12 percent more than the percentage of at-home mothers who rated themselves similarly.

Among factors that differentiate how well or poorly parents rated themselves, balancing work and family life and spending enough time with one's children were important sources of difference. Parents who described themselves as spending the right amount of time with their children were much more likely to say they were doing a very good or excellent job as parents than were those reporting that they spend too little time (77 percent versus 54 percent). Similarly, employed parents who reported that balancing responsibilities of their jobs and their family lives was not difficult for them were more likely to give themselves very good or excellent ratings as parents than were those who indicated that they had difficulty balancing work and family (77 percent versus 66 percent) (Pew Research Center 2013b). We look more closely at work and family in Chapter 11.

### How Child Effects Influence Parents

Although this chapter focuses more on how parents shape their children, children are also socializers in their own right. When an infant cries to be picked up and held, to have a diaper changed, or to be burped, or when he or she smiles when being played with, fed, or cuddled, the parents are being socialized. The child is creating strong bonds with the parents. Although the infant's actions are not at first consciously directed toward reinforcing parental behavior, they nevertheless can have that effect. In this sense, even very young children can be viewed as participants in creating their own environment, influencing the parenting they receive, and in contributing to their further development.

A variety of child effects influence how mothers and fathers experience the transition to and reality of parenting. A child's gender, temperament, and developmental or behavioral problems can all affect parents and shape parenting. Having a daughter, especially when the birth is unplanned, has been associated with new parents experiencing a larger drop in marital quality than experienced by new parents who have a son. This decline has been linked to fathers' behavior. Evidence shows fathers' greater participation in hands-on parenting of sons than of daughters. In sharing more of the caregiving responsibilities, fathers of sons are more likely to meet their spouses' expectations of shared parenting than do lesser-involved fathers of daughters. This, in turn, can affect how new mothers feel about their marriages (Kluwer 2010).

Children's temperamental characteristics such as fussiness and unpredictability negatively affect women's experiences, which, in turn, can affect their feelings about their marriages (Kluwer 2010). Children who have developmental, emotional, and/or behavioral problems all have been shown to put parents under greater stress and to negatively impact parents' psychological functioning and well-being (Umberson, Pudrovska, and Reczek 2010). In examining such child effects as mentioned previously, at least some researchers have come to the conclusion that "child effects" on parenting may be as strong or stronger than parenting effects (Umberson, Pudrovska, and Reczek 2010).

## ◼ Strategies and Styles of Child Rearing

Twentieth-century parenting was shaped by child-rearing advice from such notable authorities as Benjamin Spock, T. Berry Brazelton, and Penelope Leach. These three authors sold well over 40 million copies of their books advising parents, especially mothers, as to the best ways to raise their children. Building on psychological theories of development, they stressed the importance of parents understanding their child's cognitive and emotional development.

So what did these experts advocate as effective parenting? Sharon Hays (1996) suggests that they all advocated the ideology of intensive mothering, discussed earlier in this chapter. Aside from the belief in the special nurturing capacities of mothers, this ideology contains the following assumptions about what children need from parents:

- Raising children is and should be an emotionally absorbing experience characterized by affectionate nurture. Emotional attachment is essential for healthy development; parental unconditional love and loving nurture are seen as critical to children, no less essential, Spock asserts, than "vitamins and calories" (Spock and Rothenberg 1985, quoted in Hays 1996).

- It is the mother's job to respond to the needs and wants of her child. Parents should follow the cues given by their children, and this requires knowledge of children's needs and developmental phases as well as great parental sensitivity.

- Parents must develop sensitivity to the particular needs of their children. This includes, for example, recognizing the different meanings of a child's crying, and understanding the unique and individual developmental pattern of each child.

- Physical punishment is frowned upon. Instead, setting limits, providing a good example of what parents expect from their child, and giving the child

One of the most important things parents can do is show their children that they are loved and that their company is desired.

Asiaselects/Alamy Stock Photo

lots of love are preferred ways to convince the child to internalize and act on parents' standards. Punishment consists of "carefully managed temporary withdrawal of loving attention," a labor-intensive, emotionally absorbing method of discipline. Once a child can question, parents are urged to reason with the child, negotiate, and discuss motives and alternative ways of acting. This strategy obviously involves more time and effort than spanking.

## Contemporary Child-Rearing Strategies

One of the most challenging aspects of child rearing is knowing how to change, stop, encourage, or otherwise influence children's behavior. We can request, reason, command, cajole, compromise, yell, or threaten with physical punishment or the suspension of privileges; alternatively, we can just get down on our knees and beg. Some of these approaches may be appropriate at certain times; others clearly are never appropriate. The techniques of child rearing that educators, psychologists, and others involved with child development currently teach or endorse differ somewhat in their emphasis but share most of the tenets that follow:

- *Respect.* Mutual respect between children and parents must be fostered for growth and change to occur. One important way to teach respect is through modeling—treating the child and others respectfully.

- *Consistency and clarity.* Consistency is crucial in child rearing. Without it, children become hopelessly confused, and parents become hopelessly frustrated. Patience and teamwork (maintaining a united front when there are two parents) on the parents' part help ensure consistency. Parents should beware of making promises or threats they won't be able to keep, and a child needs to know the rules and the consequences for breaking them.

- *Logical consequences.* One of the most effective ways to learn is by experiencing the logical consequences of our actions. Some of these consequences occur naturally—if you forget your umbrella on a rainy day, you are likely to get wet. Sometimes parents need to devise consequences appropriate to their child's misbehavior.

- *Open communication.* The lines of communication between parents and children must be kept open. Numerous techniques exist for fostering communication. Among these are active listening and the use of "I" messages. In *active listening,* the parent verbally reflects the child's communications to confirm that they have a mutual *understanding.* "I" *messages* (e.g., "I am disappointed that you aren't being nicer to your sister") are important because they impart facts without placing blame and are less likely to promote rebellion in children than are *"you" messages* (e.g., "You are such a bully to your sister"). In addition, regular weekly family meetings provide an opportunity to be together and air gripes, solve problems, and plan activities.

- *Behavior modification.* Effective types of discipline use some form of behavior modification. Rewards (hugs, stickers, or special activities) are given for good behavior, and privileges are taken away when misbehavior is involved. Good behavior can be kept track of on a simple chart listing one or several of the desired behaviors. Time-outs—sending a child to his or her room or to a "boring" place for a short time or until the misbehavior stops—are useful for particularly disruptive behaviors. They also give the parent an opportunity to cool off. Of course, some parents choose other means such as spankings in response to children's misbehavior. The following "Public Policies, Private Lives" feature examines the controversial question of whether or not to spank.

## Styles of Child Rearing

A parent's approach to training, teaching, nurturing, and helping a child will vary according to cultural influences, the parent's personality, the parent's basic attitude toward children and child rearing, and the role model that the parent presents to the child.

Efforts have been made to identify and differentiate between styles of child rearing. The most influential formulations contrast styles of child rearing based upon the extent of demandingness and responsiveness shown by parents, or—said differently—the amount of control exercised and support shown by parents. The four types that result from the differing combinations of control and support are briefly described in the following material. All four can be found among parents. Parents who practice **authoritarian child rearing** typically require absolute obedience. They are high in their level of demandingness and are not responsive. The parents' ability to maintain control is of primary importance. "Because I said so" is a typical response to a child's questioning of parental authority, and physical force may be used to ensure obedience.

There is little opportunity for children to discuss, question, or suggest behavioral rules or expectations. Authoritarian parents expect compliance.

*Permissive* or **indulgent child rearing** is one in which parents are lenient, showing more responsive and less demanding behavior toward their child. The child's freedom of expression and autonomy are valued. Permissive parents rely on reasoning and explanations, and make few demands that their children take on household responsibilities.

Parents who favor **authoritative child rearing** rely on positive reinforcement and infrequent use of punishment. This style is both demanding and responsive. Parents direct the child in a manner that shows awareness of his or her feelings and capabilities, and they encourage the development of the child's autonomy within reasonable limits. They foster an atmosphere of give-and-take in parent–child communication. Parental support is a crucial ingredient in child socialization. It is positively related to cognitive development, self-control, self-esteem, moral behavior, conformity to adult standards, and academic achievement (Gecas and Seff 1991). Control is exercised in conjunction with support by authoritative parents.

Finally, **uninvolved parenting** refers to parents who are neither responsive to their children's needs nor demanding of them in their behavioral expectations. Children and adolescents of uninvolved parents may suffer consequences in social competence, academic performance, psychosocial development, and problem behavior (Darling 1999).

Although the authoritative style is often presented as the most desirable and effective, and seems to be favored in much child-rearing advice literature, the evidence is both somewhat modest and more mixed. The authoritative style may be most attractive to and effective for white, middle-class parents. In other economic and cultural contexts, such as among lower-income and working-class parents and among certain ethnic and racial minorities, authoritarian parenting may actually be favored and effective (Cancian 2002). With the exception of uninvolved parenting, research results have been fairly modest in pointing to one style as having significantly different outcomes. When differences are found, they tend to be relatively weak (Cancian 2002).

In discussing styles of parenting, we should reiterate a point made in Chapter 3. There appear to be considerable differences between middle-class parents and those whose economic status is lower (i.e., parents who are working-class and/or poor).

"This is going to hurt me more than it's going to hurt you." So say many parents as they prepare to spank their child. Of course, the reality is that a spanking is intended to hurt the child who is spanked, whether the "hurt" it gives the parent to spank is more or less than the hurt to the child, or the parents wouldn't waste their time.

Many physicians, psychologists, and sociologists have become harsh and vocal critics of physical punishment. Both the American Psychological Association and the American Medical Association oppose physical punishment of children. Many sociologists, most notably scholars who study family violence, such as Murray Straus, oppose corporal punishment; they note that it is related to later aggressive behavior from children, including later perpetration of spousal violence (Straus and Yodanis 1996). However, such punishment is used widely. Straus and Paschall (2009) estimate that more than 90 percent of parents of toddlers use corporal punishment in the United States. Psychologist Elizabeth Gershoff (2008) contends that 80 percent of U.S. 5-year-olds have ever been spanked; by high school, the percentage reaches 85 percent. Critics contend that spanking may "work" in the short run by stopping undesirable behavior; its long-range results—anger, resentment, fear, hatred, aggressiveness, and family violence—may be extremely problematic (McLoyd and Smith 2002; Straus and Yodanis 1996). Proponents contend that if done "right," spankings can, with more immediacy, correct the misbehavior and establish expectations for likely consequences of future misbehaviors.

In January 2007, California State Assemblywoman Sally Lieber introduced legislation (California AB 755) to make California the first state in the United States to make spanking a child under 3 years of age a crime. The proposed legislation would have made spanking a misdemeanor, punishable by up to a $1,000 fine and a year in jail. Successful passage of the bill would have put California in the same company as at least 15 other, mostly European, countries that have laws banning corporal punishment. Within a month, Lieber had withdrawn her legislation after being inundated with calls and emails from proponents of spanking, and realizing the bill could not pass. In 2008, Lieber introduced a new bill (California AB 2943) that would make it a crime to spank a child with "an implement, including, but not limited to, a stick, a rod, a switch, an electrical cord, an extension cord, a belt, a broom, or a shoe." Again, there was considerable negative reaction to the legislation, which many critics felt limited parents' right to decide how to discipline their children.

In 2012, the state of Delaware began an effort that soon may lead to the criminalization of spanking in that state. Senate Bill 234, sponsored by State Senate Majority Leader Patricia Blevins, along with numerous cosponsors, does not explicitly mention banning spanking. Instead, the proposed law creates a multitiered definition of child abuse. The elements most pertinent to the issue of spanking are as follows:

- "Child abuse in the third degree" is defined as a misdemeanor offense in which an individual "recklessly or intentionally causes physical injury to a child through an act of abuse and/or neglect of such child."
- Physical injury is defined as "any impairment of physical condition or pain."

As opponents and proponents alike agree, potentially, this could turn spankings into misdemeanor offenses. The bill was unanimously passed by the Delaware Senate and then approved, on June 27, 2012, by the state's House of Representatives by a vote of 34 to 7. The bill was then sent to Delaware Governor Jack Markell to be signed into law, which he did on September 12, 2012, making Delaware the first state to make spanking a crime. As of March 2015, forty four countries ban corporal punishment of children but none of the fifty U.S. states explicitly and specifically do so (Glum, 2015).

Proponents of spanking believe it to be an effective method of child discipline, and one that is advocated by both the Old and New Testaments. Opponents of spanking assert that spanking has harmful effects on the long-term well-being of children, in addition to the short-term effect of subjecting children to pain. The effects of spanking may depend upon the wider cultural context in which it occurs, and may vary between religious and racial or ethnic groups (Ellison, Musick, and Holden 2011). Effects such as increased likelihood of anxiety, depression, and social or behavioral problems, such as lying, arguing, and more frequently getting into fights, are also shaped by the frequency and harshness of the spankings (Ellison, Musick, and Holden 2011).

Like many of the issues covered in this book, corporal punishment elicits controversy rather than consensus. There is an absence of agreement that is not likely to be resolved, especially given that opponents and proponents base their positions on entirely different grounds. Proponents cite religious beliefs and ideas about parental rights and responsibilities as reasons to allow, if not actually advocate, spanking. Opponents argue more from the vantage point of empirical evidence of the effects spanking have as well as matters of children's rights. We are not likely to see agreement emerge anytime soon. ●

According to sociologist Melvin Kohn, parents' social class and work experiences influenced the qualities they desired in their children and the methods they used to achieve them. Kohn contended that working-class parents were more likely to value obedience, conformity, and being respectful as desirable qualities for their children. They were also more likely to punish strictly and use corporal punishment more frequently. Middle-class parents, whose success at work more often required creativity and allowed greater autonomy, will more likely value curiosity and self-reliance (Kohn 1977).

Annette Lareau's study, *Unequal Childhoods*, introduced in Chapter 3, contrasted the child-oriented, activity-heavy, middle-class style of concerted cultivation with the style more typical of the working-class and poor families she studied, which she called the accomplishment of natural growth. To working-class and poor parents, children grow naturally, from free play in the neighborhood, close contact with extended family, and doing what their parents tell them to do. Both styles have positive aspects to them and offer children beneficial experiences. However, the middle-class style of concerted cultivation is more highly valued, accepted, and recommended by educators (Lareau 2003).

Finally, on the subject of styles of parenting, in recent years we have seen much popular discourse about certain types of parents, usually directed at and about mothers, and whether their style is just different or has potentially negative effects on their children.

## What Do Children Need?

Parents often want to know what they can do to raise healthy children. Are there specific parental behaviors or amounts of behaviors (say 12 hugs, three smiles, a kiss, and a half hour of conversation per day) that all children need to grow up healthy? Of course not. Apart from saying that basic physical needs (adequate food, shelter, clothing, and so on) must be met, along with some basic psychological ones, experts cannot give parents such detailed instructions.

Noted physician Melvin Konner (1991) listed the following needs for optimal child development—which, he wrote, "parents, teachers, doctors, and child development experts with many different perspectives can fairly well agree on":

- Adequate prenatal nutrition and care
- Appropriate stimulation and care of newborns

- Formation of at least one close attachment during the first five years
- Support for the family "under pressure from an uncaring world," including child care when a parent or parents must work
- Protection from illness
- Freedom from physical and sexual abuse
- Supportive friends, both adults and children
- Respect for the child's individuality and the presentation of appropriate challenges leading to competence
- Safe, nurturing, and challenging schooling
- Adolescence "free of pressure to grow up too fast, yet respectful of natural biological transformations"
- Protection from premature parenthood

In today's society, especially in the absence of adequate health care and schools in so many lower-income communities, it is difficult to see how even these minimal needs can be met. Even when the necessary social supports are present, parents may find themselves confused, discouraged, or guilty because they do not live up to their own expectations of perfection.

Yet children have more resiliency and resourcefulness than we may ordinarily think. They can adapt to and overcome many difficult situations. A mother can lose her temper and scream at her child, and the child will most likely survive, especially if the mother later apologizes and shares her feelings with the child. A father can turn his child away with a grunt because he is too tired to listen, and the child will not necessarily grow up neurotic, especially if the father spends some "special time" with the child later.

### The Importance of Self-Esteem

High self-esteem—what Erik Erikson called "an optimal sense of identity"—is essential for growth in relationships, creativity, and productivity in the world at large. Low self-esteem is a disability that afflicts children (and the adults they grow up to be) with feelings of powerlessness, poor ability to cope, low tolerance for differences and difficulties, inability to accept responsibility, and impaired emotional responsiveness. Self-esteem has been shown to be more significant than intelligence in predicting scholastic performance.

Parents can foster high self-esteem in their children by (1) having high self-esteem themselves, (2) accepting their children as they are, (3) enforcing clearly defined limits, (4) respecting individuality within the limits that have been set, and (5) responding to their child with sincere thoughts and feelings. It is also important to single out the child's behavior—not the whole child—for criticism (Kutner 1988). Children (and adults) can benefit from specific information about how well they've performed a task. "You did a lousy job" not only makes us feel bad but also gives us no useful information about what would constitute a good job.

Misusing the concept of self-esteem with superficial praise is probably the most common way parents mishandle the issue. Children notice when praise is insincere. If, for instance, Molly refuses to comb her hair yet we continually tell her how good it looks, Molly quickly realizes that we either have low expectations or do not have a clue about hair care. Instead, parents can accomplish more by giving children timely, honest, specific feedback. For example, "I like the way you discussed Benjamin Franklin's inventions in your essay" is more effective than "You're wonderful!" Each time parents treat their child like an intelligent, capable person, they increase the child's self-esteem.

### What Do *Parents* Need?

Although children meet some needs of parents, parents have other needs. Important needs of parents during the child-rearing years are personal developmental needs (such as social contacts, privacy, and outside interests) and the need to maintain marital satisfaction. Yet so much is expected of parents that they often neglect these needs. Parents may feel varying degrees of guilt if their child is not happy or has some "defect," an unpleasant personality, or even a runny nose.

However, many forces affect a child's development and behavior. Accepting our limitations as parents (and as humans) and accepting our lives as they are (even if they haven't turned out exactly as planned) can help us cope with some of the many stresses of child rearing in an already stressful world. Contemporary parents need to guard against the "burnout syndrome" of emotional and physical overload. Parents' careers and children's school activities, organized sports, Scouts, and music, art, or dance lessons compete for the parents' energy and rob them of the unstructured (and energizing) time that should be spent with others, with their children, or simply alone.

Parents would also benefit from implementation of a number of workplace policies, discussed in Chapter 11, that would increase the time they can spend with their children and with their spouse or partner.

Available social supports, whether kin, friends, neighbors, or hired help can also reduce the difficulty inherent in the parenting circumstances that so many fathers and mothers face. Single parents and dual-earner couples, in particular, may require the help of others in meeting child care responsibilities (Hansen 2005).

## Diversity in Parent–Child Relationships

The diversity of family forms in our country creates a variety of parenting experiences, needs, and possibilities as well as a range of parent–child relationships. The problems and strengths of single-parent and stepfamilies are discussed in more detail in Chapter 14, but are touched on here, along with the influences of ethnicity, sexuality (i.e., lesbian and gay parenthood), and aging.

### Effects of Parents' Marital Status

Chapter 14 will consider in more detail how single parents and stepparents experience parenting. For now, it is enough to note that parental marital status is a factor in children's upbringing and well-being. Consistently, researchers have found that—whether because of economic advantage, social resources, amount and kind of parental attention and commitment, or some other factors—children who live with both of their biological parents have been found to benefit in a variety of ways when compared to peers in single-parent households, remarried parent or stepparent households, and cohabiting-parent households.

Sociologist Yongmin Sun notes that children in stepfamilies and single-parent families are more likely than children living with their married biological parents to have behavior and drug problems, show lower rates of graduation from high school, report lower levels of self-esteem, and perform worse on standardized tests (Sun 2003). On a few measures, such as levels of delinquency and academic achievement, teens in married stepfamilies are somewhat advantaged compared to teens from cohabiting stepfamilies (i.e., unmarried couples with one partner functioning as a stepparent) (Manning and Lamb 2003).

On matters such as economic well-being, children of cohabiting parents benefit when compared to children of single parents but are less well off when compared to children of married parents. However, these outcomes vary by race. White children benefit more than either African American or Hispanic children from their biological parents being married (Manning and Brown 2006). Interestingly, data from the 2003–2004 American Time Use Survey show that mothers' time with children does not differ significantly between married and cohabiting mothers in terms of either amount or type of time spent with children (e.g., routine child care, interactive child care, total child care). Single mothers spend less total time with children than either married or cohabiting mothers (Kendig and Bianchi 2008), perhaps because they are solely responsible for all household tasks, stretching them thin.

In accounting for differences that surface between married and cohabiting stepfamilies and among families with two biological parents, single-parent families, and stepfamilies, economic factors (e.g., family income and parents' level of education) are especially important. Compared to married mothers, cohabiting mothers tend to have lower levels of education and lower earnings and are more likely to be unemployed (Reed 2006). Economic disadvantages that single mothers as well as stepfamilies face may explain why children in such households do less well (Sun 2003).

Not only can parental cohabitation affect children but, according to interesting qualitative interview research with a sample of "parent cohabitors," children can alter the meanings that cohabiting couples attach to their relationship as well (Reed 2006). In interviewing 44 cohabiting couples with children, sociologist Joanna Reed found that pregnancy or the birth of a child is the factor that leads some couples to what she calls "shotgun cohabitations." Furthermore, the presence of children "prompts couples to construct their relationship in a new way and orient them around the child" (Reed 2006, 1128). Pregnancy and pending parenthood make the relationship more serious, increase a couple's commitment to their relationship, and signal to them that they ought to seriously entertain the idea of marriage. At minimum, Reed's research reveals the different meanings that cohabitors bring to their relationships and how those meanings are differently constructed for those who are or are about to become parents.

## Ethnicity and Parenting

A child's ethnic background can affect how he or she is socialized. According to some researchers, minority families socialize their children to more highly value obligation, cooperation, and interdependence (Demo and Cox 2000). Racial and ethnic minority parents also have as part of their child-rearing responsibilities the task of race/ethnic socialization. This entails helping their children to understand, accept, and value their backgrounds as well as teaching them about discrimination and how to deal with it (Crosnoe and Cavanagh 2010). Parenting styles that characterize minority families may be unfavorably assessed when compared against the dominant cultural patterns. For example, certain groups such as Mexican Americans and Chinese Americans may appear either too controlling or not sufficiently engaged with their children (Crosnoe and Cavanagh 2010). It has been suggested that Mexican American parents tend to value cooperation and family unity more than individualism and competition. Asian Americans and Latinos traditionally stress the authority of the father in the family. In both groups, parents command considerable respect from their children, even when the children become adults. Older siblings, especially brothers, have authority over younger siblings and are expected to set a good example. Many Asian Americans tend to discourage aggression in children and expect them to sacrifice their personal desires or interests out of loyalty to their elders and to family authority more generally (Demo and Cox 2000). In disciplining their children, Asian parents tend to rely on compliance based on the desire for love and respect.

African Americans, too, may have group-specific emphases in the ways they socialize their children. As reported in Chapters 3 and 4, African American parents tend to socialize their children into less rigid, more flexible gender roles. They reinforce certain traits, such as assertiveness and independence, in both their sons and their daughters. They also seek to promote such values as pride, closeness to other African Americans, and racial awareness (Demo and Cox 2000).

African American men's experiences as fathers are entangled in their socioeconomic and occupational statuses. In light of race differences in rates of stable employment, many African American men can less easily live up to the provider role that has been a major aspect of fatherhood. Research that indicates that white men attach greater importance to their roles as fathers than do black men can, in part, be seen in this light. When experiencing difficulty meeting its mandates, African American men may place somewhat less priority on fatherhood. However, it is also true that fatherhood can be a source of self-worth for some African American men that work is not (Tichenor et al. 2011).

Groups with minority status in the United States may be different from one another in some key ways, but they also have much in common. Such groups often emphasize education as the means for the children to achieve success. Studies show that immigrant children tend to excel as students until they become acculturated and discover that it's not "cool." Minority groups are often dual-worker families, meaning that the children may have considerable exposure to television while the parents are away from home. This may be viewed as a mixed blessing: On the one hand, television may help children who need to acquire English-language skills; on the other, it can promote fear, violence, and negative stereotypes of women and minority-status groups. Some U.S. children are raised with a strong positive sense of ethnic identification; however, that can also result in a sense of separateness that is imposed by the greater society.

Discrimination and prejudice shape the lives of many U.S. children. According to Mary Kay DeGenova (1997), to reduce an environment of racism, it is important for us to identify the similarities among various cultures. These include people's hopes, aspirations, desire to survive, search for love, and need for family—to name just a few. Although we may be superficially dissimilar, the essence of being human is very much the same for all of us.

## Gay and Lesbian Parents and Their Children

Estimates of the number of families of same-sex couples and their children range widely. Using the U.S. Census Bureau's 2012 American Community Survey, Krista Payne (2014b) provides the following profile of same-sex households with children. The census estimates do not include gay men or lesbians who are raising children as single parents.

Almost 236,000 children under age 18 live in same-sex couple-headed households. This represents a fifth of all same-sex households, though female-female households are twice as likely (at 28 percent) as male-male households (13 percent) to include a minor child. Of same-sex households with minor children, roughly half have only one child, with male-male couples (at 33 percent) slightly more likely than female-female couples to include two children, and roughly the same proportions (19 percent and 18 percent, respectively) of male-male and female-female couple households including three or more children. As shown in Table 10.8, there

**TABLE 10.8 Percent of Same Sex Couples With Children, by Race**

| Race | Male-Male Couples | Female-Female Couples |
|------|-------------------|------------------------|
| White | 11% | 25% |
| Black | 25% | 44% |
| Asian | 20% | 38% |
| Hispanic | 22% | 37% |

SOURCE: U.S. Census Bureau, American Community Survey, 1-year Estimates, (2012).

are some racial variations, with white same-sex couples least likely and black same-sex couples most likely to include children.

Payne reports that there are also differences by educational attainment, with male-male and female-female couple households headed by those with less than a high school education most likely to contain children. The educational difference is more notable among male-male households, as 29 percent of those headed by men with less than a high school education containing minor children but only 9 percent of those with a bachelor degree or more have children living in the household. Among female-female households, education differences are smaller, ranging from 40 percent among households headed by a woman with less than a high school degree to 24 percent among those with at least a college degree (Payne 2014b).

As sociologist Nancy Mezey (2015) points out, LGBT parents are diverse. One source of difference is that some were in heterosexual relationships and later came out as LGBT, while others first came out as LGBT and then later chose to become parents. In the latter case, unlike heterosexuals for whom pregnancies and births might have been unplanned, near zero percent of pregnancies are unplanned among LGBT individuals. As parents, LGBT individuals construct a variety of family forms, including stepfamilies, families with a combination of multiple parents, and families Mezey calls heterogay, in which heterosexual and gay parents intentionally form a family together.

Mezey also notes that the parental experiences of LGBT parents bear both similarities and differences in comparison to heterosexual parents. Regardless of their sexual identities or preferred partners, parents have to manage child care, attend to the "minutia of parenting and household chores that accompany parenting," and make the big and not-so-big decisions that will shape their children's lives (Mezey 2015, 94). As to

Consider the following account by Abigail Garner, author of Families Like Mine: Children of Gay Parents Tell It Like It Is (2005) and creator of the website FamiliesLikeMine.com:

When I was 5, my father came out as gay to his family and friends and moved in with another man. By the time I entered elementary school, I was learning about the cruelty of homophobia. "Faggot" was the favorite put-down among the boys in my class. I didn't know what it meant until my parents explained that it was a mean way of saying someone was gay. Since my classmates seemed to be so hostile about gay people, I decided I should keep quiet about my family.

I remember when I was about 8, I was walking down the street between my father and his partner and holding both of their hands. It felt dangerous, because by standing as a link between them I was "outing" them. What would happen if others realized my dad was gay? Would he lose his job? Get beaten up? Be declared an unfit parent?

Fortunately, my mother (who is heterosexual) made no attempt to limit my father's custody rights. If she had,

she probably would have gained full custody. Our courts have a history of favoring straight parents over gay ones in custody battles.

. . . [I]t wasn't having a gay father that made growing up a challenge, it was navigating a society that did not accept him and, by extension, me.

Twenty-something Elizabeth Wall was adopted by her fathers at 5 days old. Talking to ABC News, although she acknowledged that growing up with gay parents "can be isolating and challenging" because of how the wider society treats households headed by gay men or lesbians, she also believes that she is more tolerant of differences and a stronger person because of her upbringing.

Cathy Renna, a lesbian and mother of a 4-year-old, says of her daughter, "This kid could not be more loved or more wanted." Looking down the road, Renna speculates optimistically about what the future may bring: "This generation of kids growing up will look back and wonder what all the fuss was about," she told ABCNews.com.

---

differences, both gay male and lesbian families tend to have more egalitarian divisions of labor, including more sharing of child care. Lesbian families tend to be more egalitarian than gay

**False**

**9.** Children of gay or lesbian parents are not more likely than children of heterosexual parents to be gay themselves.

male families, though it would be a mistake to assume that this is true that all lesbian families divide parenting equally. Gay male and lesbian parents do confront certain issues that heterosexual parents don't. The most obvious is that same-sex parents have to deal with homophobia and heterosexism in the wider society, which means that, like racial or ethnic minority parents, they must prepare their children for possible encounters or confrontations with hostility, negative judgments, and discrimination. Nancy Mezey notes that same-sex parents may need to work more closely with teachers and administrators in the schools to assist their children with any such difficulties (Mezey 2015).

In summarizing the research on children of gay and lesbian parents as they compare with children of heterosexual parents, researchers typically have reported few if any differences. For example, psychologist Charlotte Patterson (2000, 2005) notes that there are no significant differences in their gender identities, gender-role behaviors, self-concepts, moral judgment, intelligence, success with peer relations, behavioral problems, or successful relations with adults of both genders. Research

Families headed by lesbians or gay men generally experience the same joys and pains as those headed by heterosexuals, but they are also likely to face insensitivity or discrimination from society.

DGLimages/iStock/Getty Images

commonly finds children of gay males and lesbians to be well adjusted and no more likely to be gay as adults (Flaks et al. 1995; Goleman 1992; Kantrowitz 1996).

One should note that comparisons of children of same- and opposite-sex parents can be presented in ways that downplay or exaggerate differences, depending on the intention of those making the comparisons. Fifteen years ago, sociologists Judith Stacey and Timothy Biblarz (2001) suggested that there may be some defensiveness on the part of researchers studying such differences, especially from those sympathetic to gay and lesbian parents. Aware of the social stigma and lack of support gay and lesbian families face, Stacey and Biblarz suggested that there may be a tendency to minimize differences. In so doing, some differences that might even be strengths of gay and lesbian families may go underemphasized along with any less-than-positive comparisons.

More recently, the question of differences between children of same-sex and heterosexual parents surfaced loudly, visibly, and controversially with the July 2012 publication of a study by sociologist Mark Regnerus, in the journal *Social Science Research*. Regnerus conducted what he called the New Family Structures Study (NFSS), a survey of a large, random sample of 18- to 39-year-olds that purports to identify "numerous, consistent differences, especially between the children of women who have had a lesbian relationship (LM) and those with still-married (heterosexual) parents" (Regnerus 2012, 752). Although the study had data comparing children from across eight different family backgrounds (including also those who had been adopted, whose parents divorced in the respondent's adulthood, who were in stepfamilies, who lived with single parents, and "all others"), Regnerus focused his attention on differences that surfaced between children whose parent(s) had had a same-sex sexual relationship and those from what he called an "intact biological family (IBF)." Challenging head-on the long-standing finding of "no differences" or "very few differences" between children of married heterosexual parents and those with lesbian mothers (LMs) and/or gay fathers (GFs), Regnerus reported finding a number of significant differences. He reported that

> when compared with children who grew up in biologically (still) intact mother–father families, the children of women who reported a same-sex relationship look markedly different on numerous outcomes, including many that are obviously suboptimal (such as education, depression, employment status, or marijuana use) (p. 764).

Regnerus also found significant differences in the relationship and sexual experiences of IBF children and those of women who reported having had a same-sex relationship. The study also revealed differences between young adult children of IBFs and those with fathers who had had a same-sex relationship (GFs), but Regnerus claimed to identify "many more differences by any method of analysis" between children of LMs and IBFs. Although he acknowledged that his data do not and cannot establish or prove causal relationships between having a parent who has had same-sex sexual experiences and the "suboptimal outcomes" he identified, he suggested "the empirical claim that no notable differences exist must go" (p. 766). In interviews, Regnerus went further, suggesting that because "children of same-sex parents experience greater household instability than others . . . it could be too much of a social gamble to support this new (but tiny) family form" (Frank 2012).

The Regnerus study was met with open arms by conservatives who were opposed to same-sex couples raising children and those who wished to block marriage equality but generated a firestorm of well-deserved criticism from the academic community, much of which pointed to potentially serious methodological problems and, to a lesser extent, raised ethical questions about the sponsorship and peer-review process associated with this research. The biggest methodological deficiency was the way Regnerus operationalized parents' sexual orientation. In other words, critics pointed out that the study failed to accurately define gay and lesbian parents. In including as GFs and LMs *any parent* who had been involved in a same-sex relationship, the Regnerus study wasn't really a study of gay and lesbian parents, in which one or both parents self-identifies as gay or lesbian and has had a meaningful and committed same-sex relationship, of which their children were aware. As historian Nathaniel Frank noted, the Regnerus study "does not actually compare children raised by same-sex couples with those raised by different-sex couples. . . . In fact, only a small proportion of its sample spent more than a few years living in a household headed by a same-sex couple" (Frank 2012). Furthermore, most of the sample of adult children with parents who had had same-sex relationships came from families in which parents had split up, thus confounding the comparisons. Critics pointed out that, at best, what the study showed is that children from households where parents broke up fare worse than children from households where parents remain married.

Despite the fact that individuals, groups, and organizations opposed to same-sex marriage seized upon the Regnerus study to justify and solidify their continued opposition, we really learned no more about how growing up in a stable household with same-sex parents compared with growing up in a stable household headed by heterosexual parents than we knew before Regnerus conducted the New Family Structures Study.

Ultimately, the quality of parenting and the stability and harmony within the family—not the sexuality of the parents—matter most to children. Like children of heterosexual parents, children whose gay or lesbian parents are in "warm and caring relationships," experiencing less stress and conflict, and receiving more support from partners (as well as from other family members), tend to fare better.

## What about Nonparental Households?

Yet another way to see the influence of parents on children is to examine the experiences of children in households with *no* biological parents. As indicated in Table 10.9, in 2014, close to 3 million children—almost 4 percent of the more than 73.6 million U.S. children under 18 years of age—lived in households with *neither* biological parent (ChildStats.gov 2011). Among the 2.8 million children not living with a parent in 2014, 56 percent (1.6 million) lived with grandparents, 24 percent lived with other relatives only, and 20 percent lived with nonrelatives. Of children in nonrelatives' homes, 38 percent (214,000) lived with foster parents. Of those living with relatives, 80 percent lived with a grandparent.

Older children (ages 15 to 17) are more likely than children under age 5 to live in one of these nonparental households. As shown in Table 10.9, ethnicity also makes a difference in the percentage of children who live with neither parent as well as the percentage who live in two-parent households. Generally, research has documented that children in nonparental households suffer when compared to children who live with at least one parent. Comparisons of children in foster care, albeit only one type of nonparental care, show negative effects in areas ranging from children's mental health, academic achievement, drug use, and behavioral problems (Sun 2003). Likewise, children in nonparental "kinship care" have been found to have poorer health, mental health, and school achievement than children in "parent-present" families, whether single- or two-parent families (Sun 2003). Sociologist Yongmin Sun suggests that it is likely that the absence of a mother results in the greatest impact. In accounting for the observed effects in **nonparental households**, Sun argues that the differences result mainly from the resource differences between these family structures and those with at least one parent. Key resources include income and parents' education, parents' expectations for their children's education, frequency of conversations between parents and children about school, involvement of parents with the schools and with other parents, and children's experiences of various cultural activities. No differences of note existed between kinship care and nonrelative care, and no differences were observed between girls and boys in how they fare in nonparental environments (Sun 2003).

# ■ Parenting and Caregiving throughout Life

As parents and their adult children age, the relationship between them undergoes a number of changes. The needs of adult children are different than the needs of younger, dependent sons and daughters. At the same time, the aging of parents often introduces a need for someone to provide them with or oversee their care.

## Parenting Adult Children

Many years ago, a Miami Beach couple reported their son missing (Treas and Bengtson 1987). Joseph Horowitz still doesn't understand why his mother became so upset. He wasn't "missing" from their home in Miami Beach: He had just decided to go

**TABLE 10.9** Percent of U.S. Children Living with Two Parents and with No Parents, by Race, 2011 (numbers in thousands)

| Group | Total Children | % with Two Parents | % with No Parents |
|---|---|---|---|
| All | 73,623 | 69.2% | 3.9% |
| White, non-Hispanic | 38,08440,616 | 77.5% | 2.90% |
| African American | 11,091 | 38.6% | 7.7% |
| Asian American | 3,736 | 85.7% | 1.7% |
| Hispanic | 17,981 | 67.0% | 4.1% |

SOURCE: U.S. Census Bureau, America's Families and Living Arrangements (2015, Table C-3).

north for the winter. Etta Horowitz, however, called authorities. Social worker Mike Weston finally located Joseph in Monticello, New York, where he was visiting friends. Etta, 102, and her husband, Solomon, 96, had feared that harm had befallen their son Joseph, 75. As the Horowitz story reminds us, parenting does not end when children grow up.

By some measures, children are "growing up" later than at any time in the past. They lack the means to be financially independent and are delaying entry into marriage, parenthood, and independent living, away from their families. In one study that compares 1960 census data to 2000 census data, researchers noted that there has been a significant decrease in the percentage of young adults who, by age 20 or 30, have completed all of the following five traditionally defined major adult transitions: leaving the parental home, completing their schooling, achieving financial independence (being in the labor force and/or—for women—being married and a mother), marrying, and becoming a parent. In 1960, more than three-fourths of women and two-thirds of men had reached all five of these markers by age 30, yet in 2000, less than half of women and less than a third of men had achieved all five of these (Furstenberg et al. 2004). The recession of the last decade has likely further reduced the numbers who'd met all milestones. As a consequence, parents are being asked to provide support or assistance sometimes well into their children's young adulthood. Attitude surveys suggest growing acceptance of this reality. According to a 2012 Pew Social and Demographic Trends Report ("Young, Underemployed, and Optimistic: Coming of Age Slowly in a Tough Economy"), nearly a third (31 percent) of parents surveyed believed that young adults shouldn't have to be on their own until they were 25 years old or older. Two decades earlier, in a 1993 *Newsweek* magazine poll, 80 percent of parents said children should be financially independent by age 22 (Taylor et al. 2012).

As we noted in Chapter 8, young adult children are increasingly "staying in" or returning to their parents' households, creating what earlier we called "accordion families," using anthropologist Katherine Newman's clever term. According to a 2015 report by the Pew Research Center (Fry 2015a), between 2010 and 2015, the percentage of 18- to 34-year-olds who lived independently of their parents dropped from 69 percent to 67 percent. In 2010, 24 percent of young adults lived with their parents, whereas in the first few months of 2015, that percentage had increased to 26 percent.

In 2007, only 22 percent lived with their parents (Fry 2015a). The recent recession may have made moving in or remaining in one's parents' household more necessary and accepted; the percentage of young adults doing so has actually increased despite an economic recovery. Unemployment of 18- to 34-year-olds peaked at 12.4 percent in 2010; by 2015 the rate was down to 7.7 percent, still higher than the prerecession rate of 6.2 percent in 2007, but trending downward while the percentage of young adults living with parents rose.

Living with parents is more common among those younger adults; nevertheless, 15 percent of 25- to 34-year-olds were living with their parents. According to U.S. Census data, over half of 18- to 24-year-olds live with parents. Other characteristics to note among those young adults living with parents are as follows:

- Demographically, males are more likely to live with parents than are females: Of those 18 to 34, 42.8 percent of men and 36.4 percent of women currently live with parents (or other relatives).

- Though males are more likely than females to live with family, the percentage of women living with parents is the highest it has been in more than 70 years. This is partly a reflection of the fact that young women are more likely to go to college and are less likely to be married than were young women previously.

- Young adults who are attending college are more likely than those not attending college to live with family. In 2014, 45 percent of female college students compared to 33 percent of young women who weren't in college lived with family.

- There are educational differences in the percentage of young adults living on their own, as opposed to living with family. In 2015, 86 percent of those 25- to 34-year-olds with at least a bachelor degree lived independently of their families, compared with 79 percent of those with some college and 75 percent of those with high school or less education. Of note, among both the more highly educated and the less educated, the percentages of 25- to 34-year-olds living independently declined slightly since 2010.

- As might be expected, research that has examined reasons young people give for moving back home or remaining in their parents' household find economic reasons to be primary. For example, research by economists Lisa Dettling and Joanne Hsu suggests that student loan debt factors prominently,

**TABLE 10.10** Relationship Effects of Living with Parents

| Age | Bad | Good | No Effect |
|-----|-----|------|-----------|
| 18–24 | 12% | 41% | 47% |
| 25–34 | 25% | 24% | 48% |
| All | 18% | 34% | 47% |

SOURCE: Parker (2012b).

explaining perhaps as much as 30 percent of the increase in young adults returning to the parents' household (Dettling and Hsu 2014). Other research cites reasons such as "to save money," "inability to find a job," and "needing my parents' help" (Payne and Copp 2013). Krista Payne and Jennifer Copp also show that among those who have remained (as opposed to returned) in the parental household, needing to save money was mentioned by more than 90 percent of those surveyed (Payne and Copp 2013).

- As can be seen in Table 10.10, Pew Research Center data suggests that 18- to 25-year-olds are more likely than 26- to 34-year-olds to report that living with parents has been good for their relationships (Parker 2012b).

Most parents with adult children still feel themselves to be parents even when their "children" are middle-aged. However, their parental role is considerably less important in their daily lives. They generally have some kind of regular contact with their adult children, usually by phone calls, text messages, or emails. It is estimated that almost half (46 percent) of U.S. adults with one or more grown children living outside of their home are in daily contact with their children. Another 39 percent say they are in contact at least once a week. Perhaps unsurprisingly, more mothers (52) than fathers (38 percent) report maintaining daily contact, most commonly via phone. Sixty percent of U.S. parents indicate that they also typically communicate with their adult children via text messaging (Pew Research Center 2015c).

Parents and adult children also visit each other fairly frequently and often celebrate holidays and birthdays together. Financially, they may make loans, give gifts, or pay bills for their children. According to Pew Research Center survey data, more than 60 percent of U.S. parents with grown children have given an adult child some financial support (Pew Research Center 2015c). Further assistance may come in the form of shopping, house care, child care (for grandchildren), and transportation and help in times of illness.

Parents tend to assist those whom they perceive to be in need, especially children who are single or divorced. Parents perceive their single children as being "needy" when they have not yet established themselves in occupational and family roles. These children may need financial assistance and may lack intimate ties; parents may provide both until the children are more firmly established. Parents often assist divorced children, especially if grandchildren are involved, by providing financial and emotional support. They may also provide child care and housekeeping services.

Parents tend to be deeply affected by the circumstances in which their adult children find themselves. Adult children who seem well adjusted and who have fulfilled the expected life stages (becoming independent, starting a family, and so on) provide their aging parents with a vicarious gratification as well as satisfaction that their children are "doing well" at what they are supposed to do (Mitchell 2010). In addition, sociologist Barbara Mitchell identifies the belief that one's children are "good people" of high morals and good judgment, along with satisfaction in the quality of one's relationship with them, as sources of parents' happiness in their parental roles (Mitchell 2010). On the other hand, adult children who experience "developmental delays," failing to achieve independent adulthood or to take on adult responsibilities, are sources of parents' unhappiness in their parental roles (Mitchell 2010). Similarly, children who have stress-related or chronic problems (e.g., with alcohol) cause higher levels of parental depression (Allen, Blieszner, and Roberto 2000). The same is likely true for children's marital problems and divorces, especially for mothers (Kalmijn and De Graaf 2012). Family researchers Emily Greenfield and Nadine Marks reviewed national survey data (National Survey of Midlife in the United States, http://midus.wisc.edu) to examine the effects of adult children's problems on their parents' psychological well-being. Parents who reported that their adult children had a greater number and more types of problems experienced poorer well-being, suffering negative outcomes in areas such as self-acceptance (e.g., liking oneself and being satisfied with one's life), positive affect (e.g., feeling cheerful, happy, calm, and peaceful), and parent–child relationship quality. They also reported more family relationship strain (e.g., feeling as if family members other than one's spouse demand too much, let them down, or get on their nerves) and more negative affect (e.g., feeling sad, hopeless, worthless, or nervous). Both mothers and fathers were affected. Parents' marital status made a difference. Single parents suffered greater negative impact on their positive affect, while married parents experienced more negative consequences in their parent–child relationships (Greenfield and Marks 2006).

Midlife parental unhappiness is also increased when the parent–child relationship is perceived by the parent as weak or is characterized by personality conflicts. Children who are perceived to expect "way too much," have a sense of entitlement about them, or are seen as "a drain," and are also sources of parental unhappiness (Mitchell 2010). In addition to such qualitative assessments of their children or relationships, parents experiencing financial problems, poor health, or excessive demands from their work or caregiving responsibilities are at greater risk for reduced parental well-being (Mitchell 2010).

Some elderly parents never cease being parents because they provide home care for children who are severely limited either physically or mentally. Many elderly parents, like middle-aged parents, are taking on parental roles again as children return home for financial or emotional reasons. Although we don't know how elderly parents "parent," presumably they are less involved in traditional parenting roles.

## Grandparenting

There are approximately 64 million grandparents in the United States (Stykes, Manning, and Brown 2014). An estimated 60 percent of adults 50 years old or older, and more than 80 percent of those over 65 are grandparents). Two-thirds of those who are grandparents report having at least four grandchildren (Krogstad 2015).

In the not-too-distant past, a discussion of grandparents might well have brought to mind an image of a lonely, frail grandmother in a rocking chair. Such an image, never fully representative of grandmothers, is certainly well out of sync with current reality. Grandparents are often neither old nor lonely, and they are certainly not absent in contemporary American family life. Grandparents are often a very active and meaningful presence in the family lives of young children as well as young adults. As reflected in a profile from the National Center for Family and Marriage Research, the percentage of children who lived with at least one

grandparent in 2010 was the highest it had been in at least 70 years, and had more than doubled between 1970 and 2010 (Wilson 2013).

In 2014, close to a million children (874,000) lived with both their grandmother and grandfather and both of their parents. Another 1.5 million lived with either a grandmother (1.2 million) or a grandfather and both of their parents. Among those children who lived with their mother without their father, 3 million also lived with at least one grandparent. For those 2.8 million children living with their father but not their mother, more than 400,000 live with at least one grandparent. And, finally, among the 2.8 million children who live with neither of their parents, as was mentioned earlier, more than half live with at least one grandparent.

Based on 2013 data, of the more than 7 million children who lived in a household that included at least one grandparent, a majority of those children, almost two-thirds, lived in a grandparent's household. This amounts to an estimated 5.7 million grandchildren under 18 years of age who lived in a grandparent-headed household in 2013. Almost half, 47 percent (2.7 million) of those grandchildren were younger than 6.

Grandparenting has expanded considerably in recent years, leading grandparents to take on roles that relatively few Americans played a few generations back. Grandparents play important emotional roles in American families; the majority appears to establish strong bonds with their grandchildren, and grandparents frequently provide needed child care for their grandchildren. They may play an important role in helping achieve family cohesiveness by conveying family history, stories, and customs. Grandparents also influence their grandchildren directly as they provide direct, practical assistance and act as caretakers, playmates, and mentors. They influence indirectly when they provide psychological and material support to parents, who may consequently have more resources for parenting.

Grandparents seem to take on greater importance in single-parent and stepparent families and among certain ethnic groups. Children of single mothers are considerably more likely than children of married parents to live in the same household as a grandparent. Of children who live in a household with a grandparent, a third (34 percent) also lived with two parents in a three generational household. Approximately 40 percent of children in households with grandparents also had their mothers but not fathers in the

Grandparents are important to their grandchildren as caregivers, playmates, and mentors.

Plush Studios/Bill Reitzel/Blend Images/Getty Images

household. A fifth of children living in a household with grandparents lived with neither parent. Only 5 percent of children living with a grandparent also had their fathers without their mothers in the household (Ellis and Simmons 2014).

Among those households containing coresident grandparents, there is some diversity in the role grandparents play. In just under 40 percent of such households, grandparents are caregivers. This is more often the case among blacks, where 50 percent of coresident grandparents are caregivers, with whites (44 percent) and Hispanics (43 percent) not too far behind. Among Asians, 19 percent of coresident grandparents were caregivers for their grandchildren (Wilson 2012).

Researchers Rachel Dunifon and Lori Kowaleski-Jones note that there are many ways that living with a grandparent could be beneficial to children. Grandparents might reduce the stress felt by single parents, provide additional income, help monitor the children, and engage in activities with grandchildren that benefit them. Dunifon and Kowaleski-Jones found that for white children, living in a single-mother household with a grandparent present was associated with greater cognitive stimulation when compared with children in a single-mother household without a grandparent. For African American children, such household structure is associated with less cognitive stimulation. These different outcomes may reflect different skills

that grandparents bring with them to the household as well as different circumstances that lead to sharing a household (Dunifon and Kowaleski-Jones 2007).

Researchers have found that the experiences of grandparent caregivers vary depending on whether they are in "skipped-generation" households with no parent present, or acting as co-parents along with their adult children. When no parent is present in the household, grandparent caregivers are more at risk of negative outcomes in their health and well-being (Hughes et al. 2007). This is less true among African Americans. African American custodial grandmothers fare better than African American co-parent grandmothers (Umberson, Pudrovska, and Reczek 2010).

Single parenting and remarriage have made grandparenthood more painful and problematic for many grandparents. Stepfamilies have created stepgrandparents who are often confused about their grandparenting role. Are they really grandparents?

The grandparents whose sons or daughters do not have custody often express concern about their future grandparenting roles (Goetting 1990). Although research indicates that children in stepfamilies tend to do better if they continue to have contact with both sets of grandparents, it is not uncommon for the parents of the noncustodial parent to lose contact with their grandchildren. Nineteen percent of grandparents in the United States reported that they have contact with their grandchildren less than once a month to never. However, still on the subject of contact, 20 percent of U.S. grandparents reported that they communicate daily with a grandchild, and another 40 percent said they are in touch weekly.

Among adults with living grandparents, 6 percent report being in daily contact, 22 percent—a little more than a fifth—say they are in contact weekly, and 36 percent report being in touch at least monthly. Only 2 percent report having no contact with their grandparent(s) (the remainder are in contact less than monthly). Most such contact is via phone (Pew Research Center 2015c).

A variety of circumstances may lead to situations in which the grandparent role and the relationship with grandchildren are strained, if not disrupted. Divorce and single parenthood may be the most prominent of such circumstances, but death of a spouse, distance, or estrangement between parents and children can all impede grandparent–grandchild relationships (Keith and Wacker 2002). Over the past 40 years, grandparent visitation statutes have been enacted in all 50 states, and grandparents' visitation rights have been increased. Generally, courts have not wanted to expand grandparents' rights at the expense of parents' rights, especially parents' rights to control the custody of their children (Keith and Wacker 2002).

## Children Caring for Parents

Parent–child relationships do not flow just in one direction. A common experience many American families face is the need to provide care for aging or ill parents. The previously noted idea of the **sandwich generation** captures the experience of many adults, sandwiched between the simultaneous demands of raising their own children and caring for their parents. Certain circumstances create **parentified children**—children forced to become caregivers for their parents well before adulthood (Winton 2003). In situations of *parentification*, children may be pressed into taking care of parents who have become chronically ill, chemically dependent, mentally ill, incapacitated after a divorce or widowhood, or socially isolated or incapacitated (Winton 2003).

Much of the psychological and sociological literature depicts parentification as problematic. Taking on caregiving responsibilities for a parent or parents while still a child or adolescent can disrupt normal developmental processes. Sociologists tend to focus on the nonnormative nature of children being responsible for their parents. However, definitions of normative and nonnormative vary by culture. Among many populations other than white, middle-class European Americans, parentification is expected and obligatory. Additionally, under certain circumstances, parentification may be beneficial for the development of certain personality traits, the maintenance of certain family relationships, and the acquisition of particular skills. Chester Winton (2003) suggests that parentification may be a normative part of childhood in many contemporary U.S. families. *Destructive parentification* occurs when the circumstances become extreme and long term and the responsibilities that children carry are age inappropriate (Jurkovic 1997, cited in Winton 2003). Parentification can also delay one's own entry into marriage, lead to the acquisition or development of certain personality traits or tendencies, and later relationship issues that result from seeking partners that need caregiving. Having experienced caring for parents may also lead them into careers such as nursing, social work, medicine, or teaching.

## Adults and Aging Parents

There are strong norms of **filial responsibility**—to see that one's aging parents have support, assistance, company, and are well cared for—in the United States (Gans and Silverstein 2006). Based on survey estimates, three-fourths of U.S. adults view providing financial assistance to aging parents in need as their responsibility (Pew Research Center 2015c). Acceptance of these norms neither reflects nor guarantees that one will provide such support, though it is logical to assume that support is more likely among those who accept such norms. Earlier research suggests that such norms are more strongly felt by women than by men, those with no college education, and those who have had experience providing assistance or support to their parents (Gans and Silverstein 2006).

Surveys conducted by the Pew Research Center found that in the United States, almost 60 percent of adults say they provided assistance to their aging parents in the prior 12 months, with such things as running errands, housework, and home repairs. Almost 30 percent say they have helped financially, and 14 percent report having assisted a parent with such personal care needs as dressing or bathing. Roughly a third say that assisting an aging parent is stressful. Nearly 90 percent of U.S. adults surveyed who are providing some help to an aging parent say that doing so is rewarding (Pew Research Center 2015c).

One factor that affects actual caregiving behavior is the adult child's marital status. Although most Americans want to and do maintain relationships with their aging parents, those relationships are often constrained by the "more pressing demands of marriage" (Sarkisian and Gerstel 2008, 373). Married people are less likely than never-married or divorced adults to share a household with their parents; to have frequent contact by letter, phone, or in person; to give and receive emotional support or practical help; or to give financial help to or receive financial help from their parents. Other factors affecting the level of involvement between adults and their parents include parental marital status (unmarried parents receive more assistance and contact from adult sons and daughters), degree and nature of parents' needs, and gender.

Most actual elder care is provided by women, generally daughters or daughters-in-law. Women do more "kinkeeping" than men, staying in touch with and providing care and assistance to parents and in-laws

> **True** **10.** Married men and women are less likely than their never-married or divorced counterparts to have frequent contact with their aging parents.

(Sarkisian and Gerstel 2008). Elder caregiving seems to affect men and women differently, with women reporting greater distress, greater decline in happiness, more hostility, less autonomy, and more depression from caregiving than do men (Marks, Lambert, and Choi 2002). Interestingly, when caring for a parent out of the household, many women feel a *caregiver gain*, a greater sense of purpose in life than might be felt by noncaregiving women (Marks, Lambert, and Choi 2002).

Having and raising children are among the most fulfilling and satisfying activities in which most people eventually take part. They are also among the most frustrating and stressful. Some of the stress and frustration is inevitable given the breadth, depth, and length of the commitment one makes when raising children. However, in keeping with one of the wider themes that underlie much of this textbook, a good portion of the stresses of parenting results from the wider social context in which families live and the need to balance or juggle parenting with other roles that one plays, paid and unpaid work roles. Chapter 11 turns to those issues.

## Summary

- There were just under 4 million births in the United States in 2014, the first increase in percentage since 2007, when a record number of 4.3 million births were recorded. Birth and fertility rates vary by the race and ethnicity, age, education, income, and marital status of mothers.

- Approximately 20 percent of 40- to 44-year-old women have not had children, either because they couldn't or chose not to become mothers.

- The birthrate among teenagers dropped 10 percent between 2012 and 2013, and 57 percent since 1991. The rate has declined among both non-Hispanic white and black teens as well as among Hispanics and Asian Americans. There were nearly 6.4 million pregnancies in 2009, the outcomes of which varied according to mothers' marital status and race. More than a third of pregnancies are unplanned, being either mistimed or unwanted. Babies born from unplanned pregnancies face greater health and social risks, and their mothers have greater risks of postpartum depression.

- Critics, including especially feminist critics, have alleged that the *medicalization of childbirth*—making this natural process into a medical "problem"—has caused an overdependence on medical professionals, an excessive use of technology, and an alienation of women from their bodies and feelings. Research with new mothers documented mixed reactions to the treatment they received, the care they were given, and the experiences of childbirth.

- Miscarriages, which typically occur during the first trimester of pregnancy, are the most common form of pregnancy loss. Infant mortality rates in the United States are higher than in other industrialized nations and vary by race and ethnicity.

- According to the U.S. Census Bureau, more than 1.8 million children under age 18 were adopted in the United States in 2008, 2 percent of all children living with a parent. The trend has been toward open adoption, where there is contact between the adoptive parents and the birth mother or parents during pregnancy and/or continuing after birth.

- The transition to parenthood is unlike other role transitions. It is irreversible and sudden, and it comes with little preparation. Reduced sexual desire and depression during the *postpartum period* are among the potential problematic reactions to childbirth.

- Although there is no concrete evidence of a biological maternal drive, it is clear that socialization for motherhood, as well as an *ideology of intensive mothering*, affect women's expectations and intentions to have children.

- The traditional expectations of fathers are being supplemented and perhaps supplanted by enlarged and more expressive ones. This may be truer of our beliefs about fathers (the *culture of fatherhood*) than of fathers' real behavior (the *conduct of fathers*).

- Parenthood affects many areas of men's and women's lives, including their social activities, intergenerational family ties, occupational behavior, and free time. Research indicates that marital satisfaction declines when couples become parents, though it also declines among nonparents.

- Most hands-on child care is done by mothers. Fathers are less engaged with and less accessible to their children than are mothers.

- Parents differ in terms of their styles of parenting. Four styles are *authoritarian, permissive* (or *indulgent*), *authoritative,* and *uninvolved*. Of these, most research portrays the authoritative as most effective.

- Parents' marital status, ethnicity, and sexuality all influence parenting and child socialization. Economic, cultural, and political institutions have neglected to adopt policies that would allow parents and children deeper and more frequent contact with each other.

- Most gay and lesbian parents are or have been married. Studies indicate that children of both lesbians and gay men are as well adjusted as children of heterosexual parents.

- Increasingly, older parents provide financial and emotional support to their adult children; they often take active roles in child care and housekeeping for their daughters who are single parents.

- When parents are chronically ill, chemically dependent, mentally ill, or incapacitated after a divorce or widowhood, children often become caregivers to their parents, a process known as *parentification*.

- Family caregiving activities often begin when an a*ged parent* becomes infirm or dependent. Caregiving responsibilities may create conflicts involving previous unresolved problems, the caregiver's inability to accept the parent's dependence, conflicting loyalties, resentment, anger, and money or inheritance conflicts.

## Key Terms

authoritarian child rearing  390
authoritative child rearing  390
child-free  360
conduct of fatherhood  384
culture of fatherhood  384
filial responsibility  403
generativity  386
ideology of intensive mothering  381
impaired fecundity  360
indulgent child rearing  390
infant mortality  372
medicalization of childbirth  370
nonparental households  398
open adoption  376
parentified children  403
permissive child rearing  390
postpartum period  376
sandwich generation  403
sudden infant death syndrome (SIDS)  373
uninvolved parenting  390

# 11 Marriage, Work, and Economics

Regine Mahaux/Photographer's Choice/Getty Images

*Are the following statements True or False? You may be surprised by the answers as you read this chapter.*

**T  F  1.** A majority of couples with children are dual-earner couples.

**T  F  2.** The United States is among the most generous countries in the world in terms of offering workers time off with pay for vacations and holidays.

**T  F  3.** Work–family tensions are greater for employed parents than for employed women and men without children.

**T  F  4.** Because the cultural expectations of fathers have broadened, fathers in two-earner households experience more work–family role conflict.

**T  F  5.** In most families with preschool-age children, fathers are the sole wage earners.

**T  F  6.** Compared to young men, young women today place as much or more importance on achieving career success.

**T  F  7.** The biggest factor determining how much housework and child care men do is their gender ideology, what they think they *ought to do.*

**T  F  8.** Couples who work different shifts have more satisfying and stable marriages.

**T  F  9.** Women in the United States currently make 90 cents for every dollar that men earn.

**T  F  10.** When companies offer policies to reduce work–family conflict, most women and men make full use of them.

# Chapter Outline

Imagine yourself at a party put on by your school's alumni association. As you float around the room, trying to meet and mingle with some people who graduated in recent years, you overhear the following exchanges among some of the other guests. Each snippet of conversation illustrates some unspoken assumptions people have about work and family. Can you recognize the assumptions and identify what is wrong in each exchange?

- *Exchange No. 1.* A trio of women is in a corner. "What do you do?" one of the women inquires politely while being introduced by a second woman to the third. "I don't work; I'm a stay-at-home mom," the third responds. "Oh, that's . . . nice," the first woman replies, seeming to lose interest and turning toward the woman handling the introduction.

- *Exchange No. 2.* A bearded man is talking to a couple. "So, what do you two do?" the man asks. "I'm a doctor," the woman responds as she picks up her son, who is impatiently tugging on her. "And I'm an accountant, well, I was, but like a lot of other people, I was laid off. So, in the meantime, I'm home with them," her husband says, nodding toward their fidgeting son while feeding their second child with a bottle.

Although they are subtle, we can observe the following assumptions being made and attitudes being displayed. In the first exchange, both women ignore that the woman who identified herself as a stay-at-home mother does, indeed, work. Despite being unpaid, her caregiving activities do in fact constitute work. They also appear to devalue such unpaid work in comparison with paid work. Finally, those very undervalued activities are essential components of her family's economic stability, as her performance of them allows her spouse to invest more time and energy into paid work and wage earning.

In the second exchange, the woman identifies herself as a physician without acknowledging that she is also a parent. Her husband defines himself in terms of a job he no longer has while downplaying his role as a full-time caregiver for their children. In fact, as husband and wife, father and mother, both the physician and the out-of-work accountant are unpaid family workers making important—but generally unrecognized—contributions to the family's economy.

Because "family work" is unpaid and perhaps because it is done mostly by women, family work is often ignored and looked on as less important than paid work, regardless of how difficult, time consuming, creative, rewarding, and essential it is for our lives and future as humans. This is not surprising because, despite claims to the contrary, employment in the United States tends to take precedence over family.

Understanding the role of work in families may require reconsidering how one defines "work" and what one thinks families do. People ordinarily think of families in terms of relationships and feelings—the family as an emotional unit. But families are also economic units that

Often, when we think about "work," we fail to include the domestic work and child care that families need.

Myrleen Pearson/PhotoEdit

happen to be bound by emotional ties (Ross, Mirowsky, and Goldsteen 1991). Both paid work and unpaid family work, as well as the economy itself, profoundly affect the way we live in and as families. Our most intimate relationships vary according to how we participate in, divide, and share paid work and family work.

One's paid work helps shape the quality of family life: It affects time, roles, incomes, spending, leisure, and even individual identities. Whatever time individuals have for one another, for fun, for their children, and even for sex is time not taken up by paid work. Work regulates the family, and for most families, as in the past, a woman's work molds itself to her family, whereas a man's family molds itself to his work (Ross, Mirowsky, and Goldsteen 1991). Work roles and family roles must both be fulfilled despite the difficulties and complexity involved in meeting the demands they collectively impose. These facts are the focus of this chapter.

**True** **1. A majority of couples with children are dual-earner couples.**

## Workplace and Family Linkages

Outside of sleeping, the activity to which most employed men and women devote the most time is their jobs. Paid employment, or lack thereof, can and does affect family life in numerous ways.

## It's about Time

Of the 146.3 million Americans employed in February 2015 (U.S. Bureau of Labor Statistics 2015a), 118.7 million were employed full-time. Of the 27.6 million people employed part-time, 5.9 million were is such positions for economic reasons such as an inability to obtain full-time position or having their full-time position cut back to part-time by their employers. Another 20.2 million were employed part-time by choice (U.S. Bureau of Labor Statistics 2015b).

The 118.7 million women and men employed full-time worked an average of 42.5 hours per week, with men averaging close to three hours more per week than women, 43.6 to 40.9, among men and women 16 years of age and older (U.S. Bureau of Labor Statistics 2015c). Hours spent at work also vary by race and by marital status, as illustrated in Figure 11.1.

If one looks specifically at employment patterns among married couples with children, in 2014, 60 percent were dual-earner couples. Nearly a third of couples with children had fathers employed but mothers not employed, 6 percent had mothers employed but fathers not employed, and neither parent was employed in just over 3 percent of married couples with

Figure 11.1 Average Hours of Work per Week for Those Age 16 and Over Employed Full-Time, 2015

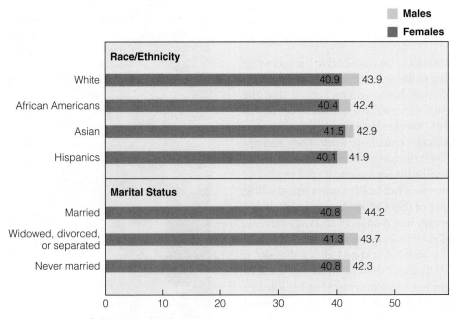

SOURCE: U.S. Bureau of Labor Statistics (2015c).

**TABLE 11.1** Employment Status of Parents among Married Couples with Children, by Age of Youngest Child, 2014 Annual Average

| | Children Under 18 | Children 6–17, None Younger | Youngest Child 6 or Under |
|---|---|---|---|
| **One or both parents** | | | |
| Employed | 96.6% | 96.3% | 96.9% |
| Both employed | 60.2% | 63.9% | 55.3% |
| Father only employed | 30.8% | 25.9% | 37.1% |
| Mother only employed | 5.6% | 6.6% | 4.4% |
| Neither parent employed | 3.4% | 3.7% | 3.1% |

SOURCE: U.S. Bureau of Labor Statistics (2015d).

children. Table 11.1 shows how this varied depending on the age of the youngest child.

The American Time Use Survey provides a more detailed picture of what mothers and fathers who are employed full-time do on an average day. The differences between mothers and fathers with regard to work and work-related activities and leisure and sports activities were very small. When one looks at housework and primary child care, the picture changes and larger differences emerge. Much larger percentages of employed mothers (74 percent) than employed fathers (55 percent) reported participating in child care activities on an "average day." Among those married, full-time workers with children under 18, mothers were also more likely than fathers to do housework (88 versus 66 percent).

Data reveal that, in contrast to declines throughout much of Europe, at least until recently, Americans were increasing their hours spent at work (Jacobs and Gerson 2004). Although U.S. workers and workers elsewhere may face similar "time dilemmas," the societal responses to these pressures are often vastly different. Jerry Jacobs and Kathleen Gerson (2004, 124) asserted the following:

Several European countries, especially those in northern Europe, have made sustained, highly publicized, and well-organized efforts to reduce working time as a strategy for reducing unemployment, increasing family time, and reducing gender inequalities in the market and at home.

One indicator of the different approaches to working time is evident in how countries approach time off from work. When it comes to vacations and paid holidays, "the United States is the only advanced economy in the world that does not guarantee its workers paid vacation" or holidays (Ray, Sanes, and Schmitt 2013).

Economists Rebecca Ray, Milla Sanes, and John Schmitt note that although European workers are guaranteed at least 20 paid vacation days and an average of 6 paid holidays, workers in the United States are not legally guaranteed any paid holidays (Ray, Sanes, and Schmitt 2013).

According to data gathered by Gallup in their 2013 and 2014 Work and Education polls, fully half of the 1,271 adults employed full-time that were interviewed say they work more than 40 hours "in a typical week." In fact, almost 40 percent reported that they worked 50 or more hours a week, with almost one out of five responding that they work 60 or more hours. Lydia Saad, a senior editor at Gallup, shows that since 2002, respondents to Gallup polls indicate that the average workweek has fluctuated only slightly, at between 45 to 47 hours, and only 42 percent of respondents who work full-time said they work a 40-hour work week (Saad 2014). These Gallup results are higher than those found in U.S. National Health Interview data, where in 2010, almost a fifth of working adults reported working 48 hour or more per week and over 7 percent worked 60 hours or more (www.cdc.gov /niosh/topics/workschedules/). Of course, it goes without saying that the more hours individuals work, the less time they have for their families, friends, communities, and themselves.

Overwork carries consequences. Table 11.2 reveals the comparison of those employed women and men who experience high levels of overwork to those who experience low levels.

It bears mentioning that not all workers experience high levels of overwork. Some categories of workers—for example, professionals, executives, managers, and small business owners—are more likely to experience the greatest time demands. Others are underemployed and would prefer to work more, while still others are unemployed and

**False** **2.** The United States is among the least generous countries in the world when it comes to vacations and holidays.

TABLE 11.2  Levels of Overwork

|  | High | Low |
|---|---|---|
| Feel very angry toward employers | 39% | 1% |
| Often/very often resent coworkers | 34% | 12% |
| Make a lot of mistakes at work | 20% | 0% |
| Feel highly stressed | 36% | 6% |
| Report their health as good | 52% | 65% |
| Feel like they take good care of themselves | 41% | 68% |

SOURCE: Galinsky et al. (2005).

would prefer any work at all (Jacobs and Gerson 2004; Perry-Jenkins, Repetti, and Crouter 2000). This **bifurcation of working time**, wherein some work longer and longer days and weeks while others work fewer hours than they need or want, is revealed by findings from the National Study of the Changing Workforce. Well before the recent recession in the U.S. economy, 60 percent of both men and women said that they would prefer to work less; however, 19.3 percent men and 18.5 percent of women (that's about one in five) said that they would prefer to work more hours than they were working at the time (Jacobs and Gerson 2004).

Time spent at one's job directly affects time spent with one's family.

Terry Vine/Stockbyte/Getty Images

## Time Strains

Whether a person loves, loathes, or merely learns to live with it, one's job structures the time that can be spent with one's family (Hochschild 1997). The time demands faced at work can create a feeling of **time strain**, in which individuals feel that they do not have or spend enough time in certain roles and relationships. This is especially true for parents, more than half of whom, according to a 2015 Pew Research Center report, say it is somewhat or very difficult to balance the responsibilities of their jobs with those of their families. Specifically, more than half (52 percent) of employed fathers and 60 percent of employed mothers feel this way (Pew Research Center 2015d). "Balance" refers here to when one's participation in both domains, paid work and family, is considered effective (Milkie et al. 2010).

Sociologists Kei Nomaguchi, Melissa Milkie, and Suzanne Bianchi (2005) found interesting gendered patterns in their investigation into the psychological effects of time strains:

- More fathers than mothers reported feeling that they did not have enough time with their children or their spouses. More mothers than fathers felt they had too little time for themselves.

- Life satisfaction was significantly reduced *for mothers but not for fathers* when they felt that they had or spent "too little time with children."

- Feelings of insufficient time with a spouse were associated with significantly higher levels of distress *for women but not for men*.

- Feelings of insufficient time for oneself were associated with reduced levels of family and life satisfaction and with increased feelings of distress *for men but not for women*.

- Fathers articulated feeling strained for time both with their spouses and their children, but these feelings did not affect them as much psychologically as they did women.

A decade later, survey respondents to a Pew Research Center survey of working parents echoed some of these the same patterns. Half of fathers employed full-time and 39 percent of full-time employed mothers said they spend too little time with their children. More than 40 percent of married or cohabiting mothers employed full-time report spending too little time with their partners. Almost 60 percent of full-time employed mothers and 53 percent of full-time employed fathers say they don't have enough leisure time—that is, time away from their children to get together with friends or to pursue hobbies (Pew Research Center 2015d).

Melissa Milkie and her colleagues (2010) found that parents' time with children was associated with feeling successful in having achieved "balance" between work and family, though more than quantity of time was important. Parents' own subjective assessments of whether they were spending the "right" amount of time with their children mattered for feeling successful at reaching a balance between the domains of work and family. Also, more parental time in *more interactive* activities, such as playing with, teaching, or helping children, gave parents more of a sense of "balancing" work and family, whereas time spent in routine care activities did not. This was especially true for mothers more than for fathers, and for parents without college degrees than for college-educated parents (Milkie et al. 2010).

Generally, despite the fact that mothers and fathers have increased the time they spend with children since the 1960s, their perceptions often don't match that reality. They may continue to feel squeezed for time and often perceive that they aren't spending enough time with their children. In the opposite direction, despite spending no greater amount of time with their children, when given greater schedule control of their work hours, a sample of employed mothers felt less of a sense that they weren't spending enough time with their children (Hill, Tranby, Kelly, and Moen 2013). Unfortunately, an estimated 17 percent of the workforce is employed in jobs with irregular, on-call, or rotating shifts. Thus, many employed parents find themselves in irregular and unpredictable work schedules in which they have little control and even little notice of what their work schedules will be for the coming week. Unsurprisingly, such circumstances heighten stress and work–family conflict, and have been found to affect children as well as parent–child relationships (Golden 2015; Morsy and Rothstein 2015a, 2015b).

Jennifer Keene-Reid and John Reynolds (2005) argue that workers feel more success in juggling work and family when they have some control in scheduling their time at work. Furthermore, because family demands and needs can and do arise unexpectedly, the ability of employed parents to adjust their schedules accordingly is a useful and important family-friendly benefit. Later in this chapter, we will look at workplace policies and benefits that could lessen the extent of time strains, work–family conflict, and resulting family stress.

## Work and Family Spillover

In addition to the time we have available to our families, paid work affects home life in other ways. We can call this **work spillover**—the effect that work has on individuals and families, absorbing their time and energy and impinging on their psychological states. It links our home lives to our workplace (Small and Riley 1990). Work is as much a part of our marriages and home lives as love is. What happens at work—frustration or worry, a rude customer, an unreasonable boss, or inattentive students—has the potential to affect our moods, perhaps making us irritable or depressed. Often, we take such moods home with us, affecting the emotional quality of our relationships.

Not surprisingly, research demonstrates that work-induced energy depletion, fatigue, or, in more extreme cases, exhaustion can affect the quality of our family relationships. Fatigue and exhaustion can make us angry, anxious, less cheerful, and more likely to complain and can cause us to experience more difficulty interacting and communicating in positive ways. Yet, according to one study, although both stress and exhaustion from work affect marital relationships, "stress is far more toxic" (Roberts and Levenson 2001, 1065). These researchers suggest that although common, job stress can seriously and negatively affect marital happiness, creating dynamics that may even contribute to divorce when left unchecked.

Scholars have increasingly looked at how and how often negative spillover affects us. Although negative work spillover occurs neither every day nor to everyone, it is accurate to consider it fairly common (Roehling, Jarvis, and Swope 2005). This is revealed in Figure 11.2, based on data from the 1997 National Study of the Changing Workforce.

In general, women report more negative spillover than do men, though college-educated men are the most likely to indicate that work interferes with their family lives (Ammons and Kelly 2008; Sloan Work and Family Research Network 2008). Such work–family tensions are greater for employed parents than they are for employed women and men without children. Furthermore, the effects seem to be greater on mothers than on fathers, just as the differences

> **True**
> **3.** Work–family tensions are greater for employed parents than for employed women and men without children.

**Figure 11.2** Work-to-Family Spillover

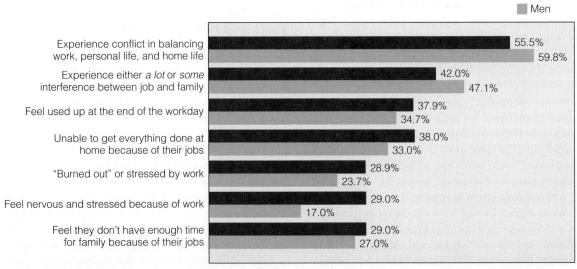

SOURCE: Jacobs and Gerson (2004, 85).

between parents and nonparents are greater among women than among men. Studying young adults ages 21 to 31 who were part of the Youth Development Study in St. Paul, Minnesota, sociologists Samantha Ammons and Erin Kelly report that although the percentages themselves are small, twice as many young women as young men reported that work affected their childbearing decisions (7 percent of women versus 3.6 percent of men). Among those with children, mothers were 1.5 times more likely than were fathers to say that work interfered with their relationships with their children (23 percent versus 14 percent) (Ammons and Kelly 2008).

Sociologists Jerry Jacobs and Kathleen Gerson note that children's ages make little difference in parents' experiences of work–family stress (Jacobs and Gerson 2004). Workplace stress often causes us to focus on our problems at work rather than on our families, even when we are home with our families. It can lead to fatigue, stomach ailments, and poorer health as well as depression, anxiety, increased drug use, and problem drinking (Crouter and Manke 1994; Roehling, Jarvis, and Swope 2005).

### Family-to-Work Spillover

As many employed parents can attest, the relationship between paid work and family life cuts both ways. The emotional climate in our homes can affect our morale and performance in our jobs. Positively, family can help alleviate some workplace stress. More research has focused on how the demands of our home lives may impinge on our concentration, energy, or availability at work (Jacobs and Gerson 2004). Similarly, negative marital interactions have been found to be associated with lower work satisfaction (Sandberg et al. 2013).

The research by Samantha Ammons and Erin Kelly (2008) showed family-to-work spillover to be an even more common problem than work-to-family spillover. Furthermore, young adults with less than college educations were more likely to report experiencing family-to-work spillover than were those with BA degrees or higher (50 percent versus 34 percent). Women reported higher levels of family-to-work spillover, with large differences between women with (41 percent) and without (62 percent) college degrees. Because women more often than men face or anticipate the intrusion of their family responsibilities into their work lives, they are forced to make more work-related adjustments because of family needs (Keene-Reid and Reynolds 2005). Additionally, higher levels of family-to-work spillover have been found among parents compared to nonparents (Roehling, Jarvis, and Swope 2005).

Meeting family demands such as assuming more household and child care responsibility often comes with hidden or unanticipated work-related financial costs. Regardless of gender, those who carry responsibility for traditionally female housework chores are likely to suffer reduced wages. This is probably the result of having less effort and energy available to spend on paid work activities (Noonan 2001).

## Role Conflict, Role Strain, and Role Overload

Two-parent families in which both partners are employed face more severe work-related problems than do nonparents. Being an employed parent usually means performing at least three demanding roles simultaneously: worker, parent, and spouse or partner. In juggling these roles, we might experience role conflict, role strain, role overload, or a combination of these.

When the multiple social statuses or positions that we occupy (e.g., spouse, parent, and worker) present us with competing, contradictory, or simultaneous role expectations, we experience **role conflict.** In such instances, the successful performance of either role is made difficult *because of* the demands of the other role (Nomaguchi 2009). When the role demands attached to any particular status (e.g., mother, husband, or employee) are contradictory or incompatible, we experience **role strain**. Finally, when the various roles we play require us to do more than we can comfortably or adequately handle or when we feel that we have so much to do that we will never "catch up" or have enough time for ourselves, we experience **role overload** (Crouter et al. 2001; MacDermid 2006).

In the specific case of family and paid work roles, when we feel torn between spending time with our spouses or children and finishing work-related tasks, we experience role conflict. We cannot be in two places at once. Fathers in two-earner households experienced an increase in work–family role conflict in recent decades as the cultural expectations of fathers broadened to include sharing more of the domestic responsibilities and spending greater amounts of time with one's children. Fathers and mothers both have set higher standards of how much time they should spend with their children. Meanwhile, at the same time, demands at work remain high and the pace and activities of family life have become faster and busier. All of these likely contribute to a sense of increased work–family role conflict (Nomaguchi 2009).

As far as role strain is concerned, men who see themselves as traditional providers may experience role strain when pressed into higher levels of housework or child care. Employed wives, exhausted by their combination of paid work, housework, and child care,

> **True**
> **4.** Because the cultural expectations of fathers have broadened, fathers in two-earner households experience more work–family role conflict.

may also experience role strain, finding less time and energy for enjoying sexual intimacy with their spouses.

Professor of family studies and human development Ann Crouter and colleagues (2001) found that both fathers' role overload and the amount of hours they worked affected the quality of their relationships with their adolescent children. When fathers worked long hours but did not experience overload, their relationships with their adolescents did not seem to suffer. It appears that for fathers and children, the combination of hours and overload have the greatest effect (Crouter et al. 2001).

Parental "availability" to children is affected by the levels of stress that parents experience. Particularly stressful days at work may be followed by parents being withdrawn at home. This may sometimes prove beneficial because, by withdrawing, less negative emotion is brought into the relationships (Perry-Jenkins, Repetti, and Crouter 2000).

However, the more important sources of role conflict and overload are not within the person but rather *within the person's role responsibilities*. Men experience role conflict when trying to balance their family and work roles. Because men are expected to give priority to their jobs over their families, it is not easy for men to be as involved in their families as they may like. A study examining role conflict among men (O'Neil and Greenberger 1994; see also Greenberger 1994; Marks 1994) found that men with the least role conflict fell into two groups: One group consisted of men who were highly committed to both work and family roles. They were determined to succeed at both. The other group consisted of men who put their family commitments above their job commitments. They were willing to work at less demanding or more flexible jobs, spend less time at work, and put their family needs first. In both instances, however, the men received strong encouragement and support from their spouses.

The various issues surrounding spillover, role conflict, role overload, and role strain vary depending on the household structure and division of labor. Single-parent households with full-time working parents are easily susceptible to role overload and role conflict. Two-parent, dual-earner households also face versions of work-to-family spillover different from those of households with one provider and a partner at home full-time.

Comparing levels of expressed work–family interference from two large survey sources, the Quality of

Employment Survey in 1977 and the National Study of the Changing Workforce in 1997, sociologist Sarah Winslow (2005) offered the following conclusions about work–family conflict. Compared with respondents in 1977, respondents in 1997 reported greater difficulty balancing work and family. This was greatest among parents regardless of whether they were in dual-earner or single-earner households. In addition, women and men reported similar levels of work–family interference.

### Crossover

Yet another aspect of the work–family relationship is conveyed by the concept of **crossover**. Crossover refers specifically to the effects of one's job-related feelings on one's spouse or intimate partner. It is therefore different from though related to the previously discussed idea of *spillover*. Spillover occurs when work comes home with the worker, affecting the quality and quantity of his or her interactions with family members. It is a within-person/between domains (e.g., home and workplace) phenomenon. *Crossover* occurs when a worker's work-related stress (negative) or engagement (positive) affects one's partner in similar ways. According to the *spillover-crossover model* of organizational psychologist Arnold Bakker and colleagues (2008, 2009), workplace experiences first spill over into the household and then cross over to one's partner. *Direct crossover* occurs through the empathy of one spouse for what the other is experiencing. When one spouse feels a lot of pressure or overload at work, the other spouse may begin to feel depressed or overloaded as well. *Indirect crossover* results more from the conflict between spouses that might result from negative spillover. Job demands can create work–family conflict, which then leads to marital conflict from which one's spouse's well-being may then suffer.

Anxiety, stress, depression, and burnout can cross over, as can engagement, vigor, enthusiasm, life satisfaction, and well-being (Bakker, Westman, and van Emmerik 2009; Demerouti, Bakker, and Schaufeli 2005). Examining a Netherlands sample of dual-earner couples with children, psychologists Evangelia Demerouti, Arnold Bakker, and Wilmar Schaufeli (2005) found that wives' job-induced exhaustion crossed over to their husbands, leading to husbands' exhaustion, while husbands' job-derived life satisfaction crossed over to wives' life satisfaction. Mina Westman and colleagues found that the unemployment of a spouse can also lead to crossover, as the anxiety created by unemployment felt by the out-of-work spouse crosses over to the other, still-employed spouse

Crossover refers to when one's emotional state comes home with one from work and affects one's spouse's or partner's mood.

The Image Bank/Getty Images

(Westman, Etzion, and Horovitz 2004). Less clear is how much "crosses over" from parents to children or whether parent–child relationships are affected in similar ways as marriages.

## The Familial Division of Labor

Families divide their labor in a number of ways. Some follow more traditional male–female patterns, the majority share wage earning, and a small number reverse roles. Even within a single family, there will likely be a number of divisions of labor over time as family members move through various life stages.

### The Traditional Pattern

In what we often consider the "traditional" division of labor in the family, work roles are complementary: The husband is expected to work outside the

**Industrialization "Creates" the Traditional Family**

I n the 19th century, industrialization transformed the face of the United States. It also transformed U.S. families from self-sufficient farm families to wage-earning urban families. As factories began producing farm machinery such as harvesters, combines, and tractors, significantly fewer farmworkers were needed. Workers migrated to the cities, where they found employment in the ever-expanding factories and businesses.

Because goods were now bought rather than made in the home, the family began to shift from being primarily a production unit to being a consumer and service-oriented unit. With this shift, a radically new division of labor arose in the family. Men began working outside the home in factories or offices for wages to purchase the family's necessities and other goods. Men became identified as the family's sole providers or "breadwinners." Their work began to be identified as

"real" work and was given higher status than women's work because it was paid in wages.

Industrialization also created the housewife, the woman who remained at home attending to household duties and caring for children. Because industrialization meant that much of what the family needed had to be purchased with the husband's earnings, the wife's contribution in terms of unpaid work and services went unrecognized, much as it continues today (Ferree 1991).

In earlier times, the necessities of family-centered work gave marriage and family a strong center based on economic need. The emotional qualities of a marriage mattered little as long as the marriage produced an effective working partnership. Without its productive center, however, the family focused on the relationships between husband and wife and between parent and child. Affection, love, and emotion became the defining qualities of a good marriage (Coontz 2005). ●

home for wages, and the wife is expected to remain at home caring for children and maintaining the household. In this pattern, a man's family role is satisfied by his performance of the provider role, whereas even when employed outside the home, a woman's employment role is secondary to her family role (Blair 1993). As the next Exploring Diversity feature reveals, this pattern was a product of industrialization.

In 2014, among married-couple families with children, fathers were the sole wage earners in the following percentages based on the age of children:

**Married-Couple Families with Males as Sole Wage Earners**
Families with children under 18: 30.8%
Families with children 6 to 17 years old: 25.9%
Families with children under 6: 37.1%

Overall, then, the traditional division of labor is evident in one-fourth of married-couple families with school-age children and less than 40 percent of such families with preschool-age children. However, this does not mean that less than 40 percent of families with young children desire the traditional pattern, nor does it mean that they prefer this pattern

**False** **5.** Fathers are the sole wage earners in less than half of families with preschool-age children.

(U.S. Bureau of Labor Statistics 2015d, Table 4). Sociologist Arlie Hochschild's classic study, *The Second Shift*, illustrated that the actual division of roles and responsibilities for some couples may not reflect their preferred role arrangement or their gender ideologies. One cannot assume that what couples do is an expression of their **gender ideologies**, their beliefs about what women and men *ought* to do (Hochschild 1989). Some male breadwinner–female homemaker heterosexual couples might possess more egalitarian attitudes than their household arrangement suggests. Many two-earner couples might desire an arrangement more in keeping with the traditional pattern but be unable to afford it financially.

Whether a reflection of choice or circumstance, the difference in primary roles between men and women in traditional households profoundly affects the most basic family tasks, such as who cleans the toilet, mops the floors, does the laundry or grocery shopping, and mows the lawn. Whether or not they are employed outside the home, women remain primarily responsible for most day-to-day household tasks, though of the previously mentioned tasks, men are most likely to participate in and bear primary responsibility for mowing the lawn.

The division of family roles along stereotypical gender lines varies by race and class. It is more characteristic of Caucasian families than of African American families. African American women, for example, are less likely than Caucasian women to be exclusively responsible for household tasks. Latino and Asian families are more likely to be closer to the traditional than are African Americans or Caucasians (Rubin 1994).

Class differences are somewhat ambiguous. Among middle-class couples, greater ideological weight is given to sharing and fairness. Working-class couples, although less ideologically traditional than in the past, are still not as openly enthusiastic about more egalitarian divisions of labor. However, in terms of *who does what*, working-class families are more likely than middle-class families to piece together work-shift arrangements that allow parents to take turns caring for the children and working outside the home. Such "opposite-shift" arrangements may force couples to depart from tradition, even if they neither believe they should nor boast that they do.

## Men's Traditional Family Work

The husband's traditional role as provider has been perhaps his most fundamental role in marriage. As sociologists Barbara Arrighi and David Maume (2000, 470) put it, "It is the activity in which they spend most of their time and depend on most for their identity." In the traditional equation, as long as the male was a good provider, he was considered a good husband and a good father (Bernard 1981). Conversely, if a man failed as a provider, in his eyes as well as the eyes of others (including his family), he was not living up to expectations held for husbands and fathers.

In the traditional model, men are expected to contribute to family work by providing household maintenance. Such maintenance consists primarily of repairs, light construction, mowing the lawn, and other activities consistent with instrumental male norms. Although men often and increasingly contribute to housework and child care, their contribution may not be notable in terms of the total amount of work to be done. Men, especially those in traditional households, tend to see their role in housekeeping or child care as "helping" their partner, not as assuming equal responsibility for such work. Husbands become more equal partners in family work when they, their wives, or both have egalitarian views of family work or when such a role

is pressed on them by either circumstantial necessity or ultimatum (Greenstein 1996; Hochschild 1989). Men who believe that they should act as traditional providers may resist performing more housework or do so only when necessary (though perhaps reluctantly), whether or not their wives are employed outside the home. If both spouses share a traditional gender ideology (traditional beliefs about what each should contribute to paid and family work), men's low level of household participation is not problematic.

Looking at marriages in which spouses are "mutually dependent"—with wives earning between 40 percent and 59 percent of the family income—such couples increased nearly 300 percent between 1970 and 2001. As many as 30 percent of dual-earner couples and 20 percent of all married couples fit such a pattern. In 29 percent of dual-earner couples in 2013, wives outearned their husbands. If one includes couples where wives are sole wage earners along with those where both spouses are employed, in 2013, 38 percent of married women "outearned" their husbands (U.S. Bureau of Labor Statistics, 2015g).

## Women's Traditional Family Work

Although the majority of women now earn salaries as paid employees, contributing more than 40 percent of family income in dual-earner households, neither traditional women nor their partners regard employment as a woman's fundamental role (Coontz 1997). For those with traditional gender ideologies, women are not duty bound to provide, though their incomes are likely to be essential in maintaining a livable lifestyle. They are, however, expected to perform household tasks (Thompson and Walker 1991).

No matter what kind of work the woman does outside the home, there is seldom equality when it comes to housework. Although this may cause tension and distress among more egalitarian-minded couples, traditional couples do not expect housework to be divided equally. Women's family work is considerably more diverse than that of men, permeating every aspect of the family. It ranges from housekeeping to child care, maintaining kin relationships to organizing recreation, socializing children to caring for aged parents, and cooking to managing the family finances. Ironically, family work is often invisible, even to the women who do most of it (Brayfield 1992).

Sociologist Ann Oakley (1985) described four primary aspects of the **homemaker role**:

- Exclusive allocation to women rather than to adults of both sexes
- Association with economic dependence
- Status as nonwork, which is distinct from "real," economically productive paid employment
- Primacy to women—that is, having priority over other women's roles

Many stay-at-home wives are dissatisfied with housework, perceiving it to be routine, monotonous, unpleasant, unappreciated, and unstimulating. However, such dissatisfaction doesn't necessarily mean that they are dissatisfied with all aspects of the stay-at-home role, which offers them some degree of autonomy. Young women, for example, may find increasing pleasure as they experience a sense of mastery over cooking, entertaining, or rearing happy children. If homemakers have formed a network among other women—such as friends, neighbors, or relatives—they may share many of their responsibilities. They discuss ideas and feelings and give one another support. They may share tasks as well as problems.

Full-time male houseworkers may call themselves *househusbands*, but they are less likely to do so than full-time female homemakers are to call themselves *housewives*. Instead, they may identify themselves as retired, unemployed, laid off, or disabled (Bird and Ross 1993). Increasingly, if they have children at home, they may call themselves "stay-at-home dads" (Smith 2009).

## ◼ Women in the Labor Force

Women have always worked outside the home. Like many of today's families, early American families were **co-provider families**—families that were economic partnerships dependent on the efforts of both the husband and the wife. Although women may have lacked the economic rights that men enjoyed, they worked with or alongside men in the tasks necessary for family survival (Coontz 1997). Beginning in the early 19th century, "work" and "family work" were separated. Men were assigned the responsibility for the wage-earning labor that increasingly occurred away from the home in factories and other centralized workplaces.

Women stayed within the home, tending to household tasks and child rearing. But this gendered division of labor was never total. Single women have traditionally been members of the paid labor force. There have also been large numbers of employed mothers, especially among lower-income and working-class families, African Americans, and many other ethnic minorities. Economic and other societal circumstances (e.g., World Wars I and II) also influence the numbers of employed mothers.

In 2014, of the 128.2 million women 16 years of age or older, 73.3 million (57.3 percent) were in the labor force, and 68.6 million (53.6 percent) were employed. In comparison, 82.2 million, or 68.8 percent of the over 119 million males 16 and over were in the labor force, and 63.8 percent of males age 16 and over were employed (U.S. Bureau of Labor Statistics 2015e).

Data for 2014 reveal that among those age 16 and older, African American women were more likely to be in the labor force (59.2 percent) than were Caucasian (56.7 percent), Hispanic (56.0 percent) and Asian (55.8 percent) women. However, they were less likely to be employed (52.9 percent) than were white women (53.7 percent) or Asian women (53.3 percent) (U.S. Bureau of Labor Statistics 2015f).

The most dramatic changes in women's labor force participation have occurred since 1960, resulting in the emergence of a family model in which both husbands and wives are employed outside the home. Between 1960 and 2013, the percentage of married women in the labor force rose from 32 percent to almost 59 percent. Nearly 64 percent of married women with spouses present were employed. Among married women, almost 75 percent of those with children under 18 were employed in 2013. There is some variation in maternal labor force participation and employment based on children's ages and mothers' marital status. The 2013 rates of maternal labor force participation (i.e., meaning that one was employed or actively seeking employment) were as follows:

- Among married mothers of 6- to 17-year-olds, 72.8 percent participated in the labor force.
- Among mothers whose youngest child was age 6 or younger, 64.7 percent were in the labor force.
- Among mothers of children under 3, 62.1 percent were in the labor force.
- As to employment, 69.6 percent of married mothers of children 6 to 17 years old, 58.6 percent of those with children under 6, and almost 56.3 percent of those with children under 3 years of age were employed.

- Of unmarried mothers (including never-married, divorced, separated, or widowed mothers) whose youngest child was under 18 years of age, 75.4 percent were in the labor force in 2013. Nearly two-thirds of unmarried mothers (65.4 percent) were employed (U.S. Bureau of Labor Statistics 2014b).

In 2013, wives' incomes exceeded husbands' incomes in 29 percent of dual-earner families, up from 16 percent in 1981. If one also adds in those married households where men have no earnings, close to 40 percent of married women were the major or sole wage earners (U.S. Bureau of Labor Statistics 2015g). In 60.2 percent of married couples with children in 2014, both parents were employed. In just under 6 percent of all married-couple families, women were sole wage earners compared to almost 31 percent in which men were the only ones who were employed. In 2014, there were 8.6 million families with children under 18 headed by single women. In over two-thirds of those households, mothers were employed. In over 60 percent of the mother-headed families with preschool-age children, mothers were employed. This amounted to slightly more than two million single mothers with children under 6 years of age (U.S. Bureau of Labor Statistics 2015d).

Couples in which wives outearn husbands have been found to experience certain negative outcomes, including lower levels of happiness, more marital conflict, and a higher likelihood of divorce. In addition, the gap between husbands' and wives' housework participation is greater when the wife outearns her husband (Bertrand, Kamenica, and Pan 2015). It has also been found that more economically dependent men, like men who earn significantly more than their wives, were more likely to cheat on their spouses, whereas economically dependent wives were less likely to cheat (Munsch 2015).

## Why Did Women's Employment Increase?

Looking at the wider societal and cultural context, a number of social and economic trends contributed to the steady increase in women's employment that occurred from 1960 to 2000 (Cotter, England, and Hermsen 2007). What accounts for this increase? Sociologists David Cotter, Paula England, and Joan Hermsen point to each of the following:

- Increases in the numbers of single mothers, resulting from increasing divorce rates and births to unmarried women

- Increases in women's educational attainment
- Pro-employment messages and the equal opportunity emphasis of the women's movement
- Better job opportunities for women
- Decline in men's wages and the reduced ability to support families on one income

As a result of these factors, more women sought, valued, and needed employment, and more opportunities awaited women in the paid labor force. After the 1970s, social norms changed, making it more than acceptable for mothers to hold a job. In fact, one might contend that by the social norms of today, women are expected to be employed outside the home and to return to employment should they become mothers and leave to have and raise children.

As far as individual women's motives for seeking employment, at least a couple are worth noting. Among the factors that lead women to seek employment, the most prominent factor is the same as it is for men—economic need. For unmarried women and single mothers, their income from employment is likely to be the only income on which they live. The incomes of married women may be primary or secondary to their husband's income, but couples increasingly depend on and require income from both partners to achieve and maintain a decent if not desirable standard of living.

Economic pressures traditionally have been powerful influences on African American women. Many married women and/or mothers of all races entered into the labor force or increased working hours to compensate for their husband's loss in earning power because of inflation or unemployment. In addition, the social status of a husband's employment often influences the level of employment a

**Women seek the same gratifications from paid work that men seek. These include, but go beyond, wages.**

Kzenon/Shutterstock.com

wife chooses (Smits, Ultee, and Lammers 1996).

Yet as is also true of men, women have other reasons for working outside the home. Among the psychological reasons for employment are an increase in a woman's self-esteem and sense of control. Employment and wage earning can enhance a woman's stature in her household, adding greater weight to her need for more involvement from her spouse or partner in household labor and child care.

Additionally, employed women may find social support, recognition, and appreciation at work that they don't get or feel at home (Hochschild 1997). As a consequence, married women with children have more positive and less negative emotional experiences at work. In contrast, men report feeling more positive and less negative at home, thus creating a situation wherein spouses live in "different realities" and have different emotional reactions to being at work and being at home (Wilhelm and Perrez 2004).

## Attitudes of and about Employed Women

The wider social and cultural context surrounding women's employment has changed in the direction of greater acceptance and support. Survey data gathered by the Pew Research Center indicates that a majority of the population perceives wider female employment in a positive light and deems dual-earner marriages as more satisfying than traditional male breadwinner–female homemaker marriages. Roughly three-quarters of those surveyed responded that the trend toward wider female employment has been for the better, and more than 60 percent identify marriages where both spouses work outside and share the tasks of housework and child care as more attractive (Pew Research Center 2009). At the same time, attitudes about women's employment are somewhat mixed. In a 2013 Pew survey, two-thirds of the 1,003 adults surveyed said that the increasing numbers of women employed outside the home had made it easier (versus 28 percent saying harder) for families to earn enough to live comfortably. However, 50 percent of respondents felt such increases make it harder to make marriages successful, as opposed to 35 percent who thought such increases made it easier. Almost three-fourths of respondents (74 percent) thought increases in women's employment made it harder for parents to raise children, while only 19 percent thought the increases made it easier. In the same survey, 51 percent of respondents

**True** **6.** Compared to young men, young women today place as much or more importance on achieving career success.

said that children are better off if their mother is home and doesn't hold a job, as compared to 34 percent saying children are just as well off if their mother works (and 13 percent saying it depends on circumstances). Only 8 percent said that children are better off if their father is at home and doesn't hold a job, whereas 76 percent said children are just as well off if their father works (Wang, Parker, and Taylor 2013).

There is also evidence, as shown in Table 11.3, that young women now place greater importance on achieving career success than did young women 15 years ago. Perhaps more interesting, now young women even more than young men report that achieving success in a career is very important in their lives.

As also shown in Table 11.3, the attitudes of older women changed much more than the attitudes of older men, reaching a point where, among 35- to 64-year-olds, the genders were equally likely to put high priority on career success.

However, there is also evidence that people continue to perceive financial success and being a good provider in somewhat gendered ways. More than twice as many people consider providing a good income to be a very important quality in making a man a good spouse or partner (41 percent) as thought it to be important in making a woman a good spouse or partner (19 percent) (Pew Research Center 2010). Two-thirds say that, to be ready to marry, a man should be able to provide financially. Only a third say the same about women.

Race appears to make a difference in such attitudes; 35 percent of whites responded that the ability to provide a good income is a very important quality in making a man a good spouse or partner, compared to 67 percent of blacks and 50 percent of Hispanics. As indicated in Table 11.4, African Americans and

**TABLE 11.3** Percentage of Women and Men Saying Success in a High-Paying Career is "One of the Most Important Things" or Very Important in Their Lives

| Year | 18- to 34-year-olds | | 35- to 64-year-olds | |
| --- | --- | --- | --- | --- |
| | % Women | % Men | % Women | % Men |
| 1997 | 56% | 58% | 26% | 41% |
| 2010–2011 | 66% | 59% | 42% | 43% |

SOURCE: Patten and Parker (2011).

**TABLE 11.4** Importance of Ability to Provide Financially in Readiness to Marry

|  | White | African American | Hispanic |
|---|---|---|---|
| Importance for a man | 62% | 88% | 77% |
| Importance for a woman | 28% | 50% | 47% |

SOURCE: Pew Social and Demographic Trends, "The Decline of Marriage and Rise of New Families," Pew Research Center (2010).

Hispanics attach greater importance than do whites to the ability to be able to provide financially as a measure of both genders' readiness to marry. Like whites, they attach substantially more importance to a man's than to a woman's ability to provide financially.

## Women's Employment Patterns

The employment of women has generally followed a pattern that reflects their family and child care responsibilities. Because of the family demands they face, women must consider the number of hours they can work, what time of day to work, and whether adequate child care is available. Traditionally, women's employment rates dropped during their prime childbearing years from 20 to 34 years. But this is no longer true; a significant number of women with children are in the labor force regardless of age of child, marital status, and racial or ethnic affiliation.

Women no longer automatically leave the job market when they become mothers. Among first-time mothers, more than half return to their jobs within six months of giving birth, and two-thirds have returned by the time their child celebrates his or her first birthday. Looking only at women who worked during their pregnancies, close to 60 percent returned to work within three months of giving birth, almost three-fourths returned within six months, and 79 percent returned within 12 months. In other words, of those women who worked at their jobs while pregnant, just over 20 percent stayed at home for the entire first year of motherhood. Among women who did not work during their pregnancies, a quarter of women were employed at six months and a third within a year of giving birth (Laughlin 2011). For those who returned to work for the same employer as before childbirth, around 70 percent returned to a job at the same pay, skill level, and hours worked per week (Laughlin 2011).

Because of family responsibilities, many employed women work part-time or work shifts other than the nine-to-five workday. Furthermore, when family demands increase, wives, not husbands, are more likely to cut back their job commitments, and, because of family commitments, women tend to interrupt their job and career lives far more often than do men. Asked what they would prefer, a majority of employed mothers said they would prefer part-time employment. Employed fathers overwhelmingly preferred full-time work to part-time (see Figure 11.3). When nonemployed parents were surveyed about reasons for being home, 27 percent of women but only 3 percent of men identify family needs or child care. Women were also four times as likely as men (14 percent versus 3 percent) to say they are home because their spouse or family doesn't want them to work. Nonemployed fathers were likely to attribute their being at home to having been laid off (37 percent) or unable to find a job (51 percent).

**Figure 11.3** What Working Situation Would Be Ideal for You?

Considering everything, what would be the ideal situation for you—working full-time, working part-time, or not working at all outside the home?

| | Ideal Situation Would Be | | | | |
|---|---|---|---|---|---|
| | Not working | Part-time work | Full-time work | Don't know | N |
| | % | % | % | % | |
| **Have childern under 18** | | | | | |
| Fathers | 16 | 12 | 72 | *=100 | 343 |
| Mothers | 29 | 50 | 20 | 1=100 | 414 |
| **Mothers with childern under 18** | | | | | |
| Employed full-time | 21 | 49 | 29 | 1=100 | 184 |
| Employed part-time | 15 | 80 | 5 | *=100 | 75 |
| Not employed | 48 | 33 | 16 | 3=100 | 153 |

SOURCE: Pew Research Center, Social and Demographic Trends Survey, July 20 to August 2, 2009. www.pewsocialtrends.org/2009/10/01/the-harried-life-of-the-working-mother/.

# Dual-Earner and Dual-Career Families

Since the 1970s, inflation, a dramatic decline in real wages, the flight of manufacturing, and the rise of a low-paying service economy have altered the economic landscape. These economic changes have reverberated through families, altering the division of household roles and responsibilities. In 2014, 60.2 percent of married-couple families with children under 18 years of age were two-earner families. This includes 63.9 percent of two-parent families with children ages 6 to 17, and 55.3 percent of families with children under age 6 (U.S. Bureau of Labor Statistics 2015d).

In 2014, the median income among married couples who depended solely on the wages of a full-time, year-round male breadwinner was $70,573. Families in which both husbands and wives were employed full-time and year-round had median incomes of $115,902. Families in which wives worked full-time and year-round and husbands didn't work had median incomes of $68,834. Looking at married-couple families with one or more children under age 18, the median income of those who had only a male wage earner was $63,562. The median income of dual-earner couples with one or more children, in which both spouses worked full-time year-round, was $115,144. Finally, those married couples with children in which only wives were employed full-time and year-round had a median income of $65,017 (www.census.gov/hhes/www/income/data/historical/families/).

Economic changes have led to a significant increase in dual-earner marriages. Many employed women still find themselves in lower-paying, lower-status jobs—administrative assistants, clerks, child care workers, factory workers, waitresses, and so on. Economic factors, such as rising prices and declining wages, as well as increasing cultural expectations about women and employment, pushed most of them into the job market.

**Dual-career families** are a subcategory of dual-earner families. They differ from other dual-earner families in that both husband and wife have high-achievement orientations, a greater emphasis on gender equality, and a stronger desire to exercise their capabilities. Unfortunately, these couples may find it difficult to achieve both their professional and their family goals. Often, they have to compromise one goal to achieve the other because the work world generally is still not structured to meet the family needs of its employees.

## Typical Dual Earners

Although we traditionally separate housework, such as mopping and cleaning, from child care, the two are inseparable in reality (Thompson 1991). Although fathers have increased their participation in housework and child care, they have made smaller increases in the frequency with which they swing a mop or scrub a toilet. If we continue to separate the two domains, men will take the more pleasant child care tasks of playing with the baby or taking the children to the playground, and women will take the more unpleasant duties of washing floors, cleaning ovens, and ironing. Furthermore, someone must do behind-the-scenes dirty work for the more pleasant tasks to be performed. Alan Hawkins and Tomi-Ann Roberts (1992) note the following:

> Bathing a young child and feeding him/her a bottle before bedtime is preceded by scrubbing the bathroom and sterilizing the bottle. If fathers want to romp with their children on the living room carpet, it is important that they be willing to vacuum regularly. . . . Along with dressing their babies in the morning and putting them to bed at night comes willingness to launder jumper suits and crib sheets.

If we are to develop a more equitable division of domestic labor, we need to see housework and child care as different aspects of the same thing: domestic labor that keeps the family running.

## Housework

Using data from the American Time Use Survey, one can see the persistence of a very gendered pattern of household activities and responsibilities. On an average day, 83 percent of women and 65 percent of men spent some time doing household activities such as housework, cooking, lawn care, or financial and other household management. Over two-thirds of women (69 percent) and 43 percent of men reported engaging in food preparation or cleanup. Nearly half of women (49 percent) but only a fifth of men reported doing housework tasks such as cleaning or laundry. Looking at *how much time* women and men spend in household activities reveals that, in 2014, women spent an average of 2.6 hours versus 2.1 hours men spent (Table A-1 2014).

The American Time Use Survey also sheds light on the time *married* women and men with full-time employment spend on housework. Married men reported spending an average 1.2 hours a day on household activities compared to 1.9 hours a day reported by married women. This doesn't include time spent "purchasing goods and services" or "caring for household members," both of which are activities in which married women also spend more time than married men spend. If one adds these to the time spent in household activities, married women spend an average 0.58 more hours a day than married men do on such tasks. Married men spend more time than married women on working (6.1 hours versus 5.3) and leisure and sports activities (3.6 hours versus 2.9) (U.S. Bureau of Labor Statistics 2013a, Table A-6).

Research on the division of responsibility for housework among married couples tells a somewhat mixed story. Housework still falls more heavily on women's shoulders regardless of whether they are employed outside the home. Evidence continues to show that although men do "pitch in" much more often than in the past, they are not yet near sharing the burden of housework equally.

We should also note that when asked how they divide such things as housework and child care, mothers and fathers in two-parent households differ in what they perceive and/or report. For example, in a recent study of how working parents "share the load," fathers were more likely than mothers to see the load as shared. In other words, half of mothers say that they do more of the daily household chores and responsibilities, while only a third (32 percent) of fathers report the same. Fathers are more likely than mothers (56 percent versus 46 percent) to say household chores and responsibilities are shared equally.

In the same survey, more than 60 percent of mothers said that they did more of the management of their children's schedules and activities (64 percent) and taking care of them when they were sick (62 percent), and fathers were less likely to say mothers did more in these areas (53 percent versus 42 percent). In the opposite direction, fathers were more likely than mothers to report housework and child care as shared equally (Pew Research Center 2015d).

In a briefing paper for the Council on Contemporary Families Online Symposium on Housework, Gender, and Parenthood, sociologists Oriel Sullivan, Jonathan Gershuny, and John Robinson reviewed 50 years of time use data from 14 countries. Across all 14, the amount of time women reported spending on such household tasks as cooking, cleaning, and laundry showed striking declines. Over this same period of time, the amount of time men reported spending in such activities increased, though the reported increase in men's time was more modest than women's reported decrease (Sullivan, Gershuny, and Robinson 2015). More specifically, they report that data from the 1960s for most of the 14 countries studied showed women spending over four hours a day on such household tasks; in the United States, women reported spending close to four hours daily. By the 2000s, in most sample countries, women were spending less than two and a half hours a day. Over that same time span, men's reported time spent in such core housework activities increased, from less than an hour in the 1960s to between an hour and a half per day in the 2000s.

As indicated in Chapter 9, there are differences between cohabiting couples and married couples in the division of labor. One such difference is that cohabiting couples have tended to display somewhat

Data suggest that men's share of housework has increased both in the United States and in other industrialized societies.

more equitable divisions of household labor in comparison to married couples (Baxter 2005). Cohabiting women also do significantly less housework than married women do. It seems that marriage, rather than living with a man, transforms a woman into a homemaker (Baxter 2005). Marriage seems to change the house from a space to keep clean to a home to care for.

Much as cohabiting couples have displayed more equitable divisions of housework than married couples, survey research suggests that so, too, do same-sex couples display more task sharing than do heterosexual couples. As addressed in Chapter 9, absent a gender difference, such couples must negotiate and construct their division of labor. Although such active negotiation may push them in the direction of more sharing, as sociologist Carla Pfeffer notes, the situation is actually more complex. Egalitarianism may be exaggerated by couples who are aware of the assumed desirability of such arrangements. In some ways, they may use family myths and gender strategies in much the same fashion as heterosexual couples, and "create the semblance of egalitarianism in their relationships" (Pfeffer 2010, 168). Pfeffer also notes that equal divisions of housework and child care may not be equally valued among all lesbian couples. In their review of a decade's worth of research on same-sex couples, Timothy Biblarz and Evren Savci point out that an emphasis on equal sharing of housework, paid work, and child care may be more characteristic of white, middle-class lesbian couples. Among many lesbian and gay male couples, choice, ability, and fairness are the criteria that shape their division of work and family caregiving (Biblarz and Savci 2010).

Among married heterosexuals, various factors seem to affect men's participation in housework. Men appear to share more when their female partners are employed more hours, earn more money, and have more years of education (Sullivan and Coltrane 2008). Earnings appear to be an especially important factor in the allocation of housework. Sociologist Sanjiv Gupta reports that how much women earn and not how their earnings compare to those of their partners is what affects how much housework they do. Using data from the National Survey of Families and Households, Gupta (2007) reported that every additional $7,500 in married women's incomes was associated with one less hour of household work.

Men tend to contribute more when their hours and their wives' hours at work do not overlap (see the discussion of shift work later in this chapter). As their income rises, wives report more participation by their husbands in household tasks; increased income and job status motivate women to try to ensure that their husbands share tasks. However, research suggests that men who are economically dependent on their wives do less housework (Arrighi and Maume 2000; Carlson et al. 2014). Men whose wives earn the same or greater amounts of income may attempt to restore their masculinity by avoiding housework. Thus, the benefit of greater sharing of housework associated with women's earnings has its limits (Bittman et al. 2003; Tichenor 2005a and b).

Other factors that appear to influence men's involvement in housework include the following:

- *Men's socialization experience and modeling of parents.* Although it does not seem to influence women's participation in those same tasks, early parental division of labor acts as a strong predictor of men's involvement in the "female tasks" of housework (Cunningham 2001).

- *Men's status in the workplace.* Men who have their "masculinity challenged" at work may reduce their involvement in housework as a way of avoiding feminine behavior (Arrighi and Maume 2000).

- *Men's age and generation.* Older men do less housework than younger men do. Arrighi and Maume (2000) speculate that this may be a reflection of generational change, with younger men having been socialized toward more participation than older men were.

Whether a couple has children is a factor affecting men's and women's participation in household labor. Even though the presence of young children increases women's and men's housework, as we saw last chapter, it also skews the division of housework in even more traditional directions. Men tend to work more hours in their paid jobs, and women tend to work fewer hours at paid work and more in the home. Women then end up with a larger share of housework than before the arrival of children.

One factor that turns out not to be as strong a determinant as one might predict is the husband's gender ideology—what he believes he *ought to do* as a husband

**False** 7. Gender ideology—or what men think they ought to do—is not the biggest factor in determining how much housework and child care they actually do.

and how paid and unpaid work should be divided. As Arlie Hochschild's (1989) research showed, even traditional men can become more egalitarian if wives successfully use direct and indirect gender strategies. In some instances, repeated requests might be enough. In other cases, ultimatums may be necessary. Aside from these direct strategies, more indirect strategies—helplessness, withholding sexual intimacy, and so on—may work with husbands who otherwise would not do more.

Furthermore, necessity may create more male involvement. Wives with particularly demanding jobs or who work unusual hours (described later) may force their husbands to share household work more, simply because they are not available (Gerson 1993; Rubin 1994). Women appeared to be more satisfied if their husbands shared traditional women's chores (such as laundry) rather than limiting their participation to traditional male tasks (such as mowing the lawn). African Americans are less likely to divide household tasks along gender lines than Caucasians.

We might assume that the stresses and inequalities of juggling paid work and domestic work among employed mothers undermines women's well-being, but research on consequences related to marital, mental, and physical health tells a much different story. Analyzing more than a quarter century of General Social Survey data, sociologist Jason Schnittker finds that "women who are employed, regardless of the number of hours they work or how they combine work with family obligations, report better health than do those who are unemployed" (American Sociological Association 2004).

Employed women in dual-earner families also appear to have better mental health than women who are at home full-time. In juggling multiple roles, they suffer less depression, experience more variety, interact with a wider social circle, and have less dependency on their marital or familial roles to provide all their needed gratification. These psychological benefits accrue despite the unequal division of labor. As historian Stephanie Coontz wrote in *The New York Times*, "At all income levels, stay-at-home mothers report more sadness, anger, and episodes of diagnosed depression than their employed counterparts" (Coontz 2013).

## Emotion Work

Although we might not typically think about them as "work" or include them in a discussion of "family work," there are other tasks that need to be performed to generate and maintain successful and satisfying relationships and families. Such tasks are often referred to as **emotion work** (Erickson 2005; Hochschild 2003; Umberson, Thomeer, and Lodge 2015). It can include the following:

- Confiding innermost feelings
- Trying to bring our partner out of a bad mood
- Praising our partner
- Suggesting solutions to relationship problems
- Raising relationship problems for consideration and discussion
- Taking initiative to begin the process of "talking things over"
- Monitoring the relationship and sensing when our partner is disturbed about something (Stevens, Kiger, and Riley 2001)

Although these might not cleanly fit many people's notions of "tasks," they may be experienced as work by those who feel unevenly burdened by them. Women do more of the emotion work in their relationships and report being less than satisfied with how these "responsibilities" are divided (Erickson 2005; Stevens, Kiger, and Riley 2001; Umberson, Thomeer, and Lodge 2015). This has important consequences because both men's and women's satisfaction with the division of emotion work in their relationships was significantly and positively associated with their marital satisfaction (Stevens, Kiger, and Riley 2001).

## Caring for Children

As noted in the previous chapter, men increasingly believe that they should be more involved as fathers than men have been in the past. Men have substantially increased the time they spend caring for children, though they are not usually fully equal caregivers in two-parent, dual-earning households. When looking at time spent in child care, men have experienced more substantial increases than we saw earlier regarding their time in daily household tasks. In the United States, from the 1960s to the first decade of this century, fathers have dramatically increased their time spent caring for children. High school–educated fathers have nearly doubled the time spent in daily child care, while college-educated fathers have more than tripled theirs (Sullivan, Gershuny, and Robinson 2015).

Child care responsibility varies according to the marital status of parents and their employment roles and schedules. In a two-parent family, care for children remains more the responsibility of mothers than fathers

(Yeung et al. 2001). When we examine data on actual involvement in tasks associated with child care or time spent with children, mothers are more involved in such tasks than fathers. In making such comparisons, it is helpful to differentiate between **engagement** with children—time spent in *direct interaction* with a child across any number of different activities—and **accessibility**—or *availability to a child,* when the parent is at the same location but not in direct interaction (Yeung et al. 2001). Even though fathers' proportional involvement with children has increased, it is estimated that fathers' engagement with children lags farther behind their accessibility to children. Fathers estimated accessibility to their children is closer to mothers' estimated accessibility than is their actual hands-on engagement.

Circumstances affect how much time fathers spend with children. One study, based on analyses of data from the Panel Study of Income Dynamics (Yeung et al. 2001), noted that a child's direct engagement with biological fathers in two-parent homes ranged from a daily average of 1 hour and 13 minutes on weekdays to 2 hours and 29 minutes on weekends. The total time (engagement plus accessibility) these fathers are involved with their children 12 years and younger is roughly 2.5 hours a day on weekdays and 6.5 hours a day on weekends.

## Active Child Care

Active, or "hands-on," child care is more often "in the hands" of mothers than fathers. Mothers take care of and think about their children more than fathers do (Walzer 1998). In most two-parent households, mothers' child care responsibility and involvement greatly exceed fathers' involvement (Aldous and Mulligan 2002).

What do mothers and fathers do in the time they spend with children? Research from the latter 20th century suggested that fathers spent more time in interactive activities, such as play or helping with homework, whereas mothers spent more time doing custodial child care, such as feeding and cleaning (Yeung et al. 2001). Mothers still perform most of such custodial care.

In addition, research has indicated that fathers are more involved with sons than with daughters, with younger children more than with older children, and with firstborn more than with later-born children (Pleck 1997, cited in Doherty, Kouneski, and Erickson 1998). Fathers' time with children has been found to be positively associated with life satisfaction and with their living healthier lives (Schindler 2010). Psychologist Holly Schindler found that fathers' engagement

with their children as well as their contribution of financial support both contributed to improved psychological well-being. Both their relationships with their children and their success as providers proved beneficial to fathers' mental health (Schindler 2010).

There is suggestive evidence that when it comes to assessing the time they spend with their children and rating themselves as parents, employed fathers are more critical of themselves than employed mothers. In a 2012 Pew Research Center survey of employed parents, nearly half of the 249 employed fathers compared to a quarter of the 229 employed mothers reported spending "too little time" with their children. Two-thirds of the employed mothers said they spent "the right amount of time" with their children. Among fathers who said that they spent "too little time" with their children, only half (49 percent) rated their performance as a parent as excellent or good. More than 80 percent of fathers who said they spent the right amount of time with their children rated their parenting as "excellent" or "good" (Parker 2015).

Both mothers' and fathers' own time in employment can influence fathers' parental engagement. When mothers are employed, fathers are more likely to increase their involvement in activities associated with parenting. They may even cut back some on their own time at work. Fathers' time in employment has the opposite effect on their participation in child care activities (Schindler 2010). Research suggests that fathers who work more hours and who have prestigious, time-demanding occupations tend to be less engaged in child rearing (NICHD Early Child Care Research Network 2000). On weekends, fathers become somewhat more equal caregivers, and their involvement is greater when mothers contribute a "substantial" portion of the family income (Yeung et al. 2001). Although fathers "help" mothers with the caregiving work and supervision involved in raising teenaged children, most fathers do less of such work than most mothers (Kurz 2002).

## Mental Child Care

Responsibility for child care doesn't consist only of what we *do* with and for our children. In her book *Thinking about the Baby: Gender and Transitions into Parenthood,* sociologist Susan Walzer (1998) examined the division of responsibility for infants in 25 two-parent households. Her focus was less on "who did what" with their children than on "who thought what and how often" about their children. Walzer identifies this "invisible" parenting as **mental labor**—the process of worrying about the baby, seeking and processing information about infants and their needs, and managing the division of infant care in the household (i.e., seeking the "assistance" of their spouse). It might be thought of as similar to the "emotion work" that surrounds marriage.

Sociologist Demie Kurz reports similar kinds of mental labor among mothers of teenaged children. Fearful for their adolescents' safety and especially fearful about the sexual vulnerability of their teenaged daughters, mothers worry (Kurz 2002). Thus, mothers continue to worry as children grow. One key to understanding this mental labor at both the earliest and the late adolescent or young adult stages is that mothers feel responsible for and judged by what happens to their children in ways that most fathers do not.

## How the Division of Household Labor Affects Couples

Aside from the obvious matters of unequal responsibilities and differences in the numbers of tasks and roles that spouses and partners play, how couples divide household labor affects their relationship. The balance and distribution of power, the amount of satisfaction partners feel with their relationship, their levels of sexual intimacy, and the likelihood of their remaining together are all influenced by how they divide and allocate household labor.

### Marital Power

An important consequence of women's employment is a shift in the decision-making patterns in a marriage. As noted earlier in Chapter 7, decision-making power in a family is not based solely on economic resources, but economics influence how couples make decisions and divide responsibilities. Employed wives exert greater power in the home than that exerted by nonemployed wives (Blair and Lichter 1991; Schwartz 1994). Marital decision-making power is greater among women employed full-time than among those employed part-time. Conversely, full-time stay-at-home wives may find themselves taken for granted and, because of their economic dependency on their husbands, relatively powerless (Schwartz 1994).

Some researchers are puzzled about why many employed wives, if they do have more power, do not demand greater participation in household work on the part of their husbands. There are many possible reasons for women's apparent reluctance to insist on their husbands' equal participation in housework. These include (1) cultural norms that housework is the woman's responsibility, (2) fears that demands for increased participation will lead to conflict, (3) the belief that husbands are not competent, and (4) the concern that men's participation comes with "strings attached," and that they will use that participation to gain other advantage over their spouses.

### Satisfaction, Sex, and Stability

How do patterns of employment and the division of family work affect the quality of marital relationships? Does the division of labor affect a couple's level of marital satisfaction? Oriel Sullivan and Scott Coltrane (2008) point out that when men share more of the housework, women's marital satisfaction increases and the frequency of marital conflict decreases. In 2008, a number of media sources publicized a report by psychologist Joshua Coleman of the Council on Contemporary Families suggesting a connection between the amount of housework men do and the amount of sex in marriage. Coleman asserts that women are more prone to depression, more likely to contemplate divorce, and less interested in sexual intimacy with their husbands when they do a disproportionate amount of the housework. He asserted that wives are "more sexually interested" in husbands who do more housework (Sullivan and Coltrane 2008). Thus, the "takeaway" message, picked up on by various media, seemed to be that husbands who do more housework would have more active sexual relationships with their wives (Elliott and Umberson 2008; Gager and Yabiku 2010).

Other research has failed to bear this out. In fact, a study by Sabino Kornrich, Julie Brines, and Katrina Leupp (2013) found that sexual frequency was greater among more traditional couples. Looking at men's participation in traditionally male and traditionally female tasks, they found that both husbands and wives reported higher levels of sexual frequency in couples who maintained more conventional divisions of labor. However, the data upon which this analysis was drawn were collected in 1987 and 1992. Finally, using more recent data (from 2006), Daniel Carlson, Amanda Miller, Sharon Sassler, and Sarah Hanson (2014) found small and conditional differences in sexual frequency between conventional and egalitarian couples, with the latter having more frequent and more satisfying sexual relationships. They further found that couples in which men performed the bulk of routine housework reported less sex, more dissatisfaction with the amount of sex, and lower-quality sexual relationships (Carlson et al. 2014).

What can we say about the effect of women's employment on the likelihood of divorce? A number of older studies suggested that employed women were more likely to divorce. Because employed women are less likely to conform to traditional gender roles, it was thought that this could potentially cause tension and conflict in the marriage. They are also more likely to be less economically dependent and do not have to tolerate unhappy or unsatisfactory marriages for economic reasons. However, in such instances, it is the unsatisfying marriage—not the woman's employment—that is leading them out of marriage. Employment just makes it more possible for a single woman to survive. Of importance, employed women are not more likely to leave satisfying or happy marriages, nor are their husbands. As to men, they are more likely to leave marriages when they, themselves, are *un*employed. The same is true of women married to unemployed men. Men's employment tends to be expected, thus unemployment is a violation of gender expectations. Women's employment increases their ability to leave, but they are no more likely to leave if they find their marriages meeting their expectations (Sayer et al. 2011).

Overall, despite an increased divorce rate, the overall effect of wives' employment on marital satisfaction in recent years has shifted from a negative effect to no effect to even a positive effect. The effect of a wife's full-time employment on a couple's marital satisfaction is affected by such variables as social class, the

How dual-earner couples divide housework and child care is associated with their levels of marital conflict, marital satisfaction, and physical intimacy and their risk of divorce.

Ron Chapple studios/Hemera/Getty Images

presence of children, and the husband's and wife's attitudes and commitment to her working. Thus, the more the wife is satisfied with her employment, the higher their marital satisfaction will be. In addition, the higher the husband's approval of his wife's employment and the greater his participation in household tasks, the higher the wife's marital satisfaction.

Data on the effects of the division of domestic labor on marital satisfaction consistently indicate a positive effect. Couples who share report themselves as happier and are less at risk of divorce than couples in which men do little of the family work. This appears to be true regardless of whether couples' gender ideologies are traditional, egalitarian, or transitional (somewhere between the other two) (Hochschild 1989). In addition, the fewer hours women spend on household tasks, the more time they can spend in "status-enhancement" activities and the greater their marital satisfaction (Stevens, Kiger, and Riley 2001). Daphne Stevens, Gary Kiger, and Pamela Riley report that marital satisfaction is affected by the way couples divide each of the three dimensions of domestic labor: domestic work, emotion work, and "status-enhancement" work (helping a partner's career development by building goodwill with the partner's clients or coworkers, ensuring that the partner has the needed time to commit to work, and so on). Women felt more resentment and less marital satisfaction when they do the majority of both domestic and emotion work. Only among women with traditional gender ideologies did this differ. For them, marital satisfaction was positively influenced by feeling that they have fulfilled their marital obligations. In the case of status-enhancement work, the more of such tasks women do, the more satisfied they and their husbands report themselves to be with their marriages. Nevertheless, the division of emotion work was most related to marital satisfaction, and the performance of status enhancement activities was least related (Stevens, Kiger, and Riley 2001).

## Atypical Dual Earners: Shift Couples and Peer Marriages

Despite sharing in common the features of having both spouses employed and the higher incomes that two wage earners can provide, there are some interesting lifestyle variations among dual-earner couples. Couples with these lifestyles differ from more common two-earner couples in one of two ways. Shift couples have household arrangements in which the parents work opposite, mostly nonoverlapping shifts and thus take turns working outside the home and caring for children. Peer or postgender couples have consciously adopted a belief in equality and fairness into how they divide domestic responsibilities. As a result of either of these differences, such atypical couples show much higher rates of male participation in child care and housework than among more typical dual earners. Here, we briefly consider each of these types.

## Shift Work and Family Life

It is estimated that one-fifth of all employed Americans work a nonstandard shift or schedule. By nonstandard, researchers include night (e.g., midnight to 8 A.M.), evening (e.g., 3 P.M. to 11 P.M.), rotating, variable, and weekend shifts. A third of dual-earner couples with kids are estimated to have one parent working such a schedule. A 2011 paper by sociologists Harriet Presser and Brian Ward found that by age 39, almost 90 percent of their representative sample had had some experience with nonstandard schedules, and 70 percent with night or evening schedules. Clearly, such nonstandard shifts and schedules are becoming more and more commonplace.

A 2007 report by the Bureau of Labor Statistics provided a snapshot of shift work in the United States in 2004. Approximately 18 percent of wage and salary workers worked hours other than the typical nine-to-five or 8:30-to-4:30 daytime shifts. This amounts to more than 21 million workers. Fifteen percent of full-time workers and 30 percent of part-time workers worked alternate shifts. Nearly 7 percent of all wage and salary workers were employed in jobs where they worked between 2:00 p.m. and midnight, making this the most common alternate shift (McMenamin 2007).

Shift work varies by gender, ethnicity, and occupation. Men are more likely than women to work an alternate shift (19 percent versus 16 percent). African Americans are more likely than either whites, Asians, or Hispanics to work alternate shifts. Almost one-fourth (23.2 percent) of African American workers worked an alternate shift in 2004, compared to 17 percent of whites and 18 percent of Asians and Hispanics. Among those employed full-time, 20.8 percent of African Americans, 13.7 percent of whites, 15.7 percent of Asian Americans, and 16 percent of Hispanic Americans worked some alternate shift in 2004 (McMenamin 2007).

Shift work is more prevalent in certain occupations and nearly nonexistent in others. Service work, especially protective services (e.g., police, guards, and firefighters), nursing, and food preparation and serving, make up the occupations where shift work is most common. In addition, transportation (e.g., truck or bus drivers) and material moving (e.g., truck drivers), warehousing, and mining have large portions of employees working night or alternate shifts. On the other hand, few managerial, professional, business, or educational positions are conducive to alternate shifts (McMenamin 2007; Presser 2003).

The late sociologist Harriet Presser (2003), a leading authority on shift work and its consequences for individuals and families, identified three macro-level changes that have contributed to the continuing increase in nonstandard work schedules—changes in the economy, demographics, and technology:

- *Changes in the economy.* There has been a substantial increase in the service sector of the economy, which has a high prevalence of nonstandard schedules. Simultaneously, women's labor force participation doubled between 1975 and 2000—from one-third to two-thirds of all adult women—which necessitates child care for more families and some share this role.

- *Changes in demographics.* Both delayed age at marriage (by nearly three years between 1960 and 2000) and sizable increases in dual-earner couples have contributed to an increased demand for entertainment and recreation at night and over weekends (Presser 2003). In addition, as the U.S. population has aged, there has been a need for medical services available to people 24 hours a day, seven days a week.

- *Changes in technology.* Computers, faxes, overnight mailing, and other communications technology have made round-the-clock offices a norm for many multinational corporations.

Although such large-scale changes have expanded the opportunity to work atypical schedules, *why* do individuals choose to do so? More than 60 percent of individuals working nonstandard schedules identify job demands or constraints as the driving force behind their work schedules. These include such reasons as "they could not get any other job, the hours were mandated by the employer, or the nature of the job required nonstandard hours" (Presser 2003, 20). Only among mothers of children under 5 years of age do we find as many as 43.8 percent identifying "caregiving" needs as a reason for their employment schedule. Looking specifically at child care arrangements, 35 percent of mothers and 7.6 percent of fathers of children under 5 years of age identify "better child care arrangements" as a main reason for their nonstandard shifts.

Couples in which one spouse works such a nonstandard shift and the other remains in a more typical shift are sometimes referred to as *opposite-shift, split-shift,* or simply *shift couples.* **Shift couples** may have to structure their home and work lives into a turn-taking, alternating system of paid work and family work. When one is at work, the other is at home. When the at-work partner returns home, the at-home partner departs for work, giving them a kind of "hello, good-bye" lifestyle. Research indicates that nearly 30 percent of dual-earner couples have at least one spouse working an atypical schedule (Presser 2003).

When this lifestyle is the product of choice, shift couples may perceive it as a reasonable trade-off. Through it, they stress the importance of child rearing over the importance of marital relations. In fact, in the absence of children in the household, shift couples can cope with the stresses imposed by their work schedules. Spouses may not see each other much, but they communicate frequently and maintain sufficient flexibility to allow them to find or make time for each other.

Significantly, in households with children, the outcomes associated with shift work are mixed and depend on the specific shift combination. If wives work second-shift (late afternoon through midnight) or third-shift (late night through morning) jobs, husbands must feed children dinner or breakfast; see that they do their homework, take baths, go to bed, or get up for school; make lunches to take to school; and so on. For the household to function, men are pressed to do a greater share of domestic work and especially child care than among either traditional couples or more typical dual earners (Presser 2003).

Psychologists Rosalind Chait Barnett, Karen Gareis, and Robert Brennan note that among shift couples in which wives are employed from 3:00 P.M. to 11:00 P.M. or from 4:00 P.M. to midnight, fathers typically "act like mothers" (Barnett, Gareis, and Brennan 2008). The more hours that husbands are at home with children while wives are at work, the less traditional the couple becomes, as men do housework and child care that women traditionally do. Conversely, the more hours that wives are home while husbands are at work, the more time wives spend on traditional female tasks, thus reinforcing traditional gender stereotypes (Barnett, Gareis, and Brennan 2008).

**False** 8. Couples who work different shifts have less satisfying and stable marriages.

### TABLE 11.5 Family–Life Satisfaction of Paid Workers

| Shift | Percentage "Extremely" or "Very" Satisfied with Family Life |
|---|---|
| Day | 71% |
| Evening | 56% |
| Night | 54% |
| Rotating | 63% |
| Split | 67% |
| Flexible | 74% |
| Total (all shifts) | 69% |

SOURCE: Grosswald (2004).

Aside from parental involvement, what does shift work do to workers and their family lives? Much research is pessimistic, linking nonstandard work to lower job satisfaction, lower marital and sexual satisfaction, higher levels of anxiety and irritability, higher blood pressure and poorer health, greater sleep difficulties, and increased risk of depression and alcohol problems. Shift workers who work such nonstandard shifts have higher levels of negative work–family spillover and experience more work–family conflict. Shift workers suffer more distress, greater dissatisfaction, and higher risks of divorce. Professor of social work Blanche Grosswald (2004) notes that shift workers have been found to have lower levels of marital satisfaction, more disagreements and marital and sexual difficulties, higher divorce rates, and more problematic relationships with their children. Grosswald observed that 69 percent of respondents to the Families and Work Institute's 1997 National Study of the Changing Workforce reported themselves to be satisfied with their family lives; however, among shift workers, the results were as shown in Table 11.5.

There are also some positive familial outcomes that can result from shift work, such as the abilities to take turns, to have a parent home with children when the other is at work, and to increase father–child closeness. Shift couples save money on child care as well as reduce some of whatever stress parents might feel about outside caregivers. Additional

In opposite-shift arrangements, as one spouse returns home from work, the other is getting ready to leave.

Annika Erickson/Blend Images/Alamy Stock Photo

economic benefits might include the opportunity to earn potentially higher wages and the flexibility to work a second job (Grosswald 2004). However, Presser (2003) found that those who work nonstandard schedules are more likely to be economically disadvantaged than those who work more typical schedules. Sociologist Hiu Liu and her colleagues found that the benefits of nonstandard schedules were more likely for married parents than for cohabiting parents. The latter, being less likely to pool their money, less committed to each other, less likely to receive social support from others, and more likely to be living with their partner's children, experience a negative relationship between their schedules and their well-being (Liu et al. 2011).

Although research has suggested that shift work can be detrimental to the quality of marital relationships, it appears that this is truest of couples in which a spouse works a night shift and/or rotating shifts. Furthermore, in couples in which wives work evening shifts, men develop more rewarding relationships with their children than do men whose wives work day shifts (Barnett, Gareis, and Brennan 2008).

## Peer and Postgender Marriages

Among some dual-earner couples, there is explicit agreement that household tasks will be divided along principles of fairness. Many couples believe their family's division of labor is fair. Among those couples who can afford household help, questions of fairness become mostly irrelevant; husbands are likely excused entirely from many household chores, such as cleaning and mopping. Because of their incomes, they are allowed to "hire" substitutes to do their share of housework (Perry-Jenkins and Folk 1994).

It is important to note that an equitable division is not the same as an equal division. Relatively few couples divide housework exactly 50/50. For women, a fair division of household work is more important than both spouses putting in an equal number of hours. There is no standard of fairness, however (Thompson 1991). Because most women work fewer hours than men spend in paid work and because wives tend to work more hours in the home, some women believe that the household labor should be divided proportionately to hours worked outside the home. Other women believe that it is equitable for higher-earning husbands to have fewer household responsibilities. Still others believe that the traditional division of labor is equitable, wherein household work is women's work by definition.

Middle-class women are more likely to demand equity; equity is less important for working-class women, who tend to be more traditional in their gender-role expectations (Perry-Jenkins and Folk 1994; Rubin 1994).

*Peer marriages* (or *postgender marriages*, to use Barbara Risman and Danette Johnson-Sumerford's [1998] term) take concerns for fairness and sharing to heart in how they structure each facet of their relationships. Rarer than shift couples, they too depart from the model of typical dual earners described previously. Whereas shift arrangements may be the result of choice, necessity, or circumstance, peer relationships typically emerge from egalitarian values and conscious intent. Peer or postgender couples base their relationships on principles of deep friendship, fairness, and sharing. Hence, they monitor each other's level of commitment and involvement, maintain equally valued investments in their paid work, and share household tasks and child care.

Research by Pepper Schwartz (1994) and Barbara Risman and Danette Johnson-Sumerford (1998) indicated that such relationships avoid many of the trappings that often accompany more traditional divisions of labor, including female powerlessness and resentment and male ingratitude and lack of respect. Furthermore, children receive attention and care from both parents, and men develop deeper relationships with their children than commonly found. Although such couples are rare, they show that the inequities in either the traditional or the more typical dual-earner household are not inevitabilities. Indeed, couples can—and some do—commit themselves to "doing it fairly" (Risman and Johnson-Sumerford 1998).

## Coping in Dual-Earner Marriages

Dual-earner marriages are here to stay. They remain stressful today because society has not pursued ways to alleviate the resulting work–family conflict. Inflexible workplaces, inadequate access to quality child care, and inequities in the household can all exacerbate stress and conflict.

The three greatest social needs in dual-earner marriages are (1) redefining gender roles to eliminate role overload for women, (2) providing adequate child care facilities for working parents, and (3) restructuring the workplace to recognize the special needs of parents and families. Coping strategies include reorganizing the family system and reevaluating household expectations.

The goal for most dual-earner families is to manage their family relationships and their paid work to achieve a reasonable balance that allows their families to thrive rather than merely survive. Achieving such balance will continue to be a struggle until society and the workplace adapt to the needs of dual-earner marriages and families.

# At-Home Fathers and Breadwinning Mothers

An additional departure from both the typical dual-earner and the traditional family is the family type in which spouses switch places or reverse roles. Such households include a wage-earning woman, a man at home full-time, and one or more children. Labels such as "role reversal" conjure images of a chosen lifestyle, and likely overstate the extent to which men who stay home are just like women who stay home with children. Even the terms to refer to the men who stay home may imply a subtle but meaningful difference. The label "stay-at-home dads," like the label "role reversal," suggests a chosen lifestyle. The label "at-home dads" seems more inclusive of all the many reasons men might have for being at home. As we noted last chapter, these include unemployment, disability, being in school, and choice.

Of the 21.5 million married-couple families with children under age 15 in 2014, 6.2 percent of them (1.25 million) had fathers who stayed home (U.S. Census Bureau 2014a). The reasons men give for staying home do not typically identify "to care for home and family" (only 24 percent of the slightly more than 1.25 million at-home men stated this as their reason). Most men are home for other reasons, such as disability, unemployment, retirement, school, or something else, but while home, they can and often do provide care for their children. In contrast, 85.6 percent of the 6.8 million mothers who are out of the labor force and home with children under age 15 cite "to care for home and family" as their reason (U.S. Census Bureau 2014a).

What happens to such couples? We can point to five areas in which couples experience some impact from having switched (traditional) places:

- *Economic impact.* Such couples live on less money but also spend less on child care. Hence, though they get by on less income, the decline may not be as dramatic, especially if women's careers are enhanced and men's occupations were not high paying. Men gain an opportunity to take a "time-out," refocus, and try new career possibilities. They do, however, surrender the provider status and confront the reality of economic dependency.

- *Social impact.* Socially, men experience some isolation as they lose the primary source of social interaction—the workplace. In addition, couples may become the targets of curiosity or even criticism for their choices. Men, however, may also receive supportive responses, especially from women, "put on pedestals," while their wives often receive envious reactions, especially from coworkers. In general, at-home fathers become visible in their domestic role in contrast to the invisibility that more traditionally befalls housewives.

- *Marital impact.* This lifestyle leads to higher levels of male involvement in housework and child care. Although men don't take over everything to the same extent that stay-at-home mothers do, they are likely to share more or do most domestic work. In addition, couple relationships change. Whereas Russell (1987) found the changes to be negative, Cohen and Durst (2000) found high levels of communication, empathy, and appreciation among the couples they studied. In some ways, men who are home full-time, like traditional men before them, benefit from having wives. Wives, in particular, know what it takes to care for households and children. When women are the ones at home (while husbands are employed), they are often married to men who lack such understanding and appreciation. Women are also aware that their spouses have taken risks and made sacrifices by staying home and may support them in ways that breadwinning husbands probably don't go out of their way to support nonemployed wives.

- *Parental impact.* Perhaps the most noticeable area of impact is the enlarged relationship between fathers and children. Fathers get to know their children in ways that are not otherwise likely and may not even be possible. Children see fathers in nontraditional ways. Mothers maintain the same sorts of relationships as other employed mothers do with their children, but they have greater peace of mind. Children are not in day care, at the sitter, or home alone. They are home with dad.

- *Personal impact.* Being an at-home father can change the ways men look at their lives, resulting in a reshuffling of priorities and the construction of a new social identity. Breadwinning mothers may also enlarge their sense of themselves as providers, take advantage of the at-home resource, and make work a larger component of their own identities.

A 2012 study of 31 at-home fathers by Brad Harrington, Fred Van Deusen, and Iyar Mazar looked at a sample of men who chose to be stay-at-home parents. About their sample they note: There is no one type of person, no necessary socialization experience, shared by—or needed to become—at-home dads. Those men who do become stay-at-home dads find themselves confronting challenges and obstacles at the same time that they reap the benefits of active, engaged relationships with their children, and their partner gains advantages in work and careers. Among the challenges and obstacles, Harrington, Van Deusen, and Mazar mention social isolation, potential stigma, and disrupted career possibilities (Harrington, Van Deusen, and Mazar 2012).

A 2009 book by journalist Jeremy Adam Smith (2009) illustrated how "reverse-traditional couples" contribute to the possibilities for contemporary families. Such couples reveal the following:

- Even when couples consist of a sole wage-earning partner while the other stays home and cares for children and the household, they need not fall into a pattern of one dominant and one submissive partner.

- Gender need not define the roles that heterosexual partners play in families and households. Such couples show that men are capable of taking care of children, that women are capable of supporting families, and that circumstances rather than gender can dictate what roles each partner plays at any given time.

- One need not choose between having a career and staying at home to raise children, as many women are told they must. For some women and men, homemaking and caregiving can be a stage of family life (perhaps followed by full-time wage earning), for others a lifelong "career" choice, and for still others something they do along with some kind of paid work.

- Caregiving creates stronger attachments between men and their children and strengthens men's abilities to provide care.

The increase in both actual involvement and social visibility of at-home fathers can be seen in a variety of ways and places. A Pew Research Center survey of "at-home dads" noted that men had become a sixth of all at-home parents (Livingston 2014b). The percentage who said they were home to care for their families had increased fourfold between 1989 and 2012, from 5 percent of those at home to 21 percent of men at home.

There are now a variety of organizations, websites, blogs, newsletters, and books (such as Smith's *Daddy Shift: How Stay-at-Home Dads, Breadwinning Moms, and Shared Parenting Are Transforming the American Family* or Peter Baylies's *The Stay-at-Home Dad Handbook*) focusing on or catering to the needs of at-home fathers. There is good reason to think that the number of men with this lifestyle will increase in coming years, but it is difficult to know by how much.

## ◼ Family Issues in the Workplace

Many workplace issues, such as economic discrimination against women, occupational stratification, adequate child care, and an inflexible work environment, directly affect families. They are more than economic issues—they are also family issues.

### Discrimination Against Women

A woman's earnings significantly affect family well-being regardless of whether the woman is the primary or secondary contributor to a dual-earner family or the

## TABLE 11.6 Median Weekly Earnings, Full-Time Employees by Gender and Race

| Group | Male | Female | Female as % of Male |
|---|---|---|---|
| Total | $871 | $719 | 82.5% |
| White | $897 | $734 | 81.8% |
| African American | $680 | $611 | 89.9% |
| Asian American | $1080 | $841 | 77.9% |
| Hispanic | $616 | $548 | 89% |

SOURCE: U.S. Bureau of Labor Statistics (2015h).

sole provider in a single-parent family. Furthermore, as we have seen, women's family responsibilities significantly affect their earnings. Given the importance, however, of women's wage contributions to their families, we need to consider—at least briefly—economic discrimination against women and sexual harassment. By affecting women's employment status and experiences in their jobs, these become important family issues as well as economic ones.

### Economic Discrimination

The effects of economic discrimination can be devastating for women. Data from 2015 showed that women who worked full-time had median weekly earnings of $719, 82.5 percent of men's median of $871 per week. This overall gap varies by race and ethnicity, as shown in Table 11.6.

The differences in women's and men's incomes translate into more women than men experiencing poverty and in need of state or federal assistance. Wage differentials are especially important to single women.

Although employment and pay discrimination are prohibited by Title VII of the 1964 Civil Rights Act, the law did not end the pay discrepancy between men and women. Much of the earnings gap is the result of occupational differences, gender segregation, and women's tendency to interrupt their employment for family reasons and to take jobs that do not interfere extensively with their family lives. Earnings are about 30 percent to 50 percent higher in traditionally male

**False**

**9.** Women in the United States currently make only 82 cents for every dollar that men make.

occupations, such as truck driver or corporate executive, than in predominantly female or sexually integrated occupations, such as administrative assistant or schoolteacher. The more women dominate an occupation, the less it pays.

### Sexual Harassment

Sexual harassment is a mixture of sex and power, with power often functioning as the dominant element. **Sexual harassment** can be defined as two distinct types of harassment: (1) the **abuse of power** for sexual ends, and (2) the creation of a hostile environment. In abuse of power, sexual harassment consists of unwelcome sexual advances, requests for sexual favors, or other verbal or physical conduct of a sexual nature as a condition of instruction or employment. Only a person with power over another can commit the first kind of harassment. In a **hostile environment**, someone acts in sexual ways to interfere with a person's performance by creating a hostile or offensive learning or work environment. Sexual harassment is illegal.

In fiscal year 2011, 11,364 reports of sexual harassment were filed with the Equal Employment Opportunity Commission (EEOC) and the state and local Fair Employment Practices agencies around the country. This is 350 fewer than reported in 2010, and represented the smallest number of complaints in the last 15 years. Of these charges, women brought 84 percent (U.S. Equal Employment Opportunity Commission 2012). These numbers need to be approached with some caution. According to a 2011 ABC/*Washington Post* poll, 25 percent of women and 10 percent of men said they had been harassed. Of the women acknowledging having been harassed, only 41 percent said they had reported it to their employers. The implication of this is that "official reports" are likely just a portion of the actual harassment experienced in the workplace. Thus, there are likely more—perhaps considerably more—women harassed than the EEOC data would indicate. More recent data, which includes only a count of cases filed with the EEOC, show 6,862 cases filed, down some each year from the 7,944 filed in 2010 to 2014 (U.S. Equal Employment Opportunity Commission 2014).

Some estimate that as many as half of employed women are harassed during their working years. Experiencing sexual harassment can lead to a variety of serious consequences. Some people quit their jobs, and others may be dismissed as part of their harassment. Victims also often report feeling depressed, anxious, ashamed, humiliated, and angry (Paludi 1990).

## The Need for Adequate Child Care

As we saw in the previous chapter, even though mothers continue to enter the workforce in ever-increasing numbers, high-quality, affordable child care remains an important but highly uncertain support (Gould and Cooke 2015). For many women, especially for those with younger children and for single mothers, the availability of child care is critical to their employment. With 59.7 percent of married and single mothers of children under the age of 6 employed, it is unsurprising that the demand for affordable and trusted child care is high (U.S. Bureau of Labor Statistics 2015d). The 2015 report from the National Association of Child Care Resource and Referral Agencies, "Parents and the High Cost of Child Care," indicates that nearly 11 million children under 5 years of age are in some sort of child care setting, with a third in multiple arrangements.

For most employed mothers with children 5 to 14 years old, school attendance is their primary day care solution. Women with preschool children, however, do not have that option. Instead, they tend to use in-home care by their child's other parent, grandparents or other relatives, day care centers, family day care (care provided in someone else's home), and preschools or nursery schools as their most important resources. Perhaps as many as two-fifths of preschool-age children experience on a regular basis a combination of multiple arrangements. Psychologist Taryn Morrissey points out that the use of multiple arrangements was more common among mothers with older preschoolers than among mothers of infants. Also, single mothers, mothers who worked less than full-time, and those who used informal arrangements were more likely to utilize multiple care arrangements (Bianchi and Milkie 2010). Some may decide to use multiple arrangements to expose children to a variety of situations and people and enrich them in the process. Others decide out of necessity because of practical constraints, such as work schedules and limited availability of options. Morrissey also notes that at least some research casts the use of multiple arrangements in a somewhat negative light in terms of its impact on children's social adjustment and the stress it induces for parents who have to coordinate

schedules and arrange for transportation from one care arrangement to another. This, in turn, can affect parents' performance at work (Morrissey 2008).

Frustration is one of the most common experiences in finding or maintaining adequate child care. Changing family situations, such as unemployed fathers' finding work or grandparents' becoming ill or overburdened, may lead to these relatives being unable to care for the children. Family day care homes and child care centers may close because of low wages or lack of funding. Where one lives and one's socioeconomic status both affect one's ability to find quality care. Higher-income communities and metropolitan areas offer parents greater access to center-based care. Meanwhile, without sufficient income, one will have greater difficulty obtaining high-quality child care (Bianchi and Milkie 2010). Parents, especially mothers, may face such problems as having to cut back or restrict their hours at work, having to change child care arrangements, or being placed on a waiting list to obtain child care. Studies show that younger parents, poor parents, and never-married parents experience more of these sorts of constraints, and nonrelative child care has been found to produce more child care problems (Laughlin 2007).

Furthermore, child care is expensive. Data on the costs of child care reveal that whether in a center-based or an in-home (family) child care setting, child care is costly. Costs range throughout the United States, depending on where one lives and the type of care one uses. Across the 50 states, full-time infant care in an in-home family setting ranges from an annual average of $3,972 in Mississippi to a high cost of $10,666 in Massachusetts. In the nation's capital, the cost is even higher: $16,006 for full-time infant care in an in-home setting. Center-based infant care ranges from a low average yearly cost of $4,822 in Mississippi to highs of $17,062 in Massachusetts and $17,842 in Washington, DC. In 37 states, the average yearly cost of full-time, center-based care for an infant is more than 10 percent of the state's median income.

Costs are less for older preschoolers but again cover a wide range. The average cost of full-time, in-home family child care for a 4-year-old ranges from $3,675 per year in Mississippi to $13,668 in Washington, DC. Average costs of center-based care ranges from $3,997 in Mississippi to $17,842 in Washington, DC (Child Care Aware® of America 2015).

Of course, many employed parents have more than one child who requires care. The estimated costs of care for two preschool-age children range between $8,819 and $40,437 for center-based care. In the Northeast and Midwest, the costs associated with caring for two children are the highest single household expense, while the costs for child care for two children in the West and South are exceeded only by the cost of housing. In all regions, the average cost of child care in a center for an infant is greater than what families spend each month on food (National Association of Child Care Resource and Referral Agencies 2015). In 2014, the average annual cost for an infant in center-based care exceeded the cost of a year of tuition and fees at a four-year, public college in 28 of the 50 states. Costs for a 4-year-old were higher than public college tuition and fees in 19 states. Of note, parents pay roughly 60 percent of child care costs but less than a quarter of public college costs, as the balance is subsidized by state and federal funds (Fraga, Dobbins, and McCready 2015).

Despite the clear need for more affordable, dependable, and accessible quality child care, the United States compares poorly to many European countries. Take, for example, France, where child care is publicly funded as part of early education (Clawson and Gerstel 2002). Nearly all 3- to 5-year-olds are enrolled in full-day programs taught by well-paid teachers.

What is the effect of early outside child care on children? The results of research are mixed. In evaluating such data, it is important to keep in mind the family's education, the personalities involved, and the family interests—key factors that play a part in which parent chooses to return to work and which must return to work once a child is born (Crouter and McHale 1993). Furthermore, a child's personality, the child's age when the custodial parent reenters the workforce, the involvement of the other parent in the home, the quantity of time spent working or with the child, the nature of the work, and the quality of care all contribute to how child care affects a child.

Evidence is mixed as to the effects of maternal employment on mother–child attachment (Hazen et al. 2015). Although earlier research suggested that infants of employed mothers developed an insecure attachment to their mothers, later research indicates that instead of whether the mother is employed—and therefore away more hours from her infant—it is her sensitivity to her infant's needs when together that affects mother–infant attachment (Owen 2003). Recent research by Nancy Hazen and colleagues found that excessive amounts of nonmaternal care at 6 to 8 months of age was associated with disorganized maternal attachment. Very long hours of nonmaternal care when accompanied by certain maternal characteristics (e.g., lower maternal sensitivity), have also been associated with insecure infant attachment (Hazen et al. 2015). Other subsequent consequences of child care—such as behavior problems, lowered cognitive performance, distractibility, and inability to focus attention—also have been noted. These negative effects are not necessarily the consequence of being cared for by outside caregivers. Rather, they may be the result of *poor-quality* child care. It has been noted that high-quality care, given by sensitive, responsive, and stimulating caregivers in a safe environment with a low teacher-to-student ratio, can actually facilitate the development of positive social qualities, consideration, and independence (Field 1991). In school-age and adolescent children, maternal employment is associated with self-confidence and independence, especially for girls whose mothers become role models of competence (Hoffman 1979).

National concern periodically is focused on day care by revelations of sexual abuse of children by their caregivers or by stories of children being injured or killed while in nonparental care. For example, in 2012, a 3-year-old in Dallas, Texas, died after being left in a day care van as temperatures climbed to 105 degrees (Sakmari and Guerra 2012). However, there are also a number of similar stories of children dying when parents left them in cars as temperatures soared. As to abuse, children have a far greater likelihood of being sexually abused by a father, stepfather, or other relative than by a day care worker. Furthermore, parents with children in child care should take some degree of comfort from the evidence demonstrating that children in outside, especially in organized, child care facilities are safe. Overall, all types of child care are safer than care within children's own families (Finkelhor and Ormrod 2001; Wrigley and Dreby 2005).

## Older Children, School-Age Child Care, and Self-Care

Although there are particularly acute needs when children are young, employed parents' child care needs are not restricted to families with preschoolers; parents of children in middle school also have child care needs. A number of terms are used to refer to caregiving for these older children, including *after-school*, *around-school*, *out-of-school*, and *school-age care* (Polatnik 2002).

Many children express strong opposition to after-school programs, seeing them as geared toward "little kids," but they find activities such as sports or other recreational or artistic programs more appealing (Polatnik 2002). Unfortunately, even when the programs and activities are free or when the costs are affordable, they are neither consistent nor continuous enough to cover the whole time children are out of school before parents return from work. Many parents of these children feel pressed to allow them to stay home alone. In 2011, approximately 4.4 million 5- to 14-year-olds were in **self-care**—that is, caring for themselves without supervision by an adult or older adolescent while their mothers are at work (Laughlin 2013). Around 1.3 million of these children are between 5 and 11 years old.

Although it may be unsettling to consider 5-year-olds fending for themselves, that is also uncommon. The U.S. Census Bureau data on child care arrangements reveal that self-care is very rare for young children (5 to 8 years old). For example, Figure 11.4 shows the percentages of children in self-care in 2010. In the spring of 2011, roughly 2 percent of 5- to 8-year-old children were in self-care. Regardless of race, mother's education, mother's employment status, or household income, few children of this age were taking care of themselves. Considering older children, 10 percent of 9- to 11-year-olds and 33 percent of 12- to 14-year-olds were in self-care during the spring of 2011 (Laughlin 2013). Census data also indicate that whites, those with some college, and those with the highest incomes were most likely to have had their children in self-care during the spring of 2011 (Laughlin 2013).

The shortage of affordable, high-quality child care is an obstacle for many employed parents.

Susan Chiang/E+/Getty Images

As with a number of critical services in our society, those who most need supplementary child care are those who can least afford it. The United States is one of the few industrialized nations without a comprehensive national day care policy. In fact, beginning in 1981, the federal government dramatically cut federal contributions to day care; many state governments followed suit.

## Effect on Employment and Educational Opportunities

The lack of affordable child care or inadequate child care has the following consequences:

- It prevents many mothers from taking paid jobs.
- It keeps many women in part-time jobs, most often with low pay, few or no benefits, and little career mobility.

**Figure 11.4  Children Home Alone: Percent of 5- to 14-Year-Olds in Self-Care, Spring 2010**

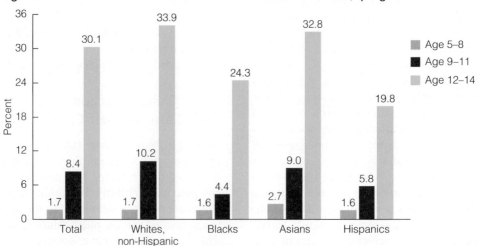

SOURCE: U.S. Census Bureau (2010).

- It keeps many women in jobs for which they are overqualified and prevents them from seeking or taking job promotions or training necessary for advancement.

- It sometimes conflicts with women's ability to perform their work.

- It restricts women from participating in education programs.

For women, cost, lack, or inadequacy of child care are major barriers to equal employment opportunity (DeSilver 2014). Many women who want and need to work are unable to find or to afford adequate child care. Child care issues may also play a significant role in women's choices concerning work schedules, especially among women who work part-time.

## Inflexible Work Environments, Stressful Households, and the Time Bind

In dual-worker families, the effects of the work environment stem from not only one workplace but two. Although many companies and unions are developing programs that are more responsive to family situations, the workplace in general has failed to recognize that the family has been radically altered during the past 50 years. Most businesses are run as if every worker were male with a full-time wife at home to attend to his needs and those of his children, or were wealthy enough to have paid domestic help to fill the needs of their families. But the reality is that women make up a significant part of the workforce, and they do not have wives at home. Nor do the vast majority of Americans have nannies for their children or domestic help to assist them in their households. Allowances are not made in the American workplace for flexibility in work schedules, day care, emergency time off to look after sick children, and so on. Many parents would reduce their work schedules to minimize work–family conflict. Unfortunately, many do not have that option.

More than 20 years ago, Carol Mertensmeyer and Marilyn Coleman (1987) contended that our society provides little evidence that it esteems parenting. It appears that little has changed. This seems to be especially true in the workplace, where corporate needs are placed high above family needs. Mertensmeyer and Coleman suggested that family policy makers encourage employers to be more responsive in providing

**False** **10.** When companies offer policies to reduce work–family conflict, most women and men do not make full use of them.

parents with alternatives that alleviate forced choices that are incongruent with parents' values. For example, corporate-sponsored child care may offset the conflict a mother feels because she is not at home with her child. Flextime and paid maternal and paternal leaves are additional benefits that employers could provide employees. These benefits would help parents fulfill their own and family expectations and would give parents evidence that our nation views parenting as a valuable role.

Unfortunately, the presence of policies cannot guarantee that employees will use them. In her book *The Time Bind: When Work Becomes Home and Home Becomes Work*, sociologist Arlie Hochschild (1997) described the official policies and corporate culture at a large corporation that she called Amerco (so as to protect its anonymity). At Amerco, workers had access to a number of family-friendly time-enhancing policies, including job sharing, part-time work, parental leave, flextime, and "flexplace," now more widely referred to as "telecommuting" (where workers could work from home). Despite the apparent availability of such opportunities, Hochschild notes that Amerco employees rarely made use of them.

Hochschild noted that Amerco employees are typical of employees at other large corporations. Fortune 500 manufacturing companies tended to offer family-friendly policies, yet few employees used them. This lack of use was especially puzzling given that Amerco employees acknowledged not having enough family time.

In accounting for the lack of utilization of workplace policies, Hochschild considered a variety of explanations. Can employees afford to work fewer hours? Do they fear being laid off? Do employees even know about policies? Do they have insensitive and insincere supervisors?

These explanations had partial validity. Some hourly employees feared potential layoffs or reduced wages. Some supervisors seemed reluctant to embrace and resentful at having to accommodate family-friendly policies. But Hochschild contended that the biggest reason employees made little use of potential family time-enhancing initiatives is because they would rather be at work.

Hochschild maintained that with the dramatic changes in the division of labor and the growth of dual-earner families, home life has become more stressful and tightly scheduled. There is too much to do, too little time to do it, and not enough appreciation or recognition for what is done. On the other end of the work–family divide, many workplaces in the United States have implemented "humanistic management" policies designed to enhance worker morale and productivity and to reduce turnover. Thus, at work, people can find social support, appreciation, and a sense of control and competence, which makes them feel better about themselves. In other words, for some, work had become homelike, and home often felt like a job (Hochschild 1997).

There is little to suggest that dual-earner home life has since become stress-free. Indeed, it is likely as stressful as when Hochschild authored *The Time Bind*. But because Hochschild (1997) studied only one company, we cannot say how commonly experienced her findings may be among dual-earner families. In fact, other researchers failed to support Hochschild's conclusions, at least to the same extent. For example, a study by Susan Brown and Alan Booth (2002b) that used the National Survey of Families and Households and is based on more than 1,500 dual-earner couples with children indicates that Hochschild's findings may not be generalizable.

Job status seems to be an important determinant of whether individuals see their jobs as more satisfying than their home lives. Brown and Booth claim that this is true only among workers in positions of lower occupational status. In addition, respondents who have high satisfaction with work and low satisfaction at home do not necessarily work significantly more hours at work. Only those who are satisfied with work, unsatisfied with home, and have adolescent children work more hours (Brown and Booth 2002b).

Another study by K. Jill Kiecolt (2003), based on General Social Survey data from 1973 to 1994, challenged several of Hochschild's (1997) conclusions. She argued, for example, that a "cultural reversal" in favor of work over home had not taken place and that employed parents with children under age 6 actually are more likely to find home rather than work to be a haven.

Even if Hochschild's findings are somewhat limited, her study was clear evidence that the presence of policies does not guarantee their use (see also Blair-Loy and Wharton 2002). People must take advantage of policies. This suggests that people's values must be directed more toward home and family. Furthermore, cultural support for using family-friendly policies must be more widespread and reflected in *company* "cultures." If the message people learn at work is that their commitment to their job is being measured by how much time they put in, then reducing work time for family needs makes them appear undercommitted.

Change must also occur at home. Dual-earner family life must be made less stressful. One way in which this can occur is by men doing more of the second-shift work discussed earlier, thereby reducing the overload and time drain that their wife more consistently feels.

Employees who feel supported by their employer with respect to their family responsibilities are less likely to experience work–family conflict. A model corporation would provide *and support* the use of family-oriented policies that would benefit both its employees and itself, such as flexible work schedules, job-sharing alternatives, extended maternity and/or paternity leaves and benefits, and child care programs or subsidies. Such policies could increase employee satisfaction, morale, and commitment.

## Living without Work: Unemployment and Families

In the latter years of the last decade, the U.S. economy suffered what came to be called "the Great Recession." Beginning in December 2007 and continuing even after the official end of the recession in June 2009, the effects of the recession were profound. One immediate and visible impact was a dramatic increase in unemployment. In June 2007, the U.S. unemployment rate stood at 4.5 percent. Seven million Americans were counted as unemployed. By June 2009, the unemployment rate had more than doubled, reaching 9.5 percent. More than 14.5 million people were unemployed, and 5 million of them were considered "long-term unemployed," having been jobless for 27 or more weeks. By July 2012, unemployment stood at 8.3 percent; nearly 13 million people were unemployed, with "long termers" making up a little over 40 percent of all unemployed women and men. Two years later, the unemployment rate had fallen to 6.2 percent, with more than 9.6 million unemployed, more than 3 million long-term. As these words are being typed in December 2015, unemployment stands at 5 percent, still a half percent higher than before the recession. Almost 8 million people are unemployed; more than 2 million people, over a quarter of those unemployed, have been unemployed for at least 27 weeks (U.S. Bureau of Labor Statistics, The Employment Situation, December 5, 2015).

The real employment situation is always worse than the official numbers. To be counted among the "unemployed," one must have looked for a job within the four weeks prior to being surveyed. As a consequence of this measurement criterion, another 1.7 million people who were not in the labor force, who wanted and were available for work, and who had looked for a job within the prior 12 months but not within the most recent four-week period, were not included in the count of the unemployed. Instead, they were classified as "marginally attached" to the labor force.

Unemployment rates vary across lines of age, gender, race, and ethnicity. This is evident in Figure 11.5. The biggest variations are by ethnic and racial background and by age. Among teens, nearly one out of four are unemployed. Although 4.3 percent of whites were unemployed in November 2015, 6.4 percent of Hispanics and 9.4 percent of African Americans were unemployed (see Figure 11.5).

Unemployment is a major source of stress for individuals, with its consequences spilling over into their families (Cottle 2001). In 2014, 6.5 million families, 8.0 percent of all families, had an unemployed member. Blacks (14.1 percent) and Hispanics (10.8 percent) were more likely than whites (7.0 percent) or Asians (7.5 percent) to have an unemployed family member. Even employed workers suffer anxiety about possible job loss caused by economic restructuring and downsizing (Pugh 2015). Job insecurity leads to uncertainty that affects the well-being of both workers and their spouse. They feel anxious, depressed, and unappreciated. For some, the uncertainty before losing a job causes as much or even more emotional and physical upset than the actual job loss.

## Families in Distress

In times of such hardship, economic strain increases; the rates of infant mortality, alcoholism, family abuse, homicide, suicide, and admissions to psychiatric institutions and prisons also sharply increase. Those who find themselves unemployed and facing income loss make other adjustments in family living. Some may postpone childbearing; others may move in with relatives or have relatives or boarders join the household (Voydanoff 1991).

The emotional and financial cost of unemployment to workers and their families is high. A common public policy assumption, however, is that unemployment is primarily an economic problem. Joblessness also seriously affects health and the family's well-being.

The families of the unemployed experience considerably more stress than that experienced by those of the employed. Whether women or men experience unemployment, families will suffer economically. With employed women's earnings in dual-earner households having become a greater share of total family income, the economic impact of women's job loss has increased. For many, it will approach or even exceed the economic impact of men's job loss. Families may find themselves pushed to or even over the edge when a wage earner loses his or her job. The resulting economic strain can generate increased stress and conflict. Unemployment has been found to be associated with reduced marital quality.

As families struggle to adapt to unemployment, family roles and routines are likely to change. In cases of either male or female unemployment, one area that changes is the division of household labor.

**Figure 11.5** U.S. Unemployment, June 2007, July 2009, July 2012

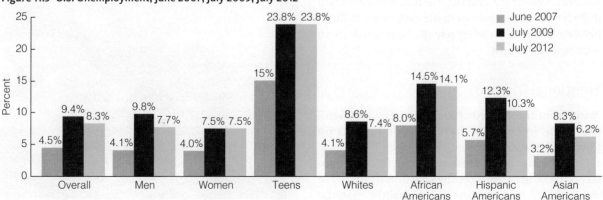

SOURCE: U.S. Bureau of Labor Statistics (2012).

Unemployed spouses increase the share of housework they shoulder, spending increased time in housework activities compared to the time they spent when they were employed. Meanwhile, spouses of unemployed individuals decrease the time they spend in housework (Gough and Killewald 2011). Researchers Margaret Gough and Alexandra Killewald found that though both men and women increased their time spent in housework upon becoming unemployed, the increase in women's housework was twice as large as the increase in men's. Regardless of whether women or men became unemployed, one consequence was an increase in the total time spent in housework (Gough and Killewald 2011).

After the first few months of their husbands' unemployment, wives of unemployed men begin to feel emotional strain, depression, anxiety, and sensitiveness in marital interactions. Children of the unemployed are likely to avoid social interactions and tend to be distrustful; they report more problems at home than do children in families with employed fathers. Families seem to achieve stable but sometimes dysfunctional patterns around new roles and responsibilities after six or seven months. If unemployment persists beyond a year, dysfunctional families become highly vulnerable to marital separation and divorce; family violence may also begin or increase at this time (Teachman et al. 1994).

The types of families hardest hit by unemployment are single-parent families headed by women, African American and Latino families, and young families. Wage earners in African American, Latino, and female-headed, single-parent families tend to remain unemployed longer than other types of families. Because of discrimination and the resultant poverty, they may not have important education and employment skills. Young families with preschool children often lack the seniority, experience, and skills to regain employment quickly. Therefore, families in the early years of childbearing and child rearing pay the largest toll in an economic downturn.

## Emotional Distress

Aside from the obvious economic effect of unemployment, job loss can have profound effects on how family members see each other and themselves (see the Real Families feature in this chapter). This in turn can alter the emotional climate of the family as much as lost wages alter the material conditions. Men are particularly affected by unemployment because wage earning is still a major way men satisfy their family responsibilities. Thus, when men fail as workers, they may feel they failed as a husband, father, and a man (Cottle 2001). As Lillian Rubin (1994, 103) poignantly conveyed in her book *Families on the Fault Line*, when men lose their jobs, "it's like you lose a part of yourself." Psychologist Thomas Cottle (2001, 21) suggests that a man's unemployment can strip "away his pride, his self-definition, his sense of personal power, and his belief in his ability to control outcomes, not the least of which is his own personal destiny." Unemployed men may display a variety of psychological and relationship consequences, including emotional withdrawal, spousal abuse, marital distress, increased alcohol intake, and diminished self-identity (Cottle 2001; Rubin 1994). Long-term unemployment can bring with it feelings of despair, desperation, confusion, impotency, and powerlessness. Katherine Newman (1988) suggested that

Unemployed men may suffer losses in self-esteem, and feel as though they are failing in their roles as spouses and parents.

Thinkstock/Stockbyte/Getty Images

It was christened the "man-cession" to highlight the fact that the U.S. recession of last decade took a greater toll on men's employment than on women's employment. Economist Heather Boushey pointed out that men experienced three out of every four job losses and that, as of July 2009, 15.6 percent of employed married women had husbands who were not employed. Among employed wives with children under age 6, 5.9 percent had unemployed husbands. Given that, on average, women in two-earner households bring in 37 percent of the household income, losing the man's income can drive families into dire economic straits. Worsening the problem is that health insurance in two-earner households is usually obtained through the man's job (Boushey 2009). Thus, when men lose their jobs, the economic impact can be severe, but other problems also arise. Families often have to shuffle responsibilities, downsize their standards of living, and watch nervously as whatever savings they accumulated diminish.

*New York Times* journalist Jennifer Steinhauer described 41-year-old Phil Winkler's transition from "primary breadwinner" to family "bus driver, disciplinarian, schedule organizer, and head chef" (Steinhauer 2009). A former production worker for Nestlé, he had never before experienced job loss, yet Winkler found himself out of work for a year. To manage, he sold his favorite car, canceled the family cable television package, and started scavenging junkyards for auto parts he could resell on e-Bay. As a consequence of his job loss, Winkler experienced wounds to his self-esteem, felt strains in his marriage, and had to impose painful sacrifices on his family. No summer swimming lessons for his daughters, no "shopping at random," no more eating out on Friday nights—these are just some of the adjustments his family had to make.

They were not alone. In fact, as Steinhauer reported, in the Winkler's neighborhood of Beth Court, a cul-de-sac in Moreno Valley, California, hardships such as these became part of life, as "job losses swept the neighborhood like an unstoppable illness." As he saw neighbor after neighbor lose their jobs, Mr. Winkler said that where once he'd seen men going off to work, instead dads began picking kids up at school, dropping them at practice, overseeing homework, and depending on their wives' jobs and incomes as well as unemployment benefits. There were some silver linings in this cloudy economic situation. Being home, unemployed men can (theoretically) take on enlarged roles in the household and with their children. Boushey (2009) noted that as difficult as families may find it, today's couples are less likely locked into traditional notions of men as the family breadwinners. Thus, a man's unemployment wouldn't necessarily trigger "life-altering role reversal" (Goodman 2009). Some men may have seized the opportunity to become primary parents. Some, however, continued to desperately search for work, seeing it as more than just their economic contribution to their families.

Although men were the ones to face harsher job loss in the recession, they have also experienced greater gains after the recession ended. Between the end of the recession and July 2011, the number of jobs held by men increased by close to 770,000, whereas the number held by women actually dropped by over 218,000. In a report for the Pew Research Center, economist Rakesh Kochhar described these gender swings:

> At the start of the recession, in December 2007, men held 3.4 million more jobs than women. In the recession, job losses for men were more severe than for women and by the end, in June 2009, men held only 223,000 more jobs than women. This gap stretched out to 1.2 million in May 2011, two years into the economic recovery. (Kochhar 2012)

As journalist Catherine Rampell noted, we moved from a man-cession into a "he-covery." As more formerly unemployed men returned to work more quickly and widely than did women, such economic fluctuations once again carried the potential to alter how families on the Winklers' cul-de-sac and elsewhere experienced their day-to-day life (Rampell 2011). ●

when families suffer downward mobility as a result of male unemployment, relations between spouses or between fathers and children are likely to be strained. Although children and spouses may be initially supportive, their support may wear thin or run out if joblessness persists and other resources are unavailable, thus preventing families from maintaining their previous economic lifestyle.

Of course, women also suffer nonmaterial losses when they lose their jobs, but those losses are different in degree and kind from those that men are likely to suffer. Men have more of their self-identities—and especially their gendered identities—tied up in working; success at work comes to define successful masculinity and symbolizes having satisfied a central element of one's roles as a husband and father (Arrighi and Maume 2000). Women have other acceptable ways of maintaining or achieving adult status (e.g., as a mother). Thus, although both women and men will suffer from lost work relationships, lost gratification, and even lost structure and purpose to their day, women have not put as many of their "identity eggs" into the "work basket" as have most men.

## Coping with Unemployment

Economic distress does not necessarily lead to family disruption. In the face of unemployment, some families may even experience increased closeness. Families with serious problems, however, may disintegrate. Individuals and families use a number of coping resources and behaviors to deal with economic distress. Coping resources include an individual's psychological disposition, such as optimism; a strong sense of self-esteem; and a feeling of mastery. Family coping resources include a family system that encourages adaptation and cohesion in the face of problems and flexible family roles that encourage problem solving. In addition, social networks of friends and family may provide important support, such as financial assistance, understanding, and willingness to listen.

Several important coping behaviors assist families in economic distress caused by unemployment. These include the following:

- *Defining the meaning of the problem.* Unemployment means not only joblessness but also diminished self-esteem if the person feels that the job loss was his or her fault. If a worker is unemployed because of layoffs or plant closings, the individual and family need to define the unemployment in terms of market failure, not personal failure. For example, during the period of time between December 2007 and June 2009, some 4.1 million people filed unemployment claims as a result of the nearly 40,000 mass layoffs that occurred (U.S. Bureau of Labor Statistics 2009). Though these were in large part products of the Great Recession, individuals may still have wrestled with whether they were at fault

for their job loss. An ability to see the larger reality will lessen such concerns.

- *Problem solving.* An unemployed person needs to attack the problem by beginning the search for another job, dealing with the consequences of unemployment (e.g., by seeking unemployment benefits and cutting expenses), or improving the situation (e.g., by changing occupations or seeking job training or more schooling). Spouses and adolescents can assist by increasing their paid work efforts.

- *Managing emotions.* Individuals and families need to understand that stress may create roller-coaster emotions, anger, self-pity, and depression. Family members need to talk with one another about their feelings; they need to support and encourage one another. They also need to seek individual or family counseling services to cope with problems before they get out of hand.

# ■ Reducing Work–Family Conflict

Before ending this chapter, it is worthwhile to consider ways in which many of the problems resulting from the conflict between work and family might be reduced. Although individuals and/or couples can craft personal strategies for managing their busy work and family lives—such as the opposite-shift arrangement discussed earlier—more external support from employers and the government is needed. Individuals alone cannot design their jobs in such a way as to best fit with their family needs.

**Family policy** is a set of objectives concerning family well-being and the specific government measures designed to achieve those objectives. Given the host of issues raised in this chapter, we might argue that if families were truly the national priority we claim them to be, we would entertain and enact policies to reduce the complexities associated with combining work and family life.

Of the following policies, some emphasize increasing people's time and availability to meet their family needs by making their work time or place flexible; others are designed to provide assistance in securing safe and affordable alternative, nonparental care arrangements. Still others are aimed at guaranteeing a certain economic standard and equal opportunity so that employed women and men and their families can meet their economic needs. Consider the following examples of different policies:

- *Policies designed to make work more flexible.* There are a number of different ways to increase flexibility in the scheduling, amount, or location of work. Most familiar to us is **flextime**, or flexible work schedules, where workers can choose and change their start and quit times to meet their personal and familial needs. A 2008 study by the National Study of the Changing Workforce found that workers with higher levels of flexibility were more likely to report low interference between job and family life than workers with low levels of flexibility (44 percent to 23 percent for low-wage workers, and 26 percent to 19 percent for high-wage workers). Those with highly flexible work arrangements also reported higher levels of job satisfaction than those with low levels of workplace flexibility (low-wage workers reporting a difference of 62 percent to 15 percent, and high-wage workers reporting a difference of 62 percent to 27 percent) (Matos and Galinsky, 2011).

At present, although nearly 80 percent of workers would like and would use more flexible options as long as they experienced no negative consequences in potential advancement, less than 30 percent of workers work flexible schedules. Forty percent believe that their chances of advancing would suffer if they used a flexible option.

According to data reported by the Families and Work Institute, 2012 National Study of Employers, three-fourths (77 percent) of employers allow at least some employees to use flextime and periodically change their start and quit times—an increase over the two-thirds who allowed it in 2005. In addition, almost 90 percent (up from 77 percent in 2005) allow employees to take time during the workday to tend to family or personal affairs without any lost wages or dock in pay.

In addition to flexible daily work schedules, however, each of the following is also a means of increasing the flexibility of work, and thus enabling workers to more easily fit work and home life together:

- *Job sharing.* Job-sharing policies allow two employees to share a full-time job, usually splitting the benefits between the two. In 2007, 20 percent of employers offered employees this alternative. According to a survey of more than 500 companies by the Society for Human Resource Management, by 2010, that number had dropped to 13 percent (Shellenbarger 2010).

- *Compressed workweeks.* Roughly a third of employers, up from 26 percent in 2008, allow employees to work shorter weeks by working longer days (Tahmincioglu 2011).

- *Telecommuting.* Telecommuting refers to working from home, typically using phone and computer connection to one's workplace. Almost two-thirds of companies allow workers to do at least some work from home. Data on telecommuting are problematic because although many employees might work from home at least occasionally, fewer do so on a regularly scheduled and/or formally arranged basis. Nevertheless, it is estimated that 24 percent of those who are employed regularly work some hours at home. Between 1997 and 2010, the percentage of the workforce working at least one day a week from home increased from 7 to 9.4 percent, representing an increase from 9.2 million to 13.4 million workers. Of those, approximately 4 million work from home occasionally, making them what the Census Bureau calls "mixed" workers. Mixed workers tend to be highly educated, better paid, and work more hours. Those who work from home face "the flexibility stigma," and compensate by working ever harder and ever crazier hours" (Rosen 2015). Research suggests that telecommuting may add to one's work hours, that it "relocates" rather than reduces one's work time. Although it has definite advantages—it could cut commuting time and expense, lessen traffic congestion, and allow those with caregiving responsibilities to achieve greater work–family balance—it has so far minimally reduced the hours workers spend on-site (Noonan and Glass 2012).

- *Policies to provide, support, or assist working parents in locating alternative nonparental child care arrangements for children of employees.* Although fewer than 10 percent of employers with 50 or more employees provide on-site child care (Bond et al. 2005), other policies assist employed parents in their search for and access to affordable and dependable child care. These include employer-provided child care resource and referral services as well as assistance or reimbursement programs for some of the expenses incurred for child care.

- *Family leave policies.* These are used for pregnancy, for childbirth, and/or to care for sick children or time off to meet other familial emergencies, and paid personal days for parenting and other family responsibilities.

- *Policies to ensure that employees receive adequate wages to provide for their families and to protect them from discrimination (by gender, ethnicity, sexual orientation, family status, or disability) in hiring, advancement, or pay.* Such initiatives include increasing the minimum wage so that workers can support their families, policies to ensure fair employment, measures to ensure pay equity between men and women for the same or comparable jobs, and affirmative action programs for women and ethnic groups.

In 2014, *Working Mother* magazine published a list of work–family and work–life benefits, identifying what percentage of U.S. companies provided such benefits and what percentage of companies they consider the "100 Best Companies" for working mothers offered those benefits. As seen in Table 11.7, companies in the United States still have far to go in the quest to create family-friendlier workplaces.

To reduce work–family conflict, lessen the stress of juggling jobs and families, and improve the quality of family life for working parents and their children, policies such as these need to be more widely adopted. At the same time, policies such as these must then be supplemented by sincere cultural support for families and children. People must believe that if they commit themselves to their families and take advantage of such measures, they will not suffer unfair economic consequences. This may be harder to convey and carry out than are most specific workplace policies.

Our marriages and families are not simply emotional relationships—they are also work relationships in which we divide or share many household and child-rearing tasks, ranging from changing diapers, washing dishes, cooking, and fixing running toilets and leaky faucets to planning a budget and paying the monthly bills. These household tasks are critical to maintaining the well-being of our families. They are also unpaid and insufficiently honored. In addition to household work and child rearing, we have our employment, the work we do for pay. Our jobs usually take us out of our homes from 20 to 80 hours a week. They not only are sources of income, but they also help our self-esteem and provide status. They may be sources of work and family conflict as well.

Now a decade into the 21st century, we still need to rethink the relationship between our work and our families. Too often, household work, child rearing, and employment are sources of conflict within our relationships. We need to reevaluate how we divide household and child-rearing tasks so that our relationships reflect greater mutuality. For many, poverty and chronic unemployment lead to distressed and unhappy families. Because most individuals do not have direct control over the number of hours they spend at work or the timing of those hours, they need outside help from either the government or the private sector to reduce the extent of work–family conflict. Designing and supporting policies that help reduce the conflict between paid employment and families will, in return, build stronger families.

**TABLE 11.7** *Working Mother's* **100 "Best Companies" and U.S. Companies, Work–Family Benefits Provided**

| Benefit | Working Mother's 100 "Best Companies" | U.S. Companies |
|---|---|---|
| Fully paid maternity leave | 100% | 5%* |
| Child care resource and referral service | 96% | 9% |
| Paid adoption leave | 93% | 17% |
| Adoption assistance | 93% | 7% |
| Paid paternity leave | 90% | 17% |
| Backup child care | 89% | 4% |
| Telecommuting | 100% | 60% |
| Flextime | 100% | 54% |
| Compressed work schedule | 94% | 31% |
| Job-sharing | 82% | 10% |

NOTE: National numbers are based on the 2015 benefits survey of Society for Human Resource Management members.
*2014 National Study of Employers, Family, and Work Institute
SOURCE: Working Mother (2015).

In the United States, the passage of the **Family and Medical Leave Act of 1993** (FMLA) provided employees with unpaid, job-protected leave of up to 12 weeks to care for an ill family member or to take time off after childbirth or adopting a child. The job protection was an important provision, as it meant that employees could not be let go just because they took time off at the birth or adoption of a child or for a family medical emergency. However, because the leave is unpaid, many workers cannot afford to lose the income they would have to sacrifice for three months and therefore are unable to use it. In addition, such leaves are guaranteed only if one works in a workplace with 50 or more employees. As a consequence, as many as half of U.S. workers are left unprotected by the policy. In addition, to be eligible for the FMLA guaranteed leave, one must have been employed for 12 months by the same company, during which time they must have worked 1,250 hours (Brown 2009).

The FMLA specifies only the minimum required for parental (and family) leaves. Roughly 16 percent of companies with at least 100 employees go beyond what is required by providing fully paid maternity leave. This is down from 1998, when 27 percent of companies provided fully paid leaves (Brown 2009). Some companies' policies are considerably more generous than the federally guaranteed leave. For example, Ernst & Young offers new mothers 39 weeks of leave with full pay. Bank of America offers 12 weeks of paid parental leave for mothers, fathers, or adoptive parents. PricewaterhouseCoopers offers new mothers, fathers, or adoptive parents 12 to 14 weeks of paid leave. Thus, the FMLA sets the minimum that a company or organization must offer employees, assuming of course that the other criteria are met.

In 2004, California became the first state in the United States to enact its own paid family leave policy, providing up to six weeks of leave per year, during which employees are eligible to earn 55 percent of their regular wages, up to $728 per week. California's leave policy does not impose a minimum workplace size criterion for eligibility as does the FMLA, nor are employees prevented from taking the state leave simultaneously with an FMLA leave, thus leading to 12 weeks of leave with 55 percent of one's salary for half the period that provides employees paid maternity leave (Gonzalez 2006). Currently, at least 11 states have enacted their own policies, some extending them to smaller workplaces than the federal law indicates, others extending the benefit to include domestic or civil union partners, in-laws, and grandparents. A 2009–2010 survey of over 250 California employers found that 87 percent of them responded that the paid leave policy did not result in increased costs; 9 percent actually responded that the policy saved money. Ninety percent or more reported that paid leave had either a "positive effect" or "no noticeable effect" on employee turnover (96 percent), productivity (89 percent), profitability/performance (91 percent), and employee morale (99 percent) (Farrell and Glynn 2013).

On a global scale, the family leave policies in the United States compare unfavorably to the leave policies in most other developed nations. For example, the United Nations reports that out of 185 countries and territories about which they have information, all but three—Oman, Papua New Guinea, and the United States—offer women guaranteed leave with pay associated with childbirth (Kim 2015). Nearly 60 percent of these countries offer paid leaves of 14 or more weeks. In 2013 Senator Kirsten Gillibrand and Representative Rosa DeLauro introduced a bill that would, if passed by both the Senate and the House of Representatives, remedy this situation. The bill (HR 1439/S. 786) known as the Family and Medical Insurance Leave Act, or The FAMILY Act, would guarantee that individuals had some income during their family or medical leave. It would apply to workers in all companies, regardless of their size, and be managed by a new Office of Paid Family and Medical Leave, which would be housed within the Social Security Administration. Workers would be provided up to 12 weeks of partial income, capped at 66 percent of their monthly wages.

In March 2015, the bill was assigned to a congressional committee that will decide whether to send it on to the full Senate or the House of Representatives. Unfortunately, the bill is not seen as having much chance of being enacted (govtrack.us). Assuming that pessimistic assumption is accurate, the United States will remain in very select company, though not the kind to which one aspires, in being one of three countries without paid leave. ●

# Summary

- Families are economic units bound together by emotional ties. Families are involved in two types of work: *paid work* at the workplace and unpaid *family work* in the household.

- The *bifurcation of work time* refers to the fact that although many feel overworked, others are underemployed or lacking work.

- Families are economic units bound together by emotional ties. Families are involved in two types of work: *paid work* at the workplace and unpaid *family work* in the household.

- The *bifurcation of work time* refers to the fact that although many feel overworked, others are underemployed or lacking work.

- *Work-to-family spillover* is the effect that employment has on the time, energy, and psychological well-being of workers and their families at home. *Family-to-work spillover* is when the demands from home life reduce the time and energy available to succeed at work.

- *Role strain* refers to difficulties that individuals have in carrying out the multiple responsibilities attached to a particular role; *role overload* occurs when the total prescribed activities for one or more roles are greater than an individual can handle; and *role conflict* occurs when roles conflict with one another. *Crossover* refers to when one's feelings about work affect one's spouse's or partner's feelings.

- The traditional division of familial labor is complementary: Husbands work outside the home for wages, and wives work inside the home without wages. A minority of U.S. families displays a traditional allocation of paid work and family roles.

- The level of women's participation in the paid labor force increased as a result of social and economic changes. Individual women enter the workforce for economic, social, and psychological reasons. Employed women tend to have better physical and emotional health than do nonemployed women.

- Women perform a majority of the daily housework and carry more responsibility for managing the division of housework. Women's household tasks tend to include the daily chores (such as cooking, shopping, cleaning, and so on) and child care. Men's household tasks tend to be more occasional and often outdoors.

- Men's involvement in routine housework is less affected by their gender-role attitudes than by the more immediate circumstances in which they find themselves. Their involvement is also influenced by their upbringing, their experiences and status at work, and their age.

- A variety of indicators point to increases in men's share of housework and decreases in the amount of time women spend on housework.

- The division of paid and unpaid labor and the allocation of housework affect marital power, marital satisfaction, sexual intimacy, and marital stability (i.e., the risk of divorce).

- Mothers' caregiving responsibilities include doing more of the *mental labor* of child care, including worrying about the children, gathering information, and managing the father's involvement as well as monitoring where their children go, who they are with, and what they do. Fathers' involvement in caregiving is influenced by the age and gender of children, age and gender attitudes of the fathers, and the fathers' occupations and earnings.

- Supplementary child care outside the home is a necessity for many families. Most children who receive outside care are in child care centers. Overall, child care is safe, and center-based care is safer than "family day care" or paid care by others in the child's home.

- Nonstandard shift work has increased because of changes in the economy, demographic changes, and technological changes. It affects family experiences in both negative and positive ways.

- Approximately 1.25 million fathers of children under 15 years stay home full-time. In such households, we can identify marital, parental, economic, and social consequences that follow from this arrangement.

- Among the problems women encounter in the labor force are economic discrimination and *sexual harassment*. Families suffer from lack of adequate child care and an inflexible work environment.

- Unemployment can cause both economic and emotional distress. Unemployment most often affects female-headed single-parent families, African American and Latino families, and young families.

- Family policy is a set of objectives concerning family well-being and the specific government measures designed to achieve those objectives.

## Key Terms

abuse of power  433
accessibility  425
bifurcation of working time  410
co-provider families  417
crossover  414
dual-career families  421
emotion work  424
engagement  425
Family and Medical Leave Act of 1993  445
family policy  442
flextime  443
gender ideology  415
homemaker role  417
hostile environment  433
mental labor  425
role conflict  413
role overload  413
role strain  413
self-care  436
sexual harassment  433
shift couples  429
time strain  410
work spillover  411

# Intimate Violence and Sexual Abuse

Joel Sartore/National Geographic/Getty Images

**What Do You Think?**

*Are the following statements True or False? You may be surprised by the answers as you read this chapter.*

T F **1.** Intimate relationships of any kind increase the likelihood of violence.

T F **2.** Although abuse may be nonviolent, all family violence is considered abusive and unacceptable.

T F **3.** The value placed on family privacy contributes to a reluctance to report suspected family violence.

T F **4.** Intimate violence occurs at similar rates among all social classes.

T F **5.** Marital rape is a crime in all 50 states of the United States.

T F **6.** The most common form of child maltreatment is neglect.

T F **7.** Sibling violence is the most widespread form of family violence.

T F **8.** Violence by children against parents has been the most neglected form of family violence.

T F **9.** Under mandatory arrest policies, every state now requires police to make an arrest when responding to a domestic violence report.

T F **10.** Sexual and other forms of physical abuse occur more frequently in households with stepparents than in households with only biological parents.

# Chapter Outline

Like most Americans, you might feel as though you are safest once you've returned home at the end of your day, locked your door, and are home for the night. It is then that you might believe you have protected yourself from the threat of violence by locking out any would-be intruders. Unfortunately, the sad reality for many people is that they also *lock in* violence once they close and lock their doors to the outside world. It may seem a cruel irony, but the relationships people most value are also the relationships that may well become the most violent. The people one loves and lives with are often the people one is most likely to be hurt or assaulted by, or the people one is most likely to assault or hurt. It is an unhappy fact that intimacy or relatedness increases the likelihood of experiencing abuse, violence, sexual abuse, or even homicide.

Consider these examples: Twenty-four year-old Emily Young; 10-year-old Luis Fuentes and his younger brothers, Juan and Alexander, ages 9 and 8; 34-year-old Mary Fa'anunu; and 43-year-old Valmir Keco were all victims of 2015 U.S. homicides committed by their spouse, partner, or parent. The following were also victims and perpetrators: In October 2015, 27-year-old Glenn Baxter of Arizona, killed himself, his 25-year-old wife, Danica, and their three children (ages 3, 2, and 1 year old), a month after 45-year-old Brian Short of Minnesota shot and killed his wife Karen, his three children, and then himself. On September 28, 2015, 33-year-old Jennifer Berry threw her newborn daughter out of a seventh-floor, Bronx, New York– apartment window, minutes after giving birth at home. This was one of three such incidents in New York City over a three-month period in 2015 (Rojas and Rosenberg 2015). Brothers Robert and Michael Bever, 18 and 16, stabbed and killed their parents and three siblings—brothers Daniel and Christopher and sister Victoria—on July 22, 2015. Assistant District Attorney of Tulsa, Oklahoma, Sarah McAmis called 2015, "an awful year for child abuse and child abuse murder cases" in Oklahoma, as two late 2015 deaths—of 16-month-old Sawyer Paige Jefferson, and one-year-old Nevaeh Brookens-Roldan, brought the number of child abuse murder cases in the state to nine (Vincent 2015a, 2015b). On March 30, 2010, 37-year-old Annamarie Rintala was beaten and strangled by her spouse, Cara Rintala. Cara Rintala is the first person in Massachusetts to stand trial for murdering her same-sex spouse. She is awaiting her third trial, scheduled to begin in September 2016, after two previous trials resulted in hung juries (Everette 2016).

> **True**
> **1.** Intimate relationships of any kind increase the likelihood of violence.

Although not commonplace, crimes such as these are not unique. For example, in the United States, husbands and intimate partners are estimated to commit a third of homicides against women. Worldwide, according to the World Health Organization (2013), almost 40 percent of homicides against women are committed by spouses or intimate partners. Also, over 1,500 children died as a result of abuse or neglect in 2013, nearly 80 percent of them victimized by one or both parents (U.S. Department of Health and Human Services 2015b). Of course, deaths at the hands of partners, spouses, parents, siblings, or other relatives are neither typical of intimate violence and abuse nor representative of most homicides in the United States. However, they are chilling reminders of the worst-case outcomes of the broader phenomenon of violence among family members.

Now, consider also the following items:

- Nearly one out of every two women and one out of every five men experience sexual violence victimization other than rape in their lives (Walters, Chen, and Breiding 2013).

- Teens and young women experience the highest rate of relationship violence, with an estimated one in ten high school females indicating they had suffered.

- In 2010, an estimated 1,182 women and 304 men were killed by a current or former intimate partner (Catalano 2013).

- Each year between 2002 and 2010, an average of more than 800,000 females, and nearly 174,000 males, age 12 and older, experienced nonfatal violence (rape/sexual assault, robbery, or aggravated or simple assault) at the hands of a current or former spouse, boyfriend, or girlfriend (Catalano 2013).

- Globally, according to a World Health Organization study and report on 79 countries and 2 territories, it is estimated that 30 percent of ever-partnered women have experienced intimate partner violence (World Health Organization 2013).

- Based on a nationwide study of adolescent relationship abuse among a sample of 12- to 18-year-olds, almost a fifth of both males and females reported that they had been subjected to physical and/or sexual abuse, and more than 60 percent of both females and males had been victims and perpetrators of psychological abuse (Taylor and Mumford 2014).

- In a 2013 survey by the Centers for Disease Control, over 10 percent of teenagers in a dating relationship reported having been hit, slammed into something, or injured with a weapon or object on purpose by a partner in the 12 months before the survey (Kann et al. 2014).

- Each year, more than 2 million older Americans, typically women, are subjected to some form of abuse or neglect; 90 percent of the perpetrators are family members (National Coalition Against Domestic Violence 2008b).

- Almost a million parents are physically assaulted by their adolescents or younger children every year.

- The prevalence of same-sex intimate partner violence is estimated to be at least about the same (estimated as between 21 percent and 35 percent) as the prevalence of intimate violence experienced by heterosexual women and men. In 2011, there were 19 LGBTQH (lesbian, gay, bisexual, transgender, queer, and HIV-affected) intimate partner homicides (Mezey 2015).

In addition, as many as 90 percent of American parents spank their children. Although spankings are clearly different from beatings, assaults, physical and sexual abuse, or homicide, they are still acts of violence and therefore merit attention and consideration in this chapter.

Think for a moment about who our society "permits" us to shove, hit, kick, or spank. If we assault a stranger, push a coworker or employer, or spank or slap a fellow student or professor, we would run great risk of being arrested, fired, suspended, or expelled. With our intimates, though, we are "allowed" to do such things.

Those closest to us are the ones we are most likely to slap, punch, kick, bite, burn, stab, or shoot. And our intimates are the most likely to do these things to us (Gelles and Cornell 1990). Furthermore, living together provides people more opportunity to disagree, get angry at one another, and hurt one another. The intensity of emotions and the cultural expectations surrounding family privacy need to be added to the mix. In effect, families and households can be very dangerous places.

To understand family and intimate violence and abuse requires consideration of a range of behaviors and examination of the various factors—social, psychological, and cultural—that shed light on why it is that

Relationships can become violent or abusive regardless of the gender of partners.

Ghislain & Marie David de Lossy/The Image Bank/Getty Images

people often hurt the ones they most love. In this chapter, we look at violence and abuse between husbands and wives (including marital rape), between cohabiting partners, between gay and lesbian partners, between dating partners (including acquaintance rape), and between siblings as well as violence and abuse committed against children by parents and against parents by grown children. We look, too, at the various models researchers use in studying intimate violence, and we discuss the dynamics of battering relationships. We also discuss prevention and treatment strategies. In the last section of the chapter, we discuss child sexual abuse and focus on the types, perpetrators, and victims.

## ■ Abuse, Intimate Partner Violence, and Family Violence: Definitions and Prevalence

In exploring the violent and abusive underside of families and intimate relationships, researchers have used different and changing terminology, trying to keep pace with increasing knowledge about the phenomenon (McHugh, Livingston, and Ford 2005). Commonly, we find researchers using the terms **intimate partner violence** or **intimate partner abuse**, instead of marital violence or spousal abuse. Such terms are preferred because they more fully capture the broad scope of violence among intimate couples, of which marital or dating violence are types. Intimate partner violence (IPV) includes violence between young teenagers in middle school or high school dating relationships, unmarried cohabiting adults, married women and men, and applies to both heterosexual and same-sex couples. Other forms of family violence, such as those acts of violence between siblings or between parents and children, still most often fall under the broader umbrella term *family violence,* which itself may be officially counted as a category of abuse or maltreatment/mistreatment. The term *child maltreatment* has come to refer to the many forms of neglectful and abusive treatment to which children are subjected. They are addressed later in this chapter.

One immediate distinction to draw is between *violence* and *abuse.* For the purpose of this book, we use the definition of **violence** offered by sociologists Richard Gelles and Claire Pedrick Cornell (1990): "an act carried out with the intention or perceived intention of causing physical pain or injury to another person." Abuse includes acts such as neglect and emotional abuse, including verbal abuse, that are not violent. Thus, familial abuse is broader than and encompasses more than family violence.

*Violence* may best be understood along a continuum, with "normal" and "routine" violence at one end and lethal violence at the other extreme (Gelles and Straus 1988). Thus, family violence ranges from spanking to homicide. If one wishes to understand acts that fall anywhere between those extremes, it is necessary to examine the full continuum: "families who shoot and stab each other as well as those who spank and shove, . . . [as] one cannot be understood without considering the other" (Straus, Gelles, and Steinmetz 1980). In this chapter, we focus most of our attention on physical violence and sexual abuse that occurs between intimate partners and between family members.

Abuse can take multiple forms. Whether looking at relationships between spouses, intimate partners, parents and children, siblings, or elderly women and men and their familial caregivers, abuse can occur without any physical violence. This is certainly true of such categories as emotional or psychological abuse, financial abuse, and neglect. Emotional abuse consists of verbally insulting, embarrassing, and threatening someone, scaring them, calling them names, or threatening to destroy their property and/or acting on such threats. Financial abuse consists of the misuse of someone's money, property, or assets without their permission or comprehension. Neglect consists of failing to meet the needs of someone whose care one is responsible for, such as one's child or an ill or elderly relative, whether that is one's spouse, sibling, or parent. Failing to meet their needs for food, housing, or medical care would constitute neglect.

It may be less clear where to draw the line between physical abuse and violence; you might even find yourself wondering as you read this: Aren't those the same? Interestingly and importantly, they are not the same. Although the category of physical abuse (or maltreatment) may consist of such violent acts as hitting, kicking, pushing, slapping, striking with an object, beating, and biting, some violent acts are not considered abusive. The clearest and most controversial of such violence is corporal punishment, which most commonly consists of spanking but which can include being struck with objects. Spankings *are* violent acts, intended to inflict pain on the recipient, even if such pain is perceived to be part of an effort to teach the recipient about consequences,

> **False**
> 2. Although abuse may be violent or nonviolent, not all family violence is considered abusive and unacceptable.

differentiate acceptable from unacceptable behavior, or to protect the recipient from potential harm by punishing behavior that is seen as dangerous. Parents may mean it when they say to a child they are about to spank, "This is going to hurt me more than it is going to hurt you," but the reality is that they wouldn't bother spanking at all if it didn't hurt the child. Pain is intended. We don't allow anyone except one's family—primarily parents (and much less often teachers in certain school settings)—the right to spank, and if they do so, we subject them to a range of possible civil and/or criminal consequences.

Typically, researchers also separate sexual violence from physical abuse. That is not to ignore or minimize the violent nature of sexual abuse but rather a way to differentiate violence with a sexual element involved from those acts of violence where sexual aspects are absent. As used in typologies of abuse, sexual abuse or sexual violence entails forcing a person (e.g., one's child, an elderly parent, one's intimate partner) to engage in a sexual act when such an act is unwanted and/or inappropriate. Sexual abuse can also occur without any physical contact between perpetrator and recipient, as in the case of exposing one's "intimate parts" to a child, forcing a child to view pornographic material, or engaging in voyeuristic acts, such as secretly watching a child or family member undress or bathe.

Aware of multiple types of abuse, researchers increasingly speak of **poly-victimization** or multiple victimizations, to draw attention to the reality that the various types of abuse often co-occur (Finkelhor, Ormrod, and Turner 2007). For example, sociologists David Finkelhor, Richard Ormrod, and Heather Turner found that 22 percent of a nationally representative sample of over 2,000 2- to 17-year-olds experienced multiple forms of victimization, and that such multiple victimizations were associated with greater trauma for victims. Negative consequences of victimization are notably worse for those who suffer from higher levels of concurrent abuse. Sociologist Kristin Anderson points out that when researchers study only a single type of abuse, some of the variation in how victims respond may actually be due to their having also experienced other forms of mistreatment (Anderson 2010).

## Types of Intimate Partner Violence

Even limiting the discussion to violence in intimate couple relationships leaves one with a range of behaviors that require some kind of differentiation.

Michael Johnson and Kathleen Ferraro (2000) offered the following highly influential and widely used typology of partner violence:

- **Situational couple violence** (also sometimes called **common couple violence**) is violence that erupts during an argument when one partner strikes the other in the heat of the moment. Such violence is not part of a wider relationship pattern; it is as likely to come from a woman as a man or to be mutual. It less often escalates, and it is less likely to lead to serious injury or fatality. Situational couple violence is also not likely to be "passed on" from one generation to the next (Anderson 2010).

- **Intimate terrorism** occurs in relationships where one partner tries to dominate and control the other. Violent episodes that escalate and emotional abuse are two common traits. Victims are left "demoralized and trapped" as their sense of self and their place in the world are greatly diminished by their partner's dominance. The violence in intimate terrorism is likely to recur, escalate, and lead to injury. It is also less likely to be mutual but more likely to be transmitted intergenerationally than is situational couple violence (Anderson 2010).

- **Violent resistance** refers to violence used to resist intimate terrorism (Johnson 2006). Such violence can be self-defensive but "can also function primarily as an expression of anger or resistance even if the resistor expects that it may actually provoke greater violence from the controlling partner" (Johnson 2006, 4). In that it is largely a resistance to what is a predominantly male pattern of violence, it is more often perpetrated by women than by men.

- **Mutual violent control** refers to relationships in which both partners are violently trying to control each other and the relationship.

Distinctions such as these are important if one is to make sense of the data on who commits violence against a partner or spouse, particularly the distinction between situational couple violence and intimate terrorism. Of the four types, situational couple violence is the form that most displays a pattern of **gender symmetry**, a controversial term that refers to the similarity in survey research estimates of male-on-female and female-on-male intimate partner violence. Intimate terrorism is usually perpetrated by men, and violent resistance is typically committed by women (Johnson and Ferraro 2000).

Such a typology helps differentiate among different motives and outcomes of violence. Not all intimate violence is an attempt to control a partner, and injuries and fatalities do not occur equally in all types. Other outcomes—economic, psychological, and health related—also differ by the type of violence. For example, post-traumatic stress disorder is more likely suffered by a victim of intimate terrorism than by a victim of situational couple violence. However, the attempt to control a partner need not be attempted or achieved violently. Sociologist Kristin Anderson points out that some people who suffer the extreme subordination and control associated with intimate terrorism but without the violence suffer similar emotional, psychological, and even physical consequences. Thus, not all acts of intimate partner abuse that are control driven may be captured within the category of intimate terrorism (Anderson 2008).

Tension and conflict are normal features in intimate relationships but can escalate into situational or common couple violence.

RelaXimages/Alloy/Corbis

## Prevalence of Intimate Violence

To put it bluntly, it is impossible to know exactly how much violence there is in families and relationships in the United States. Part of the difficulty stems from the socially unacceptable, frequently illegal nature of the behavior; some stems from methodological limitations in the various data we gather. Depending on *how* we gather the information, estimates of *how much* there is, *who does it*, and *where it happens* will vary. You might think that there must be some kind of "official statistics" we could use, such as arrest records or emergency room visits. Yet so much family violence is unreported that such "official data" are incomplete (U.S. Bureau of Justice Statistics 1998). Perhaps as much as three-quarters of all physical assaults, four-fifths of all rapes, and half of all stalking incidents suffered by females at the hands of their intimate partners are never reported to police (National Institute of Justice and the Centers for Disease Control 2000). Aside from underreporting by victims, some people are better positioned to hide their abusive behavior from authorities. Upper- and middle-class abusers are more likely to live in their own private homes, where their conflict may remain more hidden. They also are more likely to be given more credibility by police as well as medical and social service professionals. People who can afford to use nonhospital medical resources (such as personal physicians and family doctors to treat injuries) may avoid suspicion because the incident won't show up in hospital admitting records, and their accounts of what happened are more likely to be believed.

Data from domestic violence shelters are even more severely limited because most victims don't seek out a shelter. In addition, most women who use shelters are typically from lower economic backgrounds and most commonly have been victims of the severest forms of mistreatment (Cunradi, Caetano, and Schafer 2002). Thus, the information about shelter populations does not reflect the extent or distribution of the wider problem.

Most research on intimate and family violence is based on survey data. Many discussions of intimate violence rely on surveys of large random samples drawn from the wider U.S. population. Such studies include the National Intimate Partner and Sexual Violence Survey, the National Family Violence Survey, the National Survey of Families and Households, the National Violence Against Women Survey, and the National Longitudinal Couples Survey. In addition, broader studies of crime and victimization, such as the National Crime Victimization Survey, the FBI's Supplementary Homicide Reports, and the Study of

Injured Victims of Violence, have been used to better estimate the prevalence of intimate violence and to understand the influence of social and economic factors (Field and Caetano 2005).

Of course, reports and estimates based on survey data are themselves prone to problems. In asking people to admit to family violence, researchers are likely to receive underreports. Even in anonymous surveys, individuals may downplay their involvement—as offenders or victims—in socially undesirable behavior. Nevertheless, the estimates from such large-scale national surveys are the ones that give us our best ideas of the frequency and spread of family violence. Most estimates in this chapter are based on such data. Keep in mind, however, that it is probable that these estimates are underrepresentative of the real prevalence of violence and victimization. In other words, reality is most likely worse than the data you will now see.

## What the Data Reveal

Based on survey data from large nationally representative samples of heterosexual couples in the United States, the rate of intimate partner violence ranges from 17 percent to 39 percent of couples each year (Caetano, Vaeth, and Ramisetty-Mikler 2008). Roughly one out of five couples in the general population report having experienced intimate partner violence according to 25 years of survey data summarized by Craig Field and Raul Caetano (2005).

Using multiple sources of data, Catalano (2013) produced a report on violence between intimates. Key findings are as follows:

- Between 2002 and 2011, 22 percent of nonfatal violent victimizations of females over age 12 were committed by intimate partners. Among males 12 and older, intimate partners were responsible for 4 percent of nonfatal violent victimizations.

- Between 1993 and 2010, on average 34 percent of female murder victims and 3 percent of male murder victims were killed by an intimate.

With support from the National Institute of Justice and the Department of Defense, in 2011, the Centers for Disease Control and Prevention's National Center for Injury Prevention and Control conducted a large-scale study, the National Intimate Partner and Sexual Violence Survey (NISVS), in which more than 12,727 people completed interviews about their experiences of violence, stalking, and sexual violence in their relationships. Results shown in Tables 12.1 and 12.2 illustrate the percentage of the sample of women and men who had ever experienced the forms of intimate

**TABLE 12.1** U.S. Women, NISVS 2011, Intimate Partner Violence Victimization Experienced in Lifetime and Prior 12 Months

| | Lifetime % | Estimated Number of People | % Estimated in the Prior 12 months | Estimated Number of Individuals Experiencing |
|---|---|---|---|---|
| **Slapped, pushed, or shoved** | 29.7 | 35,872,000 | 3.7 | 4,447,000 |
| Slapped | 18.9 | 22,864,000 | 1.7 | 2,056,000 |
| Pushed or shoved | 27.3 | 32,955,000 | 3.1 | 3,736,000 |
| **Any severe physical violence** | 22.3 | 26,928,000 | 2.3 | 2,752,000 |
| Hurt by pulling hair | 9.4 | 11,397,000 | 0.9 | 1,088,000 |
| Hit with a fist or something hard | 13.2 | 15,881,000 | 1.2 | 1,471,000 |
| Kicked | 6.7 | 8,033,000 | 0.4 | 494,000 |
| Slammed against something | 15.4 | 18,638,000 | 1.3 | 1,614,000 |
| Tried to hurt by choking or suffocating | 9.2 | 11,120,000 | 0.7 | 896,000 |
| Beaten | 10.5 | 12,719,000 | 0.7 | 896,000 |
| Burned on purpose | 1.2 | 1,423,000 | * | * |
| Threatened with a knife or gun | 4.2 | 5,101,000 | * | * |

*Estimate not reported.
SOURCE: Breiding et al. (2014).

**TABLE 12.2** U.S. Men, NISVS 2011, Behavior Experienced in Lifetime and Prior 12 Months

| | Lifetime % | Estimated Number of People | % Estimated in the Prior 12 months | Estimated Number of Individuals Experiencing |
|---|---|---|---|---|
| **Slapped, pushed, or shoved** | 25.5 | 28,992,000 | 4.4 | 4,983,000 |
| Slapped | 19.5 | 22, 216,000 | 2.7 | 3,072,000 |
| Pushed or shoved | 18.3 | 20,849,000 | 3.2 | 3,641,000 |
| **Any severe physical violence** | 14.0 | 15,985,000 | 2.1 | 2,374,000 |
| Hurt by pulling hair | 2.6 | 3,014,000 | * | * |
| Hit with fist or something hard | 10.1 | 11,506,000 | 1.5 | 1,695,000 |
| Kicked | 4.6 | 5,190,000 | 0.5 | 555,000 |
| Slammed against something | 2.5 | 2,836,000 | 0.4 | 455,000 |
| Tried to hurt by choking or suffocating | 0.7 | 814,000 | * | * |
| Beaten | 2.3 | 2,654,000 | * | * |
| Burned on purpose | 0.3 | 384,000 | * | * |
| Threatened with a knife or gun | 2.3 | 2,661,000 | * | * |

*Estimate not reported
SOURCE: Breiding et al. (2014).

partner violence indicated, and the percentage who had experienced such violence in the more immediate 12 months prior to the survey. Also shown is the prevalence of such experiences in the U.S. population that can be estimated from the sample data.

The NISVS data suggest that nearly one in three women and one in four men has been slapped, pushed, or shoved by an intimate partner, and nearly one in four women and one in seven men has suffered severe physical violence by an intimate partner in their lifetime. Although the data for the prior 12 months are lower, they do indicate that 3.7 percent of women and 4.4 percent of men had been slapped, pushed, or shoved during the 12 months prior to being interviewed in 2011, and 2.3 percent of women and 2.1 percent of men had experienced severe physical violence by an intimate partner during that same time frame. The NISVS also revealed that for more than one in five women (23.1 percent) and nearly one in seven men (14 percent) who ever experienced rape, physical violence, and/or stalking by an intimate partner, their first experience of some form of intimate partner violence occurred when they were between 11 and 17 years of age.

# Why Families Are Violent: Models of Family Violence

All families have their ups and downs, and all family members experience anger toward one another at times. But why does violence erupt more often and with more severe consequences in some families than in others? To better understand violence within the family, we must look at its place in the larger sociocultural environment. The principal models used in understanding family violence are discussed in the following sections.

## Individualistic Explanations

An individualistic approach emphasizes how a perpetrator's violence is related to a personality disorder, mental or emotional illness, or alcohol or drug misuse (O'Leary 1993). Sociologists Richard Gelles and Claire Pedrick Cornell (1990) suggested that individualistic explanations may be especially appealing to abusers. If they can attribute the violence and abuse they inflict and the hurt they cause as due to an aberration or illness, then abusers can believe that their acts are not deliberately hurtful or abusive. They can neutralize

the stigma attached to being an abuser by using such explanations as excuses for what they've done. Such individualistic thinking might be comforting to the wider population as well, as it suggests that no sane or sober person would do such things. Unfortunately, to more fully understand intimate violence and abuse, we must step back and look at the bigger picture—at the family and society that influence abusers.

Although the use of alcohol and other drugs has been found to be associated with family violence, as sociologists Richard Gelles and Mary Cavanaugh remind us, association is not the same as causation (Gelles and Cavanaugh 2005). The claim that alcohol acts as a disinhibitor and therefore consumption can cause violence to occur needs to be put into a cultural context. The chemical properties in alcohol do not produce or guarantee any given behavior. In some cultures, alcohol makes people more passive. That we believe alcohol will unleash aggressiveness, including violence toward family members, is more important in explaining the observed associations between alcohol use and family violence.

Psychological explanations are a different matter. Although there is no single mental illness that is unique to and common among those who abuse, psychopathology may well be a factor in the repeated and most extreme instances of domestic and intimate partner violence and abuse. In such instances, offenders may be more likely to fit profiles as either psychopathic batterers or having a borderline and antisocial personality (Dutton and Bodnarchuk 2005). However, it is less useful if and when employed to account for the more common, more routine slaps, shoves, pushes, and spankings that characterize family experience for many in the United States.

The idea that people are violent only because they are disturbed or drunk is not supported by the facts. Furthermore, such thinking carries the implication that perpetrators are not really responsible for their behavior.

## Ecological Model

The ecological model uses a systems perspective to explore child abuse. Psychologist James Garbarino (1982) suggested that cultural approval of physical punishment of children combines with lack of community support for the family to increase the risk of violence within families. Under this model, children who don't "match" well with their parents (such as children with emotional or developmental disabilities), and families under stress (such as from unemployment or poor health) or that have little community support (such as child care or medical care), can be at increased risk for child abuse.

## Feminist Model

The feminist model stresses the role of gender inequalities, gendered power and powerlessness, and cultural concepts of masculinity as causes of violence. Using a historical perspective, this approach holds that most social systems have traditionally placed women in a subordinate position to men, thus supporting male dominance even when that includes violence (Toews, Catlett, and McKenry 2005; Ÿllo 1993).

There is no doubt that violence against women and children—and indeed violence in general—has had an integral place in most societies throughout history. Feminist theory must be credited for advancing our understanding of domestic violence by insisting that the patriarchal roots of domestic relations be taken into account. However, the patriarchy model alone does not adequately explain the variations in degrees of violence among families in the same society, or the phenomenon of female violence (Ÿllo 1993). Women are sometimes violent toward their husbands and partners. More mothers are implicated in child abuse than fathers. Although the latter fact has much to do with women's greater responsibility for and time with children, it does illustrate that the capacity for violence is not limited to men or completely explained by concepts such as masculinity or patriarchy. Finally and most telling, rates of violence between lesbian partners, though difficult and problematic to measure, are estimated to be comparable to rates among heterosexual partners (Mezey 2015). Furthermore, like heterosexual partner violence, when same-gender partner violence occurs, it is more likely to be a recurrent feature of the relationship than a one-time event. The feminist approach may have its greatest use in efforts to understand and explain male-on-female intimate partner violence among heterosexual couples.

## Social Stress and Social Learning Models

The two social models discussed here can be related to the ecological and feminist models in that they view violence as originating in the social structure.

First, the social *stress* model views family violence as arising from two main factors: (1) structural stress such as low income or illness, and (2) cultural norms such as the "spare the rod and spoil the child" ethic (Gelles and Cornell 1990). Groups with few resources, such as the poor, are seen to be at greater risk for family violence.

Second, the social *learning* model holds that people learn to be violent from society and their families (Wareham, Boots, and Chavez 2009). The core premise is that children, especially boys, learn to become violent when they are a victim of or witness to violence and abuse (Bevan and Higgins 2002). This is even more likely if the child experiences positive reinforcement for displaying violence. Although it is true that many perpetrators of family violence were abused as children, it is also true that many victims of childhood violence do not become violent parents. These theories do not account for this discrepancy.

### Resource Model

William Goode's (1971) resource theory can be applied to family violence. This model assumes that social systems are based on force or the threat of force. A person acquires power by mustering personal, social, and economic resources. Thus, according to Goode, the person with the most resources is the least likely to resort to overt force. Although family violence occurs among all income levels, Gelles and Cornell (1990) describe the typical situation: "A husband who wants to be the dominant person in the family but has little education, has a job low in prestige and income, and lacks interpersonal skills may choose to use violence to maintain the dominant position."

### Exchange–Social Control Model

Richard Gelles (Gelles 1993; Gelles and Cornell 1990) posits the two-part exchange–social control theory

Research indicates that males, like females, are victims of violence and abuse too, though the causes and consequences of such victimization may be different.

Cultura RM Exclusive/Howard Kingsnorth/Cultura Exclusive/Getty Images

of family violence. The first part, exchange theory, holds that in our interactions, we constantly weigh the perceived rewards against the costs. When Gelles says that "people hit and abuse family members because they can," he is applying exchange theory, contending that the risks and costs associated with being violent are deemed less than the reward of getting what one wants from, or doing what one wants to, one's partner.

**True** **3.** The value placed on family privacy contributes to a reluctance to report suspected family violence.

The expectation is that "people will only use violence toward family members when the costs of being violent do not outweigh the rewards." The possible rewards of violence might be getting their own way, exerting superiority, working off anger or stress, or exacting revenge. Costs could include being hit back, being arrested, being jailed, losing social status, or dissolving the family. Three characteristics of families that may reduce those costs of violence and thus reduce social control are the following:

- *Inequality.* Men are stronger than women and often have more economic power and social status. Adults are more powerful than children.

- *Private nature of the family.* People are reluctant to look outside the family for help, and outsiders (e.g., the police or neighbors) may hesitate to intervene in private matters. The likelihood of family violence decreases as the number of nearby friends and relatives increases (Gelles and Cornell 1990).

- *"Real man" image.* In some American subcultures, aggressive male behavior brings approval.

A violent man may gain status among his peers for asserting his "authority." The exchange–social control model is useful for looking at treatment and prevention strategies for family violence, discussed later in this chapter.

Although these various emphases and approaches each attempt to account for family and intimate violence, researchers in the first decade of the 21st century increasingly looked at what *follows from* intimate and family violence. As summarized by sociologist Kristin Anderson, an increasing number of researchers began looking at IPV as an independent variable, examining how it might explain relationship formation and dissolution and other individual and couple experiences (Anderson 2010). Intimate violence was shown to be connected to higher risk of separation and relationship dissolution, a reduced likelihood of marriage, a higher likelihood of teen pregnancy, and economic instability and poverty.

## The Importance of Gender, Power, Stress, and Intimacy

Each of the previously mentioned perspectives has valuable insight to offer concerning some aspects of a complex problem with no easy or single solution. Looking across the theories, several factors surface repeatedly.

### Gender

Although there is female-on-male violence and female-on-female violence (discussed later), violence by males tends to be more extreme, often has different causes (power and control versus self-defense), and typically results in different consequences (in terms of both physical injuries and domination). Thus, although certain survey data may indicate a kind of gender symmetry in experiencing and expressing some kinds of intimate violence, gender matters greatly when it comes to fully understanding the reality of intimate and family violence.

### Power

Power is a central motive in much intimate violence, especially the long-term and extreme forms of spousal violence that Michael Johnson and Janel Leone (2005) called intimate terrorism. Power and control surface in much partner violence whether between dating, cohabiting, or married couples. In addition, powerlessness can be linked to violence when those who feel dominated and unable to legitimately assert their rights may turn to violence as a last resort.

### Stress

As individuals are subjected to a variety of stresses (such as unemployment, underemployment, illness, pregnancy, work-related relocations, and difficult or disabled children), tensions among family members may rise. Stress-based explanations help account for the greater prevalence of violence among lower-income families and households facing unemployment, or for the eruption of violence at certain times in the sharing of family life, but stress alone cannot account for the breadth and depth of family violence (McCaghy, Capron, and Jamieson 2000; Straus, Gelles, and Steinmetz 1980). Stress may raise the likelihood of violence, but it is not the cause. Somewhere, the individual must have learned that acting violently toward loved ones is appropriate and acceptable (Gelles and Straus 1988).

## Intimacy

The heightened emotions and long-term commitments that characterize family relationships are among the qualities we most value about those relationships. Those same qualities contribute to a greater likelihood that we will have disagreements and that when we find ourselves disagreeing those disagreements will be more emotional. Furthermore, cultural beliefs promote the idea that we have the right to influence our loved one's behavior. Some abusive men explain that they assault their spouses "because they love them." This indicates that the cultural expectations surrounding love and intimacy—one's sense of what he or she can and should do within their closest relationships—contribute to the worst aspects of those same relationships.

In addition, as discussed in Chapter 3, members in contemporary families are granted and expect to enjoy a level of privacy, occasionally even secrecy, in experiencing intimate and family relationships. Even when family conflict is in a public setting, others are reluctant to intervene in such "domestic disputes." In some ways, our society thus legitimizes violence and force within families and then compels us to turn the other way and "mind our own business" when they occur.

## Women and Men as Victims and Perpetrators

It has long been fairly common to hear the phrase "battered women" used to refer to women who have been subjected to abuse from their spouses or partners. **Battering**, as used in the literature on family violence, includes slapping, punching, knocking down, choking, kicking, hitting with objects, threatening with weapons, stabbing, and shooting. In fact, the term *battering* does not specify the gender of the batterer, yet most people likely assume that the batterer is male and that the victim is female. Interestingly, survey research has found that the number of women who report expressing any violence toward their male partners is the *same as or greater than* the number of men who report expressing violence toward their

female partners. This is true of research on spousal, cohabiting, and dating relationships. This was described earlier as *gender symmetry*.

Ignored or rejected by many researchers or interpreted as signs of "self-defensive" or reactive violence by female victims, we now know that women use violence with male partners, perhaps about as often as men do with female partners (Frieze 2005). A 2010 article by sociologist Murray Straus summarizes findings from more than 200 studies in which researchers found gender symmetry in the perpetration of violence in marital or dating relationships (Straus 2010).

However, even when the rates of violence are similar for males and females, the motives and outcomes of male-on-female and female-on-male violence may be quite different. Most violence perpetrated by women on men (as well as most male-on-female partner violence) is of the more situational, routine, and relatively more minor variety. It is not the sort of violence that typically leads to hospitals or shelters. It does not typically include the subjugation and domination that more extreme forms of male violence and abuse might. The less common and more extreme violence that escalates and causes serious injury or even death is usually committed by men against women (Johnson 1995).

There is also reason to suspect that women and men use violence for different reasons. As Maureen McHugh, Nicole Livingston, and Amy Ford (2005) assert, men's violence tends to be instrumental: They use violence to get what they want and to assert control and gain power over their partners. Women's motives include self-defense, retaliation, expression of anger, attention seeking, stress or frustration, jealousy, depression, and loss of self-control.

Historically and culturally, women have unfortunately been considered "appropriate" victims of domestic violence (Gelles and Cornell 1990). Many mistakenly accept the misogynistic idea that women sometimes need to be "put in their place" by men, thus providing a disturbing cultural basis for the physical and sexual abuse of women. There is no comparable cultural justification for the physical or sexual abuse of men.

As far as outcomes are concerned, more female victims than male victims are injured from partner violence, and their injuries tend to be more severe than those received by male victims. Even the same acts are not really the same: A slap that breaks a victim's jaw is not the same as a slap that reddens a victim's face. In other words, men's slaps (or punches, shoves, kicks, and so on) are not identical to those of women (McHugh, Livingston, and Ford 2005). Thus, even if women and men had identical rates of expressing intimate partner violence, differences in the causes, context, and consequences make women's and men's experiences something other than symmetrical.

## Female Victims and Male Perpetrators

No one knows with certainty exactly how many women are victims of partner violence each year, but, as shown earlier, the data we have paint a less-than-encouraging picture. Consider also these facts from the Bureau of Justice Statistics:

- Of all violent crime women experienced between 2002 and 2011, almost 26 percent was from an intimate partner (current or former spouse, boyfriend, or girlfriend). During that same period, intimate partner violence accounted for nearly 5 percent of nonfatal violence against men (Catalano 2013).

- Females between the ages of 20 and 24 were at the greatest risk of nonfatal intimate partner violence. Males between the ages of 12 and 15 or those over age 65 experienced the lowest rates of nonfatal partner violence (Catalano et al. 2009).

- Between 2002 and 2011, half the female victims of nonfatal intimate partner violence suffered an injury, with 13 percent suffering a serious injury (e.g., sexual violence, stab or gunshot wound, broken bones, unconsciousness, or internal injury) (Catalano 2013).

- Unmarried women have higher rates of intimate partner violence victimization than do married women. Separated women experienced the highest rates compared to females of all other marital statuses (Catalano et al. 2009).

In addition to these facts, data illustrate the vulnerability and victimization of pregnant women due to abuse from their spouses or intimate partners (National Coalition Against Domestic Violence 2008c). For example, it is estimated that almost 10 percent of females are subjected to intimate partner violence during pregnancy. Most women (70 percent) who are abused before pregnancy continue to experience abuse during pregnancy, and violence tends to intensify *after* the offender learns of the woman's pregnancy. Women abused during pregnancy have been found to have a greater likelihood of miscarriages, low-birth-weight

babies, high blood pressure, vaginal bleeding, and kidney or urinary tract infections, and they are much more likely to delay prenatal care. Finally, homicide is one of the leading causes of death among pregnant women (National Coalition Against Domestic Violence 2008c).

In the midst of all the disturbing data and discussion, it is worth noting that there is more encouraging evidence to add. According to the Bureau of Justice Statistics, from 1994 to 2012, there were substantial declines in the rate of serious intimate partner violence (Truman and Morgan 2014). Such violence declined 72 percent for females and 64 percent for males (Catalano 2013).

### Which Women Are Victimized?

Women of all races, ages, and socioeconomic statuses experience intimate violence, although they are not victimized equally. Younger women (ages 20 to 24), Native American and African American women, and lower-income women are more frequent victims of partner violence. Black women suffered higher rates of nonlethal violence (5 per 1,000 people 12 or older) than did Hispanic (4.3), white (4.0), or Asian (1.4) women. The rate of violence Native American women experienced (11.1) was more than twice the rate African American women experienced.

Socioeconomic differences can also be noted. As income increases, the rate of female victimization decreases (Catalano et al. 2009). Although no social class is immune to partner violence, it is more likely to occur in lower-income, low-status families (Aaltonen et al. 2012).

Although early studies of battering relationships seemed to indicate a cluster of personality characteristics constituting a typical battered woman, more recent studies have not borne out this viewpoint. Factors such as low self-esteem or childhood experiences of violence do not appear to be necessarily associated with a woman being in an assaultive relationship (Hotaling and Sugarman 1990). Two characteristics, however, do appear to be highly correlated with wife assault. First, a number of studies have found that wife abuse is more common and more severe in families of lower socioeconomic status. However, keep in mind that higher-income adults have greater privacy and thus greater ability to conceal domestic violence (Fineman and Mykitiuk 1994). Second, marital conflict—and the inability to resolve conflict—is a factor in many battering relationships.

Battered women's shelters provide safe havens for women in abusive relationships. Shelters provide counseling and emotional support, as well as temporary lodging, meals, and other necessities for women and their children.

*Stephan Gladieu/Getty Images News/Getty Images*

## Characteristics of Male Perpetrators

A heterosexual man who systematically inflicts violence on his wife or girlfriend is likely to have at least some of the following traits:

- He believes in the "traditional" home, family, and gender-role stereotypes and in the moral rightness of his violence (although he may acknowledge "accidentally" going too far).

- He has low self-esteem and may use violence as a means of demonstrating power or adequacy.

- He may be sadistic, pathologically jealous, or passive-aggressive and may use sex as an act of aggression (Gelles and Cornell 1990; Margolin, Sibner, and Gleberman 1988; Vaselle-Augenstein and Ehrlich 1992; Walker 2009).

The Centers for Disease Control (CDC) identify a number of individual factors that contribute to perpetrating intimate violence, including low self-esteem of the abuser, depression, antisocial or borderline personality traits, emotional dependency and insecurity, and heavy drug and alcohol use. In addition to such characteristics, the CDC adds being unemployed, being socially isolated (i.e., having few friends), and having a history of experiencing poor parenting, physical discipline, or physical or psychological abuse as a child (Centers for Disease Control 2015c). Psychologist Maureen McHugh and colleagues (2005) note that in addition to perpetrating violence, violent men are likely to be the target of violence, either in the present or in their past. In other words, they are either victims of mutual violence or have histories of being abused themselves.

We often read or hear that a major factor in predicting a man's violence is if he experienced family violence as a child. According to research, a childhood troubled by parental violence accounts for only one percent of adult dating violence and approximately the same proportion of violence in marriage or marriage-like relationships (see review by Johnson and Ferraro 2000). Although it is true that sons of the *most* violent parents have a 1,000 percent greater rate of wife beating than sons of nonviolent parents, the majority of these sons, including sons of the most violent parents, are not violent (Johnson and Ferraro 2000).

## Female Perpetrators and Male Victims

The incidence and experiences of males victimized by violence committed by their wives or girlfriends are poorly understood. Although it is undoubtedly true that some heterosexual men are injured in attacks by wives or girlfriends, most injured victims of severe intimate partner violence are women. The males most likely to be injured or killed as a result of intimate partner violence are gay men, who are assaulted or killed by their same-gender partner (Greenwood et al. 2002). Thus, we may not consider violence by women as significant as that committed by heterosexual men (Straus 1993). Often, even if a woman attempts to inflict damage on a man in self-defense or retaliation, her chances of prevailing in hand-to-hand combat with a man are slim. A woman may be severely injured simply trying to defend herself. Remember, though, that when we combine common couple violence and violent resistance, about the same rate of female-on-male acts of violence occur.

Suzanne Steinmetz (1987) long ago suggested that some scholars downplay the extent and importance of women's violence against male partners. As such, there is a "conspiracy of silence [that] fails to recognize that family violence is never inconsequential." Sociologist Murray Straus (1993) offered four reasons to take the study of female violence seriously:

- Assaulting a spouse—either a wife or a husband—is an "intrinsic moral wrong."

- Not doing so unintentionally validates cultural norms that condone a certain amount of violence between spouses.

- There is always the danger of escalation. A violent act—whether committed by a man or a woman—may lead to increased violence.

- Spousal assault is a model of violent behavior for children. Children are affected as strongly by viewing the violent behavior of their mothers as by viewing that of their fathers.

## Familial and Social Risk Factors

In addition to the kinds of characteristics noted previously, the relationship dynamics, the wider social environment, and the society at large also affect rates of family violence. As far as relationship factors are concerned, the presence of persistent tension, conflict, and fighting between spouses creates a context that heightens the likelihood of intimate violence. If the family is experiencing economic stress, the risk of violence also increases.

If one is living within an environment where there are weak sanctions against violence, where neighbors are unwilling to intervene even when they witness acts of violence, and where the relationships and interactions within the surrounding community are weak and infrequent, the risk of intimate partner violence grows. Traditional gender norms along with beliefs about family privacy, the legitimacy of violence as a means to solve problems, and the right of partners to influence or control each other are additional cultural or societal risk factors (Centers for Disease Control, National Center for Injury Prevention and Control 2008b).

## ■ Socioeconomic Class and Race

We often hear about how "democratic" intimate violence is, occurring among all groups, regardless of economic status, race, or sexual orientation. Indeed, there is truth to that statement: Intimate partner violence *can* be found among all ethnic and economic groups; however, the amount of violence varies greatly. More than three decades of research demonstrates an association between socioeconomic status and partner violence.

**False** | **4.** Rates of intimate violence vary greatly among social classes.

### Socioeconomic Class

Using data from the 1995 National Alcohol and Family Violence Survey, Carol Cunradi, Raul Caetano, and John Schafer (2002) found that household income had the greatest influence on intimate partner violence across racial and ethnic lines. Both females and

Abuse and violence do exist among all social classes, but they tend to be more common among those with lower incomes and greater economic stress.

Ingram Publishing/Getty Images

males of higher-income households were found to experience less partner violence than their counterparts from lower-earning households. Noteworthy, too, are the following facts: Within each income level, females were at higher risk than males of victimization, and low-income men were victimized at lower rates than high-income women. Hyunkag Cho notes too that previous research has found low socioeconomic status to be a risk factor for experiencing IPV. In his analysis of data from the Collaborative Psychiatric Epidemiology Surveys (CPES), Cho found that "sound financial conditions" were associated with lower risks of IPV, and suggests that the elevated rate of IPV he found among blacks and Latinas was likely due to their having the lowest levels of financial security of those groups sampled (Cho 2012).

Although there are consistent and strong associations between low economic status and violence, research reveals partner violence and abuse among high-status couples as well (Weitzman 2000). Their economic position may even create unique problems for women who are victimized. They may find other people less sympathetic to their circumstances, more skeptical of their allegations, and less willing to help because they perceive the women as having considerable means with which to help themselves.

### Race

Using data from the 2011 National Intimate Partner and Sexual Violence Survey, the prevalence of intimate partner violence varies across lines of race. American

Indian/Alaska Native and multiracial women and men reported the highest rates of victimization; Asian or Pacific Islander women and men reported the lowest rates (Breiding et al. 2014). Just over half of American Indian/Alaska Native (51.7 percent) and multiracial women (51.3 percent) reported having experienced physical violence by an intimate partner in their lifetimes. These rates were followed by 41.2 percent of non-Hispanic Black women, 30.5 percent of non-Hispanic white women, 29.7 percent of Hispanic women, and 15.3 percent of Asian or Pacific Islander women.

Rates among men were lower than those among women, but showed vary similar racial difference. In order, the reported prevalence of lifetime victimization ranged as follows: an estimated 43 percent of American Indian/Alaska Native men, 39.3 percent of multiracial men, 36.3 percent of non-Hispanic Black men, 27.1 percent of Hispanic men, 26.6 percent of non-Hispanic white men, and 11.5 percent of Asian or Pacific Islander men reported ever having experienced intimate partner violence (Breiding, Smith, Basile, et al. 2014).

Psychological aggression, such as name calling, insulting or humiliating a partner, showed a similar pattern of victimization across race for women. The highest reported lifetime prevalence was reportred by American Indian/Alaska Native women, followed by multiracial women, non Hispanic Black women, non Hispanic white women, Hispanic women, and Asian or Pacific Islander women. With regard to men's experience of psychological aggression, the pattern differed from that found for intimate violence. Multiracial men reported the highest rate of having ever experienced psychological aggression, followed by non-Hispanic Black men, Hispanic men, American Indian/Alaska Native men, white non-Hispanic men, and Asian or Pacific Islander men.

In addition to these reported differences, other research shows that there is within group variation, especially among Hispanic and Asian American populations, though as Hyunkag Cho notes, it is difficult to ensure sufficient representation of specific Asian or Hispanic ethnic groups in national surveys in order to draw such comparisons. Cho also contends that in analyzing data from the Collaborative Psychiatric Epidemiology Surveys, race differences diminish greatly or disappear when one controls for other variables, including income and employment status. This suggests that at least some reported racial differences are products of socioeconomic disadvantage (Cho 2012).

## LGBT Experience of Intimate Violence

Until fairly recently, little was known about violence in lesbian and gay relationships. One reason is that such relationships had not been given the same social status as those of heterosexuals. In addition, long-term same-sex relationships have been less common than long-term heterosexual relationships. Finally, some gay or lesbian individuals are likely to be reluctant to identify their sexuality for fear of resulting stigma or mistreatment. However, understanding violence in same-sex relationships is important for at least two reasons: People are being victimized, and their victimization is mostly invisible and unaddressed. Relationships between gay men or lesbians obviously lack the gender differences that otherwise reflect male dominance and female subordination. Clearly, neither male dominance nor male socialization toward dominance, aggressiveness, or violence can account for physical abuse in lesbian relationships.

The available research estimates on LGBT intimate violence is fraught with additional problems. Methodological limitations include the use of nonrepresentative samples, absence of consistent definitions of behavior to be measured, and failure to examine differences among the subpopulations involved (lesbians, gay males, bisexuals, transgender people). Perhaps most problematic, data often fail to indicate whether past experience of violence or abuse was within a same-sex relationship or a prior heterosexual relationship (Mezey 2015).

Recent research indicates that the rate of abuse in gay and lesbian relationships is somewhat greater than that in heterosexual relationships. Figure 12.1 shows the lifetime prevalence of violence by an intimate partner for heterosexuals, lesbians and gay men, and for bisexual women and men as found by the 2010 NISVS.

One estimate placed the range between 25 percent and 50 percent for lesbian couples (McClennen, Summers, and Daly 2002, in Frieze 2005). A study by Kimberly Balsam and Dawn Szymanski (2005) found that of the 272 lesbian and bisexual women in their sample, 40 percent reported being violent, and 44 percent reported being victims of violence within relationships with female partners (Frieze 2005). In their four-city study mentioned earlier, psychologist Gregory Greenwood and colleagues found nearly 40 percent of their sample of 2,881 gay or bisexual men had experienced some form of abuse. Most common was psychological abuse (34 percent), but 22 percent experienced physical

## Figure 12.1 Lifetime Prevalence of Intimate Partner Violence, by Gender and Sexual Orientation

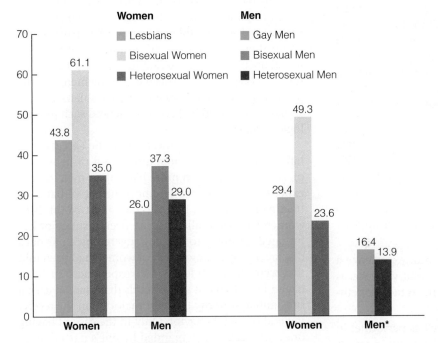

**Lifetime prevalence of rape, physical violence, and/or stalking**

**Lifetime prevalence of severe violence (e.g., hit with fist or something hard, slammed against something, or beaten)**

Women: Lesbians, Bisexual Women, Heterosexual Women

Men: Gay Men, Bisexual Men, Heterosexual Men

Left chart (rape, physical violence, and/or stalking):
Women — Lesbians 43.8, Bisexual Women 61.1, Heterosexual Women 35.0
Men — Gay Men 26.0, Bisexual Men 37.3, Heterosexual Men 29.0

Right chart (severe violence):
Women — Lesbians 29.4, Bisexual Women 49.3, Heterosexual Women 23.6
Men* — Gay Men 16.4, Heterosexual Men 13.9

*For bisexual men, numbers too small to report.

SOURCE: Walters, Chen, and Breiding (2013).

abuse, 5 percent indicated that they had been sexually abused by their partner, and 18 percent suffered multiple forms of abuse (Greenwood et al. 2002).

The 2010 National Intimate Partner and Sexual Violence Survey reported lifetime prevalence rates of violence by intimate partners experienced by women and men of different sexual orientations. Figure 12.1 shows the percentages of lesbians, bisexual women, heterosexual women, gay men, bisexual men, and heterosexual men who reported ever having experienced rape, physical violence, and/or stalking by an intimate partner, and the percentages reporting ever having experienced severe physical violence (being punched, hit with something hard, slammed against something, or beaten). As can be seen from the data, according to the NISVS, among women, a higher percentage of bisexual women had ever experienced rape, physical violence, and/or stalking by an intimate partner, followed by lesbians and then heterosexual women. Among men, too, bisexual men had the highest prevalence of having experienced rape, physical violence, and/or

stalking. Heterosexual and gay men had similar and lower levels of prevalence. In the case of severe physical violence, among women, again, a higher percentage of bisexual women reported having experienced such violence, followed by lesbians and then heterosexual women. Among men, similar percentages of gay men and heterosexual men reported having experienced such violence. It is important to note, however, that by looking at "lifetime prevalence," some past victimization experienced by gay men, lesbians, and, especially, bisexual men and women, was violence perpetrated by opposite-sex partners. The NISVS estimates that roughly 90 percent of bisexual women who reported past experience of intimate partner victimization report only male perpetrators of such violence. Nearly 80 percent of bisexual men reported only female perpetrators of the violence they'd experienced. Among gay men who had experienced intimate partner violence, more than 90 percent reported only male perpetrators. Among lesbian women, two-thirds (67.4 percent) reported only female perpetrators of the intimate violence to which they'd been subjected (Walters, Chen, and Breiding 2013).

When violence occurs among same-sex couples, it is rarely a one-time event; once violence occurs, it is likely to recur. It also appears to be as serious as violence in heterosexual relationships, including physical, psychological, and/or financial abuse. Michael Johnson and Kathleen Ferraro (2000) note that intimate terrorism can be found among lesbian couples. One additional form of abuse, unique to same-sex couples, is the threat of "outing" (revealing another's gay orientation without consent). Threatening to out a partner to coworkers, employers, or family may be used as a form of psychological abuse in same-sex relationships.

Sociologist Nancy Mezey summarizes a number of research findings that consistently indicate higher rates of IPV victimization experienced by bisexuals than by either gay males, lesbians, or heterosexuals (Mezey 2015). Citing research findings on dating relationships, as well as sexual, emotional and financial IPV, Mezey reports that results indicate bisexuals suffer higher levels of such mistreatment at the hands of intimate partners. Additionally, available research indicates that transgender people experience "serious levels of IPV," much of which is directed at them because they are transgender or gender nonconforming (Mezey 2015, 146).

For LGBT battered partners, there are likely fewer sources of support. Services are often nonexistent or uninformed about the multifaceted issues that such victims face. Researchers have identified several policy issues that must be addressed among service providers and domestic violence agencies:

- Consider how homophobia and heterosexism shape how LGBT victims of abuse experience IPV. Included in this is the fact that among LGBT people who have not publicly identified as such, the threat of being "outed" by an abusive partner may keep some individuals from seeking help or from leaving the relationship.

- Recognize that IPV victims of color experience a triple jeopardy: as victims of domestic violence, as stigmatized sexual minorities, and as racial or ethnic minorities.

- Law enforcement officials as well as personnel in IPV support services may lack training in issues specific to LGBT people. Some services, especially shelters, are designed to assist biologically born females who are being victimized by male partners. Thus, broader training is needed.

- Address the issue of gay men and lesbians as both batterers and victims who may seek services at the same time from the same agency. (Mezey 2015; Renzetti 1995)

# Emotional and Psychological Abuse

Despite the childhood rhyme about sticks and stones, words can and do hurt. The 2010 National Intimate Partner and Sexual Violence Survey asked participants about their experiences with a number of kinds of verbal or emotional abuse. Tables 12.3 and 12.4 show the percentages of heterosexual and gay men, and the percentages of heterosexual women, lesbians, and bisexual women who had ever experienced such psychological aggression.

As the percentages indicate, with the exception of having one's whereabouts monitored or kept track of, women were much more likely than men to have experienced the forms of emotional or verbal mistreatment indicated. Between half and two-thirds of women had been called names, been insulted or humiliated, or made to feel endangered by a partner's display of anger. Almost half of women had been told they were failures or losers. Men experienced such behavior less often but still at levels that place at least a substantial minority to slight majority on the receiving end of such behavior. Only in the case of having one's whereabouts demanded by one's partner were men roughly on par with women.

In addition to the previous information, the 2010 NISVS found that 45.5 percent of women had physical threats made against them, 44 percent had partners try to prevent them from talking to friends or family, 40 percent had partners destroy things of importance to them, and 36 percent were prevented from leaving the house when they wanted to leave. On these,

TABLE 12.3 **Percentage of Heterosexual and Gay Men Who Report Ever Experiencing Types of Psychological Aggression, NISVS 2010**

| Type | % of heterosexual men | % of gay men |
|---|---|---|
| Being called fat, crazy, or stupid | 51% | 65% |
| Seeing partner act in an angry fashion that seemed dangerous | 39% | 48% |
| Being insulted, humiliated, or made fun of | 39.5% | 39.4% |
| Being kept track of, partner demanded to know whereabouts | 63.5% | 49.3% |
| Told not good enough, called a loser or failure | 42.3% | 61.6% |

SOURCE: Walters, Chen, and Breiding (2013).

**TABLE 12.4  Percentage of Heterosexual Women, Lesbians, and Bisexual Women Who Report Ever Experiencing Types of Psychological Aggression, NISVS 2010**

| Type | % of heterosexual Women | % of lesbians | % of bisexual women |
|---|---|---|---|
| Being called fat, crazy, or stupid | 63.8% | 59.6% | 75.7% |
| Seeing partner act in an angry fashion that seemed dangerous | 57.7% | 63.5% | 69.7% |
| Being insulted, humiliated, or made fun of | 58.1% | 55.9% | 64.2% |
| Being kept track of, partner demanded to know whereabouts | 61.4% | 60.7% | 66.3% |
| Told not good enough, called a loser or failure | 48.8% | 42.5% | 56.4% |

SOURCE: Walters, Chen, and Breiding (2013).

men were much less likely to have experienced such mistreatment by a partner. Twenty percent had been threatened with harm, 19.4 percent had been prevented from leaving the house, and 29 percent had things of theirs destroyed by an angry partner. Men were also less likely (28.3 percent) to have their partners try to keep them from seeing or talking to family or friends. Even with this gender difference, it is noteworthy that a fifth or more of men acknowledged experiencing these forms of psychological aggression.

## Spousal and Intimate Partner Sexual Violence

Sexual violence, including rape, is among the most serious form of intimate violence. Spousal or intimate partner rape is a form of battering that is inflicted by intimate partners, often as part of a pattern of intimate terrorism. **Rape** may involve vaginal, oral, or anal penetration, and it may involve the insertion of objects other than the penis. It may be perpetrated by force, threat of harm, or when the victim [is] intoxicated, and it may be committed by males or females and against males or females.

According to the National Violence Against Women Survey, 7.7 percent of the 8,000 women age 18 and older who were sampled had been raped by an intimate partner. This translates to almost 8 million women having been raped by a boyfriend, ex-boyfriend, husband, or ex-husband at some point in their lives. Of those women who had been raped, 20.2 percent were raped by a current spouse or ex-spouse, 4.3 percent by a current or former cohabiting partner, and 21.5 percent had experienced rape by a current or former date, boyfriend, or girlfriend. Of all female rape victims, 43 percent had been raped by a former or current intimate partner (Tjaden and Thoennes 2006). Based on the same survey, more than 200,000 women endure more than 320,000 rapes by an intimate partner each year. This leads to an estimated rate of 3.2 intimate partner rapes per 1,000 women in the United States.

Divorced or separated women reported having experienced more intimate partner violence than married women. Regarding rape, specifically, 69 percent of those women who reported having been raped by a former spouse or partner were raped before the relationship ended. However, a fourth of those who had experienced marital/partner rape were raped both before and after the relationship ended, and 6.3 percent reported having been raped only after they had ended their relationship with their former spouse/partner. Rape cases involving intimates, though no less likely to be reported to police, were less likely to be prosecuted and, once prosecuted, less likely to result in convictions than were cases of rape by nonintimates (Tjaden and Thoennes 2006).

The data from the National Violence Against Women Survey were collected more than 20 years ago, between 1995 and 1996. Using more recent data from the 2010 National Intimate Partner and Sexual Violence Survey (NISVS), almost 10 percent of women in the United States have experienced intimate partner rape. An estimated 11.2 million women have been raped by an intimate partner, 6.6 percent having experienced forced penetration by an intimate partner, 2.5 percent attempted forced penetration, and 3.4 percent experienced drug- or alcohol-facilitated penetration.

In addition, nearly 19 million, or one in six women (and one in 12 men or 8 million men), have experienced sexual violence other than rape from an intimate partner. The data also indicate that an estimated 3.4 million women had experienced rape (686,000) or other sexual violence (2,747,000) by an intimate partner in the year prior to their participation in the NISVS (Breiding, Chen, and Black 2014).

Historically, heterosexual legal marriage was regarded as giving husbands unlimited sexual access to their wives. Beginning in the late 1970s, most states enacted legislation to make at least some forms of marital rape illegal. On July 5, 1993, marital rape became a crime in all 50 states. Throughout the United States, a husband can be prosecuted for raping his wife, although many states limit the conditions, such as requiring extraordinary force or violence, especially if the spouses were still living together. The precise definition of marital rape differs from state to state, however.

**True**
**5.** Marital rape is a crime in all 50 states of the United States.

## ◼ Dating Violence and Date Rape

In the past few decades, researchers have grown increasingly aware that violence and sexual assault can take place in all forms of intimate relationships. Violence between intimates is not restricted to spouses or family members. Even casual or dating relationships can be marred by violence or rape. In fact, although all 50 states have laws that prohibit the kinds of behaviors that make up dating violence (e.g., stalking, sexual assault, and assault), evidence has long indicated that the level of dating violence exceeds the level of marital violence.

### Tweens, Teens, and Young Adults: Dating Violence and Abuse

The incidence of physical violence and emotional or verbal abuse in dating relationships, including those of teenagers, is alarming. It is also hard to pin down with precision. The National Coalition Against Domestic Violence (NCADV) compiled the following statistics in its "Dating Abuse and Teen Violence" Fact Sheet (NCADV 2015):

- More than 20 percent of female high school students and more than 13 percent of male high school students report having been sexually or physically abused by a dating partner.
  - In a study of 10th graders, more than a third (35 percent) had been either physically or verbally abused, and nearly a third (31 percent) had abused a dating partner.
- Over a quarter of teens in dating relationships were victims of cyberdating abuse, with females twice as likely as males to have experienced such abuse.
- More than 40 percent of surveyed college women who were in dating relationships reported having experienced abusive behavior from their partners.

Meanwhile, research that has looked into the percentages of teens who *know someone* who has been victimized by dating violence indicates that between 50 to 60 percent of teens surveyed knew someone who has been the victim of physical, sexual, or verbal abuse from a dating partner (Liz Claiborne Inc. 2005; NCADV 2015).

Using data from the National Survey on Teen Relationships and Intimate Violence (STRiV), researchers Bruce Taylor and Elizabeth Mumford examined victimization and perpetration of adolescent relationship abuse (ARA) among a nationally representative sample of 12- to 18-year-old females and males. Findings indicated that 69 percent reported being victims, and 63 percent reported perpetrating relationship abuse. Fifty-eight percent reported both ARA victimization and perpetration. Of the different forms of ARA (physical, sexual, and psychological), psychological abuse was most common among sampled adolescents, with two-thirds reporting victimization and 62 percent reporting having been perpetrators of such abuse. Females age 15 to 18 reported the highest levels of both victimization (73 percent) and perpetration (66 percent) of ARA, whereas 12- to 14-year-old females reported the lowest rates of both victimization (53 percent) and perpetration (56 percent). Approximately a fifth of 14- to 18-year-old females reported having been victims of sexual abuse (21 percent) and perpetrators of physical abuse (17 percent). More than a quarter (27 percent) of males age 12 to 15 reported physical abuse victimization, and 15 percent of males age 15 to 18 reported having perpetrated sexual abuse. Although the study found differences by both age and gender, there were not meaningful differences for other variables such as ethnicity, education, household income, and region (Taylor and Mumford 2014).

Technology has made certain aspects of couple relationships easier. Partners can stay in touch, make plans, and express affection via talk or text. Facebook posts of appreciation, Instagram pictures, and affectionate texts can supplement and support the closeness partners feel for each other. But so, too, can the technology turn troubling and become tools of abusive partners, allowing them to harass, torment, and attempt to control their relationship partners. According to recent research, such cyber-abuse affects many teens who are in dating relationships, especially females.

One study of over 5,600 middle and high school students from New York, New Jersey, and Pennsylvania, found that more than a quarter (26 percent) had, in the prior year, experienced some form of cyberdating abuse, with females twice as likely as males to report being victims (Zweig et al. 2013). Twelve percent of respondents said that they had perpetrated such abuse during the prior 12 months, with females reporting higher levels of perpetrating nonsexual cyberdating abuse. Regarding such sexual cyberdating abuse, males were significantly more likely to report perpetrating such abuse in the prior year.

Of note, for both the victims and perpetrators, cyberdating abuse tended to occur in relationships that were fraught with other forms of dating abuse. For example, more than 80 percent of cyberdating abuse victims reported that they also experienced psychological dating abuse, and more than half of cyber-abuse victims reported experiencing physical dating violence. A third of cyber-abuse victims reported having experienced sexual coercion. Perpetrators of cyberdating abuse were likely to commit other forms of relationship abuse; almost three-fourths reported committing psychological dating abuse, 52 percent physical dating violence, and 11 percent acknowledged having engaged in sexual coercion. Much higher rates of cyberdating abuse

victimization and perpetration were reported by LGBTQ youth than heterosexual youth (Zweig et al. 2013).

According to an earlier national online survey of 615 13- to 18-year-olds and 414 parents with teens in that age group (Picard 2007):

- Of teens who responded, 25 percent reported having been called names or were harassed by their partners in text messages or cell phone calls.
- Of respondents, 22 percent have been pressed via cell phone or Internet contact to have sex.
- Of surveyed teens who were in dating relationships, 24 percent communicated with their dating partner via cell phone every hour between midnight and 5:00 A.M.
- Of those teens who responded, 30 percent said that they receive frequent (between 10 and 30) text messages per hour by partners asking where they are, who they are with, and what they are doing.
- Of teens, 68 percent acknowledged that boyfriends or girlfriends sharing embarrassing or private pictures or videos of them with others through cell phones and computers is a serious problem. A slightly higher percentage report that spreading rumors about a boyfriend or girlfriend via social networking sites (e.g., MySpace or Facebook) or cell phones was a serious problem.

In addition to the data from teens, information from parents reveals their lack of awareness of the magnitude of the problem. For example, 82 percent of parents of teens who had been emailed or texted 30 times per hour were unaware that their sons or daughters were experiencing such harassment. Additionally, 67 percent of parents were unaware that their teen was being pressured via email, text message, or cell phone to have sex, while 71 percent did not know that their teen was afraid of what their partners might do if they failed to respond to a text, an instant message, or a cell phone call. ●

SOURCES: Picard (2007); Zweig et al. (2013).

Dating violence and abuse may be found at very young ages, in fact as soon as young people begin relationships. In one survey, among "tweens" (11- to 14-year-olds), 62 percent knew friends who had suffered verbal abuse by a girlfriend or boyfriend.

As these young people move into high school and then for those who continue on to college, levels of dating violence victimization increase. One study of relationships among a sample of 572 college students found that 21 percent had engaged in "physically

aggressive" behavior, acts that included throwing something; pushing, grabbing, or hitting; slapping; kicking, biting, or punching; beating up; choking; and threatening to or using a gun or a knife on a partner.

In two studies of undergraduate couples in ongoing relationships (18 to 25 years old), Jennifer Katz and colleagues found that a third to nearly half of the students were in relationships in which their partners had acted violently toward them. In both studies, rates at which men and women were victimized were similar, although men experienced higher levels of moderate violence (Katz, Kuffel, and Coblentz 2002). Other research suggested that about a third of college students report experiencing dating violence in a previous relationship, while 21 percent indicated that they had been victimized by a current partner (Sellers and Bromley 1996).

Some of the issues involved in dating violence appear to be different than those generally involved in spousal violence. Whereas marital violence may surface over domestic issues such as housekeeping and child rearing, dating violence is far more likely to be precipitated by jealousy or rejection, or center around issues of control.

Although females and males may sustain dating violence at comparable levels, they do not appear to suffer comparable consequences. As in the case of marital violence, women react with more distress than men do to relationship violence, even within mutually violent relationships (Katz, Kuffel, and Coblentz 2002). They also sustain more physical injuries from dating violence. More surprising is the finding that "partner violence generally is unrelated to decreased relationship satisfaction" (Katz, Kuffel, and Coblentz 2002, 250). Many teen victims remain in relationships even after experiencing violence from a partner, even in the absence of a legal tie or shared residence.

Those who experience dating violence are at greater risk for a variety of health consequences, including increased risk of injury, attempted suicide, binge drinking, and physical fights. They also have higher rates of alcohol, tobacco, and illegal drug use. Teen victims are also more likely than nonvictims to engage in unhealthy and unsafe sexual practices, putting themselves at greater risk of sexually transmitted infections, HIV, and unintended pregnancy (Centers for Disease Control 2009b).

Many women do leave a dating relationship after one violent incident; others stay through repeated episodes. Women who have "romantic" attitudes about jealousy and possessiveness and who have witnessed physical violence between their own parents may be more likely to stay in such relationships (Follingstad et al. 1992). Women who leave violent partners cite the following factors in making the decision to break up: a series of broken promises that the man will end the violence, an improved self-image ("I deserve better"), escalation of the violence, and physical and emotional help from family and friends (Lloyd and Emery 1990). Apparently, counselors, physicians, and law enforcement agencies are not widely used by victims of dating violence.

## Date Rape and Coercive Sex

Sexual intercourse with a dating partner that occurs against his or her will with force or the threat of force—often referred to as **date rape**—is the most common form of rape. Date rape is also known as **acquaintance rape**. One study found that women were more likely than men to define date rape as a crime. Disturbingly, date rape was considered less serious when the woman was African American (Foley et al. 1995).

Date rapes are usually not planned. Two researchers (Bechhofer and Parrot 1991) describe a typical date rape. He plans the evening with the intent of sex, but if the date does not progress as planned and his date does not comply, he becomes angry and takes what he feels is his right—sex. Afterward, the victim feels raped but the assailant believes that he has done nothing wrong. He may even ask the victim out on another date.

Alcohol or drugs are often involved. When both people are drinking, they are viewed as more sexual. Men who believe in rape myths are more likely to see drinking as a sign that females are sexually available (Abbey and Harnish 1995). In many instances, either the female rape victim or her assailant had been drinking or taking drugs before the rape occurred (Caponera 1998). There are also high levels of alcohol and drug use among middle school and high school students who have unwanted sex (Erickson and Rapkin 1991).

In recent years, certain "date rape drugs," most often either gamma-hydroxybutyrate (GHB) or Rohypnol (flunitrazepam, popularly known as "roofies"), have surfaced as major public safety concerns. Both drugs have sedative effects, especially when combined with alcohol. They may reduce inhibitions, and they affect memory. Both are used by some men to sedate and later victimize women, many of whom wake up unaware of where they are, how they got there, or what they have done. Samantha Reid, a 15-year-old, died as a result of drinking a soft drink that had been laced with GHB. Knowing only that the drink tasted funny, she died just hours later. Her friend, Melanie Sindone, recovered after entering a coma that lasted less than a day. In Reid's death, three men were convicted of involuntary manslaughter, punishable by 15 years in prison (Bradsher 2000). In 2000, then President Bill Clinton signed into law the Hillory J. Farias and Samantha Reid Date-Rape Drug Prohibition Act of 2000, named for Reid and another teenage victim who died after unknowingly drinking a beverage mixed with GHB. It is a federal crime, punishable by up to 20 years in prison, to manufacture, distribute, or possess GHB.

## Affirmative Consent: Only "Yes" Means "Yes"

There has long been considerable confusion and argument about sexual consent. Much sexual communication is done nonverbally and ambiguously. In the absence of verbal consent, individuals often tried to gauge sexual interest through nonverbal signs. However, as we saw in Chapter 6, nonverbal communication is imprecise. It can be misinterpreted easily if it is not reinforced verbally. For example, a woman's friendliness might have suggested to some men that she was sexually interested in them, even when she was merely being friendly (Johnson, Stockdale, and Saal 1991; Stockdale 1993). Similarly, a woman's cuddling, kissing, and fondling might have indicated that she wished to go further, such as wishing to engage in sexual intercourse, even though she may have only wanted to cuddle, kiss, or fondle (Gillen and Muncher 1995; Muehlenhard 1988; Muehlenhard and Linton 1987). Our sexual scripts often assumed a "yes" unless a "no" was directly stated. This left much ambiguity, and the assumption of consent—that a woman was interested unless and until she firmly and clearly said no—put women at a definite disadvantage.

To avoid the confusion that some claimed made them read a woman's "no," as not truly "NO!" but more as "maybe," or even "yes, if you keep trying," many colleges and universities have adopted a standard called **affirmative consent**. Following in the footsteps of California, New York colleges and universities have instituted policies in which "yes," and only "yes," means "yes, I am interested in engaging in sexual relations with you." There must be a clear, affirmative agreement between partners regarding mutual interest in further sexual intimacy. As the new legislation signed into California law by Governor Jerry Brown specifies, consent cannot be given if one is intoxicated, incapacitated by drugs, or asleep. Additionally, neither silence nor lack of resistance amount to consent. Affirmative consent must be evident throughout a sexual encounter (Chappell 2014). Though California and New York were among the earliest to adopt such policies into law, other states are considering similar legislation (Medina 2015). As of January 2016, an estimated 1,500 colleges and universities had adopted affirmative consent policies

(Bennett 2016). In California, high school students are required to be taught the concept and implications of affirmative consent.

There are a variety of suggestions made to women about how to protect themselves from date rape. These suggestions include examining how one communicates one's desires (e.g., one should be clear and unambiguous about one's intent and consent), and avoiding using drugs and alcohol (and, if one does drink, be careful about who one accepts a drink from and where one puts it down). Beyond these strategies and suggestions, however, is an important and unpleasant reality. As with avoidance of stranger rapes, a woman can do everything right and still be victimized.

# When and Why Some Women Stay in Violent Relationships

Violence in relationships generally develops a continuing pattern of abuse over time. We know from systems theory that all relationships have some degree of mutual dependence, and battering relationships are certainly no different. Despite the mistreatment they receive, some women stay in or return to violent situations for many reasons. However, we need to be careful not to overstate the tendency for abuse victims to stay with their abusers. As Michael Johnson and Kathleen Ferraro (2000) point out, the focus is often misplaced on answering "why women stay," even when a study finds that two-thirds of women have left a violent relationship. They suggest that it would be more appropriate to ask questions such as "how and why women leave." For those women who do stay in violent or abusive situations, their reasons include the following:

- *Economic dependence.* Even if a woman is financially secure, she may not perceive herself as being able to cope with economic matters. For low-income or poor families, the threat of losing the man's support—if he is incarcerated, for example—may be a real barrier against change.

- *Religious pressure.* She may feel that the teachings of her religion require her to keep the family together at all costs, to submit to her husband's will, and to try harder.

- *Children's need for a father.* She may believe that even a father who beats the mother is better than no father. If the abusing husband also assaults the children, the woman may be motivated to seek help (but this is not always the case).

- *Fear of being alone.* She may have no meaningful relationships outside her marriage. Her husband may have systematically cut off her ties to other family members, friends, and potential support sources. She may have no one to go to for any real perspective on her situation.

- *Belief in the American dream.* The woman may have accepted without question the myth of the perfect woman and happy household. Even though her existence belies this, she continues to believe that it is how it should (and can) be.

- *Guilt, pity, and shame.* She feels that it is her own fault if her marriage isn't working. She worries about who else will take care of her husband. If she leaves, she believes, everyone will know that she is a failure, or her husband might kill himself.

- *Duty and responsibility.* She feels that she must keep her marriage vows "till death us do part."

- *Fear for her life.* She believes that she may be killed if she tries to escape.

- *Love.* She loves him; he loves her. Love may make one want to believe that one's partner will change, that he really is a good person, and so on.

- *Cultural reasons.* Certain minority women may face greater obstacles to leaving a relationship. She may not speak English, may not know where to go for help, and may fear she will not be understood. She often fears that her husband will lose his job, retaliate against her, or take the children back to the country of origin. Recent immigrants may be especially fearful that their revelations will reflect badly on the family and community.

- *Nowhere else to go.* She may have no alternative place to live. Shelter space is limited and temporary. Relatives and friends may be unable or unwilling to house a woman who has left, especially if she brings children with her.

- *Learned helplessness.* Lenore Walker (1979, 1993) theorized that a woman stays in a battering relationship as a result of **learned helplessness.** After being repeatedly battered, she develops a low self-concept and comes to feel that she cannot control the battering or the events that surround it. Through a process of behavioral reinforcement, she "learns" to become helpless and feels that she has no control over the circumstances of her life.

Michael Johnson and Kathleen Ferraro's (2000) distinction between common couple violence and intimate terrorism is important to add here. Women subjected to situational violence are less likely to leave than victims of intimate terrorism. Victims of intimate terrorism leave their partners more often, most commonly seeking friends and relatives for help, and look for destinations that are safe and secret (Johnson and Leone 2005).

## ■ The Costs and Consequences of Intimate Violence

The cumulative financial costs associated with intimate violence are considerable. The Centers for Disease Control reports an estimated cost in excess of $8 billion per year (Max et al. 2004). This includes both costs for direct medical and mental health services for victims of partner violence, rape, assault, and stalking; millions of dollars worth of broken or stolen property; and the wages lost to victims due to time out of work. The "bottom line" is indeed steep.

Then there are the nonfinancial costs. These include the actual health and mental health effects with which victims of violence must cope. DeMaris (2001) reports that thousands of women and men are treated in emergency rooms each year for injuries suffered in partner violence. Victims of intimate partner violence also suffer twice the rate of depression and four times the rate of post-traumatic stress disorder as nonvictims (Zink and Putnam 2005).

The 2010 NISVS also included questions about the impact of such behavior on the recipients. As shown in Figure 12.2, pronounced gender differences surfaced. Women were *five times more likely* to say that the behavior made them fearful (25.7 percent versus 5.2 percent), concerned for their safety (22.2 percent versus 4.5 percent), or caused them to suffer such symptoms of post-traumatic stress disorder as nightmares, feeling detached or numb, or being constantly on guard (22.3 percent versus 4.7 percent). They were also 3½ times as likely to suffer an injury (14.8 percent versus 4.0 percent), and five times as likely to need medical care (7.9 percent versus 1.6 percent).

According to the Centers for Disease Control and Prevention (2003), victims of severe intimate violence lose a cumulative total of nearly 8 million days of paid work—the equivalent of more than 32,000 full-time jobs—and almost 5.6 million days of household productivity each year.

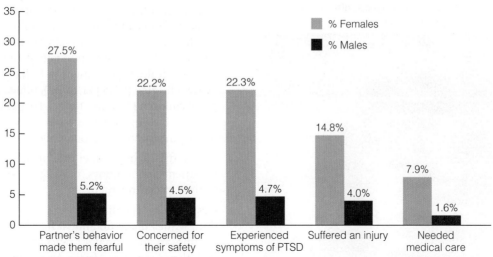

**Figure 12.2** Reported Consequences of Intimate Partner Violence and Abuse, 2010

SOURCE: Black et al. (2011); Breiding et al. (2014).

# Children as Victims: Child Abuse and Neglect

Although it is an all-too-familiar concept today, child abuse was not recognized as a serious problem in the United States until the early 1960s. At that time, C. H. Kempe and his colleagues (1962) coined the medical term **battered child syndrome** to describe the patterns of injuries commonly observed in physically abused children.

Looking at the variety of forms of potential abuse and adding the issue of neglect, the U.S. Centers for Disease Control (2009a) provided the following information regarding various aspects of **child**

**True**   **6.** The most common form of child maltreatment is neglect.

**maltreatment** in the United States. Child maltreatment includes the following:

- *Neglect.* Failing to meet a child's basic needs for such things as food, housing, clothing, education, and access to medical care.

- *Physical abuse.* Actions such as hitting, kicking, shaking, or burning a child, resulting in the child sustaining an injury or dying.

- *Sexual abuse.* Such actions as fondling, raping, or exposing a child to other sexual activities.

- *Emotional abuse.* Subjecting a child to such behaviors as name-calling, threatening, withholding affection, and shaming, all of which can harm the child's emotional well-being and sense of self-worth (see Table 12.5).

**TABLE 12.5** Forms of Emotional Child Abuse

Based on the work of psychologists Stuart Hart and Marla Brassard, all of the following six categories of psychological maltreatment (or emotional abuse) convey to the child victim that she or he is unloved, unwanted, endangered, worthless, or flawed:

- *Spurning.* Ridiculing or belittling a child

- *Terrorizing.* Threatening a child or placing the child in a dangerous situation

- *Isolating.* Denying a child the opportunity to interact with others, confining a child, or imposing unreasonable limitations on a child's freedom of movement

- *Exploiting or corrupting.* Permitting the child to drink alcohol while underage or to use tobacco or illegal drugs; encouraging a child to engage in prostitution or other criminal activities; or exposing a child to criminal activities

- *Denying emotional responsiveness.* Refusing or failing to express affection or ignoring a child's attempts to interact

- *Neglect of a child's mental or medical health or educational needs*

SOURCE: Hart, Gelardo, and Brassard (1986).

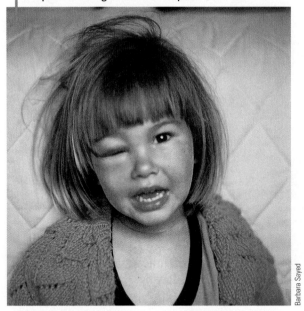

Children are the least protected members of our society. Much physical abuse is camouflaged as discipline or as the parent "losing" his or her temper.

Barbara Sayed

## Prevalence of Child Maltreatment

Based on data collected in the National Child Abuse and Neglect Data System, a federal government effort to collect annual data on child abuse and neglect, the following picture of child maltreatment in the United States emerges:

- In 2013, 3.5 million referrals to children's protective services (CPS) agencies were made, involving nearly almost 6.5 million children. Of these, 2.1 million reports, involving 3.2 million children, received a CPS response, of either an investigation or an alternative response. CPS determined that 679,000 children were in fact victims of maltreatment.

- Most (61.6 percent) reports to children's protective services of abuse or neglect were made by professionals who had contact with the victim through their jobs as social services staff (11.0 percent), teachers and other educational personnel (17.5 percent), or police officers or legal personnel (17.5 percent). Friends, relatives, neighbors, or coaches made the remainder, almost a fifth (18.6 percent), of reports. Twenty percent were made anonymously, or by "other" or "unknown" sources.

- Although child abuse occurs at every economic level, among all religions, and across racial and ethnic lines, rates of victimization appear to vary by race and gender, with African American, multiracial children, and American Indian and Alaska Native children at greatest risk. African American children had the highest victimization rate, at 14.6 per 1,000 children, followed by American Indian and Alaska Native children (12.5 per 1,000), biracial or multiracial children (10.6 per 1,000), Hispanic children (8.5 per 1,000), whites (8.1 per 1,000), and Asian American children (1.7 per 1,000).

- Girls suffer slightly higher risk for all forms of maltreatment than do boys, making up 51 percent of victims of maltreatment.

- Despite the racial differences in rates of victimization, whites made up almost half (44 percent) of victims; an additional 22 percent were Hispanic, and 21 percent were African American.

- At 23.1 per 1,000, children under 1 year of age had the highest rate of victimization (see Figure 12.3). More than a quarter of all victims (27 percent) were under 3 years of age. Another 20 percent were between 3 and 5 years of age (see Figure 12.4).

- The most common form of child maltreatment was neglect, which was found in 80 percent of the cases of substantiated maltreatment. Of the other forms of maltreatment, the child suffered physical abuse in 18 percent of the cases, 9 percent experienced sexual abuse, and 9 percent were victims of psychological maltreatment. An additional 10 percent experienced some "other" form of maltreatment (e.g., threats to harm child, abandonment, or congenital drug addiction). Two percent experienced medical neglect. Some children were poly-victims, so the percentages do not add up to 100 percent. In 2013, an estimated 1,520 children died from abuse and/or neglect.

- Over 80 percent of the perpetrators of child maltreatment were parents, 13 percent were nonparents, and 5 percent were unknown. Women represented 56.5 percent of perpetrators. Perpetrators were relatively young; 75 percent were under 40 years of age.

- Of parent perpetrators, almost 90 percent were biological parents. Mothers acted as the sole perpetrator in 40.7 percent of the cases. In 22.5 percent of the cases, mothers and fathers were both perpetrators, and in nearly 7 percent, mothers and "others" together were perpetrators. Fathers were the sole perpetrators in 20.3 percent of the cases and acted with

**Figure 12.3  Rate per 1,000 of Victimizations, by Age of Child**

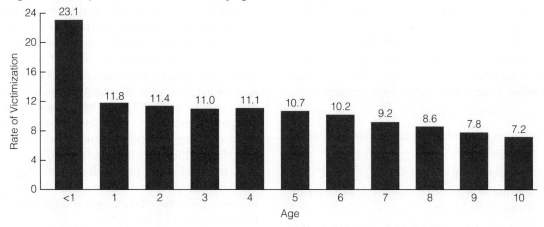

SOURCE: U.S. Department of Health and Human Services. "Child Maltreatment 2013." Administration for Children and Families, Administration on Children, Youth and Families, Children's Bureau, 2015a. www.acf.hhs.gov/programs/cb/research-data-technology/statistics-research/child-maltreatment.

**Figure 12.4  Percentage of Victims by Age**

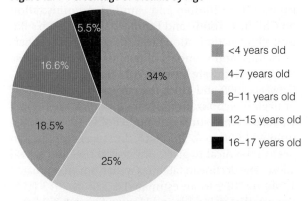

- <4 years old
- 4–7 years old
- 8–11 years old
- 12–15 years old
- 16–17 years old

SOURCE: U.S. Department of Health and Human Services. "Child Maltreatment 2013." Administration for Children and Families, Administration on Children, Youth and Families, Children's Bureau, 2015a. www.acf.hhs.gov/programs/cb /research-data-technology/statistics-research/child-maltreatment.

Data indicate that children commonly experience psychological aggression.

someone other than the child's mother in an additional 1 percent of the cases of maltreatment (U.S. Department of Health and Human Services 2015a).

Considering the state of U.S. children in 2012, the Children's Defense Fund (2014) reported the following:

- Every 47 seconds, a child suffers from abuse or neglect, with infants and toddlers most at risk.
- Each day, 1,837 children are confirmed to have been abused or neglected.
- Each day, more than four children die as a result of abuse or neglect, an average of a child every five and a half hours.

Like intimate partner violence, the real prevalence of child abuse may be beyond our ability to accurately measure. It is widely assumed that official reports on the numbers of abused or neglected children do not include all actual cases of abuse and neglect, and thus do not capture the full scope of the problem of child maltreatment (Committee on Child Maltreatment Research, Policy, and Practice for the Next Decade, et al. 2014). As is also true of partner relationships, children are subjected to nonphysical forms of mistreatment by parents. In examining the national prevalence of **psychological aggression** by parents, Murray Straus and Carolyn Field (2003) found that verbal attacks on children are so common as to be "just about universal."

Based on nearly 1,000 interviews with a nationally representative sample of households with at least one child under 18 years old living at home, Straus and Field explored the prevalence of psychological aggression. They defined psychological aggression as consisting of the following kinds of behaviors, with the latter three constituting "more severe" psychological aggression:

- Shouting, yelling, or screaming at one's child
- Threatening to spank or hit one's child but not actually doing it
- Swearing or cursing at one's child
- Threatening to send one's child away or kick him or her out of the house
- Calling one's child dumb or lazy, or making some other disparaging comment

Of the sample parents, 89 percent reported having committed at least one of the five kinds of psychological aggression, and 33 percent reported at least one instance of the more severe forms. The prevalence of the various forms of psychological aggression is illustrated in Table 12.6.

Use of psychological aggression varies with the age of the child. A total of 43 percent of parents of infants reported using psychological aggression, and nearly 90 percent of parents of 2-year-olds use some form of psychological aggression. The percentage peaks at 98 percent at age 7, and as late as age 17, the rate still remains as high as 90 percent.

Conversely, research on corporal (physical) punishment shows it declining with the age of the child; only 12 percent of parents of 17-year-olds report still using corporal punishment (Straus and Field 2003).

TABLE 12.6 **Prevalence of Psychological Aggression**

| Prevalence | Measure (% in last year) |
| --- | --- |
| Overall | 88.6 |
| Severe | 33.4 |
| Shouting, yelling, screaming | 74.7 |
| Threatening to spank or hit | 53.6 |
| Swearing or cursing | 24.3 |
| Name-calling | 17.5 |
| Threatening to kick out of house | 6.0 |

SOURCE: Straus and Field (2003).

However, more than 90 percent of toddlers in the United States are reportedly spanked (Straus and Field 2003). Most child-rearing experts currently advise that parents use alternative disciplinary measures.

Parents' ages matter, too. Younger parents (ages 18 to 29) reported the most frequent use of psychological aggression (22 times in the past 12 months) compared to parents 30 to 39 (19 times in the past 12 months) and parents over 40 (15 times in the past 12 months). Aside from age differences, there was "a lack of demographic differences in use of psychological aggression; this means that nearly all parents, regardless of sociodemographic characteristics, used at least some psychological aggression as a disciplinary tactic" (Straus and Filed 2003, 805).

In the midst of all these distressing facts, is there any good news to report? If we look at the five years of data in the report, *Child Maltreatment: 2013,* from the U.S. Department of Health and Human Services, Administration for Children and Families, Administration on Children, Youth, and Families, and Children's Bureau, child victimization rates, the estimated numbers of victims, and numbers of fatalities linked to abuse and/or neglect were lower in 2013 than in recent years. Between 2009 and 2013, the rate of child abuse and/or neglect victimization declined from 9.3 to 9.1 per 1,000, and the estimated number of victims declined from 693,000 to 679,000. Declines were reported in neglect, medical neglect, physical abuse, and sexual abuse. The decline in fatalities went from an estimated 1,740 in 2009 to an estimated 1,520 in 2013 (U.S. Department of Health and Human Services 2015a).

Writing a decade ago, sociologist David Finkelhor and psychologist Lisa Jones noted that a variety of forms of maltreatment and abuse of children had declined. Finkelhor and Jones (2006) reported the following:

- In the early 1990s, sexual abuse began to decline after a 15-year-long period of increases. From 1990 through 2004, substantiated sexual abuse was down 49 percent.
- From 1992 through 2004, physical abuse was down 43 percent. Thirty-eight states experienced declines.

In accounting for the declines in most forms of child abuse, Finkelhor and Jones (2006) pointed to a number of factors, including increases in the numbers of police and other agents of social control and intervention, a decrease in the numbers of unwanted children, a fairly robust economy, changing cultural norms, and the arrival of new psychopharmaceuticals, such as Prozac. Unfortunately, in the midst of the U.S. economic downturn, a number of state and local agencies around the United States (e.g., Arizona, New York, Georgia, Washington, California, Ohio, and Massachusetts) reported increases in reports to children's protective services, calls to hotlines, hospital emergency room visits, and arrests. This suggests that economic matters may have been large factors in the decreases that Finkelhor and Jones described as well as in more recent decreases that happened in the post-recession years.

## Families at Risk

Research has established that the following sets of factors put families at risk for child abuse and neglect: (1) parental characteristics, (2) child characteristics, (3) family factors, and (4) the family ecosystem—that is, the family system's interaction with the larger environment. The characteristics described in the next sections are likely to be present in abusive families (Goldman et al. 2003; Straus, Gelles, and Steinmetz 1980).

### Parental Characteristics

Some or all of the following characteristics are likely to be present in parents who abuse their children:

- The abusive parent was physically punished by his or her parents, and his or her father physically abused his or her mother. However, it is important to emphasize that a history of childhood maltreatment does not guarantee that one will become a maltreating parent. Likewise, many parents who abuse or neglect their children were not themselves abused.
- The parents believe in corporal discipline of children.
- Parents have unrealistic expectations of their children and are less understanding of age-appropriate behaviors. Either of these can lead to parental frustration or disappointment, both of which may lead them to take frustrations out on their child.
- Certain personality traits or characteristics are often identified among abusers. These include low self-esteem, poor impulse control, anxiety, or depression.
- The marital relationship itself may not be valued by the parents. There may be spousal violence.
- The parents appear unconcerned about the seriousness of a child's injury, responding, "Oh well, accidents happen."
- A parent may engage in substance abuse.

### Child Characteristics

Who are the battered children? Are they any different from other children? Surprisingly, the answer is often yes; they are different in some way or at least are perceived to be so by their parents. Children who are abused are often labeled by their parents as "unsatisfactory," a term that may describe any of the following:

- A "normal" child who is the product of a difficult or unplanned pregnancy, is of the "wrong" sex, or is born outside of marriage.
- An "abnormal" child, one who was premature or of low birth weight, possibly with congenital defects or illness. Children with disabilities suffer rates of maltreatment that may be almost twice as high as children without disabilities.
- A "difficult" child, one who shows such traits as fussiness or hyperactivity. Researchers note that all too often, a child's perceived difficulties are a result (rather than a cause) of abuse and neglect.

In addition to these, the child's age has been found to be associated with certain increased risks. Young children, especially infants to toddlers, are at greatest risk for neglect. They are also more vulnerable to "shaken baby syndrome." Teens are at greater risk of sexual abuse.

### Family Characteristics

In addition to the aforementioned characteristics of adult offenders and child victims, the following family characteristics are associated with greater risks of child abuse and maltreatment:

- Children in single-parent homes may have higher risk of victimization, although it may be less a

Children need to have someone, such as a teacher who they trust, in whom they can confide about their suffering.

Yellow Dog Productions/Photodisc/Getty Images

matter of the number of parents than of other characteristics that often accompany single parenthood (e.g., lower income and increased stress). For example, the elevated rate of abusive violence found in homes headed by single mothers is likely a function of the poverty that characterizes such families.

- Marital conflict, especially if it becomes frequent, intense, and violent, is often accompanied by child maltreatment.
- Families experiencing high levels of stress (e.g., from unemployment, serious illness, or death of a family member) may be susceptible to higher risks of child abuse or neglect.
- The kind and quality of parent–child interaction may be a factor leading to harsh discipline and more use of corporal punishment or verbal aggression. Parents who abuse or neglect their children are typically less affectionate, playful, or supportive, and focus more on children's negative behaviors than on their positive behaviors.

## Family Ecosystem

As discussed earlier in this chapter, the community and the family's relation to it may be relevant to the existence of domestic violence. The following characteristics may be found in families that experience child abuse:

- The family experiences poverty and/or unemployment.
- The family is socially isolated, with few or no close contacts with relatives, friends, or groups.
- The family has a low level of income, creating economic stress.
- The family lives in an unsafe neighborhood that is characterized by higher-than-average levels of violence.
- The home is crowded, hazardous, dirty, or unhealthy.

Notice the clustering of such socioeconomic characteristics as unemployment, low income, neighborhood, and housing. This combination tells an important story. Like spousal or partner violence, the mistreatment of children can be found across the socioeconomic spectrum. But, like spousal violence, it happens more often at the lower economic levels. As noted earlier, the culprit in these associations is most likely stress.

Children who experience abuse or neglect often endure long-lasting adverse effects, including physical, emotional, cognitive, and social effects. They are more likely than children who are not maltreated to experience poor physical health, including such outcomes as hypertension and chronic fatigue; poor emotional and mental health, including depression, anxiety, or suicidal thoughts; social relationship difficulties; and cognitive deficits. They are also more likely to engage in more high-risk behaviors, such as onset of sexual activity at a younger age; to have more sexual partners; to become pregnant as a teen; and to engage in substance use and to display such behavioral problems as aggression, delinquency, or violent behavior as adults (Wang and Holton 2007).

Children also suffer from witnessing violence between family members. As revealed in the National Survey of Children's Exposure to Violence (NatSCEV), one in 12 young people under 18 years of age indicated that they had witnessed some form of family violence in the preceding 12-month period, with over 17 percent seeing interparental violence. More than one-fourth (20.3 percent) of the more than 4,500 young people, age 17 and younger, had past exposure to some violence between family members. Furthermore, in their lifetimes, 14 percent saw a parent pushed, 15 percent saw "displaced aggression," wherein someone threw something or broke something while in the course of conflict, 12 percent saw a parent assaulted, and 10 percent witnessed hitting or slapping between parents. One out of 20 saw severe interparental violence, as in seeing a parent kicked, choked, or beaten up.

Exposure to IPV as a child is associated with such consequences as depression, anxiety, post-traumatic stress disorder, as well as an elevated risk of criminal offending as an adult. In this way, children who are not direct victims are nevertheless clearly victimized (Finkelhor et al. 2015; Hamby et al. 2011).

## ■ Hidden Victims of Family Violence: Siblings, Parents, and the Elderly

Most studies of, and much of the public discussion and concern about, family violence have focused on violence between spouses and on parental violence toward children. There is, however, considerable violence between siblings, between teenage children and their parents, and between adult children and their aging parents. These are the "hidden victims" of family violence (Gelles and Cornell 1990).

## Sibling Violence

More than three decades of research illustrates that violence between siblings is by far the most common form of family violence (Hoffman, Kiecolt, and Edwards 2005; Straus, Gelles, and Steinmetz 1980). Estimates in different studies have ranged from 40 percent to more than 90 percent of children under age 18 experiencing sibling violence or abuse.

Estimates of childhood exposure to sibling violence suggest greater victimization from siblings than from peers, whereas among adolescents, more experience peer victimization than sibling victimization (Tucker et al. 2014). Drawing from data from the National Survey of Children's Exposure to Violence and its large, representative sample of children 3 to 17, Corinna Jenkins Tucker, David Finkelhor, Heather Turner, and Anne Shattuck reported levels of peer and sibling victimization. They found that in the year prior to the study, among sample children (respondents 3 to 9 years of age), whereas 40 percent reported no victimization from siblings or peers, a third experienced sibling violence, 12 percent peer violence, and 15 percent were victimized both by siblings and peers. Among those 10 to 17, they found 49 percent reporting no victimization, 14 percent reported victimization by siblings only, 22 percent reported only peer victimization, and once again, 15 percent experienced victimization by both peers and siblings in the prior year (Tucker et al. 2014).

Among junior high school/middle school–age children, 88 percent of males and 94 percent of females reported ever being victims of sibling violence, and 85 percent of males and 96 percent of females admitted ever assaulting a sibling (Caspi 2008). Although violence declines as children age, no less than two-thirds of teenagers annually commit an act of violence— pushing, slapping, throwing, or hitting with an object or something more severe—against a sibling. In addition, despite decreasing prevalence from childhood to adolescence, there is evidence that the seriousness of sibling abuse increases; adolescent sibling violence is more likely to result in injury or include weapons.

Studies of college-age samples demonstrate that such behavior can persist at least into young adulthood. One study of 651 college undergraduates found that nearly 70 percent acknowledged having acted violently toward their closest-age sibling while seniors in high school. The violence most commonly consisted of hitting with a hand or object, pushing or shoving, and throwing things but often included slapping, punching, and pulling hair (Hoffman, Kiecolt, and Edwards 2005). Another study of college students found that 65 percent of the sample had experienced severe physical violence (e.g., kicking, biting, and choking) within the prior 12-month period (Wilson and Fromuth 1997).

> **True** **7.** Sibling violence is the most widespread form of family violence.

This type of sibling interaction is often simply taken for granted by our culture—"You know how kids are!" However, in 2002, siblings were responsible for 6 percent of all intrafamilial murders in the United States (Kiselica and Morrill-Richards 2007). Thus, sibling violence, like other forms of familial violence, can reach extreme levels.

Also like other forms of intrafamilial violence, sibling violence and abuse can take varying forms: It can be physical, psychological, or sexual. However, unlike most other forms, it is often seen as normal, harmless, and socially acceptable (Caspi 2011). In reality, suffering violence from a sibling, being consistently targeted and subjected to humiliation and/or aggression, can be at least as, if not more, damaging when compared to being bullied by one's peers or classmates. It is well beyond normal and far from harmless (O'Connor 2013). Chronic exposure to verbal or physical abuse from a sibling, especially for those who have been attacked or intimidated by a sibling, is associated with higher levels of depression, anger, and anxiety.

Counseling psychologists Mark Kiselica and Mandy Morrill-Richards review the literature on sibling abuse from a wide range of sources. In their review, they note the following:

- A study of 150 adult survivors of sibling abuse found that 78 percent had experienced emotional abuse.

- A national study of family violence reported that 80 percent of 3- to 17-year-olds had experienced sibling violence and that more than 50 percent had experienced severe violence—such as acts as stabbing, striking with an object, punching, or kicking.

- The rates of physical violence tended to be higher for males and also to decline with age.

- As to sexual abuse, research suggests that as many as 2.3 percent of women have been sexually victimized by a sibling.

Despite the fact that sibling violence and abuse remain the most common form of intrafamilial violence, they

Nixzmary Brown was 7 years old when she died at the hands of her abusive stepfather in 2006. Details of her sad life and tragic death reveal abuses that included many nights tied to a chair, being forced to use a litter box rather than the toilet, and frequent beatings from her stepfather. On January 11, 2006, her stepfather beat her to death for "stealing a cup of yogurt" from the refrigerator. According to prosecutor Ama Dwimoh, "After being beaten, battered, broken, and thrown naked onto a cold wooden floor, the last words of 7-year-old Nixzmary Brown—moaning in pain, gasping for air—were, 'Mommy, Mommy, Mommy.'" Yet her mother, Nixzaliz Santiago, did not protect or help her daughter and, in fact, may have helped initiate some of the prolonged abuse the child endured. She was convicted of manslaughter and sentenced to more than 40 years in prison. The stepfather, Cesar Rodriguez, was sentenced to 29 years in prison. As details of this case unfolded, it was learned that the system designed to protect children had failed Nixzmary. A social worker at her school had made numerous pleas with the Administration for Children's Services (ACS) to save Nixzmary and had even tried to visit Nixzmary at home. The school principal had expressed concern for Nixzmary's safety. Despite such expressions of concern to ACS, caseworkers for the city agency failed to act.

On September 12, 2009, the New York state legislature sent a bill, Nixzmary's Law, on to New York Governor David Paterson. The law made those who torture or intentionally kill a child in an "especially cruel and wanton manner" eligible for life sentences without parole.

The case and the legal legacy it leaves are reminiscent of the 1995 murder of 6-year-old Elisa Izquierdo by her mother, Awilda Lopez. In addition to sharing similar kinds of horrific abuse, the two girls were cases that somehow slipped through the cracks in the agency intended to protect children and prevent such tragedies. After the outrage surrounding Elisa's death, New York passed Elisa's Law, which—among other things—was designed to increase disclosure of child abuse allegations made by requiring "authorities to disclose this information in several situations, among them: when a person is charged with child abuse and when an abused child dies. It thus seeks to make investigators more accountable for their mistakes" (Hernandez 1996, B6). The law also mandated that records of investigated but unsubstantiated allegations of abuse be kept, thus enabling authorities to more effectively detail the history of alleged abuse victimization. Unfortunately, neither Elisa's Law nor Nixzmary's Law—nor any statutory change for that matter—can perfectly protect children and prevent abuse. However, these laws can facilitate a more effective reaction from those whose job it is to investigate complaints and/or punish perpetrators. More recently, through the efforts of Erin Merryn, a survivor of child sexual abuse, 26 states have passed legislation that has been come to be known as "Erin's Law." It requires all public schools to implement child sexual abuse prevention programs. This includes providing age-appropriate strategies and techniques to students from prekindergarten through 12th grade to enable them to recognize child sexual abuse and to tell an adult they trust. It also educates school personnel, parents, and guardians about the problem of child sexual abuse, the warning signs to watch for, and referral or resource information to assist child sexual abuse victims or their families. ●

SOURCES: Blain (2009); Haberman (2009); Hernandez (1996); and Shifrel (2008); http://erinslaw.org/about/what-is-erins-law/.

have received surprisingly less attention and seem to generate considerably less concern than other forms of family abuse.

## Parents as Victims

Many people might find it difficult to imagine children attacking their parents because it so profoundly violates our image of parent–child relations. Parents possess the authority and power in the family hierarchy. Furthermore, there is greater social disapproval of a child striking a parent than of a parent striking a child; it is the parent—not the child—who has the "right" to hit.

Indeed, child–parent violence (CPV) may well be the least understood, most understudied form of family violence, even though some estimates suggest it to be more prevalent than spousal abuse and child abuse (Walsh and Krienert 2007). Researcher Amanda Holt (2012) calls it "the most underresearched form of family abuse." Criminal justice researchers Jeffrey Walsh and Jessie Krienert report that past estimates have ranged widely. Studies using survey data have suggested that 5 to 10 percent of adolescents strike their parents each

> **True**  **8.** Violence by children against parents has been the most neglected form of family violence.

year, that child violence toward a parent is experienced in close to one in five two-parent and more than one in four single-parent families, that 20 percent of children strike a parent each year, and that 10 percent commit violence with high risk of injury to the parent(s) (Walsh and Krienert 2007). Paul Robinson, Leah Davidson, and Michael Drebot stated that most estimates ranged between one in ten and one in three parents of 10- to 23-year-olds report being abused by their children (Robinson, Davidson, and Drebot 2004). Estimates vary depending on the ages of the children studied, the length of time one inquires about (e.g., the previous year or ever), the type of violence one includes, and the nature of the sample itself (i.e., studies of delinquent youth produce higher estimates) (Ulman and Straus 2003).

After reviewing 11 years of data from the FBI's National Incident-Based Reporting System, they analyzed all criminal events in which a child between 7 and 21 years of age committed violence against a parent. Between 1995 and 2005, there were over 108,000 such incidents of child–parent violence (CPV) and 79 cases of **parricide**, the murder of a parent by a child (Walsh and Kreinert 2009).

Of the 108,231 CPV cases, offenders were most likely to be white males between 14 and 16 years of age who victimized females, typically their biological mothers. In addition, the following was also noted:

- Males committed 63 percent, females 37 percent.
- Victims and offenders were primarily white (78 and 76 percent, respectively).
- The peak age of offenders was 14 to 16; more than half of victims were 35 to 44.

Parricide offenders and victims tended to be somewhat older; almost two-thirds of offenders were 17 to 21, and 60 percent of the victims were 45 years old or older. Both offenders and victims were typically white males. Males committed 85 percent of the parricides. In just under 80 percent of the cases, the victims were biological parents, though there was a higher percentage of stepparents (stepfathers) than found among CPV victims. Walsh and Krienert (2009) note that parricide may be the result of an escalation process. What often starts as verbal abuse may progress over time, becoming both more frequent and more intense, escalating to emotional and physical abuse "when the intended effect is no longer achieved" (p. 23). For some, primarily males, this may further increase in both frequency and intensity, reaching a point where "parricide is the culminating action" (p. 23).

## Elder Abuse

Of all the forms of hidden family violence, only the abuse of elderly parents by their grown children (or, in some cases, by their grandchildren) has received considerable public attention. Elder mistreatment may be an act of commission (abuse) or omission (neglect) (Wolf 1995). It is estimated that more than 2 million older people (over 50) are physically abused annually (American Psychological Association 2005). Perhaps as many as 10 percent of the elderly population has been abused (Centers for Disease Control and Prevention 2016; Lachs and Pillemer 2004). In approximately 90 percent of elder abuse cases, the victim is abused by a family member—a spouse or partner, adult children and grandchildren, or other family members (National Coalition Against Domestic Violence 2008a).

Like the abuse of children and intimate partners, elder abuse can be physical, psychological, sexual, verbal, or financial. Also like child abuse, neglect is one of the more common forms of elder maltreatment, consisting less of acts against them than actions withheld, resulting in the failure to provide sufficient food, clothing, housing, or medical care. Sexual abuse of elders includes unwanted sexual contact as well as forced viewing of pornography or forced listening to sexual accounts. It most commonly involves a male caregiver as perpetrator and a female victim, over 70 years of age, who is either functioning at a low level or totally dependent on her caregiver(s).

Financial abuse encompasses a range of acts, including coercing or deceiving an elder into signing wills, contracts, or other similar documents; taking advantage of an elder suffering from dementia by taking control over his or her money or financial decisions; and forcing a victim to part with resources or property. Physical and verbal forms of elder abuse are much like physical and verbal abuse of other family members.

Physical abuse includes any acts of physical contact intended to cause pain or injury to the victim. Signs of physical abuse include bruises, abrasions, burns, fractures, and dislocations, along with any injuries that are either unexplained or whose explanations don't fit the injury. Verbal abuse would encompass such acts as name-calling, embarrassing or scaring a victim, along with other acts that damage the emotional well-being or self-worth of the recipient (Centers for Disease Control and Prevention 2008a; National Coalition Against Domestic Violence 2008a). Most abuse of the elderly goes unnoticed, unrecognized, and unreported. Even though mandatory reporting of suspected cases of elder

The abuse of the elderly by children, grandchildren, and others can be physical, psychological, sexual, or financial.

John Birdsall/The Image Works

abuse is the law in at least 44 of the 50 states (Wei and Herbers 2004), sociologists Karl Pillemer and David Finkelhor estimate that only 1 of every 14 cases is reported to authorities (Pillemer and Finkelhor 1998). Elderly people are often confined to a bed or a wheelchair, and many do not report their mistreatment out of fear of institutionalization or other reprisal.

Among the factors that can protect elderly men and women from potential victimization is having a number of strong relationships with people of different social statuses, thus diminishing the likelihood of any single caregiver or contact being able to mistreat someone without such mistreatment being discovered. The more contacts one has, the less dependent one is, and the greater chance for someone to prevent the abuse or protect the potential victim (National Coalition Against Domestic Violence 2008b).

The most likely victims—in most cases, women—of elder abuse are suffering from physical or mental impairments, especially Alzheimer's disease or other forms of dementia. Their advanced age renders them dependent on their caregivers for many if not all of their daily needs. Their dependency may increase their likelihood of being abused. Other research indicates that many abusers are financially dependent on their elderly parents; they may resort to violence out of feelings of powerlessness.

## The Economic Costs of Family Violence

The economic impact of family violence is staggering. Whether one assesses the effects in terms of dollars and cents, impaired work performance, lost time and wages in one's job, or some other measure, the costs are great.

Other economic "costs" can be seen in the following information on job performance and workplace experiences:

- The Centers for Disease Control and Prevention and Bureau of Labor Statistics Survey of Workplace Violence Prevention estimate that victims of domestic violence collectively lose almost 8 million days of paid work, an amount that translates to more than 32,000 full-time jobs, and over 5.5 million days of household productivity (Centers for Disease Control and Prevention 2015c; U.S. Bureau of Labor Statistics 2006)

- Nearly a third of women killed in the workplace between 2003 and 2008 were killed by a current or former intimate partner.

- Of the 1 million women who are victims of stalking, about 250,000 report that stalking led them to miss work; they missed an average of 11 days of work per year. Seven percent never returned to their jobs.

- Between 2005 and 2006, approximately 130,000 women were fired or asked to leave their jobs because of stalking by a current or former intimate partner or spouse.

- Between 35 percent and 56 percent of employed abused women were harassed by an intimate partner while at work.

Attitudes and opinions of corporate leaders are especially instructive. More than 90 percent of Fortune 1000 corporate leaders believe that both the private and the working lives of employees are affected by domestic violence: 56 percent are aware of employees within their companies who are affected by intimate violence against women; 32 percent of corporate leaders contend that violence against women has damaged their company's "bottom line," while 66 percent believe that their company's performance would be better if they could address violence against women; and 48 percent believe that worker productivity in their organization has been negatively affected by intimate violence against women, while 42 percent believe

## Working the Front Line in the Fight against Child Abuse

In 2010, almost 1,600 children died from abuse or neglect by their parents or guardians. Some cases remain relatively unknown to the wider public, reported in small articles in mostly local newspapers if reported at all. Others become major news stories, the focus of not only local but also wider regional or even national attention. Both kinds of cases can be seen in the following list of cases that occurred over the past 25 years. The list includes Eli Creekmore, age 3, beaten to death by his father in 1986; Elizabeth "Lisa" Steinberg, age 6, beaten to death by her adopted father in 1987; Joseph Wallace, age 3, hanged by his mother in 1993; Nadine Lockwood, age 4, intentionally starved to death by her mother in 1996; and Serenity Deal, age 5, beaten to death by her father in 2011. In just a three-month period, between late 2005 and early 2006, Sierra Roberts, age 7; Dahquay Gillians, age 16 months; and Joziah Bunch, age 1; died at the hands of their parents.

This is but a partial list of child abuse homicides, selected because in each instance some agency or individuals in a position to intervene didn't—despite what in retrospect looked like clear and unambiguous evidence of severe abuse. Many of these cases were met by public outcry and led to changes in the policies used by the relevant protective agencies. Add to them the three-quarters of a million children or youth treated in emergency rooms due to violence. Confronted with this kind of information, the most extreme outrage is typically and understandably expressed at the parent perpetrators. However, there is also often intense anger and blame directed at the agency or caseworkers who failed to rescue the child from his or her abusive, lethal surroundings. "How could they miss this?" many wonder. "How could

they let this happen?" The next time you hear someone or feel yourself wondering such things, try to keep the following two things in mind:

- Child Protective Services caseworkers are asked to do what sometimes feels like an impossible job, given the constraints of their numbers and the numbers of their cases. In 2013, the national average caseload for CPS caseworkers was 68, though ten states had average caseloads that reached into triple digits. Utah, with an average caseload per caseworker of 199 was the highest, followed by Indiana (175), Massachusetts (163), Louisiana (147), and New Hampshire (143). The smallest average caseload per caseworker was Connecticut (25). Each case puts the caseworker in a position of determining whether a child is or isn't being mistreated, whether the mistreatment warrants removal from the home, and whether some alternative plan or intervention can preserve the family. The consequences of "guessing wrong" include subjecting a child to further, worsening, perhaps fatal mistreatment, or removing a child and thus breaking up a family that could otherwise be preserved.
- Compounding this situation is the double bind–like message that is reflected in some of our cultural values about families, specifically the tension between "hands off," or our expectation of family privacy, versus "hands on," our expectation that the most vulnerable among us will be protected from mistreatment by one's family when need be. Each and every day, caseworkers walk that fine line between those positions, knowing all along that their choices carry life-altering—perhaps even life-and-death—consequences. ●

that such violence has contributed to high levels of employee turnover (Roper Starch Worldwide 2002).

The Children's Defense Fund estimated that child maltreatment led to an estimated $80 billion a year in direct costs and lost productivity. Others put the estimated costs higher. Researchers Xiangming Fang, Derek S. Brown, Curtis S. Florence, and James A. Mercy of the Centers for Disease Control and Prevention estimated that the dollar costs associated with child abuse and neglect in the United States in 2008 reached over $121 billion. For each victim of nonfatal child maltreatment, they project lifetime cost of $210,000. For each child maltreatment fatality, they estimate an average cost of $1.3 million. They also acknowledge

that these estimates do not include all possible negative impacts resulting from abuse and their associated costs. Therefore, they acknowledge that the estimates are likely underestimates (Fang et al. 2012).

## Responding to Intimate and Family Violence

Based on the foregoing evidence, you may by now have concluded that the U.S. family is well on its way to extinction as family members bash, thrash, cut, shoot, and otherwise wipe themselves out of existence.

Statistically, the safest family homes are those with one or no children in which the husband and wife experience little life stress and in which decisions are made democratically. By this definition, most of us probably do not live in homes that are particularly safe. What can we do to protect ourselves (and our posterity) from ourselves?

Professionals who deal with domestic violence have long debated the most appropriate strategy: control and deterrence versus compassion (Mederer and Gelles 1989). Both approaches have their place. Controlling measures such as arrest, prosecution, and imprisonment, as well as compassionate measures such as shelters, education, counseling, and support groups, have been shown to be successful to varying degrees. Used together, these interventions may be quite effective. Helen Mederer and Richard Gelles (1989) suggest that controlling measures may be effective in motivating perpetrators to take part in treatment programs.

**False** **9.** Mandatory arrest policies now exist in 20 out of the 50 states.

## Intervention and Prevention

The goals of intervention in domestic violence include protecting victims, punishing offenders, and perhaps rehabilitating offenders through therapeutic intervention, and assisting and strengthening families. In dealing with intimate partner and family violence, especially with child abuse, professionals and government agencies may be called on to provide medical care, counseling, and services such as day care, child care education, telephone crisis lines, and temporary foster care. Many of these services are costly, and many of those who require them cannot afford to pay. Our system does not currently provide the human and financial resources necessary to deal with these problems.

Prevention strategies usually take one of the following paths: (1) eliminating social stress; (2) educating about the causes, costs, and consequences of violence and abuse; or (3) strengthening families. Family violence experts make the following general recommendations:

- Reduce societal sources of stress, such as poverty, racism and inequality, unemployment, and inadequate health care.
- Reduce sexism, and provide employment and educational opportunities for women and men.
- Furnish adequate day care.

- Promote sex education and family planning to prevent unplanned and unwanted pregnancies.
- End social isolation, and explore means of establishing supportive networks that include relatives, friends, and community.
- Break the family cycle of violence, eliminate corporal punishment, promote education about disciplinary alternatives, and support parent education classes to deal with inevitable parent–child conflict.
- Address the cultural norms that legitimize and glorify violence.

## Intimate Partner Violence and the Law

Early family violence studies and feminist pressure spurred a movement toward the implementation of stricter policies for dealing with domestic offenders. Once long ignored, intimate violence has become a top concern for legislators and law enforcement agencies throughout the country in the past two decades (Wilson 1997).

Today, many of the largest U.S. police forces have implemented **mandatory arrest** policies in which much discretion is taken from police officers responding to a call about intimate partner violence. Under such policies, "if an officer finds probable cause that a crime occurred, he or she must arrest" (Goodman and Epstein 2005, 480). As of 2007, mandatory arrest policies existed in Washington, DC, and at least 20 states of the United States: Alaska, Arizona, Colorado, Connecticut, Iowa, Kansas, Louisiana, Maine, Mississippi, Nevada, New Jersey, New York, Ohio, Oregon, Rhode Island, South Carolina, South Dakota, Utah, Virginia, and Washington. Thirty-three states also mandate arrests for those who violate restraining orders.

Looking at the impact of mandatory arrest shows that the rates of arrest are higher in states with mandatory arrest policies. However, research also indicates that the presence of such policies—accompanied specifically by knowing that calling the police is initiating a process that will lead to the arrest of one's intimate partner and could result in one's own arrest—may discourage victims from reporting their victimization to police (Peralta and Novisky 2015). By not calling police, one remains at risk of further, even escalating violence to a potentially fatal level (Stop Abusive and Violent Environments [SAVE] 2010).

Police practices under the heading of preferred or **pro-arrest policy** are an alternative to mandatory arrest. In such instances, officers retain the discretion of whether or not to make an arrest. However, in such preferred arrest contexts, the lack of an arrest requires officers to justify their decisions with detailed reports. In recognition of some of the unintended outcomes associated with mandatory arrest, the 2005 reauthorization of the Violence Against Women Act recommended pro-arrest rather than mandatory arrest policies.

In addition to police policy, the adoption of **no-drop prosecution** policies compels prosecutors to proceed in the prosecution of an intimate partner violence case as long as evidence exists, regardless of a victim's expressed wishes (Goodman and Epstein 2005).

For police to play any effective role in combating intimate partner violence, they must first *know of the violence*. According to a "Fact Sheet on Intimate Partner Violence" put out by the National Center for Injury Prevention and Control, only about 20 percent of rapes or sexual assaults by a partner, 25 percent of physical assaults, and 50 percent of the incidents of stalking directed toward women are reported (Centers for Disease Control and Prevention 2008b). The rate at which men report their victimization is even less.

Even when incidents are reported, we have reason to question whether police are sufficiently and consistently committed to becoming involved in domestic disputes. This has long been a complaint of women who are victimized and who may find police reluctant to intervene, even under mandatory arrest policies. Male victims of female perpetrators find that police are often dismissive of their concerns (Migliaccio 2002).

Aside from the sincerity of the commitment of criminal justice personnel, the innovations in policy have potentially mixed consequences. Lisa Goodman and Deborah Epstein (2005) note some unintended and potentially unavoidable outcomes of no-drop policies, including a loss of a spouse's earnings due to his or her arrest and subsequent prosecution or suffering future victimization from a spouse who has been provoked into retaliatory abuse. As noted previously, research indicates that mandatory arrest policies have contributed to increased levels of intimate partner homicide, perhaps by as much as 60 percent in states that have adopted them. In part, mandatory arrest policies may reduce women's likelihood of reporting violence that she experiences because of not wanting the offender arrested. Such policies may also exacerbate tensions between parties that fester and remain after the arrest, possible prosecution, and potential punishment or penalty. Finally, mandatory arrest and no-drop prosecution have been criticized

because although they take the burden off of victims as to whether to press charges, such policies also take away victims' desires about whether and how to proceed, and limit their abilities to decide.

Some states, including but not limited to California, Florida, Massachusetts, Michigan, and Tennessee, have what are known as preferred arrest or pro-arrest policies. Pro-arrest policies do not compel police to make an arrest but they do put the burden of accounting for a decision not to arrest on the police who respond to a domestic violence call.

## Working with Offenders: Abuser Programs

Treatment services for abusers provide one important component of a coordinated response to domestic violence. Such services might include psychotherapy, group discussion, stress management, or communication skills classes, all of which may be available through mental health agencies, women's crisis programs, or various self-help groups. Such interventions can also be made mandatory as part of the sanctions imposed on convicted offenders.

The most common intervention combines a criminal justice response (i.e., jail) with a mandated group intervention program. The extent to which attending batterers' groups changes violent behavior of abusing men is difficult to measure, but based on evaluations of many different programs, effects associated with different intervention strategies are "typically small," with reoffending rates ranging from 21 percent to 35 percent. One review of the effects of mandatory offender programs concluded that, based on victim reports, the average effect was zero (Day, O'Leary, and Carson 2009). A more optimistic assessment from a National Institute of Justice report was that intervention strategies—especially coordinated, multiagency interventions—yield "modest but statistically significant reductions in recidivism (repeat offending)" among those who participate (Healey and Smith 1998).

Strategies of intervention are shaped by how one attempts to explain or account for the problem in the first place. Rehabilitative interventions tend to be based more on individualistic and psychological approaches than on more structural explanations of violence. Perhaps the most widely used program is based on the **Duluth model**, the curriculum of which emphasizes helping batterers develop critical thinking skills around such themes as nonviolence, respect, partnership, and negotiation. The program proceeds in stages from intake and assessment to orientation and group sessions to program completion. To successfully complete the intervention, many programs require batterers to meet specific criteria—such as writing a "responsibility letter" acknowledging one's behavior that is read before the whole group.

What has become apparent is the ineffectiveness of the "one-size-fits-all" approach and the need to adopt a more sophisticated understanding of an individual's violent behaviors (Day, O'Leary, and Carson 2009). In addition, coordinated community response that includes proactive police and criminal justice strategies, advocacy and services for victims and their children, and responses by other community institutions that promote safety for victims and sanctions for those who batter are necessary interventions (Tolman 1995).

Yet, as Michael Johnson and Janel Leone (2005) warn, failure to differentiate types of violence can be problematic. For example, such common interventions as couples counseling or mediation may work effectively with more situational couple violence. However, given the more extreme nature of intimate terrorism, a woman who has been the victim of such abuse may be at considerable risk of injury or worse if she continues to live with her abuser while they partake in such common interventions.

## Confronting Child and Elder Abuse

The first step in treating child abuse is locating the children who are threatened. Mandatory reporting of suspected child abuse is now required of professionals such as teachers, doctors, and counselors in all 50 states. Reported incidents of child abuse have increased greatly since mandatory reporting went into effect, but the actual number of incidents appears to have decreased. This is good news as far as it goes. Still, levels of violence against children remain unacceptably high, and not nearly enough resources are available to assist children. As addressed in the Real Families feature, child welfare workers are notoriously overburdened with cases, and adequate foster placement is often difficult to find (Gelles and Cornell 1990).

Society must address this tragedy of continued child abuse from a variety of levels:

- Parents must learn how to deal more positively and effectively with their children.
- Children need to be taught skills to recognize and report abuse as soon as it occurs.

- Professionals working with children and families should be required to receive adequate training in child abuse and neglect and to be sensitive to cultural norms.
- Agencies should coordinate their efforts for preventing and investigating abuse.
- Public awareness of child abuse needs to be created and broadened by methods such as posters and public service announcements.

While researchers continue to sort out the whys and wherefores of elder abuse, battered older people have a number of pressing needs. Two decades ago, Karl Pillemer and Jill Suitor (1988) recommended the following services for elders and their caregiving families:

- Housing services, including temporary respite care to give caregivers a break and permanent housing (such as rest homes, group housing, and nursing homes)
- Health services, including home health care, adult day care centers, and occupational, physical, and speech therapy
- Housekeeping services, including shopping and meal preparation
- Support services, such as visitor programs and recreation
- Guardianship and financial management

# Child Sexual Abuse

The 2011 cases of child sexual abuse at Penn State University brought such mistreatment to the attention of many who had previously been unaware, or who had underestimated or ignored such behavior. It is estimated that between 100,000 and 500,000 children are sexually abused annually. By the time they reach age 18, between 15 percent and 25 percent of children will likely have been sexually abused; among females, the likelihood is greater—30 percent to 40 percent—whereas the percentage is between 10 percent and 15 percent for males (Bahroo 2003).

Whether committed by relatives or nonrelatives, **child sexual abuse** is defined as any sexual interaction (including fondling, erotic kissing, or oral sex as well as genital penetration) between an adult or older adolescent and a prepubertal child. It does not matter whether the adult perceives the child as freely engaging in the sexual activity. Because of the child's age, he or she cannot legally give consent; the activity can be considered only as self-serving to the adult.

For a variety of reasons, as the American Psychological Association (APA) reports, definitive statistics "are difficult to collect because of problems of underreporting and the lack of one definition of what constitutes such abuse." In lieu of specific statistics, the APA states that child sexual abuse is "not uncommon and is a serious problem in the United States" (www.apa.org/pubs/info/brochures/sex-abuse.aspx#).

Child sexual abuse is generally categorized in terms of the involvement or noninvolvement of kin. **Extrafamilial sexual abuse** is conducted by nonrelated individuals. **Intrafamilial sexual abuse** is conducted by related individuals, including step-relatives. The child's victimization may include force or the threat of force, pressure, or taking advantage of trust or innocence. The most serious forms of sexual abuse include actual or attempted penile–vaginal penetration, fellatio, cunnilingus, and anal sex with or without the use of force. Other serious forms range from forced digital penetration of the vagina to fondling of the breasts (unclothed) or simulated intercourse without force. Sexual abuse can also include acts ranging from kissing to intentional sexual touching of the clothed genitals, breasts, or other body parts with or without the use of force.

## Children at Risk

Typically, girls are the more common victims of child sexual abuse, making up between 70 percent and 89 percent of abuse survivors (Snyder 2000). More of the abuse that boys experience is extrafamilial. Lower-income children are at greater risk (www.unh.edu/ccrc/factsheet/pdf/childhoodSexualAbuseFactSheet.pdf). Males are the more common perpetrators, perhaps as many as 90 percent or more, though research estimates on the percentage of female perpetrators range between 5 percent and 20 percent of child sexual abuse.

At higher risk appear to be children who have poor relationships with their parents (especially mothers) or whose parents are absent or unavailable and have high levels of marital conflict. A child in such a family may be less well supervised and, as a result, more vulnerable to manipulation and exploitation by an adult. Finally, children with stepfathers are at greater risk for sexual abuse. The higher risk may result from the weaker incest taboo in stepfamily relationships and because stepfathers have not built inhibitions

> **True** **10.** Sexual and other forms of physical abuse occur more frequently in households with stepparents than in households with only biological parents.

resulting from parent–child bonding beginning from infancy. As a result, stepfathers may be more likely to view their stepdaughters sexually.

## Forms of Intrafamilial Child Sexual Abuse

The incest taboo, though in varying forms, is nearly universal in human societies. It prohibits sexual activities between closely related individuals. **Incest** is generally defined as sexual intercourse between people too closely related to marry legally (usually interpreted to mean father–daughter, mother–son, or brother–sister). Sexual abuse in families can involve blood relatives (most commonly uncles and grandfathers) and step-relatives (most often stepfathers and stepbrothers).

There is general agreement that the most traumatic form of sexual victimization is father–daughter abuse, including that committed by stepfathers. Some factors contributing to the severity of reactions to father–daughter sexual relations include fathers being more likely to engage in penile–vaginal penetration than other relatives, fathers sexually abusing their daughters more frequently, and fathers being more likely to use force or violence. The presence of a stepfather increases the risk of sexual abuse for girls, making them twice as likely to be abused as girls who live with their fathers (Putnam 2003). In fact, comparative data from a number of countries indicates that sexual and other forms of abuse occur substantially more frequently in households with stepparents (stepmothers or stepfathers) than in households with only biological parents, a pattern psychologists Martin Daly and Margo Wilson call "the Cinderella effect" (Daly and Wilson 1998, 2002, 2009). Sexual abuse by a stepfather represents a violation of the basic parent–child relationship.

## Sibling Sexual Abuse

One resource (Psychpage.com) defines all brother–sister (or cousin) sexual interaction as *abuse,* largely in terms of the use of force or coercion and when penetration occurs or injury results. Vernon Wiehe (1997) suggests that official estimates of sibling sexual abuse are severely underestimated because most goes undetected by parents and unreported to authorities. The abuse is usually committed by an older brother (or sister) on a younger sibling.

The criteria of force and/or coercion may be the aspect most highly associated with negative outcomes, regardless of the specific sexual behavior (e.g., kissing, fondling, simulated intercourse, or exhibition). Studies assert that the circumstances, characteristics, and potential outcomes of brother–sister incest are as serious as, if not more than, those of father–daughter incest (Cyr et al. 2002; Rudd and Herzberger 1998). Furthermore, according to the U.S. Department of Health and Human Services (2002), sibling sexual abuse is more common than abuse committed by an adult relative.

## Effects of Child Sexual Abuse

There is extensive research indicating that potential "profound, long-term consequences for an adult's sexual behavior and intimate relationships" can result from child sexual abuse (Cherlin et al. 2004, 770). Among the numerous well-documented consequences of child sexual abuse are both initial and long-term consequences. Many abused children experience symptoms of post-traumatic stress disorder (McLeer et al. 1992). Other initial effects include emotional disturbances, such as fear, anxiety, guilt, and shame; social disturbances, such as running away or truancy; physical consequences, such as changes in eating or sleeping; and sexual disturbances, such as open masturbation and sexual preoccupation (Perez-Fuentes et al. 2013).

### Long-Term Effects of Sexual Abuse

Although the initial effects of child sexual abuse can subside to some extent, the abuse may leave lasting scars on adult survivors who often have significantly higher incidences of psychological, physical, and sexual problems than the general population. Cherlin et al. (2004) list such outcomes as feelings of betrayal, lack of trust, feelings of powerlessness, low self-image, depression, and a lack of clear boundaries between self and others. Abuse as a child may predispose some women to early onset of sexual involvement, more involvement in risky sexual behavior, multiple partners, and sexually abusive dating relationships (Cate and Lloyd 1992; Cherlin et al. 2004). Cherlin and colleagues also identify the following:

- Greater anxiety and less pleasure from sex
- Behaviors such as using drugs and/or alcohol with sex that increase risk of sexually transmitted disease or HIV infection
- More frequent sexual encounters

As Cherlin et al. (2004, 771) point out, childhood sexual abuse victimization may also affect the ability to maintain long-term intimate relationships in adulthood: "Overall, the relationship difficulties associated with childhood sexual abuse would seem to be more consistent with frequent, short-term unions than with long-term unions." Psychiatrist Gabriela Perez-Fuentes and colleagues report that couples in which one partner is a survivor of childhood sexual abuse are at risk for a number of possible relationship or mental health problems. They suggest that couples therapists become familiar with specific interventions for working with such clients and become more aware of the long-term impact of child sexual abuse (Perez-Fuentes et al. 2013).

**Traumatic sexualization** refers to a process in which a sexually abused child's sexuality develops inappropriately and the child becomes interpersonally dysfunctional. Sexually traumatized children learn inappropriate sexual behaviors (such as manipulating an adult's genitals for affection), are confused about their sexuality, and inappropriately associate certain emotions—such as loving and caring—with sexual activities.

As adults, sexual issues may become especially important. Survivors may suffer flashbacks, sexual dysfunctions, and negative feelings about their bodies. They may also be confused about sexual norms and standards. A fairly common confusion is the belief that sex may be traded for affection. Some women label themselves as "promiscuous," but this label may be a result more of their negative self-image than of their actual behavior.

Children feel betrayed when they discover that someone on whom they have been dependent has manipulated, used, or harmed them. Children may also feel betrayed by other family members, especially mothers, for not protecting them from abuse. As adults, survivors may experience depression, a sense of distrust, hostility and anger, or social isolation and avoidance of intimate relationships. Anger may express a need for revenge or retaliation.

Children experience a basic kind of powerlessness when their bodies and personal spaces are invaded against their will. In adulthood, such powerlessness may be experienced as fear or anxiety; a person feels unable to control events. It may also be related to increased vulnerability or revictimization through rape or marital violence; survivors may feel unable to prevent subsequent victimization. Other survivors, however, may attempt to cope with their earlier powerlessness by an excessive need to control or dominate others. Ideas about being a bad person as well as feelings of guilt and shame about sexual abuse are transmitted to abused children and then internalized by them. Stigmatization is communicated in numerous ways. The abuser conveys it by blaming the child or, through secrecy, communicating a sense of shame. As adults, survivors may feel extreme guilt or shame about having been sexually abused. They may also feel different from others because they mistakenly believe that they alone have been abused.

Obviously, the violence and abuse discussed in this chapter are complex phenomena. They are products of individual characteristics of perpetrators and victims, relationship dynamics, and certain social and cultural factors. Not every home becomes a center of violence and abuse, and most families are not embattled. We need to realize that those families and relationships that are violent or abusive are products of a blend of qualities and are affected on multiple levels. This understanding is important if we hope to reduce the prevalence of violence and abuse and if we care to help those who are most at risk or already victimized.

## Summary

- Abuse and violence are separate though related and overlapping phenomena. Not all abuse is violent, and some intimate violence is considered appropriate and not abusive.

- Violence between intimate partners ranges from routine to extreme, from *common couple violence*, which is typically less severe, to *intimate terrorism*, which is a more severe, most often male-on-female form of violence and abuse in which power and domination are key motives.

- Different perspectives are used to study sources of family violence: (1) individualistic explanations, (2) the feminist model, (3) the social situational model, (4) the social learning model, (5) the resource model, and (6) the exchange–social control model.

- Researchers have stressed the roles played by gender, power and control, stress, and intimacy in explaining intimate violence.

- It is difficult to know exactly how much violence there is in intimate relationships. Either the use of official records and/or survey data give us underestimates of how much intimate violence there is in the United States.

- Two characteristics that correlate highly with intimate violence against women in the United States are low socioeconomic status and a high degree of marital conflict.

- *Gender symmetry* refers to the survey data findings of similarity in both expressing and experiencing violence between the genders. However, the context and consequences of partner violence are not the same for men and women.

- Age, race, and social class all factor into domestic violence. Research on male victims shows both similarities and differences with what research has revealed about male perpetrators and female victims.

- Violence among same-sex couples is similar to the levels of violence among heterosexuals. Because such relationships lack the social supports that heterosexual couples can draw upon, the experience of victimization may be worse.

- Dating violence, including verbal abuse, physical violence, and coercive sex, is often precipitated by jealousy or rejection.

- Reasons some women may stay in or return to abusive relationships include economic dependency, religious pressure or beliefs, the perceived need for a father for the children, a sense of duty, fear, love, and reasons pertaining to their particular culture.

- Most child abuse cases are unreported. Parental violence is one of the five leading causes of childhood death.

- Families at risk for child abuse often have specific parental, child, and family ecosystem characteristics.

- Nearly 90 percent of parents acknowledged using some form of psychological aggression with at least one child during the prior 12-month period, with younger parents using such aggression more often.

- Intimate violence generates high costs in terms of time lost at work, mental health, and medical expenses for injuries or trauma sustained.

- The hidden victims of family violence include siblings (who have the highest rate of violent interaction), parents assaulted by their adolescent or youthful children, and elderly parents assaulted by their middle-aged children.

- Recommendations for reducing family violence include reducing sources of societal stress, such as poverty and racism; establishing supportive networks; breaking the family cycle of violence; and eliminating the legitimization and glorification of violence.

- Recent legal innovations such as *mandatory arrest* and *no-drop prosecution* have had mixed results. In some ways, they raise the costs for victims reporting the violence.

- Mandatory reporting of suspected child abuse may be helping to decrease the number of abused children in the United States. Early intervention and education also may help reduce abuse.

- Sexual victimization of children may include incest, but it can also involve other family members and other sexual activities.

- Children most at risk for sexual abuse include females, preadolescents, children with absent or unavailable parents, children with poor parental relationships, children with parents in conflict, and children living with a stepfather.

## Key Terms

# Coming Apart: Separation and Divorce

# 13

Gladskikh Tatiana/Shutterstock.com

**What Do You Think?**

*Are the following statements True or False? You may be surprised by the answers as you read this chapter.*

**T F** 1. Annulment, like divorce, is a way to terminate a marriage in which both spouses feel that something has gone wrong in their marriage.

**T F** 2. The U.S. divorce rate is at an all-time high and continues to climb.

**T F** 3. College-educated women and men are the most likely to marry and the least likely to divorce.

**T F** 4. Age at marriage is the best predictor of the likelihood of divorce.

**T F** 5. Couples in remarriages benefit from past experience and are less likely to divorce than couples in first marriages.

**T F** 6. The critical emotional event when a marriage breaks down is separation rather than divorce.

**T F** 7. Most divorces are uncontested.

**T F** 8. Many of the problems children of divorced parents experience are present before the marital disruption.

**T F** 9. Nowadays, after a divorce, children are as likely to live with their father as they are to live with their mother.

**T F** 10. Increasingly, couples who are in the process of divorcing consult first with mediators whose job is to help them resolve their problems and save their relationships.

s one woman told sociologist Joseph Hopper (2001), there was nothing she and her husband could do:

> It's something that had to happen, and it wasn't something that either one of us really controlled. It was just an awful situation that we had to get out of, and I recognized it and he didn't.

A second person offered the following:

> I had wanted that forever—the white picket fence and the whole dream. But it didn't come true. But I was at least smart enough to realize it wasn't happening and no matter what I did, it wasn't going to.

In reading those words, some people will feel empathy for the speakers, sense their despair, and understand their decisions. Others are likely to question such assumptions as "it had to happen" or "no matter what I did, it wasn't going to work." To some, marriage is supposed to bring us our deepest sense of fulfillment and intimacy, and if those feelings are absent, we ought to move on and try to find them elsewhere. To others, marriage is a commitment that is for life, even if it means enduring periods of disappointment or, worse yet, never finding or feeling all of what one hoped to find or feel.

As we posed in Chapter 8, this raises such questions as the following: Are most Americans pro-marriage? Are we too soft on divorce? Do we believe in the importance of marriage and the commitment we make when we exchange wedding vows? Or when we say "I do," are we really adding, perhaps not under our breath but silently in our heads, "at least for now"?

This chapter addresses many aspects of divorce, including the relative risk factors associated with divorce; the process of separating; the experiences and outcomes of divorce for men, women, and children; and such policy matters as joint custody, no-fault divorce, mediation, and covenant marriage. This exploration will help you better understand what parents, children, and families experience and how they cope with what increasingly has become part of our marriage system—divorce.

## The Meaning of Divorce

Some scholars suggest that divorce reflects an idealization of marriage, not a devaluation or rejection. Though initially such a notion might raise an eyebrow, they reason that many individuals who divorce would not divorce if they did not have such high expectations of marriage—so much hope about finding the fulfillment of their various needs in their spouses. According to Frank Furstenberg and Graham Spanier (1987), divorce may well be a critical part of our contemporary marriage system, which emphasizes emotional fulfillment and personal satisfaction. And, as sociologist Andrew Cherlin notes in his book, *The Marriage-Go-Round: The State of Marriage and the Family in America Today* (2009), Americans simultaneously highly value both marriage and individual fulfillment. Failing to find the individual fulfillment they desire, expect, and seek in their marriage, they divorce and keep looking.

A high divorce rate also tells us that many people may no longer believe in the permanence of marriage. Norval Glenn (1991) suggested that there has been a "decline in the ideal of marital permanence and . . . in the expectation that marriages will last until one of the spouses dies." Instead, marriages disintegrate when love goes or a potentially better partner comes along. Divorce is a persistent fact of American marital and family life and one of the most important forces affecting and changing American lives today (Furstenberg and Cherlin 1991).

### The Legal Meaning of Divorce

Aside from the more debatable question about the meaning of divorce within the culture and family system in the United States, there is a more straightforward question about the legal meaning of divorce. Just what does divorce represent? Obviously, divorce is a means of terminating one's marriage. Where it once meant that the actions of one of the spouses had permanently damaged the marriage to such a degree that the marriage could not be fixed or endured, now it more often means that the marriage suffers from something that cannot be fixed. This reflects the distinction between **fault-based divorce** and **no-fault divorce**. In a fault-based divorce, one spouse files for divorce, alleging that his or her spouse is responsible for the failed marriage through such actions as adultery, cruel and inhumane treatment, mental cruelty, habitual drunkenness, and desertion. In a no-fault divorce, the couple can divorce without either having to accuse the other or prove the other responsible for the failure of their marriage. They can simply claim that *irreconcilable differences* make it impossible for them to continue as married. Some states permit fault-based divorce, but, with New York's passage of no-fault divorce in 2010, every state now allows for no-fault divorce. A later section considers the issue of no-fault divorce in more detail.

At present, however, it is important to note that divorce is not the only way a marriage can be ended. Annulment and desertion are two other means of ending or exiting an existing marriage. The difference between annulments and divorces are many, but perhaps the most important is what each says about the terminated marriage. In a divorce, spouses agree that the relationship has failed, or one spouse proves that the other committed an act that caused the marriage to fail. Typically, an annulment is granted when it is determined that the marriage never met the legal requirements of marriage. For example, in instances where one married when underage, entered marriage already married, or married incestuously, an annulment is likely. In addition, if it can be demonstrated that at the time of marriage one lacked the mental capacity to understand the commitment they were making—or perhaps was under the influence of drugs or alcohol—the marriage is likely to be annulled. The literal legal meaning of an annulment is that the marriage has been nullified because it never really existed as a legitimate marriage. In other words, it invalidates a marriage because the marriage is considered to have never been a valid one.

In addition to divorce and annulment, some marriages end when one spouse simply leaves. Such acts of desertion leave the other spouse in legal limbo because he or she technically remains married to the absent spouse. In the latter decades

> **False** **1.** Annulment, unlike divorce, is granted when it is determined a marriage never met the legal requirements of marriage.

of the 19th century as well as during the Depression, desertion occurred more often. Throughout the latter decades of the 20th century and the first decade of the 21st century, divorce is the more common termination of marriage (Faust and McKibben 1999). In most states today, desertion (sometimes called abandonment) is, itself, grounds for divorce in states that still allow fault-based divorces. However, until the marriage is officially terminated, the deserted spouse lives with considerable uncertainty and ambiguity. As difficult as divorce is to experience, at least it conveys a clearer message about one's marital status and one's eligibility for reentering marriage.

## The Multiple Realities of Divorce

Yet another way of considering the "meaning" of divorce is to recognize the complex and multidimensional nature of the divorce experience in the lives of those who divorce. This is what the late anthropologist Paul Bohannan had in mind by his concept of the **stations of divorce**. Bohannan (1970) developed a descriptive model of the divorce process that focused on the multiple experiences people have as their marriages end. He emphasized the emotional, legal, economic, co-parental, community, and psychic stations of divorce, noting that as people divorce they experience multiple transitions and transformations. These "stations" neither have a particular order nor begin and end simultaneously. Some occur before one is legally divorced (e.g., the emotional station), and others can stretch on for years after one's divorce is final (e.g., the co-parental station). The level of intensity of these different divorces varies at different times and for different couples. In no specific order, the stations are the following:

- The *emotional divorce*, when one spouse (or both) begins to disengage from the marriage and starts to feel that "something isn't quite right," begins well before the legal divorce. But even as divorce papers are filed, one or both of the partners may find themselves feeling ambivalent. Because the emotional divorce is not complete, they may try to reconcile.

- The *legal divorce* is the court-ordered termination of a marriage. By the time a couple is legally divorced, much has happened. The legal decree permits divorced spouses to remarry and conduct themselves in a way that is legally independent of each other. Of course, legal "independence" may be an overstatement, as a divorce decree sets the terms for the division of property and child custody, issues that may lead to bitterly contested and/or recurring divorce battles. Many other unresolved issues surrounding a divorce, such as feelings of hurt and betrayal, may be acted out during the legal divorce. No-fault divorce was, in part, intended to minimize these issues.

- The *economic divorce* makes the economic aspect of marriage painfully apparent. Property acquired during a marriage is considered joint marital property and must somehow be divided between the divorcing spouses. The settlement is based on the assumption that each spouse contributes to the estate. This contribution may be nonmonetary, as in the case of traditional homemakers whose support and practical assistance enabled their husbands to work outside the home. As part of the economic divorce, child support and, less often, alimony may be ordered by the court. As the partners go their own ways, husbands and wives often experience different consequences in their standards of living as they set up separate households and no longer pool their resources. Women usually experience a decline in their standard of living, whereas men sometimes see theirs increase.

- The *co-parental divorce* is experienced when a couple has children. Parenthood and parental responsibility persist even after marriages end. Even those parents who never see their children remain fathers and mothers. The co-parental divorce is among the most complicated aspect of divorce because it also gives rise to single-parent families and, in most cases, stepfamilies, both of which are considered in more detail in Chapter 14. As parents divorce, issues of child custody, visitation, and support must be addressed. With divorce, new ways of relating to the children and former spouses must be developed, ideally keeping the children's best interest foremost in mind.

- The *community divorce* means that when people divorce, their social world changes. In-laws become ex-laws; often they lose (or stop) contact. (This is particularly troublesome when in-laws are also grandparents.) Friends may choose sides or, not wanting to be caught in the middle, drop out; they may not be as supportive as desired. New friends may replace old ones as divorced men and women begin dating again. They may enter the singles subculture, where activities center on dating. Single parents may feel isolated from such activities because child rearing often leaves them little or no time or leisure and diminished income leaves them little or no money.

- The *psychic divorce* is accomplished when one once again feels like a separate individual and no longer feels like part of a couple. A former spouse becomes irrelevant to one's sense of self and emotional well-being. As part of the psychic divorce, former spouses develop a sense of independence, completeness, and stability, and reconstruct their identities. Navigating through the psychic station may be more difficult and take a good deal longer than it does to experience most of the other stations of divorce.

## Divorce in the United States

Before 1974, the view of marriage as lasting "till death do us part" reflected marital reality for most Americans. However, a surge in divorce rates began in the mid-1960s, and in 1974, a watershed in U.S. history was reached when more marriages ended by divorce than by death. Divorce rates peaked in 1981, subsequently leveling off in the later 1980s. Today, approximately 35 percent to 45 percent of all new marriages are likely to end in divorce (Stevenson and Wolfers 2007).

### Measuring Divorce: How Do We Know How Much Divorce There Is?

How common is divorce, and how likely are we to experience it, should we marry? According to the U.S. Census Bureau, there were nearly 25.3 million divorced people age 18 and older in the United States in 2014, representing almost 11 percent of the population. This included 14.6 million women and 10.7 million men. Race and gender patterns are illustrated in Figure 13.1. As to how many divorces occur in any given year, what numbers we know turn out to be incomplete ones, as they are based on less than full reporting of all 50 U.S. states. For example, the 2014 estimate of some 813,862 divorces and annulments does not include data from California, Georgia, Hawaii, Indiana, or Minnesota. The number of

**Figure 13.1 Percentage of Population 18 and Over Currently Divorced, by Race/Ethnicity and Gender, 2014**

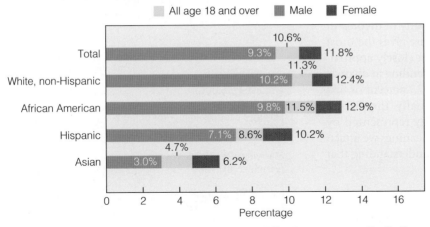

SOURCE: U.S. Census Bureau (2015). "Families and Living Arrangements," Table MS-1 www.census.gov/hhes/families/data/marital.html.

marriages, 2,140,272, doesn't include data from the state of Georgia (Centers for Disease Control and Prevention 2015d). In part because of such incomplete data, some researchers are somewhat skeptical of the widely stated decline in divorce that has occurred since the latter years of the last century (Kennedy and Ruggles 2014).

There are a variety of ways to measure and represent the prevalence of divorce in the United States. The most common measures are discussed next.

## Ratio Measure of Divorces to Marriages

It is likely that many of you have heard the gloomy—and potentially misleading—statement that *one out of two marriages ends in divorce*. What exactly does that statistic mean? On what is it based? The prevalence of divorce is typically reported using one of the following statistics. The **ratio measure of divorce** is calculated by taking the number of divorces and the number of marriages in a given year and producing a ratio to represent how often divorce occurs relative to marriage. As mentioned previously, the needed data for such a calculation are incomplete. In 2014, there were a reported 2,140,272 marriages and 813, 682 divorces. Though this amounts to less than half as many divorces as marriages, the data reflect divorces in 45 states and marriages in 49. Calculating and comparing the rates of marriage and divorce shows a marriage rate of 6.9 and a divorce rate of 3.2. But recognize that even if there were data from all 50 states that showed twice as many marriages in a year than

divorces in that same year, the ratio measure is still potentially misleading. Those who divorce in a given year are not likely drawn from those that married in that same year. What the ratio measure truly reflects is the relative popularity or commonality of marriage and divorce in a given year.

## Crude Divorce Rate

The **crude divorce rate** represents the number of divorces in a given year for every 1,000 people in the population. As mentioned, the crude divorce rate in 2014 was 3.2 per 1,000 people. There were also 6.9 marriages per 1,000 people in the population, in this instance being slightly more that the ratio of one divorce for every two marriages (CDC/NCHS National Vital Statistics System).

Crude divorce or marriage rates have certain problems. Obviously, when calculating the crude divorce rate, counting every 1,000 people in the population means including many unmarried people, children, the elderly, the already divorced, and so on. All of you reading this book would be counted among that statistic even though the vast majority of you are not even married. Unmarried people cannot become divorced. It is therefore a statistic that is highly susceptible to the age distribution, proportions of married and single people in the population, and changes in such population characteristics.

## Refined Divorce Rate

Considered the most useful measure of divorce, the **refined divorce rate** measures the number of divorces that occur in a given year for every 1,000 marriages (as measured by married women age 15 and older). In 2013, the refined divorce rate in the United States was 18.5 per 1,000 marriages (Payne 2014a).

A 2011 report by the National Center for Marriage and Family Research offers a look at the refined rate of *first divorces* per every 1,000 women in their *first marriages*. The report, by Larry Gibbs and Krista Payne, shows a 2010 refined first divorce rate of 17.5 per 1,000 women in first marriages. Gibbs and Payne (2011) also show that there is considerable variation in divorce, by such factors as race, educational attainment, and age.

Note that the range of available statistics produces different impressions about the reality of divorce in the United States. The ratio measure gives the most alarming impression, the one most closely approximating "one out of two marriages" ending in divorce. When we use the refined rate of 1.75 percent of first marriages ending in divorce annually, the picture seems much less bleak. The reality represented by each statistic is the same, but the meanings we attach to each statistic and therefore the understanding that each creates vary significantly.

## Predicting Divorce

Another divorce statistic worth mentioning is the **predictive divorce rate**. This calculation (too complicated for our purposes) allows researchers to estimate how many new marriages will likely end in divorce. Sociologist Paul Amato notes that at the end of the 20th century, it was predicted that between 43 to 46 percent of marriages would end in dissolution (Amato 2010). The prevailing estimate is that somewhere between 35 percent and 45 percent of marriages entered into in a year are likely to end in divorce, though some continue to estimate as many as 50 percent.

Estimating future trends is a tricky business. This estimate is based on the experiences of prior birth cohorts (people born between specific years) and therefore we cannot be confident that current and future cohorts will make the same choices or face the same circumstances as their predecessors.

But even these predictions need to be more carefully assessed. As we show in subsequent sections describing factors associated with divorce, not everyone faces the same risk of divorce. As explained by Barbara Dafoe Whitehead and David Popenoe (2004), the risk of divorce is greatly affected—either increased or diminished—by the background characteristics that spouses bring with them into marriage. They go on to report the decreases in vulnerability to divorce during the first ten years of marriage that are shown in Table 13.1.

Ultimately, a 2011 report from The National Marriage Project offers the following more reassuring assessment of the risk of divorce: "So if you are a reasonably well-educated person with a decent income, come from an intact family and are religious, and

**False** **2.** The U.S. divorce rate peaked in the early 1980s.

**TABLE 13.1** **Vulnerability to Divorce in First Ten Years of Marriage**

| Divorce | Percentage Reduction in Risk of Divorce |
|---|---|
| Annual income over $50,000 (vs. under $25,000) | –30 |
| Having a baby seven months or more after marriage (vs. before marriage) | –24 |
| Marrying after 25 years of age (vs. under 18) | –24 |
| Own family of origin intact (vs. divorced parents) | –14 |
| Religious affiliation (vs. none) | –14 |
| College (vs. high school dropout) | –25% |

SOURCE: National Marriage Project (2011).

marry after age 25 without having a baby first, your chances of divorce are low indeed."

## Divorce Trends in the United States

If we look at long-term divorce trends, the unmistakable conclusion is that from the start to the end of the 20th century, the United States saw dramatic increases in marital breakups. If we look, instead, over the past 30 years, a different picture emerges. In more recent decades, though, the divorce rate has dropped, returning to levels not seen since the early 1970s (see Table 13.2). As we show shortly, this did not occur equally for all groups.

Both marriage and divorce rates have declined. The 2,140,272 marriages in 2014 illustrate a slight increase over the 2,081,301 in 2013 but are less than the 2,157,000 in 2008. As to divorce, we can see that fewer divorces were granted in 2014 than had been granted since before 1980. Again, however, the data are incomplete, especially regarding divorces. In recent years, the divorce rate per 1,000 people in the population has hovered between 3.2 to 3.6 divorces per 1,000, a rate lower than we've seen since the 1970s.

| TABLE 13.2 | Divorce and Marriage through the 20th Century and Beyond | | | |
|---|---|---|---|---|
| Year | Marriages | Rate per 1,000 | Divorces | Rate per 1,000 |
| 1900 | 709,000 | 9.3 | 55,751 | 0.7 |
| 1920 | 1,274,476 | 12.0 | 170,506 | 1.6 |
| 1940 | 1,595,879 | 12.1 | 264,000 | 2.0 |
| 1960 | 1,523,000 | 8.5 | 393,000 | 2.2 |
| 1970 | 2,158,802 | 10.6 | 708,000 | 3.5 |
| 1980 | 2,406,708 | 10.6 | 1,189,000 | 5.2 |
| 1985 | 2,413,000 | 10.2 | 1,178,000 | 5.0 |
| 1990 | 2,448,000 | 9.8 | 1,182,000 | 4.7 |
| 1995 | 2,336,000 | 8.9 | 1,169,000 | 4.4 |
| 2000 | 2,315,000 | 8.2 | 944,000 | 4.0 |
| 2005 | 2,249,000 | 7.6 | 847,000 | 3.6 |
| 2010 | 2,096,000 | 6.8 | 872,000 | 3.6 |
| 2014 | 2,140,272 | 6.9 | 813,862 | 3.2 |

SOURCE: Centers for Disease Control and Prevention (2015d).

# Factors Affecting Divorce

Sometimes it is easy to point to the cause of a particular divorce. Perhaps one spouse was unfaithful or abusive, and the marriage was brought to a quick end. In other instances, even the divorcing parties cannot identify the exact cause or causes that led to divorce. Furthermore, every instance of divorce brings a unique combination of such causes. Researchers have looked at factors affecting both wider societal divorce rates and individual divorce decisions.

Some analyses address the complex sets of changes that make divorce rates hard to predict. For example, Heaton (2002) notes that there have been increases in the prevalence of premarital sex, premarital births, cohabitation, and both racial and religious intermarriage. All these tend to be associated with higher likelihood of marital instability, especially divorce. Yet there also have been increases in age at marriage and in educational attainment that tend to be associated with higher rates of stable marriage. In this section, we

look at both the larger societal or demographic factors and the individual and couple characteristics that may be related to the likelihood of divorce.

## Societal Factors

As seen earlier, even the reduced divorce rates especially evident in and since the 1990s were considerably higher than the rates early in the 20th century. For example, the 2014 crude divorce rate was *four and a half times* the rate at the beginning of the 20th century. It was 45 percent higher than the rates in 1960. In addition, divorce rates in the United States are higher than rates elsewhere in the industrialized world.

Comparing divorce rates internationally illustrates that they range widely. Furthermore, as can be gleaned from the following Table 13.3, many countries had divorce rates less than half of the 2010 U.S. rate of 3.6.

| TABLE 13.3 | International Variation in Crude Divorce Rates, 2010 |
|---|---|
| Country | Divorces per 1,000 |
| Ireland | 0.6 |
| Italy | 0.9 |
| Croatia, Greece | 1.3 |
| Bulgaria, Iceland (2011), Romania, Turkey | 1.6 |
| Poland | 1.7 |
| China, Japan | 1.8 |
| Austria, France, Iran, Luxembourg, Norway, United Kingdom | 2.0 |
| Netherlands | 2.1 |
| Germany, Hungary, Spain | 2.2 |
| Korea | 2.3 |
| Cyprus, Estonia, Finland, Portugal | 2.4 |
| Sweden | 2.5 |
| Denmark | 2.6 |
| Switzerland | 2.8 |
| Czech Republic | 2.9 |
| Belgium, Lithuania | 3.0 |
| Latvia | 3.6 |

SOURCE: Eurostat Statistics Explained (2010). http://epp.eurostat.ec.europa.eu /statistics_explained/index.php.

Other countries with relatively low crude divorce rates include Chile (0.3), Mexico (0.6), Jamaica (0.9), Singapore (1.4), Iran (1.7), and Japan (2.0). Very few countries—the Russian Federation (4.5), Aruba (4.0) and Belarus (3.9)—all had rates that exceeded the U.S. rate (United Nations Statistics Division 2012; OECD Family Database 2011). Also not shown in the table but worth noting is the fact that rates throughout Africa, Asia, South and Central America, and Canada all tend to be lower than U.S. rates.

## Changed Nature of the Family

In the United States, the shift from an agricultural society to an industrial one undermined many of the family's traditional functions. Schools, the media, and peers are now important sources of child socialization and child care. Hospitals and nursing homes manage birth and care for the sick and aged. Because the family pays cash for goods and services rather than producing or providing them itself, its members are no longer interdependent.

As a result of losing many of its social and economic underpinnings, the family is less of a necessity. It is now simply one of many choices we have. We may choose singlehood, cohabitation, marriage, or divorce—and if we choose to divorce, we enter the cycle of choices again: singlehood, cohabitation, or marriage and possibly divorce for a second time. A second divorce leads to our entering the cycle for a third time and so on.

## Social Integration

**Social integration**—the degree of interaction between individuals and the larger community—is a potentially important factor related to the incidence of divorce. The social integration approach regards such factors as urban residence, church membership, and population change as especially important in explaining divorce rates.

Among African Americans, the lowest divorce rate is found among those born and raised in the South; African Americans born and raised in the North and West have higher divorce rates. Similarly, those who live in urban areas, where the divorce rate is higher than in rural areas, are less likely to be subject to the community's social or moral pressures. They are more independent and have greater freedom of personal choice.

An interesting aspect of the social integration–divorce relationship can be seen in the effects of divorce on members of one's social network. According to research by Rose McDermott, James Fowler,

and Nicholas Christakis (2009, 2013), people who get divorced can influence others in their wider networks to divorce. In that way, divorce is not dissimilar from other family behaviors that are subject to peer influence, such as fertility decisions and contraceptive use. Using the idea of a *contagion*, they contend that "divorce can spread through a network like a rumor, affecting friends up to two degrees removed" (p. 11). In other words, people who get a divorce may not only influence their friends but also friends of their friends as well. Whether by demonstrating that divorce can be survived or even be experienced as beneficial, or by being available to lend support and offer empathy, divorced individuals may promote divorce among others in their networks (McDermott, Fowler, and Christakis 2013). In their research, McDermott, Fowler, and Christakis found that both women and men are "equally susceptible to divorce if their friends do it" (McDermott, Fowler, and Christakis 2009). Thus, it appears as though social ties can protect or provoke divorce.

## Individualistic or Collectivist Cultural Values

American culture has traditionally been individualistic. We highly value individual rights, we cherish images of an individual battling nature, and we believe in individual responsibility. It should not be surprising that many view the individual as having priority over the family when the two conflict. As marriage and the family lost many of their earlier social and economic functions, their meaning shifted. Marriage and family are viewed as paths to *individual* happiness and fulfillment. We marry for love and then expect marriage and our partners to bring us happiness. When individual needs conflict with family demands, however, we no longer automatically submerge our needs to those of the family. We often struggle to balance individual and family needs. But if we are unable to do so, divorce has emerged as an alternative to an unhappy or unfulfilling marriage and as an escape from a mean-spirited or violent marriage. What about more collectivist cultures? How do they view divorce? The Exploring Diversity feature describes the climate and culture surrounding divorce in three collectivist societies—Iran, India, and the Philippines.

## Demographic Factors

A number of demographic factors appear to have a correlation with divorce, including employment status, income, education level, ethnicity, and religion.

**Divorcing in Iran and India, but NOT the Philippines**

A sign of increasing "godlessness"? "A national threat"? Corrupting consequences of exposure to Western media? According to journalist William Yong, these are just some of the ways more conservative commentators describe and try to account for the increasing divorce rate in Iran. Yong notes that although the prevalence of divorce in Iran pales in comparison to divorce in the United States, "divorce is skyrocketing in Iran," and he refers to the trend as "striking" with "few signs of slowing down."

Against the conservative interpretation of Iran's divorce trend, other more liberal analysts look to changes in Iranian society—increasing unemployment, the rapid pace and extent of urbanization, and especially, changes in women's attitudes and behavior. Yong offers this comment from an interview with Iranian psychiatrist Azardokht Mofidi, who suggests that increasingly, Iranian women ". . . are no longer prepared to put up with hardships in marriage, and their expectations have risen to include equality in marriage."

More than 1,500 miles separate the Iranian city of Tehran from Delhi, India, where one can observe similar phenomena—increasing divorce within a society that has traditionally strongly condemned it. Journalists Saher Mahmood and Somini Sengupta note that the adjustment is ongoing in Indian society. They comment, "Marriage is still, by and large, socially compulsory. But in a measure of the slow churning of Indian social mores, divorce and remarriage are slowly gaining acceptability."

Yuvraj Raina is a businessman in New Delhi, India, and the founder of the Aastha Center for Remarriage, a matrimonial agency for women and men seeking to reenter marriage after a divorce or the death of a spouse.

On his list of 5,000 prospective brides and grooms, one finds many divorced women and men. Raina himself is remarried after a divorce. His business and his personal experience illustrate the changing Indian attitudes toward divorce and remarriage. As he told Mahmood and Sengupta, "In general, it's no taboo these days."

A six-plus hour flight would take you the almost 3,000 miles from Delhi to Manila, the Philippines. As of January 2015, the Philippines was the only country in the world in which divorce was illegal for most of its citizens. A strongly Catholic society, the Philippines maintains the prohibition against divorce, even though other strongly Catholic societies have legalized divorce: Italy (in 1974), Brazil (in 1977), Spain (in 1981), Argentina (1987), Ireland (in 1997), Chile (in 2004), and most recently Malta (in 2011). The state allows Muslims, who are just over 10 percent of the population, the right to divorce. Among the vast Catholic majority, the only recourse for ending a bad marriage is a church or civil annulment. To annul their marriage, a couple must prove that there were conditions or circumstances present at the beginning of the marriage that made it fail to meet the requirements of marriage (e.g., either spouse was underage, one spouse was already married, or either spouse was psychologically incapacitated). The process is described as long, expensive, and traumatic, in that it often forces couples who would split up more amicably to create or fake adversarial relations, so as to be granted their desired annulment (Hundley and Santos 2015). As reported by journalists Tom Hundley and Ana Santos, in the country of more than 100 million people, only about 10,000 annulments are granted annually. For the majority of other failing Catholic marriages, couples just split up (Hundley and Santos 2015). ●

## Employment Status

Low status, low pay occupations, unemployment, and poverty have all been identified as factors that increase one's susceptibility to divorce (Amato 2010). They subject couples and households to economic distress that contributes to marital stress. Marital stress is then related to decisions to divorce.

Studies conflict as to whether employed wives are more likely than nonemployed wives to divorce; an argument can be made for both possibilities. Certainly, employed women are likely to have more financial resources with which to move on from an unhappy or failed marriage. Employment likely reduces some financial dependence that they might

otherwise feel if they had no separate individual earnings of their own. As we saw in Chapter 11, dual-earner couples face time strains. The hours wives spend at work can lessen their or their spouse's happiness with their marriages. In addition to the question of *how many hours* people work is the issue of *which hours*. Sociologist Harriet Presser (2000) estimated that among men married less than five years and with young children, working night shifts increased their likelihood of divorce or separation by six times compared to men with similar families who worked days. Women with similar families who work nights face three times the likelihood of separation and divorce compared to those who work days. In the absence of children, the same effects are not found.

When women are employed, they may be unhappy with either the expectations on them at work or the responsibilities that await them at home. Wives' employment creates more situations in which husbands will be asked or expected to share more of the housework and child care. Discussions of what's "fair" and "right" become more common and contested. If women feel as though their partner is not contributing, such feelings can lower their satisfaction with that partner and the relationship they share.

At the same time, when both spouses are employed, even though the complexity of "who does what" becomes greater, the increase in their financial well-being might protect them by reducing economic distress.

Employment also creates more opportunities for spouses to meet someone else and to embark on an extramarital sexual relationship. The presence and numbers of attractive alternative partners positively influences the risk of divorce. Scott South, Katherine Trent, and Yang Shen (2001) call this the *macrostructural opportunity perspective*, calling attention to the importance of opportunities for spouses to form potentially destabilizing opposite-sex relationships that are embedded within macrosocial structures, such as the workplace.

Research by sociologists Liana Sayer, Paula England, Paul Allison, and Nicole Kangas used data from three waves of the National Survey of Families and Households, and showed that employment and satisfaction both factor into shaping the fate of marriages. When women experience above-average marital satisfaction, their employment affects neither their likelihood of leaving nor their husband's. When wives experience below-average marital satisfaction, their being employed increases the likelihood that they will leave the marriage. Wives' employment had no effect on men's likelihood of leaving their marriages. Sayer and colleagues also looked at the effects of men's employment status and found that when men are not employed, either spouse is more likely

to leave than in marriages where men are employed (Sayer et al. 2011).

## Income

The higher the family income, the lower the divorce rate for both Caucasians and African Americans (Clarke-Stewart and Brentano 2007).

Each spouse's income alone does not explain divorce. Some research suggests that the probability of a divorce is heightened by arrangements in which a wife outearns her husband or in couples with a male breadwinner (Cooke 2006). Stacy Rogers (2004) found that the highest risk of divorce occurred in marriages in which wives contributed between 50 percent and 60 percent of the family's resources if spouses were at low or moderate levels of happiness. However, "happier spouses have little incentive to divorce, irrespective of spouses' relative economic contributions" (Rogers 2004, 71). Thus, neither higher-earning wives nor lower-earning husbands are automatically prone to divorce.

## Educational Level

The rate of divorce in first marriages for women in the United States varies by education. As shown in Figure 13.2, the rate is highest among those who have obtained some education beyond high school (i.e., have "some college") but have not earned a bachelor's degree. Among such women, the rate of divorce is 23.3 per 1,000. Overall, the education–divorce association is such that the highest-educated and least-educated women are those with the lowest divorce rates (in first marriages), with college-educated women at 14.6 and those with less than high school (having neither graduated nor earned a GED)

**True** **3.** College-educated women and men are the most likely to marry and the least likely to divorce.

Figure 13.2 Educational Differences in First Divorce Rates per 1,000 Women, 2010

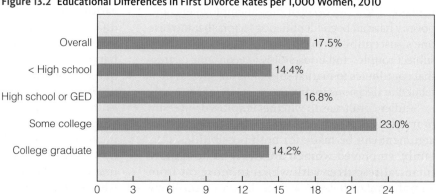

SOURCE: Stykes, Gibbs, and Payne (2014).

at 13.2 per 1,000. Among whites, blacks, Hispanic, and Asian Americans, a similar education pattern exists (Stykes, Gibbs, and Payne 2014).

Of course, educational attainment is usually linked with other factors that affect marital success. For example, men and women pursuing higher education tend to delay marriage and children until they're older. In addition, increased education may lead to acquiring values more conducive to marital success (Heaton 2002). One study concluded that college graduates had the most restrictive attitudes toward divorce, believing that "it should be more difficult to obtain a divorce than it is now" (Martin and Parashar 2006). Women who haven't completed high school tended to have the least restrictive attitudes.

A Pew Research Center report offers another way of seeing education effects on the likelihood of divorce. Where around half of marriages in the United States are expected to last at least 20 years, among college-educated women, it is estimated that nearly 80 percent will make it to at least their 20th anniversary. The Pew data suggest that 49 percent of women with "some college" and 40 percent of women with high school educations or less are estimated to see their marriages last at least 20 years (Wang 2015b).

## Ethnicity

As illustrated in a topical profile by the National Center for Family and Marriage Research, 2012 divorce rates in first marriages varied by race and ethnicity in the following ways. Using a refined rate of first divorce (number per 1,000 women in a first marriage), Asian women had the lowest first divorce rate (10.9 per 1,000 women in a first marriage), and U.S.-born Hispanic women had the highest rate (26.8). Black women's rate (25.4) was the next highest. White women's rate was 16.1. These rates represent declines for all racial/ethnic groups in recent years (since 2008) (Stykes, Gibbs, and Payne 2014).

Pew Research Center researcher, Wendy Wang, estimates that nearly 70 percent of Asian women who married for the first time between 2006 and 2010 are likely to see their marriages last at least 20 years. This is higher than the estimated likelihood for Hispanic women (54 percent), white women (53 percent), or black women (37 percent). Among men, Hispanic men (at 62 percent) had the highest predicted likelihood of remaining married at least 20 years, followed by white (54 percent) and black (53 percent) men (the sample had too few Asian men to be considered nationally representative) (Wang 2015b).

## Religion

Both *religiosity* (strength of religious commitment and participation) and religious affiliation have been linked to risk of divorce. Couples who are more religious are less at risk of divorce because such couples tend to have higher levels of marital satisfaction as well as reduced likelihood of domestic violence. Conversely, couples in which wives are but husbands aren't fundamentalists or evangelicals have an elevated risk of divorce.

Frequency of attendance at religious services (not necessarily the depth of beliefs) tends to be negatively associated with the divorce rate. That is, the greater the involvement in religious activities, the less the likelihood of divorce. Frequent attendees may also benefit from the community of "like-minded" congregants who can serve as social supports for remaining strongly committed to one's marriage, and from the religious education classes, pastoral counseling, seminars, and other such formal and informal marriage reinforcement available to them. But, interestingly, a difference between spouses in frequency of attendance is a risk factor, too. Marriages in which wives attend services weekly and husbands don't attend have a greater risk of divorce than even those marriages in which neither spouse attends religious services. The lowest risk is found among couples in which both spouses attend services regularly (Call and Heaton 1997; Vaaler, Ellison, and Powers 2009).

Because all major religions discourage divorce, highly religious men and women are less likely to accept divorce because it violates their values. It may also be that a shared religion and participation in organized religious life affirms the couple relationship (Call and Heaton 1997; Guttman 1993; Wineberg 1994). Religiosity even seems to influence the likelihood of divorce when marital problems arise, suggesting that religion plays a role in the decision of whether to seek a divorce (Lowenstein 2005).

## Life Course Factors

Among those women (15 and older) who divorced in 2012, the median length of time of their marriages before divorcing was 12 years. Despite the popular notion of a "seven=year itch," only a third of marriages that ended in 2012 were within their first seven years (with a sixth divorcing within 0 to 4 years, 40 percent within the first nine years). Almost a third of women who divorced in 2012 had been married between 10 and 19 years. Around one percent even divorced after 50 years of marriage (Spangler and Payne 2014).

Different aspects of the life course may affect the probability of divorce. These include age at time of marriage, premarital pregnancy and childbirth, cohabitation, remarriage, and intergenerational transmission.

## Age at Time of Marriage

The age at which people marry is perhaps "the most consistent predictor of marital stability identified in social science research" (Heaton 2002). Young, especially teenage, marriages are more likely to end in divorce than are marriages that take place when people are in their 20s or older. Close to 50 percent of those who marry before age 18 and 40 percent of those who marry before turning 20 years old divorce. Younger partners are less likely to be emotionally mature, younger marriages may be more likely to involve premarital pregnancy, and marrying "young" may be associated with curtailment of education, which has economic consequences that can undermine marital stability. Only 25 percent of those who marry when older than 25 end up divorced. The effect of age at marriage is not the same for all ethnic groups, however. Marrying in their teens has a "destabilizing effect" on Caucasian and African American marriages but not on Mexican American marriages (Phillips and Sweeney 2005).

**True** **4.** Age at marriage is the best predictor of the likelihood of divorce.

Sociologist Nicholas Wolfinger (2015) suggests that age risks are not limited to couples who are "too young." Wolfinger notes that it is still true that someone who marries at 25 is over 50 percent less likely to get divorced than is someone who weds at age 20. He contends that for each year that one waits to marry up until around age 31 to 32, one reduces the likelihood of eventually divorcing. However, he warns that beyond that point, each additional year one waits to enter a first marriage, one's odds of divorce increase by around 5 percent. He places the optimal age to marry in one's late 20s.

One other age-related divorce trend to note is the increase in what has been coined "gray divorce," referring to divorce among those over 50 years old, now at a record high (Brown and Lin 2013). According to Susan Brown and I-Fen Lin, married people in their 50s are twice as likely to experience a divorce as did their 50-plus counterparts of 1990. Among those over 65, the risk of divorce has more than doubled. One out of four individuals who experience a divorce is over 50, and one in ten is over 64. Though they see some of this as a product of failing remarriages, there

Marrying young, especially in one's teens, significantly increases one's risk of divorce.

Kichigin/Shutterstock.com

has also been an increase of divorce among over-50s who are in their first marriages. Strikingly, more than half of such gray divorces occur among couples married at least 20 years. Furthermore, unlike with younger women and men, the divorce risk among older couples does not vary by education. Brown and Lin also point out that the consequences of divorce are likely to be different for those who divorce later in life. They may face especially difficult financial consequences, particularly women who may have been at-home moms and out of the workforce. In comparison both to same-age married or widowed counterparts, "gray divorceds" may struggle more to make their resources meet their needs (Brown and Lin 2014). Given the increasing trend and potential consequences such as these, society may need to develop and implement ways of providing care and support to an aged population that will contain increasing numbers of unmarried elders needing such services (Brown and Lin 2013).

## Cohabitation

As shown in Chapter 9, premarital cohabitation had been associated with an elevated risk of a later divorce, though this is no longer evident. The potential negative impact that cohabitation was found to have had varied depending on the nature of one's cohabiting relationship, much depended on the expectations and intentions of couples when they entered cohabiting relationships. Those who entered with an understanding that they are eventually going to marry are unlikely to suffer the so-called "cohabitation effect."

On the other hand, those who seemingly "drifted" into marriage as a result of their having accumulated constraints to otherwise ending their relationship were those that were most likely to eventually end the marriage they enter together. For such couples, the lifestyles that they created by living together may have made marriage more likely, in part by making ending their relationship seem too costly. They find themselves "sliding into marriage," more than deciding to marry (Stanley, Rhoades, and Markman 2006).

In addition, as we noted in Chapter 9, certain characteristics of cohabitors may well have made them more likely to choose to divorce should their marriage become troubled. Having become so common and normative as it is today, cohabitation with one's eventual spouse is not likely to increase one's marital vulnerability.

## Remarriage

It might seem reasonable to expect that having been married and divorced at least once would make people better at making a subsequent marriage succeed. One might think that formerly married people would know what to avoid in reentering marriage and that they would reenter with greater commitment and determination for success this time around. Although this may sound logical, the reality is that the divorce rate among those who remarry is *higher* than it is for those who enter first marriages.

> **False** **5.** Couples in remarriages have a higher divorce rate than those in first marriages.

It is not entirely clear why there is a higher divorce rate in remarriages. Some researchers suggest that the cause may lie in a "kinds of people" explanation. The probability factors associated with the kinds of people who divorced in first marriages—everything from low levels of education to unwillingness to settle for unsatisfactory marriages—are present in subsequent marriages, increasing their likelihood of divorce. Similarly, people bring their same personality problems to any new relationship. Others argue that the unique dynamics of subsequent marriages, especially the presence of stepchildren, increase the chances of divorce. In fact, subsequent marriages that involve stepchildren have twice the likelihood of divorce as first marriages (Schoen 2002/2003). Additional factors that may account for an elevated risk of divorce in remarriages include different levels of commitment and greater acceptance of divorce among those who have already divorced once or more (Whitton et al. 2013). Those who have already gone through a divorce are more likely to take steps toward divorce when encountering marital distress. In that regard, it isn't remarriage per se that is a risk. Remarried women and men who are happy and who experience low levels of conflict seem no more prone to divorce than did those in first marriages of high quality (Whitton et al. 2013).

## Intergenerational Transmission

Those whose parents divorce are subject to **intergenerational transmission**—the increased likelihood that divorce will later occur to them (Amato 1996; Cavanagh and Sullivan 2009; Li and Wu 2008). Daughters of divorced parents appear at somewhat greater risk of themselves experiencing divorce, especially early in marriage (McDermott et al. 2013). If both the husband's and the wife's parents have been divorced, the odds of divorce increase dramatically. Sociologist Nicholas Wolfinger (2005) found that when both spouses are children of divorces, their risk of divorce is three times greater than if neither are children of divorced parents. How can we explain this intergenerational cycle?

Paul Amato (1996) noted that children of divorced parents are more likely to marry younger, cohabit, and experience higher levels of economic hardship. They become more pessimistic about lifelong marriage and develop more liberal attitudes toward divorce. In addition, females whose parents divorce develop less traditional attitudes about women's family roles, value self-sufficiency, and possess stronger attachments to paid employment. Also, children of divorce report both less commitment to their marriages and more positive and accepting attitudes about divorce (Whitton et al. 2013). All of these could raise susceptibility to divorce. Interestingly, parental "marital discord" in the absence of divorce has been found to have little consequence for their children's risk of divorce. Furthermore, high-discord marriages that ended in divorce only minimally raised their children's risk of divorce. However, where low-discord marriages ended in divorce, the children were especially vulnerable to divorce themselves (Amato and De Boer 2001).

Keep in mind that, as with intergenerational cycles of family violence, this relationship is neither automatic nor inevitable. It is, however, an important factor that can undermine marital success. Children of divorced parents, especially daughters of divorced parents, are more likely to possess pro-divorce attitudes (Kapinus 2004). Research that examined the effect of parents' attitudes on more than 400 children of divorce (Kapinus 2004) offered the following conclusions:

- There appears to be a "critical period," namely, the late teens, when parents' attitudes toward divorce have special influence on their children. This is the period of time when one is likely to be dating and forming expectations about relationships.

- Parental divorce affects sons' and daughters' attitudes toward divorce differently. Daughters of divorce are more likely to express "pro-divorce attitudes" than are sons of divorce.

- Diminished relationships with fathers after divorce and continued postdivorce conflict between parents may lead sons toward negative attitudes toward divorce. Yet postdivorce conflict between parents does not have the same effect on daughters.

## Family Processes

Among other factors that one might expect to influence one's likelihood of divorce are family processes that affect couples' levels of happiness and their experience of problems. The actual day-to-day marital processes of communication—handling conflict, showing affection, and other marital interactions—may be among the most important factors affecting marital outcomes (Gottman 1994).

### Marital Happiness

Although it seems reasonable that there would be a strong link between marital happiness (or, rather, the lack of happiness) and divorce, this is true only during the earliest years of marriage. Low levels of liking and trusting a partner are associated with long-term outcomes such as reduced satisfaction and elevated risk of divorce. The strength of the relationship between low marital happiness and divorce decreases in later stages of marriage, however (White and Booth 1991).

Eventually, alternatives to marriage and barriers to divorce appear to influence divorce decisions more strongly than does marital happiness. With nothing better to leave for or if there are too many obstacles to overcome in leaving, a couple might stay married even if unhappy. Although the opposite is also true—even if happy, one partner might leave for a more attractive alternative—it is probably less common. The presence of alternatives to a spouse has an effect on marital stability that can be observed among both high- and

The more people one is exposed to who could be attractive alternatives to one's spouse, the greater the risk of divorce.

Fancy/Photolibrary

low-risk couples (i.e., among those with other predisposing factors and those without).

The importance of the availability of attractive alternatives to a spouse has sometimes been overlooked as a factor accounting for divorce. Scott South, Katherine Trent, and Yang Shen (2001, 753) note that "satisfied and dissatisfied spouses alike remain, consciously or not, in the marriage market." This is further addressed in the following Issues and Insights feature.

### Children

Although 60 percent of divorces involve children, couples with children divorce somewhat less often than couples without children. The birth of the first child reduces the chance of divorce to almost nil in the year following the birth. Furthermore, couples with two children divorce less often than couples with one child or no children (Diekmann and Schmidheiny 2004). This does not mean that having children will spare parents from a divorce or that troubled spouses should become parents so that their troubles will disappear. It may well be that troubled spouses hold off having children or, if they have a child, resist having more because of their troubles. Thus, the quality of the marriage may lead to childbearing as much as or more than vice versa. Of course, having children raises the potential emotional, social, and financial costs of a divorce. One might well assess all such costs in making decisions about one's continuing marital future. Not having children together may make individuals more likely to act solely on the basis of the rewards or benefits one was experiencing in their marriage.

There are also some situations in which the presence of children may be related to higher divorce rates.

## Ending a "Not Quite Good Enough" Marriage

The decision to seek a divorce is not an easy one, and the process of getting divorced is an uncomfortable one for the individuals involved. For all the difficulty and discomfort experienced, a divorce may well bring relief for one or both spouses and even for children when it brings an end to a high-conflict, unhappy marriage. After all, as has been observed previously in this text and studied at length by many researchers, high-conflict, unhappy marriages may be worse—for adults and their children—than getting and being divorced, at least when one compares the well-being of those adults and children living within high conflict versus the well-being of those who have ended such marriages.

It may seem like a "no-brainer" to most people that divorce is a choice made by women and men in unhappy, conflict-ridden marriages, and that these former spouses will be better off as a result. Be careful when it comes to such seemingly obvious assumptions. Sociologists Paul Amato and Bryndl Hohmann-Marriott looked at what transpires when high- and low-distress marriages end. "High-distress" marriages were characterized by frequent conflict, physical aggression, and low levels of both marital happiness and interaction between spouses. They also had frequent thoughts about divorce. "Low-distress" marriages reported less conflict, fewer arguments, little aggression, and few thoughts of divorce. They also claimed moderate levels of both marital happiness and interaction, yet they ended their marriages.

In comparing the "high-distress" and "low-distress" couples, Amato and Hohmann-Marriott noted that both groups possessed certain "risk factors" associated with divorce, including being less religious, being in a second or subsequent marriage, and growing up in divorced families. Both groups were also more likely than continuously married couples to have cohabited prior to marrying, to have cohabited with a partner other than their spouse, to have stepchildren, and to possess less conservative values and more liberal attitudes toward divorce. Generally speaking, high-distress and low-distress couples were more similar than different in their backgrounds. Where they most differed was in terms of the perceived quality of the marriages that they chose to end. To those in "high-distress" marriages, divorce was a way out of a bad marriage. Those who ended their "low-distress" marriages left relationships that they might have characterized as "not bad" but also as "not quite good enough."

Amato and Hohmann-Marriott put this in the context of social exchange theory. High-distress couples who divorce do so primarily because they see their marriages as costly and unsatisfying. Low-distress couples who divorce do so for other reasons. Amato and Hohmann-Marriott also suggest that low-distress couples may represent as much as half of all divorces, meaning that ". . . only about half of all divorces fit the stereotype of a wretched couple continuously fighting and driven to divorce by the need to escape an aversive relationship" (2007, 636).

Looked at from the outside, the divorces of low-distress couples may well be unexpected and seem puzzling. One might be tempted to think that there must have been troubles that were hidden and kept private. In fact, it may be that low-distress couples who divorce "have been influenced by the belief that a good marriage should provide a high level of personal growth and self-actualization, criteria that many, perhaps most, marriages cannot meet" (p. 637). Looking at the outcome of divorce for the low-distress versus high-distress couples revealed that women and men who left high-distress marriages tended to experience an increase in life happiness after divorce, whereas those who ended low-distress marriages reported lower levels of happiness, leading Amato and Hohmann-Marriott to question whether such couples "overestimated the rewards of alternatives to the current marriage" (p. 634). ●

---

Children conceived during adolescence and children with physical or mental challenges are associated with divorce, as are children from prior marriages or relationships. For example, research indicates that parents whose children were diagnosed with attention deficit/hyperactivity disorder (ADHD) were more likely to have divorced than were parents of children without ADHD. The issues surrounding parenting a child with ADHD elevated the stress and marital conflict, which increased the likelihood of divorce (Wymbs et al. 2008).

In addition, the elevated risk of divorce posed by not having children may, itself, vary based on the reasons for childlessness. For example, a Danish study of 47,515 women in fertility treatment found that those who sought fertility treatment that failed were at greater risk of divorce for up to 12 years after first seeking help compared to those who sought help and had a child. Childless couples, including "child-free" couples, may divorce because they are at odds about how they feel about their childlessness.

## Marital Problems

If you ask divorced people to give the reasons for their divorce, they are not likely to say, "I blame the changing nature of the family" or "It was demographics." They are more likely to respond, "She was on my case all the time" or "He just didn't understand me"; if they are charitable, they might say, "We just weren't right for each other." Personal characteristics leading to conflicts are important factors in the dissolution of relationships.

Studies of divorced men and women cite such problems as alcoholism, drug abuse, domestic violence, marital infidelity, sexual incompatibility, and conflicts about gender roles as relationship factors leading to their divorces (Amato 2010). They also often cite external events—problems with in-laws or the effect of jobs (Amato and Previti 2003). Paul Amato and Denise Previti (2003) found the most common reasons given by their sample were infidelity, incompatibility, alcohol or drug use, growing apart, personality problems, lack of communication, and abuse (physical or mental).

Gender differences in reasons for divorce indicate that, in general, women cite emotional or relationship reasons, incompatibility, infidelity, unhappiness, and insufficient love as well as aspects of their former husband's personality or behaviors (such as abusiveness, neglect of children or home, and substance use). They are less likely to blame themselves. Men more often cite external factors or claim ignorance—they say they do not know what happened (Amato and Previti 2003).

People of high socioeconomic status are more likely to stress communication problems, incompatibility of or changes in values or interests, and their former spouse's self-centeredness. People of low socioeconomic status more often mention such things as financial problems, physical abuse, going out frequently with "the boys" or "with the girls," employment problems, neglect of home responsibilities, and drinking.

We know from studying enduring marriages that marriages often continue in the face of such problems. Based on interviews with almost 2,000 people, Paul Amato and Stacy Rogers (1997) found the following:

- Although men's and women's reports differed in the particular problems they emphasized, both predicted divorce equally well.
- Certain problems such as jealousy, moodiness, anger, poor communication, and drinking increased the odds

> **True** **6.** The critical emotional event when a marriage breaks down is separation rather than divorce.

of later divorce; sexual infidelity was an especially strong predictor of divorce.

- People who later divorce report a higher number of problems as early as 9 to 12 years before their divorce. Thus, their assessments of problems are not after-the-fact justifications concocted to account for or justify their divorce.
- Marital problems are **proximal causes** of later divorce. They are those things experienced within the daily life of married couples—such as high levels of conflict—that directly raise the probability of divorce. There are also background characteristics, such as age at marriage, prior cohabitation, education, income, church attendance, and parental divorce, that operate as more **distal causes**. These are brought by each spouse to the relationship and raise the likelihood that marital problems will later arise.

### No-Fault Divorce

Since 1970, beginning with California's Family Law Act, all 50 states have adopted no-fault divorce—the legal dissolution of a marriage in which guilt or fault by one or both spouses does not have to be established. It is unclear exactly how or how much no-fault divorce has affected divorce rates. Some contend that liberalization of divorce law led to increases in the divorce rate in both the United States and in other countries (e.g., Scotland, England, and Wales) (Lowenstein 2005). It is debatable that this has, by itself, affected the divorce rate. Unambiguously, however, liberalization of divorce law has altered the process of divorce by decreasing the time involved in the legal process, and it has altered the grounds for determining postdivorce financial responsibility.

## ▶ Uncoupling: The Process of Separation

Divorce is not caused by a single event. You don't wake up one morning and say, "I'm getting a divorce," and then leave. It's a far more complicated process (Kitson and Morgan 1991). It may start with little things that at first you hardly notice—a rude remark, thoughtlessness, an unreasonable act, or a "closed-ness." Whatever the particulars may be, they begin to add up. Other times, however, the sources of unhappiness are more blatant—yelling, threatening, or battering. For whatever reasons, the marriage eventually becomes unsatisfactory; one or both partners become unhappy.

Perhaps the crucial event in a marital breakdown is the act of separation. Although separation generally precedes divorce, not all separations lead to divorce. Furthermore, those that do may first involve attempts at reconciliation, in that about one-third of divorced women become divorced after attempting at least one marital reconciliation (Wineberg 1999). A statistic now more than a decade old indicates that perhaps one in ten marriages experiences a separation and reconciliation (Wineberg and McCarthy 1993). Those who reconcile may have separated to dramatize their complaints, create emotional distance, or dissipate their anger (Kitson 1985).

## Initiators and Partners

People gradually move apart through a set of fairly predictable stages. Sociologist Diane Vaughan (1990) calls this process *uncoupling*. The process appears to be the same for married or unmarried heterosexual couples and for gay or lesbian relationships. The length of time together does not seem to affect the process.

"Uncoupling begins," Vaughan observes, "as a quiet, unilateral process." Usually one person, the **initiator**, is unhappy or dissatisfied but keeps such feelings to himself or herself. Because the dissatisfied partner is unable to find satisfaction within the relationship, he or she begins turning elsewhere. This is not a malicious or intentional turning away; it is done to find self-validation without leaving the relationship. In doing so, however, the dissatisfied partner

In the early phases of the process of separation, estrangement can grow before both parties are fully aware of what has happened.

Walter Hodges/The Image Bank/Getty Images

"creates a small territory independent of the coupled identity" (Vaughan 1990).

Eventually, the initiator decides that he or she can no longer go on. He or she may go through a process of mourning the demise of what is still an intact marriage (Emery 1994, cited in Amato 2000). After the relationship ends, initiators have better adjustment to divorce and carry less postdivorce attachment to their former spouses (Wang and Amato 2000).

Uncoupling does not end when the end of a relationship is announced or even when the couple physically separates. Acknowledging that the relationship cannot be saved represents the beginning of the last stage of uncoupling. Diane Vaughan (1990) describes the process:

> Partners begin to put the relationship behind them. They acknowledge that the relationship is unsaveable. Through the process of mourning they, too, eventually arrive at an account that explains this unexpected denouement. "Getting over" a relationship does not mean relinquishing that part of our lives that we shared with another but rather coming to some conclusion that allows us to accept and understand its altered significance. Once we develop such an account, we can incorporate it into our lives and go on.

## The New Self: Separation Distress and Postdivorce Identity

Examining the experiences of those who divorce may be as good a way as any to see how much our married self becomes part of our deepest self. When people separate or divorce, many feel as if they have "lost an arm or a leg." This analogy, as well as the traditional marriage rite in which a man and a woman are pronounced "one," reveals an important truth of marriage: The constant association of both partners makes each almost a physical part of the other. This dynamic is true even if two people are locked in conflict; they, too, are attached to each other (Masheter 1991).

Almost everyone suffers **separation distress** when a marriage breaks up. The distress is real, results from the absence of one's spouse, but, fortunately, does not last forever (although it may seem so). The distress is situational and is modified by numerous external factors. About the only men and women who do not experience distress are those whose marriages were riddled by high levels of conflict. In these cases, one or both partners may view the separation with relief (Raschke 1987).

During separation distress, almost all attention is centered on the missing partner and is accompanied by apprehensiveness, anxiety, fear, and often panic. "What am I going to do?" "What is he or she doing?" "I need him. . . . I need her. . . . I hate him. . . . I love him. . . . I hate her. . . . I love her."

Social support is positively correlated with lower distress and positive adjustment. Additionally, as with other stressors in a person's life, it is often the individual's perception of the event, not the stress itself, that influences how a person adjusts to change. If those experiencing separation and divorce can begin to view and accept their changing circumstances as presenting new challenges and opportunities, there is a greater likelihood that the physiological and psychological symptoms of stress that follow divorce can be reduced.

## Establishing a Postdivorce Identity

A person goes through two distinct phases in establishing a new identity following marital separation: *transition* and *recovery* (Weiss 1975). The transition period begins with the separation and is characterized by separation distress and then loneliness. In this period's later stages, most people begin functioning in an orderly way again, although they still may experience bouts of upset and turmoil. The transition period generally ends within the first year. During this time, individuals have already begun making decisions that provide the framework for new selves. They have entered the role of single parent or noncustodial parent, have found a new place to live, have made important career and financial decisions, and have begun to date. Their new lives are taking shape.

The recovery period usually begins in the second year and lasts between one and three years. By this time, the separated or divorced individual has already created a reasonably stable pattern of life. The marriage is becoming more of a distant memory, and the former spouse does not arouse the intense passions he or she once did. Mood swings are not as extreme, and periods of depression are fewer. Yet the individual still has self-doubts that lie just beneath the surface. A sudden reversal, a bad time with the children, or doubts about a romantic involvement can suddenly destroy a divorced person's confidence. By the end of the recovery period, the distress has passed. It may take some people longer than others to recover because each person experiences the process in his or her own way. But most are surprised by how long the

recovery takes—they forget that they are undergoing a major discontinuity in their lives.

## Dating Again

A new partner reduces much of the distress caused by separation. A new relationship prevents the loneliness caused by emotional isolation. It also reinforces a person's sense of self-worth. It will not necessarily eliminate separation distress caused by the disruption of intimate personal relations with the former partner, children, friends, and relatives, but it "often produces a decline in depression, health complaints, and visits to the doctor, and an increase in self-esteem. When someone loves you and values you, you begin thinking that you are worth caring about" (Hetherington and Kelly 2002, 78–79).

Initiating this process may be stressful. A first date after years of marriage and subsequent months of singlehood might evoke some of the same emotions inexperienced adolescents feel.

For many divorced men and women, the greatest problem is how to meet other unmarried people. They believe that marriage has put them "out of circulation," and many are not sure how to get back in. Because of the marriage squeeze (the unequal ratio of available women to available men), separated and divorced men in their 20s and 30s are at a particular disadvantage: Considerably fewer women are available than men. The squeeze reverses itself at age 40, when significantly fewer single men are available. The problem of meeting others is most acute for single mothers who are full-time parents in the home because they lack opportunities to meet potential partners. Divorced men, having fewer child care responsibilities and more income than divorced women, tend to have more active social lives.

Several features of dating following separation and divorce differ from premarital dating. First, dating does not seem to be a leisurely matter. Divorced people are often too pressed for time to waste it on a first date that might not go well. Second, dating may be less spontaneous if a divorced woman or man has primary responsibility for children. Parents must make arrangements about child care; they may wish not to involve the children in dating. Third, finances may be strained; divorced mothers may have income only from low-paying or part-time jobs or payments from Temporary Assistance for Needy Family benefits and have many child care expenses. In some cases, a father's finances may be strained by paying alimony or child support.

# Making Personal Trouble Public: Sharing One's Divorce Online or in Print

The Sunday *New York Times* has long carried wedding announcements, sometimes including among them longer feature stories on particular couples, detailing how they met, how and when the relationship intensified, when and where they married. But in 2013, the paper began periodically carrying something else, in the same Sunday Styles section where one expects to find smiling photos of newly married women and men, the *Times* published stories under the heading "Unhitched." These features are described by the *Times* as follows: "Longtime couples tell the story of their relationships, from romance to vows to divorce to life afterward." Mostly authored by journalist Louise Rafkin, the stories featured long-married couples who divorced. With headlines such as, "Lessons Learned When Its All Over" (Rafkin 2013a), "Disappointments in 'Happily Ever After'" (Rafkin 2013b), and "Unraveling in Slow Motion" (Rafkin 2014), each feature told a couple's story from both points of view. Among the details such articles included were such things as: where they each grew up, how they met, what drew them to each other, why they married, when trouble surfaced, who asked for the split, and how life has moved on since their divorce. These provide an unusually candid look inside a relationship that ended.

Going public with one's separation or divorce is not limited to newspaper stories. Many people watched the online videos, uncomfortable as they were, when Tricia Walsh-Smith, 52, being divorced by her husband, 76-year-old theater executive Philip Smith, turned her marital crisis into a public spectacle by airing a series of self-made videos on YouTube. In one, she reported that though she and Philip were not physically intimate, he had a supply of condoms, pornography, and Viagra. Attempting to garner attention, sympathy, and support from the public and to generate revenue by selling music videos online, Walsh-Smith lost her case and had to vacate the New York penthouse she lived in and settle, instead, for the $750,000 settlement prescribed by the prenuptial agreement she and Philip entered.

Although the first and most well-known case of such video/Internet disclosure, numerous blogs and podcasts from divorcing spouses can be found online. Still others have taken to Facebook and other sites to announce their separations and divorces. Hannah Seligson described such efforts as "Facebook's Last Taboo: The Unhappy Marriage" (Seligson 2014), noting that the social media site that was more typically swimming in happy photos and postings about couples in love, now also increasingly featured divorce announcements, divorce selfies, and postings about failed relationships and former partners (Seligson 2014).

Although divorce has been and remains a deeply personal crisis, it has become increasingly public via the Internet and television.

And then there are Jon and Kate Gosselin, formerly of the television show *Jon & Kate Plus 8*, whose episode announcing the separation and divorce that ended their ten-year marriage became the most widely watched of any of their episodes throughout their five-season program. In that episode, they expressed sentiments that are widely shared among people who undergo divorces: Jon states, "People think I've changed, and I have changed. But I'm now the person I know I am." Among Kate's comments, she reflects, "We can't go back now. We can only go forward. And that's what we're going to do, and we're going to learn a lot going forward. And I know that we'll all come out of this on the other side, hopefully stronger, better, wiser."

Whether such public airing of disturbing details of divorce are motivated by revenge or revenues—via increased attention, more readers/viewers/listeners, and increased advertising income—they represent a clear departure from the days when divorce was kept more private and personal, except, of course, for those who appeared on *Divorce Court*. ●

SOURCES: Kaufman (2008); Starr (2009).

Sexual relationships are often an important component in the lives of separated and divorced men and women. Engaging in sexual relations for the first time following separation may help people accept their newly acquired single status. Because sexual fidelity is an important element in marriage, becoming sexually active with someone other than an ex-spouse is a dramatic symbol that the old marriage vows are no longer valid.

# Consequences of Divorce

Most divorces are not contested; according to the American Bar Association, close to 95 percent of divorces are uncontested and therefore avoid a trial (public.findlaw.com). Instead, they are settled out of court through negotiations between spouses, often with assistance from a mediator, or by their lawyers. But divorce, whether or not it is amicable, is a complex legal process involving highly charged feelings about custody, property, and children (who are sometimes treated by angry partners as property to be fought over).

**True**

**7.** Most divorces are uncontested.

## Economic Consequences of Divorce

One of the most damaging consequences of the no-fault divorce laws is that they systematically impoverish divorced women and their children. Following divorce, women are primarily responsible for both child rearing *and* economic support. As a result, women have been at a greater risk for economic distress than they were during their marriage. Even if a woman is not driven into poverty, she often experiences a dramatic downward turn in her economic condition (McKeever and Wolfinger 2001).

Husbands typically enhance their earning capacity during marriage. In contrast, wives historically decreased their earning capacity because they either quit or limited their participation in the workforce to fulfill family roles. Withdrawal from full participation limits one's earning capacity when one attempts to reenter the workforce. Women who were stay-at-home mothers may find their experience outdated, their skills lacking, and a lack of seniority that causes them to be vulnerable at work. Thus, they may not be "equal" to their former husbands at the point of divorce. Furthermore, many may never catch up.

Another factor contributing to women's economic slide is lack of child support. When marriage ends, many women face the triple consequences of gender, ethnic, and age discrimination as they seek to support themselves and their children. Because the workplace favors men in terms of opportunity and income, separation and divorce do not affect them as adversely.

Efforts to make the economic consequences less severe, such as changes in child support laws and an increase in the average amount of payments, along with the greater likelihood that women have remained employed during marriage, have helped lessen some of the economic impact confronting divorced women (McKeever and Wolfinger 2001). Add to these the slow but real narrowing of the wage gap and the increase in women's educational attainment. Cumulatively, these help account for a reduction in the economic cost of divorce for women. In fact, as sociologist Nicholas Wolfinger (2007, F16) describes, "divorced women are faring better financially than ever before."

Although it is often claimed that men experience enhanced financial well-being following divorce, truly this outcome depends on the division of wage earning that characterized the failed marriage. For white men who contributed less than 60 percent of their marital standard of living, divorce precipitates a decline in their living standards. On the other hand, men whose share of the household income was greater than 80 percent experience significant increases in their living standards after their marriages end (McManus and DiPrete 2001).

Another economic consequence of divorce that mostly affects women is the possible loss of health insurance. Roughly a quarter of women under age 65 receive dependent health coverage through a spouse or family member. According to research by Bridget Lavelle and Pamela Smock (2012), approximately 115,000 women lose private health insurance annually as a result of divorce. More than half (65,000) become uninsured, and the rate of insurance coverage doesn't improve for more than two years. As Lavelle and Smock point out, given the connection between having health coverage and access to health care, the loss of coverage was likely to affect women's health, their lifestyles, and other noneconomic consequences of divorce (Lavelle and Smock 2012). This research was conducted prior to the Affordable Care Act, and so individuals who lose spousal coverage upon divorcing have additional available options for coverage, though they may find themselves having to pay more and/or change providers that they had used before divorcing.

## Alimony and Child Support

**Alimony**, or *spousal support,* is monetary payment a former spouse makes to the other to meet his or her economic needs. It is *not* intended to be punitive (to the spouse required to pay). It is instead designed to address the economic vulnerability that a spouse may find himself or herself in after the end of the marriage. In some instances, alimony might have been awarded for an indefinite period of time, perhaps until the receiving spouse remarried or died. Death of the paying spouse may not even bring an end to alimony obligations, however. The deceased's estate may be required to continue to honor the alimony decision even after the paying spouse dies (Ho and Johnson 2004). As reviewed by law professor J. Thomas Oldham (2008/2009), such indefinite awards became less common in the latter decades of the 20th century, as did the awarding of spousal support more generally. In specific instances where a long-term marriage ends and the former spouses' incomes differ greatly, there appears to be somewhat greater judicial willingness to award indefinite alimony (Oldham 2008/2009). Typically, in addition to the length of marriage and income discrepancy, such factors as employability and future earnings of divorcing spouses, the ability to pay alimony, and the length of time support is deemed to be needed shape decisions about whether and how much alimony is to be mandated.

A number of states have implemented restrictions on the amount or duration of spousal support. For example, in 2011, Massachusetts Governor Deval Patrick signed into law a piece of legislation that sets limits on and provides guidelines for determining alimony in his state. The law ties the length of time spouses are to receive postmarital support to the length of their marriages. It also stipulates that the spouse paying alimony can stop upon retirement from employment. Proponents of the new law hope it achieves greater consistency in alimony judgments and provides more equitable outcomes for both divorcing spouses (Bidgood 2011; Powers 2011). Texas generally restricts spousal support to divorcing couples married at least ten years. New York and Pennsylvania are among the states that have developed formulas for calculating alimony amounts. Still other states attach a maximum term for receipt of alimony regardless of how long a couple was married (Oldham 2008/2009).

Alimony is spousal support and, as such, is different from **child support**—the monetary payments made by the noncustodial spouse to the custodial spouse to assist in child-rearing expenses. However, child support is not an automatic outcome of divorce; only about half of custodial parents are awarded child support (Grall 2011). Because women are the parent with physical custody in the vast majority (82 percent) of instances, we typically think of child support in the context of single mothers. In fact, child support is awarded to single custodial fathers, too, just not in the same proportion as mothers.

More critically, being awarded support does not guarantee receiving it. In 2011, close to half of all custodial parents had either legal or informal child support agreements, with mothers (53.4 percent) almost twice as likely as fathers (28.8 percent) to have such agreements. Mothers made up nine of every ten custodial parents who were due to receive child support. Of those with agreements, only 43 percent of custodial parents received the full amount of child support they were granted. All told, 74.1 percent received at least partial support amounts, while almost a quarter received *none* of what they were awarded (Grall 2013). For those custodial parents living in poverty, 40 percent received all that they were due. Custodial parents who had joint custody arrangements were more likely to receive the full amount they were awarded. At the opposite end, in instances where there was no contact between the noncustodial parent and the child(ren), less than a third (30.7 percent) received the full amount of child support they'd been awarded (Grall 2013).

In 2011, child support amounts awarded or agreed to informally averaged $6,050 a year, approximately $500 per month. The average amount actually received by custodial parents due child support was $3,770, roughly $315 per month. A little over a quarter of custodial parents due child support received $5,000 or more in annual support payments. In total, $23.6 billion of child support was received, representing 62 percent of the almost $38 billion due (Grall 2013).

The legal criteria around child support have undergone some notable changes in the past few decades. The Child Support Enforcement Amendments, passed in 1984, and the Family Support Act of 1988 require states to deduct delinquent support from fathers' paychecks, authorize judges to use their discretion when support agreements cannot be met, and mandate periodic reviews of award levels to keep up with the rate of inflation. In addition, all states implemented systems to conduct automatic wage withholding of child support in 1994. Chien-Chung Huang, Ronald Mincy, and Irwin Garfinkel contend that nearly every year for the past two decades, Congress has passed new laws designed to strengthen child support enforcement. Furthermore, spending by both state and federal governments on child support enforcement increased from less than $1 billion a year in 1978 to $5.2 billion in 2002 (Huang, Mincy, and Garfinkel 2005). For those noncustodial parents who fail to pay child support, a variety of consequences can result. They can lose their driver's licenses (as well as certain professional, occupational, and recreational licenses), be denied passports, and have their bank accounts frozen and/or seized. In addition, the Office of Child Support Enforcement also assists in measures designed to establish paternity and to locate noncustodial parents (Office of Child Support Enforcement 2011).

People are generally more approving, at least in principle, of child support than they are of alimony. In the past, alimony represented the continuation of the husband's responsibility to support his wife. Currently, laws determine that alimony be awarded on the basis of need to those women or men who would otherwise be indigent. At the same time, some assert that alimony represents the return of a woman's "investment" in marriage. Sociologist Lenore Weitzman (1985) argued that a woman's homemaking and child care activities must be considered important contributions to her husband's present and future earnings. If divorce rules do not give a wife a share of her husband's enhanced earning capacity, then the "investment" she made in her spouse's future earnings is discounted. According to Weitzman, alimony and child support awards should be made to divorced women in recognition of the wife's primary child care responsibilities and her contribution to her ex-husband's work or career. Such awards can help raise some divorced women and their children above the level of poverty to which they have been cast as a result of no-fault divorce's specious equality.

Before leaving the subject of child support, it is important to note that most (60 percent) custodial parents receive some form of nonmonetary support from

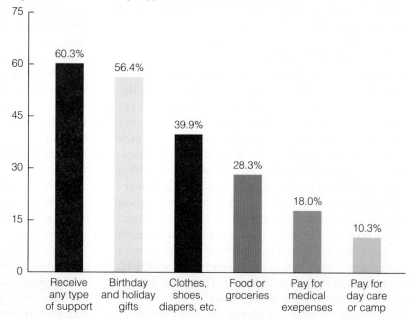

**Figure 13.3 Nonmonetary Support to Custodial Parents, 2010**

SOURCE: Grall (2011).

their former spouse for their children, with husbands more likely than wives to receive such support. Such noncash support includes things like birthday and holiday presents, clothing, food or groceries, medical expenses other than insurance, full or partial payment for summer camps, and assistance paying for child care. As shown in Figure 13.3, of these, birthday and holiday gifts were the most common form of noncash support (Grall 2011).

## Employment

The economic effect of divorce on women with children is especially difficult because their employment opportunities are often constrained by the necessity of caring for children (Maccoby et al. 1993). Child care costs may consume a third or more of a single mother's low income. Women may work fewer hours because of the need to care for their children.

Separation and divorce can dramatically change many mothers' employment patterns. If a mother was not employed before separation, she is likely to seek a job following the split. The reason is simple: If most women and their children relied on alimony and child support alone, they would soon find themselves on the street. Most employed single mothers are not far from being on the verge of financial disaster, however. In 2009, 30 percent of single custodial mothers were in poverty. This poverty rate was even worse for custodial mothers who had not completed high school (59.1 percent), were in public assistance programs (58.7 percent), or had three or more children (51.5 percent). Custodial mothers with full-time, year-round jobs were much less likely to be poor (9.5 percent), thus illustrating the importance of women's employment. The general problems of women's lower earning capacity and lack of adequate child care are particularly severe for single mothers. Gender discrimination in employment and lack of societal support for child care condemn millions of single mothers and their children to poverty.

## Noneconomic Consequences of Divorce

In comparison to married people, the picture of divorced individuals is fairly bleak. Reviewing the research literature of the 1990s, Paul Amato (2000) notes the following. Compared with married people, divorced individuals experience more psychological distress, poorer self-concepts, lower levels of psychological well-being, lower levels of happiness, more social isolation, less satisfying sex lives, and more negative life events. They also have greater risks of mortality and report more health problems. Compared to married women and men, major depression is three times higher for separated or divorced women and *nine times higher* for separated or divorced men. British data reveal a similar story. Marital separation is accompanied by significant increases in heavy drinking during the period of separation (Power, Rogers, and Hope 1999). In addition, Terrance Wade and David Pevalin (2004) found that for those exiting a marriage through separation and divorce, there is a much higher prevalence of mental health problems. Of note, they also found that such problems are evident before the marital disruption, indicating that the relationship between mental health and divorce goes both ways. Sociologist Paul Amato reports that mental health consequences are comparable for divorced women and men (Amato 2010).

Some have questioned whether the mental health and well-being of individuals who divorce results from or precedes the divorce. Similarly, sociologist Linda Waite and journalist Maggie Gallagher, in their book *The Case for Marriage* (2000), looked at the question of whether being married makes people happier or whether happier people get married and *stay* married. Citing research that compared the emotional health of a sample of people over time—some who married and stayed married, some who never married or remained divorced, and others who married and

divorced—they report the following: When people married, their mental health substantially improved. When people separated and divorced, they suffered declines in their emotional and mental well-being.

Waite and Gallagher (2000) also noted that compared to married people, divorced (and widowed) women and men were three times as likely to commit suicide. Among the divorced, as among the general population, more men than women commit suicide. However, among women, those who are divorced are "the most likely to commit suicide, followed by widowed, never-married, and married, in that order." As parents, divorced individuals have more difficulty raising children. They display more role strain, whether they are custodial or noncustodial parents, and they display less authoritative parenting styles (Amato 2000).

Despite the stark picture that surfaces, divorce is associated with positive consequences for some people. These include higher levels of personal growth, greater autonomy, and—for some women—improvements in self-confidence, career opportunities, social lives, and happiness as well as a stronger sense of control (Amato 2000). In addition, it would be remiss not to stress the fact that evidence indicates that remaining unhappily married is worse than divorcing. People who find themselves in "long-term low-quality marriages" are less happy than those who divorce and remarry. They also have lower overall life satisfaction, lower overall health, and lower self-esteem than those who divorce and remain single (Hawkins and Booth 2005).

# Children and Divorce

Divorce not only ends marriages and breaks up families, but it can also create new family forms from the old ones. It can lead one or both former spouses into remarriages (which, as Chapter 14 will show, are different from first marriages). If children are involved, it gives birth to single-parent families, stepfamilies, and the *binuclear family*, the family created when parents divorce, making their children now members of two separate households. Today, more than one out of every four U.S. families is a single-parent family, and half of all children will become stepchildren (Stewart 2007). Within the singles subculture is an immense pool of divorced men and women (most of whom are on their way to remarriage). Consider also the increasing numbers of marriages that are truly remarriages for one or both spouses. As noted earlier, such marriages carry increased risk of divorce.

The greatest concern that social scientists express about divorce is its effect on children. But even in studies of the children of divorce, the research may be distorted by traditional assumptions about marriage in which divorce is clearly frowned upon (Amato 1991). For example, problems that children experience may be attributed to divorce rather than to other causes, such as personality traits. Although some effects are caused by the disruption of the family itself, others may be linked to the family climate that preceded and led to the divorce. Then there is the matter of the new social environment—most notably poverty and parental stress—into which children are thrust by their parents' divorce. Slightly more than half of all divorces involve children. Popular images of divorce depict "broken homes," but it is important to remember that an intact nuclear family, merely because it is intact, does not guarantee children an advantage over children in a single-parent family or a stepfamily. An intact family wracked with frequent and intense conflict between spouses, spousal violence, sexual or physical abuse of children, alcoholism, neglect, or psychopathology creates a destructive environment likely to inhibit children's healthy development.

Although living in a two-parent family with severe marital conflict is often more harmful to children than living in a tranquil single-parent family or stepfamily, what about what researchers consider low-discord marriages? In such relationships, parents avoid overt conflict, cooperate with each other, behave in a civil—even respectful—manner toward each other, but feel as though their marriages are lacking in some way. Researchers contend that children in such households may be better off if their parents stay together, even if only until they are grown and out of the household and even if parents are less than fulfilled by their marriage. By keeping conflict limited and away from the children, the risk of children having developmental and emotional problems is low. Children suffer more interpersonal and psychological problems when their parents end such "good-enough" marriages (Amato 2003). For them, divorce may represent "an unexpected, unwelcome, and uncontrollable event." They face the loss of one parent, the emotional distress of the remaining parent, and perhaps a decline in standard of living (Booth and Amato 2001).

Unmistakably, children living in happy, two-parent families appear to be the best adjusted, and those from conflict-ridden two-parent families appear to be the worst adjusted. Children from stable, well-functioning single-parent families are in the middle (Coontz 1997).

## How Children Are Told

Telling children that their parents are separating is one of the most difficult and unhappy events in life. Whether or not the parents are relieved about the separation, they often feel extremely guilty about how the separation will affect their children. Many children may not even be aware of the extent of parental discord, especially in low-discord marriages (Amato 2006; Furstenberg and Cherlin 1991). Even those that are may be upset by the separation, but their distress may not be immediately apparent.

Qualitative research by Heather Westberg, Thorana Nelson, and Kathleen Piercy (2002) indicated that children's reaction is influenced by how the news is disclosed and is shaped by the perception that life will be relatively better or relatively worse afterward. For those to whom the news was disclosed long before the divorce occurred, it was experienced as relief by the time it "finally" happened. For those who had lived with frequent and intense conflict between parents, divorce may also be welcome. For those who fear that they will have to move or that they won't see their noncustodial parent, the news will cause distress.

As the late psychologist Judith Wallerstein suggested in her book *Second Chances* (Wallerstein and Blakeslee 1989), divorce is differently experienced within the family. For at least one of the divorcing spouses, divorce is welcomed as an escape from an unpleasant or unfulfilling relationship. Both spouses may come to appreciate the "second chance" they receive with divorce—the opportunity to make a better choice and build themselves a better relationship. Children may not see the breakup of their parents' marriage as an "opportunity." However, under certain circumstances—most notably when parental conflict has been long-term, overt, and unresolved—children are at risk of developing emotional and developmental problems so long as their parents stay together (Booth and Amato 2001). For such children, divorce may, indeed, come as a relief.

Lisa Strohschein (2005) found that children's antisocial behaviors such as bullying and lying were reduced after the divorce of parents who had been experiencing high levels of dysfunction. The stress relief that comes with divorce may, however, become apparent only after enough time passes (Strohschein 2005).

Notifying children of a decision to divorce is difficult for both the parents and children.

## The Three Stages of Divorce for Children

Part of the difficulty in determining the effect of divorce on children is a failure to recognize that, just as it is for adults, divorce is a process as opposed to a single event. Divorce contains a series of events and changes in life circumstances. Many studies focus on only one part of the process and identify that part with divorce itself. Yet at different points in the process, children are confronted with different tasks and adopt different coping strategies. Furthermore, the diversity of children's responses to divorce is the result, in part, of differences in temperament, gender, age, and past experiences.

A study by psychologist Judith Wallerstein and Deborah Resnikoff (1997) found that children from divorced families suffered both emotionally and developmentally. Young children fared worse than older children. Depending on the point in the process, boys tend to do less well than girls. In the "crisis period" of the two years following separation, boys' suffering is especially evident. This may be because they must internalize different gendered styles of reacting to distress. It is also the case, however, that after separation, most boys live with their mothers and not their fathers. This, too, can exacerbate their suffering, as they suffer diminished contact with their fathers (Furstenberg and Cherlin 1991).

According to Wallerstein (1983), children experience divorce as a three-stage process. Studying 60 California families during a five-year period, she argued that divorce consisted of the initial, transition, and restabilization stages:

- *Initial crisis.* The initial stage, following the decision to separate, was extremely stressful; conflict escalated, and unhappiness was endemic. The children's aggressive responses were magnified by the parents' inability to cope because of the crisis in their own lives.

- *Transition and short-term aftermath.* The transition stage began about a year after the separation, when the extreme emotional responses of the children had diminished or disappeared. The period was characterized by restructuring of the family and by economic and social changes: living with only one parent and visiting the other, moving, making new friends and losing old ones, financial stress, and so on. The transition period lasted between two and three years for half the families in the study.

- *Restabilization stage.* Families had reached the restabilization stage by the end of five years. Economic and social changes had been incorporated into daily living. The postdivorce family, usually a single-parent family or stepfamily, had been formed.

## Children's Responses to Divorce

Decisive in children's responses to divorce are their age and developmental stage (Guttman 1993). A child's age affects how one responds to one parent leaving home, changes (usually downward) in socioeconomic status, moving from one home to another, transferring schools, making new friends, and so on.

Children react differently to divorce depending on their age. Most feel sad, but the eventual outcome for children depends on many factors, including having a competent and caring custodial parent, siblings, and friends, as well as their own resiliency. The postdivorce relationship between parents and the custodial parent's economic situation are also important factors.

## Developmental Tasks of Divorce

Judith Wallerstein (1983) suggested that children must undertake six developmental tasks when their parents divorce. The first two tasks need to be resolved during the first year. The other tasks may be worked on later; often they may need to be reworked because the issues may recur. How children resolve these tasks differs by age and social development. The tasks are as follows:

> **True**
> **8.** Many of the problems children of divorced parents experience are present before the marital disruption.

- *Acknowledging parental separation.* Children often feel overwhelmed by feelings of rejection, sadness, anger, and abandonment. They may try to cope with them by denying that their parents are "really" separating. They need to accept their parents' separation and face their fears.

- *Disengaging from parental conflicts.* Children need to psychologically distance themselves from their parents' conflicts and problems. They require such distance so that they can continue to function in their everyday activities without being overwhelmed by their parents' crisis.

- *Resolving loss.* Children lose not only their familiar parental relationship but also their everyday routines and structures. They need to accept these losses and focus on building new relationships, friends, and routines.

- *Resolving anger and self-blame.* Children, especially young ones, often blame themselves for the divorce. They are angry with their parents for disturbing their world. Many often "wish" their parents would divorce, and when their parents do, they feel responsible and guilty for "causing" it.

- *Accepting the finality of divorce.* Children need to realize that their parents will probably not get back together. Younger children hold "fairy tale" wishes that their parents will reunite and "live happily ever after." The older the child is, the easier it is for him or her to accept the divorce.

- *Achieving realistic expectations for later relationship success.* Children need to understand that their parents' divorce does not condemn them to unsuccessful relationships as adults. They are not damaged by witnessing their parents' marriage; they can have fulfilling relationships themselves.

**Younger Children.** Younger children react to the initial news of a parental breakup in many different ways.

Feelings range from guilt to anger and from sorrow to relief, often vacillating among all of these. Preadolescent children, who seem to experience a deep sadness and anxiety about the future, are usually the most upset. Some may regress to immature behavior, wetting their beds or becoming excessively possessive. Most children, regardless of their age, are angry because of the separation. Very young children tend to have more temper tantrums. Slightly older children become aggressive in their play, games, and fantasies—for example, pretending to hit one of their parents.

Data collected over a 12-year period examined parent–child relationships before and after divorce. Researchers found that marital discord may exacerbate children's behavior problems, making them more difficult to manage (Amato and Booth 1996). Because discord between parents often preoccupies and distracts them from the tasks of parenting, they appear unavailable and unable to deal with their children's needs. This study reinforced a growing body of evidence showing that many problems assumed to be caused by divorce are present before marital disruption.

School-age children may blame one parent and direct their anger toward him or her, believing the other one innocent. But even in these cases, the reactions are varied. If the father moves out of the house, the children may blame the mother for making him go, or they may be angry at the father for abandoning them regardless of reality. Younger schoolchildren who blame the mother often mix anger with placating behavior, fearing that she will leave them. Preschool children often blame themselves, feeling that they drove their parents apart by being naughty or messy. They beg their parents to stay, promising to be better. It is heartbreaking to hear a child say, "Mommy, tell Daddy I'll be good. Tell him to come back. I'll be good. He won't be mad at me anymore." A study of 121 white children between the ages of 6 and 12 found that about 33 percent initially blamed themselves for their parents' divorce. After a year, the figure dropped to 20 percent (Healy, Stewart, and Copeland 1993). The largest factor in self-blaming was being caught in the middle of parental conflict. Children who blamed themselves displayed more psychological symptoms and behavior problems than those who did not blame themselves.

When parents separate, children want to know with whom they are going to live. If they feel strong bonds with the parent who leaves, they want to know when they can see him or her. If they have brothers or sisters, they want to know if they will remain with their siblings. They want to know what will happen to them if the parent they are living with dies. Will they go to their grandparents, their other parent, an aunt or uncle, or a foster home? These are practical questions, and children have a right to answers. They need to know what lies ahead for them amid the turmoil of a family split so that they can prepare for the changes.

Some parents report that their children seemed to do better psychologically than they themselves did after a split. Children often have more strength and inner resources than parents realize. The outcome of separation for children depends on several factors related to the children's age. Young children need a competent and loving parent to take care of them; they tend to do poorly when a parenting adult becomes enmeshed in constant turmoil, depression, and worry. With older, preadolescent children, the presence of brothers and sisters helps because the children have others to play with and rely on in addition to the single parent. If they have good friends or do well in school, this contributes to their self-esteem. Regardless of the child's age, it is important that the absent parent continue to play a role in the child's life. The children need to know that they have not been abandoned and that the absent parent still cares (Wallerstein and Kelly 1980). They need continuity and security, even if the old parental relationship has radically changed.

**Adolescents.** Many adolescents find parental separation traumatic. Studies indicate that much of what appear to be negative results of divorce (personal changes, parental loss, economic hardships, and psychological adjustments) are often more likely the result of parental conflict that precedes and surrounds the divorce (Amato and Booth 1996; Amato and Keith 1991; Morrison and Cherlin 1995). A study by Yongmin Sun (2001) found that such problems as poor psychological well-being, academic difficulties, and behavioral problems are present among adolescents from divorced families *at least a year before* the divorce.

Adolescents may try to protect themselves from the conflict preceding separation by distancing themselves. Although they usually experience immense turmoil within, they may outwardly appear cool and detached. Unlike younger children, they rarely blame themselves for the conflict. Rather, they are likely to be angry with both parents, blaming them for upsetting their lives. Adolescents may be particularly bothered

by their parents beginning to date again. Some are shocked to realize that their parents are sexual beings, especially when they see a separated parent kiss someone or bring someone home for the night. The situation may add greater confusion to the adolescents' emerging sexual life. Some may take the attitude that if their mother or father sleeps with a date, why can't they? Others may condemn their parents for acting "immorally."

Kathleen Boyce Rodgers and Hilary Rose (2002) assert that the negative effects of divorce on adolescents can be tempered. They suggest that strong peer support, a strong attachment to school, and high levels of support and monitoring by parents can lessen the negative consequences adolescents otherwise encounter.

## How to Help Children Adjust

There are certain steps parents can attempt to take to ease their children's adjustment after divorce. Talking openly with children on a level they will be able to understand about having decided to separate and/or divorce and what changes and potential problems to expect can be helpful (Raschke 1987). In addition, as is strongly and widely advocated, a **cooperative co-parenting** strategy is thought to be the optimal arrangement parents can attempt to construct to assist their children's adjustment. In co-operative co-parenting, parents cooperate, communicate, and avoid or minimize conflict. In other words, they have infrequent disagreements about caregiving, communicate frequently about their children's caregiving needs, and engage in few to no attempts to undermine each other's parenting. Also important is to see that children maintain relationships with their nonresidential parent, and to ensure that the quality of parenting, especially by the custodial parent, is high.

## Betwixt and Between: Children Caught in the Middle

Recently divorced women and men often suffer from a lack of self-esteem and a sense of failure. One means of dealing with the feelings caused by divorce is to blame the other person. To prevent further hurt or to get revenge, divorced parents may try to control each other through their children. One consequence of this for children is the sense of being caught in the middle, forced to choose sides, and pulled in different directions by their parents. Some have even suggested that feeling caught between parents may be one of the factors that differentiate children's reactions to divorce, explaining why some do better and some do worse. Such feelings may also lead to adolescent depression and deviant behavior. Evidence indicates that older adolescents are more likely than younger adolescents and children to feel caught. In addition, such feelings may extend well into adulthood, although reduced contact with both parents may lessen the intensity of such feelings.

When caught in the middle, children may opt for one of three strategies: try to maintain positive relationships with both parents, form an alliance with one parent over and against the other, or reject both parents. Trying to remain close to two embattled parents may exact costs that outweigh the benefits of such relationships. Choosing sides comes at the expense of a relationship with one parent and can trigger guilt toward the abandoned parent and resentment toward the custodial parent. Rejecting both parents means losing closeness to both—a steep price to pay. Paul Amato and Tamara Afifi (2006) also found that parents put more pressure on daughters than on sons to take sides in their disputes, and that feeling caught in the middle is of more negative consequence for mothers and daughters than for mothers and sons.

## Perspectives on the Long-Term Effects of Divorce on Children

There are multiple perspectives on how and why divorce affects children (Amato 1993). Specified outcomes range from negative through neutral to positive (Coontz 1997; Whitehead 1996). There is enough divergent information that we could selectively cite research to make either a more pessimistic or a more optimistic generalization. We review some of these mixed findings here.

A variety of studies reviewed by Barbara Dafoe Whitehead in her strongly antidivorce book, *The Divorce Culture* (1997), suggest multiple ways in which children suffer after their parents divorce. First, across racial lines, children of divorce suffer substantial reduction in family income as a direct result of divorce. Second, most children experience a weakening of ties with their fathers, suffering damage when and after fathers leave. She suggests that separation and later divorce induce a "downward spiral" in father–child relationships wherein distance between them grows and children eventually lose their fathers' "love, support, and substantial involvement." Third, children suffer a loss of "residential stability," often having to move from the family home because of drops in their economic standing.

Whitehead (1997) goes on to detail other measurable ways in which children suffer: reduced school performance, increased likelihood of dropping out, worsened and increased behavioral problems, and a greater likelihood of becoming teen parents. Many of these same outcomes were identified as among the "risks and problems associated with stepfamily life" (Whitehead 1996).

In her more optimistic book, *The Way We Really Are: Coming to Terms with America's Changing Families*, Stephanie Coontz (1997) tempers some of this distressing news. While acknowledging the "agonizing process" that accompanies divorce and the ways in which children, especially, can be hurt by divorce, Coontz qualifies the more pessimistic interpretations. In a subtle but important comparison, she notes that research shows "*not* that children in divorced families have more problems but that *more* children of divorced parents have problems" (Coontz 1997, 99, emphasis in original). In other words, all children of divorce do not suffer the negative consequences identified by researchers and reported by people such as Whitehead. Coontz reminds us that although more children in divorced homes drop out or become pregnant than do children whose parents stay married, "divorce does not account for the majority of such social problems as high school dropout rates and unwed teen motherhood" (Coontz 1997, 100). Finally, Coontz went even further, by noting that there are some measures on which large proportions of children of divorced homes score higher than do children from homes with two parents. She reports that children of single parents (usually single mothers) spend more time talking with their custodial parent, receive more praise for their academic successes, and face fewer pressures toward conventional gender roles. Thus, she argues, single-parent households may in some ways be beneficial environments within which to be raised (Coontz 1997). Furthermore, comparisons of children of divorced parents to children of intact parental marriages are as potentially flawed as comparing divorced adults to married adults. A more nuanced, three-way comparison that looks at both parental marital stability and marital quality is needed. Children of divorced parents are children whose parents were unhappily married. Unless they are compared against children whose parents are unhappily married, we cannot draw the kinds of conclusions we desire. Clearly, the optimal arrangement for children would be to live with happily married parents.

## Just How Bad Are the Long-Term Consequences of Divorce?

The message about the long-term consequences varies according to the research examined. Influential longitudinal research conducted by Judith Wallerstein highlights fairly extensive, long-term trauma and distress that stays with and affects children of divorce well into adulthood. Beginning with *Surviving the Breakup: How Children and Parents Cope with Divorce* (Wallerstein and Kelly 1980) through *Second Chances: Men, Women, and Children a Decade after Divorce* (Wallerstein and Blakeslee 1989) and culminating with *The Unexpected Legacy of Divorce: A 25-Year Landmark Study* (Wallerstein, Lewis, and Blakeslee 2000), Wallerstein has followed a sample of (originally) 60 families, with 131 children among them, as they divorced and went through the subsequent adjustment processes at 18 months, 5 years, 10 years, 15 years, and ultimately 25 years. Seventy-five percent of the original families, and 71 percent of the 131 children were studied for all three books.

Wallerstein found that at the five-year mark, more than a third of the children were struggling in school and experiencing depression, had difficulty with friendships, and continued to long for a parental reconciliation. At the ten-year follow-up, she indicated that almost half the children carried lingering problems and that they had become worried, sometimes angry, underachieving young adults. Three-fifths of the children of divorce retained a persistent sense of rejection by one or both parents and suffered especially poor relationships with their fathers. Finally, at the 25-year point, Wallerstein asserted that the effects of divorce on children reached their peak in adulthood, where the ability to form and maintain committed intimate relationships was negatively affected (see Amato 2003).

A more moderate view of the long-term effects of divorce emerges from other studies (Amato 2003; Hetherington and Kelly 2002). E. Mavis Hetherington undertook the Virginia Longitudinal Study of Divorce and Remarriage, which initially consisted of following a sample of 144 families with a 4-year-old "target child." Half the sample families were divorced, and half were married. Initially, they were to be followed and restudied at two years to compare how those who divorced fared in comparison to those who did not. Eventually, the sample was expanded, and subsequent research was conducted at 2, 6, 11, and 20 years after divorce. As the "target children" (i.e., the

initial 4-year-olds) married, had a child, or cohabited for more than six months, they were further studied (Hetherington and Kelly 2002). Meanwhile, families were added to the sample at each wave to reach a final sample of 450, evenly split between nondivorced, divorced, and remarried families.

Throughout the research, a variety of qualitative and quantitative data were collected on personalities of parents and children, adjustment, and relationships within and outside the family (Hetherington 2003). The impression that Hetherington's research leaves is more encouraging than the one received from Wallerstein's studies. For example, most adults and children adapt to the divorce within two to three years. Although 70 percent of the divorced parents were wrestling with animosity, loneliness, persistent attachment, and doubts about the divorce at the one-year mark, by six years, most were moving toward building new lives. More than 75 percent of the sample said that the divorce had been a good thing, more than 50 percent of the women and 70 percent of the men had remarried, and most had embarked on the postdivorce paths they would continue to take (Hetherington 2003).

In considering the effects of divorce on children, Hetherington reports that 20 percent of her sample of youths from divorced and remarried families was troubled and displayed a range of problems, including depression and irresponsible, antisocial behavior. They had the highest dropout rate, had the highest divorce rate (as they themselves married), and were the most likely to be struggling economically. But perhaps more important, "80 percent of children from divorced homes eventually are able to adapt to their new life and become reasonably well adjusted" (Hetherington and Kelly 2002, 228). Given that 10 percent of youths from nondivorced homes also were struggling, the difference for children from divorced as opposed to nondivorced homes was fairly small (10 percent).

As Hetherington points out, the optimal outcome for adults and their children is to be in a happily married household. Nevertheless, her research indicates that we may overstate the risks and fail to recognize the resilience of men, women, and children of divorce.

Paul Amato (2003) suggests that much of the divorce research supports Wallerstein's claims that divorce is "disruptive and disturbing" in the lives of children, but he fails to find the same strength and pervasiveness of the supposed effects. Using still other longitudinal data gathered as part of the Marital

As the preceding discussion illustrates, there are multiple viewpoints about whether, how, and for how long children are affected by a parental divorce. There is no simple singular message to draw from that research. Moreover, when in the midst of a deteriorating or failed marriage, emotions run high, making decisions that much harder to make. Put yourself into each of the following four scenarios:

1. You are married, in your fourth year of marriage, and find yourself unfulfilled by your marriage. There is no abuse, no infidelity that you know about, and no irresponsible behavior by your spouse to fuel your unhappiness, but things just don't feel right. You've tried counseling. You believe marriage ought to be happy and, increasingly, you have come to doubt that your marriage will ever make you feel that way. You and your spouse have no children. What, if anything, are you going to do?

2. Now, consider this exact same scenario but add a 3-year-old into your family. You love your child, as does your spouse. Neither you nor your spouse can imagine what it would be like—feel like—to live without your child in his/her day-to-day life. What are you going to do?

Now, let's tweak the situation and complicate it even more:

3. Let's imagine that you are six years into a marriage in which you and your spouse fight a lot, even over little things. Often the fights are very heated, bordering on but not yet reaching a point of violence between you. Although it has yet to happen, you have to admit that you can envision a situation where either or both of you could become violent. You feel like you have tried to make the marriage better, the relationship stronger, and yourself and your spouse happier. It hasn't happened. You have no children. What, if anything, are you going to do?

4. Add a 4½-year-old daughter and a 1-year-old son into this very same situation. You love your children, your spouse loves your children, your spouse is a good parent; it's just that the two of you cannot get along. The conflict has gotten more constant and is more frequently intense. What, if anything, would you do?

You may well have somewhat different answers for the four situations. How easily were you able to come to those answers? How easily do you think such answers are to make in real life?

Instability Over the Life Course Study, Amato reports that 90 percent of children with divorced parents achieve the same level of adult well-being as children of "continuously married parents" (Amato 2003). Amato further suggests that children who experience multiple family transitions (parental divorce, remarriages, subsequent divorces, and so on) are the ones who most suffer. He found that children who experienced only a single parental divorce (without any additional parental transitions) were no different in their psychological well-being than children of continuously married parents.

**False** **9.** Physical custody of children is given to the mother in the majority of divorces.

## Child Custody

Of all the issues surrounding separation and divorce, custody issues are particularly poignant because they represent continued versus strained or even severed ties between one parent and his or her children. Historically, until the mid-19th century, child custody tended to go to fathers. The pendulum then swung in the direction of granting mothers custody of children, unless they were considered unfit parents. Where mothers were typically awarded custody, fathers were granted visitation. Beginning then in the 1970s, custody decisions were made on a more gender-neutral basis, taking into account the "best interests of the child" (Elrod and Dale 2008).

Despite the more recent trend of gender-neutral custody considerations and assignments, physical custody of the children is given to the mother in over 80 percent of the cases. This is unsurprising. After all, women and men both often believe that children will be better off living with their mother. Additionally, giving custody to the mother is in keeping with our cultural beliefs about intensive mothering and children's developmental needs. Furthermore, in past decades, there may well have been a judicial bias that favored mothers even when fathers sought custody with equal determination. This latter reason is much less applicable to present custody disputes and decisions.

Along with gender, sexual orientation has been a traditional basis for awarding custody (Baggett 1992; Beck and Heinzerling 1993). In the past, a parent's homosexuality has been sufficient grounds for denying custody, but courts are increasingly determining custody on the basis of parenting ability rather than sexual orientation. Interviews with children whose parents are gay or lesbian testify to the children's acceptance of their parents' orientation without negative consequences.

## Types of Custody

The major types of custody are sole, joint, and split. In **sole custody**, the child lives with one parent who has sole responsibility for physically raising the child and making all decisions regarding his or her upbringing. Until the 1970s, sole custody was the norm (Elrod and Dale 2008). In **split custody**, the children in the family are divided between the divorcing parents, with each parent receiving physical custody of at least one of the children. Split custody can have especially harsh effects on relationships between siblings. For the most part, courts are reluctant to use split custody without compelling reasons to do otherwise (Oliphant and ver Steegh 2007).

To keep both parents involved with children after a divorce, custody decisions moved toward shared or joint custody. If and when parents can cooperate and effectively co-parent, joint custody is the preferred form of custody arrangements in the United States today. There are two forms of **joint custody:** legal and physical. In **joint legal custody,** the children live primarily with one parent, but both parents share jointly in decisions about their children's education, religious training, and general upbringing. In **joint physical custody,** the children actually live with both parents, dividing time between the two households.

Even though joint custody is not a guarantee that the child's time will be evenly divided between parents, it does give children the chance for a more normal and realistic relationship with each parent (Arditti and Keith 1993). Joint physical custody, however, requires considerable energy from the parents in working out both the logistics of the arrangement and their feelings about each other. Joint physical custody is particularly problematic in instances where conflict is high or domestic violence has occurred (Elrod and Dale 2008).

Any custody arrangement has both benefits and drawbacks, and joint custody is no exception. It has the obvious appeal of keeping both parents engaged and involved with their children; that is at the heart of joint custody. It appears as though its effectiveness may be dependent on the kind and quality of relationships between ex-spouses. Parental conflict can cause suffering among children in joint custody arrangements. Furthermore, imposed joint custody, over the strong objections of one of the parents, may be more harmful to the children and their relationships with parents than would sole custody (Elrod and Dale 2008). Joint custody may force two parents to interact (*cooperate* is too benign a word) when they would rather never see each other again, and the resulting conflict and ill will may be detrimental to the children. Parental hostility may make joint custody the worst form of custody (Opie 1993).

It is often assumed that children in joint custody/shared parenting arrangements might experience more stress, moving as they do between parental households on a regular basis. Researchers in Sweden looked at national data on nearly 150,000 sixth- through ninth-graders and their experiences of such stress-related health problems as difficulty sleeping, loss of appetite, feelings of sadness or tension, and headaches. Children living in households with both parents had the fewest problems. However, children who lived with both of their separated parents—that is, were in joint custody arrangements—fared better than children who lived with a single parent (Oaklander 2015). Researcher Malin Bergstrom suggests that the everyday contact with both parents and the greater resources two parents can provide offer great benefit to children. Indeed, internationally and domestically, family and child development experts recommend, where possible, joint or shared custody. Many states in the United States are considering or have enacted legislation favoring shared custody (National Parents Organization 2015).

Parental satisfaction with court-imposed custody arrangements depends on many factors. These include how hostile the divorce was, whether the parent without physical custody perceives visitation as lengthy and frequent enough, and how close that parent feels to his or her children. In addition, the amount of support payments affects satisfaction. If parents feel that they are paying too much or were "cheated" in the property settlement, they are also likely to feel that the custody arrangements are unfair. Unfortunately, custodial satisfaction is not necessarily related to the best interests of the child.

## Noncustodial Parents

Much public attention has been directed at "deadbeat" parents who are depicted as absent, uncaring, and failing to honor and maintain their child support obligations. In fact, noncustodial parents demonstrate varying degrees of involvement. Noncustodial parent involvement exists on a continuum in terms of caregiving, decision making, and parent–child interaction, from highly involved to completely removed. Involvement also changes depending on whether the custodial family is a single-parent family or a stepfamily (Bray and Berger 1990).

Typically, one hears more about "deadbeat dads" than either deadbeat parents or "deadbeat moms." The label hardly captures the reality for most noncustodial fathers. Many noncustodial fathers suffer grievously from the disruption or disappearance of their father role following divorce. They feel depressed, anxious, and guilt ridden; they feel a lack of self-esteem (Arditti 1990). The change in status from full-time father to noncustodial parent leaves fathers bewildered about how they are to act; there are no norms for an involved noncustodial parent.

Some men act irresponsibly after a divorce, failing to pay child support and possibly becoming infrequent parts of their children's lives. This lack of norms makes it especially difficult if the relationship between the former spouses is bitter. Without adequate norms, fathers may become "Disneyland dads" who interact with their children only during weekends, when they provide treats such as movies and pizza, or they may become "disappearing dads," absenting themselves from all contact with their children. For many concerned noncustodial fathers, the question is simple but painful: "How can I be a father if I'm not a father anymore?"

Noncustodial parents often weigh the costs of continued involvement with their children, such as emotional pain and role confusion, against the benefits, such as emotional bonding. Those who perceive that they have a say in the decisions affecting their children and who are stably employed and thus have the means to contribute tend to stay involved and to pay support (Braver et al. 1993a, 1993b). However, a lack of sufficient income is one significant source of nonsupport, prompting social work professor Ronald Mincy and economist Elaine Sorensen to refer to them as "turnips" in recognition of the cliché that "you can't get blood from a turnip" (Mincy and Sorensen 1998).

Children may eventually have little contact with their nonresidential parents. This reduced contact seems to weaken the bonds of affection. Divorced fathers are less likely to consider their children sources of support in times of need (Amato 1994; Cooney 1994). Although perhaps better than Frank

Furstenberg and Christine Nord's (1985) claim of almost three decades ago, that "marital dissolution involves either a complete cessation of contact between the nonresidential parent and child or a relationship that is tantamount to a ritual form of parenthood," noncustodial parents certainly see their relationships suffer considerably.

Factors that affect visitation by noncustodial parents include geographic proximity, feelings about one's former spouse, and the remarriage of either parent. Noncustodial mothers in many ways behave similarly to noncustodial fathers, indulging their children during visitations and reducing involvement in children's daily lives. Research indicates that compared to noncustodial fathers, noncustodial mothers do speak more often with their children by phone, write them more letters, and have more frequent extended visits. However, they experience the same difficulty maintaining consistent visitation and, like noncustodial fathers, blame the other parent (Pollack and Mason 2004).

Of course, not all noncustodial parents behave the same way. J. Bart Stykes, a researcher at the National Center for Family and Marriage Research, looked at nationally representative data on visitation patterns of nonresident fathers. Using data from the 2006–2010 National Survey of Family Growth, he separated the

It is usually important for a child's postdivorce adjustment that he or she has continuing contact with the noncustodial parent. Noncustodial parents are involved with their children in varying degrees.

Steve Skjold/Alamy Stock Photo

fathers into one of three categories based on how often they saw their children: frequent visitation, moderate visitation, and infrequent visitation. *Frequent visitation* meant seeing one's children several times a week, a pattern displayed by 25 percent of the fathers. Almost a third of the men (31 percent) described what Stykes refers to as *moderate visitation,* and saw their children one to three times a month. Finally, *infrequent visitation* referred to having seen one's nonresident child(ren) "several times last year or less." Roughly two-fifths (44 percent) of the fathers fit this pattern (Stykes 2012). Stykes points out that there were notable differences among fathers based on race, education, and fathers' marital/relationship status. African American men were more likely than whites or Hispanics (36, 25, and 20 percent, respectively) to report frequent visitation. They were also as likely as white fathers (35 and 34 percent, respectively) to report infrequent visitation. As to education, frequent and infrequent visitation were affected by father's education; men with at least a bachelor's degree were most likely to report frequent visitation (39 percent) and least likely to report infrequent visitation (29 percent). Infrequent visitation was most commonly reported by men with less than a high school education (51 percent), and least likely to be reported by men with college degrees (29 percent). Fathers who remarried or were cohabiting with someone were more likely than single fathers to report infrequent visitation: More than half of remarried men (54 percent) and almost half (48 percent) of men in cohabiting relationships reported infrequent visitation, compared to a third of the men not in relationships. Less than a fifth of men who were remarried or living with someone (17 percent each) reported frequent visitation, compared to 37 percent of men not in relationships (Stykes 2012).

A recommendation about visitation worth considering is one made by social work professors Daniel Pollack and Susan Mason. They advocate **mandatory visitation**, suggesting that rather than *award* or *grant* visitation rights to noncustodial parents, we should treat visitation as an obligation of parents and an expectation to be enforced by authorities. If children gain socially, emotionally, and psychologically from continued contact with and involvement by both parents, then much could be gained by mandating contact and visitation, as long as there is sufficient cooperation between ex-spouses and an absence of high levels of ongoing conflict (Pollack and Mason 2004).

## Divorce Mediation

The courts are supposed to act in the best interests of the child, but they often victimize children by their emphasis on legal criteria rather than on the children's psychological well-being and emotional development (Schwartz 1994). There is increasing support for the idea that children are better served by those with psychological training than by those with legal backgrounds (Miller 1993). Growing concern about the effect of litigation on children's well-being has led to the development of divorce mediation as an alternative to legal proceedings.

**Divorce mediation** is the process in which a mediator attempts to assist divorcing couples in resolving personal, legal, and parenting issues in a cooperative manner. More than two-thirds of U.S. states offer or require mediation through the courts over such legal issues as custody and visitation. Mediators act as facilitators to help couples arrive at mutually agreed-on solutions. Although some mediators are attorneys by profession, they neither act as lawyers for nor give advice to either party in the divorce process. Mediators can be either private or court ordered. Mediators generally come from marriage counseling, family therapy, and social work backgrounds, although increasing numbers are coming from other backgrounds and are seeking training in divorce mediation (DeWitt 1994).

Mediation has many goals. A primary goal is to encourage divorcing parents to reduce anxiety about and to see shared parenting as a viable alternative. Their role is not to save the marriage but to see that couples exit the marriage with less conflict, feeling that their interests were represented. In that way, they also differ from attorneys who represent individual spouses in the process of dissolving their marriage. Mediators are supposed to be neutral parties whose job is to help couples arrive at a mutually agreeable arrangement on matters of property, support, children, and contact. Some mediators are also attorneys. They will play very different roles, however, depending on the capacity in which they are employed in any particular divorce.

Data on satisfaction indicate that those who use mediators as part of their divorce process have greater

**False** **10.** Although increasing numbers of couples consult mediators during the process of divorce, the goal of mediation is not to save the marriage.

levels of satisfaction than those who divorce through adversarial means. They also spend less to end their marriages because divorce mediation is less financially costly than divorce that relies on litigation (through lawyers) alone (www.divorceinfo.com/doesmediationwork.htm). Research suggests that mediation is associated with more postdivorce contact and communication between parents and children, as well as less conflict and more communication between former spouses (Emery, Sbarra, and Grover 2005).

When mediation is court mandated, topics are generally limited to custody and visitation issues. Divorcing parents often find mediation helpful for these issues. In contrast to court settings, mediation provides an informal setting to work out volatile issues. Both men and women report that mediation is more successful at validating their perceptions and feelings than is litigation. Some courts order parents to participate in seminars covering the children's experience of divorce as well as problem solving and building co-parent relationships. Karen Blaisure and Margie Geasler (2006) found that half of all court systems in the United States had divorcing parents attend such parent education courses. Programs vary in duration and substance as well as in who is used to provide instruction. The research that has been done on the success of such programs has been largely positive, finding such results as less conflict between former spouses, less psychological distress, and better family functioning and divorce adjustment (Amato 2010). Parents also report that these seminars help them become more aware of their children's reactions and give them more options for resolving child-related disputes.

## ■ What to Do about Divorce

As the previous pages have illustrated, divorcing is a painful process for those involved, and it leaves families and individuals changed forever. Most people will agree that we would be better off reducing the rate of divorce, but how can that goal be achieved? First, we must decide on the most important causes of the high divorce rates in the United States.

If we believe that divorce rates rose partly because we made it easier and more acceptable to divorce, should we restigmatize divorce? Make exiting a marriage more difficult? If divorce rates rose with the increasing economic independence of women, how can we reduce divorce? Do we need to encourage employed women to stay home? How, then, do their families

survive without their incomes (see Chapter 11)? If part of the explanation for rising divorce rates is in the increasing importance given to self-fulfillment and the decline of both familistic self-sacrifice and religious constraints, does reducing divorce require changing people's values? Finally, if increases in divorce result from the weakening of all but the emotional function of marriage and the reduction of the family's economic role, can *anything* be done about divorce?

Part of the dilemma has to do with how we perceive divorce. Is divorce the *problem,* or is it a *solution* to other problems? Do we want to impose restrictions on divorce that require people to remain in unfulfilling, possibly dangerous relationships? The societal reactions to reducing divorce have been largely of two kinds: cultural and legal. From a cultural perspective, some commentators bemoan the popular cultural attitudes about marriage (Popenoe 1993; Whitehead 1993, 1997). They suggest that we "dismantle the divorce culture" that we have constructed by more consistently championing and effectively demonstrating the benefits of stable, lifelong marriage. Instead of celebrating "family diversity" and glorifying single-parent households, they believe that we should consistently reiterate the idea that marriage is a lifelong commitment involving considerable sacrifice. If that means that we must "restigmatize divorce," then that is what we should do (Whitehead 1997).

The other emphasis has been a legal one. Believing that marriage was weakened and divorce increased by no-fault divorce legislation, some have argued that we make divorce *harder to obtain.* Some states have contemplated repealing no-fault divorce legislation or raising marriage ages. Some states have enacted a two-tiered system of marriage in which couples are allowed and encouraged to consider *covenant marriage*—marriage under laws that require couples to undergo premarital counseling, swear to the lifelong commitment of marriage, and promise to divorce only under extraordinary circumstances and only after seeking marriage counseling (see the Public Policies, Private Lives feature on covenant marriages in this chapter). Still relatively new, the covenant marriage system has appealed to both those who wish to reduce divorce and those who wish to establish a more traditional, even religious, understanding of marriage commitments.

The difficulty behind both cultural and legal efforts is that in attempting to make divorce harder or less attractive, they do little to make staying married easier. This, too, could be done. It might entail enacting some work–family policy initiatives to ease the stress and

**Covenant Marriage as a Response to Divorce**

In 1996, as a way of trying to strengthen marriage and reduce divorce rates, Louisiana became the first state in the United States to establish a two-tiered system of marriage.

Marrying couples could choose either a "standard marriage" or a covenant marriage (Hewlett and West 1998). Following Louisiana's lead, two other states have enacted their own covenant marriage legislation, and at least 20 others have legislation awaiting approval by their state legislatures (National Healthy Marriage Resource Center 2010). Regardless of the state in question or the status of pending legislation, covenant marriage usually consists of something close to the following, which is drawn from the Louisiana law:

> We do solemnly declare that marriage is a covenant between a man and a woman who agree to live together as husband and wife for so long as they both may live. We have chosen each other carefully and disclosed to one another everything which could adversely affect the decision to enter into this marriage.
>
> We have received premarital counseling on the nature, purposes, and responsibilities of marriage. We have read the Covenant Marriage Act, and we understand that a Covenant Marriage is for life. If we experience marital difficulties, we commit ourselves to take all reasonable efforts to preserve our marriage, including marital counseling.
>
> With full knowledge of what this commitment means, we do hereby declare that our marriage will be bound by Louisiana law on Covenant Marriages and we promise to love, honor, and care for one another as husband and wife for the rest of our lives.

This is supplemented by an affidavit by the parties that they have discussed with a religious representative or counselor their intent to enter a covenant marriage.

Included is their agreement to seek marital counseling in times of marital difficulties and their agreement to the grounds for terminating the marriage. As pointed out by family law attorney Mark Baer, covenant marriages are "essentially reinstating fault-based divorce" (Baer 2014), for which a couple can divorce from a covenant marriage: adultery, spousal commission of a felony, or physical or sexual abuse by a spouse of the other spouse or children of either spouse. Fault-based divorces cost more and potentially lead to greater suffering by children of the divorcing couple (Baer 2014).

We cannot say whether covenant marriage will "work" to reduce the prevalence of divorce. It may have no effect because the people who elect to enter such a marriage may already perceive marriage as a relationship to keep "till death do us part."

Research indicates that covenant marriage likely has and will have modest effects on divorce rates. Few couples appear to choose a covenant over a standard marriage. Of the more than 370,000 Louisiana marriages between 2000 and 2010, 1 percent were covenant marriages. In Arizona, even fewer marriages—0.025 percent—have been covenant marriages (National Healthy Marriage Resource Center 2010). Those who do opt for covenant marriages are more conservative and more religious and tend to see the covenant marriage as a public expression of their conviction of the sacred commitment that one makes in marrying. Thus, these same people would be less likely to divorce even in the absence of covenant marriage statutes. For those women and men who have concerns about inequalities in traditional marriage or who worry about women's rights in families, covenant marriage will be unappealing.

To do more than "preach to the choir"—appealing to those who already share the covenant marriage philosophy—will be more difficult for proponents of such reform. However, sociologists Steven Nock, Laura Sanchez, and James Wright (2008) argue that noncovenant marriages could benefit from the same kind of mandatory premarital counseling or education that covenant marriage requires. Furthermore, they suggested that if nothing else, there is a symbolic value to covenant marriage, independent of religious symbolism, in that it reflects a society-wide concern about the meaning and importance of marriage. •

strain two-earner households face. On the subject of financial resources, because we know that divorce hits hardest at lower- and working-class levels, bolstering the economic stability and security of low-income families might also lead to less divorce.

If we cannot reduce or eliminate divorce, we should at least do what we can to protect those who go through divorce, especially children (Coontz 1997; Furstenberg and Cherlin 1991). We should devote resources that will help custodial parents raise their children more effectively. This means, among other things, ensuring their access to quality child care when parents are at work, guaranteeing their receipt of financial obligations (such as child support and alimony) from their former spouses, and helping them avoid the devastating plunge into poverty. In addition, ex-spouses must be instructed in how to display more amicable relationships with each other and should be expected to do so. Because at least some effects of divorce are tied to the level of postdivorce conflict and adjustment, taking steps to reduce conflict and ensure more effective adjustment will benefit children and their parents. Early and aggressive intervention into the postdivorce family (such as teaching anger management or instructing fathers about the vital roles they can still play) constitutes such intervention (Coontz 1997; Furstenberg and Cherlin 1991).

There is no denying that separation or divorce are typically filled with pain for all involved—spouses, children, and even one's wider network of kin and friends. Furthermore, as we have seen, both the process and its outcomes are often different for husbands and wives and for parents and children. It is hoped that this chapter has increased your understanding of how much divorce there is, the multiple factors that have led to shifts in the divorce rate and that expose individuals to greater or lesser risk of divorce, and the different perspectives on what we can and should do about divorce. Keep in mind that as one family ends, new family forms emerge. These include new relationships and possibilities, new circumstances and responsibilities, and new families with unique relationships: the single parent or the stepfamily. These are the families that we explore in the next chapter.

## Summary

- Divorce is an integral part of the contemporary U.S. marriage system, which values individualism and emotional gratification.

- Among the statistics researchers use to measure divorce are the *ratio* of marriages to divorces, the *crude divorce rate* per 1,000 people in a population, the *refined divorce rate* per 1,000 marriages, and the *predictive rate* of the future likelihood of divorce within a cohort. The trend in divorce has been downward since the 1980s.

- Compared to other countries, the divorce rate in the United States is among the highest.

- A variety of societal, demographic, and life course factors can affect the likelihood of divorce.

- *No-fault divorce* revolutionized divorce by eliminating fault finding and the adversarial process and by treating husbands and wives as equals. An unintended consequence of no-fault divorce is the growing poverty of divorced women with children.

- *Uncoupling,* the process by which couples drift apart in predictable stages, is differently experienced by the initiator and his or her partner.

- Women frequently experience downward mobility after divorce. The economic effect on men is more mixed and depends on what proportion of the marital income they were responsible for before the divorce.

- *Child support* often goes unpaid despite a number of legal initiatives to increase compliance by parents who owe support. A major determinant of compliance is what percentage of the parent's income is expected in support.

- Although psychological distress, reduced self-esteem, less happiness, more isolation, and less satisfying sex lives are among noneconomic consequences of divorce, the consequences of divorce for some are more positive than negative. Remaining in an unhappy marriage reduces life satisfaction, mental and physical health, and self-esteem.

- Children are typically told about the divorce by mothers. Children's overall reactions are usually negative, though for some, news of the divorce may be experienced as relief.

- Children in the divorce process go through stages: Turmoil is greatest in the initial stage. By the restabilization stage, changes have been integrated into the children's lives.

- A significant factor affecting the responses of children to divorce is their age. Younger children tend to act out and blame themselves, whereas adolescents tend to remain aloof and angry at both parents for disrupting their lives.

- Many problems assumed to be caused by divorce are present before marital disruption. Although divorce has been said to put children in the middle of parental conflict, this seems to occur more in intact, high-conflict parental marriages.

- Longitudinal studies following children of divorce over decades have shown different conclusions about how bad the long-term consequences of divorce are and how long they last.

- Custody is generally based on one of two standards: the best interests of the child or the least detrimental of the available alternatives. *Physical custody* is generally awarded to the mother. *Joint custody* has become more popular because men are becoming increasingly involved in parenting.

- Noncustodial parent involvement exists on a continuum from absent to intimately and regularly involved.

- *Divorce mediation* is a process in which a mediator attempts to assist divorcing couples in resolving personal, legal, and parenting issues in a cooperative manner.

- Recent legislative initiatives such as covenant marriage are attempts to reduce the divorce rate by strengthening the marriage commitment.

## Key Terms

# 14

# New Beginnings: Single-Parent Families, Remarriages, and Blended Families

*Are the following statements True or False? You may be surprised by the answers as you read this chapter.*

**T** **F** **1.** It is expected that in their lifetimes a majority of Americans will be connected with a stepfamily.

**T** **F** **2.** Single-parent families today are as likely to be headed by a father as by a mother.

**T** **F** **3.** Children tend to have greater power in single-parent families than in traditional nuclear families.

**T** **F** **4.** In most married couples, both spouses are in their first marriages.

**T** **F** **5.** Women are more likely than men are to remarry.

**T** **F** **6.** Second marriages are more likely than first marriages to end in divorce.

**T** **F** **7.** Becoming a stepfamily is a process that can take years to fully complete.

**T** **F** **8.** Stepmothers generally experience less stress in stepfamilies than stepfathers because stepmothers can find fulfillment in nurturing their stepchildren.

**T** **F** **9.** Stepfathers tend to build relationships with stepchildren that are as warm and affectionate as those between biological fathers and children.

**T** **F** **10.** Compared to children in single-parent families, children in stepfamilies are less likely to have emotional and psychological problems.

Blend Images/Getty Images

When Paige was 6 and Daniel was 8, their parents separated and divorced. The children continued to live with their mother, Sophia, in a single-parent household while spending weekends and holidays with their father, David. After a year, David began living with Jane, a single mother who had a 5-year-old daughter, Lisa. Three years after the divorce, Sophia married John, who had joint physical custody of his two daughters, Sally and Mary, aged 7 and 9. Some eight years after their parents divorced, Paige and Daniel's family included two biological parents, two stepparents, three stepsisters, one stepbrother, and two half brothers. In addition, they had assorted grandparents, step-grandparents, biological and step aunts, uncles, and cousins.

Today's families mark a definitive shift from the traditional family system, based on lifetime marriage and the intact nuclear family, to a pluralistic family system, including families created by divorce, remarriage, and births to single women. This new pluralistic family system consists of three major types of families: (1) intact nuclear families, (2) single-parent families (either never married or formerly married), and (3) stepfamilies. As we discussed in Chapter 9, cohabiting couples who may or may not have children are also part of this wider family system.

**Single-parent families** are families consisting of one parent and one or more children; the parent can be divorced, widowed, or never married. **Stepfamilies** are families in which one or both partners have children from a previous marriage or relationship. Stepfamilies are sometimes referred to as *blended families*.

Although nearly three-fifths (58 percent) of recent marriages are first marriages for both spouses, over 40 percent involved one or both spouses remarrying. In fact, a fifth (21 percent) involve both spouses marrying for at least the second time (Lewis and Kreider 2015). More than 60 percent (63 percent) of remarriages include stepchildren (Stykes and Guzzo 2015). If one considers all possible types of stepfamilies, it is expected that a majority of Americans are or will be connected with a stepfamily (Stewart 2007).

To better understand the world that children such as Paige and David live in, one needs to examine some major patterns in our continually evolving pluralistic family system. This chapter examines single-parent families, binuclear families, remarriage, and stepfamilies. Given the wide diversity that characterizes families today, researchers can no longer consider such families to be unusual family forms (Stewart 2007). Shifting attention from family structure or form to function, the important question becomes whether a specific family—regardless of whether it is a two-parent nuclear family, a single-parent family, or a stepfamily— succeeds in performing its functions. In a practical sense, as long as a family is fulfilling its functions and meeting its needs, it is one kind of normal family. This chapter considers these versions of normal families.

## Single-Parent Families

In the United States, as throughout the world, single-parent families have increased and continue to grow in number. Although no other family type has increased in number as rapidly, single-parent families are not always accurately or adequately understood. They may still be treated negatively in the popular imagination, considered to be "broken homes" as opposed to "intact families." Single-parent families created by unmarried motherhood, especially those that are headed by teenagers or young adults, are stereotyped as well, with the image being one of young women who casually bear children "out of wedlock." These images are clearly inadequate, based on ideas and stereotypes that misdirect us from a more accurate understanding. The "broken home" image is based on the ideal of the "happy" intact family; the assumed irresponsibility of single mothers is based on misunderstandings and

**True**
**1.** It is expected that in their lifetimes a majority of Americans will be connected with a stepfamily.

moralism, occasionally mixed with racism, condemning women for bearing children outside of marriage; and the "promiscuous teenage mother" stereotype ignores the reality that most births to single mothers are to women older than 20. Finally, although roughly 85 percent of single, custodial parents are female, these images overlook the situations and experiences of single fathers.

In the four-plus decades from 1970 to 2014, the percentage of children living in single-parent families doubled, increasing from 13 percent to nearly 28 percent. More than 20 million children lived with an unmarried parent in 2014 (U.S. Census Bureau 2014a).

In previous generations, the life pattern most women expected and experienced was (1) marriage, (2) motherhood, and (3) widowhood. Single-parent families existed in the past, but they were typically the result of widowhood more than either divorce or births to unmarried women. Significant numbers were headed by men. But a new marriage and family pattern later emerged. Its greatest effect has been on women and their children. Divorce and births to unmarried mothers are the key factors creating today's single-parent family.

The life pattern many married women today experience is (1) marriage, (2) motherhood, (3) divorce, (4) single parenting, (5) remarriage, and (6) widowhood. For those who are not married at the time of their child's birth, the pattern may be (1) dating or cohabitation, (2) motherhood, (3) single parenting with the later possibility but no certainty of (4) marriage, and (5) widowhood. Finally, some who marry, divorce, and remarry may experience subsequent divorces and or remarriages; they embody the characteristics that make up *serial monogamy*—having a number of monogamous marriages across one's life.

## Characteristics of Single-Parent Families

Single-parent families share a number of characteristics, including the following: They are created by either divorce, births to unmarried mothers, or the death of a spouse. They are usually female headed. They are characterized by a diversity of living arrangements. Some single-parent families are created intentionally through planned pregnancy, artificial insemination, and/or adoption. Others are headed by lesbians and gay men. Many single-parent households contain two cohabiting adults and are therefore not *single-adult* households, as many might assume them to be (Fields 2003). The prevalence and experiences of

single parents vary across racial or ethnic lines. Finally, although there is considerable economic diversity among single-parent families, they are often economically disadvantaged, perhaps even impoverished.

### Creation by Divorce or Births to Unmarried Women

Single-parent families today are usually created by marital separation, divorce, or births to unmarried women rather than by widowhood. Throughout the world, including the United States, single-parent families created through births to unmarried women have increased at a higher rate than single-parent families created through divorce (Gonzalez 2005). In 2014, just over 40 percent of U.S. births were to unmarried women (Hamilton et al. 2015). The number of children living with an unmarried couple more than tripled between 1980 and 2014. In 2014,  20 million children under age 18 lived in 11.8 million families headed by either the mother only or the father only (U.S. Census Bureau 2015a, Table C9).

In 2014, 51 percent of all single mothers were previously married. This is even more the case among white single mothers, two-thirds of whom were divorced or separated. In contrast, almost half of Hispanic single mothers and a third of African American single mothers were previously married (Mather 2010).

In comparison to the much less common single parenting by widows, single parenting by divorced or never-married mothers has tended to receive considerably less social support. Widowed mothers often receive social support from their husband's relatives. A divorced mother may receive some assistance from her own kin but typically receives considerably less (or none) from her former partner's relatives. Many in the wider society still display ambivalence about divorce and consider divorce-induced, single-parent families as less certain environments for protecting children's well-being.

### Headed by Mothers (and Sometimes Fathers)

More than 80 percent of single-parent families are headed by women (Kreider and Ellis 2011b). This has important economic ramifications because of gender discrimination in wages and job opportunities, as discussed in Chapter 11. Still, more than 2.8 million children live in a single-father household (U.S. Census Bureau 2014a). Like women, men take a range of paths to single parenthood. They are either divorced or separated from their children's mothers, widowed,

Although there is a tendency to assume that single-parent household means single-mother household, many single fathers are raising children.

Antenna/fStop Images GmbH/Alamy Stock Photo

or they raise children from relationships in which they were never married. They are less likely to have never been married than are single mothers. In 2015, 61.6 percent of single-father families were the result of divorce or marital separation. Among single mother–headed families, 47.5 percent were divorced or separated, while 49.1 percent were never married (U.S. Census Bureau 2015a, Table C3). Single-father families are also less likely than single-mother families to be poor (Livingston 2013).

**False** **2.** More than 80 percent of single-parent families are headed by women.

## Significance of Ethnicity

As can be seen in Figure 14.1, ethnicity remains an important demographic factor in single-parent families. In 2014, among non-Hispanic white children, 19.8 percent lived in single-parent families; among African American children, 54.9 percent lived in such families; among Hispanics, 30.7 percent lived in single-parent families; and among Asian American children, 11.2 percent lived in such households (U.S. Census Bureau 2015a, Table C9). White single mothers were more likely

to be divorced than their African American or Latino counterparts, who were more likely to be unmarried at the time of the birth, or widowed.

## Poverty

Married women tend to experience a drop—frequently steep—in their income when they separate or divorce (as discussed in Chapter 13). Among unmarried single mothers, poverty and motherhood often go hand in hand. In 2014, 39.8 percent of single mother–headed families with children under 18 were living below the poverty level, and almost 60 percent of poor children lived in female-headed families.

Because they are women, because they are often young, and because they are often from ethnic minorities, single mothers often have few financial resources. They may find themselves under constant economic stress in trying to make ends meet, working for low wages, experiencing unemployment, or both. They are unable to plan because of their constant financial uncertainty. They move more often than two-parent families as economic and living situations change, uprooting themselves and their children. They accept material support from kin but often at the price of receiving unsolicited "free advice," especially from their mothers.

Looking specifically at families with children under age 18, both mother-only and father-only families are more likely to be poor than are two-parent families (see Figure 14.2). Where nearly 40 percent of single-mother families with children under 18 were in

**Figure 14.1 Percentage of U.S. Children Living in Single-Parent Households, by Race and Ethnicity, 2014**

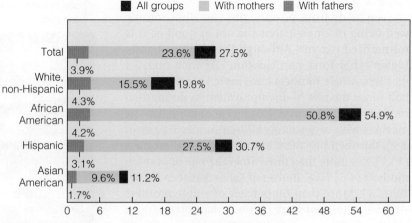

■ All groups ■ With mothers ■ With fathers

| | |
|---|---|
| Total | 23.6%  27.5%  3.9% |
| White, non-Hispanic | 15.5%  19.8%  4.3% |
| African American | 50.8%  54.9%  4.2% |
| Hispanic | 27.5%  30.7%  3.1% |
| Asian American | 9.6%  11.2%  1.7% |

SOURCE: U.S. Census Bureau, Families and Living Arrangements (2015a), Table C9.

**Figure 14.2** Percentage of People in Families with Related Children under 18 in Poverty, by Family Structure and Race, 2010

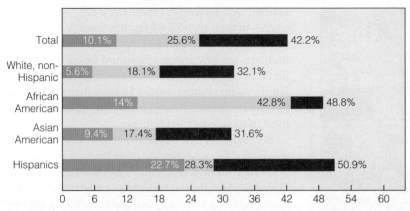

■ Single, female-headed households
▨ Single, male-headed households
▨ Married-couple households

| | | | |
|---|---|---|---|
| Total | 10.1% | 25.6% | 42.2% |
| White, non-Hispanic | 5.6% | 18.1% | 32.1% |
| African American | 14% | 42.8% | 48.8% |
| Asian American | 9.4% | 17.4% | 31.6% |
| Hispanics | 22.7% | 28.3% | 50.9% |

0  6  12  18  24  30  36  42  48  54  60

SOURCE: U.S. Census Bureau, Current Population Survey (2011b), Annual Social and Economic Supplement, Table POV03.

poverty in 2014, among male-headed single-parent families with children, 22 percent were in poverty, as were 8.2 percent of married couple–headed families with children (U.S. Census Bureau 2015c, Historical Poverty Tables, Table 4). Overall, in 2014, 6.2 percent of married couple–headed families lived in poverty compared to 15.7 percent of male-headed families and 30.6 percent of female-headed families (DeNavas-Walt and Proctor 2015).

Clearly, the association between single parenthood and poverty is greater for mothers than for fathers (Zhan and Pandey 2004). Among fathers, single fathers are less well-off compared to married fathers. They tend to be younger, less educated, less likely to have jobs, and more likely to receive public assistance and to live in poverty.

Even aside from poverty, the overall economic well-being of single parents is not as good as it is for married parents. Although they have a slightly higher labor force participation rate than married mothers, single mothers have more than two and a half times the rate of unemployment as do married mothers (10.3 percent to 4.0 percent in 2014. Among mothers with very young children (under 3 years of age), the unemployment rate for unmarried mothers (15.1) was more than three times the rate of married mothers (4.5) (U.S. Bureau of Labor Statistics 2015d, Table 4). More than 60 percent of single mothers who are employed work in lower-wage occupations (Mather 2010).

If one considers those in *low-income* households (defined as having household incomes under 200 percent of the poverty threshold), children in single-mother households are the most likely to be either poor or low income. If one looks at extreme poverty (defined as incomes less than half of the poverty threshold), 52 percent of poor single-mothers are in extreme poverty (singlemotherguide.com/single-mother-statistics/). Compared to all single mothers, low-income single mothers tend to be younger, have less education, and a greater likelihood of being unemployed than higher-income single mothers. Among single mothers, never-married single mothers are more economically disadvantaged than divorced single mothers (Mather 2010).

## Diversity of Living Arrangements

There are many different kinds of single-parent households. Although 27.5 percent of all children under age 18 live in families headed by a single parent, 23.6 percent live in mother-headed households, and only 3.9 percent live in households headed by an unmarried father (U.S. Census Bureau 2014a). Single-parent families need greater flexibility in managing child care and housing with limited resources.

The economic well-being of single parents, especially single mothers, is much less than that of two-parent families.

Aleph Studio/Shutterstock.com

In doing so, they rely on a greater variety of household arrangements than is suggested by the umbrella heading "single-parent household." For example, many young African American mothers live with their own mothers in a three-generation setting.

It is worth pointing out that, in fact, "single-parent households" may actually contain the parent and his or, more often, her unmarried partner. In 2014, for example, 22 percent of children living with single fathers and 11 percent of children living with single mothers were living with their parent and their parent's cohabiting partner (www.childstats.gov/americaschildren/family1.asp). Thus, though unmarried parents headed such households, those parents were not necessarily alone in the tasks of child rearing (Fields 2004; Kreider and Ellis 2011b).

Even in the absence of parents' *live-in* partners, parents' romantic partners may play important roles in their children's lives. For example, many children of single mothers and nonresidential biological fathers have a **social father**—a male relative, family associate, or mother's partner—"who demonstrates parental behaviors and is like a father to the child" (Jayakody and Kalil 2002).

Along these same lines, single parents, especially mothers, often rely on a combination of state or federal assistance and **private safety nets**: support from their social networks on which they can fall back in times of economic need (Hamer and Marchioro 2002; Harknett 2006). Social support, whether from family or friends, can lead to enhanced well-being and self-esteem among economically disadvantaged single mothers. These, in return, may lead to more effective parenting, even under difficult and highly stressful conditions. Without such support, mothers raising children on their own in economically distressed, potentially dangerous, urban neighborhoods are more likely to experience psychological distress, which then negatively affects their parenting behavior (Kotchick, Dorsey, and Heller 2005).

## Transitional Form

Single parenting is usually a transitional state. Though less so today, a single mother has strong motivation to marry or remarry because of cultural expectations, economic stress, role overload, and a need for emotional security and intimacy. The increasing presence of social fathers, including mothers' live-in romantic partners, may be part of the reason low-income families increasingly cohabit rather than marry. The presence of such men can reduce the various pushes toward marriage or remarriage (Jayakody and Kalil 2002).

## Intentional Single-Parent Families

For many single women in their 30s and 40s, single parenting has become a more accepted, intentional, and less transitional lifestyle (Seltzer 2000). Some older women choose unmarried single parenting because they have not found a suitable partner and are concerned about declining fertility. They may plan their pregnancies or choose donor insemination or adoption. If their pregnancies are unplanned, they decide to bear and rear the child. Others choose single parenting because they do not want their lives and careers encumbered by the compromises necessary in marriage. Still others choose it because they don't want a husband but do want a child.

## Lesbian and Gay Single Parents

Although estimates from the research literature were provided in Chapter 10, it is in fact nearly impossible to confidently assert how many children in the United States are being raised by gay or lesbian couples. Some of these parents were married either before they were aware of their sexual orientation or to "fit in" with social or familial expectations. Such men and women become single parents as a result of divorce. Others were always aware of being lesbian or gay; they chose adoption or donor insemination to have children. The long-standing inability to marry also forced many lesbian or gay parents to live as unmarried parents, with or without co-parenting partners. Many states also imposed restrictions on adoption by lesbians or gay men (as of 2016 only Mississippi retains such restriction, and it is under court challenge) or second-parent adoption by a gay or lesbian parent's partner, and again the result is an inflated number of unmarried parents among the LGB population.

The American Academy of Pediatrics supports "statutory and legal means to enable children to be adopted by the second parent or co-parent in families headed by gay and lesbian couples." About half of the U.S. states either explicitly allow or have approved petitions for second-parent adoptions. Second-parent adoption allows each partner in a gay or lesbian couple to adopt a child without the other partner/parent losing parental rights. Second-parent adoptions allow both partners to claim parental rights and allow an adopted child of a gay or lesbian couple the right to have two legally recognized parents (Human Rights Campaign 2012).

Limited marriage and adoption rights make same-gender families with children into legal single-parent families.

## Children in Single-Parent Families

Children born outside of marriage tend to suffer economic disadvantages that may then lead to other educational, social, and behavioral outcomes. Their disadvantages tend to be worse than those experienced by children of divorced parents or by children in two-parent, married households (Seltzer 2000). They are more likely to engage in high-risk, "health-compromising" behaviors, such as cigarette smoking, drug and alcohol use, and unprotected sex; are less likely to graduate from high school and college; are more likely to have a child outside of marriage and/or during their teens; are more likely to be "idle" (out of school and out of work), have lower earnings, and suffer lower levels of psychological well-being; and are more vulnerable to divorce and marital instability as adults (King, Harris, and Heard 2004).

Research examining the effects divorced, single-parent households have on children often points to some negative outcomes in areas such as behavioral problems, academic performance, social and psychological adjustment, and health. The gaps between children in such households, usually headed by single mothers, and those whose parents remain continuously married are relatively small but consistent. As Paul Amato (2000) reported, especially when exposed to a series of associated negative life events such as having to move or change schools, the effects of living in a divorced, single-parent home can create particular adjustment difficulties.

> **True**
> **3.** Children tend to have greater power in single-parent families than in traditional nuclear families.

The consequences appear to be linked to the lack of economic resources but also to the reduced money, attention, guidance, and social connections—what researchers call **social capital**—that two parents provide.

### Parental Stability and Loneliness

After a divorce, single parents are usually glad to have the children with them. Everything else may seem to have fallen apart, but as long as divorced parents have their children, they retain their parental function. Their children's need for them reassures them of their own importance. A custodial parent's success as a parent becomes even more important to counteract the feelings of low self-esteem that result from divorce.

Feeling depressed, single parents know that they must bounce back for the children. Yet after a short period, they come to realize that their children do not fill the void left by their former spouse or partner. As much pleasure as children bring, raising them is also difficult, and custodial single parents may resent being constantly tied down by their children's needs. Thus, minor incidents with the children—a child's refusal to eat or a temper tantrum—may be blown out of proportion. A major disappointment for many new single parents is the discovery that they are still lonely. It seems almost paradoxical. How can a person be lonely amid all the noise and bustle that accompany children? However, children do not ordinarily function as attachment figures; they can never be potential partners. Any attempt to make them so is harmful to both parent and child. Yet children remain the central figures in the lives of single parents. This situation leads to a second paradox: Although children do not completely fulfill a person, they rank higher in most single mothers' priorities than anything else.

### Changed Family Structure

A single-parent family is not the same as a two-parent family with one parent temporarily absent. The permanent absence of one parent dramatically changes the way in which the parenting adult relates to the children. Generally, the mother becomes closer and more responsive to her children. Her authority role changes, too.

A greater distinction between parents and children exists in two-parent homes. Both mothers and fathers develop rules. Parents generally have an implicit understanding to back each other up in child-rearing matters and to enforce mutually agreed-on rules. In a single-parent family, no other partner is available to help maintain such agreements; as a result, the children may find themselves in a more egalitarian situation with more power to negotiate rules. They can be more stubborn, cry more often and louder, whine, pout, and throw temper tantrums. Any parent who has tried to convince children to do something they do not want to do knows how soon an adult can be worn down.

In Chapter 10, we saw the mixed effects that parenthood has been shown to have on parents' well-being. The situation is more difficult for single parents than married or coupled parents. Compared with married parents, unmarried parents, especially unmarried mothers, are at increased risk of suffering higher levels of distress, and much of the distress results from elevated levels of stress to which they are exposed. Single mothers face greater financial and caregiving stress as well as greater difficulty balancing employment and family life (Umberson, Pudrovska, and Reczek 2010). The distress that single custodial parents experience can affect the tone of the whole household. For example, single mothers with higher levels of life stresses and less time for themselves are more likely to be anxious and may transmit their anxiety to their children. Repeated experiences of transmitted anxiety from mother to child can lead to chronic distress in children (Larson and Gillman 1999). Parental depression, especially among custodial mothers, can affect their abilities to parent effectively and thus exposes their children to more "adjustment problems" (Amato 2000). Without a second parent in the household, there may be no way to buffer children from the stresses and hardships that their mothers are experiencing.

At the same time, there may be benefits or advantages to being a child of a single parent. Children in single-parent homes may learn more responsibility, spend more time talking with their custodial parent, and face less pressure to conform to more traditional gender roles (Coontz 1997). They may learn to help with kitchen chores, to clean up their messes, or to be more considerate. In the single-parent setting, children are encouraged to recognize the work their mother does, assist in household chores, and understand the importance of cooperation (Stewart 2007).

Although most single parents continue to demonstrate love and creativity in the face of adversity, research on their children reveals some negative long-term consequences. In adolescence and young adulthood, children from single-parent families had fewer years of education and were more likely to drop out of high school. They had lower earnings and were more likely to be poor. They were more likely to initiate sex earlier, become pregnant in their teens, and cohabitate but not marry earlier (Furstenberg and Teitler 1994). Furthermore, they were more likely to divorce. These conclusions are consistent for Caucasians, African Americans, Latinos, and Asian Americans. The reviewers note that socioeconomic status accounts for much but not all of the effects. Some effects are attributed to family structure.

Harriette Pipes McAdoo (1988, 1996, 1998) attributed the cause to poverty, not to single parenthood. She also noted that African American families are able to meet their children's needs in a variety of structures. "The major problem arising from female-headed families is poverty," she asserted (McAdoo 1988). "The impoverishment of black families has been more detrimental than the actual structural arrangement."

## Successful Single Parenting

Single parenting is difficult, but the problems are manageable for many single parents. Almost two-thirds of divorced single parents found that single parenting grows easier over time (Richards and Schmiege 1993). Thus, it is important to note that many of the characteristics of successful single parents and their families are shared by all successful families.

### Characteristics of Successful Single Parents

In-depth interviews with successful single parents found certain themes running through their lives:

- *Acceptance of responsibilities and challenges of single parenthood.* Successful single parents saw themselves as primarily responsible for their families; they were determined to do the best they could under varying circumstances.

- *Parenting as first priority.* In balancing family and work roles, their parenting role ranked highest. Romantic relationships were balanced with family needs.

- *Consistent, nonpunitive discipline.* Successful single parents realized that their children's development required discipline. They adopted an authoritative style of discipline that respected their children and helped them develop autonomy.

- *Emphasis on open communication.* They valued and encouraged expression of their children's feelings and ideas. Parents similarly expressed their feelings.

- *Fostering individuality supported by the family.* Children were encouraged to develop their own interests and goals; the family valued differences.

- *Recognition of the need for self-nurturance.* Successful single parents realized that they needed time for themselves. They needed to maintain an independent self that they achieved through other activities, such as dating, music, dancing, reading, classes, and trips.

- *Dedication to rituals and traditions.* Successful single parents maintained or developed family rituals and traditions, such as bedtime stories, family prayer or meditation, sit-down family dinners at least once a week, picnics on Sundays, visits to Grandma's, or watching television or going for walks together. (Olson and Haynes 1993)

### Single-Parent Family Strengths

Although most studies use a **deficit approach** emphasizing the stresses and difficulties single parents and their children face, some studies note the potential that single parents have to build strength and confidence. This may be especially true for women (Amato 2000; Coontz 1997).

Qualitative interview research has uncovered ways in which both single mothers and single fathers display and develop strengths that enable them to effectively and successfully raise their children. For example, family researchers Leslie Richards and Cynthia Schmiege (1993) conducted a study of 60 white single mothers and 11 white single fathers (most of whom were divorced) that identified five family strengths associated with successful single parenting:

- *Parenting skills.* Successful single parents have the ability to take on both expressive and instrumental roles and traits. Single mothers may teach their children household repairs or car maintenance; single fathers may become more expressive and involved in their children's daily lives.

- *Personal growth.* Developing a positive attitude toward the changes that have taken place in their lives helps single parents, as does feeling success and pride in overcoming obstacles.

- *Communication.* Through good communication, single parents can develop trust and a sense of honesty with their children as well as an ability to convey their ideas and feelings clearly to their children and friends.

- *Family management.* Successful single parents develop the ability to coordinate family, school, and work activities and to schedule meals, appointments, family time, and alone time.

- *Financial support.* Developing the ability to become financially self-supporting and independent is important to single parents.

Among the single parents in the study, more than 60 percent identified parenting skills as one of their family strengths. In addition, 40 percent identified family management as a strength in their families (Richards and Schmiege 1993). About 25 percent identified personal growth and communication among their family strengths.

Sociologist Barbara Risman's (1986) research on custodial single fathers showed their abilities to be attentive, nurturing caregivers to their children. Rather than relying on paid help or female social supports, men became the nurturers in their children's lives. They were involved in their personal, social, and academic lives and saw to it that their emotional and physical needs were met. To Risman, they affirmatively answer the question in her article title, "Can Men Mother?"

## ◼ Binuclear Families

One of the most complex and ambiguous relationships in contemporary United States is what some researchers call the *binuclear family*—a postdivorce family system with children (Ahrons and Rodgers 1987; Ganong and Coleman 1994). It is the original nuclear family divided in two. The binuclear family consists of two nuclear families—the maternal nuclear family headed by the mother and the paternal one headed by the father. Both single-parent families and stepfamilies are forms of binuclear families.

Divorce ends a marriage but not a family. It dissolves the husband–wife relationship but not necessarily the father–mother, mother–child, or father–child relationship. The family reorganizes itself into a binuclear family. In this new family, an ex-husband and ex-wife may continue to relate to each other and to their children, although in substantially altered ways. The significance of the maternal and paternal components of the binuclear family varies. In families with joint physical custody, the maternal and paternal families may be equally important to the children. In single-parent families, especially those headed by women, the former spouse's family component may be minimal.

## Subsystems of the Binuclear Family

To clarify the different relationships, researchers Constance Ahrons and Roy Rodgers (1987) divided the binuclear family into five subsystems: former spouse, remarried couple, parent–child, sibling (step-siblings and half siblings), and mother/stepmother–father/stepfather.

### Former Spouse Subsystem

Divorce may sever the marital relationship between two people, but it doesn't bring an end to their parenting responsibilities. Thus, their relationship as co-parents endures. As noted in the previous chapter, children tend to benefit when they maintain relationships with both parents. However, achieving this requires that former spouses work through the following:

- Anger and hostility they may feel toward each other as a consequence of their previous marriage, the separation, and whatever postdivorce arrangements have been reached regarding such matters as custody, child support, visitation, and spousal support
- Conflict over different parenting styles, values, and aspirations concerning the children
- Shifting roles and relationships between former spouses when one or both remarries
- The incorporating of others as stepparents, step-siblings and step-grandparents as one spouse or both former spouses remarry

In essence, former spouses must separate their marital and personal issues from their mutual desire to raise their children effectively.

### Remarried Couple Subsystems

As we shall soon see, remarried life is complicated by the need to manage such issues as the exchange of money, sharing decision-making power, time, and

children with former spouses. Those with physical custody must provide their former spouse with access to the children. As a consequence of these complexities, the remarried couple experiences many typical marital issues that become more complicated just by the presence or involvement of the former spouse.

### Parent–Child Subsystem

Such matters force both biological and stepparents to make adjustments. Former single parents must now make room for the presence and involvement of a second parent in decision making and child-rearing practices. Stepparents who assume that their role will be similar to the parent role have the greatest and hardest adjustment to make. Stepmothers tend to experience greater stress than stepfathers.

### Sibling, Step-Sibling, and Half-Sibling Subsystem

A parent's remarriage may introduce "instant" sisters or brothers into one's life. They likely differ, perhaps considerably, in their temperaments. They must compete and contend with and accommodate each other's needs for parental affection, attention, space, and so on. Sharing one's parent with new step-siblings may be even harder than sharing one's parent with a new stepparent. Visiting biological children compete with stepchildren who are now living with the visiting children's biological parent. This may affect older children who are out of the house (or on their way) as well as children who must adapt to the feeling of being out of place in their own parent's home.

### Mother/Stepmother–Father/Stepfather Subsystems

The relationship between new spouses and former spouses often influences the remarried family. The former spouse can be an intruder in the new marriage and a source of conflict between the remarried couple. Other times, the former spouse is a handy scapegoat for displacing problems. Much of current spouse–former spouse interaction depends on how the ex-spouses feel about each other.

## Recoupling: Relationship Development in Repartnering

Certain norms governing relationship development before first marriage are fairly well understood. As the relationship progresses and intensifies, individuals spend more time together; at the same time, their family and friends limit time and energy demands because "they're in love." Norms for second and subsequent marriages or cohabiting relationships, however, are not so clear (Ganong and Coleman 1994; Rodgers and Conrad 1986).

For example, when is it acceptable for formerly married (and presumably sexually experienced) men and women to become sexually involved? What type of commitment validates "premarital" sex among postmarital men and women? When should a parent's new partner be introduced to his or her children? How long should couples wait before making a commitment to marriage or a decision to move in together? Without clear norms, the progression of relationships following divorce can be plagued by uncertainty about what to expect.

Remarriage courtships tend to be short unless preceded by cohabitation. Research on how postdivorce cohabitation affects the timing of remarriage shows that postdivorce cohabitation tends to lead to a longer waiting time until remarriage than is experienced by those who don't cohabit before remarrying (Xu, Hudspeth, and Bartkowski 2006).

As noted earlier, some divorced individuals re-enter marriage within a year of their divorces. This may indicate, however, that they knew their future partners before they were divorced. If neither partner has children, progression into remarriage may resemble what couples experienced before the first marriage, with one major exception: The memory of the earlier marriage exists as a model for the second marriage. The discussion and consideration of remarrying (or cohabiting) may trigger old fears, regrets, habits of relating, wounds, or doubts. At the same time, having experienced the day-to-day living of marriage, the partners may have more realistic expectations. Their courtship may be complicated if one or both are noncustodial parents. In that event, continued visitation with their children presents an additional element that not only further complicates courtship but also elevates the risk of a later divorce (Sweeney 2010).

## Cohabitation

Increases in the rates of cohabitation in the United States include many divorced women and men who cohabit before or instead of remarrying. As great an increase as has occurred in premarital cohabitation, *postdivorce* cohabitation has also greatly

increased (Xu, Hudspeth, and Bartkowski 2006). Thus, although remarriage rates have declined in recent years, "recoupling" through cohabitation remains common (Coleman, Ganong, and Fine 2000). In fact, cohabitation might be considered "the primary way people prepare for remarriage" (Ganong and Coleman 1994). However, couples who lived together before remarriage, like those who cohabited prior to engagement in first marriages, face an elevated risk of divorce. With postmarital cohabitation prior to remarriage, the effect of cohabitation on remarriage success did not depend upon the timing relative to getting engaged (Stanley et al. 2010).

## Repartnering and Children

The process leading up to a remarriage differs considerably from that preceding a first marriage if one or both members in the dating relationship are custodial parents. Single parents are not often a part of the singles world because such participation requires leisure time and money, which single parents generally lack. Children rapidly consume both of these resources.

Although single parents may wish to find a new partner, their children usually remain the central figures in their lives. This creates a number of new problems. First, the single parent's decision to go out at night may lead to guilt feelings about the children. If a single mother works and her children are in day care, for example, should she go out in the evening or stay at home with them? Second, a single parent must look at a potential partner as a potential parent. A person may be a good companion and listener and be fun to be with, but if he or she does not want to assume parental responsibilities, the relationship will often stagnate or be broken off. A single parent's new companion may be interested in assuming parental responsibilities, but the children may regard him or her as an intruder and try to sabotage the new relationship.

A single parent may also have to decide whether to permit a romantic partner to spend the night when children are in the home. This is often an important symbolic act. For one thing, it brings the children into the parent's new relationship. If the couple has no commitment, the parent may fear the consequences of the children's emotional involvement with the romantic partner; if the couple breaks up, the children may be adversely affected.

# Remarriage

The 18th-century writer Samuel Johnson described **remarriage**—a marriage in which one or both partners have been previously married—as "the triumph of hope over experience." Americans are a hopeful people. Many newly divorced men and women express great wariness about marrying again, yet find themselves actively searching for mates. Women often view their divorced time as important for their development as individuals, and they tend to have a wider network of friends and, thus, more social support. Men may seek to reenter marriage in search of the emotional and social support they hope marriage will provide (Ganong, Coleman, and Hans 2006).

## Rates and Patterns of Remarriage

Nearly two-fifths of recent marriages are remarriages for at least one of the spouses (Stykes and Guzzo 2015). A recent report on remarriage in the United States from the U.S. Census Bureau examined marriage and remarriage patterns for 2008 to 2012 (Lewis and Kreider 2015). Approximately 15 percent of currently married couples involve a wife (7 percent) or husband (8 percent) in a second marriage whose spouses are in their first marriage. Nine percent of currently married couples involve couples where both spouses are married for a second time. Five percent of currently married men and 4.5 percent of currently married women are in at least their third marriage. In 1.4 percent of marriages, both have been married three or more times. Overall, although 30.7 percent of women and men 15 and older had never married, 52.3 percent had married once, 13.5 percent had married twice, and 3.6 percent had married at least three times (Lewis and Kreider 2015).

Looking at changes in marriage and remarriage shows that remarriage rates in the United States have declined in recent decades. In 1990, for example, the rate of remarriage was 50 per 1,000 previously married individuals. In the mid 1990s, more than three-quarters of men and more

than two-thirds of women remarried after a divorce. By 2013, the remarriage rate had dropped to 28 per 1,000 divorced people 18 and older. Men were still more likely to remarry than women, but the rates had fallen to 40 per 1,000 among men and 21 per 1,000 among women. (Payne 2015). The decline may be partly the result of the desire on the part of divorced men and women to avoid the legal responsibilities accompanying marriage. Instead of remarrying, many are choosing to cohabit

Additional points about remarriage include the following: Remarriage is less common among widowed women and men, as only about 5 percent of women and 12 percent of men reenter marriage after losing their spouse (Sweeney 2010). Racially, the highest rate of remarriage among men is the 58 per 1,000 among foreign-born Hispanic men, while the lowest rate is 39 per 1,000 among black men. Among women, native-born Hispanic women had the highest rate of remarriage (31 per 1,000), while black women had the lowest (16 per 1,000). Figure 14.3 illustrates 2012 remarriage rates for women and men of different ethnic and racial backgrounds.

Women and men with at least some college education (or college degrees) have higher rates of remarriage than those with less education. Among those with at least some college, the remarriage rate is 51 per 1,000 among men and 29 per 1,000 among women. Those with less than a high school education tend to have the lowest rates of remarriage—29 per 1,000 men

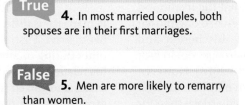

**True** **4.** In most married couples, both spouses are in their first marriages.

**False** **5.** Men are more likely to remarry than women.

**Figure 14.3  Remarriage Rates per 1,000 Men and Women by Race and Ethnicity**

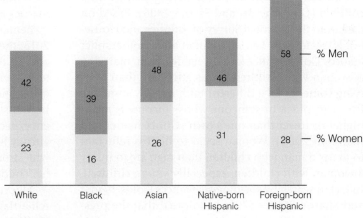

SOURCE: Lamidi and Cruz (2014).

and 12 per 1,000 among women (Lamidi and Cruz 2014). Remarriages are more unstable than first marriages, failing at higher rates and ending sooner than failed first marriages. Among women under 45, whereas one in five first marriages ends in divorce within the first five years of marriage, the rate for remarriages is one in three (31 percent). The duration of first marriages among women and men who divorced in 2012 was 13 years. Remarriages that ended in divorce lasted around 10 years.

## Gender

More men remarry than women for a number of reasons. First, divorced women tend to be older than never-married women. Given the tendency for men to marry women younger than themselves and the fact that older women may be seen as less attractive and therefore less desirable as spouses, women face more competition and possess fewer "resources" to bring to a remarriage. They are also more likely to have custody of children, which can reduce both the ease with which they socialize or date and, as shown later, their appeal as potential spouses.

## Presence of Children

Although research has indicated that the likelihood of remarriage is affected by the presence of children (de Graaf and Kalmjin 2003; Goldscheider and Sassler 2006), the majority of remarriages do, in fact, include children (Stykes and Guzzo 2015). However, the presence of dependent children may be a stronger obstacle for women's chances of remarriage than for men's chances, especially if she has many children or very young children (Stewart, Manning, and Smock 2003). The presence of children may especially affect the likelihood of women marrying men without children (Goldscheider and Sassler 2006). Based on data from the National Survey of Families and Households, sociologist Gayle Kaufman and demographer Frances Goldscheider found that, for men, marrying a woman with children is less attractive than marrying someone of a different faith or race, someone five years older, or someone whose income is much higher or lower than one's own (Goldscheider and Kaufman 2006). Women seem to be less reluctant to marry a man with children than men are to marry a woman with children, especially young children. In fact, research by sociologists Frances Goldscheider and Sharon Sassler (2006) indicated that the presence of children might even positively affect men's

establishing new cohabiting or married relationships. Men who have children seem less hesitant to remarry a woman who is also a mother, and even some men without children may regard children as a "benefit" in the form of a ready-made family (Ganong and Coleman 1994). Women and men who were once themselves children in stepfamilies express a greater willingness to marry a partner who has children (Sweeney 2010).

## Initiator Status

Research suggests that initiators of divorce will be more likely to remarry than noninitiators. In making their decisions about seeking a divorce, initiators may have factored in the prospect for reentering marriage. They also may be "better prepared emotionally" than noninitiators to remarry. The advantage initiators have over noninitiators may be temporary because noninitiators lag behind initiators in the process of adjusting to and accepting the ending of their marriages (Sweeney 2002). Indeed, Megan Sweeney found that initiators enter new relationships substantially more quickly than noninitiators.

## Characteristics of Remarriage

Remarriage is different from first marriage in a number of ways. First, the new partners get to know each other during or after a time of significant changes in life relationships, which may include confusion, guilt, stress, and mixed feelings about the past. They have great hope that they will not repeat past mistakes, but there is also often some fear that the hurts of the previous marriage will recur (McGoldrick and Carter 1989). The past is still part of the present. A Talmudic scholar once commented, "When a divorced man marries a divorced woman, four go to bed."

Remarriages occur later than first marriages. In 2013, the median age at remarriage was 45.5 for men and 42.8 for women (Payne 2015). Thus those entering remarriages are at different stages in their life cycles and may have different goals than those entering a first marriage. Divorced people may have different expectations of their new marriage. A woman who already has had children may enter a second marriage with stronger career goals. In her first marriage, raising her children may have been more important.

In reviewing a decade's worth of family research on remarriage and stepfamilies, sociologist Megan Sweeney (2010), and sociologists Deborah Carr and Kristen

Springer (2010) noted the following characteristics of remarried couples:

- Remarried couples display patterns of communication and interaction that differentiate them some from people in a first marriage. They are both less positive and less negative in their communication. Not unexpectedly then, they are less likely to engage and more likely to withdraw from discussions.

- Remarried couples endorse more autonomous financial and parenting decisions, likely due to the financial circumstances (resources and obligations) and the children that they bring with them from their prior marriage.

- Those who remarry derive benefits from reentering marriage, in terms of improved mental and physical health, but those benefits are smaller than those associated with a first marriage. The remarried report less distress, less alcohol use, and fewer symptoms of depression than those who were previously married but remain unmarried (Carr and Springer 2010).

Deborah Carr and Kristen Springer characterize the remarriage benefits as "more modest" than those associated with a first marriage. They also appear to be shorter-term benefits, mostly experienced in the early years of remarriage (2010, 749).

Economically, remarried women are less advantaged than continuously married women but potentially better off than formerly married women. Being remarried can "offset" some of the financial distress accompanying divorce (Sweeney 2010).

## Marital Satisfaction and Stability in Remarriage

According to various studies, remarried people are about as satisfied or happy in a remarriage as they were in their first marriage. Sociologist Susan Stewart notes that *newly* remarried couples actually report greater marital satisfaction than *newly* married first-marriage couples. As in first marriages, marital satisfaction then appears to decline with the passage of time and decline more rapidly in remarriages (Stewart 2007). Once "stabilized," marital happiness and satisfaction in remarriages are similar to marital quality in first marriages. However, remarried couples are more likely to divorce (Sweeney 2010). As Coleman,

Ganong, and Fine (2000) note, "Serial remarriages are increasingly common." How do we account for this paradox? Researchers have suggested several reasons for the higher divorce rate in remarriage.

*First*, as we saw last chapter, people who remarry after divorce often have a different outlook on marital stability and are more likely to use divorce to resolve an unhappy marriage (Whitton et al. 2013). As Frank Furstenberg and Graham Spanier (1987) remarked roughly 30 years ago, the willingness of remarried individuals to dissolve unhappy marriages somewhat surprised them: "Regardless of how unattractive they thought this eventuality, most indicated that after having endured a first marriage to the breaking point, they were unwilling to be miserable again simply for the sake of preserving the union."

*Second*, despite its prevalence, remarriage remains what sociologist Andrew Cherlin called an "incomplete institution" (Cherlin 1981). This concept has greatly influenced thinking about remarriages, especially those with children, and continues to do so today (Stewart 2007). What is it that makes remarriage an incomplete institution? Cherlin pointed out that society has not evolved norms, customs, and traditions to guide couples in their second or subsequent marriages. There are neither norms nor laws, for example, defining a stepfather's responsibility to a child: Is he a friend, a father, a sort of uncle, or what? What should one call one's stepmother? Nor are there rules establishing the relationship between an individual's former spouse and his or her present partner: Are they friends, acquaintances, rivals, or strangers? Remarriages don't usually receive the same family and kin support as do first marriages (Goldenberg and Goldenberg 1994).

*Third*, remarriages are subject to stresses that are not present in first marriages. The vulnerability of remarriage to divorce is especially real if children from a prior marriage or relationship are in the home (Booth and Edwards 1992). Children can make the formation of the husband–wife relationship more difficult both because they compete for their parents' love, energy, and attention and because they may resent and blame their parents' new partners for their parents' failed marriage. Additionally, in such families, time together alone becomes a precious and all-too-rare commodity for remarried spouses. Furthermore, although children have little influence in selecting a parent's new husband or wife, they have immense power in "deselecting" them by acting in ways that bring out differences

> True **6.** Second marriages are more likely than first marriages to end in divorce.

Remarriages are more likely to be unhappy and/or end in divorce than are first marriages.

SpeedKingz/Shutterstock.com

in parenting styles and values or create tension between siblings and step-siblings. The relationships between parents and children have a stronger effect on the happiness and marital quality in stepfamilies than in intact families (Stewart 2007).

The divorce proneness of remarriage seems to lessen and become more like that of a first marriage as people age. People who enter remarriage after turning 40 may even face a lower divorce likelihood than that found among first marriages (Coleman, Ganong, and Fine 2000). As mentioned earlier in the chapter, those who cohabit before remarrying face greater risk of divorce (Xu, Hudspeth, and Bartkowski 2006).

## Remarried Families

A majority of remarriages involve children, and those remarriages that include children are vastly different from those that do not. The families that emerge from remarriages with children are traditionally known as *stepfamilies*. Social scientists also sometimes call them *reconstituted, restructured,* or *remarried families*—names that emphasize their structural differences from other families. In popular discourse, we often consider them "blended families," as the new spouses and their children attempt to blend into a single functioning family. Attempting to focus more on the positive aspect of blending (and striving to steer clear of the negative connotations that sometimes accompany "steps," as in "evil stepmother"), some even refer to their new stepchildren or stepparents as "bonus" children or "bonus" parents. A website for Bonus Families (www.bonusfamilies.com), a nonprofit organization whose goal is to promote "peaceful coexistence between

divorced or separated parents and their new families," suggests that different terms may be more appropriate or acceptable at different phases.

Whatever terms one decides to use, there soon may be more stepfamilies in the United States than any other family form. It is not a simple task to provide a count of the number of stepfamilies. Sociologist Susan Stewart points out that there are few official counts of stepfamilies, the "official" estimates that we have come from multiple sources and leave us with a number of different statistics, and official data are incomplete, as we saw in Chapter 13 with the data on divorce. Nonetheless, Stewart also suggests that if we broaden the definition, take into account the full range of stepfamilies including the otherwise "invisible" stepfamilies and "hidden" stepfamily members—such as remarried couples with children, cohabiting couples with stepchildren, stepfamilies that are created through nonmarital childbearing and subsequent marriage or cohabitation, and gay or lesbian stepfamilies—it is probable that most Americans will at some point live in a stepfamily. In fact, Stewart (2007) points out that although the model of two married heterosexual parents and their biological children is still treated as the normative family type, this creates a mismatch between our cultural values and our lived experiences in real families. In fact, almost half of remarried men and 9 percent or remarried women are full-time residential stepparents (Stykes and Guzzo 2015). If we care about *families* rather than some idealized *family*, we need to understand and support stepfamilies.

Even with declines in the remarriage rate, in approximately 40 percent of recent marriages, one or both spouses had been married once before. More than 60 percent of remarriages include stepchildren. Data from 2014 indicated that almost 6 percent of children, at least 4.2 million children under 18, were living with at least one stepparent. Of these children, 911,000 lived with a biological father and a stepmother, while at least 2.9 million lived with their biological mother and a stepfather.

Over 9 million children lived in a household with a step-sibling or half sibling. That amounts to 12.5 percent of U.S. children under 18 living with a step-sibling (1.7 percent) or half sibling (10.8 percent) in 2009 (Kreider and Ellis 2011b). Stepfamilies in which there are two sets of children living in the same household face a greater likelihood of a parental divorce (Stewart 2007). Further complexity is revealed by the fact that more than half of coresidential

I n her book, *We're Still Family: What Grown Children Have to Say about Their Parents' Divorce*, psychologist Constance Ahrons tells the story of a wedding she attended. The bride was her late ex-husband's daughter by his second wife. Looking for ways to describe or explain whose wedding she was flying off to attend, she told some people it was the wedding of her daughters' sister. Confused by this designation (she was really their half sister, but they avoided such a potentially condescending qualifier—as far as her daughters were concerned, the bride was their real sister), people were similarly perplexed that Ahrons would travel across the country to attend such a wedding or that she had even been invited.

During the wedding, standing alongside her late ex-husband's widow, she caused further puzzlement when someone from the groom's side asked about her connection to the bride's mother, and they replied, "We used to be married to the same man" (Ahrons 2004, 5). If this was true, the questioner wondered, why was Ahrons at this wedding? How were the two women getting along? This is not the way people assume or expect it to be.

Constance Ahrons has spent more than two decades researching and writing about divorce and the binuclear family. She has followed a sample of divorced families over time, beginning in 1979, with 98 pairs of ex-spouses with children and most recently interviewing 173 of the 204 children from these families. As Ahrons describes her own "extended family rearranged by divorce," she remarks, "My family, and the many others like mine, don't fit the ideal images we have about families. . . . [T]hey're not tidy. There are extra people and relationships that don't exist in nuclear families and are awkward to describe because we don't have familiar and socially defined kinship terms to do so" (Ahrons 2004, 6).

As 20-something-year-old Margaret described to sociologist Ted Cohen, the series of transitions that produced a complicated set of relationships that she called her "wild world of steps and ex-steps," it was the term "ex-steps" that made Cohen most curious. Like Constance Ahrons, sociologist Susan Stewart has written, "there are no specific kinship terms for ex-stepfamily members," and not a lot is known about relations

between them (Stewart 2007, 135). When Cohen asked Margaret to elaborate, she offered the following details. At 12, Margaret's parents divorced after a 15-year-long marriage that produced Margaret, her two brothers, and a sister. Initially after the divorce, the four children lived with their mother, which as we've seen, is typical. Meanwhile, Margaret's father remarried a younger woman who, herself, had a 4-year-old daughter from an earlier marriage. Margaret's mom also got engaged and subsequently married a man with two children from his previous marriage, a 16-year-old daughter and a 12-year-old son. Margaret's dad and his wife—her stepmother—had a child together. Margaret's mom married, leaving Margaret and her siblings with a half brother, a stepbrother, and two stepsisters. Her mother's marriage to her stepdad did not last and her mom began the process of getting a divorce. This caused Margaret and her siblings to "lose (our) stepdad and stepbrother and stepsister, hence, now, the 'ex-steps' who we don't talk to ever."

In recounting her experiences, Margaret observed that "adjusting to gaining 'steps' was a difficult process—we never actually were able to blend—and then (we) got rid of them via divorce, as *if they were never there*" (emphasis added). In part, Margaret's experiences and Ahrons's account both illustrate what sociologists refer to as *boundary ambiguity*, a lack of clarity about who's in and who's out, who is and who's not (or is no longer) a member of one's family. Some people may have considerable difficulty fully incorporating stepparents and stepsiblings into their perception of their family, whereas others may continue to consider former stepparents and stepsiblings as at least partial members of their family even after any official connection has been severed (Sweeney 2010).

Families, composed of people rearranged by divorce and expanded by remarriages, are increasingly common. Ahrons advocates that we reconceptualize what we mean by family; recognize that our nuclear family bias no longer describes the family lives of millions of men, women, and children; and extend the idea of family to include all of those hard-to-describe relationships among those who consider themselves family. ●

stepparents also have children living outside of their homes (Sweeney 2010).

The picture drawn from these estimates probably understates the extent of stepfamily life in the United States. Research by the Pew Research Center of a

nationally representative sample of 2,691 women and men offers us additional data and reinforces the fact that stepfamily experience has become quite common:

- Projecting from the survey sample, an estimated 95.5 million adults in the United States, 42 percent of the adult population, have at least one step-relative, either a stepparent, stepchild, or step-sibling.

- Thirty percent indicated that they have a step-sibling or half sibling, 18 percent a stepparent, and 13 percent at least one stepchild.

- The likelihood of having a step-relative and complex family and sibling relationships is greater among blacks (60 percent) and Hispanics (46 percent) than among whites (39 percent). It is also higher among younger people, and those without a college degree.

- Of those younger than 30, 52 percent report having a step-relative, and 44 percent report having a step-sibling or half sibling, compared to those 30 to 49 years old, of which 45 percent have a step-relative and 35 percent a step-sibling.

- Of those without a college degree, 45 percent have a step-relative, and 34 percent have a step-sibling or half-sibling, compared to 33 percent and 21 percent, respectively, of those with a college degree.

## A Different Kind of Family

When we enter a stepfamily, many of us expect to re-create a family identical to an intact family. The intact nuclear family becomes the model against which we judge our successes and failures. But, as nicely expressed by family counselor Susan Wisdom and author Jennifer Green (2002, 32), "Evaluating a stepfamily by standards you'd apply to a nuclear one is like grading an algebra exam with the key to a geometry final." Indeed, researchers believe that blended families are significantly different from what are typically called intact families in multiple ways. As expressed in a Facts for Families Guide on "Stepfamily Problems," produced by the American Academy of Child and Adolescent Psychiatry (2015), at the time of its formation, members of stepfamilies have neither shared family histories nor shared routines or ways of doing things. Indeed, they may have very different religious, educational, or ethnic backgrounds. Bonds among members, especially among resident stepparents and stepchildren, must be built, and the process of such building may take a good deal of time (American Academy of Child and Adolescent Psychiatry 2015).

Furthermore, these differences are ample enough to leave people experiencing a sense of culture shock, akin to traveling to and living in a foreign country (Stewart 2007). If one tries, expects, and attempts to

Although adults enter into it with enthusiasm, blending a family is a complex and difficult process in which new relationships and roles must be constructed.

Blend Images/Getty Images

make his or her feelings and relationships in a stepfamily identical to those of an intact family, one is bound to fail. But by recognizing that stepfamilies work differently and provide different satisfactions and challenges, one can appreciate the richness they bring and have a greater likelihood of building a successful stepfamily.

### Structural Differences

Six structural characteristics make the stepfamily different from the traditional first-marriage family (Visher and Visher 1991). Each one is laden with potential difficulties:

1. *Almost all the members in a stepfamily have lost an important primary relationship.* The children may mourn the loss of a parent or parents, and the spouses may mourn the loss of their former mate and their marriage. Anger and hostility may be displaced onto the new stepparent.

2. *One biological parent typically lives outside the current family.* In stepfamilies that form after divorce, the absent former spouse may either support or interfere with the new family. Power struggles may occur between the absent parent and the custodial parent, and there may be jealousy between the absent parent and the stepparent.

3. *The relationship between a parent and his or her children predates the relationship between the new partners.* Children have often spent considerable time in a single-parent family structure. They have formed close and different bonds with the parent. A new husband or wife may seem to be an interloper in

A Pew Research Center survey ("A Portrait of Stepfamilies," 2011) inquired into people's attitudes about their stepfamily lives and their feelings of responsibility for assisting family members in need. First, comparing people with and without stepfamilies revealed that although both groups were largely very satisfied with their family lives, those without step-relatives were somewhat more likely to feel that way. Of those with at least one step-relative (stepparent, step-sibling, or stepchild), 70 percent said they were very satisfied with their family lives compared with 78 percent who don't have any step-relatives.

When asked questions about their sense of obligation to help their family members, clear differences emerged in what people said about their step-relatives and their biological relatives:

- Among those with both a living parent and a living stepparent, 85 percent said they would feel "very

obligated" to help out their parent, but only 56 percent said the same about a stepparent.
- Among those with both a grown biological and stepchild, 78 percent said they would feel "very obligated" to help their biological child in need. In comparison, 62 percent said the same about a grown stepchild.
- For those with both biological and step-siblings, 64 percent would feel obligated to help their biological brother or sister, whereas 42 percent said they would feel the same way about helping a stepbrother or stepsister.

As these data reveal, there is less of a sense of obligation to step-relatives than to biological relatives. This may not be surprising but it is telling. The Pew data also showed that step-relatives did fare better than friends. Compared to the previous percentages, 39 percent said they would feel obligated to assist their best friend in a situation where the friend was facing a serious problem. ●

the children's special relationship with the parent. A new stepparent may find that he or she must compete with the children for the parent's attention. The stepparent may even be excluded from the parent–child system.

4. *Stepparent roles are ill defined.* No one knows quite what he or she is supposed to do as a stepparent. Ambiguity and ambivalence are common. Remarried families may try to model themselves after traditional nuclear families, so stepparents often expect that their role will be similar to the parent role. However, some are reluctant to assume an active parenting role, and some attempt to assume such a role too quickly. Children may resist the efforts stepparents make to become involved in their lives. Most stepparents try role after role (e.g., parent, friend, big brother, or big sister) until they find one that fits. Researchers estimate that this process may stretch over a decade or longer (Papernow 1993).

5. *Many children in stepfamilies are also members of a noncustodial parent's household.* Each home may have differing rules and expectations. When conflict arises, children may try to play one household against the other. They may feel a conflict of loyalties and a sense of lost closeness with their newly married parent (Visher and Visher 1979).

6. *Children in stepfamilies have at least one extra pair of grandparents.* Children gain a new set of step-grandparents, but the roles these new grandparents are to play are far from clear.

In addition to these structural differences, there are differences in the perceptions people have about the meaning of their step-relationships, especially surrounding the question of family obligation. This is discussed in the following Issues and Insights box.

Numerous researchers have found that children in stepfamilies exhibit at least about the same number of adjustment problems as children in single-parent families and more problems than children in original, two-parent families (Coleman, Ganong, and Fine 2000; Nicholson, Fergusson, and Horwood 1999; Sweeney 2007). Others suggest that stepfamily life actually may be *more* difficult for children than living in a single-parent household. As reported by sociologist Megan Sweeney, recent research indicates that, for children, living with a stepparent rather than a single parent is associated with poorer emotional well-being but better well-being in areas such as health and behavioral outcomes (Sweeney 2010).

In addition, research reveals that relations between stepparents and their stepchildren are often of "low quality," characterized by less frequent activities together than between biological parents and children, less warmth and support from stepparents to stepchildren, and less involvement by stepparents in monitoring and controlling their stepchildren's activities (Stewart 2007). At the same time, a new partner is "a second pair of eyes and hands" who can share in the various, often burdensome, tasks of child rearing. Likewise, new partners can be sources of emotional and social support, strengthening the mother's authority in the household, assisting her with difficult decisions, comforting her when parenting is stressful, and potentially inhibiting her from acting in negative or hurtful ways toward her children. Certainly, these effects will be for the better for children (Thomson et al. 2001).

## The Developmental Stages of Stepfamilies

Individuals and families blend into and become a stepfamily through a process—a stepfamily cycle—of experiencing a series of developmental stages. Each person—the biological parent, the stepparent, and the stepchild (or children)—experiences the process differently. For family members, it involves seven stages, according to psychologist Patricia Papernow (1993), whose model of the stepfamily cycle was drawn, in part, from interviews with stepfamilies in her clinical practice. The early stages are fantasy, immersion, and awareness; the middle stages are mobilization and action; and the later stages are contact and resolution.

**True**

**7.** Becoming a stepfamily is a process that can take years to fully complete.

It may take stepfamilies about seven years to complete the developmental process. Some may complete it in four years, and others take many, many years. Some only go through a few of the stages and become stuck. Others split up with divorce. But many are successful. Becoming a stepfamily is a slow process that moves in small ways to transform strangers into family members.

### Early Stages: Fantasy, Immersion, and Awareness

The early stages in becoming a stepfamily include the courtship and early period of remarriage, when each individual has his or her fantasy of their new family. It is a time when the adults (and sometimes the children) hope for an "instant" nuclear family that will fulfill their dreams of how families should be. They have not yet realized that stepfamilies are different from nuclear families.

**Fantasy Stage.** During the fantasy stage, biological parents hope that the new partner will be a better spouse and parent than the previous partner. They want their children to be loved, adored, and cared for by their new partners. They expect their children to love the new parent as much as they do.

New stepparents fantasize that they will be loving parents who are accepted and loved by their new stepchildren. They believe that they can ease the load of the new spouse, who may have been a single parent for years. One stepmother recalled her fantasy: "I would meet the children and they would gradually get to know me and think I was wonderful. . . . I just knew they would love me to pieces. I mean, how could they not?" (Papernow 1993). Of course, they did not.

The children, meanwhile, may have quite different fantasies. They may still feel the loss of their original family. Their fantasies are often that their parents will get back together. Others fear that they may "lose" their parent to an interloper, the new stepparent. Some fear that their new family may "fail" again. Still others are concerned about upheavals in their lives, such as moving, going to new schools, and so on.

**Immersion Stage.** The immersion stage is the "sink-or-swim" stage in a stepfamily. Reality replaces fantasy. "We thought we would just add the kids to this wonderful relationship we'd developed. Instead we spent three years in a sort of cold war over them," recalled one stepparent (Papernow 1993).

For children, a man's transformation from "mom's date" to stepfather may be the equivalent of the transformation from Dr. Jekyll to Mr. Hyde. Suddenly, an outsider becomes an insider—with authority, as described by one 12-year-old (whose new stepmother also had children): "In the beginning it's fun. Then you realize that your whole life is going to change. Everything changes. . . . [N]ow there's all these new people and new rules" (Papernow 1993). Children may also feel disloyal to their absent biological parent if they show affection to a stepparent. (Biological parents can make a difference: They can let their children know it's okay to love a stepparent.)

**Awareness Stage.** The awareness stage in stepfamily development is reached when family members "map" the territory. This stage involves individual and joint family tasks. The individual task is for each member to identify and name the feelings that he or she experiences in being in the new stepfamily. A key feeling for stepparents to acknowledge is feeling like an outsider. They need to become aware of feelings of aloneness, they must discover their own needs, and they must set some distance between themselves and their stepchildren. They need to understand why their stepchildren are not warmly welcoming them as they had expected.

Biological parents need to become aware of unresolved feelings from their earlier marriages and from being single parents. They may feel pulled from the multiple demands of their children and their partners. Biological parents may feel resentment toward their children, their partners, or both (Papernow 1993).

Children in the awareness stage often feel "bumped" from their close relationship with the single parent. They miss cuddling in bed in the morning, the bedtime story, and the wholehearted attention. When a new stepparent moves in, their feelings of loss over their parents' divorce are often rekindled. Loyalty issues resurface. If they are not pressured into feeling "wonderful" about their new family, however, they can slowly learn to appreciate the benefits of an added parent and friend who will play with them or take them places.

## Middle Stages: Mobilization and Action

In the middle stages of stepfamily development, family members are clearer about their feelings and relationships with one another. They have given up many of their fantasies. They understand more of their own needs. They have mapped the new territory. The family, however, remains biologically oriented. Parent–child relationships are central. In this stage, changes involve the emotional structure of the family.

**Mobilization Stage.** In the mobilization stage, family members recognize differences. Conflict becomes more open. Members mobilize around their unmet needs. A stepmother described this change: "I started realizing that I'm different than Jim [the husband] is, and I'm going to be a different person than he is. I spent years trying to be just like him and be sweet and always gentle with his daughter. But I'm not always that way. I think I made a decision that what I was seeing was right" (Papernow 1993). The challenge in this stage is to resolve differences while building the stepfamily's sense of family.

Stepparents begin to take a stand. They stop trying to be the ideal parent. They no longer are satisfied with being outsiders. Instead, they want their needs met.

They begin to make demands on their stepchildren: to pick up their clothes, be polite, and do the dishes. Similarly, they make demands on their partner to be consulted; they often take positions regarding their partner's former spouse. Because stepparents make their presence known in this stage, the family begins to change. The family begins to integrate the stepparent into its functioning. In doing so, the stepparent ceases being an outsider, and the family increasingly becomes a real stepfamily.

For biological parents, the mobilization stage can be frightening. A stepparent's desire for change leaves a biological parent torn. Biological parents feel that they must protect their children and yet satisfy the needs of their partner.

Children often attempt to resolve loyalty issues at this stage. They have been tugged and pulled in opposite directions by angry parents too long. Often the adults paid no attention to them. Finally, the children have had enough and can articulate their feelings. After hearing her parents squabble one time too many, one girl reflected, "I thought, this stinks. It's horrible. After the 50 millionth time I said, 'That's your problem. Talk to each other about it,' and they didn't do it again" (Papernow 1993).

**Action Stage.** In the action stage, the family begins to take major steps in reorganizing itself as a stepfamily. It creates new norms and family rituals. Although members have different feelings and needs, they begin to accept each other. Most important, stepfamily members develop shared, realistic expectations and act on them.

Step-couples begin to develop their own relationship independent from the children. They also begin working together as a parental team. Stepparents begin to take on disciplinary and decision-making roles; they are supported by the biological parents. Stepparents begin to develop relationships with their stepchildren independent of the biological parents. Stepparent–stepchild bonds are strengthened.

## Later Stages: Contact and Resolution

The later stages in stepfamily development involve solidifying the stepfamily. Much of the hard work has been accomplished in the middle stages.

**Contact Stage.** In the contact stage, stepfamily members make intimate contact with one another. Their relationships become genuine. They communicate with a sense of ease and intimacy. The couple relationship becomes a sanctuary from everyday family life. The stepparent becomes an "intimate outsider" with whom stepchildren can talk about things "too hot" for their biological parents, such as sex, drugs, their feelings about the divorce, and religion.

For the stepparent, a clear role finally emerges—what is now called the **stepparent role**. The role varies from stepparent to stepparent and from stepfamily to stepfamily because, as shown earlier, it is undefined in our society. It is mutually suitable to both the individual and the different family members.

**Resolution Stage.** The stepfamily is solid in its resolution stage. It no longer requires the close attention and work of the middle stages. Family members feel that earlier issues have been resolved. The stepparent becomes "an intimate outsider" (Stewart 2007).

Not all relationships in a stepfamily are necessarily the same; they may differ according to the personalities of each individual. Some relationships develop more closely than others. But in any case, there is a sense of acceptance. The stepfamily has made it and has benefited from the effort.

# Stepparenting

To become a stepparent, one must be willing to bring into a relationship children from a prior relationship or to enter a relationship with someone who has children from a previous relationship. Many people, especially many men, are unwilling or reluctant to do so. Evidence of this can be seen in the survey data (from the National Survey of Families and Households) that indicate that although women did not express a strong desire to marry someone who already had children, their opposition to the idea was not as great as what men expressed (Goldscheider and Kaufman 2003; Stewart 2007).

## Problems of Women and Men in Stepfamilies

Most people go into stepfamily relationships expecting to re-create the traditional nuclear family: They are full of love, hope, and energy. However, parenting stepchildren appears to be less satisfying than parenting one's biological children.

Although women and men may enter stepfamilies equally hopeful, they do not experience the same things. Women who are stepmothers report feeling a range of feelings. They report positive reactions to the challenges that stepfamily life creates, the personal growth that it engenders, and the opportunity to play the maternal role that it offers. Negative reactions include feeling isolated, resentful, and guilty (Stewart 2007).

Garrett Pace, Kevin Shafer, Todd Jensen, and Jeffry Larson looked at the relationship between relationship quality and stepparenting. As one might imagine, stepparenting issues and stresses can have negative impact on couples who form stepfamilies together. They stress the importance of clear communication between partners as protective of couple relationships and for individual spouses dealing with the stresses of stepparenting (Pace et al. 2013). Jensen, Shafer, and Larson (2014) also found that stepparents who favor or prioritize their own biological children, see their stepchildren as interfering with their romantic relationship, and who expect their stepchildren to be obedient have greater difficulty, more stepparenting issues, and lower relationship quality (Jensen, Shafer, and Larson 2014).

## Women in Stepfamilies

Most research indicates that stepfamily life is more difficult for women than for men. Some factors accounting for the difficulty include the cultural expectations for women as parents, the nurturing responsibilities that fall more heavily on women, and the ongoing involvement of biological mothers in raising their children (Stewart 2007). To various degrees, women enter stepfamilies with high hopes, expecting to do the following:

- Make up to the children for the divorce or provide children whose mothers have died with a maternal figure
- Create a happy, close-knit family and a new nuclear family
- Keep everyone happy
- Prove that they are not wicked stepmothers
- Love the stepchild instantly and as much as their biological children
- Receive instant love from their stepchildren (Visher and Visher 1979, 1991)

Needless to say, most women are disappointed. Expectations of total love, happiness, and the like would be unrealistic in any

> **False** 8. Stepmothers experience more stress in stepfamilies than do stepfathers.

kind of family, be it a traditional family or a stepfamily. The warmer a woman is to her stepchildren, the more hostile they may become to her because they feel that she is trying to replace their "real" mother. If a stepmother tries to meet everyone's needs—especially her stepchildren's, which are often contradictory, excessive, and distancing—she is likely to exhaust herself emotionally and physically. It takes time for her and her children to become emotionally integrated as a family.

One thing that makes stepmothering more difficult than stepfathering is the role that women typically play in child rearing. Women are expected to and expect to become nurturing, primary caregivers, although their stepchildren may not adequately acknowledge or appreciate this role. Consequently, they have more opportunities to encounter stress and experience conflict with their stepchildren, and thus poorer relationships with their stepchildren may occur.

Stepchildren tend to view relationships with stepmothers as more stressful than relationships with stepfathers. If their biological mother is still living, they may feel that their stepmother threatens their relationship with their birth mother (Hetherington and Stanley-Hagan 1999). Bitter custody fights may leave children emotionally troubled and hostile to a stepmother. In other instances, children (especially adolescents) may have moved into their father's home because their mother could no longer handle them. In either case, the stepmother may be required to parent children who have special needs or problems. Stepmothers may find these relationships especially difficult. Typically, stepmother–stepdaughter relationships are the most problematic (Stewart 2007). Relationships become even more difficult when a stepmother never intended to become a full-time stepparent.

## Men in Stepfamilies

Different expectations are placed on men in stepfamilies. Because men are generally less involved in child rearing, they usually have few "cruel stepparent" myths to counter. Nevertheless, men entering stepparenting roles may find certain areas particularly difficult at first (Visher and Visher 1991). A critical factor in a man's stepparenting is whether he has children of his own. If he does, they are more likely to live with his ex-wife. In this case, the stepfather may experience guilt and confusion in his stepparenting because he feels that he should be parenting his own children. When his children visit, he may try to be "superdad," spending all his time with them and taking them to special places. His wife and stepchildren may feel excluded and angry.

A stepfather usually joins an already established single-parent family. He may find himself having to squeeze into it. The longer a single-parent family has been functioning, the more difficult it usually is to reorganize it. The children may resent what they perceive as "interfering" with their relationship with their mother. His ways of handling the children may be different from his wife's, resulting in conflict with her or with her children (Marsiglio 2004b; Wallerstein and Kelly 1980).

Working out rules of family behavior is often the area in which a stepfamily encounters its first real difficulties. Although the mother usually wants help with discipline, she often feels protective if the stepfather's style is different from hers. Allowing a stepparent to discipline a child requires the biological parent to trust and let go. Disciplining often elicits a child's testing response: "You're not my real father. I don't have to do what you tell me." Nevertheless, disciplining establishes legitimacy because only a parent or a parent figure is expected to discipline in our culture. Disciplining may be the first step toward family integration because it establishes the stepparent's presence and authority in the family.

In comparison to birth parents, stepfathers tend to have more limited and less positive relationships with their stepchildren. They communicate less, display less warmth and affection, and are typically less involved. Conversely, some research also indicates that among divorced, noncustodial fathers, remarriage and stepfathering may lead to development of closer relationships with stepchildren than with their biological children.

However, the process of **paternal claiming**, embracing stepchildren as if they were biological children and becoming involved in the processes of nurturing, providing for, and protecting them, is a two-way process. Stepfathers must build an appropriate identity, but both birth mothers and the stepchildren also help create or hinder the development of a sense of familial "we-ness" (Marsiglio 2004b). The complex role that a stepfather brings to his family often creates role ambiguity and confusion that takes time to work out.

**False** **9.** Stepfathers have less positive relationships with their stepchildren than those between biological fathers and children.

Sometimes I feel like I'm on the outside looking in because—sometimes I wish she was mine. I guess because we're just that close. . . . [I]n my heart, I feel like I'm her father. . . . I know in reality, I'm not but, I'm going to give her all the benefit that a father should. I'm going to make sure she gets those benefits. Even though her dad is giving them to her, she is given a little extra, and I figure that extra goes a long way. . . .

Sociologist William Marsiglio conducted interview research with a diverse group of 36 stepfathers, including the 35-year-old stepfather just quoted. Of the men, 25 were married, seven more cohabited with their female partners, and four lived apart from their partners. They ranged in age from 20 to 54, with an average of 36 years of age. Educationally, 16 of the men were college graduates, 12 more graduated high school and attended some college, and eight had either just completed or failed to complete high school. Racially, 27 of the men were Caucasian, and 9 were African American. In addition, 22 men had biological children of their own, and 11 were living with at least one of "their own" biological children. Marsiglio (2004b, 34) wanted to uncover men's experiences of "claiming stepchildren" and identified 10 properties of the claiming process.

Among these properties is the *degree of deliberativeness*— how much thought men give to their relationships with their stepchildren and how conscious and deliberate they

are in coming to orient themselves to their stepchildren as "their" children. Although some men experience the paternal claiming process gradually, as events unfold that may include key turning points, some men, like 41-year-old Terry, decide at the outset that the relationship is to be "all or nothing." As he told Marsiglio, "It was like, if I'm coming into this relationship, then I'm coming in a hundred percent. I'm either going to be an all husband and an all father or nothing at all. I can't have like half a relationship. I can't be half a father. Where do you draw the line? . . . If I'm going to love you, I'm going to be your father. I'm going to be there all the way."

Other properties include the degree to which they have and use opportunities to be involved across a range of paternal behaviors; the extent to which they find themselves thinking about, mindful of, or daydreaming about their stepchildren in ways that biological fathers do; and the degree to which they seek and are publicly acknowledged as a father figure by others: schoolteachers, coaches, neighbors, and—in the case of adoptions—the law.

Marsiglio (2004b) also identified five conditions that encourage men to perceive their stepchildren as their own: the stepfather's identification with the stepchild, the stepfather's personality, the birth mother's involvement, the stepchildren's perceptions and reactions, and the biological father's presence and involvement.

Here is how the first condition—the degree to which stepfathers identify with their stepchildren, seeing

However, the potential for deep, mutually gratifying, and meaningful relationships between a stepfather and stepchildren is there, as illustrated in the Issues and Insights feature in this chapter.

One can see, even in the terms children and adolescents use to describe or refer to their stepfathers, variations in their sense of closeness to their stepfather and their having "claimed" their stepfather as family. For example, the decision to refer to a stepfather as "Ted," "my stepfather," or "my mother's husband" is one that reflects different kinds and degrees of felt connection between child and stepfather (Thorsen and King 2015). Calling a stepfather "dad" (or even "my stepdad") signals a closer bond than calling a stepfather "my mother's husband." According to research by Maggie Thorsen and Valarie King, "my stepfather"

was used by the majority of their sample. It was more likely to be used by adolescents who had been part of a stepfamily for a longer period of time, had stepsiblings in the household, had less close relationships with their biological fathers, and had fewer other father figures in their lives. Those who were closer to their biological father, less close with their mother, had been in a stepfamily for a shorter period of time, and had fewer siblings in the household were the ones more likely to use the term "my mother's husband." Thorsen and King stress that the choice of label one uses is more heavily influenced by the felt closeness to one's nonresidential biological father and the degree of closeness felt toward one's mother than to the closeness felt toward one's stepfather (Thorsen and King 2015).

similarities in personalities, interests, or personalities—was expressed by a man whom Marsiglio calls Thomas:

> They're my kids. I look at them like they're my boys, I tell everybody, they're my boys. And I don't want to take nothing away from [Danny's] dad, but I've raised them for so long now, I mean . . . you have a child in your home for the amount of time that I have, you feed them and long enough, they'll start acting and looking just like you, you know what I'm saying? They just do. They just call me, call me "dad."

For men who have biological children, perceiving that stepchildren are their own may mean coming to feel similarly toward their stepchildren and biological children. Such reactions may not be common, but neither are they aberrations. Some stepfather–stepchild ties become quite powerful, becoming the equivalent of relationships between biological parents and children. Thus, when we read or hear generalizations about distance or deficiency in stepparent–stepchild relationships, we would do well to remember the words and sentiments expressed by Marsiglio's interviewees. We would also be well advised to consider some factors that might enhance or facilitate the paternal claiming process.

It is unsurprising that research shows that the claiming process can go in both directions. In other words, stepchildren may well "claim" their stepparents or stepsiblings, coming to see them as family (Coleman et al. 2015). Looking specifically at relationships between former stepparents and stepchildren, Coleman and colleagues found that roughly half of their sample had at an earlier point claimed stepparents as family. The breakup of the marriage between one's parent and stepparent led roughly half of those who had claimed their stepparent as family (around a quarter of their overall sample) to continue to claim these former stepparents as family. This included referring to them as such, making an effort to maintain a familial relationship, maintaining reciprocal relationships in which support and or resources flow between them, and spending time together. This was more likely when and if one's biological parent either approved or were neutral of such efforts, and occurred more often found within relationships in which stepparents had been involved in their stepchildren's lives since early childhood. But half of those who had once claimed stepparents as family eventually *disclaimed* them. Upon learning from their parents some of the details about the reasons their remarriage failed, they reevaluated what they thought and felt about their former stepparent.

Keep in mind that Coleman et al. found that for half of their sample, such claiming of stepparents as family never occurred. Either the child's age at the time that stepparents entered their lives, unfavorable impressions or experiences with the stepparent, or having been through other parental marital transitions can all be inhibitors to claiming stepparents as family. ●

## Children in Stepfamilies

The research into effects on children of being in a stepfamily reveal a number of negative outcomes when compared with being in families headed by married biological parents. These include academic, psychological, and behavioral outcomes. Academically, attendance in school, school performance, and overall educational attainment tend to be worse among children in stepfamilies. Psychologically, children in stepfamilies are more likely to have needed and/or received psychological help, to be sad or anxious, and to be less sociable. Behaviorally, they are more likely to smoke cigarettes, drink, use illegal drugs, and engage in early sexual behavior and childbearing. In addition to these, longer-term effects have been identified, including leaving home at an earlier age and maintaining less adult contact with siblings (Stewart 2007). Lower levels of stress felt by children in stepfamilies have been associated with closeness with resident stepparents and resident biological parents. However, greater closeness to non-residential biological parents has been associated with slightly elevated stress levels, perhaps out of felt loyalty conflicts (Jensen, Shafer, and Holmes 2015).

Some of the adverse outcomes are reduced when children feel a strong sense of belonging in their stepfamilies. In biologically based two-parent families, children are related to both parents and all siblings, which can foster a sense of belonging to and in this unit. The perceived quality of mother–child and stepfather–child relationships has been associated with fostering a greater sense of belonging and, thus, potentially less negative impacts of divorce and stepfamily life such as negative academic behaviors, substance use, and earlier onset of sexual activity (King, Boyd, and Thorsen 2015).

Of course, any comparisons with children in intact two-parent homes are complicated by other differences between stepfamilies and families headed by married biological parents. In part because of some of the economic factors that contribute to divorce, in part because of postdivorce economic consequences, and in part because of the economic context in which nonmarital childbearing is most common, stepfamilies are often at a disadvantage. Sociologist Susan Stewart (2007) notes that stepfamilies have less savings and lower incomes and are less likely to be homeowners. These socioeconomic differences could be associated with some of the outcomes that have been reported. For example, the lower earnings and education of stepfathers in comparison to biological fathers is likely among the causes of the lesser academic achievement of children in stepfamilies (Hofferth 2006).

Children's adjustment to and needs in stepfamilies vary by age of the child. Preschoolers (2 to 5 years old) may fantasize that their original family will one day be reunited, see remarriage and the presence of stepparents as threats to that end, and feel guilty or responsible for their parents' divorce. They also may fear abandonment by the remarried parent. They need reassurance from their biological parents that they are loved and that their parents will continue to be a major part of their lives. Young children may also feel guilty about developing close, loving relationships with stepparents and need to be assured that one can love both one's biological parents and stepparents.

Children of elementary school age may also feel guilt about having caused their parents' divorce. They may feel as though their lives are out of control and need to be given opportunities to exercise some control in their daily lives (e.g., decorating their rooms or selecting their clothes). They may grieve upon learning of their parents' remarrying, as it signals an end to their fantasy of reunification.

Preteens (11 to 12 years old) typically begin to pull away from their families and assert more independence. They may focus their resentment of authority on the stepparent. Without a history together, there is less fear that one will lose the love of the stepparent. Practitioners suggest that what these preteens need is reassurance that they will always find support whenever they need it.

Teens may resent the changes in the household that accompany remarriage. They may not welcome shifts in responsibilities, especially if they are surrendering more adult responsibilities that they acquired in their single-parent households. Involving them in discussions and allowing them to make decisions about the tasks they want to retain and the time they may want to spend with the nonresidential parent may be important for their adjustment (Rubinyi, Tralle, and Lee 2012).

Stepfamilies formed during adolescence are more susceptible to strained relationships, especially between adolescents and a stepfather. Children who are younger when their parent remarries may have an easier time coming to see the stepparent in a parental light, are more likely to refer to their stepparents as dad or mom, and are more likely to still consider the stepparent a parent when they reach young adulthood (Thorsen and King 2015).

Further complicating assessments of outcomes is the fact that outcomes are not the same for all children or all types of stepfamilies. In addition to the aforementioned age differences, such factors as the age of the stepparents, the child's gender, whether one or both parents have previous children, and other socioeconomic or demographic characteristics may make a difference in what children need and how well they respond. For example, girls are reported to have more difficulties with stepfamilies than do boys. They are particularly likely to experience poorer relationships with and negative consequences from relationships with stepmothers. Older stepchildren experience different outcomes when their parent remarries or cohabits with a stepparent. They leave their parents' household at earlier ages, are less likely to live with their parents as adults, they marry earlier, and are less likely to go away to college. When their parent has young stepchildren or new biological children, they may feel cut off financially, receive less to no help for their schooling, and become resentful of their younger step-siblings or half siblings (Stewart 2007).

Three other points about effects on children bear mentioning. First, the more relevant comparison may be between children in stepfamilies and children in single-parent families, because children rarely move from households with married birth parents to

**False** **10.** Children in stepfamilies are more likely to have emotional and psychological problems than children in single-parent families.

stepfamily households. The existing evidence, as noted earlier, shows that compared to children in single-parent homes, children with a stepparent are typically better off economically and gain an additional adult role model. The presence of a stepparent (typically stepfather) may also lessen the burden of child care or, conversely, lessen other responsibilities otherwise shouldered by a single parent and, therefore, allow the formerly single parent to devote more time to parenting (Sweeney 2010). However, compared to children in single-parent households, children in stepfamilies may be at greater risk for psychological or emotional problems.

Second, the greater risks for adverse outcomes for children are found in situations of multiple transitions. In other words, when children not only experience the divorce of their parents and parental remarriage by one or both parents but also acquire stepsiblings, have to move, have to change schools, and so on, the outcomes will be more problematic. Finally, when considering differences between stepchildren and biological children, the differences tend to be small ones (Anderson and Greene 2013; Stewart 2007).

## Conflict in Stepfamilies

Conflict takes place in all families: traditional nuclear families, single-parent families, and stepfamilies. If some family members do not like each other, they will bicker, argue, tease, and fight. Sometimes they have no better reason for disruptive behavior than that they are bored or frustrated and want to take it out on someone. These are fundamentally personal conflicts. Other conflicts are about definite issues, such as dating, use of the car, manners, television, or friends. These conflicts can be between partners, between parents and children, or among the children themselves. Certain types of stepfamily conflicts, however, are of a frequency, intensity, or nature that distinguishes them from conflicts in traditional nuclear families. Recent research on how conflict affects children in stepfamily households found that parental conflict does not account for children's lower level of well-being (Hanson, McLanahan, and Thomson 1996). These conflicts are about favoritism, divided loyalties, discipline, and money, goods, and services.

### Favoritism

Favoritism exists in families of first marriages as well as in stepfamilies. In stepfamilies, however, the favoritism often takes a different form. Whereas a parent may favor a child in a biological family on the basis of age, sex, or personality, favoritism in stepfamilies tends to run along kinship lines. A parent favors a child because he or she is the parent's biological child. If a new child is born to the remarried couple, they may favor him or her as a child of their joint love. In American culture, where parents are expected to treat children equally, favoritism based on kinship seems particularly unfair.

### Divided Loyalties

"How can you stand that lousy, low-down, sneaky, nasty mother (or father) of yours?" demands a hostile parent. It is one of the most painful questions children can confront because it forces them to take sides against someone they love. One study (Lutz 1983) found that about half the adolescents studied confronted situations in which one divorced parent talked negatively about the other. Almost half the adolescents felt themselves "caught in the middle." Three-quarters found such talk stressful.

Divided loyalties put children in no-win situations, forcing them not only to choose between parents but also to reject new stepparents. Children feel disloyal to one parent for loving the other parent or stepparent. But, as shown in the previous chapter, divided loyalties, like favoritism, can exist in traditional nuclear families as well. This is especially true of conflict-ridden families in which warring parents seek their children as allies.

### Discipline

Researchers generally agree that discipline issues are among the most important causes of conflict among remarried families (Ihinger-Tallman and Pasley 1987). Discipline is especially difficult to deal with if the child is not the person's biological child. Disciplining a stepchild often gives rise to conflicting feelings within a stepparent. Stepparents may feel that they are overreacting to the child's behavior, that their feelings are out of control, and that they are being censured by the child's biological parent. Compensating for fears of unfairness, the stepparent may become overly tolerant.

The specific discipline problems vary from family to family, but a common problem is interference by the biological parent with the stepparent (Mills 1984). The biological parent may feel resentful or overreact to the stepparent's disciplining if he or she has been reluctant to give the stepparent authority. As one

## Inconsistent to Nonexistent: Lack of Legal Policies about Stepfamilies

Sociologist Susan Stewart tells the following story of a stepfather's unsuccessful attempt to gain custody of his stepchild after the child's mother died. Married to the child's mother, the stepfather helped raise his stepchild from the time the child was 16 months old until the mother's death when the child was 9. Despite the stated preferences of the child and expert testimony of a psychiatrist about the child remaining with the stepfather being in the child's best interests, the stepfather's attempt to gain custody was denied. Custody was instead given to the child's nonresident biological father.

What does a stepparent owe a stepchild? What rights and responsibilities do members of stepfamilies have? If the biological parent dies, can a stepparent gain legal custody? Speaking of custody, can a stepparent seek custody in a divorce? What about visitation rights? Must she or he pay child support? Despite legal questions that might arise around stepfamily responsibilities and relationships, as Stewart notes, when it comes to stepfamilies, there are no clear and consistent legal guidelines. For the most part, stepparents remain legal strangers to their stepchildren. In that vein, they are not guaranteed custody or visitation and may not even be able to include coverage of their stepchildren in health insurance policies (Stewart 2007).

Like other areas of family law, such as divorce law and rights to marry, *where they exist*, laws and policies concerning stepfamilies vary from state to state. Thus, there is no consistency to the answers to questions posed such as these. In some states, stepparents can seek and obtain visitation of stepchildren after a divorce or custody of stepchildren after the biological parent dies. In most states, unless one adopts one's stepchild—which requires a biological parent to sign away his or her parental rights—there is no legal standard as to what is expected of one or what one is entitled to as a stepparent. Even the weighty issues of financial support or authority to approve medical treatment are not uniformly treated throughout the United States (Engel 2000).

Thus, what one finds in examining the legal rights and responsibilities of stepfamily members is policy that is inconsistent at best (varying from state to state). As Stewart (2007) describes, inconsistency is not the only problem. In general, there is a deficit of stepfamily law. Given how many men, women, and children are or will be members of stepfamilies, it seems crucial for states to clarify how far stepparent rights extend and how wide stepparent responsibilities range.

---

biological mother who believed that she had a good remarriage stated:

> Sometimes I feel he is too harsh in disciplining, or he doesn't have the patience to explain why he is punishing and to carry through in a calm manner, which causes me to have to step into the matter (which I probably shouldn't do). . . . I do realize that it was probably hard for my husband to enter marriage and the responsibility of a family instantly . . . but this has remained a problem. (Ihinger-Tallman and Pasley 1987)

As a result of interference, the biological parent implies that the stepparent is wrong and undermines his or her status in the family. Over time, the stepparent may decrease involvement in the family as a parent figure.

### Money, Goods, and Services

Problems of allocating money, goods, and services exist in all families, but they can be especially difficult in stepfamilies. In first marriages, husbands and wives form an economic unit in which one or both may produce income for the family, the husband and wife being interdependent. Following divorce, the binuclear family consists of two economic units: the custodial family and the noncustodial family. Both must provide separate housing, dramatically increasing their basic expenses. Despite their separation, the two households may nevertheless continue to be extremely interdependent. The mother in the custodial single-parent family, for example, probably has reduced income. She may be employed but still dependent on child support payments or welfare, such as Temporary Assistance for Needy Families. She may have to rely more extensively on child care, draining her resources dramatically. The father in the noncustodial family may make child support payments or contribute to medical or school expenses, depleting his income. Both households may have to deal with financial instability. Custodial parents cannot count on always receiving their child support payments, making it difficult to undertake financial planning.

When one or both of the former partners remarry, their financial situation may be altered significantly. On remarriage, the mother receives less income from her former partner or lower welfare benefits. Instead, her new partner becomes an important contributor to the family income. At this point, a major problem in stepfamilies arises. What responsibility does the stepfather have in supporting his stepchildren? Should he or the biological father provide financial support? Because there are no norms, each family must work out its own solution.

Stepfamilies typically have resolved the problem of distributing their economic resources by using a one-pot or two-pot pattern (Fishman 1983). In the *one-pot* (or "common pot") pattern, families pool their resources and distribute them according to need rather than biological relationship. It doesn't matter whether the child is a biological child or a stepchild. One-pot families typically have relatively limited resources and consistently fail to receive child support from the noncustodial biological parent. On the plus side, by sharing their resources, one-pot families increase the likelihood of family cohesion and are more likely than two-pot families to report higher levels of family satisfaction (American Psychological Association 2016).

In *two-pot* families, resources are distributed by biological relationship; need is secondary. These families tend to have a higher income, and one or both parents have former spouses who regularly contribute to the support of their biological children. Expenses relating to children are generally handled separately; usually, there are no shared checking or savings accounts. Two-pot families maintain strong bonds among members of the first family. For these families, a major problem is achieving cohesion in the stepfamily while maintaining separate checking accounts. Some families are still better characterized as three-pot families, "yours," "mine," and "ours" (Deal and Olson 2010).

Just as economic resources need to be redistributed following divorce and remarriage, so do goods and services. Whereas a two-bedroom home or apartment may have provided plenty of space for a single-parent family with two children, a stepfamily with additional residing or visiting step-siblings can experience instant overcrowding. Rooms, bicycles, and toys, for example, need to be shared; larger quarters may have to be found. Time becomes a precious commodity for harried parents and stepparents in a stepfamily. When visiting stepchildren arrive, duties are doubled. Stepchildren compete with parents and other children for time and affection.

It may appear that remarried families are confronted with many difficulties, but traditional nuclear families also encounter financial, loyalty, and discipline problems. We need to put these problems in perspective. Neither first marriages nor remarriages are problem free. Although clearly different, both traditional "intact" nuclear families and stepfamilies have strengths as well as struggles.

## Strengths of Stepfamilies

Because we have traditionally viewed stepfamilies as less than ideal, we have often ignored their strengths. Instead, we have seen only their problems. We end this chapter by focusing on the strengths of blended families.

### Family Functioning

Although traditional nuclear families may be structurally less complicated than stepfamilies, stepfamilies are nevertheless able to fulfill traditional family functions. A binuclear, single-parent, custodial, or noncustodial family may provide more companionship, love, and security than the particular traditional nuclear family it replaces. If the nuclear family was ravaged by conflict or violence, for example, the single-parent family or stepfamily that replaces it may be considerably better, and because children now see happy parents, they have positive role models of marriage partners (Rutter 1994).

New partners may have greater objectivity regarding old problems or relationships. Opportunity presents itself for flexibility and patience. As family boundaries expand, individuals grow and adapt to new personalities and ways of being. In addition, new partners are sometimes able to intervene between former spouses to resolve long-standing disagreements, such as custody or child care arrangements.

### Benefits for Children

As illustrated earlier, stepfamilies are often associated with a number of problematic outcomes for children. But potentially, blended families can offer children benefits that can compensate for the negative consequences of divorce and of living with a single parent. Remember the notion of "bonus families" introduced earlier? Here are some ways in which stepfamilies could offer children some bonuses:

- Children gain additional role models from which to choose. Instead of having only one mother or father after whom to model themselves, children may have two mothers or fathers—the biological parents and the stepparents.

- Children gain greater flexibility. They may be introduced to new ideas, different values, or alternative politics. For example, biological parents may be unable to encourage certain interests, such as music or model airplanes, whereas a stepparent may play the piano or be a die-hard modeler. In such cases, that stepparent can assist the stepchildren in pursuing their development. In addition, children often have alternative living arrangements that enlarge their perspectives.

- Stepparents may act as a sounding board for their children's concerns. They may be a source of support or information in areas in which the biological parents feel unknowledgeable or uncomfortable.

- Children may gain additional siblings, either as step-siblings or half siblings. Although there is evidence that living with step-siblings or half siblings is associated with problematic outcomes, especially educational outcomes, there are other potential outcomes that can be seen more positively. With step-siblings, children gain additional experience in interacting, cooperating, and learning to settle disputes among peers.

- Children gain an additional extended kin network, which may become at least as important and loving as their original kin network.

- A child's economic situation is often improved, especially if a single mother remarried.

- Children may gain parents who are happily married. Most research indicates that children are significantly better adjusted in happily remarried families than in conflict-ridden nuclear families.

As sociologist Susan Stewart points out, "There are both risks and protective factors present in *all* families." She goes on to state the following, which seems an especially useful way to close:

> In the end, the structure of a child's family (intact, single parent, stepparent, etc.) is probably less important than what goes on *within* a child's particular family environment. (Stewart 2007)

If anything, it should be clear by now that the American family is no longer what it was through most of the 20th century. Remember, though, that families are dynamic and diverse—they change and they differ—and thus the rise of the single-parent family and stepfamily does not imply an end to the family, nor does divorce or cohabitation. Rather, these forms provide different paths that contemporary families take as they strive to fulfill the hopes, needs, and desires of their members.

## Summary

- Many of today's families, including single-parent families and stepfamilies, depart from the traditional family system based on lifetime marriage and the intact nuclear family.

- Single-parent families tend to be created by divorce or births to unmarried women, and are generally headed by women. Single, custodial fathers typically take different paths to single parenthood, obtaining custody because mothers are financially unable to provide adequate care for children, are physically or psychologically unfit, or do not want full-time responsibility for raising children.

- Both mother-only and father-only families are more likely to be poor than are two-parent families, but because of gender discrimination and inequality in wages or job opportunities, many female-headed families face more extensive economic hardship.

- Many "single-parent households" actually contain the parent and his or her unmarried partner. Even in the absence of parents' live-in partners, parents' romantic partners, relatives, or family associates may play important roles in children's lives.

- Single parents, especially mothers, often come to rely on a combination of state or federal assistance and *private safety nets*—support from their social networks on which they can fall back in times of economic need.

- Because of a lack of resources (such as money, attention, and guidance), children of single parents are more likely to engage in high-risk, "health-compromising" behaviors and to suffer a variety of educational, economic, and personal costs.

- The inability to marry and/or the inability to adopt the children of one's partner forces many gay or lesbian parents into legal single parenthood while denying others the opportunity to develop and maintain ties with their partners' children.

- The *binuclear family* is a postdivorce family system with children. It consists of two nuclear families: the mother-headed family and the father-headed family.

- Courtship for second marriage lacks clear norms. Courtship is complicated by the presence of children because remarriage involves the formation of a stepfamily.

- Cohabitation is more common in the "courtship" process leading to remarriage. As with cohabitation in first marriages, cohabitation before remarriage leads to higher rates of marital instability.

- Remarried couples are more likely to divorce than couples in their first marriage.

- Traditionally, researchers viewed stepfamilies from a "deficit" perspective, assuming that stepfamilies are very different from traditional nuclear families. More recently, stepfamilies have been viewed as normal families.

- Research in the United States and a number of other countries reveals some hazards of stepfamilies for children, including academic, behavioral, and health-related difficulties. Some research indicates that girls adjust less well than boys to stepfamily life.

- Relations between stepparents and their stepchildren have been characterized as "disengaged."

- Becoming a stepfamily is a process that is experienced differently by each person involved—the biological parent, the stepparent, and the stepchild (or children). Stepmothers tend to experience greater stress in stepfamilies than do stepfathers. The warmer a woman is to her stepchildren, the more hostile they may become to her because they feel that she is trying to replace their "real" mother. Men are generally less involved in child rearing; they usually have few "cruel stepparent" myths to counter.

- A stepfather usually joins an already established single-parent family. The longer a single-parent family has been functioning, the more difficult it usually is to reorganize it.

- Conflict in stepfamilies is often over favoritism, divided loyalties, discipline, and money, goods, and services. The addition of a new baby into a stepfamily neither solidifies nor divides the family. A key issue for stepfamilies is family solidarity—the feeling of oneness with the family.

- Stepfamily strengths may include improved family functioning and reduced conflict between former spouses. Children may gain multiple role models, more flexibility, concerned stepparents, additional siblings, additional kin, an improved economic situation, and happily married parents.

## Key Terms

deficit approach  534
paternal claiming  547
private safety nets  531
remarriage  537
single-parent families  527
social capital  532
social father  531
stepfamilies  527
stepparent role  546

# Glossary

## A

**abstinence** Refraining from sexual intercourse, often on religious or moral grounds.

**abuse of power** Sexual harassment consisting of unwelcome sexual advances, requests for sexual favors, or other verbal or physical conduct of a sexual nature as a condition of instruction or employment.

**accepters** Individuals who perceive family trends as either good or making no difference in society.

**accessibility** Availability of a parent or caregiver to a child, when the parent is in the same location but not in direct interaction.

**acquaintance rape** Rape in which the assailant is personally known to the victim, usually in the context of a dating relationship. Also known as *date rape*.

**acquired immunodeficiency syndrome (AIDS)** An infection caused by the human immunodeficiency virus (HIV), which suppresses and weakens the immune system, leaving it unable to fight opportunistic infections.

**adaptation** In ecological theory, the processes of responding to the circumstances imposed by the physical, social, cultural, and economic environments within which we live.

**adolescence** The social and psychological state occurring during puberty.

**agents of socialization** Individuals, groups, and organizations that shape our self identities, our ideas about gender differences, and our understanding of expected and acceptable behavior for each gender.

**affiliated kin** Unrelated individuals who are treated as if they were related.

**affirmative consent** Policy in which there must be clear, affirmative agreement between partners regarding mutual interest in further sexual intimacy.

**alimony** Court-ordered monetary support to a spouse or former spouse following separation or divorce.

**ambiguous loss** A situation of uncertainty and unclear loss, resulting from confusion about a family's boundaries, and from not knowing who is in or out of a particular family.

**anal eroticism** Sexual activities involving the anus.

**anal intercourse** Penetration of the anus by the penis.

**androgyny** Displaying both masculine and feminine qualities.

**anonymity** A state or condition requiring that no one, including the researcher, can connect particular responses to the individuals who provided them.

**antigay prejudice** Strong dislike, fear, or hatred of gay men and lesbians because of their homosexuality. See also *homophobia*.

**applied research** The focus of such research is more practical than theoretical. It tends to be less concerned with formulating theories, generating concepts, or testing hypotheses. Data are gathered in an effort to solve problems, evaluate policies or programs, or estimate the outcome of some proposed future change in policy.

**asexual** One who lacks sexual attraction to anyone.

**assisted reproductive technologies** Medical procedures, such as in vitro fertilization, used in the treatment of infertility.

**attachment theory of love** A theory maintaining that the degree and quality of an infant's attachment to his or her primary caregiver is reflected in his or her love relationships as an adult.

**attributions** The ways individuals account for and explain relationship failures. Such attributions influence the level of distress felt after a breakup.

**authoritarian child rearing** A parenting style characterized by the demand for absolute obedience.

**authoritative child rearing** A parenting style that recognizes the parent's legitimate power and also stresses the child's feelings, individuality, and need to develop autonomy.

**autoeroticism** Erotic behavior involving only the self; usually refers to masturbation but also includes erotic dreams and fantasies.

## B

**battered child syndrome** Medical term to describe patterns of injuries commonly found in physically abused children.

**battering** A violent act directed against another, such as hitting, slapping, beating, stabbing, shooting, or threatening with weapons.

**bias** A personal leaning or inclination.

**bifurcation of working time** Situation wherein some work longer and longer days and weeks while others work fewer hours than they want or need.

**binuclear family** A postdivorce family with children, consisting of the original nuclear family divided into two families, one headed by the mother, the other by the father; the two "new" families may be either single-parent or stepfamilies.

**biphobia** Negative attitudes toward bisexuals that is separate and distinct from homophobia.

**bisexuality** Sexual involvement with both sexes, usually sequentially rather than during the same time period.

**blended family** A family in which one or both partners have a child or children from an earlier marriage or relationship; a stepfamily. See also *binuclear family*.

**boomerang generation** Individuals who, as adults, return to their family home and live with their parents.

**bundling** A colonial Puritan courtship custom in which a couple slept together with a board separating them.

## C

**caballerismo** Gender concept in Latino culture, emphasizing male qualities such as chivalry, and protecting and respecting women's honor.

**case-study method** In clinical research, the in-depth examination of an individual or small group in some form of psychological treatment in order to gather data and formulate hypotheses.

**centrists** These people share aspects of both conservative and liberal positions. Like liberals, they identify wider social changes (e.g., economic or demographic) as major determinants of the changes in family life, but like conservatives, they believe that some familial changes have had negative consequences.

**child-free** A term used to refer to partners who choose not to have children.

**child maltreatment** Instances of the victimization of children from neglect, physical abuse, sexual abuse, and/or emotional abuse.

**child sexual abuse** Any sexual interaction, including fondling, erotic kissing, oral sex, or genital penetration, that occurs between an adult (or older adolescent) and a prepubertal child.

**child support** Court-ordered financial support by the noncustodial parent to pay or assist in paying child-rearing expenses incurred by the custodial parent.

**churning** Relationships characterized by breakups followed by reconciliation, and sexual intimacy between ex-partners. See also *relationship cycling*.

**cisgender** A term used to refer to those whose gender identities match their biological sex and sex of assignment.

**clan** A group of families related along matrilineal or patrilineal descent lines, regarded as the basic family unit in some cultures.

**clinical research** The in-depth examination of an individual or small group in clinical treatment in order to gather data and formulate hypotheses. See also *case-study method*.

**closed field** A setting in which potential partners may meet, characterized by a small number of people who are likely to interact, such as at a class, dormitory, or party. Compare to *open field*.

**cognition** The mental processes, such as thought and reflection, that occur between the moment we receive a stimulus and the moment we respond to it.

**cognitive development theory** A theory of socialization associated with Swiss psychologist Jean Piaget in which the emphasis was placed on the child's developing abilities to understand and interpret their surroundings.

**cohabitation** The sharing of living quarters by two heterosexual, gay, or lesbian individuals who are involved in an ongoing emotional and sexual relationship. The couple may or may not be married.

**cohabitation effect** The cohabitation effect refers to the fact that people who cohabit before marriage appear to have a greater risk of divorce than those couples who don't cohabit.

**coitus** The insertion of the penis into the vagina and subsequent stimulation; sexual intercourse.

**coming out** For gay, lesbian, and bisexual individuals, the process of publicly acknowledging one's sexual orientation.

**commitment** Factors that help maintain a relationship, for better or worse, including love, obligation, and social pressure.

**common couple violence** Sociologist Michael Johnson's term for the more routine forms of partner violence that result from disputes and disagreements, and for which there is a high degree of gender symmetry.

**common law marriage** A legal marriage resulting from two people living together as though married, and presenting themselves as married, even though their relationship was never formalized by religious or secular ceremony.

**companionate love** A form of love emphasizing intimacy and commitment.

**companionate marriage** A marriage characterized by shared decision making and emotional and sexual expressiveness.

**complementary needs theory** A theory of mate selection suggesting that we select partners whose needs are different from and/or complement our own needs.

**concepts** Abstract ideas that we use to represent the reality in which we are interested, and to focus our research and organize our data.

**conceptualization** The specification and definition of concepts used by the researcher.

**conduct of fatherhood** Men's actual participation in raising their children.

**confidentiality** An ethical rule according to which the researcher knows the identities of participants and can connect what was said to who said it but promises not to reveal such information publicly.

**conflict-habituated marriages** Relationships in which tension, arguing, and conflict permeate the relationship.

**conflict theory** A social theory that views individuals and groups as being basically in competition with one another. Power is seen as the decisive factor in interactions.

**conjugal family** A family consisting of husband, wife, and children. See also *nuclear family*.

**conjugal relationship** A relationship formed by marriage.

**consanguineous relationship** A relationship formed by common blood ties.

**consummate love** In Robert Sternberg's theory, a love that includes passion, intimacy, and commitment. It is idealized, sought after, yet difficult to find and sustain.

**contempt** Verbal or nonverbal communication with another that conveys that the recipient is undesirable. Contempt can be displayed verbally through insults, sarcasm, and mockery, or nonverbally through such expressions as rolling one's eyes.

**conservatives** People who tend to believe that cultural values have shifted from individual self-sacrifice for their families toward personal self-fulfillment. Conservatives believe that as a result of such changes, today's families are weaker and less able to meet the needs of children, adults, or the wider society.

**cooperative co-parenting** In co-operative co-parenting, parents cooperate, communicate, and avoid or minimize conflict about caregiving and what their children need.

**co-provider families** Families that are dependent on economic activity from both men and women.

**co-rumination** Self-disclosure consisting of excessive discussion of personal problems.

**crossover** A situation in which one's job-related emotional state affects one's partner in the same way.

**crude divorce rate** A statistical measure of divorce calculated on the basis of the number of divorces per 1,000 people in the population.

**cultural lag** The outcome of rapid social change, when part of the culture changes more rapidly than another part.

**culture of fatherhood** Ralph LaRossa's term for the beliefs we have about the roles, responsibilities, and involvement of fathers in raising their children. LaRossa noted that these beliefs have changed more dramatically than has the conduct of fatherhood.

**culture of poverty** A concept that depicts problems faced by poor and especially poor black families as due to their having internalized values and adopted practices that end up keeping them poor.

**cunnilingus** Oral stimulation for female genitals.

# D

**date rape** Rape in which the assailant is personally known to the victim, usually in the context of a dating relationship. Also known as *acquaintance rape*.

**deductive research** Research designed to test hypotheses and examine causal relationships between variables.

**deficit approach** An approach to studying single-parent families that emphasizes the stresses and difficulties parents and children face.

**deinstitutionalization of marriage** The weakening of social norms that define people's behavior in marriage.

**Defense of Marriage Act** Federal legislation signed into law by President Clinton denying recognition to same-sex couples, should any state legalize same-sex marriage.

**demand–withdraw communication** A communication pattern in which one person makes an effort to engage the other person in a discussion of some issue of importance. The one raising the issue may criticize, complain, or suggest a need for change in the other's behavior or in the relationship. In response, the other party withdraws by either leaving the discussion, failing to reply, or changing the subject.

**demography** "the study of the size, distribution, and composition of the population, changes therein, and the demographic processes underlying demographic change, including fertility, mortality, and migration" (Lichter 2013, F2).

**dependent variable** A variable that is observed or measured in an experiment and may be affected by another variable. See *independent variable*.

**devitalized marriages** A marriage type that begins with high levels of emotional intensity that dwindles over time.

**distal causes** Background characteristics (e.g., age at marriage, prior cohabitation, parental divorce) brought by spouses into marriage that raise the likelihood of later marital problems.

**division of labor** The interdependence of people with specialized tasks and abilities. Within the family, labor is traditionally divided along gender lines. See also *complementary marriage model*.

**divorce mediation** The process in which a mediator (counselor) assists a divorcing couple in resolving personal, legal, and parenting concerns in a cooperative manner.

**domestic partnership** Cohabitation of heterosexual or same-gender couples who can enter legal relationships as domestic partners, wherein they gain many of the legal rights otherwise accorded to married couples.

**double standard of aging** The devaluation of women in contrast to men in terms of attractiveness as they age.

**dual-career families** Subcategory of dual-earner families in which both partners or spouses have high achievement orientations, a greater commitment to equality, and a stronger desire to exercise their capabilities through their work.

**Duluth Model** A multistage rehabilitative approach for batterers, which emphasizes helping batterers develop critical thinking skills around themes of nonviolence, respect, partnership, and negotiation.

**duration-of-marriage effect** The accumulation over time of various factors, such as poor communication, unresolved conflicts, role overload, heavy work schedules, and child-rearing responsibilities, that negatively affect marital satisfaction.

**dyspareunia** Painful sexual intercourse.

# E

**egalitarian** As used to describe both societies and social groups, instances in which women and men possess similar amounts of power and neither dominates economically or politically.

**egocentric fallacy** The mistaken belief that one's own personal experience and values are those of others in general.

**emotion work** Activities that are designed to manage the emotional climate within families and relationships, such as providing emotional support, eliciting and sharing feelings, and monitoring the relationship for signs of trouble or problems.

**empty nest** The experience of parents when the last grown child has left home. The "empty nest syndrome," in which the mother becomes depressed after the children have gone, is believed to be more of a myth than a reality.

**endogamy** Marriage within a particular group. Compare to *exogamy*.

**engagement** A pledge to marry.

**environment** A central concept in ecological theory, referring to the physical, social, cultural, and economic situations and circumstances within which individuals and families live.

**environmental influences** The wider context and external influences on families that are the focus of family ecological theory.

**equity** In social exchange theory, the result of exchanges that are fair and balanced.

**erectile dysfunction** Inability or difficulty in achieving erection.

**ethical guidelines** Standards agreed upon by professional researchers. These guidelines protect the privacy and safety of individuals who provide information in a research setting.

**ethnic group** A large group of people distinct from others because of cultural characteristics, such as language, religion, and customs, transmitted from one generation to another. See also *minority group* and *racial group*.

**ethnocentric fallacy** (also ethnocentrism) The belief that one's own ethnic group, nation, or culture is inherently superior to others. See also *racism*.

**exogamy** Norm requiring one to marry outside certain groups (e.g., outside one's family).

**experimental research** A research method involving the isolation of specific factors (variables) under controlled circumstances to determine the effects of each factor.

**expressive displays** Ways of communicating feelings of love, largely through verbal expression.

**expressive trait** A supportive or emotional personality trait or characteristic.

**extended family** The family unit of parent(s), child(ren), and other kin, such as grandparents, uncles, aunts, and cousins.

**extended household** A household composed of several different families.

**extensive mothering** A version of mothering that stresses self-sacrificing, labor-intensive, all-encompassing responsibility for one's child.

**extrafamilial sexual abuse** Child sexual abuse that is perpetrated by nonrelated individuals. Compare to *intrafamilial sexual abuse*.

**extramarital sex** Sexual activities, especially sexual intercourse, occurring outside the marital relationship.

# F

**fallacy** A fundamental error in reasoning that affects our understanding of a subject.

**familism** A pattern of social organization in which family loyalty and strong feelings for the family are important.

**families of choice** Family-like relationships, with fluid boundaries crossing multiple households, constructed by gay men and lesbians, consisting of those who can be counted upon to provide emotional, financial, and practical support.

**family** A unit of two or more people, of which one or more may be children who are related by blood, marriage, or affiliation and who cooperate economically and may share a common dwelling place.

**Family and Medical Leave Act of 1993** U.S. policy requiring employers to provide employees 12 weeks of job-protected, unpaid leave from work to meet family or medical needs. Such a leave is required of employers who employ 50 or more employees. To qualify for such a leave, an employee must have been employed for 12 months by the same company and worked a minimum of 1,250 hours.

**family development theory** A micro-level perspective that emphasizes the patterned changes that occur in families through stages and across time.

**family ecology theory** A macro-level theory, emphasizing how families are influenced by and in turn influence the wider environment.

**family life cycle** A developmental approach to studying families, emphasizing the family's changing roles and relationships at various stages, beginning with marriage and ending when both spouses have died.

**family of orientation** The family in which a person is reared as a child. Compare to *family of procreation*.

**family of procreation** The family formed by a couple and their child or children. See also *family of cohabitation*.

**family policy** A set of objectives concerning family well-being and specific measures initiated by government to achieve them.

**family systems theory** A theory viewing family structure as created by the pattern of interactions between its various subsystems, and individual actions as being strongly influenced by the family context.

**fault-based divorce** Adversarial process of divorce in which one spouse alleges that the other has committed an act that has caused the marriage to break down or has failed to act in ways that would preserve the marriage.

**fellatio** Oral stimulation of the male genitals.

**feminist perspectives** A variety of perspectives that stress the importance of gender and gender inequality in shaping social and familial experience.

**feminization of love** The idea that our cultural construction of love is based on mostly expressive qualities, more compatible with women's earlier socialization. More instrumental displays of love tend not to be recognized as love.

**feminization of poverty** The shift of poverty to females, primarily as a result of high divorce rates and births to unmarried women.

**fictive kin ties** The extension of kinship-like attributes to nonblood relationships (such as friends or neighbors) to demonstrate their importance and to symbolize the mutual reciprocity found within them.

**filial responsibility** Responsibility to see that one's aging parents have support, assistance, and company, and are well cared for.

**filter theories** These theories suggest that the choice of a partner or spouse is based on a number of different factors that become more or less important as relationships develop, grow, and change.

**flextime** A policy allowing workers to choose and change their work schedules to meet family needs.

**foreplay** Erotic activity prior to coitus, such as kissing, caressing, sex talk, and oral/genital contact; petting.

## G

**gatekeeping** A method of exerting control over and exercising power in such familial situations as caring for children by making the decisions that determine how.

**gender** The division into male and female, often in a social sense; sex.

**gender binary** Term referring to a dichotomous understanding of gender, in which individuals are either male or female.

**gender identity** The psychological sense of whether one is male or female.

**gender ideology** Arlie Hochschild's term for what individuals believe they ought to do as husbands or wives, and how they believe paid and unpaid work should be divided.

**gender polarization** The cultural belief in the genders as truly opposites, fundamentally different from each other.

**gender-rebellion feminism** Versions of feminism that emphasize the interconnectedness between multiple inequalities (race, class, sexual orientation, age, and gender), and see gender inequality as only one aspect of wider social inequality.

**gender-reform feminism** Versions of feminism that stress how similar women and men are and emphasize the need for equal rights and opportunities for both genders.

**gender-resistant feminism** Versions of feminism that advocate separatist strategies, wherein women establish women-only social institutions and settings.

**gender role** Expectations attached to being identified as male or female that guide one's behavior.

**gender socialization** Refers to the process by which we come to learn what behaviors, demeanor, and temperament is expected of us, by virtue of our assigned gender.

**gender spectrum** A more nuanced understanding of gender as consisting of a wide range of identities and ways of expressing oneself.

**gendered role** The culturally assigned role that a person is expected to perform based on male or female gender.

**gender stratification** Inequality between women and men, wherein one gender occupies a privileged position relative to the other gender in access to economic resources, political power, and societal influence.

**gender symmetry** The similarity in survey estimates of male-on-female and female-on-male intimate partner violence.

**gender theory** A theory in which gender is viewed as the basis of hierarchal social relations that justify greater power to males.

**generativity** A commitment to guiding or nurturing others.

**grounded theory** Theory that emerges from inductive research and is rooted in repeated observations.

## H

**halo effect** The tendency to infer positive characteristics or traits based on a person's physical attractiveness.

**hegemonic masculinity and emphasized femininity** Constructions of gender that become dominant within a society at a particular point in time and come to be seen as the normative way of being male and female.

**heterogamy** Marriage between those with different social or personal characteristics. Compare to *homogamy*.

**heterosexism** Actions that deny, denigrate, or stigmatize nonheterosexual behavior, relationships, identity, or community.

**heterosexuality** Sexual orientation toward members of the opposite sex.

**hierarchy of needs** Abraham Maslow's theoretical construct in which human needs are ranked from most basic to higher-level needs. The most basic needs are physiological (e.g., water, oxygen, food), followed by safety needs. Once these are met, the need for intimacy and belonging is most fundamental.

**homemaker role** A family role usually allocated to women, in which they are primarily responsible for home management, child rearing, and the maintenance of kin relationships. Traditionally the role is associated with economic dependency and has primacy over other female roles.

**homeostasis** A social group's tendency to maintain internal stability or balance and to resist change.

**homogamy** Marriage between those with similar social or personal characteristics. Compare to *heterogamy*.

**homophobia** Irrational or phobic fear of gay men and lesbians.

**homosexuality** Sexual orientation toward members of the same sex. See also *gay male* and *lesbian*.

**honeymoon effect** The tendency of newly married couples to overlook problems, including communication problems.

**hooking up** Relationship pattern more common among college students in which a male and female pair off after a party or evening at a bar with the expectation of some physical intimacy to follow.

**hostile environment** An environment created through sexual harassment in which the harassed person's ability to learn or work is negatively influenced by the harasser's actions.

**household** As defined by the U.S. Census Bureau, one or more people; everyone living in a housing unit makes up a household.

**human immunodeficiency virus (HIV)** The virus causing AIDS.

**hypergamy** A marriage in which one's spouse is of a higher social class or rank.

**hypogamy** A marriage in which one's spouse is from a lower social standing.

**hypothesis** An unproven theory or proposition tentatively accepted to explain a collection of facts.

## I

**ideology of intensive mothering** Ideology of intensive mothering is a belief system that portrays mothering as emotionally absorbing, labor-intensive, guided by experts, and financially expensive.

**impaired fecundity** The experience of difficulty conceiving or carrying a pregnancy to term.

**incest** Sexual intercourse between individuals too closely related to marry, usually interpreted to mean father/daughter, mother/son, or brother/sister. See also *intrafamilial sexual abuse.*

**independent variable** A variable that may be changed or manipulated in an experiment.

**individualized marriage** Marriage type where the emphasis is on personal self-fulfillment more than marital commitment.

**inductive research** Unlike deductive research, inductive research does not begin with hypotheses to test. Research begins with a topic of interest and some concepts that the researcher explores. As data are collected, concepts are refined, patterns are identified, and hypotheses are generated.

**indulgent child rearing** Permissive childrearing.

**infant mortality rate** The number of deaths for every 1,000 live births.

**initiator** The spouse or partner whose unhappiness or dissatisfaction leads them to consider divorce or a breakup and sets in motion the process of "uncoupling."

**instrumental displays** Displaying love more by what one does than what one says; tasks done out of and displaying love.

**instrumental trait** A practical or task-oriented personality trait or characteristic.

**intensive mothering ideology** The belief that children need full-time, unconditional attention from mothers to develop into healthy, well-adjusted people.

**interaction** In communication, a reciprocal act that takes place between at least two people.

**intergenerational transmission** The increased likelihood that children of divorced parents will themselves later divorce.

**intersectionality** An approach to studying gender that pays additional attention to race, ethnicity, and class.

**intersexed** Condition of having been born with reproductive or sexual anatomy that doesn't seem to fit the typical definitions of female or male.

**intervening variable** A variable that is affected by the independent variable and in turn affects the dependent variable.

**intimacy** Closeness between two people, can be expressed physically (through sexual relations or displays of affection), or emotionally (by sharing ourselves, our feelings, with another).

**intimate partner abuse and intimate partner violence** Terms currently used to address the full scope of violence and/or abuse among intimate couples, regardless of gender, marital status, or sexual orientation.

**intimate terrorism** Attempts by one partner to dominate and control the other through escalating violence, threats of further violence, and emotional abuse. Such violence is most often committed by men, is likely to recur, escalate, and lead to injury or death.

**intimate zone** Spatial distance (0 to 18 inches in the United States) typically reserved for one's most intimate relationships.

**intrafamilial sexual abuse** Child sexual abuse that is perpetrated by related individuals, including step-relatives. See also *incest*; compare to *extrafamilial sexual abuse.*

## J

**jealousy** An aversive response occurring because of a partner's or other significant person's real, imagined, or likely involvement with or interest in another person.

**joint custody** Custody arrangement in which both parents are responsible for the care of the child. Joint custody takes two forms: *joint legal custody* and *joint physical custody.* See also *sole custody* and *split custody.*

**joint legal custody** Joint custody in which the child lives primarily with one parent but both parents jointly share in important decisions regarding the child's education, religious training, and general upbringing.

**joint physical custody** Joint custody in which the child lives with both parents in separate households and spends more or less equal time with each parent.

## K

**kinship system** The social organization of the family conferring rights and obligations based on an individual's status.

## L

**learned helplessness** Term to describe the state of mind experienced by a female victim of repeated battering, resulting in her developing a low self-concept and a feeling that she cannot control the circumstances of her life.

**liberals** People who tend to believe that the changes in family patterns are products of and adaptations to wider social and economic changes rather than a shift in cultural values.

**life chances** Opportunities to enjoy a healthy and fulfilling life that are affected by one's social class standing.

**linked lives** Concept central to the developmental/life course perspective, emphasizing how the shape of one's life course is influenced by the shape of the life courses of others.

## M

**machismo** Traditional and exaggerated Latino gender role expectations for males, emphasizing male superiority and dominance over women.

**macro-level theories** Theories that focus on the family as a social institution: family ecology theory, conflict theory, structural functional theory, and feminist perspectives are all macro-level theories.

**mandatory arrest** A criminal justice policy requiring police who respond to a report of domestic violence to make an arrest if they determine that an incident of domestic abuse has occurred.

**mandatory visitation** Recommended policy wherein visitation by noncustodial parents is considered an obligation that will be enforced by authorities.

**marianismo** Female gender ideal among Latinas, emphasizing qualities such as sexual purity, submissiveness, selflessness, and self-sacrifice.

**marital commitment** Factors that keep people married, including the personal, moral, and structural commitments.

**marital history homogamy** Marriage between people of similar marital backgrounds, such as two formerly married people, two widowed people, or two never-married people.

**marital paradigm** A set of images about how marriage ought to be done, "for better or worse."

**marital sanctification** A process through which one's marriage is believed to be sacred, have divine character, and in which God is believed to be an active partner in the relationship.

**marketplace of relationships** A concept that portrays relationship formation as an exchange, wherein we are evaluated by and evaluate others on the basis of the qualities they possess or the resources they bring.

**marriage** The legally recognized union between a man and woman in which economic cooperation, legitimate sexual interactions, and the rearing of children may take place.

**marriage debate** Disagreement about the health and future of marriage.

**marriage squeeze** The phenomenon in which there are greater numbers of marriageable women than marriageable men, particularly among older women and African American women. See also *mating gradient*.

**masturbation** Manual or mechanical stimulation of the genitals by self or partner; a form of autoeroticism.

**mating gradient** The tendency for women to marry men of higher status.

**matriarchy** A form of social organization in which the mother or eldest female is recognized as the head of the family, kinship group, or tribe, and descent is traced through her. Compare to *patriarchy*.

**matrilineal** Descent or kinship traced through the mother. Compare to *patrilineal*.

**matrimania** The glorification of marriage as the ultimate source of happiness and fulfillment with the implication that unmarried people can never achieve that same fulfillment.

**medicalization of childbirth** The process wherein women receive impersonal care during childbirth, without having the opportunity to exercise control or have input over their experiences, and where the birth process entails routine but potentially unnecessary use of medical interventions.

**menopause** Cessation of menses for at least one year as a result of aging.

**mental labor** The process of worrying about one's baby, seeking and processing information on child care, and managing the division of infant care in the household.

**micro-level theories** Theories that emphasize what happens within families, looking at everyday behavior, interaction between family members, patterns of communication, and so on. Symbolic interaction theory, social exchange theory, family developmental theory, and family systems theory are all examples of micro-level theories.

**middle class** Often subdivided into lower- and upper-middle class, the middle class is the large group of people who are between the upper classes and the working class in terms of wealth, income, education, and occupational status.

**minority group** A social category composed of people whose status places them at economic, social, and political disadvantage. Compare to *majority group*; see also *ethnic group*.

**minority status** Social rank having unequal access to economic and political power.

**modeling** Learning gender role expectations by imitation.

**modified extended family** Extended families in which members share contact, care, and support even though they don't share a residence.

**monogamy** (1) The practice of having only one husband or wife at a time; and (2) sexual exclusiveness (colloquial).

**mutual violent control** Relationships in which both partners are using violence in an effort to control each other and their relationship.

## N

**no-drop prosecution** A criminal justice policy requiring prosecutors to proceed in prosecuting a case of alleged intimate violence as long as evidence exists, regardless of the expressed wishes of the victim.

**no-fault divorce** The dissolution of marriage because of irreconcilable differences for which neither party is held responsible.

**nonmarital sex** Sexual activities, especially sexual intercourse, that take place among older single individuals. Compare to *premarital sex* and *extramarital sex*.

**nonparental households** Households in which children live with no biological parents.

**nonverbal communication** Communication of emotion by means other than words, such as touch, body movement, and facial expression.

**normative age-graded influences** Biological or social influences on family experience that are correlated with age.

**normative history-graded influences** Influences that are common to a particular generation, such as the political and economic influences of wars and economic depressions, and that are similar for individuals in a particular age group.

**nuclear family** The basic family building block, consisting of a mother, father, and at least one child; in popular usage, used interchangeably with *traditional family*. Some anthropologists argue that the basic nuclear family is the mother and child dyad.

## O

**objective statement** A factual statement presenting information based on scientifically measured findings, not on opinions or personal values.

**objectivity** Suspending the beliefs, biases, or prejudices we have about a subject until we have really understood what is being said.

**observational research** Research method using unobtrusive, direct observation.

**open adoption** A form of adoption in which the birth mother has an active part in choosing the adoptive parents; there is a certain amount of information exchanged between the birth mother and the adoptive parents, and there may be some form of continuing contact between the birth mother, the child, and the adoptive family following adoption.

**open field** A setting in which potential partners may not be likely to meet, characterized by large numbers of people who do not ordinarily interact, such as at a beach, shopping mall, or large university campus. Compare to *closed field*.

**operationalization** The identification and/or development of research strategies to observe or measure concepts.

**opinion** An unsubstantiated belief or conclusion based on personal values or biases.

**outing** The act of publicly disclosing the sexual orientation of gays, lesbians, or bisexuals.

## P

**palimony** Financial support paid to an ex-partner, like alimony, but with reference to former cohabitants.

**pansexual** Someone who is physically or romantically attracted to others, regardless of their gender identity or biological sex.

**parental image theory** A theory of mate selection suggesting that we select partners similar to our opposite-sex parents.

**parentified children** Children who find themselves in a position of caregiving for a parent.

**parricide** The killing of a parent by one's child.

**passionate love** Intense, impassioned love. Compare to *companionate love*.

**Passive-congenial marriages** Relationships that begin without the emotional spark or intensity contained in our romantic idealizations of marriage.

**paternal claiming** Process by which stepfathers come to see stepchildren as their own children.

**patriarchy** A form of social organization in which the father or eldest male is recognized as the head of the family, kinship group, or tribe, and descent is traced through him. Compare to *matriarchy*.

**patrilineal** Descent or kinship traced through the father. Compare to *matrilineal*.

**peer** A person of equal status, as in age, class, position, or rank.

**performance anxiety** The fear of failing in sexual encounters. It is probably the most important immediate cause of erectile dysfunctions in men and, to a lesser extent, of orgasmic dysfunctions in women.

**permissive child rearing** A parenting style stressing the child's autonomy and freedom of expression, often over the needs of the parents.

**personal space** More common distance (1.5 to 4 feet in the United States) at which people interact. One's personal space is the spatial zone just beyond the intimate zone, within which one can access a variety of sensory information (e.g., scent, touch).

**phenotype** A set of genetically determined anatomical and physical characteristics, such as skin and hair color and facial structure.

**pleasuring** The giving and receiving of sensual pleasure through nongenital touching.

**polyandry** The practice of having more than one husband at the same time. See also *polygamy*; compare to *polygyny*.

**polygamy** The practice of having more than one husband or wife at the same time; plural marriage. See also *polyandry, polygyny,* and *consanguineous relationship*.

**polygyny** The practice of having more than one wife at the same time. See also *polygamy*; compare to *polyandry*.

**poly-victimization** Refers to the reality that many victims of intimate partner violence and abuse suffer from multiple forms of violence and abuse.

**positive assortative mating** Another term to describe patterns of homogamy.

**postgender relationships** Relationships lived outside the constraints of gender expectations and that emphasize such relationship characteristics as sharing, equity, and equality.

**postpartum period** A period of about three months following childbirth during which critical family adjustments are made.

**power** The ability to exert one's will, influence, or control over another person or group.

**precursor to marriage** Type of cohabiting relationship in which partners share in the expectation that they will eventually marry.

**predictive divorce rate** A statistical calculation of the expected divorce rate of people who enter marriage in a given year.

**premarital sex** Sexual activities, especially sexual intercourse, prior to marriage, especially among young, never-married individuals.

**principle of least interest** A theory of power in which the person less interested in sustaining a relationship has the greater power.

**private safety nets** Social support from one's social network on which one can fall back in times of economic need.

**pro-arrest policy** An alternative to mandatory arrest, wherein officers retain the discretion of whether or not to make an arrest. However, in such preferred arrest contexts, the lack of an arrest requires officers to justify their decisions with detailed reports.

**prototype** In psychology, concepts organized into a mental model.

**proximal causes** Experiences in the daily life of married couples that directly raise their probability of divorce.

**proximity** Nearness to another in terms of both physical space and time.

**psychological aggression** Nonphysical form of mistreatment, typically of children by parents, consisting of the following kinds of behavior: shouting; threatening to hit; cursing at one's child; threatening to send child away; or calling one's child dumb or lazy.

## Q

**qualitative research** Small groups or individuals are studied in an in-depth fashion.

**quantitative research** Samples taken from a large number of subjects.

**questionnaire** Documents consisting of a series of questions or statements to which a researcher wants research participants to answer or react.

## R

**race** A race (or racial group) is a group of people, such as whites, blacks, and Asians, classified according to their phenotype— their anatomical and physical characteristics. Members of a race share common phenotypical characteristics, such as skin color and facial structure. We perceive or identify ourselves and others within racial classifications, and we may be treated or act toward others on the basis of race.

**racial socialization** Efforts undertaken by racial or ethnic minority parents to prepare their children to live in a society in which they are likely to experience discrimination or racism.

**racial group** A large group of people defined as distinct because of their phenotype (genetically transmitted anatomical and physical characteristics, especially facial structure and skin color). Compare to *ethnic group*.

**rape** Sexual act against a person's will or consent as defined by law, usually including sexual penetration by the penis or other object; it may not, however, necessarily include penile penetration of the vagina. Also known as *sexual assault*. See also *acquaintance rape* and *marital rape*.

**ratio measure of divorce** A statistical calculation reflecting the ratio of the number of divorces in a given year to the number of marriages in that same year.

**reactive jealousy** Jealousy that occurs when a partner's past, present, or anticipated involvement with another is revealed. Compare to *suspicious jealousy*.

**refined divorce rate** A statistic reflecting the number of divorces in a given year for every thousand married couples.

**reflection hypothesis** How media content reflects the values and ideals of what people want to see or what already exists.

**rejection sensitivity** The tendency to anticipate and overreact to rejection.

**rejecters** Individuals who perceive changing family patterns as bad for society.

**relationship cycling** Much the same as churning, relationship cycling refers to on-again off-again relationships.

**relative love and need theory** A theory of power in which the person gaining the most from a relationship is the most dependent.

**remarriage** A marriage in which one or both partners have been previously married.

**residential propinquity** A pattern in which the chances of two people marrying are greater the closer they live to each other.

**resource theory of power** A theory of marital power that suggests that the person who is the sole or larger source of financial resources gains and exercises decision-making power.

**retaliation thesis** The viewpoint that divorce may lead to more violence as men try to intimidate their ex-spouses into reconciliation or regain and exercise control over their former wives.

**role conflict** Role conflict occurs when individuals face competing and incompatible demands from two or more roles that they play.

**role overload** The experience of having more prescribed activities in one or more roles than can be comfortably or adequately performed. See also *role strain*.

**role strain** Difficulties, tensions, or contradictions experienced in performing a role, often because of multiple role demands.

**role theory** Theory of spousal choice in which individuals choose spouses who share the same expectations of spousal roles.

## S

**safety-valve thesis** The viewpoint that divorce reduces the risk of violence by reducing the amount of physical contact between spouses.

**sandwich generation** Individuals and families who care for both their own children and their aging parents at the same time.

**scientific method** A method of investigation in which a hypothesis is formed on the basis of impartially gathered data and is then tested empirically.

**second shift** Arlie Hochschild's term for the domestic responsibilities awaiting employed women after their paid work hours are completed.

**secondary data analysis** Use of research gathered by public sources of information.

**self-care** Children under age 14 caring for themselves at home without supervision by an adult or older adolescent.

**self-disclosure** The revelation of deeply personal information about oneself to another.

**self-silencing** Act of keeping quiet, not expressing one's feelings or venting one's frustrations.

**separation distress** A psychological state following separation that may be characterized by depression, anxiety, intense loneliness, or feelings of loss.

**serial cohabitation** The experience of cohabiting with a sequence of partners.

**serial monogamy** A practice in which one person may have several spouses over his or her lifetime although no more than one at any given time.

**sexual dysfunction** Recurring problems in sexual functioning that cause distress to the individual or partner; may have a physiological or psychological basis.

**sexual enhancement** Any means of improving a sexual relationship, including developing communication skills, fostering a positive attitude, giving a partner accurate and adequate information, and increasing self-awareness.

**sexual harassment** Deliberate or repeated unsolicited verbal comments, gestures, or physical contact that is sexual in nature and unwelcomed by the recipient. Two types of sexual harassment involve (1) the abuse of power and (2) the creation of a hostile environment. See also *hostile environment*.

**sexual identity** How one perceives oneself sexually.

**sexual intercourse** Coitus; heterosexual penile/vaginal penetration and stimulation.

**sexual orientation** The direction of one's sexual attraction.

**sexual script** A culturally approved set of expectations as to how one should behave sexually as male or female and as heterosexual, gay, or lesbian.

**shift couples** Two-earner households in which spouses work different, often nonoverlapping shifts, so that one partner is home while the other is at work.

**SIDS** See *sudden infant death syndrome*.

**single-parent family** A family with children, created by divorce or unmarried motherhood, in which only one parent is present. A family consisting of one parent and one or more children.

**singlism** Prejudice, stereotyping, and discrimination experienced by unmarried women and men.

**situational couple violence** A common form of violence wherein the violence erupts during an argument. It is not part of a wider pattern, is as likely to come from a woman as from a man, rarely escalates, and is less likely to result in serious injury or fatality.

**skeptics** Individuals who are relatively tolerant of changes in family lifestyles but are somewhat concerned about their impact on society.

**social capital** The guidance, attention, and social connections that can enhance the quality of one's life and/or enlarge one's opportunities.

**social causation** A perspective that explains the greater happiness, health, and well-being found among married people compared to unmarried people as being the result of marriage.

**social classes** Groupings of people who share a common economic position by virtue of their wealth, income, power, and prestige, and thus have similar social and familial experiences.

**social desirability bias** The tendency for participants in research to answer questions with more acceptable answers. Such a bias may be especially prominent in research about sexual behavior.

**social exchange theory** A theory that emphasizes the process of mutual giving and receiving of rewards in social relationships, such as love or sexual intimacy, calculated by the equation Reward − Cost = Outcome.

**social father** A male relative, family associate, or mother's partner who acts like a father to her children.

**social institution** The organized pattern of statuses and structures, roles, and rules by which a society attempts to meet certain of its most basic needs.

**social integration** The degree of interaction between individuals and the larger community.

**social learning theory** A theory of human development that emphasizes the role of cognition (thought processes) in learning.

**social mobility** A term used to refer to movement up or down the socioeconomic ladder, which can occur within a person's lifetime or between generations.

**social role** A socially established pattern of behavior that exists independently of any particular person, such as the husband or wife role or the stepparent role.

**social selection** A perspective that explains the greater happiness, health, and well-being found among married people compared to unmarried people as a consequence of the types of people who marry.

**socialization** The shaping of individual behavior to conform to social or cultural norms.

**socioeconomic status** A term used to refer to the combined effects of income, occupational prestige, wealth, education, and income on a person's lifestyle and opportunities.

**sole custody** Child custody arrangement in which only one parent has both legal and physical custody of the child. See also *joint custody* and *split custody*.

**split custody** Custody arrangement when there are two or more children in which custody is divided between the parents, the mother generally receiving the girls and the father receiving the boys.

**stations of divorce** The multiple experiences people have as their marriages end.

**stepfamily** A family in which one or both partners have a child or children from an earlier marriage or relationship. Also known as a *blended family*; see also *binuclear family*.

**stepparent role** The role a stepparent forges for herself or himself within the stepfamily as there is no such role clearly defined by society.

**stereotype** A rigidly held, simplistic, and overgeneralized view of individuals, groups, or ideas that fails to allow for individual differences and is based on personal opinion and bias rather than critical judgment.

**stimulus–value–role theory** A three-stage theory of romantic development proposed by Bernard Murstein: (1) stimulus brings people together; (2) value refers to the compatibility of basic values; and (3) role has to do with each person's expectations of how the other should fulfill his or her roles.

**structural functionalism theory** A sociological theory that examines how society is organized and maintained by examining the functions performed by its different structures. In marriage and family studies, structural functionalism examines the functions the family performs for society, the functions the individual performs for the family, and the functions the family performs for its members.

**structural mobility** When large segments of a population experience upward or downward mobility resulting from changes in the society and economy.

**subsystem** A system that is part of a larger system, such as family; for example, religious and economic systems are subsystems of society, and the parent/child system is a subsystem of the family.

**sudden infant death syndrome** The sudden and unexplained death of a seemingly healthy infant, often during sleep. See also *SIDS*.

**survey research** A research method using questionnaires or interviews to gather information from small, representative groups and to infer conclusions that are valid for larger populations.

**suspicious jealousy** Jealousy that occurs when there is either no reason for suspicion or only ambiguous evidence that a partner is involved with another. Compare to *reactive jealousy*.

**symbolic interaction theory** A theory that focuses on the subjective meanings of acts and how these meanings are communicated through interactions and roles to give shared meaning.

### T

**theory** A set of general principles or concepts used to explain a phenomenon and to make predictions that may be tested and verified experimentally.

**time strain** Situation in which individuals feel as though they do not have enough time or spend enough time in certain roles or relationships.

**time use survey/time use diary** Self-administered survey in which people record their activities at designated points in time and/or report how much time they spend in various activities.

**total marriages** Rare type of marriage relationship in which characteristics of vital relationships are present but to an even wider and deeper degree.

**traditionalization** Process whereby couples become more traditional in their performance of household and family roles once they become parents.

**transgender** A broad category describing individuals who stand outside and apart from societal conventions regarding gender, neither identifying with nor acting within societal conventions for males or females. Some may undergo sex reassignment procedures, whereas others alter their social but not necessarily their physical characteristics, presenting themselves as of the opposite sex.

**traumatic sexualization** The process of developing inappropriate or dysfunctional sexual attitudes, behaviors, and feelings by a sexually abused child.

**trial marriage** Cohabitation with the purpose of determining compatibility prior to marriage.

**triangular theory of love** A theory developed by Robert Sternberg emphasizing the dynamic quality of love as expressed by the interrelationship of three elements: intimacy, passion, and decision/commitment.

**triangulation** The use of multiple data collection techniques in a single study.

**two-person career** An arrangement in which it takes the efforts of two spouses to ensure the career success of one. One spouse, typically the husband, can devote himself or herself fully to career pursuits because of the help and assistance received from his or her spouse. This help and assistance includes taking care of all family and domestic needs, but also often includes unpaid supportive roles (such as entertaining business colleagues).

### U

**uninvolved parenting** Parents who are neither responsive to their children's needs nor demanding of them in their expectations.

**upper-middle class** A socioeconomic class consisting of college-educated, highly paid professionals (for example, lawyers, doctors, engineers) who have annual incomes that may reach into the hundreds of thousands of dollars.

### V

**value judgment** An evaluation based on ethics or morality rather than on objective observation.

**value theory** The theory that we choose spouses based on similarity of values.

**variable** In experimental research, a factor, such as a situation or behavior, that may be manipulated. See also *independent variable* and *dependent variable*.

**violence** An act carried out with the intention of causing physical pain or injury to another.

**violent resistance** Violence used by victims—more often women—as self-defense.

**virginity** The state of not having engaged in sexual intercourse.

**vital marriages** Marriages that appeal more to our romantic notions of marriage because they begin and continue with high levels of emotional intensity.

### W

**work spillover** The effect that employment has on time, energy, activities, and psychological functioning of workers and their families.

**working class** A socioeconomic class comprised of skilled laborers with high school or vocational educations. The working class lives somewhat precariously, with little savings and few liquid assets should illness or job loss occur.

# Bibliography

Aaltonen, Mikko, Janne Kivivuori Mikkom, Pekka Martikainen, and Venla Salmi. "Socio-Economic Status and Criminality as Predictors of Male Violence: Does Victim's Gender or Place of Occurrence Matter?" *British Journal of Criminology* 52,6 (2012): 1192–1211.

Abbey, A., and R. J. Harnish. "Perception of Sexual Intent: The Role of Gender, Alcohol Consumption, and Rape Supportive Attitudes." *Sex Roles* 32, 5–6 (March 1995): 297–313.

Abbott, Elizabeth. "A History of Marriage: From Same-Sex Unions to Private Vows and Common Law, the Surprising Diversity of a Tradition." New York: Seven Stories Press, 2010.

Absi-Semaan, Nada, Gail Crombie, and Corinne Freeman. "Masculinity and Femininity in Middle Childhood: Developmental and Factor Analyses." *Sex Roles: A Journal of Research* 28 (1993): 187–207.

Acevedo, B., and A. Aron. "Does a Long-Term Relationship Kill Romantic Love?" *Review of General Psychology* 13, 1 (2009): 59–65.

Ackerman, Diane. *The Natural History of Love.* New York: Random House, 1994.

Acs, Gregory. "Downward Mobility from the Middle Class: Waking Up from the American Dream." Pew Charitable Trusts: Economic Mobility Project. Washington, DC, 2011.

Adler, Nancy, Susan Hendrick, and Clyde Hendrick. "Male Sexual Preference and Attitudes toward Love and Sexuality." *Journal of Sex Education and Therapy* 12, 2 (September 1996): 27–30.

Adler, Nancy, Susan Hendrick, and Clyde Hendrick. "Male Sexual Preference and Attitudes toward Love and Sexuality." *Journal of Sex Education and Therapy* 12 (1989): 27–30.

Administration on Aging, U.S. Department of Health and Human Services. "A Profile of Older Americans: 2011." www.aoa.gov/Aging _Statistics/Profile/2011/docs/2011profile.pdf.

————. "Marital Status," 2014. www.aoa.acl.gov/Aging_Statistics /Profile/2014/5.aspx.

Ahrons, Constance *The Good Divorce: Keeping Your Family Together When Your Marriage Comes Apart.* New York: HarperCollins, 1994.

Ahrons, C. *We're Still Family: What Grown Children Have to Say about Their Parents' Divorce.* New York: HarperCollins, 2004.

Ahrons, Constance, and Roy Rodgers. *Divorced Families: A Multidisciplinary View.* New York: Norton, 1987.

Ainsworth, Mary D., M. D. Blehar, E. Waters, and S. Wall. *Patterns of Attachment: A Psychological Study of the Strange Situation.* Hillsdale, NJ: Lawrence Erlbaum, 1978.

Alapack, Richard. "The Adolescent First Kiss." *Humanistic Psychologist* 19, 1 (March 1991): 48–67.

Aldous, Joan. *Family Careers: Developmental Change in Families.* New York: Wiley, 1978.

————. "Perspectives on Family Change." *Journal of Marriage and the Family* 52, 3 (August 1990): 571–583.

————. *Family Careers: Rethinking the Developmental Perspective.* Thousand Oaks, CA: Sage, 1996.

Aldous, Joan, and Gail M. Mulligan. "Fathers' Child Care and Children's Behavior Problems: A Longitudinal Study." *Journal of Family Issues* 23, 5 (July 2002): 624–647.

Alfano, Sean. "Poll: Women Strive to Find Balance." CBSNews.com, 2009.

Ali, Nojoud, and Delphine Minoui. ( 2010). *I Am Nujood, Age 10 and Divorced.* New York: Three Rivers Press.

Allen, K. R. "Ambiguous Loss After Lesbian Couples with Children Break Up: A Case for Same-Sex Divorce." *Family Relations* 56 (2007a): 174–182.

Allen, Katherine, Rosemary Bleiszner, and Karen Roberto. "Families in the Middle and Later Years: A Review and Critique of Research in the 1990s." *Journal of Marriage and the Family* 62, 4 (November 2000): 911–926.

Allen, Katherine R., David H. Demo, and Mark A. Fine, *Handbook of Family Diversity.* New York: Oxford University Press, 2000.

Allen, K., E. Husser, D. Stone, and C. Jordal. "Agency and Error in Young Adults' Stories of Sexual Decision Making." *Family Relations* 57, 4 (2008): 517–529.

Allen, William, and David Olson. "Five Types of African American Marriages." *Journal of Marital and Family Therapy* 27, 3 (July 2001): 301–314.

Altman, I., and M. Chemers. *Culture and Environment.* Belmont, CA: Wadsworth, 1984.

Amato, Paul. "President's Report: Studying Divorce and Couple Relationships—Difficult Times for Family Scholars." *National Council on Family Relations Report* 60.1 (Spring 2015).

————. "Hypotheses Are Many Splendored Things." *National Council on Family Relations Report* 59.4 (Winter 2014): 3–6.b.

Amato, P. "Parental Absence during Childhood and Depression in Later Life." *The Sociological Quarterly* 32 (1991): 543–556.

————. "Children's Adjustment to Divorce: Theories, Hypotheses, and Empirical Support." *Journal of Marriage and the Family* 55 1 (February 1993): 23–32.

Amato, P. R. "Life-Span Adjustment of Children to Their Parents' Divorce." *The Future of Children* 4 (1994): 143–164. www.princeton .edu/futureofchildren/publications/docs/04_01_08.pdf.

————. "Reconciling Divergent Perspectives: Judith Wallerstein, Qualitative Family Research, and Children of Divorce." *Family Relations* 52 (2003): 332–339.

Amato, Paul. "Who Cares for Children in Public Places? Naturalistic Observation of Male and Female Caretakers." *Journal of Marriage and the Family* 51 (November 1989): 981–990.

Amato, Paul. "Explaining the Intergenerational Transmission of Divorce." *Journal of Marriage and the Family* 58 (1996): 628–640.

————. "The Consequences of Divorce for Adults and Children," *Journal of Marriage and the Family* 62, 4 (November 2000): 1269–1288.

————. "Tension between Institutional and Individual Views of Marriage." *Journal of Marriage and Family* 66, 4 (November 2004a): 959–965.

————. (2004b). "The Future of Marriage. Vision 2004: What Is the Future of Marriage." Minneapolis, MN: National Council on Family Relations, pp. 99–102.

————. "For the Sake of the Children." Council on Contemporary Families, 2006.

————. "Research on Divorce: Continuing Trends and New Developments." *Journal of Marriage and Family* 72 (June 2010): 650–666.

Amato, Paul R., and Tamara D. Afifi. "Feeling Caught between Parents: Adult Children's Relations with Parents and Subjective Well-Being." *Journal of Marriage and Family* 68, 1 (February 2006): 222–235.

Amato, Paul R., and Alan Booth. "A Prospective Study of Divorce and Parent-Child Relationships." *Journal of Marriage and the Family* 58, 2 (May 1996): 356–365.

Amato, P., A. Booth, D. Johnson, and S. Rogers. *Alone Together: How Marriage in America Is Changing.* Cambridge, MA: Harvard University Press, 2007.

Amato, Paul R., and Danelle De Boer. "The Transmission of Marital Instability Across Generations: Relationship Skills or Commitment to Marriage?" *Journal of Marriage and Family* 63 (November 2001): 1038–1051.

Amato, Paul, and Bryndl Hohmann-Marriott. "A Comparison of High- and Low-Distress Marriages that End in Divorce." *Journal of Marriage and Family* 69, 3 (August 2007): 621–638.

Amato, Paul R., and Bruce Keith. "Parental Divorce and the Well-Being of Children: A Meta-Analysis." *Psychological Bulletin* 110 (1991): 26–46.

Amato, Paul R., and Denise Previti. "People's Reasons for Divorcing: Gender, Social Class, the Life Course, and Adjustment." *Journal of Family Issues* 24, 5 (July 2003): 602–626.

Amato, Paul, and Stacy Rogers. "A Longitudinal Study of Marital Problems and Subsequent Divorce." *Journal of Marriage and the Family* 59 (August 1997): 612–624.

American Academy of Child and Adolescent Psychiatry. "Stepfamily Problems." Facts for Families Guide, 2015. www.aacap.org/AACAP/Families_and_Youth?Facts_for_Families_Guide.

American Psychological Association. "Elder Abuse," 2005. www.apa.org/pi/prevent-violence/resources/elder-abuse.aspx.

————. "Making Stepfamilies Work," 2016. www.apa.org/helpcenter/stepfamily.aspx.

American Sociological Association. "24/7 Economy's Work Schedules Are Family Unfriendly and Suggest Needed Policy Changes." ASA News, May 2004. www.asanet.org.

Ames, B., W. A. Brosi, and K. M. Damiano-Teixeira. "I'm Just Glad My Three Jobs Could be during the Day: Women and Work in a Rural Community." *Family Relations* 55, 1 (January 2006): 119–131.

Ammons, S., and P. Edgell. "Religious Influences on Work-Family Tradeoffs." *Journal of Family Issues* 28, 6 (June 2007): 794–826.

Ammons, S., and E. Kelly. "Social Class and the Experience of Work-Family Conflict during the Transition to Adulthood." *New Directions for Child and Adolescent Development* 119 (2008): 71–84.

Andersen, J. E., P. A. Andersen, and M. W. Lustig. "Opposite-Sex Touch Avoidance: A National Replication and Extension." *Journal of Nonverbal Behavior*, 11 (1987): 89–109.

Anderson, Edward, and Shannon Greene. "Beyond Divorce: Research on Children in Repartnered and Remarried Families." *Family Court Review* 51, 1 (January 2013): 119–130.

Anderson, Monica. 2015. "A Rising Share of the U.S. Black Population Is Foreign Born; 9 Percent Are Immigrants; and While Most Are from the Caribbean, Africans Drive Recent Growth." Pew Research Center, Washington, DC: April.

Anderson, K. L. "Is Partner Violence Worse in the Context of Control?" *Journal of Marriage and Family* 70, 5 (2008): 1157–1168.

Anderson, Kristin. "Conflict, Power, and Violence in Families." *Journal of Marriage and Family* 72 (June 2010): 726–742.

Andersen, P., L. Guerrero, and S. Jones. "Nonverbal Behavior in Intimate Interactions and Intimate Relationships." in *The Sage Handbook of Nonverbal Communication*, edited by V. Manusov and M. Patterson. Thousand Oaks, CA: Sage, 2006.

Aponte, Robert, with Bruce Beal and Michelle Jiles. "Ethnic Variation in the Family: The Elusive Trend toward Convergence." In *Handbook of Marriage and the Family*, 2nd ed., edited by M. Sussman, S. Steinmetz, and G. Peterson. New York: Plenum, 1999.

Archer, J. "Sex Differences in Aggression between Heterosexual Partners: A Metaanalytic Review." *Psychological Bulletin* 126, 5 (2000): 651–680.

Arditti, Joyce A. "Noncustodial Fathers: An Overview of Policy and Resources." *Family Relations* 39, 4 (October 1990): 460–465.

Arditti, Joyce, and T. Keith. "Visitation Frequency, Child Support Payment, and the Father–Child Relationship Post-Divorce." *Journal of Marriage and the Family* 55 (1993): 699–712.

Aries, Phillipe. *Centuries of Childhood.* New York: Vintage, 1962.

Aron, Arthur, and Elaine Aron. "Love and Sexuality." In *Sexuality in Close Relationships*, edited by K. McKinney and S. Sprecher. Hillsdale, NJ: Lawrence Erlbaum, 1991.

Arrighi, Barbara A., and David J. Maume Jr. "Workplace Subordination and Men's Avoidance of Housework." *Journal of Family Issues* 21, 4 (May 2000): 464–487.

Asian Pacific American Legal Center and Asian American Justice Center. "A Community of Contrasts: Asian Americans in the United States," 2011. www.advancingjustice.org/publication/community-contrasts-asian-americans-us-2011.

Associated Press-NORC Center for Public Affairs Research. "Same-Sex Marriage and Gay Rights: A Shift in Americans' Attitudes." Issue Brief, 2015. www.apnorc.org/PDFs/SameSexStudy/LGBT%20issues_D5_FINAL.pdf.

Atkins, D., and D. Kessel. "Religiousness and Infidelity: Attendance, but Not Faith and Prayer, Predict Marital Fidelity." *Journal of Marriage and Family* 70, 2 (May 2008): 407–418.

Atkinson, Maxine P., and Stephen P. Blackwelder. "Fathering in the 20th Century." *Journal of Marriage and the Family* 55, 4 (November 1993): 975–986.

Atwood, J. D., and M. Seifer. "Extramarital Affairs and Constructed Meanings: A Social Constructionist Therapeutic Approach." *The American Journal of Family Therapy* 25 (1997): 55–75.

Aubrey, Jennifer Stevens, Elizabeth Behm-Morawitz, and Kyung Bo Kim. "Understanding the Effects of MTV's *16 and Pregnant* on Adolescent Girls' Beliefs, Attitudes, and Behavioral Intentions toward Teen Pregnancy." *Journal of Health Communication: International Perspectives* 19, 10 (2014): 124–41.

Austen, Ian. "Afghan Family, Led by Father Who Called Girls a Disgrace, Is Guilty of Murder." *New York Times*, January 29, 2012.

Avellar, Sarah, and Pamela Smock. "The Economic Consequences of the Dissolution of Cohabiting Unions." *Journal of Marriage and Family* 67, 2 (May 2005): 315–327.

Babbie, E. *The Practice of Social Research*, 11th ed. Belmont, CA: Wadsworth, 2007.

Baer, Mark. "The Problem with Covenant Marriages and Fault-Based Divorce," Blog post, November 6, 2014. www.huffingtonpost.com/mark-baer/the-problem-with-covenant_b_6110600.html.

Baggett, Courtney R. "Sexual Orientation: Should It Affect Child Custody Rulings?" *Law and Psychology Review* 16 (1992): 189–200.

Bahroo, B. "Special Issue: Child Protection in the 21st Century: Pedophilia: Psychiatric Insights." *Family Court Review* 41, 4 (October 2003): 497–507.

Baird, Julia. "Neither Female nor Male." *New York Times*, April 6, 2014. www.nytimes.com/2014/04/07/opinion/neither-female-nor-male.html.

Baker-Sperry, L., and L. Grauerholz. "The Pervasiveness and Persistence of the Feminine Beauty Ideal in Children's Fairy Tales." *Gender and Society* 17, 5 (2003): 711–726.

Bakker, A. B., E. Demerouti, E. and R. Burke. "Workaholism and Relationship Quality: A Spillover-Crossover Perspective." *Journal of Occupational Health Psychology*, 14, 1 (2008): 23–33.

Bakker, A. B., M. Westman, and I. J. H. van Emmerik. "Advancements in Crossover Theory." *Journal of Managerial Psychology* 24, 3 (2009): 206–219.

Balsam, K., T. Beauchaine, E. Rothblum, and S. Solomon. "Three-Year Follow-Up of Same-Sex Couples Who Had Civil Unions in Vermont, Same-Sex Couples Not in Civil Unions, and Heterosexual Married Couples." *Developmental Psychology* 44 (2008): 102–116.

Balsam, Kimberly F., and Dawn M. Szymanski. "Relationship Quality and Domestic Violence in Women's Same-Sex Relationships: The Role of Minority Stress." *Psychology of Women Quarterly* 29, 3 (September 2005): 258–269.

Bankston, Carl, and Jacques Henry. "Endogamy among Louisiana Cajuns: A Social Class Explanation." *Social Forces* 77, 4 (1999): 1317–1338.

Barnes, Medora W. "Our Family Functions: Functions of Traditional Weddings for Modern Brides and Postmodern Families." *Qualitative Sociology Review* 10 (2): 60–78.

Barnett, R., K. Gareis, and R. Brennan. "Wives' Shift Work Schedules and Husbands' and Wives' Well-Being in Dual-Earner Couples with Children: A Within-Couple Analysis." *Journal of Family Issues* 29, 3 (2008): 396–422.

Bartholomew, Kim. "Avoidance of Intimacy: An Attachment Perspective." *Journal of Social and Personal Relationships* 7, 2 (1990): 147–178.

Basu, Moni. "Gender Identity: In Conservative Nepal, It's OK to Be an 'Other.'" CNN, June 6, 2015. www.cnn.com/2015/06/05/world/nepal-transgender-rights.

Batson, C. D., Z-C. Qian, and D. T. Lichter. "Interracial and Intraracial Patterns of Mate Selection among America's Diverse Black Populations." *Journal of Marriage and Family* 68 (2006): 658–672.

Baucom, Brian, Pamela McFarland, and Andrew Christensen. "Gender, Topic, and Time in Observed Demand-Withdraw Interaction in Cross- and Same-Sex Couples." *Journal of Family Psychology* 24, 3 (2010): 233–242.

Baxter, Janeen. "To Marry or Not to Marry: Marital Status and the Household Division of Labor." *Journal of Family Issues* 26, 3 (April 2005): 300–321.

BBC News. "Young Saudi Girl's Marriage Ended." BBC News, April 30, 2009. http://news.bbc.co.uk/2/hi/middle_east/8026545.stm.

Beatie, T. *Labor of Love: The Story of One Man's Extraordinary Pregnancy.* Berkeley, CA: Seal Press, 2008.

Bechhofer, L., and L. Parrot. "What Is Acquaintance Rape?" In *Acquaintance Rape: The Hidden Crime,* edited by A. Parrott and L. Bechhofer. New York: Wiley, 1991.

Beck, Joyce W., and Barbara M. Heinzerling. "Gay Clients Involved in Child Custody Cases: Legal and Counseling Issues." *Psychotherapy in Private Practice* 12, 1 (1993): 29–41.

Bell, Robin. "Homosexual Men and Women." *British Medical Journal* 318, 7181 (February 13, 1999): 452–455.

Bell, B., R. Lawton, and H. Dittmar. "The Impact of Thin Models in Music Videos on Adolescent Girls' Body Dissatisfaction." *Body Image* 4, 2 (June 2007): 137–145.

Bellah, Robert, et al. *Habits of the Heart.* Berkeley, CA: University of California Press, 1985.

Belluck, Pam. "Maternal Depression Often Starts before Giving Birth, Study Says." *New York Times,* February 2, 2015. well.blogs.nytimes .com/2015/02/02/maternal-depression-often-starts-before-giving-birth -study-says/?_.

Belsky, Jay, and John Kelly. *The Transition to Parenthood: How a First Child Changes a Marriage.* Delacorte Press, 1994.

Belsky, Jay, and Michael Rovine. "Patterns of Marital Change across the Transition to Parenthood." *Journal of Marriage and the Family* 52 (February 1990): 5–19.

Bem, S. *The Lenses of Gender: Transforming the Debate on Sexual Inequality.* New Haven, CT: Yale University Press, 1993.

Benenson, Joyce, and Athena Christakos. "The Greater Fragility of Females' Versus Males' Closest Same-Sex Friendships." *Child Development* 74, 4 (July 2003): 1123–1129.

Bengtson, V. L., A. C. Acock, K. R. Allen, P. Dilworth-Anderson, and D. M. Klein. (eds.). *Sourcebook of Family Theory and Research.* Thousand Oaks, CA: SAGE Publications, Inc., 2005.

Benner, Aprile, and Su Yeong Kim. "Intergenerational Experience of Discrimination in Chinese American Families: Influence of Socialization and Stress." *Journal of Marriage and Family* 71, 4 (November 2009): 862–877.

Bennett, Jessica. "Campus Sex … With a Syllabus", *New York Times,* January 9, 2016.

Benokraitis, Nijole V., ed. *Feuds about Families: Conservative, Centrist, Liberal, and Feminist Perspectives."* Upper Saddle River, NJ: Prentice Hall, 2000.

Bernard, Jessie. *The Future of Marriage,* 2nd ed. New York: Columbia University Press, 1982.

———. "The Good-Provider Role: Its Rise and Fall." *American Psychologist* 36, 1 (January 1981): 1–12.

Bernard, Tara Siegel. "Fate of Domestic Partner Benefits in Question after Marriage Ruling." *New York Times,* June 28, 2015. www.nytimes .com/2015/06/29/your-money/fate-of-domestic-partner-benefits-in -question-after-marriage-ruling.html.

Bernstein, Elizabeth. "When It's Just Another Fight, and When It's Over." *Wall Street Journal,* April 3, 2012. www.wsj.com/articles/SB100014240 5270230402350457731962333618562.

Bersamin, M., M. Todd, D. A. Fisher, et al. "Parenting Practices and Adolescent Sexual Behavior: A Longitudinal Study." *Journal of Marriage and Family* 70 (February 2008): 97–112.

Bertrand Marianne, Emir Kamenica and Jessica Pan. "Gender Identity and Relative Income within Households." *Quarterly Journal of Economics* 130: 571–614.

Betchen, Stephen. "Male Masturbation as a Vehicle for the Pursuer /Distancer Relationship in Marriage." *Journal of Sex and Marital Therapy* 17, 4 (December 1991): 269–278.

Bevan, Emma, and Daryl Higgins. "Is Domestic Violence Learned? The Contribution of Five Forms of Child Maltreatment to Men's Violence and Adjustment." *Journal of Family Violence* 17, 3 (September 2002): 223–245.

Bianchi, S., and L. Casper. "American Families." *Population Bulletin* 55, 4 (2000): 1–43.

Bianchi S., J. Robinson, and M. Milkie. *Changing Rhythms of American Family Life.* New York: Russell Sage, 2006.

Bianchi, Suzanne. "Family Change and Time Allocation in American Families." *The ANNALS of the American Academy of Political and Social Science* 638, 1 (November 2011): 21–24.

Bianchi, Suzanne, and Melissa Milkie. "Work and Family Research in the First Decade of the 21st Century." *Journal of Marriage and Family* 72, 3 (June 2010): 705–725.

Biblarz, Timothy, and Even Savci. "Lesbian, Gay, Bisexual, and Transgender Families." *Journal of Marriage and Family* 72, 3 (2010): 480–497.

Bidgood, Jess. "Alimony in Massachusetts Gets Overhaul, with Limits." *New York Times,* September 26, 2011.

Bird, Chloe E., and Catherine E. Ross. "Houseworkers and Paid Workers: Qualities of the Work and Effects on Personal Control." *Journal of Marriage and the Family* 55, 4 (November 1993): 913–925.

Bisin, A., G. Topa, and T. Verdier. "Religious Intermarriage and Socialization in the United States." *Journal of Political Economy* 112, 3 (June 2004): 615–664.

Bittman, M., P. England, L. Sayer, N. Folbre, and G. Matheson. "When Does Gender Trump Money? Bargaining and Time in Household Work." *American Journal of Sociology* 109, 1 (July 2003): 186–214.

Black, Michelle, Kathleen Basile, Matthew Breiding, Sharon Smith, Mikel Walters, Melissa Merrick, Jieru Chen, and Mark Stevens. "The National Intimate Partner and Sexual Violence Survey (NISVS): 2010 Summary Report." Atlanta, GA: National Center for Injury Prevention and Control, Centers for Disease Control and Prevention, 2011.

Blackwell, Debra. "Marital Homogamy in the United States: The Influence of Individual and Paternal Education." *Social Science Research* 27, 2 (1998): 159–164.

Blackwell, Debra, and Daniel Lichter. "Homogamy among Dating, Cohabiting, and Married Couples. *The Sociological Quarterly* 45, 4 (2004): 719–737.

Blain, G. "Gov. Paterson Signs Nixzmary's Law." *New York Daily News,* October 10, 2009.

Blair, S., and Daniel Lichter. "Measuring the Division of Household Labor: Gender Segregation of Housework among American Couples." *Journal of Family Issues* 12 (1991): 91–113.

Blair, Sampson Lee. "Employment, Family, and Perceptions of Marital Quality among Husbands and Wives." *Journal of Family Issues* 14, 2 (1993): 189–212.

Blair-Loy, M. *Competing Devotions: Career and Family among Women Executives.* Cambridge, MA: Harvard University Press, 2003.

Blair-Loy, Mary and Amy Wharton. "Employees Use of Work-Family Policies and the Workplace Social Context." *Social Forces* 80, 3 (March 2002): 813–845.

Blaisure, Karen, and Margie Geasler. "Educational Interventions for Separating and Divorcing Parents and Their Children." In *Handbook of Divorce and Relationship Dissolution,* edited Mark Fine and John Harvey (pp. 575–602). Hillsdale, NJ: Erlbaum, 2006.

Blood, Robert, and Donald Wolfe. *Husbands and Wives.* Glencoe, IL: Free Press, 1960.

Blumer, Markie, Mary Green, Sarah Knowles, and April Williams. "Shedding Light on Thirteen Years of Darkness: Content Analysis of Articles Pertaining to Transgender Issues in Marriage/Couple and Family Therapy Journals." *Journal of Marital and Family Therapy* 38, s1 (2012): 244–256.

Blumstein, Philip, and Pepper Schwartz. *American Couples.* New York: McGraw-Hill, 1983.

Bogle, K. "The Shift from Dating to Hooking Up: What Scholars Have Missed." Paper presented at the annual meeting of the American Sociological Association, Philadelphia, 2005. www.allacademic.com /meta/p23315_index.html.

———. *Hooking Up: Sex, Dating, and Relationships on Campus.* New York: New York University Press, 2008.

Bohannan, Paul, ed. *Divorce and After.* New York: Doubleday, 1970.

Boivin J, L., G. Bunting, J. Collins, and K. Nygren. "International Estimates of Infertility Prevalence and Treatment-Seeking: Potential Need and Demand for Infertility Medical Care." *Human Reproduction* 22 (2007): 1506–1512.

Bond, James, Erin Galinsky, Kim Stacy, and Kim Erin Brownfield. *National Study of Employers.* New York: Families and Work Institute, 2005.

Booth, A., and J. N. Edwards. "Starting Over: Why Remarriages Are More Unstable. *Journal of Family Issues* 13, 2 (June 1992): 179–194.

Booth, Alan, and Paul Amato. "Parental Predivorce Relations and Offspring Postdivorce Well-Being." *Journal of Marriage and Family* 63 (February 2001): 197–212.

Boren, Cindy. "Adrian Peterson Released on Bond in Child-Abuse Case, but NFL's Problems Continue." *The Washington Post,* September 13, 2014. www.washingtonpost.com/news/early-lead/wp/2014/09/13 /adrian-peterson-released-on-bond-in-child-abuse-case-but-nfls -problems-continue/.

Boss, P. "Rethinking Adoptions that Dissolve: What We Need to Know about Research, Practice, Policies and Attitudes." *National Council on Family Relations Report: Family Focus on Adoption,* Issue FF39 (September 2008): F5–7, 19.

Boss, Pauline. "Ambiguous Loss Research, Theory, and Practice: Reflections after 9/11." *Journal of Marriage & Family* 66, 3 (August 2004): 551–566.

Boss, Pauline. *Loss, Trauma, and Resilience: Therapeutic Work with Ambiguous Loss.* New York: W.W. Norton, 2006.

Boushey, H. "Women Breadwinners, Men Unemployed." Center for American Progress, 2009. www.americanprogress.org/issues /economy/news/2009/07/20/6314/women-breadwinners-men -unemployed/.

Bowe, J. "Gay Donor or Gay Dad? Gay Men and Lesbians Are Having Babies and Redefining Fatherhood, Commitment and What a Family Can Be." *New York Times Magazine*, November 19, 2006, 66.

Bowen, Gary, Roderick Rose, Joelle Powers, and Elizabeth Glennie. "The Joint Effects of Neighborhoods, Schools, Peers, and Families on Changes in the School Success of Middle School Students." *Family Relations*, 57, 4 (October 2008): 504–516.

Bowlby, John. *Attachment and Loss.* New York: Basic Books, 1969.

Bowlby, J. *Attachment and Loss. Vol. 2: Separation: Anxiety and Anger.* New York: Basic Books, 1973 (reissued in 1999).

Bowlby, John. *Attachment and Loss.* 3 vols. New York: Basic Books, 1980.

Boyd, Monica, and Anne Li. "May–December: Canadians in Age-Discrepant Relationships." *Canadian Social Trends* (Autumn 2003, Statistics Canada): 29–33.

Bradbury, T. N., and B. R. Karney. "Understanding and Altering the Longitudinal Course of Marriage." *Journal of Marriage and Family* 66, 4 (November 2004): 862–879.

Bradshaw, Matt, and Christopher G. Ellison. "The Nature-Nurture Debate is Over and Both Sides Lost: Reflections on the Study of Gender and Religion." *Journal for the Scientific Study of Religion* 48 (2009): 241–251.

Bradsher, Keith. "3 Guilty of Manslaughter in Slipping Drug to Girl," *New York Times*, March 15, 2000.

Bratter, J., and R. King. "But Will It Last?": Duration of Interracial Unions Compared to Similar Race Relationships." *Family Relations* 57 (2008): 160–171.

Braver, Sanford, et al. "A Social Exchange Model of Nonresidential Parent Involvement." In *Nonresidential Parenting: New Vistas in Family Living*, edited by C. E. Depner and J. H. Bray. Newbury Park, CA: Sage Publications, 1993a.

———. "A Longitudinal Study of Noncustodial Parents: Parents without Children." *Journal of Family Psychology* 7, 1 (June 1993b): 9–23.

Bray, J. H., and S. H. Berger. "Noncustodial Parent and Grandparent Relationships in Stepfamilies." *Family Relations* 39, 4 (1990): 414–419.

Brayfield, April A. "Employment Resources and Housework in Canada." *Journal of Marriage and the Family* 54, 1 (February 1992): 19–30.

Breiding, M. J., J. Chen, and M. C. Black. "Intimate Partner Violence in the United States — 2010." Atlanta, GA: National Center for Injury Prevention and Control, Centers for Disease Control and Prevention, 2014.

Breiding, Matthew, Sharon Smith, Kathleen Basile, Mikel Walters, Jieru Chen, and Melissa Merrick. "Prevalence and Characteristics of Sexual Violence, Stalking, and Intimate Partner Violence Victimization— National Intimate Partner and Sexual Violence Survey, United States, 2011." Centers for Disease Control and Prevention, Surveillance Summaries 63, SS08 (September 5, 2014): 1–18.

Bretthauer, B., T. Schindler Zillerman, and J. Banning. "A Feminist Analysis of Popular Music: Power over, Objectification of, and Violence against Women." *Journal of Feminist Family Therapy* 18, 4 (2007): 29–51.

Brewster, Karin, and Irene Padavic. "Change in Gender-Ideology, 1977–1996: The Contributions of Intracohort Change and Population Turnover." *Journal of Marriage and the Family* 62, 2 (May 2000): 477–488.

Bringle, Robert G., and Glenda J. Bagby. "Self-Esteem and Perceived Quality of Romantic and Family Relationships in Young Adults." *Journal of Research in Personality* 26, 4 (1992): 340–356.

Bringle, Robert, and Bram Buunk. "Jealousy and Social Behavior: A Review of Person, Relationship, and Situational Determinants." In *Review of Personality and Social Psychology, Vol. 6, Self, Situation, and Social Behavior*, edited by P. Shaver. Newbury Park, CA: Sage, 1985.

———. "Extradyadic Relationships and Sexual Jealousy." In *Sexuality in Close Relationships*, edited by K. McKinney and S. Sprecher. Hillsdale, NJ: Lawrence Erlbaum, 1991.

Bronfenbrenner, Urie, and P. A. Morris. "The Ecology of Developmental Processes." In *Handbook of Child Psychology*, 5th ed., vol. 1: Theoretical Models of Human Development, edited by W. Damon and R. M. Lerner (pp. 993–1023). New York: Wiley, 1998.

Brown, Susuan, and I-Fen Lin. "The Gray Divorce Revolution." National Council on Family Relations, NCFR Report, Family Focus on Families and Demography 58.2 (Summer 2013): F4–F5.

Brown, Susan, and I-Fen Lin. "Gray Divorce: A Growing Risk Regardless of Class or Education." Council on Contemporary Families, October 8, 2014. https://contemporaryfamilies.org/growing-risk-brief-report/.

Brown, Susan L., and Sayaka K. Shinohara. "Dating Relationships in Older Adulthood: A National Portrait." *Journal of Marriage and Family* 75 (2013):1194–1202.

Brown, Erin. J. CREW Pushes Transgendered Child Propaganda. Media Research Center, April 8, 2011. www.mrc.org/articles/jcrew-pushes -transgendered-child-propaganda.

Brown, H. "U.S. Maternity Leave Benefits Are Still Dismal." *Forbes.com*, May 4, 2009.

Brown, J. D., K. L. L'Engle, C. J. Pardun, G. Guo, K. Kenneavy, and C. Jackson. "Sexy Media Matter: Exposure to Sexual Content in Music, Movies, Television and Magazines Predicts Black and White Adolescents' Sexual Behavior." *Pediatrics* 117, 4 (2006): 1018–1027.

Brown, Susan, and Alan Booth. "Cohabitation versus Marriage: A Comparison of Relationship Quality," *Journal of Marriage and the Family* 58 (August) 1996: 668–678.

———. "Bending the Time Bind: Rejoinder to Hochschild and Goodman." *Social Science Quarterly* 83, 4 (December 2002b): 941–946.

Brown, Tonya Ballard. "Did You Know It's Legal in Most States to Discriminate against LGBT People?" National Public Radio, April 28, 2015.

Brown, S. L., J. R. Bulanda, and G. R. Lee. "Transitions Into and Out of Cohabitation in Later Life." *Journal of Marriage and Family* 74 (2012): 774–793.

Brown, Susan L., Wendy D. Manning, and Krista K. Payne. "Relationship Quality among Cohabiting versus Married Couples." *National Center for Family and Marriage Research Working Paper Series*, WP-14-03. Bowling Green, OH: Bowling Green State University, National Center for Family and Marriage Research, 2014.

Bryant, A. "Changes in Attitudes toward Women's Roles: Predicting Gender Role Traditionalism among College Students." *Sex Roles* 48, 3/4 (February 2003): 131–143.

Bryant, Chalandra, K. A. S. Wickrama, John Boland, Barlynda Bryant, Carolyn Cutrona, and Christine Stanik. "Race Matters, Even in Marriage: Identifying Factors Linked to Marital Outcomes for African Americans." *Journal of Family Theory & Review* 2, 3 (September 2010): 157–174.

Budig, M. "Feminism and the Family." In *The Blackwell Companion to the Sociology of Families*, edited by J. Scott Treas and M. Richards. Malden, MA: Blackwell, 2004.

Buffum J. "Prescription Drugs and Sexual Function." *Psychiatric Medicine* 10 (1992): 181–198.

Bulanda, Jennifer. "Doing Family, Doing Gender, Doing Religion: Structured Ambivalence and the Religion-Family Connection." *Journal of Family Theory & Review* 3, 3 (September 2011): 179–197.

Bulanda, Jennifer, and Susan Brown. "Race-Ethnic Differences in Marital Quality and Divorce." Working Paper Series 06-08, Center for Family and Demographic Research, Bowling Green State University, 2004.

Bullock, Melinda, Jana Hackathorn, Eddie M. Clark, and Brent A. Mattingly. "Can We Be (and Stay) Friends? Remaining Friends after Dissolution of a Romantic Relationship." *The Journal of Social Psychology* 151, 5 (2011): 662–666.

Burgess, Ernest. "The Family as a Unity of Interacting Personalities." *The Family* 7, 1 (March 1926): 3–9.

Burgess, M., S. Stermer, and S. Burgess. "Sex, Lies and Video Games: The Portrayal of Male and Female Characters on Video Game Covers." *Sex Roles* 57, 5–6 (September 2007): 419–433.

Burgoyne, Sarah. "Relationship Quality among Married and Cohabiting Couples" (FP-12-12). National Center for Family and Marriage Research, 2012. www .bgsu.edu/content/dam/BGSU/college-of-arts-and-sciences/NCFMR /documents/FP/FP-12-12.pdf.

Burleigh, Nina. "Sexting, Shame, and Suicide." *Rolling Stone*, September 17, 2013, RS1192. www.rollingstone.com/culture/news/sexting-shame -and-suicide-20130917.

Burleson, Brant, and Wayne Denton. "The Relationship between Communication Skill and Marital Satisfaction: Some Moderating Effects." *Journal of Marriage and the Family* 59, 4 (November 1997): 884–902.

Burr, Wesley, Loren Marks, and R. Day. *Sacred Matters: Religion and Spirituality in Families.* New York: Routledge, 2012.

Burton Linda. "Child Adultification in Economically Disadvantaged Families: A Conceptual Model." *Family Relations* 56 (2007): 329–345.

Burton, Linda, Eduardo Bonilla-Silva, Victor Ray, Rose Buckelew, and Elizabeth Hordge Freeman. "Critical Race Theories, Colorism, and the Decade's Research on Families of Color." *Journal of Marriage and Family* 72 (June 2010): 440–459.

Butler, Amy C. "Gender Differences in the Prevalence of Same-Sex Sexual Partnering: 1988–2002." *Social Forces* 84, 1 (September 2005): 421–449.

Buunk, Bram, and Pieternel Dijkstra. "Gender Differences in Rival Characteristics that Evoke Jealousy in Response to Emotional versus Sexual Infidelity." *Personal Relationships* 11 (2004): 395–408.

Buunk, B. P., P. Dijkstra, D. Fetchenhauer, and D. T. Kenrick. "Age and Gender Differences in Mate Selection Criteria for Various Involvement Levels." *Personal Relationships* 9 (2002): 271–278.

Buunk, Bram, and Ralph Hupka. "Cross-Cultural Differences in the Elicitation of Sexual Jealousy." *Journal of Sex Research* 23, 1 (February 1987): 12–22.

Byers, E. S., H. A. Sears, and A. D. Weaver. "Parents' Reports of Sexual Communication with Children in Kindergarten to Grade 8." *Journal of Marriage and Family* 70 (2008): 86–96.

Byers, E. Sandra. "Beyond the *Birds and the Bees* and *Was It Good for You*?: Thirty Years of Research on Sexual Communication." *Canadian Psychology/Psychologie canadienne* 52, 1 (February 2011): 20–28.

Byrne, Michael, Alan Carr, and Marie Clark. "Power in Relationships of Women with Depression." *Journal of Family Therapy* 26, 4 (November 2004): 407–429.

Cabrera, N., J. Fagan, and D. Farrie. "Explaining the Long Reach of Fathers' Prenatal Involvement on Later Paternal Engagement." *Journal of Marriage and Family* 70, 5 (2008): 1094–1107.

Caetano, R., P. A. C. Vaeth, and S. Ramisetty-Mikler. *Intimate Partner Violence Victim and Perpetrator Characteristics among Couples in the United States.* New York: Springer Science + Business Media, 2008.

Cahn, Naomi R., and June Carbone. "Red Families v. Blue Families" (August 16, 2007). University of Florida *Journal of Law & Public Policy*; GWU Law School Public Law Research Paper No. 343; GWU Legal Studies Research Paper No. 343. Available at SSRN: http://ssrn.com /abstract=1008544.

Cahn, Naomi, and June Carbone. *Red Families v. Blue Families: Legal Polarization and the Creation of Culture.* New York: Oxford, 2010.

Call, Vaughn R. A., and Tim B. Heaton. "Religious Influence on Marital Stability." *Journal for the Scientific Study of Religion* 36, 3 (September 1997): 382–392.

Calmes, C. A., and J. E. Roberts. "Rumination in Interpersonal Relationships: Does Co-Rumination Explain Gender Differences in Emotional Distress and Relationship Satisfaction among College Students?" *Cognitive Therapy and Research* 32 (2008): 577–590.

Campa, M., and J. J. Eckenrode. "Pathways to Intergenerational Adolescent Childbearing in a High-Risk Sample." *Journal of Marriage and Family* 68, 3 (2006): 558–572.

Campbell, L., R. Martin, and J. Ward. "An Observational Study of Humor Use while Resolving Conflict in Dating Couples." *Personal Relationships* 15 (2008): 41–55.

Campos, Belinda, Anthony Graesch, Rena Repetti, Thomas Bradbury, and Elinor Ochs. "Opportunity for Interaction? A Naturalistic Observation Study of Dual-Earner Families after Work and School." *Journal of Family Psychology* 23, 6 (2009): 798–807.

Cancian, Francesca. "Gender Politics: Love and Power in the Private and Public Spheres." In *Gender and the Life Course*, edited by A. Rossi. Hawthorne, NY: Aldine, 1985: 253–262.

Cancian, Francesca. "Defining 'Good' Child Care: Hegemonic and Democratic Standards." In *Child Care and Inequality: Re-Thinking Carework for Children and Youth*, edited by Francesca Cancian, Demie Kurz, Andrew London, Rebecca Reviere, and Mary Tuominen, pp. 65–80. New York: Routledge, 2002.

Cannon, H. Brevy. "'Whose Turn to Pay?' Can Be a Real Deal-Breaker for Cohabiting Couples." UVa Today, 2009. www.virginia.edu/uvatoday /newsRelease.php?id=8573.

Caponera, B. (1998). *The Nature of Sexual Assault in New Mexico I (1995–1996).* Albuquerque, NM: New Mexico Clearinghouse on Sexual Abuse and Assault Services, 1998.

Carlson, Daniel L., Amanda J. Miller, Sharon Sassler, and Sarah Hanson. "The Gendered Division of Housework and Couples' Sexual Relationships. A Re-Examination." *Sociology Faculty Publications.* Paper 2, 2014.http://scholarworks.gsu.edu/sociology_facpub/2.

Carpenter, Laura. "Gender and the Meaning and Experience of Virginity Loss in the Contemporary United States." *Gender and Society* 16, 3 (June 2002): 345–365.

Carpenter, Laura. *Virginity Lost: An Intimate Portrait of First Sexual Experiences.* New York: New York University Press, 2005.

Carr, Deborah. "The Desire to Date and Remarry among Older Widows and Widowers." *Journal of Marriage and Family* 66, 4 (November 2004): 1051–1068.

Carr, Deborah, and Kristen Springer. "Advances in Family and Health Research in the 21st Century." *Journal of Marriage and Family* 72 (June 2010): 743–761.

Carr, Deborah, Kristen Springer, and Philip Cowan. "The 100th Anniversary of Father's Day: A Council on Contemporary Families Media Advisory." Council on Contemporary Families. 2010. contemporaryfamilies.org/the-100th-anniversary-of-fathers-day/.

Casper, L., and L. Sayer. "Cohabitation Transitions: Different Purposes and Goals, Different Paths." Paper presented at the meeting of the Population Association of America, Los Angeles, 2002.

Casper, Lynne, and Suzanne Bianchi. *Continuity and Change in the American Family.* Thousand Oaks, CA: Sage Publications, 2002.

Caspi, Jonathan. "Building a Sibling Aggression Treatment Model: Design and Development Research in Action." *Research on Social Work Practice* 18, 6 (November 2008): 575–585. http://rsw.sagepub.com/content/18/6/575. abstract.

Caspi, Jonathan. *Sibling Aggression: Assessment and Treatment.* New York: Springer, 2011.

Catalano, Shannon. "Intimate Partner Violence: Attributes of Victimization, 1993–2011." Bureau of Justice Statistics, Washington, DC, 2013. www.bjs.gov/index.cfm?ty=pbdetail&iid=4801.

Catalano, Shannan. "Intimate Partner Violence: Attributes of Victimization, 1993–2011." U.S. Department of Justice Office of Justice Programs, Bureau of Justice Statistics. BJS Special Report, November 2013, NcJ 243300.

Catalano, Shannan, Erica Smith, Howard Snyder, and Michael Rand. "Female Victims of Violence." Bureau of Justice Statistics, Selected Findings, NCJ 228356, September 2009. http://bjs.ojp.usdoj.gov /content/pub/pdf/fvv.pdf.

Cate, Rodney M., and Sally A. Lloyd. *Courtship.* Newbury Park, CA: Sage, 1992.

Caughlin, John, and Erin Basinger. "Completely Open and Honest Communication: Is That What We Really Want?" *National Council on Family Relations*, NCFR Family Focus on Effective Communication FF64 (Summer 2015): F1–F3.

Caughlin, J. P., and R. S. Malis. "Demand/Withdraw Communication between Parents and Adolescents as a Correlate of Relational Satisfaction." *Communication Reports* (June 22, 2004): 59–71.

Caughlin, J. P., and T. L. Huston. "The Flourishing Field of Flourishing Relationships." *Journal of Family Theory & Review* 2 (2010): 25–35.

Caughlin, J., and A. Scott. "Toward a Communication Theory of the Demand /Withdraw Pattern of Interaction in Interpersonal Relationships." In *New Directions in Interpersonal Communication Research*, edited by S. Smith and S. Wilson. Thousand Oaks, CA: Sage, 2010.

Cavanagh, S., and K. Sullivan. "Family Instability, Childhood Relationship Skills and Romance in Adolescence." National Center for Family and Marriage, Working Paper Series, WP-09-11, September 2009.

Centers for Disease Control and Prevention. "Intimate Partner Violence: Consequences," 2003. www.cdc.gov/violenceprevention /intimatepartnerviolence/consequences.html.

————. "Fact Sheet: Understanding Elder Maltreatment," 2008a.

————. "Intimate Partner Violence Prevention, Scientific Information: Risk and Protective Factors." National Center for Injury Prevention and Control, 2008b.

————. "Child Maltreatment: Facts at a Glance," 2009a. http:// indianaparentinginstitute.org/pdf/maltreatment1.pdf.

————. "Intimate Partner Violence: Dating Violence Fact Sheet," 2009b. www.cdc.gov/violenceprevention/intimatepartnerviolence /teen_dating_violence.html.

————. "Health, United States," 2011. www.cdc.gov/nchs/data/hus /hus11.pdf#022.

————. "Infertility FAQs: Reproductive Health," 2012a. Centers for Disease Control and Prevention. www.cdc.gov/reproductivehealth/infertility/.

_____. "Reproductive Health: Depression among Women of Reproductive Age and Postpartum Depression," 2012b. Centers for Disease Control and Prevention. www.cdc.gov/reproductivehealth/depression/.

_____. "Sexually Transmitted Diseases (STDs)," 2012c. www.cdc.gov/std/stats/.

_____. "STDs in Racial and Ethnic Minorities," 2013. www.cdc.gov/std/stats13/minorities.htm.

_____. "HIV among Women: Fast Facts," 2015a.www.cdc.gov/hiv/pdf/risk_women.pdf.

_____. "HIV in the United States: At a Glance," 2015b. www.cdc.gov/hiv/statistics/overview/ataglance.html.

_____. "Intimate Partner Violence: Consequences." National Center for Injury Prevention and Control, Division of Violence Prevention, 2015c. www.cdc.gov/violenceprevention/intimatepartnerviolence/consequences.html.

_____. "National Marriage and Divorce Rate Trends, Provisional Number of Marriages and Marriage Rate: United States, 2000–2014," 2015d. www.cdc.gov/nchs/nvss/marriage_divorce_tables.htm.

_____. "Sudden Unexpected Infant Deaths and Sudden Infant Death Syndrome," Data and Statistics, 2015e. www.cdc.gov/sids/data.htm.

_____. "Understanding Elder Abuse." Division of Violence Prevention, 2016a. www.cdc.gov/violenceprevention/pdf/em-factsheet-a.pdf.

_____. "Understanding Teen Dating Violence," 2016b. www.cdc.gov/violenceprevention/pdf/teen-dating-violence-factsheet-a.pdf.

Central Intelligence Agency. "Country Comparison: Infant Mortality Rate." The World Factbook, 2015. https://www.cia.gov/library/publications/the-world-factbook/rankorder/2091rank.html.

Chambers, Anthony, and Aliza Kravitz. "Understanding the Disproportionately Low Marriage Rate among African Americans: An Amalgam of Sociological and Psychological Constraints." *Family Relations* 60 (December 2011): 648–660.

Chandra, A., C. E. Copen, and E. H. Stephen. "Infertility and Impaired Fecundity in the United States, 1982–2010: Data from the National Survey of Family Growth. National for Health Statistics Reports 67. Hyattsville, MD: National Center for Health Statistics. 2013.

Chandra, A., W. D. Mosher, C. Copen, and C. Sionean. "Sexual Behavior, Sexual Attraction, and Sexual Identity in the United States: Data from the 2006–2008 National Survey of Family Growth." *National Health Statistics Reports* 36. Hyattsville, MD: National Center for Health Statistics, Table 6, 2011.

Chappell, Bill. "California Enacts 'Yes Means Yes' Law, Defining Sexual Consent." NPR, The Two-Way: Breaking News from NPR, September 29, 2014.

Cherlin, Andrew. *Marriage, Divorce, Remarriage.* Cambridge, MA: Harvard University Press, 1981.

_____. "On Single Mothers 'Doing' Family." *Journal of Marriage and Family* 68 (2006): 800–803.

_____. *The Marriage-Go-Round: The State of Marriage and the Family in America Today.* New York: Knopf, 2009.

_____. "Demographic Trends in the United States: A Review of Research in the 2000s." *Journal of Marriage and Family* 72 (June 2010): 403–419.

Cherlin, Andrew. *Labor's Love Lost: The Rise and Fall of the Working-Class Family in America.* New York: Russell Sage Foundation, 2014.

Cherlin, Andrew, Linda M. Burton, Tera R. Hurt, and Diane M. Purvin. "The Influence of Physical and Sexual Abuse on Marriage and Cohabitation." *American Sociological Review* 69 (2004): 768–789.

Cherlin, Andrew J. "The Deinstitutionalization of American Marriage." *Journal of Marriage and Family* 66, 4 (November 2004): 848–861.

Chesser, Barbara Jo. "Analysis of Wedding Rituals: An Attempt to Make Weddings More Meaningful." *Family Relations* 29, 2 (April 1980).

Chetty, Raj, and Nathaniel Hendren. "The Impacts of Neighborhoods on Intergenerational Mobility: Childhood Exposure Effects and County-Level Estimates." Harvard University and NBER, 2015. http://scholar.harvard.edu/files/hendren/files/nbhds_paper.pdf?m=1430722623

Child Care Aware® of America. "Parents and the High Cost of Child Care," 2015. http://usa.childcareaware.org

Child Welfare Information Gateway. "Fact Sheet for Families: Impact of Adoption on Adoptive Parents," November 2010. www.childwelfare.gov/pubs/factsheets/impact_parent/impactparent.pdf.

_____. "Costs of Adopting" (fact sheet). Washington, DC: U.S. Department of Health and Human Services, Children's Bureau, 2011a.

_____. "How Many Children Were Adopted in 2007 and 2008?" Washington, DC: U.S. Department of Health and Human Services, Children's Bureau, 2011b.

_____. "Adoption Disruption and Dissolution." U.S. Department of Health and Human Services, 2012. www.childwelfare.gov/pubs/s_disrup.cfm.

Children's Defense Fund. *The State of America's Children: 2014 Report.* Washington, DC, 2014.

ChildStats.gov. Forum on Child and Family Statistics, America's Children: Key National Indicators of Well-Being, 2011. www.childstats.gov/americaschildren/index3.asp.

Chinitz, J., and R. Brown. "Religious Homogamy, Marital Conflict, and Stability in Same-Faith and Interfaith Jewish Marriages." *Journal for the Scientific Study of Religion* 40, 4 (December 2001): 723–733.

Cho, H. "Racial Differences in the Prevalence of Intimate Partner Violence against Women and Associated Factors." *Journal of Interpersonal Violence* 27, 2 (2012): 344–363.

Chodorow, Nancy. *The Reproduction of Mothering: Psycho-Analysis and the Sociology of Gender.* Berkeley: University of California Press, 1978.

Choi, H., and N. F. Marks, "Marital Conflict, Depressive Symptoms, and Functional Impairment." *Journal of Marriage and Family,* 70 (2008): 377–390.

Christensen, A., and C. L. Heavey, "Gender Differences in Marital Conflict: The Demand–Withdraw Interaction Pattern." In *Gender Issues in Contemporary Society,* edited by S. Oskamp, and M. Costanzo, pp. 113–141. Newbury Park, CA: Sage, 1993.

Christensen, A., and L. Pasch. "The Sequence of Marital Conflict: An Analysis of Seven Phases of Marital Conflict in Distressed and Nondistressed Couples." *Clinical Psychology Review* 13 (1993): 3–14.

Christensen, A., B. D. Doss, and D. C. Atkins, "A Science of Couple Therapy: For What Should We Seek Empirical Support?" In *Family Psychology: The Art of the Science,* edited by W. M. Pinsof and J. Lebow. Oxford/New York: Oxford University Press, 2006.

Christopher, F. Scott, and Susan Sprecher. "Sexuality in Marriage, Dating, and Other Relationships: A Decade Review." *Journal of Marriage and the Family* 62, 4 (November 2000): 999–1017.

Christopher, Karen. "Extensive Mothering: Employed Mothers' Construction of the Good Mother." *Gender and Society* 26, 1 (January 2012): 73–96.

Ciabattari, T. "Cohabitation and Housework: The Effects of Marital Intentions." *Journal of Marriage and Family* 66 (2004): 118–125.

Clarke-Stewart, A., and C. Brentano. *Divorce: Causes and Consequences.* New Haven, CT: Yale University Press, 2007.

Clatterbaugh, Kenneth. *Contemporary Perspectives on Masculinity: Men, Women, and Politics in Modern Society,* 2nd ed. Boulder, CO: Westview, 1997.

Clawson, Dan, and Naomi Gerstel. "Caring for Our Young: Child Care in Europe and the United States." *Contexts* 1, 4 (Fall/Winter 2002): 28–35.

Claxton, A., and M. Perry-Jenkins. "No Fun Anymore: Leisure and Marital Quality across the Transition to Parenthood." *Journal of Marriage and Family* 70, 1 (February 2008): 28–43.

Clearfield, M., and N. M. Nelson. "Sex Differences in Mothers' Speech and Play Behavior with 6-, 9-, and 14-Month-Old Infants." *Sex Roles* 54, 1/2 (January 2006): 127–137.

Clements, Marie, Scott M. Stanley, and Howard J. Markman. "Before They Said 'I Do': Discriminating among Marital Outcomes over 13 Years." *Journal of Marriage and Family* 66, 3 (August 2004): 613–626.

Cohen, Oma, Yael Geron, and Alva Farchi. "A Typology of Marital Quality of Enduring Marriages in Israel." *Journal Of Family Issues* 31, 6 (2010): 727–774.

Cohen, Philip. "Has the Recession Increased Family Violence?" Family Equality (blog), 2011. http://familyinequality.wordpress.com.

Cohen, Phillip. "Divorce and the Recession." Council on Contemporary Families, October 8, 2014. https://contemporaryfamilies.org/divorce-recession-brief-report.

Cohen, T. F. "What Do Fathers Provide? Reconsidering the Economic and Nurturant Dimensions of Men as Parents." In *Men, Work and Family,* edited by J. C. Hood. Newbury Park, CA: Sage, 1993.

Cohen, Theodore (ed.). *Men and Masculinity: A Text Reader.* Belmont CA: Wadsworth/Thompson Learning, 2001.

Cohen, Theodore, and John C. Durst. "Leaving Work and Staying Home: The Impact on Men of Terminating the Male Economic Provider Role." In *Men and Masculinity: A Text-Reader.* Belmont, CA: Wadsworth, 2001.

Cohn, D'Vera. "Divorce and the Great Recession." Pew Research Center, 2012. pewsocialtrends.org.

Cohn, D'Vera. "Love and Marriage." Pew Research Center, 2013. www.pewsocialtrends.org/2013/02/13/love-and-marriage.

_____. "American Indian and White, but Not Multiracial." Pew Research Center, 2015. www.pewresearch.org/fact-tank/2015/06/11/american-indian-and-white-but-not-multiracial/.

Cohn, D'Vera, Gretchen Livingston, and Wendy Wang. "After Decades of Decline, a Rise in Stay-at-Home Mothers." Washington, DC: Pew Research Center's Social & Demographic Trends project, April 2014.

Cohn, D'Vera, Jeffrey S. Passel, Wendy Wang, and Gretchen Livingston. "Barely Half of U.S. Adults Are Married—A Record Low." Pew Research Center, 2011. www.pewsocialtrends.org/2011/12/14/barely-half-of-u-s-adults-are-married-a-record-low/.

Colby, Sandra L., and Jennifer M. Ortman. "Projections of the Size and Composition of the U.S. Population: 2014 to 2060." Current Population Reports, P25-1143, U.S. Census Bureau, Washington, DC, 2014.

Cole, Robert. "Mental Illness and the Family." In *Vision 2010: Families and Health Care*, edited by B. A. Elliott. Minneapolis: National Council on Family Relations (1993): 18–19.

Coleman, Marilyn, Lawrence Ganong, and Mark Fine. "Reinvestigating Remarriage: Another Decade of Progress." *Journal of Marriage and the Family* 62, 4 (November 2000): 1288–1307.

Coleman, Marilyn, Lawrence Ganong, Luke Russell, and Nick Frye-Cox. "Stepchildren's Views about Former Step Relationships Following Stepfamily Dissolution." *Journal of Marriage and Family* 77, 3 (June 2015): 775–790.

Coleman, Thomas. "A Week for Singles but Greeting Cards Can't Be Found." Column One: Eye on Unmarried America, unmarriedamerica.org, September 2005.

College Board. "Trends in College Pricing 2015." College Board, 2015. http://trends.collegeboard.org/sites/default/files/trends-college-pricing-web-final-508-2.pdf.

Collins, Rebecca, Steven Martino, Marc Elliott, and Angie Miu. "Relationships between Adolescent Sexual Outcomes and Exposure to Sex in Media: Robustness to Propensity Analysis." *Developmental Psychology* 47, 2 (2011): 585–591.

Collins, R., M. N. Elliott, S. H. Berry, D. E. Kanouse, D. Kunkel, S. B. Hunter, and A. Miu. "Watching Sex on Television Predicts Adolescent Initiation of Sexual Behavior." *Pediatrics* 114, 3 (September 2004): e280–e289.

Collins, Rebecca L., Steven C. Martino, Marc N. Elliott, and Angela Miu. "Relationships between Adolescent Sexual Outcomes and Exposure to Sex in Media: Robustness to Propensity-Based Analysis." *Developmental Psychology*, 47, 2 (March 2011): 585–591.

Coltrane, Scott. "Research on Household Labor: Modeling and Measuring the Social Embeddedness of Routine Family Work." *Journal of Marriage and the Family* 62, 4 (November 2000): 1208–1233.

Committee on Child Maltreatment Research, Policy, and Practice for the Next Decade: Phase II; Board on Children, Youth, and Families; Committee on Law and Justice; Institute of Medicine; National Research Council; Petersen AC, Joseph J, Feit M, eds. "Describing the Problem." *New Directions in Child Abuse and Neglect Research*, Mar 25, 2014. Washington, DC: National Academies Press. www.ncbi.nlm.nih.gov/books/NBK195982/.

Conger, Rand, Katherine Conger, and Monica Martin. "Socioeconomic Status, Family Processes, and Individual Development." *Journal of Marriage and Family* 72 (June 2010): 685–704.

Conlin, Jennifer. "The Freedom to Choose Your Pronoun." *New York Times*, September 30, 2011.

Connell, Robert. *Gender and Power: Society, the Person, and Sexual Politics.* Stanford, CA: Stanford University Press, 1987.

_____. *Masculinities.* Berkeley: University of California Press, 1995.

Connolly, Lisa. "Anti-Gay Bullying in Schools—Are Anti-Bullying Statutes the Solution?" *Law Review* 248 (April 2012), 87 N.Y.U.

Connidis, I. A. "Intimate Relationships: Learning from Later Life Experience." In *Age Matters*, edited by T. Calasanti and K. Slevin. New York: Routledge, 2006.

Connidis, Ingrid, Arnet. "Theoretical Directions for Studying Family Ties and Aging." In *Handbook of Families and Aging*, 2nd ed., edited by R. Blieszner and V. H. Bedford. Santa Barbara, CA: Praeger, 2012.

Contrera, Jessica. "Target Will Stop Separating Toys and Bedding into Girls' and Boys' Sections." *Washington Post*, August 9, 2015. www.washingtonpost.com/news/arts-and-entertainment/wp/2015/08/09/target-will-stop-separating-toys-and-bedding-into-girls-and-boys-sections/.

Cooke, L. "'Doing' Gender in Context: Household Bargaining and Risk of Divorce in Germany and the United States." *American Journal of Sociology* 112 (2006): 442–472.

Cooke, L. P. "'Traditional' Marriages Now Less Stable than Ones Where Couples Share Work and Household Chores," 2008. www.contemporaryfamilies.org.

Cooney, Teresa M. "Young Adults' Relations with Parents: The Influence of Recent Parental Divorce." *Journal of Marriage and the Family* 56, 1 (February 1994): 45–56.

Coontz, Stephanie. *The Way We Really Are: Coming to Terms with America's Changing Families.* New York: Basic Books, 1997.

_____. "Divorcing Reality: Other Researchers Question Wallerstein's Conclusions," *Children's Advocate*, Action Alliance for Children, January–February, 1998.

_____. "The World Historical Transformation of Marriage." *Journal of Marriage and Family* 66, 4 (November 2004): 974–979.

_____. *Marriage, A History: From Obedience to Intimacy, or How Love Conquered Marriage.* New York: Viking, 2005.

_____. "The Moynihan Family Circus: How a Fifty-Year-Old Report on the Black Family and Poverty Continues to Distort American Social Policy." BookForum, June/July/August 2015. www.bookforum.com/inprint/022_02/14579.

_____. "The Triumph of the Working Mother." *New York Times*, June 1, 2013. www.nytimes.com/2013/06/02/opinion/sunday/coontz-the-triumph-of-the-working-mother.html?_r=0.

Cordova, J. V., C. B. Gee, and L. Z. Warren. "Emotional Skillfulness in Marriage: Intimacy as a Mediator of the Relationship between Emotional Skillfulness and Marital Satisfaction." *Journal of Social and Clinical Psychology* 24 (2005): 218–235.

Cotter, D., P. England, and J. Hermsen. "Moms and Jobs: Trends in Mothers' Employment and Which Mothers Stay Home," 2007. Council on Contemporary Families. https://contemporaryfamilies.org/moms-jobs-trends-mothers-employment-mothers-stay-home/.

Cotter, David, Joan M. Hermsen, and Reeve Vanneman. "The End of the Gender Revolution? Gender Role Attitudes from 1977 to 2008." *American Journal of Sociology* 117.1 (2011): 259–289.

Cotter, David, Joan Hermsen, and Reeve Vanneman. "Brief: Back on Track? The Stall and Rebound in Support for Women's New Roles in Work and Politics, 1977–2012." Council on Contemporary Families, July 30, 2014. https://contemporaryfamilies.org/gender-revolution-rebound-brief-back-on-track/.

Cottle, T. *Hardest Times: The Trauma of Long Term Unemployment.* Westport, CT: Praeger, 2001.

Cowan, C. P., and P. A. Cowan. *When Partners Become Parents: The Big Life Change for Couples.* New York: Basic Books, 1992.

Cowan, Carolyn Pope, and Philip Cowan. *When Partners Become Parents: The Big Life Change for Couples.* Mahwah, NJ: Lawrence Erlbaum, 2000.

Cowan, Philip, and Carolyn Pope Cowan. "After the Baby: Keeping the Couple Relationship Alive." *National Council on Family Relations Report: Family Focus on Transition to Parenthood, FF 49* (2011).

Cowan, Philip, and Carolyn Cowan. "Becoming a Family: Research and Intervention." In *Methods of Family Research: Biographies of Research Projects*, edited by I. Sigel and G. Brody. Hillsdale, NJ: Lawrence Erlbaum, 1990.

Cowdery, Randi, Norma Scarborough, Carmen Knudson-Martin, Gita Seshadri, Monique Lewis, and Anne Rankin Mahoney. "Gendered Power in Cultural Contexts: Part II. Middle Class African American Heterosexual Couples with Young Children." *Family Process* 48, 1 (March 2009): 25–39.

Coyne, Sarah, Brad Bushman, and Amy Nathanson. "Media and the Family: A Note from the Guest Editors." *Family Relations* 61 (3): 359–362.

Crabb, P. B., and Marciano, D. L. "Representations of Material Culture and Gender in Award-Winning Children's Books: A Twenty-Year Follow-Up." *Journal of Research in Childhood Education* 25 (2011): 390–398.

Cramer, R. E., R. E. Lipinski, J. D. Meteer, and J. A. Houska. "Sex Differences in Subjective Distress to Unfaithfulness: Testing Competing Evolutionary and Violation of Infidelity Expectations Hypotheses." *Journal of Social Psychology* 148, 4 (2008), 389–406.

Cramer, Robert E., William T. Abraham, Lesley M. Johnson, and Barbara Manning-Ryan. "Gender Differences in Subjective Distress to Emotional and Sexual Infidelity: Evolutionary or Logical Inference Explanation." *Current Psychology* 20, 4 (Winter, 2001–2002): 327–336.

Crissey, S. R. "Race/Ethnic Differences in the Marital Expectations of Adolescents: The Role of Romantic Relationships." *Journal of Marriage and Family* 67 (2005): 697–709.

Crohan, Susan E. "Marital Quality and Conflict across the Transition to Parenthood in African American and White Couples." *Journal of Marriage and Family* 58, 4 (November 1996): 933–944.

Crosnoe, Robert, and Shannon Cavanagh. "Families with Children and Adolescents: A Review, Critique, and Future Agenda." *Journal of Marriage and Family* 72 (June 2010): 594–611.

Crouter, A., M. Bumpus, M. Head, and S. McHale. "Implications of Overwork and Overload for the Quality of Men's Family Relationships." *Journal of Marriage and Family* 63, 2 (2001): 404–416.

Crouter, Ann C., and Beth Manke. "The Changing American Workplace: Implications for Individuals and Families." *Family Relations* 43, 2 (April 1994): 117–124.

Crouter, Ann C., and Susan M. McHale. "The Long Arm of the Job: Influences of Parental Work on Child Rearing." In *Parenting: An Ecological Perspective,* edited by T. Luster and L. Okagaki. Hillsdale, NJ: Lawrence Erlbaum, 1993.

Cubanski, J., G. Casillas, and A. Damico. "Poverty among Seniors: An Updated Analysis of National and State Level Poverty Rates under the Official and Supplemental Poverty Measures." Kaiser Family Foundation, 2015. http://kff.org/report-section/poverty-among-seniors-issue-brief/

Cuber, John F., and Peggy Harroff. *The Significant Americans, a Study of Sexual Behavior among the Affluent.* NY: Appleton-Century, 1965.

Cui, Ming, and M. Brent Donnellan. "Trajectories of Conflict over Raising Adolescent Children and Marital Satisfaction." *Journal of Marriage and Family* 71 (2009): 478–494.

Cummings, E. Mark, Marcie C. Goeke-Morey, and Lauren M. Papp. "Children's Responses to Everyday Marital Conflict Tactics in the Home." *Child Development* 74, 6 (November 2003): 1918–1929.

Cunningham, Mick. "Parental Influences of the Gendered Division of Housework." *American Sociological Review* 66 (April 2001): 184–203.

Cunradi, Carol, Raul Caetano, and John Schafer. "Socio-Economic Predictors of Intimate Partner Violence among White, Black, and Hispanic Couples in the United States." *Journal of Family Violence* 17, 4 (December 2002): 377–389.

Curry, Tim, Robert Jiobu, and Kent Schwirian. *Sociology for the 21st Century.* Prentice Hall, 2002.

Curtin, S. C., J. C. Abma, S. J. Ventura, and S. K. Henshaw. "Pregnancy Rates for U.S. Women Continue to Drop." NCHS data brief, no 136. Hyattsville, MD: National Center for Health Statistics, 2013. www.cdc.gov/nchs/data/databriefs/db136.pdf.

Cyr, Mireille, John Wright, Pierre McDuff, and Alain Perron. "Intrafamilial Sexual Abuse: Brother-Sister Incest Does Not Differ from Father-Daughter and Stepfather-Stepdaughter Incest." *Child Abuse & Neglect* 26, 9 (September 2002): 957–974.

Dahlhamer, J. M., A. M. Galinsky, S. S. Joestl, and B. W. Ward. "Sexual Orientation in the 2013 National Health Interview Survey: A Quality Assessment." *Vital Health Statistics* 2, 169 (2014).

*Daily Mail Reporter.* "Corpse Bride: French Woman Marries Her DEAD Boyfriend." *Daily Mail,* June 24, 2011. www.dailymail.co.uk/news/article-2007803.

Dalla, R. "I Fell Off (the Mothering) Track: Barriers to 'Effective Mothering' among Prostituted Women." *Family Relations* 53 (March 2004): 190–200.

Daly, M., and M. Wilson. *The Truth about Cinderella.* New Haven, CT: Yale University Press, 1998.

———. "The Cinderella Effect: Parental Discrimination against Stepchildren." *Samfundsokonomen* 4 (2002): 39–46.

———. "Happiness, Unhappiness, and Suicide: An Empirical Assessment." *Journal of the European Economic Association,* April 2009.

Darling, Nancy. "Parenting Style and Its Correlates." ERIC Digest Champaign: ERIC Clearinghouse on Elementary and Early Childhood Education (ED 427 896), University of Illinois, 1999. www.ericdigests.org/1999-4/parenting.htm.

Dasgupta, Nilanjana, and Luis Rivera. "When Social Context Matters: The Influence of Long-Term Contact and Short-Term Exposure to Admired Outgroup Members on Implicit Attitudes and Behavioral Intentions." *Social Cognition* 26, 1 (2008): 112–123.

Davidson, Helen. "Third Gender Must Be Recognised by NSW after Norrie Wins Legal Battle." *The Guardian,* April 1, 2014. www.theguardian.com/world/2014/apr/02/third-gender-must-be-recognised-by-nsw-after-norrie-wins-legal-battle.

Davies, Lizzy. "French Woman Marries Dead Partner." *The Guardian,* November 17, 2009.

Davis, Shannon, and Theodore Greenstein. "Why Study Housework? Cleaning as a Window into Power in Couples." *Journal of Family Theory & Review,* Special Issue on Household Labor 5, 2 (June 2013): 63–71.

Day, A., D. Chung, P. O'Leary, and E. Carson. "Programs for Men Who Perpetrate Domestic Violence: An Examination of the Issues Underlying the Effectiveness of Intervention Programs." *Journal of Family Violence* 24, 3 (2009): 203–212.

De Coulanges, N. D. F. *The Ancient City: A Study on the Religion, Laws, and Institutions of Greece and Rome.* Baltimore, MD: Johns Hopkins University, 1980.

de Graaf, P. M., and M. Kalmijn. "Alternative Routes in the Remarriage Market: Competing-Risk Analyses of Union Formation after Divorce." *Social Forces* 81, 4 (June 2003): 1459–1498.

De Judicibus, Margaret A., and Marita P. McCabe. "Psychological Factors and the Sexuality of Pregnant and Postpartum Women." *Journal of Sex Research* 39, 2 (May 2002): 94–83.

Deal, Ron, and David Olson. *The Remarriage Checkup: Tools to Help Your Marriage Last a Lifetime.* Bloomington, MN: Bethany House, 2010.

Declercq, E. R., C. Sakala, M. P. Corry, and S. Applebaum. *Listening to Mothers II: Report of the Second National U.S. Survey of Women's Childbearing Experiences.* New York: Childbirth Connection, 2006.

Declercq, E. R., C. Sakala, M. P. Corry, S. Applebaum, and A. Herrlich. *Listening to Mothers III: Pregnancy and Birth.* New York: Childbirth Connection, May 2013. http://transform.childbirthconnection.org/wp-content/uploads/2013/06/LTM-III_Pregnancy-and-Birth.pdf.

DeGenova, Mary Kay. *Families in Cultural Context: Strengths and Challenges in Diversity.* Mountain View, CA: Mayfield, 1997.

Degler, Carl. *Women and the Family in America from the Revolution to the Present.* New York: Oxford University Press, 1980.

DeMaris, Alfred. "The Influence of Intimate Violence on Transitions out of Cohabitation." *Journal of Marriage and Family* 63 (February 2001): 235–246.

DeMaris, Al. "Burning the Candle at Both Ends: Extramarital Sex as a Precursor of Marital Disruption." *Journal of Family Issues* 34, 11 (January 4, 2013): 1474–1499.

Demerouti, E., A. Bakker, and W. Schaufeli. "Spillover and Crossover of Exhaustion and Life Satisfaction among Dual-Earner Parents." *Journal of Vocational Behavior* 67 (2005): 266–289.

Demo, David, and Martha Cox. "Families with Young Children: A Review of Research in the 1990s." *Journal of Marriage and the Family,* 62, 4 (November 2000): 876–895.

Demos, John. *A Little Commonwealth.* New York: Oxford University Press, 1970.

Demos, Vasilikie. "Black Family Studies in the *Journal of Marriage and the Family* and the Issue of Distortion: A Trend Analysis." *Journal of Marriage and the Family* 52, 3 (August 1990): 603–612.

DeNavas-Walt, Carmen, and Bernadette D. Proctor. "U.S. Census Bureau, Current Population Reports, P60-249." Income and Poverty in the United States: 2013, U.S. Government Printing Office, Washington, DC, 2014, Table B-1.

———. "U.S. Census Bureau, Current Population Reports, P60-249." Income and Poverty in the United States: 2013, U.S. Government Printing Office, Washington, DC, 2014.

———. "Current Population Reports, P60-252, Income and Poverty in the United States: 2014." U.S. Census Bureau, U.S. Government Printing Office, Washington, DC, 2015.

DeNavas-Walt, Carmen, Bernadette Proctor, and Jessica Smith. U.S. Census Bureau, Current Population Reports, P60-239, Income, Poverty, and Health Insurance Coverage in the United States: 2010, U.S. Government Printing Office, Washington, DC, 2011.

DePaulo, B. *Singled Out: How Singles Are Stereotyped, Stigmatized, and Ignored, and Still Live Happily Ever After.* New York: St. Martin's Press, 2006.

DeSilver, Drew. "After recession, more children living with Grandma or Grandpa. Pew Research Center September 18, 2013. www.pewresearch.org/fact-tank/2013/09/18/after-recession-more-children-living-with-grandma-or-grandpa/.

———. "Rising Cost of Child Care May Help Explain Recent Increase in Stay-at-Home Moms." Pew Research Center, 2014. www.pewresearch.org/fact-tank/2014/04/08/rising-cost-of-child-care-may-help-explain-increase-in-stay-at-home-moms/.

Dettling, Lisa, and Joanne Hsu. "Returning to the Nest: Debt and Parental Co-residence Among Young Adults." Finance and Economics Discussion Series No.2014-80. Washington: Board of Governors of the Federal Reserve System, September 2014.

Dew, J. P. "The Many Interfaces between Money and Marriage." NCFR Report, FF63 (2015): F11–F12.

Dew, J. P., S. Britt, and S. J. Huston. "Examining the Relationship between Financial Issues and Divorce." *Family Relations* 61 (2012): 615–628.

Dew, Jeffrey. "Bank on It: Thrifty Couples are the Happiest." In the State of Our Unions, 2009: Money and Marriage National Marriage Project and the Institute for American Values (pp. 23–30).

———. "Financial Issues and Relationship Outcomes among Cohabiting Individuals." *Family Relations* 60 (April 2011): 178–190.

Dew, Jeffery, and Bradley Wilcox. "If Momma Ain't Happy: Explaining Declines in Marital Satisfaction among New Mothers." *Journal of Marriage and Family* 73 (February 2011): 1–12.

Dewey, Caitlin. "Confused by Facebook's New Gender Options? Here's What They Mean." *Washington Post*, February 14, 2014. www.washingtonpost.com/news/arts-and-entertainment/wp/2014/02/14/confused-by-facebooks-new-gender-options-heres-what-they-mean/.

DeWitt, P. M. "Breaking Up Is Hard To Do." *American Demographics*, reprint package (1994): 14–16.

Diamond, Lisa. "Emerging Perspectives on Distinctions between Romantic Love and Sexual Desire." *Current Directions in Psychological Science* 13, 3 (2004): 116–119.

Dibiase, Rosemarie, and Jaime Gunnoe. "Gender and Culture Differences in Touching Behavior." *Journal of Social Psychology* 144, 1 (February 2004): 49–62.

Diekmann, Andreas, and Kurt Schmidheiny. "Do Parents of Girls Have a Higher Risk of Divorce? An Eighteen-Country Study." *Journal of Marriage & Family* 66, 3 (August 2004): 651–660.

Digest of Education Statistics, 2014. https://nces.ed.gov/programs/digest/2014menu_tables.asp.

Dilorio, C., M. Kelley, and M. Hockenberry-Eaton. "Communication about Sexual Issues: Mothers, Fathers, and Friends." *Journal of Adolescent Health* 23, (1999):181–189.

Dilworth-Anderson, Peggye, and Harriette Pipes McAdoo. "The Study of Ethnic Minority Families: Implications for Practitioners and Policymakers." *Family Relations* 37, 3 (July 1988): 265–267.

Dion, Karen, et al. "What Is Beautiful Is Good." *Journal of Personality and Social Psychology* 24 (1972): 285–290.

Dittus, P., J. Jaccard, and V. Gordon. "Direct and Indirect Communication of Maternal Beliefs to Adolescents: Adolescent Motivations for Premarital Sexual Activity." *Journal of Applied Social Psychology* 29 (1999): 1927–1963.

Do, K. Anh, and Yan R. Xia. "Cultural Influences on Dating and Courtship." *Family Focus on Dating and Mate Selection, National Council on Family Relations* (Winter 2014): F16–F17.

Dobner, Jennifer. "Polygamous Family Launches Challenge of Utah Law." Fox News/Associated Press, December 30, 2011.

Dodson, Lisa, and Jillian Dickert. "Girls' Family Labor in Low-Income Households: A Decade of Qualitative Research." *Journal of Marriage & the Family* 66, 2 (May 2004): 218–332.

Doherty, William J., Edward Kouneski, and Martha Erickson. "Responsible Fathering: An Overview and Conceptual Framework." *Journal of Marriage and the Family* 60, 2 (1998): 277–292.

Dolgin, Kim. "Men's Friendships: Mismeasured, Demeaned, and Misunderstood?" In *Men and Masculinity: A Text Reader*, edited by T. Cohen. Belmont, CA: Wadsworth, 2001.

Donnelly, D., and E. Burgess. "The Decision to Remain in an Involuntarily Celibate Relationship." *Journal of Marriage and Family* 70, 2 (May 2008): 519–535.

Dovidio, J. F., Hebl, M., Richeson, J., & Shelton, J. N. "Nonverbal Communication, Race, and Intergroup Interaction." In *Handbook of Nonverbal Communication*, edited by V. Manusov, and M. L. Patterson (pp. 481–500). Thousand Oaks, CA: Sage, 2006.

Downey, D. B., and D. J. Condron. "Playing Well with Others in Kindergarten: The Benefits of Siblings at Home." *Journal of Marriage and Family* 66 (2004): 333–350.

Downey, Geraldine, C. Bonica, and C. Rincon. "Rejection Sensitivity and Conflict in Adolescent Romantic Relationships." In *Adolescent Romantic Relationships*, edited by W. Furman, B. Brown, and C. Feiring. New York: Cambridge University Press, 1999.

Downey, Geraldine, and Scott I. Feldman. "Implications of Rejection Sensitivity for Intimate Relationships." *Journal of Personality and Social Psychology* 70, 6 (1996): 1327–1343.

Doyle, James. *The Male Experience* (3rd ed.). Dubuque, Iowa: W. C. Brown, 1994.

Driver, Janice, and John Gottman. "Daily Marital Interactions and Positive Affect During Marital Conflict among Newlywed Couples." *Family Process* 43, 3 (September 2004): 301–314.

Drouin, Michelle, and Carly Landgraff. "Texting, Sexting, Attachment, and Intimacy in College Students' Romantic Relationships." *Computers in Human Behavior* 28 (2012): 444–449.

Duck, Steve, ed. *Dynamics of Relationships.* Thousand Oaks, CA: Sage, 1994.

Dunifon, R., and L. Kowaleski-Jones. "The Influence of Grandparents in Single-Mother Families." *Journal of Marriage and Family* 69 (2007): 465–481.

du Plessis, K., and D. Clarke. "Couples' Helpful, Unhelpful and Ideal Conflict Resolution Strategies: Secure and Insecure Attachment Differences and Similarities." *Interpersona* 2, 1 (2008): 65–88.

Dutton, Donald, and Mark Bodnarchuk. "Through a Psychological Lens: Personality Disorder and Spousal Assault." In *Current Controversies on Family Violence,* 2nd ed. edited by Donileen Loseke, Richard Gelles, and Mary Cavanaugh (pp. 5–18). Thousand Oaks, CA: Sage, 2005.

East, Patricia L. "Children's Provision of Family Caregiving: Benefit or Burden?" *Child Development Perspectives* 4, 1 (2010): 55–61.

Eastwick, P. W., and E. J. Finkel. "Sex Differences in Mate Preferences Revisited: Do People Know What They Initially Desire in a Romantic Partner?" *Journal of Personality and Social Psychology* 94 (2008): 245–264.

Eaton, Asia, and Suzanna Rose. "Has Dating Become More Egalitarian? A 35-Year Review Using *Sex Roles*." *Sex Roles* 64, 11–12 (2011): 843–862.

Eckman-Himanek, Celeste. "Transitioning to Parenthood: The Experience of Infertility." Family Focus, Issue FF49, Transitioning to Parenthood, National Council on Family Relations, 2011, pp. F9–F11.

Economides, Anastasia. "From Florida to Brooklyn to Care for Her Ailing Sister." *New York Times*, December 11, 2011.

*Economist.* "China's Corpse Brides: Wet Goods and Dry Goods. A Lucrative, Grisly Market for Grave Robbers and Murderers." *Economist*, July 26, 2007. www.economist.com/node/9558423.

Edin, K., and M. Kefalas. *Promises I Can Keep: Why Poor Women Put Motherhood before Marriage.* Berkeley: University of California Press, 2005.

Edin, Kathryn, Maria Kefalas, and Joanna Reed. "A Peek Inside the Black Box: What Marriage Means for Poor Unmarried Parents." *Journal of Marriage and Family* 66, 4 (November 2004): 1007–1014.

Eggebeen, David, and Chris Knoester. "Does Fatherhood Matter for Men?" *Journal of Marriage and Family* 63 (May 2001): 381–393.

Einolf, Christopher J., and Deborah M. Philbrick. "Generous or Greedy Marriage? Evidence from a Longitudinal Survey of Volunteering and Charitable Giving." *Journal of Marriage and Family,* 76 (June 2014): 573–586.

Ela, Elizabeth "Adoption Motivations among U.S. Parents" (FP-11-06), National Center for Family and Marriage Research, 2011. www.bgsu.edu/content/dam/BGSU/college-of-arts-and-sciences/NCFMR/documents/FP/FP-11-06.pdf.

Elliott, S., and D. Umberson. "The Performance of Desire: Gender and Sexual Negotiation in Long-Term Marriages." *Journal of Marriage and Family* 70 (2008): 391–406.

Ellis, Renee R., and Tavia Simmons. "Coresident Grandparents and Their Grandchildren: 2012." Current Population Reports, P20-576, U.S. Census Bureau, Washington, DC, 2014.

Ellison, Christopher, Amy Burdette, and W. Bradford Wilcox. "The Couple that Prays Together: Race and Ethnicity, Religion, and Relationship Quality among Working-Age Adults." *Journal of Marriage and Family* 72, 4 (August 2010): 963–975.

Ellison, Christopher G., Andrea K. Henderson, Norval D. Glenn, and Kristine E. Harkrider. "Sanctification, Stress, and Marital Quality." *Family Relations* 60 (2011): 404–420.

Ellison, Christopher, Marc Musick, and George Holden. "Does Conservative Protestantism Moderate the Association between Corporal Punishment and Child Outcomes?" *Journal of Marriage and Family* 73, 5 (October 2011): 946–961.

Elrod, L., and M. Dale. "Paradigm Shifts and Pendulum Swings in Child Custody: The Interests of Children in the Balance." *Family Law Quarterly* 42, 3 (2008): 381.

Ember, Melvin, Carol Ember, and Bobbi Low. "Comparing Explanations of Polygyny." *Cross Cultural Research* 41, 4 (November 2007): 428–440.

Emery, R. *Renegotiating Family Relationships, Divorce, Child Custody, and Mediation.* New York: The Guildford Press, 1994.

Emery, R. E., D. Sbarra, and T. Grover. "Divorce Mediation: Research and Reflections." *Family Court Review,* 43 (January 2005): 22–37.

Engel, M. "Do I Have an Obligation to Support My Stepchildren?" National Stepfamily Resource Center, 2000. www.stepfamilies.info/articles/do-i-have-an-obligation-to-support-my-stepchildren.php.

English, Bella. "Gay Spouse Murder Case Puts Light on Long-Hidden Problem." Boston Globe, February 24, 2012.

Erickson, Rebecca. "Why Emotion Work Matters: Sex, Gender, and the Division of Household Labor." *Journal of Marriage and Family* 67, 2 (May 2005): 337–351.

Erickson P. I., and A. Rapkin. "Unwanted Sexual Experiences among Middle and High School Youth." *Journal of Adolescent Health* 12 (1991): 319–325.

Equal Employment Opportunity Commission. "Charges Alleging Sexual Harassment FY 2010–FY 2014," 2014. www.eeoc.gov/eeoc/statistics /enforcement/sexual_harassment_new.cfm.

Eshelman, J. Ross. *The Family*, 8th ed. Needham Heights, MA: Allyn and Bacon, 1997.

Evenson, Ranae J., and Robin W. Simon. "Clarifying the Relationship between Parenthood and Depression." *Journal of Health & Social Behavior* 46, 4 (December 2005): 341–358.

Everett, Rebecca. "Cara Rintala's Third Murder Trial Scheduled for September." *Mass Live*, January 27, 2016.

Fagundes, Christopher. "Getting Over You: Contributions of Attachment Theory for Post-Breakup Emotional Adjustment." *Personal Relationships* 19, 1 (March 2012): 37–50.

Fallon, Stephen J. "It's the Best Time to Get HIV, and the Worst: While the Virus Finds Its Way Around." Test Positive Aware Network, 2005. www.thebody.com/content/art1019.html.

Fang, Xiangming, Derek Brown, Curtis Florence, and James Mercy. "The Economic Burden of Child Maltreatment in the United States and Implications for Prevention." *Child Abuse & Neglect* 36, 2 (February 2012): 156–165.

Farrell, Jane, and Sarah Jane Glynn. "The FAMILY Act: Facts and Frequently Asked Questions." Center for American Progress, December 12, 2013. www.americanprogress.org/issues/labor/report/2013/12/12/81037 /the-family-act-facts-and-frequently asked-questions.

Farrington, Brendan. "Florida Supreme Court Rules on Gay Parental Rights." The Boston Globe, November 8, 2013. www.bostonglobe .com/news/nation/2013/11/08/fla-supreme-court-settles-lesbian -custody-battle/UWSsMngQiYqt1k4sxikFJN/story.html.

Faulkner, Rhonda A., Maureen Davey, and Adam Davey. "Gender-Related Predictors of Change in Marital Satisfaction and Marital Conflict." *American Journal of Family Therapy* 33, 1 (January–February 2005): 61–83.

Faust, K. A., and J. N. McKibben. "Marital Dissolution: Divorce, Separation, Annulment, and Widowhood." In *Handbook of Marriage and the Family*, edited by M. B. Sussman, S. K. Steinmetz, and G. W. Peterson. New York: Plenum, 1999.

Fehr, Beverly. "Prototype Analysis of the Concepts of Love and Commitment." *Journal of Personality and Social Psychology* 55, 4 (1988): 557–579.

Ferree, Myra Marx. "Beyond Separate Spheres: Feminism and Family Research." In *Contemporary Families: Looking Forward, Looking Back*, edited by A. Booth. Minneapolis: National Council on Family Relations, 1991.

Field, Craig, and Raul Caetano. "Intimate Partner Violence in the U.S. General Population: Progress and Future Directions." *Journal of Interpersonal Violence* 20, 4 (April 2005): 463–469.

Field, Tiffany. "Quality Infant Day Care and Grade School Behavior and Performance." *Child Development* 62 (1991): 863–870.

Fields, J. "Children's Living Arrangements and Characteristics: March 2002." U.S. Census Bureau. Current Population Reports (P20-547), 2003.

Fields, J. "America's Families and Living Arrangements: 2003." *Current Population Reports*, P20-553. Washington, DC: U.S. Census Bureau, 2004.

Fields, Jason. "The Living Arrangements of Children," 2005. www.census .gov/population/pop-profile/dynamic/LivArrChildren.pdf.

Fields, Jason, and Lynne Casper. "America's Families and Living Arrangements: March 2000." *Current Population Reports* (Series P20, No. 537). Washington, DC: U.S. Census Bureau, 2001.

Filene, Peter. *Him/Her/Self: Sex Roles in Modern America*, 2nd ed. Baltimore: Johns Hopkins University Press, 1986.

Fincham, Frank, and Steven Beach. "Marriage in the New Millennium: A Decade in Review." *Journal of Marriage and Family* 72 (June 2010): 630–649.

Fincham, Frank D., and Steven R. H. Beach. "Conflict in Marriage: Implications for Working with Couples." *Annual Review of Psychology* 50, 1 (1999): 47–77.

Fincham, F. D., and S. R. Beach. "Forgiveness in Marriage: Implications for Psychological Aggression and Constructive Communication." *Personal Relationships* 9 (2002): 239–251.

Fine, M., and F. D. Fincham, (eds.). *Handbook of Family Theories: A Content-Based Approach*. New York: Routledge, 2013.

Fineman, Martha Albertson, and Roxanne Mykitiuk. *The Public Nature of Private Violence: The Discovery of Domestic Abuse*. New York: Routledge, 1994.

Finkel, Eli. "The All-or-Nothing Marriage." *New York Times*, February 16, 2014: Sunday Review, 1, 6.

Finkelhor, D. and L. Jones. "Why Have Child Maltreatment and Child Victimization Declined?" *Journal of Social Issues* 62, 4 (2006): 685–716.

Finkelhor, D., R. K. Ormrod, and H. A. Turner. "Poly-Victimization: A Neglected Component in Child Victimization Trauma." *Child Abuse & Neglect* 31 (2007): 7–26.

Finkelhor, David, and Richard Ormrod. "Factors in the Underreporting of Crimes against Juveniles." *Child Maltreatment* 6, 3 (2001): 219–229.

Finkelhor, D., H. Turner, A. Shattuck, S. Hamby, and K. Kracke. "Children's Exposure to Violence, Crime, and Abuse: An Update." *Juvenile Justice Bulletin: National Survey of Children's Exposure to Violence*, U.S. Department of Justice, September 2015. www.ojjdp .gov/pubs/248547.pdf.

Fisher, C., and R. Lerner. "Sex Differences." In *Encyclopedia of Applied Developmental Science*, vol. 2. Newbury Park, CA: Sage, 2005.

Fishman, Barbara. "The Economic Behavior of Stepfamilies." *Family Relations* 32 (July 1983): 356–366.

Flaks, David K., Ilda Ficher, Frank Masterpasqua, and G. Joseph. "Lesbians Choosing Motherhood: A Comparative Study of Lesbians and Heterosexual Parents and Their Children." *Developmental Psychology* 31, 1 (January 1995): 105–114.

Foley, L., et al. "Date Rape: Effects of Race of Assailant and Victim and Gender of Subjects." *Journal of Black Psychology* 21, 1 (February 1995): 6–18.

Follingstad, D. R., E. S. Hause, L. L. Rutledge, and D. S. Polek. "Effects of Battered Women's Early Responses on Later Abuse Patterns." *Violence and Victims* 7 (1992): 109–128.

Foote, Jeffrey, Carrie Wilkens, Nicole Kosanke, and Stephanie Higgs. *Beyond Addiction: How Science and Kindness Help People Change*. New York: Scribner, 2014.

Foran, H. M., and K. D. O'Leary. "Alcohol and Intimate Partner Violence: A Meta-Analytic Review." *Clinical Psychology Review* 28 (2008): 1222–1234.

Fowers, Blaine, and David Olson. "Four Types of Premarital Couples: An Empirical Typology Based on PREPARE." *Journal of Family Psychology* 6, 1 (1992): 10–21.

Fox, Bonnie. *When Couples Become Parents: The Creation of Gender in the Transition to Parenthood*. University of Toronto Press, 2009.

Fox, Greer, and Velma McBride Murry. "Gender and Families: Feminist Perspectives and Family Research." *Journal of Marriage and the Family* 62, 4 (November 2000): 1160–1172.

Fraga, Lynette, Dionne Dobbins, and Michelle McCready. "Parents and the High Cost of Child Care: 2015 Report." Arlington, VA: Child Care Aware® of America, 2015. www.usa.childcareaware.org.

Frank, Nathaniel. "Dad and Dad vs. Mom and Dad." *Los Angeles Times*, June 13, 2012.

Fraser, Antonia. *The Weaker Vessel*, 2nd ed. New York: Knobpf, 1984.

Fraser, Julie, Eleanor Maticka-Tyndale, and Lisa Smylie. "Sexuality of Canadian Women at Midlife." *Canadian Journal of Human Sexuality* 13, 3–4 (2004): 171–187.

Friedan, Betty. *The Feminine Mystique*. New York: Dell, 1963.

Friedman, H. L. *Playing to Win: Raising Children in a Competitive Culture*. Berkeley: University of California Press, 2013.

Frieze, Irene H. "Female Violence against Intimate Partners: An Introduction." *Psychology of Women Quarterly* 29 (2005): 229–237.

Fry, Richard. "More Millennials Living with Family Despite Improved Job Market." Pew Research Center, Washington, DC, July 2015a.

———. "Record Share of Young Women Are Living with Their Parents, Relatives." Pew Research Center, 2015b. www.pewresearch.org/fact -tank/2015/11/11/record-share-of-young-women-are-living-with -their-parents-relatives/.

Fry, Richard, and D'Vera Cohn. "Living Together: The Economics of Cohabitation." Pew Research Center, Social and Demographic Trends. Washington, DC, 2011. pewsocialtrends.org.

Fuchs, Dale. "Spanish Socialists' Proposals Opposed by Church." *New York Times*, May 30, 2004.

Fugate, Maria. "No Sex for Me, Please! Ex-Transexual Australian, Norrie May-Welby Is First Legally Genderless Person." *New York Daily News*, March 6, 2010.

Fuhrman, R. W., Flannagan, D., and Matamoros, M. (2009). "Behavior Expectations in Cross-Sex Friendships, Same-Sex Friendships, and Romantic Relationships." *Personal Relationships* 16: 1350–4126.

Fulwiler, Michael. "The Six Things that Predict Divorce." The Gottman Relationship Blog, October 10, 2014. www.gottmanblog.com /archives/2014/10/31/the-6-things-that-predict-divorce.

Furstenberg, Frank F., Jr., and A. J. Cherlin. *Divided Families: What Happens to Children When Parents Part.* Cambridge, MA: Harvard University Press, 1991.

Furstenberg, Frank F., Jr., Sheela Kennedy, Vonnie C. Mcloyd, Rubén G. Rumbaut, and Richard A. Settersten Jr. "Growing Up Is Harder To Do." *Contexts* 3, 3 (Summer 2004): 33–41.

Furstenberg, Frank F., Jr., and Christine Nord. "Parenting Apart: Patterns in Childrearing after Marital Disruption." *Journal of Marriage and the Family* 47, 4 (November 1985): 893–904.

Furstenberg, Frank F., Jr., and Graham Spanier, eds. *Recycling the Family— Remarriage after Divorce,* rev. ed. Newbury Park, CA: Sage, 1987.

Furstenberg, Frank F., Jr., and J. O. Teitler. "Reconsidering the Effects of Marital Disruption: What Happens to Children of Divorce in Early Adulthood?" *Journal of Family Issues* 15, 2 (June 1994): 173–190.

Gager, C., T. M. Cooney, and K. T. Call. "The Effects of Family Characteristics and Time Use on Teenage Girls' and Boys' Household Labor." *Journal of Marriage and Family* 61 (1999): 982–994.

Gager, C. T., and S. T. Yabiku. "Who Has the Time? The Relationship between Household Labor Time and Sexual Frequency." *Journal of Family Issues* 31 (2010): 135–163.

Galinsky, E., J. T. Bond, S. S. Kim, L. Backon, E. Brownfield, and K. Sakai. *Overwork in America: When the Way We Work Becomes Too Much.* New York, NY: Families and Work Institute, 2005.

Gallagher, James. "UK Approves Three-Person Babies." News, February 24 2015. www.bbc.com/news/health-31594856.

Ganong, L., M. Coleman, and J. Hans. "Divorce as Prelude to Stepfamily Living and the Consequences of Re-Divorce." In *Handbook of Divorce,* edited by M. Fine and J. Harvey. Hillsdale, NJ: Lawrence Erlbaum, 2006.

Ganong, Lawrence, and Marilyn Coleman. *Remarried Family Relation*ships. Newbury Park, CA: Sage Publications, 1994.

Gans, D., and M. Silverstein. "Norms of Filial Responsibility for Aging Parents across Time and Generations." *Journal of Marriage and Family* 68 (2006): 961–986.

Gans, Herbert. "Symbolic Ethnicity: The Future of Ethnic Groups and Cultures in America." In *On the Making of Americans,* edited by H. Gans. Philadelphia: University of Pennsylvania, 1979.

Gans, Herbert. "From 'Underclass' to 'Undercaste': Some Observations about the Future of the Postindustrial Economy and its Major Victims." *International Journal of Urban and Regional Research* 17, 3 (September 1993): 327–335.

Gao, George. "Most Americans Now Say Learning Their Child Is Gay Wouldn't Upset Them." Fact Tank: News in the Numbers, Pew Research Center, June 29, 2015. www.pewresearch.org/fact -tank/2015/06/29/most-americans-now-say-learning-their-child-is -gay-wouldnt-upset-them/.

Garbarino, James. *Children and Families in the Social Environment.* Hawthorne, NY: Aldine De Gruyter, 1982.

Garner, Abigail. *Families Like Mine: Children of Gay Parents Tell It Like It Is.* New York: HarperCollins, 2005.

Gates, Gary. "How Many People are Lesbian, Gay, Bisexual and Transgender?" The Williams Institute, 2011. williamsinstitute .law.ucla.edu/research/census-lgbt-demographics-studies /how-many-people-are-lesbian-gay-bisexual-and-transgender/.

———. "Same-Sex Couples in Census 2010: Race and Ethnicity." The Williams Institute, 2012.

———. "Demographics of Married and Unmarried Same-Sex Couples: Analyses of the 2013 American Community Survey." Los Angeles, CA: Williams Institute, UCLA School of Law, 2015a.

———. "Lesbian, Gay, Bisexual, and Transgender Demographics." In *Gay, Lesbian, Bisexual, and Transgender Civil Rights: A Public Policy Agenda for Uniting a Divided America,* edited by Wallace Swan (pp. 1–20). Boca Raton, FL: CRC Press/Taylor & Francis Group, 2015b.

———. "Lesbian, Gay, Bisexual, and Transgender Family Formation and Demographics." In *Gay, Lesbian, Bisexual, and Transgender Civil Rights: A Public Policy Agenda for Uniting a Divided America* edited by Wallace Swan (pp. 21–34). Boca Raton, FL: CRC Press/Taylor & Francis Group, 2015c.

———. "Marriage and Family: LGBT Individuals and Same-Sex Couples." *The Future of Children* 25, 2 (Fall 2015d). Brookings. www .futureofchildren.org.

Gates, Gary, and Taylor N. T. Brown. *Marriage and Same-Sex Couples after Obergefell.* Los Angeles, CA: Williams Institute, UCLA School of Law, 2015.

Gates, Gary, and Frank Newport. "An Estimated 780,000 Americans in Same-Sex Marriages," Gallup, Social Issues, April 2015. www.gallup .com/poll/182837/estimated-780-000-americans-sex-marriages.aspx.

Gecas, Viktor, and Monica Seff. "Families and Adolescents." In *Contemporary Families: Looking Forward, Looking Back,* edited by A. Booth. Minneapolis: National Council on *Family Relations,* 1991.

Gelles, Richard J. "Constraints against Family Violence: How Well Do They Work?" *The American Behavioral Scientist* 36 (May/June 1993): 575–586.

Gelles, Richard, and Mary Cavanaugh. "Association Is Not Causation: Alcohol and Other Drugs Do Not Cause Violence." In *Current Controversies on Family Violence,* 2nd ed., edited by Donileen Loseke, Richard Gelles, and Mary Cavanaugh (pp. 178–189). Thousand Oaks, CA: Sage, 2005.

Gelles, Richard J., and Claire Pedrick Cornell. *Intimate Violence in Families,* 2nd ed. Newbury Park, CA: Sage, 1990.

Gelles, Richard J., and Murray Straus. *Intimate Violence: The Definitive Study of the Causes and Consequences of Abuse in the American Family.* New York: Simon and Schuster, 1988.

Gershoff, E. T. *Report on Physical Punishment in the United States: What Research Tells Us about Its Effects on Children.* Columbus, OH: Center for Effective Discipline, 2008.

Gerson, Kathleen. *Hard Choices: How Women Decide about Work, Career, and Motherhood.* Berkeley: University of California Press, 1985.

———. *No Man's Land: Men's Changing Commitments to Family and Work.* New York: Basic Books, 1993.

———. *The Unfinished Revolution: How a New Generation is Reshaping Family, Work, and Gender in America.* Oxford University Press, 2010.

Gerstel, N., and N. Sarkisian. "Marriage: The Good, the Bad, and the Greedy." *Contexts* 5, 4 (November 2006): 16–21.

Gibbs, Larry, and Krista Payne. "First Divorce Rate, 2010" (FP-11-09). National Center for Family and Marriage Research, Bowling Green, OH, 2011. www.bgsu.edu/content/dam/BGSU/college-of-arts-and -sciences/NCFMR/documents/FP/FP-11-09.pdf.

Gibson-Davis, Christina, Kathryn Edin, and Sara McLanahan. "High Hopes but Even Higher Expectations: The Retreat from Marriage among Low-Income Couples." *Journal of Marriage and Family* 67, 3 (December 2005): 1301–1312.

Gilbert, Dennis. *The American Class Structure in an Age of Growing Inequality,* 8th ed. Los Angeles: Pine Forge Press, 2011.

———. *The American Class Structure in an Age of Growing Inequality.* SAGE Publications, 2014.

Giles, H., and B. LePoire. "Introduction: The Ubiquity and Social Meaningfulness of Nonverbal Communication." In *The Sage Handbook of Nonverbal Communication,* edited by V. Manusov and M. Patterson. Thousand Oaks, CA: Sage, 2006.

Gillen, K., and S. J. Muncher. "Sex Differences in the Perceived Casual Structure of Date Rape: A Preliminary Report." *Aggressive Behavior* 21, 2 (1995): 101–112.

Gillespie, Rosemary. "Childfree and Feminine: Understanding the Gender Identity of Voluntarily Childless Women." *Gender & Society* 17, 1 (February 2003): 122–136.

Gilmore, David. *Manhood in the Making.* New Haven, CT: Yale University Press, 1990.

Glenn, Norval. "The Recent Trend in Marital Success in the United States," *Journal of Marriage and the Family* 53 (2) May 1991: 261–270.

———. "Who's Who in the Family Wars: A Characterization of the Major Ideological Factions." In *Feuds about Families: Conservative, Centrist, Liberal, and Feminist Perspectives,* edited by Nijle Benokraitis. Upper Saddle River, NJ: Prentice Hall, 2000.

Glum, Julia. "Spanking Children In America: Corporal Punishment Forbidden In Many Countries, but U.S. Ban Is Unlikely," *International Business Times,* March 6, 2015. http://www.ibtimes.com/spanking -children-america-corporal-punishment-forbidden-many-countries -us-ban-1837856.

Gneezy, U., and A. Rustichini. "Gender and Competition at a Young Age." *American Economic Review Papers and Proceedings,* May 2004, 377–381.

Godbeer, Richard. "Courtship and Sexual Freedom in Eighteenth-Century America." *OAH Magazine of History* 18, 4 (2004): 9–13.

Goeke-Morey, M. C., E. M. Cummings, and L. M. Papp. "Children and Marital Conflict Resolution: Implications for Emotional Security and Adjustment." *Journal of Family Psychology* 21, 4 (2007): 744–753.

Goetting, Ann. "The Six Stages of Remarriage: Developmental Tasks of Remarriage after Divorce." *Family Relations* 31 (April 1982): 213–222.

––––––. "Patterns of Support among In-Laws in the United States: A Review of Research." *Journal of Family Issues* 11, 1 (1990): 67–90.

Goldberg, Abbie. "'Doing' and 'Undoing' Gender: The Meaning and Division of Housework in Same-Sex Couples." *Journal of Family Theory & Review* 5 (June 2013): 85–104.

Goldberg, A. E., and A. G. Sayer. Lesbian Couples' Relationship Quality across the Transition to Parenthood. *Journal of Marriage and Family* 68 (2006): 87–100.

Golden, Lonnie. "Irregular Work Scheduling and Its Consequences." *Economic Policy Institute Report*, Briefing Paper 394, April 9, 2015.

Goldenberg, H., and I. Goldenberg. *Counseling Today's Families*, 2nd ed. Pacific Grove, CA: Brooks/Cole, 1994.

Goldman, J., M. K. Salus, D. Wolcott, and K. Y. Kennedy. *A Coordinated Response to Child Abuse and Neglect: The Foundation for Practice.* Washington, DC: U.S. Department of Health and Human Services, Administration for Children and Families, Administration on Children, Youth and Families, Children's Bureau, Office on Child Abuse and Neglect, 2003. www.childwelfare.gov/pubs/usermanuals /foundation.

Goldscheider, F., and G. Kaufman. "Willingness to Stepparent: Attitudes about Partners Who Already Have Children." *Journal of Family Issues* 27, 10 (2006): 1415–1436.

Goldscheider, F., and S. Sassler. "Creating Stepfamilies: Integrating Children into the Study of Union Formation." *Journal of Marriage and Family* 68 (2006): 275–291.

Goldscheider, Frances K., and Gayle Kaufman. "Willingness to Stepparent: Men and Women's Attitudes toward Marrying Someone with Children." Paper presented at the annual meetings of the American Sociological Association, Atlanta, August 2003.

Goleman, Daniel. "Gay Parents Called No Disadvantage." *New York Times*, March 11, 1992.

Gonzalez, Libertad. "The Determinants of the Prevalence of Single Mothers: A Cross-Country Analysis." Discussion Paper No. 1677, Institute for the Study of Labor, July 2005.

Gonzalez, N. "Beyond the FMLA: California's Paid Family Leave Is a National First." *National Council on Family Relations Report: Family Focus on Families and Work-Life* 51, 2 (June 2006): 11.

Goode, William. "Force and Violence in the Family." *Journal of Marriage and the Family* 33 (November 1971): 624–636.

Goode, William J. "Family Cycle and Theory Construction." In *The Family Life Cycle in European Societies*. The Hague: Mouton, 1977.

––––––, ed. *The Family*, 2nd ed. Englewood Cliffs, NJ: Prentice Hall, 1982.

Goodman, Lisa, and Deborah Epstein. "Refocusing on Women: A New Direction for Policy and Research on Intimate Partner Violence." *Journal of Interpersonal Violence* 20, 4 (April 2005): 479–487.

Goodman, M. "What Laid-Off Dads Want This Father's Day. Think about What's Best for That Unemployed Dad," 2009. www.abcnews.go.com.

Goodman, M., D. Dollahite, L. Marks, and E. Layton. "Religious Faith and Transformational Processes in Marriage." *Family Relations* 62 (December 2013): 808–823.

Gootman, Elissa. "The Officiant among Us." *New York Times*, March 9, 2012.

Gottman, J. M., R. W. Levenson, J. Gross, B. Fredrickson, K. McCoy, L. Rosenthal, A. Ruel, and D. Yoshimoto. "Correlates of Gay and Lesbian Couples' Relationship Satisfaction and Relationship Dissolution." *Journal of Homosexuality* 45, 1 (2003): 23–43.

Gottman, J., K. Ryan, S. Carrere, and A. Erley. "Toward a Scientifically Based Marital Therapy." In *Family Psychology: Science-Based Interventions*, edited by H. Liddle, D. Sanstistban, et al. (pp. 147–174). Washington, DC: American Psychological Association, 2002.

Gottman, John. *Marital Interaction: Experimental Investigations*. New York: Academic Press, 1979.

Gottman, John M. *What Predicts Divorce? The Relationship between Marital Processes and Marital Outcomes*. Hillsdale, NJ: Lawrence Erlbaum, 1994.

––––––. *Why Marriages Succeed or Fail and How You Can Make Yours Work*. New York: Simon and Schuster, 1995.

Gottman, John, James Coan, Sybil Carrere, and Catherine Swanson. "Predicting Marital Happiness and Stability from Newlywed Interactions." *Journal of Marriage and the Family* 60, 1 (February 1998): 5–22.

Gottman, John, and Robert Levenson. "Marital Processes Predictive of Later Dissolution Behavior, Physiology, and Health." *Journal of Personality and Social Psychology* 63, 2 (August 1992): 221–233.

Gottman Institute. "The Four Horsemen: The Antidote." Gottman Relationship Blog, April 26, 2013. www.gottman.com/blog /the-four-horsemen-the-antidotes/.

Gough, Kathleen. "Is the Family Universal: The Nayer Case." In *A Modern Introduction to the Family*, edited by N. Bell and E. Vogel. New York: Free Press, 1968.

Gough, Margaret, and Alexandra Killewald. "Unemployment in Families: The Case of Housework." *Journal of Marriage and Family* 73 (2011): 1085–1100.

Gould, Elise, and Tanyell Cooke. "High-Quality Child Care Is Out of Reach for Working Families." *Economic Policy Institute*, October 6, 2015. www.epi.org/publication/child-care-affordability/.

Graham, J. L., E. Keneski, and T. J. Loving. "The Physiological and Health -Relevant Consequences of Nonmarital Relationship Dissolution." *National Family Relations Council Report*, 2014.

Graham, Jamie, Elizabeth Keneski, and Timothy Loving. "Mental and Physical Health Correlates of Nonmarital Relationship Dissolution." *Family Focus on Dating and Mate Selection, National Council on Family Relations* (Winter 2014): F7–F9.

Graham-Kevan, Nicola, and John Archer. "Investigating Three Explanations of Women's Relationship Aggression." *Psychology of Women Quarterly* 29 (2005): 270–277.

Grall, Timothy. "Custodial Mothers and Fathers and their Child Support: 2009." U.S. Census Bureau, Current Population Reports P60-240, 2011.

––––––. "Custodial Mothers and Fathers and Their Child Support: 2011." Current Population Reports, 2013. U.S. Census Bureau, October 2013. P60-246.

Grant, J.M., J.L. Herman, and M. Keisling. "Injustice at Every Turn: A Report of the National Transgender Discrimination Survey." Washington, DC: National Center for Transgender Equality and National Gay and Lesbian Task Force, 2011.

Greeff, Abraham P., and Tanya de Bruyne. "Conflict Management Style and Marital Satisfaction." *Journal of Sex and Marital Therapy* 26, 4 (October 2000): 321–334.

Green, P. "Your Mother Is Moving In? That's Great." *New York Times*, January 15, 2009, D1.

Greenberger, Ellen. "Explaining Role Strain: Intrapersona." *Journal of Marriage and the Family* 52, 1 (February 1994): 115–118.

Greene, K., and S. L. Faulkner. "Gender Belief in the Sexual Double Standard, and Sexual Talk in Heterosexual Dating Relationships." *Sex Roles* 53, 3/4 (August 2005): 239–251.

Greenfield, E. A., and N. F. Marks. "Linked Lives: Adult Children's Problems and Their Parents' Psychological and Relational Well-Being." *Journal of Marriage and Family* 68 (2006): 442–454.

Greenstein, Theodore N. "Husbands' Participation in Domestic Labor: Interactive Effects of Wives' and Husbands' Gender Ideologies." *Journal of Marriage and the Family* 58, 3 (August 1996): 585–595.

Greenwood, G., M. V. Relf, B. Huang, L. M. Pollack, J. A. Canchola, and J. A. Catania. "Battering Victimization among a Probability-Based Sample of Men Who Have Sex with Men." *American Journal of Public Health* 92, 12 (December 2002): 1964–1969.

Gregory, Elizabeth. "Myths about Later Motherhood: A Fact Sheet Prepared for the Council on Contemporary Families." Fact sheet, 2012. contemporaryfamilies.org/wp-content/uploads/2013/10/2012 _Factsheet_Gregory_Myths-about-later-motherhood.pdf.

Gross, J. "Boomerang Parents." *New York Times*, November 18, 2008.

Grossman, C. L., and I. Yoo. "Civil Marriage on Rise across USA." *USA Today*, October 7, 2003, 1A.

Grosswald, Blanche. "The Effects of Shift Work on Family Satisfaction." *Families in Society* 85, 3 (July–September 2004): 413–423.

Grotevant, H. "Open Adoption: What Is It and How Is It Working?" *National Council on Family Relations Report: Family Focus on Adoption*, Issue FF39 (September 2008): F1, 2, 17–19.

Grotevant, Harold, and Julie Kohler. "Adoptive Families." In *Parenting and Child Development in "Nontraditional" Families*, edited by M. E. Lamb. Mahwah, NJ: Lawrence Erlbaum, 1999.

Grov, C., D. Bimbi, J. E. Nanin, and J. T. Parsons. "Coming-Out Process among Gay, Lesbian, and Bisexual Individuals." *Journal of Sex Research* 43, 2 (May 2006): 115–121.

Guadagno, Rosanna, and Brad Sagarin. "Sex Differences in Jealousy: An Evolutionary Perspective on Online Infidelity." *Journal of Applied Social Psychology* 40, 10 (2010): 2636–2655.

Guerrero, L. K., and Anderson, P. A. "The Waxing and Waning of Relational Intimacy: Touch as a Function of Relational Stage, Gender, and Touch Avoidance." *Journal of Social and Personal Relationships* 8 (1991): 147–165.

Gupta, S. "Autonomy, Dependence, or Display? The Relationship between Married Women's Earnings and Housework." *Journal of Marriage and Family* 69 (2007): 399–417.

Gupta, Pooja. "Television Shows, Gay Characters and the Origin of Younger Americans' Support for LGBTQ Rights." Journalist's Resource, 2014. journalistsresource.org/studies/society/gender-society/popular-media-and-the-origin-of-younger-americans-greater-support-for-gay-rights.

Gurian, Michael, and Kathy Stevens. *The Minds of Boys: Saving Our Sons from Falling Behind in School and Life.* San Francisco: Jossey-Bass, 2005.

Guttman, Herbert. *The Black Family: From Slavery to Freedom.* New York: Pantheon, 1976.

Guttman, Joseph. *Divorce in Psychosocial Perspective: Theory and Research.* Hillsdale, NJ: Lawrence Erlbaum, 1993.

Guzzo, Karen Benjamin. "Trends in Cohabitation Outcomes: Compositional Changes and Engagement among Never-Married Young Adults." *Journal of Marriage and Family* 76 (2014): 826–842.

Haas, A., M. Eliason, V. Mays, R. Mathy, S. Cochran, A. D'Angelli, and P. Clayton. "Suicide and Suicide Risk in Lesbian, Gay, Bisexual, and Transgender Populations: Review and Recommendations." *Journal of Homosexuality* 58, 1 (2011): 10–51.

Haberman, M. "Senate OKs Tough 'Nixzmary' Law." *New York Post,* September 12, 2009.

Haberman, Clyde. "Beyond Caitlyn Jenner Lies a Long Struggle by Transgender People." *New York Times,* June 14, 2015. www.nytimes.com/2015/06/15/us/beyond-caitlyn-jenner-lies-a-long-struggle-by-transgender-people.html.

Hall, E. *The Hidden Dimension.* Garden City, NY: Doubleday, 1966.

Hamby, Sherry, David Finkelhor, Heather Turner, and Randy Ormrod. "Children's Exposure to Intimate Partner Violence and Other Family Violence." *Juvenile Justice Bulletin,* U.S. Department of Justice, October 2011.

Hamer, J., and K. Marchioro. "Becoming Custodial Dads: Exploring Parenting among Low-Income and Working-Class African American Fathers." *Journal of Marriage and Family* 64 (2002): 116–129.

Hamilton, B. E., J. A. Martin, M. J. K. Osterman, and S. C. Curtin. "Births: Preliminary data for 2014." *National Vital Statistics Reports* 64, 6 (2015a). Hyattsville, MD: National Center for Health Statistics.

Hamilton, B. E., J. A. Martin, M. J. K. Osterman, et al. "Births: Final Data for 2014." *National Vital Statistics Reports* 64, 12 (2015b). Hyattsville, MD: National Center for Health Statistics.

Hamilton, B. E., J. A. Martin, S. J. Ventura. "Births: Preliminary Data for 2009." National Vital Statistics Reports Volume 59, Number 3, December 21, 2010.

Hamplová, D., C. Le Bourdais, and E. Lapierre-Adamcyk. "Is the Cohabitation-Marriage Gap in Money Pooling Universal?" *Journal of Marriage and Family* 76 (2014): 983–997.

Hansen, Karen. *Not So Nuclear Families: Class Gender and Networks of Care.* New Brunswick, NJ: Rutgers University Press, 2005.

Hanson, Thomas L., Sara S. McLanahan, and Elizabeth Thomson. "Double Jeopardy: Parental Conflict and Stepfamily Outcomes for Children." *Journal of Marriage and the Family,* 58, 1 (February 1996): 141–154.

Harknett, K. "The Relationship between Private Safety Nets and Economic Outcomes among Single Mothers." *Journal of Marriage and Family* 68, 1 (2006): 172–191.

Harms, Roger. "How Does Paternal Age Affect a Baby's Health?," 2012. www.mayoclinic.com/health/paternal-age/AN02180.

Harrington, Michael. *The Other America: Poverty in the United States.* New York: Macmillan, 1962.

Harrington, Brad, Fred Van Deusen, and Iyar Mazar. "The New Dad: Right at Home, Boston College Center for Work and Family." Carroll School of Management, 2012. www.bc.edu/content/dam/files/centers/cwf/pdf/The%20New%20Dad%20Right%20at%20Home%20BCCWF%202012.pdf.

Harris, David, and Hiromi Ono. "How Many Interracial Marriages Would There Be If All Groups Were of Equal Size in All Places? A New Look at National Estimates of Interracial Marriage." *Social Science Research* 34, 1 (2005): 236–251.

Hart, Betsy. "'Living Apart Together' Relationship Ultimately Selfish." *Deseret News* (Salt Lake City), May 14, 2006.

Hart, S. N., M. Gelardo, and M. Brassard. "Psychological Maltreatment." In *Psychiatric Sequelae of Child Abuse,* edited by J. J. Jacobson. Springfield, IL: Charles C. Thomas, Publisher, 1986.

Hartwell-Walker, M. "Legal Issues for Cohabiting Couples," 2008. http://psychcentral.com/lib/2008/legal-issues-for-cohabiting-couples.

Harwood, Lori. "UA Researcher Finds Negative Impact of MTV Show *16 and Pregnant.*" UA News, University of Arizona, March 25, 2014.

https://uanews.arizona.edu/story/ua-researcher-finds-negative-impact-of-mtv-show-16-and-pregnant.

Haselschwerdt, M. L. "Who Cares about the Rich Folk? An Argument for More Research on Affluent Families and Communities." National Council on Family Relations Report: Family Focus On . . . Open Mic, 57.3 (FF54) (Fall 2012): 14–16.

Hassebrauck, Manfred, and Beverly Fehr. "Dimensions of Relationship Quality." *Personal Relationships* 9 (2002): 253–270.

Hatch, L., and K. Bulcroft. "Does Long-Term Marriage Bring Less Frequent Disagreements? Five Explanatory Frameworks." *Journal of Family Issues* 25, 4 (2004): 465–495.

Hatfield, Elaine, and Richard L Rapson. "Love and intimacy." In *Encyclopedia of Human Behavior* 3, edited by V. S. Ramachandran in New York: Academic Press (1994): 1145–1149.

———. *Love, Sex, and Intimacy: The Psychology, Biology, and History.* New York: Harper-Collins, 1993.

———. "Love." In *The Concise Corsini Encyclopedia of Psychology and Behavioral Science,* edited by W. E. Craighead and C. B. Nemeroff (pp. 898–901). New York: John Wiley & Sons, 2000.

Hatfield, Elaine, Lisamarie Bensman, and Richard Rapson. "A Brief History of Social Scientists' Attempts to Measure Passionate Love." *Journal of Social and Personal Relationships* 29, 2 (2012): 143–164.

Hatfield, Elaine, and Susan Sprecher. *Mirror, Mirror: The Importance of Looks in Everyday Life.* New York: State University of New York, 1986.

Hatfield, Elaine, and G. William Walster. *A New Look at Love.* Reading, MA: Addison-Wesley, 1981.

Hattery, Angela, and Earl Smith. *African American Families Today: Myths and Realities.* Lanham, Maryland: Rowman and Littlefield, 2012.

Hawkins, Alan J., and Tomi-Ann Roberts. "Designing a Primary Intervention to Help Dual-Earner Couples Share Housework and Childcare." *Family Relations* 41, 2 (April 1992): 169–177.

Hawkins, Alan, and David Dollahite, eds. *Generative Fathering: Beyond Deficit Perspectives.* Vol. 3, *Current Issues in the Family.* Thousand Oaks, CA: Sage, 1997.

Hawkins, D., and Alan Booth. "'Unhappily Ever After': Effects of Long-Term Low-Quality Marriages on Well-Being." *Social Forces* 84, 1 (September 2005): 451–471.

Hays, Sharon. *Cultural Contradictions of Motherhood.* New Haven, CT: Yale University Press, 1996.

Hazan, Cindy, and Philip Shaver. "Romantic Love Conceptualized as an Attachment Process." *Journal of Personality and Social Psychology* 52 (March 1987): 511–524.

Hazen, N. L., S. D. Allen, C. H. Christopher, T. Umemura, and D. B. Jacobvitz. "Very Extensive Nonmaternal Care Predicts Mother–Infant Attachment Disorganization: Convergent Evidence from Two Samples." *Development and Psychopathology* 27, 3 (2015): 649–661.

Healey, K. M., and C. Smith. "Batterer Programs: What Criminal Justice Agencies Need to Know." National Institute of Justice: Research in Action, U.S. Department of Justice, 1998.

Healy, Joseph M., Abigail J. Stewart, and Anne P. Copeland. "The Role of Self-Blame in Children's Adjustment to Parental Separation." *Personality and Social Psychology Bulletin* 19, 3 (1993): 279–289.

Heard, Holly. "Fathers, Mothers, and Family Structure: Family Trajectories, Parent Gender, and Adolescent Schooling." *Journal of Marriage and Family* 69 (May 2007): 435–450.

Heaton, T. "Factors Contributing to Increasing Marital Stability in the United States." *Journal of Family Issues* 23, 3 (2002): 392–409.

Helms, H. M., A. J. Supple, and C. M. Proulx. "Mexican-Origin Couples in the Early Years of Parenthood: Marital Well-Being in Ecological Context." *Journal of Family Theory and Review* 3 (2011): 67–95.

Helms-Erikson, Heather. "Marital Quality Ten Years after the Transition to Parenthood: Implications of the Timing of Parenthood and the Division of Housework." *Journal of Marriage and Family* 63 (November 2001): 1099–1110.

Henderson, Stephen. "Weddings: Vows—Rakhi Dhanoa and Ranjeet Purewal." *New York Times,* August 18, 2002.

Hendrick, Clyde, and Susan S. Hendrick. "Research on Love: Does It Measure Up?" *Journal of Personality and Social Psychology* 56, 5 (May 1989): 784–794.

Hendrick, Clyde, and Susan Hendrick. "Attachment Theory and Close Adult Relationships." *Psychological Inquiry* 5, 1 (1994): 38–41.

Hendrick, S. S., and C. Hendrick. "Gender differences and similarities in sex and love." *Personal Relationships* 2 (1995): 55–65.

Hendrick, Susan. "Self-Disclosure and Marital Satisfaction." *Journal of Personality and Social Psychology* 40 (1981): 1150–1159.

Henley, N. *Body Politics: Power, Sex, and Nonverbal Communication.* Englewood Cliffs, NJ: Prentice Hall, 1977.

Hennigan, Randi, and Linda Ladd. "Conflict Resolution Methods in Heterosexual and Same-Sex Relationships." *NCFR Family Focus on Effective Communication* (Summer 2015): F7–F9.

Henslin, J. *Essentials of Sociology: A Down-to-Earth Approach.* Boston: Allyn and Bacon, 2006.

Henslin, James. *Essentials of Sociology: A Down to Earth Approach,* 11th ed. Boston: Allyn & Bacon, 2015.

Herbenick, Debby, Michael Reece, Vanessa Schick, Stephanie A. Sanders, Brian Dodge, and J. Dennis Fortenberry. "Sexual Behaviors, Relationships, and Perceived Health Status among Adult Women in the United States: Results from a National Probability Sample." *Journal of Sexual Medicine* 7, 5 (supplement) (2010): 277–290.

Herek, G. M., Chopp, R., and Strohl, D. "Sexual Stigma: Putting Sexual Minority Health Issues in Context." In *The Health of Sexual Minorities: Public Health Perspectives on Lesbian, Gay, Bisexual, and Transgender Populations,* edited by I. Meyer and M. Northridge (pp. 171–208). New York: Springer, 2007.

Herek, G. M., J. R. Gillis, and J. C. Cogan. "Psychological Sequelae of Hate-Crime Victimization among Lesbian, Gay, and Bisexual Adults." *Journal of Consulting and Clinical Psychology* 67 (1999): 945–951.

Herek, Gregory M. "Heterosexuals' Attitudes toward Bisexual Men and Women in the United States." *Journal of Sex Research* 39, 4 (November 2002): 264–275.

Hernandez, R. "Law to Ease Disclosures on Child Abuse." *New York Times,* February 13, 1996, B6.

Hertlein, Katherine. "Digital Dwelling: Technology in Couple and Family Relationships." *Family Relations* 61 (July 2012): 374–387.

Hesketh, Therese, Li Lu, and Zhu Wei Xing. "The Consequences of Son Preference and Sex-Selective Abortion in China and Other Asian Countries." *Canadian Medical Association Journal* 183 (September 6, 2011): 1374–1377; published ahead of print March 14, 2011.

Heslin, R., and Alper, T. "Touch: A Bonding Issue." In *Nonverbal Communication,* edited by J. M. Weimann and R. P. Harrison. Beverly Hills, CA: Sage, 1983.

Hetherington, E. Mavis. "Intimate Pathways: Changing Patterns in Close Personal Relationships across Time." *Family Relations* 52, 4 (October 2003): 318–331.

Hetherington, E. Mavis, and John Kelly. *For Better or Worse: Divorce Reconsidered.* New York: W.W. Norton, 2002.

Hetherington, E. Mavis, and Margaret Stanley-Hagan. "Stepfamilies." In *Parenting and Child Development in 'Nontraditional' Families* edited by M. Lamb. Mahwah, pp. 137–160. New Jersey: Lawrence Erlbaum, 1999.

Heuveline, Patrick, and Jeffrey Timberlake. "The Role of Cohabitation in Family Formation: The United States in Comparative Perspective." *Journal of Marriage & Family* 66, 5 (December 2004): 1214–1230.

Hewlett, Sylvia, and Cornel West. *The War against Parents: What We Can Do for America's Beleaguered Moms and Dads.* New York: Houghton Mifflin, 1998.

Hill, Shirley. "Teaching and Doing Gender in African American Families." *Sex Roles* 47, 11/12 (December 2002).

Hill Collins, Patricia. "Intersections of Race, Class, Gender, and Nation: Some Implications for Black Family Studies." *Journal of Comparative Family Studies* 29 (1) 1998: 27–36.

Hill Collins, Patricia. *Black Feminist Thought: Knowledge, Consciousness, and the Politics of Empowerment,* 2nd ed. New York: Routledge, 2000.

Hill, Rachelle, Eric Tranby, Erin Kelly, and Phyllis Moen. "Relieving the Time Squeeze? Effects of a White-Collar Workplace Change on Parents." *Journal of Marriage and Family* 75 (August 2013): 1014–1029.

Hirschl, Thomas A., Joyce Altobelli, and Mark R. Rank. "Does Marriage Increase the Odds of Affluence? Exploring the Life Course Probabilities." *Journal of Marriage and Family* 65, 4 (November 2003): 927–938.

Ho, V., and J. Johnson. "Overview of Florida Alimony." *Florida Bar Journal* 71 (October 2004).

Hochschild, Arlie. *The Second Shift: Working Parents and the Revolution at Home.* New York: Viking Press, 1989.

————. *The Time Bind: When Work Becomes Home and Home Becomes Work.* New York: Holt, 1997.

————. *The Managed Heart: Commercialization of Human Feeling.* Berkeley, University of California Press, 2003.

Hoeffel, Elizabeth, Sonya Rastogi, Myoung Ouk Kim, and Hasan Shahid. The Asian Population: 2010. U.S. Census Bureau, 2010 Census Briefs, 2012.

Hoesly, Dusty. "'Need a Minister? How about Your Brother?': The Universal Life Church between Religion and Non-Religion." *Secularism and Nonreligion* 4 (2015). www.secularismandnonreligion.org/articles/10.5334/snr.be/.

Hoff, C. C., and S. C. Beougher. "Sexual Agreements among Gay Male Couples." *Archives of Sexual Behavior* 39 (2010): 774–787.

Hofferth, S. "Residential Father Family Type and Child Well-Being: Investment versus Selection." *Demography* 43 (2006): 53–77.

Hoffman, Kristi L., K. Jill Kiecolt, and John N. Edwards. "Physical Violence between Siblings: A Theoretical and Empirical Analysis." *Journal of Family Issues* 26, 8 (November 2005): 1103–1130.

Hoffman, Lois Wladis. "Maternal Employment: 1979." *American Psychologist* 34 (1979): 859–865.

Hogh-Olesen, Henrik. "Human Spatial Behaviour: The Spacing of People, Objects and Animals in Six Cross-Cultural Samples." *Journal of Cognition and Culture* 8 (2008), 245–280.

Holley, Sarah, Claudia Haase, and Robert Levenson. "Age-Related Changes in Demand Withdraw Communication Behaviors." *Journal of Marriage and Family* 75 (August 2013): 822–836.

Holley, Sarah, Virginia Sturm, and Robert Levenson. "Exploring the Basis for Gender Differences in the Demand–Withdraw Pattern." *Journal of Homosexuality* 57, 5 (May 2010): 666–684.

Holmes, J., and S. Murray. "Conflict in Close Relationships." In *Social Psychology: Handbook of Basic Principles,* edited by E. T. Higgins and A. W. Kruglanski. New York, NY: Guilford Press, 1996: 622–654.

Holt, Amanda. (2012). "Researching Parent Abuse: A Critical Review of the Methods." *Social Policy and Society* 11: 289–298.

Hook, J. L., and S. Chalasani. "Gendered Expectations? Reconsidering Single Fathers' Childcare Time." *Journal of Marriage and Family* 70, 4 (November 2008): 835–1091.

Hook, Misty, Lawrence Gerstein, Lacy Detterich, and Betty Gridley. "How Close Are We? Measuring Intimacy and Examining Gender Differences." *Journal of Counseling and Development* 81 (Fall 2003): 462–472.

Hopkins, Christopher Dean. "Jenner: 'For All Intents and Purposes, I Am a Woman.'" April 24, 2015. www.npr.org/sections/thetwoway/2015/04/24/402069597/jenner-for-all-intents-and-purposes-i-am-a-woman.

Hopper, Joseph. "The Symbolic Origins of Conflict in Divorce." *Journal of Marriage and Family* 63 (May 2001): 430–445.

Hopper, Tristin. "Genderless Passports 'Under Review' in Canada." *National Post,* May 8, 2012. http://news.nationalpost.com/2012/05/08/genderless-passports-under-review-in-canada/.

Hotaling, Gerald T., and David B. Sugarman. "A Risk Marker Analysis of Assaulted Wives." *Journal of Family Violence* 5, 1 (March 1990): 1–14.

Houts, Leslie A. "But Was It Wanted? Young Women's First Voluntary Sexual Intercourse." *Journal of Family Issues* 26, 8 (November 2005): 1082–1102.

Hsueh, Annie, Kristen Rahbar Morrison, and Brian Doss. "Qualitative Reports of Problems in Cohabiting Relationships: Comparisons to Married and Dating Relationships." *Journal of Family Psychology* 23, 2 (2009): 236–246.

Huang, Chien-Chung, Ronald B. Mincy, and Irwin Garfinkel. "Child Support Obligations and Low-Income Fathers." *Journal of Marriage and Family* 67, 5 (December 2005): 1213–1225.

Hughes, M., K. Morrison, and K. J. K. Asada. "What's Love Got to Do with It? Exploring the Impact of Maintenance Rules, Love Attitudes, and Network Support on Friends with Benefits Relationships." *Western Journal of Communication* 69, 1 (January 2005): 49–66.

Hughes, Diane, Emilie Smith, Howard Stevenson, James Rodriguez, Deborah Johnson, and Paul Spicer. "Parents' Ethnic-Racial Socialization Practices: A Review of Research and Directions for Future Study." *Developmental Psychology* 42, 5 (2006): 747–770.

Hughes, Mary Elizabeth, Linda J. Waite, Tracey A. LaPierre, and Ye Luo. "All in the Family: The Impact of Caring for Grandchildren on Grandparents' Health." *Journal of Gerontology, Social Sciences* 60 (2007): S108–S119.

Human Rights Campaign, 2012. www.hrc.org/.

Humble, A., C. R. Solomon, K. R. Allen, K. R. Blaisure, and M. P. Johnson. "Feminism and Mentoring of Graduate Students." *Family Relations* 55 (January 2006): 2–15.

Hundley, Tom, and Ana Santos. "The Last Country in the World Where Divorce Is Illegal." *Foreign Policy,* January 19, 2015. http://foreignpolicy.com/2015/01/19/the-last-country-in-the-world-where-divorce-is-illegal-philippines-catholic-church/.

Hummer, Robert, and Erin Hamilton. "Race and Ethnicity in Fragile Families." *Future of Children* 20, 2 (Fall 2010): 113–131.

Hurtado, Aida, and Mrinal Sinha. "More than Men: Latino Feminist Masculinities and Intersectionality." *Sex Roles* 59 (special issue on "Gender: An Intersectionality Perspective") (2008): 337–349.

Huston, T. L. "What's Love Got To Do with It? Why Some Marriages Succeed and Others Fail." *Personal Relationships* 16 (2009): 301–327.

Huston, Ted, and Heidi Melz. "The Case for (Promoting) Marriage: The Devil Is in the Details." *Journal of Marriage and Family* 66, 4 (November 2004): 943–958.

Huston, Ted, S. M. McHale, and A. C. Crouter. "When the Honeymoon's Over: Changes in the Marriage Relationship over the First Year." In *The Emerging Field of Personal Relationships*, edited by S. Duck and R. Gilmour. Hillsdale, NJ: Lawrence Erlbaum, 1986.

Huston, Ted. "The Social Ecology of Marriage and Other Intimate Unions." *Journal of Marriage and the Family* 62, 2 (May 2000): 298–321.

Hyde, J. S. "The Gender Similarities Hypothesis." *American Psychologist* 60, 6 (2005): 581–592.

Hynie, Michaela, John E. Lydon, Sylvana Cote, and Seth Wiener. "Relational Sexual Scripts and Women's Condom Use: The Importance of Internalized Norms." *Journal of Sex Research* 35, 4 (November 1998): 370–380.

IANS. "China Crackdown on Illegal Abortions." *The South Asian Times*, September 4, 2014. www.thesouthasiantimes.info/news-China_crackdown_on_illegal_abortions-61810-International-17.html.

Ihinger-Tallman, Marilyn, and Kay Pasley. *Remarriage*. Newbury Park, CA: Sage, 1987.

Isaacs, J. "Economic Mobility of Black and White Families," 2007. www.brookings.edu/papers/2007/11_blackwhite_isaacs.aspx.

Ishii-Kuntz, Masako. "Japanese American Families." In *Families in Cultural Context*, edited by M. K. DeGenova. Mountain View, CA: Mayfield, 1997.

Jacobs, Jerry, and Kathleen Gerson. *The Time Divide: Work, Family, and Gender Inequality*. Cambridge, MA: Harvard University Press, 2004.

Jacobson, Neal, and John Gottman. *When Men Batter Women: New Insights into Ending Abusive Relationships*. NY: Simon & Schuster, 2007.

Jadva, V., T. Freeman, E. Tranfield, and S. Golombok. "'Friendly Allies in Raising a Child': A Survey of Men and Women Seeking Elective Co-Parenting Arrangements via an Online Connection Website." *Human Reproduction* 30, 8 (2015): 1896–1906.

Jaffe, Ina. "Older Americans' Breakups Are Causing a 'Graying' Divorce Trend." National Public Radio, February 24, 2014. www.npr.org/2014/02/24/282105022/older-americans-breakups-are-causing-a-graying-divorce-trend.

James, Scott. "Many Successful Gay Marriages Share an Open Secret." *New York Times*, January 28, 2010.

Jankowiak, W., and E. Fischer. "A Cross-Cultural Perspective on Romantic Love," *Ethnology* 31 (1992): 149–1906.

Jankowiak, W., M. Sudakov, and B. Wilreker. "Co-Wife Conflict and Co-Operation." *Ethnology* 44, 1 (Winter 2005): 81–98.

Jankowiak, William, Shelly Volsche, and Justin Garcia. "Is the Romantic-Sexual Kiss a Near-Human Universal?" *American Anthropologist* 117, 3 (September 2015).

Jaschik, Scott. "The Deceptive Data on Asians," *Inside Higher Ed*, June 7, 2013. www.insidehighered.com/news/2013/06/07/report-calls-end-grouping-asian-american-students-one-category.

Jayakody, R., and A. Kalil. "Social Fathering in Low-Income, African American Families with Preschool Children." *Journal of Marriage and Family* 64, 2 (2002): 504–516.

Jayson, Sharon. "Birthrate for U.S. Teens is Lowest in History." *USA Today*, April 10, 2012.

Jensen, Todd, Kevin Shafer, and Erin Holmes. "Transitioning to Stepfamily Life: The Influence of Closeness with Biological Parents and Stepparents on Children's Stress." *Child & Family Social Work* (April 2015).

Jensen, Todd, Kevin Shafer, and Jeffry Larson. "(Step)Parenting Attitudes and Expectations: Implications for Stepfamily Functioning and Clinical Intervention." *Families in Society: The Journal of Contemporary Social Services* 95, 3 (2014): 213–220.

Jewkes, R. "Intimate Partner Violence: Causes and Prevention." *Lancet* 359, 9315 (2002): 1423–1429.

Johnson, Judith. "Why NOT to Have a Friend or Family Member Officiate at Your Wedding," 2013. www.huffingtonpost.com/judith-johnson/why-not-to-have-a-friend-_b_3860883.html.

Johnson, Kevin. "Domestic Violence Rises in Sluggish Economy, Police Report." *USA Today*, April 30, 2012.

Johnson, Catherine B., Margaret S. Stockdale, and Frank E. Saal. "Persistence of Men's Misperceptions of Friendly Cues across a Variety of Interpersonal Encounters." *Psychology of Women Quarterly* 15, 3 (September 1991): 463–475.

Johnson, Kevin. "Domestic Violence Rises in Sluggish Economy, Police Report." *USA Today*, April 30, 2012.

Johnson, M. "An Exploration of Men's Experience and Role at Childbirth" *Journal of Men's Studies* 10, 2 (Winter 2002).

Johnson, M., J. Caughlin, and T. Huston. "The Tripartite Nature of Marital Commitment: Personal, Moral, and Structural Reasons to Stay Married." *Journal of Marriage and Family* 61 (1999): 160–177.

Johnson, Michael, and Janel Leone. "The Differential Effects of Intimate Terrorism and Situational Couple Violence: Findings from the National Violence against Women Survey." *Journal of Family Issues* 26, 3 (April 2005): 322–349.

Johnson, Michael, and Kathleen Ferraro. "Research on Domestic Violence in the 1990s: Making Distinctions," *Journal of Marriage and the Family*, 62, 4 (November 2000): 948–963.

Johnson, Michael. "Patriarchal Terrorism and Common Couple Violence: Two Forms of Violence against Women." *Journal of Marriage and the Family* 57 (2) (May 1995): 283–294.

Johnson, Michael P. "A 'General' Theory of Intimate Partner Violence: A Working Paper." Paper presented at the Theory Construction and Research Methodology Pre-Conference Workshop, National Council on Family Relations annual meeting. Minneapolis, Minnesota, November 2006.

Johnston, Alexandra. "Gatekeeping in the Family: How Family Members Position One Another as Decision Makers." In *Family Talk: Discourse and Identity in Four American Families*, edited by Deborah Tannen, Shari Kendall, and Cynthia Gordon. New York: Oxford University Press, 2007.

Jones, N. A., and J. J. Bullock. "Understanding Who Reported Multiple Races in the U.S. Decennial Census: Results From Census 2000 and the 2010 Census." *Family Relations* 62 (2013): 5–16.

Judge, Abigail M. "'Sexting' among U.S. Adolescents: Psychological and Legal Perspectives." *Harvard Review of Psychiatry* 20, 2 (2012).

Jurkovic, G. *Lost Childhoods: The Plight of the Parentified Child*. New York: Brunner/Mazell, 1997.

Kachadourian, Lorig K., Frank Fincham, and Joanne Davila. "The Tendency to Forgive in Dating and Married Couples: The Role of Attachment and Relationship Satisfaction." *Personal Relationships* 11, 3 (September 2004): 373–393.

Kahlor, L. A., and D. Morrison. "Television Viewing and Rape Myth Acceptance among College Women." *Sex Roles* 56, 11–12 (June 2007): 729–739.

Kaiser Family Foundation. "Teens Say Sex On TV Influences Behavior of Peers: Some Positive Effects Seen," 2002. www.popline.org/node/253992.

Kaiser Family Foundation. "National Survey of Adolescents and Young Adults: Sexual Health, Knowledge and Experience," 2003. www.kff.org.

Kaiser Family Foundation. "HIV/AIDS Epidemic in the United States," 2014. http://kff.org/hivaids/fact-sheet/the-hivaids-epidemic-in-the-united-states/.

Kalmijn, M. "Educational Inequality and Extended Family Relationships." Paper presented at the Ross Colloquium Series, University of California, Los Angeles (UCLA), April 7, 2004.

Kalmijn, Matthijs. "Intermarriage and Homogamy: Causes, Patterns, Trends." *Annual Review of Sociology* 24 (1998): 395–421.

Kalmijn, Matthijs, and Paul de Graaf. "Life Course Changes of Children and Well-Being of Parents." *Journal of Marriage and Family* 74, 2 (April 2012): 269–280.

Kalmijn, Matthijs, and Henk Flap. "Assortative Meeting and Mating: Unintended Consequences of Organized Settings for Partner Choices." *Social Forces* 79, 4 (June 2001): 1289–1312.

Kane, Emily. "'I Wanted a Soul Mate': Gendered Anticipation and Frameworks of Accountability in Parents' Preferences for Sons and Daughters." *Symbolic Interaction* 32, 4 (Fall 2009): 372–389.

Kane, Emily. *The Gender Trap*. New York: NYU Press, 2012.

Kann, L., S. Kinchen, S. Shanklin, K. Flint, J. Hawkins, et al. "Youth Risk Behavior Surveillance—United States, 2013." Centers for Disease Control and Prevention. Surveillance Summaries 63, SS04 (June 13, 2014): 1–168. www.cdc.gov/mmwr/preview/mmwrhtml/ss6304a1.htm?s_cid=ss6304a1_w.

Kann, L., S. Kinchen, S. L. Shanklin, K. H. Flint, J. Kawkins, W. A. Harris, R. Lowry, E. O. Olsen, T. McManus, D. Chyen, L. Whittle, E. Taylor, Z. Demissie, N. Brener, J. Thornton, J. Moore, and S. Zaza. "Youth Risk Behavior Surveillance—United States, 2013." Centers for Disease Control and Prevention (CDC). MMWR Surveillance Summaries 63, 4 (June 13, 2014): 1–168.

Kann, L., S. Kinchen, S. Shanklin, et al. "Youth Risk Behavior Surveillance—United States, 2013." Morbidity and Mortality Weekly Report 63, 4 (June 13, 2014), Table 63.

Kann, L., S. Kinchen, S. Shanklin, et al. "Youth Risk Behavior Surveillance—United States, 2013." Morbidity and Mortality Weekly Report 63, 4 (June 13, 2014), Table 65.

Kantrowitz, Barbara. "Gay Families Come Out." Newsweek, November 4, 1996, 51–57.

Kapinus, Carolyn. "The Effects of Parents' Attitudes toward Divorce on Offspring's Attitudes: Gender and Parental Divorce as Mediating Factors." Journal of Family Issues 25,1 (January 2004): 112–135.

Kaplan, Helen Singer. Disorders of Desire. New York: Simon and Schuster, 1979.

Kaplan, Helen Singer. Evaluation of Sexual Disorders: Psychological and Medical Aspects. Levittown, PA: Taylor & Francis, 1983.

Kaplan, Lori. "A Couplehood Typology for Spouses of Institutionalized Persons with Alzheimer's Disease: Perceptions of "We"–"I.".* Family Relations 50, 1 (January 2001): 87–98.

Karp, D. The Burden of Sympathy. New York: Oxford University Press, 2001.

Kassop, Mark. "Salvador Minuchin: A Sociological Analysis of His Family Therapy Theory." Clinical Sociological Review 5 (1987): 158–167.

Katz, J. N. "'Homosexual' and 'Heterosexual': Questioning the Terms." In Sexualities: Identities, Behaviors and Society, edited by Michael Kimmel and Rebecca Plante. New York: Oxford University Press, 2004.

Katz, J. E., and M. Aakhus. "Conclusion: Meaning Making of Mobiles—A Theory of Apparatgeist. In Perpetual Contact: Mobile Communication, Private Talk, Public Performance, edited by J. E. Katz and M. Aakhus (pp. 301–320). Cambridge, England: Cambridge University Press, 2002.

Katz, Jennifer, Stephanie Washington Kuffel, and Amy Coblentz. "Are There Gender Differences in Sustaining Dating Violence? An Examination of Frequency, Severity, and Relationship Satisfaction." Journal of Family Violence 17, 3 (September 2002): 247–271.

Katz, Lynn Fainsilber, and Erica M. Woodin. "Hostility, Hostile Detachment, and Conflict Engagement in Marriages: Effects on Child and Family Functioning." Child Development 73, 2 (March 2002): 636–690.

Kaufman, L. "When the Ex Writes a Blog, Dirty Laundry Is Aired." New York Times, April 18, 2008.

Kawamoto, Walter T., and Tamara C. Cheshire. "American Indian Families." In Families in Cultural Context, edited by M. K. DeGenova. Mountain View, CA: Mayfield, 1997.

Kearney, Melissa, and Phillip Levine. "Media Influences on Social Outcomes: The Impact of MTV's 16 & Pregnant on Teen Childbearing." NBER Working Paper, No. 19795, January 2014.

Keene-Reid, Jennifer, and John Reynolds. "Gender Differences in the Job Consequences of Work-to-Family Spillover." Journal of Family Issues 26, 3 (2005): 275–299.

Keith, Pat, and Robbyn Wacker. "Grandparent Visitation Rights: An Inappropriate Intrusion or Appropriate Protection?" International Journal of Aging and Human Development 54, 3 (2002): 191–204.

Kempe, C. H., F. Silverman, B. Steele, W. Droegmueller, and H. Silver. "The Battered Child Syndrome." Journal of the American Medical Association 181 (1962): 17–24.

Kendig, S., and S. M. Bianchi. "Single, Cohabitating, and Married Mothers' Time with Children." Journal of Marriage and Family 70, 5 (2008): 1228–1240.

Kennedy, Sheela, and Steven Ruggles. "Breaking Up Is Hard to Count: The Rise of Divorce in the United States, 1980–2010." Demography 51, 2 (2014): 587–598.

Kent, Erin, and Amani El Alayli. "Public and Private Physical Affection Differences between Same-Sex and Different-Sex Couples: The Role of Perceived Marginalization." Interpersona 5, 2 (2011): 149–167.

Kerpelman, Jennifer. "Healthy Adolescent Romantic Relationships. Family Focus on Dating and Mate Selection." National Council on Family Relations Report 59, 4 (Winter 2014): F2.

Kerr, David, Deborah Capaldi, Lee Owen, Margit Wiesner, and Katherine C. Pear. "Changes in At-Risk American Men's Crime and Substance Use Trajectories Following Fatherhood." Journal of Marriage and Family 73, 5 (October 2011): 1101–1116.

Kett, Joseph. "Reflections on the History of Adolescence in America." The History of the Family 8 (2003): 355–373.

Khan, Sharful, and Mohammed Hussain. "Living on the Extreme Margin: Social Exclusion of the Transgender Population (Hijra) in Bangladesh." Journal of Health, Population, and Nutrition 27, 4 (August 2009): 441–451.

Kiecolt, K. Jill. "Satisfaction with Work and Family Life: No Evidence of a Cultural Reversal." Journal of Marriage and Family 65 (February 2003): 23–35.

Kikumura, Akemi, and Harry Kitano. "The Japanese American Family." In Ethnic Families in America: Patterns and Variations, 3rd ed., edited by C. Mindel et al. New York: Elsevier North Holland, 1988.

Kilmartin, Christopher. The Masculine Self. New York: Macmillan, 1994.

Kim, Janna L., and L. Monique Ward. "Pleasure Reading: Associations between Young Women's Sexual Attitudes and Their Reading of Contemporary Women's Magazines." Psychology of Women Quarterly 28, 1 (March 2004): 48.

Kim, Jungsik, and Elaine Hatfield. "Love Types and Subjective Well-Being: A Cross-Cultural Study." Social Behavior and Personality 32, 2 (2004): 173–182.

Kim, Susanna. "US Is Only Industrialized Nation Without Paid Maternity Leave," ABC News, May 6, 2015. http://abcnews.go.com/Business/us-industrialized-nation-paid-maternity-leave/story?id=30852419

Kim, Young Shin, and Bennett Leventhal. "Bullying and Suicide: A Review." International Journal of Adolescent Medicine and Health 20, 2 (April/June 2008): 133–154.

Kimmel, Michael. The Gendered Society, 4th ed. New York: Oxford University Press, 2011.

Kimmel, Michael, and Amy Aronson. Sociology Now, Census Update. Upper Saddle River, NJ: Prentice Hall, 2011.

Kimmel, M., and M. Messner. Men's Lives, 7th ed. Boston: Pearson/Allyn and Bacon, 2007.

Kimmel, Michael, and Michael Messner, eds. Men's Lives, 8th ed. Boston: Pearson/Allyn and Bacon, 2010.

Kimmel, Michael, and Rebecca Plante. "The Gender of Desire: The Sexual Fantasies of College Women and Men." In Sexualities: Identities, Behaviors, and Society, edited by Michael Kimmel and Rebecca Plante. New York: Oxford University Press, 2004.

King, Valerie, Lisa Boyd, and Maggie Thorsen. "Adolescents Perceptions of Family Belonging in Stepfamilies." Journal of Marriage and Family 77 (June 2015): 761–774.

King, V., K. M. Harris, and H. E. Heard. "Racial and Ethnic Diversity in Nonresident Father Involvement." Journal of Marriage and Family 66 (2004): 1–21.

King, Valarie, and Mindy E. Scott. "A Comparison of Cohabiting Relationships among Older and Younger Adults." Journal of Marriage and Family 67, 2 (May 2005): 271–285.

Kiselica, M. S., and M. Morrill-Richards. Sibling Maltreatment: The Forgotten Abuse." Journal of Counseling and Development 25, 2 (2007): 148.

Kitano, K., and H. Kitano. The Japanese-American Family." In Ethnic Families in America: Patterns and Variations, 4th ed., edited by C. Mindel, R. Habenstein, and R. Wright, Jr. (pp. 311–330). Upper Saddle River, NJ: Prentice Hall, 1998.

Kito, M. 2005. "Self-Disclosure in Romantic Relationships and Friendships among American and Japanese College Students." The Journal of Social Psychology 145, 2 (2005): 127–140.

Kitson, G. C., R. D. Clark, N. B. Rushforth, P. M. Brinich, H. S. Sudak, and S. J. Zyzanski. "Research on Difficult Family Topics: Helping New and Experienced Researchers Cope with Research on Loss." Family Relations, 45 (1996): 183–188.

Kitson, Gay. "Marital Discord and Marital Separation: A County Survey." Journal of Marriage and the Family 47 (August 1985): 693–700.

Kitson, Gay, and Leslie Morgan. "The Multiple Consequences of Divorce: A Decade Review." In Contemporary Families: Looking Forward, Looking Back, edited by A. Booth. Minneapolis: National Council on Family Relations, 1991.

Kitzinger, Sheila. *The Complete Book of Pregnancy and Childbirth*. New York: Knopf, 1989.

Klein, Fritz. "The Need to View Sexual Orientation as a Multivariate Dynamic Process: A Theoretical Perspective." In *Homosexuality /Heterosexuality: Concepts of Sexual Orientation*, edited by David McWhirter, Stephanie Sandovers, and June Machover Reinisch. New York: Oxford University Press, 1990.

Klinenberg, Eric. *Going Solo: The Extraordinary Rise and Surprising Appeal of Living Alone*. New York: Penguin, 2012.

Klinetob, Nadya, and David Smith. "Demand-Withdraw Communication in Marital Interaction: Tests of Interspousal Contingency and Gender Role Hypothesis." *Journal of Marriage and the Family*, 58, 4 (November 1996): 945–957.

Kluwer, Esther. "From Partnership to Parenthood: A Review of Marital Change across the Transition to Parenthood." *Journal of Family Theory and Review* 2, 2 (June 2010): 105–125.

Kluwer, E. S., J. A. Heesink, E. Van De Vliert. "The Marital Dynamics of Conflict over the Division of Labor." *Journal of Marriage and the Family* 59 (1997): 635–653.

Knafo, D., and Y. Jaffe. "Sexual Fantasizing in Males and Females." *Journal of Research in Personality* 18 (1984): 451–462.

Knox, David. "'TEXT ME': Impact of Technology on Today's Romantic Relationships." *NCFR Reports, Family Focus on Dating and Mate Selection* (Winter 2014): F13–14.

Knox, David, and Caroline Schacht. *Choices in Relationships: An Introduction to Marriage and the Family*, 6th ed. Belmont, CA: Wadsworth Pub Co, 2000.

Knudson-Martin, Carmen, and Anne Rankin Mahoney. "Language and Processes in the Construction of Equality in New Marriages." *Family Relations* 47, 1 (January 1998): 81–91.

Kochhar, Rakesh. "The Demographics of the jobs recovery: Employment Gains by Race, Ethnicity, Gender, and Nativity." Pew Hispanic Center, March 21, 2012.

Kochhar, Rakesh, Richard Fry, and Paul Taylor. "Wealth Gaps Rise to Record Highs between Whites, Blacks, Hispanics: Twenty-to-One." Pew Research Center, 2011. www.pewsocialtrends.org/2011/07/26 /wealth-gaps-rise-to-record-highs-between-whites-blacks-hispanics/.

Koerner, Ascan, and Mary Anne Fitzpatrick. "Nonverbal Communication and Marital Adjustment and Satisfaction: The Role of Decoding Relationship Relevant and Relationship Irrelevant Affect." *Nonverbal Communications Monographs* 69, 1 (March 2002): 33–51.

Kohlberg, Lawrence. "The Cognitive-Development Approach to Socialization." In *Handbook of Socialization Theory and Research*, edited by A. Goslin. Chicago: Rand McNally, 1969.

Kohn, Melvin. *Class and Conformity: A Study in Values, with a Reassessment*. Chicago: University of Chicago Press, 1977.

Kohn, M. "Social Structure and Personality: A Quintessentially Sociological Approach to Social Psychology." *Social Forces* 68 (September 1989): 26–33.

Konner, Melvin. *Childhood*. Boston: Little, Brown, 1991.

Kornhaber, Spencer. "The *Modern Family* Effect: Pop Culture's Role in the Gay-Marriage Revolution." *The Atlantic*, June 26, 2015. www .theatlantic.com/entertainment/archive/2015/06/gay-marriage -legalized-modern-family-pop-culture/397013/.

Kornrich, S., J. Brines, and K. Leupp. "Egalitarianism, Housework, and Sexual Frequency in Marriage." *American Sociological Review* 78 (2013): 26–50.

Kosciw, Joseph G., et al. "The Effect of Negative School Climate on Academic Outcomes for LGBT Youth and the Role of In-School Supports." *Journal of School Violence* 12, 1 (2013): 45–63.

Kotchick, B. A., S. Dorsey, and L. Heller. "Predictors of Parenting among African American Single Mothers: Personal and Contextual Factors." *Journal of Marriage and Family* 67 (2005): 448–460.

Kowal, A. K., and L. Blinn-Pike. "Sibling Influences on Adolescents' Attitudes toward Safe Sex Practices." *Family Relations* 53, 4 (2004): 377–384.

Kranichfeld, Marion. "Rethinking Family Power." *Journal of Family Issues* 8, 1 (March 1987): 42–56.

Kreider, Rose M., and Renee Ellis, "Living Arrangements of Children: 2009." Current Population Reports, P70-126, U.S. Census Bureau, Washington, DC, 2011a.

Kreider, Rose, and Renee Ellis, "Number, Timing, and Duration of Marriages and Divorces: 2009." Current Population Reports, P70-125, U.S. Census Bureau, Washington, DC, 2011b.

Kreider, Rose M., and Daphne A. Lofquist. "Adopted Children and Stepchildren: 2010." Current Population Reports, P20-572, U.S. Census Bureau, Washington, DC, 2014.

Kristoff, Nicholas. "Divorced before Puberty." *New York Times*, March 3, 2010.

Krogstad, Jens Manuel. "Five Facts about Today's Grandparents." Fact Tank Pew Research Center, September 13, 2015.

Kulu, Hill, and Paul Boyle. "Premarital Cohabitation and Divorce: Support for the 'Trial Marriage' Theory?" *Demographic Research* 23, 31 (2010). www.demographic-research.org.

Kunkel, D., K. Eyal, K. Finnerty, and E. Donnderstein. *Sex on TV5: A Biennial Report to the Kaiser Family Foundation*. Menlo Park, CA: Kaiser Family Foundation, 2005.

Kuperberg, Arielle. "Does Premarital Cohabitation Raise Your Risk of Divorce?" Council on Contemporary Families, March 10, 2014. https://contemporaryfamilies.org/cohabitation-divorce-brief-report/.

Kurdek, L. A. "Conflict Resolution Styles in Gay, Lesbian, Heterosexual Nonparent, and Heterosexual Parent Couples." *Journal of Marriage and Family* 56, 3 (August 1994): 705–722.

Kuroki, Masanori. "Opposite-Sex Coworkers and Marital Infidelity." *Economics Letters*, 118, 1 (*2013*): 71–73.

Kurz, Demie. "Caring for Teenage Children." *Journal of Family Issues* 23, 6 (September 2002): 748–767.

Kutner, L. "Parent and Child." *New York Times*, January 28, 1988.

Lachs, Mark S., and Karl Pillemer. "Elder Abuse." *The Lancet*, 364 (2004): 1192–1263.

Lakoff, Robin. *Language and Women's Place*. New York, 1975.

Lamanna, Mary Ann, and Agnes Riedmann. *Marriages and Families: Making Choices in a Diverse Society*. Belmont, CA: Wadsworth, 1997.

Lamb, Michael. "Book Review." *Journal of Marriage and the Family* 55, 4 (November 1993): 1047–1049.

Lamidi, Esther. "Single, Cohabiting, and Married Households, 1995 –2012" (FP-14-1). National Center for Family & Marriage Research, 2014. http://ncfmr.bgsu.edu/pdf/family_profiles/file141218.pdf

Lamidi, E., and J. Cruz. "Remarriage Rate in the U.S., 2012" (FP-14-10). National Center for Family and Marriage Research, 2014. www .bgsu.edu/content/dam/BGSU/college-of-arts-and-sciences/NCFMR /documents/FP/FP-14-10-remarriage-rate-2012.pdf.

Lamidi, Esther. "Trends in Cohabitation: The Never Married and Previously Married, 1995–2014" (FP-15-21). National Center for Family and Marriage Research, 2015. http://bgsu.edu/ncfmr /resources/data/family-profiles/lamidi-cohab-trends-never -previously-married-fp-15-21.

Lamidi, Esther, Susan Brown, and Wendy Manning. "Assortative Mating: Age Heterogamy in U.S. Marriages, 1964–2014" (FP-15-14). National Center for Family and Marriage Research, 2015a.

Lamidi, Esther, Susan Brown, and Wendy Manning. "Assortative Mating: Educational Homogamy in U.S. Marriages, 1964–2014" (FP-15-15). National Center for Family and Marriage Research, 2015b.

Lamidi, Esther, and Krista Payne. Marital Status in the U.S., 2012 (FP-14-07). National Center for Family & Marriage Research, 2014.

Laner, Mary R., and N. Ventrone. "Dating Scripts Revisited." *Journal of Family Issues* 21, 4 (May 2000): 488–500.

Lang, M. M., and B. Risman. "A Stalled Revolution or a Still Unfolding One? The Continuing Convergence of Men's and Women's Roles." Paper prepared for the 10th Anniversary Conference of the Council on Contemporary Families, May 4–5, 2007, Chicago, IL.

Langman, L. "Social Stratification." In *Handbook of Marriage and the Family*, edited by M. G. Sussman and S. K. Steinmetz (pp. 211–246). New York: Plenum Press, 1987.

Lannutti, Pamela J., and Kenzie A. Cameron. "Beyond the Breakup: Heterosexual and Homosexual Post-Dissolutional Relationships." *Communication Quarterly* 50, 2 (2002), 153–170.

Lantz, Herman. "Family and Kin as Revealed in the Narratives of Ex-Slaves." *Social Science Quarterly* 60, 4 (March 1980): 667–674.

Lareau, Annette. *Unequal Childhoods*. Berkeley, CA: University of California Press, 2003.

LaRossa, Ralph, and Maureen Mulligan LaRossa. *The Transition to Parenthood: How Infants Change Families*. Beverly Hills, CA: Sage, 1981.

LaRossa, Ralph. *The Modernization of Fatherhood: A Social and Political History*. University of Chicago, 1997.

LaRossa, Ralph. *Of War and Men: World War II in the Lives of Fathers and Their Families*. University of Chicago Press, 2011.

Larson, Reed, and Sally Gillman. "Transmission of Emotions in the Daily Interactions of Single-Mother Families." *Journal of Marriage & Family* 61, 1 (February 1999): 21–373.

Larson, Jeffrey, and Rachel Hickman. "Are College Marriage Textbooks Teaching Students the Premarital Predictors of Marital Quality?" *Family Relations* 53, 4 (July 2004): 385–392.

LaSala, Michael. "Extradyadic Sex and Gay Male Couples: Comparing Monogamous and Nonmonogamous Relationships." *Families in Society: The Journal of Contemporary Social Services* 85, 3 (2004): 405–412.

Lauer, Jeanette, and Robert Lauer. *'Til Death Do Us Part: How Couples Stay Together.* New York: Haworth Press, 1986.

Laughlin, Lynda. "Child Care Constraints among America's Families." U.S. Census Bureau, Housing and Household Economic Statistics Division, 2007. www.census.gov/content/dam/Census/library/working-papers/2007/demo/SEHSD-WP2007-05.pdf.

————. "Maternity Leave and Employment Patterns of First-Time Mothers: 1961–2008. U.S. Census Bureau, 2011.

————. Who's "Minding the Kids? Child Care Arrangements: Spring 2011." *Current Population Reports,* P70-135. U.S. Census Bureau, Washington, DC, 2013.

Laumann, E., D. Glasser, R. Neves, and E. Moreira Jr. "A Population-Based Survey of Sexual Activity, Sexual Problems and Associated Help-Seeking Behavior Patterns in Mature Adults in the United States of America." *International Journal of Impotence Research* 21 (2009): 171–178.

Laumann, Edward, John Gagnon, Robert Michael, and Stuart Michaels. *The Social Organization of Sexuality: Sexual Practices in the United States.* Chicago: University of Chicago Press, 1994.

Lavee, Y., and D. H. Olson. "Seven Types of Marriage: Empirical Typology Based on ENRICH." *Journal of Marital and Family Therapy* 19 (1993): 325–340.

Lavelle, B., & P. J. Smock. "Divorce and Women's Risk of Health Insurance Loss." *Journal of Health and Social Behavior* 53, 4 (2012): 413–431. http://doi.org/10.1177/0022146512465758.

Layne, Linda. "Breaking the Silence: An Agenda for a Feminist Discourse of Pregnancy Loss." *Feminist Studies* 23, 2 (1997): 289–315.

Lederer, William, and Don Jackson. *Mirages of Marriage.* New York: Norton, 1968.

Lee, John A. *The Color of Love.* Toronto: New Press, 1973.

————. "Love Styles." In *The Psychology of Love,* edited by R. Sternberg and M. Barnes. New Haven, CT: Yale University Press, 1988.

Lee, Y., and L. Waite. "Husbands and Wives Time Spent on Housework: A Comparison of Measures." *National Council on Family Relations Report: Family Focus on Families and Work-Life* 51, 2 (June 2006): F1.

Lehr, Sally, Alice Demi, Colleen DiIorio, and Jeffrey Facteau. Predictors of father–son communication about sexuality. *Journal of Sex Research* 42 (2005): 119–129.

Liefbroer, Aart, Anne-Rigt Poortman, and Judith A. Seltzer. "Why Do Intimate Partners Live Apart? Evidence on LAT Relationships across Europe." *Demographic Research* 32 (Jan–Jun 2015): 251–286.

Leonard, R., and S. Elias. *Family Law Dictionary.* Berkeley, CA: Nolo Press, 1990.

Lerner, Adam. "The Supreme Court's Most Memorable Quotes on Gay Marriage." www.politico.com, June 26, 2015.

Leslie, Leigh, and Bethany Letiecq. "Marital Quality of African American and White Partners in Interracial Couples." *Personal Relationships* 11 (2004): 559–574.

Lev, Arlene Istar. "Resilience in Lesbian and Gay Couples." Chapter 3 in K. Skerrett and K. Fergus (eds.), *Couple Resilience:* 45–61.

Levaro, L. B. "Living Together or Living Apart Together: New Choices for Old Lovers." *National Council on Family Relations Report: Family Focus on Cohabitation* 54, 2 (Summer 2009): F9–10.

Levenson, Robert W., Laura L. Carstensen, and John M. Gottman. "Long-Term Marriage: Age, Gender, and Satisfaction." *Psychology and Aging* 8, 2 (1993): 301–313.

————. "The Influence of Age and Gender on Affect, Physiology, and Their Interrelations: A Study of Long-Term Marriages." *Journal of Personality & Social Psychology* 67, 1 (July 1994): 56–68.

Levine, J., C. Emery, and H. Pollack. "Well-Being of Children Born to Teen Mothers." *Journal of Marriage and Family* 69, 1 (February 2007): 105–122.

Levitt, H., R. Swanger, and J. Butler. "Male Perpetrators' Perspectives on Intimate Partner Violence, Religion, and Masculinity." *Sex Roles.* 58, 5–6 (March, 2008): 435–448.

Lewin, Tamar. "Financially Set, Grandparents Help Keep Families Afloat, Too." *New York Times,* July 14, 2005, A1.

Lewin, Tamar. "Chicago Court Gives Woman Frozen Embryos Despite Ex-Boyfriend's Objections." *New York Times,* June 13, 2015.

Lewis, Oscar. "The Culture of Poverty." *Scientific American,* 215, 4 (1966): 19–25.

Lewis, Jamie M., and Rose M. Kreider, "Remarriage in the United States." American Community Survey Reports, ACS-30, U.S. Census Bureau, Washington, DC, 2015.

Li, J., and L. L. Wu. "No Trend in the Intergenerational Transmission of Divorce." *Demography* 45 (2008): 875–83.

Lichter, Daniel T. "Studying Families: Why Demography Matters." *Family Focus on Families and Demography National Council on Family Relations Report* 58, 2 (Summer 2013).

Lichter, D., and Z. Qian. "Serial Cohabitation and the Marital Life Course." *Journal of Marriage and Family* 70 (2008): 861–878.

Lieberson, Stanley, and Mary Waters. *From Many Strands: Ethnic and Racial Groups in Contemporary America.* New York: Russell Sage Foundation, 1988.

Liebow, Elliot. *Tally's Corner: A Study Of Negro Streetcorner Men.* Boston, Little, Brown, 1967.

Liefbroer, Aart C., Anne-Rigt Poortman, and Judith Seltzer. "Why Do Intimate Partners Live Apart? Evidence on LAT Relationships across Europe." *Demographic Research* 32 (June–July 2015): 251–286.

Lin, Chin-Yau Cindy, and Victoria Fu. "A Comparison of Child-Rearing Practices among Chinese, Immigrant Chinese, and Caucasian-American Parents." *Child Development* 61, 2 (April 1990): 429–434.

Lin, I-Fen, and Susan Brown. "Unmarried Boomers Confront Old Age: A National Portrait." *The Gerontologist* 52, 2 (2012): 1–13.

Linden, M. "The Cost of Doing Nothing: The Economic Impact of Recession-Induced Child Poverty, First Focus: Making Children and Families the Priority," 2008. www.firstfocus.net/Download/CostNothing.pdf.

Lindsey, E. W., Y. Caldera, and L. Tankersley. "Marital Conflict and the Quality of Young Children's Peer Play Behavior: The Mediating and Moderating Role of Parent-Child Emotional Reciprocity and Attachment Security." *Journal of Family Psychology* 23, 2 (2009): 130–145.

Lindsey, Linda. *Gender Roles: A Sociological Perspective,* 3rd ed. Upper Saddle River, NJ: Prentice Hall, 1997.

Linn, Allison. "Women Who Try To Have It All Will Likely Pay a Mommy Penalty." CNBC, 2013. www.cnbc.com/2013/11/14/study-finds-women-with-young-kids-earn-less-than-men.html.

Lino, Mark. "Expenditures on Children by Families, 2012." U.S. Department of Agriculture, Center for Nutrition Policy and Promotion (Miscellaneous Publication No. 1528-2012), 2013. www.cnpp.usda.gov/sites/default/files/expenditures_on_children_by_families/crc2012.pdf.

Lippa, Richard. *Gender, Nature, and Nurture,* 2nd ed. Mahwah, NJ: Lawrence Earlbaum Associates, 2005.

Lippman, Laura, and W. Bradford Wilcox. *World Family Map 2014.* Washington, DC: Child Trends, 2014.

Lips, Hilary. *Sex and Gender,* 3rd ed. Mountain View, CA: Mayfield, 1997.

Liu, H., and C. Reczek. "Cohabitation and U.S. Adult Mortality: An Examination by Gender and Race." *Journal of Marriage and Family* 74 (2012): 794–811.

Liu, Hui, Qiu Wang, Venessa Keesler, and Barbara Schneider. 2011. "Nonstandard Work Schedules, Work-Family Conflict and Parental Well-Being: A Comparison of Married and Cohabiting Unions." *Social Science Research* (2011) 40: 473–484.

Livingston, Gretchen. "The Rise of Single Fathers: A Ninefold Increase Since 1960." Pew Research Center, 2013. www.pewsocialtrends.org/2013/07/02/the-rise-of-single-fathers/.

Livingston, Gretchen. "Four-in-Ten Couples are Saying 'I Do,' Again: Growing Number of Adults Have Remarried." Pew Research Center, November 14, 2014. www.pewsocialtrends.org/2014/11/14/four-in-ten-couples-are-saying-i-do-again.

————. "Growing Number of Dads Home with the Kids: Biggest Increase among Those Caring for Family." Washington, DC: Pew Research Center's Social and Demographic Trends project, June 2014b.

————. "Childlessness Falls, Family Size Grows among Highly Educated Women." Pew Research Center, Pew Social and Demographic Trends, May 7, 2015. www.pewsocialtrends.org/2015/05/07/childlessness-falls-family-size-grows-among-highly-educated-women/.

Liz Claiborne Inc. "Omnibuzz® Topline Findings–Teen Relationship Abuse Research." Teenage Research Unlimited, 2005. http://nrhs.nred.org/www/nred_nrhs/site/hosting/Resources4GuidanceSocialWork/GuidanceSocialWork/HandoutsSocialWork/Facts_TeenDatingViolence.pdf.

Lloyd, Sally A., and Beth C. Emery. "The Dynamics of Courtship Violence." Paper presented at the annual meeting of the National Council on Family Relations, Seattle, WA, November 1990.

Lo, Shao-Kang. "The Nonverbal Communication Functions of Emoticons in Computer-Mediated Communication." *Cyberpsychology and Behavior* 11, 5 (2008).

Lofquist, Daphne. "Multigenerational Households: 2009–2011." American Community Survey Briefs, U.S. Census Bureau, Washington, DC, 2012.

LoPiccolo, J. "Counseling and Therapy for Sexual Problems in the Elderly." *Clinics in Geriatric Medicine* 7 (1991): 161–179.

Lorber, Judith. *Paradoxes of Gender*. New Haven, CT: Yale University Press, 1994.

———. *Gender Inequality: Feminist Theories and Politics*. Los Angeles: Roxbury, 1998.

"Louisiana Civil Code Art. 98. Mutual Duties of Married Persons," marriagedebate.com, 1999.

Lounsbury, Kaitlin, Kimberly Mitchell, and David Finkelhor. "The True Prevalence of 'Sexting.'" Crimes Against Children Research Center Fact Sheet, University of New Hampshire, April 2011.

Lowenstein, Ludwig. "Causes and Associated Features of Divorce as Seen by Recent Research." *Journal of Divorce & Remarriage* 42, 3/4 (2005): 153–171.

Lundquist, Jennifer Hickes, Michelle J. Budig, and Anna Curtis (graduate student). "Race and Childlessness in America: Similarities in Trends and Pathways to Childlessness among Black and White Women, 1988–2002." *Journal of Marriage and the Family* 71, 3 (2009): 741–55.

Luster, Tom, Desiree Qin, Laura Bates, Deborah Johnson, and Meenal Rana. "The Lost Boys of Sudan: Ambiguous Loss, Search for a Family, and Reestablishing Relationships with Family Members." *Family Relations* 57 (October 2008): 444–456.

Lutz, Patricia. "The Stepfamily: An Adolescent Perspective." *Family Relations* 32, 3 (July 1983): 367–375.

Lynley, Matt. "This Dating App that Just Raised $22 Million is about to Take Over the World." *Business Insider*, May 13, 2012. www.businessinsider.com/this-dating-app-that-just-raised-22-million-is-about-to-take-over-the-world-2012-5#ixzz25XE3vSls.

Lyssens-Danneboom, V., and D. Mortelmans. "Living Apart Together and Money: New Partnerships, Traditional Gender Roles." *Journal of Marriage and Family* 76 (2014): 949–966.

Maccoby, E. E., C. M. Buchanan, R. H. Mnookin, and S. M. Dornbusch. "Postdivorce Roles of Mothers and Fathers in the Lives of Their Children." *Journal of Family Psychology* 7 (1993): 24–38.

MacDermid, S. "Work-Family Research: Learning from the 'Best of the Best.'" *National Council on Family Relations Report: Family Focus on Families and Work-Life* 51, 2 (June 2006): F15–16.

MacDorman, M. F., and T. J. Mathews. *Recent Trends in Infant Mortality in the United States*. NCHS Data Brief, no. 9. Hyattsville, MD: National Center for Health Statistics, 2008.

MacDorman, Marian, and Sharon Kirmeyer. "The Challenge of Fetal Mortality." NCHS Data Brief #16, 2009. www.cdc.gov/nchs/data/databriefs/db16.pdf.

Mace, David, and Vera Mace. "Enriching Marriage." In *Family Strengths*, edited by N. Stinnet et al. Lincoln: University of Nebraska Press, 1979.

Maciel, Jose, Zanetta Van Putten, and Carmen Knudson-Martin. "Gendered Power in Cultural Contexts: Part 1. Immigrant Couples." *Family Process* 48, 1 (March 2009): 9–23.

Madden, Mary, and Amanda Lenhart. "On-line Dating." Pew Internet and American Life Project, www.pewinternet.org, March 5, 2006.

Mahoney, Annette. "Religion in Families, 1999–2009: A Relational Spirituality Framework." *Journal of Marriage and Family* 72 (2010): 805–827.

Manning, Wendy. "Children and the Stability of Cohabiting Couples." *Journal of Marriage & Family* 66, 3 (August 2004): 674–689.

Manning, Wendy. "Trends in Cohabitation: Twenty Years of Change, 1987–2008" (FP-10-07). National Center for Family and Marriage Research, 2010. www.bgsu.edu/content/dam/BGSU/college-of-arts-and-sciences/NCFMR/documents/FP/FP-10-07.pdf.

Manning, W. D., and S. L. Brown. "Children's Economic Well-Being in Married and Cohabiting Parent Families." *Journal of Marriage and Family* 68 (2006): 345–362.

Manning, W. D., and J. A. Cohen. "Premarital Cohabitation and Marital Dissolution: An Examination of Recent Marriages." *Journal of Marriage and Family* 74 (2012): 377–387.

Manning, Wendy D., and Kathleen Lamb. "Adolescent Well-Being in Cohabiting, Married, and Single-Parent Families." *Journal of Marriage & Family* 65, 4 (November 2003): 876–893.

Manning, W., M. A. Longmore, and P. C. Giordano. "The Changing Institution of Marriage: Adolescents' Expectations to Cohabit and to Marry." *Journal of Marriage and Family* 69 (August 2007): 559–575.

Manning, Wendy, and Bart Stykes. "Twenty-Five Years of Change in Cohabitation in the U.S., 1987–2013" (FP-15-01), National Center for Family and Marriage Research, 2015. www.bgsu.edu/content/dam/BGSU/college-of-arts-and-sciences/NCFMR/documents/FP/lamidi-25yr-change-marriage-fp-15-17.pdf.

Mapes, Diane. "Should We Call It Quits? A New Kind of Couples Counseling." *Today*, May 3, 2012. www.today.com/health/should-we-call-it-quits-new-kind-couples-counseling-750599.

Margolin, Gayla, Linda Gorin Sibner, and Lisa Gleberman. "Wife Battering." In *Handbook of Family Violence*, edited by V. B. Van Hasselt et al. New York: Plenum Press, 1988.

Mark, Kristen, Erick Janssen, and Robin Milhausen. "Infidelity in Heterosexual Couples: Demographic, Interpersonal, and Personality-Related Predictors of Extradyadic Sex." *Archives of Sexual Behavior* 40, 5 (October 2011): 971–82.

Markey, Patrick, Charlotte Markey, Christopher Nave, and Kristin August. "Interpersonal Problems and Relationship Quality: An Examination of Gay and Lesbian Romantic Couples." *Journal of Research in Personality* 51 (2014): 1–8.

Markman, Howard, et al. "The Prediction and Prevention of Marital Distress: A Longitudinal Investigation." In *Understanding Major Mental Disorders: The Contribution of Family Interaction Research*, edited by K. Hahlweg and M. Goldstein. New York: Family Process Press, 1987.

Markman, Howard. "Prediction of Marital Distress: A Five-Year Follow-Up." *Journal of Consulting and Clinical Psychology* 49 (1981): 760–761.

Markman, Howard, Galena Rhoades, Scott Stanley, Erica Ragan, and Sarah Whitton. "The Premarital Communication Roots of Marital Stress and Divorce: The First Five Years of Marriage." *Journal of Family Psychology* 24, 3 (2010): 289–298.

Marks, Nadine, James Lambert, and Heejeong Choi. "Transitions to Caregiving, Gender and Psychological Well-Being: A Prospective U.S. National Study," *Journal of Marriage and the Family* 64, 3 (August 2002): 657–667.

Marks, Stephen. "What Is a Pattern of Commitment?" *Journal of Marriage and the Family* 52, 1 (February 1994): 112–115.

Marks, Stephen. *Three Corners: Integrating Marriage & The Self*. Lexington, MA: Lexington Books, 1986.

Marsiglio, William. "Paternal Engagement Activities with Minor Children." *Journal of Marriage & Family* 53, 4 (November 1991): 973–986.

———. *Procreative Man*. New York: New York University, 1998.

———. *Stepdads: Stories of Love, Hope, and Repair*. Lanham, MD: Rowman and Littlefield, 2004a.

Marsiglio, W. "When Stepfathers Claim Stepchildren: A Conceptual Analysis." *Journal of Marriage and Family* 66 (2004b): 22–39.

Marsiglio, William, and Kevin Roy. *Nurturing Dads: Social Initiatives for Contemporary Fatherhood*. ASA Rose Monograph Series. New York: Russell Sage Foundation, 2012.

Martin J. A., B. E. Hamilton, S. J. Ventura, et al. "Births: Final Data for 2009." National Vital Statistics Reports, vol. 60, no. 1. Hyattsville, MD: National Center for Health Statistics, 2011.

Martin, J, B. Hamilton, S. Ventura, et al. "Births: Final Data for 2010." National Vital Statistics Reports, vol. 61, no. 1. Hyattsville, MD: National Center for Health Statistics, 2012.

Martin, J. A., B. E. Hamilton, M. J. K. Osterman, et al. "Births: Final Data for 2013." National Vital Statistics Reports, vol. 64, no. 1. Hyattsville, MD: National Center for Health Statistics, 2015.

Martin, Karin A. "Giving Birth Like a Girl," *Gender & Society* 17, 1 (February 2003): 54–72.

Martin, S., and S. Parashar. "Women's Changing Attitudes towards Divorce, 1974–2002: Evidence for an Educational Crossover." *Journal of Marriage and Family* 68, 1 (2006): 29–40.

Martinez, G. M., and J. C. Abma. "Sexual Activity, Contraceptive Use, and Childbearing of Teenagers Aged 15–19 in the United States." NCHS data brief, no 209. Hyattsville, MD.

Martinez, Gladys, Casey Copen, and Joyce Abma. "Teenagers in the United States: Sexual Activity, Contraceptive Use, and Childbearing, 2006–2010." National Survey of Family Growth. U.S. Department of Health and Human Services. *Vital and Health Statistics* 23, 31 (October 2011).

Martinez, Gladys, Kimberly Daniels, and Anjani Chandra. "Fertility of Men and Women Aged 15–44 Years in the United States: National Survey of Family Growth, 2006–2010." National Health Statistics Reports, no. 51. Hyattsville, MD: National Center for Health Statistics. Division of Vital Statistics, 2012.

Martinez-Prather, Kathy, and Donna M. Vandiver. "Sexting among Teenagers in the United States: A Retrospective Analysis of Identifying Motivating Factors, Potential Targets, and the Role of a Capable Guardian." International Journal of Cyber Criminology 8, 1 (January–June 2014): 21–35.

Mascarenhas, M. N., S. R. Flaxman, T. Boerma, S. Vanderpoel, and G. A. Stevens. "National, Regional, and Global Trends in Infertility Prevalence Since 1990: A Systematic Analysis of 277 Health Surveys." PLoS Med 9, 12 (2012): e1001356.

Masheter, Carol. "Postdivorce Relationships between Ex-Spouses: The Roles of Attachment and Interpersonal Conflict." Journal of Marriage and the Family 53, 1 (April 1991): 103–110.

Maslow, Abraham. Motivation and Personality, 2nd ed. New York, Harper & Row, 1970.

Mason, M. A., M. Fine, and S. Carnochan. "Family Law in the New Millennium: For Whose Families?" Journal of Family Issues 22, 7 (October 2001): 859–881.

Masters, William, and Virginia Johnson. Human Sexual Inadequacy. Boston: Little, Brown, 1970.

Masterson, John. "Wife Swap—It Should be Compulsory for Couples." Irish Independent, August 24, 2008.

Mastropasqua, Kristina. "Global Discrimination against LGBT Persons: 2015 United Nations Report." Journalist's Resource, June 8, 2015. http://journalistsresource.org/studies/international/human-rights/global-discrimination-against-lgbt-persons-2015-united-nations-report.

Mather, Mark. "Population Reference Bureau Data Brief: U.S. Children in Single-Mother Families," May 2010. www.prb.org.

Mathews T. J., and B. E. Hamilton. "Mean Age of Mother, 1970–2000." National Vital Statistics Reports, vol. 51, no 1. Hyattsville, Maryland: National Center for Health Statistics, 2002.

Mathews, T. J., M. F. MacDorman, and M. E. Thoma. "Infant Mortality Statistics from the 2013 Period Linked Birth/Infant Death Data Set." National Vital Statistics Reports 64, 9 (2015). Hyattsville, MD: National Center for Health Statistics.

Matos, K., and E. Galinsky. "Workplace Flexibility in the United States: A Status Report," Table 3: Effects of Offering Flexible Work Options. Families and Work Institute, 2011. http://familiesandwork.org/downloads/WorkplaceFlexibilityinUS.pdf.

Matsumoto, D. "Culture and Nonverbal Behavior." In The Sage Handbook of Nonverbal Communication, edited by Valerie Manusov and Miles Patterson. Thousand Oaks, CA: Sage, 2006.

Max, Wendy, Dorothy P. Rice, Eric Finkelstein, Robert A. Bardwell, Steven Leadbetter. "The Economic Toll of Intimate Partner Violence against Women in the United States." Violence and Victims 19, 3 (2004): 259–272.

Maza, Carlos. " Five Things News Outlets Should Know about the Newest Same-Sex Parenting Study." Media Matters for America, June 12, 2012. http://mediamatters.org /blog/2012/06/12/five-things-news-outlets-should-know-about-the/168813.

McAdoo, Harriette Pipes. "Changes in the Formation and Structure of Black Families: The Impact on Black Women." Working paper no. 182, Center for Research on Women, Wellesley College, Wellesley, MA, 1988.

McAdoo, H. P. Black Families, 3rd ed. Thousand Oaks, CA: Sage, 1996.

McAdoo, H. P. "African American Families: Strength and Realities." In Resiliency in Ethnic Minority Families, Vol. 2: African American Families, edited by H. I. McCubbin, E. A. Thompson, A. I. Thompson, and J. A. Futrell (pp. 17–30). Thousand Oaks, CA: Sage, 1998.

Macartney, S., A. Bishaw, and K. Fontenot. "Poverty Rates for Selected Detailed Race and Hispanic Groups by State and Place: 2007–2011." U.S. Census Bureau, 2013.

McCabe, Janice. "What's in a Label? The Relationship between Feminist Self Identification and 'Feminist' Attitudes among U.S. Women and Men." Gender & Society 19, 4 (2005): 480–505.

McCabe, K. "Chinese Immigrants in the United States." Migration Information Source (ISSN No. 1946-40370). Washington, DC: Migration Policy Institute, 2012.

McCaghy, Charles, Timothy Capron, and J. D. Jamieson. Deviant Behavior: Crime, Conflict, and Interest Groups, 5th ed. Needham Heights, MA: Allyn and Bacon, 2000.

McCarthy, Justin. "Gallup Social Issues: Record-High 60% of Americans Support Same-Sex Marriage." Gallup, May 19, 2015. www.gallup.com/poll/183272/record-high-americans-support-sex-marriage.aspx.

McClennen, J. C., A. B. Summers, and J. G. Daley. "The Lesbian Partner Abuse Scale." Research on Social Work Practice 12 (2002): 277–292.

McCracken, Allison. "Domestic Reality TV." FlowTV, January 21, 2005. http://flowtv.org/2005/01/domestic-reality-tv/.

McDermott, Rose, James H. Fowler, and Nicholas A. Christakis. "Breaking Up is Hard to Do, Unless Everyone Else is Doing It Too: Social Network Effects on Divorce in a Longitudinal Sample Followed for 32 Years" (October 18, 2009). http://ssrn.com/abstract=1490708.

———. "Breaking Up is Hard to Do, Unless Everyone Else Is Doing it Too: Social Network Effects on Divorce in a Longitudinal Sample." Social Forces; a Scientific Medium of Social Study and Interpretation 92, 2 (2013), 491–519.

McGoldrick, Monica, and B. Carter, eds. The Changing Family Life Cycle, 2nd ed. Boston: Allyn and Bacon, 1989.

McHugh, Maureen, Nicole Livingston, and Amy Ford. "A Postmodern Approach to Women's Use of Violence: Developing Multiple and Complex Conceptualizations." Psychology of Women Quarterly 29, 4 (December 2005): 323–336.

McKeever, M., and N. Wolfinger. "Reexamining the Economic Costs of Marital Disruption for Women." Social Science Quarterly 82, 1 (2001): 202–217.

McLanahan, Sara, and Karen Booth. "Mother-Only Families: Problems, Prospects, and Politics." Journal of Marriage and the Family 51 (1989): 557–580.

McLeer, S. V., et al. "Sexually Abused Children at High Risk for Post-Traumatic Stress Disorder." Journal of the American Academy of Child and Adolescent Psychiatry 31, 5 (September 1992): 875–879.

McLoyd, Vonnie, Ana Marie Cauce, David Takeuchi, and Leon Wilson. "Marital Processes and Parental Socialization in Families of Color: A Decade Review of Research." Journal of Marriage and the Family 62, 4 (November 2000): 1070–1093.

McLoyd, Vonnie, and Julia Smith. "Physical Discipline and Behavior Problems in African American, European American, and Hispanic Children: Emotional Support as a Moderator," Journal of Marriage and the Family 64, 1 (February 2002): 40–53.

McManus, Patricia, and Thomas DiPrete. "Losers and Winners: The Financial Consequences of Separation and Divorce for Men." American Sociological Review 66 (April 2001): 246–268.

McMenamin, T. "A Time to Work: Recent Trends in Shift Work and Flexible Schedules." Monthly Labor Review 130, 12 (December 2007): 3–15.

McNamara, Mary. "Review, ABC's 'black-ish' Gamely Takes on Racial Identity." L.A. Times, 2014. www.latimes.com/entertainment/tv/la-et-st-blackish-review-20140924-column.html.

Mead, Margaret. Male and Female. New York: Morrow, 1975.

Mederer, H., and R. J. Gelles. "Compassion or Control: Intervention in Cases of Wife Abuse." Journal of Interpersonal Violence 4, 1 (March 1989): 25–43.

Medina, Jennifer. "Sex Ed Lesson: 'Yes Means Yes,' but It's Tricky." New York Times, October 14, 2015. www.nytimes.com/2015/10/15/us/california-high-schools-sexual-consent-classes.html?_r=0.

Medora, Nilufer P., Jeffry H. Larson, Nuran Hortaçsu, and Parul Dave. "Perceived Attitudes towards Romanticism: A Cross-Cultural Study of American, Asian-Indian, and Turkish Young Adults." Journal of Comparative Family Studies 33, 2 (Spring 2002): 155–178.

Mertensmeyer, C., and Coleman, M. "Correlates of Inter-Role Conflict in Young Rural and Urban Parents." Family Relations: An Interdisciplinary Journal of Applied Family Studies 36, 4 (October 1987): 425–429.

Messner, Michael. Politics of Masculinities: Men in Movements. Lanham, Maryland: AltaMira, 1997.

Metts, Sandra, and William Cupach. "The Role of Communication in Human Sexuality." In Human Sexuality: The Social and Interpersonal Context, edited by K. McKinney and S. Sprecher. Norwood, NJ: Ablex, 1989.

Metz, Michael E., and Norman Epstein. "Assessing the Role of Relationship Conflict in Sexual Dysfunction." Journal of Sex and Marital Therapy 28, 2 (March 2002): 139–164.

Mezey, Nancy. LGBT Families. Thousand Oaks, CA: Sage Publications, 2015.

Michael, Robert, John Gagnon, Edward Laumann, and Gina Kolata. Sex in America: The Definitive Survey. Boston: Little, Brown, 1994.

Migliaccio, Todd. "Abused Husbands: A Narrative Analysis." Journal of Family Issues 23, 1 (January 2002): 26–52.

Mikelson, K. S. "He Said, She Said: Comparing Mother and Father Reports of Father Involvement." *Journal of Marriage and Family* 70, 3 (2008): 613–624.

Milan, A., and A. Peters. "Couples Living Apart." *Canadian Social Trends* (Summer 2003). Statistics Canada, Catalogue 11–008: Ottawa, Ontario.

Milkie, Melissa, Alex Bierman, and Scott Schieman. "How Adult Children Influence Older Parents' Mental Health: Integrating Stress-Process and Life-Course Perspectives." *Social Psychology Quarterly* 71, 1 (2008): 86–105.

Milkie, Melissa, Sarah Kendig, Kei Nomaguchi, and Kathleen Denny. "Time with Children, Children's Well-Being, and Work-Family Balance among Employed Parents." *Journal of Marriage and Family,* 272 (2010): 1329–1343.

Miller, Glenn. "The Psychological Best Interests of the Child." *Journal of Divorce and Remarriage* 19, 1–2 (1993): 21–36.

Miller, R., C. Hollist, J. Olsen, and D. Law. "Marital Quality and Health Over 20 Years: A Growth Curve Analysis." *Journal of Marriage and the Family* 75 (June 2013): 667–680.

Miller, Kim, Martin L. Levin, Daniel J. Whitaker, and Xiaohe Xu. "Patterns of Condom Use among Adolescents: The Impact of Mother–Adolescent Communication." *American Journal of Public Health* 88, 10 (October 1998).

Miller, Alan, and Rodney Stark. "Gender and Religiousness: Can Socialization Explanations Be Saved?" *American Journal of Sociology* 107, 6 (2002):1399–1423.

Miller, T. *Making Sense of Fatherhood: Gender, Caring and Work.* Cambridge: Cambridge University Press, 2011.

Miller-Ott, Aimee, Lynne Kelly, and Robert Duran. "The Effects of Cell Phone Usage Rules on Satisfaction in Romantic Relationships." *Communication Quarterly* 60, 1 (2012): 17–34. doi:10.1080/0146337 3.2012.642263.

Mills, C. Wright. *The Sociological Imagination.* New York, Oxford University Press, 1959.

Mills, David. "A Model for Stepparent Development." *Family Relations* 33 (1984): 365–372.

Min, Pyong Gap. "Changes in Korean Immigrants' Gender Role and Social Status, and Their Marital Conflicts." *Sociological Forum* 16, 2 (2001): 301–320.

Mincy, R. B., and E. J. Sorensen. "Deadbeats and Turnips in Child Support Reform." *Journal of Policy Analysis and Management* 17, 1 (1998): 44–51.

Mintz, Steven. *Huck's Raft: A History of American Childhood.* Cambridge, MA: Belknap Press of Harvard University, 2004.

Mintz, Steven, and Susan Kellogg. *Domestic Revolutions: A Social History of American Family Life.* New York: Free Press, 1988.

Minuchin, Salvador. *Family Therapy Techniques.* Cambridge, MA: Harvard University Press, 1981.

Miracle, Tina, A. Miracle, and R. Baumeister, *Human Sexuality: Meeting Your Basic Needs.* Upper Saddle River, NJ: Prentice Hall, 2003.

Mirowsky, John. "Parenthood and Health: The Pivotal and Optimal Age at First Birth." *Social Forces* 81, 1 (2002): 315.

Mishori, Ranit. "Feeling Her Pain: Intensity of Childbirth Experience Leaves Some Men Feeling Ill-Prepared." *Washington Post,* July 4, 2006.

Mitchell, V. "Lesbian Family Life, Like the Fingers of a Hand: Under-Discussed and Controversial Topics." *Journal of Lesbian Studies* 12, 2/3 (2008): 119–125.

Mitchell, Barbara. "Happiness in Midlife Parental Roles: A Mixed Methods Analysis." *Journal of Family Relations* 59, 3 (2010): 326–339.

Mollborn, S. "Making the Best of a Bad Situation: Material Resources and Teenage Parenthood." *Journal of Marriage and Family* 69, 1 (2007): 92–104.

Mongeau, P. A., K. Knight, J. Williams, J. Eden, and C. Shaw. "Identifying and Explicating Variation among Friends with Benefits Relationships." *Journal of Sex Research* 50, 1 (2013): 37–47.

Monte, Lindsay M., and Renee R. Ellis. "Fertility of Women in the United States: 2012." Population Characteristics, P20-575, U.S. Census Bureau, Washington, DC, 2014. www.census.gov/content/dam /Census/library/publications/2014/demo/p20-575.pdf.

Moore-Hirschl, S., Parra, L., Weis, D. L., and Laflin, M. T. "Attitudes of College Females toward Marital Exclusivity over a Nine-Year Period." *Journal of Psychology & Human Sexuality* 7, 3 (1995): 61–75.

Morabito, Andrea. "All Three *Married at First Sight* Couples Are Splitting Up." *New York Post,* June 16, 2015. nypost.com/2015/06/16 /all-three-married-at-first-sight-couples-are-divorcing/.

Morin, Ashley. "Use It or Lose It: The Enforcement of Polygamy Laws in America." *Rutgers Law Review,* 497 (Winter 2014): 497–530. www .rutgerslawreview.com/wp-content/uploads/2015/02/MORIN-Final -Macro.pdf.

Morin, Rich. "The Public Renders a Split Verdict on Changes in Family Structure." Pew Social and Demographic Trends, Pew Research Center, 2011. www.pewsocialtrends.org/2011/02/16/ .

———. "New Academic Study Links Rising Income Inequality to 'Assortative Mating.'" Pew Research Center, 2014. www.pewresearch .org/fact-tank/2014/01/29/new-academic-study-links-rising-income -inequality-to-assortive-mating/.

Morin, R., and D. Cohn. "Women Call the Shots at Home: Public Mixed on Gender Roles in Jobs." Pew Research Center, 2008. http:// pewresearch.org/pubs/967/gender-power.

Morrissey, Taryn. "Familial Factors Associated with the Use of Multiple Child-Care Arrangements." *Journal of Marriage and Family* 70, 2 (May 2008): 549–563.

Morrison, Donna R., and Andres J. Cherlin. "The Divorce Process and Young Children's Well-Being: A Prospective Analysis." *Journal of Marriage and the Family* 57, 3 (August 1995): 800–812.

Morse, Katherine A., and Steven L. Neuberg. "How Do Holidays Influence Relationship Processes and Outcomes? Examining the Instigating and Catalytic Effects of Valentine's Day." *Personal Relationships* 11, 4 (December 2004): 509–527.

Morsy, Leila, and Richard Rothstein. "Five Social Disadvantages That Depress Student Performance." Washington, DC Economic Policy Institute, 2015a.

———. "Parents' Non-Standard Work Schedules Make Adequate Childrearing Difficult: Reforming Labor Market Practices Can Improve Children's Cognitive and Behavioral Outcomes." Washington, DC, 2015b.

Mosher, W. D., A. Chandra, and J. Jones. "Sexual Behavior and Selected Health Measures: Men and Women 15–44 Years of Age, United States, 2002." *Vital and Health Statistics,* no. 362. Hyattsville, MD: National Center for Health Statistics, 2005.

Mosher William, Jo Jones, and Joyce Abma. "Intended and Unintended Births in the United States: 1982–2010." National Health Statistics Reports, no 55. Hyattsville, MD: National Center for Health Statistics, 2012.

Moss, Hilary. "Fox News Goes After J. Crew's Jenna Lyons For Painting Son's Toenails Pink." Huffington Post, April 12, 2011. www .huffingtonpost.com/2011/04/12/fox-news-jcrew-jenna-lyons-_n _848152.html.

Motel, Seth, and Eileen Patten. "The 10 Largest Hispanic Origin Groups: Characteristics, Rankings, Top Counties." Washington, DC: Pew Hispanic Center, 2012.

Moultrup, D. J. *Husbands, Wives and Lovers: The Emotional System of the Extramarital Affair.* New York: Guilford, 1990.

Moynihan, Daniel Patrick. *The Negro Family: The Case for National Action.* Washington, DC: U.S. Government Printing Office, 1965.

Muehlenhard, Charlene. "Misinterpreted Dating Behaviors and the Risk of Date Rape." *Journal of Social and Clinical Psychology* 9, 1 (1988): 20–37.

Muehlenhard, Charlene L., and M. Linton. "Date Rape and Sexual Aggression in Dating Situations." *Journal of Consulting Psychology* 34 (April 1987): 186–196.

Mullen, Paul E. "The Crime of Passion and the Changing Cultural Construction of Jealousy." *Criminal Behavior and Mental Health* 3, 1 (1993): 1–11.

Mundy, Liza. *The Richer Sex: How the New Majority of Female Breadwinners Is Transforming Sex, Love and Family.* New York: Simon and Schuster, 2012.

Munguia, Hayley. "What Makes Someone Identify as Multiracial?" FiveThirtyEight, 2015. fivethirtyeight.com/datalab/ what-makes-someone-identify-as-multiracial.

Munsch, Cristin. "Her Support, His Support: Money, Masculinity, and Marital Infidelity." *American Sociological Review* 80, 3 (2015): 469–495.

Munson, M. L., and P. D. Sutton. "Births, Marriages, Divorces, and Deaths: Provisional Data for 2004." *National Vital Statistics Reports,* vol. 53, no. 21. Hyattsville, MD: National Center for Health Statistics, 2005.

Murphey, David, Tawana Bandy, Kristin Anderson Moore, and Mae Cooper. "Do Parents Feel More Aggravated These Days? *Child Trends* (#2014-14), 2014. childtrends.org.

Murphy, Caryle "Fact Tank—Our Lives in Numbers: Interfaith Marriage Is Common in U.S., Particularly among the Recently Wed." Pew Research Center, Fact Tank, June 2, 2015. www.pewresearch.org /fact-tank/2015/06/02/interfaith-marriage.

Murphy-Berman, Virginia, Helen Levesque, and John Berman. "U.N. Convention on the Rights of the Child." *American Psychologist* 51, 12 (December 1996): 1257–1262.

Murstein, Bernard. *Paths to Marriage: Family Studies Text Series*, vol. 5. Beverly Hills, CA: Sage, 1986.

Musick, Kelly, and Larry Bumpass. "Re-Examining the Case for Marriage: Union Formation and Changes in Well-Being." *Journal of Marriage and Family* 74, 1 (2012): 1–18.

Myers, Jane, and Matthew Shurts. "Measuring Positive Emotionality: A Review of Instruments Assessing Love." *Measurement and Evaluation in Counseling and Development* 34 (January 2002): 238–254.

Myers, Scott. "Religious Homogamy and Marital Quality: Historical and Generational Patterns, 1980–1997." *Journal of Marriage and Family* 68, 2 (May 2006): 292–304.

Nadal, K. L., J. Sriken, K. C. Davidoff, Y. Wong, and K. McLean. "Microaggressions within Families: Experiences of Multiracial People." *Family Relations* 62 (2013): 190–201.

Nanda, Serena. *Neither Man nor Woman: The Hijras of India*. Belmont. CA: Wadsworth, 1990.

Nandi, Jacinta. "Germany Got It Right by Offering a Third Gender Option on Birth Certificates." *The Guardian*, November 10, 2013. www.theguardian.com/commentisfree/2013/nov/10/germany-third -gender-birth-certificate.

National Coalition Against Domestic Violence. "NCADV Fact Sheet, Abuse in Later Life," 2008a. Denver, CO. www.ncadv.org.

———. "NCADV Fact Sheet, International Violence against Women," 2008b.

———. "NCADV Fact Sheet: Reproductive Health and Pregnancy," 2008c.

———. "Facts about Dating Abuse and Teen Violence," 2015. www .ncadv.org.

National Conference of State Legislatures. Common Law Marriage. ncsl. org. National Domestic Violence Hotline. "Increased Financial Stress Affects Domestic Violence Victims," 2009. www.ndvh.org/2009/01 /increased-financial-stress-affects-domestic-violence-victims-2.

National Healthy Marriage Resource Center. "Covenant Marriage: A Fact Sheet." NHMRC Marriage Facts and Research, February 2010. www .healthymarriageinfo.org/research-and-policy/marriage-facts/index .aspx.

National Institute of Justice and the Centers for Disease Control. "Extent, Nature and Consequences of Intimate Partner Violence: Findings from the National Violence Against Women Survey," 2000.

National Marriage Project. "Social Indicators of Marital Health and Well-Being: Trends of the Past Five Decades: Your Chances of Divorce May be Much Lower than You Think." From The State of Our Unions: Marriage in America, pp. 73–74. National Marriage Project and Institute for American Values, 2011.

National Parents Organization. "Shared Parenting: A Solution to the Nation's Divorce Rate," October 20, 2015. www .nationalparentsorganization.org/2015-07-16-13-55-58/news -releases/22637-shared-parenting-a-solution-to-the-nation-s -divorce-rate.

National Survey of Sexual Health and Behavior, 2010. www .nationalsexstudy.indiana.edu/.

Nauert, Rick. "Boys May Find Talking about Problems a 'Waste of Time.'" PsychCentral, August 23, 2011. Psychcentral.com/news/2011/08/23 /males-have-a-problem-discussing-problems/28844.html.

Nelson, Margaret. "Caring for Anna." *Contexts* 5, 4 (November 2006).

Nelson, S. Katherine, Kostadin Kushlev, and Sonja Lyubomirsky. "The Pain and Pleasures of Parenting: When, Why, and How Is Parenthood Associated with More or Less Well-Being?" *Psychological Bulletin* 140, 3 (2014): 846–895.

Neuman, W. L. *Basics of Social Research: Qualitative and Quantitative Approaches*, 2nd ed. Boston: Allyn and Bacon, 2006.

Newcomer, Susan, and Richard Udry. "Oral Sex in an Adolescent Population." *Archives of Sexual Behavior* 14, 1 (February 1985): 41–46.

Newman, Katherine. *Falling from Grace*. New York: Free Press, 1988.

———. *The Accordion Family: Boomerang Kids, Anxious Parents, and the Private Toll of Global Competition*. Boston: Beacon Press, 2012.

Newman, W. Lawrence. *Basics of Social Research: Quantitative and Qualitative Approaches*. Boston, MA: Pearson, 2004.

Newport, Frank. "In U.S., 87% Approve of Black-White Marriage, vs. 4% in 1958." Gallup, July 25, 2013. www.gallup.com/poll/163697 /approve-marriage-blacks-whites.aspx.

Newport, Frank, and Joy Wilke 2013. "Most in U.S. Want Marriage, but Its Importance Has Dropped." Gallup, August 2, 2013. www.gallup .com/poll/163802/marriage-importance-dropped.aspx.

Newton, Paula. "Boy or a Girl? It's a Secret—and an International Controversy." CNN, May 27, 2011. www.cnn.com/2011/WORLD /americas/05/27/canada.gender.storm/.

NICHD Early Child Care Research Network, The Eunice Kennedy Shriver National Institute of Child Health and Human Development, 2000.

Nicholson, Jan M., David Fergusson, and L. John Horwood. "Effects on Later Adjustment of Living in a Stepfamily During Childhood and Adolescence." *Journal of Child Psychology* 40, 3 (1999): 405–416.

Nielsen Company, "Nielsen Estimates 116.3 Million TV Homes in the U.S., Up 0.4%." Media and Entertainment, 2014 (May). www.nielsen .com/us/en/insights/news/2014/nielsen-estimates-116-3-million-tv -homes-in-the-us.html.

Nielsen Company. *Total Audience Report 2015*. www.nielsen.com/content /dam/corporate/us/en/reports-downloads/2015-reports/total- audience-report-q1-2015.pdf.

Nock, S., L. Sanchez, and J. Wright. *Covenant Marriage: The Movement to Reclaim Tradition in America*. New Brunswick, NJ: Rutgers University Press, 2008.

Noller, Patricia, and Mary Anne Fitzpatrick. "Marital Communication." In *Contemporary Families: Looking Forward, Looking Back*, edited by A. Booth. Minneapolis: National Council on Family Relations, 1991.

Noller, Patricia. "Couple Communication: Conflict Styles." NCFR Reports, National Council on Family Relations, Family Focus on Effective Communication, Issue FF64 (Summer 2015): F3–F5.

Nomaguchi, K. M., and M. A. Milkie. "Costs and Rewards of Children: The Effects of Becoming a Parent on Adults' Lives." *Journal of Marriage and Family* 65 (2003): 356–374.

Nomaguchi, Kei. "Change in Work-Family Conflict among Employed Parents between 1977 and 1997." *Journal of Marriage and Family* 71, 1 (February 2009): 15–32.

Nomaguchi, Kei, Melissa Milkie, and Suzanne Bianchi. "Time Strains and Psychological Well-Being: Do Dual-Earner Mothers and Fathers Differ?" *Journal of Family Issues* (September 2005): 756–791.

Noonan, Mary. "The Impact of Domestic Work on Men's and Women's Wages." *Journal of Marriage and Family* 63 (November 2001) 1134–1145.

Noonan, Mary, and Jennifer Glass. "The Hard Truth about Telecommuting." *Monthly Labor Review*, June 2012.

Norris, Tina, Paula Vines, and Elizabeth Hoeffel. "The American Indian and Alaska Native Population: 2010 Census Briefs," 2012. www .census.gov/prod/cen2010/briefs/c2010br-10.pdf.

Norton, Arthur J. "Family Life Cycle: 1980." *Journal of Marriage and the Family* 45 (1983): 267–275.

Notman, Malkah T., and Eva P. Lester. "Pregnancy: Theoretical Considerations." *Psychoanalytic Inquiry* 8, 2 (1988): 139–159.

Nowak, Nicole, Glenn E. Weisfeld, Olcay Imamoğlu, Carol C. Weisfeld, Marina Butovskaya, and Jiliang Shen. "Attractiveness and Spousal Infidelity as Predictors of Sexual Fulfillment without the Marriage Partner in Couples from Five Cultures." *Human Ethology Bulletin* 29, 1 (2014): 18–38.

NPR. "Is '16 and Pregnant' An Effective Form of Birth Control?" Television, January 13, 2014. www.npr.org/2014/01/13/262175399 /is-16-and-pregnant-an-effective-form-of-birth-control.

Oaklander, Mandy. "This Divorce Arrangement Stresses Kids Out Most." *Time*, April 27, 2015.

Oakley, Ann, ed. *Sex, Gender, and Society*, rev. ed. New York: Harper and Row, 1985.

O'Connor, Anahad. "When the Bully is a Sibling," *New York Times*, June 17, 2013.

Office of Child Support Enforcement. "2010 Preliminary Report." Department of Health and Human Services, Administration for Children and Families, Office of Child Support Enforcement, May 2011. www.acf.hhs.gov/programs/cse/pubs/2011/reports /preliminary_report_fy2010/.

Oldham, J. T. "What If the Beckhams Move to L.A. and Divorce? Marital Property Rights of Mobile Spouses When They Divorce in the United States." *Family Law Quarterly* 42 (2008/2009): 263.

Oliphant, R., and N. ver Steegh. *Family Law: Examples and Explanations*, 2nd ed. New York: Aspen Publishers, 2007.

Olson, M. R., and J. A. Haynes. "Successful Single Parents." *Families in Society: Journal of Contemporary Human Services* (1993): 259–267.

Ono, Hiromi. "Marital History Homogamy between the Divorced and the Never Married among non-Hispanic Whites." *Social Science Research* 34, 2 (2005): 333.

Opie, Anne. "Ideologies of Joint Custody." *Family and Conciliation Courts Review* 31, 3 (1993): 313–326.

Orbuch, T., F. Bauermeister, W. Brown, and McKinley. "Early Family Ties and Marital Stability over 16 years: The Context of Race and Gender." *Family Relations* 62 (April 2013): 255–268.

Organization for Economic Co-operation and Development (OECD). Families and Children, OECD Family Database, OECD, Paris, 2011.

Oropesa, R. S., and Nancy S. Landale. "The Future of Marriage and Hispanics." *Journal of Marriage and Family* 66, 4 (November 2004): 901–920.

Over, Ray, and Gabriel Phillips. "Differences between Men and Women in Age Preferences for a Same-Sex Partner." *Behavioral and Brain Sciences* 20 (1997): 137–143.

O'Leary, K. Daniel. "Through a Psychological Lens: Personality Traits, Personality Disorders, and Levels of Violence." In *Current Controversies in Family Violence*, edited by R. Gelles and D. Loseke. Newbury Park, CA: Sage, 1993.

O'Neil, Robin, and Ellen Greenberger. "Patterns of Commitment to Work and Parenting: Implications for Role Strain." *Journal of Marriage and the Family* 52, 1 (February 1994): 101–115.

Online Dating Magazine. "Online Dating Statistics," 2012. www .onlinedatingmagazine.com/onlinedatingstatistics.html.

O'Sullivan, L., B. Jaramillo, D. Moreau, and H. Meyer Bahlburg. "Mother-Daughter Communication about Sexuality in a Clinical Sample of Hispanic Adolescent Girls." *Hispanic Journal of Behavioral Sciences* 21, 4 (1999): 447–469.

Owen, Pamela. "100,000 Women Undergo Brutal Genital Mutilation Illegally in Britain." MailOnline, April 22, 2012. www.dailymail. co.uk/news/article-2133427.

Owen, Margaret Tresch. "Working Parents," *Encyclopedia of Clinical Child and Pediatric Psychology*, edited by in Thomas Ollendick and Carolyn Schroeder. New York: Plenum, 2003.

Owen, J., and F. Fincham. "Effects of Gender and Psychosocial Factors on 'Friends with Benefits' Relationships among Young Adults." *Archives of Sexual Behavior* 40, 2 (2011): 311–320.

Pace, Garret, Kevin Shafer, Todd Jensen, and Jeffry Larson. "Stepparenting Issues and Relationship Quality: The Role of Clear Communication." *Journal of Social Work* 15, 1 (2015): 24–44.

Palkovitz, Robert. "Reconstructing 'Involvement': Expanding Conceptualizations of Men's Caring in Contemporary Families." In *Generative Fathering: Beyond Deficit Perspectives*, Vol. 3, *Current Issues in the Family*, edited by A. Hawkins and D. Dollahite (pp. 200–216). Thousand Oaks, CA: Sage, 1997.

Paludi, M. A. "Sociopsychological and Structural Factors Related to Women's Vocational Development." *Annals of New York Academy of Sciences*, 602 (1990): 157–168.

Paolisso, Michael, and Raymond Hames. "Time Diary versus Instantaneous Sampling: A Comparison of Two Behavioral Research Methods." *Field Methods* 22, 4 (2010): 357–377.

Papanek, Hannah. "Men, Women, and Work: Reflections on the Two-Person Career." *American Journal of Sociology* 78, 4 (January 1973): 90–110.

Papernow, Patricia L. *Becoming a Stepfamily*. San Francisco: Jossey-Bass, 1993.

Papp, Lauren, M., E. M. Cummings, and M. C. Goeke-Morey. "For Richer, for Poorer: Money as a Topic of Marital Conflict in the Home." *Family Relations* 58, 1 (2009): 91–103.

Papp, Lauren, Marcie Goeke Morey, and E. Mark Cummings. "Let's Talk about Sex: A Diary Investigation of Couples' Intimacy Conflicts in the Home." *Couple Family Psychology* 2, 1 (March 2013).

Papp, Lauren, Chrytyna Kouros, and E. Mark Cummings. "Demand-Withdraw Patterns in Marital Conflict in the Home." *Personal Relationships* 16 (2009): 285–300.

Pardun, C. J., K. L. L'Engle, and J. D. Brown. "Linking Exposure to Outcomes: Early Adolescents' Consumption of Sexual Content in Six Media." *Mass Communication & Society* 8, 2 (2005): 75–91.

Park, Kristin. "Stigma Management among the Voluntarily Childless." *Sociological Perspectives* 45, 1 (Spring 2002): 21–45.

Parker, Kim. "Women, Work and Motherhood: A Sampler of Recent Pew Research Survey Findings." Pew Research Center, Pew Social and Demographic Trends, April 13, 2012a.

Parker, Kim. "The Boomerang Generation: Feeling OK about Living with Mom and Dad." Pew Social and Demographic Trends, Pew Research Center, 2012b.

———. "Working-Mom Guilt? Many Dads Feel It Too." Pew Research Center, April 1, 2015.

Parker, Kim, and Wendy Wang. "Modern Parenthood: Roles of Moms and Dads Converge as They Balance Work and Family." Pew Research Center, March 4, 2013. www.pewsocialtrends.org/files/2013/03 /FINAL_modern_parenthood_03-2013.pdf.

Parker-Pope, Tara. "When Silence Isn't Golden." *New York Times*, October 2, 2007.

Parkin, R. *Kinship: An Introduction to Basic Concepts* (p. 30). Oxford: Blackwell, 1997.

Pascoe, C. J. *Dude You're a Fag: Masculinity and Sexuality in High School*. Berkeley: University of California Press, 2007.

Pasquesoone, Valentine. "7 Countries Giving Transgender People Fundamental Rights the U.S. Still Won't." Mic.com, April 9, 2014. http://mic.com/articles/87149/7-countries-giving-transgender -people-fundamental-rights-the-u-s-still-won-t.

Patterson, Charlotte. "Sexual Orientation and Family Life: A Decade Review." *Journal of Marriage and the Family* 62 (2000): 1052–1069.

Patterson, Charlotte. "Lesbian and Gay Parents and Their Children: Summary of Research Findings." In *Lesbian and Gay Parenting American Psychological Association*, 2005.

Patten, Eileen, and Jens Manuel Krogstad. "Black Child Poverty Rate Holds Steady, Even as Other Groups See Declines." Pew Research Center, 2015. www.pewresearch.org/fact-tank/2015/07/14/black -child-poverty-rate-holds-steady-even-as-other-groups-see-declines/.

Patten, Eileen, and Kim Parker. "A Gender Reversal on Career Aspirations: Young Women Now Top Young Men in Valuing a High Paying Career." Pew Social and Demographic Trends, 2011.

Paul, P. *Parenting, Inc*. New York: Henry Holt, 2008.

Paul, Pamela. "The Un-Divorced." *New York Times*, July 30, 2010.

Paulson, Sharon, and Cheryl Somers. "Students' Perceptions of Parent -Adolescent Closeness and Communication about Sexuality: Relations with Sexual Knowledge, Attitudes, and Behaviors." *Journal of Adolescence* 23, 5 (October 2000): 629–644.

Payne, Krista. "The Divorce Rate in the U.S.: 2013" (FP-14-17). National Center for Family and Marriage Research, 2014a. www.bgsu.edu /content/dam/BGSU/college-of-arts-and-sciences/NCFMR /documents/FP/FP-14-17-divorce-rate-2013.pdf.

———. "Marriage Rate in the U.S., 2013" (FP-14-15). National Center for Family and Marriage Research, 2014b. www.bgsu.edu/content /dam/BGSU/college-of-arts-and-sciences/NCFMR/documents/FP/FP -14-15-marriage-rate-2013.pdf.

———. "The Remarriage Rate: Geographic Variation, 2013." (FP-15-08) National Center for Family and Marriage Research, 2015. www .bgsu.edu/ncfmr/resources/data/family-profiles/payne-remarriage -rate-fp-15-08.html.

Payne, Krista, Wendy Manning, and Susan Brown. "Unmarried Births to Cohabiting and Single Mothers, 2005–2010" (FP-12-06). National Center for Family and Marriage Research, 2012. www.bgsu.edu /content/dam/BGSU/college-of-arts-and-sciences/NCFMR /documents/FP/FP-12-06.pdf.

Payne, K. K., and J. Copp. "Young Adults in the Parental Home and the Great Recession." National Center for Family and Marriage Research (FP-13-07), 2013. www.bgsu.edu/content/dam/BGSU/college-of-arts -and-sciences/NCFMR/documents/FP/FP-13-07.pdf.

Peltola, P., M. Milkie, and S. Presser. "The 'Feminist' Mystique Feminist Identity in Three Generations of Women." *Gender and Society* 18, 1 (February 2004): 122–144.

Peoples, J., and G. Bailey. *Humanity: An Introduction to Cultural Anthropology*, 7th ed. Belmont, CA: Thomson Wadsworth, 2006.

Peoples, James, and Garrick Bailey. *Humanity: An Introduction to Cultural Anthropology*, 10th ed. Belmont, CA: Thomson-Wadsworth, 2014.

Peplau, L. A. "Rethinking Women's Sexual Orientation: An Interdisciplinary,Relationship-Focused Approach." *Personal Relationships* 8 (2001): 1–19.

Peplau, L. A., and A. W. Fingerhut. "The Close Relationships of Lesbians and Gay Men." *Annual Review of Psychology* 58 (2007): 405–424.

Peplau, Letitia Anne, Rosemary Veniegas, and Susan Miller Campbell. "Gay and Lesbian Relationships," (pp. 200–215). In *Sexualities: Identities, Behaviors, and Society*, edited by Michael Kimmel, Rebecca Plante. New York: Oxford University, 2004.

Peplau, Letitia, and Susan Cochran. "Value Orientations in the Intimate Relationships of Gay Men." *Gay Relationships*, edited by J. DeCecco. New York: Haworth Press, 1988.

Peralta, Robert, and Meghan Novisky. "When Women Tell: Intimate Partner Violence and the Factors Related to Police Notification." *Violence Against Women* 21, 1 (January 2015): 65–86.

Perez-Fuentes, Gabriela, Mark Olfson, Laura Villegas, Carmen Morcillo, Shuai Wang, and Carlos Blanco. "Prevalence and Correlates of Child Sexual Abuse: A National Study." *Comprehensive Psychiatry* 54, 1 (2013): 16–27.

Perry-Jenkins, Maureen, and Karen Folk. "Class, Couples, and Conflict: Effects of the Division of Labor on Assessments of Marriage in

Dual-Earner Marriages." *Journal of Marriage and the Family* 56, 1 (February 1994): 165–180.

Perry-Jenkins, Maureen, Rena Repetti, and Ann Crouter. "Work and Family in the 1990s." *Journal of Marriage and the Family*, 62, 4 (November 2000): 981–998.

Peterson, Richard. "Employment, Unemployment, and Rates of Intimate Partner Violence: Evidence from the National Crime Victim Surveys." In *Economic Crisis and Crime* (Sociology of Crime Law and Deviance, Volume 16), edited by Mathieu Deflem (pp. 171–193). Emerald Group Publishing Limited, 2011.

Pettigrew, J. "Text Messaging and Connectedness Within Close Interpersonal Relationships." *Marriage & Family Review* 45, 6 (2009): 697–716.

Petts, R., and C. Knoester, "Parents' Religious Heterogamy and Children's Well-Being." *Journal for the Scientific Study of Religion* 46, 3 (2007): 373–389.

Pew Research Center. "A Social and Demographic Trends Report. As Marriage and Parenthood Drift Apart, Public Is Concerned about Social Impact: Generation Gap in Values, Behavior." Washington, DC: Pew Research Center, 2007a. pewresearch.org/pubs/526 /marriage-parenthood.

———. "Modern Marriage: 'I Like Hugs. I Like Kisses. But What I Really Love is Help with the Dishes.'" http://pewresearch .org/pubs/542 /modern-marriage, July 18, 2007b.

———. "The Harried Life of the Working Mother." Pew Social and Demographic Trends, October 2009. www.pewsocialtrends. org/2009/10/01/the-harried-life-of-the-working-mother/.

———. "The Decline of Marriage and Rise of New Families," Pew Social and Demographic Trends, 2010. www.pewsocialtrends.org /2010/11/18/the-decline-of-marriage-and-rise-of-new-families /3/#iii-marriage.

———. "A Portrait of Stepfamilies." Pew Social and Demographic Trends, January 13, 2011.

———. "Social and Demographic Trends Project." The Rise of Intermarriage, 2012. www.pewsocialtrends.org/2012/02/16 /the-rise-of-intermarriage/.

———. "Young, Underemployed, and Optimistic: Coming of Age Slowly in a Tough Economy." *Social and Demographic Trends Report*, 2012. www.pewsocialtrends.org/files/2012/02/young -underemployed-and-optimistic.pdf.

———. "Modern Parenthood." *Social and Demographic Trends*, 2013a.

———. "The Rise of Asian Americans." *Social Trends*, 2013b. www .pewsocialtrends.org/files/2013/04/Asian-Americans-new-full -report-04-2013.pdf.

———. "Couples, the Internet, and Social Media." Pew, February 2014. pewinternet.org/Reports/2014/Couples-and-the-internet.aspx.

———. "America's Changing Religious Landscape," May 12, 2015a.

———. "Childlessness Falls, Family Size Grows among Highly Educated Women." Washington DC, May 2015b.

———. "Family Support in Graying Societies: How Americans, Germans and Italians Are Coping with an Aging Population," May 2015c. www.pewsocialtrends.org/files/2015/05/2015-05-21_family -support-relations_FINAL.pdf.

———. "Raising Kids and Running a Household: How Working Parents Share the Load," November 2015d. www.pewsocialtrends. org/2015/11/04/raising-kids-and-running-a-household-how-working -parents-share-the-load/st_2015-11-04_working-parents-15/.

———. "Support for Same-Sex Marriage at Record High, but Key Segments Remain Opposed," June 2015e. www.people-press.org /files/2015/06/6-8-15-Same-sex-marriage-release1.pdf.

Pfeffer, Carla. "'Women's Work'? Women Partners of Transgender Men Doing Housework and Emotion Work." *Journal of Marriage and Family*, 72 (February 2010): 165–183.

Phillips, Julie A., and Megan M. Sweeney. "Premarital Cohabitation and Marital Disruption among White, Black, and Mexican American Women." *Journal of Marriage and Family* 67, 2 (May 2005): 296–314.

Picard, P. *Research Topline: Tech Abuse in Teen Relationships Study*. Northbrook, IL: Teenage Research Unlimited, 2007.

Picker, Leslie, and Alex Sherman. "Grindr Said to Explore Sale of Gay Men's Dating App." *Bloomberg*, May 8, 2015. www.bloomberg.com /news/articles/2015-05-08/grindr-said-to-explore-sale-of-gay-men-s -dating-app.

Pillemer, K., and D. Finkelhor. "The Prevalence of Elder Abuse: A Random Sample Survey." *Violence Against Women* 4, 5 (1998): 559–571.

Pillemer, Karl, and J. Jill Suitor. "Elder Abuse." In *Handbook of Family Violence*, edited by V. B. Van Hasselt et al. New York: Plenum Press, 1988.

Piña-Watson, B., L. G. Castillo, E. Jung, L. Ojeda, and R. Castillo-Reyes. "The Marianismo Beliefs Scale: Validation with Mexican American Adolescent Girls and Boys." *Journal of Latina/o Psychology 2*, 2 (2014): 113–130.

Pistole, M. Carole. "Attachment in Adult Romantic Relationships: Style of Conflict Resolution and Relationship Satisfaction." *Journal of Social and Personal Relationships* 6 (1989): 505–510.

Pistole, M. C., E. M. Clark, and A. L. Tubbs, Jr. "Adult attachment and the investment model." *Journal of Mental Health Counseling* 17 (1995): 199–209.

Pleck, J. H. "Paternal Involvement: Levels, Origins, and Consequences." In *The Role of the Father in Child Development*, 3rd ed., edited by M. E. Lamb (pp. 66–103). New York: Wiley, 1997.

Poehlmann, J., R. Shlafer, E. Maes, and A. Hanneman. "Factors Associated with Young Children's Opportunities for Maintaining Family Relationships during Maternal Incarceration." *Family Relations* 57 (July 2008): 267–280.

Pogarsky, G., T. P. Thornberry, and A. J. Lizotte. "Developmental Outcomes for Children of Young Mothers." *Journal of Marriage and Family* 68 (2006): 332–334.

Poisson, Jayme. Parents keep child's gender secret: Parents believe children can make meaningful decisions on their own. Toronto Star, May 21, 2011. www.thestar.com/life/parent/2011/05/21/parents _keep_childs_gender_secret.html.

Polatnik, M. Rivka. "Too Old for Child Care? Too Young for Self Care? Negotiating After-School Arrangements for Middle School." *Journal of Family Issues* 23, 6 (September 2002): 728–747.

Pollack, D., and S. Mason. "Mandatory Visitation: In the Best Interests of the Child." *Family Court Review* 42, 1 (2004): 74–84.

Pollack, William. *Real Boys: Rescuing Our Sons from the Myths of Boyhood*. New York: Random House, 1998.

Pollard, Michael, and Kathleen Mullan Harris. "Nonmarital Cohabitation, Marriage, and Health among Adolescents and Young Adults" (May 24, 2013). RAND Working Paper Series WR-997.

Popenoe, David. "American Family Decline, 1960–1990: A Review and Appraisal." *Journal of Marriage and the Family* 55 (August 1993): 527–542.

Postpartum Depression: Action Towards Causes and Treatment (PACT) Consortium. "Heterogeneity of Postpartum Depression: A Latent Class Analysis." *The Lancet Psychiatry* 2, 1 (January 2015): 59–67.

Power, C., Rodgers, B., Hope, S. "Heavy Alcohol Consumption and Marital Status: Disentangling the Relationship in a National Study of Young Adults." *Addiction* 94 (1999): 1477–1487.

Powers, Martine. "Legislation Overhauls Bay State Alimony Law. Patrick Expected to Sign Measure Today." *Boston Globe*, September 26, 2011.

Powledge, Tabitha. "Where Are the Missing Females? Do Skewed Sex Ratios in China, Elsewhere Lead to Social Problems?" Genetic Literacy Project, 2015. www.geneticliteracyproject.org/2015/01/27 /where-are-the-missing-females-do-skewed-sex-ratios-in-china -elsewhere-lead-to-social-problems.

Presser, Harriet. "Nonstandard Work Schedules and Marital Instability." *Journal of Marriage and the Family*, 62, 1 (February 2000): 93–110.

Presser, Harriet. *Working in a 24/7 Economy: Challenges for American Families*. New York: Russell Sage Foundation, 2003.

Price, Joseph, and Gordon Dahl. "Using Natural Experiments to Study the Impact of Media on the Family." *Family Relations* 61 (July 2012): 363–373.

Primack, B., E. Douglas, M. Fine, and M. Dalton. "Association of Favorite Music and Sexual Behavior." *American Journal of Preventive Medicine* 36 (2009): 317–323.

Pugh, Allison. *The Tumbleweed Society: Working and Caring in an Age of Insecurity*. New York: Oxford University Press, 2015.

Putnam, F. "Ten-Year Research Update Review: Child Sexual Abuse." *Journal of the American Academy of Child and Adolescent Psychiatry* 42, 3 (March 2003): 269–278.

Pyke, Karen. "Women's Employment as a Gift or Burden? Marital Power across Marriage, Divorce, and Remarriage." *Gender & Society* 8 (1994): 73–91.

Pyke, Karen, and Michele Adams. "What's Age Got To Do with It? Power and Gender in Husband-Older Marriages." *Journal of Family Issues* 31 (2010): 748–777.

Qian, Zhenchao, and Daniel Lichter. "Changing Patterns of Interracial Marriage in a Multiracial Society." *Journal of Marriage and Family* 73, 5 (October 2011): 1065–1084.

Raffaelli, Marcela, and Lenna Ontai. "Gender Socialization in Latino/a Families: Results from Two Retrospective Studies." *Sex Roles: A Journal of Research* 50, 5/6 (March 2004): 287–299.

Rafkin, Louise. "Lessons Learned When Its All Over." *New York Times*, August 9, 2013a. www.nytimes.com/2013/08/11/booming/lessons-learned-when-its-all-over.html.

———. "Disappointments in 'Happily Ever After.'" *New York Times*, September 9, 2013b. www.nytimes.com/2013/09/22/booming/disappointments-in-happily-ever-after.html.

———. "Unraveling in Slow Motion." *New York Times*, November 7, 2014. www.nytimes.com/2014/11/09/fashion/weddings/unraveling-in-slow-motion.html.

Raley, R. K., and M. M. Sweeney. "What Explains Race and Ethnic Variations in Cohabitation, Marriage, Divorce, and Nonmarital Fertility?" California Center for Population Research Online Working Paper Series, University of California, Los Angeles, September 2007.

Rampell, Catherine. "Mancession to He-covery." *New York Times*, July 6, 2011.

Randall, Ashley, and Guy Bodenmann. "The Role of Stress on Close Relationships and Marital Satisfaction." *Clinical Psychology Review* 29 (2009): 105–115.

Raschke, Helen. "Divorce." In *Handbook of Marriage and the Family*, edited by M. Sussman and S. Steinmetz. New York: Plenum Press, 1987.

Rastogi, Sonya, Tallese Johnson, Elizabeth Hoeffel, and Malcom Drewery Jr. "The Black Population 2010," U.S. Census Bureau, 2010 Census Briefs, September 2011.

Ray, Rebecca, Milla Sanes, and John Schmitt. "No-Vacation Nation Revisited." Center for Economic and Policy Research, Washington, DC, May 2013. http://cepr.net/documents/no-vacation-update-2014-04.pdf.

Raymond, Victor. "Another Divide: Bisexuality in U.S. Politics." In *Gay, Lesbian, Bisexual, and Transgender Civil Rights: A Public Policy Agenda for Uniting a Divided America*, edited by Wallace Swan (pp. 139–158). Boca Raton, FL: CDC Press 2015.

Reed, J. "Not Crossing the 'Extra Line': How Cohabitors with Children View Their Unions." *Journal of Marriage and Family* 68 (2006): 1117–1131.

Regan, P. C. "The Role of Sexual Desire and Sexual Activity in Dating Relationships." *Social Behavior and Personality* 28 (2000): 51–60.

Regan, Pamela. *The Mating Game: A Primer on Love, Sex, and Marriage*. Thousand Oaks, CA: Sage, 2003.

Regan, P., E. Kocan, and T. Whitlock. "Ain't Love Grand! A Prototype Analysis of the Concept of Romantic Love." *Journal of Social and Personal Relationships* 15, 3 (1998): 411–420.

Regnerus, Mark. "How Different are the Adult Children of Parents Who Have Same-Sex Relationships? Findings from the New Family Structures Study." *Social Science Research* 41, 4 (July 2012): 752–770.

Rehel, Erin, and Emily Baxter. "Men, Fathers, and Work-Family Balance." Center for American Progress, 2015. www.americanprogress.org/issues/women/report/2015/02/04/105983/men-fathers-and-work-family-balance/.

Rehman, U.S., and A. Holtzworth-Munroe. "A Cross-Cultural Analysis of the Demand–Withdraw Marital Interaction: Observing Couples from a Developing Country." *Journal of Consulting and Clinical Psychology* 74 (2006): 755–766.

Reid, A., and N. Purcell, "Pathways to Feminist Identification." *Sex Roles* 50 11/12 (2004): 759–769.

Reimondos, Anna, Ann Evans, and Edith Gray. "Living Apart Together (LAT) Relationships in Australia." *Family Matters, Australian Institute of Family Studies* 87 (2011): 43–55.

Reiss, Ira. *Family Systems in America*, 3rd ed. New York: Holt, Rinehart, and Winston, 1980.

Rendall, Michael. "Breakup of New Orleans Households after Hurricane Katrina." *Journal of Marriage and Family* 73 (June 2011): 654–668.

Renwick, Trudi. "Child Poverty Down—Income of Families with Children Up." Random Samplings: The Official Blog of the U.S. Census Bureau, September 16, 2014. http://blogs.census.gov/2014/09/16/child-poverty-down-income-of-families-with-children-up.

Renzetti, C., and D. Curran. *Women, Men and Society*, 4th ed. Boston: Allyn and Bacon, 1999.

———. *Women, Men, and Society*, 5th ed. Needham Heights, MA: Allyn and Bacon, 2003.

Renzetti, Claire. "Violence in Gay and Lesbian Relationships." In *Vision 2010: Families and Violence, Abuse and Neglect*, edited by R. J. Gelles. Minneapolis: National Council on *Family Relations*, 1995.

Rhoades, Galena, Scott Stanley, and Howard Markman. "The Pre-Engagement Cohabitation Effect: A Replication and Extension of Previous Findings." *Journal of Family Psychology* 23, 1 (February 2009): 107–111.

Rice, F. Phillip, and Kim Gale Dolgin. *The Adolescent: Development, Relationships and Culture*, 10th ed. Boston: Allyn and Bacon, 2002.

Richards, Leslie N., and Cynthia J. Schmiege. "Problems and Strengths of Single-Parent Families: Implications for Practice and Policy." *Family Relations* 42, 3 (July 1993): 277–285.

Rideout, Victoria, Ulla G. Foehr, and Donald F. Roberts. "GENERATION M2: Media in the Lives of 8- to 18-Year-Olds." Melo Park, CA: The Henry J. Kaiser Family Foundation, 2010.

Ridley, Jane, and Michael Crowe. "The Behavioural-Systems Approach to the Treatment of Couples." *Sexual and Marital Therapy* 7, 2 (1992): 125–140.

Riffkin, Rebecca. "New Record Highs in Moral Acceptability." Gallup Poll, May 30, 2014. www.gallup.com/poll/170789/new-record-highs-moral-acceptability.aspx.

Risman, Barbara. "Can Men Mother? Life as a Single Father." *Family Relations* 35 (1986): 95–102.

———. "Intimate Relationships from a Microstructural Perspective: Men Who Mother." *Gender & Society* 1, 1 (March 1987): 6–32.

———. "Can Men Mother? Life as a Single Father." In *Gender in Intimate Relationships*, edited by B. Risman and P. Schwartz. Belmont, CA: Wadsworth, 1989.

———. *Gender Vertigo: American Families in Transition*. New Haven, CT: Yale University Press, 1998.

Risman, Barbara, and Danette Johnson-Sumerford. "Doing It Fairly: A Study of Post Gender Marriages." *Journal of Marriage and the Family* 60 (February 1998): 23–40.

Rizzo K. M., H. H. Schiffrin, and M. Liss. "Insight into the Parenthood Paradox: Mental Health Outcomes of Intensive Mothering." *Journal of Child and Family Studies*, 2012. doi: 10.1007/s10826-012-9615-z.

Robbins, Cynthia, Vanessa Schick, Michael Reece, Debra Herbenick, Stephanie Sanders, Brian Dodge. "Prevalence, Frequency, and Associations of Masturbation with Partnered Sexual Behaviors among U.S. Adolescents." *Archives of Pediatric Adolescent Medicine* 165, 12 (December 2011): 1087–1093.

Roberts, Michelle. "3,200 Still Missing After Katrina." Associated Press, January 19, 2006. http://staugustine.com/stories/011906/nat_3585867.shtml.

Roberts, Nicole, and Robert Levenson. "The Remains of the Workday: Impact of Job Stress and Exhaustion on Marital Interaction in Police Couples." *Journal of Marriage and Family* 63, 4 (November 2001): 1052–1067.

Roberts, Anabel, and Marian Smith. "Horrific Taboo: Female Circumcision on the Rise in U.S." NBC News, March 30, 2014. www.nbcnews.com/news/world/horrific-taboo-female-circumcision-rise-u-s-n66226.

Robins, Simon. "Ambiguous Loss in a Non-Western Context: Families of the Disappeared in Postconflict Nepal." *Family Relations* 59 (July 2010): 253–268.

Robinson, Paul, Leah J. Davidson, and Michael E. Drebot. "Parent Abuse on the Rise: A Historical Review." American Association of Behavioral Social Science Online Journal. *AABSS Online Journal*, 2004.

Robinson, John. "The Time Diary Method: Structure and Uses. In *Time Use Research in the Social Sciences*, edited by Wendy E. Pentland, Andrew Harvey, M. P. Lawton, and Mary Ann McColl. New York: Kluwer Academic Publishers Group, 1999.

Robinson, John, and Steven Martin. "Changes in American Daily Life: 1965–2005." *Social Indicators Research* 93, 1 (2009): 47–56.

Rodgers, Kathleen Boyce, and Hilary A. Rose. "Risk and Resiliency Factors among Adolescents Who Experience Marital Transitions." *Journal of Marriage and Family* 64, 4 (November 2002): 1024–1037.

Rodgers, Roy, and L. Conrad. "Courtship for Remarriage: Influences on Family Reorganization after Divorce." *Journal of Marriage and the Family* 48 (1986): 767–775.

Rodgers, Roy. *Family Interaction and Transaction: The Developmental Approach*. Englewood Cliffs, NJ: Prentice Hall, 1973.

Roehling, Patricia V., Lorna Hernandez Jarvis, and Heather Swope. "Variations in Negative Work-Family Spillover among White, Black, and Hispanic American Men and Women: Does Ethnicity Matter?" *Journal of Family Issues* 26, 6 (September 2005): 840–865.

Roenrich, L., and B. N. Kinder. "Alcohol Expectancies and Male Sexuality: Review and Implications for Sex Therapy." *Journal of Sex and Marital Therapy* 17 (1991): 45–54.

Rogers, Stacy. "Dollars, Dependency, and Divorce: Four Perspectives on the Role of Wives' Income." *Journal of Marriage and the Family* 66, 1 (February 2004): 59–74.

Rojas, Rick, and Eli Rosenberg. "Mother Is Accused of Throwing Newborn to Death from 7th Floor Window in Bronx." *New York Times*, September 29, 2015.

Rolf, Karen, and Joseph Ferrie. "The May-December Relationship Since 1850: Age Homogamy in the U.S." Paper presented at the Population Association of America Annual Conference, New Orleans, 2008.

Rollins A., and A. G. Hunter. "Racial Socialization of Biracial Youth: Maternal Messages and Approaches to Address Discrimination." *Family Relations* 62, 1 (2013): 140–53.

Romance Writers of America. "Romance Reader Statistics," 2013. www.rwa.org/p/cm/ld/fid=582.

Roper Starch Worldwide. Liz Claiborne Inc. Study of Fortune 1000 Senior Executives, 2002.

Roscoe, Will. *Changing Ones: Third and Fourth Genders in Native North America.* New York: St. Martin's Press, 2000.

Rose, Amanda. "Co-Rumination in the Friendships of Girls and Boys." *Child Development* 73, 6 (December 2002): 1830–1843.

Rose, Amanda, Wendy Carlson, and Erika Waller. "Prospective Associations of Co-Rumination with Friendship and Emotional Adjustment: Considering the Socioemotional Trade-Offs of Co-Rumination." *Developmental Psychology* 43, 4 (2007): 1019–1031.

Rosen, Rebecca. "The Hidden Cost of a Flexible Job." *The Atlantic,* February 2015. www.theatlantic.com/business/archive/2015/02/the-hidden-cost-of-a-flexible-job/385170/.

Rosen, L., N. Cheever, C. Cummings, and J. Felt. "The Impact of Emotionality and Self-Disclosure on Online Dating versus Traditional Dating." *Computers in Human Behavior* 24 (2008): 2124–2157.

Rosen, M. P., et al. "Cigarette Smoking: An Independent Risk Factor for Atherosclerosis in the Hypogastric-Cavernous Arterial Bed of Men with Arteriogenic Impotence." *Journal of Urology* 145, 4 (April 1991): 759–776.

Rosenblum, Constance. "Living Apart Together." *New York Times,* September 13, 2013.

Roseneil, Sasha. "On Not Living with a Partner: Unpicking Coupledom and Cohabitation." *Sociological Research Online* 11, 3 (2006). www.socresonline.org.uk/11/3/roseneil.html.

Rosenbloom, Stephanie. "Love, Lies and What They Learned." *New York Times,* November 12, 2011.

Rosenfeld, Michael. "Couple Longevity in the Era of Same-Sex Marriage in the United States." *Journal of Marriage and the Family* 76 (October 2014): 905–918.

Ross, C. E., J. Mirowsky, and K. Goldsteen. "The Impact of the Family on Health." In *Contemporary Families: Looking Forward, Looking Back,* edited by A. Booth (pp. 341–360). Minneapolis: National Council on Family Relations, 1991.

Rossi, Alice. "Transition to Parenthood." *Journal of Marriage and the Family* 30 (1) February 1968: 26–39.

Rostosky, S. S., E. D. B. Riggle, D. Pascale-Hague, and L. E. McCants. "The Positive Aspects of a Bisexual Self-Identification." *Psychology & Sexuality* 1 (2010): 131–144.

Roy, R., Weibust, K., and Miller, C. "Effects of Stereotypes about Feminists on Feminist Self-Identification." *Psychology of Women Quarterly* 31 (2007): 146–156.

Ruane, J., and K. Cerulo. *Second Thoughts: Seeing Conventional Wisdom through the Sociological Eye,* 3rd ed. Thousand Oaks, CA: Pine Forge Press, 2004.

Rubin, Lillian. *Intimate Strangers: Men and Women Together.* New York: Perennial, 1983.

———. *Just Friends: The Role of Friendship in Our Lives.* New York: Harper and Row, 1985.

———. *Families on the Faultline: America's Working Class Speaks about the Family, the Economy, Race, and Ethnicity.* New York: HarperCollins, 1994.

Rubinyi, Wendy, Minnell Tralle, and Heather Lee. "How Age Affects Children's Adjustment to Stepfamilies." Parents Forever, University of Minnesota Extension, January 2012. www.extension.umn.edu.

Rudd, J., and S. Herzberger. "Brother-Sister Incest—Father-daughter Incest: A Comparison of Characteristics and Consequences." *Child Abuse and Neglect* 23, 9 (September 1999): 915–928.

Ruggles, Steven. "Marriage, Family Systems, and Economic Opportunity in the United States Since 1850." University of Minnesota Department of History (Working Paper No. 2014-11), December 2014.

Runyan, William. "In Defense of the Case Study Method." *American Journal of Orthopsychiatry* 52, 3 (July 1982): 440–446.

Ruppaner, Leah. "Conflict and Housework: Does Country Context Matter?" *European Sociological Review* 26, 5 (2010): 557–570.

Russell, Graeme. "Problems in Role-Reversed Families." In *Reassessing Fatherhood,* edited by C. Lewis and M. O'Brien (pp. 161–179). London: Sage, 1987.

Rust, Paula. "Two Many and Not Enough: The Meanings of Bisexual Identities." In *Sexualities: Identities, Behaviors, and Society,* edited by Michael Kimmel and Rebecca Plante. New York: Oxford University Press, 2004.

Rutter, V. "Lessons from Stepfamilies." *Psychology Today* (May 1994): 30–33, 60.

Saad, Lydia. U.S. "Acceptance of Gay/Lesbian Relations Is the New Normal." Gallup Politics, May 14, 2012. www.gallup.com/poll/154634/Acceptance-Gay-Lesbian-Relations-New-Normal.aspx.

Saad, Lydia. 2014. "The '40-Hour' Workweek Is Actually Longer by Seven Hours." Gallup, August 29, 2014. www.gallup.com/poll/175286/hour-workweek-actually-longer-seven-hours.aspx.

Sack, Kevin. "Health Benefits Inspire Rush To Marry or Divorce." *New York Times,* August 12, 2008.

Sadker, Myra, and David Sadker. *Failing at Fairness: How American Schools Cheat Girls.* New York: C. Scribner's Sons, 1994.

Safilios-Rothschild, Constantina. "Family Sociology or Wives' Sociology? A Cross-Cultural Examination of Decision Making." *Journal of Marriage and the Family* 38 (1976): 355–362.

Sagrestano, Lynda, Christopher Heavey, and Andrew Christensen. "Perceived Power and Physical Violence in Marital Conflict." *Journal of Social Issues* 55, 1 (Spring 1999): 65–79.

Sakmari, Elvira, and Amanda Guerra. "Child Found in Day Care Van Has Died." NBCDFW.com, July 21, 2012.

Sandberg, Jonathan, James Harper, E. Jeffrey Hill, Richard Miller, Jeremy Yorgason, and Randal Day. "What Happens at Home Does Not Necessarily Stay at Home: The Relationship of Observed Negative Couple Interaction with Physical Health, Mental Health, and Work Satisfaction." *Journal of Marriage and Family* 75 (August 2013): 808–821.

Sanford, Keith. "Problem–Solving Conversations in Marriage: Does It Matter What Topics Couples Discuss?" *Personal Relationships* 10, 1 (March 2003): 97–112.

Santrock, John. *Life Span Development.* Madison, WI: WCB Brown & Benchmark Publishers, 1995.

Sarkisian, N., and N. Gerstel. "Till Marriage Do Us Part: Adult Children's Relationships with Parents." *Journal of Marriage and Family* 70, 2 (May 2008): 360–376.

Sarkisian, N., M. Gerena, and N. Gerstel. "Extended Family Integration among Euro and Mexican Americans: Ethnicity, Gender, and Class." *Journal of Marriage and Family* 69, 1 (February 2007): 40–54.

Sarkisian, Natalia, and Naomi Gerstel. *Nuclear Family Values, Extended Family Lives: The Power of Race, Class, and Gender.* New York: Routledge, 2012.

Sassler, Sharon. "The Process of Entering into Cohabiting Unions." *Journal of Marriage and Family* 66, 2 (May 2004): 491–505.

Sassler, Sharon. "Partnering across the Life Course: Sex, Relationships, and Mate Selection." *Journal of Marriage and Family* 72, 3 (June 2010): 557–575.

Sassler, Sharon. "But How Do Relationships Get Started?" Brief Reports: Experts Respond to: "Does Premarital Cohabitation Raise Your Risk for Divorce?" Council on Contemporary Families, 2015. https://contemporaryfamilies.org/cohabitation-divorce-commentaries/#Sassler.

Sassler, Sharon, and Amanda J. Miller. "Class Differences in Cohabitation Processes." *Family Relations* 60, 2 (2011a): 163–177.

Sassler, Sharon, and Amanda J. Miller. "Waiting to Be Asked: Gender, Power, and Relationship Progression among Cohabiting Couples." *Journal of Family Issues* 32, 4 (2011b): 482–506.

Sassler, Sharon, Amanda J. Miller, and Tamara Green. "Social Class Differences in Relationship Negotiation among Cohabiting Couples." Paper presented at Population Association of America, 2011 meeting, Washington DC.

Sayer, Liana. "The Complexities of Interpreting Changing Household Patterns." Council on Contemporary Families, May 7, 2015. contemporaryfamilies.org/complexities-brief-report.

Sayer, Liana, Paula England, Paul Allison, and Nicole Kangas. "She Left, He Left: How Employment and Satisfaction Affect Men's and Women's Decisions to Leave Marriages." *American Journal of Sociology* 116, 6 (May 2011): 1982–2018.

Schade, Lori Cluffa, Jonathan Sandberga, Roy Beana, Dean Busbya, and Sarah Coynea. "Using Technology to Connect in Romantic Relationships: Effects on Attachment, Relationship Satisfaction, and Stability in Emerging Adults." *Journal of Couple & Relationship Therapy: Innovations in Clinical and Educational Interventions* 12, 4 (2013): 314–338.

Scanzoni, J. "Social Exchange and Behavioral Interdependence." In *Social Exchange in Developing Relationships*, edited by R. Burgess and T. Huston. New York: Academic Press, 1979.

Scanzoni, J., K. Polonko, J. Teachman, and L. Thompson. *The Sexual Bond: Rethinking Families and Close Relationships*. Newbury Park, CA: Sage, 1989.

Schaap, Cas, Bram Buunk, and Ada Kerkstra. "Marital Conflict Resolutions." In *Perspectives on Marital Interaction*, edited by P. Noller and M. A. Fitzpatrick. Philadelphia: Multilingual Matters, 1988.

Scelfo, Julie. "A University Recognizes a Third Gender: Neutral." *New York Times*, February 3, 2015. www.nytimes.com/2015/02/03 /education/edlife/a-university-recognizes-a-third-gender-neutral.html.

Scelfo, Julie. "A Gender-Neutral Glossary." *New York Times*, February 8, 2015.

Schilt, Kristen. *Just One of the Guys? Transgender Men and the Persistence of Gender Inequality*. University of Chicago Press, 2010.

Schilt, Kristen, and Laurel Westbrook. "Doing Gender, Doing Heteronormativity: 'Gender Normals,' Transgender People, and the Social Maintenance of Heterosexuality." *Gender and Society* 23, 4 (August 2009): 440–464.

Schindler, Holly. "The Importance of Parenting and Financial Contributions in Promoting Fathers' Psychological Health." *Journal of Marriage and Family* 72 (April 2010): 318–332.

Schlabach S. "The Importance of Family, Race, and Gender for Multiracial Adolescent Well-being." *Family Relations* 62, 1 (2013): 154–74.

Schoen, Robert, and Yen-Hsin Alice Cheng. "Partner Choice and the Differential Retreat from Marriage." *Journal of Marriage and Family* 68, 1 (February 2006): 1–10.

Schoen, Robert. "Union Disruption in the United States." *International Journal of Sociology* 32, 4 (Winter 2002/2003): 36–50.

Schoenborn, C. A. "Marital Status and Health: United States, 1999–2002." *Vital and Health Statistics*, no. 351. Hyattsville, MD: National Center for Health Statistics, 2004.

Schwartz, Christine, and Nikki L. Graf. "Assortative Matching among Same-Sex and Different-Sex Couples in the United States, 1990–2000." *Demographic Research* 21, 28 (2009): 843–878.

Schwartz, C., and R. Mare. "Trends in Educational Assortative Marriage from 1940 to 2003." California Center for Population Research, 2005.

Schwartz, John. "Highlights From the Supreme Court Decision on Same-Sex Marriage." *New York Times*. www.nytimes.com/ interactive/2015/us/2014-term-supreme-court-decision-same-sex-marriage.html.

Schwartz, Pepper. *Peer Marriage: How Love between Equals Really Works*. New York: Free Press, 1994.

Scislowska, Monika. "Poland OK's Edict on Domestic Violence." Associated Press, February 7, 2015. www.bostonglobe.com/news /world/2015/02/07/poland-oks-law-against-domestic-violence-after -fiery-debate/7Kl7uCqhVAWiFzjBhX8AL3I/story.html.

Seccombe, Karen. "Families in Poverty in the 1990s: Trends, Causes, Consequences, and Lessons Learned." *Journal of Marriage and the Family* 62, 4 (November 2000): 1094–1113.

Sedikides, C., Oliver, M. B., and Campbell, W. K. "Perceived Benefits and Costs of Romantic Relationships for Women and Men: Implications for Exchange Theory." *Personal Relationships* 1 (1994): 5–21.

Seligson, Hannah. "Facebook's Last Taboo: The Unhappy Marriage." *New York Times*, December 26, 2014. www.nytimes.com/2014/12/28 /fashion/facebook-last-taboo-the-unhappy-marriage.html.

Sellers, C., and M. Bromley. "Violent Behavior in College Student Dating Relationships." *Journal of Contemporary Criminal Justice* 12, 1 (1996): 1–27.

Seltzer, Judith. "Families Formed Outside of Marriage." *Journal of Marriage and the Family*, 62, 4 (November 2000): 1247–1268.

Sennett, Richard, and Jonathan Cobb. *The Hidden Injuries of Class*. New York: Vintage, 1972.

Serewicz, M. C. M., and E. Gale. "First-Date Scripts: Gender Roles, Context, and Relationship." *Sex Roles* 58 (2008): 149–164.

Sharpsteen, Don J. "Romantic Jealousy as an Emotion Concept: A Prototype Analysis." *Journal of Social and Personal Relationships* 10, 1 (1993): 69–82.

Shaver, Phillip, Cindy Hazan, and D. Bradshaw. "Love as Attachment: The Integration of Three Behavioral Systems." In *The Psychology of Love*, edited by R. Sternberg and M. Barnes. New Haven, CT: Yale University Press, 1988.

Shellenbarger, S. "When 20-Somethings Move Back Home, It Isn't All Bad." *Wall Street Journal*, May 21, 2008.

————. "How Growing Up in a Recession Can Shape a Child's Future." *Wall Street Journal*, February 18, 2009, D1.

————. "Job-Sharing: Appealing, But Little-Used." *Wall Street Journal*, WSJ blogs, *The Juggle*, August 2, 2010. http://blogs.wsj.com/juggle /2010/08/02/job-sharing-appealing-but-little-used/.

Shelton, Beth A., and Daphne John. "Does Marital Status Make a Difference? Housework among Married and Cohabiting Men and Women." *Journal of Family Issues* 14, 3 (September 1993): 401–420.

Sherkat, Darren, Kylan Mattias de Vries, and Stacia Creek. "Race, Religion, and Opposition to Same-Sex Marriage." *Social Science Quarterly* 91, 1 (March 2010): 80–98.

Shibley-Hyde, J., and S. Jaffee. "Becoming a Heterosexual Adult: The Experiences of Young Women." *Journal of Social Issues* 56, 2 (2000): 283–296.

Shields, B. *Down Came the Rain: My Journey through Postpartum Depression*. New York: Hyperion, 2005.

Shifrel, Scott. "Nixzmary Brown's Last Words Open Mother's Murder Trial." *New York Daily News*, September 18, 2008.

Shifren, Jan, Brigitta Monz, Patricia Russo, Anthony Segreti, and Catherine Johannes. "Sexual Problems and Distress in United States Women." *Obstetrics and Gynecology* 112, 5 (November 2008): 970–978.

Siegel, Deborah, and Susan Livingston Smith. *Openness in Adoption: From Secrecy and Stigma to Knowledge and Connections*. Evan B. Donaldson Adoption Institute, New York, March 2012.

Silverschanz, P., J. Konik, L. M. Cortina, and V. J. Magley. "Slurs, Snubs and Queer Jokes: Incidence and Impact of Heterosexist Harassment in Academica." *Sex Roles* 58 (2008): 179–191.

Simon, R. "The Joys of Parenthood, Reconsidered." *Contexts: Understanding People in Their Social Worlds* 7, 2 (Spring 2008): 40–45.

Sims, Paul. "A Father's Presence during Childbirth 'Makes Labour Longer and Harder and Could Damage Mother and Child's Health.'" *MailOnline*, October 19, 2009.

Sloan Work and Family Research Network. *Questions and Answers about Spillover: Negative Impacts*. Boston: Boston College Graduate School of Social Work, 2008. www.bc.edu/wfnetwork.

Slotter, Erica, Wendi Gardner, and Eli Finkel. "Who Am I without You? The Influence of Romantic Breakup on the Self-Concept." *Personality and Social Psychology Bulletin* 36 (2010): 147–60.

Small, S. "Bridging Research and Practice In the Family and Human Sciences." *Family Relations* 54 (April 2005): 320–334.

Small, S. A., and Riley, D. "Toward a Multidimensional Assessment of Work Spillover into Family Life." *Journal of Marriage and the Family* 52 (1990): 51–61.

Smith, Aaron, and Maeve Duggan, "Online Dating and Relationships," Pew Research Center, October 21, *2013*.

Smith, Craig S. "A Love that Transcends Death Is Blessed by the State." *New York Times*, February 20, 2004.

Smith, Joann Seeman, David Vogel, Stephanie Madon, and Sarah Edwards. "The Power of Touch: Nonverbal Communication within Married Dyads." *The Counseling Psychologist* 39, 5 (2011): 764–787.

Smith, J. A. *The Daddy Shift: How Stay-at-Home Dads, Breadwinning Moms, and Shared Parenting Are Transforming American Families*. Boston: Beacon Press, 2009.

Smith, Suzanne, and Raeann Hamon. *Exploring Family Theories*. Oxford University Press, 2012.

Smith, Tom, Peter V. Marsden, and Michael Hout. "General Social Surveys, 1972–2014." Chicago: National Opinion Research Center, 2015. https://sda.berkeley.edu/archive.htm.

Smits, Jeroen, Wout Ultee, and Jan Lammers. "Effects of Occupational Status Differences between Spouses on the Wife's Labor Force Participation and Occupational Achievement: Findings from 12 European Countries." *Journal of Marriage and the Family* 58, 1 (February 1996): 101–115.

Smock, P. J., L. M. Casper, and J. Wyse. "Nonmarital Cohabitation: Current Knowledge and Future Directions for Research." Population Studies Center Research Report, Institute for Social Research, University of Michigan, Ann Arbor, MI, July 2008.

Smock, Pamela J. "Cohabitation in the United States: An Appraisal of Research Themes, Findings, and Implications." *Annual Review of Sociology* 26 (Summer 2000).

Smock, Pamela J. "The Wax and Wane of Marriage: Prospects for Marriage in the 21st Century." *Journal of Marriage and Family* 66 (2004): 966–73.

Snarey, John, et al. "The Role of Parenting in Men's Psychosocial Development." *Developmental Psychology* 23, 4 (July 1987): 593–603.

Snyder, H. *Sexual Assault of Young Children as Reported to Law Enforcement: Victim, Incident, and Offender Characteristics.* Washington, DC: Bureau of Justice Statistics, U.S. Department of Justice, 2000.

Snyder, Karrie. "A Vocabulary of Motives: How Parents Understand Quality Time. *Journal of Marriage and Family* 69 (2007): 320–340.

Snyder, T. D., and S. A. Dillow. Digest of Education Statistics 2010 (NCES 2011-015). National Center for Education Statistics, Institute of Education Sciences, U.S. Department of Education. Washington, DC, 2011.

Solomon, S. E., E. D. Rothblum, and K. F. Balsam. "Money, Housework, Sex and Conflict: Same-Sex Couples in Civil Unions, Those Not in Civil Unions, and Heterosexual Married Siblings." *Sex Roles* 52 (2005): 561–575.

Solot, Dorian, and Marshall Miller. "Common Law Marriage Fact Sheet," August 2005. www.unmarried.org/common.html.

South, Scott, Katherine Trent, and Yang Shen. "Changing Partners: Toward a Macrostructural-Opportunity Theory of Marital Dissolution." *Journal of Marriage and Family* 63, 3 (August 2001): 743–754.

Spangler, Ashley, and Krista Payne. "Marital Duration at Divorce, 2012" (FP-14-11). National Center for Family and Marriage Research, 2014. www.bgsu.edu/content/dam/BGSU/college-of-arts-and-sciences/NCFMR/documents/FP/FP-14-11-marital-duration-2012.pdf.

Spitalnick, Josh, and Lily McNair. "Couples Therapy with Gay and Lesbian Clients: An Analysis of Important Clinical Issues." *Journal of Sex & Marital Therapy* 31, 1 (January/February 2005): 43–56.

Spock, B., and M. Rothenberg. *Dr. Spock's Baby and Child Care: 40th Anniversary Edition.* New York: Pocket Books, 1985.

Sprecher, S., and D. Felmlee. "The Balance of Power in Romantic Heterosexual Couples over Time from 'His' and 'Her' Perspectives." *Sex Roles* 37, 5/6 (1997): 361–379.

Sprecher, Susan, and Elaine Hatfield. "Matching Hypothesis." In H. Reis and S. Sprecher (eds.), *Encyclopedia of Human Relationships.* New York: Sage, 2009.

Sprecher, Susan, and Susan Hendrick. "Self-Disclosure in Intimate Relationships: Associations with Individual and Relationship Characteristics over Time." *Journal of Social and Clinical Psychology* 23 (2004): 836–856.

Sprecher, Susan, and Pamela Regan. "Liking Some Things (in Some People) More than Others: Partner Preferences in Romantic Relationships and Friendships." *Journal of Social and Personal Relationships* 19 (2002): 436–481.

Stacey, Judith, and Timothy Biblarz. "(How) Does the Sexual Orientation of Parents Matter?" *American Sociological Review* 66, 2 (2001): 159–183.

Stack, Carol B. *All Our Kin: Strategies for Survival in a Black Community.* New York: Harper and Row, 1974.

Stanley, S., and G. Rhoades. "'Sliding vs. Deciding': Understanding a Mystery." *National Council on Family Relations Report: Family Focus on Cohabitation* 54, 2 (Summer 2009): F1–4.

Stanley S. M., G. K. Rhoades, and H. J. Markman. "Sliding vs. Deciding: Inertia and the Premarital Cohabitation Effect." *Family Relations* 55 (2006): 499–509.

Stanley, S. M., G. K. Rhoades, P. A. Olmos-Gallo, and H. J. Markman. "Mechanisms of Change in a Cognitive Behavioral Couples Prevention Program: Does Being Naughty or Nice Matter?" *Prevention Science* 8 (2007): 227–239.

Stanley, Scott, Galena Rhoades, Paul Amato, Howard J. Markman, and Christine A. Johnson. "The Timing of Cohabitation and Engagement: Impact on First and Second Marriages." *Journal of Marriage and Family* 72 (August 2010): 906–918.

Starr, Colleen. "'Jon and Kate Plus 8' Announcement: We are Getting Divorced." Examiner.com, June 22, 2009.

Steinhauer, J. "A Cul-de-Sac of Lost Dreams, and New Ones." *New York Times,* August 22, 2009.

Steinmetz, Suzanne, Sylvia Clavan, and K. Stein. *Marriage and Family Realities.* New York: Harper and Row, 1990.

Steinmetz, Suzanne. "Family Violence." In *Handbook of Marriage and the Family,* edited by M. Sussman and S. Steinmetz. New York: Plenum Press, 1987.

Sternberg, R. "Triangulating Love." in *The Altruism Reader,* edited by Thomas Oord. West Conshocken, PA, (2007): 331–347.

Stevens, Daphne, Gary Kiger, and Pamela J. Riley. "Working Hard and Hardly Working: Domestic Labor and Marital Satisfaction among Dual-Earner Couples." *Journal of Marriage and Family* 63, 2 (May 2001): 514–526.

Stevenson, Betsey, and Justin Wolfers. "Marriage and Divorce: Changes and Their Driving Forces." American Economic Association, *Journal of Economic Perspectives* 21, 2 (Spring 2007): 27–52.

Stewart, A. J., and C. McDermott. "Gender in Psychology." *Annual Review of Psychology* 55 (2004): 519–544.

Stewart, S., W. D. Manning, and P. J. Smock. "Union Formation among Men in the U.S.: Does Having Prior Children Matter?" *Journal of Marriage and Family* 65, 1 (February 2003): 90–104.

Stewart, Susan. *Brave New Stepfamilies: Diverse Paths toward Stepfamily Living.* Thousand Oaks, CA: Sage, 2007.

Stimpson, J. P., and F. A. Wilson. "Cholesterol Screening by Marital Status and Sex in the United States." *Preventing Chronic Disease* 6, 2 (2009). www.ncbi.nlm.nih.gov/pmc/articles/PMC2687861/pdf/PCD62A55.pdf.

Stockdale, M. S. "The Role of Sexual Misperceptions of Women's Friendliness in an Emerging Theory of Sexual Harassment." *Journal of Vocational Behavior* 42, 1 (February 1993): 84–101.

Stokes, Charles E., and Christopher G. Ellison. "Religion and Attitudes toward Divorce. Laws among U.S. Adults." *Journal of Family Issues* 31, 10 (2010): 1279–1304.

Stolzenberg, L., and S. J. D' Alessio. "The Effect of Divorce on Domestic Crime." *Crime and Delinquency* 52, 2 (April 2007): 281–302.

Stop Abusive and Violent Environments (SAVE). "Arrest Policies for Domestic Violence" Special Report, November 2010.

Stout, Hilary. "Technologies Help Adult Children Monitor Aging Parents." *New York Times,* July 29, 2010.

Strasburger, V. "Adolescents, Sex and the Media: Ooooo Baby, Baby: A Q and A." *Adolescent Medicine Clinics* 16 (2005): 269–288.

Straus, M. "Prevalence of Violence against Dating Partners by Male and Female University Students Worldwide." *Violence against Women* 10, 7 (July 2004): 790–811.

Straus, M. "Dominance and Symmetry in Partner Violence by Male and Female University Students in 32 Nations." *Children and Youth Services Review* 30 (2008): 252–275.

Straus, Murray A. "Physical Assaults by Wives: A Major Social Problem." In *Current Controversies in Family Violence,* edited by R. Gelles and D. Loseke. Newbury Park, CA: Sage, 1993.

Straus, Murray. 2010. "Thirty Years of Denying the Evidence on Gender Symmetry in Partner Violence: Implications for Prevention and Treatment." *Partner Abuse* 1, 3 (2010): 332–362.

Straus, Murray, and Carolyn Field. "Psychological Aggression by American Parents: National Data on Prevalence, Chronicity, and Severity." *Journal of Marriage and the Family* 65, 4 (November 2003): 795–808.

Straus, Murray, Richard Gelles, and Suzanne Steinmetz. *Behind Closed Doors.* Garden City, NY: Anchor Books, 1980.

Straus, Murray A., and Mallie J. Paschall. "Corporal Punishment by Mothers and Development of Children's Cognitive Ability: A Longitudinal Study of Two Nationally Representative Age Cohorts." *Journal of Aggression, Maltreatment, and Trauma* 18, 5 (2009): 459–483.

Straus, Murray, and Carrie Yodanis. "Corporal Punishment in Adolescence and Physical Assaults on Spouses Later in Life: What Accounts for the Link?" *Journal of Marriage and the Family* 58, 4 (November 1996): 825–841.

Streib, Jessi. "Marrying across Class Lines." *Contexts: Understanding People in Their Social Worlds,* vol. 14, no. 2 (Spring 2015):40–45.

Strohmaier, Heidi, Megan Murphy, and David DeMatteo. "Youth Sexting: Prevalence Rates, Driving Motivations, and the Deterrent Effect of Legal Consequences." *Sexuality Research and Social Policy: Journal of NSRC* 11, 3 (September 2014): 245–255.

Strohschein, Lisa. "Parental Divorce and Child Mental Health Trajectories." *Journal of Marriage and Family* 67, 5 (December 2005): 1286–1300.

Strong, Bryan, and Christine DeVault. *Human Sexuality: Diversity in Contemporary America,* 2nd ed. Mountain View, CA: Mayfield, 1997.

Stutzman, Rene. "Both Lesbian Moms Have Parental Rights, Daytona Court Rules in Custody Dispute." *Orlando Sentinel,* December 29, 2011.

Stykes, B., L. Gibbs, and K. K. Payne. "First Divorce Rate, 2012," (FP-14-09). National Center for Family and Marriage Research, 2014. www.bgsu.edu/content/dam/BGSU/college-of-arts-and-sciences/NCFMR/documents/FP/FP-14-09-divorce-rate-2012.pdf.

Stykes, Bart, and Karen Benjamin Guzzo. "Remarriage and Stepfamilies" (FP-15-10). National Center for Family and Marriage Research, 2015.

Stykes, James. "Fatherhood in the U.S.: Number of Children, 1987–2010" (FP-11-10). *National Center for Family and Marriage Research,* 2011.

Stykes, Bart. "Who are Nonresident Fathers? Demographic Characteristics of Nonresident Fathers" (FP-12-08). National Center for Family and Marriage Research, Bowling Green State University, 2012. www.bgsu.edu/content/dam/BGSU/college-of-arts-and-sciences/NCFMR/documents/FP/FP-12-08.pdf.

Stykes, B., K. Payne, and L. Gibbs. "First Marriage Rate in the U.S., 2012" (FP-14-08). National Center for Family and Marriage Research, 2014. www.bgsu.edu/content/dam/BGSU/college-of-arts-and-sciences/NCFMR/documents/FP/FP-14-08-marriage-rate-2012.pdf.

Stykes, Bart, Wendy Manning, and Susan Brown. "Grandparenthood in the U.S: Prevalence of Grandparenthood among Adults Aged 50+." National Center for Family and Marriage Research (FP-14-14), 2014. www.bgsu.edu/content/dam/BGSU/college-of-arts-and-sciences/NCFMR/documents/FP/FP-14-14-grandparent-50-plus.pdf.

Suicide Prevention Resource Center. "Issues Brief: Suicide and Bullying." SPRC, March 2011. www.sprc.org/sites/sprc.org/files/library/Suicide_Bullying_Issue_Brief.pdf.

Sullivan, P. "The Rising Cost of Youth Sports, in Money and Emotion. *New York Times*, January 15, 2015.

Sullivan, Oriel, Jonathan Gershuny, and John Robinson. "The Continuing 'Gender Revolution' in Housework and Care." Council on Contemporary Families, The Society Pages, 2015. thesocietypages.org/ccf/2015/06/16/the-continuing-gender-revolution-in-housework-and-care/.

Sullivan, O., and S. Coltrane. "Men's Changing Contribution to Housework and Childcare: A Discussion Paper on Changing Family Roles." Paper presented at the 11th Annual Conference of the Council on Contemporary Families, University of Illinois, Chicago, April 25–26, 2008.

Sun, Shirley Hsiao-Li. "The 'Final Say' Is Not the Last Word: Gendered Patterns, Perceptions, and Processes in Household Decision Making among Chinese Immigrant Couples in Canada." *Journal of Comparative Research in Anthropology and Sociology* 1, 1 (2010): 91–105.

Sun, Yongmin. "Family Environment and Adolescents' Well-Being Before and After Parents' Marital Disruption: A Longitudinal Analysis." *Journal of Marriage and Family* 63, 3 (August 2001): 697–713.

Sun, Yongmin. "The Well-Being of Adolescents in Households with No Biological Parents." *Journal of Marriage & Family* 65, 4 (November 2003): 894–909.

Surra, C. A., T. M. J. Boettcher-Burke, N. R. Cottle, A. R. West, and C. R. Gray. "The Treatment of Relationship Status in Research on Dating and Mate Selection." *Journal of Marriage and Family* 69 (2007): 207–221.

Sutfin, E., M. Fulcher, R. P. Bowles, and C. J. Patterson. "From 1997 to 2007: Fewer Mothers Prefer Full-Time Work," 2007. http://pewresearch.org.

Swain, Scott. "Covert Intimacy: Closeness in Men's Friendships." In *Gender in Intimate Relationships: A Microstructural Approach,* edited by B. Risman and P. Schwartz. Belmont, CA: Wadsworth, 1989.

Sweeney, Megan M. "Remarriage and the Nature of Divorce: Does It Matter Which Spouse Chose to Leave?" *Journal of Family Issues* 23, 3 (April 2002): 410–440.

Sweeney, Megan. "Remarriage and Stepfamilies: Strategic Sites for Family Scholarship in the First Decade of the 21st Century." *Journal of Marriage and Family* 72 (June 2010): 667–684.

Szinovacz, Maximiliane. "Family Power." In *Handbook of Marriage and the Family,* edited by M. Sussman and S. Steinmetz. New York: Plenum Press, 1987.

Tach, L., and S. Halpern-Meekin. "How Does Premarital Cohabitation Affect Trajectories of Marital Quality?" *Journal of Marriage and Family* 71 (2009): 298–317.

Tahmincioglu, Eve. "Employers Rethinking Five-Day Workweek. Some See Improved Productivity, but Old Habits Die Hard." Careers on NBC News.com, May 8, 2011. www.msnbc.msn.com/id/42918666/ns/business-careers/t/employers-rethinking-five-day-workweek/#.UCvFEqPJDIV.

Tannen, Deborah. "Sex, Lies and Conversation: Why Is It So Hard for Men and Women to Talk to Each Other?" *Washington Post,* June 24, 1990.

Tashiro, Ty, and Patricia Frazier. "'I'll Never Be in a Relationship Like That Again': Personal Growth Following Romantic Relationship Breakups." *Personal Relationships* 10, 1 (March 2003): 113–128.

Tavernise, Sabrina. "Black Children in U.S. Are Much More Likely to Live in Poverty, Study Finds." *New York Times,* July 14, 2015.

Taylor, Bruce G., and Elizabeth A. Mumford. "A National Descriptive Portrait of Adolescent Relationship Abuse Results from the National Survey on Teen Relationships and Intimate Violence." *Journal of Interpersonal Violence* (2014): 0886260514564070.

Taylor, Paul, Kim Parker, Rakesh Kochhar, Richard Fry, Cary Funk, Eileen Patten, and Seth Motel. "Young, Underemployed and Optimistic: Coming of Age, Slowly, in a Tough Economy." Pew Research Center, 2012.

Taylor, Paul, Rich Morin, D'Vera Cohn, April Clark, and Wendy Wang. "Men or Women: Who's the Better Leader? A Paradox in Public Attitudes." Pew Research Center Reports, August 25, 2008.

Taylor, Robert J. "Minority Families in America." In *Minority Families in the United States: A Multicultural Perspective,* edited by R. L. Taylor. Englewood Cliffs, NJ: Prentice Hall, 1994.

Taylor, Robert J., Linda M. Chatters, Belinda Tucker, and Edith Lewis. "Developments in Research on Black Families." In *Contemporary Families: Looking Forward, Looking Back,* edited by A. Booth. Minneapolis: National Council on Family Relations, 1991.

Taylor, Robert Joseph, Linda M. Chatters, Amanda Toler Woodward, and Edna Brown. "Racial and Ethnic Differences in Extended Family, Friendship, Fictive Kin and Congregational Informal Support Networks." *Family Relations* 62, 4 (October 1, 2013): 609–624.

Teachman, Jay D., R. Vaughn, A. Call, and Karen P. Carver. "Marital Status and Duration of Joblessness among White Men." *Journal of Marriage and the Family* 56, 2 (May 1994): 415–428.

Tejada-Vera, B., and P. D. Sutton. "Births, Marriages, Divorces, and Deaths: Provisional Data for 2008." *National Vital Statistics Reports,* vol. 57, no. 19. Hyattsville, MD: National Center for Health Statistics, 2009.

Tessina, Tina. *Gay Relationships: For Men and Women. How to Find Them, How to Improve Them, How to Make Them Last.* Los Angeles: Jeremy P. Tarcher, 1989.

Thompson, Anthony. "Emotional and Sexual Components of Extramarital Relations." *Journal of Marriage and the Family* 46, 1 (February 1984): 35–42.

Thompson, Linda, and Alexis Walker. "Gender in Families." In *Contemporary Families,* edited by Alan Booth (pp. 76–102). Minneapolis, MN: National Council on Family Relations, 1991.

Thompson, Linda, and Alexis J. Walker. "The Place of Feminism in Family Studies." *Journal of Marriage & Family* 57, 4 (November 1995): 847–865.

Thompson, Linda. "Family Work: Women's Sense of Fairness," *Journal of Family Issues* 12, 2 (June 1991): 181–196.

———. "Conceptualizing Gender in Marriage: The Case of Marital Care." *Journal of Marriage and the Family* 55, 3 (August 1993): 557–569.

Thomson, E., J. Mosley, T. L. Hanson, and S. S. McLanahan. "Remarriage, Cohabitation, and Changes in Mothering Behavior." *Journal of Marriage and Family* 63 (2001): 370–380.

Thorsen, Maggie, and Valarie King. "My Mother's Husband: Factors Associated with How Adolescents Label Their Stepfathers." *Journal of Social and Personal Relationships.* August 13, 2015.

Tichenor, V. *Earning More and Getting Less: Why Successful Wives Can't Buy Equality.* New Brunswick, NJ: Rutgers University Press, 2005a.

———. "Maintaining Men's Dominance: Negotiating Identity and Power When She Earns More." *Sex Roles* 53, 3/4 (August 2005b).

Tichenor, Veronica, Julia McQuillan, Arthur Greil, Raleigh Contreras, and Karina Shreffler. "The Importance of Fatherhood to U.S. Married and Cohabiting Men." *Fathering* 9, 3 (Fall 2011): 232–251.

Tjaden, Patricia, and Nancy Thoennes. "Extent, Nature, and Consequences of Rape Victimization: Findings from the National Violence against Women Survey." Research Report, January 2006. Washington, DC: U.S. Department of Justice, National Institute of Justice, NCJ 210346.

Toews, M. L., B. S. Catlett, and P. C. McKenry. "Women's Use of Aggression during Marital Separation." *Journal of Divorce and Remarriage* 42 (2005): 1–14.

Tolman, Richard M. "Treatment Program for Men Who Batter." In *Vision 2010: Families and Violence, Abuse and Neglect,* edited by R. J. Gelles. Minneapolis: National Council on Family Relations, 1995.

Topham, Glade L., Jeffry H. Larson, and Thomas B. Holman. "Family-of-Origin Predictors of Hostile Conflict in Early Marriage." *Contemporary Family Therapy* 27, 1 (March 2005): 101–121.

Townsend, Tiffany. "Protecting Our Daughters: Intersection of Race, Class, and Gender in African American Mothers' Socialization of Their Daughters' Heterosexuality." *Sex Roles* 59 (2008): 429–442.

Tracy, Kathleen. *The Secret Story of Polygamy.* Naperville, IL: Sourcebooks, 2002.

Tran, Than Van. "The Vietnamese American Family." In *Ethnic Families in America: Patterns and Variations*, 3rd ed., edited by C. H. Mindel et al. New York: Elsevier, 1988.

Treas, Judith, and Deirdre Giesen, "Sexual Infidelity among Married and Cohabiting Americans." *Journal of Marriage and the Family* 62, 1 (2000): 48–60.

Treas, Judith, and Vern L. Bengtson. "The Family in Later Years." In *Handbook of Marriage and the Family*, edited by M. Sussman and S. Steinmetz. New York: Plenum Press, 1987.

Treichler, Paula. "Feminism, Medicine, and the Meaning of Childbirth." In *Body/Politics: Women and the Discourses of Science*, edited by Mary Jacobus, Evelyn Fox Keller, and Sally Shuttleworth (pp. 113–138). New York: Routledge, 1990.

Troiden, Richard. *Gay and Lesbian Identity: A Sociological Analysis.* New York: General Hall, 1988.

Truman, Jennifer, and Rachel Morgan. "Nonfatal Domestic Violence, 2003–2012." U.S. Department of Justice, Bureau of Justice Statistics, April 2014.

Tucker, Corinna Jenkins, David Finkelhor, Heather Turner, and Anne Smith. "Sibling and Peer Victimization in Childhood and Adolescence." *Child Abuse & Neglect* 38 (2014): 1599–1606.

Tucker, Raymond K., M. G. Marvin, and B. Vivian. "What Constitutes a Romantic Act." *Psychological Reports* 89, 2 (October 1991): 651–654.

Tudge, Jonathan, Irina Mokrova, Bridget Hatfield, and Rachel Karnick. "Uses and Misuses of Brofenbrenner's Bioecological Theory of Human Development." *Journal of Family Theory & Review* 1 (December 2009): 198–210.

Twenge, Jean, Ryne Sherman, and Brooke Wells. "Changes in American Adults' Sexual Behaviors and Attitudes, 1972–2012." *Archives of Sexual Behavior* (May 2015): 1–13.

Tyre, Peg. "The Trouble With Boys." *Newsweek* (January 30, 2006): 44–51.

Umberson, Debra, Tetyana Pudrovska, and Corinne Reczek. "Parenthood, Childlessness, and Well-Being: A Life Course Perspective." *Journal of Marriage and Family* 72 (June 2010): 612–629.

Umberson, D., M. B. Thomeer, and A. C. Lodge. "Intimacy and Emotion Work in Lesbian, Gay, and Heterosexual Relationships." *Journal of Marriage and Family* 77 (2015): 542–556.

Umland, Beth. "How Will the Same-Sex Marriage Decision Affect DP Benefits?" Mercer Signal, July 13, 2015. http://ushealthnews.mercer.com/article/424/how-will-the-same-sex-marriage-decision-affect-dp-benefits-.

UNAIDS World AIDS Day Report. "Joint United Nations Programme on HIV/AIDS (UNAIDS)." UNAIDS World AIDS Day report, 2011. Geneva, Switzerland: Joint United Nations Programme on HIV/AIDS (UNAIDS). www.unaids.org/sites/default/files/en/media/unaids/contentassets/documents/factsheet/2011/20111121_FS_WAD2011_global_en.pdf.

UN Human Rights Council. *Discrimination and Violence against Individuals Based on Their Sexual Orientation and Gender Identity*, May 4, 2015, A/HRC/29/23. www.refworld.org/docid/5571577c4.html.

United Nations Statistics Division. "Demographic Yearbook, Table 25," 2006. http://unstats.un.org/unsd/demographic/products/dyb/dyb2006/Table25.pdf.

United Nations Statistics Division. "Demographic Yearbook," 2012. http://unstats.un.org/unsd/demographic/products/dyb/dyb2012.htm.

USlegal.com. "Cohabitation," 2010. http://cohabitation.uslegal.com/.

U.S. Bureau of Justice Statistics, *Violence by Intimates: Analysis of Data on Crimes by Current or Former Spouses, Boyfriends, and Girlfriends.* Washington, DC: Department of Justice, 1998.

U.S. Bureau of Justice Statistics. "News: Survey of Workplace Violence Prevention, 2005."

U.S. Bureau of Labor Statistics, 2006. www.bls.gov/iif/oshwc/osnr0026.pdf.

———. "Employment Situation Summary," March 6, 2009.

———. "Labor Force Statistics from the Current Population Survey," Table 2, Employment Status of the Civilian Noninstitutional Population 16 Years and Over by Sex, 1971 to Date, 2012. www.bls.gov/cps/cpsaat02.htm.

———. "American Time Use Survey," Table A-6: "Time Spent in Primary Activities 1 and the Percent of Married Mothers and Fathers who Did the Activities on an Average Day by Employment Status and Age of Youngest Own Household Child, Average for the Combined Years 2009–13," 2013a. www.bls.gov/tus/tables/a6_0913.pdf.

———. "Highlights of Women's Earnings in 2012." Bureau of Labor Statistics, October 2013b, Report 1045. www.bls.gov.

———. "Labor Force Characteristics by Race and Ethnicity," *BLS Reports*, August 2014a.

———. "Women in the Labor Force: A Databook," Table 6: "Employment Status of Women by Presence and Age of Youngest Child, Marital Status, Race, and Hispanic or Latino Ethnicity," December 2014b. www.bls.gov/opub/reports/cps/women-in-the-labor-force-a-databook-2014.pdf.

———. "Labor Force Statistics from the Current Population Survey," Table 1, "Employment Status of the Civilian Noninstitutional Population, 1944 to Date," 2015a. www.bls.gov/cps/cpsaat01.htm.

———. "Labor Force Statistics from the Current Population Survey," Table 8: "Employed and Unemployed Full- and Part-Time Workers by Age, Sex, Race, and Hispanic or Latino Ethnicity," 2015b. www.bls.gov/cps/cpsaat08.htm.

———. "Labor Force Statistics from the Current Population Survey," Table 22: "Persons at Work in Nonagricultural Industries by Age, Sex, Race, Hispanic or Latino Ethnicity, Marital Status, and Usual Full- or Part-Time Status," 2015c. www.bls.gov/cps/cpsaat22.htm.

———. "Economic News Release." Table 4: "Families with Own Children: Employment Status of Parents by Age of Youngest Child and Family Type, 2013-2014 Annual Averages," 2015d. www.bls.gov/news.release/famee.t04.htm.

———. "Labor Force Statistics from the Current Population Survey," Table 2, "Employment Status of the Civilian Noninstitutional Population 16 Years and Over by Sex, 1974 to Date," 2015e. www.bls.gov/cps/cpsaat02.htm.

———. "Labor Force Statistics from the Current Population Survey," Table 3. "Employment Status of the Civilian Noninstitutional Population by Age, Sex, and Race," 2015f. www.bls.gov/cps/cpsaat03.htm.

———. "Labor Force Statistics from the Current Population Survey," Wives Who Earn More than Their Husbands 1987–2013, June 2015g. www.bls.gov/cps/wives_earn_more.htm.

———. "Median Weekly Earnings of Full-Time Wage and Salary Workers by Selected Characteristics," Table 37, 2015h. www.bls.gov/cps/cpsaat37.htm.

———. "A Profile of the Working Poor, 2013," *BLS Reports*, August 2015i.

U.S. Census Bureau. "Current Population Survey Reports, America's Families and Living Arrangements," 2008.

———. "America's Families and Living Arrangements: 2010," Table 1A. Marital Status of People 15 Years and Over, by Age, Sex, Personal Earnings, Race, and Hispanic Origin, 2010.

———. "America's Families and Living Arrangements: 2011," Table A1. Marital Status of People 15 Years and Over, by Age, Sex, Personal Earnings, Race, and Hispanic Origin, 2011a. www.census.gov/hhes/families/data/cps2011.html.

———. "March 2010 Current Population Survey, America's Families and Living Arrangements: 2010," Table C9, Children by Presence and Type of Parents, Race, and Hispanic Origin, 2011b.

———. "Who's Minding the Kids, Child Care Arrangements," Spring 2010, Table 4, 2011c. www.census.gov.

———. "Facts for Features, Asian/Pacific American Heritage Month: May 2012." CB12-FF.09, March 21, 2012a.

———. "Facts for Features, Unmarried and Single Americans Week Sept. 16–22, 2012." CB12-FF.18, July 31, 2012b.

———. "America's Families and Living Arrangements: 2013," Table A1: Adults, 2013. www.census.gov/hhes/families/data/cps2013A.html.

———. "America's Families and Living Arrangements: 2014," 2014a.

———. "Facts for Features: Unmarried and Signle Americans Week," 2014b. http://www.census.gov/newsroom/facts-for-features/2014/cb14-ff21.html

———. "Current Population Survey (CPS), Definitions, 2015a. www.census.gov/cps/about/cpsdef.html.

———. "Facts for Features: Fathers' Day: June 21," 2015b, CB15-FF-11. www.census.gov/newsroom/facts-for-features/2015/cb15-ff110.html.

———. "Historical Poverty Tables," Table 4, 2015c. www.census.gov/hhes/www/poverty/data/historical/families.html.

U.S. Department of Health and Human Services, Administration for Children and Families. "Child Maltreatment 2002," 2002. www.acf.hhs.gov/programs/cb/pubs/cm02/index.htm.

U.S. Department of Health and Human Services. "Child Maltreatment 2013." Administration for Children and Families, Administration on Children, Youth and Families, Children's Bureau, 2015a. www.acf.hhs.gov/programs/cb/research-data-technology/statistics-research/child-maltreatment.

_____. "Fewer Child Abuse and Neglect Victims for Seventh Consecutive Year." Administration on Children and Families, 2015b. www.acf.hhs.gov/media/press/2015/fewer-child-abuse-and-neglect-victims-for-seventh-consecutive-year.

U.S. Food and Drug Administration. "Vaccines, Blood and Biologics," 2015a. www.fda.gov/BiologicsBloodVaccines/default.htm.

_____. "HIV Testing," 2015b. www.fda.gov/forpatients/illness/hivaids/prevention/ucm117922.htm.

Uecker, J. E., and C. E. Stokes. "Early Marriage in the United States." _Journal of Marriage and Family_ 70 (2008): 835–846.

Ulman, A., and M. Straus. "Violence by Children against Mothers in Relation to Violence between Parents and Corporal Punishment by Parents." _Journal of Comparative Family Studies_ 34 (2003): 41–60.

Urbina, Ian. "Philadelphia to Handle Domestic Violence Calls Differently." _New York Times_, December 30, 2009.

Vaaler, Margaret, Christopher Ellison, and Daniel Powers. "Religious Influences on the Risk of Marital Dissolution. _Journal of Marriage and Family_, 71, 4 (November 2009): 917–934.

Van Damme, Elke. "Gender and Sexual Scripts in Popular American Teen Series. A Study on the Gendered Discourses in _One Tree Hill_ and _Gossip Girl_." _Catalan Journal of Communication and Cultural studies_ 2, 1 (2010): 77–92.

Van Laningham, J., D. Johnson, and P. Amato. "Marital Happiness, Marital Duration, and the U-Shaped Curve: Evidence from a Five-Wave Panel Study." _Social Forces_ 79, 4 (June 2001): 1313–1341.

Vandivere, Sharon, Karin Malm, and Laura Radel. _Adoption USA: A Chartbook Based on the 2007 National Survey of Adoptive Parents._ Washington, DC: The U.S. Department of Health and Human Services, Office of the Assistant Secretary for Planning and Evaluation, 2009.

Vaquera, Elizabeth, and Grace Kao. "Private and Public Displays of Affection among Interracial and Intra-Racial Adolescent Couples." _Social Science Quarterly_ 86, 2 (June 2005): 484–508.

Vaselle-Augenstein, Renata, and Annette Ehrlich. "Male Batterers: Evidence for Psychopathology." In _Intimate Violence: Interdisciplinary Perspectives_, edited by E. C. Viano. Washington, DC: Hemisphere, 1992.

Vasquez-Nuttall, E., et al. "Sex Roles and Perceptions of Femininity and Masculinity of Hispanic Women: A Review of the Literature." _Psychology of Women Quarterly_ 11 (1987): 409–426.

Vaughan, Diane. _Uncoupling: Turning Points in Intimate Relationships._ New York: Vintage Reprint Edition, 1990.

Vennum, Amber, and Matthew Johnson. "The Impact of Premarital Cycling on Early Marriage." _Family Relations_ 63 (October 2014): 439–452.

Vermont Guide to Civil Unions, 2005. www.vermontjudiciary.org/gtc/family/SharedDocuments/Pamphlet%2035.pdf.

Vespa, Jonathan, Jamie M. Lewis, and Rose M. Kreider. "America's Families and Living Arrangements: 2012, Current Population Reports, P20-570," U.S. Census Bureau, Washington, DC, 2013.

Vincent, Samantha. "2015: 'An awful Year for Child Abuse and Child Abuse Murder Cases.'" Tulsa World, December 8, 2015. www.tulsaworld.com/news/crimewatch/an-awful-year-for-child-abuse-and-child-abuse-murder/article_b8200f57-9ece-531f-9314-7712bca752c1.html.

Vincent, Samantha. "Broken Arrow Killings: Autopsy Reports Reveal Bever Parents Stabbed Dozens of Times." Tulsa World, November 10, 2015. www.tulsaworld.com/news/local/broken-arrow-killings-autopsy-reports-reveal-bever-parents-stabbed-dozens/article_0c5ae34d-95e2-55f6-8f9c-c7218f614bf9.html.

Visher, Emily B., and John S. Visher. _Stepfamilies: A Guide to Working with Stepparents and Stepchildren._ New York: Brunner/Mazel, 1979.

_____. _How to Win as a Stepfamily._ New York: Brunner/Mazel, 1991.

Vogel, David L., Stephen R. Wester, and Martin Heesacker. "Dating Relationships and the Demand–Withdraw Pattern of Communication." _Sex Roles_ 41, 3–4 (1999): 297–306.

Voydanoff, Patricia. "Economic Distress and Family Relations: A Review of the Eighties." In _Contemporary Families: Looking Forward, Looking Back_, edited by A. Booth. Minneapolis: National Council on Family Relations, 1991.

Vredevelt, P. _Empty Arms: Emotional Support for Those Who Have Suffered Miscarriage or Stillbirth._ Sisters, OR: Questar, 1994.

Wade, Terrance J., and David J. Pevalin. "Marital Transitions and Mental Health." _Journal of Health and Social Behavior_ 45, 2 (June 2004): 155–170.

Wagner, Cynthia G. "Homosexual Relationships." _Futurist_ 40, 3 (May/June 2006): 6.

Waite, Linda, and Maggie Gallagher. _The Case for Marriage: Why Married People are Happier, Healthier, and Better off Financially._ New York: Doubleday, 2000; Broadway Books, 2001.

Walker, Karen. "Men, Women, and Friendship: What They Say, What They Do," _Gender and Society_ 8, 2 (June 1994).

Walker, Lenore. _The Battered Woman Syndrome._ New York: Harper Colophon, 1979.

Walker, Lenore. _The Battered Woman Syndrome_, 3rd ed. New York: Springer Publishing, 2009.

Walker, L. "The Battered Woman Syndrome Is a Psychological Consequence of Abuse." In _Current Controversies in Family Violence_, edited by R. Gelles and D. Loseke. Newbury Park, CA: Sage Publications, 1993.

Waller, Willard, and Reuben Hill. _The Family: A Dynamic Interpretation._ New York: Dryden Press, 1951.

Wallerstein, Judith. _Surviving the Breakup, How Children and Parents Cope with Divorce._ New York: Basic Books, 1980.

Wallerstein, J. "Children of Divorce: The Psychological Tasks of the Child." _American Journal of Orthopsychiatry_ 53, 2 (April 1983): 230–243.

Wallerstein, Judith (with S. Blakeslee). _Second Chances: Men, Women and Children a Decade After Divorce._ Boston: Ticknor & Fields, 1989.

Wallerstein, J., and Deborah Resnikoff. "Parental Divorce and Developmental Progression: An Inquiry into Their Relationship." _International Journal of Psycho-Analysis_ 78 (1997): 135–154.

Wallerstein, Judith (with Julia Lewis, Sandra Blakeslee). _The Unexpected Legacy of Divorce: A 25 Year Landmark Study._ New York : Hyperion, 2000.

Wallerstein, Judith, and Joan Kelly. _Surviving the Breakup: How Children and Parents Cope with Divorce._ New York: Basic Books, 1980.

Wallis, Cara. "Performing Gender: A Content Analysis of Gender Display in Music." _Sex Roles_ 64, 3–4 (2011): 160–172.

Walls, Jill, Heather Helms, and Joseph Grzywacz. "Intensive Mothering Beliefs among Full-Time Employed Mothers of Infants." _Journal of Family Issues_ (2014): 245–69.

Walsh, Marie-Therese. "Postpartum Depression Can Begin during Pregnancy," 2015. www.sciguru.org/newsitem/18241/postpartum-depression-can-begin-during-pregnancy.

Walsh, J. A., and J. L. Krienert. "Child-Parent Violence: An Empirical Analysis of Offender, Victim, and Event Characteristics in a National Sample of Reported Incidents." _Journal of Family Violence_, 22 (2007): 563–574.

Walsh, J., and M. Ward. "Adolescent Gender Role Portrayals in the Media: 1950 to the Present." in _The Changing Portrayal of Adolescents in the Media since 1950_, edited by P. Jamieson and D. Romer. New York: Oxford University Press, 2008.

Walters, M. L., J. Chen, and M. Breiding. _The National Intimate Partner and Sexual Violence Survey (NISVS): 2010 Findings on Victimization by Sexual Orientation._ Atlanta, GA: National Center for Injury Prevention and Control, Centers for Disease Control and Prevention, 2013.

Walzer, Susan. _Thinking about the Baby: Gender and Transitions into Parenthood._ Philadelphia: Temple University, 1998.

Wang, C., and J. Holton. _Economic Impact Study: Total Estimated Cost of Child Abuse and Neglect in the United States._ Chicago, IL: Prevent Child Abuse America, 2007.

Wang, H., and P. Amato. "Predictors of Divorce Adjustment: Stressors, Resources, and Definitions." _Journal of Marriage and the Family_ 62 (August 2000): 655–668.

Wang, T., W. Parish, E. Laumann, and Y. Luo. "Partner Violence and Sexual Jealousy in China: A Population-Based Survey." _Violence Against Women_ 15, 7 (2009): 774–798.

Wang, Wendy. "The Rise of Intermarriage: Rates, Characteristics Vary by Race, Gender." Pew Research Center, February 16, 2012.

_____. "For Young Adults, the Ideal Marriage Meets Reality." Pew Research Center, July 10, 2013.

_____. "Interracial Marriage: Who Is 'Marrying Out'?" Pew Research Center, June 12, 2015.

_____. "The Link between a College Education and a Lasting Marriage." Pew Research Center, December 4, 2015b. www.pewresearch.org/fact-tank/2015/12/04/education-and-marriage/.

Wang, Wendy, and Kim Parker. "Record Share of Americans Have Never Married: As Values, Economics and Gender Patterns Change." Washington, DC: Pew Research Center's Social & Demographic Trends project, September 2014.

Wang, Wendy, Kim Parker, and Paul Taylor. "Breadwinner Moms." Pew Research Center, May 29, 2013.

Wang, Wendy. "The Rise of Intermarriage Rates, Characteristics Vary by Race and Gender." Pew Research Center, Social and Demographic Trends, 2012.

Ward, M., and K. Friedman. "Using TV as a Guide: Associations between Television Viewing and Adolescents' Sexual Attitudes and Behavior." *Journal of Research on Adolescence* 16, 1 (March 2006): 105–131.

Wareham, J., D. P. Boots, and J. Chavez. "A Test of Social Learning and Intergenerational Transmission among Batterers." *Journal of Criminal Justice* 37, 2 (March–April 2009): 163–173.

Warren-Gash, Charlotte. "Worldwide Infertility Rates Unchanged in Twenty Years Says World Health Organization." *BioNews* 687 (January 7, 2013). www.bionews.org.uk/page_232839.asp.

Wei, G. S., and J. E. Herbers. "Reporting Elder Abuse: A Medical, Legal, and Ethical Overview." *Journal of the American Medical Women's Association* 59, 4 (2004): 248–254.

Weinberg, Martin S., C. J. Williams, and Douglas W. Pryor. *Dual Attraction: Understanding Bisexuality.* New York: Oxford University Press, 1994.

Weiss, Robert. *Marital Separation.* New York: Basic Books, 1975.

Weisshaar, K. "Earnings Equality and Relationship Stability for Same-Sex and Heterosexual Couples." *Social Forces* 93, 1 (2014): 93–123.

Weitzman, Lenore. *The Marriage Contract: Spouses, Lovers, and the Law.* New York: Macmillan, 1981.

———. *The Divorce Revolution: The Unexpected Social and Economic Consequences for Women and Children in America.* New York: Free Press, 1985.

Weitzman, Susan. *Not to People Like Us: Hidden Abuse in Upscale Marriages.* New York: Basic Books, 2000.

Wells, Georgia. "The Dating Business: Love on the Rocks." *The Wall Street Journal*, June 10, 2015.

Werner, Carol M., Barbara B. Brown, Irwin Altman, and Brenda Staples. "Close Relationships in Their Physical and Social Contexts: A Transactional Perspective." *Journal of Social and Personal Relationships* 9, 3 (1992): 411–431.

West, Candace, and Don Zimmerman. "Doing Gender." *Gender and Society* 1 (1987): 125–151.

Westberg, Heather, Thorana S. Nelson, and Kathleen W. Piercy. "Disclosure of Divorce Plans to Children: What the Children Have To Say." *Contemporary Family Therapy* 24, 4 (December 2002): 525–542.

Westman, M., D. Etzion, and S. Horovitz. "The Toll of Unemployment Does Not Stop with the Unemployed." *Human Relations* 57, 7 (July 2004): 823–844.

White, James, and D. Klein. *Family Theories*, 2nd ed. Thousand Oaks, CA: Sage, 2002.

White, James, and David Klein. *Family Theories*, 3rd ed. Thousand Oaks, CA: Sage, 2008.

White, James, and David Klein. *Family Theories*, 4th ed. Thousand Oaks, CA: Sage, 2014.

White, Lynn, and Alan Booth. "Divorce over the Life Course: The Role of Marital Happiness." *Journal of Family Issues* 12, 1 (March 1991): 5–22.

Whitehead, Barbara Dafoe. "Dan Quayle Was Right," *Atlantic Monthly.* April 1993: 47–84.

———. *The Divorce Culture: Rethinking Poor Commitments to Marriage and the Family.* New York: Knopf Publishing Group, 1996.

———. *The Divorce Culture.* New York: Knopf, 1997.

Whitehead, Andrew. "Gender Ideology and Religion: Does a Masculine Image of God Matter?" *Review of Religious Research* 54 (2012):139–56.

Whitehead, B., and D. Popenoe, "The State of Our Unions: 2001." New Brunswick, NJ: National Marriage Project at Rutgers University, June 27, 2001.

Whitehead, Barbara Dafoe, and David Popenoe. "Social Indicators of Marital Health and Well Being." The State of Our Unions, 2004: The Social Health of Marriage in America. The National Marriage Project. New Brunswick, NJ: Rutgers University, 2004.

Whitehead, Barbara Dafoe, and David Popenoe. "Social Indicators of Marital Health and Well Being." The State of Our Unions, 2005: The Social Health of Marriage in America. The National Marriage Project. New Brunswick, NJ: Rutgers University, 2005.

Whitehurst, Lindsay. "Judge Rules for TV Polygamists, Strikes Down Part of Utah Ban." *The Seattle Times*, August 14, 2014a. www.seattletimes.com/nation-world/judge-rules-for-tv-polygamists-strikes-down-part-of-utah-ban/.

———. "Governor: Utah Should Defend Anti-Polygamy Law." Associated Press, August 30, 2014b. gazette.com/governor-utah-should-defend-anti-polygamy-law/article/1536555.

Whiteman, S., S. McHale, and A. Crouter. "Longitudinal Changes in Marital Relationships: The Role of Offspring's Pubertal Development." *Journal of Marriage and Family* 69 (2007): 1005–1020.

Whiteman, S. D., S. M. McHale, and A. Soli. "Theoretical Perspectives on Sibling Relationships." *Journal of Family Theory and Review, 3* (2011): 124–139.

"Who's Who in the Family Wars: A Characterization of the Major Ideological Factions," In *Feuds About Families: Conservative, Centrist, Liberal, and Feminist Perspectives,* edited by N. Benokraitis (pp. 2–13). Upper Saddle River, NJ: Prentice Hall, 2000.

Whitton, S. W., S. M. Stanley, H. J. Markman, and C. A. Johnson. "Attitudes toward Divorce, Commitment, and Divorce Proneness in First Marriages and Remarriages." *Journal of Marriage and the Family* 75, 2 (2013): 276–287.

Widmer, Eric D., Judith Treas, and Robert Newcomb. "Attitudes toward Nonmarital Sex in 24 Countries." *Journal of Sex Research* 35 (November 1998): 349–358.

Wiehe, Vernon. "Sibling Abuse: Hidden Physical, Emotional, and Sexual Trauma." Thousand Oaks, CA: Sage, 1997.

Wilcox, W. Bradford. "A Red Family Advantage? Marriage and Family Stability in Red and Blue America." *Institute for Family Studies,* July 1, 2015. http://family-studies.org/a-red-family-advantage-marriage-and-family-stability-in-red-and-blue-america/.

Wilcox, W. Bradford and Nicholas Zill. "Red State Families: Better Than We Knew." *Institute for Family Studies,* June 11, 2015. http://family-studies.org/red-state-families-better-than-we-knew/.

Wildsmith, E., and R. K. Raley. "Race-Ethnic Differences in Nonmarital Fertility: A Focus on Mexican American Women." *Journal of Marriage and Family* 68, 2 (2006): 491–508.

Wildsmith, Ph.D., Nicole R. Steward-Streng, M.A., and Jennifer Manlove. "Childbearing Outside of Marriage: Estimates and Trends in the United States." Child Trends, 2011. www.childtrends.org/Files/Child_Trends-2011_11_01_RB_NonmaritalCB.pdf.

Wilhelm, Peter, and Meinrad Perrez. "How Is My Partner Feeling in Different Daily-Life Settings? Accuracy of Spouses' Judgments about Their Partner's Feelings at Work and at Home." *Social Indicators Research* 67 (2004): 183–246.

Wilkinson, Doris Y. "American Families of African Descent." In *Families in Cultural Context,* edited by M. K. DeGenova. Mountain View, CA: Mayfield, 1997.

Willetts, Marion. "An Exploratory Investigation of Heterosexual Licensed Domestic Partners." *Journal of Marriage and the Family* 65, 4 (November 2003): 939–952.

Williams, Rhiannon. "Facebook's 71 gender options come to UK users." Telegraph, June 27, 2014. www.telegraph.co.uk/technology/facebook/10930654/Facebooks-71-gender-options-come-to-UK-users.html.

Williams, Alex. "Just Wait until Your Mother Gets Home." *New York Times,* August 10, 2012.

Williams, Dmitri, Nicole Martins, Mia Consalvo, and James Ivory. "The Virtual Census: Representations of Gender, Race and Age in Video Games." *New Media & Society* 11, 5 (August 2009): 815–834.

Willing, Richard. "Research Downplays Risk of Cousin Marriages." *USA Today* April 4, 2002.

Willoughby, Brian, Adam Farero, and Dean Busby. "Exploring the Effects of Sexual Desire Discrepancy among Married Couples." *Archives of Sexual Behavior* 43 (2014): 551–562.

Wilson, Breana. "Coresident Grandparents: Caregivers versus Non-Caregivers" (FP-12-18). National Center for Family and Marriage Research, 2012. www.bgsu.edu/content/dam/BGSU/college-of-arts-and-sciences/NCFMR/documents/FP/FP-12-18.pdf.

Wilson, Breana. "Grandchildren: Living in a Grandparent-Headed Household" (FP-13-03). National Center for Family and Marriage Research, 2013. http://scholarworks.bgsu.edu/cgi/viewcontent.cgi?article=1020&context=ncfmr_family_profiles.

Wilson, K. J. *When Violence Begins at Home: A Comprehensive Guide to Understanding and Ending Abuse.* Salt Lake City, UT: Publishers Press, 1997.

Wilson, L. "Sexting and Minors: Child Pornography, Not Child's Play." 2008. http://blog.lawinfo.com/ 2008/10/08/sexting-and-minors-child-pornography-not-childs-play.

Wilson, William Julius. "Another Look at the Truly Disadvantaged." *Political Science Quarterly* 106, 4 (1991): 639–656.

Wilson, C. D., and M. E. Fromuth. "Characteristics of Abusive Sibling Relationships and Correlations with Later Relationships." Chicago: Presentation at the American Psychological Association Conference, 1997.

Wilson, A. C., and T. L. Huston. "Premarital Roots of Divorce: The Importance of Shared Reality and Grounded Feelings." *Journal of Marriage and Family* 75 (2013): 681–96.

Wilson, William Julius. *The Truly Disadvantaged: The Inner City, the Underclass, and Public Policy.* Chicago: University of Chicago Press, 1987.

Wineberg, H. "Marital Reconciliation in the United States: Which Couples Are Successful?" *Journal of Marriage and the Family* 56, 1 (February 1994): 80–88.

Wineberg, H. "The Timing of Remarriage among Women Who Have a Failed Marital Reconciliation in the First Marriage." *Journal of Divorce and Remarriage* 30, 3/4 (June 1999): 57–69.

Wineberg, H., and J. McCarthy. "Separation and Reconciliation in American Marriages." *Journal of Divorce & Remarriage* 20, 1–2 (1993): 21–42.

Winslow, Sarah. "Work–Family Conflict, Gender, and Parenthood, 1977–1997." *Journal of Family Issues* 26, 6 (2005): 727–755.

Winton, Chester. *Frameworks for Studying Families.* Guilford, CT: Dushkin, 1995.

————. *Children as Caregivers: Parental and Parentified Children.* Boston: Pearson Allyn and Bacon, 2003.

Wisdom, S., and J. Green. *Stepcoupling: Creating and Sustaining a Strong Marriage in Today's Blended Family.* New York: Crown Publishing, 2002.

Wise, T. N., S. Epstein, and R. Ross. "Sexual Issues in the Medically Ill and Aging." *Psychiatric Medicine* 10 (1992): 169–180.

Witters, Dan, and Lindsey Sharpe. "Women's Well-Being Suffers More when Marriage Ends." Gallup, October 15, 2014. www.gallup.com /poll/178553/women-suffers-marriage-ends.aspx.

Wlodarski, Rafael, and Robin Dunbar. "What's in a Kiss? The Effect of Romantic Kissing on Mating Desirability." *Evolutionary Psychology* 12, 1 (2014): 178–199.

Wohl, M., and A. McGrath. "The Perception of Time Heals All Wounds: Temporal Distance Affects Willingness to Forgive Following an Interpersonal Transgression." *Personality and Social Psychology Bulletin* 33, 7 (2007): 1023–1035.

Wolf, Rosalie S. "Abuse and Neglect of the Elderly." In *Vision 2010: Families and Violence, Abuse and Neglect,* edited by R. J. Gelles. Minneapolis: National Council on Family Relations, 1995.

Wolfinger, N. *Understanding the Divorce Cycle: The Children of Divorce in Their Own Marriages.* New York: Cambridge University Press, 2005.

————. "Hello and Goodbye to Divorce Reform." *National Council on Family Relations Report: Family Focus on Divorce and Relationship Dissolution,* Issue FF36 (December 2007): F15–16.

Wolfinger, Nicholas. "Want to Avoid Divorce? Wait to Get Married, But Not Too Long." Family Studies: The Blog of The Institute for Family Studies, July 16, 2015. http://family-studies.org/want-to-avoid-divorce -wait-to-get-married-but-not-too-long/.

Wolfinger, Nicholas H. "Parental Divorce and Offspring Marriage: Early or Late?" *Social Forces* 82, 1 (September 2003): 337–353.

Women of Color Policy Network. "Income and Poverty in Communities of Color: A Reflection on 2010 U.S. Census Bureau Data. Policy Brief, September 2011.

World Health Organization. Global Health Observatory, "HIV/AIDS," February 2012a. www.who.int/mediacentre/factsheets/fs241/en/.

World Health Organization. Media Centre, Fact Sheet no. 241, "Female Genital Mutilation," February 2012. www.who.int/gho/hiv/en/index .html.

————. "Global and Regional Estimates of Violence against Women: Prevalence and Health Effects of Intimate Partner Violence and Nonpartner Sexual Violence." World Health Organization, Department of Reproductive Health and Research, London School of Hygiene and Tropical Medicine, South African Medical Research Council, 2013. www.who.int/reproductivehealth/publications/ violence/9789241564625/en/.

Worthen, Meredith. "An Argument for Separate Analyses of Attitudes toward Lesbian, Gay, Bisexual Men, Bisexual Women, MtF and FtM Transgender Individuals." *Sex Roles,* published online, April 4, 2012.

Wright, Britanny, and Timothy Loving. "Health Implications of Conflict in Close Relationships." *Social and Personality Psychology Compass* 5/8 (2011): 552–562.

Wrigley, Julia, and Dreby, Joanna. "Fatalities and the Organization of Childcare." *American Sociological Review* 70, 5 (October 2005): 729–757.

Wu, Zheng, and Randy Hart. "Union Disruption in Canada." *International Journal of Sociology* 32, 4 (Winter 2002/2003): 51–75.

Wu, Zheng. "The Future of Cohabitation." In D. Cheal (ed.), *Family: Critical Concepts in Sociology* (pp. 164–193) London: Routledge.

Wu, Zheng, Margaret J. Penning, Michael S. Pollard, and Randy Hart. "In Sickness and in Health." *Journal of Family Issues* 24, 6 (September 2003): 811–838.

Wymbs, Brian T., William E. Pelham Jr., Brooke S. G. Molina, Elizabeth M. Gnagy, Tracey K. Wilson, and Joel B. Greenhouse. "Rate and Predictors of Divorce among Parents of Youths with ADHD." *Journal of Consulting and Clinical Psychology,* 76, 5 (October 2008): 735–744.

Xu, Xiaohe, John Bartkowski, and Kimberly Dalton. "The Role of Cohabitation in Remarriage: A Replication." *International Review of Sociology* 21, 3 (2011): 549–564.

Xu, X., C. D. Hudspeth, and J. P. Bartkowski. "The Timing of First Marriage: Are There Religious Variations?" *Journal of Family Issues* 26, 5 (July 2005): 584–618.

Xu, Xiaohe, Clark D. Hudspeth, and John P. Bartkowski. "The Role of Cohabitation in Remarriage." *Journal of Marriage and Family* 68, 2 (May 2006): 261–274.

Yabiku, Scott, and Constance Gager. "Sexual Frequency and the Stability of Marital and Cohabiting Unions." *Journal of Marriage and Family* 71 (November 2009): 983–1000.

Yardley, J. "Dead Bachelors in Remote China Still Find Wives." *New York Times,* October 5, 2006.

Yeager, Erica Owens. "High-Conflict Couple Interaction and the Role of Relative Power." *Sociology Compass* 3/4 (2009): 672–688.

Yee, Barbara, Barbara DeBaryshe, Sylvia Yuen, Su Yeong Kim, and Hamilton McCubbin, H. "Asian American and Pacific Islander Families: Resiliency and Life-Span Socialization in a Cultural Context. In *Handbook of Asian American Psychology,* 2nd ed., edited by F. T. Leong, A. G. Inman, A. Ebreo, L. H. Yang, M. Kinoshita, and M. Fu, pp. 69–86. Thousand Oaks, CA: Sage, 2007.

Yellowbird, Michael, and C. Matthew Snipp. "American Indian Families." In *Minority Families in the United States: A Multicultural Perspective,* edited by R. L. Taylor. Englewood Cliffs, NJ: Prentice Hall, 1994.

Yeung, W. Jean, John Sandberg, Pamela Davis-Kean, and Sandra Hofferth. "Children's Time with Fathers in Intact Families." *Journal of Marriage and Family* 63 (February 2001): 136–154.

Ÿllo, Kersti. "Through a Feminist Lens: Gender, Power, and Violence." In *Current Controversies in Family Violence,* edited by R. Gelles and D. Loseke. Newbury Park, CA: Sage, 1993.

Young, Robin. "Polygamy in America." Public Radio International, February 12, 2010.

Yuki, M., Maddux, W. W., and T. Masuda. "Are the Windows to the Soul the Same in the East and West? Cultural Differences in Using the Eyes and Mouth as Cues to Recognize Emotions in Japan and the United States." *Journal of Experimental Social Psychology* 43 (2007): 303–311.

Zhang, Y., and J. Van Hook. "Marital Dissolution among Interracial Couples." *Journal of Marriage and the Family* 71, 1 (February 2009): 95–107.

Zhan, M., and S. Pandey. "Economic Well-Being of Single Mothers: Work First or Postsecondary Education?" *Journal of Sociology and Social Welfare* 31, 3 (2004): 87–112.

Zilbergeld, B. *The New Male Sexuality: The Truth about Men, Sex, and Pleasure.* New York: Bantam, 1993.

Zimmerman, Kevin. "Effective Couple Communication." *NCFR Reports: Family Focus on Effective Communication* (Summer 2015): F13–F14.

Zimmerman, Julie. "Obergefell, Arthur Wed on Plane in MD." *Cincinnati Enquirer,* April 18 2014. www.cincinnati.com.

Zink, Therese, and Frank Putnam. "Intimate Partner Violence Research in the Health Care Setting: What are Appropriate and Feasible Methodological Standards?" *Journal of Interpersonal Violence* 20, 4 (April 2005): 365–372.

Ziv, Stav. "How Jim Obergefell's Fight for His Dying Spouse Legalized Gay Marriage in America." Newsweek, June 26, 2015. www .newsweek.com/jim-obergefell-man-behind-supreme-courts-same -sex-marriage-decision-347314.

Zucker, Alyssa. "Disavowing Social Identities: What It Means When Women Say 'I'm Not a Feminist But. . . .'" *Psychology of Women Quarterly* 28 (December 2004): 423–435.

Zvonkovic, Anisa N., Kathleen M. Greaves, Cynthia J. Schmiege, and Leslie D. Hall. "The Marital Construction of Gender through Work and Family Decisions: A Qualitative Analysis." *Journal of Marriage and the Family* 58, 1 (February 1996): 91–100.

Zweig, J. M., M. Dank, P. Lachman, and J. Yahner. (2013). "Technology, Teen Dating Violence and Abuse, and Bullying." https://ncjrs.gov /pdffiles1/nij/grants/243296.pdf.

# Name Index

Walker, Alexis J., 45, 416
Walker, Alice, 214
Walker, Karen, 161
Walker, Lenore, 460, 470
Wall, Elizabeth, 396
Wallace, Joseph, 481
Waller, Erika, 160, 161
Waller, Willard, 256
Wallerstein, Judith, 58, 512, 513, 514, 515, 517, 547
Wallis, Cara, 138
Walls, Jill, 145, 382
Walsh, J. A., 478
Walsh, Jeffrey, 478–479
Walsh, Jennifer, 138
Walsh, Marie-Therese, 378
Walsh, Seth, 137
Walsh-Smith, Tricia, 507
Walster, G. William, 184
Walster, William, 188–189
Walters, Barbara, 118
Walters, Mikel, 454, 462, 471
Walters, M. L., 449, 463, 464, 465
Walzer, Susan, 425
Wang, C., 476
Wang, H., 505
Wang, Qiu, 430
Wang, Shuai, 486
Wang, T., 184
Wang, Wendy, 77, 94, 108, 109, 144, 147, 277, 278, 290, 291, 292, 295, 313, 325, 328, 382–383, 419, 499
Ward, Brian, 428
Ward, B. W., 214
Ward, J., 262
Ward, Monique, 138, 199
Wareham, J., 456
Warren, L. Z., 304
Warren-Gash, Charlotte, 359
Waters, Frances Cudjoe, 99
Waters, Mary, 108
Weaver, A. D., 195
Wei, G. S., 480
Weibust, K., 149
Weinberg, Martin S., 216
Weis, D. L., 227
Weisfeld, Carol C., 226

Weisfeld, Glenn E., 226
Weiss, Robert, 506
Weisshaar, K., 338
Weitzman, Lenore, 308, 510
Weitzman, Susan, 461
Wells, Brooke, 201
Wells, Georgia, 178
Wendi Gardner, 186
Werner, Carol M., 307
West, A. R., 175
West, Candace, 124
West, Cornel, 33, 523
Westberg, Heather, 512
Westbrook, Laurel, 116
Wester, Stephen, 180
Westman, Mina, 414
Weston, Kath, 351
Weston, Mike, 399
Wharton, Amy, 438
Whitaker, Daniel J., 195
White, James, 39, 42, 50, 51, 52, 53, 54
White, Lynn, 502
Whitehead, Andrew, 138–139
Whitehead, Barbara Dafoe, 299, 494, 516, 522
Whitehurst, Lindsay, 4
Whiteman, S., 314
Whiteman, S. D., 41
Whitlock, T., 166, 167
Whitton, Sarah, 248
Whitton, S. W., 501, 539
Wickrama, K. A. S., 97, 98, 99, 101
Widmer, Eric, 226
Wiehe, Vernon, 485
Wiener, Seth, 192
Wiesner, Margit, 385
Wilcox, W. Bradley, 6, 7, 16, 24, 141, 283, 284, 285, 293
Wildsmith, E., 95
Wilhelm, Peter, 419
Wilke, Joy, 278, 327
Wilkens, Carrie, 253
Willetts, Marion, 339, 346, 347, 348
William Cupach, 176
Williams, April, 148

Williams, C. J., 216
Williams, Dmitri, 138
Williams, J., 161
Williams, Rhiannon, 116
Willing, Richard, 286, 287
Willoughby, Brian, 265
Wilreker, B., 10
Wilson, April, 304
Wilson, Breana, 401
Wilson, C. D., 477
Wilson, F. A., 300
Wilson, Karen, 150
Wilson, K. J., 482
Wilson, L., 98, 99, 103, 143, 146, 205
Wilson, Margo, 485
Wilson, Tracey K., 503
Wilson, William Julius, 85, 100
Wineberg, H., 499, 505
Winkler, Phil, 441
Winslow, Sarah, 414
Winton, Chester, 51, 403
Wisdom, Susan, 542
Wise, T. N., 230
Witterick, Kathy, 133
Witters, Dan, 300
Wlodarski, Rafael, 163
Wohl, Michael, 60
Wolcott, D., 475
Wolf, Rosalie S., 479
Wolfe, Donald, 256
Wolfers, Justin, 121, 492
Wolfinger, Nicholas, 296, 500, 501, 508
Wong, Y., 110
Woodward, Amanda Toler, 12, 97
Working Mother, 444
World Health Organization, 120, 234, 235, 359, 448, 449
World Values Survey, 279
Worthen, Meredith, 213, 216
Wright, Brittany, 259, 265, 268–269
Wright, James, 523
Wright, John, 486
Wrigley, Julia, 435
Wu, L. L., 501
Wu, Zheng, 15, 300, 339–340

Wymbs, Brian T., 503
Wyse, J., 332, 333, 336, 337, 343, 346

**X**

Xia, Yan R., 179
Xiaohe Xu, 195
Xu, X., 283, 284, 335–336, 536, 540

**Y**

Yabiku, Scott, 50, 426
Yahner, J., 467
Yahya, Tooba, 118
Yardley, J., 8
Yeager, Erica Owens, 268, 269
Yee, Barbara, 133
Yellowbird, Michael, 12, 107
Yeung, W. Jean, 425
Ÿllo, Kersti, 456
Yodanis, Carrie, 391
Yoffe, Emily, 34
Yong, William, 497
Yoo, I., 307
Yorgason, Jeremy, 412
Yoshimoto, D., 350
Young, Emily, 448
Young, Robin, 4
Yuen, Sylvia, 133
Yuhua, Guo, 8
Yuki, Masaki, 245

**Z**

Zhan, M., 530
Zhang, Y., 292
Zhu Wei Xing, 118
Zilbergeld, Bernie, 229
Zill, Nicholas, 24
Zillerman, T. Schindler, 138
Zimmerman, Don, 124
Zimmerman, Julie, 9
Zimmerman, Kevin, 253, 254
Zink, Therese, 470
Ziv, Stav, 9
Zucker, Alyssa, 149
Zvonkovic, Anisa N., 45
Zweig, J. M., 467
Zyzanski, S. J., 35, 58

# Subject Index

Biphobia, 216
Birthrates, 356–358
Births, trends in, 75, 76
Bisexual identities, 207–216
Bisexual invisibility, 216
Bisexuality, 207, 214, 216
Bisexuality erasure, 216
Black feminism, 45
*Blackish*, 99
Blended families, 77, 527, 541. *See also* Stepfamilies
Boomerang generation, 314–315
Border work, 136
Boundaries, establishing in early marriage, 309–311
Boundary ambiguity, 541
Boundary maintaining, 136
Breaking up, 184–188
*Brokeback Mountain*, 154
Buckeye Singles Council, 326
Bullying, 135–136, 137
Bureau of Labor Statistics Survey of Workplace Violence Prevention, 480

## C

Caballerismo, 133, 143
Caldecott Medal, 130–131
California's Family Law Act, 504
Caregiver gain, 404
Caregiving
  aging parents, 403–404
  children caring for parents, 403
  cyber caregiving, 19
  family and, 14
  grandparenting, 401–403
  parenting adult children, 398–401
*The Case for Marriage* (Waite and Gallagher), 511
Case-study method, 58
Catholic Church, 20
Celibate marriages, 226
Centers for Disease Control and Prevention, 214, 232, 234, 235, 300, 360, 364, 366, 373, 377, 449, 453, 460, 470, 480, 481
Centrists, 22
Channeling, 131
Cherokee, 106
Child abuse and neglect
  by age, 473
  Child Protective Services, 481
  forms of emotional child abuse, 471
  intervention and prevention, 484
  prevalence of, 472–475
Childbearing, 93, 201
Childbirth
  giving birth, 371
  infant mortality, 372–373
  loss, coping with, 373–374
  medicalization of, 370–371
  men and, 372
  mother's mean age at, 365–366
Child care, 72, 424–426, 434–437
Child-centered parents, 48
Child custody, 4, 268, 518–522
Child-free, 360
Childhood, 67, 69–70, 91, 129–130
"Childless by choice," 360
Childlessness, 359–361
Child maltreatment, 471
*Child Maltreatment: 2013*, 474
Child–parent violence (CPV), 478–479

Child rearing
  African Americans, 100
  Asian Americans, 105
  children's needs, 392–393
  cohabitation and, 77
  contemporary strategies, 389–390
  fathers, 80
  introduction to, 388–389
  Native Americans, 65–66
  parents' needs, 393
  recession and, 93
  spanking, 391
  strategies and styles of, 388–393
  styles of, 390, 392
  women and, 69
Children
  caught in the middle of divorce, 515–516
  class and, 88–89
  demographics, 1970–2014, 76
  developmental tasks of divorce, 514–515
  divorce and, 502–503, 511–518
  explaining divorce to, 512
  living with grandparents, 94
  long-term effects of divorce, 516–518
  Native Americans, 65–66
  poverty and, 86
  recession, long-term effects of, 93–94
  remarriage and, 538
  in single-parent families, 532–533
  slavery and, 68, 70
  in stepfamilies, 549–551
  three stages of divorce for, 513
Children's Bureau, 474
Children's Defense Fund, 481
Children's literature, 130–131
Children's protective services (CPS) agencies, 472, 478, 481
Child sexual abuse
  effects of, 486–487
  forms of intrafamilial child sexual abuse, 485
  introduction to, 484–485
  as risk children, 485
  sibling sexual abuse, 485–486
Child support, 508, 509–510
Child Support Enforcement Amendments, 509
Child Welfare Information Gateway, 376
Child well-being, 269
China, 8, 115, 118
Chinese Americans, 105–106
Chlamydia, 232–234
Christianity, 7
Church of Spiritual Humanism, 298
"Cinderella effect," 485
Cisgender, 114, 116
Civil Rights Act of 1964, 433
Civil unions, 287–288
Class. *See* Social class
Clinical psychologists, 272
Clinical research, 58
Closed fields, 176
"Closeted" gays or lesbians, 211
Coercion, 262, 270
Coercive sex, 468–469
Coexistence, 270
Cognition, 126, 128
Cognitive development theory, 128–129

Cohabitation
  African Americans and, 20
  as alternative way of entering marriage, 306
  binuclear family, 536
  conflict and, 260
  contemporary state of, 76–77
  demographics, 1970-2014, 76
  divorce and, 500–501
  educational attainment, 320, 332–333
  effect of on later marriage, 341, 343–346
  engagement and, 305–306
  gay and lesbian, 349–352
  housework, 423
  in late 20th century, 75
  legal issues, 342
  vs. marriage, 9, 336–341
  masturbation and, 219
  meaning of, 335
  parenthood and, 394
  race and ethnicity, 330–332
  redefining sex in, 217
  remarriage and, 335–336
  rise of, 330–333
  sexual expression, 224
  single-parent families, 531
  types of, 333–335
Cohabitation effect, 32, 341
Coitus, 220–221
Collaboration, 263
Collaborative Psychiatric Epidemiology Surveys (CPES), 461
Collectivist cultures, 158
College, 140–141
College Board, 362
Colonial families, 66–67
Coming out, 208–209, 210–212
Commission on the Status of Women, 149
Commitment
  changes in over time, 188
  defining, 159
  forging in adolescence, 201
  love and, 171
  marital commitments, 312
  marriage vs. cohabitation, 336–337
  technology and, 242
  in triangular theory of love, 169
  verbal and nonverbal, 241–246
Common couple violence, 451
Common law marriages, 346–349
Common sense explanations, 31–32
Communal expressiveness, 304
Communication
  about sexuality, 195
  barriers to, 252–253
  demand-withdraw, 249–251
  difficult topics, 252
  gender differences, 246–247
  introduction to, 240–241
  marital power, sources of, 255–256
  overcommunication, 251
  parenting and, 534
  patterns of in marriage, 247–251
  positive strategies, 253–254
  power and intimacy, 254–255
  problems in, 251–253
  sexual, 248–249
  topic-related difficulty, 251–252
Communication privacy management theory, 251
Community divorce, 492

"A Community of Contrasts: Asian Americans in the United States, 2011," 104
Companionate family, 72
Companionate love, 168
Companionate marriage, 157
Competition, 263
Complementary needs theory, 297
Compliance, 263
Compressed workweeks, 443
Compromise, 263
Concepts, 37
Conceptualization, 37
Concerted cultivation, 88, 91
Condoms, 236
Conduct of fatherhood, 384
Confidentiality, 55
Conflict
  anger, dealing with, 260
  beneficial aspects, 270
  common areas of, 264–268
  consequences of, 268–270
  decision to stay or leave, 272
  determinants of how couples handle, 262–263
  experience of, 259–260
  gender and marital conflict, 261
  gender differences, 260–261
  housework, 266–268
  intimacy and, 259–268
  management and avoidance, 264
  money, 265–266
  sex, 265
  sources of, 44
Conflict-avoiding couples, 319
Conflicted couples, 319
Conflict engagement, 263
Conflict-habituated marriages, 318
Conflict resolution, 262–264, 270–273
Conflict Resolution Styles Inventory (CRSI), 263
Conflict Tactics Scales (CTS2), 61
Conflict theory, 44–45, 53
Conjugal family, 157, 159
Conjugal relationships, 18
Consanguineous relationships, 18
Conservatives, 22
Consummate love, 188
Contact stage, 545–546
Contagion, 496
Contemporary sexual scripts, 194
Contempt, 243, 252–253, 305
*Contexts* (Streib), 82
Contraception, 75, 79, 224, 237
Cooperative co-parenting, 515
Cooperative exchanges, 49–50
Cooperativeness, 263
Co-parental divorce, 492
Co-parenting, 353
Co-provider families, 417
Coresidential dating, cohabitation as, 334, 336–337
Corporal punishment, 100, 391
Correlational studies, 60
Co-rumination, 160–161
*CosmoGirl*, 204
*Cosmopolitan*, 197, 199
"Costs of Adopting" fact sheet, 376
Council on Contemporary Families, 422, 426
Counseling, 271–273
Courtships, 302
Covenant marriage, 523
Criticism, 243, 252, 305
Crossover, 414
Cross-sex friendships, 161
Crude divorce rate, 493, 495–496

Cultivators, 132
Cultural constructions of love, 158
Cultural lag, 320
Culture of fatherhood, 384
Cultures of poverty, 81, 95–96
Cunnilingus, 220
Current Population Survey, 17
Cyber caregiving, 19
Cyberdating abuse, 467

 **D**

*Daddy Shift* (Smith), 432
Date rape, 468–469
Dating
  after a divorce, 506, 508
  vs. arranged marriages, 285
  breaking up, 184–188
  cultural differences, 179
  gendered roles, 179
  gender roles, 175
  hooking up, 181–182
  jealousy, 182–184
  meeting face-to-face, 175–176
  meeting online, 176–178
  in older adulthood, 180–181
  overview of, 178–179
  physical attractiveness, 173–175
  problems in, 179–181
  sexual expression, 224
"Dating Abuse and Teen Violence"
  Fact Sheet, 466
Dating violence, 61, 466–469
Daughters' household
  responsibilities, 127
Deductive research, 37–38
Defense of Marriage Act, 9, 287, 288
Defensiveness, 243, 252–253, 305
Deficit approach, 534
Degree of deliberativeness,
  548–549
Delayed orgasm, 230
Demand-withdraw
  communications, 249–251,
  260, 262
Demography, 58
Dependent variables, 38
Desertion, 491
Destructive parentification, 403
Devitalized couples, 319
Devitalized marriages, 318
Direct crossover, 414
Direct interaction, 425
Disasters, conceptualizing the
  effects of, 40–41
Discernment counseling, 272
Discipline, 551–552
Discrimination
  about LGBT individuals, 212–214
  against bisexuals, 214–216
  children and, 395
  against single people, 330
  against women in the workplace,
  432–434
Disengaging from parental
  conflicts, 514
Distal causes, 504
Divided loyalties, 551
Division of labor, 68. *See also*
  Housework
  men's traditional family work, 416
  traditional pattern, 414–416
  women's traditional family work,
  416–417
Divorce
  age at time of marriage, 500
  alimony and child support,
  509–510

child custody, 4, 518–522
children and, 502–503, 511–518
cohabitation and, 500–501
conflict and, 268
consequences of, 508–511
contemporary state of, 77
crude divorce rate, 493
dating following, 506, 508
decision to stay or leave, 503
demographic factors, 496–499
developmental tasks of, 514–515
economic consequences,
  508–511
educational level, 498–499
employment and, 497–498,
  510–511
ethnicity and, 499
factors affecting, 495–504
family processes, 502–504
helping children adjust, 515
how children are told, 512
income and, 498
increased rate of, 79
individualistic vs. collectivist
  values, 496
intergenerational transmission,
  501–502
internationally, 495–496, 497
intimate violence and, 453
introduction to, 490
legal meaning of, 490–491
life course factors, 499–502
long-term effects on children,
  516–518
marital happiness, 502
marital problems, 504
meaning of, 490–492
mediation, 521–522
multiple realities of, 491–492
no-fault divorce, 504
noneconomic consequences
  of, 511
postdivorce identity, 505–506
poverty and, 86
predicting, 494
prevalence of, 492–494
public aspects of, 507
by race, ethnicity, and gender,
  492–493
ratio measure of divorces to
  marriages, 493
recession and, 93
refined divorce rate, 493–494
religion and, 499
remarriage and, 501
separation, process of, 504–508
societal factors, 495–496
three stages of for children, 513
trends, 75, 76, 494–495
what to do about, 522–524
*The Divorce Culture* (Whitehead),
  516
Divorce mediation, 521–522
Divorce rate, 277, 327
Domestic partnerships, 347–349
Domestic violence. *See* Intimate
  violence/abuse
Double standard of aging, 175
Downward mobility, 92
*Dr. Phil*, 34
Dual-earner families
  atypical dual earners, 427–431
  caring for children, 424–426
  changing gender roles, 80, 144
  coping in, 431
  effects of division of household
  labor, 426–427
  emotion work, 424

at-home fathers and
  breadwinning mothers,
  431–432
housework, 421–424
increases in, 75, 78
marriage and, 166
peer and postgender marriages,
  430–431
shift work and family life,
  428–430
typical dual earners, 421
Duluth model, 483–484
Duration-of-marriage effect, 308
Dyspareunia, 229

**E**

Early ejaculation, 230
Early marriage
  effects of parenthood on,
  312–313
  establishing boundaries, 309–311
  establishing marital roles,
  308–309
  marital commitments, 312
  social context and social stress,
  311–312
Ecological niche, 40–41
Ecology theory, 39–42, 53
Economic cooperation, 13
Economic dependence, 469
Economic discrimination, 433
Economic divorce, 492
Economic prosperity, 74
Economic resources, 300–301
Economics, and gender, 121
Education, 70
Educational attainment, 104, 135
  childless, 360–361
  cohabitation and, 332–333
  fertility rates and birthrates,
  357–358
  intermarriage, 290–291
  marriage and, 281, 294, 320
  remarriage and, 537–538
Educational homogamy, 294
Egalitarianism, 115, 423
Egocentric fallacy, 35
Elder abuse, 479–480, 484
Elective co-parenting, 353
Elisa's Law, 478
Emotional abuse, 450, 464–465,
  471
Emotional divorce, 491
Emotions, expressing nonverbally,
  243
Emotion work, 424
Empathy stage, 253
Emphasized femininity, 123
Employment patterns, 408–410
Empty nest, 314
Endogamy, 289
Enduring marriages, 318–320
Engagement, 305–306, 425
Environment, 39
Equal Employment Opportunity
  Commission (EEOC), 433
Equal Pay Act of 1963, 149
Equity, 49
Erectile dysfunction, 230
Erin's Law, 478
Eros, 167–168
Ethical guidelines, 55
Ethnic group, 95. *See also* Race and
  ethnicity; *specific groups*
Ethnic identity, 15, 107–108
Ethnic neighborhoods, 107–108

Ethnic socialization, 100
Ethnocentric fallacies, 35, 96
European ethnic families, 107–108
Exchange-social control model,
  456–457
Exogamy, 289
Exosystem, 40
Expectations, 49
Experimental research, 60
Expressive displays, 159
Expressive traits, 43, 125
Extended families, 16–18, 84, 89,
  102, 107
Extended households, 100
Extensive mothering, 145, 381–382
Extrafamilial sexual abuse, 485
Extramarital sex, 223
Extramarital sexuality, 226–229
Eye contact, 245

**F**

Facebook, 116, 177, 507
"Facebook's Last Taboo: The
  Unhappy Marriage" (Seligson),
  507
Facial expressions, 245
Facts for Families Guide, 542
"Fact Sheet on Intimate Partner
  Violence," 483
Failed repair attempts, 305
Fallacies, 35
Familial well-being, 269
Families and Work Institute, 443
*Families Like Mine* (Garner), 396
Families of choice, 351–352
*Families on the Fault Line* (Rubin),
  440
*Families We Choose* (Weston), 351
Familism, 80
Family
  in the 20th century, 71–74
  advantages of living in, 16
  attitudes toward changes in,
  22–23, 25
  caregiving, 14
  choosing and redesigning,
  351–352
  conservative, liberal, and centrist
  perspectives, 22, 24
  defining, 10–12
  disagreement among family
  scientists, 23, 25
  diversity of, 2, 26–27
  division of labor, 414–417
  dynamism of, 26
  economic cooperation, 13
  experience vs. expertise, 2
  extended families and kinship,
  16–18
  friends as, 352–353
  functions of, 12–15
  gender and, 26–27, 121–122,
  142–147
  vs. household, 10–11
  industrialization, 415
  interdependence with wider
  society, 27–28
  intimate relations and, 12–13
  introduction to, 2
  *Modern Family* effect, 25
  multiple viewpoints of, 19–25
  optimism about, 21–22
  outside influences on, 27
  race and ethnicity, 26
  reasons for living in, 15–16
  reducing work-family conflict,
  442–445